W9-AMC-888

THE INTEL MICROPROCESSORS

8086/8088, 80186/80188, 80286, 80386, 80486, Pentium, Pentium Pro Processor, Pentium II, Pentium III, and Pentium 4

Architecture, Programming, and Interfacing

Seventh Edition

BARRY B. BREY
DeVry University

Upper Saddle River, New Jersey
Columbus, Ohio

Library of Congress Cataloging-in-Publication Data
Brey, Barry B.
The Intel microprocessors : 8086/8088, 80186/80188, 80286, 80386, 80486, Pentium, Pentium Pro processor, Pentium II, Pentium III, and Pentium 4 : architecture, programming, and interfacing / Barry B. Brey—7th ed.
p. cm.
Includes index.
ISBN 0-13-119506-9
1. Intel 80xxx series microprocessors. 2. Pentium (Microprocessor) I. Title.

QA76.8.I292B75 2005
004.165—dc22

2005001757

Assistant Vice President and Publisher: Charles E. Stewart, Jr.
Production Editor: Alexandrina Benedicto Wolf
Design Coordinator: Diane Ernsberger
Cover Designer: Terry Rohrbach
Cover Art: Index Stock
Production Manager: Matt Ottenweller
Marketing Manager: Ben Leonard

This book was set in Times Roman by The GTS Companies/York, PA Campus. It was printed and bound by Hamilton Printing. The cover was printed by Coral Graphic Services, Inc.

Pearson Education Ltd.
Pearson Education Singapore Pte. Ltd.
Pearson Education Canada, Ltd.
Pearson Education—Japan
Pearson Education Australia Pty. Limited
Pearson Education North Asia Ltd.
Pearson Educación de Mexico, S. A. de C.V.
Pearson Education Malaysia Pte. Ltd.

10 9 8 7 6 5 4 3 2 1
ISBN: 0-13-119506-9

This text is dedicated to my progeny, Brenda (the programmer) and Gary (the veterinary technician).

PREFACE

This practical reference text is written for students who require a thorough knowledge of programming and interfacing of the Intel family of microprocessors. Today, anyone studying or working in a field that uses computers must understand assembly language programming and interfacing. Intel microprocessors have gained wide, and at times exclusive, application in many areas of electronics, communications, and control systems, particularly in desktop computer systems. A major addition to this seventh edition explains how to interface C/C++ with assembly language for both the older DOS and the Windows environments. Many applications include Visual C++ as a basis for learning assembly language using the inline assembler. Updated sections detailing new events in the fields of microprocessors and microprocessor interfacing have been added.

ORGANIZATION AND COVERAGE

To cultivate a comprehensive approach to learning, each chapter begins with a set of objectives that briefly define its content. Chapters contain many programming applications and examples that illustrate the main topics. Each chapter ends with a numerical summary, which doubles as a study guide and reviews the information just presented. Questions and problems are provided for reinforcement and practice, including research paper suggestions.

This text contains many example programs using the Microsoft Macro Assembler program and the inline assembler in the Visual C++ environment, which provide a learning opportunity to program the Intel family of microprocessors. Operation of the programming environment includes the linker, library, macros, DOS function, BIOS functions, and Visual C/C++ program development. The inline assembler (C/C++) is illustrated for both the 16- and 32-bit programming environments of various versions of Visual C++. The text is written to use Visual Studio.NET 2003 as a development environment, but Visual Studio 6.0 can also be used with almost no change.

This text also provides a thorough description of family members, memory systems, and various I/O systems that include disk memory, ADC and DAC, 16550 UART, PIAs, timers, keyboard/display controllers, arithmetic coprocessors, and video display systems. Also discussed are the personal computer system buses (AGP, ISA, PCI, USB, serial ports, and parallel port). Through these systems, a practical approach to microprocessor interfacing can be learned.

APPROACH

Because the Intel family of microprocessors is quite diverse, this text initially concentrates on real mode programming, which is compatible with all versions of the Intel family of microprocessors. Instructions for each family member, which includes the 80386, 80486, Pentium, Pentium Pro,

Pentium II, Pentium III, and Pentium 4 processors, are compared and contrasted with those for the 8086/8088 microprocessors. This entire series of microprocessors is very similar, which allows more advanced versions and their instructions to be learned with the basic 8086/8088. Please note that the 8086/8088 are still used in embedded systems along with their updated counterparts, the 80186/80188 and 80386EX embedded microprocessors.

This text also explains the programming and operation of the numeric coprocessor, the MMX extension, and the SIMD extension, which function in a system to provide access to floating-point calculations that are important in control systems, video graphics, and computer-aided design (CAD) applications. The numeric coprocessor allows a program to access complex arithmetic operations that are otherwise difficult to achieve with normal microprocessor programming. The MMX and SIMD instructions allow both integer and floating-point data to be manipulated in parallel at very high speed.

This text also describes the pin-outs and function of the 8086–80486 and all versions of the Pentium microprocessor. First, interfacing is explained using the 8086/8088 with some of the more common peripheral components. After the basics are explained, a more advanced emphasis is placed on the 80186/80188, 80386, 80486, and Pentium through Pentium 4 microprocessors. Coverage of the 80286, because of its similarity to the 8086 and 80386, is minimized so the 80386, 80486, and Pentium versions can be covered in complete detail.

Through this approach, the operation of the microprocessor and programming with the advanced family members, along with interfacing all family members, provide a working and practical background of the Intel family of microprocessors. Upon completing a course using this text, you will be able to:

1. Develop software to control an application interface microprocessor. Generally, the software developed will also function on all versions of the microprocessor. This software also includes DOS-based and Windows-based applications. The main emphasis is on developing inline assembly and C++ mixed language programs in the Windows environment.
2. Program using MFC controls, handlers, and functions to use the keyboard, video display system, and disk memory in assembly language and C++.
3. Develop software that uses macro sequences, procedures, conditional assembly, and flow control assembler directives that are linked to a Visual C++ program.
4. Develop software for code conversions using lookup tables and algorithms.
5. Program the numeric coprocessor to solve complex equations.
6. Develop software for the MMX and SIMD extensions.
7. Explain the differences between the family members and highlight the features of each member.
8. Describe and use real and protected mode operation of the microprocessor.
9. Interface memory and I/O systems to the microprocessor.
10. Provide a detailed and comprehensive comparison of all family members and their software and hardware interfaces.
11. Explain the function of the real-time operating system in an embedded application.
12. Explain the operation of disk and video systems.
13. Interface small systems to the ISA, PCI, serial ports, parallel port, and USB bus in a personal computer system.

CONTENT OVERVIEW

Chapter 1 introduces the Intel family of microprocessors with an emphasis on the microprocessor-based computer system: its history, operation, and the methods used to store data in a microprocessor-based system. Number systems and conversions are also included. Chapter 2 explores the programming model of the microprocessor and system architecture. Both real and protected mode operations are explained.

Once an understanding of the basic machine is grasped, Chapters 3 through 6 explain how each instruction functions with the Intel family of microprocessors. As instructions are explained, simple applications are presented to illustrate the operation of the instructions and develop basic programming concepts.

Chapter 7 introduces the use of Visual C/C++ with the inline assembler and separate assembly language programming modules. It also explains how to configure a simple Visual C++ program with assembly application.

After the basis for programming is developed, Chapter 8 provides applications using Visual C++ with the inline assembler program. These applications include programming using the keyboard and mouse through message handlers in the Windows environment. Disk files are explained using CFile, as well as keyboard and video operations on a personal computer system through Windows. This chapter provides the tools required to develop virtually any program on a personal computer system.

Chapter 9 introduces the 8086/8088 family as a basis for learning basic memory and I/O interfacing, which follow in later chapters. This chapter shows the buffered system as well as the system timing.

Chapter 10 explains memory interface using both integrated decoders and programmable logic devices using VHDL. The 8-, 16-, 32-, and 64-bit memory systems are provided so the 8086–80486 and the Pentium through Pentium 4 microprocessors can be interfaced to memory.

Chapter 11 provides a detailed look at basic I/O interfacing, including PIAs, timers, the 16550 UART, and ADC/DAC. It also describes the interface of both DC and stepper motors.

Once these basic I/O components and their interface to the microprocessor is understood, Chapters 12 and 13 provide detail on advanced I/O techniques that include interrupts and direct memory access (DMA). Applications include a printer interface, real-time clock, disk memory, and video systems.

Chapter 14 details the operation and programming for the 8087–Pentium 4 family of arithmetic coprocessors, as well as MMX and SIMD instructions. Today few applications function efficiently without the power of the arithmetic coprocessor. Remember that all Intel microprocessors since the 80486 contain a coprocessor; since the Pentium, an MMX unit; and since the Pentium II, an SIMD unit.

Chapter 15 shows how to interface small systems to the personal computer through the use of the parallel port, serial ports, and the ISA and PCI bus interfaces.

Chapters 16 and 17 cover the advanced 80186/80188–80486 microprocessors and explore their differences with the 8086/8088, as well as their enhancements and features. Cache memory, interleaved memory, and burst memory are described with the 80386 and 80486 microprocessors. Chapter 16 also covers real-time operating systems (RTOS), and Chapter 17 also describes memory management and memory paging.

Chapter 18 details the Pentium and Pentium Pro microprocessors. These microprocessors are based upon the original 8086/8088.

Chapter 19 introduces the Pentium II, Pentium III, and Pentium 4 microprocessors. It covers some of the new features, package styles, and the instructions that are added to the original instruction set.

Appendices are included to enhance the text. Appendix A provides an abbreviated listing of the DOS INT 21H function calls. It also details the use of the assembler program and the Windows Visual C++ interface. A complete listing of all 8086–Pentium 4 instructions, including many example instructions and machine coding in hexadecimal as well as clock timing information, is found in Appendix B. Appendix C provides a compact list of all the instructions that change the flag bits. Answers for the even-numbered questions and problems are provided in Appendix D.

ACKNOWLEDGMENTS

I greatly appreciate the feedback from the following reviewers: James Archibald, Brigham Young University; William Evans, University of Toledo; Issac Ghansah, California State University; and William Murray, Broome Community College.

STAY IN TOUCH

We can stay in touch through the Internet. My Internet site contains information about all of my textbooks and many important links that are specific to the personal computer, microprocessors, hardware, and software. Also available is a weekly lesson that details many of the aspects of the personal computer. Of particular interest is the "Technical Section," which presents many notes on topics that are not covered in this text. Please feel free to contact me at bbrey@ee.net if you need any type of assistance. I answer all of my email within 24 hours.

My website at **http://members.ee.net/brey**

DeVry University at **http://199.218.238.2/facstaff/bbrey/**

CONTENTS

CHAPTER 1

Introduction to the Microprocessor and Computer

INTRODUCTION

This chapter provides an overview of the Intel family of microprocessors. Included is a discussion of the history of computers and the function of the microprocessor in the microprocessor-based computer system. Also introduced are terms and jargon used in the computer field, so you will understand *computerese* in the discussion of microprocessors and computers.

The block diagram and a description of the function of each block detail the operation of a computer system. Blocks in the block diagram show how the memory and input/output (I/O) system of the personal computer interconnect. The chapter details how data are stored in the memory so each data type can be used as software is developed. Numeric data are stored as integers, floating-point, and binary-coded decimal (BCD); alphanumeric data are stored by using the ASCII (American Standard Code for Information Interchange) code.

CHAPTER OBJECTIVES

Upon completion of this chapter, you will be able to:

1. Converse by using appropriate computer terminology such as bit, byte, data, real memory system, protected mode memory system, Windows, DOS, I/O, and so forth
2. Briefly detail the history of the computer and list applications performed by computer systems
3. Provide an overview of the various 80X86 and Pentium family members
4. Draw the block diagram of a computer system and explain the purpose of each block
5. Describe the function of the microprocessor and detail its basic operation
6. Define the contents of the memory system in the personal computer
7. Convert between binary, decimal, and hexadecimal numbers
8. Differentiate and represent numeric and alphabetic information as integers, floating-point, BCD, and ASCII data

1–1 A HISTORICAL BACKGROUND

This first section outlines the historical events leading to the development of the microprocessor and, specifically, the extremely powerful and current 80X86,[1] Pentium, Pentium Pro, Pentium III, and Pentium 4[2] microprocessors. Although a study of history is not essential to understand the microprocessor, it furnishes interesting reading and provides a historical perspective of the fast-paced evolution of the computer.

The Mechanical Age

The idea of a computing system is not new—it was around long before modern electrical and electronic devices were developed. The idea of calculating with a machine dates to 500 BC when the Babylonians invented the **abacus**, the first mechanical calculator. The abacus has strings of beads with which to perform calculations. It was used by the Babylonian priests to keep track of their vast storehouses of grain. Still in use today, the abacus was not improved until 1642, when mathematician Blaise Pascal invented a calculator that was constructed of gears and wheels. Each gear contained 10 teeth that, when moved one complete revolution, advanced a second gear one place. This is the same principal that is used in the automobile's odometer mechanism and is the basis of all mechanical calculators. Incidentally, the PASCAL programming language is named in honor of Blaise Pascal for his pioneering work in mathematics and with the mechanical calculator.

The arrival of the first practical geared mechanical machines used to automatically compute information dates to the early 1800s. This is before humans invented the lightbulb or before much was known about electricity. In this dawn of the computer age, humans dreamed of mechanical machines that could compute numerical facts with a program—not merely calculate facts, as with a calculator.

In 1937 it was discovered through plans and journals that one early pioneer of mechanical computing machinery was Charles Babbage, aided by Augusta Ada Byron, the Countess of Lovelace. Babbage was commissioned in 1823 by the Royal Astronomical Society of Great Britain to produce a programmable calculating machine. This machine was to generate navigational tables for the Royal Navy. He accepted the challenge and began to create what he called his **Analytical Engine**. This engine was a steam-powered mechanical computer that stored 1000 twenty-digit decimal numbers and a variable program that could modify the function of the machine to perform various calculating tasks. Input to his engine was through punched cards, much as computers in the 1950s and 1960s used punched cards. It is assumed that he obtained the idea of using punched cards from Joseph Jacquard, a Frenchman who used punched cards as input to a weaving machine he invented in 1801, which is today called *Jacquard's loom*. Jacquard's loom used punched cards to create intricate weaving patterns in the cloth that it produced. The punched cards programmed the loom.

After many years of work, Babbage's dream began to fade when he realized that the machinists of his day were unable to create the mechanical parts needed to complete his work. The Analytical Engine required more than 50,000 machined parts, which could not be made with enough precision to allow his engine to function reliably.

The Electrical Age

The 1800s saw the advent of the electric motor (conceived by Michael Faraday); with it came a multitude of motor-driven adding machines, all based on the mechanical calculator developed by

---

[1]80X86 is an accepted acronym for 8086, 8088, 80186, 80188, 80286, 80386, and 80486 microprocessors and also includes the Pentium series.

[2]Pentium, Pentium Pro, Pentium II, Pentium III, and Pentium 4 are registered trademarks of Intel Corporation.

Blaise Pascal. These electrically driven mechanical calculators were common pieces of office equipment until well into the early 1970s, when the small hand-held electronic calculator, first introduced by Bomar Corporation and called the **Bomar Brain**, appeared. Monroe was also a leading pioneer of electronic calculators, but its machines were desktop, four-function models the size of cash registers.

In 1889, Herman Hollerith developed the punched card for storing data. Like Babbage, he too apparently borrowed the idea of a punched card from Jacquard. He also developed a mechanical machine—driven by one of the new electric motors—that counted, sorted, and collated information stored on punched cards. The idea of calculating by machinery intrigued the U.S. government so much that Hollerith was commissioned to use his punched-card system to store and tabulate information for the 1890 census.

In 1896, Hollerith formed the Tabulating Machine Company, which developed a line of machines that used punched cards for tabulation. After a number of mergers, the Tabulating Machine Company became the International Business Machines Corporation, now referred to more commonly as IBM, Inc. The punched cards used in computer systems are often called **Hollerith cards**, in honor of Herman Hollerith. The 12-bit code used on a punched card is called the **Hollerith code**.

Mechanical machines driven by electric motors continued to dominate the information processing world until the construction of the first electronic calculating machine in 1941. A German inventor named Konrad Zuse, who worked as an engineer for the Henschel Aircraft Company in Berlin, invented the first modern computer. In 1936 Zuse constructed a mechanical version of his system and later in 1939 he constructed his first electromechanical computer system, called the Z2. His Z3 calculating computer, as pictured in Figure 1–1, was probably used in aircraft and missile design during World War II for the German war effort. The Z3 was a relay logic machine that

FIGURE 1–1 The Z3 computer developed by Konrad Zuse uses a 5.33 hertz clocking frequency. (Photo courtesy of Horst Zuse, the son of Konrad.)

was clocked at 5.33 Hz (far slower than the latest multiple GHz microprocessors). Had Zuse been given adequate funding by the German government, he most likely would have developed a much more powerful computer system. Zuse is today finally receiving some belated honor for his pioneering work in the area of digital electronics and for his Z3 computer system.

It has recently been discovered (through the declassification of British military documents) that the first electronic computer was placed into operation in 1943 to break secret German military codes. This first electronic computing system, which used vacuum tubes, was invented by Alan Turing. Turing called his machine **Colossus**, probably because of its size. A problem with Colossus was that although its design allowed it to break secret German military codes generated by the mechanical **Enigma machine**, it could not solve other problems. Colossus was not programmable—it was a fixed-program computer system, which today is often called a **special-purpose computer**.

The first general-purpose, programmable electronic computer system was developed in 1946 at the University of Pennsylvania. This first modern computer was called the ENIAC (**Electronic Numerical Integrator and Calculator**). The ENIAC was a huge machine, containing over 17,000 vacuum tubes and over 500 miles of wires. This massive machine weighed over 30 tons, yet performed only about 100,000 operations per second. The ENIAC thrust the world into the age of electronic computers. The ENIAC was programmed by rewiring its circuits—a process that took many workers several days to accomplish. The workers changed the electrical connections on plug-boards that looked like early telephone switchboards. Another problem with the ENIAC was the life of the vacuum tube components, which required frequent maintenance.

Breakthroughs that followed were the development of the transistor on December 23, 1947 at Bell Labs by John Bardeen, William Shockley, and Walter Brattain. This was followed by the 1958 invention of the integrated circuit by Jack Kilby of Texas Instruments. The integrated circuit led to the development of digital integrated circuits (RTL, or resistor-to-transistor logic) in the 1960s and the first microprocessor at Intel Corporation in 1971. At that time, Intel engineers Federico Faggin, Ted Hoff, and Stan Mazor developed the 4004 microprocessor (U.S. Patent 3,821,715)—the device that started the microprocessor revolution that continues today at an ever-accelerating pace.

Programming Advancements

Now that programmable machines had been developed, programs and programming languages began to appear. As mentioned earlier, the first programmable electronic computer system was programmed by rewiring its circuits. Because this proved too cumbersome for practical application, early in the evolution of computer systems, computer languages began to appear in order to control the computer. The first such language, **machine language**, was constructed of ones and zeros using binary codes that were stored in the computer memory system as groups of instructions called a program. This was more efficient than rewiring a machine to program it, but it was still extremely time-consuming to develop a program because of the sheer number of codes that were required. Mathematician John von Neumann was the first modern person to develop a system that accepted instructions and stored them in memory. Computers are often called **von Neumann machines** in honor of John von Neumann. (Recall that Babbage also had developed the concept long before von Neumann.)

Once computer systems such as the UNIVAC became available in the early 1950s, **assembly language** was used to simplify the chore of entering binary code into a computer as its instructions. The assembler allows the programmer to use mnemonic codes, such as ADD for addition, in place of a binary number such as 0100 0111. Although assembly language was an aid to programming, it wasn't until 1957, when Grace Hopper developed the first high-level programming language, called **FLOWMATIC**, that computers became easier to program. In the same year, IBM developed

FORTRAN (**FORmula TRANslator**) for its computer systems. The FORTRAN language allowed programmers to develop programs that used formulas to solve mathematical problems. FORTRAN is still used by some scientists for computer programming. Another similar language, introduced about a year after FORTRAN, was ALGOL (**ALGOrithmic Language**).

The first truly successful and widespread programming language for business applications was COBOL (**COmputer Business Oriented Language**). Although COBOL usage has diminished somewhat in recent years, it is still a major player in many large business systems. Another once-popular business language is RPG (**Report Program Generator**), which allows programming by specifying the form of the input, output, and calculations.

Since these early days of programming, additional languages have appeared. Some of the more common are BASIC, Java, C#, C/C++, PASCAL, and ADA. The BASIC and PASCAL languages were both designed as teaching languages, but have escaped the classroom and are used in many computer systems. The BASIC language is probably the easiest of all to learn. Some estimates indicate that the BASIC language is used in the personal computer for 80% of the programs written by users. In the past decade, a new version of BASIC, Visual BASIC, has made programming in the Windows environment easier. The Visual BASIC language may eventually supplant C/C++ and PASCAL as a scientific language, but it is doubtful.

In the scientific community, C/C++ and occasionally PASCAL and FORTRAN appear as control programs. These languages, especially C/C++, allow the programmer almost complete control over the programming environment and computer system. In many cases, C/C++ is replacing some of the low-level machine control software or drivers normally reserved for assembly language. Even so, assembly language still plays an important role in programming. Most video games written for the personal computer are written almost exclusively in assembly language. Assembly language is also interspersed with C/C++ and PASCAL to perform machine control functions efficiently. Some of the newer parallel instructions (SIMD) found on the newest Pentium microprocessors are only programmable in assembly language.

The ADA language is used heavily by the Department of Defense. The ADA language was named in honor of Augusta Ada Byron, Countess of Lovelace. The countess worked with Charles Babbage in the early 1800s in the development of software for his Analytical Engine.

The Microprocessor Age

The world's first microprocessor, the Intel 4004, was a 4-bit microprocessor—a programmable controller on a chip. It addressed a mere 4096 4-bit-wide memory locations. (A **bit** is a binary digit with a value of one or zero. A 4-bit-wide memory location is often called a **nibble**.) The 4004 instruction set contained only 45 instructions. It was fabricated with the then-current state-of-the-art P-channel MOSFET technology that only allowed it to execute instructions at the slow rate of 50 KIPs (**kilo-instructions per second**). This was slow when compared to the 100,000 instructions executed per second by the 30-ton ENIAC computer in 1946. The main difference was that the 4004 weighed much less than an ounce.

At first, applications abounded for this device. The 4-bit microprocessor debuted in early video game systems and small microprocessor-based control systems. One such early video game, a shuffleboard game, was produced by Bailey. The main problems with this early microprocessor were its speed, word width, and memory size. The evolution of the 4-bit microprocessor ended when Intel released the 4040, an updated version of the earlier 4004. The 4040 operated at a higher speed, although it lacked improvements in word width and memory size. Other companies, particularly Texas Instruments (TMS-1000), also produced 4-bit microprocessors. The 4-bit microprocessor still survives in low-end applications such as microwave ovens and small control systems and is still available from some microprocessor manufacturers. Most calculators are still based on 4-bit microprocessors that process 4-bit BCD (**binary-coded decimal**) codes.

TABLE 1–1 Early 8-bit microprocessors.

Manufacturer	*Part Number*
Fairchild	F-8
Intel	8080
MOS Technology	6502
Motorola	MC6800
National Semiconductor	IMP-8
Rockwell International	PPS-8
Zilog	Z-8

Later in 1971, realizing that the microprocessor was a commercially viable product, Intel Corporation released the 8008—an extended 8-bit version of the 4004 microprocessor. The 8008 addressed an expanded memory size (16K bytes) and contained additional instructions (a total of 48) that provided an opportunity for its application in more advanced systems. (A **byte** is generally an 8-bit-wide binary number and a **K** is 1024. Often, memory size is specified in K bytes.)

As engineers developed more demanding uses for the 8008 microprocessor, they discovered that its somewhat small memory size, slow speed, and instruction set limited its usefulness. Intel recognized these limitations and introduced the 8080 microprocessor in 1973—the first of the modem 8-bit microprocessors. About six months after Intel released the 8080 microprocessor, Motorola Corporation introduced its MC6800 microprocessor. The floodgates opened and the 8080—and, to a lesser degree, the MC6800—ushered in the age of the microprocessor. Soon, other companies began to introduce their own versions of the 8-bit microprocessor. Table 1–1 lists several of these early microprocessors and their manufacturers. Of these early microprocessor producers, only Intel and Motorola (IBM also produces Motorola-style microprocessors) continue successfully to create newer and improved versions of the microprocessor. Zilog still manufactures microprocessors, but remains in the background, concentrating on microcontrollers and embedded controllers instead of general-purpose microprocessors. Rockwell has all but abandoned microprocessor development in favor of modem circuitry. Motorola has declined from having nearly a 50% share of the microprocessor market to a much smaller share.

What Was Special about the 8080? Not only could the 8080 address more memory and execute additional instructions, but it executed them 10 times faster than the 8008. An addition that took 20 μs (50,000 instructions per second) on an 8008-based system required only 2.0 μs (500,000 instructions per second) on an 8080-based system. Also, the 8080 was compatible with TTL (transistor-transistor logic), whereas the 8008 was not directly compatible. This made interfacing much easier and less expensive. The 8080 also addressed four times more memory (64K bytes) than the 8008 (16K bytes). These improvements are responsible for ushering in the era of the 8080 and the continuing saga of the microprocessor. Incidentally, the first personal computer, the MITS Altair 8800, was released in 1974. (Note that the number 8800 was probably chosen to avoid copyright violations with Intel.) The BASIC language interpreter, written for the Altair 8800 computer, was developed in 1975 by Bill Gates and Paul Allen, the founders of Microsoft Corporation. The assembler program for the Altair 8800 was written by Digital Research Corporation, which once produced DR-DOS for the personal computer.

The 8085 Microprocessor. In 1977, Intel Corporation introduced an updated version of the 8080—the 8085. The 8085 was to be the last 8-bit, general-purpose microprocessor developed by Intel. Although only slightly more advanced than an 8080 microprocessor, the 8085 executed software at an even higher speed. An addition that took 2.0 μs (500,000 instructions per second on the 8080) required only 1.3 μs (769,230 instructions per second) on the 8085. The main advantages of the 8085 were its internal clock generator, internal system controller, and higher clock frequency. This higher level of component integration reduced the 8085's cost and increased its usefulness. Intel has managed to sell well over 100 million copies of the 8085 microprocessor, its most successful 8-bit general-purpose microprocessor. Because the 8085 is also manufactured (second-sourced) by many other companies, there are over 200 million of these microprocessors in existence. Applications that contain the 8085 will likely continue to be popular. Another company that sold 500 million 8-bit microprocessors is Zilog Corporation, which produced the Z-80 microprocessor. The Z-80 is machine language–compatible with the 8085, which means that there are over 700 million microprocessors that execute 8085/Z-80 compatible code!

The Modern Microprocessor

In 1978, Intel released the 8086 microprocessor; a year or so later, it released the 8088. Both devices were 16-bit microprocessors, which executed instructions in as little as 400 ns (2.5 MIPs, or 2.5 **millions of instructions per second**). This represented a major improvement over the execution speed of the 8085. In addition, the 8086 and 8088 addressed 1M byte of memory, which was 16 times more memory than the 8085. (A **1M-byte memory** contains 1024K byte-sized memory locations or 1,048,576 bytes.) This higher execution speed and larger memory size allowed the 8086 and 8088 to replace smaller minicomputers in many applications. One other feature found in the 8086/8088 was a small 4- or 6-byte instruction cache or queue that prefetched a few instructions before they were executed. The queue sped the operation of many sequences of instructions and proved to be the basis for the much larger instruction caches found in modern microprocessors.

The increased memory size and additional instructions in the 8086 and 8088 have led to many sophisticated applications for microprocessors. Improvements to the instruction set included multiply and divide instructions, which were missing on earlier microprocessors. In addition, the number of instructions increased from 45 on the 4004, to 246 on the 8085, to well over 20,000 variations on the 8086 and 8088 microprocessors. Note that these microprocessors are called CISC (**complex instruction set computers**) because of the number and complexity of instructions. The additional instructions eased the task of developing efficient and sophisticated applications, even though the number of instructions are at first overwhelming and time-consuming to learn. The 16-bit microprocessor also provided more internal register storage space than the 8-bit microprocessor. The additional registers allowed software to be written more efficiently.

The 16-bit microprocessor evolved mainly because of the need for larger memory systems. The popularity of the Intel family was ensured in 1981, when IBM Corporation decided to use the 8088 microprocessor in its personal computer. Applications such as spreadsheets, word processors, spelling checkers, and computer-based thesauruses were memory-intensive and required more than the 64K bytes of memory found in 8-bit microprocessors to execute efficiently. The 16-bit 8086 and 8088 provided 1M byte of memory for these applications. Soon, even the 1M-byte memory system proved limiting for large databases and other applications. This led Intel to introduce the 80286 microprocessor, an updated 8086, in 1983.

The 80286 Microprocessor. The 80286 microprocessor (also a 16-bit architecture microprocessor) was almost identical to the 8086 and 8088, except it addressed a 16M-byte memory system instead of a 1M-byte system, and it had a few additional instructions that managed the extra 15M bytes of memory. The clock speed of the 80286 was increased, so it executed some instructions in as little as 250 ns (4.0 MIPs) with the original release 8.0 MHz version. Some changes also occurred to the internal execution of the instructions, which led to an eightfold increase in speed for many instructions when compared to 8086/8088 instructions.

The 32-Bit Microprocessor. Applications began to demand faster microprocessor speeds, more memory, and wider data paths. This led to the 1986 arrival of the 80386 by Intel Corporation. The 80386 represented a major overhaul of the 16-bit 8086–80286 architecture. The 80386 was Intel's first practical 32-bit microprocessor that contained a 32-bit data bus and a 32-bit memory address. (Note that Intel produced an earlier, although unsuccessful, 32-bit microprocessor called the iapx-432.) Through these 32-bit buses, the 80386 addressed up to 4G bytes of memory. (**1G** of memory contains 1024M, or 1,073,741,824 locations.) A 4G-byte memory can store an astounding 1,000,000 typewritten, double-spaced pages of ASCII text data. The 80386 was available in a few modified versions such as the 80386SX, which addressed 16M bytes of memory through a 16-bit data and 24-bit address bus, and the 80386SL/80386SLC, which addressed 32M bytes of memory through a 16-bit data and 25-bit address bus. An 80386SLC version contained an internal cache memory that allowed it to process data at even higher rates. In 1995, Intel released the 80386EX microprocessor. The 80386EX microprocessor is called an **embedded PC** because it contains all the components of the AT class personal computer on a single integrated circuit. The 80386EX also contains 24 lines for input/output data, a 26-bit address bus, a 16-bit data bus, a DRAM refresh controller, and programmable chip selection logic.

Applications that require higher microprocessor speeds and large memory systems include software systems that use a GUI, or **graphical user interface**. Modern graphical displays often contain 256,000 or more picture elements (**pixels**, or **pels**). The least sophisticated VGA (**variable graphics array**) video display has a resolution of 640 pixels per scanning line with 480 scanning lines (this is the resolution used when the computer boots and displays the boot screen). To display one screen of information, each picture element must be changed, which requires a high-speed microprocessor. Virtually all new software packages use this type of video interface. These GUI-based packages require high microprocessor speeds and accelerated video adapters for quick and efficient manipulation of video text and graphical data. The most striking system, which requires high-speed computing for its graphical display interface, is Microsoft Corporation's Windows.[3] We often call a GUI a WYSIWYG (**what you see is what you get**) display.

The 32-bit microprocessor is needed because of the size of its data bus, which transfers real (single-precision floating-point) numbers that require 32-bit-wide memory. In order to efficiently process 32-bit real numbers, the microprocessor must efficiently pass them between itself and memory. If the numbers pass through an 8-bit data bus, it takes four read or write cycles; when passed through a 32-bit data bus, however, only one read or write cycle is required. This significantly increases the speed of any program that manipulates real numbers. Most high-level languages, spreadsheets, and database management systems use real numbers for data storage. Real numbers are also used in graphical design packages that use vectors to plot images on the video screen. These include such CAD (**computer-aided drafting/design**) systems as AUTOCAD, ORCAD, and so forth.

Besides providing higher clocking speeds, the 80386 included a memory management unit that allowed memory resources to be allocated and managed by the operating system. Earlier microprocessors left memory management completely to the software. The 80386 included hardware circuitry for memory management and memory assignment, which improved its efficiency and reduced software overhead.

The instruction set of the 80386 microprocessor was upward-compatible with the earlier 8086, 8088, and 80286 microprocessors. Additional instructions referenced the 32-bit registers and managed the memory system. Note that memory management instructions and techniques used by the 80286 are also compatible with the 80386 microprocessor. These features allowed older, 16-bit software to operate on the 80386 microprocessor.

The 80486 Microprocessor. In 1989, Intel released the 80486 microprocessor, which incorporated an 80386-like microprocessor, an 80387-like numeric coprocessor, and an 8K-byte cache memory system into one integrated package. Although the 80486 microprocessor was not radically different from the 80386, it did include one substantial change. The internal structure of the 80486 was modified from the 80386 so that about half of its instructions executed in one clock instead of two clocks. Because the 80486 was available in a 50 MHz version, about half of the instructions executed in 25 ns (50 MIPs). The average speed improvement for a typical mix of instructions was about 50% over the 80386 that operated at the same clock speed. Later versions

[3]Windows is a registered trademark of Microsoft Corporation and is currently available as Windows 98, Windows 2000, Windows ME, and Windows XP.

of the 80486 executed instructions at even higher speeds with a 66 MHz double-clocked version (80486DX2). The double-clocked 66 MHz version executed instructions at the rate of 66 MHz, with memory transfers executing at the rate of 33 MHz. (This is why it was called a double-clocked microprocessor.) A triple-clocked version from Intel, the 80486DX4, improved the internal execution speed to 100 MHz with memory transfers at 33 MHz. Note that the 80486DX4 microprocessor executed instructions at about the same speed as the 60 MHz Pentium. It also contained an expanded 16K-byte cache in place of the standard 8K-byte cache found on earlier 80486 microprocessors. Advanced Micro Devices (AMD) has produced a triple-clocked version that runs with a bus speed of 40 MHz and a clock speed of 120 MHz. The future promises to bring microprocessors that internally execute instructions at rates of up to 10 GHz or higher.

Other versions of the 80486 were called OverDrive[4] processors. The OverDrive processor was actually a double-clocked version of the 80486DX that replaced an 80486SX or slower-speed 80486DX. When the OverDrive processor was plugged into its socket, it disabled or replaced the 80486SX or 80486DX, and functioned as a doubled-clocked version of the microprocessor. For example, if an 80486SX, operating at 25 MHz, was replaced with an OverDrive microprocessor, it functioned as an 80486DX2 50 MHz microprocessor using a memory transfer rate of 25 MHz.

Table 1–2 lists many microprocessors produced by Intel and Motorola with information about their word and memory sizes. Other companies produce microprocessors, but none have attained the success of Intel and, to a lesser degree, Motorola.

The Pentium Microprocessor. The Pentium, introduced in 1993, was similar to the 80386 and 80486 microprocessors. This microprocessor was originally labeled the P5 or 80586, but Intel decided not to use a number because it appeared to be impossible to copyright a number. The two introductory versions of the Pentium operated with a clocking frequency of 60 MHz and 66 MHz, and a speed of 110 MIPs, with a higher-frequency 100 MHz one and one-half clocked version that operated at 150 MIPs. The double-clocked Pentium, operating at 120 MHz and 133 MHz, was also available, as were higher-speed versions. (The fastest version produced by Intel is the 233 MHz Pentium, which is a three and one-half clocked version.) Another difference was that the cache size was increased to 16K bytes from the 8K cache found in the basic version of the 80486. The Pentium contained an 8K-byte instruction cache and an 8K-byte data cache, which allowed a program that transfers a large amount of memory data to still benefit from a cache. The memory system contained up to 4G bytes, with the data bus width increased from the 32 bits found in the 80386 and 80486 to a full 64 bits. The data bus transfer speed was either 60 MHz or 66 MHz, depending on the version of the Pentium. (Recall that the bus speed of the 80486 was 33 MHz.) This wider data bus width accommodated double-precision floating-point numbers used for modern high-speed, vector-generated graphical displays. These higher bus speeds should allow virtual reality software and video to operate at more realistic rates on current and future Pentium-based platforms. The widened data bus and higher execution speed of the Pentium allow full-frame video displays to operate at scan rates of 30 Hz or higher—comparable to commercial television. Recent versions of the Pentium also included additional instructions, called multimedia extensions, or MMX instructions. Although Intel hoped that the MMX instructions would be widely used, it appears that few software companies have used them. The main reason is that there is no high-level language support for these instructions.

Intel had also released the long-awaited Pentium OverDrive (P24T) for older 80486 systems that operate at either 63 MHz or 83 MHz clock. The 63 MHz version upgrades older 80486DX2 50 MHz systems; the 83 MHz version upgrades the 80486DX2 66 MHz systems. The upgraded 83 MHz system performs at a rate somewhere between a 66 MHz Pentium and a 75 MHz Pentium. If older VESA local bus video and disk-caching controllers seem too expensive to toss out, the Pentium OverDrive represents an ideal upgrade path from the 80486 to the Pentium.

---

[4]OverDrive is a registered trademark of Intel Corporation.

TABLE 1–2 Many modern Intel and Motorola microprocessors.

Manufacturer	*Part Number*	*Data Bus Width*	*Memory Size*
Intel	8048	8	2K internal
	8051	8	8K internal
	8085A	8	64K
	8086	16	1M
	8088	8	1M
	8096	16	8K internal
	80186	16	1M
	80188	8	1M
	80251	8	16K internal
	80286	16	16M
	80386EX	16	64M
	80386DX	32	4G
	80386SL	16	32M
	80386SLC	16	32M + 8K cache
	80386SX	16	16M
	80486DX/DX2	32	4G + 8K cache
	80486SX	32	4G + 8K cache
	80486DX4	32	4G + 16 cache
	Pentium	64	4G + 16K cache
	Pentium OverDrive	32	4G + 16K cache
	Pentium Pro	64	64G + 16K L1 cache + 256K L2 cache
	Pentium II	64	64G + 32K L1 cache + 256K L2 cache
	Pentium III	64	64G + 32K L1 cache + 256K L2 cache
	Pentium 4	64	64G + 8K L1 cache + 512K L2 cache (or larger)
Motorola	6800	8	64K
	6805	8	2K
	6809	8	64K
	68000	16	16M
	68008D	8	4M
	68008Q	8	1M
	68010	16	16M
	68020	32	4G
	68030	32	4G + 256 cache
	68040	32	4G + 8K cache
	68050	32	Proposed, but never released
	68060	64	4G + 16K cache
	PowerPC	64	4G + 32K cache

Probably the most ingenious feature of the Pentium is its dual integer processors. The Pentium executes two instructions, which are not dependent on each other, simultaneously because it contains two independent internal integer processors called superscaler technology. This allows the Pentium to be able to execute two instructions per clocking period. Another feature that enhances performance is a jump prediction technology that speeds the execution of program

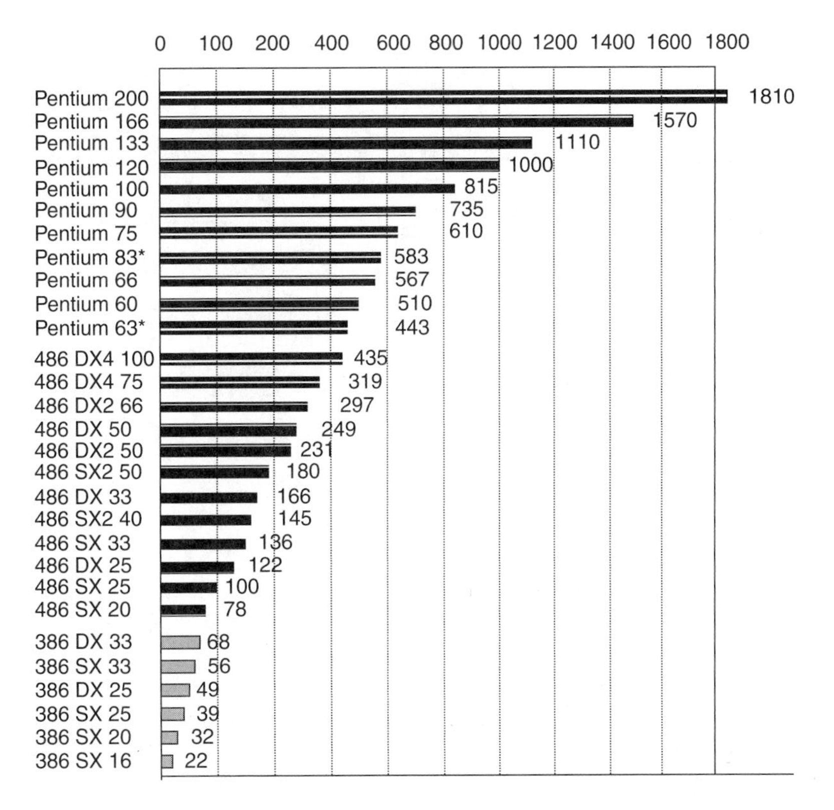

FIGURE 1–2 The Intel iCOMP index.

loops. As with the 80486, the Pentium also employs an internal floating-point coprocessor to handle floating-point data, albeit at five times the speed improvement. These features portend continued success for the Intel family of microprocessors. Intel also may allow the Pentium to replace some of the RISC (**reduced instruction set computer**) machines that currently execute one instruction per clock. Note that some newer RISC processors execute more than one instruction per clock through the introduction of superscaler technology. Motorola, Apple, and IBM produce the PowerPC, a RISC microprocessor that has two integer units and a floating-point unit. The PowerPC certainly boosts the performance of the Apple Macintosh, but at present is slow to efficiently emulate the Intel family of microprocessors. Tests indicate that the current emulation software executes DOS and Windows applications at speeds slower than the 80486DX 25 MHz microprocessor. Because of this, the Intel family should survive for many years in personal computer systems. Note that there are currently 6 million Apple Macintosh[5] systems and well over 260 million personal computers based on Intel microprocessors. In 1998, reports stated that 96% of all PCs were shipped with the Windows operating system.

In order to compare the speeds of various microprocessors, Intel devised the iCOMP-rating index. This index is a composite of SPEC92, ZD Bench, and Power Meter. The iCOMP1 rating index is used to rate the speed of all Intel microprocessors through the Pentium. Figure 1–2 shows relative speeds of the 80386DX 25 MHz version at the low end to the Pentium 233 MHz version at the high end of the spectrum.

Since the release of the Pentium Pro and Pentium II, Intel has switched to the iCOMP2 index, which is scaled by a factor of 10 from the iCOMP1 index. A microprocessor with an index

---

[5]Macintosh is a registered trademark of Apple Computer Corporation.

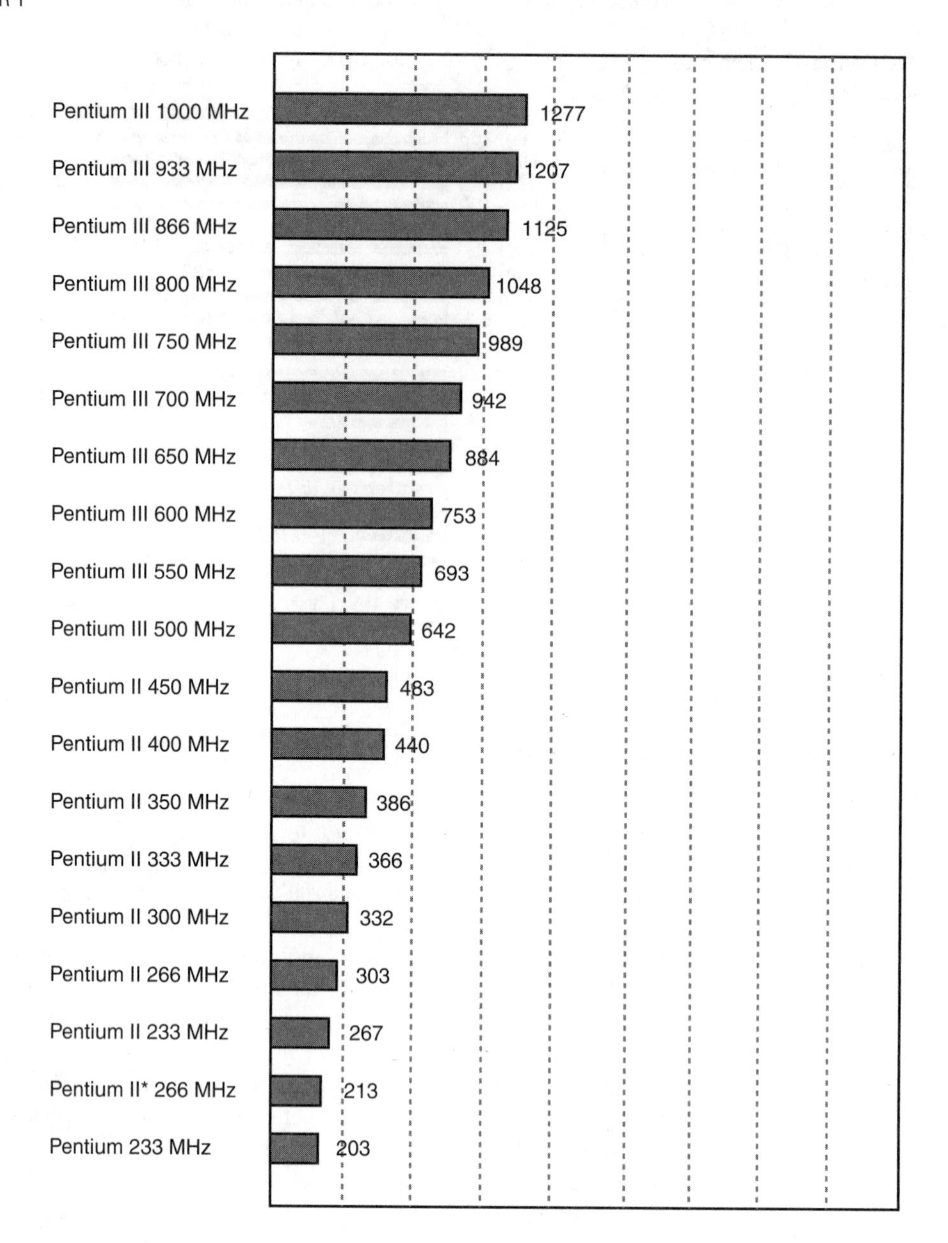

FIGURE 1–3 The Intel iCOMP2 index.

of 1000 using iCOMP1 is rated as 100 using iCOMP2. Another difference is the benchmarks used for the scores. Figure 1–3 shows the iCOMP2 index listing the Pentium III at speeds up to 1000 MHz. Figure 1–4 shows SYSmark 2002 for the Pentium III and Pentium 4.

Pentium Pro Processor. A recent entry from Intel is the Pentium Pro processor, formerly named the P6 microprocessor. The Pentium Pro processor contains 21 million transistors, three integer units, and a floating-point unit to increase the performance of most software. The basic clock frequency was 150 MHz and 166 MHz in the initial offering made available in late 1995. In addition to the internal 16K level-one (L1) cache (8K for data and 8K for instructions), the Pentium Pro processor also contains a 256K level-two (L2) cache. One other significant change is that the Pentium Pro processor uses three execution engines, so it can execute up to

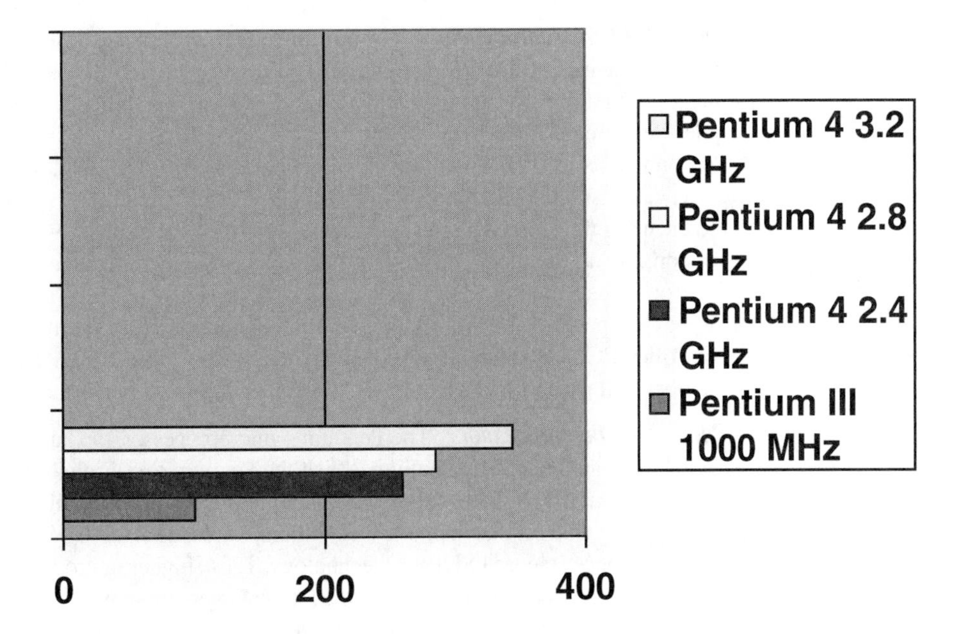

FIGURE 1–4 Intel microprocessor performance using SYSmark 2002.

three instructions at a time, which can conflict and still execute in parallel. This represents a change from the Pentium, which executes two instructions simultaneously as long as they do not conflict. The Pentium Pro microprocessor has been optimized to efficiently execute 32-bit code; for this reason, it was often bundled with Windows NT rather than with normal versions of Windows 95. Intel launched the Pentium Pro processor for the server market. Still another change is that the Pentium Pro can address either a 4G-byte memory system or a 64G-byte memory system. The Pentium Pro has a 36-bit address bus if configured for a 64G memory system.

Pentium II and Xeon Microprocessors. The Pentium II microprocessor (released in 1997) represents a new direction for Intel. Instead of being an integrated circuit as with prior versions of the microprocessor, the Pentium II is placed on a small circuit board. The main reason for the change is that the L2 cache found on the main circuit board of the Pentium was not fast enough to function properly with the Pentium II. On the Pentium system, the L2 cache operates at the system bus speed of 60 MHz or 66 MHz. The L2 cache and microprocessor are on a circuit board called the **Pentium II module**. This onboard L2 cache operates at a speed of 133 MHz and stores 512K bytes of information. The microprocessor on the Pentium II module is actually a Pentium Pro with MMX extensions.

In 1998, Intel changed the bus speed of the Pentium II. Because the 266 MHz through the 333 MHz Pentium II microprocessors used an external bus speed of 66 MHz, there was a bottleneck, so the newer Pentium II microprocessors use a 100 MHz bus speed. The Pentium II microprocessors rated at 350 MHz, 400 MHz, and 450 MHz all use this higher 100 MHz memory bus speed. The higher speed memory bus requires the use of 8 ns SDRAM in place of the 10 ns SDRAM found in use with the 66 MHz bus speed.

In mid-1998 Intel announced a new version of the Pentium II called the Xeon,[6] which was specifically designed for high-end workstation and server applications. The main difference between the Pentium II and the Pentium II Xeon is that the Xeon is available with an L1 cache size of 32K bytes and an L2 cache size of either 512K, 1M, or 2M bytes. The Xeon functions with the 440GX chip set. The Xeon is also designed to function with four Xeons in the same system, which is similar to the Pentium Pro. This newer product represents a change in Intel's strategy: Intel now produces a professional version and a home/business version of the Pentium II microprocessor.

[6]Xeon is a registered trademark of Intel Corporation.

Pentium III Microprocessor. The Pentium III microprocessor uses a faster core than the Pentium II, but it is still a P6 or Pentium Pro processor. It is also available in the slot 1 version mounted on a plastic cartridge and a socket 370 version called a flip-chip, which looks like the older Pentium package. Intel claims the flip-chip version costs less. Another difference is that the Pentium III is available with clock frequencies of up to 1 GHz. The slot 1 version contains a 512K cache and the flip-chip version contains a 256K cache. The speeds are comparable because the cache in the slot 1 version runs at one-half the clock speed, while the cache in the flip-chip version runs at the clock speed. Both versions use a memory bus speed of 100 MHz, while the Celeron[7] uses a memory bus clock speed of 66 MHz.

The speed of the front side bus, the connection from the microprocessor to the memory controller, PCI controller, and AGP controller, is now either 100 MHz or 133 MHz. Although the memory still runs at 100 MHz, this change has improved performance.

Pentium 4 Microprocessor. The Pentium 4 microprocessor, the latest Intel release, was first made available in late 2000. The Pentium 4, like the Pentium Pro through the Pentium III, uses the Intel P-6 architecture. The main difference is that the Pentium 4 is available in speeds to 3.2 GHz and faster and the chip set that supports the Pentium 4 uses the RAMBUS or DDR memory technologies in place of once standard SDRAM technology. These higher microprocessor speeds are made available by an improvement in the size of the internal integration, which at present is 0.095 micron technology. It is also interesting to note that Intel has changed the level-one cache size from 32K to 8K bytes. Research must have shown that this size is large enough for the initial release version of the microprocessor, with future versions possibly containing a 32K L1 cache. The L2 cache remains at 256K bytes as in the Pentium coppermine version with the latest versions containing a 512K cache. The Pentium 4 Extreme Edition contains a 2M L2 cache and the Pentium 4e contains a 1M L2 cache.

Another change likely to occur is a shift from aluminum to copper interconnections inside the microprocessor. Because copper is a better conductor, it should allow increased clock frequencies for the microprocessor in the future. This is especially true now that a method for using copper has surfaced at IBM Corporation. Another event to look for is a change in the speed of the front side bus, which will likely increase beyond the current maximum 200 MHz.

Table 1–3 shows the various Intel P numbers and the microprocessors that belong to each class. The P versions show what internal core microprocessor is found in each of the Intel microprocessors. Notice that all of the microprocessors since the Pentium Pro use the same basic microprocessor core.

The Future of Microprocessors. No one can really make accurate predictions, but the success of the Intel family should continue for quite a few years. What may occur is a change to RISC technology, but more likely are improvements to a new joint technology by Intel and Hewlett-Packard called hyper-threading technology. Even this new technology embodies the CISC

TABLE 1–3 Intel microprocessor core (P) versions.

Core (P) Version	*Microprocessor*
P1	8086, 8088, 80186, and 80188
P2	80286
P3	80386
P4	80486
P5	Pentium
P6	Pentium Pro, Pentium II, Pentium III, and Pentium 4

[7]Celeron is a trademark of Intel Corporation.

instruction set of the 80X86 family of microprocessors, so that software for the system will survive. The basic premise behind this technology is that many microprocessors communicate directly with each other, allowing parallel processing without any change to the instruction set or program. Currently, the superscaler technology uses many microprocessors, but they all share the same register set. This new technology contains many microprocessors, each containing its own register set that is linked with the other microprocessors' registers. This technology offers true parallel processing without writing any special program.

The hyper-threading technology should continue into the future, bringing even more parallel processors (at present two processors). There are suggestions that Intel may also incorporate the chip set into the microprocessor package.

In 2002, Intel released a new microprocessor architecture that is 64 bits in width and has a 128-bit data bus. This new architecture, named the Itanium,[8] is a joint venture called EPIC (Explicitly Parallel Instruction Computing) by Intel and Hewlett-Packard. The Itanium architecture allows greater parallelism than traditional architectures, such as the Pentium III or Pentium 4. These changes include 128 general-purpose integer registers, 128 floating-point registers, 64 predicate registers, and many execution units to ensure enough hardware resources for software. The Itanium is designed for the server market and may or may not trickle down to the home/business market in the future.

Figure 1–5 is a conceptual view, comparing the 80486 through Pentium 4 microprocessors. Each view shows the internal structure of these microprocessors: the CPU, coprocessor, and cache memory. This illustration shows the complexity and level of integration in each version of the microprocessor.

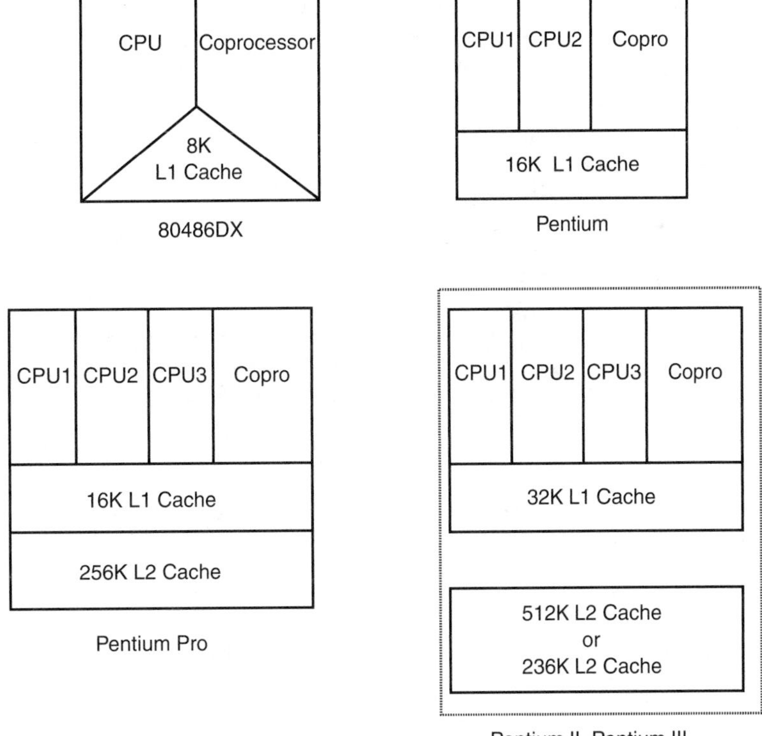

FIGURE 1–5 Conceptual views of the 80486, Pentium Pro, Pentium II, Pentium III, and Pentium 4 microprocessors.

[8]Itanium is a trademark of Intel Corporation.

1–2 THE MICROPROCESSOR-BASED PERSONAL COMPUTER SYSTEM

Computer systems have undergone many changes recently. Machines that once filled large areas have been reduced to small desktop computer systems because of the microprocessor. Although these desktop computers are compact, they possess computing power that was only dreamed of a few years ago. Million-dollar mainframe computer systems, developed in the early 1980s, are not as powerful as the Pentium 4–based computers of today. In fact, many smaller companies have replaced their mainframe computers with microprocessor-based systems. Companies such as DEC (Digital Equipment Corporation, now owned by Hewlett-Packard Company) have stopped producing mainframe computer systems in order to concentrate their resources on microprocessor-based computer systems.

This section shows the structure of the microprocessor-based personal computer system. This structure includes information about the memory and operating system used in many microprocessor-based computer systems.

See Figure 1–6 for the block diagram of the personal computer. This diagram also applies to any computer system, from the early mainframe computers to the latest microprocessor-based systems. The block diagram is composed of three blocks that are interconnected by buses. (A **bus** is set of common connections that carry the same type of information. For example, the address bus, which contains 20 or more connections, conveys the memory address to the memory.) These blocks and their function in a personal computer are outlined in this section of the text.

The Memory and I/O System

The memory structures of all Intel-based personal computers are similar. This includes the first personal computers based on the 8088, introduced in 1981 by IBM, to the most powerful high-speed versions of today, based on the Pentium 4. Figure 1–7 illustrates the memory map of a personal computer system. This map applies to any IBM personal computer or to any of the many IBM-compatible clones that are in existence.

The memory system is divided into three main parts: TPA (**transient program area**), system area, and XMS (**extended memory system**). The type of microprocessor in your computer

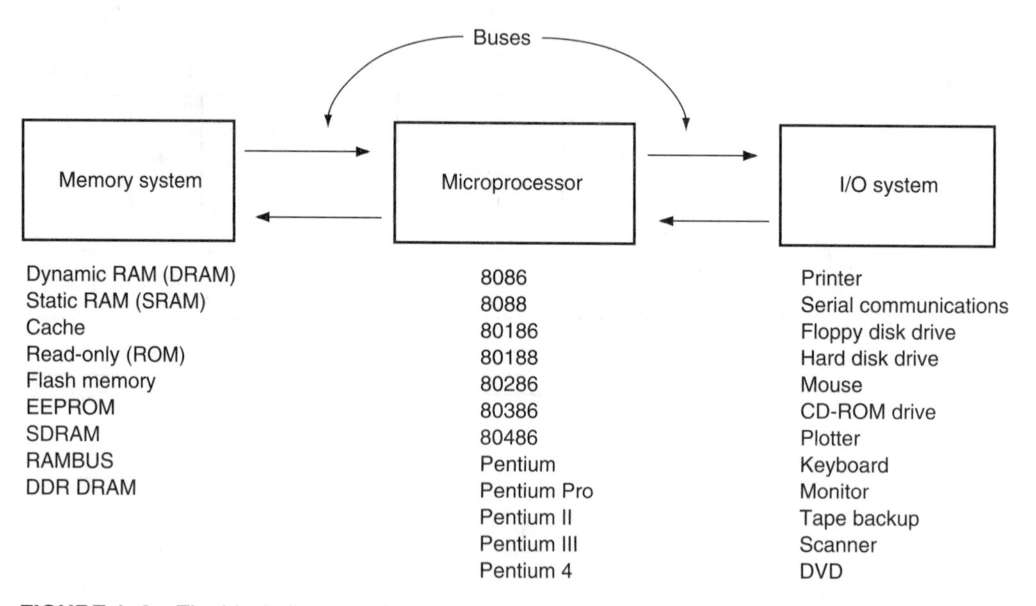

FIGURE 1–6 The block diagram of a microprocessor-based computer system.

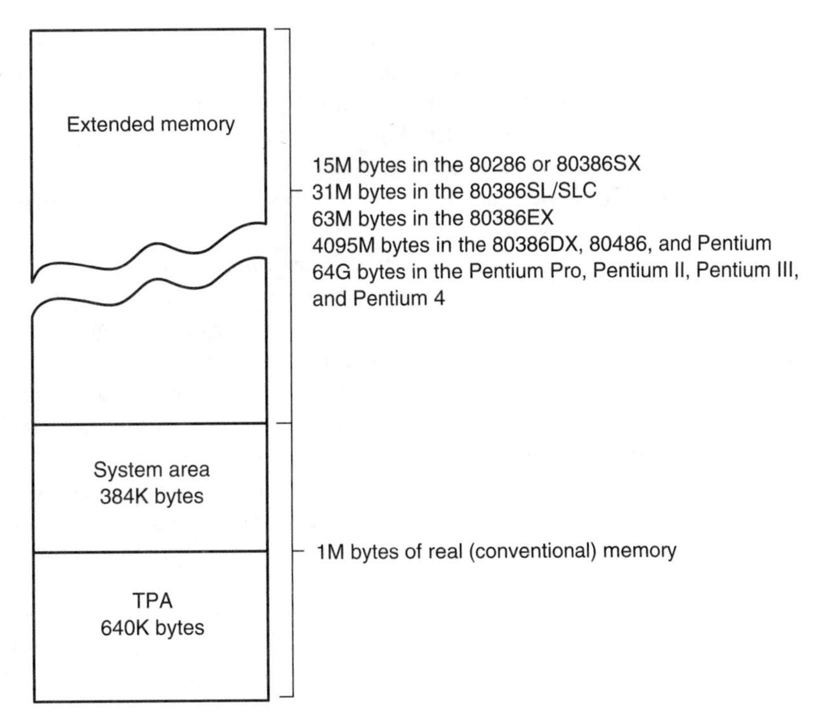

FIGURE 1–7 The memory map of the personal computer.

determines whether an extended memory system exists. If the computer is based upon a really old 8086 or 8088 (a PC or XT), the TPA and systems area exist, but there is no extended memory area. The PC and XT computers contain 640K bytes of TPA and 384K bytes of system memory, for a total memory size of 1M bytes. We often call the first 1M byte of memory the real or conventional memory system because each Intel microprocessor is designed to function in this area by using its real mode of operation.

Computer systems based on the 80286 through the Pentium 4 not only contain the TPA (640K bytes) and **system area** (384K bytes), they also contain extended memory. These machines are often called AT class machines. The PS/l and PS/2, produced by IBM, are other versions of the same basic memory design. Sometimes, these machines are also referred to as ISA (**industry standard architecture**) or EISA (**extended ISA**) machines. The PS/2 is referred to as a micro-channel architecture system, or ISA system, depending on the model number.

A change beginning with the introduction of the Pentium microprocessor and the ATX class machine is the addition of a bus called the PCI (**peripheral component interconnect**) bus, now being used in all Pentium–Pentium 4 systems. Extended memory contains up to 15M bytes in the 80286 and 80386SX-based computers, and up to 4095M bytes in the 80386DX, 80486, and Pentium microprocessors, in addition to the first 1M byte of real or conventional memory. The Pentium Pro through Pentium 4 computer systems have up to 1M less than 4G or 1M less than 64G of extended memory. Servers tend to use the larger 64G memory map, while home/business computers use the 4G-byte memory map. The ISA machine contains an 8-bit peripheral bus that is used to interface 8-bit devices to the computer in the 8086/8088-based PC or XT computer system. The AT class machine, also called an ISA machine, uses a 16-bit peripheral bus for interface and may contain the 80286 or above microprocessor. The EISA bus is a 32-bit peripheral interface bus found in a few older 80386DX- and 80486-based systems. Note that each of these buses is compatible with the earlier versions. That is, the 8-bit interface card functions in the 8-bit ISA, 16-bit ISA, or 32-bit EISA bus system. Likewise, a 16-bit interface card functions in the 16-bit ISA or 32-bit EISA system.

Another bus type found in many 80486-based personal computers is called the **VESA** local bus, or VL bus. The **local bus** interfaces disk and video to the microprocessor at the local bus level, which allows 32-bit interfaces to function at the same clocking speed as the microprocessor. A recent modification to the VESA local bus supports the 64-bit data bus of the Pentium microprocessor and competes directly with the PCI bus, although it has generated little, if any, interest. The ISA and EISA standards function at only 8 MHz, which reduces the performance of the disk and video interfaces using these standards. The PCI bus is either a 32- or 64-bit bus that is specifically designed to function with the Pentium through Pentium 4 microprocessors at a bus speed of 33 MHz.

Three newer buses have appeared in ATX class systems. The first to appear was the USB (**universal serial bus**). The universal serial bus is intended to connect peripheral devices such as keyboards, a mouse, modems, and sound cards to the microprocessor through a serial data path and a twisted pair of wires. The main idea is to reduce system cost by reducing the number of wires. Another advantage is that the sound system can have a separate power supply from the PC, which means much less noise. The data transfer rates through the USB are 10 Mbps at present for USB1; they increase to 480 Mbps in USB2.

The second newer bus is the AGP (**advanced graphics port**) for video cards. The advanced graphics port transfers data between the video card and the microprocessor at higher speeds (66 MHz, with a 64-bit data path, or 533M bytes per second) than were possible through any other bus or connection. The latest AGP speed is 8X or 2G bytes per second. This video subsystem change has been made to accommodate the new DVD players for the PC.

The latest new bus to appear is the serial ATA interface (**SATA**) for hard disk drives, which transfers data from the PC to the hard disk drive at rates of 150M bytes per second. The serial ATA standard will eventually reach speeds of 450M bytes per second.

The TPA. The transient program area (TPA) holds the DOS (**disk operating system**) operating system and other programs that control the computer system. The TPA is a DOS concept and not really applicable in Windows. The TPA also stores any currently active or inactive DOS application programs. The length of the TPA is 640K bytes. As mentioned, this area of memory holds the DOS operating system, which requires a portion of the TPA to function. In practice, the amount of memory remaining for application software is about 628K bytes if MSDOS[9] version 7.x is used as an operating system. Earlier versions of DOS required more of the TPA area and often left only 530K bytes or less for application programs. Figure 1–8 shows the organization of the TPA in a computer system running DOS.

The DOS memory map shows how the many areas of the TPA are used for system programs, data, and drivers. It also shows a large area of memory available for application programs. To the left of each area is a hexadecimal number that represents the memory addresses that begin and end each data area. Hexadecimal **memory addresses** or **memory locations** are used to number each byte of the memory system. (A **hexadecimal number** is a number represented in radix 16 or base 16, with each digit representing a value from 0 to 9 and A to F. We often end a hexadecimal number with an H to indicate that it is a hexadecimal value. For example, 1234H is 1234 hexadecimal. We also represent hexadecimal data as 0x1234 for a 1234 hexadecimal.)

The Interrupt vectors access various features of the DOS, BIOS (**basic I/O system**), and applications. The system BIOS is a collection of programs stored in either a read-only (ROM) or flash memory that operates many of the I/O devices connected to your computer system. The system BIOS and DOS communications areas contain transient data used by programs to access I/O devices and the internal features of the computer system. These are stored in the TPA so they can be changed as DOS operates.

The IO.SYS is a program that loads into the TPA from the disk whenever an MSDOS system is started. The IO.SYS contains programs that allow DOS to use the keyboard, video

[9]MSDOS is a trademark of Microsoft Corporation and version 7.x is supplied with Windows XP.

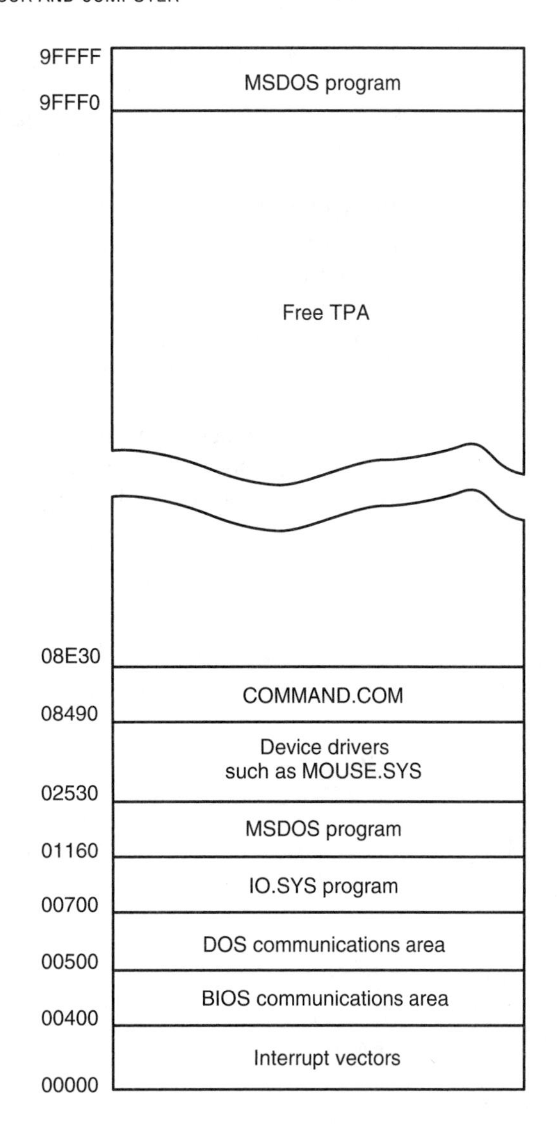

FIGURE 1–8 The memory map of the TPA in a personal computer. (Note that this map will vary between systems.)

display, printer, and other I/O devices often found in the computer system. The IO.SYS program links DOS to the programs stored on the system BIOS ROM.

The size of the driver area and number of drivers changes from one computer to another. **Drivers** are programs that control installable I/O devices such as a mouse, disk cache, hand scanner, CD-ROM memory (**Compact Disk Read-Only Memory**), DVD (**Digital Versatile Disk**), or installable devices, as well as programs. Installable drivers are programs that control or drive devices or programs that are added to the computer system. DOS drivers are normally files that have an extension of .SYS, such as MOUSE.SYS; in DOS version 3.2 and later, the files have an extension of .EXE, such as EMM386.EXE. Note that even though these files are not used by Windows, they are still used to execute DOS applications, even with Windows XP. Windows uses a file called SYSTEM.INI to load drivers used by Windows. In newer versions of Windows such as Windows XP, a registry is added to contain information about the system and the drivers used by the system. You can view the registry with the REGEDIT program.

The COMMAND.COM program (**command processor**) controls the operation of the computer from the keyboard when operated in the DOS mode. The COMMAND.COM program processes the DOS commands as they are typed from the keyboard. For example, if DIR is typed, the COMMAND.COM program displays a directory of the disk files in the current disk directory. If the COMMAND.COM program is erased, the computer cannot be used from the keyboard in DOS mode. Never erase COMMAND.COM, IO.SYS, or MSDOS.SYS to make room for other software, or your computer will not function.

The System Area. The DOS system area, although smaller than the TPA, is just as important. The **system area** contains programs on either a read-only memory (ROM) or flash memory, and areas of read/write (RAM) memory for data storage. Figure 1–9 shows the system area of a typical computer system. As with the map of the TPA, this map also includes the hexadecimal memory addresses of the various areas.

The first area of the system space contains video display RAM and video control programs on ROM or flash memory. This area starts at location A0000H and extends to location C7FFFH. The size and amount of memory used depends on the type of video display adapter attached to the system. Display adapters generally have their video RAM located at A0000H–AFFFFH, which stores graphical or bit-mapped data, and the memory at B0000H–BFFFFH stores text data. The video BIOS, located on a ROM or flash memory, is at locations C0000H–C7FFFH and contains programs that control the DOS video display.

The area at locations C8000H–DFFFFH is often open or free. This area is used for the expanded memory system (EMS) in a PC or XT system, or for the upper memory system in an AT system. Its use depends on the system and its configuration. The expanded memory system allows a 64K-byte page frame of memory to be used by application programs. This 64K-byte page

FIGURE 1–9 The system area of a typical personal computer.

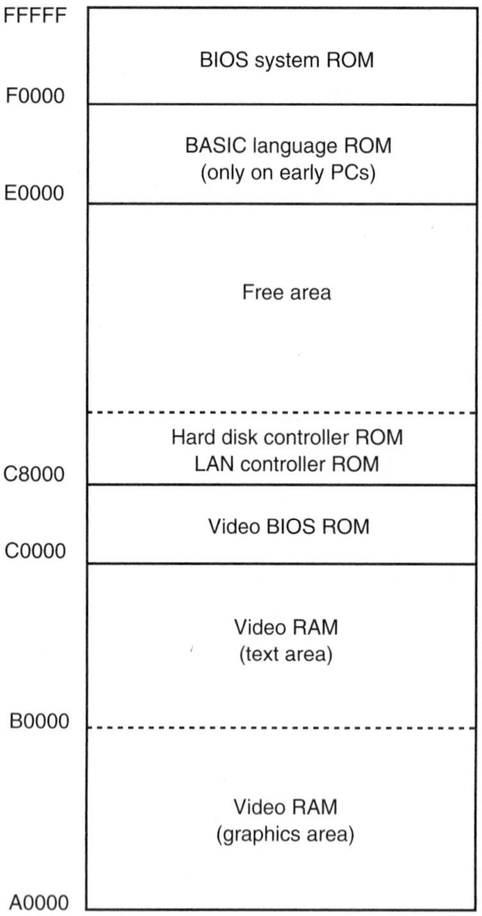

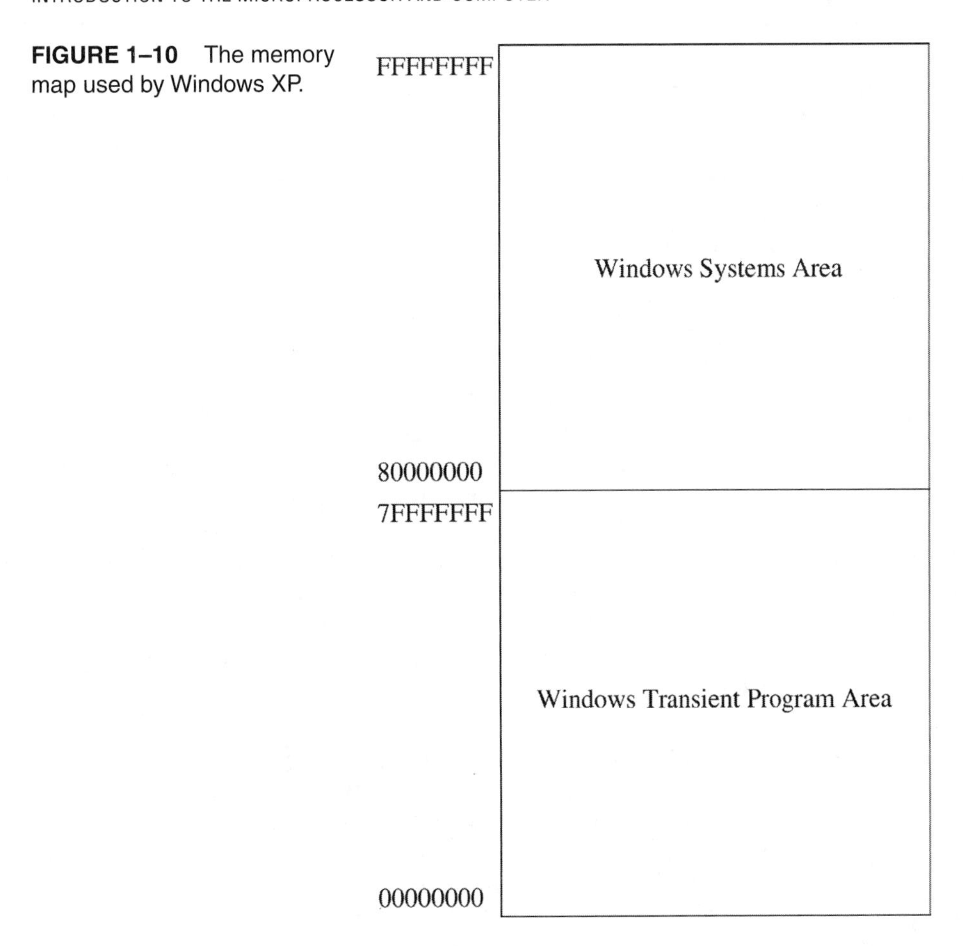

FIGURE 1–10 The memory map used by Windows XP.

frame (usually locations D0000H through DFFFFH) is used to expand the memory system by switching pages of memory from the EMS into this range of memory addresses.

Memory locations E0000H–EFFFFH contain the cassette BASIC language on ROM found in early IBM personal computer systems. This area is often open or free in newer computer systems.

Finally, the system BIOS ROM is located in the top 64K bytes of the system area (F0000H–FFFFFH). This ROM controls the operation of the basic I/O devices connected to the computer system. It does not control the operation of the video system, which has its own BIOS ROM at location C0000H. The first part of the system BIOS (F0000H–F7FFFH) often contains programs that set up the computer; the second part contains procedures that control the basic I/O system.

Windows Systems. Modern computers use a different memory map with Windows than the DOS memory maps of Figures 1–8 and 1–9. The Windows memory map appears in Figure 1–10 and has two main areas, a TPA and a system area. The difference between it and the DOS memory map are the sizes and locations of these areas.

The Windows TPA is the first 2G bytes of the memory system from locations 00000000H to 7FFFFFFFH. The Windows system area is the last 2G bytes of memory from locations 80000000H to FFFFFFFFH. It appears that the same idea used to construct the DOS memory map was also used in a modern Windows-based system. The system area is where the system BIOS and the video memory is located. Also located in the system area is the actual Windows program and drivers. Every program that is written for Windows can use up to 2G bytes of memory located at linear addresses 00000000H through 7FFFFFFFH.

Does this mean that my program will begin at physical address 00000000H? No, the memory system physical map is much different for the linear programming model shown in Figure 1–10. Every process in a Windows XP or Windows 2000 system has its own set of page tables, which define where in the physical memory each 4K-byte page of the process is located. This means that the process can be located anywhere in the memory, even in noncontiguous pages. Page tables and the paging structure of the microprocessor are discussed later in this chapter and are beyond the scope of the text at this point. As far as an application is concerned, you will always have 2G bytes of memory even if the computer has less memory. The operating system (Windows) handles assigning physical memory to the application and if not enough physical memory exists, it uses the hard disk drive for any that is not available.

I/O Space. The I/O (input/output) space in a computer system extends from I/O port 0000H to port FFFFH. (An **I/O port address** is similar to a memory address, except that instead of addressing memory, it addresses an I/O device.) The I/O devices allow the microprocessor to communicate between itself and the outside world. The I/O space allows the computer to access up to 64K different 8-bit I/O devices, 32K different 16-bit devices, or 16K different 32-bit devices. A great number of these locations are available for expansion in most computer systems. Figure 1–11 shows the I/O map found in many personal computer systems. To view the map on your computer in Windows XP, go to the Control Panel, Performance and Maintenance, System, Hardware tab, Device Manager, View tab, then select resources by type and click on the plus next to Input/Output (I/O).

The I/O area contains two major sections. The area below I/O location 0400H is considered reserved for system devices; many are depicted in Figure 1–10. The remaining area is available I/O space for expansion that extends from I/O port 0400H through FFFFH. Generally, I/O addresses between 0000H and 00FFH address components on the main board of the computer, while addresses between 0100H and 03FFH address devices located on plug-in cards (or on the main board). Note that the limitation of I/O addresses between 0000 and 03FFH comes from the original PC standard, as specified by IBM. When using the ISA bus, you must only use addresses between 0000H and 03FFH. The PCI bus uses I/O addresses between 0400H and FFFFH.

Various I/O devices that control the operation of the system are usually not directly addressed. Instead, the system BIOS ROM addresses these basic devices, which can vary slightly in location and function from one computer to the next. Access to most I/O devices should always be made through Windows, DOS, or BIOS function calls to maintain compatibility from one computer system to another. The map shown in Figure 1–11 is provided as a guide to illustrate the I/O space in the system.

The Microprocessor

At the heart of the microprocessor-based computer system is the microprocessor integrated circuit. The microprocessor, sometimes referred to as the CPU (**central processing unit**), is the controlling element in a computer system. The microprocessor controls memory and I/O through a series of connections called buses. The buses select an I/O or memory device, transfer data between an I/O device or memory and the microprocessor, and control the I/O and memory system. Memory and I/O are controlled through instructions that are stored in the memory and executed by the microprocessor.

The microprocessor performs three main tasks for the computer system: (1) data transfer between itself and the memory or I/O systems, (2) simple arithmetic and logic operations, and (3) program flow via simple decisions. Although these are simple tasks, it is through them that the microprocessor performs virtually any series of operations or tasks.

The power of the microprocessor is in its capability to execute billions of millions of instructions per second from a program or software (**group of instructions**) stored in the memory

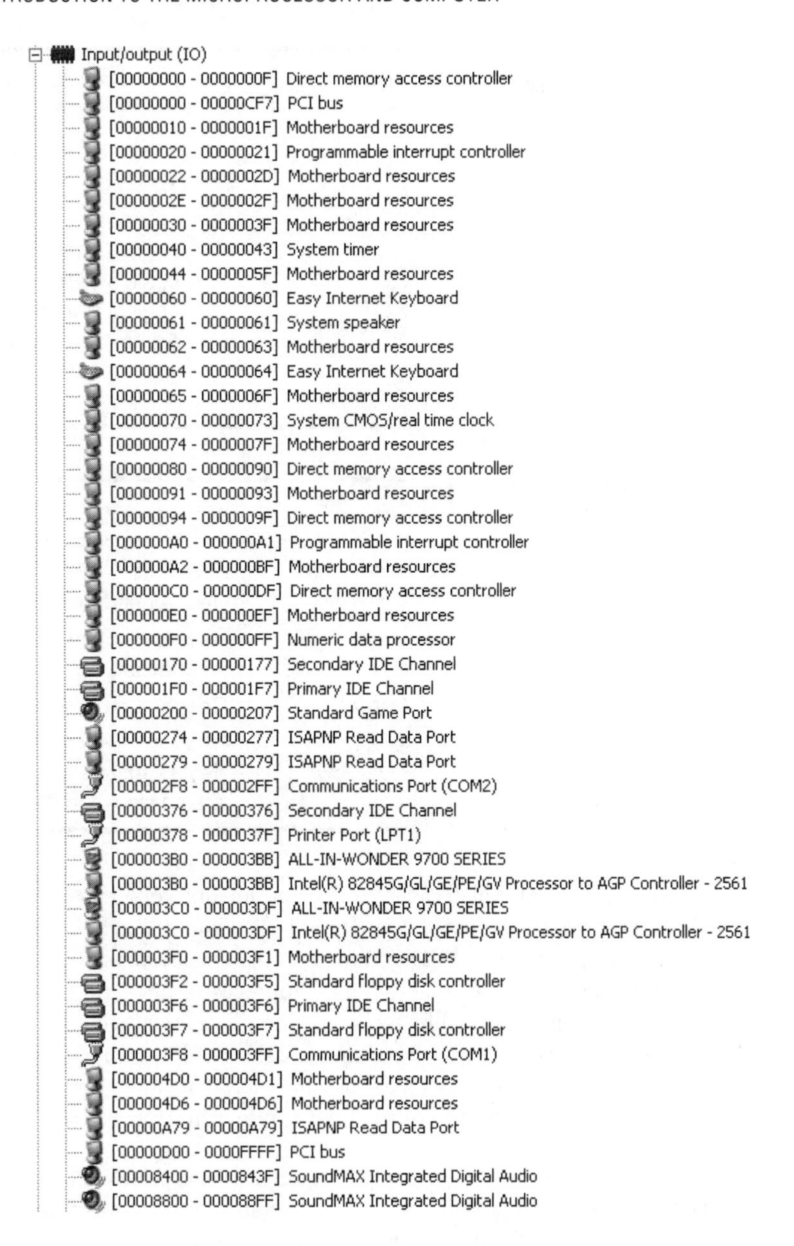

FIGURE 1–11 Some I/O locations in a typical personal computer.

system. This stored program concept has made the microprocessor and computer system very powerful devices. (Recall that Babbage also wanted to use the stored program concept in his Analytical Engine.)

Table 1–4 shows the arithmetic and logic operations executed by the Intel family of microprocessors. These operations are very basic, but through them, very complex problems are solved. Data are operated on from the memory system or internal registers. Data widths are variable and include a **byte** (8 bits), **word** (16 bits), and **doubleword** (32 bits). Note that only the 80386 through the Pentium 4 directly manipulate 8-, 16-, and 32-bit numbers. The earlier 8086–80286 directly manipulated 8- and 16-bit numbers, but not 32-bit numbers. Beginning with the 80486, the microprocessor contained a numeric coprocessor that allowed it to perform complex arithmetic using floating-point arithmetic. The numeric coprocessor, which is similar to a calculator chip, was an additional component in the 8086- through the 80386-based personal computers. The

TABLE 1–4 Simple arithmetic and logic operations.

Operation	Comment
Addition	
Subtraction	
Multiplication	
Division	
AND	Logical multiplication
OR	Logic addition
NOT	Logical inversion
NEG	Arithmetic inversion
Shift	
Rotate	

TABLE 1–5 Decisions found in the 8086–Pentium 4 microprocessors.

Decision	Comment
Zero	Test a number for zero or not-zero
Sign	Test a number for positive or negative
Carry	Test for a carry after addition or a borrow after subtraction
Parity	Test a number for an even or an odd number of ones
Overflow	Test for an overflow that indicates an invalid result after a signed addition or a signed subtraction

numeric coprocessor is also capable of performing integer operations on **quadwords** (64 bits). The MMX and SIMD units inside the Pentium–Pentium 4 function with integers and floating-point numbers in parallel. The SIMD unit requires numbers stored as **octalwords** (128 bits).

Another feature that makes the microprocessor powerful is its ability to make simple decisions based upon numerical facts. For example, a microprocessor can decide if a number is zero, if it is positive, and so forth. These simple decisions allow the microprocessor to modify the program flow, so that programs appear to "think through" these simple decisions. Table 1–5 lists the decision-making capabilities of the Intel family of microprocessors.

Buses. A **bus** is a common group of wires that interconnect components in a computer system. The buses that interconnect the sections of a computer system transfer address, data, and control information between the microprocessor and its memory and I/O systems. In the microprocessor-based computer system, three buses exist for this transfer of information: address, data, and control. Figure 1–12 shows how these buses interconnect various system components such as the microprocessor, read/write memory (RAM), read-only memory (ROM or flash), and a few I/O devices.

The address bus requests a memory location from the memory or an I/O location from the I/O devices. If I/O is addressed, the address bus contains a 16-bit I/O address from 0000H through FFFFH. The 16-bit I/O address, or port number, selects one of 64K different I/O devices. If memory is addressed, the address bus contains a memory address, which varies in width with the different versions of the microprocessor. The 8086 and 8088 address 1M byte of memory using a 20-bit address that selects locations 00000H–FFFFFH. The 80286 and 80386SX address 16M bytes of memory using a 24-bit address that selects locations 000000H–FFFFFFH. The 80386SL, 80386SLC, and 80386EX address 32M bytes of memory using a 25-bit address that selects locations 0000000H–1FFFFFFH. The 80386DX, 80486SX, and 80486DX address 4G bytes of memory using a 32-bit address that selects locations 00000000H–FFFFFFFFH. The Pentium also addresses 4G bytes of memory, but it uses a 64-bit data bus to access up to 8 bytes

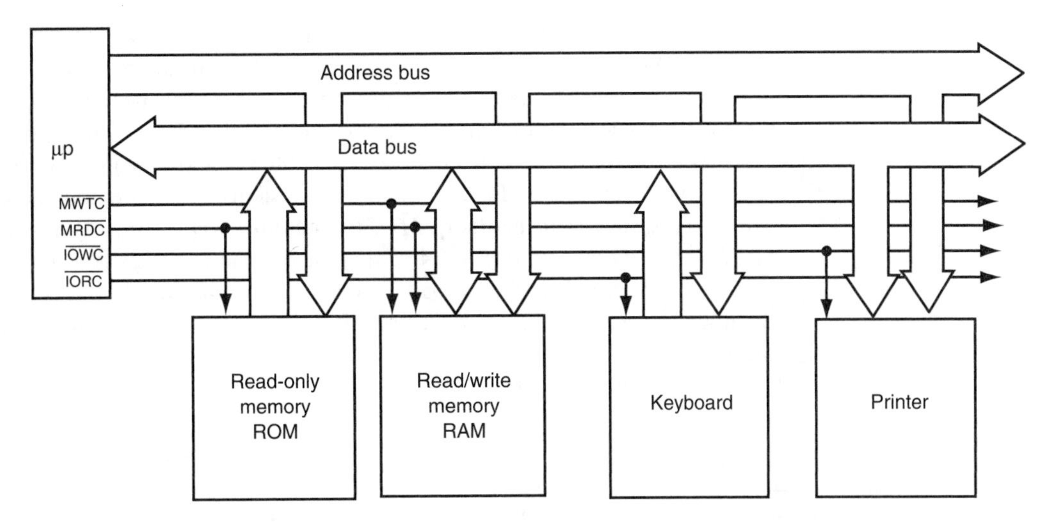

FIGURE 1–12 The block diagram of a computer system showing the address, data, and control bus structure.

of memory at a time. The Pentium Pro through Pentium 4 microprocessors have a 64-bit data bus and a 32-bit address bus that address 4G of memory from locations 00000000H–FFFFFFFFH, or a 36-bit address bus that addresses 64G of memory at locations 000000000H–FFFFFFFFFH, depending on their configuration. Refer to Table 1–6 for a complete listing of bus and memory sizes of the Intel family of microprocessors.

The data bus transfers information between the microprocessor and its memory and I/O address space. Data transfers vary in size, from 8 bits wide to 64 bits wide in various members of the Intel microprocessor family. For example, the 8088 has an 8-bit data bus that transfers 8 bits of data at a time. The 8086, 80286, 80386SL, 80386SX, and 80386EX transfer 16 bits of data through their data buses; the 80386DX, 80486SX, and 80486DX transfer 32 bits of data; and the Pentium through Pentium 4 microprocessors transfer 64 bits of data. The advantage of a wider data bus is speed in applications that use wide data. For example, if a 32-bit number is stored in memory, it takes the 8088 microprocessor four transfer operations to complete because its data bus is only 8 bits wide. The 80486 accomplishes the same task with one transfer because its data bus is 32 bits

TABLE 1–6 The Intel family of microprocessor bus and memory sizes.

Microprocessor	*Data Bus Width*	*Address Bus Width*	*Memory Size*
8086	16	20	1M
8088	8	20	1M
80186	16	20	1M
80188	8	20	1M
80286	16	24	16M
80386SX	16	24	16M
80386DX	32	32	4G
80386EX	16	26	64M
80486	32	32	4G
Pentium	64	32	4G
Pentium Pro–Pentium 4	64	32	4G
Pentium Pro–Pentium 4 (if extended addressing is enabled)	64	36	64G

wide. Figure 1–13 shows the memory widths and sizes of the 8086–80486 and Pentium–Pentium 4 microprocessors. Notice how the memory sizes and organizations differ between various members of the Intel microprocessor family. In all family members, the memory is numbered by byte. Notice that the Pentium–Pentium 4 microprocessors all contain a 64-bit-wide data bus.

The control bus contains lines that select the memory or I/O and cause them to perform a read or write operation. In most computer systems, there are four control bus connections: $\overline{MRDC}$ (**memory read control**), $\overline{MWTC}$ (**memory write control**), $\overline{IORC}$ (**I/O read control**), and $\overline{IOWC}$ (**I/O write control**). Note that the overbar indicates that the control signal is active-low; that is, it is active when a logic zero appears on the control line. For example, $\overline{IOWC} = 0$, the microprocessor is writing data from the data bus to an I/O device whose address appears on the address bus.

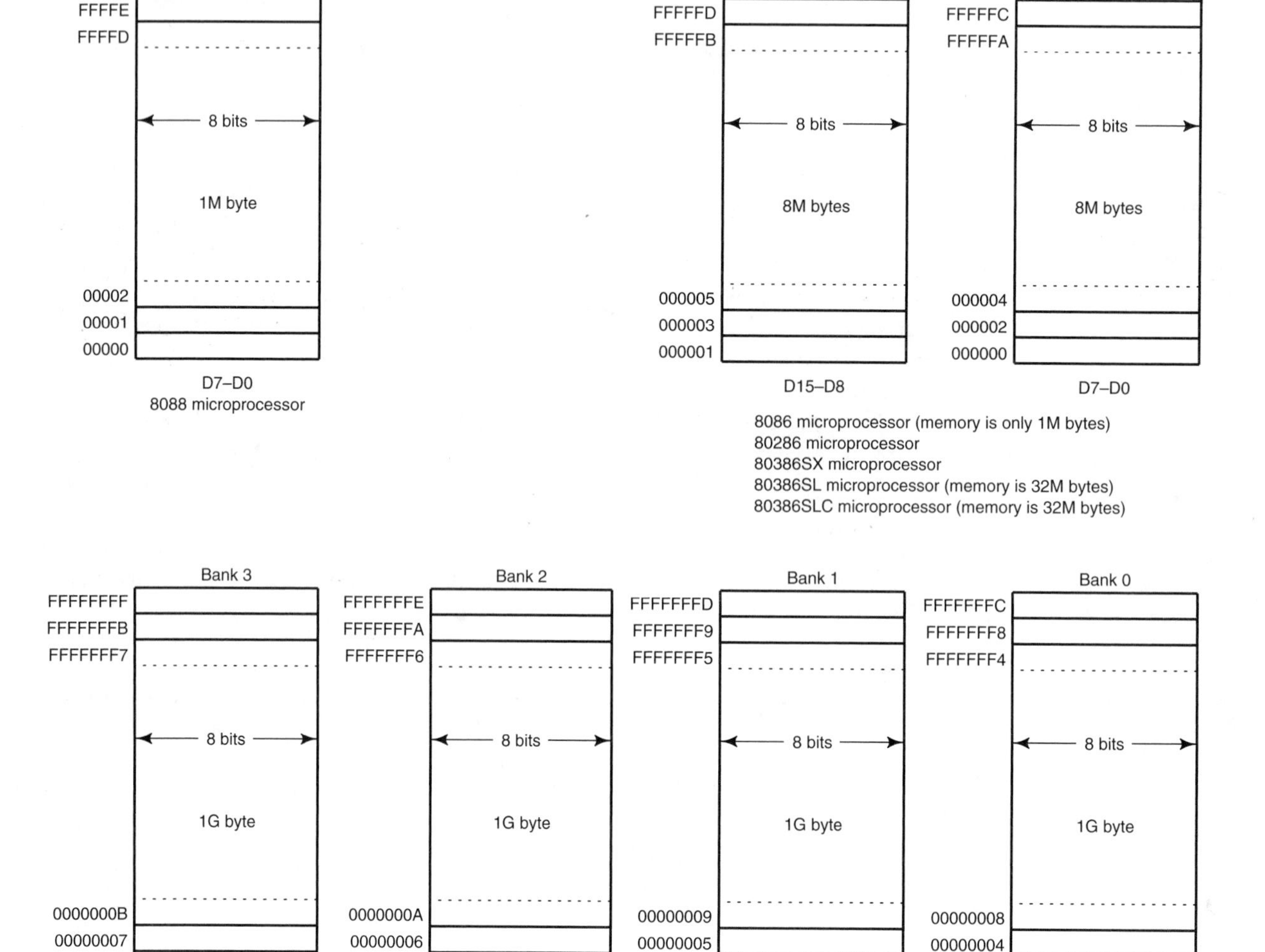

FIGURE 1–13 The physical memory systems of the 8086 through the Pentium 4 microprocessors.

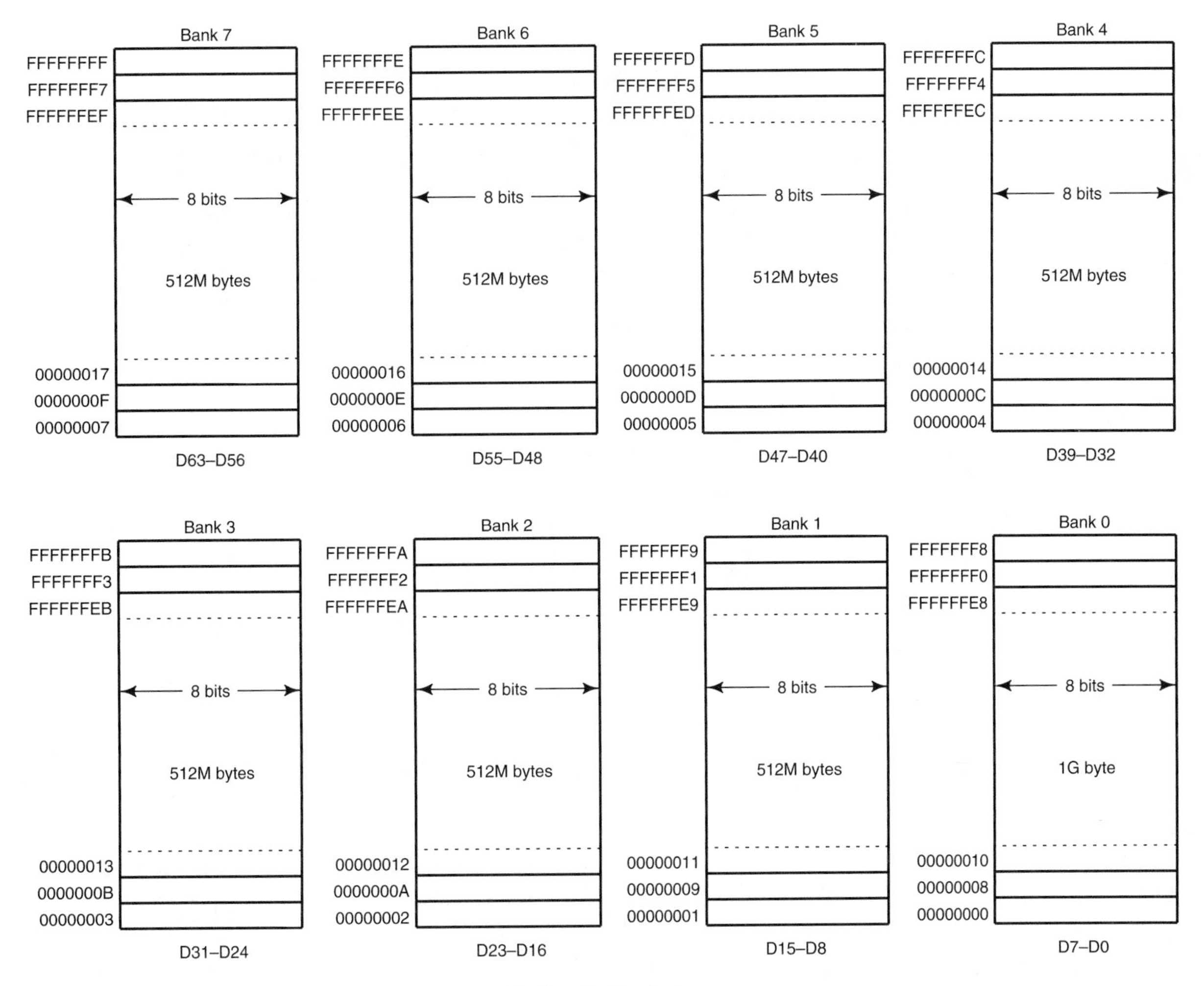

FIGURE 1–13 *(continued)*

The microprocessor reads the contents of a memory location by sending the memory an address through the address bus. Next, it sends the memory read control signal ($\overline{MRDC}$) to cause the memory to read data. Finally, the data read from the memory are passed to the microprocessor through the data bus. Whenever a memory write, I/O write, or I/O read occurs, the same sequence ensues, except that different control signals are issued and the data flow out of the microprocessor through its data bus for a write operation.

1–3 NUMBER SYSTEMS

The use of the microprocessor requires a working knowledge of binary, decimal, and hexadecimal numbering systems. This section of the text provides a background for those who are unfamiliar with these numbering systems. Conversions between decimal and binary, decimal and hexadecimal, and binary and hexadecimal are described.

Digits

Before numbers are converted from one number base to another, the digits of a number system must be understood. Early in our education, we learned that a decimal (base 10) number is constructed with 10 digits: 0 through 9. The first digit in any numbering system is always zero. For example, a base 8 (**octal**) number contains 8 digits: 0 through 7; a base 2 (**binary**) number contains 2 digits: 0 and 1. If the base of a number exceeds 10, the additional digits use the letters of the alphabet, beginning with an A. For example, a base 12 number contains 12 digits: 0 through 9, followed by A for 10 and B for 11. Note that a base 10 number does contain a *10* digit, just as a base 8 number does not contain an *8* digit. The most common numbering systems used with computers are decimal, binary, and hexadecimal (base 16). (Many years ago octal numbers were popular.) Each of these number systems is described and used in this section of the chapter.

Positional Notation

Once the digits of a number system are understood, larger numbers are constructed by using positional notation. In grade school, we learned that the position to the left of the units position is the tens position, the position to the left of the tens position is the hundreds position, and forth. (An example is the decimal number 132: This number has 1 hundred, 3 tens, and 2 units.) What we probably did not learn was the exponential value of each position: The units position has a weight of 10^0, or 1; the tens position has a weight of 10^1, or 10; and the hundreds position has a weight of 10^2, or 100. The exponential powers of the positions are critical for understanding numbers in other numbering systems. The position to the left of the radix (**number base**) point, called a decimal point only in the decimal system, is always the units position in any number system. For example, the position to the left of the binary point is always 2^0, or 1; the position the left of the octal point is 8^0, or 1. In any case, any number raised to its zero power is always 1, or the units position.

The position to the left of the units position is always the number base raised to the first power; in a decimal system, this is 10^1, or 10. In a binary system, it is 2^1, or 2; and in an octal system, it is 8^1, or 8. Therefore, an 11 decimal has a different value from an 11 binary. The decimal number is composed of 1 ten plus 1 unit, and has a value of 11 units; the binary number 11 is composed of 1 two plus 1 unit, for a value of 3 decimal units. The 11 octal has a value of 9 decimal units.

In the decimal system, positions to the right of the decimal point have negative powers. The first digit to the right of the decimal point has a value of 10^{-1}, or 0.1. In the binary system the first digit to the right of the binary point has a value of 2^{-1}, or 0.5. In general, the principles that apply to decimal numbers also apply to numbers in any other number system.

Example 1–1 shows 110.101 in binary (often written as 110.101_2). It also shows the power and weight or value of each digit position. To convert a binary number to decimal, add weights of each digit to form its decimal equivalent. The 110.101_2 is equivalent to 6.625 in decimal (4 + 2 + 0.5 + 0.125). Notice that this is the sum of 2^2 (or 4) plus 2^1 (or 2), but 2^0 (or 1) is not added because there are no digits under this position. The fraction part is composed of 2^{-1} (.5) plus 2^{-3} (or .125), but there is no digit under the 2^{-2} (or .25) so .25 is not added.

EXAMPLE 1–1

Power	2^2		2^1		2^0		2^{-1}		2^{-2}		2^{-3}		
Weight	4		2		1		.5		.25		.125		
Number	1		1		0	.	1		0		1		
Numeric Value	4	+	2	+	0	+	.5	+	0	+	.125	=	6.625

Suppose that the conversion technique is applied to a base 6 number, such as 25.2_6. Example 1–2 shows this number placed under the powers and weights of each position. In the example, there is a 2 under 6^1, which has a value of 12_{10} (2 × 6), and a 5 under 6^0, which has a value of 5 (5 × 1). The whole number portion has a decimal value of 12 + 5, or 17. The number to the right of the hex point is a 2 under 6^{-1}, which has a value of .333 (2 × .167). The number 25.2_6, therefore, has a value of 17.333.

EXAMPLE 1–2

Power	6^1	6^0	6^{-1}
Weight	6	1	.167
Number	2	5	.2
Numeric Value	12 +	5 +	.333 = 17.333

Conversion to Decimal

The prior examples have shown that to convert from any number base to decimal, determine the weights or values of each position of the number, and then sum the weights to form the decimal equivalent. Suppose that 125.7_8 octal is converted to decimal. To accomplish this conversion, first write down the weights of each position of the number. This appears in Example 1–3. The value of 125.7_8 is 85.875 decimal, or 1 × 64 plus 2 × 8 plus 5 × 1 plus 7 × .125.

EXAMPLE 1–3

Power	8^2	8^1	8^0	8^{-1}
Weight	64	8	1	.125
Number	1	2	5	.7
Numeric Value	64 +	16 +	5 +	.875 = 85.875

Notice that the weight of the position to the left of the units position is 8. This is 8 times 1. Then notice that the weight of the next position is 64, or 8 times 8. If another position existed, it would be 64 times 8, or 512. To find the weight of the next higher-order position, multiply the weight of the current position by the number base (or 8, in this example). To calculate the weights of position to the right of the radix point, divide by the number base. In the octal system, the position immediately to the right of the octal point is $^1/_8$, or .125. The next position is $^{.125}/_8$, or .015625, which can also be written as $^1/_{64}$. Also note that the number in Example 1–3 can also be written as the decimal number $85^7/_8$.

Example 1–4 shows the binary number 11011.0111 written with the weights and powers of each position. If these weights are summed, the value of the binary number converted to decimal is 27.4375.

EXAMPLE 1–4

Power	2^4	2^3	2^2	2^1	2^0	2^{-1}	2^{-2}	2^{-3}	2^{-4}
Weight	16	8	4	2	1	.5	.25	.125	.0625
Number	1	1	0	1	1 .	0	1	1	1
Numeric Value	16 +	8 +	0 +	2 +	1 +	0 +	.25 +	.125 +	.0625 = 27.4375

It is interesting to note that 2^{-1} is also $^1/_2$, 2^{-2} is $^1/_4$, and so forth. It is also interesting to note that 2^{-4} is $^1/_{16}$, or .0625. The fractional part of this number is $^7/_{16}$ or .4375 decimal. Notice that 0111 is a 7 in binary code for the numerator and the rightmost one is in the $^1/_{16}$ position for the denominator. Other examples: The binary fraction of .101 is $^5/_8$ and the binary fraction of .001101 is $^{13}/_{64}$.

Hexadecimal numbers are often used with computers. A 6A.CH (**H** for hexadecimal) is illustrated with its weights in Example 1–5. The sum of its digits is 106.75, or $106^3/_4$. The whole number part is represented with 6 × 16 plus 10 (A) × 1. The fraction part is 12 (C) as a numerator and 16 (16^{-1}) as the denominator, or $^{12}/_{16}$, which is reduced to $^3/_4$.

EXAMPLE 1–5

```
Power           16^1    16^0    16^-1
Weight          16      1       .0625
Number          6       A   .   C
Number Value    96  +   10  +   .75   =  106.75
```

Conversion from Decimal

Conversions from decimal to other number systems are more difficult to accomplish than conversion to decimal. To convert the whole number portion of a number to decimal, divide by the radix. To convert the fractional portion, multiply by the radix.

Whole Number Conversion from Decimal. To convert a decimal whole number to another number system, divide by the radix and save the remainders as significant digits of the result. An algorithm for this conversion is as follows:

1. Divide the decimal number by the radix (number base).
2. Save the remainder (first remainder is the least significant digit).
3. Repeat steps 1 and 2 until the quotient is zero.

For example, to convert 10 decimal to binary, divide it by 2. The result is 5, with a remainder of 0. The first remainder is the units position of the result (in this example, a 0). Next divide the 5 by 2. The result is 2, with a remainder of 1. The 1 is the value of the twos (2^1) position. Continue the division until the quotient is a zero. Example 1–6 shows this conversion process. The result is written as 1010_2 from the bottom to the top.

EXAMPLE 1–6

```
2) 10      remainder = 0
 2) 5      remainder = 1
  2) 2     remainder = 0
   2) 1    remainder = 1           result = 1010
      0
```

To convert 10 decimal into base 8, divide by 8, as shown in Example 1–7. The number 10 decimal is 12 octal.

EXAMPLE 1–7

```
8) 10      remainder = 2
 8) 1      remainder = 1           result = 12
    0
```

Conversion from decimal to hexadecimal is accomplished by dividing by 16. The remainders will range in value from 0 through 15. Any remainder of 10 through 15 is then converted to the letters A through F for the hexadecimal number. Example 1–8 shows the decimal number 109 converted to 6DH.

EXAMPLE 1–8

```
16) 109    remainder = 13 (D)
 16)  6    remainder = 6           result = 6D
      0
```

Converting from a Decimal Fraction. Conversion from a decimal fraction to another number base is accomplished with multiplication by the radix. For example, to convert a decimal fraction into binary, multiply by 2. After the multiplication, the whole number portion of the result is saved as a significant digit of the result, and the fractional remainder is again multiplied by the radix. When the fraction remainder is zero, multiplication ends. Note that some numbers are never-ending (repetend). That is, a zero is never a remainder. An algorithm for conversion from a decimal fraction is as follows:

1. Multiply the decimal fraction by the radix (number base).
2. Save the whole number portion of the result (even if zero) as a digit. Note that the first result is written immediately to the right of the radix point.
3. Repeat steps 1 and 2, using the fractional part of step 2 until the fractional part of step 2 is zero.

Suppose that the number .125 decimal is converted to binary. This is accomplished with multiplications by 2, as illustrated in Example 1–9. Notice that the multiplication continues until the fractional remainder is zero. The whole number portions are written as a binary fraction (0.001) in this example.

EXAMPLE 1–9

```
   .125
x     2
   0.25       digit is 0

   .25
x    2
   0.5        digit is 0

   .5
x   2
  1.0         digit is 1         result = 0.001₂
```

This same technique is used to convert a decimal fraction into any number base. Example 1–10 shows the same decimal fraction of .125 from Example 1–9 converted to octal by multiplying by 8.

EXAMPLE 1–10

```
   .125
x     8
    1.0       digit is 1         result = 0.1₈
```

Conversion to a hexadecimal fraction appears in Example 1–11. Here, the decimal .046875 is converted to hexadecimal by multiplying by 16. Note that .046875 is 0.0CH.

EXAMPLE 1–11

```
   .046875
x       16
      0.75  digit is 0

   .75
x   16
  12.0        digit is 12(C)     result = 0.0C₁₆
```

Binary-Coded Hexadecimal

Binary-coded hexadecimal (BCH) is used to represent hexadecimal data in binary code. A binary-coded hexadecimal number is a hexadecimal number written so that each digit is represented by a 4-bit binary number. The values for the BCH digits appear in Table 1–7.

Hexadecimal numbers are represented in BCH code by converting each digit to BCH code with a space between each coded digit. Example 1–12 shows 2AC converted to BCH code. Note that each BCH digit is separated by a space.

TABLE 1–7 Binary-coded hexadecimal (BCH) code.

Hexadecimal Digit	*BCH Code*
0	0000
1	0001
2	0010
3	0011
4	0100
5	0101
6	0110
7	0111
8	1000
9	1001
A	1010
B	1011
C	1100
D	1101
E	1110
F	1111

EXAMPLE 1–12

```
2AC = 0010 1010 1100
```

The purpose of BCH code is to allow a binary version of a hexadecimal number to be written in a form that can be easily converted between BCH and hexadecimal. Example 1–13 shows a BCH coded number converted back to hexadecimal code.

EXAMPLE 1–13

```
1000 0011 1101 . 1110 = 83D.E
```

Complements

At times, data are stored in complement form to represent negative numbers. Two systems are used to represent negative data: **radix** and **radix –1** complements. The earliest system was the radix –1 complement, in which each digit of the number is subtracted from the radix –1 to generate the radix –1 complement to represent a negative number.

Example 1–14 shows how the 8-bit binary number 01001100 is one's (radix –1) complemented to represent it as a negative value. Notice that each digit of the number is subtracted from one to generate the radix –1 (one's) complement. In this example, the negative of 01001100 is 10110011. The same technique can be applied to any number system, as illustrated in Example 1–15, in which the fifteen's (radix –1) complement of a 5CD hexadecimal is computed by subtracting each digit from a 15.

EXAMPLE 1–14

```
  1111 1111
- 0100 1100
  1011 0011
```

EXAMPLE 1–15

```
  15 15 15
-  5  C  D
   A  3  2
```

Today, the radix –1 complement is not used by itself; it is used as a step for finding the radix complement. The radix complement is used to represent negative numbers in modem computer systems. (The radix –1 complement was used in the early days of computer technology.) The main problem with the radix –1 complement is that a negative or a positive zero exists; in the radix complement system, only a positive zero can exist.

To form the radix complement, first find the radix –1 complement, and then add a one to the result. Example 1–16 shows how the number 0100 1000 is converted to a negative value by two's (radix) complementing it.

EXAMPLE 1–16

```
  1111 1111
- 0100 1000
  1011 0111 (one's complement)
+         1
  1011 1000 (two's complement)
```

To prove that 0100 1000 is the inverse (negative) of 1011 1000, add the two together to form an 8-digit result. The ninth digit is dropped and the result is zero because 0100 1000 is a positive 72, while 1011 1000 is a negative 72. The same technique applies to any number system. Example 1–17 shows how the inverse of 345 hexadecimal is found by first fifteen's complementing the number, and then by adding one to the result to form the sixteen's complement. As before, if the original 3-digit number 345 is added to the inverse of CBB, the result is a 3-digit 000. As before, the fourth bit (carry) is dropped. This proves that 345 is the inverse of CBB. Additional information about one's and two's complements is presented with signed numbers in the next section of the text.

EXAMPLE 1–17

```
  15 15 15
-  3  4  5
   C  B  A  (fifteen's complement)
+        1
   C  B  B  (sixteen's complement)
```

1–4 COMPUTER DATA FORMATS

Successful programming requires a precise understanding of data formats. In this section, many common computer data formats are described as they are used with the Intel family of microprocessors. Commonly, data appear as ASCII, Unicode, BCD, signed and unsigned integers, and floating-point numbers (real numbers). Other forms are available, but are not presented here because they are not commonly found.

ASCII and Unicode Data

ASCII (**American Standard Code for Information Interchange**) data represent alphanumeric characters in the memory of a computer system (see Table 1–8). The standard ASCII code is a 7-bit code, with the eighth and most significant bit used to hold parity in some antiquated systems. If ASCII data are used with a printer, the most significant bits are 0 for alphanumeric printing and 1 for graphics printing. In the personal computer, an extended ASCII character set is selected by placing a 1 in the leftmost bit. Table 1–9 shows the extended ASCII character set, using codes 80H–FFH. The extended ASCII characters store some foreign letters and punctuation, Greek characters,

TABLE 1–8 ASCII code.

								Second								
	X0	X1	X2	X3	X4	X5	X6	X7	X8	X9	XA	XB	XC	XD	XE	XF
First																
0X	NUL	SOH	STX	ETX	EOT	ENQ	ACK	BEL	BS	HT	LF	VT	FF	CR	SO	SI
1X	DLE	DC1	DC2	DC3	DC4	NAK	SYN	ETB	CAN	EMS	SUB	ESC	FS	GS	RS	US
2X	SP	!	"	#	$	%	&	'	(	)	*	+	,	-	.	/
3X	0	1	2	3	4	5	6	7	8	9	:	;	<	=	>	?
4X	@	A	B	C	D	E	F	G	H	I	J	K	L	M	N	O
5X	P	Q	R	S	T	U	V	W	X	Y	Z	[	\	]	^	_
6X	`	a	b	c	d	e	f	g	h	i	j	k	l	m	n	o
7X	p	q	r	s	t	u	v	w	x	y	z	{	|	}	~	:::

TABLE 1–9 Extended ASCII code, as printed by the IBM ProPrinter.

First						Second										
	X0	X1	X2	X3	X4	X5	X6	X7	X8	X9	XA	XB	XC	XD	XE	XF
0X		☺	☻	♥	♦	♣	♠	•	◘	○	◙	♂	♀	♪	♫	☼
1X	►	◄	↕	‼	¶	§	▬	↨	↑	↓	→	←	∟	↔	▲	▼
8X	Ç	ü	é	â	ä	à	å	ç	ê	ë	è	ï	î	ì	Ä	Å
9X	É	æ	Æ	ô	ö	ò	û	ù	ÿ	Ö	Ü	¢	£	¥	Pt	ƒ
AX	á	í	ó	ú	ñ	Ñ	ª	º	¿	⌐	¬	½	¼	¡	«	»
BX	░	▒	▓	│	┤	╡	╢	╖	╕	╣	║	╗	╝	╜	╛	┐
CX	└	┴	┬	├	─	┼	╞	╟	╚	╔	╩	╦	╠	═	╬	╧
DX	╨	╤	╥	╙	╘	╒	╓	╫	╪	┘	┌	█	▄	▌	▐	▀
EX	α	β	Γ	π	Σ	σ	µ	γ	Φ	Θ	Ω	δ	∞	∮	∈	∩
FX	≡	±	≥	≤	⌠	⌡	÷	≈	°	•	.	√	n	2	■	

mathematical characters, box-drawing characters, and other special characters. Note that extended characters can vary from one printer to another. The list provided is designed to be used with the IBM ProPrinter, which also matches the special character set found in most word processors.

The ASCII control characters, also listed in Table 1–8, perform control functions in a computer system, including clear screen, backspace, line feed, and so on. To enter the control codes through the computer keyboard, hold down the Control key while typing a letter. To obtain the control code 01H, type Control-A; 02H is obtained by Control-B, and so on. Note that the control codes appear on the screen, from the DOS prompt, as ^A for Control-A, ^B for Control-B, and so forth. Also note that the carriage return code (CR) is the Enter key on most modern keyboards. The purpose of CR is to return the cursor or print head to the left margin. Another code that appears in many programs is the line feed code (LF), which moves the cursor down one line.

To use Tables 1–8 or 1–9 for converting alphanumeric or control characters into ASCII characters, first locate the alphanumeric code for conversion. Next, find the first digit of the hexadecimal ASCII code. Then find the second digit. For example, the capital letter A is ASCII code 41H, and the lowercase letter a is ASCII code 61H. Many Windows-based applications use the **Unicode** system to store alphanumeric data. This system stores each character as 16-bit data. The codes 0000H–00FFH are the same as standard ASCII code. The remaining codes, 0100H–FFFFH, are used to store all special characters from many worldwide character sets. This allows software written for the Windows environment to be used in many countries in the world.

ASCII data are most often stored in memory by using a special directive to the assembler program called define byte(s), or DB. (The assembler is a program that is used to program a computer in its native binary machine language.) An alternative to DB is the word BYTE. The DB and BYTE directives, and several examples of their usage with ASCII-coded character strings, are listed in Example 1–18. Notice how each character string is surrounded by apostrophes (')—never use the quote ("). Also notice that the assembler lists the ASCII-coded value for each character to the left of the character string. To the far left is the hexadecimal memory location where the character string is first stored in the memory system. For example, the character string WHAT is stored beginning at memory address 001D. The first letter is stored as 57 (W), followed by 68 (H), and so forth. Example 1–19 shows the same three strings defined as CString character strings for use with Visual C++. Note that Visual C++ uses quotes to surround strings.

EXAMPLE 1–18

```
0000  42 61 72 72 79   NAMES  DB   'Barry B. Brey'
      20 42 2E 20 42
      72 65 79
000D  57 68 65 20 63   MESS   DB   'Where can it be?'
      20 63 61 6E 20
      69 74 20 62 65
      3F
001D  57 69 20 74 20   WHAT   DB   'What is on first.'
      69 73 20 6F 6E
      20 66 69 72 73
      74 2E
```

EXAMPLE 1–19

```
CString NAMES = "Barry B. Brey"

CString MESS = "Where can it be?"

CString WHAT = "What is on first."
```

BCD (Binary-Coded Decimal) Data

Binary-coded decimal (BCD) information is stored in either packed or unpacked forms. **Packed BCD** data are stored as two digits per byte and **unpacked BCD** data are stored as one digit per byte. The range of a BCD digit extends from 0000_2 to 1001_2, or 0–9 decimal. Unpacked BCD data are returned from a keypad or keyboard. Packed BCD data are used for some of the instructions included for BCD addition and subtraction in the instruction set of the microprocessor.

Table 1–10 shows some decimal numbers converted to both the packed and unpacked BCD forms. Applications that require BCD data are point-of-sale terminals and almost any device that performs a minimal amount of simple arithmetic. If a system requires complex arithmetic, BCD

TABLE 1–10 Packed and unpacked BCD data.

Decimal	*Packed*		*Unpacked*		
12	0001 0010		0000 0001	0000 0010	
623	0000 0110	0010 0011	0000 0110	0000 0010	0000 0011
910	0000 1001	0001 0000	0000 1001	0000 0001	0000 0000

data are seldom used because there is no simple and efficient method of performing complex BCD arithmetic.

Example 1–20 shows how to use the assembler to define both packed and unpacked BCD data. Example 1–21 shows how to do this using Visual C++ and char or bytes. In all cases, the convention of storing the least-significant data first is followed. This means that to store 83 in memory, the 3 is stored first, followed by the 8. Also note that with packed BCD data, the letter H (hexadecimal) follows the number to ensure that the assembler stores the BCD value rather than a decimal value for packed BCD data. Notice how the numbers are stored in memory as unpacked, one digit per byte; or packed, as two digits per byte.

EXAMPLE 1–20

```
                    ;Unpacked BCD data (least-significant data first)
                    ;
0000  03 04 05      NUMB1 DB    3,4,5        ;defines number 543
0003  07 08         NUMB2 DB    7,8          ;defines number 87
                    ;
                    ;Packed BCD data (least-significant data first)
                    ;
0005  37 34         NUMB3 DB    37H,34H      ;defines number 3437
0007  03 45         NUMB4 DB    3,45H        ;defines number 4503
```

EXAMPLE 1–21

```
//Unpacked BCD data (least-significant data first)
//
char Numb1 = 3,4,5;      ;defines number 543
char Numb2 = 7,8         ;defines number 87
//
//Packed BCD data (least-significant data first)
//
char Numb3 = 0x37,0x34   ;defines number 3437
char Numb4 = 3,0x45      ;defines number 4503
```

Byte-Sized Data

Byte-sized data are stored as *unsigned* and *signed* integers. Figure 1–14 illustrates both the unsigned and signed forms of the byte-sized integer. The difference in these forms is the weight of the leftmost bit position. Its value is 128 for the unsigned integer and minus 128 for

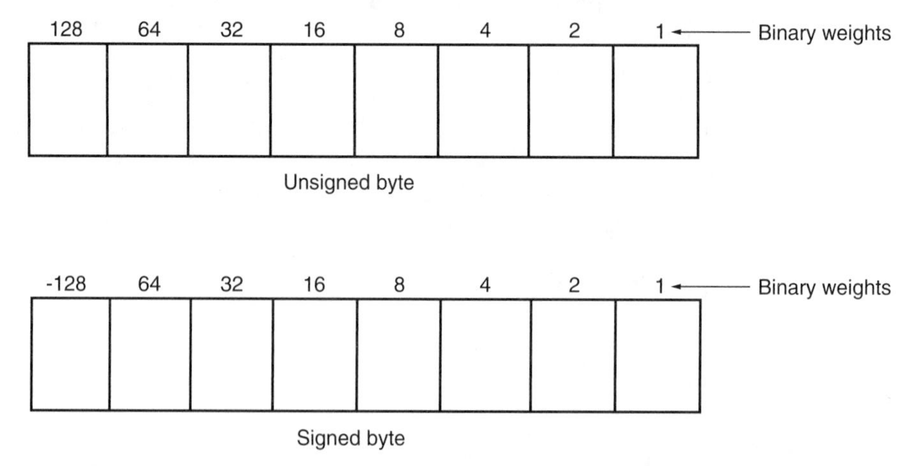

FIGURE 1–14 The unsigned and signed bytes illustrating the weights of each binary-bit position.

the signed integer. In the signed integer format, the leftmost bit represents the sign bit of the number, as well as a weight of minus 128. For example, 80H represents a value of 128 as an unsigned number; as a signed number, it represents a value of minus 128. Unsigned integers range in value from 00H to FFH (0–255). Signed integers range in value from –128 to 0 to +127.

Although negative signed numbers are represented in this way, they are stored in the two's complement form. The method of evaluating a signed number by using the weights of each bit position is much easier than the act of two's complementing a number to find its value. This is especially true in the world of calculators designed for programmers.

Whenever a number is two's complemented, its sign changes from negative to positive or positive to negative. For example, the number 00001000 is a +8. Its negative value (–8) is found by two's complementing the +8. To form a two's complement, first one's complement the number. To one's complement a number, invert each bit of a number from zero to one or from one to zero. Once the one's complement is formed, the two's complement is found by adding a one to the one's complement. Example 1–22 shows how numbers are two's complemented using this technique.

EXAMPLE 1–22

```
+8 = 00001000
     11110111 (one's complement)
+           1
-8 = 11111000 (two's complement)
```

Another, and probably simpler, technique for two's complementing a number starts with the rightmost digit. Start writing down the number from right to left. Write the number exactly as it appears until the first one. Write down the first one, and then invert all bits to its left. Example 1–23 shows this technique with the same number as in Example 1–22.

EXAMPLE 1–23

```
+8 = 00001000
         1000 (write number to first 1)
     1111   (invert the remaining bits)
-8 = 11111000
```

To store 8-bit data in memory using the assembler program, use the DB directive as in prior examples or char as in Visual C++ examples. Example 1–24 lists many forms of 8-bit numbers stored in memory using the assembler program. Notice in the example that a hexadecimal number is defined with the letter H following the number, and that a decimal number is written as is, without anything special. Example 1–25 shows the same byte data defined for use with a Visual C++ program.

EXAMPLE 1–24

```
             ;Unsigned byte-sized data
             ;
0000 FE      DATA1 DB     254    ;define 254 decimal
0001 87      DATA2 DB     87H    ;define 87 hexadecimal
0002 47      DATA3 DB     71     ;define 71 decimal
             ;
             ;Signed byte-sized data
             ;
0003 9C      DATA4 DB     -100   ;define -100 decimal
0004 64      DATA5 DB     +100   ;define +100 decimal
0005 FF      DATA6 DB     -1     ;define -1 decimal
0006 38      DATA7 DB     56     ;define 56 decimal
```

EXAMPLE 1–25

```
//Unsigned byte-sized data
//
unsigned char Data1 = 254;      //define 254 decimal
unsigned char Data2 = 0x87;     //define 87 hexadecimal
unsigned char Data3 = 71        //define 71 decimal
//
//Signed byte-sized data
//
char Data4 = -100;              //define -100 decimal
char Data5 = +100;              //define +100 decimal
char Data6 = -1;                //define -1 decimal
char Data7 = 56;                //define 56 decimal
```

Word-Sized Data

A word (16-bits) is formed with two bytes of data. The least significant byte is always stored in the lowest-numbered memory location, and the most significant byte is stored in the highest. This method of storing a number is called the **little endian** format. An alternate method, not used with the Intel family of microprocessors, is called the **big endian** format. In the big endian format, numbers are stored with the lowest location containing the most significant data. The big endian format is used with the Motorola family of microprocessors. Figure 1–15(a) shows the weights of each bit position in a word of data, and Figure 1–15(b) shows how the number 1234H appears when stored in the memory locations 3000H and 3001H. The only difference between a signed and an unsigned word is the leftmost bit position. In the unsigned form, the leftmost bit is unsigned and has a weight of 32,768; in the signed form, its weight is –32,768. As with byte-sized signed data, the signed word is in two's complement form when representing a negative number. Also, notice that the low-order byte is stored in the lowest-numbered memory location (3000H) and the high-order byte is stored in the highest-numbered location (3001H).

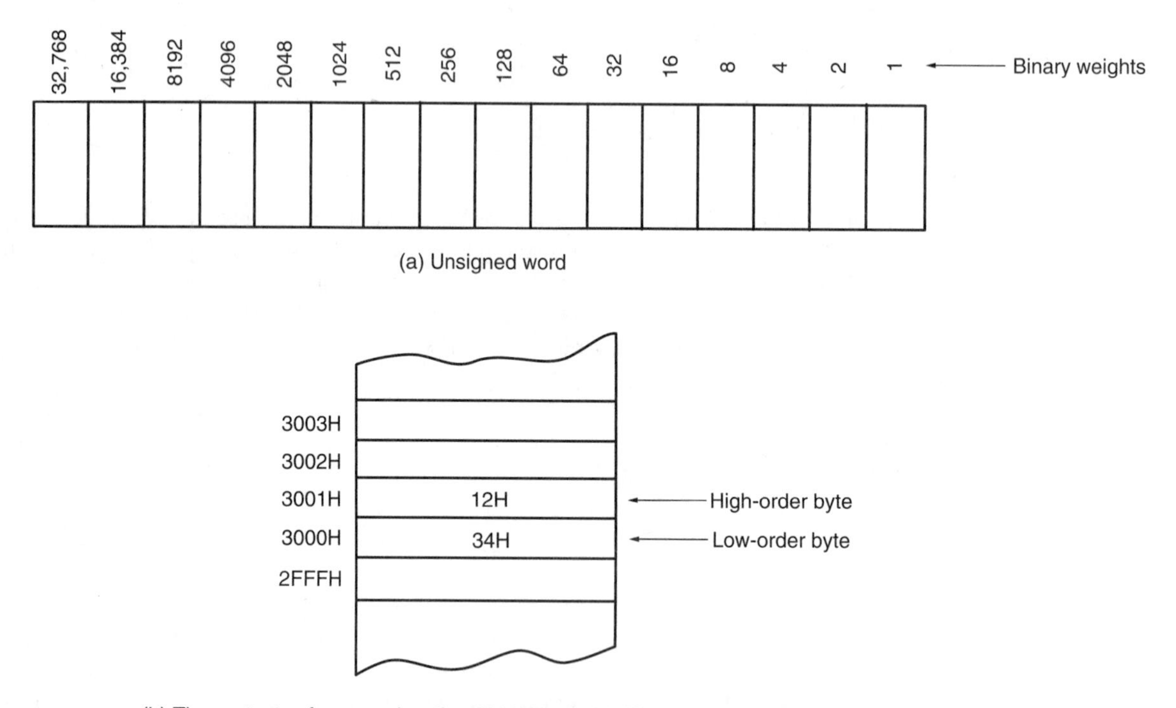

FIGURE 1–15 The storage format for a 16-bit word in (a) a register and (b) two bytes of memory.

Example 1–26 shows several signed and unsigned word-sized data stored in memory using the assembler program. Example 1–27 shows how to store the same numbers in a Visual C++ program (assuming version 5.0 or newer), which uses the **short** directive to store a 16-bit integer. Notice that the **define word(s)** directive, or **DW**, causes the assembler to store words in the memory instead of bytes, as in prior examples. The WORD directive can also be used to define a word. Notice that the word *data* is displayed by the assembler in the same form as entered. For example, 1000H is displayed by the assembler as 1000. This is for our convenience because the number is actually stored in the memory as 00 10 in two consecutive memory bytes.

EXAMPLE 1–26

```
                ;Unsigned word-sized data
                ;
0000  09F0      DATA1 DW    2544          ;define 2544 decimal
0002  87AC      DATA2 DW    87ACH         ;define 87AC hexadecimal
0004  02C6      DATA3 DW    710           ;define 710 decimal
                ;
                ;Signed word-sized data
                ;
0006  CBA8      DATA4 DW    -13400        ;define -13400 decimal
0008  00C6      DATA5 DW    +198          ;define +198 decimal
000A  FFFF      DATA6 DW    -1            ;define -1 decimal
```

EXAMPLE 1–27

```
//Unsigned word-sized data
//
unsigned short Data1 = 2544;        //define 2544 decimal
unsigned short Data2 = 0x87AC       //define 87AC hexadecimal
unsigned short Data3 = 710;         //define 710 decimal
//
//Signed word-sized data
//
short Data4 = -13400;               //define -13400 decimal
short Data5 = +198;                 //define +198 decimal
short Data6 = -1;                   //define -1 decimal
```

Doubleword-Sized Data

Doubleword-sized data requires four bytes of memory because it is a 32-bit number. Doubleword data appear as a product after a multiplication and also as a dividend before a division. In the 80386 through the Pentium 4, memory and registers are also 32 bits in width. Figure 1–16 shows the form used to store doublewords in the memory and the binary weights of each bit position.

When a doubleword is stored in memory, its least significant byte is stored in the lowest numbered memory location, and its most significant byte is stored in the highest-numbered memory location using the little endian format. Recall that this is also true for word-sized data. For example, 12345678H that is stored in memory locations 00100H–00103H is stored with the 78H in memory location 00100H, the 56H in location 00101H, the 34H in location 00102H, and the 12H in location 00103H.

To define doubleword-sized data, use the assembler directive **define doubleword(s)**, or **DD**. (You can also use the DWORD directive in place of DD.) Example 1–28 shows both signed and unsigned numbers stored in memory using the DD directive. Example 1–29 shows how to define the same doublewords in Visual C++ using the **int** directive.

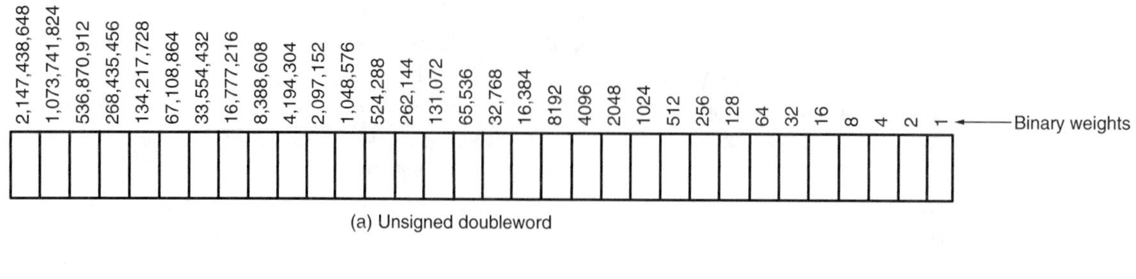

(a) Unsigned doubleword

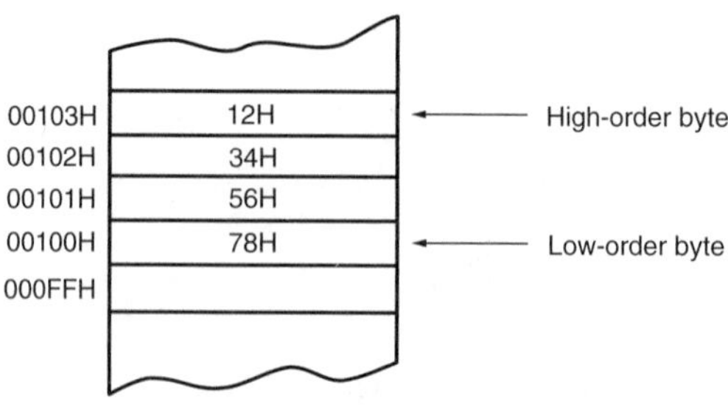

(b) The contents of memory location 00100H–00103H are the doubleword 12345678H.

FIGURE 1–16 The storage format for a 32-bit word in (a) a register and (b) four bytes of memory.

EXAMPLE 1–28

```
                        ;Unsigned doubleword-sized data
                        ;
0000    0003E1C0        DATA1 DD    254400      ;define 254400 decimal
0004    87AC1234        DATA2 DD    87AC1234H   ;define 87AC1234 hexadecimal
0008    00000046        DATA3 DD    70          ;define 70 decimal
                        ;
                        ;Signed doubleword-sized data
                        ;
000C    FFEB8058        DATA4 DD    -1343400    ;define -1343400 decimal
0010    000000C6        DATA5 DD    +198        ;define +198 decimal
0014    FFFFFFFF        DATA6 DD    -1          ;define -1 decimal
```

EXAMPLE 1–29

```
//Unsigned doubleword-sized data
//
unsigned int Data1 = 254400;            //define 254400 decimal
unsigned int Data2 = 0x87AC1234;        //define 87AC1234 hexadecimal
unsigned int Data3 = 70;                //define 70 decimal
//
//Signed doubleword-sized data
//
int Data4 = -1343400;                   //define -1342400 decimal
int Data5 = +198;                       //define +198 decimal
int Data6 = -1;                         //define -1 decimal
```

Integers may also be stored in memory that is of any width. The forms listed here are standard forms, but that doesn't mean that a 256-byte-wide integer can't be stored in the memory. The microprocessor is flexible enough to allow any size of data in assembly language. When nonstandard-width numbers are stored in memory, the DB directive is normally used to store them. For example, the 24-bit number 123456H is stored using a DB 56H,34H,12H directive. Note that this conforms to the little endian format. This could also be done in Visual C++ using the char directive.

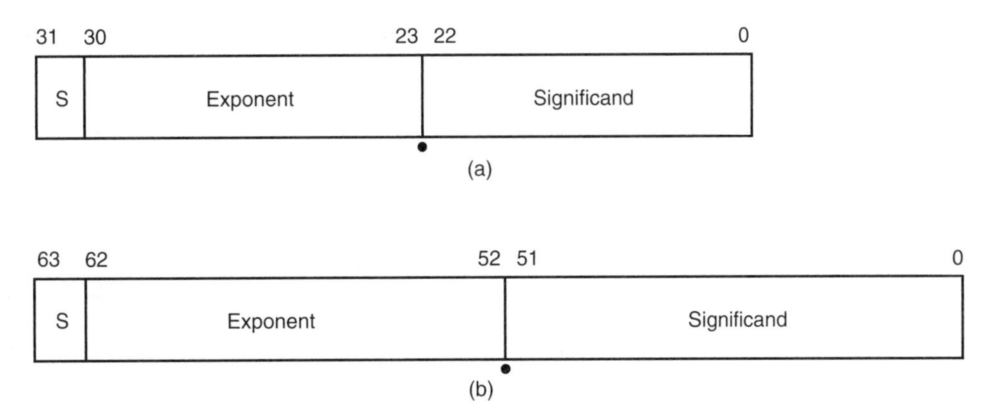

FIGURE 1–17 The floating-point numbers in (a) single-precision using a bias of 7FH and (b) double-precision using a bias of 3FFH.

Real Numbers

Because many high-level languages use the Intel family of microprocessors, real numbers are often encountered. A real number, or a **floating-point number**, as it is often called, contains two parts: a mantissa, significand, or fraction; and an exponent. Figure 1–17 depicts both the 4- and 8-byte forms of real numbers as they are stored in any Intel system. Note that the 4-byte number is called **single-precision** and the 8-byte form is called **double-precision**. The form presented here is the same form specified by the IEEE[10] standard, IEEE-754, version 10.0. The standard has been adopted as the standard form of real numbers in virtually all programming languages and many applications packages. The standard also applies the data manipulated by the numeric coprocessor in the personal computer. Figure 1–17(a) shows the single-precision form that contains a sign-bit, an 8-bit exponent, and a 24-bit fraction (mantissa). Note that because applications often require double-precision floating-point numbers (see Figure 1–17(b)), the Pentium–Pentium 4 with their 64-bit data bus perform memory transfers at twice the speed of the 80386/80486 microprocessors.

Simple arithmetic indicates that it should take 33 bits to store all three pieces of data. Not true—the 24-bit mantissa contains an **implied** (hidden) one-bit that allows the mantissa to represent 24 bits while being stored in only 23 bits. The hidden bit is the first bit of the normalized real number. When normalizing a number, it is adjusted so that its value is at least 1, but less than 2. For example, if 12 is converted to binary (1100_2), it is normalized and the result is 1.1×2^3. The whole number 1 is not stored in the 23-bit mantissa portion of the number; the 1 is the hidden one-bit. Table 1–11 shows the single-precision form of this number and others.

TABLE 1–11 Single-precision real numbers.

Decimal	*Binary*	*Normalized*	*Sign*	*Biased Exponent*	*Mantissa*
+12	1100	1.1×2^3	0	10000010	10000000 00000000 00000000
−12	1100	1.1×2^3	1	10000010	10000000 00000000 00000000
+100	1100100	1.1001×2^6	0	10000101	10010000 00000000 00000000
−1.75	1.11	1.11×2^0	1	01111111	11000000 00000000 00000000
+0.25	0.01	1.0×2^{-2}	0	01111101	00000000 00000000 00000000
+0.0	0	0	0	00000000	00000000 00000000 00000000

[10]IEEE is the Institute of Electrical and Electronic Engineers.

The exponent is stored as a **biased exponent**. With the single-precision form of the real number, the bias is 127 (7FH) and with the double-precision form, it is 1023 (3FFH). The bias and exponent are added before being stored in the exponent portion of the floating-point number. In the previous example, there is an exponent of 2^3, represented as a biased exponent of 127 + 3 or 130 (82H) in the single-precision form, or as 1026 (402H) in the double-precision form.

There are two exceptions to the rules for floating-point numbers. The number 0.0 is stored as all zeros. The number infinity is stored as all ones in the exponent and all zeros in the mantissa. The sign-bit indicates either a positive or a negative infinity.

As with other data types, the assembler can be used to define real numbers in both single- and double-precision forms. Because single-precision numbers are 32-bit numbers, use the DD directive or use the **define quadwords(s)**, or DQ, directive to define 64-bit double-precision real numbers. Optional directives for real numbers are REAL4, REAL8, and REAL10 for defining single-, double-, and extended precision real numbers. Example 1–30 shows numbers defined in real number format for the assembler. If using the inline assembler in Visual C++ single-precision numbers are defined as **float** and double-precision numbers are defined as **double** as shown in Example 1–31. There is no way to define the extended-precision floating-point number for use in Visual C++.

EXAMPLE 1–30

```
                                ;single-precision real numbers
                                ;
0000 3F9DF3B6                   NUMB1    DD     1.234        ;define 1.234
0004 C1BB3333                   NUMB2    DD     -23.4        ;define -23.4
0008 43D20000                   NUMB3    REAL4  4.2E2        ;define 420
                                ;
                                ;Double-precision real numbers
                                ;
000C 405ED9999999999A           NUMB4    DQ     123.4        ;define 123.4
0014 C1BB333333333333           NUMB5    REAL8  -23.4        ;define -23.4
                                ;
                                ;Extended-precision real numbers
                                ;
001C 4005F6CCCCCCCCCCCCCD       NUMB6    REAL10  123.4       ;define 123.4
```

EXAMPLE 1–31

```
//Single-precision real numbers
//
float Numb1 = 1.234;
float Numb2 = -23.4;
float Numb3 = 4.3e2;
//
//Double-precision real numbers
double Numb4 = 123.4;
double Numb5 = -23.4;
```

1–5 SUMMARY

1. The mechanical computer age began with the advent of the abacus in 500 BC. This first mechanical calculator remained unchanged until 1642, when Blaise Pascal improved it. An early mechanical computer system was the Analytical Engine developed by Charles Babbage in 1823. Unfortunately, this machine never functioned because of the inability to create the necessary machine parts.

2. The first electronic calculating machine was developed during World War II by Konrad Zuse, an early pioneer of digital electronics. His computer, the Z3, was used in aircraft and missile design for the German war effort.
3. The first electronic computer, which used vacuum tubes, was placed into operation in 1943 to break secret German military codes. This first electronic computer system, the Colossus, was invented by Alan Turing. Its only problem was that the program was fixed and could not be changed.
4. The first general-purpose, programmable electronic computer system was developed in 1946 at the University of Pennsylvania. This first modern computer was called the ENIAC (Electronics Numerical Integrator and Calculator).
5. The first high-level programming language, called FLOWMATIC, was developed for the UNIVAC I computer by Grace Hopper in the early 1950s. This led to FORTRAN and other early programming languages such as COBOL.
6. The world's first microprocessor, the Intel 4004, was a 4-bit microprocessor—a programmable controller on a chip—that was meager by today's standards. It addressed a mere 4096 4-bit memory location. Its instruction set contained only 45 different instructions.
7. Microprocessors that are common today include the 8086/8088, which were the first 16-bit microprocessors. Following these early 16-bit machines were the 80286, 80386, 80486, Pentium, Pentium Pro, Pentium II, Pentium III, and Pentium 4 processors. The architecture has changed from 16 bits to 32 bits and, with the Itanium, to 64 bits. With each newer version, improvements followed that increased the processor's speed and performance. From all indications, this process of speed and performance improvement will continue, although the performance increases may not always come from an increased clock frequency.
8. The DOS-based personal computers contain memory systems that include three main areas: TPA (transient program area), system area, and extended memory. The TPA holds application programs, the operating system, and drivers. The system area contains memory used for video display cards, disk drives, and the BIOS ROM. The extended memory area is only available to the 80286 through the Pentium 4 microprocessor in an AT-style or ATX-style personal computer system. The Windows-based personal computers contain memory systems that include two main areas: TPA and systems area.
9. The 8086/8088 address 1M byte of memory from locations 00000H–FFFFFH. The 80286 and 80386SX address 16M bytes of memory from locations 000000H–FFFFFFH. The 80386SL addresses 32M bytes of memory from locations 0000000H 1FFFFFFH. The 80386DX through the Pentium 4 address 4G bytes of memory from locations 00000000H–FFFFFFFFH. In addition, the Pentium Pro through the Pentium 4 can operate with a 36-bit address and access up to 64G bytes of memory from locations 000000000H–FFFFFFFFFH.
10. All versions of the 8086 through the Pentium 4 microprocessors address 64K bytes of I/O address space. These I/O ports are numbered from 0000H to FFFFH with I/O ports 0000H–03FFH reserved for use by the personal computer system. The PCI bus allows ports 0400H–FFFFH.
11. The operating system in early personal computers was either MSDOS (Microsoft disk operating system) or PCDOS (personal computer disk operating system from IBM). The operating system performs the task of operating or controlling the computer system, along with its I/O devices. Modern computers use Microsoft Windows in place of DOS as an operating system.
12. The microprocessor is the controlling element in a computer system. The microprocessor performs data transfers, does simple arithmetic and logic operations, and makes simple decisions. The microprocessor executes programs stored in the memory system to perform complex operations in short periods of time.
13. All computer systems contain three buses to control memory and I/O. The address bus is used to request a memory location or I/O device. The data bus transfers data between the microprocessor and its memory and I/O spaces. The control bus controls the memory and I/O, and requests

reading or writing of data. Control is accomplished with $\overline{IORC}$ (I/O read control), $\overline{IOWC}$ (I/O write control), $\overline{MRDC}$ (memory read control), and $\overline{MWTC}$ (memory write control).

14. Numbers are converted from any number base to decimal by noting the weights of each position. The weight of the position to the left of the radix point is always the units position in any number system. The position to the left of the units position is always the radix times one. Succeeding positions are determined by multiplying by the radix. The weight of the position to the right of the radix point is always determined by dividing by the radix.
15. Conversion from a whole decimal number to any other base is accomplished by dividing by the radix. Conversion from a fractional decimal number is accomplished by multiplying by the radix.
16. Hexadecimal data are represented in hexadecimal form or in a code called binary-coded hexadecimal (BCH). A binary-coded hexadecimal number is one that is written with a 4-bit binary number that represents each hexadecimal digit.
17. The ASCII code is used to store alphabetic or numeric data. The ASCII code is a 7-bit code; it can have an eighth bit that is used to extend the character set from 128 codes to 256 codes. The carriage return (Enter) code returns the print head or cursor to the left margin. The line feed code moves the cursor or print head down one line. Most modern applications use Unicode, which contains ASCII at codes 0000H–00FFH.
18. Binary-coded decimal (BCD) data are sometimes used in a computer system to store decimal data. These data are stored either in packed (two digits per byte) or unpacked (one digit per byte) form.
19. Binary data are stored as a byte (8 bits), word (16 bits), or doubleword (32 bits) in a computer system. These data may be unsigned or signed. Signed negative data are always stored in the two's complement form. Data that are wider than 8 bits are always stored using the little endian format. In 32-bit Visual C++ these data are represented with char (8 bits), short (16 bits), and int (32 bits).
20. Floating-point data are used in computer systems to store whole, mixed, and fractional numbers. A floating-point number is composed of a sign, a mantissa, and an exponent.
21. The assembler directives DB or BYTE define bytes; DW or WORD define words; DD or DWORD define doublewords; and DQ or QWORD define quadwords.

1–6 QUESTIONS AND PROBLEMS

1. Who developed the Analytical Engine?
2. The 1890 census used a new device called a punched card. Who developed the punched card?
3. Who was the founder of IBM Corporation?
4. Who developed the first electronic calculator?
5. The first electronic computer system was developed for what purpose?
6. The first general-purpose, programmable computer was called the ____________.
7. The world's first microprocessor was developed in 1971 by ____________.
8. Who was the Countess of Lovelace?
9. Who developed the first high-level programming language called FLOWMATIC?
10. What is a von Neumann machine?
11. Which 8-bit microprocessor ushered in the age of the microprocessor?
12. The 8085 microprocessor, introduced in 1977, has sold ____________ copies.
13. Which Intel microprocessor was the first to address 1M byte of memory?

14. The 80286 addresses ____________ bytes of memory.
15. How much memory is available to the 80486 microprocessor?
16. When did Intel introduce the Pentium microprocessor?
17. When did Intel introduce the Pentium Pro processor?
18. When did Intel introduce the Pentium 4 microprocessor?
19. Which Intel microprocessors address 64G of memory?
20. What is the acronym MIPs?
21. What is the acronym CISC?
22. A binary bit stores a(n) ____________ or a(n) ____________.
23. A computer K (pronounced kay) is equal to ____________ bytes.
24. A computer M (pronounced meg) is equal to ____________ K bytes.
25. A computer G (pronounced gig) is equal to ____________ M bytes.
26. How many typewritten pages of information are stored in a 4G-byte memory?
27. The first 1M byte of memory in a DOS-based computer system contains a(n)____________ and a(n) ____________ area.
28. How large is the Windows application programming area?
29. How much memory is found in the DOS transient program area?
30. How much memory is found in the Windows systems area?
31. The 8086 microprocessor addresses ____________ bytes of memory.
32. The Pentium 4 microprocessor addresses ____________ bytes of memory.
33. Which microprocessors address 4G bytes of memory?
34. Memory above the first 1M byte is called ____________ memory.
35. What is the system BIOS?
36. What is DOS?
37. What is the difference between an XT and an AT computer system?
38. What is the VESA local bus?
39. The ISA bus holds ____________-bit interface cards.
40. What is the USB?
41. What is the AGP?
42. What is the XMS?
43. A driver is stored in the ____________ area.
44. The personal computer system addresses ____________ bytes of I/O space.
45. What is the purpose of the BIOS?
46. Draw the block diagram of a computer system.
47. What is the purpose of the microprocessor in a microprocessor-based computer?
48. List the three buses found in all computer systems.
49. Which bus transfers the memory address to the I/O device or to the memory?
50. Which control signal causes the memory to perform a read operation?
51. What is the purpose of the $\overline{IORC}$ signal?
52. If the $\overline{MRDC}$ signal is a logic 0, which operation is performed by the microprocessor?
53. Define the purpose of the following assembler directives:
    (a) DB
    (b) DQ
    (c) DW
    (d) DD
54. Define the purpose of the following 32-bit Visual C++ directives:
    (a) char
    (b) short
    (c) int
    (d) float
    (e) double

55. Convert the following binary numbers into decimal:
    (a) 1101.01
    (b) 111001.0011
    (c) 101011.0101
    (d) 111.0001
56. Convert the following octal numbers into decimal:
    (a) 234.5
    (b) 12.3
    (c) 7767.07
    (d) 123.45
    (e) 72.72
57. Convert the following hexadecimal numbers into decimal:
    (a) A3.3
    (b) 129.C
    (c) AC.DC
    (d) FAB.3
    (e) BB8.0D
58. Convert the following decimal integers into binary, octal, and hexadecimal:
    (a) 23
    (b) 107
    (c) 1238
    (d) 92
    (e) 173
59. Convert the following decimal numbers into binary, octal, and hexadecimal:
    (a) 0.625
    (b) .00390625
    (c) .62890625
    (d) 0.75
    (e) .9375
60. Convert the following hexadecimal numbers into binary-coded hexadecimal code (BCH):
    (a) 23
    (b) AD4
    (c) 34.AD
    (d) BD32
    (e) 234.3
61. Convert the following binary-coded hexadecimal numbers into hexadecimal:
    (a) 1100 0010
    (b) 0001 0000 1111 1101
    (c) 1011 1100
    (d) 0001 0000
    (e) 1000 1011 1010
62. Convert the following binary numbers to the one's complement form:
    (a) 1000 1000
    (b) 0101 1010
    (c) 0111 0111
    (d) 1000 0000
63. Convert the following binary numbers to the two's complement form:
    (a) 1000 0001
    (b) 1010 1100
    (c) 1010 1111
    (d) 1000 0000

64. Define byte, word, and doubleword.
65. Convert the following words into ASCII-coded character strings:
    (a) FROG
    (b) Arc
    (c) Water
    (d) Well
66. What is the ASCII code for the Enter key and what is its purpose?
67. What is the Unicode?
68. Use an assembler directive to store the ASCII-character string 'What time is it?' in memory.
69. Convert the following decimal numbers into 8-bit signed binary numbers:
    (a) +32
    (b) –12
    (c) +100
    (d) –92
70. Convert the following decimal numbers into signed binary words:
    (a) +1000
    (b) –120
    (c) +800
    (d) –3212
71. Use an assembler directive to store byte-sized –34 into memory.
72. Create a byte-sized variable called Fred1 and store –34 in it in Visual C++.
73. Show how the following 16-bit hexadecimal numbers are stored in the memory system (use the standard Intel little endian format):
    (a) 1234H
    (b) A122H
    (c) B100H
74. What is the difference between the big endian and little endian formats for storing numbers that are larger than 8 bits in width?
75. Use an assembler directive to store 123A hexadecimal into memory.
76. Convert the following decimal numbers into both packed and unpacked BCD forms:
    (a) 102
    (b) 44
    (c) 301
    (d) 1000
77. Convert the following binary numbers into signed decimal numbers:
    (a) 10000000
    (b) 00110011
    (c) 10010010
    (d) 10001001
78. Convert the following BCD numbers (assume that these are packed numbers) to decimal numbers:
    (a) 10001001
    (b) 00001001
    (c) 00110010
    (d) 00000001
79. Convert the following decimal numbers into single-precision floating-point numbers:
    (a) +1.5
    (b) –10.625
    (c) +100.25
    (d) –1200

80. Convert the following single-precision floating-point numbers into decimal numbers:
    (a) 0 10000000 11000000000000000000000
    (b) 1 01111111 00000000000000000000000
    (c) 0 10000010 10010000000000000000000
81. Use the Internet to write a short report about any one of the following computer pioneers:
    (a) Charles Babbage
    (b) Konrad Zuse
    (c) Joseph Jacquard
    (d) Herman Hollerith
82. Use the Internet to write a short report about any one of the following computer languages:
    (a) COBOL
    (b) ALGOL
    (c) FORTRAN
    (d) PASCAL
83. Use the Internet to write a short report detailing the features of the Itanium 2 microprocessor.
84. Use the Internet to detail the Intel 90 nm (nanometer) fabrication technology.

CHAPTER 2

The Microprocessor and Its Architecture

INTRODUCTION

This chapter presents the microprocessor as a programmable device by first looking at its internal programming model and then at how it addresses its memory space. The architecture of the entire family of Intel microprocessors is presented simultaneously, as are the ways that the family members address the memory system.

The addressing modes for this powerful family of microprocessors are described for both the real and protected modes of operation. Real mode memory (DOS memory) exists at locations 00000H–FFFFFH—the first 1M byte of the memory system—and is present on all versions of the microprocessor. Protected mode memory (Windows memory) exists at any location in the entire memory system, but is available only to the 80286–Pentium 4, not to the earlier 8086 or 8088 microprocessors. Protected mode memory for the 80286 contains 16M bytes; for the 80386–Pentium, 4G bytes; and for the Pentium Pro through the Pentium 4, either 4G or 64G bytes.

CHAPTER OBJECTIVES

Upon completion of this chapter, you will be able to:

1. Describe the function and purpose of each program-visible register in the 8086–Pentium 4 microprocessors
2. Detail the flag register and the purpose of each flag bit
3. Describe how memory is accessed using real mode memory-addressing techniques
4. Describe how memory is accessed using protected mode memory-addressing techniques
5. Describe the program-invisible registers found within the 80286 through Pentium 4 microprocessors
6. Detail the operation of the memory-paging mechanism

2–1 INTERNAL MICROPROCESSOR ARCHITECTURE

Before a program is written or any instruction investigated, the internal configuration of the microprocessor must be known. This section of the chapter details the program-visible internal architecture of the 8086–Pentium 4 microprocessors. Also detailed are the function and purpose of each of these internal registers.

The Programming Model

The programming model of the 8086 through the Pentium 4 is considered to be **program visible** because its registers are used during application programming and are specified by the instructions. Other registers, detailed later in this chapter, are considered to be **program invisible** because they are not addressable directly during application programming, but may be used indirectly during system programming. Only the 80286 and above contain the program-invisible registers used to control and operate the protected memory system and other features.

Figure 2–1 illustrates the programming model of the 8086 through the Pentium 4 microprocessors. The earlier 8086, 8088, and 80286 contain **16-bit** internal architectures, a subset of the registers shown in Figure 2–1. The 80386 through the Pentium 4 microprocessors contain full **32-bit** internal architectures. The architectures of the earlier 8086 through the 80286 are fully upward-compatible to the 80386 through the Pentium 4. The shaded areas in this illustration

FIGURE 2–1 The programming model of the Intel 8086 through the Pentium 4.

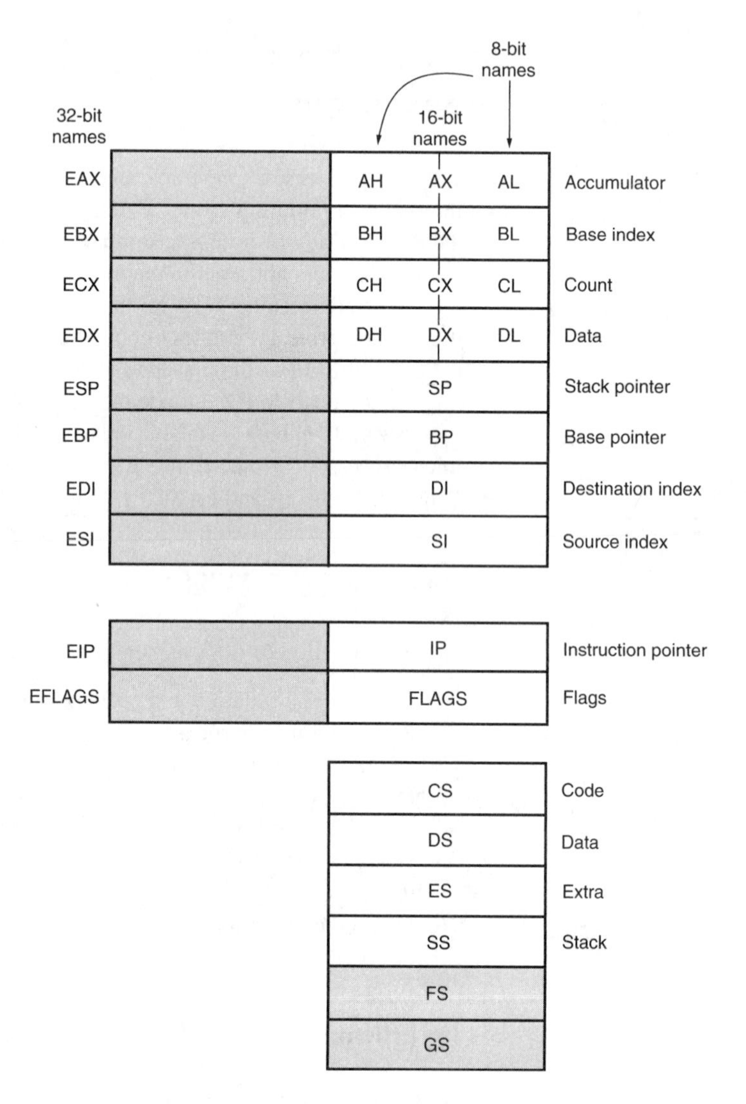

represent registers that are not found in the 8086, 8088, or 80286 microprocessors and are enhancements provided on the 80386–Pentium 4 microprocessors. Refer to Chapter 19 for the 64-bit architecture of some Pentium 4s.

The programming model contains 8-, 16-, and 32-bit registers. The 8-bit registers are AH, AL, BH, BL, CH, CL, DH, and DL and are referred to when an instruction is formed using these two-letter designations. For example, an ADD AL,AH instruction adds the 8-bit contents of AH to AL. (Only AL changes due to this instruction.) The 16-bit registers are AX, BX, CX, DX, SP, BP, DI, SI, IP, FLAGS, CS, DS, ES, SS, FS, and GS. These registers are also referenced with the two-letter designations. For example, an ADD DX,CX instruction adds the 16-bit contents of CX to DX. (Only DX changes due to this instruction.) The extended 32-bit registers are EAX, EBX, ECX, EDX, ESP, EBP, EDI, ESI, EIP, and EFLAGS. These 32-bit extended registers, and 16-bit registers FS and GS, are available only in the 80386 and above. These registers are referenced by the designations FS or GS for the two new 16-bit registers, and by a three-letter designation for the 32-bit registers. For example, an ADD ECX,EBX instruction adds the 32-bit contents of EBX to ECX. (Only ECX changes due to this instruction.)

Some registers are general-purpose or multipurpose registers, while some have special purposes. The **multipurpose registers** include EAX, EBX, ECX, EDX, EBP, EDI, and ESI. These registers hold various data sizes (bytes, words, or doublewords) and are used for almost any purpose, as dictated by a program.

Multipurpose Registers.

EAX (accumulator) — EAX is referenced as a 32-bit register (EAX), as a 16-bit register (AX), or as either of two 8-bit registers (AH and AL). Note that if an 8- or 16-bit register is addressed, only that portion of the 32-bit register changes without affecting the remaining bits. The accumulator is used for instructions such as multiplication, division, and some of the adjustment instructions. For these instructions, the accumulator has a special purpose, but is generally considered to be a multipurpose register. In the 80386 and above, the EAX register may also hold the offset address of a location in the memory system.

EBX (base index) — EBX is addressable as EBX, BX, BH, or BL. The BX register sometimes holds the offset address of a location in the memory system in all versions of the microprocessor. In the 80386 and above, EBX also can address memory data.

ECX (count) — ECX is a general-purpose register that also holds the count for various instructions. In the 80386 and above, the ECX register also can hold the offset address of memory data. Instructions that use a count are the repeated string instructions (REP/REPE/REPNE); and shift, rotate, and LOOP/LOOPD instructions. The shift and rotate instructions use CL as the count, the repeated string instructions use CX, and the LOOP/LOOPD instructions use either CX or ECX.

EDX (data) — EDX is a general-purpose register that holds a part of the result from a multiplication or part of the dividend before a division. In the 80386 and above, this register can also address memory data.

EBP (base pointer) — EBP points to a memory location in all versions of the microprocessor for memory data transfers. This register is addressed as either BP or EBP.

EDI (destination index) — EDI often addresses string destination data for the string instructions. It also functions as either a 32-bit (EDI) or 16-bit (DI) general-purpose register.

ESI (source index) ESI is used as either ESI or SI. The source index register often addresses source string data for the string instructions. Like EDI, ESI also functions as a general-purpose register. As a 16-bit register it is addressed as SI; as a 32-bit register, it is addressed as ESI.

Special-Purpose Registers. The special-purpose registers include EIP, ESP, and EFLAGS; and the segment registers CS, DS, ES, SS, FS, and GS.

EIP (instruction pointer) EIP addresses the next instruction in a section of memory defined as a code segment. This register is IP (16 bits) when the microprocessor operates in the real mode and EIP (32 bits) when the 80386 and above operate in the protected mode. Note that the 8086, 8088, and 80286 do not contain an EIP register and only the 80286 and above operate in the protected mode. The instruction pointer, which points to the next instruction in a program, is used by the microprocessor to find the next sequential instruction in a program located within the code segment. The instruction pointer can be modified with a jump or a call instruction.

ESP (stack pointer) ESP addresses an area of memory called the stack. The stack memory stores data through this pointer and is explained later in the text with the instructions that address stack data. This register is referred to as SP if used as a 16-bit register and ESP if referred to as a 32-bit register.

EFLAGS EFLAGS indicate the condition of the microprocessor and control its operation. Figure 2–2 shows the flag registers of all versions of the microprocessor. Note that the flags are upward-compatible from the 8086/8088 through the Pentium 4 microprocessors. The 8086–80286 contain a FLAG register (16 bits) and the 80386 and above contain an EFLAG register (32-bit extended flag register).

The rightmost five flag bits and the overflow flag change after many arithmetic and logic instructions execute. The flags never change for any data transfer or program control operation. Some of the flags are also used to control features found in the microprocessor. Following is a list of each flag bit, with a brief description of their function. As instructions are introduced in subsequent chapters, additional detail on the flag bits is provided. The rightmost five flags and the overflow flag is changed by most arithmetic and logic operations, although data transfers do not affect them.

C (carry) Carry holds the carry after addition or the borrow after subtraction. The carry flag also indicates error conditions, as dictated by some programs and procedures. This is especially true of the DOS function calls.

P (parity) Parity is a logic 0 for odd parity and a logic 1 for even parity. Parity is the count of ones in a number expressed as even or odd. For example, if a number contains three binary one bits, it has odd parity. If a

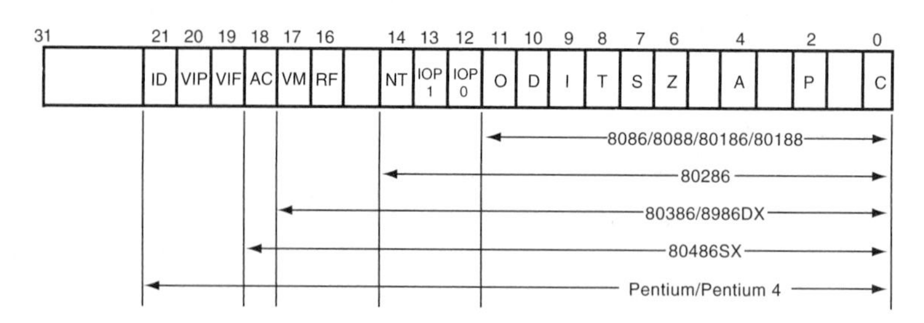

FIGURE 2–2 The EFLAG and FLAG register counts for the entire 80X86 and Pentium microprocessor family.

number contains no one bits, it has even parity. The parity flag finds little application in modern programming; it was implemented in early Intel microprocessors for checking data in data communications environments. Today parity checking is often accomplished by the data communications equipment instead of the microprocessor.

A (auxiliary carry) The auxiliary carry holds the carry (half-carry) after addition or the borrow after subtraction between bit positions 3 and 4 of the result. This highly specialized flag bit is tested by the DAA and DAS instructions to adjust the value of AL after a BCD addition or subtraction. Otherwise, the A flag bit is not used by the microprocessor or any other instructions.

Z (zero) The zero flag shows that the result of an arithmetic or logic operation is zero. If Z = 1, the result is zero; if Z = 0, the result is not zero. This may be confusing, but that is how Intel decided to name this flag.

S (sign) The sign flag holds the arithmetic sign of the result after an arithmetic or logic instruction executes. If S = 1, the sign bit (leftmost bit of a number) is set or negative; if S = 0, the sign bit is cleared or positive.

T (trap) The trap flag enables trapping through an on-chip debugging feature. (A program is debugged to find an error or bug.) If the T flag is enabled (1), the microprocessor interrupts the flow of the program on conditions as indicated by the debug registers and control registers. If the T flag is a logic 0, the trapping (debugging) feature is disabled. The Visual C++ debugging tool uses the trap feature and debug registers to debug faulty software.

I (interrupt) The interrupt flag controls the operation of the INTR (interrupt request) input pin. If I = 1, the INTR pin is enabled; if I = 0, the INTR pin is disabled. The state of the I flag bit is controlled by the STI (**set I flag**) and CLI (**clear I flag**) instructions.

D (direction) The direction flag selects either the increment or decrement mode for the DI and/or SI registers during string instructions. If D = 1, the registers are automatically decremented; if D = 0, the registers are automatically incremented. The D flag is set with the STD (**set direction**) and cleared with the CLD (**clear direction**) instructions.

O (overflow) Overflows occur when signed numbers are added or subtracted. An overflow indicates that the result has exceeded the capacity of the machine. For example, if 7FH (+127) is added—using an 8-bit addition—to 01H (+1), the result is 80H (–128). This result represents an overflow condition indicated by the overflow flag for signed addition. For unsigned operations, the overflow flag is ignored.

IOPL (I/O privilege level) IOPL is used in protected mode operation to select the privilege level for I/O devices. If the current privilege level is higher or more trusted than the IOPL, I/O executes without hindrance. If the IOPL is lower than the current privilege level, an interrupt occurs, causing execution to suspend. Note that an IOPL of 00 is the highest or most trusted and an IOPL of 11 is the lowest or least trusted.

NT (nested task) The nested task flag indicates that the current task is nested within another task in protected mode operation. This flag is set when the task is nested by software.

RF (resume) The resume flag is used with debugging to control the resumption of execution after the next instruction.

VM (virtual mode) The VM flag bit selects virtual mode operation in a protected mode system. A virtual mode system allows multiple DOS memory partitions that are 1M byte in length to coexist in the memory system. Essentially, this allows the system program to execute multiple DOS programs. VM is used to simulate DOS in the modern Windows environment.

AC (alignment check) The alignment check flag bit activates if a word or doubleword is addressed on a non-word or non-doubleword boundary. Only the 80486SX microprocessor contains the alignment check bit that is primarily used by its companion numeric coprocessor, the 80487SX, for synchronization.

VIF (virtual interrupt) The VIF is a copy of the interrupt flag bit available to the Pentium–Pentium 4 microprocessors.

VIP (virtual interrupt pending) VIP provides information about a virtual mode interrupt for the Pentium–Pentium 4 microprocessors. This is used in multitasking environments to provide the operating system with virtual interrupt flags and interrupt pending information.

ID (identification) The ID flag indicates that the Pentium–Pentium 4 microprocessors support the CPUID instruction. The CPUID instruction provides the system with information about the Pentium microprocessor, such as its version number and manufacturer.

Segment Registers. Additional registers, called segment registers, generate memory addresses when combined with other registers in the microprocessor. There are either four or six segment registers in various versions of the microprocessor. A segment register functions differently in the real mode when compared to the protected mode operation of the microprocessor. Details on their function in real and protected mode are provided later in this chapter. Following is a list of each segment register, along with its function in the system:

CS (code) The code segment is a section of memory that holds the code (programs and procedures) used by the microprocessor. The code segment register defines the starting address of the section of memory holding code. In real mode operation, it defines the start of a 64K-byte section of memory; in protected mode, it selects a descriptor that describes the starting address and length of a section of memory holding code. The code segment is limited to 64K bytes in the 8088–80286, and 4G bytes in the 80386 and above when these microprocessors operate in the protected mode.

DS (data) The data segment is a section of memory that contains most data used by a program. Data are accessed in the data segment by an offset address or the contents of other registers that hold the offset address. As with the code segment and other segments, the length is limited to 64K bytes in the 8086–80286, and 4G bytes in the 80386 and above.

ES (extra) The extra segment is an additional data segment that is used by some of the string instructions to hold destination data.

SS (stack) The stack segment defines the area of memory used for the stack. The stack entry point is determined by the stack segment and stack pointer registers. The BP register also addresses data within the stack segment.

FS and GS The FS and GS segments are supplemental segment registers available in the 80386–Pentium 4 microprocessors to allow two additional memory segments for access by programs. Windows uses these segments for internal operations, but no definition of their usage is available.

2–2 REAL MODE MEMORY ADDRESSING

The 80286 and above operate in either the real or protected mode. Only the 8086 and 8088 operate exclusively in the real mode. This section of the text details the operation of the microprocessor in the real mode. **Real mode operation** allows the microprocessor to address only the first 1M byte of memory space—even if it is the Pentium 4 microprocessor. Note that the first 1M byte of memory is called the **real memory**, **conventional memory**, or **DOS memory** system. The DOS operating system requires the microprocessor to operate in the real mode. Windows does not use the real mode. Real mode operation allows application software written for the 8086/8088, which only contains 1M byte of memory, to function in the 80286 and above without changing the software. The upward compatibility of software is partially responsible for the continuing success of the Intel family of microprocessors. In all cases, each of these microprocessors begins operation in the real mode by default whenever power is applied or the microprocessor is reset.

Segments and Offsets

A combination of a segment address and an offset address accesses a memory location in the real mode. All real mode memory addresses must consist of a segment address plus an offset address. The **segment address**, located within one of the segment registers, defines the beginning address of any 64K-byte memory segment. The **offset address** selects any location within the 64K-byte memory segment. Segments in the real mode always have a length of 64K bytes. Figure 2–3 shows how the **segment plus offset** addressing scheme selects a memory location. This illustration shows a memory segment that begins at location 10000H and ends at location IFFFFH—64K bytes in length. It also shows how an offset address, sometimes called a **displacement**, of

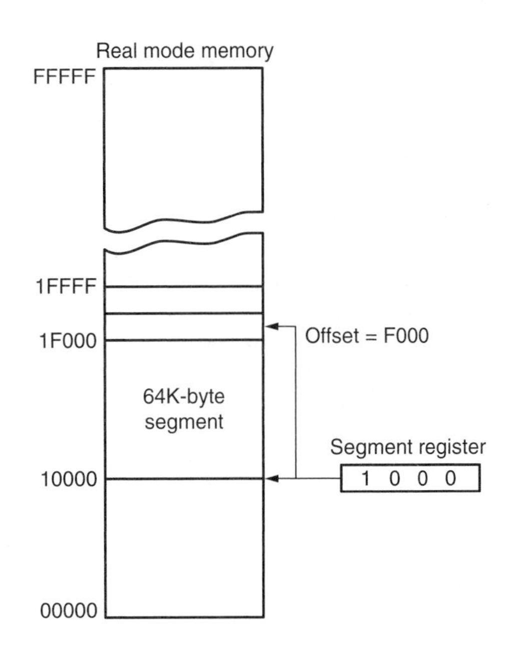

FIGURE 2–3 The real mode memory-addressing scheme, using a segment address plus an offset.

TABLE 2–1 Example of real mode segment addresses.

Segment Register	*Starting Address*	*Ending Address*
2000H	20000H	2FFFFH
2001H	20010H	3000FH
2100H	21000H	30FFFH
AB00H	AB000H	BAFFFH
1234H	12340H	2233FH

F000H selects location 1F000H in the memory system. Note that the offset or displacement is the distance above the start of the segment, as shown in Figure 2–3.

The segment register in Figure 2–3 contains 1000H, yet it addresses a starting segment at location 10000H. In the real mode, each segment register is internally appended with a **0H** on its rightmost end. This forms a 20-bit memory address, allowing it to access the start of a segment. The microprocessor must generate a 20-bit memory address to access a location within the first 1M of memory. For example, when a segment register contains 1200H, it addresses a 64K-byte memory segment beginning at location 12000H. Likewise, if a segment register contains 1201H, it addresses a memory segment beginning at location 12010H. Because of the internally appended 0H, real mode segments can begin only at a 16-byte boundary in the memory system. This 16-byte boundary is often called a **paragraph**.

Because a real mode segment of memory is 64K in length, once the beginning address is known, the **ending address** is found by adding FFFFH. For example, if a segment register contains 3000H, the first address of the segment is 30000H, and the last address is 30000H + FFFFH or 3FFFFH. Table 2–1 shows several examples of segment register contents and the starting and ending addresses of the memory segments selected by each segment address.

The offset address, which is a part of the address, is added to the start of the segment to address a memory location within the memory segment. For example, if the segment address is 1000H and the offset address is 2000H, the microprocessor addresses memory location 12000H. The offset address is always added to the starting address of the segment to locate the data. The segment and offset address is sometimes written as 1000:2000 for a segment address of 1000H with an offset of 2000H.

In the 80286 (with special external circuitry) and the 80386 through the Pentium 4, an extra 64K minus 16 bytes of memory is addressable when the segment address is FFFFH and the HIMEM.SYS driver for DOS is installed in the system. This area of memory (0FFFF0H–10FFEFH) is referred to as **high memory**. When an address is generated using a segment address of FFFFH, the A20 address pin is enabled (if supported) when an offset is added. For example, if the segment address is FFFFH and the offset address is 4000H, the machine addresses memory location FFFF0H + 4000H or 103FF0H. Notice that the A20 address line is the one in address 103FF0H. If A20 is not supported, the address is generated as 03FF0H because A20 remains a logic zero.

Some addressing modes combine more than one register and an offset value to form an offset address. When this occurs, the sum of these values may exceed FFFFH. For example, the address accessed in a segment whose segment address is 4000H and whose offset address is specified as the sum of F000H plus 3000H will access memory location 42000H instead of location 52000H. When F000H and 3000H are added, they form a 16-bit (**modulo 16**) sum of 2000H used as the offset address, not 12000H, the true sum. Note that the carry of 1 (F000H + 3000H = 12000H) is dropped for this addition to form the offset address of 2000H. The address is generated as 4000:2000 or 42000H.

Default Segment and Offset Registers

The microprocessor has a set of rules that apply to segments whenever memory is addressed. These rules, which apply in the real and protected mode, define the segment register and offset register combination. For example, the code segment register is always used with the instruction

TABLE 2–2 Default 16-bit segment and offset combinations.

Segment	Offset	Special Purpose
CS	IP	Instruction address
SS	SP or BP	Stack address
DS	BX, DI, SI, an 8- or 16-bit number	Data address
ES	DI for string instructions	String destination address

pointer to address the next instruction in a program. This combination is **CS:IP** or **CS:EIP**, depending upon the microprocessor's mode of operation. The **code segment** register defines the start of the code segment and the **instruction pointer** locates the next instruction within the code segment. This combination (CS:IP or CS:EIP) locates the next instruction executed by the microprocessor. For example, if CS = 1400H and IP/EIP = 1200H, the microprocessor fetches its next instruction from memory location 14000H + 1200H or 15200H.

Another of the default combinations is the **stack**. Stack data are referenced through the stack segment at the memory location addressed by either the stack pointer (SP/ESP) or the pointer (BP/EBP). These combinations are referred to as SS:SP (SS:ESP) or SS:BP (SS:EBP). For example, if SS = 2000H and BP = 3000H, the microprocessor addresses memory location 23000H for the stack segment memory location. Note that in real mode, only the rightmost 16 bits of the extended register address a location within the memory segment. In the 80386–Pentium 4, never place a number larger than FFFFH into an offset register if the microprocessor is operated in the real mode. This causes the system to halt and indicate an addressing error.

Other defaults are shown in Table 2–2 for addressing memory using any Intel microprocessor with 16-bit registers. Table 2–3 shows the defaults assumed in the 80386 and above using 32-bit registers. Note that the 80386 and above have a far greater selection of segment/offset address combinations than do the 8086 through the 80286 microprocessors.

The 8086–80286 microprocessors allow four memory segments and the 80386–Pentium 4 microprocessors allow six memory segments. Figure 2–4 shows a system that contains four memory segments. Note that a memory segment can touch or even overlap if 64K bytes of memory are not required for a segment. Think of segments as windows that can be moved over any area of memory to access data or code. Also note that a program can have more than four or six segments, but can only access four or six segments at a time.

Suppose that an application program requires 1000H bytes of memory for its code, 190H bytes of memory for its data, and 200H bytes of memory for its stack. This application does not require an extra segment. When this program is placed in the memory system by DOS, it is loaded in the TPA at the first available area of memory above the drivers and other TPA programs. This area is indicated by a **free-pointer** that is maintained by DOS. Program loading is

TABLE 2–3 Default 32-bit segment and offset combinations.

Segment	Offset	Special Purpose
CS	EIP	Instruction address
SS	ESP or EBP	Stack address
DS	EAX, EBX, ECX, EDX, ESI, EDI, an 8- or 32-bit number	Data address
ES	EDI for string instructions	String destination address
FS	No default	General address
GS	No default	General address

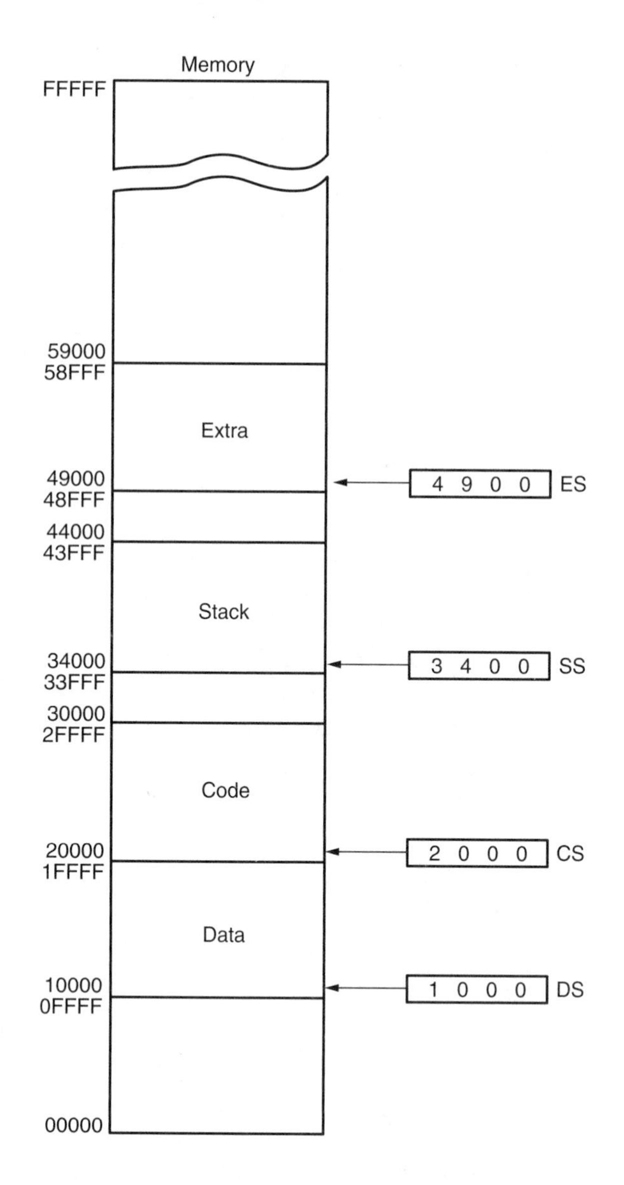

FIGURE 2–4 A memory system showing the placement of four memory segments.

handled automatically by the **program loader** located within DOS. Figure 2–5 shows how an application is stored in the memory system. The segments show an overlap because the amount of data in them does not require 64K bytes of memory. The side view of the segments clearly shows the overlap. It also shows how segments can be moved over any area of memory by changing the segment starting address. Fortunately, the DOS program loader calculates and assigns segment starting addresses.

Segment and Offset Addressing Scheme Allows Relocation

The segment and offset addressing scheme seems unduly complicated. It is complicated, but it also affords an advantage to the system. This complicated scheme of segment plus offset addressing allows DOS programs to be relocated in the memory system. It also allows programs written to function in the real mode to operate in a protected mode system. A **relocatable program** is one that can be placed into any area of memory and executed without change. **Relocatable data** are data that can be placed in any area of memory and used without any change to the program. The segment and offset addressing scheme allows both programs and data to be relocated without changing a thing in a program or data. This is ideal for use in a general-purpose computer system in which not all

FIGURE 2–5 An application program containing a code, data, and stack segment loaded into a DOS system memory.

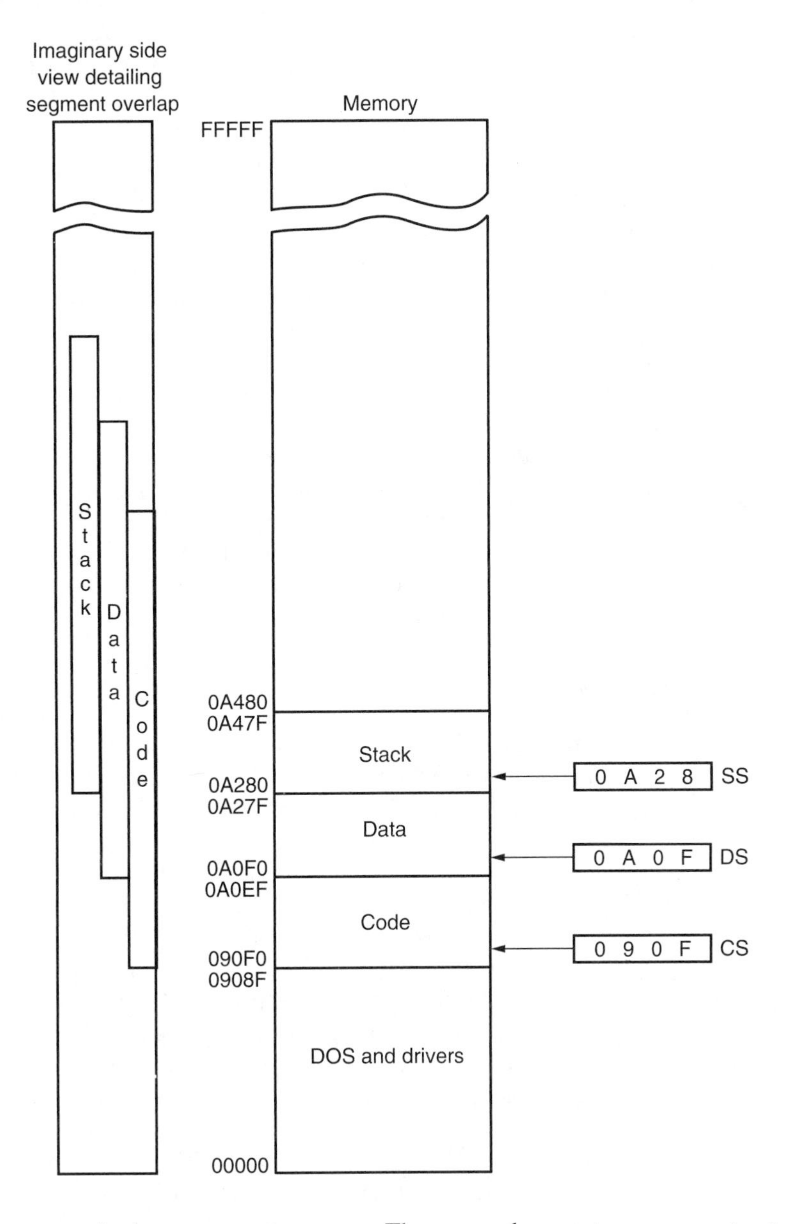

machines contain the same memory areas. The personal computer memory structure is different from machine to machine, requiring relocatable software and data.

Because memory is addressed within a segment by an offset address, the memory segment can be moved to any place in the memory system without changing any of the offset addresses. This is accomplished by moving the entire program, as a block, to a new area and then changing only the contents of the segment registers. If an instruction is four bytes above the start of the segment, its offset address is 4. If the entire program is moved to a new area of memory, this offset address of 4 still points to four bytes above the start of the segment. Only the contents of the segment register must be changed to address the program in the new area of memory. Without this feature, a program would have to be extensively rewritten or altered before it is moved. This would require additional time or many versions of a program for the many different configurations of computer systems. This concept also applies to programs written to execute in the protected mode for Windows. In the Windows environment all programs are written assuming that the first 2G of memory are available for code and data. When the program is loaded it is placed in the actual memory, which may be anywhere and a portion may be located on the disk in the form of a swap file.

2–3 INTRODUCTION TO PROTECTED MODE MEMORY ADDRESSING

Protected mode memory addressing (80286 and above) allows access to data and programs located above the first 1M byte of memory, as well as within the first 1M byte of memory. **Protected mode** is where Windows operates. Addressing this extended section of the memory system requires a change to the segment plus a offset addressing scheme used with real mode memory addressing. When data and programs are addressed in extended memory, the offset address is still used to access information located within the memory segment. One difference is that the segment address, as discussed with real mode memory addressing, is no longer present in the protected mode. In place of the segment address, the segment register contains a **selector** that selects a descriptor from a descriptor table. The **descriptor** describes the memory segment's location, length, and access rights. Because the segment register and offset address still access memory, protected mode instructions are identical to real mode instructions. In fact, most programs written to function in the real mode will function without change in the protected mode. The difference between modes is in the way that the segment register is interpreted by the microprocessor to access the memory segment. Another difference, in the 80386 and above, is that the offset address can be a 32-bit number instead of a 16-bit number in the protected mode. A 32-bit offset address allows the microprocessor to access data within a segment that can be up to 4G bytes in length.

Selectors and Descriptors

The selector, located in the segment register, selects one of 8192 descriptors from one of two tables of descriptors. The **descriptor** describes the location, length, and access rights of the segment of memory. Indirectly, the segment register still selects a memory segment, but not directly as in the real mode. For example, in the real mode, if CS = 0008H, the code segment begins at location 00080H. In the protected mode, this segment number can address any memory location in the entire system for the code segment, as explained shortly.

There are two descriptor tables used with the segment registers: one contains global descriptors and the other contains local descriptors. The **global descriptors** contain segment definitions that apply to all programs, whereas the **local descriptors** are usually unique to an application. You might call a global descriptor a **system descriptor** and call a local descriptor an **application descriptor**. Each descriptor table contains 8192 descriptors, so a total of 16,384 descriptors are available to an application at any time. Because the descriptor describes a memory segment, this allows up to 16,384 memory segments to be described for each application. Since a memory segment can be up to 4G bytes in length, this means that an application could have access to 4G × 16,384 bytes of memory or 64T bytes.

Figure 2–6 shows the format of a descriptor for the 80286 through the Pentium 4. Note that each descriptor is eight bytes in length, so the global and local descriptor tables are each a

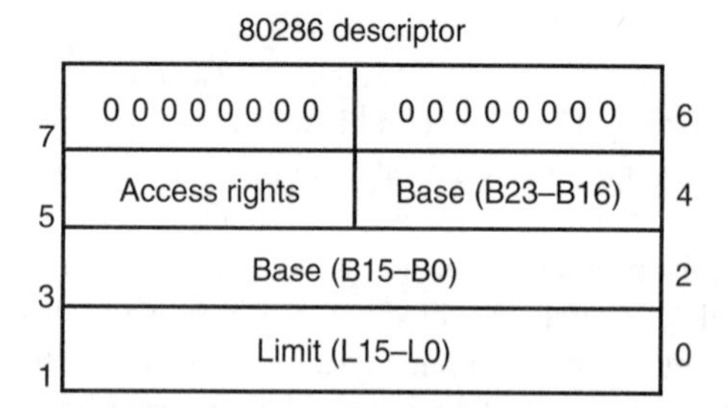

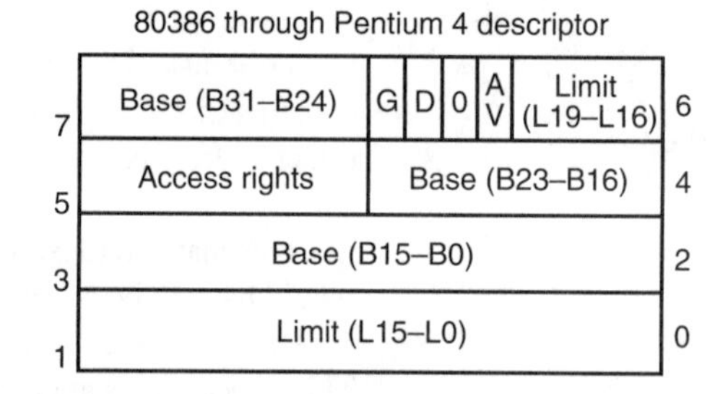

FIGURE 2–6 The descriptor formats for the 80286 and 80386 through Pentium 4 microprocessors.

maximum of 64K bytes in length. Descriptors for the 80286 and the 80386–Pentium 4 differ slightly, but the 80286 descriptor is upward-compatible.

The **base address** portion of the descriptor indicates the starting location of the memory segment. For the 80286 microprocessor, the base address is a 24-bit address, so segments begin at any location in its 16M bytes of memory. Note that the paragraph boundary limitation is removed in these microprocessors when operated in the protected mode so segments may begin at any address. The 80386 and above use a 32-bit base address that allows segments to begin at any location in its 4G bytes of memory. Notice how the 80286 descriptor's base address is upward-compatible to the 80386 through the Pentium 4 descriptor because its most-significant 16 bits are 0000H. Refer to Chapters 18 and 19 for additional detail on the 64G memory space provided by the Pentium Pro through the Pentium 4.

The segment limit contains the last offset address found in a segment. For example, if a segment begins at memory location F00000H and ends at location F000FFH, the base address is F00000H and the limit is FFH. For the 80286 microprocessor, the base address is F00000H and the limit is 00FFH. For the 80386 and above, the base address is 00F00000H and the limit is 000FFH. Notice that the 80286 has a 16-bit limit and the 80386 through the Pentium 4 have a 20-bit limit. An 80286 accesses memory segments that are between 1 and 64K bytes in length. The 80386 and above access memory segments that are between 1 and 1M byte, or 4K and 4G bytes in length.

There is another feature found in the 80386 through the Pentium 4 descriptor that is not found in the 80286 descriptor: the G bit, or **granularity bit**. If G = 0, the limit specifies a segment limit of 00000H to FFFFFH. If G = 1, the value of the limit is multiplied by 4K bytes (appended with FFFH). The limit is then 00000FFFFH to FFFFFFFFH, if G = 1. This allows a segment length of 4K to 4G bytes in steps of 4K bytes. The reason that the segment length is 64K bytes in the 80286 is that the offset address is always 16 bits because of its 16-bit internal architecture. The 80386 and above use a 32-bit architecture that allows an offset address, in the protected mode operation, of the 32 bits. This 32-bit offset address allows segment lengths of 4G bytes and the 16-bit offset address allows segment lengths of 64K bytes. Operating systems operate in a 16- or 32-bit environment. For example, DOS uses a 16-bit environment, while most Windows applications use a 32-bit environment called **WIN32**.

Example 2–1 shows the segment start and end if the base address is 10000000H, the limit is 001FFH, and the G bit = 0.

EXAMPLE 2–1

```
Base = Start = 10000000H
G = 0
End = Base + Limit = 10000000H + 001FFH = 100001FFH
```

Example 2–2 uses the same data as Example 2–1, except that the G bit = 1. Notice that the limit is appended with FFFH to determine the ending segment address.

EXAMPLE 2–2

```
Base = Start = 10000000H
G = 1
End = Base + Limit = 10000000H + 001FFFFFH = 101FFFFFH
```

The AV bit, in the 80386 and above descriptor, is used by some operating systems to indicate that the segment is available (AV = 1) or not available (AV = 0). The D bit indicates how the 80386 through the Pentium 4 instructions access register and memory data in the protected or real mode. If D = 0, the instructions are 16-bit instructions, compatible with the 8086–80286 microprocessors. This means that the instructions use 16-bit offset addresses and 16-bit registers by default. This mode is often called the 16-bit instruction mode or DOS mode. If D = 1, the instructions are 32-bit

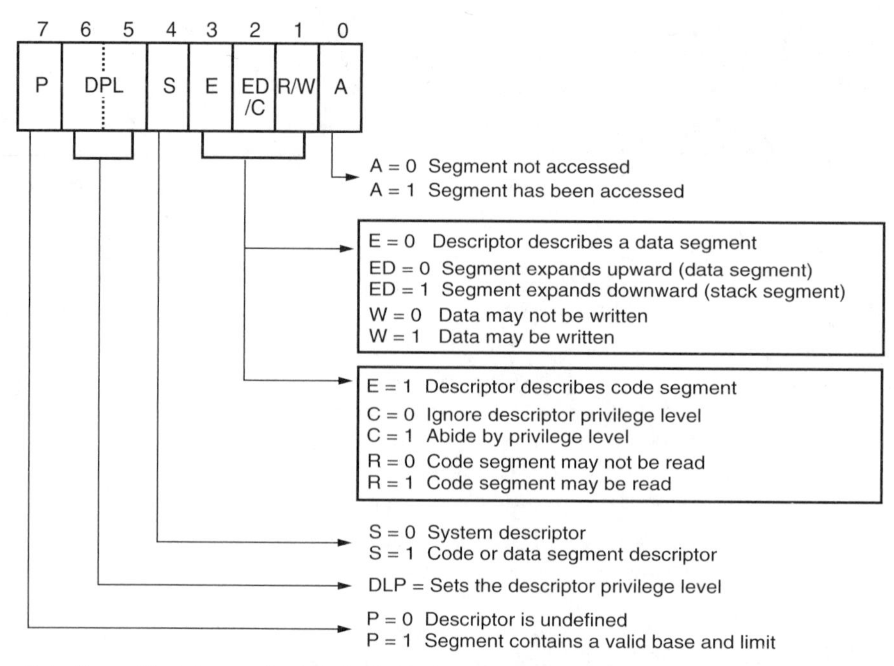

Note: Some of the letters used to describe the bits in the access rights bytes vary in Intel documentation.

FIGURE 2–7 The access rights byte for the 80286 through Pentium 4 descriptor.

instructions. By default, the 32-bit instruction mode assumes that all offset addresses and all registers are 32 bits. Note that the default for register size and offset address is overridden in both the 16- and 32-bit instruction modes. Both the MSDOS and PCDOS operating systems require that the instructions always be used in the 16-bit instruction mode. Windows 3.1, and any application that was written for it, also require that the 16-bit instruction mode be selected. Note that the instruction mode is accessible only in a protected mode system such as Windows XP. More detail on these modes and their application to the instruction set appears in Chapters 3 and 4.

The **access rights byte** (see Figure 2–7) controls access to the protected mode segment. This byte describes how the segment functions in the system. The access rights byte allows complete control over the segment. If the segment is a data segment, the direction of growth is specified. If the segment grows beyond its limit, the microprocessor's operating system program is interrupted, indicating a general protection fault. You can even specify whether a data segment can be written or is write-protected. The code segment is also controlled in a similar fashion and can have reading inhibited to protect software.

Descriptors are chosen from the descriptor table by the segment register. Figure 2–8 shows how the segment register functions in the protected mode system. The segment register contains a 13-bit selector field, a table selector bit, and a requested privilege level field. The 13-bit **selector** chooses one of the 8192 descriptors from the descriptor table. The **TI bit** selects either the global descriptor table (TI = 0) or the local descriptor table (TI = 1). The **requested privilege level** (RPL) requests the access privilege level of a memory segment. The highest privilege level is 00 and the lowest is 11. If the requested privilege level matches or is higher in priority than the privilege level set by the access rights byte, access is granted. For example, if the requested privilege level is 10 and the access rights byte sets the segment privilege level at 11, access is granted because 10 is higher in priority than privilege level 11. Privilege levels are used in multiuser environments. Windows uses privilege level 00 (**ring 0**) for the kernel and driver programs and level 11 (**ring 3**) for applications. Windows does not use levels 01 or 10. If privilege levels are violated, the system normally indicates an application or privilege level violation.

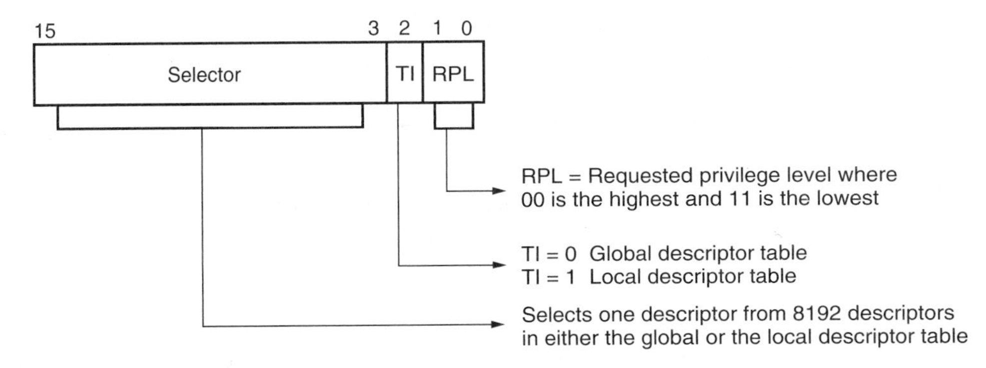

FIGURE 2–8 The contents of a segment register during protected mode operation of the 80286 through Pentium 4 microprocessors.

Figure 2–9 shows how the segment register, containing a selector, chooses a descriptor from the global descriptor table. The entry in the global descriptor table selects a segment in the memory system. In this illustration, DS contains 0008H, which accesses the descriptor number 1 from the global descriptor table using a requested privilege level of 00. Descriptor number 1 contains a descriptor that defines the base address as 00100000H with a segment limit of 000FFH. This means that a value of 0008H loaded into DS causes the microprocessor to use memory locations 00100000H–001000FFH for the data segment with this example descriptor

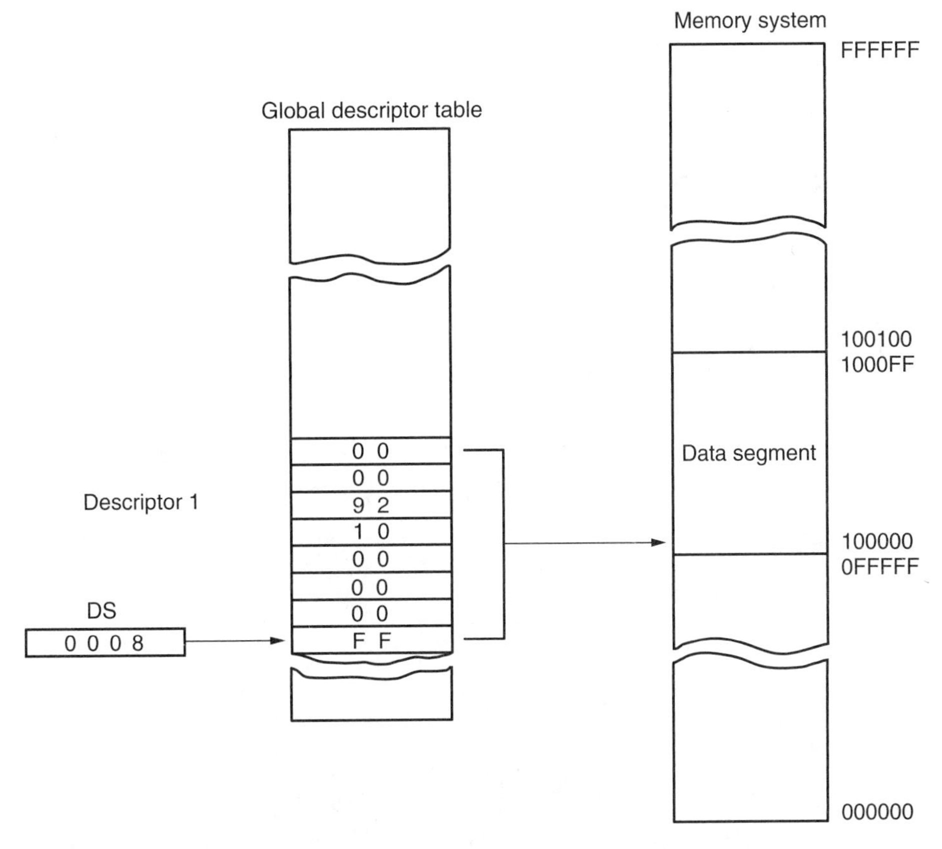

FIGURE 2–9 Using the DS register to select a description from the global descriptor table. In this example, the DS register accesses memory locations 100000H–1000FFH as a data segment.

table. Note that descriptor zero is called the null descriptor, must contain all zeros, and may not be used for accessing memory.

Program-Invisible Registers

The global and local descriptor tables are found in the memory system. In order to access and specify the address of these tables, the 80286–Pentium 4 contain program-invisible registers. The program-invisible registers are not directly addressed by software so they are given this name (although some of these registers are accessed by the system software). Figure 2–10 illustrates the program-invisible registers as they appear in the 80286 through the Pentium 4. These registers control the microprocessor when operated in protected mode.

Each of the segment registers contains a program-invisible portion used in the protected mode. The program-invisible portion of these registers is often called cache memory because cache is any memory that stores information. This cache is not to be confused with the level 1 or level 2 caches found with the microprocessor. The program-invisible portion of the segment register is loaded with the base address, limit, and access rights each time the number segment register is changed. When a new segment number is placed in a segment register, the microprocessor accesses a descriptor table and loads the descriptor into the program-invisible portion of the segment register. It is held there and used to access the memory segment until the segment number is again changed. This allows the microprocessor to repeatedly access a memory segment without referring to the descriptor table (hence the term cache).

The GDTR (**global descriptor table register**) and IDTR (**interrupt descriptor table register**) contain the base address of the descriptor table and its limit. The limit of each descriptor table is 16 bits because the maximum table length is 64K bytes. When the protected mode operation is desired, the address of the global descriptor table and its limit are loaded into the GDTR.

Before using the protected mode, the interrupt descriptor table and the IDTR must also be initialized. More detail is provided on protected mode operation later in the text. At this point programming and additional description of these registers are impossible.

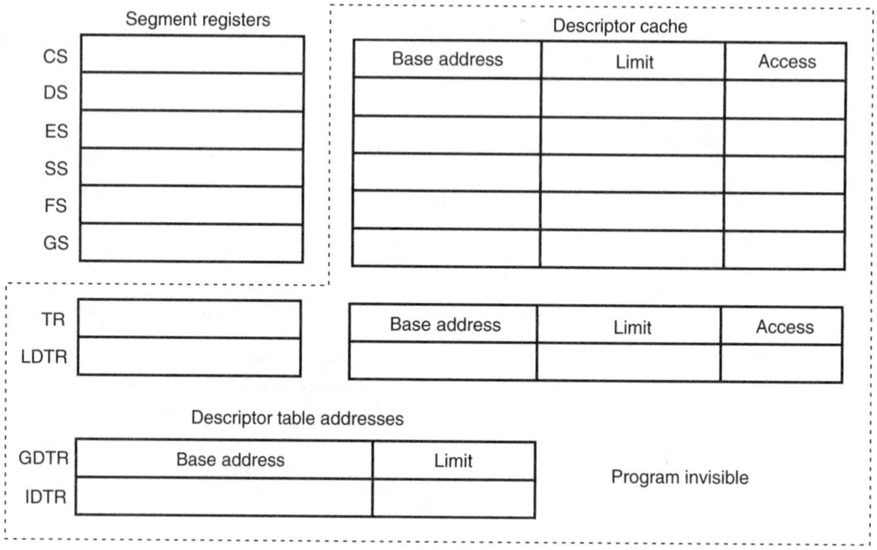

Notes:
1. The 80286 does not contain FS and GS nor the program-invisible portions of these registers.
2. The 80286 contains a base address that is 24-bits and a limit that is 16-bits.
3. The 80386/80486/Pentium/Pentium Pro contain a base address that is 32-bits and a limit that is 20-bits.
4. The access rights are 8-bits in the 80286 and 12-bits in the 80386/80486/Pentium–Pentium 4.

FIGURE 2–10 The program-invisible register within the 80286–Pentium 4 microprocessors.

The location of the local descriptor table is selected from the global descriptor table. One of the global descriptors is set up to address the local descriptor table. To access the local descriptor table, the LDTR (**local descriptor table register**) is loaded with a selector, just as a segment register is loaded with a selector. This selector accesses the global descriptor table and loads the address, limit, and access rights of the local descriptor table into the cache portion of the LDTR.

The TR (**task register**) holds a selector, which accesses a descriptor that defines a task. A task is most often a procedure or application program. The descriptor for the procedure or application program is stored in the global descriptor table, so access can be controlled through the privilege levels. The task register allows a context or task switch in about 17 μs. Task switching allows the microprocessor to switch between tasks in a fairly short amount of time. The task switch allows multitasking systems to switch from one task to another in a simple and orderly fashion.

2–4 MEMORY PAGING

The **memory paging mechanism** located within the 80386 and above allows any physical memory location to be assigned to any linear address. The **linear address** is defined as the address generated by a program. The **physical address** is the actual memory location accessed by a program. With the memory paging unit, the linear address is invisibly translated to any physical address, which allows an application written to function at a specific address to be relocated through the paging mechanism. It also allows memory to be placed into areas where no memory exists. An example is the upper memory blocks provided by EMM386.EXE in a DOS system.

The EMM386.EXE program reassigns extended memory, in 4K blocks, to the system memory between the video BIOS and the system BIOS ROMS for upper memory blocks. Without the paging mechanism, the use of this area of memory is impossible.

In Windows each application is allowed a 2G linear address space from locations 00000000H–7FFFFFFFH even though there may not be enough memory or memory available at these addresses. Through paging to the hard disk drive and paging to the memory through the memory paging unit any Windows application can be executed.

Paging Registers

The paging unit is controlled by the contents of the microprocessor's control registers. See Figure 2–11 for the contents of control registers CR0 through CR4. Note that these registers are available to the 80386 through the Pentium microprocessors. Beginning with the Pentium, an additional control register labeled CR4 controls extensions to the basic architecture provided in the Pentium or newer microprocessors. One of these features is a 4M-byte page that is enabled by setting bit position 4 of CR4. Refer to Chapters 18 and 19 for additional details on the 4M-byte memory paging system.

The registers important to the paging unit are CR0 and CR3. The leftmost bit (PG) position of CR0 selects paging when placed at a logic 1 level. If the PG bit is cleared (0), the linear address generated by the program becomes the physical address used to access memory. If the PG bit is set (1), the linear address is converted to a physical address through the paging mechanism. The paging mechanism functions in both the real and protected modes.

CR3 contains the page directory base or root address and the PCD and PWT bits. The PCD and PWT bits control the operation of the PCD and PWT pins on the microprocessor. If PCD is set (1), the PCD pin becomes a logic one during bus cycles that are not paged. This allows the external hardware to control the level 2 cache memory. (Note that the level 2 cache memory is an internal [on modern versions of the Pentium] high-speed memory that functions as a buffer between the microprocessor and the main DRAM memory system.) The PWT bit also appears on the PWT pin during bus cycles that are not paged to control the write-through cache in the system. The page directory base address locates the directory for the page translation unit. Note

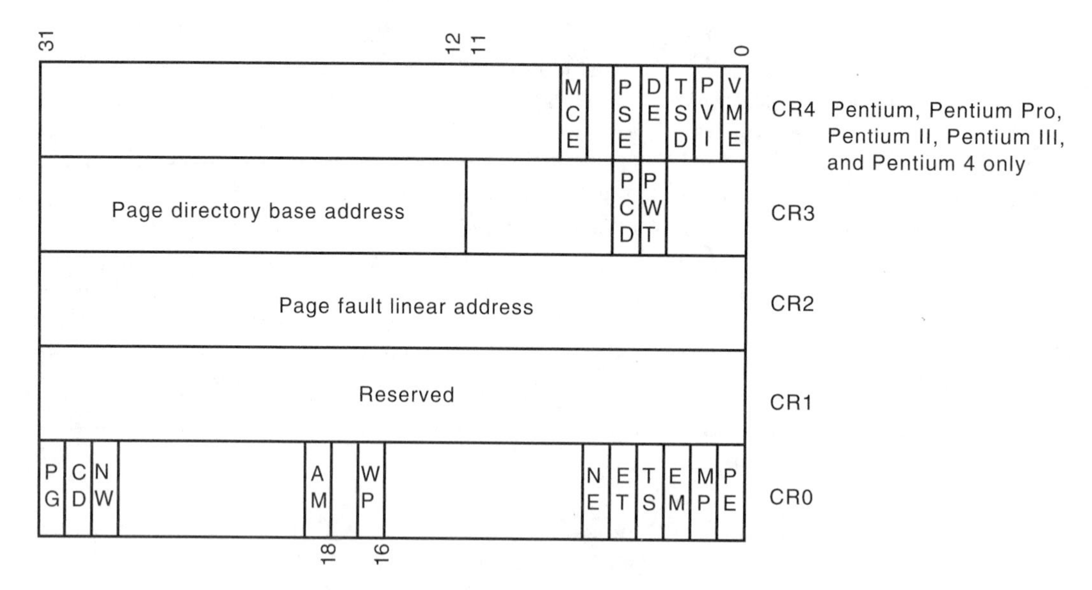

FIGURE 2–11 The control register structure of the microprocessor.

that this address locates the page directory at any 4K boundary in the memory system because it is appended internally with 000H. The page directory contains 1024 directory entries of four bytes each. Each page directory entry addresses a page table that contains 1024 entries.

The linear address, as it is generated by the software, is broken into three sections that are used to access the **page directory entry**, **page table entry**, and **memory page offset address**. Figure 2–12 shows the linear address and its makeup for paging. Notice how the leftmost 10 bits address an entry in the page directory. For linear address 00000000H–003FFFFFH, the first page directory is accessed. Each page directory entry represents or repages a 4M section of the memory system. The contents of the page directory select a page table that is indexed by the next 10 bits of

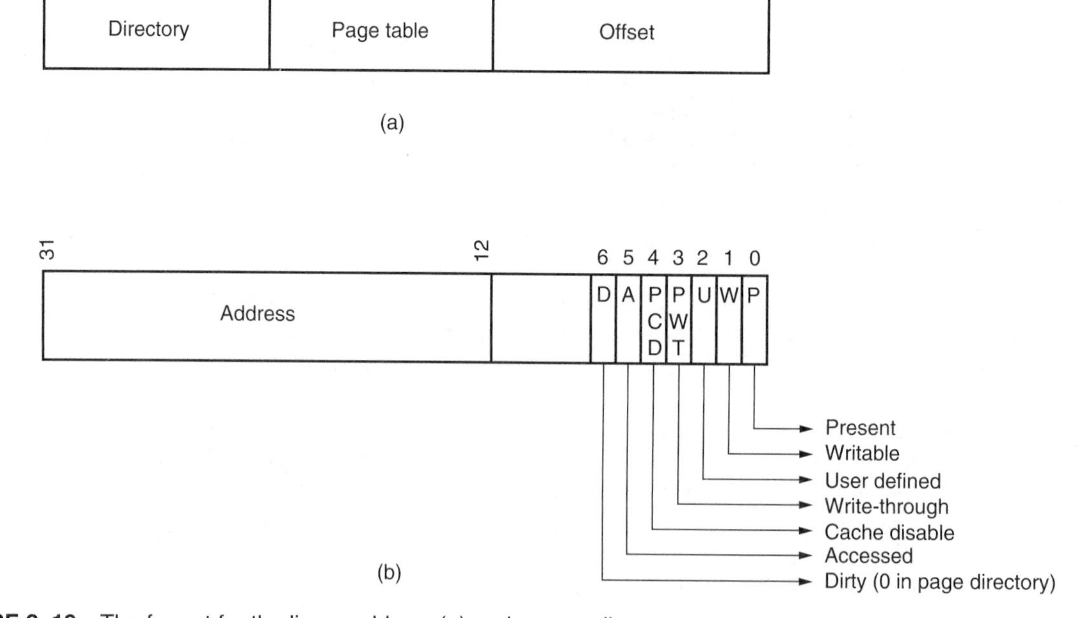

FIGURE 2–12 The format for the linear address (a) and a page directory or page table entry (b).

the linear address (bit positions 12–21). This means that address 00000000H–00000FFFH selects page directory entry 0 and page table entry 0. Notice this is a 4K-byte address range. The offset part of the linear address (bit positions 0–11) next selects a byte in the 4K-byte memory page. In Figure 2–12, if the page table entry 0 contains address 00100000H, then the physical address is 00100000H–00100FFFH for linear address 00000000H–00000FFFH. This means that when the program accesses a location between 00000000H and 00000FFFH, the microprocessor physically addresses location 00100000H–00100FFFH.

Because the act of repaging a 4K-byte section of memory requires access to the page directory and a page table, which are both located in memory, Intel has incorporated a special type of cache called the TLB (**translation look-aside buffer**). In the 80486 microprocessor, the cache holds the 32 most recent page translation addresses. This means that the last 32 page table translations are stored in the TLB, so if the same area of memory is accessed, the address is already present in the TLB, and access to the page directory and page tables is not required. This speeds program execution. If a translation is not in the TLB, the page directory and page table must be accessed, which requires additional execution time. The Pentium–Pentium 4 microprocessors contain separate TLBs for each of their instruction and data caches.

The Page Directory and Page Table

Figure 2–13 shows the page directory, a few page tables, and some memory pages. There is only one page directory in the system. The page directory contains 1024 doubleword addresses that locate up to 1024 page tables. The page directory and each page table are 4K bytes in length. If the entire 4G byte of memory is paged, the system must allocate 4K bytes of memory for the page directory, and 4K times 1024 or 4M bytes for the 1024 page tables. This represents a considerable investment in memory resources.

The DOS system and EMM386.EXE use page tables to redefine the area of memory between locations C8000H–EFFFFH as upper memory blocks. This is done by repaging

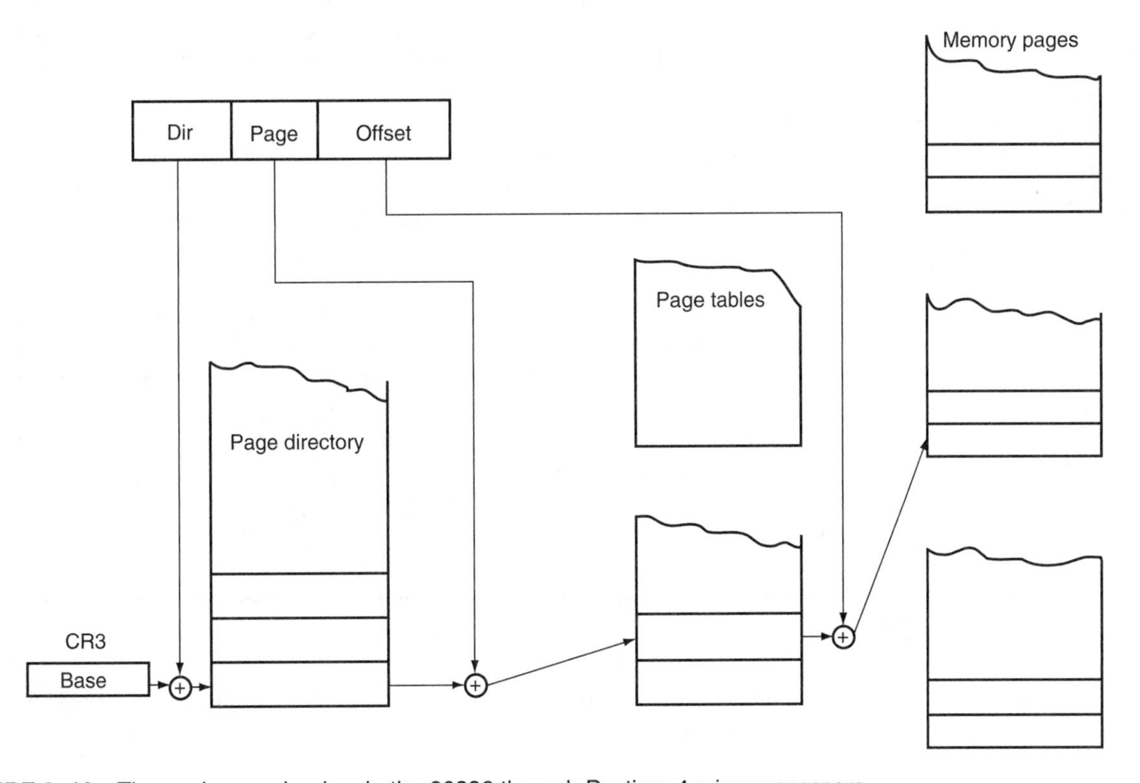

FIGURE 2–13 The paging mechanism in the 80386 through Pentium 4 microprocessors.

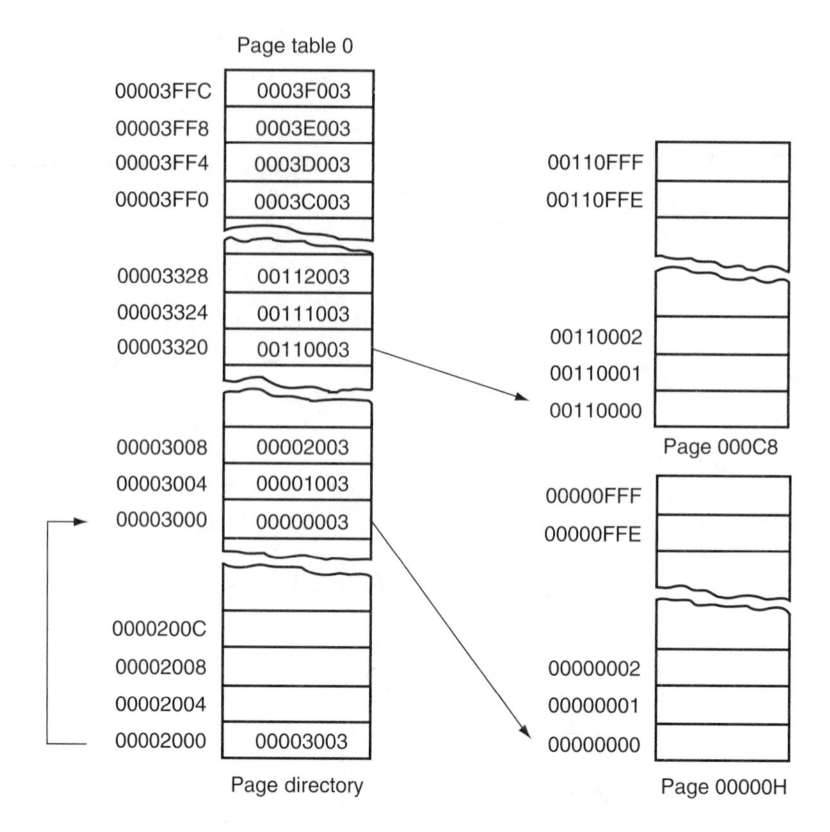

FIGURE 2–14 The page directory, page table 0, and two memory pages. Note how the address of page 000C8000–000C9000 has been moved to 00110000–00110FFF.

extended memory to backfill this part of the conventional memory system to allow DOS access to additional memory. Suppose that the EMM386.EXE program allows access to 16M bytes of extended and conventional memory through paging and locations C8000H–EFFFFH must be repaged to locations 110000–138000H, with all other areas of memory paged to their normal locations. Such a scheme is depicted in Figure 2–14.

Here, the page directory contains four entries. Recall that each entry in the page directory corresponds to 4M bytes of physical memory. The system also contains four page tables with 1024 entries each. Recall that each entry in the page table repages 4K bytes of physical memory. This scheme requires a total of 16K of memory for the four page tables and 16 bytes of memory for the page directory.

As with DOS, the Windows program also repages the memory system. At present, Windows version 3.11 supports paging for only 16M bytes of memory because of the amount of memory required to store the page tables. Newer versions of Windows repage the entire memory system. On the Pentium–Pentium 4 microprocessors, pages can be either 4K bytes in length or 4M bytes in length. Although no software currently supports the 4M-byte pages, as the Pentium 4 and more advanced versions pervade the personal computer arena, operating systems of the future will undoubtedly begin to support 4M-byte memory pages.

2–5 SUMMARY

1. The programming model of the 8086 through 80286 contains 8- and 16-bit registers. The programming model of the 80386 and above contains 8-, 16-, and 32-bit extended registers as well as two additional 16-bit segment registers: FS and GS.

2. The 8-bit registers are AH, AL, BH, BL, CH, CL, DH, and DL. The 16-bit registers are AX, BX, CX, DX, SP, BP, DI, and SI. The segment registers are CS, DS, ES, SS, FS, and GS. The 32-bit extended registers are EAX, EBX, ECX, EDX, ESP, EBP, EDI, and ESI. In addition, the microprocessor contains an instruction pointer (IP/EIP) and flag register (FLAGS or EFLAGS).
3. All real mode memory addresses are a combination of a segment address plus an offset address. The starting location of a segment is defined by the 16-bit number in the segment register that is appended with a hexadecimal zero at its rightmost end. The offset address is a 16-bit number added to the 20-bit segment address to form the real mode memory address.
4. All instructions (code) are accessed by the combination of CS (segment address) plus IP or EIP (offset address).
5. Data are normally referenced through a combination of the DS (data segment) and either an offset address or the contents of a register that contains the offset address. The 8086–Pentium 4 use BX, DI, and SI as default offset registers for data if 16-bit registers are selected. The 80386 and above can use the 32-bit registers EAX, EBX, ECX, EDX, EDI, and ESI as default offset registers for data.
6. Protected mode operation allows memory above the first 1M byte to be accessed by the 80286 through the Pentium 4 microprocessors. This extended memory system (XMS) is accessed via a segment address plus an offset address, just as in the real mode. The difference is that the segment address is not held in the segment register. In the protected mode, the segment starting address is stored in a descriptor that is selected by the segment register.
7. A protected mode descriptor contains a base address, limit, and access rights byte. The base address locates the starting address of the memory segment; the limit defines the last location of the segment. The access rights byte defines how the memory segment is accessed via a program. The 80286 microprocessor allows a memory segment to start at any of its 16M bytes of memory using a 24-bit base address. The 80386 and above allow a memory segment to begin at any of its 4G bytes of memory using a 32-bit base address. The limit is a 16-bit number in the 80286 and a 20-bit number in the 80386 and above. This allows an 80286 memory segment limit of 64K bytes, and an 80386 and above memory segment limit of either 1M byte (G = 0) or 4G bytes (G = 1).
8. The segment register contains three fields of information in the protected mode. The leftmost 13 bits of the segment register address one of 8192 descriptors from a descriptor table. The TI bit accesses either the global descriptor table (TI = 0) or the local descriptor table (TI = 1). The rightmost 2 bits of the segment register select the requested priority level for the memory segment access.
9. The program-invisible registers are used by the 80286 and above to access the descriptor tables. Each segment register contains a cache portion that is used in protected mode to hold the base address, limit, and access rights acquired from a descriptor. The cache allows the microprocessor to access the memory segment without again referring to the descriptor table until the segment register's contents are changed.
10. A memory page is 4K bytes in length. The linear address, as generated by a program, can be mapped to any physical address through the paging mechanism found within the 80386 through the Pentium 4 microprocessors.
11. Memory paging is accomplished through control registers CR0 and CR3. The PG bit of CR0 enables paging, and the contents of CR3 addresses the page directory. The page directory contains up to 1024 page table addresses that are used to access paging tables. The page table contains 1024 entries that locate the physical address of a 4K-byte memory page.
12. The TLB (translation look-aside buffer) caches the 32 most recent page table translations. This precludes page table translation if the translation resides in the TLB, speeding the execution of the software.

2–6 QUESTIONS AND PROBLEMS

1. What are program-visible registers?
2. The 80286 addresses registers that are 8 and _________ bits wide.
3. The extended registers are addressable by which microprocessors?
4. The extended BX register is addressed as _________.
5. Which register holds a count for some instructions?
6. What is the purpose of the IP/EIP register?
7. The carry flag bit is not modified by which arithmetic operations?
8. Will an overflow occur if a signed FFH is added to a signed 01H?
9. A number that contains 3 one bits is said to have _________ parity.
10. Which flag bit controls the INTR pin on the microprocessor?
11. Which microprocessors contain an FS segment register?
12. What is the purpose of a segment register in the real mode operation of the microprocessor?
13. In the real mode, show the starting and ending addresses of each segment located by the following segment register values:
    (a) 1000H
    (b) 1234H
    (c) 2300H
    (d) E000H
    (e) AB00H
14. Find the memory address of the next instruction executed by the microprocessor, when operated in the real mode, for the following CS:IP combinations:
    (a) CS = 1000H and IP = 2000H
    (b) CS = 2000H and IP = 1000H
    (c) CS = 2300H and IP = 1A00H
    (d) CS = 1A00H and IP = B000H
    (e) CS = 3456H and IP = ABCDH
15. Real mode memory addresses allow access to memory below which memory address?
16. Which register or registers are used as an offset address for the string instruction destination in the microprocessor?
17. Which 32-bit register or registers is/are used to hold an offset address for data segment data in the Pentium 4 microprocessor?
18. The stack memory is addressed by a combination of the _________ segment plus _________ offset.
19. If the base pointer (BP) addresses memory, the _________ segment contains the data.
20. Determine the memory location addressed by the following real mode 80286 register combinations:
    (a) DS = 1000H and DI = 2000H
    (b) DS = 2000H and SI = 1002H
    (c) SS = 2300H and BP = 3200H
    (d) DS = A000H and BX = 1000H
    (e) SS = 2900H and SP = 3A00H
21. Determine the memory location addressed by the following real mode Pentium 4 register combinations:
    (a) DS = 2000H and EAX = 00003000H
    (b) DS = 1A00H and ECX = 00002000H
    (c) DS = C000H and ESI = 0000A000H
    (d) SS = 8000H and ESP = 00009000H
    (e) DS = 1239H and EDX = 0000A900H

22. Protected mode memory addressing allows access to which area of the memory in the 80286 microprocessor?
23. Protected mode memory addressing allows access to which area of the memory in the Pentium 4 microprocessor?
24. What is the purpose of the segment register in protected mode memory addressing?
25. How many descriptors are accessible in the global descriptor table in the protected mode?
26. For an 80286 descriptor that contains a base address of A00000H and a limit of 1000H, what starting and ending locations are addressed by this descriptor?
27. For a Pentium 4 descriptor that contains a base address of 01000000H, a limit of 0FFFFH, and G = 0, what starting and ending locations are addressed by this descriptor?
28. For a Pentium 4 descriptor that contains a base address of 00280000H, a limit of 00010H, and G = 1, what starting and ending locations are addressed by this descriptor?
29. If the DS register contains 0020H in a protected mode system, which global descriptor table entry is accessed?
30. If DS = 0103H in a protected mode system, the requested privilege level is ________.
31. If DS = 0105H in a protected mode system, which entry, table, and requested privilege level are selected?
32. What is the maximum length of the global descriptor table in the Pentium 4 microprocessor?
33. Code a descriptor that describes a memory segment that begins at location 210000H and ends at location 21001FH. This memory segment is a code segment that can be read. The descriptor is for an 80286 microprocessor.
34. Code a descriptor that describes a memory segment that begins at location 03000000H and ends at location 05FFFFFFH. This memory segment is a data segment that grows upward in the memory system and can be written. The descriptor is for a Pentium 4 microprocessor.
35. Which register locates the global descriptor table?
36. How is the local descriptor table addressed in the memory system?
37. Describe what happens when a new number is loaded into a segment register when the microprocessor is operated in the protected mode.
38. What are the program-invisible registers?
39. What is the purpose of the GDTR?
40. How many bytes are found in a memory page?
41. What register is used to enable the paging mechanism in the 80386, 80486, Pentium, Pentium Pro, and Pentium 4 microprocessors?
42. How many 32-bit addresses are stored in the page directory?
43. Each entry in the page directory translates how much linear memory into physical memory?
44. If the microprocessor sends linear address 00200000H to the paging mechanism, which paging directory entry is accessed, and which page table entry is accessed?
45. What value is placed in the page table to redirect linear address 20000000H to physical address 30000000H?
46. What is the purpose of the TLB located within the Pentium class microprocessor?
47. Using the Internet, write a short report that details the TLB. Hint: You might want to go to the Intel Web site and search for information.
48. Locate articles about paging on the Internet and write a report detailing how paging is used in a variety of systems.

CHAPTER 3

Addressing Modes

INTRODUCTION

Efficient software development for the microprocessor requires a complete familiarity with the addressing modes employed by each instruction. In this chapter, the MOV (**move data**) instruction is used to describe the data-addressing modes. The MOV instruction transfers bytes or words of data between registers, or between registers and memory in the 8086 through the 80286. Bytes, words, or doublewords are transferred in the 80386 and above by a MOV. In describing the program memory-addressing modes, the CALL and JUMP instructions show how to modify the flow of the program.

The data-addressing modes include register, immediate, direct, register indirect, base-plus-index, register-relative, and base relative-plus-index in the 8086 through the 80286 microprocessors. The 80386 and above also include a scaled-index mode of addressing memory data. The program memory-addressing modes include program relative, direct, and indirect. The chapter explains the operation of the stack memory so that the PUSH and POP instructions and other stack operations will be understood.

CHAPTER OBJECTIVES

Upon completion of this chapter, you will be able to:

1. Explain the operation of each data-addressing mode
2. Use the data-addressing modes to form assembly language statements
3. Explain the operation of each program memory-addressing mode
4. Use the program memory-addressing modes to form assembly and machine language statements
5. Select the appropriate addressing mode to accomplish a given task
6. Detail the difference between addressing memory data using real mode and protected mode operation
7. Describe the sequence of events that place data onto the stack or remove data from the stack
8. Explain how a data structure is placed in memory and used with software

3–1 DATA-ADDRESSING MODES

Because the MOV instruction is a very common and flexible instruction, it provides a basis for the explanation of the data-addressing modes. Figure 3–1 illustrates the MOV instruction and defines

FIGURE 3–1 The MOV instruction showing the source, destination, and direction of data flow.

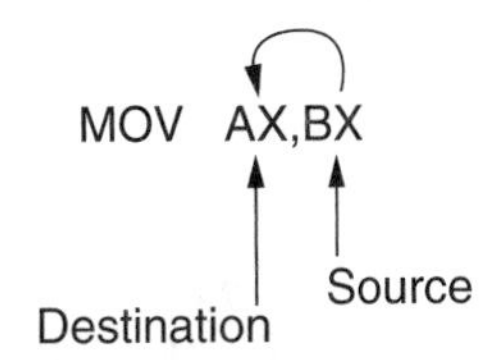

the direction of data flow. The **source** is to the right and the **destination** is to the left, next to the opcode MOV. (An **opcode**, or operation code, tells the microprocessor which operation to perform.) This direction of flow, which is applied to all instructions, is awkward at first. We naturally assume that things move from left to right, whereas here they move from right to left. Notice that a comma always separates the destination from the source in an instruction. Also, note that memory-to-memory transfers are *not* allowed by any instruction except for the MOVS instruction.

In Figure 3–1, the MOV AX,BX instruction transfers the word contents of the source register (BX) into the destination register (AX). The source never changes, but the destination always changes.[1] It is crucial to remember that a MOV instruction always *copies* the source data into the destination. The MOV never actually picks up the data and moves it. Also, note that the flag register remains unaffected by most data transfer instructions. The source and destination are often called **operands**.

Figure 3–2 shows all possible variations of the data-addressing modes using the MOV instruction. This illustration helps to show how each data-addressing mode is formulated with the MOV instruction and also serves as a reference on data-addressing modes. Note that these are the same data-addressing modes found with all versions of the Intel microprocessor, except for the scaled-index addressing mode, which is found only in the 80386 through the Pentium 4. The data-addressing modes are as follows:

Register addressing Register addressing transfers a copy of a byte or word from the source register or contents of a memory location to the destination register or memory location. (Example: The MOV CX,DX instruction copies the word-sized contents of register DX into register CX.) In the 80386 and above, a doubleword can be transferred from the source register or memory location to the destination register or memory location. (Example: The MOV ECX,EDX instruction copies the doubleword-sized contents of register EDX in register ECX.)

Immediate addressing Immediate addressing transfers the source, an immediate byte or word data, into the destination register or memory location. (Example: The MOV AL,22H instruction copies a byte-sized 22H into register AL.) In the 80386 and above, a doubleword of immediate data can be transferred into a register or memory location. (Example: The MOV EBX,12345678H instruction copies a doubleword-sized 12345678H into the 32-bit-wide EBX register.)

Direct addressing Direct addressing moves a byte or word between a memory location and a register. The instruction set does not support a memory-to-memory transfer, except with the MOVS instruction. (Example: The MOV CX,LIST instruction copies the word-sized contents of memory location LIST into register CX.) In the 80386 and above, a doubleword-sized memory location can also be addressed. (Example: The MOV ESI,LIST instruction copies a 32-bit number, stored in four consecutive bytes of memory, from location LIST into register ESI.)

[1]The exceptions are the CMP and TEST instructions, which never change the destination. These instructions are described in later chapters.

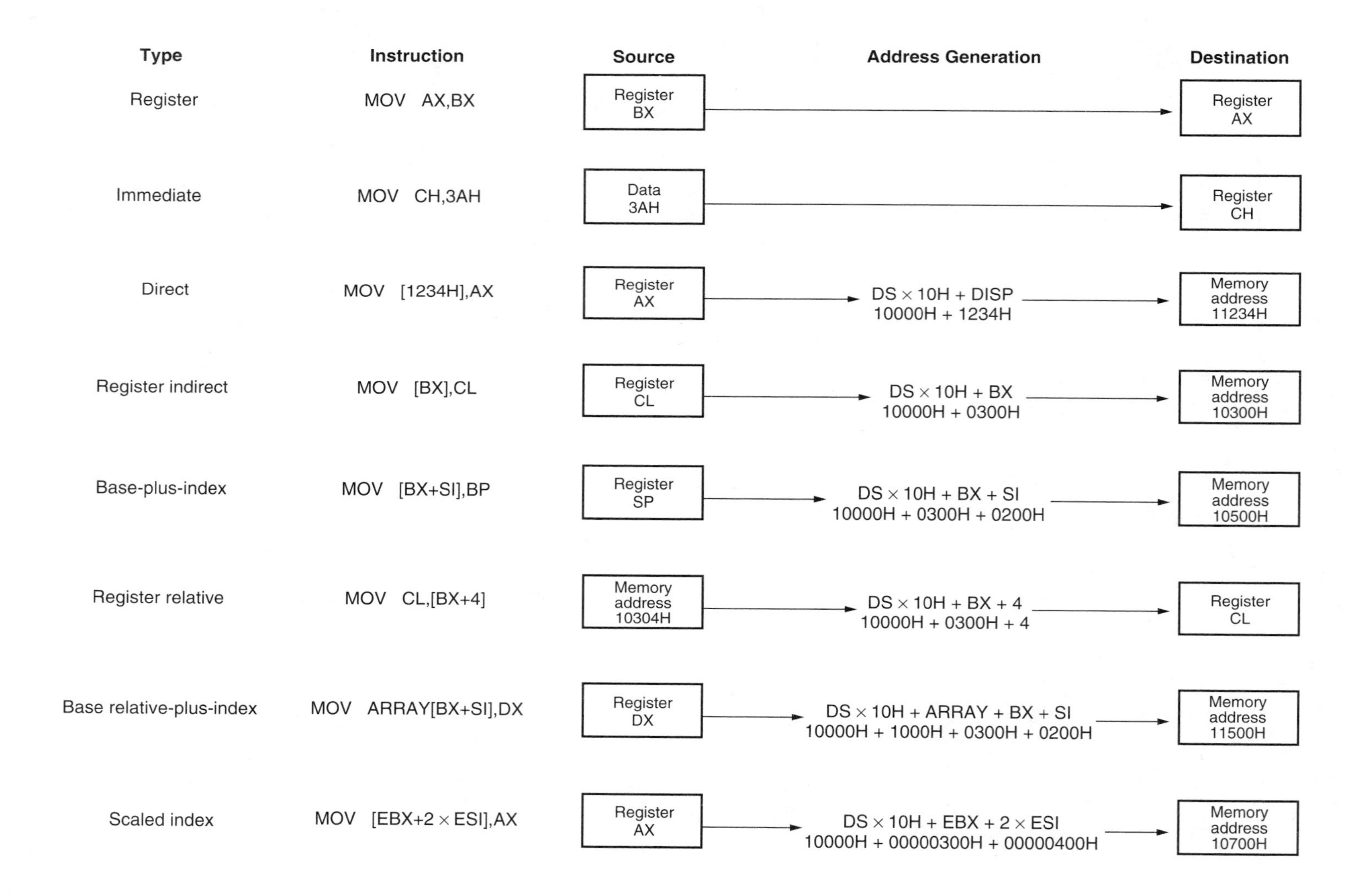

FIGURE 3–2 8086–Pentium 4 data-addressing modes.

Register indirect addressing

Register indirect addressing transfers a byte or word between a register and a memory location addressed by an index or base register. The index and base registers are BP, BX, DI, and SI. (Example: The MOV AX,[BX] instruction copies the word-sized data from the data segment offset address indexed by BX into register AX.) In the 80386 and above, a byte, word, or doubleword is transferred between a register and a memory location addressed by any register: EAX, EBX, ECX, EDX, EBP, EDI, or ESI. (Example: The MOV AL,[ECX] instruction loads AL from the data segment offset address selected by the contents of ECX.)

Base-plus-index addressing

Base-plus-index addressing transfers a byte or word between a register and the memory location addressed by a base register (BP or BX) plus an index register (DI or SI). (Example: The MOV [BX+DI],CL instruction copies the byte-sized contents of register CL into the data segment memory location addressed by BX plus DI.) In the 80386 and above, any two registers (EAX, EBX, ECX, EDX, EBP, EDI, or ESI) may be combined to generate the memory address. (Example: The MOV [EAX+EBX],CL instruction copies the byte-sized contents of register CL into the data segment memory location addressed by EAX plus EBX.)

Register relative addressing

Register relative addressing moves a byte or word between a register and the memory location addressed by an index or base register plus a displacement. (Example: MOV AX,[BX+4] or MOV AX,ARRAY[BX]. The first instruction loads AX from the data segment address formed by BX plus 4. The second instruction loads AX from the data segment memory location in ARRAY plus the contents of BX.) The 80386 and above use any 32-bit register except ESP to address memory. (Example: MOV AX,[ECX+4] or MOV AX,ARRAY[EBX]. The first instruction loads AX from the data segment address formed by ECX plus 4. The second instruction loads AX from the data segment memory location ARRAY plus the contents of EBX.)

Base relative-plus-index addressing

Base relative-plus-index addressing transfers a byte or word between a register and the memory location addressed by a base and an index register plus a displacement. (Example: MOV AX,ARRAY[BX+DI] or MOV AX,[BX+DI+4]. These instructions load AX from a data segment memory location. The first instruction uses an address formed by adding ARRAY, BX, and DI and the second by adding BX, DI, and 4.) In the 80386 and above, MOV EAX,ARRAY[EBX+ECX] loads EAX from the data segment memory location accessed by the sum of ARRAY, EBX, and ECX.

Scaled-index addressing

Scaled-index addressing is available only in the 80386 through the Pentium 4 microprocessors. The second register of a pair of registers is modified by the scale factor of 2×, 4×, or 8× to generate the operand memory address. (Example: A MOV EDX,[EAX+4*EBX] instruction loads EDX from the data segment memory location addressed by EAX plus four times EBX.) Scaling allows access to word (2×), doubleword (4×), or quadword (8×) memory array data. Note that a scaling factor of 1× also exists, but it is normally implied and does not appear explicitly in the instruction. The MOV AL,[EBX+ECX] is an example in which the scaling factor is a one. Alternately, the instruction can be rewritten as MOV AL,[EBX+1*ECX]. Another example is a MOV AL,[2*EBX] instruction, which uses only one scaled register to address memory.

Register Addressing

Register addressing is the most common form of data addressing and, once the register names are learned, is the easiest to apply. The microprocessor contains the following 8-bit register names used with register addressing: AH, AL, BH, BL, CH, CL, DH, and DL. Also present are the following 16-bit register names: AX, BX, CX, DX, SP, BP, SI, and DI. In the 80386 and above, the extended 32-bit register names are: EAX, EBX, ECX, EDX, ESP, EBP, EDI, and ESI. With register addressing, some MOV instructions and the PUSH and POP instructions also use the 16-bit segment register names (CS, ES, DS, SS, FS, and GS). It is important for instructions to use registers that are the same size. *Never* mix an 8-bit register with a 16-bit register, an 8-bit register with 32-bit register, or a 16-bit register with 32-bit register because this is not allowed by the microprocessor and results in an error when assembled. This is even true when a MOV AX,AL (MOV EAX,AL) instruction may seem to make sense. Of course, the MOV AX,AL or MOV EAX,AL instruction is *not* allowed because these registers are of different sizes. Note that a few instructions, such as SHL DX,CL, are exceptions to this rule, as indicated in later chapters. It is also important to note that *none* of the MOV instructions affects the flag bits.

Table 3–1 shows many variations of register move instructions. It is impossible to show all combinations because there are too many. For example, just the 8-bit subset of the MOV instruction has 64 different variations. A segment-to-segment register MOV instruction is about the only type of register MOV instruction *not* allowed. Note that the code segment register is *not* normally changed by a MOV instruction because the address of the next instruction is found by both IP/EIP and CS. If only CS were changed, the address of the next instruction would be unpredictable. Therefore, changing the CS register with a MOV instruction is not allowed.

Figure 3–3 shows the operation of the MOV BX,CX instruction. Note that the source register's contents do not change, but the destination register's contents do change. This instruction moves (*copies*) a 1234H from register CX into register BX. This *erases* the old contents (76AFH) of register BX, but the contents of CX remain unchanged. The contents of the destination register or destination memory location change for all instructions except the CMP and TEST instructions. Note that the MOV BX,CX instruction does not affect the leftmost 16 bits of register EBX.

Example 3–1 shows a sequence of assembled instructions that copy various data between 8-, 16-, and 32-bit registers. As mentioned, the act of moving data from one register to another changes

TABLE 3–1 Examples of register-addressed instructions.

Assembly Language	*Size*	*Operation*
MOV AL,BL	8 bits	Copies BL into AL
MOV CH,CL	8 bits	Copies CL into CH
MOV AX,CX	16 bits	Copies CX into AX
MOV SP,BP	16 bits	Copies BP into SP
MOV DS,AX	16 bits	Copies AX into DS
MOV SI,DI	16 bits	Copies DI into SI
MOV BX,ES	16 bits	Copies ES into BX
MOV ECX,EBX	32 bits	Copies EBX into ECX
MOV ESP,EDX	32 bits	Copies EDX into ESP
MOV DS,CX	16 bits	Copies CX into DS
MOV ES,DS	—	Not allowed (segment to segment)
MOV BL,DX	—	Not allowed (mixed sizes)
MOV CS,AX	—	Not allowed (the code segment register may not be the destination register)

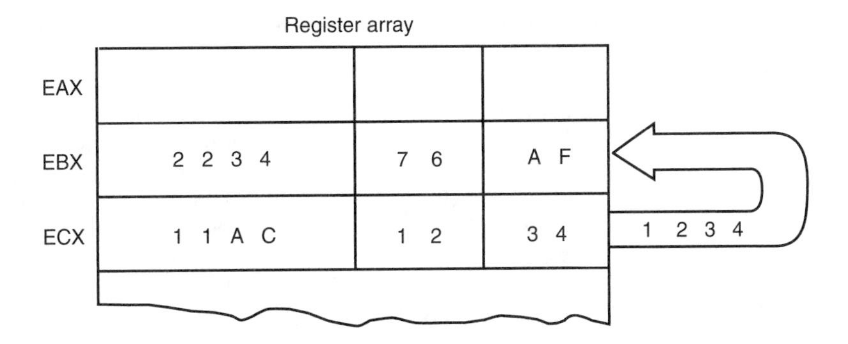

FIGURE 3–3 The effect of executing the MOV BX, CX instruction at the point just before the BX register changes. Note that only the rightmost 16 bits of register EBX change.

only the destination register, never the source. The last instruction in this example (MOV CS,AX) assembles without error, but causes problems if executed. If only the contents of CS change without changing IP, the next step in the program is unknown and therefore causes the program to go awry.

EXAMPLE 3–1

```
0000 8B C3          MOV AX,BX     ;copy contents of BX into AX
0002 8A CE          MOV CL,DH     ;copy contents of DH into CL
0004 8A CD          MOV CL,CH     ;copy contents of CH into CL
0006 66|8B C3       MOV EAX,EBX   ;copy contents of EBX into EAX
0009 66|8B D8       MOV EBX,EAX   ;copy contents of EAX into EBX
000C 66|8B C8       MOV ECX,EAX   ;copy contents of EAX into ECX
000F 66|8B D0       MOV EDX,EAX   ;copy contents of EAX into EDX
0012 8C C8          MOV AX,CS     ;copy CS into DS (two steps)
0014 8E D8          MOV DS,AX
0016 8E C8          MOV CS,AX     ;copy AX into CS (causes problems)
```

Immediate Addressing

Another data-addressing mode is immediate addressing. The term *immediate* implies that the data immediately follow the hexadecimal opcode in the memory. Also note that immediate data are **constant data**, whereas the data transferred from a register or memory location are **variable data**. Immediate addressing operates upon a byte or word of data. In the 80386 through the Pentium 4 microprocessors, immediate addressing also operates on doubleword data. The MOV immediate instruction transfers a copy of the immediate data into a register or a memory location. Figure 3–4 shows the operation of a MOV EAX,13456H instruction. This instruction copies the 13456H from the instruction, located in the memory immediately following the hexadecimal opcode, into register EAX. As with the MOV instruction illustrated in Figure 3–3, the source data overwrites the destination data.

In symbolic assembly language, the symbol # precedes immediate data in some assemblers. The MOV AX,#3456H instruction is an example. Most assemblers do not use the # symbol, but represent immediate data as in the MOV AX,3456H instruction. In this text, the # symbol is not used for immediate data. The most common assemblers—Intel ASM, Microsoft

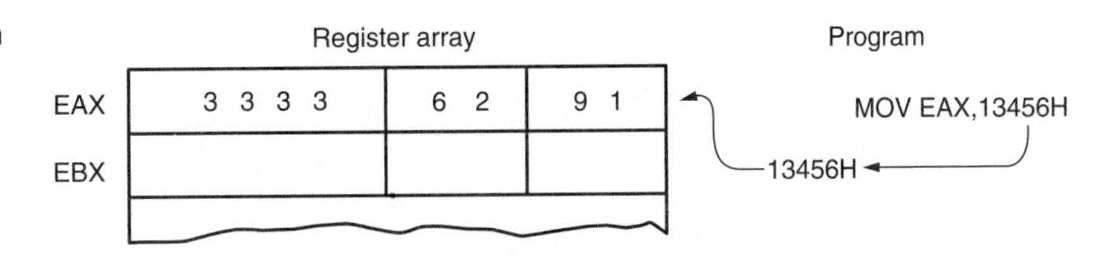

FIGURE 3–4 The operation of the MOV EAX,3456H instruction. This instruction copies the immediate data (13456H) into EAX.

TABLE 3–2 Examples of immediate addressing using the MOV instruction.

Assembly Language	*Size*	*Operation*
MOV BL,44	8 bits	Copies 44 decimal (2CH) into BL
MOV AX,44H	16 bits	Copies 0044H into AX
MOV SI,0	16 bits	Copies 0000H into SI
MOV CH,100	8 bits	Copies 100 decimal (64H) into CH
MOV AL, 'A'	8 bits	Copies ASCII A into AL
MOV AX, 'AB'	16 bits	Copies ASCII BA* into AX
MOV CL,11001110B	8 bits	Copies 11001110 binary into CL
MOV EBX,12340000H	32 bits	Copies 12340000H into EBX
MOV ESI,12	32 bits	Copies 12 decimal into ESI
MOV EAX,100B	32 bits	Copies 100 binary into EAX

*This is not an error. The ASCII characters are stored as BA, so exercise care when using word-sized pairs of ASCII characters.

MASM,[2] and Borland TASM[3]—do not use the # symbol for immediate data, but an older assembler used with some Hewlett-Packard logic development system does, as may others.

The symbolic assembler portrays immediate data in many ways. The letter H appends hexadecimal data. If hexadecimal data begin with a letter, the assembler requires that the data start with a 0. For example, to represent hexadecimal F2, 0F2H is used in assembly language. In some assemblers (though not in MASM, TASM, or this text), hexadecimal data are represented with 'h, as in MOV AX,#'h1234. Decimal data are represented as is and require no special codes or adjustments. (An example is the 100 decimal in the MOV AL,100 instruction.) An ASCII-coded character or characters may be depicted in the immediate form if the ASCII data are enclosed in apostrophes. (An example is the MOV BH,'A' instruction, which moves an ASCII-coded letter A [41H] into register BH.) Be careful to use the apostrophe (') for ASCII data and not the single quotation mark ('). Binary data are represented if the binary number is followed by the letter B, or, in some assemblers, the letter Y. Table 3–2 shows many different variations of MOV instructions that apply immediate data.

Example 3–2 shows various immediate instructions in a short assembly language program that places 0000H into the 16-bit registers AX, BX, and CX. This is followed by instructions that use register addressing to copy the contents of AX into registers SI, DI, and BP. This is a complete program that uses programming models for assembly and execution with MASM. The .MODEL TINY statement directs the assembler to assemble the program into a single code segment. The .CODE statement or directive indicates the start of the code segment; the .STARTUP statement indicates the starting instruction in the program; and the .EXIT statement causes the program to exit to DOS. The END statement indicates the end of the program file. This program is assembled with MASM and executed with CodeView[4] (CV) to view its execution. Note that the most recent version of TASM will also accept MASM code without any changes. To store the program into the system use the DOS EDIT program, Windows NotePad,[5] or Programmer's WorkBench[6] (PWB). Note that a TINY program always assembles as a command (.COM) program.

EXAMPLE 3–2

```
                    .MODEL TINY         ;choose single segment model
0000                .CODE               ;start of code segment
                    .STARTUP            ;start of program
```

[2]MASM (MACRO assembler) is a trademark of Microsoft Corporation.

[3]TASM (Turbo assembler) is a trademark of Borland Corporation.

[4]CodeView is a registered trademark of Microsoft Corporation.

[5]NotePad is a registered trademark of Microsoft Corporation.

[6]Programmer's WorkBench is a registered trademark of Microsoft Corporation.

```
0100  B8 0000         MOV AX,0                ;place 0000H into AX
0103  BB 0000         MOV BX,0                ;place 0000H into BX
0106  B9 0000         MOV CX,0                ;place 0000H into CX

0109  8B F0           MOV SI,AX               ;copy AX into SI
010B  8B F8           MOV DI,AX               ;copy AX into DI
010D  8B E8           MOV BP,AX               ;copy AX into BP

                      .EXIT                   ;exit to DOS
                      END                     ;end of program
```

Each statement in an assembly language program consists of four parts or fields, as illustrated in Example 3–3. The leftmost field is called the *label*. It is used to store a symbolic name for the memory location that it represents. All labels must begin with a letter or one of the following special characters: @, $, -, or ?. A label may be of any length from 1 to 35 characters. The label appears in a program to identify the name of a memory location for storing data and for other purposes that are explained as they appear. The next field to the right is called the *opcode field*; it is designed to hold the instruction, or opcode. The MOV part of the move data instruction is an example of an opcode. To the right of the opcode field is the *operand field*, which contains information used by the opcode. For example, the MOV AL,BL instruction has the opcode MOV and operands AL and BL. Note that some instructions contain between zero and three operands. The final field, the *comment field*, contains a comment about an instruction or a group of instructions. A comment always begins with a semicolon (;).

EXAMPLE 3–3

```
LABEL      OPCODE  OPERAND    COMMENT

DATA1      DB      23H        ;define DATA1 as a byte of 23H
DATA2      DW      1000H      ;define DATA2 as a word of 1000H

START:     MOV     AL,BL      ;copy BL into AL
           MOV     BH,AL      ;copy AL into BH
           MOV     CX,200     ;copy 200 into CX
```

When the program is assembled and the list (.LST) file is viewed, it appears as the program listed in Example 3–2. The hexadecimal number at the far left is the offset address of the instruction or data. This number is generated by the assembler. The number or numbers to the right of the offset address are the machine-coded instructions or data that are also generated by the assembler. For example, if the instruction MOV AX,0 appears in a file and it is assembled, it appears in offset memory location 0100 in Example 3–2. Its hexadecimal machine language form is B8 0000. The B8 is the opcode in machine language and the 0000 is the 16-bit-wide data with a value of zero. When the program was written, only the MOV AX,0 was typed into the editor; the assembler generated the machine code and addresses, and stored the program in a file with the extension .LST. Note that all programs shown in this text are in the form generated by the assembler.

EXAMPLE 3–4

```
int MyFunction(int temp)
{
        _asm
        {
                mov   eax,temp
                add   eax,20h
                mov   temp,eax
        }
        return temp;
}
```

Programs are also written using the inline assembler in some Visual C++ programs. Example 3–4 shows a function in a Visual C++ program that includes some code written with the inline assembler. This function adds 20H to the number returned by the function. Notice that the assembly code accesses C++ variable temp and all of the assembly code is placed in an **_asm** code block. Many examples in this text are written using the inline assembler within a C++ program.

Direct Data Addressing

Most instructions can use the direct data-addressing mode. In fact, direct data addressing is applied to many instructions in a typical program. There are two basic forms of direct data addressing: (1) **direct addressing**, which applies to a MOV between a memory location and AL, AX, or EAX, and (2) **displacement addressing**, which applies to almost any instruction in the instruction set. In either case, the address is formed by adding the displacement to the default data segment address or an alternate segment address.

Direct Addressing. Direct addressing with a MOV instruction transfers data between a memory location, located within the data segment, and the AL (8-bit), AX (16-bit), or EAX (32-bit) register. A MOV instruction using this type of addressing is usually a three-byte-long instruction. (In the 80386 and above, a register size prefix may appear before the instruction, causing it to exceed three bytes in length.)

The MOV AL,DATA instruction, as represented by most assemblers, loads AL from the data segment memory location DATA (1234H). Memory location DATA is a symbolic memory location, and 1234H is the actual hexadecimal location. With many assemblers, this instruction is represented as a MOV AL,[1234H][7] instruction. The [1234H] is an absolute memory location that is not allowed by all assembler programs. Note that this may need to be formed as MOV AL,DS:[1234H] with some assemblers, to show that the address is in the data segment. Figure 3–5 shows how this instruction transfers a copy of the byte-sized contents of memory location 11234H into AL. The effective address is formed by adding 1234H (the offset address) and 10000H (the data segment address of 1000H times 10H) in a system operating in the real mode.

Table 3–3 lists the three direct-addressed instructions. These instructions often appear in programs, so Intel decided to make them special three-byte-long instructions to reduce the length of programs. All other instructions that move data from a memory location to a register, called **displacement-addressed instructions**, require four or more bytes of memory for storage in a program.

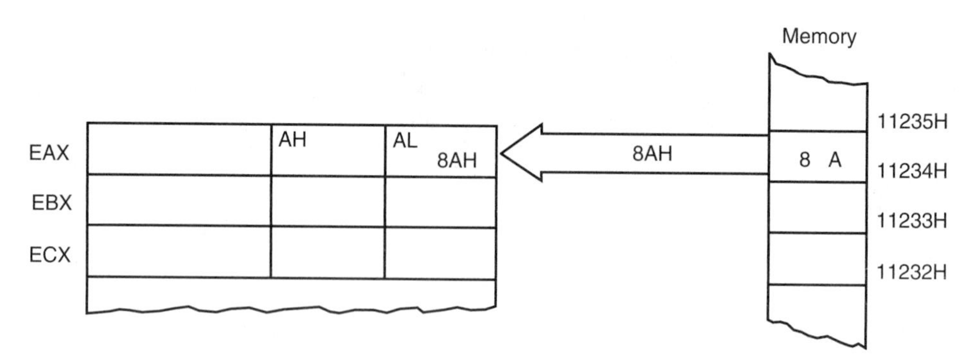

FIGURE 3–5 The operation of the MOV AL,[1234H] instruction when DS = 1000H.

[7]This form may be typed into a MASM program, but it most often appears when the debugging tool is executed.

TABLE 3–3 Direct-addressed instructions using EAX, AX, and AL.

Assembly Language	*Size*	*Operation*
MOV AL,NUMBER	8 bits	Copies the byte contents of data segment memory location NUMBER into AL
MOV AX,COW	16 bits	Copies the word contents of data segment memory location COW into AX
MOV EAX,WATER*	32 bits	Copies the doubleword contents of data segment location WATER into EAX
MOV NEWS,AL	8 bits	Copies AL into byte memory location NEWS
MOV THERE,AX	16 bits	Copies AX into word memory location THERE
MOV HOME,EAX*	32 bits	Copies EAX into doubleword memory location HOME
MOV ES:[2000H],AL	8 bits	Copies AL into extra segment memory at offset address 2000H

*The 80386–Pentium 4 at times use more than three bytes of memory for 32-bit instructions.

Displacement Addressing. Displacement addressing is almost identical to direct addressing, except that the instruction is four bytes wide instead of three. In the 80386 through the Pentium 4, this instruction can be up to seven bytes wide if both a 32-bit register and a 32-bit displacement are specified. This type of direct data addressing is much more flexible because most instructions use it.

If the operation of the MOV CL,DS:[1234H] instruction is compared to that of the MOV AL,DS:[1234H] instruction of Figure 3–5, we see that both basically perform the same operation except for the destination register (CL versus AL). Another difference only becomes apparent upon examining the assembled versions of these two instructions. The MOV AL,DS:[1234H] instruction is three bytes long and the MOV CL,DS:[1234H] instruction is four bytes long, as illustrated in Example 3–5. This example shows how the assembler converts these two instructions into hexadecimal machine language. You must include the segment register DS: in this example, before the [offset] part of the instruction. You may use any segment register, but, in most cases, data are stored in the data segment, so this example uses DS:[1234H].

EXAMPLE 3–5

```
0000 A0 1234 R          MOV AL,DS:[1234H]
0003 BA 0E 1234 R       MOV CL,DS:[1234H]
```

Table 3–4 lists some MOV instructions using the displacement form of direct addressing. Not all variations are listed because there are many MOV instructions of this type. The segment registers can be stored or loaded from memory.

Example 3–6 shows a short program using models that address information in the data segment. Note that the **data segment** begins with a .DATA statement to inform the assembler where the data segment begins. The model size is adjusted from TINY, as shown in Example 3–3, to SMALL so that a data segment can be included. The **SMALL model** allows one data segment and one code segment. The SMALL model is often used whenever memory data are required for a program. A SMALL model program assembles as an execute (.EXE) program file. Notice how this example allocates memory locations in the data segment by using the DB and DW directives. Here the .STARTUP statement not only indicates the start of the code, but it also loads the data segment register with the segment address of the data segment. If this

TABLE 3–4 Examples of direct data addressing using a displacement.

Assembly Language	*Size*	*Operation*
MOV CH,DOG	8 bits	Copies the byte contents of data segment memory location DOG into CH
MOV CH,DS:[1000H]*	8 bits	Copies the byte contents of data segment memory offset address 1000H into CH
MOV ES,DATA6	16 bits	Copies the word contents of data segment memory location DATA6 into ES
MOV DATA7,BP	16 bits	Copies BP into data segment memory location DATA7
MOV NUMBER,SP	16 bits	Copies SP into data segment memory location NUMBER
MOV DATA1,EAX	32 bits	Copies EAX into data segment memory location DATA1
MOV EDI,SUM1	32 bits	Copies the doubleword contents of data segment memory location SUM1 into EDI

*This form of addressing is seldom used with most assemblers because an actual numeric offset address is rarely accessed.

program is assembled and executed with CodeView, the instructions can be viewed as they execute and change registers and memory locations.

EXAMPLE 3–6

```
                                 .MODEL SMALL       ;choose small model
0000                             .DATA              ;start data segment

0000 10                DATA1 DB  10H                ;place 10H into DATA1
0001 00                DATA2 DB  0                  ;place 00H into DATA2
0002 0000              DATA3 DW  0                  ;place 0000H into DATA3
0004 AAAA              DATA4 DW  0AAAAH             ;place AAAAH into DATA4

0000                             .CODE              ;start code segment
                                 .STARTUP           ;start program

0017 A0 0000 R                   MOV  AL,DATA1      ;copy DATA1 into AL
001A 8A 26 0001 R                MOV  AH,DATA2      ;copy DATA2 into AH
001E A3 0002 R                   MOV  DATA3,AX      ;copy AX into DATA3
0021 8B 1E 0004 R                MOV  BX,DATA4      ;copy DATA4 into BX

                                 .EXIT              ;exit to DOS
                                 END                ;end program listing
```

Register Indirect Addressing

Register indirect addressing allows data to be addressed at any memory location through an offset address held in any of the following registers: BP, BX, DI, and SI. For example, if register BX contains 1000H and the MOV AX,[BX] instruction executes, the word contents of data segment offset address 1000H are copied into register AX. If the microprocessor is operated in the real mode and DS = 0100H, this instruction addresses a word stored at memory bytes 2000H and 2001H and transfers it into register AX (see Figure 3–6). Note that the contents of 2000H are moved into AL and the contents of 2001H are moved into AH. The [ ] symbols denote indirect addressing in assembly language. In addition to using the BP, BX, DI, and SI registers to indirectly address memory, the 80386 and above allow register indirect

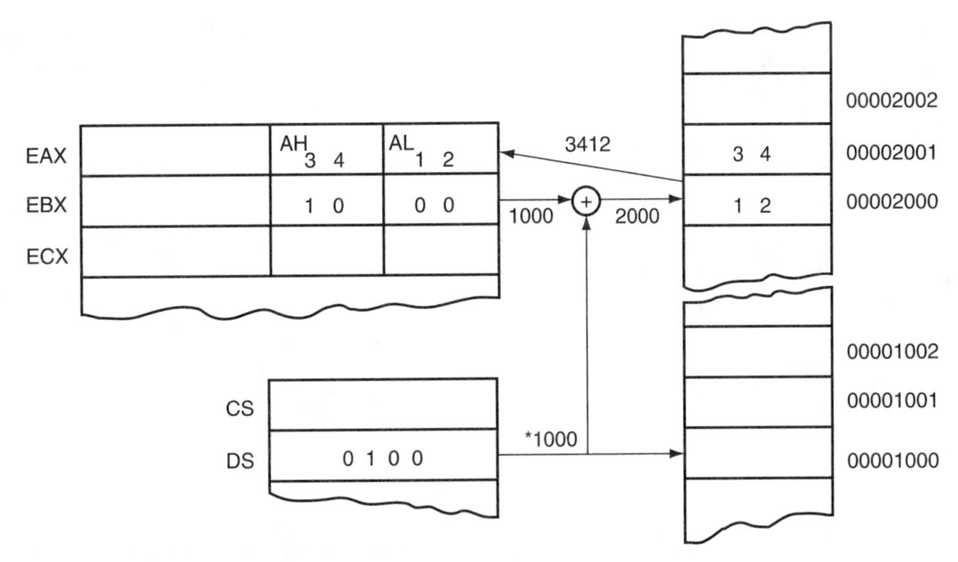

*After DS is appended with a 0.

FIGURE 3–6 The operation of the MOV AX,[BX] instruction when BX = 1000H and DS = 0100H. Note that this instruction is shown after the contents of memory are transferred to AX.

addressing with any extended register except ESP. Some typical instructions using indirect addressing appear in Table 3–5.

The **data segment** is used by default with register indirect addressing or any other addressing mode that uses BX, DI, or SI to address memory. If the BP register addresses memory, the **stack segment** is used by default. These settings are considered the default for these four index and base registers. For the 80386 and above, EBP addresses memory in the stack segment by default; EAX, EBX, ECX, EDX, EDI, and ESI address memory in the data segment by default. When using a 32-bit register to address memory in the real mode, the contents of the 32-bit register must never exceed 0000FFFFH. In the protected mode, any value can be used in a 32-bit

TABLE 3–5 Examples of register indirect addressing.

Assembly Language	*Size*	*Operation*
MOV CX,[BX]	16 bits	Copies the word contents of the data segment memory location addressed by BX into CX
MOV [BP],DL*	8 bits	Copies DL into the stack segment memory location addressed by BP
MOV [DI],BH	8 bits	Copies BH into the data segment memory location addressed by DI
MOV [DI],[BX]	—	Memory-to-memory transfers are not allowed except with string instructions
MOV AL,[EDX]	8 bits	Copies the byte contents of the data segment memory location addressed by EDX into AL
MOV ECX,[EBX]	32 bits	Copies the doubleword contents of the data segment memory location addressed by EBX into ECX

*Data addressed by BP or EBP are in the stack segment by default, while other indirect addressed instructions use the data segment by default.

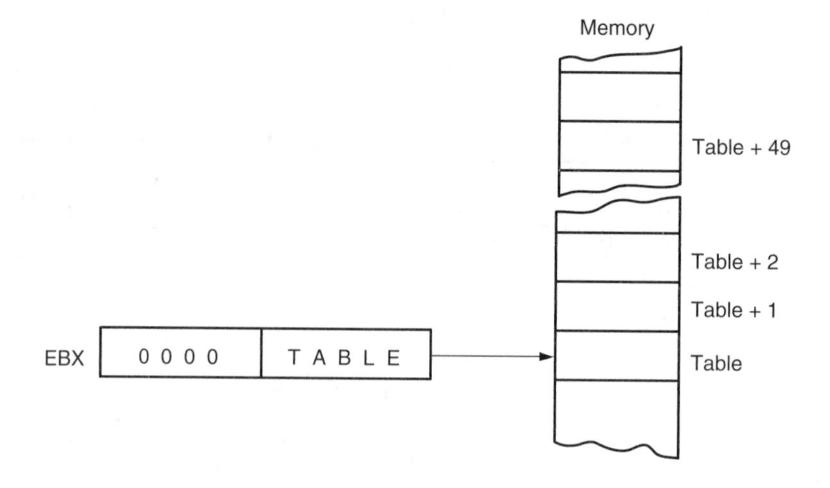

FIGURE 3–7 An array (TABLE) containing 50 bytes that are indirectly addressed through register BX.

register that is used to indirectly address memory, as long as it does not access a location outside of the segment, as dictated by the access rights byte. An example 80386–Pentium 4 instruction is MOV EAX,[EBX]. This instruction loads EAX with the doubleword-sized number stored at the data segment offset address indexed by EBX.

In some cases, indirect addressing requires specifying the size of the data. The size is specified by the **special assembler directive** BYTE PTR, WORD PTR, or DWORD PTR. These directives indicate the size of the memory data addressed by the memory pointer (PTR). For example, the MOV AL,[DI] instruction is clearly a byte-sized move instruction, but the MOV [DI],10H instruction is ambiguous. Does the MOV [DI],10H instruction address a byte-, word-, or doubleword-sized memory location? The assembler can't determine the size of the 10H. The instruction MOV BYTE PTR [DI],10H clearly designates the location addressed by DI as a byte-sized memory location. Likewise, MOV DWORD PTR [DI],10H clearly identifies the memory location as doubleword-sized. The BYTE PTR, WORD PTR, and DWORD PTR directives are used only with instructions that address a memory location through a pointer or index register with immediate data, and for a few other instructions that are described in subsequent chapters. Another directive that is occasionally used is QWORD PTR, where a QWORD is a quadword (64 bits). If programs are using the SIMD instructions, the OWORD PTR is also used to represent a 128-bit-wide number.

Indirect addressing often allows a program to refer to tabular data located in the memory system. For example, suppose that you must create a table of information that contains 50 samples taken from memory location 0000:046C. Location 0000:046C contains a counter in DOS that is maintained by the personal computer's real-time clock. Figure 3–7 shows the table and the BX register used to sequentially address each location in the table. To accomplish this task, load the starting location of the table into the BX register with a MOV immediate instruction. After initializing the starting address of the table, use register indirect addressing to store the 50 samples sequentially.

The sequence shown in Example 3–7 loads register BX with the starting address of the table and initializes the count, located in register CX, to 50. The **OFFSET** directive tells the assembler to load BX with the offset address of memory location TABLE, not the contents of TABLE. For example, the MOV BX,DATAS instruction copies the contents of memory location DATAS into BX, while the MOV BX,OFFSET DATAS instruction copies the offset address DATAS into BX. When the OFFSET directive is used with the MOV instruction, the assembler calculates the offset address and then uses a MOV immediate instruction to load the address in the specified 16-bit register.

EXAMPLE 3–7

```
                                .MODEL SMALL             ;select small model
0000                            .DATA                    ;start data segment

0000 0032 [            DATAS   DW   50 DUP(?) ;setup array of 50 words
          0000
               ]
0000                            .CODE                    ;start code segment
                                .STARTUP                 ;start program
0017 B8 0000                    MOV  AX,0
001A 8E C0                      MOV  ES,AX               ;address segment 0000 with ES
001C B8 0000 R                  MOV  BX,OFFSET DATAS     ;address DATAS array with BX
001F B9 0032                    MOV  CX,50               ;load counter with 50
0022                   AGAIN:
0022 26:A1 046C                 MOV  AX,ES:[046CH]       ;get clock value
0026 89 07                      MOV  [BX],AX             ;save clock value in DATAS
0028 43                         INC  BX                  ;increment BX to next element
0029 43                         INC  BX
002A E2 F6                      LOOP AGAIN               ;repeat 50 times

                                .EXIT                    ;exit to DOS
                                END                      ;end program listing
```

Once the counter and pointer are initialized, a repeat-until CX = 0 loop executes. Here data are read from extra segment memory location 46CH with the MOV AX,ES:[046CH] instruction and stored in memory that is indirectly addressed by the offset address located in register BX. Next, BX is incremented (1 is added to BX) twice to address the next word in the table. Finally, the LOOP instruction repeats the LOOP 50 times. The LOOP instruction decrements (subtracts 1 from) the counter (CX); if CX is not zero, LOOP causes a jump to memory location AGAIN. If CX becomes zero, no jump occurs and this sequence of instructions ends. This example copies the most recent 50 values from the clock into the memory array DATAS. This program will often show the same data in each location because the contents of the clock are changed only 18.2 times per second. To view the program and its execution, use the CodeView program. To use CodeView, type CV XXXX.EXE, where XXXX.EXE is the name of the program that is being debugged. You can also access it as DEBUG from the Programmer's WorkBench program under the RUN menu. Note that CodeView functions only with .EXE or .COM files. Some useful CodeView switches are /50 for a 50-line display and /S for use of high-resolution video displays in an application. To debug the file TEST.COM with 50 lines, type CV /50 /S TEST.COM at the DOS prompt.

Base-Plus-Index Addressing

Base-plus-index addressing is similar to indirect addressing because it indirectly addresses memory data. In the 8086 through the 80286, this type of addressing uses one base register (BP or BX) and one index register (DI or SI) to indirectly address memory. The base register often holds the beginning location of a memory array, whereas the index register holds the relative position of an element in the array. Remember that whenever BP addresses memory data, both the stack segment register and BP generate the effective address.

In the 80386 and above, this type of addressing allows the combination of any two 32-bit extended registers except ESP. For example, the MOV DL,[EAX+EBX] instruction is an example using EAX (as the base) plus EBX (as the index). If the EBP register is used, the data are located in the stack segment instead of in the data segment.

Locating Data with Base-Plus-Index Addressing. Figure 3–8 shows how data are addressed by the MOV DX,[BX+DI] instruction when the microprocessor operates in the real mode. In this

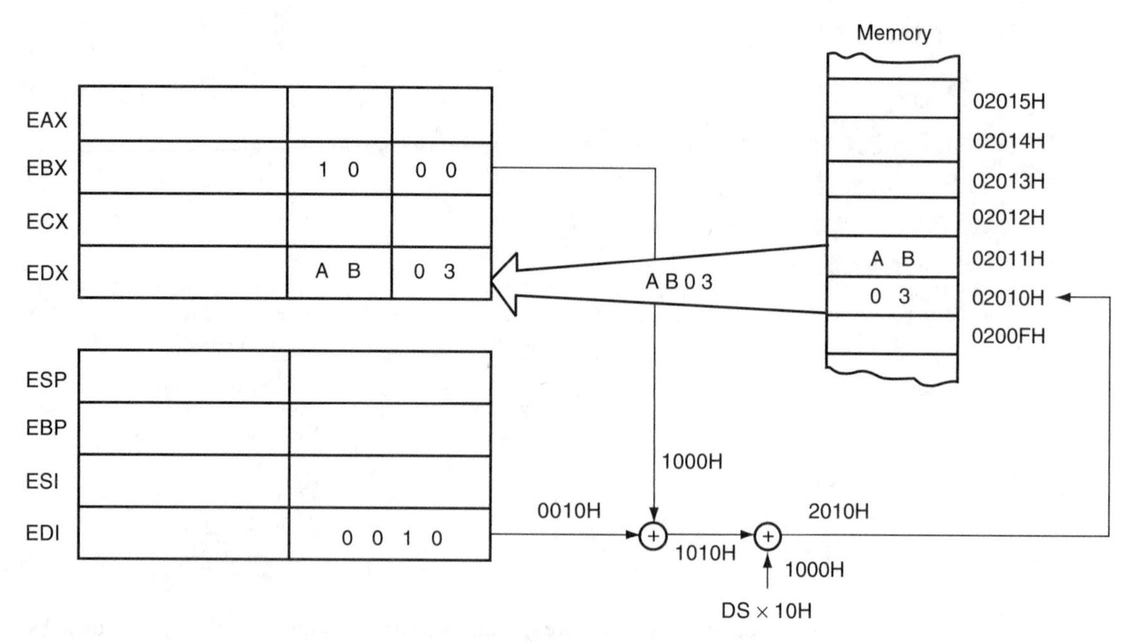

FIGURE 3–8 An example showing how the base-plus-index addressing mode functions for the MOV DX,[BX+DI] instruction. Notice that memory address 02010H is accessed because DS = 0100H, BX = 100H, and DI = 0010H.

example, BX = 1000H, DI = 0010H, and DS = 0100H, which translate into memory address 02010H. This instruction transfers a copy of the word from location 02010H into the DX register. Table 3–6 lists some instructions used for base-plus-index addressing. Note that the Intel assembler requires that this addressing mode appear as [BX][DI] instead of [BX+DI]. The MOV DX,[BX+DI] instruction is MOV DX,[BX][DI] for a program written for the Intel ASM assembler. This text uses the first form in all example programs, but the second form can be used in many assemblers, including MASM from Microsoft. Instructions like MOV DI,[BX+DI] will assemble, but will not execute correctly.

TABLE 3–6 Examples of base-plus-index addressing.

Assembly Language	*Size*	*Operation*
MOV CX,[BX+DI]	16 bits	Copies the word contents of the data segment memory location addressed by BX plus DI into CX
MOV CH,[BP+SI]	8 bits	Copies the byte contents of the stack segment memory location addressed by BP plus SI into CH
MOV [BX+SI],SP	16 bits	Copies SP into the data segment memory location addressed by BX plus SI
MOV [BP+DI],AH	8 bits	Copies AH into the stack segment memory location addressed by BP plus DI
MOV CL,[EDX+EDI]	8 bits	Copies the byte contents of the data segment memory location addressed by EDX plus EDI into CL
MOV [EAX+EBX],ECX	32 bits	Copies ECX into the data segment memory location addressed by EAX plus EBX

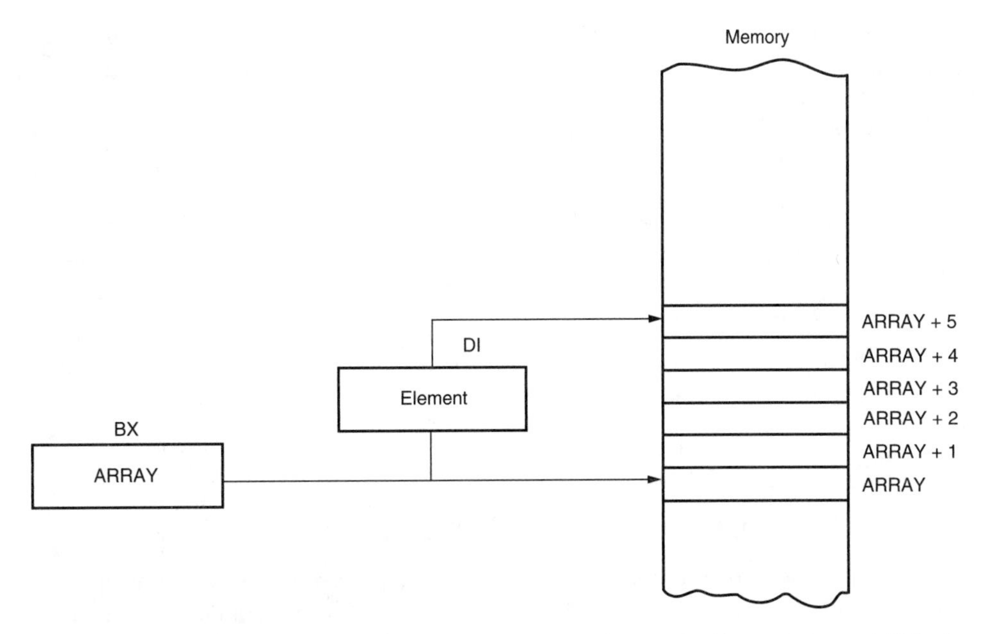

FIGURE 3–9 An example of the base-plus-index addressing mode. Here an element (DI) of an ARRAY (BX) is addressed.

Locating Array Data Using Base-Plus-Index Addressing. A major use of the base-plus-index addressing mode is to address elements in a memory array. Suppose that the elements in an array located in the data segment at memory location ARRAY must be accessed. To accomplish this, load the BX register (base) with the beginning address of the array and the DI register (index) with the element number to be accessed. Figure 3–9 shows the use of BX and DI to access an element in an array of data. A short program, listed in Example 3–8, moves array element 10H into array element 20H. Notice that the array element number, loaded into the DI register, addresses the array element. Also notice how the contents of the ARRAY have been initialized so that element 10H contains 29H.

EXAMPLE 3–8

```
                                .MODEL  SMALL          ;select small model
0000                            .DATA                  ;start data segment
0000 0010 [            ARRAY    DB  16  DUP(?)         ;setup array of 16 bytes
            00
               ]
0010 29                         DB  29H                ;element 10H
0011 001E [                     DB  20  dup(?)
            00
               ]
0000                            .CODE                  ;start code segment
                                .STARTUP

0017 B8 0000 R                  MOV  BX,OFFSET ARRAY   ;address ARRAY
001A BF 0010                    MOV  DI,10H            ;address element 10H
001D 8A 01                      MOV  AL,[BX+DI]        ;get element 10H
001F BF 0020                    MOV  DI,20H            ;address element 20H
0022 88 01                      MOV  [BX+DI],AL        ;save in element 20H

                                .EXIT                  ;exit to DOS
                                END                    ;end program
```

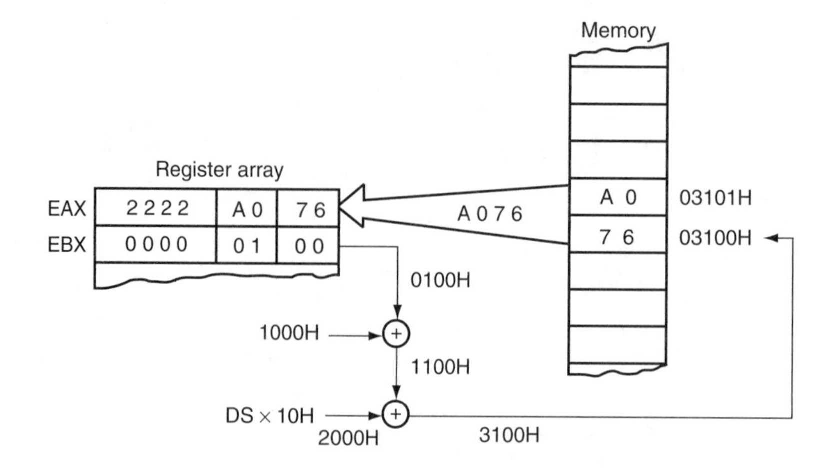

FIGURE 3–10 The operation of the MOV AX,[BX+1000H] instructon, when BX = 0100H and DS = 0200H.

Register Relative Addressing

Register relative addressing is similar to base-plus-index addressing and displacement addressing. In register relative addressing, the data in a segment of memory are addressed by adding the displacement to the contents of a base or an index register (BP, BX, DI, or SI). Figure 3–10 shows the operation of the MOV AX,[BX+1000H] instruction. In this example, BX = 0100H and DS = 0200H, so the address generated is the sum of DS × 0H, BX, and the displacement of 1000H, which addresses location 03100H. Remember that BX, DI, or SI addresses the data segment and BP addresses the stack segment. In the 80386 and above, the displacement can be a 32-bit number and the register can be any 32-bit register except the ESP register. Remember that the size of a real mode segment is 64K bytes long. Table 3–7 lists a few instructions that use register relative addressing.

The displacement is a number added to the register within the [ ], as in the MOV AL,[DI+2] instruction, or it can be a displacement subtracted from the register, as in MOV AL,[SI-l]. A displacement also can be an offset address appended to the front of the [ ], as in MOV AL,DATA[DI]. Both forms of displacements also can appear simultaneously, as in the MOV AL,DATA[DI+3] instruction. In all cases, both forms of the displacement add to the base or base and index register within the [ ]. In the 8086–80286 microprocessors, the value of the displacement is limited to a 16-bit signed number with a value ranging between +32,767 (7FFFH)

TABLE 3–7 Examples of register relative addressing.

Assembly Language	*Size*	*Operation*
MOV AX,[DI+100H]	16 bits	Copies the word contents of the data segment memory location addressed by DI plus 100H into AX
MOV ARRAY[SI],BL	8 bits	Copies BL into the data segment memory location addressed by ARRAY plus SI
MOV LIST[SI+2],CL	8 bits	Copies CL into the data segment memory location addressed by the sum of LIST, SI, and 2
MOV DI,SET_IT[BX]	16 bits	Copies the word contents of the data segment memory location addressed by SET_IT plus BX into DI
MOV DI,[EAX+10H]	16 bits	Copies the word contents of the data segment location addressed by EAX plus 10H into DI
MOV ARRAY[EBX],EAX	32 bits	Copies EAX into the data segment memory location addressed by ARRAY plus EBX

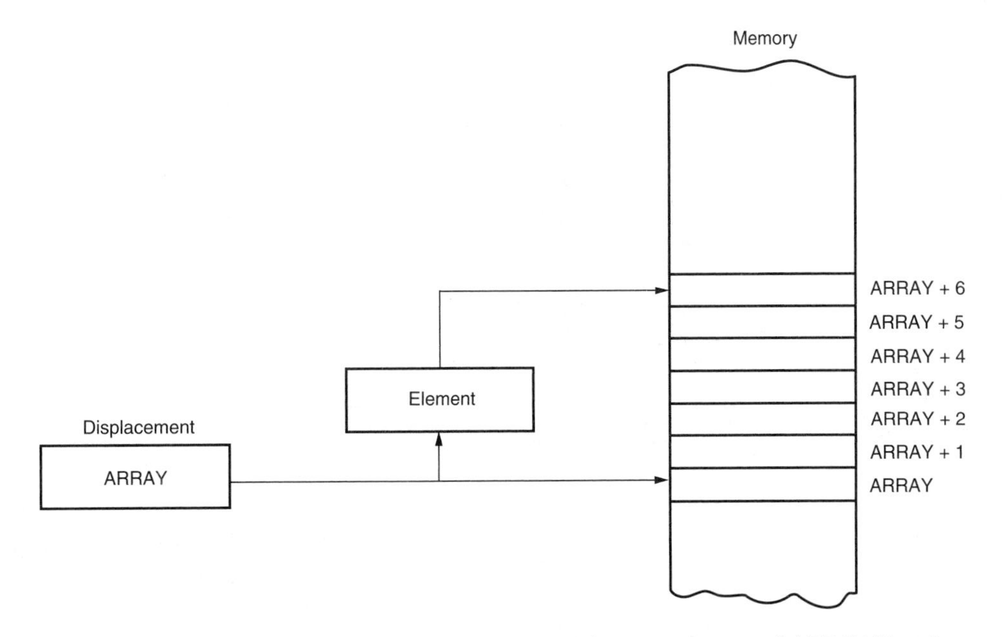

FIGURE 3–11 Register relative addressing used to address an element of ARRAY. The displacement addresses the start of ARRAY, and DI accesses an element.

and –32,768 (8000H); in the 80386 and above, a 32-bit displacement is allowed with a value ranging between +2,147,483,647 (7FFFFFFFH) and –2,147,483,648 (80000000H).

Addressing Array Data with Register Relative. It is possible to address array data with register relative addressing, such as one does with base-plus-index addressing. In Figure 3–11, register relative addressing is illustrated with the same example as for base-plus-index addressing. This shows how the displacement ARRAY adds to index register DI to generate a reference to an array element.

Example 3–9 shows how this new addressing mode can transfer the contents of array element 10H into array element 20H. Notice the similarity between this example and Example 3–8. The main difference is that, in Example 3–9, register BX is not used to address memory ARRAY; instead, ARRAY is used as a displacement to accomplish the same task.

EXAMPLE 3–9

```
                             .MODEL SMALL          ;select small model
0000                         .DATA                 ;start data segment
0000 0010 [          ARRAY   DB  16 dup(?)         ;setup ARRAY
          00
            ]
0010 29                      DB  29                ;element 10H
0011 001E [                  DB  30 dup(?)
          00
            ]
0000                         .CODE                 ;start code segment
                             .STARTUP              ;start program
0017 BF 0010                 MOV  DI,10H           ;address element 10H
001A 8A 85 0000 R            MOV  AL,ARRAY[DI]     ;get ARRAY element 10H
001E BF 0020                 MOV  DI,20H           ;address element 20H
0021 88 85 0000 R            MOV  ARRAY[DI],AL     ;save it in element 20H
                             .EXIT                 ;exit to DOS
                             END                   ;end of program
```

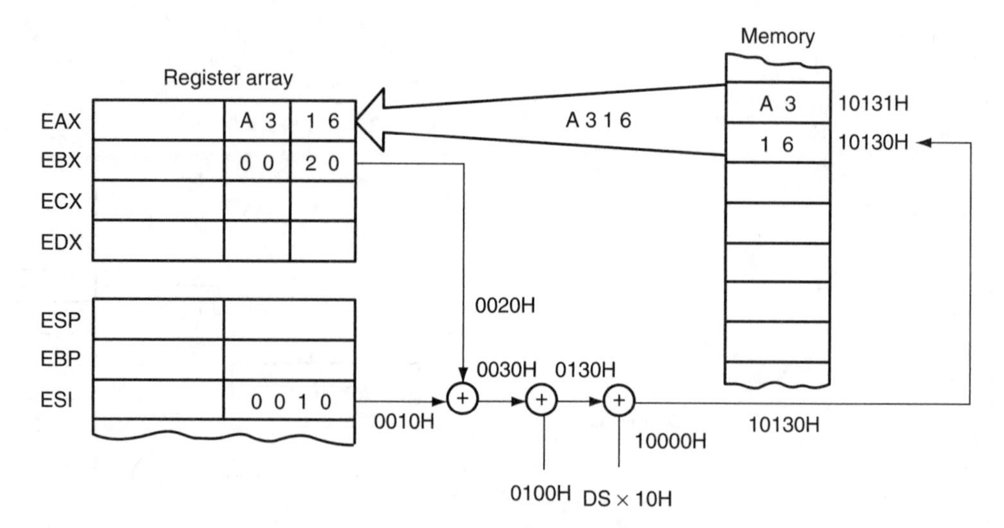

FIGURE 3–12 An example of base relative-plus-index addressing using a MOV AX,[BX+SI+100H] instruction. Note: DS = 1000H.

Base Relative-Plus-Index Addressing

The base relative-plus-index addressing mode is similar to the base-plus-index addressing, but it adds a displacement, besides using a base register and an index register, to form the memory address. This type of addressing mode often addresses a two-dimensional array of memory data.

Addressing Data with Base Relative-Plus-Index. Base relative-plus-index addressing is the least-used addressing mode. Figure 3–12 shows how data are referenced if the instruction executed by the microprocessor is MOV AX,[BX+SI+100H]. The displacement of 100H adds to BX and SI to form the offset address within the data segment. Registers BX = 0020H, SI = 0100H, and DS = 1000H, so the effective address for this instruction is 10130H—the sum of these registers plus a displacement of 100H. This addressing mode is too complex for frequent use in programming. Some typical instructions using base relative-plus-index addressing appear in Table 3–8. Note that with the 80386 and above, the effective address is generated by the sum of two 32-bit registers plus a 32-bit displacement.

TABLE 3–8 Example base relative-plus-index instructions.

Assembly Language	*Size*	*Operation*
MOV DH,[BX+DI+20H]	8 bits	Copies the byte contents of the data segment memory location addressed by the sum of BX, DI and 20H into DH
MOV AX,FILE[BX+DI]	16 bits	Copies the word contents of the data segment memory location addressed by the sum of FILE, BX and DI into AX
MOV LIST[BP+DI],CL	8 bits	Copies CL into the stack segment memory location addressed by the sum of LIST, BP, and DI
MOV LIST[BP+SI+4],DH	8 bits	Copies DH into the stack segment memory location addressed by the sum of LIST, BP, SI, and 4
MOV EAX,FILE[EBX+ECX+2]	32 bits	Copies the doubleword contents of the memory location addressed by the sum of FILE, EBX, ECX, and 2 into EAX

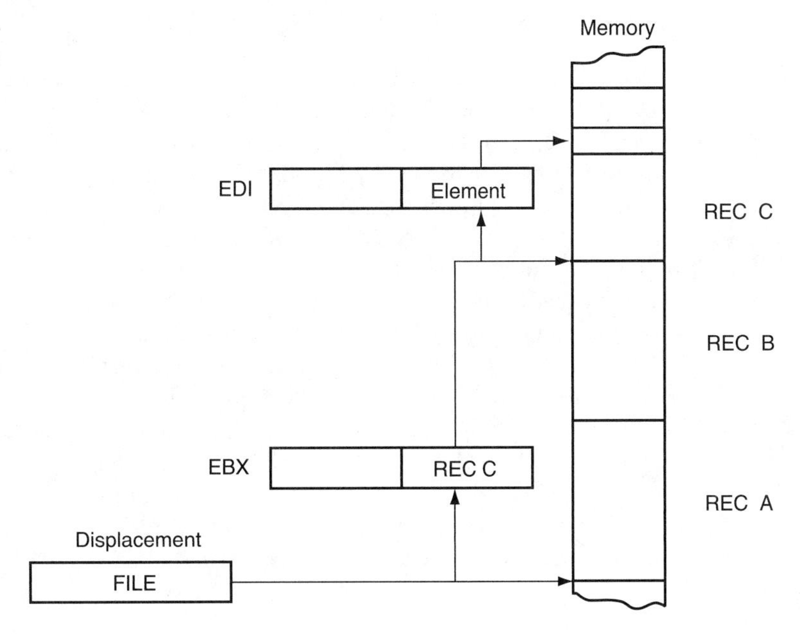

FIGURE 3–13 Base relative-plus-index addressing used to access a FILE that contains multiple records (REC).

Addressing Arrays with Base Relative-Plus-Index. Suppose that a file of many records exists in memory and each record contains many elements. The displacement addresses the file, the base register addresses a record, and the index register addresses an element of a record. Figure 3–13 illustrates this very complex form of addressing.

Example 3–10 provides a program that copies element 0 of record A into element 2 of record C by using the base relative-plus-index mode of addressing. This example FILE contains four records and each record contains 10 elements. Notice how the THIS BYTE statement is used to define the label FILE and RECA as the same memory location.

EXAMPLE 3–10

```
                                .MODEL SMALL            ;select small model
0000                            .DATA                   ;start data segment
0000 = 0000             FILE    EQU  THIS BYTE          ;assign FILE to this byte
0000 000A [             RECA    DB   10 dup(?)          ;10 bytes for record A
            00
               ]
000A 000A [             RECB    DB   10 dup(?)          ;10 bytes for record B
            00
               ]
0014 000A [             RECC    DB   10 dup(?)          ;10 bytes for record C
            00
               ]
001E 000A [             RECD    DB   10 dup(?)          ;10 bytes for record D
            00
               ]
0000                            .CODE                   ;start code segment
                                .STARTUP                ;start program
0017 BB 0000 R                  MOV  BX,OFFSET RECA     ;address record A
001A BF 0000                    MOV  DI,0               ;address element 0
001D 8A 81 0000 R               MOV  AL,FILE[BX+DI]     ;get data
0021 BB 0014 R                  MOV  BX,OFFSET RECC     ;address record C
0024 BF 0002                    MOV  DI,2               ;address element 2
0027 88 81 0000 R               MOV  FILE[BX+DI],AL     ;save data
                                .exit                   ;exit to DOS
                                end                     ;end of program
```

Scaled-Index Addressing

Scaled-index addressing is the last type of data-addressing mode discussed. This data-addressing mode is unique to the 80386 through the Pentium 4 microprocessors. Scaled-index addressing uses two 32-bit registers (a base register and an index register) to access the memory. The second register (index) is multiplied by a scaling factor. The scaling factor can be 1×, 2×, 4×, or 8×. A scaling factor of 1× is implied and need not be included in the assembly language instruction (MOV AL,[EBX+ECX]). A scaling factor of 2× is used to address word-sized memory arrays, a scaling factor of 4× is used with doubleword-sized memory arrays, and a scaling factor of 8× is used with quadword-sized memory arrays.

An example instruction is MOV AX,[EDI+2*ECX]. This instruction uses a scaling factor of 2×, which multiplies the contents of ECX by 2 before adding it to the EDI register to form the memory address. If ECX contains 00000000H, word-sized memory element 0 is addressed; if ECX contains 00000001H, word-sized memory element 1 is accessed, and so forth. This scales the index (ECX) by a factor of 2 for a word-sized memory array. Refer to Table 3–9 for some examples of scaled-index addressing. As you can imagine, there are an extremely large number of the scaled-index addressed register combinations. Scaling is also applied to instructions that use a single indirect register to access memory. The MOV EAX,[4*EDI] is a scaled-index instruction that uses one register to indirectly address memory.

Example 3–11 shows a sequence of instructions that uses scaled-index addressing to access a word-sized array of data called LIST. Note that the offset address of LIST is loaded into register EBX with the MOV EBX,OFFSET LIST instruction. Once EBX addresses array LIST, elements (located in ECX) 2, 4, and 7 of this word-wide array are added, using a scaling factor of 2 to access the elements. This program stores the 2 at element 2 into elements 4 and 7. Also notice the .386 directive to select the 80386 microprocessor. This directive must follow the .MODEL statement for the assembler to process 80386 instructions for DOS. If the 80486 is in use, the .486 directive appears after the .MODEL statement; if the Pentium is in use, then use .586; and if the Pentium Pro, Pentium II, Pentium III, or Pentium 4 is in use, the .686 directive. If the microprocessor selection directive appears before the .MODEL statement, the microprocessor executes instructions in the 32-bit protected mode, which must execute in Windows.

EXAMPLE 3–11

```
                                  .MODEL SMALL          ;select small model
                                  .386                  ;select 80386 microprocessor
0000                              .DATA                 ;start data segment
```

TABLE 3–9 Examples of scaled-index addressing.

Assembly Language	*Size*	*Operation*
MOV EAX,[EBX+4*ECX]	32 bits	Copies the doubleword contents of the data segment memory location addressed by the sum of 4 times ECX plus EBX into EAX
MOV [EAX+2*EDI+100H],CX	16 bits	Copies CX into the data segment memory location addressed by the sum of EAX, 100H, and 2 times EDI
MOV AL,[EBP+2*EDI+2]	8 bits	Copies the byte contents of the stack segment memory location addressed by the sum of EBP, 2, and 2 times EDI into AL
MOV EAX,ARRAY[4*ECX]	32 bits	Copies the doubleword contents of the data segment memory location addressed by the sum of ARRAY and 4 times ECX into EAX

```
0000 0000 0001 0002  LIST  DW  0,1,2,3,4          ;define array LIST
     0003 0004
000A 0005 0006 0007        DW  5,6,7,8,9
     0008 0009
0000                       .CODE                  ;start code segment
0010 66|BB 00000000 R      MOV EBX,OFFSET LIST    ;address array LIST
0016 66|B9 00000002        MOV ECX,2              ;address element 2
001C 67&8B 04 4B           MOV AX,EBX+2*ECX]      ;get element 2
0020 66|B9 00000004        MOV ECX,4              ;address element 4
0026 67&89 04 4B           MOV [EBX+2*ECX],AX     ;store in element 4
002A 66|B9 00000007        MOV ECX,7              ;address element 7
0030 67&89 04 4B           MOV [EBX+2*ECX],AX     ;store in element 7
                           .exit                  ;exit to DOS
                           end
```

Data Structures

A data structure is used to specify how information is stored in a memory array and can be quite useful with applications that use arrays. It is best to think of a data structure as a template for data. The start of a structure is identified with the STRUC assembly language directive and the end with the ENDS statement. A typical data structure is defined and used three times in Example 3–12. Notice that the name of the structure appears with the STRUC and with the ENDS statement. The example shows the data structure as it was typed without the assembled version.

EXAMPLE 3–12

```
;define the INFO data structure
;
INFO     STRUC

NAMES    DB  32 dup(?)   ;reserve 32 bytes for a name
STREET   DB  32 dup(?)   ;reserve 32 bytes for the street address
CITY     DB  16 dup(?)   ;reserve 16 bytes for the city
STATE    DB   2 dup(?)   ;reserve 2 bytes for the state
ZIP      DB   5 dup(?)   ;reserve 5 bytes for the zipcode

INFO     ENDS

NAME1    INFO <'Bob Smith', '123 Main Street', 'Wanda', 'OH', '44444'>
NAME2    INFO <'Steve Doe', '222 Moose Lane', 'Miller', 'PA', '18100'>
NAME3    INFO <'Jim Dover', '303 Main Street', 'Orender', 'CA', '90000'>
```

The data structure in Example 3–12 defines five fields of information. The first is 32 bytes long and holds a name; the second is 32 bytes long and holds a street address; the third is 16 bytes long for the city; the fourth is 2 bytes long for the state; the fifth is 5 bytes long for the ZIP code. Once the structure is defined (INFO), it can be filled, as illustrated, with names and addresses. Three example uses for INFO are illustrated. Note that literals are surrounded with apostrophes and the entire field is surrounded with < > symbols when the data structure is used to define data.

When data are addressed in a structure, use the structure name and the field name to select a field from the structure. For example, to address the STREET in NAME2, use the operand NAME2.STREET, where the name of the structure is first followed by a period and then by the name of the field. Likewise, use NAME3.CITY to refer to the city in structure NAME3.

A short sequence of instructions appears in Example 3–13 that clears the name field in structure NAME1, the address field in structure NAME2, and the ZIP code field in structure

NAME3. The function and operation of the instructions in this program are defined in later chapters in the text. You may wish to refer to this example once you learn these instructions.

EXAMPLE 3–13

```
                          ;clear NAMES in array NAME1
                          ;
0000 B9 0020                     MOV  CX,32
0003 B0 00                       MOV  AL,0
0005 BE 0000 R                   MOV  DI,OFFSET NAME1.NAMES
0008 F3/AA                       REP  STOSB
                          ;
                          ;clear STREET in array NAME2
                          ;
000A B9 0020                     MOV  CX,32
000D B0 00                       MOV  AL,0
000F BE 0077 R                   MOV  DI,OFFSET NAME2.STREET
0012 F3/AA                       REP  STOSB
                          ;
                          ;clear ZIP in NAME3
                          ;
0014 B9 0005                     MOV  CX,5
0017 B0 00                       MOV  AL,0
0019 BE 0100 R                   MOV  DI,OFFSET NAME3.ZIP
001C F3/AA                       REP  STOSB
```

3–2 PROGRAM MEMORY-ADDRESSING MODES

Program memory-addressing modes, used with the JMP (jump) and CALL instructions, consist of three distinct forms: direct, relative, and indirect. This section introduces these three addressing forms, using the JMP instruction to illustrate their operation.

Direct Program Memory Addressing

Direct program memory addressing is what many early microprocessors used for all jumps and calls. Direct program memory addressing is also used in high-level languages, such as the BASIC language GOTO and GOSUB instructions. The microprocessor uses this form of addressing, but not as often as relative and indirect program memory addressing are used.

The instructions for direct program memory addressing store the address with the opcode. For example, if a program jumps to memory location 10000H for the next instruction, the address (10000H) is stored following the opcode in the memory. Figure 3–14 shows the direct intersegment JMP instruction and the four bytes required to store the address 10000H. This JMP instruction loads CS with 1000H and IP with 0000H to jump to memory location 10000H for the next instruction. (An **intersegment jump** is a jump to any memory location within the entire memory system.) The direct jump is often called a *far jump* because it can jump to any memory location for the next instruction. In the real mode, a far jump accesses any location within the first 1M byte of memory by changing both CS and IP. In protected mode operation, the far jump accesses a new code segment descriptor from the descriptor table, allowing it to jump to any memory location in the entire 4G-byte address range in the 80386 through Pentium 4 microprocessors.

The only other instruction that uses direct program addressing is the intersegment or far CALL instruction. Usually, the name of a memory address, called a *label,* refers to the location

FIGURE 3–14 The 5-byte machine language version of a JMP [10000H] instruction.

Opcode	Offset (low)	Offset (high)	Segment (low)	Segment (high)
E A	0 0	0 0	0 0	1 0

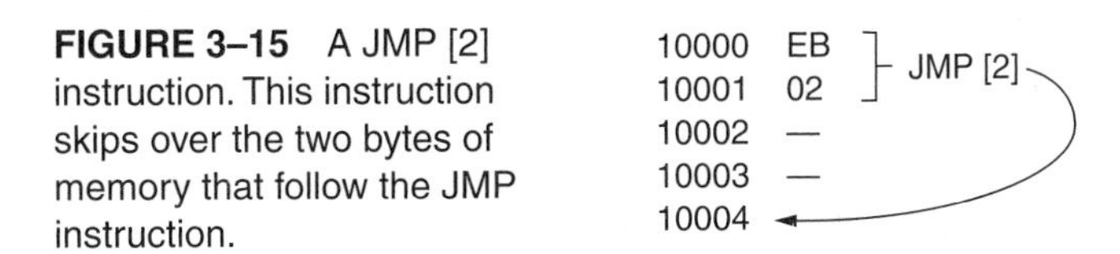

FIGURE 3–15 A JMP [2] instruction. This instruction skips over the two bytes of memory that follow the JMP instruction.

that is called or jumped to instead of the actual numeric address. When using a label with the CALL or JMP instruction, most assemblers select the best form of program addressing.

Relative Program Memory Addressing

Relative program memory addressing is not available in all early microprocessors, but it is available to this family of microprocessors. The term **relative** means "relative to the instruction pointer (IP)." For example, if a JMP instruction skips the next two bytes of memory, the address in relation to the instruction pointer is a 2 that adds to the instruction pointer. This develops the address of the next program instruction. An example of the relative JMP instruction is shown in Figure 3–15. Notice that the JMP instruction is a one-byte instruction, with a one-byte or a two-byte displacement that adds to the instruction pointer. A one-byte displacement is used in **short** jumps, and a two-byte displacement is used with **near** jumps and calls. Both types are considered to be intrasegment jumps. (An **intrasegment jump** is a jump anywhere within the current code segment.) In the 80386 and above, the displacement can also be a 32-bit value, allowing them to use relative addressing to any location within their 4G-byte code segments.

Relative JMP and CALL instructions contain either an 8-bit or a 16-bit signed displacement that allows a forward memory reference or a reverse memory reference. (The 80386 and above can have an 8-bit or 32-bit displacement.) All assemblers automatically calculate the distance for the displacement and select the proper one-, two-, or four-byte form. If the distance is too far for a two-byte displacement in an 8086 through an 80286 microprocessor, some assemblers use the direct jump. An 8-bit displacement (*short*) has a jump range of between +127 and –128 bytes from the next instruction; a 16-bit displacement (*near*) has a range of ±32K bytes. In the 80386 and above, a 32-bit displacement allows a range of ±2G bytes. The 32-bit displacement can only be used in the protected mode.

Indirect Program Memory Addressing

The microprocessor allows several forms of program indirect memory addressing for the JMP and CALL instructions. Table 3–10 lists some acceptable program indirect jump

TABLE 3–10 Examples of indirect program memory addressing.

Assembly Language	*Operation*
JMP AX	Jumps to the current code segment location addressed by the contents of AX
JMP CX	Jumps to the current code segment location addressed by the contents of CX
JMP NEAR PTR[BX]	Jumps to the current code segment location addressed by the contents of the data segment location addressed by BX
JMP NEAR PTR[DI+2]	Jumps to the current code segment location addressed by the contents of the data segment memory location addressed by DI plus 2
JMP TABLE[BX]	Jumps to the current code segment location addressed by the contents of the data segment memory location address by TABLE plus BX
JMP ECX	Jumps to the current code segment location addressed by the contents of ECX

FIGURE 3–16 A jump table that stores addresses of various programs. The exact address chosen from the TABLE is determined by an index stored with the jump instruction.

```
TABLE DW LOC0
      DW LOC1
      DW LOC2
      DW LOC3
```

instructions, which can use any 16-bit register (AX, BX, CX, DX, SP, BP, DI, or SI); any relative register ([BP], [BX], [DI], or [SI]); and any relative register with a displacement. In the 80386 and above, an extended register can also be used to hold the address or indirect address of a relative JMP or CALL. For example, JMP EAX jumps to the location address by register EAX.

If a 16-bit register holds the address of a JMP instruction, the jump is near. For example, if the BX register contains 1000H and a JMP BX instruction executes, the microprocessor jumps to offset address 1000H in the current code segment.

If a relative register holds the address, the jump is also considered to be an indirect jump. For example, JMP [BX] refers to the memory location within the data segment at the offset address contained in BX. At this offset address is a 16-bit number that is used as the offset address in the intrasegment jump. This type of jump is sometimes called an *indirect-indirect* or *double-indirect jump.*

Figure 3–16 shows a jump table that is stored, beginning at memory location TABLE. This jump table is referenced by the short program of Example 3–14. In this example, the BX register is loaded with a 4 so, when it combines in the JMP TABLE[BX] instruction with TABLE, the effective address is the contents of the second entry in the 16-bit-wide jump table.

EXAMPLE 3–14

```
                          ;Using indirect addressing for a jump
                          ;
0000 BB 0004              MOV  BX,4          ;address LOC2
0003 FF A7 23A1 R         JMP  TABLE[BX]     ;jump to LOC2
```

3–3 STACK MEMORY-ADDRESSING MODES

The stack plays an important role in all microprocessors. It holds data temporarily and stores the return addresses used by procedures. The stack memory is a LIFO (**last-in, first-out**) memory, which describes the way that data are stored and removed from the stack. Data are placed onto the stack with a **PUSH instruction** and removed with a **POP instruction**. The CALL instruction also uses the stack to hold the return address for procedures and a RET (return) instruction to remove the return address from the stack.

The stack memory is maintained by two registers: the stack pointer (SP or ESP) and the stack segment register (SS). Whenever a word of data is pushed onto the stack [see Figure 3–17(a)], the high-order 8 bits are placed in the location addressed by SP – 1. The low-order 8 bits are placed in the location addressed by SP – 2. The SP is then decremented by 2 so that the next word of data is stored in the next available stack memory location. The SP/ESP register always points to an area of memory located within the stack segment. The SP/ESP register adds to SS × 10H to form the stack memory address in the real mode. In protected mode operation, the SS register holds a selector that accesses a descriptor for the base address of the stack segment.

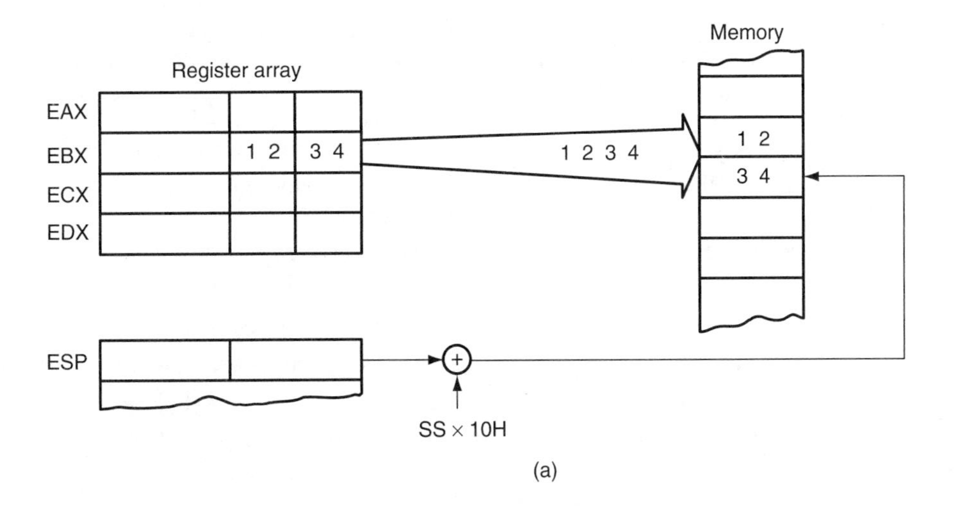

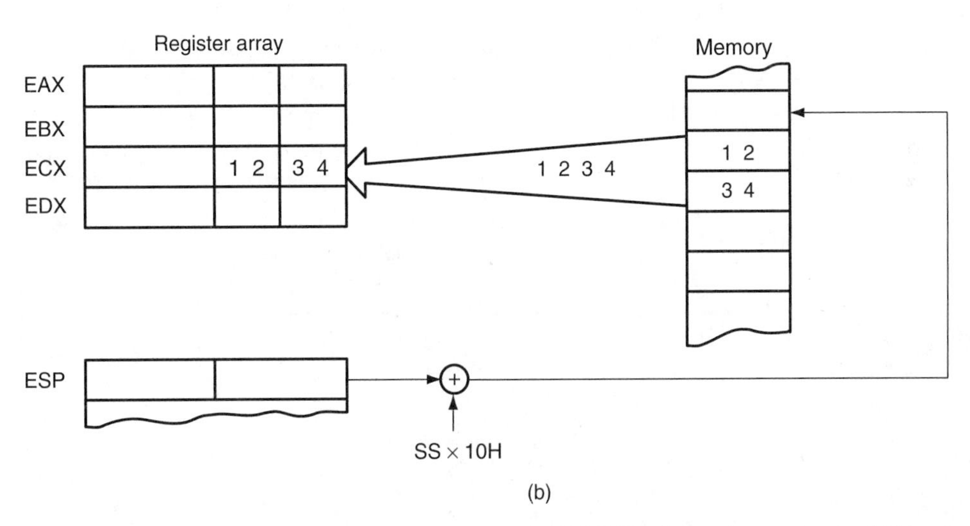

FIGURE 3–17 The PUSH and POP instructions: (a) PUSH BX places the contents of BX onto the stack; (b) POP CX removes data from the stack and places them into CX. Both instructions are shown after execution.

Whenever data are popped from the stack [see Figure 3–17(b)], the low-order 8 bits are removed from the location addressed by SP. The high-order 8 bits are removed from the location addressed by SP + 1. The SP register is then incremented by 2. Table 3–11 lists some of the PUSH and POP instructions available to the microprocessor. Note that PUSH and POP store or retrieve words of data—never bytes—in the 8086 through the 80286 microprocessors. The 80386 and above allow words or doublewords to be transferred to and from the stack. Data may be pushed onto the stack from any 16-bit register or segment register; in the 80386 and above, from any 32-bit extended register. Data may be popped off the stack into any register or any segment register except CS. The reason that data may not be popped from the stack into CS is that this only changes part of the address of the next instruction.

The PUSHA and POPA instructions either push or pop all of the registers, except segment registers, onto the stack. These instructions are not available on the early 8086/8088 processors. The push immediate instruction is also new to the 80286 through the Pentium 4 microprocessors.

TABLE 3–11 Example PUSH and POP instructions.

Assembly Language	*Operation*
POPF	Removes a word from the stack and places it into the flag register
POPFD	Removes a doubleword from the stack and places it into the EFLAG register
PUSHF	Copies the flag register to the stack
PUSHFD	Copies the EFLAG register to the stack
PUSH AX	Copies the AX register to the stack
POP BX	Removes a word from the stack and places it into the BX register
PUSH DS	Copies the DS register to the stack
PUSH 1234H	Copies a word-sized 1234H to the stack
POP CS	This instruction is illegal
PUSH WORD PTR[BX]	Copies the word contents of the data segment memory location addressed by BX onto the stack
PUSHA	Copies AX, CX, DX, BX, SP, BP, DI and SI to the stack
POPA	Removes the word contents for the following registers from the stack: SI, DI, BP, SP, BX, DX, CX, and AX
PUSHAD	Copies EAX, ECX, EDX, EBX, ESP, EBP, EDI, and ESI to the stack
POPAD	Removes the doubleword contents for the following registers from the stack: ESI, EDI, EBP, ESP, EBX, EDX, ECX, and EAX
POP EAX	Removes a doubleword from the stack and places it into the EAX register
PUSH EDI	Copies EDI to the stack

Note the examples in Table 3–11, which show the order of the registers transferred by the PUSHA and POPA instructions. The 80386 and above also allow extended registers to be pushed or popped.

Example 3–15 lists a short program that pushes the contents of AX, BX, and CX onto the stack. The first POP retrieves the value that was pushed onto the stack from CX and places it into AX. The second POP places the original value of BX into CX. The last POP places the value of AX into BX.

EXAMPLE 3–15

```
                         .MODEL TINY          ;select tiny model
0000                     .CODE                ;start code segment
                         .STARTUP             ;start program
0100 B8 1000             MOV   AX,1000H       ;load test data
0103 BB 2000             MOV   BX,2000H
0106 B9 3000             MOV   CX,3000H

0109 50                  PUSH  AX             ;1000H to stack
010A 53                  PUSH  BX             ;2000H to stack
010B 51                  PUSH  CX             ;3000H to stack

010C 58                  POP   AX             ;3000H to AX
010D 59                  POP   CX             ;2000H to CBX
010E 5B                  POP   BX             ;1000H to BX
                         .exit                ;exit to DOS
                         end                  ;end program
```

3–4 SUMMARY

1. The data-addressing modes include register, immediate, direct, register indirect, base-plus-index, register relative, and base relative-plus-index addressing. The 80386 through the Pentium 4 microprocessors have an additional addressing mode called scaled-index addressing.
2. The program memory-addressing modes include direct, relative, and indirect addressing.
3. Table 3–12 lists all real mode data-addressing modes available to the 8086 through the 80286 microprocessors. Note that the 80386 and above use these modes, plus the many

TABLE 3–12 Example real mode data-addressing modes.

Assembly Language	*Address Generation*
MOV AL,BL	8-bit register addressing
MOV AX,BX	16-bit register addressing
MOV EAX,ECX	32-bit register addressing
MOV DS,DX	Segment register addressing
MOV AL,LIST	(DS x 10H) + LIST
MOV CH,DATA1	(DS x 10H) + DATA1
MOV ES,DATA2	(DS x 10H) + DATA2
MOV AL,12	Immediate data of 12
MOV AL,[BP]	(SS x 10H) + BP
MOV AL,[BX]	(DS x 10H) + BX
MOV AL,[DI]	(DS x 10H) + DI
MOV AL,[SI]	(DS x 10H) + SI
MOV AL,[BP+2]	(SS x 10H) + BP + 2
MOV AL,[BX–4]	(DS x 10H) + BX – 4
MOV AL,[DI+1000H]	(DS x 10H) + DI + 1000H
MOV AL,[SI+300H]	(DS x 10H) + SI + 300H
MOV AL,LIST[BP]	(SS x 10H) + LIST + BP
MOV AL,LIST[BX]	(DS x 10H) + LIST + BX
MOV AL,LIST[DI]	(DS x 10H) + LIST + DI
MOV AL,LIST[SI]	(DS x 10H) + LIST + SI
MOV AL,LIST[BP+2]	(SS x 10H) + LIST + BP + 2
MOV AL,LIST[BX–6]	(DS x 10H) + LIST + BX – 6
MOV AL,LIST[DI+100H]	(DS x 10H) + LIST + DI + 100H
MOV AL,LIST[SI+200H]	(DS x 10H) + LIST + SI + 200H
MOV AL,[BP+DI]	(SS x 10H) + BP + DI
MOV AL,[BP+SI]	(SS x 10H) + BP + SI
MOV AL,[BX+DI]	(DS x 10H) + BX + DI
MOV AL,[BX+SI]	(DS x 10H) + BX + SI
MOV AL,[BP+DI+8]	(SS x 10H) + BP + DI + 8
MOV AL,[BP+SI–8]	(SS x 10H) + BP + SI – 8
MOV AL,[BX+DI+10H]	(DS x 10H) + BX + DI + 10H
MOV AL,[BX+SI–10H]	(DS x 10H) + BX + SI – 10H
MOV AL,LIST[BP+DI]	(SS x 10H) + LIST + BP + DI
MOV AL,LIST[BP+SI]	(SS x 10H) + LIST + BP + SI
MOV AL,LIST[BX+DI]	(DS x 10H) + LIST + BX + DI
MOV AL,LIST[BX+SI]	(DS x 10H) + LIST + BX + SI
MOV AL,LIST[BP+DI+2]	(SS x 10H) + LIST + BP + DI + 2
MOV AL,LIST[BP+SI–7]	(SS x 10H) + LIST + BP + SI – 7
MOV AL,LIST[BX+DI+3]	(DS x 10H) + LIST + BX + DI + 3
MOV AL,LIST[BX+SI–2]	(DS x 10H) + LIST + BX + SI – 2

defined through this chapter. In the protected mode, the function of the segment register is to address a descriptor that contains the base address of the memory segment.

4. The 80386 through Pentium 4 microprocessors have additional addressing modes that allow the extended registers EAX, EBX, ECX, EDX, EBP, EDI, and ESI to address memory. Although these addressing modes are too numerous to list in tabular form, in general, any of these registers function in the same way as those listed in Table 3–12. For example, MOV AL,TABLE[EBX+2*ECX+10H] is a valid addressing mode for the 80386–Pentium 4 microprocessors.
5. The MOV instruction copies the contents of the source operand into the destination operand. The source never changes for any instruction.
6. Register addressing specifies any 8-bit register (AH, AL, BH, BL, CH, CL, DH, or DL) or any 16-bit register (AX, BX, CX, DX, SP, BP, SI, or DI). The segment registers (CS, DS, ES, or SS) are also addressable to move data between a segment register and a 16-bit register/memory location or for PUSH and POP. In the 80386 through the Pentium 4 microprocessors, the extended registers also are used for register addressing; they consist of EAX, EBX, ECX, EDX, ESP, EBP, EDI, and ESI. Also available to the 80386 and above are the FS and GS segment registers.
7. The MOV immediate instruction transfers the byte or word that immediately follows the opcode into a register or a memory location. Immediate addressing manipulates constant data in a program. In the 80386 and above, doubleword immediate data may also be loaded into a 32-bit register or memory location.
8. The .MODEL statement is used with assembly language to identify the start of a file and the type of memory model used with the file. If the size is TINY, the program exists in one segment, the code segment, and is assembled as a command (.COM) program. If the SMALL model is used, the program uses a code and data segment and assembles as an execute (.EXE) program. Other model sizes and their attributes are listed in Appendix A.
9. Direct addressing occurs in two forms in the microprocessor: (1) direct addressing and (2) displacement addressing. Both forms of addressing are identical except that direct addressing is used to transfer data between EAX, AX, or AL and memory; displacement addressing is used with any register-memory transfer. Direct addressing requires three bytes of memory, whereas displacement addressing requires four bytes. Note that some of these instructions in the 80386 and above may require additional bytes in the form of prefixes for register and operand sizes.
10. Register indirect addressing allows data to be addressed at the memory location pointed to by either a base (BP and BX) or index (DI and SI) register. In the 80386 and above, extended registers EAX, EBX, ECX, EDX, EBP, EDI, and ESI are used to address memory data.
11. Base-plus-index addressing often addresses data in an array. The memory address for this mode is formed by adding a base register, index register, and the contents of a segment register times 10H. In the 80386 and above, the base and index registers may be any 32-bit register except EIP and ESP.
12. Register relative addressing uses a base or index register, plus a displacement to access memory data.
13. Base relative-plus-index addressing is useful for addressing a two-dimensional memory array. The address is formed by adding a base register, an index register, displacement, and the contents of a segment register times 10H.
14. Scaled-index addressing is unique to the 80386 through the Pentium 4. The second of two registers (index) is scaled by a factor of 2×, 4×, or 8× to access words, doublewords, or quadwords in memory arrays. The MOV AX,[EBX+2*ECX] and the MOV [4*ECX],EDX are examples of scaled-index instructions.
15. Data structures are templates for storing arrays of data, and are addressed by array name and field. For example, array NUMBER and field TEN of array NUMBER is addressed as NUMBER.TEN.

16. Direct program memory addressing is allowed with the JMP and CALL instructions to any location in the memory system. With this addressing mode, the offset address and segment address are stored with the instruction.
17. Relative program addressing allows a JMP or CALL instruction to branch forward or backward in the current code segment by ±32K bytes. In the 80386 and above, the 32-bit displacement allows a branch to any location in the current code segment by using a displacement value of ±2G bytes. The 32-bit displacement can be used only in protected mode.
18. Indirect program addressing allows the JMP or CALL instructions to address another portion of the program or subroutine indirectly through a register or memory location.
19. The PUSH and POP instructions transfer a word between the stack and a register or memory location. A PUSH immediate instruction is available to place immediate data on the stack. The PUSHA and POPA instructions transfer AX, CX, DX, BX, BP, SP, SI, and DI between the stack and these registers. In the 80386 and above, the extended register and extended flags can also be transferred between registers and the stack. A PUSHFD stores the EFLAGS, whereas a PUSHF stores the FLAGS.

3–5 QUESTIONS AND PROBLEMS

1. What do the following MOV instructions accomplish?
   (a) MOV AX,BX
   (b) MOV BX,AX
   (c) MOV BL,CH
   (d) MOV ESP,EBP
   (e) MOV AX,CS
2. List the 8-bit registers that are used for register addressing.
3. List the 16-bit registers that are used for register addressing.
4. List the 32-bit registers that are used for register addressing in the 80386 through the Pentium 4 microprocessors.
5. List the 16-bit segment registers used with register addressing by MOV, PUSH, and POP.
6. What is wrong with the MOV BL,CX instruction?
7. What is wrong with the MOV DS,SS instruction?
8. Select an instruction for each of the following tasks:
   (a) copy EBX into EDX
   (b) copy BL into CL
   (c) copy SI into BX
   (d) copy DS into AX
   (e) copy AL into AH
9. Select an instruction for each of the following tasks:
   (a) move 12H into AL
   (b) move 123AH into AX
   (c) move 0CDH into CL
   (d) move 1000H into SI
   (e) move 1200A2H into EBX
10. What special symbol is sometimes used to denote immediate data?
11. What is the purpose of the .MODEL TINY statement?
12. What assembly language directive indicates the start of the CODE segment?
13. What is a label?
14. The MOV instruction is placed in what field of a statement?

15. A label may begin with what characters?
16. What is the purpose of the .EXIT directive?
17. Does the .MODEL TINY statement cause a program to assemble as an execute (.EXE) program?
18. What tasks does the .STARTUP directive accomplish in the small memory model?
19. What is a displacement? How does it determine the memory address in a MOV DS:[2000H], AL instruction?
20. What do the symbols [ ] indicate?
21. Suppose that DS = 0200H, BX = 0300H, and DI = 400H. Determine the memory address accessed by each of the following instructions, assuming real mode operation:
    (a) MOV AL,[1234H]
    (b) MOV EAX,[BX]
    (c) MOV [DI],AL
22. What is wrong with a MOV [BX],[DI] instruction?
23. Choose an instruction that requires BYTE PTR.
24. Choose an instruction that requires WORD PTR.
25. Choose an instruction that requires DWORD PTR.
26. Explain the difference between the MOV BX,DATA instruction and the MOV BX,OFFSET DATA instruction.
27. Suppose that DS = 1000H, SS = 2000H, BP = 1000H, and DI = 0100H. Determine the memory address accessed by each of the following instructions, assuming real mode operation:
    (a) MOV AL,[BP+DI]
    (b) MOV CX,[DI]
    (c) MOV EDX,[BP]
28. What, if anything, is wrong with a MOV AL,[BX][SI] instruction?
29. Suppose that DS = 1200H, BX = 0100H, and SI = 0250H. Determine the address accessed by each of the following instructions, assuming real mode operation:
    (a) MOV [100H],DL
    (b) MOV [SI+100H],EAX
    (c) MOV DL,[BX+100H]
30. Suppose that DS = 1100H, BX = 0200H, LIST = 0250H, and SI = 0500H. Determine the address accessed by each of the following instructions, assuming real mode operation:
    (a) MOV LIST[SI],EDX
    (b) MOV CL,LIST[BX+SI]
    (c) MOV CH,[BX+SI]
31. Suppose that DS = 1300H, SS = 1400H, BP = 1500H, and SI = 0100H. Determine the address accessed by each of the following instructions, assuming real mode operation:
    (a) MOV EAX,[BP+200H]
    (b) MOV AL,[BP+SI–200H]
    (c) MOV AL,[SI–0100H]
32. Which base register addresses data in the stack segment?
33. Suppose that EAX = 00001000H, EBX = 00002000H, and DS = 0010H. Determine the addresses accessed by the following instructions, assuming real mode operation:
    (a) MOV ECX,[EAX+EBX]
    (b) MOV [EAX+2*EBX],CL
    (c) MOV DH,[EBX+4*EAX+1000H]
34. Develop a data structure that has five fields of one word each named Fl, F2, F3, F4, and F5 with a structure name of FIELDS.
35. Show how field F3 of the data structure constructed in question 34 is addressed in a program.
36. What are the three program memory-addressing modes?

37. How many bytes of memory store a far direct jump instruction? What is stored in each of the bytes?
38. What is the difference between an intersegment and intrasegment jump?
39. If a near jump uses a signed 16-bit displacement, how can it jump to any memory location within the current code segment?
40. The 80386 and above use a ______ -bit displacement to jump to any location within the 4G-byte code segment.
41. What is a far jump?
42. If a JMP instruction is stored at memory location 100H within the current code segment, it cannot be a ______ jump if it is jumping to memory location 200H within the current code segment.
43. Show which JMP instruction assembles (short, near, or far) if the JMP THERE instruction is stored at memory address 10000H and the address of THERE is:
    (a) 10020H
    (b) 11000H
    (c) 0FFFEH
    (d) 30000H
44. Form a JMP instruction that jumps to the address pointed to by the BX register.
45. Select a JMP instruction that jumps to the location stored in memory at the location TABLE. Assume that it is a near JMP.
46. How many bytes are stored on the stack by a PUSH AX?
47. Explain how the PUSH [DI] instruction functions.
48. What registers are placed on the stack by the PUSHA instruction? In what order?
49. What does the PUSHAD instruction accomplish?
50. Which instruction places the EFLAGS on the stack in the Pentium 4 microprocessor?

CHAPTER 4

Data Movement Instructions

INTRODUCTION

This chapter concentrates on the data movement instructions. The data movement instructions include MOV, MOVSX, MOVZX, PUSH, POP, BSWAP, XCHG, XLAT, IN, OUT, LEA, LDS, LES, LFS, LGS, LSS, LAHF, SAHF, and the string instructions MOVS, LODS, STOS, INS, and OUTS. The latest data transfer instruction implemented on the Pentium Pro through Pentium 4 is the CMOV (conditional move) instruction. The data movement instructions are presented first because they are more commonly used in programs and are easy to understand.

The microprocessor requires an assembler program, which generates machine language, because machine language instructions are too complex to generate efficiently by hand. This chapter describes the assembly language syntax and some of its directives. [This text assumes that the user is developing software on an IBM personal computer or clone. It is recommended that the Microsoft MACRO assembler (MASM) be used as the development tool, but the Intel Assembler (ASM), Borland Turbo assembler (TASM), or similar software function equally as well. The most recent version of TASM completely emulates the MASM program. This text presents information that functions with the Microsoft MASM assembler, but most programs assemble without modification with other assemblers. Appendix A explains the Microsoft assembler and provides detail on the linker program.] As a more modern alternative, the Visual C++ compiler and its inline assembler program may also be used as a development system. Both are explained in detail in the text.

CHAPTER OBJECTIVES

Upon completion of this chapter, you will be able to:

1. Explain the operation of each data movement instruction with applicable addressing modes
2. Explain the purposes of the assembly language pseudo-operations and keywords such as ALIGN, ASSUME, DB, DD, DW, END, ENDS, ENDP, EQU, .MODEL, OFFSET, ORG, PROC, PTR, SEGMENT, USEI6, USE32, and USES
3. Select the appropriate assembly language instruction to accomplish a specific data movement task
4. Determine the symbolic opcode, source, destination, and addressing mode for a hexadecimal machine language instruction
5. Use the assembler to set up a data segment, stack segment, and code segment

6. Show how to set up a procedure using PROC and ENDP
7. Explain the difference between memory models and full-segment definitions for the MASM assembler
8. Use the Visual C++ online assembler to perform data movement tasks

4–1 MOV REVISITED

The MOV instruction, introduced in Chapter 3, explains the diversity of 8086–Pentium 4 addressing modes. In this chapter, the MOV instruction introduces the machine language instructions available with various addressing modes and instructions. Machine code is introduced because occasionally it may be necessary to interpret machine language programs generated by an assembler or inline assembler of Visual C++. Interpretation of the machine's native language (**machine language**) allows debugging or modification at the machine language level. Occasionally, machine language patches are made by using the DEBUG program available with DOS and also in Visual C++ for Windows, which requires some knowledge of machine language. Conversion between machine and assembly language instructions is illustrated in Appendix B.

Machine Language

Machine language is the native binary code that the microprocessor understands and uses as its instructions to control its operation. Machine language instructions for the 8086 through the Pentium 4 vary in length from one to as many as 13 bytes. Although machine language appears complex, there is order to this microprocessor's machine language. There are well over 100,000 variations of machine language instructions, meaning that there is no complete list of these variations. Because of this, some binary bits in a machine language instruction are given, and the remaining bits are determined for each variation of the instruction.

Instructions for the 8086 through the 80286 are 16-bit mode instructions that take the form found in Figure 4–1(a). The 16-bit mode instructions are compatible with the 80386 and above if they are programmed to operate in the 16-bit instruction mode, but they may be prefixed, as shown in Figure 4–1(b). The 80386 and above assume that all instructions are 16-bit mode instructions when the machine is operated in the *real mode* (*DOS*). In the *protected mode* (*Windows*), the upper byte of the descriptor contains the D-bit that selects either the 16- or 32-bit instruction mode. At present, only Windows 95 through Windows XP and Linux operate in the 32-bit instruction mode. The 32-bit mode instructions are in the form shown in Figure 4–1(b).

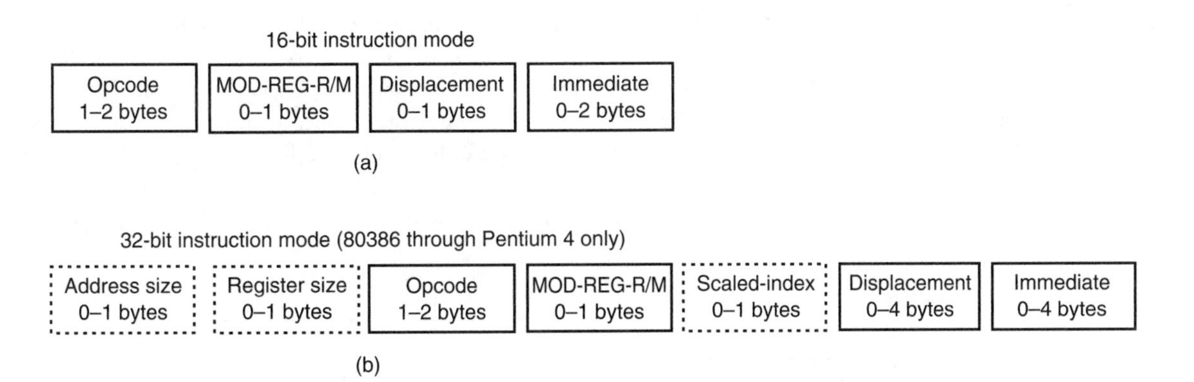

FIGURE 4–1 The formats of the 8086–Pentium 4 instructions. (a) The 16-bit form and (b) the 32-bit form.

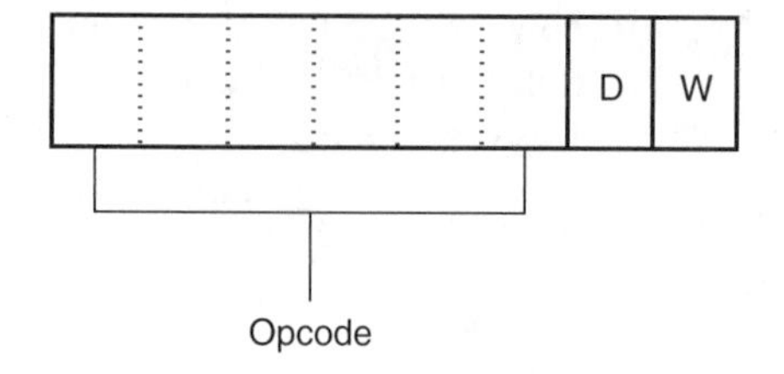

FIGURE 4–2 Byte 1 of many machine language instructions, showing the position of the D- and W-bits.

These instructions occur in the 16-bit instruction mode by the use of prefixes, which are explained later in this chapter.

The first two bytes of the 32-bit instruction mode format are called **override prefixes** because they are not always present. The first modifies the size of the operand address used by the instruction and the second modifies the register size. If the 80386 through the Pentium 4 operate as 16-bit instruction mode machines (real or protected mode) and a 32-bit register is used, the **register-size prefix** (66H) is appended to the front of the instruction. If operated in the 32-bit instruction mode (protected mode only) and a 32-bit register is used, the register-size prefix is absent. If a 16-bit register appears in an instruction in the 32-bit instruction mode, the register-size prefix is present to select a 16-bit register. The **address-size prefix** (67H) is used in a similar fashion, as explained later in this chapter. The prefixes toggle the size of the register and operand address from 16-bit to 32-bit or from 32-bit to 16-bit for the prefixed instruction. Note that the 16-bit instruction mode uses 8- and 16-bit registers and addressing modes, while the 32-bit instruction mode uses 8- and 32-bit registers and addressing modes by default. The prefixes override these defaults so that a 32-bit register can be used in the 16-bit mode or a 16-bit register can be used in the 32-bit mode. The **mode of operation** (16 or 32 bits) should be selected to function with the current application. If 8- and 32-bit data pervade the application, the 32-bit mode should be selected; likewise, if 8- and 16-bit data pervade, the 16-bit mode should be selected. Normally, mode selection is a function of the operating system. (Remember that DOS can operate only in the 16-bit mode, but Windows can operate in both modes.)

The Opcode. The **opcode** selects the operation (addition, subtraction, move, and so on) that is performed by the microprocessor. The opcode is either one or two bytes long for most machine language instructions. Figure 4–2 illustrates the general form of the first opcode byte of many, but *not* all, machine language instructions. Here, the first six bits of the first byte are the binary opcode. The remaining two bits indicate the **direction** (D)—not to be confused with the instruction mode bit (16/32) or direction flag bit (used with string instructions)—of the data flow, and indicate whether the data are a byte or a word (W). In the 80386 and above, words and doublewords are both specified when W = 1. The instruction mode and register-size prefix (66H) determine whether W represents a word or a doubleword.

If the direction bit (D) = 1, data flow *to* the register REG field from the R/M field located in the second byte of an instruction. If the D-bit = 0 in the opcode, data flow to the R/M field *from* the REG field. If the W-bit = 1, the data size is a *word* or *doubleword*; if the W-bit = 0, the data size is always a *byte*. The W-bit appears in most instructions, while the D-bit appears mainly with the MOV and some other instructions. Refer to Figure 4–3 for the binary bit pattern of the second opcode byte (reg-mod-r/m) of many instructions. Figure 4–3 shows the location of the MOD (mode), REG (register), and R/M (register/memory) fields.

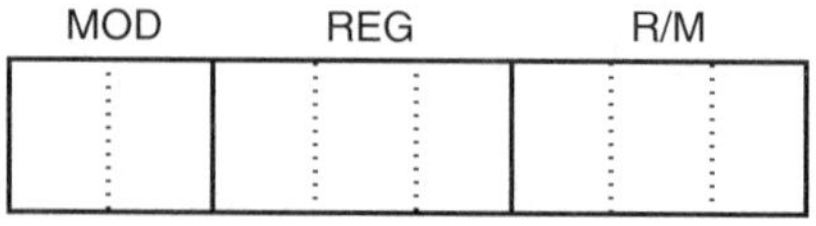

MOD REG R/M

FIGURE 4–3 Byte 2 of many machine language instructions, showing the position of the MOD, REG, and R/M fields.

TABLE 4–1 MOD field for the 16-bit instruction mode.

MOD	*Function*
00	No displacement
01	8-bit sign-extended displacement
10	16-bit signed displacement
11	R/M is a register

MOD Field. The MOD field specifies the addressing mode (MOD) for the selected instruction. The MOD field selects the type of addressing and whether a displacement is present with the selected type. Table 4–1 lists the operand forms available to the MOD field for 16-bit instruction mode, unless the operand address-size override prefix (67H) appears. If the MOD field contains 11, it selects the register-addressing mode. Register addressing uses the R/M field to specify a register instead of a memory location. If the MOD field contains 00, 01, or 10, the R/M field selects one of the data memory-addressing modes. When MOD selects a data memory-addressing mode, it indicates that the addressing mode contains no displacement (00), an 8-bit sign-extended displacement (01), or a 16-bit displacement (10). The MOV AL,[DI] instruction is an example that contains no displacement; the MOV AL,[DI+2] instruction uses an 8-bit displacement (+2); and the MOV AL,[DI+1000H] instruction uses a 16-bit displacement (+1000H).

All 8-bit displacements are sign-extended into 16-bit displacements when the microprocessor executes the instruction. If the 8-bit displacement is 00H–7FH (positive), it is sign-extended to 0000H–007FH before adding to the offset address. If the 8-bit displacement is 80H–FFH (negative), it is sign-extended to FF80H–FFFFH. To sign-extend a number, its sign-bit is copied to the next higher-order byte, which generates either a 00H or an FFH in the next higher-order byte. Some assembler programs do not use the 8-bit displacements and instead default to all 16-bit displacements.

In the 80386 through the Pentium 4 microprocessors, the MOD field may be the same as shown in Table 4–1 for the 16-bit instruction mode; if the instruction mode is 32 bits, the MOD field is as it appears in Table 4–2. The MOD field is interpreted as selected by the address-size override prefix or the operating mode of the microprocessor. This change in the interpretation of the MOD field and instruction supports many of the numerous additional addressing modes allowed in the 80386 through the Pentium 4. The main difference is that when the MOD field is a 10, this causes the 16-bit displacement to become a 32-bit displacement to allow any protected mode memory location (4G bytes) to be accessed. The 80386 and above only allow an 8- or 32-bit displacement when operated in the 32-bit instruction mode, unless the address-size override prefix appears. Note that if an 8-bit displacement is selected, it is sign-extended into a 32-bit displacement by the microprocessor.

Register Assignments. Table 4–3 lists the register assignments for the REG field and the R/M field (MOD = 11). This table contains three lists of register assignments. One is used when the W-bit = 0 (bytes), and the other two are used when the W-bit = 1 (words or doublewords). Note that doubleword registers are only available to the 80386 through the Pentium 4.

TABLE 4–2 MOD field for the 32-bit instruction mode (80386–Pentium 4 only).

MOD	*Function*
00	No displacement
01	8-bit sign-extended displacement
10	32-bit signed displacement
11	R/M is a register

TABLE 4–3 REG and R/M (when MOD = 11) assignments.

Code	*W = 0 (Byte)*	*W = 1 (Word)*	*W = 1 (Doubleword)*
000	AL	AX	EAX
001	CL	CX	ECX
010	DL	DX	EDX
011	BL	BX	EBX
100	AH	SP	ESP
101	CH	BP	EBP
110	DH	SI	ESI
111	BH	DI	EDI

Suppose that a two-byte instruction, 8BECH, appears in a machine language program. Because neither a 67H (operand address-size override prefix) nor a 66H (register-size override prefix) appears as the first byte, the first byte is the opcode. If the microprocessor is operated in the 16-bit instruction mode, this instruction is converted to binary and placed in the instruction format of bytes 1 and 2, as illustrated in Figure 4–4. The opcode is 100010. If you refer to Appendix B, which lists the machine language instructions, you will find that this is the opcode for a MOV instruction. Notice that both the D- and W-bits are a logic 1, which means that a word moves into the destination register specified in the REG field. The REG field contains 101, indicating register BP, so the MOV instruction moves data into register BP. Because the MOD field contains 11, the R/M field also indicates a register. Here, R/M = 100 (SP); therefore, this instruction moves data from SP into BP and is written in symbolic form as a MOV BP,SP instruction.

Suppose that a 668BE8H instruction appears in an 80386 or above, operated in the 16-bit instruction mode. The first byte (66H) is the register-size override prefix that selects 32-bit register operands for the 16-bit instruction mode. The remainder of the instruction indicates that the opcode is a MOV with a source operand of EAX and a destination operand of EBP. This instruction is MOV EBP,EAX. The same instruction becomes MOV BP,AX in the 80386 and above if it is operated in the 32-bit instruction mode because the register-size override prefix selects a 16-bit register. Luckily, the assembler program keeps track of the register- and address-size prefixes and the mode of operation. Recall that if the .386 switch is placed before the .MODEL statement, the 32-bit mode is selected; if it is placed after the .MODEL statement, the 16-bit mode is selected. All programs written using the inline assembler in Visual C++ are always in the 32-bit mode.

R/M Memory Addressing. If the MOD field contains 00, 01, or 10, the R/M field takes on a new meaning. Table 4–4 lists the memory-addressing modes for the R/M field when MOD is 00, 01, or 10 for the 16-bit instruction mode.

Opcode						D	W
1	0	0	0	1	0	1	1

MOD		REG			R/M		
1	1	1	0	1	1	0	0

Opcode = MOV
D = Transfer to register (REG)
W = Word
MOD = R/M is a register
REG = BP
R/M = SP

FIGURE 4–4 The 8BEC instruction placed into bytes 1 and 2 formats from Figures 4–2 and 4–3. This instruction is a MOV BP,SP.

TABLE 4–4 16-bit R/M memory-addressing modes.

R/M Code	*Addressing Mode*
000	DS:[BX+SI]
001	DS:[BX+DI]
010	SS:[BP+SI]
011	SS:[BP+DI]
100	DS:[SI]
101	DS:[DI]
110	SS:[BP]*
111	DS:[BX]

*See text section, "Special Addressing Mode."

All of the 16-bit addressing modes presented in Chapter 3 appear in Table 4–4. The displacement, discussed in Chapter 3, is defined by the MOD field. If MOD = 00 and R/M = 101, the addressing mode is [DI]. If MOD = 01 or 10, the addressing mode is [DI+33H], or LIST [DI+22H] for the 16-bit instruction mode. This example uses LIST, 33H, and 22H as arbitrary values for the displacement.

Figure 4–5 illustrates the machine language version of the 16-bit instruction MOV DL,[DI] or instruction (8AI5H). This instruction is two bytes long and has the opcode 100010, D = 1 (to REG from R/M), W = 0 (byte), MOD = 00 (no displacement), REG = 010 (DL), and R/M = 101 ([DI]). If the instruction changes to MOV DL,[DI+1], the MOD field changes to 01 for an 8-bit displacement, but the first two bytes of the instruction otherwise remain the same. The instruction now becomes 8A5501H instead of 8A15H. Notice that the 8-bit displacement appends to the first two bytes of the instruction to form a three-byte instruction instead of two bytes. If the instruction is again changed to MOV DL,[DI+1000H], the machine language form becomes 8A750010H. Here, the 16-bit displacement of 1000H (coded as 0010H) appends the opcode.

Special Addressing Mode. There is a special addressing mode that does not appear in Tables 4–2, 4–3, or 4–4. It occurs whenever memory data are referenced only by the displacement mode of addressing for 16-bit instructions. Examples are the MOV [1000H],DL and MOV NUMB,DL instructions. The first instruction moves the contents of register DL into data segment memory location 1000H. The second instruction moves register DL into symbolic data segment memory location NUMB.

Whenever an instruction has only a displacement, the MOD field is always 00 and the R/M field is always 110. As shown in the tables, the instruction contains no displacement and uses addressing mode [BP]. You *cannot* actually use addressing mode [BP] without a displacement in machine language. The assembler takes care of this by using an 8-bit displacement (MOD = 01)

Opcode D W
1 0 0 0 1 0 1 0

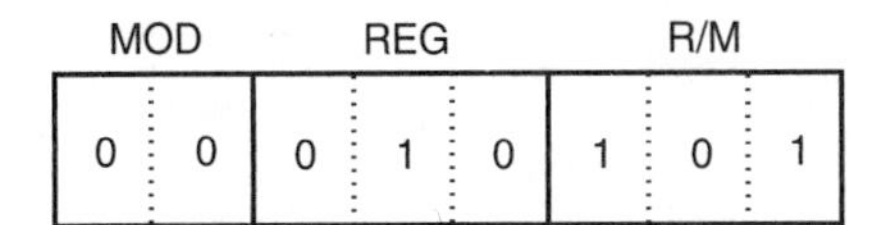

Opcode = MOV
D = Transfer to register (REG)
W = Byte
MOD = No displacement
REG = DL
R/M = DS:[DI]

FIGURE 4–5 A MOV DL,[DI] instruction converted to its machine language form.

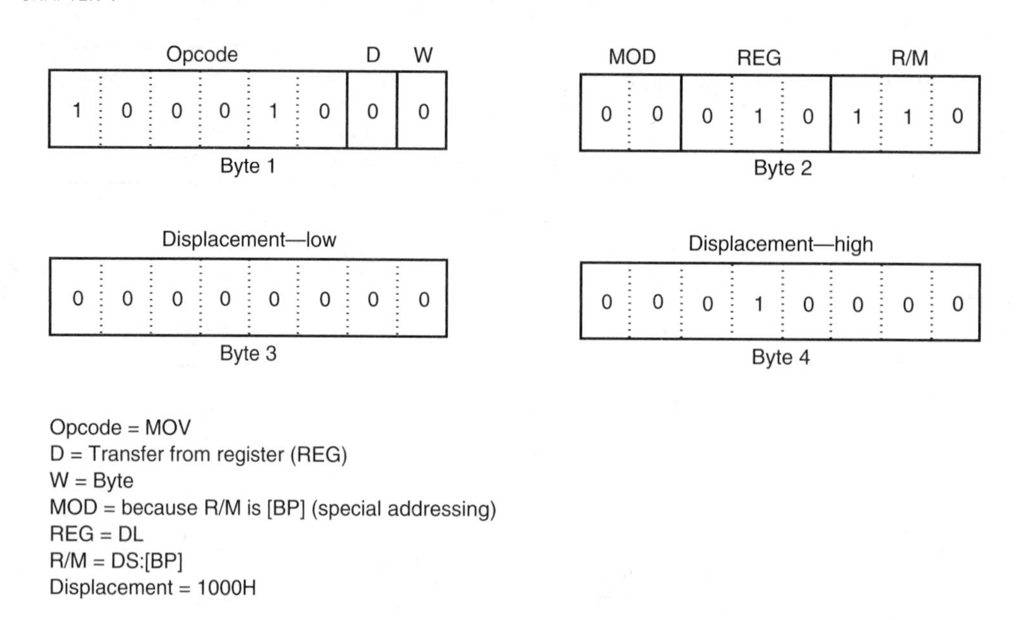

FIGURE 4–6 The MOV [1000H],Dl instruction uses the special addressing mode.

of 00H whenever the [BP] addressing mode appears in an instruction. This means that the [BP] addressing mode assembles as [BP+0], even though [BP] is used in the instruction. The same special addressing mode is also available for the 32-bit mode.

Figure 4–6 shows the binary bit pattern required to encode the MOV [1000H],DL instruction in machine language. If the individual translating this symbolic instruction into machine language does not know about the special addressing mode, the instruction would incorrectly translate to a MOV [BP],DL instruction. Figure 4–7 shows the actual form of the MOV [BP],DL instruction. Notice that this is a three-byte instruction with a displacement of 00H.

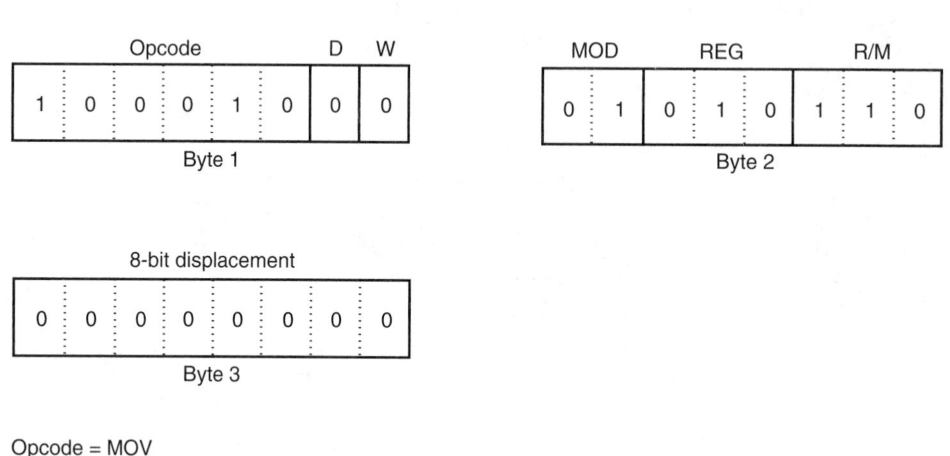

FIGURE 4–7 The MOV [BP],DL instruction converted to binary machine language.

TABLE 4–5 32-bit addressing modes selected by R/M.

R/M Code	*Function*
000	DS:[EAX]
001	DS:[ECX]
010	DS:[EDX]
011	DS:[EBX]
100	Uses scaled-index byte
101	SS:[EBP]*
110	DS:[ESI]
111	DS:[ESI]

*See text section, "Special Addressing Mode."

32-Bit Addressing Modes. The 32-bit addressing modes found in the 80386 and above are obtained by either running these machines in the 32-bit instruction mode or in the 16-bit instruction mode by using the address-size prefix 67H. Table 4–5 shows the coding for R/M used to specify the 32-bit addressing modes. Notice that when R/M = 100, an additional byte called a **scaled-index byte** appears in the instruction. The scaled-index byte indicates the additional forms of scaled-index addressing that do not appear in Table 4–5. The scaled-index byte is mainly used when two registers are added to specify the memory address in an instruction. Because the scaled-index byte is added to the instruction, there are seven bits in the opcode and eight bits in the scaled-index byte to define. This means that a scaled-index instruction has 2^{15} (32K) possible combinations. There are more than 32,000 different variations of the MOV instruction alone in the 80386 through the Pentium 4 microprocessors.

Figure 4–8 shows the format of the scaled-index byte as selected by a value of 100 in the R/M field of an instruction when the 80386 and above use a 32-bit address. The leftmost two bits select a scaling factor (multiplier) of 1×, 2×, 4×, or 8×. Note that a scaling factor of 1× is implicit if none is used in an instruction that contains two 32-bit indirect address registers. The index and base fields both contain register numbers, as indicated in Table 4–3 for 32-bit registers.

The instruction MOV EAX,[EBX+4*ECX] is encoded as 67668B048BH. Notice that both the *address size* (67H) and *register size* (66H) override prefixes appear in the instruction. This coding (67668B048BH) is used when the 80386 and above microprocessors are operated in the 16-bit instruction mode for this instruction. If the microprocessor operates in the 32-bit instruction mode, both prefixes disappear and the instruction becomes an 8B048BH instruction. The use of the prefixes depends on the mode of operation of the microprocessor. Scaled-index addressing can also use a single register multiplied by a scaling factor. An example is the MOV AL,[2*ECX] instruction. The contents of the data segment location addressed by two times ECX are copied into AL.

An Immediate Instruction. Suppose that the MOV WORD PTR [BX+1000H],1234H instruction is chosen as an example of a 16-bit instruction using immediate addressing. This instruction

FIGURE 4–8 The scaled-index byte.

s s	Index	Base

ss
00 = × 1
01 = × 2
10 = × 4
11 = × 8

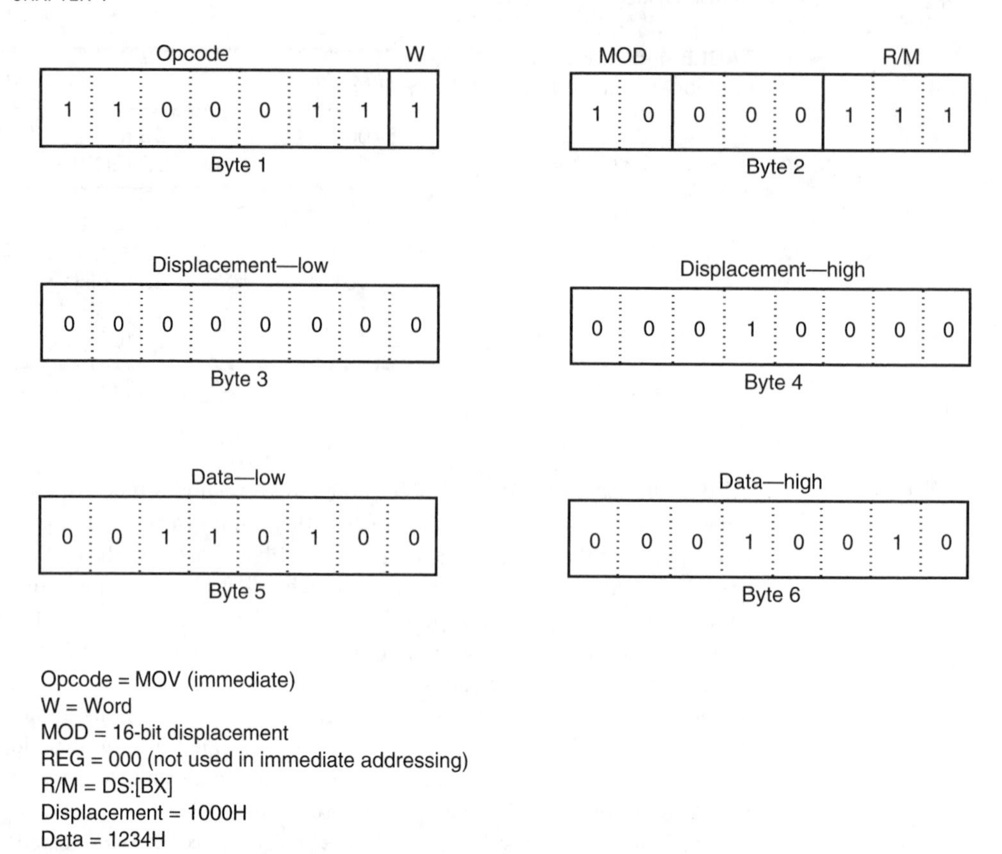

FIGURE 4–9 A MOV WORD PTR [BX+1000H], 1234H instruction converted to binary machine language.

moves 1234H into the word-sized memory location addressed by the sum of 1000H, BX, and DS × 10H. This six-byte instruction uses two bytes for the opcode, W, MOD, and R/M fields. Two of the six bytes are the data of 1234H; two of the six bytes are the displacement of 1000H. Figure 4–9 shows the binary bit pattern for each byte of this instruction.

This instruction, in symbolic form, includes WORD PTR. The WORD PTR directive indicates to the assembler that the instruction uses a word-sized memory pointer. If the instruction moves a byte of immediate data, BYTE PTR replaces WORD PTR in the instruction. Likewise, if the instruction uses a doubleword of immediate data, the DWORD PTR directive replaces BYTE PTR. Most instructions that refer to memory through a pointer do not need the BYTE PTR, WORD PTR, or DWORD PTR directives. These directives are necessary only when it is not clear whether the operation is a byte, word, or doubleword. The MOV [BX],AL instruction is clearly a byte move; the MOV [BX],9 instruction is not exact, and could therefore be a byte-, word-, or doubleword-sized move. Here, the instruction must be coded as MOV BYTE PTR [BX],9, MOV WORD PTR [BX],9, or MOV DWORD PTR [BX],9. If not, the assembler flags it as an error because it cannot determine the intent of this instruction.

Segment MOV Instructions. If the contents of a segment register are moved by the MOV, PUSH, or POP instructions, a special set of register bits (REG field) selects the segment register (see Table 4–6).

Figure 4–10 shows a MOV BX,CS instruction converted to binary. The opcode for this type of MOV instruction is different for the prior MOV instructions. Segment registers can be

TABLE 4–6 Segment register selection.

Code	*Segment Register*
000	ES
001	CS*
010	SS
011	DS
100	FS
101	GS

*MOV CS,R/M and POP CS are not allowed.

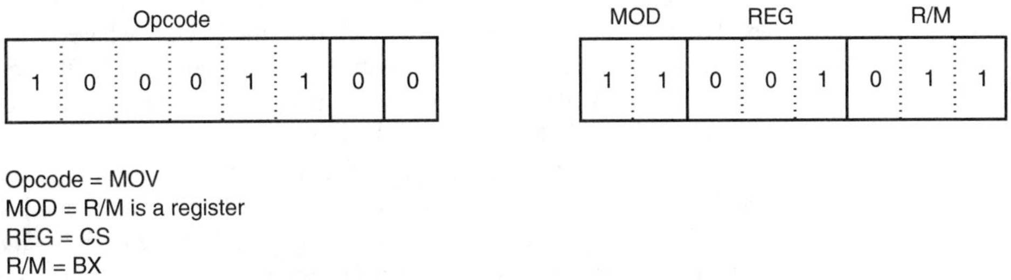

Opcode							
1	0	0	0	1	1	0	0

MOD		REG			R/M		
1	1	0	0	1	0	1	1

Opcode = MOV
MOD = R/M is a register
REG = CS
R/M = BX

FIGURE 4–10 A MOV BX,CS instruction converted to binary machine language.

moved between any 16-bit register or 16-bit memory location. For example, the MOV [DI],DS instruction stores the contents of DS into the memory location addressed by DI in the data segment. An immediate segment register MOV is not available in the instruction set. To load a segment register with immediate data, first load another register with the data and then move it to a segment register.

Although this discussion has not been a complete coverage of machine language coding, it provides enough information for machine language programming. Remember that a program written in symbolic assembly language (*assembly language*) is rarely assembled by hand into binary machine language. An assembler program converts symbolic assembly language into machine language. Since the microprocessor has more than 100,000 instruction variations, let us hope that an assembler is available for the conversion because the process is very time-consuming, although not impossible.

4–2 PUSH/POP

The PUSH and POP instructions are important instructions that *store* and *retrieve* data from the LIFO (last-in, first-out) stack memory. The microprocessor has six forms of the PUSH and POP instructions: register, memory, immediate, segment register, flags, and all registers. The PUSH and POP immediate and the PUSHA and POPA (all registers) forms are not available in the earlier 8086/8088 microprocessors, but are available to the 80286 through the Pentium 4.

Register addressing allows the contents of any 16-bit register to be transferred to or from the stack. In the 80386 and above, the 32-bit extended registers and flags (EFLAGS) can also be pushed or popped from the stack. Memory-addressing PUSH and POP instructions store the contents of a 16-bit memory location (or 32 bits in the 80386 and above) on the stack or stack data into a memory location. Immediate addressing allows immediate data to be pushed onto the stack, but not popped off the stack. Segment register addressing allows the contents of any segment register to be pushed onto the stack or removed from the stack (ES may be pushed, but data from the stack may never be popped into ES). The flags may be pushed or popped from that stack, and the contents of all the registers may be pushed or popped.

PUSH

The 8086–80286 PUSH instruction always transfers two bytes of data to the stack; the 80386 and above transfer two or four bytes, depending on the register or size of the memory location. The source of the data may be any internal 16- or 32-bit register, immediate data, any segment register, or any two bytes of memory data. There is also a PUSHA instruction that copies the contents of the internal register set, except the segment registers, to the stack. The PUSHA (**push all**) instruction copies the registers to the stack in the following order: AX, CX, DX, BX, SP, BP, SI, and DI. The value for SP that is pushed onto the stack is whatever it was before the PUSHA instruction executes. The PUSHF (**push flags**) instruction copies the contents of the flag register to the stack. The PUSHAD and POPAD instructions push and pop the contents of the 32-bit register set found in the 80386 through the Pentium 4.

Whenever data are pushed onto the stack, the first (most-significant) data byte moves to the stack segment memory location addressed by SP – 1. The second (least-significant) data byte moves into the stack segment memory location addressed by SP – 2. After the data are stored by a PUSH, the contents of the SP register decrement by 2. The same is true for a doubleword push, except that four bytes are moved to the stack memory (most-significant byte first), after which the stack pointer decrements by 4. Figure 4–11 shows the operation of the PUSH AX instruction. This instruction copies the contents of AX onto the stack where address SS:[SP – 1] = AH, SS:[SP – 2] = AL, and afterwards SP = SP – 2.

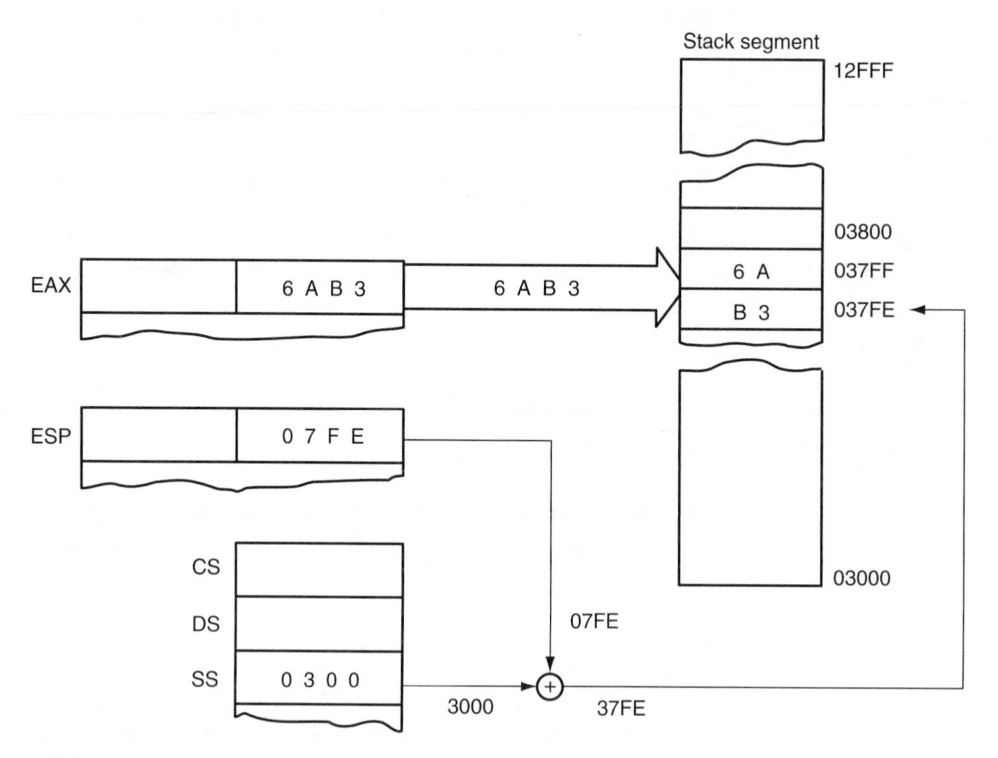

FIGURE 4–11 The effect of the PUSH AX instruction on ESP and stack memory locations 37FFH and 37FEH. This instruction is shown at the point after execution.

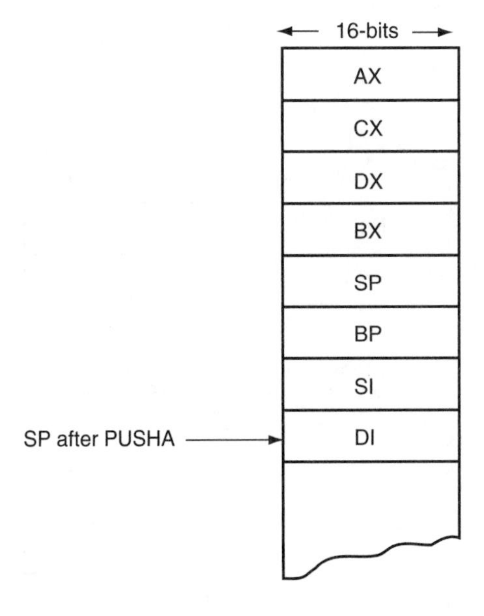

FIGURE 4–12 The operation of the PUSHA instruction, showing the location and order of stack data.

The PUSHA instruction pushes all the internal 16-bit registers onto the stack, as illustrated in Figure 4–12. This instruction requires 16 bytes of stack memory space to store all eight 16-bit registers. After all registers are pushed, the contents of the SP register are decremented by 16. The PUSHA instruction is very useful when the entire register set (microprocessor environment) of the 80286 and above must be saved during a task. The PUSHAD instruction places the 32-bit register set on the stack in the 80386 through the Pentium 4. PUSHAD requires 32 bytes of stack storage space.

The PUSH immediate data instruction has two different opcodes, but in both cases, a 16-bit immediate number moves onto the stack; if PUSHD is used, a 32-bit immediate datum is pushed. If the value of the immediate data are 00H–FFH, the opcode is 6AH; if the data are 0100H–FFFFH, the opcode is 68H. The PUSH 8 instruction, which pushes 0008H onto the stack, assembles as 6A08H. The PUSH 1000H instruction assembles as 680010H. Another example of PUSH immediate is the PUSH 'A' instruction, which pushes 0041H onto the stack. Here, the 41H is the ASCII code for the letter A.

Table 4–7 lists the forms of the PUSH instruction that include PUSHA and PUSHF. Notice how the instruction set is used to specify different data sizes with the assembler.

TABLE 4–7 The PUSH instruction.

Symbolic	*Example*	*Note*
PUSH reg16	PUSH BX	16-bit register
PUSH reg32	PUSH EDX	32-bit register
PUSH mem16	PUSH WORD PTR[BX]	16-bit pointer
PUSH mem32	PUSH DWORD PTR[EBX]	32-bit pointer
PUSH seg	PUSH DS	Segment register
PUSH imm8	PUSH 'R'	8-bit immediate
PUSH imm16	PUSH 1000H	16-bit immediate
PUSHD imm32	PUSHD 20	32-bit immediate
PUSHA	PUSHA	Save all 16-bit registers
PUSHAD	PUSHAD	Save all 32-bit registers
PUSHF	PUSHF	Save flags
PUSHFD	PUSHFD	Save EFLAGS

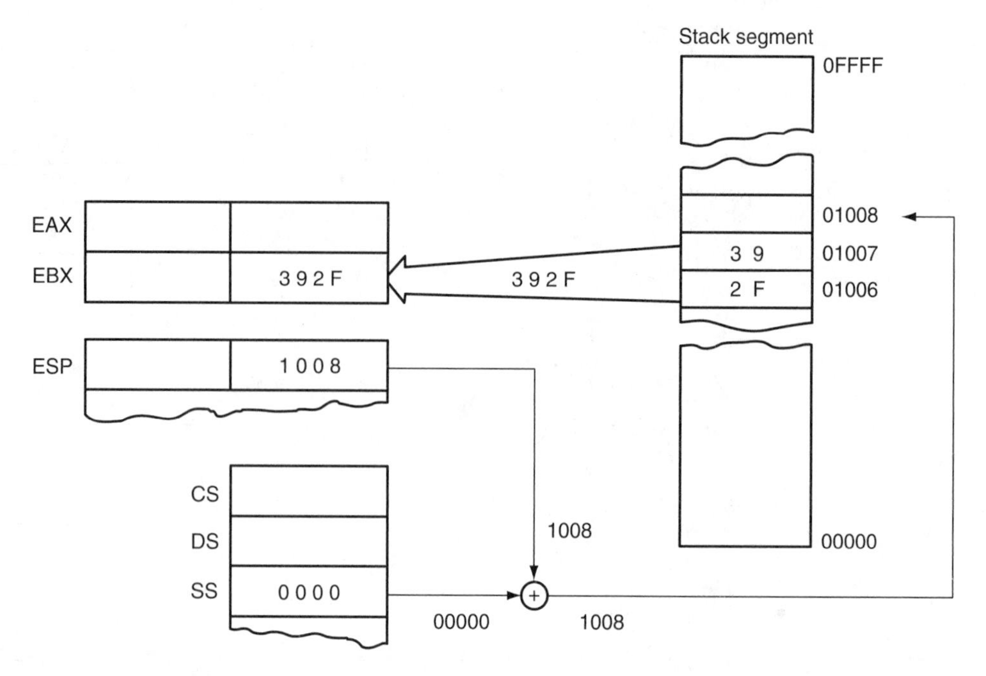

FIGURE 4–13 The POP BX instruction, showing how data are removed from the stack. This instruction is shown after execution.

POP

The POP instruction performs the inverse operation of a PUSH instruction. The POP instruction removes data from the stack and places it into the target 16-bit register, segment register, or 16-bit memory location. In the 80386 and above, a POP can also remove 32-bit data from the stack and use a 32-bit address. The POP instruction is not available as an immediate POP. The POPF (**pop flags**) instruction removes a 16-bit number from the stack and places it into the flag register; the POPFD removes a 32-bit number from the stack and places it into the extended flag register. The POPA (**pop all**) instruction removes 16 bytes of data from the stack and places them into the following registers, in the order shown: DI, SI, BP, SP, BX, DX, CX, and AX. This is the reverse order from the way they were placed on the stack by the PUSHA instruction, causing the same data to return to the same registers. In the 80386 and above, a POPAD instruction reloads the 32-bit registers from the stack.

Suppose that a POP BX instruction executes. The first byte of data removed from the stack (the memory location addressed by SP in the stack segment) moves into register BL. The second byte is removed from stack segment memory location SP + 1 and is placed into register BH. After both bytes are removed from the stack, the SP register increments by 2. Figure 4–13 shows how the POP BX instruction removes data from the stack and places them into register BX.

The opcodes used for the POP instruction and all of its variations appear in Table 4–8. Note that a POP CS instruction is not a valid instruction in the instruction set. If a POP CS instruction executes, only a portion of the address (CS) of the next instruction changes. This makes the POP CS instruction unpredictable and therefore it is not allowed.

Initializing the Stack

When the stack area is initialized, load both the stack segment (SS) register and the stack pointer (SP) register. It is normal to designate an area of memory as the stack segment by loading SS with the bottom location of the stack segment.

TABLE 4–8 The POP instructions.

Symbolic	*Example*	*Note*
POP reg16	POP CX	16-bit register
POP reg32	POP EBP	32-bit register
POP mem16	POP WORD PTR[BX+1]	16-bit pointer
POP mem32	POP DATA3	32-bit memory address
POP seg	POP FS	Segment register
POPA	POPA	Pops all 16-bit registers
POPAD	POPAD	Pops all 32-bit registers
POPF	POPF	Pop flags
POPFD	POPFD	Pop EFLAGS

For example, if the stack segment is to reside in memory locations 10000H–1FFFFH, load SS with 1000H. (Recall that the rightmost end of the stack segment register is appended with a 0H for real mode addressing.) To start the stack at the top of this 64K-byte stack segment, the stack pointer (SP) is loaded with a 0000H. Likewise, to address the top of the stack at location 10FFFH, use a value of 1000H in SP. Figure 4–14 shows how this value causes data to be pushed onto the top of the stack segment with a PUSH CX instruction. Remember that all segments are cyclic in nature—that is, the top location of a segment is contiguous with the bottom location of the segment.

In assembly language, a stack segment is set up as illustrated in Example 4–1. The first statement identifies the start of the stack segment and the last statement identifies the end of the stack segment. The assembler and linker programs place the correct stack segment address in SS and the length of the segment (top of the stack) into SP. There is no need to load these registers in your program unless you wish to change the initial values for some reason.

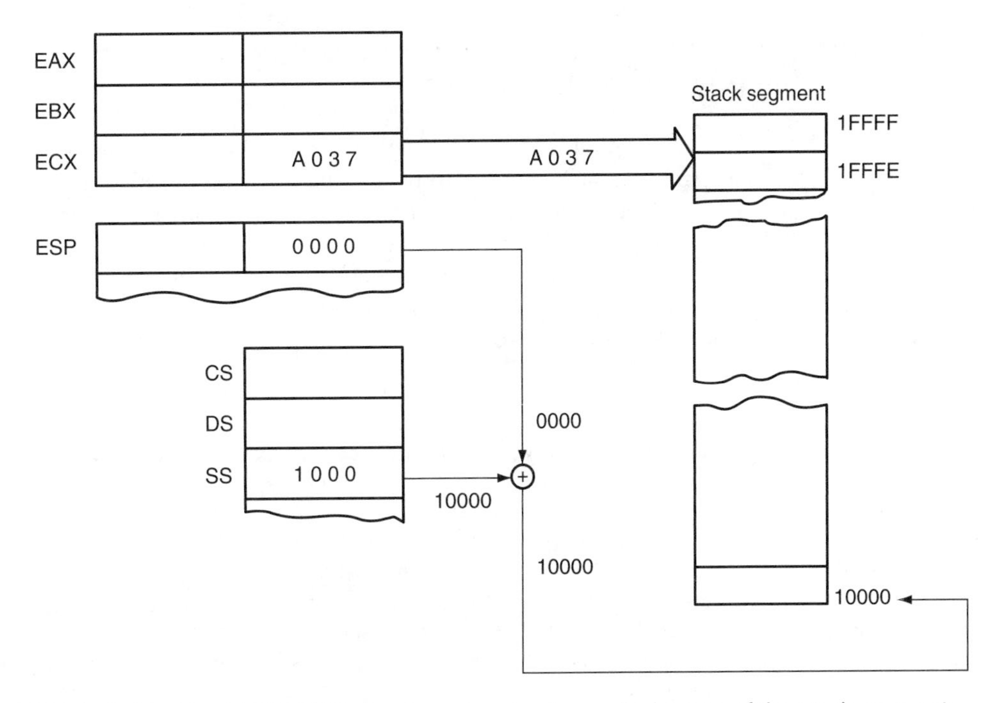

FIGURE 4–14 The PUSH CX instruction, showing the cyclical nature of the stack segment. This instruction is shown just before execution, to illustrate that the stack bottom is contiguous to the top.

EXAMPLE 4–1

```
0000                   STACK_SEG     SEGMENT  STACK

0000 0100[                           DW     100H DUP(?)
          ????
              ]
0200                   STACK_SEG     ENDS
```

An alternative method for defining the stack segment is used with one of the memory models for the MASM assembler only (refer to Appendix A). Other assemblers do not use models; if they do, the models are not exactly the same as with MASM. Here, the .STACK statement, followed by the number of bytes allocated to the stack, defines the stack area (see Example 4–2). The function is identical to Example 4–1. The .STACK statement also initializes both SS and SP. Note that this text uses memory models that are designed for the Microsoft Macro Assembler program MASM.

EXAMPLE 4–2

```
        .MODEL SMALL
        .STACK 200H     ;set stack size
```

If the stack is not specified by using either method, a warning will appear when the program is linked. The warning may be ignored if the stack size is 128 bytes or fewer. The system automatically assigns (through DOS) at least 128 bytes of memory to the stack. This memory section is located in the **program segment prefix** (PSP), which is appended to the beginning of each program file. If you use more memory for the stack, you will erase information in the PSP that is critical to the operation of your program and the computer. This error often causes the computer program to crash. If the TINY memory model is used, the stack is automatically located at the very end of the segment, which allows for a larger stack area.

4–3 LOAD-EFFECTIVE ADDRESS

There are several load-effective address instructions in the microprocessor instruction set. The LEA instruction loads any 16-bit register with the offset address, as determined by the addressing mode selected for the instruction. The LDS and LES variations load any 16-bit register with the offset address retrieved from a memory location, and then load either DS or ES with a segment address retrieved from memory. In the 80386 and above, LFS, LGS, and LSS are added to the instruction set, and a 32-bit register can be selected to receive a 32-bit offset from memory. Table 4–9 lists the load-effective address instructions.

TABLE 4–9 Load-effective address instructions.

Assembly Language	*Operation*
LEA AX,NUMB	Loads AX with the offset address of NUMB
LEA EAX,NUMB	Loads EAX with the offset address of NUMB
LDS DI,LIST	Loads DS and DI with the 32-bit contents of data segment memory location LIST
LDS EDI,LIST1	Loads the DS and EDI with the 48-bit contents of data segment memory location LIST1
LES BX,CAT	Loads ES and BX with the 32-bit contents of data segment memory location CAT
LFS DI,DATA1	Loads FS and DI with the 32-bit contents of data segment memory location DATA1
LGS SI,DATA5	Loads GS and SI with the 32-bit contents of data segment memory location DATA5
LSS SP,MEM	Loads SS and SP with the 32-bit contents of data segment memory location MEM

LEA

The LEA instruction loads a 16- or 32-bit register with the offset address of the data specified by the operand. As the first example in Table 4–9 shows, the operand address NUMB is loaded into register AX, not the contents of address NUMB.

By comparing LEA with MOV, we observe that LEA BX,[DI] loads the offset address specified by [DI] (contents of DI) into the BX register; MOV BX,[DI] loads the data stored at the memory location addressed by [DI] into register BX.

Earlier in the text, several examples were presented using the OFFSET directive. The OFFSET directive performs the same function as an LEA instruction if the operand is a displacement. For example, the MOV BX,OFFSET LIST performs the same function as LEA BX,LIST. Both instructions load the offset address of memory location LIST into the BX register. See Example 4–3 for a short program that loads SI with the address of DATA1 and DI with the address of DATA2. It then exchanges the contents of these memory locations. Note that the LEA and MOV with OFFSET instructions are both the same length (three bytes).

EXAMPLE 4–3

```
                              .MODEL SMALL              ;select small model
0000                          .DATA                     ;start data segment
0000 2000          DATA1      DW      2000H             ;define DATA1
0002 3000          DATA2      DW      3000H             ;define DATA2
0000                          .CODE                     ;start code segment
                              .STARTUP                  ;start program
0017 BE 0000 R                LES SI,DATA1              ;address DATA1 with SI
001A BF 0002 R                MOV DI,OFFSET DATA2       ;address DATA2 with DI
001D 8B 1C                    MOV BX,[SI]               ;exchange DATA1 with DATA2
001F 8B 0D                    MOV CX,[DI]
0021 89 0C                    MOV [SI],CX
0023 89 1D                    MOV [DI],BX
                              .EXIT
                               END
```

But why is the LEA instruction available if the OFFSET directive accomplishes the same task? First, OFFSET only functions with simple operands such as LIST. It may not be used for an operand such as [DI], LIST [SI], and so on. The OFFSET directive is more efficient than the LEA instruction for simple operands. It takes the microprocessor longer to execute the LEA BX,LIST instruction than the MOV BX,OFFSET LIST. The 80486 microprocessor, for example, requires two clocks to execute the LEA BX,LIST instruction and only one clock to execute MOV BX,OFFSET LIST. The reason that the MOV BX,OFFSET LIST instruction executes faster is because the assembler calculates the offset address of LIST, whereas the microprocessor calculates the address for the LEA instruction. The MOV BX,OFFSET LIST instruction is actually assembled as a move immediate instruction and is more efficient.

Suppose that the microprocessor executes an LEA BX,[DI] instruction and DI contains 1000H. Because DI contains the offset address, the microprocessor transfers a copy of DI into BX. A MOV BX,DI instruction performs this task in less time and is often preferred to the LEA BX,[DI] instruction.

Another example is LEA SI,[BX + DI]. This instruction adds BX to DI and stores the sum in the SI register. The sum generated by this instruction is a modulo-64K sum. (A **modulo-64K sum** drops any carry out of the 16-bit result.) If BX = 1000H and DI = 2000H, the offset address moved into SI is 3000H. If BX = 1000H and DI = FF00H, the offset address is 0F00H instead of 10F00H. Notice that the second result is a modulo-64K sum of 0F00H.

LDS, LES, LFS, LGS, and LSS

The LDS, LES, LFS, LGS, and LSS instructions load any 16-bit or 32-bit register with an offset address and the DS, ES, FS, GS, or SS segment register with a segment address. These instructions use any of the memory-addressing modes to access a 32-bit or 48-bit section of memory that contains both the segment and offset address. The 32-bit section of memory contains a 16-bit offset and segment address, while the 48-bit section contains a 32-bit offset and a segment address. These instructions may not use the register addressing mode (MOD = 11). Note that the LFS, LGS, and LSS instructions are only available on 80386 and above, as are the 32-bit registers.

Figure 4–15 illustrates an example LDS BX,[DI] instruction. This instruction transfers the 32-bit number addressed by DI in the data segment into the BX and DS registers. The LDS, LES, LFS, LGS, and LSS instructions obtain a new far address from memory. The offset address appears first, followed by the segment address. This format is used for storing all 32-bit memory addresses.

A far address can be stored in memory by the assembler. For example, the ADDR DD FAR PTR FROG instruction stores the offset and segment address (far address) of FROG in 32 bits of memory at location ADDR. The DD directive tells the assembler to store a doubleword (32-bit number) in memory address ADDR.

In the 80386 and above, an LDS EBX,[DI] instruction loads EBX from the four-byte section of memory addressed by DI in the data segment. Following this four-byte offset is a word that is loaded to the DS register. Notice that instead of addressing a 32-bit section of memory, the 80386 and above address a 48-bit section of the memory whenever a 32-bit offset address is loaded to a 32-bit register. The first four bytes contain the offset value loaded to the 32-bit register and the last two bytes contain the segment address.

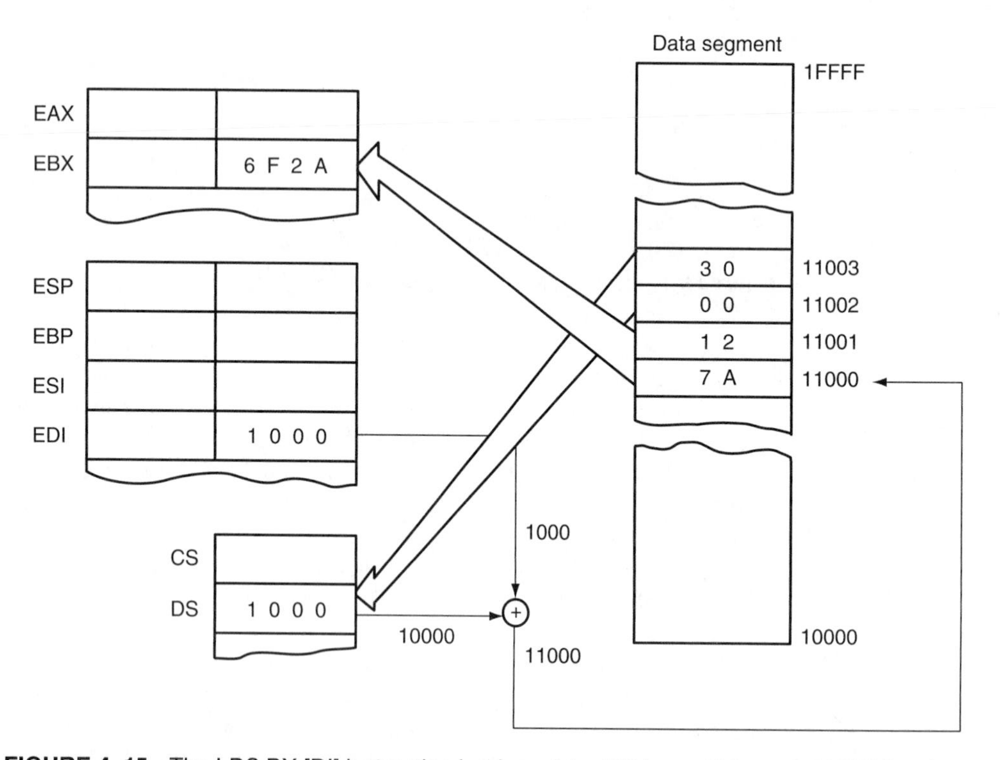

FIGURE 4–15 The LDS BX,[DI] instruction loads register BX from addresses 11000H and 11001H and register DS from locations 11002H and 11003H. This instruction is shown at the point just before DS changes to 3000H and BX changes to 127AH.

The most useful of the load instructions is the LSS instruction. Example 4–4 shows a short program that creates a new stack area after saving the address of the old stack area. After executing some dummy instructions, the old stack area is reactivated by loading both SS and SP with the LSS instruction. Note that the CLI (**disable interrupts**) and STI (**enable interrupts**) instructions must be included to disable interrupts. (This topic is discussed near the end of this chapter.) Because the LSS instruction functions in the 80386 or above, the .386 statement appears after the .MODEL statement to select the 80386 microprocessor. Notice how the WORD PTR directive is used to override the doubleword (DD) definition for the old stack memory location. If an 80386 or newer microprocessor is in use, it is suggested that the .386 switch be used to develop software for the 80386 microprocessor. This is true even if the microprocessor is a Pentium, Pentium Pro, Pentium II, Pentium III, or Pentium 4. The reason is that the 80486–Pentium 4 microprocessors add only a few additional instructions to the 80386 instruction set, which are seldom used in software development. If the need arises to use any of the CMPXCHG, CMPXCHG8 (new to the Pentium), XADD, or BSWAP instructions, select either the .486 switch for the 80486 microprocessor or the .586 switch for the Pentium. You can even specify the Pentium II–Pentium 4 using the .686 switch.

EXAMPLE 4–4

```
                                .MODEL SMALL             ;select small model
                                .386                     ;select 80386
0000                            .DATA                    ;start data segment
0000 00000000           SADDR   DD      ?                ;old stack address
0004 1000 [             SAREA   DW      1000H DUP(?)     ;new stack area
          ????
               ]
2004 = 2004             STOP    EQU THIS WORD            ;define top of new stack
0000                            .CODE                    ;start code segment
                                .STARTUP                 ;start program
0010 FA                         CLI                      ;disable interrupts
0011 8B C4                      MOV AX,SP                ;save old SP
0013 A3 0000 R                  MOV WORD PTR SADDR,AX
0016 8C D0                      MOV AX,SS                ;save old SS
0018 A3 0002 R                  MOV WORD PTR SADDR+2,AX

001B 8C D8                      MOV AX,DS                ;load new SS
001D 8E D0                      MOV SS,AX
001F B8 2004 R                  MOV AX,OFFSET STOP       ;load new SP
0022 8B E0                      MOV SP,AX
0024 FB                         STI                      ;enable interrupts

0025 8B C0                      MOV AX,AX                ;do some dummy instructions
0027 8B C0                      MOV AX,AX
0029 9F B2 26 0000 R            LSS SP,SADDR             ;get old stack
                                .EXIT                    ;exit to DOS
                                END                      ;end program listing
```

4–4 STRING DATA TRANSFERS

There are five string data transfer instructions: LODS, STOS, MOVS, INS, and OUTS. Each string instruction allows data transfers that are either a single byte, word, or doubleword (or if repeated, a block of bytes, words, or doublewords). Before the string instructions are presented, the operation of the D flag-bit (direction), DI, and SI must be understood as they apply to the string instructions.

The Direction Flag

The direction flag (D, located in the flag register) selects the auto-increment (D = 0) or the auto-decrement (D = 1) operation for the DI and SI registers during string operations. The direction flag is used only with the string instructions. The CLD instruction clears the D flag (D = 0) and the STD instruction sets it (D = I). Therefore, the CLD instruction selects the auto-increment mode (D = 0) and STD selects the auto-decrement mode (D = 1).

Whenever a string instruction transfers a byte, the contents of DI and/or SI increment or decrement by 1. If a word is transferred, the contents of DI and/or SI increment or decrement by 2. Doubleword transfers cause DI and/or SI to increment or decrement by 4. Only the actual registers used by the string instruction increment or decrement. For example, the STOSB instruction uses the DI register to address a memory location. When STOSB executes, only DI increments or decrements without affecting SI. The same is true of the LODSB instruction, which uses the SI register to address memory data. LODSB only increments or decrements SI without affecting DI.

DI and SI

During the execution of a string instruction, memory accesses occur through either or both of the DI and SI registers. The DI offset address accesses data in the extra segment for all string instructions that use it. The SI offset address accesses data, by default, in the data segment. The segment assignment of SI may be changed with a segment override prefix, as described later in this chapter. The DI segment assignment is always in the extra segment when a string instruction executes. This assignment cannot be changed. The reason that one pointer addresses data in the extra segment and the other in the data segment is so the MOVS instruction can move 64K bytes of data from one segment of memory to another.

When operating in the 32-bit mode in the 80386 or above, the EDI and ESI registers are used in place of DI and SI. This allows strings using any memory location in the entire 4G-byte protected mode address space of the microprocessor.

LODS

The LODS instruction loads AL, AX, or EAX with data stored at the data segment offset address indexed by the SI register. (Note that only the 80386 and above use EAX.) After loading AL with a byte, AX with a word, or EAX with a doubleword, the contents of SI increment, if D = 0, or decrement, if D = 1. A 1 is added to or subtracted from SI for a byte-sized LODS, 2 is added or subtracted for a word-sized LODS, and 4 is added or subtracted for a doubleword-sized LODS.

Table 4–10 lists the permissible forms of the LODS instruction. The LODSB (**loads a byte**) instruction causes a byte to be loaded into AL, the LODSW (**loads a word**) instruction

TABLE 4–10 Forms of the LODS instruction.

Assembly Language	*Operation*
LODSB	AL = DS:[SI]; SI = SI ± 1
LODSW	AX = DS:[SI]; SI = SI ± 2
LODSD	EAX = DS:[SI]; SI = SI ± 4
LODS LIST	AL = DS:[SI]; SI = SI ± 1 (if LIST is a byte)
LODS DATA1	AX = DS:[SI]; SI = SI ± 2 (if DATA1 is a word)
LODS FROG	EAX = DS:[SI]; SI = SI ± 4 (if FROG is a doubleword)

Note: The segment register can be overridden with a segment override prefix as in LODS ES:DATA4.

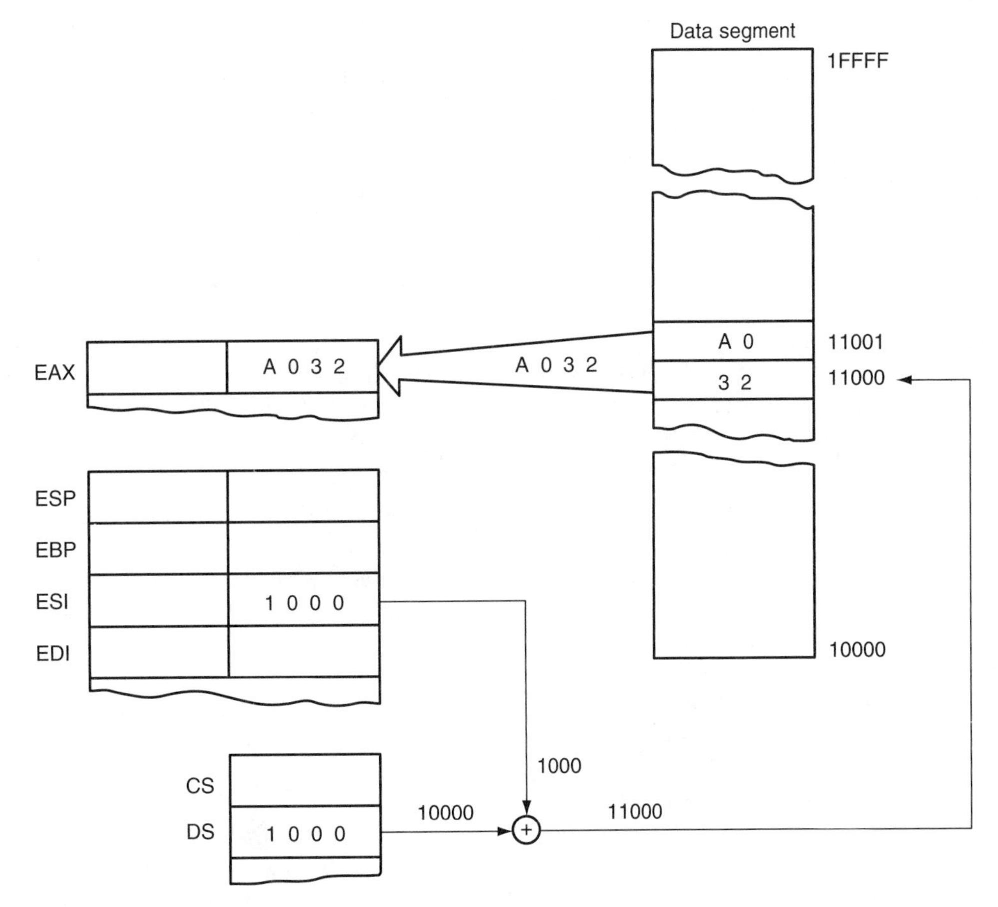

FIGURE 4–16 The operation of the LODSW instruction if DS = 1000H, D = 0, 11000H = 32, and 11001H = A0. This instruction is shown after AX is loaded from memory, but before SI increments by 2.

causes a word to be loaded into AX, and the LODSD (**loads a doubleword**) instruction causes a doubleword to be loaded into EAX. Although rare, as an alternative to LODSB, LODSW, and LODSD, the LODS instruction may be followed by a byte-, word- or doubleword-sized operand to select a byte, word, or doubleword transfer. Operands are often defined as bytes with DB, as words with DW, and as doublewords with DD. The DB pseudo-operation defines byte(s), the DW pseudo-operation defines word(s), and the DD pseudo-operations defines doubleword(s).

Figure 4–16 shows the effect of executing the LODSW instruction if the D flag = 0, SI = 1000H, and DS = 1000H. Here, a 16-bit number stored at memory locations 11000H and 11001H moves into AX. Because D = 0 and this is a word transfer, the contents of SI increment by 2 after AX loads with memory data.

STOS

The STOS instruction stores AL, AX, or EAX at the extra segment memory location addressed by the DI register. (Note that only the 80386–Pentium 4 use EAX and doublewords.) Table 4–11 lists all forms of the STOS instruction. As with LODS, a STOS instruction may be appended with a B, W, or D for byte, word, or doubleword transfers. The STOSB (**stores a byte**) instruction stores the byte in AL at the extra segment memory location addressed by DI. The STOSW

TABLE 4–11 Forms of the STOS instruction.

Assembly Language	*Operation*
STOSB	ES:[DI] = AL; DI = DI ± 1
STOSW	ES:[DI] = AX; DI = DI ± 2
STOSD	ES:[DI] = EAX; DI = DI ± 4
STOS LIST	ES:[DI] = AL; DI = DI ± 1 (if LIST is a byte)
STOS DATA3	ES:[DI] = AX; DI = DI ± 2 (if DATA3 is a word)
STOS DATA4	ES:[DI] = EAX; DI = DI ± 4 (if DATA4 is a doubleword)

(**stores a word**) instruction stores AX in the extra segment memory location addressed by DI. A doubleword is stored in the extra segment location addressed by DI with the STOSD (**stores a doubleword**) instruction. After the byte (AL), word (AX), or doubleword (EAX) is stored, the contents of DI increment or decrement.

STOS with a REP. The **repeat prefix** (REP) is added to any string data transfer instruction except the LODS instruction. It doesn't make any sense to perform a repeated LODS operation. The REP prefix causes CX to decrement by 1 each time the string instruction executes. After CX decrements, the string instruction repeats. If CX reaches a value of 0, the instruction terminates and the program continues with the next sequential instruction. Thus, if CX is loaded with 100 and a REP STOSB instruction executes, the microprocessor automatically repeats the STOSB instruction 100 times. Because the DI register is automatically incremented or decremented after each datum is stored, this instruction stores the contents of AL in a block of memory instead of a single byte of memory.

Suppose that the STOSW instruction is used to clear an area of memory called Buffer using a count called Count and the program is to function call ClearBuffer in the C++ environment using the inline assembler. (See Example 4–5). Note that both the Count and Buffer address are transferred to the function. The REP STOSW instruction clears the memory buffer called Buffer. Notice that Buffer is a pointer to the actual buffer that is cleared by this function.

EXAMPLE 4–5

```
void ClearBuffer ( int Count, short* Buffer )
{
   _asm{
         push edi              ;save registers
         push es
         push ds
         mov  ax,0
         mov  ecx, Count
         mov  edi, Buffer
         pop  es               ;load ES with DS
         rep  stosw            ;clear Buffer
         pop  es               ;restore registers
         pop  edi
   }
}
```

The operands in a program can be modified by using arithmetic or logic operators such as multiplication (*). Other operators appear in Table 4–12.

MOVS

One of the more useful string data transfer instructions is MOVS because it transfers data from one memory location to another. This is the only memory-to-memory transfer allowed in the 8086–Pentium 4 microprocessors. The MOVS instruction transfers a byte, word, or doubleword

TABLE 4–12 Common operand modifiers.

Operator	*Example*	*Comment*
+	MOV AL,6+3	Copies 9 into AL
–	MOV AL,6–3	Copies 3 into AL
*	MOV AL,4*3	Copies 12 into AL
/	MOV AX,12/5	Copies 2 into AX (remainder is lost)
MOD	MOV AX,12 MOD 7	Copies 5 into AX (quotient is lost)
AND	MOV AX,12 AND 4	Copies 4 into AX (1100 AND 0100 = 0100)
OR	MOV EAX,12 OR 1	Copies 13 into EAX (1100 OR 0001 = 1101)
NOT	MOV AL,NOT 1	Copies 254 into AL (NOT 0000 0001 = 1111 1110 or 254)

from the data segment location addressed by SI to the extra segment location addressed by SI. As with the other string instructions, the pointers then increment or decrement, as dictated by the direction flag. Table 4–13 lists all the permissible forms of the MOVS instruction. Note that only the source operand (SI), located in the data segment, may be overridden so that another segment may be used. The destination operand (DI) must always be located in the extra segment.

It is often necessary to transfer the contents of one area of memory to another. Suppose that we have two blocks of doubleword memory, BlockA and BlockB, and we need to copy BlockA into BlockB. This can be accomplished using the MOVSD instruction as illustrated in Example 4–6, which is a C++ language function written using the inline assembler. The function receives three pieces of information from the caller: BlockSize and the addresses of BlockA and BlockB. Note that all data are in the data segment in a Visual C++ program so we need to copy DS into ES, which is done using a PUSH DS followed by a POP ES. We also need to save all registers that we changed except for EAX, EBX, ECX, and EDX.

Example 4–7 shows the same function written in C++ exclusively so the two methods can be compared and contrasted. Example 4–8 shows the assembly language version of Example 4–7 for comparison to Example 4–6. Notice how much shorter the assembly language version is compared to the C++ version generated in Example 4–8. Admittedly the C++ version is a little easier to type, but if execution speed is important Example 4–6 will run much faster than Example 4–7.

TABLE 4–13 Forms of the MOVS instruction.

Assembly Language	*Operation*
MOVSB	ES:[DI] = DS:[SI]; DI = DI ± 1; SI = SI ± 1 (byte transferred)
MOVSW	ES:[DI] = DS:[SI]; DI = DI ± 2; SI = SI ± 2 (word transferred)
MOVSD	ES:[DI] = DS:[SI]; DI = DI ± 4; SI = SI ± 4 (doubleword transferred)
MOVS BYTE1, BYTE2	ES:[DI] = DS:[SI]; DI = DI ± 1; SI = SI ± 1 (byte transferred if BYTE1 and BYTE2 are bytes)
MOVS WORD1,WORD2	ES:[DI] = DS:[SI]; DI = DI ± 2; SI = SI ± 2 (word transferred if WORD1 and WORD2 are words)
MOVS TED,FRED	ES:[DI] = DS:[SI]; DI = DI ± 4; SI = SI ± 4 (doubleword transferred if TED and FRED are doublewords)

EXAMPLE 4–6

```
//Function that copies BlockA into BlockB using the inline assembler
//
void TransferBlocks (int BlockSize, int* BlockA, int* BlockB)
{
      _asm{
            push es                     ;save registers
            push edi
            push esi
            push ds                     ;copy DS into ES
            pop  es
            mov  esi, BlockA            ;address BlockA
            mov  edi, BlockB            ;address BlockB
            mov  ecx, BlockSize         ;load count
            rep  movsd                  ;move data
            pop  es                     ;restore registers
            pop  esi
            pop  edi
      }
}
```

EXAMPLE 4–7

```
//C++ version of Example 4-6
//
void TransferBlocks (int BlockSize, int* BlockA, int*BlockB)
{
      for (int a = 0; a < BlockSize; a++)
      {
            BlockA = BlockB++;
            BlockA++;
      }
}
```

EXAMPLE 4–8

```
void TransferBlocks(int BlockSize, int* BlockA, int* BlockB)
{
004136A0  push          ebp
004136A1  mov           ebp,esp
004136A3  sub           esp,0D8h
004136A9  push          ebx
004136AA  push          esi
004136AB  push          edi
004136AC  push          ecx
004136AD  lea           edi,[ebp-0D8h]
004136B3  mov           ecx,36h
004136B8  mov           eax,0CCCCCCCCh
004136BD  rep stos      dword ptr [edi]
004136BF  pop           ecx
004136C0  mov           dword ptr [ebp-8],ecx
       for( int a = 0; a < BlockSize; a++ )
004136C3  mov           dword ptr [a],0
004136CA  jmp           TransferBlocks+35h (4136D5h)
004136CC  mov           eax,dword ptr [a]
004136CF  add           eax,1
004136D2  mov           dword ptr [a],eax
004136D5  mov           eax,dword ptr [a]
004136D8  cmp           eax,dword ptr [BlockSize]
004136DB  jge           TransferBlocks+57h (4136F7h)
       {
             BlockA = BlockB++;
004136DD  mov           eax,dword ptr [BlockB]
004136E0  mov           dword ptr [BlockA],eax
004136E3  mov           ecx,dword ptr [BlockB]
004136E6  add           ecx,4
004136E9  mov           dword ptr [BlockB],ecx
```

```
                BlockA++;
004136EC  mov          eax,dword ptr [BlockA]
004136EF  add          eax,4
004136F2  mov          dword ptr [BlockA],eax
        }
004136F5  jmp          TransferBlocks+2Ch (4136CCh)
}
004136F7  pop          edi
004136F8  pop          esi
004136F9  pop          ebx
004136FA  mov          esp,ebp
004136FC  pop          ebp
004136FD  ret          0Ch
```

INS

The INS (**input string**) instruction (not available on the 8086/8088 microprocessors) transfers a byte, word, or doubleword of data from an I/O device into the extra segment memory location addressed by the DI register. The I/O address is contained in the DX register. This instruction is useful for inputting a block of data from an external I/O device directly into the memory. One application transfers data from a disk drive to memory. Disk drives are often considered and interfaced as I/O devices in a computer system.

As with the prior string instructions, there are three basic forms of the INS. The INSB instruction inputs data from an 8-bit I/O device and stores it in the byte-sized memory location indexed by SI. The INSW instruction inputs 16-bit I/O data and stores it in a word-sized memory location. The INSD instruction inputs a doubleword. These instructions can be repeated using the REP prefix, which allows an entire block of input data to be stored in the memory from an I/O device. Table 4–14 lists the various forms of the INS instruction.

Example 4–9 shows a sequence of instructions that input 50 bytes of data from an I/O device whose address is 03ACH and store the data in extra segment memory array LISTS. This software assumes that data are available from the I/O device at all times. Otherwise, the software must check to see if the I/O device is ready to transfer data precluding the use of a REP prefix.

EXAMPLE 4–9

```
                    ;Using the REP INSB to input data to a memory array
                    ;
0000 BF 0000 R            MOV DI,OFFSET LISTS     ;address array
0003 BA 03AC              MOV DX,3ACH             ;address I/O
0006 FC                   CLD                     ;auto-increment
0007 B9 0032              MOV CX,50               ;load counter
000A F3/6C                REP INSB                ;input data
```

TABLE 4–14 Forms of the INS instruction.

Assembly Language	*Operation*
INSB	ES:[DI] = [DX]; DI = DI ± 1 (byte transferred)
INSW	ES:[DI] = [DX]; DI = DI ± 2 (word transferred)
INSD	ES:[DI] = [DX]; DI = DI ± 4 (doubleword transferred)
INS LIST	ES:[DI] = [DX]; DI = DI ± 1 (if LIST is a byte)
INS DATA4	ES:[DI] = [DX]; DI = DI ± 2 (if DATA4 is a word)
INS DATA5	ES:[DI] = [DX]; DI = DI ± 4 (if DATA5 is a doubleword)

Note: [DX] indicates that DX is the I/O device address. These instructions are not available on the 8086 and 8088 microprocessors.

TABLE 4–15 Forms of the OUTS instruction.

Assembly Language	*Operation*
OUTSB	[DX] = DS:[SI]; SI = SI ± 1 (byte transferred)
OUTSW	[DX] = DS:[SI]; SI = SI ± 2 (word transferred)
OUTSD	[DX] = DS:[SI]; SI = SI ± 4 (doubleword transferred)
OUTS DATA7	[DX] = DS:[SI]; SI = SI ± 1 (if DATA7 is a byte)
OUTS DATA8	[DX] = DS:[SI]; SI = SI ± 2 (if DATA8 is a word)
OUTS DATA9	[DX] = DS:[SI]; SI = SI ± 4 (if DATA9 is a doubleword)

Note: [DX] indicates that DX is the I/O device address. These instructions are not available on the 8086 and 8088 microprocessors.

OUTS

The OUTS (**output string**) instruction (not available on the 8086/8088 microprocessors) transfers a byte, word, or doubleword of data from the data segment memory location address by SI to an I/O device. The I/O device is addressed by the DX register as it is with the INS instruction. Table 4–15 shows the variations available for the OUTS instruction.

Example 4–10 shows a short sequence of instructions that transfer data from a data segment memory array (ARRAY) to an I/O device at I/O address 3ACH. This software assumes that the I/O device is always ready for data.

EXAMPLE 4–10

```
                        ;Using the REP OUTSB to output data from a memory array
                        ;
0000 BE 0064 R              MOV SI,OFFSET ARRAY     ;address array
0003 BA 03AC                MOV DX,3ACH             ;address I/O
0006 FC                     CLD                     ;auto-increment
0007 B9 0064                MOV CX,100              ;load counter
000A F3/6E                  REP OUTSB               ;output data
```

4–5 MISCELLANEOUS DATA TRANSFER INSTRUCTIONS

Don't be fooled by the term *miscellaneous*; these instructions are used in programs. The data transfer instructions detailed in this section are XCHG, LAHF, SAHF, XLAT, IN, OUT, BSWAP, MOVSX, MOVZX, and CMOV. Because the miscellaneous instructions are not used as often as a MOV instruction, they have been grouped together and represented in this section.

XCHG

The XCHG (**exchange**) instruction exchanges the contents of a register with the contents of any other register or memory location. The XCHG instruction cannot exchange segment registers or memory-to-memory data. Exchanges are byte-, word-, or doubleword-sized (80386 and above), and they use any addressing mode discussed in Chapter 3, except immediate addressing. Table 4–16 shows some examples of the XCHG instruction.

The XCHG instruction, using the 16-bit AX register with another 16-bit register, is the most efficient exchange. This instruction occupies one byte of memory. Other XCHG instructions require two or more bytes of memory, depending on the addressing mode selected.

When using a memory-addressing mode and the assembler, it doesn't matter which operand addresses memory. The XCHG AL,[DI] instruction is identical to the XCHG [DI],AL instruction, as far as the assembler is concerned.

TABLE 4–16 Forms of the XCHG instruction.

Assembly Language	*Operation*
XCHG AL,CL	Exchanges the contents of AL with CL
XCHG CX,BP	Exchanges the contents of CX with BP
XCHG EDX,ESI	Exchanges the contents of EDX with ESI
XCHG AL,DATA2	Exchanges the contents of AL with data segment memory location DATA2

If the 80386 through the Pentium 4 microprocessor is available, the XCHG instruction can exchange doubleword data. For example, the XCHG EAX,EBX instruction exchanges the contents of the EAX register with the EBX register.

LAHF and SAHF

The LAHF and SAHF instructions are seldom used because they were designed as bridge instructions. These instructions allowed 8085 (an early 8-bit microprocessor) software to be translated into 8086 software by a translation program. Because any software that required translation was completed many years ago, these instructions have little application today. The LAHF instruction transfers the rightmost eight bits of the flag register into the AH register. The SAHF instruction transfers the AH register into the rightmost eight bits of the flag register.

At times, the SAHF instruction may find some application with the numeric coprocessor. The numeric coprocessor contains a status register that is copied into the AX register with the FSTSW AX instruction. The SAHF instruction is then used to copy from AH into the flag register. The flags are then tested for some of the conditions of the numeric coprocessor. This is detailed in Chapter 14, which explains the operation and programming of the numeric coprocessor.

XLAT

The XLAT (**translate**) instruction converts the contents of the AL register into a number stored in a memory table. This instruction performs the direct table lookup technique often used to convert one code to another. An XLAT instruction first adds the contents of AL to BX to form a memory address within the data segment. It then copies the contents of this address into AL. This is the only instruction that adds an 8-bit number to a 16-bit number.

Suppose that a seven-segment LED display lookup table is stored in memory at address TABLE. The XLAT instruction then uses the lookup table to translate the BCD number in AL to a seven-segment code in AL. Example 4–11 provides a sequence of instructions that convert from a BCD code to a seven-segment code. Figure 4–17 shows the operation of this example program if TABLE = 1000H, DS = 1000H, and the initial value of AL = 05H (5 BCD). After the translation, AL = 6DH.

EXAMPLE 4–11

```
                            TABLE   DB   3FH, 06H, 5BH, 4FH     ;lookup table
                                    DB   66H, 6DH, 7DH, 27H
                                    DB   7FH, 6FH

0017 B0 05                  LOOK:   MOV AL,5                  ;load AL with 5 (a test number)
0019 BB 1000 R                      MOV BX,OFFSET TABLE       ;address lookup table
001C D7                             XLAT                      ;convert
```

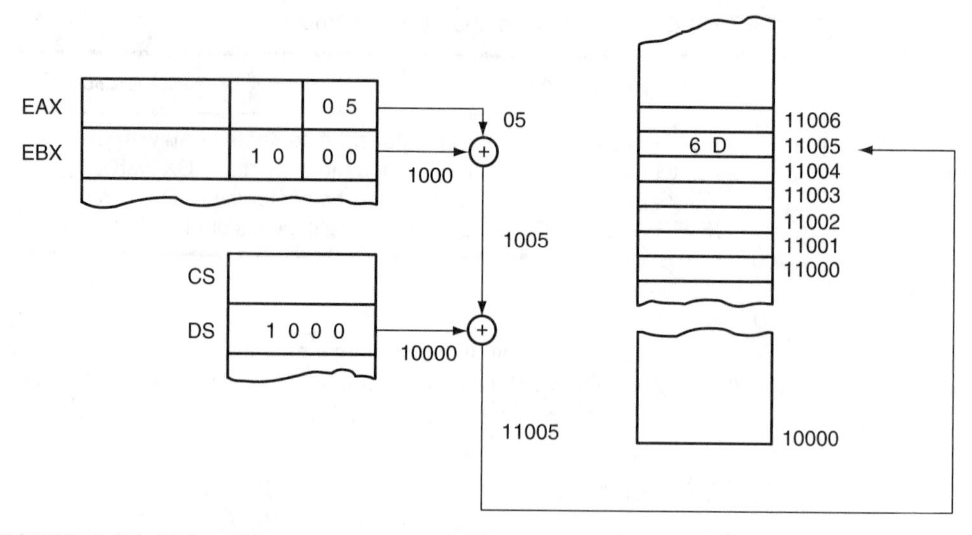

FIGURE 4–17 The operation of the XLAT instruction at the point just before 6DH is loaded into AL.

IN and OUT

Table 4–17 lists the forms of the IN and OUT instructions, which perform I/O operations. Notice that the contents of AL, AX, or EAX are transferred only between the I/O device and the microprocessor. An IN instruction transfers data from an external I/O device into AL, AX, or EAX; an OUT transfers data from AL, AX, or EAX to an external I/O device. (Note that only the 80386 and above contain EAX.)

Two forms of I/O device (port) addressing exist for IN and OUT: fixed port and variable port. *Fixed-port addressing* allows data transfer between AL, AX, or EAX using an 8-bit I/O port address. It is called fixed-port addressing because the port number follows the instruction's opcode, just as it did with immediate addressing. Often, instructions are stored in ROM. A fixed-port instruction stored in ROM has its port number permanently fixed because of the nature of read-only memory. A fixed-port address stored in RAM can be modified, but such a modification does not conform to good programming practices.

TABLE 4–17 IN and OUT instructions.

Assembly Language	*Operation*
IN AL,p8	8 bits are input to AL from I/O port p8
IN AX,p8	16 bits are input to AX from I/O port p8
IN EAX,p8	32 bits are input to EAX from I/O port p8
IN AL,DX	8 bits are input to AL from I/O port DX
IN AX,DX	16 bits are input to AX from I/O port DX
IN EAX,DX	32 bits are input to EAX from I/O port DX
OUT p8,AL	8 bits are output to I/O port p8 from AL
OUT p8,AX	16 bits are output to I/O port p8 from AX
OUT p8,EAX	32 bits are output to I/O port p8 from EAX
OUT DX,AL	8 bits are output to I/O port DX from AL
OUT DX,AX	16 bits are output to I/O port DX from AX
OUT DX,EAX	32 bits are output to I/O port DX from EAX

Note: p8 = an 8-bit I/O port number (0000H to 00FFH) and DX = the 16-bit I/O port number (0000H to FFFFH) held in register DX.

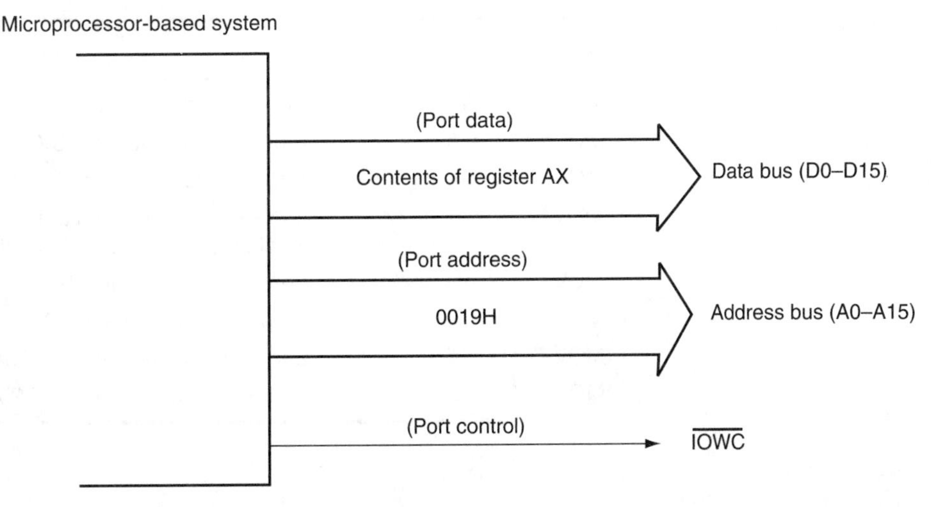

FIGURE 4–18 The signals found in the microprocessor-based system for an OUT 19H,AX instruction.

The port address appears on the address bus during an I/O operation. For the 8-bit fixed-port I/O instructions, the 8-bit port address is zero-extended into a 16-bit address. For example, if the IN AL,6AH instruction executes, data from I/O address 6AH are input to AL. The address appears as a 16-bit 006AH on pins A0–A15 of the address bus. Address bus bits A16–A19 (8086/8088), A16–A23 (80286/80386SX), A16–A24 (80386SL/80386SLC/80386EX), or A16–A31 (80386–Pentium 4) are undefined for an IN or OUT instruction. Note that Intel reserves the last 16 I/O ports (FFF0H–FFFFH) for use with some of its peripheral components.

Variable-port addressing allows data transfers between AL, AX, or EAX and a 16-bit port address. It is called *variable-port addressing* because the I/O port number is stored in register DX, which can be changed (varied) during the execution of a program. The 16-bit I/O port address appears on the address bus pin connections A0–A15. The IBM PC uses a 16-bit port address to access its I/O space. The ISA bus I/O space for a PC is located at I/O port 0000H–03FFH. Note that PCI bus cards may use I/O addresses above 03FFH.

Figure 4–18 illustrates the execution of the OUT 19H,AX instruction, which transfers the contents of AX to I/O port 19H. Notice that the I/O port number appears as a 0019H on the 16-bit address bus and that the data from AX appears on the data bus of the microprocessor. The system control signal $\overline{IOWC}$ (I/O write control) is a logic zero to enable the I/O device.

A short program that clicks the speaker in the personal computer appears in Example 4–12. The speaker (in DOS only) is controlled by accessing I/O port 61H. If the rightmost two bits of this port are set (11) and then cleared (00), a click is heard on the speaker. Note that this program uses a logical OR instruction to set these two bits and a logical AND instruction to clear them. These logic operation instructions are described in Chapter 5. The MOV CX,8000H instruction, followed by the LOOP L1 instruction, is used as a time delay. If the count is increased, the click will become longer; if shortened, the click will become shorter. To obtain a series of clicks that can be heard, the program must be modified to repeat many times.

EXAMPLE 4–12

```
                    .MODEL TINY          ;select tiny model
0000                .CODE                ;start code segment
                    .STARTUP             ;start program
0100 E4 61                 IN AL,61H     ;read I/O port 61H
0102 0C 03                 OR AL,3       ;set rightmost two bits
0104 E6 61                 OUT 61H,AL    ;speaker on
0106 B9 8000               MOV CX,8000H  ;load delay count
```

TABLE 4–18 The MOVSX and MOVZX instructions.

Assembly Language	*Operation*
MOVSX CX,BL	Sign-extends BL into CX
MOVSX ECX,AX	Sign-extends AX into ECX
MOVSX BX,DATA1	Sign-extends the byte at DATA1 into BX
MOVSX EAX,[EDI]	Sign-extends the word at the data segment memory location addressed by EDI into EAX
MOVZX DX,AL	Zero-extends AL into DX
MOVZX EBP,DI	Zero-extends DI into EBP
MOVZX DX,DATA2	Zero-extends the byte at DATA2 into DX
MOVZX EAX,DATA3	Zero-extends the word at DATA3 into EAX

```
0109                 L1:
0109 E2 FE                 LOOP L1         ;time delay
010B E4 61                 IN AL,61H       ;speaker off
010D 24 FC                 AND AL,0FCH
010F E6 61                 OUT 61H,AL
                     .EXIT
                     END
```

MOVSX and MOVZX

The MOVSX (**move and sign-extend**) and MOVZX (**move and zero-extend**) instructions are found in the 80386–Pentium 4 instruction sets. These instructions move data, and at the same time either sign- or zero-extend it. Table 4–18 illustrates these instructions with several examples of each.

When a number is zero-extended, the most-significant part fills with zeros. For example, if an 8-bit 34H is zero-extended into a 16-bit number, it becomes 0034H. Zero-extension is often used to convert unsigned 8- or 16-bit numbers into unsigned 16- or 32-bit numbers by using the MOVZX instruction.

A number is sign-extended when its sign-bit is copied into the most-significant part. For example, if an 8-bit 84H is sign-extended into a 16-bit number, it becomes FF84H. The sign-bit of an 84H is a one, which is copied into the most-significant part of the sign-extended result. Sign-extension is most often used to convert 8- or 16-bit signed numbers into 16- or 32-bit signed numbers by using the MOVSX instruction.

BSWAP

The BSWAP (**byte swap**) instruction is available only in the 80486–Pentium 4 microprocessors. This instruction takes the contents of any 32-bit register and swaps the first byte with the fourth and the second with the third. For example, the BSWAP EAX instruction with EAX = 00112233H swaps bytes in EAX, resulting in EAX = 33221100H. Notice that the order of all four bytes is reversed by this instruction. This instruction is used to convert data between the big and little endian forms.

CMOV

The CMOV (**conditional move**) class of instruction is new to the Pentium Pro–Pentium 4 instruction sets. There are many variations of the CMOV instruction. Table 4–19 lists these variations of CMOV. These instructions move the data only if the condition is true. For example, the CMOVZ instruction moves data only if the result from some prior instruction was a zero. The destination is limited only to 16- or 32-bit registers, but the source can be a 16- or 32-bit register or memory location.

Because this is a new instruction, you cannot use it with the assembler unless the .686 switch is added to the program.

TABLE 4–19 The conditional move instructions.

Assembly Language	*Flag(s) Tested*	*Operation*
CMOVB	C = 1	Move if below
CMOVAE	C = 0	Move if above or equal
CMOVBE	Z =1 or C = 1	Move if below or equal
CMOVA	Z = 0 and C = 0	Move if above
CMOVE or CMOVZ	Z = 1	Move if equal or move if zero
CMOVNE or CMOVNZ	Z = 0	Move if not equal or move if not zero
CMOVL	S != O	Move if less than
CMOVLE	Z = 1 or S != O	Move if less than or equal
CMOVG	Z = 0 and S = O	Move if greater than
CMOVGE	S = O	Move if greater than or equal
CMOVS	S = 1	Move if sign (negative)
CMOVNS	S = 0	Move if no sign (positive)
CMOVC	C = 1	Move if carry
CMOVNC	C = 0	Move if no carry
CMOVO	O = 1	Move if overflow
CMOVNO	O = 0	Move if no overflow
CMOVP or CMOVPE	P = 1	Move if parity or move if parity even
CMOVNP or CMOVPO	P = 0	Move if no parity or move if parity odd

4–6 SEGMENT OVERRIDE PREFIX

The **segment override prefix**, which may be added to almost any instruction in any memory-addressing mode, allows the programmer to deviate from the default segment. The segment override prefix is an additional byte that appends the front of an instruction to select an alternate segment register. About the only instructions that cannot be prefixed are the jump and call instructions that must use the code segment register for address generation. The segment override is also used to select the FS and GS segments in the 80386 through the Pentium 4 microprocessors.

For example, the MOV AX,[DI] instruction accesses data within the data segment by default. If required by a program, this can be changed by prefixing the instruction. Suppose that the data are in the extra segment instead of in the data segment. This instruction addresses the extra segment if changed to MOV AX,ES:[DI].

Table 4–20 shows some altered instructions that address different memory segments that are different from normal. Each time an instruction is prefixed with a segment override prefix,

TABLE 4–20 Instructions that include segment override prefixes.

Assembly Language	*Segment Accessed*	*Default Segment*
MOV AX,DS:[BP]	Data	Stack
MOV AX,ES:[BP]	Extra	Stack
MOV AX,SS:[DI]	Stack	Data
MOV AX,CS:LIST	Code	Data
MOV ES:[SI],AX	Extra	Data
LODS ES:DATA1	Extra	Data
MOV EAX,FS:DATA2	FS	Data
MOV GS:[ECX],BL	GS	Data

the instruction becomes one byte longer. Although this is not a serious change to the length of the instruction, it does add to the instruction's execution time. It is usually customary to limit the use of the segment override prefix and remain in the default segments so that shorter and more efficient software is written.

4–7 ASSEMBLER DETAIL

The assembler (MASM)[1] for the microprocessor can be used in two ways: (1) with models that are unique to a particular assembler, and (2) with full-segment definitions that allow complete control over the assembly process and are universal to all assemblers. This section of the text presents both methods and explains how to organize a program's memory space by using the assembler. It also explains the purpose and use of some of the more important directives used with this assembler. Appendix A provides additional detail about the assembler.

In most cases, the inline assembler found in Visual C++ is used for developing assembly code for use in a C++ program, but there are occasions that require separate assembly modules written using the assembler. This section of the text contrasts where possible the inline assembler and the assembler.

Directives

Before the format of an assembly language program is discussed, some details about the directives (**pseudo-operations**) that control the assembly process must be learned. Some common assembly language directives appear in Table 4–21. **Directives** indicate how an operand or section of a program is to be processed by the assembler. Some directives generate and store information in the memory; others do not. The DB (**define byte**) directive stores bytes of data in the memory, whereas the BYTE PTR directive never stores data. The **BYTE PTR** directive indicates the size of the data referenced by a pointer or index register. Note that none of the directives function in the inline assembler program that is a part of Visual C++. If you are using the inline assembler exclusively, you can skip this part of the text. Be aware that complex sections of assembly code are still written using MASM.

Note that by default the assembler accepts only 8086/8088 instructions, unless a program is preceded by the .686 or .686P directive or one of the other microprocessor selection switches. The .686 directive tells the assembler to use the Pentium Pro instruction set in the real mode, and the .686P directive tells the assembler to use the Pentium Pro protected mode instruction set. Most modern software is written assuming that the microprocessor is a Pentium Pro or newer, so the .686 switch is often used. Windows 95 was the first major operating system to use a 32-bit architecture that conforms to the 80386. Windows XP requires a Pentium class machine (.586 switch) using at least a 233MHz microprocessor.

Storing Data in a Memory Segment. The DB (**define byte**), DW (**define word**), and DD (**define doubleword**) directives, first presented in Chapter 1, are most often used with MASM to define and store memory data. If a numeric coprocessor executes software in the system, the DQ (**define quadword**) and DT (**define ten bytes**) directives are also common. These directives label a memory location with a symbolic name and indicate its size.

Example 4–13 shows a memory segment that contains various forms of data definition directives. It also shows the full-segment definition with the first SEGMENT statement to indicate the start of the segment and its symbolic name. Alternately, as in past examples in this and prior chapters, the SMALL model can be used with the .DATA statement. The last statement in this

[1]The assembler used throughout this text is the Microsoft MACRO assembler called MASM, version 6.1x.

TABLE 4–21 Common MASM directives.

Directive	*Function*
.286	Selects the 80286 instruction set
.286P	Selects the 80286 protected mode instruction set
.386	Selects the 80386 instruction set
.386P	Selects the 80386 protected mode instruction set
.486	Selects the 80486 instruction set
.486P	Selects the 80498 protected mode instruction set
.586	Selects the Pentium instruction set
.586P	Selects the Pentium protected mode instruction set
.686	Selects the Pentium Pro–Pentium 4 instruction set
.686P	Selects the Pentium Pro–Pentium 4 protected mode instruction set
.287	Selects the 80287 math coprocessor
.387	Selects the 80387 math coprocessor
.CODE	Indicates the start of the code segment (models only)
.DATA	Indicates the start of the data segment (models only)
.EXIT	Exits to DOS (models only)
.MODEL	Selects the programming model
.STACK	Selects the start of the stack segment (models only)
.STARTUP	Indicates the starting instruction in a program (models only)
ALIGN n	Align to boundary n (n = 2 for words, n = 4 for doublewords)
ASSUME	Informs the assembler to name each segment (full segments only)
BYTE	Indicates byte-sized, as in BYTE PTR
DB	Defines byte(s) (8 bits)
DD	Defines doubleword(s) (32 bits)
DQ	Defines quadwords(s) (64 bits)
DT	Defines ten byte(s) (80 bits)
DUP	Generates duplicates
DW	Defines word(s) (16 bits)
DWORD	Indicates doubleword-sized, as in DWORD PTR
END	Ends a program file
ENDM	Ends a MACRO sequence
ENDP	Ends a procedure
ENDS	Ends a segment or data structure
EQU	Equates data or a label to a label
FAR	Defines a far pointer as in FAR PTR
MACRO	Designates the start of a MACRO sequence
NEAR	Defines a near pointer as in NEAR PTR
OFFSET	Specifies an offset address
ORG	Sets the origin within a segment
OWORD	Indicates octalwords, as in OWORD PTR
PROC	Starts a procedure
PTR	Designates a pointer
SEGMENT	Starts a segment for full segments
STACK	Starts a stack segment for full segments
STRUC	Defines the start of a data structure
USES	Automatically pushes and pops registers
USE16	Uses 16-bit instruction mode
USE32	Uses 32-bit instruction mode
WORD	Indicates word-sized, as in WORD PTR

example contains the ENDS directive, which indicates the end of the segment. The name of the segment (LIST_SEG) can be anything that the programmer desires to call it. This allows a program to contain as many segments as required.

EXAMPLE 4–13

```
                    ;Using the DB, DW, and DD directives
                    ;
0000                LIST_SEG     SEGMENT

0000 01 02 03       DATA1  DB  1,2,3            ;define bytes
0003 45                    DB  45H              ;hexadecimal
0004 41                    DB  'A'              ;ASCII
0005 F0                    DB  11110000B        ;binary
0006 000C 000D      DATA2  DW  12,13            ;define words
000A 0200                  DW  LIST1            ;symbolic
000C 2345                  DW  2345H            ;hexadecimal
000E 00000300       DATA3  DD  300H             ;define doubleword
0012 4007DF3B              DD  2.123            ;real
0016 544269E1              DD  3.34E+12         ;real
001A 00             LISTA  DB  ?                ;reserve 1 byte
001B 000A[          LISTB  DB 10 DUP(?)         ;reserve 10 bytes
        ??
          ]
0025 00                    ALIGN  2             ;set word boundary
0026 0100[          LISTC  DW  100H DUP(0)      ;reserve 100H words
        0000
            ]
0226 0016[          LISTD  DD  22 DUP(?)        ;reserve 22 doublewords
        ????????
                ]
027E 0064[          SIXES  DB  100 DUP(6)       ;reserve 100 bytes
        06
          ]
02E2                LIST_SEG     ENDS
```

Example 4–13 shows various forms of data storage for bytes at DATA1. More than one byte can be defined on a line in binary, hexadecimal, decimal, or ASCII code. The DATA2 label shows how to store various forms of word data. Doublewords are stored at DATA3; they include floating-point, single-precision real numbers.

Memory is **reserved** for use in the future by using a question mark (?) as an operand for a DB, DW, or DD directive. When a ? is used in place of a numeric or ASCII value, the assembler sets aside a location and does not initialize it to any specific value. (Actually, the assembler usually stores a zero into locations specified with a ?.) The DUP (**duplicate**) directive creates an array, as shown in several ways in Example 4–12. A 10 DUP (?) reserves 10 locations of memory, but stores no specific value in any of the 10 locations. If a number appears within the ( ) part of the DUP statement, the assembler initializes the reserved section of memory with the data indicated. For example, the LIST2 DB 10 DUP (2) instruction reserves 10 bytes of memory for array LIST2 and initializes each location with a 02H.

The ALIGN directive, used in this example, makes sure that the memory arrays are stored on word boundaries. An ALIGN 2 places data on *word boundaries* and an ALIGN 4 places them on *doubleword boundaries.* In the Pentium–Pentium 4, quadword data for double-precision floating-point numbers should use ALIGN 8. It is important that word-sized data be placed at word boundaries and doubleword-sized data be placed at doubleword boundaries. If not, the microprocessor spends additional time accessing these data types. A word stored at an odd-numbered memory location takes twice as long to access as a word stored at an even-numbered memory location. Note that the ALIGN directive cannot be used with memory models because the size of the model determines the data alignment. If all doubleword data are defined first, followed by word-sized and then byte-sized data, the ALIGN statement is not necessary for aligning data correctly.

ASSUME, EQU, and ORG. The equate directive (EQU) equates a numeric, ASCII, or label to another label. Equates make a program clearer and simplify debugging. Example 4–14 shows several equate statements and a few instructions that show how they function in a program.

EXAMPLE 4–14

```
                ;Using equate directive
                ;
= 000A          TEN     EQU 10
= 0009          NINE    EQU 9

0000 B0 0A              MOV AL,TEN
0002 04 09              ADD AL,NINE
```

The THIS directive always appears as THIS BYTE, THIS WORD, or THIS DWORD. In certain cases, data must be referred to as both a byte and a word. The assembler can only assign a byte, word, or doubleword address to a label. To assign a byte label to a word, use the software listed in Example 4–15.

EXAMPLE 4–15

```
                        ;Using the THIS and ORG directives
                        ;
0000                    DATA_SEG        SEGMENT

0300                            ORG     300H

= 0300                  DATA1   EQU     THIS BYTE
0300                    DATA2   DW      ?
0302                    DATA_SEG        ENDS

0000                    CODE_SEG         SEGMENT 'CODE'
                                ASSUME CS:CODE_SEG, DS:DATA_SEG
0000 8A 1E 0300 R               MOV   BL,DATA1
0004 A1 0300 R                  MOV   AX,DATA2
0007 8A 3E 0301 R               MOV   BH,DATA1+1

000B                    CODE_SEG        ENDS
```

This example also illustrates how the ORG (**origin**) statement changes the starting offset address of the data in the data segment to location 300H. At times, the origin of data or the code must be assigned to an absolute offset address with the ORG statement. The **ASSUME** statement tells the assembler what names have been chosen for the code, data, extra, and stack segments. Without the ASSUME statement, the assembler assumes nothing and automatically uses a segment override prefix on all instructions that address memory data. The ASSUME statement is only used with full-segment definitions, as described later in this section of the text.

PROC and ENDP. The PROC and ENDP directives indicate the start and end of a procedure (**subroutine**). These directives *force structure* because the procedure is clearly defined. Note that if structure is to be violated for any reason, use the CALLF, CALLN, RETF, and RETN instructions. Both the PROC and ENDP directives require a label to indicate the name of the procedure. The PROC directive, which indicates the start of a procedure, must also be followed with a NEAR or FAR. A NEAR procedure is one that resides in the same code segment as the program. A FAR procedure may reside at any location in the memory system. Often the call NEAR procedure is considered to be *local*, and the call FAR procedure is considered to be *global*. The term *global* denotes a procedure that can be used by any program; *local* defines a procedure that is only used by the current program. Any labels that are defined within the procedure block are also defined as either local (NEAR) or global (FAR).

Example 4–16 shows a procedure that adds BX, CX, and DX and stores the sum in register AX. Although this procedure is short and may not be particularly useful, it does illustrate how to use the PROC and ENDP directives to delineate the procedure. Note that information about the operation of the procedure should appear as a grouping of comments that show the registers changed by the procedure and the result of the procedure.

EXAMPLE 4–16

```
                                    ;A procedure that adds BX, CX, and DX with the
                                    ;sum stored in AX
                                    ;
0000                                ADDEM PROC    FAR              ;start of procedure

0000 03 D9                                ADD     BX,CX
0002 03 DA                                ADD     BX,DX
0004 8B C3                                MOV     AX,BX
0006 CB                                   RET

0007                                ADDEM ENDP                     ;end of procedure
```

If version 6.x of the Microsoft MASM assembler program is available, the PROC directive specifies and automatically saves any registers used within the procedure. The USES statement indicates which registers are used by the procedure, so that the assembler can automatically save them before your procedure begins and restore them before the procedure ends with the RET instruction. For example, the ADDS PROC USES AX BX CX statement automatically pushes AX, BX, and CX on the stack before the procedure begins and pops them from the stack before the RET instruction executes at the end of the procedure. Example 4–17 illustrates a procedure written using MASM version 6.x that shows the USES statement. Note that the registers in the list are not separated by commas, but by spaces, and the PUSH and POP instructions are displayed in the procedure listing because it was assembled with the .LIST ALL directive. The instructions prefaced with an asterisk (*) are inserted by the assembler and were not typed in the source file. The USES statement appears elsewhere in this text, so if MASM version 5.10 is in use, you will need to modify the code.

EXAMPLE 4–17

```
                                          ;A procedure that includes the USES directive to
                                          ;save BX, CX, and DX on the stack and restore them
                                          ;before the return instruction.
                                          ;
0000                              ADDS    PROC  NEAR    USES BX CX DX

0000 53            *                      push  bx
0001 51            *                      push  cx
0002 52            *                      push  dx
0003 03 D8                                ADD   BX,AX
0005 03 CB                                ADD   CX,BX
0007 03 D1                                ADD   DX,CX
0009 8B C2                                MOV   AX,DX
                                          RET
000B 5A            *                      pop   dx
000C 59            *                      pop   cx
000D 5B            *                      pop   bx
000E C3            *                      ret   0000h

000F                              ADDS    ENDP
```

Memory Organization

The assembler uses two basic formats for developing software: One method uses models and the other uses full-segment definitions. Memory models, as presented in this section and briefly in earlier chapters, are unique to the MASM assembler program. The TASM assembler also uses memory models, but they differ somewhat from the MASM models. The full-segment definitions are common to most assemblers, including the Intel assembler, and are often used for software development. The models are easier to use for simple tasks. The full-segment definitions offer better control over the assembly language task and are recommended for complex programs. The model was used in earlier chapters because it is easier to understand for the beginning programmer. Models are also used with assembly language procedures that are used by high-level languages such as C/C++. Although this text fully develops and uses the memory model definitions for its programming examples, realize that full-segment definitions offer some advantages over memory models, as discussed later in this section.

Models. There are many models available to the MASM assembler, ranging in size from tiny to huge. Appendix A contains a table that lists all the models available for use with the assembler. To designate a model, use the .MODEL statement followed by the size of the memory system. The **TINY model** requires that all software and data fit into one 64K-byte memory segment; it is useful for many small programs. The **SMALL model** requires that only one data segment be used with one code segment for a total of 128K bytes of memory. Other models are available, up to the HUGE model.

Example 4–18 illustrates how the .MODEL statement defines the parameters of a short program that copies the contents of a 100-byte block of memory (LISTA) into a second 100-byte block of memory (LISTB). It also shows how to define the stack, data, and code segments. The .EXIT 0 directive returns to DOS with an error code of 0 (no error). If no parameter is added to .EXIT, it still returns to DOS, but the error code is not defined. Also note that special directives such as @DATA (see Appendix A) are used to identify various segments. If the .STARTUP directive is used (MASM version 6.x), the MOV AX,@DATA followed by MOV DS,AX statements can be eliminated. The .STARTUP directive also eliminates the need to store the starting address next to the END label. Models are important with both Microsoft Visual C++ and Borland C++ development systems if assembly language is included with C++ programs. Both development systems use inline assembly programming for adding assembly language instructions and require an understanding of programming models.

EXAMPLE 4–18

```
                            .MODEL SMALL              ;select small model
                            .STACK 100H               ;define stack
                            .DATA                     ;start data segment

0000 0064[          LISTA   DB     100 DUP(?)
          ??
            ]
0064 0064[          LISTB   DB     100 DUP(?)
          ??
            ]

                            .CODE                     ;start code segment

0000 B9 ---- ?      HERE:   MOV    AX,@DATA           ;load ES and DS
0003 8E C0                  MOV    ES,AX
0005 8E D8                  MOV    DS,AX
0007 FC                     CLD                       ;move data
0008 BE 0000 R              MOV    SI,OFFSET LISTA
```

```
000B BF 0064 R          MOV   DI,OFFSET LISTB
000E B9 0064            MOV   CX,100
0011 F3/A4              REP   MOVSB

0013                    .EXIT 0              ;exit to DOS
                        END HERE
```

Full-Segment Definitions. Example 4–19 illustrates the same program using full-segment definitions. Full-segment definitions are also used with the Borland and Microsoft C/C++ environments for procedures developed in assembly language. The program in Example 4–19 appears longer than the one pictured in Example 4–18, but it is more structured than the model method of setting up a program. The first segment defined is the STACK_SEG, which is clearly delineated with the SEGMENT and ENDS directives. Within these directives, DW 100 DUP (?) sets aside 100H words for the stack segment. Because the word STACK appears next to SEGMENT, the assembler and linker automatically load both the stack segment register (SS) and stack pointer (SP).

EXAMPLE 4–19

```
0000                  STACK_SEG      SEGMENT      'STACK'
0000 0064[                   DW      100H DUP(?)
          ????
              ]
0200                  STACK_SEG      ENDS

0000                  DATA_SEG       SEGMENT      'DATA'
0000 0064[            LISTA  DB      100 DUP(?)
          ??
            ]
0064 0064[            LISTB  DB      100 DUP(?)
          ??
            ]
00CB                  DATA_SEG       ENDS

0000                  CODE_SEG       SEGMENT      'CODE'
                             ASSUME  CS:CODE_SEG,DS:DATA_SEG
                             ASSUME  SS:STACK_SEG
0000                  MAIN   PROC    FAR
0000 B8 ---- R               MOV     AX,DATA_SEG          ;load DS and ES
0003 8E C0                   MOV     ES,AX
0005 8E D8                   MOV     DS,AX
0007 FC                      CLD                          ;save data
0008 BE 0000 R               MOV     SI,OFFSET LISTA
000B BF 0064 R               MOV     DI,OFFSET LISTB
000E B9 0064                 MOV     CX,100
0011 F3/A4                   REP     MOVSB
0013 B4 4C                   MOV     AH,4CH               ;exit to DOS
0015 CD 21                   INT     21H
0017                  MAIN   ENDP
0017                  CODE_SEG       ENDS
                             END     MAIN
```

Next, the data are defined in the DATA_SEG. Here, two arrays of data appear as LISTA and LISTB. Each array contains 100 bytes of space for the program. The names of the segments in this program can be changed to any name. Always include the group name 'DATA', so that the Microsoft program CodeView can be effectively used to symbolically debug this software. CodeView is a part of the MASM package used to debug software. To access CodeView, type CV, followed by the file name at the DOS command line; if operating from Programmer's WorkBench, select Debug under the Run menu. If the group name is not placed in a program, CodeView can still be used to debug a program, but the program will not be debugged in symbolic form. Other group names such as 'STACK', 'CODE', and so forth are listed in Appendix A. You must at least place the word 'CODE' next to the code segment SEGMENT statement if you want to view the program symbolically in CodeView.

The CODE_SEG is organized as a far procedure because most software is procedure-oriented. Before the program begins, the code segment contains the ASSUME statement. The ASSUME statement tells the assembler and linker that the name used for the code segment (CS) is CODE_SEG; it also tells the assembler and linker that the data segment is DATA_SEG and the stack segment is STACK_SEG. Notice that the group name 'CODE' is used for the code segment for use by CodeView. Other group names appear in Appendix A with the models.

After the program loads both the extra segment register and data segment register with the location of the data segment, it transfers 100 bytes from LISTA to LISTB. Following this is a sequence of two instructions that return control back to DOS (the disk operating system). Note that the program loader does not automatically initialize DS and ES. These registers must be loaded with the desired segment addresses in the program.

The last statement in the program is END MAIN. The END statement indicates the end of the program and the location of the first instruction executed. Here, we want the machine to execute the main procedure so the MAIN label follows the END directive.

In the 80386 through the Pentium 4 microprocessors, an additional directive is found attached to the code segment. The USE16 or USE32 directive tells the assembler to use either the 16- or 32-bit instruction modes for the microprocessor. Software developed for the DOS environment must use the USE16 directive for the 80386 through the Pentium 4 program to function correctly because MASM assumes that all segments are 32 bits and all instruction modes are 32 bits by default.

A Sample Program

Example 4–20 provides a sample program, using full-segment definitions, that reads a character from the keyboard and displays it on the CRT screen. Although this program is trivial, it illustrates a complete workable program that functions on any personal computer using DOS, from the earliest 8088-based system to the latest Pentium 4-based system. This program also illustrates the use of a few DOS function calls. (Appendix A lists the DOS function calls with their parameters.) The BIOS function calls allow the use of the keyboard, printer, disk drives, and everything else that is available in your computer system.

This example program uses only a code segment because there is no data. A stack segment should appear, but it has been left out because DOS automatically allocates a 128-byte stack for all programs. The only time that the stack is used in this example is for the INT 21H instructions that call a procedure in DOS. Note that when this program is linked, the linker signals that no stack segment is present. This warning may be ignored in this example because the stack is fewer than 128 bytes.

Notice that the entire program is placed into a far procedure called MAIN. It is good programming practice to write all software in procedural form, which allows the program to be used as a procedure at some future time if necessary. It is also fairly important to document register use and any parameters required for the program in the program header, which is a section of comments that appear at the start of the program.

The program uses DOS functions 06H and 4CH. The function number is placed in AH before the INT 21H instruction executes. The 06H function reads the keyboard if DL = 0FFH, or displays the ASCII contents of DL if it is not 0FFH. Upon close examination, the first section of the program moves 06H into AH and 0FFH into DL, so that a key is read from the keyboard. The INT 21H tests the keyboard; if no key is typed, it returns equal. The JE instruction tests the equal condition and jumps to MAIN if no key is typed.

When a key is typed, the program continues to the next step, which compares the contents of AL with an @ symbol. Upon return from the INT 21H, the ASCII character of the typed key is found in AL. In this program, if an @ symbol is typed, the program ends. If the @ symbol is not typed, the program continues by displaying the character typed on the keyboard with the next INT 21H instruction.

The second INT 21H instruction moves the ASCII character into DL so it can be displayed on the CRT screen. After the character is displayed, a JMP executes. This causes the program to continue at MAIN, where it repeats reading a key.

If the @ symbol is typed, the program continues at MAIN1, where it executes the DOS function code number 4CH. This causes the program to return to the DOS prompt so that the computer can be used for other tasks.

More information about the assembler and its application appears in Appendix A and in the next several chapters. Appendix A provides a complete overview of the assembler, linker, and DOS functions. It also provides a list of the BIOS (basic I/O system) functions. The information provided in the following chapters clarifies how to use the assembler for certain tasks at different levels of the text.

EXAMPLE 4–20

```
                   ;An example DOS full-segment program that reads a key and
                   ;displays it. Note that an @ key ends the program.
                   ;
0000               CODE_SEG       SEGMENT 'CODE'
                          ASSUME CS:CODE_SEG

0000               MAIN   PROC    FAR

0000 B4 06                MOV  AH,06H            ;read a key
0002 B2 FF                MOV  DL,0FFH
0004 CD 21                INT  21H
0006 74 F8                JE   MAIN              ;if no key typed
0008 3C 40                CMP  AL,'@'
000A 74 08                JE   MAIN1             ;if an @ key
000C B4 06                MOV  AH,06H            ;display key (echo)
000E 8A D0                MOV  DL,AL
0010 CD 21                INT  21H
0012 EB EC                JMP  MAIN              ;repeat
0014               MAIN1:
0014 B4 4C                MOV  AH,4CH            ;exit to DOS
0016 CD 21                INT  21H

0018               MAIN   ENDP
0018                      END  MAIN
```

EXAMPLE 4–21

```
                   ;An example DOS model program that reads a key and displays
                   ;it. Note that an @ key ends the program.
                   ;
                   .MODEL TINY
0000               .CODE
                   .STARTUP

0100               MAIN:
0100 B4 06                MOV  AH,6              ;read a key
0102 B2 FF                MOV  DL,0FFH
0104 CD 21                INT  21H
0106 74 F8                JE   MAIN              ;if no key typed
0108 3C 40                CMP  AL, '@'
010A 74 08                JE   MAIN1             ;if an @ key
010C B4 06                MOV  AH,06H            ;display key (echo)
010E 8A D0                MOV  DL,AL
0110 CD 21                INT  21H
0112 EB EC                JMP  MAIN              ;repeat
```

```
0114            MAIN1:

                .EXIT                   ;exit to DOS
                END
```

Example 4–21 shows the program listed in Example 4–20, except models are used instead of full-segment descriptions. Please compare the two programs to determine the differences. Notice how much shorter and cleaner looking the models can make a program.

4–8 SUMMARY

1. Data movement instructions transfer data between registers, a register and memory, a register and the stack, memory and the stack, the accumulator and I/O, and the flags and the stack. Memory-to-memory transfers are allowed only with the MOVS instruction.
2. Data movement instructions include MOV, PUSH, POP, XCHG, XLAT, IN, OUT, LEA, LOS, LES, LSS, LGS, LFS, LAHF, SAHF, and the following string instructions: LODS, STOS, MOVS, INS, and OUTS.
3. The first byte of an instruction contains the opcode. The opcode specifies the operation performed by the microprocessor. The opcode may be preceded by one or more override prefixes in some forms of instructions.
4. The D-bit, located in many instructions, selects the direction of data flow. If D = 0, the data flow from the REG field to the R/M field of the instruction. If D = 1, the data flow from the R/M field to the REG field.
5. The W-bit, found in most instructions, selects the size of the data transfer. If W = 0, the data are byte-sized; if W = 1, the data are word-sized. In the 80386 and above, W = 1 specifies either a word or doubleword register.
6. MOD selects the addressing mode of operation for a machine language instruction's RIM field. If MOD = 00, there is no displacement; if MOD = 01, an 8-bit sign-extended displacement appears; if MOD = 10, a 16-bit displacement occurs; and if MOD = 11, a register is used instead of a memory location. In the 80386 and above, the MOD bits also specify a 32-bit displacement.
7. A 3-bit binary register code specifies the REG and R/M fields when the MOD = 11. The 8-bit registers are AH, AL, BH, BL, CH, CL, DH, and DL. The 16-bit registers are AX, BX, CX, DX, SP, BP, DI, and SI. The 32-bit registers are EAX, EBX, ECX, EDX, ESP, EBP, EDI, and ESI.
8. When the R/M field depicts a memory mode, a 3-bit code selects one of the following modes: [BX+DI], [BX+SI], [BP+DI], [BP+SI], [BX], [BP], [DI], or [SI] for 16-bit instructions. In the 80386 and above, the R/M field specifies EAX, EBX, ECX, EDX, EBP, EDI, and ESI or one of the scaled-index modes of addressing memory data. If the scaled-index mode is selected (R/M = 100), an additional byte (scaled-index byte) is added to the instruction to specify the base register, index register, and the scaling factor.
9. By default, all memory-addressing modes address data in the data segment unless BP or EBP addresses memory. The BP or EBP register addresses data in the stack segment.
10. The segment registers are addressed only by the MOV, PUSH, or POP instructions. The MOV instruction may transfer a segment register to a 16-bit register, or vice versa. MOV CS,reg or POP CS instructions are not allowed because they change only a portion of the address. The 80386 through the Pentium 4 include two additional segment registers, FS and GS.
11. Data are transferred between a register or a memory location and the stack by the PUSH and POP instructions. Variations of these instructions allow immediate data to be pushed onto

the stack, the flags to be transferred between the stack, and all 16-bit registers to be transferred between the stack and the registers. When data are transferred to the stack, two bytes (8086–80286) always move. The most-significant byte is placed at the location addressed by SP – 1, and the least-significant byte is placed at the location addressed by SP – 2. After placing the data on the stack, SP decrements by 2. In the 80386–Pentium 4, four bytes of data from a memory location or register may also be transferred to the stack.

12. Opcodes that transfer data between the stack and the flags are PUSHF and POPF. Opcodes that transfer all the 16-bit registers between the stack and the registers are PUSHA and POPA. In the 80386 and above, PUSHFD and POPFD transfer the contents of the EFLAGS between the microprocessor and the stack, and PUSHAD and POPAD transfer all the 32-bit registers.
13. LEA, LDS, and LES instructions load a register or registers with an effective address. The LEA instruction loads any 16-bit register with an effective address; LDS and LES load any 16-bit register and either DS or ES with the effective address. In the 80386 and above, additional instructions include LFS, LGS, and LSS, which load a 16-bit register and FS, GS, or SS.
14. String data transfer instructions use either or both DI and SI to address memory. The DI offset address is located in the extra segment, and the SI offset address is located in the data segment. If the 80386–Pentium 4 are operated in protected mode, ESI and EDI are used with the string instructions.
15. The direction flag (D) chooses the auto-increment or auto-decrement mode of operation for DI and SI for string instructions. To clear D to 0, use the CLD instruction to select the auto-increment mode; to set D to 1, use the STD instruction to select the auto-decrement mode. Either or both DI and SI increment/decrement by 1 for a byte operation, by 2 for a word operation, and by 4 for a doubleword operation.
16. LODS loads AL, AX, or EAX with data from the memory location addressed by SI; STOS stores AL, AX, or EAX in the memory location addressed by DI; and MOVS transfers a byte, a word, or a doubleword from the memory location addressed by SI into the location addressed by DI.
17. INS inputs data from an I/O device addressed by DX and stores it in the memory location addressed by DI. The OUTS instruction outputs the contents of the memory location addressed by SI and sends it to the I/O device addressed by DX.
18. The REP prefix may be attached to any string instruction to repeat it. The REP prefix repeats the string instruction the number of times found in register CX.
19. Arithmetic and logic operators can be used in assembly language. An example is MOV AX,34*3, which loads AX with 102.
20. Translate (XLAT) converts the data in AL into a number stored at the memory location addressed by BX plus AL.
21. IN and OUT transfer data between AL, AX, or EAX and an external I/O device. The address of the I/O device is stored either with the instruction (fixed-port addressing) or in register DX (variable-port addressing).
22. The Pentium Pro–Pentium 4 contain a new instruction called CMOV, or conditional move. This instruction only performs the move if the condition is true.
23. The segment override prefix selects a different segment register for a memory location than the default segment. For example, the MOV AX,[BX] instruction uses the data segment, but the MOV AX,ES:[BX] instruction uses the extra segment because of the ES: override prefix. Using the segment override prefix is the only way to address the FS and GS segments in the 80386 through the Pentium 4.
24. The MOVZX (move and zero-extend) and MOVSX (move and sign-extend) instructions, found in the 80386 and above, increase the size of a byte to a word or a word to a doubleword. The zero-extend version increases the size of the number by inserting leading zeros.

The sign-extend version increases the size of the number by copying the sign-bit into the more-significant bits of the number.

25. Assembler directives DB (define byte), DW (define word), DD (define doubleword), and DUP (duplicate) store data in the memory system.
26. The EQU (equate) directive allows data or labels to be equated to labels.
27. The SEGMENT directive identifies the start of a memory segment and ENDS identifies the end of a segment when full-segment definitions are in use.
28. The ASSUME directive tells the assembler what segment names you have assigned to CS, DS, ES, and SS when full-segment definitions are in effect. In the 80386 and above, ASSUME also indicates the segment name for FS and GS.
29. The PROC and ENDP directives indicate the start and end of a procedure. The USES directive (MASM version 6.x) automatically saves and restores any number of registers on the stack if they appear with the PROC directive.
30. The assembler assumes that software is being developed for the 8086/8088 microprocessor unless the .286, .386, .486, .586, or .686 directive is used to select one of these other microprocessors. This directive follows the .MODEL statement to use the 16-bit instruction mode and precedes it for the 32-bit instruction mode.
31. Memory models can be used to shorten the program slightly, but they can cause problems for larger programs. Memory models are not compatible with all assembler programs.

4–9 QUESTIONS AND PROBLEMS

1. The first byte of an instruction is the __________, unless it contains one of the override prefixes.
2. Describe the purpose of the D- and W-bits found in some machine language instructions.
3. In a machine language instruction, what information is specified by the MOD field?
4. If the register field (REG) of an instruction contains 010 and W = 0, what register is selected, assuming that the instruction is a 16-bit mode instruction?
5. How are the 32-bit registers selected for the 80486 microprocessor?
6. What memory-addressing mode is specified by R/M = 001 with MOD = 00 for a 16-bit instruction?
7. Identify the default segment registers assigned to the following:
   (a) SP
   (b) EBX
   (c) DI
   (d) EBP
   (e) SI
8. Convert 8B07H from machine language to assembly language.
9. Convert 8B9E004CH from machine language to assembly language.
10. If a MOV SI,[BX+2] instruction appears in a program, what is its machine language equivalent?
11. If a MOV ESI,[EAX] instruction appears in a program for the Pentium 4 microprocessor operating in the 16-bit instruction mode, what is its machine language equivalent?
12. What is wrong with a MOV CS,AX instruction?
13. Form a short sequence of instructions that load the data segment register with 1000H.
14. The PUSH and POP instructions always transfer a(n) __________-bit number between the stack and a register or memory location in the 8086–80286 microprocessors.
15. What segment register may not be popped from the stack?
16. Which registers move onto the stack with the PUSHA instruction?

17. Which registers move onto the stack for a PUSHAD instruction?
18. Describe the operation of each of the following instructions:
    (a) PUSH AX
    (b) POP ESI
    (c) PUSH [BX]
    (d) PUSHFD
    (e) POP DS
    (f) PUSHD 4
19. Explain what happens when the PUSH BX instruction executes. Make sure to show where BH and BL are stored. (Assume that SP = 0100H and SS = 0200H.)
20. Repeat question 19 for the PUSH EAX instruction.
21. The 16-bit POP instruction (except for POPA) increments SP by ___________.
22. What values appear in SP and SS if the stack is addressed at memory location 02200H?
23. Compare the operation of the MOV DI,NUMB instruction with the LEA DI,NUMB instruction.
24. What is the difference between the LEA SI,NUMB instruction and the MOV SI,OFFSET NUMB instruction?
25. Which is more efficient, a MOV with an OFFSET or an LEA instruction?
26. Describe how the LDS BX,NUMB instruction operates.
27. What is the difference between the LDS and LSS instructions?
28. Develop a sequence of instructions that move the contents of data segment memory locations NUMB and NUMB+1 into BX, DX, and SI.
29. What is the purpose of the direction flag?
30. Which instructions set and clear the direction flag?
31. Which string instruction(s) use both DI and SI to address memory data?
32. Explain the operation of the LODSB instruction.
33. Explain the operation of the STOSW instruction.
34. Explain the operation of the OUTSB instruction.
35. What does the REP prefix accomplish, and what type of instruction is it used with?
36. Develop a sequence of instructions that copy 12 bytes of data from an area of memory addressed by SOURCE into an area of memory addressed by DEST.
37. Where is the I/O address (port number) stored for an INSB instruction?
38. Select an assembly language instruction that exchanges the contents of the EBX register with the ESI register.
39. Would the LAHF and SAHF instructions normally appear in software?
40. Explain how the XLAT instruction transforms the contents of the AL register.
41. Write a short program that uses the XLAT instruction to convert the BCD numbers 0–9 into ASCII-coded numbers 30H–39H. Store the ASCII-coded data in a TABLE located within the data segment.
42. Explain what the IN AL,12H instruction accomplishes.
43. Explain how the OUT DX,AX instruction operates.
44. What is a segment override prefix?
45. Select an instruction that moves a byte of data from the memory location addressed by the BX register in the extra segment into the AH register.
46. Develop a sequence of instructions that exchange the contents of AX with BX, ECX with EDX, and SI with DI.
47. What is accomplished by the CMOVNE CX,DX instruction in the Pentium 4 microprocessor?
48. What is an assembly language directive?
49. Describe the purpose of the following assembly language directives: DB, DW, and DD.
50. Select an assembly language directive that reserves 30 bytes of memory for array LIST1.

51. Describe the purpose of the EQU directive.
52. What is the purpose of the .686 directive?
53. What is the purpose of the .MODEL directive?
54. If the start of a segment is identified with .DATA, what type of memory organization is in effect?
55. If the SEGMENT directive identifies the start of a segment, what type of memory organization is in effect?
56. What does the INT 21H accomplish if AH contains 4CH?
57. What directives indicate the start and end of a procedure?
58. Explain the purpose of the USES statement as it applies to a procedure with version 6.x of MASM.
59. How is the Pentium 4 microprocessor instructed to use the 16-bit instruction mode?
60. Develop a near procedure that stores AL in four consecutive memory locations within the data segment, as addressed by the DI register.
61. Develop a far procedure that copies contents of the word-sized memory location CS:DATA4 into AX, BX, CX, DX, and SI.

CHAPTER 5

Arithmetic and Logic Instructions

INTRODUCTION

In this chapter, we examine the arithmetic and logic instructions. The arithmetic instructions include addition, subtraction, multiplication, division, comparison, negation, increment, and decrement. The logic instructions include AND, OR, Exclusive-OR, NOT, shifts, rotates, and the logical compare (TEST). The chapter also presents the 80386 through the Pentium 4 instructions XADD, SHRD, SHLD, bit tests, and bit scans. The chapter concludes with a discussion of string comparison instructions, which are used for scanning tabular data and for comparing sections of memory data. Both tasks are performed efficiently with the string scan (SCAS) and string compare (CMPS) instructions.

If you are familiar with an 8-bit microprocessor, you will recognize that the 8086 through the Pentium 4 instruction set is superior to most 8-bit microprocessors because most of the instructions have two operands instead of one. Even if this is your first microprocessor, you will quickly learn that this microprocessor possesses a powerful and easy-to-use set of arithmetic and logic instructions.

CHAPTER OBJECTIVES

Upon completion of this chapter, you will be able to:

1. Use arithmetic and logic instructions to accomplish simple binary, BCD, and ASCII arithmetic
2. Use AND, OR, and Exclusive-OR to accomplish binary bit manipulation
3. Use the shift and rotate instructions
4. Explain the operation of the 80386 through the Pentium 4 exchange and add, compare and exchange, double-precision shift, bit test, and bit scan instructions
5. Check the contents of a table for a match with the string instructions

5–1 ADDITION, SUBTRACTION, AND COMPARISON

The bulk of the arithmetic instructions found in any microprocessor include addition, subtraction, and comparison. In this section, addition, subtraction, and comparison instructions are illustrated. Also shown are their uses in manipulating register and memory data.

Addition

Addition (ADD) appears in many forms in the microprocessor. This section details the use of the ADD instruction for 8-, 16-, and 32-bit binary addition. A second form of addition, called **add-with-carry**, is introduced with the ADC instruction. Finally, the increment instruction (INC) is presented. Increment is a special type of addition that adds 1 to a number. In Section 5–3, other forms of addition are examined, such as BCD and ASCII. Also described is the XADD instruction, found in the 80486 through the Pentium 4.

Table 5–1 illustrates the addressing modes available to the ADD instruction. (These addressing modes include almost all those mentioned in Chapter 3.) However, because there are more than 32,000 variations of the ADD instruction in the instruction set, it is impossible to list them all in this table. The only types of addition *not* allowed are memory-to-memory and segment register. The segment registers can only be moved, pushed, or popped. Note that, as with all other instructions, the 32-bit registers are available only with the 80386 through the Pentium 4.

Register Addition. Example 5–1 shows a simple sequence of instructions that uses register addition to add the contents of several registers. In this example, the contents of AX, BX, CX, and DX are added to form a 16-bit result stored in the AX register.

TABLE 5–1 Example addition instructions.

Assembly Language	*Operation*
ADD AL,BL	AL = AL + BL
ADD CX,DI	CX = CX + DI
ADD EBP,EAX	EBP = EBP + EAX
ADD CL,44H	CL = CL + 44H
ADD BX,245FH	BX = BX + 245FH
ADD EDX,12345H	EDX = EDX + 12345H
ADD [BX],AL	AL adds to the byte contents of the data segment memory location addressed by BX with the sum stored in the same memory location
ADD CL,[BP]	The byte contents of the stack segment memory location addressed by BP add to CL with the sum stored in CL
ADD AL,[EBX]	The byte contents of the data segment memory location addressed by EBX add to AL with the sum stored in AL
ADD BX,[SI+2]	The word contents of the data segment memory location addressed by SI + 2 add to BX with the sum stored in BX
ADD CL,TEMP	The byte contents of data segment memory location TEMP add to CL with the sum stored in CL
ADD BX,TEMP[DI]	The word contents of the data segment memory location addressed by TEMP + DI add to BX with the sum stored in BX
ADD [BX+DI],DL	DL adds to the byte contents of the data segment memory location addressed by BX + DI with the sum stored in the same memory location
ADD BYTE PTR [DI],3	A 3 adds to the byte contents of the data segment memory location addressed by DI with the sum stored in the same location
ADD BX,[EAX+2*ECX]	The word contents of the data segment memory location addressed by EAX plus 2 times ECX add to BX with the sum stored in BX

EXAMPLE 5–1

```
0000 03 C3          ADD   AX,BX
0002 03 C1          ADD   AX,CX
0004 03 C2          ADD   AX,DX
```

Whenever arithmetic and logic instructions execute, the contents of the flag register change. Note that the contents of the interrupt, trap, and other flags do not change due to arithmetic and logic instructions. Only the flags located in the rightmost eight bits of the flag register and the overflow flag change. These rightmost flags denote the result of the arithmetic or logic operation. Any ADD instruction modifies the contents of the sign, zero, carry, auxiliary carry, parity, and overflow flags. The flag bits never change for most of the data transfer instructions presented in Chapter 4.

Immediate Addition. Immediate addition is employed whenever constant or known data are added. An 8-bit immediate addition appears in Example 5–2. In this example, DL is first loaded with 12H by using an immediate move instruction. Next, 33H is added to the 12H in DL by an immediate addition instruction. After the addition, the sum (45H) moves into register DL and the flags change, as follows:

Z = 0 (result not zero)
C = 0 (no carry)
A = 0 (no half-carry)
S = 0 (result positive)
P = 0 (odd parity)
O = 0 (no overflow)

EXAMPLE 5–2

```
0000 B2 12          MOV   DL,12H
0002 80 C2 33       ADD   DL,33H
```

Memory-to-Register Addition. Suppose that an application requires memory data to be added to the AL register. Example 5–3 shows an example that adds two consecutive bytes of data, stored at the data segment offset locations NUMB and NUMB+1, to the AL register.

EXAMPLE 5–3

```
0000 BF 0000 R      MOV   DI,OFFSET NUMB       ;address NUMB
0003 B0 00          MOV   AL,0                 ;clear sum
0005 02 05          ADD   AL,[DI]              ;add NUMB
0007 02 45 01       ADD   AL,[DI+1]            ;add NUMB+1
```

The first instruction loads the destination index register (DI) with offset address NUMB. The DI register, used in this example, addresses data in the data segment beginning at memory location NUMB. After clearing the sum to zero, the ADD AL,[DI] instruction adds the contents of memory location NUMB to AL. Finally, the ADD AL,[DI+ I] instruction adds the contents of memory location NUMB plus one byte to the AL register. After both ADD instructions execute, the result appears in the AL register as the sum of the contents of NUMB plus the contents of NUMB+1.

Array Addition. Memory arrays are sequential lists of data. Suppose that an array of data (ARRAY) contains 10 bytes, numbered from element 0 through element 9. Example 5–4 shows how to add the contents of array elements 3, 5, and 7 together.

This example first clears AL to 0, so it can be used to accumulate the sum. Next, register SI is loaded with a 3 to initially address array element 3. The ADD AL,ARRAY[SI] instruction adds the contents of array element 3 to the sum in AL. The instructions that follow add array elements 5 and 7 to the sum in AL, using a 3 in SI plus a displacement of 2 to address element 5, and a displacement of 4 to address element 7.

EXAMPLE 5–4

```
0000 B0 00             MOV  AL,0             ;clear sum
0002 BE 0003           MOV  SI,3             ;address element 3
0005 02 84 0000 R      ADD  AL,ARRAY[SI]     ;add element 3
0009 02 84 0002 R      ADD  AL,ARRAY[SI+2]   ;add element 5
000D 02 84 0004 R      ADD  AL,ARRAY[SI+4]   ;add element 7
```

Suppose that an array of data contains 16-bit numbers used to form a 16-bit sum in register AX. Example 5–5 shows a sequence of instructions written for the 80386 and above, showing the scaled-index form of addressing to add elements 3, 5, and 7 of an area of memory called ARRAY. In this example, EBX is loaded with the address ARRAY, and ECX holds the array element number. Note how the scaling factor is used to multiply the contents of the ECX register by 2 to address words of data. (Recall that words are two bytes long.)

EXAMPLE 5–5

```
0000 66|BB 00000000 R   MOV  EBX,OFFSET ARRAY    ;address ARRAY
0006 66|B9 00000003     MOV  ECX,3               ;address element 3
000C 67&8B 04 4B        MOV  AX,[EBX+2*ECX]      ;get element 3
0010 66|B9 00000005     MOV  ECX,5               ;address element 5
0016 67&03 04 4B        ADD  AX,[EBX+2*ECX]      ;add element 5
001A 66|B0 00000007     MOV  ECX,7               ;address element 7
0020 67&03 04 4B        ADD  AX,[EBX+2*ECX]      ;add element 7
```

Increment Addition. Increment addition (INC) adds 1 to a register or a memory location. The INC instruction adds 1 to any register or memory location except a segment register. Table 5–2 illustrates some of the possible forms of the increment instruction available to the 8086–Pentium 4 processors. As with other instructions presented thus far, it is impossible to show all variations of the INC instruction because of the large number available.

With indirect memory increments, the size of the data must be described by using the BYTE PTR, WORD PTR, or DWORD PTR directives. The reason is that the assembler program

TABLE 5–2 Example increment instructions.

Assembly Language	*Operation*
INC BL	BL = BL + 1
INC SP	SP = SP + 1
INC EAX	EAX = EAX + 1
INC BYTE PTR[BX]	Adds 1 to the byte contents of the data segment memory location addressed by BX
INC WORD PTR[SI]	Adds 1 to the word contents of the data segment memory location addressed by SI
INC DWORD PTR[ECX]	Adds 1 to the doubleword contents of the data segment memory location addressed by ECX
INC DATA1	Adds 1 to the contents of data segment memory location DATA1

cannot determine if, for example, the INC [DI] instruction is a byte-, word-, or doubleword-sized increment. The INC BYTE PTR [DI] instruction clearly indicates byte-sized memory data; the INC WORD PTR [DI] instruction unquestionably indicates word-sized memory data; and the INC DWORD PTR [DI] instruction indicates doubleword-sized data.

Example 5–6 shows how to modify Example 5–3 to use the increment instruction for addressing NUMB and NUMB+1. Here, an INC DI instruction changes the contents of register DI from offset address NUMB to offset address NUMB+1. Both program sequences shown in Examples 5–3 and 5–6 add the contents of NUMB and NUMB+1. The difference between them is the way that the address is formed through the contents of the DI register using the increment instruction.

EXAMPLE 5–6

```
0000 BF 0000 R       MOV  DI,OFFSET NUMB          ;address NUMB
0003 B0 00           MOV  AL,0                    ;clear sum
0005 02 05           ADD  AL,[DI]                 ;add NUMB
0007 47              INC  DI                      ;increment DI
0008 02 05           ADD  AL,[DI]                 ;add NUMB+1
```

Increment instructions affect the flag bits, as do most other arithmetic and logic operations. The difference is that increment instructions do not affect the carry flag bit. Carry doesn't change because we often use increments in programs that depend upon the contents of the carry flag. Note that increment is used to point to the next memory element in a byte-sized array of data only. If word-sized data are addressed, it is better to use an ADD DI,2 instruction to modify the DI pointer in place of two INC DI instructions. For doubleword arrays, use the ADD DI,4 instruction to modify the DI pointer. In some cases, the carry flag must be preserved, which may mean that two or four INC instructions might appear in a program to modify a pointer.

Addition-with-Carry. An addition-with-carry instruction (ADC) adds the bit in the carry flag (C) to the operand data. This instruction mainly appears in software that adds numbers that are wider than 16 bits in the 8086–80286 or wider than 32 bits in the 80386–Pentium 4.

Table 5–3 lists several add-with-carry instructions, with comments that explain their operation. Like the ADD instruction, ADC affects the flags after the addition.

Suppose that a program is written for the 8086–80286 to add the 32-bit number in BX and AX to the 32-bit number in DX and CX. Figure 5–1 illustrates this addition so that the placement and function of the carry flag can be understood. This addition cannot be easily performed without adding the carry flag bit because the 8086–80286 only adds 8- or 16-bit numbers. Example 5–7 shows how the contents of registers AX and CX add to form the least-significant 16 bits of the sum. This addition may or may not generate a carry. A carry appears in the carry

TABLE 5–3 Example add-with-carry instructions.

Assembly Language	*Operation*
ADC AL,AH	AL = AL + AH + carry
ADC CX,BX	CX = CX + BX + carry
ADC EBX,EDX	EBX = EBX + EDX + carry
ADC DH,[BX]	The byte contents of the data segment memory location addressed by BX add to DH with the sum stored in DH
ADC BX,[BP+2]	The word contents of the stack segment memory location addressed by BP plus 2 add to BX with the sum stored in BX
ADC ECX,[EBX]	The doubleword contents of the data segment memory location addressed by EBX add to ECX with the sum stored in ECX

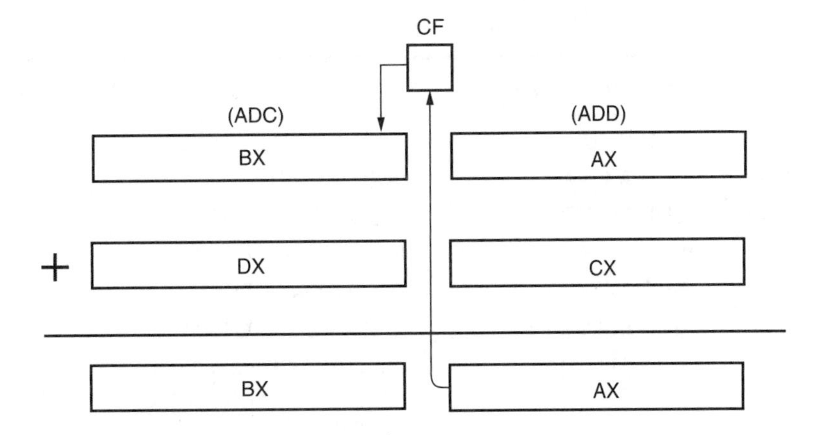

FIGURE 5–1 Addition-with-carry showing how the carry flag (C) links the two 16-bit additions into one 32-bit addition.

flag if the sum is greater than FFFFH. Because it is impossible to predict a carry, the most-significant 16 bits of this addition are added with the carry flag using the ADC instruction. The ADC instruction adds the 1 or the 0 in the carry flag to the most-significant 16 bits of the result. This program adds BX–AX to DX–CX, with the sum appearing in BX–AX.

EXAMPLE 5–7

```
0000 03 C1          ADD  AX,CX
0002 13 DA          ADC  BX,DX
```

Suppose the same software is rewritten for the 80386 through the Pentium 4, but modified to add two 64-bit numbers. The changes required for this operation are the use of the extended registers to hold the data and modifications of the instructions for the 80386 and above. These changes are shown in Example 5–8, which adds two 64-bit numbers.

EXAMPLE 5–8

```
0000 66|03 C1       ADD  EAX,ECX
0003 66|13 DA       ADC  EBX,EDX
```

Exchange and Add for the 80486–Pentium 4 Processors. A new type of addition called **exchange and add** (XADD) appears in the 80486 instruction set and continues through the Pentium 4. The XADD instruction adds the source to the destination and stores the sum in the destination, as with any addition. The difference is that after the addition takes place, the original value of the destination is copied into the source operand. This is one of the few instructions that change the source.

For example, if BL = 12H and DL = 02H, and the XADD BL,DL instruction executes, the BL register contains the sum of 14H and DL becomes 12H. The sum of 14H is generated and the original destination of 12H replaces the source. This instruction functions with any register size and any memory operand, just as with the ADD instruction.

Subtraction

Many forms of subtraction (SUB) appear in the instruction set. These forms use any addressing mode with 8-, 16-, or 32-bit data. A special form of subtraction (decrement, or DEC) subtracts 1 from any register or memory location. Section 5–3 shows how BCD and ASCII data subtract. As with addition, numbers that are wider than 16 bits or 32 bits must occasionally be subtracted. The **subtract-with-borrow** instruction (SBB) performs this type of subtraction. In the 80486 through the Pentium 4 processors, the instruction set also includes a compare and exchange instruction.

TABLE 5–4 Example subtraction instructions.

Assembly Language	*Operation*
SUB CL,BL	CL = CL – BL
SUB AX,SP	AX = AX – SP
SUB ECX,EBP	ECX = ECX – EBP
SUB DH,6FH	DH = DH – 6FH
SUB AX,0CCCCH	AX = AX – 0CCCCH
SUB ESI,2000300H	ESI = ESI – 2000300H
SUB [DI],CH	Subtracts CH from the byte contents of the data segment memory addressed by DI and stores the difference in the same memory location
SUB CH,[BP]	Subtracts the byte contents of the stack segment memory location addressed by BP from CH and stores the difference in CH
SUB AH,TEMP	Subtracts the byte contents of memory location TEMP from AH and stores the difference in AH
SUB DI,TEMP[ESI]	Subtracts the word contents of the data segment memory location addressed by TEMP plus ESI from DI and stores the difference in DI
SUB ECX,DATA1	Subtracts the doubleword contents of memory location DATA1 from ECX and stores the difference in ECX

Table 5–4 lists some of the many addressing modes allowed with the subtract instruction (SUB). There are well over 1000 possible subtraction instructions, far too many to list here. About the only types of subtraction not allowed are memory-to-memory and segment register subtractions. Like other arithmetic instructions, the subtract instruction affects the flag bits.

Register Subtraction. Example 5–9 shows a sequence of instructions that perform register subtraction. This example subtracts the 16-bit contents of registers CX and DX from the contents of register BX. After each subtraction, the microprocessor modifies the contents of the flag register. The flags change for most arithmetic and logic operations.

EXAMPLE 5–9

```
0000 2B D9          SUB  BX,CX
0002 2B DA          SUB  BX,DX
```

Immediate Subtraction. As with addition, the microprocessor also allows immediate operands for the subtraction of constant data. Example 5–10 presents a short sequence of instructions that subtract 44H from 22H. Here, we first load the 22H into CH using an immediate move instruction. Next, the SUB instruction, using immediate data 44H, subtracts 44H from the 22H. After the subtraction, the difference (0DEH) moves into the CH register. The flags change as follows for this subtraction:

Z = 0 (result not zero)
C = 1 (borrow)
A = 1 (half-borrow)
S = 1 (result negative)
P = 1 (even parity)
O = 0 (no overflow)

EXAMPLE 5–10

```
0000 B5 22          MOV  CH,22H
0002 80 ED 44       SUB  CH,44H
```

Both carry flags (C and A) hold borrows after a subtraction instead of carries, as after an addition. Notice in this example that there is no overflow. This example subtracted 44H (+68) from 22H (+34), resulting in 0DEH (–34). Because the correct 8-bit signed result is –34, there is no overflow in this example. An 8-bit overflow occurs only if the signed result is greater than +127 or less than –128.

Decrement Subtraction. Decrement subtraction (DEC) subtracts 1 from a register or the contents of a memory location. Table 5–5 lists some decrement instructions that illustrate register and memory decrements.

The decrement indirect memory data instructions require BYTE PTR, WORD PTR, or DWORD PTR because the assembler cannot distinguish a byte from a word or doubleword when an index register addresses memory. For example, DEC [SI] is vague because the assembler cannot determine whether the location addressed by SI is a byte, word, or doubleword. Using DEC BYTE PTR [SI], DEC WORD PTR [DI], or DEC DWORD PTR [SI] reveals the size of the data to the assembler.

Subtraction-with-Borrow. A subtraction-with-borrow (SBB) instruction functions as a regular subtraction, except that the carry flag (C), which holds the borrow, also subtracts from the difference. The most common use for this instruction is for subtractions that are wider than 16 bits in the 8086–80286 microprocessors or wider than 32 bits in the 80386–Pentium 4. Wide subtractions require that borrows propagate through the subtraction, just as wide additions propagate the carry.

Table 5–6 lists several SBB instructions with comments that define their operations. Like the SUB instruction, SBB affects the flags. Notice that the immediate subtract from memory instruction in this table requires a BYTE PTR, WORD PTR, or DWORD PTR directive.

When the 32-bit number held in BX and AX is subtracted from the 32-bit number held in SI and DI, the carry flag propagates the borrow between the two 16-bit subtractions. The carry flag holds the borrow for subtraction. Figure 5–2 shows how the borrow propagates through the carry flag (C) for this task. Example 5–11 shows how this subtraction is performed by a program. With wide subtraction, the least-significant 16- or 32-bit data are subtracted with the SUB instruction. All subsequent and more-significant data are subtracted by using the SBB instruction.

TABLE 5–5 Example decrement instructions.

Assembly Language	*Operation*
DEC BH	BH = BH – 1
DEC CX	CX = CX – 1
DEC EDX	EDX = EDX – 1
DEC BYTE PTR[DI]	Subtracts 1 from the byte contents of the data segment memory location addressed by DI
DEC WORD PTR[BP]	Subtracts 1 from the word contents of the stack segment memory location addressed by BP
DEC DWORD PTR[EBX]	Subtracts 1 from the doubleword contents of the data segment memory location addressed by EBX
DEC NUMB	Subtracts 1 from the contents of data segment memory location NUMB

TABLE 5–6 Example subtraction-with-borrow instructions.

Assembly Language	*Operation*
SBB AH,AL	AH = AH – AL – carry
SBB AX,BX	AX = AX – BX – carry
SBB EAX,ECX	EAX = EAX – ECX – carry
SBB CL,2	CL = CL – 2 – carry
SBB BYTE PTR[DI],3	Both 3 and carry subtract from the data segment memory location addressed by DI
SBB [DI],AL	Both AL and carry subtract from the data segment memory location addressed by DI
SBB DI,[BP+2]	Both carry and the word contents of the stack segment memory location addressed by BP plus 2 subtract from DI
SBB AL,[EBX+ECX]	Both carry and the byte contents of the data segment memory location addressed by EBX plus ECX subtract from AL

FIGURE 5–2 Subtraction-with-borrow showing how the carry flag (C) propagates the borrow.

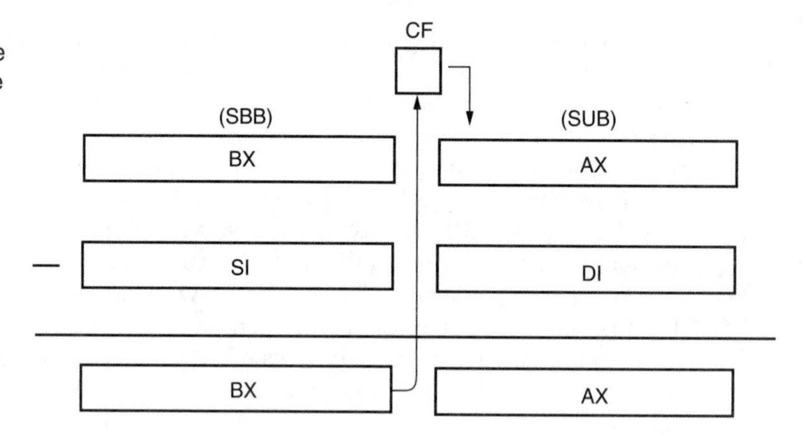

The example uses the SUB instruction to subtract DI from AX, then uses SBB to subtract-with-borrow SI from BX.

EXAMPLE 5–11

```
0000 2B C7          SUB  AX,DI
0002 1B DE          SBB  BX,SI
```

Comparison

The comparison instruction (CMP) is a subtraction that changes only the flag bits; the destination operand never changes. A comparison is useful for checking the entire contents of a register or a memory location against another value. A CMP is normally followed by a conditional jump instruction, which tests the condition of the flag bits.

Table 5–7 lists a variety of comparison instructions that use the same addressing modes as the addition and subtraction instructions already presented. Similarly, the only disallowed forms of compare are memory-to-memory and segment register compares.

Example 5–12 shows a comparison followed by a conditional jump instruction. In this example, the contents of AL are compared with 10H. Conditional jump instructions that often

TABLE 5–7 Example comparison instructions.

Assembly Language	*Operation*
CMP CL,BL	CL – BL
CMP AX,SP	AX – SP
CMP EBP,ESI	EBP – ESI
CMP AX,2000H	AX – 2000H
CMP [DI],CH	CH subtracts from the byte contents of the data segment memory location addressed by DI
CMP CL,[BP]	The byte contents of the stack segment memory location addressed by BP subtracts from CL
CMP AH,TEMP	The byte contents of data segment memory location TEMP subtracts from AH
CMP DI,TEMP[BX]	The word contents of the data segment memory location addressed by TEMP plus BX subtracts from DI
CMP AL,[EDI+ESI]	The byte contents of the data segment memory location addressed by EDI plus ESI subtracts from AL

follow the comparison are JA (**jump above**) or JB (**jump below**). If the JA follows the comparison, the jump occurs if the value in AL is above 10H. If the JB follows the comparison, the jump occurs if the value in AL is below 10H. In this example, the JAE instruction follows the comparison. This instruction causes the program to continue at memory location SUBER if the value in AL is 10H or above. There is also a JBE (**jump below or equal**) instruction that could follow the comparison to jump if the outcome is below or equal to 10H. Later chapters provide additional detail on the comparison and conditional jump instructions.

EXAMPLE 5–12

```
0000 3C 10          CMP  AL,10H          ;compare AL against 10H
0002 73 1C          JAE  SUBER           ;if AL is 10H or above
```

Compare and Exchange (80486–Pentium 4 Processors Only). The compare and exchange instruction (CMPXCHG), found only in the 80486 through the Pentium 4 instruction sets, compares the destination operand with the accumulator. If they are equal, the source operand is copied into the destination; if they are not equal, the destination operand is copied into the accumulator. This instruction functions with 8-, 16-, or 32-bit data.

The CMPXCHG CX,DX instruction is an example of the compare and exchange instruction. This instruction first compares the contents of CX with AX. If CX equals AX, DX is copied into AX; if CX is not equal to AX, CX is copied into AX. This instruction also compares AL with 8-bit data and EAX with 32-bit data if the operands are either 8- or 32-bit data.

In the Pentium–Pentium 4 processors, a CMPXCHG8B instruction is available that compares two quadwords. This is the only new data manipulation instruction provided in the Pentium–Pentium 4 when they are compared with prior versions of the microprocessor. The compare-and-exchange-8-bytes instruction compares the 64-bit value located in EDX:EAX with a 64-bit number located in memory. An example is CMPXCHG8B TEMP. If TEMP equals EDX:EAX, TEMP is replaced with the value found in ECX:EBX; if TEMP does not equal EDX:EAX, the number found in TEMP is loaded into EDX:EAX. The Z (zero) flag bit indicates that the values are equal after the comparison.

This instruction has a bug that will cause the operating system to crash. More information about this flaw can be obtained at www.intel.com.

5–2 MULTIPLICATION AND DIVISION

Only modern microprocessors contain multiplication and division instructions. Earlier 8-bit microprocessors could not multiply or divide without the use of a program that multiplied or divided by using a series of shifts and additions or subtractions. Because microprocessor manufacturers were aware of this inadequacy, they incorporated multiplication and division instructions into the instruction sets of the newer microprocessors. The Pentium–Pentium 4 processors contain special circuitry that performs a multiplication in as little as one clocking period, whereas it took over 40 clocking periods to perform the same multiplication in earlier Intel microprocessors.

Multiplication

Multiplication is performed on bytes, words, or doublewords, and can be signed integer (IMUL) or unsigned integer (MUL). Note that only the 80386 through the Pentium 4 processors multiply 32-bit doublewords. The product after a multiplication is always a double-width product. If two 8-bit numbers are multiplied, they generate a 16-bit product; if two 16-bit numbers are multiplied, they generate a 32-bit product; and if two 32-bit numbers are multiplied, a 64-bit product is generated.

Some flag bits (O [overflow] and C [carry]) change when the multiply instruction executes and produce predictable outcomes. The other flags also change, but their results are unpredictable and therefore are unused. In an 8-bit multiplication, if the most-significant 8 bits of the result are zero, both C and O flag bits equal zero. These flag bits show that the result is 8-bits wide (C = 0) or 16-bits wide (C = 1). In a 16-bit multiplication, if the most-significant 16-bits part of the product is 0, both C and O clear to zero. In a 32-bit multiplication, both C and O indicate that the most-significant 32 bits of the product are zero.

8-Bit Multiplication. With 8-bit multiplication, the multiplicand is always in the AL register, whether signed or unsigned. The multiplier can be any 8-bit register or any memory location. Immediate multiplication is not allowed unless the special signed immediate multiplication instruction, discussed later in this section, appears in a program. The multiplication instruction contains one operand because it always multiplies the operand times the contents of register AL. An example is the MUL BL instruction, which multiplies the unsigned contents of AL by the unsigned contents of BL. After the multiplication, the unsigned product is placed in AX—a double-width product. Table 5–8 illustrates some 8-bit multiplication instructions.

Suppose that BL and CL each contain two 8-bit unsigned numbers, and these numbers must be multiplied to form a 16-bit product stored in DX. This procedure cannot be accomplished by a single instruction because we can only multiply a number times the AL register for an 8-bit multiplication. Example 5–13 shows a short program that generates DX = BL × CL. This

TABLE 5–8 Example 8-bit multiplication instructions.

Assembly Language	*Operation*
MUL CL	AL is multiplied by CL; the unsigned product is in AX
IMUL DH	AL is multiplied by DH; the signed product is in AX
IMUL BYTE PTR[BX]	AL is multiplied by the byte contents of the data segment memory location addressed by BX; the signed product is in AX
MUL TEMP	AL is multiplied by the byte contents of data segment memory location TEMP; the unsigned product is in AX

example loads register BL and CL with example data 5 and 10. The product, 50, moves into DX from AX after the multiplication by using the MOV DX,AX instruction.

EXAMPLE 5–13

```
0000 B3 05          MOV  BL,5          ;load data
0002 B1 0A          MOV  CL,10
0004 8A C1          MOV  AL,CL         ;position data
0006 F6 E3          MUL  BL            ;multiply
0008 8B D0          MOV  DX,AX         ;position product
```

For signed multiplication, the product is in binary form, if positive, and in two's complement form, if negative. These are the same forms used to store all positive and negative signed numbers used by the microprocessor. If the program of Example 5–13 multiplies two signed numbers, only the MUL instruction is changed to IMUL.

16-Bit Multiplication. Word multiplication is very similar to byte multiplication. The difference is that AX contains the multiplicand instead of AL, and the 32-bit product appears in DX–AX instead of AX. The DX register always contains the most-significant 16 bits of the product, and AX contains the least-significant 16 bits. As with 8-bit multiplication, the choice of the multiplier is up to the programmer. Table 5–9 shows several different 16-bit multiplication instructions.

A Special Immediate 16-Bit Multiplication. The 8086/8088 microprocessors could not perform immediate multiplication; the 80186 through the Pentium 4 processors can do so by using a special version of the multiply instruction. Immediate multiplication must be signed multiplication, and the instruction format is different because it contains three operands. The first operand is the 16-bit destination register; the second operand is a register or memory location that contains the 16-bit multiplicand; and the third operand is either 8-bit or 16-bit immediate data used as the multiplier.

The IMUL CX,DX,12H instruction multiplies 12H times DX and leaves a 16-bit signed product in CX. If the immediate data are 8 bits, they sign-extend into a 16-bit number before the multiplication occurs. Another example is IMUL BX,NUMBER,1000H, which multiplies NUMBER times 1000H and leaves the product in BX. Both the destination and multiplicand must be 16-bit numbers. Although this is immediate multiplication, the restrictions placed on it limit its use especially the fact that it is a signed multiplication and the product is 16 bits wide.

32-Bit Multiplication. In the 80386 and above, 32-bit multiplication is allowed because these microprocessors contain 32-bit registers. As with 8- and 16-bit multiplication, 32-bit multiplication can be signed or unsigned by using the IMUL and MUL instructions. With 32-bit multiplication, the contents of EAX are multiplied by the operand specified with the instruction. The product (64 bits wide) is found in EDX–EAX, where EAX contains the least-significant 32 bits of the product. Table 5–10 lists some of the 32-bit multiplication instructions found in the 80386 and above instruction set.

TABLE 5–9 Example 16-bit multiplication instructions.

Assembly Language	*Operation*
MUL CX	AX is multiplied by CX; the unsigned product is in DX–AX
IMUL DI	AX is multiplied by DI; the signed product is in DX–AX
MUL WORD PTR[SI]	AX is multiplied by the word contents of the data segment memory location addressed by SI; the unsigned product is in DX–AX

TABLE 5–10 Example 32-bit multiplication instructions.

Assembly Language	*Operation*
MUL ECX	EAX is multiplied by ECX; the unsigned product is in EDX–EAX
IMUL EDI	EAX is multiplied by EDI; the signed product is in EDX–EAX
MUL DWORD PTR[ESI]	EAX is multiplied by the doubleword contents of the data segment memory location address by ESI; the unsigned product is in EDX–EAX

Division

As with multiplication, division occurs on 8- or 16-bit numbers in the 8086–80286 microprocessors, and on 32-bit numbers in the 80386–Pentium 4 microprocessors. These numbers are signed (IDIV) or unsigned (DIV) integers. The dividend is always a double-width dividend that is divided by the operand. This means that an 8-bit division divides a 16-bit number by an 8-bit number; a 16-bit division divides a 32-bit number by a 16-bit number; and a 32-bit division divides a 64-bit number by a 32-bit number. There is no immediate division instruction available to any microprocessor.

None of the flag bits change predictably for a division. A division can result in two different types of errors; one is an attempt to divide by zero and the other is a divide overflow. A divide overflow occurs when a small number divides into a large number. For example, suppose that AX = 3000 and that it is divided by 2. Because the quotient for an 8-bit division appears in AL, the result of 1500 causes a divide overflow because the 1500 does not fit into AL. In either case, the microprocessor generates an interrupt if a divide error occurs. In most systems, a divide error interrupt displays an error message on the video screen. The divide error interrupt and all other interrupts for the microprocessor are explained in Chapter 6.

8-Bit Division. Eight-bit division uses the AX register to store the dividend that is divided by the contents of any 8-bit register or memory location. The quotient moves into AL after the division with AH containing a whole number remainder. For a signed division, the quotient is positive or negative; the remainder always assumes the sign of the dividend and is always an integer. For example, if AX = 0010H (+16) and BL = 0FDH (–3) and the IDIV BL instruction executes, AX = 01FBH. This represents a quotient of –5 (AL) with a remainder of 1 (AH). If, on the other hand, –16 is divided by +3, the result will be a quotient of –5 (AL) with a remainder of –1 (AH). Table 5–11 lists some of the 8-bit division instructions.

With 8-bit division, the numbers are usually 8 bits wide. This means that one of them, the dividend, must be converted to a 16-bit-wide number in AX. This is accomplished differently for signed and unsigned numbers. For the unsigned number, the most-significant 8 bits must be cleared to zero (**zero-extended**). The MOVZX instruction described in Chapter 4 can be used to

TABLE 5–11 Example 8-bit division instructions.

Assembly Language	*Operation*
DIV CL	AX is divided by CL; the unsigned quotient is in AL and the unsigned remainder is in AH
IDIV BL	AX is divided by BL; the signed quotient is in AL and the signed remainder is in AH
DIV BYTE PTR[BP]	AX is divided by the byte contents of the stack segment memory location addressed by BP; the unsigned quotient is in AL and the unsigned remainder is in AH

zero-extend a number in the 80386 through the Pentium 4 processors. For signed numbers, the least-significant 8 bits are sign-extended into the most-significant 8 bits. In the microprocessor, a special instruction sign-extends AL into AH, or converts an 8-bit signed number in AL into a 16-bit signed number in AX. The CBW (**convert byte to word**) instruction performs this conversion. In the 80386 through the Pentium 4, a MOVSX instruction (see Chapter 4) sign-extends a number.

EXAMPLE 5–14

```
0000 A0 0000 R          MOV  AL,NUMB          ;get NUMB
0003 B4 00              MOV  AH,0             ;zero-extend
0005 F6 36 0002 R       DIV  NUMB1            ;divide by NUMB1
0009 A2 0003 R          MOV  ANSQ,AL          ;save quotient
000C 88 26 0004 R       MOV  ANSR,AH          ;save remainder
```

Example 5–14 illustrates a short program that divides the unsigned byte contents of memory location NUMB by the unsigned contents of memory location NUMB1. Here, the quotient is stored in location ANSQ and the remainder is stored in location ANSR. Notice how the contents of location NUMB are retrieved from memory and then zero-extended to form a 16-bit unsigned number for the dividend.

16-Bit Division. Sixteen-bit division is similar to 8-bit division, except that instead of dividing into AX, the 16-bit number is divided into DX–AX, a 32-bit dividend. The quotient appears in AX and the remainder appears in DX after a 16-bit division. Table 5–12 lists some of the 16-bit division instructions.

As with 8-bit division, numbers must often be converted to the proper form for the dividend. If a 16-bit unsigned number is placed in AX, DX must be cleared to zero. In the 80386 and above, the number is zero-extended by using the MOVZX instruction. If AX is a 16-bit signed number, the CWD (**convert word to doubleword**) instruction sign-extends it into a signed 32-bit number. If the 80386 and above is available, the MOVSX instruction can also be used to sign-extend a number.

EXAMPLE 5–15

```
0000 B8 FF9C        MOV  AX,-100        ;load a -100
0003 B9 0009        MOV  CX,9           ;load +9
0006 99             CWD                 ;sign-extend
0007 F7 F9          IDIV  CX
```

Example 5–15 shows the division of two 16-bit signed numbers. Here, –100 in AX is divided by +9 in CX. The CWD instruction converts the –100 in AX to –100 in DX–AX before the division. After the division, the results appear in DX–AX as a quotient of –11 in AX and a remainder of –1 in DX.

32-Bit Division. The 80386 through the Pentium 4 processors perform 32-bit division on signed or unsigned numbers. The 64-bit contents of EDX–EAX are divided by the operand

TABLE 5–12 Example 16-bit division instructions.

Assembly Language	*Operation*
DIV CX	DX–AX is divided by CX; the unsigned quotient is in AX and the unsigned remainder is in DX
IDIV SI	DX–AX is divided by SI; the signed quotient is in AX and the signed remainder is in DX
DIV NUMB	DX–AX is divided by the word contents of data segment memory NUMB; the unsigned quotient is in AX and the unsigned remainder is in DX

TABLE 5–13 Example 32-bit division instructions.

Assembly Language	*Operation*
DIV ECX	EDX–EAX is divided by ECX; the unsigned quotient is in EAX and the unsigned remainder is in EDX
IDIV DATA4	EDX–EAX is divided by the doubleword contents in data segment memory location DATA4; the signed quotient is in EAX and the signed remainder is in EDX
DIV DWORD PTR[EDI]	EDX–EAX is divided by the doubleword contents of the data segment memory location addressed by EDI; the unsigned quotient is in EAX and the unsigned remainder is in EDX

specified by the instruction, leaving a 32-bit quotient in EAX and a 32-bit remainder in EDX. Other than the size of the registers, this instruction functions in the same manner as the 8- and 16-bit divisions. Table 5–13 shows some 32-bit division instructions. The CDQ (**convert doubleword to quadword**) instruction is used before a signed division to convert the 32-bit contents of EAX into a 64-bit signed number in EDX–EAX.

The Remainder. What is done with the remainder after a division? There are a few possible choices. The remainder could be used to round the quotient or just dropped to truncate the quotient. If the division is unsigned, rounding requires that the remainder be compared with half the divisor to decide whether to round up the quotient. The remainder could also be converted to a fractional remainder.

EXAMPLE 5–16

```
0000 F6 F3              DIV  BL          ;divide
0002 02 E4              ADD  AH,AH       ;double remainder
0004 3A E3              CMP  AH,BL       ;test for rounding
0006 72 02              JB   NEXT        ;if OK
0008 FE C0              INC  AL          ;round
000A              NEXT:
```

Example 5–16 shows a sequence of instructions that divide AX by BL and round the unsigned result. This program doubles the remainder before comparing it with BL to decide whether to round the quotient. Here, an INC instruction rounds the contents of AL after the comparison.

Suppose that a fractional remainder is required instead of an integer remainder. A fractional remainder is obtained by saving the quotient. Next, the AL register is cleared to zero. The number remaining in AX is now divided by the original operand to generate a fractional remainder.

EXAMPLE 5–17

```
0000 B8 000D            MOV  AX,13       ;load 13
0003 B3 02              MOV  BL,2        ;load 2
0005 F6 F3              DIV  BL          ;13/2
0007 A2 0003 R          MOV  ANSQ,AL     ;save quotient
000A B0 00              MOV  AL,0        ;clear AL
000C F6 F3              DIV  BL          ;generate remainder
000E A2 0004 R          MOV  ANSR,AL     ;save remainder
```

Example 5–17 shows how 13 is divided by 2. The 8-bit quotient is saved in memory location ANSQ, and then AL is cleared. Next, the contents of AX are again divided by 2 to generate a fractional remainder. After the division, the AL register equals 80H. This is 10000000_2. If the binary point (radix) is placed before the leftmost bit of AL, the fractional remainder in AL is 0.10000000_2 or 0.5 decimal. The remainder is saved in memory location ANSR in this example.

5–3 BCD AND ASCII ARITHMETIC

The microprocessor allows arithmetic manipulation of both BCD (**binary-coded decimal**) and ASCII (**American Standard Code for Information Interchange**) data. This is accomplished by instructions that adjust the numbers for BCD and ASCII arithmetic.

The BCD operations occur in systems such as point-of-sale terminals (e.g., cash registers) and others that seldom require complex arithmetic. The ASCII operations are performed on ASCII data used by many programs. In many cases, BCD or ASCII arithmetic is rarely used today, but some of the operations can be used for other purposes.

BCD Arithmetic

Two arithmetic techniques operate with BCD data: addition and subtraction. The instruction set provides two instructions that correct the result of a BCD addition and a BCD subtraction. The DAA (**decimal adjust after addition**) instruction follows BCD addition, and the DAS (**decimal adjust after subtraction**) follows BCD subtraction. Both instructions correct the result of the addition or subtraction so that it is a BCD number.

For BCD data, the numbers always appear in the packed BCD form and are stored as two BCD digits per byte. The adjustment instructions function only with the AL register after BCD addition and subtraction.

DAA Instruction. The DAA instruction follows the ADD or ADC instruction to adjust the result into a BCD result. Suppose that DX and BX each contain four-digit packed BCD numbers. Example 5–18 provides a short sample program that adds the BCD numbers in DX and BX, and stores the result in CX.

EXAMPLE 5–18

```
0000 BA 1234        MOV  DX,1234H      ;load 1234 BCD
0003 BB 3099        MOV  BX,3099H      ;load 3099 BCD
0006 8A C3          MOV  AL,BL         ;sum BL and DL
0008 02 C2          ADD  AL,DL
000A 27             DAA
000B 8A C8          MOV  CL,AL         ;answer to CL
000D 9A C7          MOV  AL,BH         ;sum BH, DH an carry
000F 12 C6          ADC  AL,DH
0011 27             DAA
0012 8A E8          MOV  CH,AL         ;answer to CH
```

Because the DAA instruction functions only with the AL register, this addition must occur eight bits at a time. After adding the BL and DL registers, the result is adjusted with a DAA instruction before being stored in CL. Next, add BH and DH registers with carry; the result is then adjusted with DAA before being stored in CH. In this example, 1234 is added to 3099 to generate the sum 4333, which moves into CX after the addition. Note that 1234 BCD is the same as 1234H.

DAS Instruction. The DAS instruction functions as does the DAA instruction, except that it follows a subtraction instead of an addition. Example 5–19 is the same as Example 5–18, except that it subtracts instead of adds DX and BX. The main difference in these programs is that the DAA instructions change to DAS, and the ADD and ADC instructions change to SUB and SBB instructions.

EXAMPLE 5–19

```
0000 BA 1234        MOV  DX,1234H      ;load 1234 BCD
0003 BB 3099        MOV  BX,3099H      ;load 3099 BCD
0006 8A C3          MOV  AL,BL         ;subtract DL from BL
0008 2A C2          SUB  AL,DL
```

```
000A 2F              DAS
000B 8A C8           MOV  CL,AL       ;answer to CL
000D 9A C7           MOV  AL,BH       ;subtract DH
000F 1A C6           SBB  AL,DH
0011 2F              DAS
0012 8A E8           MOV  CH,AL       ;answer to CH
```

ASCII Arithmetic

The ASCII arithmetic instructions function with ASCII-coded numbers. These numbers range in value from 30H to 39H for the numbers 0–9. Four instructions are used with ASCII arithmetic operations: AAA (**ASCII adjust after addition**), AAD (**ASCII adjust before division**), AAM (**ASCII adjust after multiplication**), and AAS (**ASCII adjust after subtraction**). These instructions use register AX as the source and as the destination.

AAA Instruction. The addition of two one-digit ASCII-coded numbers will not result in any useful data. For example, if 31H and 39H are added, the result is 6AH. This ASCII addition (1 + 9) should produce a two-digit ASCII result equivalent to a 10 decimal, which is a 31H and a 30H in ASCII code. If the AAA instruction is executed after this addition, the AX register will contain 0100H. Although this is not ASCII code, it can be converted to ASCII code by adding 3030H to AX, which generates 3130H. The AAA instruction clears AH if the result is less than 10 and adds 1 to AH if the result is greater than 10.

EXAMPLE 5–20

```
0000 B8 0031         MOV  AX,31H      ;load ASCII 1
0003 04 39           ADD  AL,39H      ;add ASCII 9
0005 37              AAA              ;adjust sum
0006 05 3030         ADD  AX,3030H    ;answer to ASCII
```

Example 5–20 shows the way ASCII addition functions in the microprocessor. Please note that AH is cleared to zero before the addition by using the MOV AX,31H instruction. The operand of 0031H places 00H in AH and 31H in AL.

AAD Instruction. Unlike all other adjustment instructions, the AAD instruction appears before a division. The AAD instruction requires that the AX register contain a two-digit unpacked BCD number (not ASCII) before executing. After adjusting the AX register with AAD, it is divided by an unpacked BCD number to generate a single-digit result in AL with any remainder in AH.

Example 5–21 illustrates how 72 in unpacked BCD is divided by 9 to produce a quotient of 8. The 0702H loaded into the AX register is adjusted by the AAD instruction to 0048H. Notice that this converts a two-digit unpacked BCD number into a binary number so it can be divided with the binary division instruction (DIV). The AAD instruction converts the unpacked BCD numbers between 00 and 99 into binary.

EXAMPLE 5–21

```
0000 B3 09           MOV  BL,9        ;load divisor
0002 B8 0702         MOV  AX,702H     ;load dividend
0005 D5 0A           AAD              ;adjust
0007 F6 F3           DIV  BL          ;divide
```

AAM Instruction. The AAM instruction follows the multiplication instruction after multiplying two one-digit unpacked BCD numbers. Example 5–22 shows a short program that multiplies 5 times 5. The result after the multiplication is 0019H in the AX register. After adjusting the result with the AAM instruction, AX contains 0205H. This is an unpacked BCD result of 25. If 3030H is added to 0205H, it has an ASCII result of 3235H.

EXAMPLE 5–22

```
0000 B0 05          MOV  AL,5        ;load multiplicand
0002 B1 03          MOV  CL,3        ;load multiplier
0004 F6 E1          MUL  CL
0006 D4 0A          AAM              ;adjust
```

The AAM instruction accomplishes this conversion by dividing AX by 10. The remainder is found in AL, and the quotient is in AH. Note that the second byte of the instruction contains 0AH. If the 0AH is changed to another value, AAM divides by the new value. For example, if the second byte is changed to 0BH, the AAM instruction divides by 11. This is accomplished with DB 0D4H, 0BH in place of AAM, which forces the AMM instruction to multiply by 11.

One side benefit of the AAM instruction is that AAM converts from binary to unpacked BCD. If a binary number between 0000H and 0063H appears in the AX register, the AAM instruction converts it to BCD. For example, if AX contains 0060H before AAM, it will contain 0906H after AAM executes. This is the unpacked BCD equivalent of 96 decimal. If 3030H is added to 0906H, the result changes to ASCII code.

Example 5–23 shows how the 16-bit binary content of AX is converted to a four-digit ASCII character string by using division and the AAM instruction. Note that this works for numbers between 0 and 9999. First DX is cleared and then DX–AX is divided by 100. For example, if AX = 245_{10}, AX = 2 and DX = 45 after the division. These separate halves are converted to BCD using AAM, and then 3030H is added to convert to ASCII code.

EXAMPLE 5–23

```
0000 33 D2          XOR  DX,DX       ;clear DX
0002 B9 0064        MOV  CX,100      ;divide DX-AX by 100
0005 F7 F1          DIV  CX
0007 D4 0A          AAM              ;convert to BCD
0009 05 3030        ADD  AX,3030H    ;convert to ASCII
000C 92             XCHG AX,DX       ;repeat for remainder
000D D4 0A          AAM
000F 05 3030        ADD  AX,3030H
```

Example 5–24 uses the DOS 21H function AH = 02H to display a sample number in decimal on the video display using the AAM instruction. Notice how AAM is used to convert AL into BCD. Next, ADD AX,3030H converts the BCD code in AX into ASCII for display with DOS INT 21H. Once the data are converted to ASCII code, they are displayed by loading DL with the most-significant digit from AH. Next, the least-significant digit is displayed from AL. Note that the DOS INT 21H function calls change AL.

EXAMPLE 5–24

```
                    ;A program that displays the number in AL, loaded
                    ;with the first instruction (48H).
                    ;
                    .MODEL TINY              ;select tiny model
0000                .CODE                    ;start code segment
                    .STARTUP                 ;start program
0100 B0 48                MOV   AL,48H     ;load test data
0102 B4 00                MOV   AH,0       ;clear AH
0104 D4 0A                AAM              ;convert to BCD
0106 05 3030              ADD   AX,3030H   ;convert to ASCII
0109 8A D4                MOV   DL,AH      ;display most-significant digit
010B B4 02                MOV   AH,2
010D 50                   PUSH  AX
010E CD 21                INT   21H
0110 58                   POP   AX
0111 8A D0                MOV   DL,AL      ;display least-significant digit
0113 CD 21                INT   21H
                    .EXIT                    ;exit to DOS
                    END
```

AAS Instruction. Like other ASCII adjust instructions, AAS adjusts the AX register after an ASCII subtraction. For example, suppose that 35H is subtracted from 39H. The result will be 04H, which requires no correction. Here, AAS will modify neither AH nor AL. On the other hand, if 38H is subtracted from 37H, then AL will equal 09H and the number in AH will decrement by 1. This decrement allows multiple-digit ASCII numbers to be subtracted from each other.

5–4 BASIC LOGIC INSTRUCTIONS

The basic logic instructions include AND, OR, Exclusive-OR, and NOT. Another logic instruction is TEST, which is explained in this section of the text because the operation of the TEST instruction is a special form of the AND instruction. Also explained is the NEG instruction, which is similar to the NOT instruction.

Logic operations provide binary bit control in low-level software. The logic instructions allow bits to be set, cleared, or complemented. Low-level software appears in machine language or assembly language form and often controls the I/O devices in a system. All logic instructions affect the flag bits. Logic operations always clear the carry and overflow flags, while the other flags change to reflect the condition of the result.

When binary data are manipulated in a register or a memory location, the rightmost bit position is always numbered bit 0. Bit position numbers increase from bit 0 toward the left, to bit 7 for a byte, and to bit 15 for a word. A doubleword (32 bits) uses bit position 31 as its leftmost bit.

AND

The AND operation performs logical multiplication, as illustrated by the truth table in Figure 5–3. Here, two bits, A and B, are ANDed to produce the result X. As indicated by the truth table, X is a logic 1 only when both A and B are logic 1s. For all other input combinations of A and B, X is a logic 0. It is important to remember that 0 AND anything is always 0, and 1 AND 1 is always 1.

The AND instruction can replace discrete AND gates if the speed required is not too great, although this is normally reserved for embedded control applications. (Note that Intel has released the 80386EX embedded controller, which embodies the basic structure of the personal computer system.) With the 8086 microprocessor, the AND instruction often executes in about a microsecond. With newer versions, the execution speed is greatly increased. Take the 3.0 GHz Pentium with its clock time of 1/3 ns that can execute up to three instructions per clock (1/9 ns per AND operation). If the circuit that the AND instruction replaces operates at a much slower speed than the microprocessor, the AND instruction is a logical replacement. This replacement can save a considerable amount of money. A single AND gate integrated

A	B	T
0	0	0
0	1	0
1	0	0
1	1	1

(a)

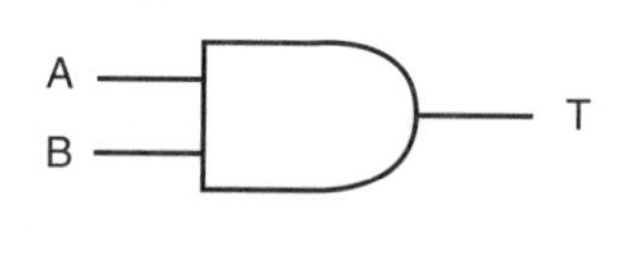

(b)

FIGURE 5–3 (a) The truth table for the AND operation and (b) the logic symbol of an AND gate.

FIGURE 5–4 The operation of the AND function showing how bits of a number are cleared to zero.

```
  x x x x  x x x x   Unknown number
• 0 0 0 0  1 1 1 1   Mask
-------------------
  0 0 0 0  x x x x   Result
```

circuit (74HC08) costs approximately 40¢, while it costs less than 1/100¢ to store the AND instruction in read-only memory. Note that a logic circuit replacement such as this only appears in control systems based on microprocessors and does not generally find application in the personal computer.

The AND operation clears bits of a binary number. The task of clearing a bit in a binary number is called **masking**. Figure 5–4 illustrates the process of masking. Notice that the leftmost four bits clear to 0 because 0 AND anything is 0. The bit positions that AND with 1s do not change. This occurs because if a 1 ANDs with a 1, a 1 results; if a 1 ANDs with a 0, a 0 results.

The AND instruction uses any addressing mode except memory-to-memory and segment register addressing. Table 5–14 lists some AND instructions and comments about their operations.

An ASCII-coded number can be converted to BCD by using the AND instruction to mask off the leftmost four binary bit positions. This converts the ASCII 30H to 39H to 0–9. Example 5–25 shows a short program that converts the ASCII contents of BX into BCD. The AND instruction in this example converts two digits from ASCII to BCD simultaneously.

EXAMPLE 5–25

```
0000 BB 3135              MOV  BX,3135H        ;load ASCII
0003 81 E3 0F0F           AND  BX,0F0FH        ;mask BX
```

OR

The **OR operation** performs logical addition and is often called the *Inclusive-OR* function. The OR function generates a logic 1 output if any inputs are 1. A 0 appears at the output only when all inputs are 0. The truth table for the OR function appears in Figure 5–5. Here, the inputs A and B OR together to produce the X output. It is important to remember that 1 ORed with anything yields 1.

TABLE 5–14 Example AND instructions.

Assembly Language	*Operation*
AND AL,BL	AL = AL and BL
AND CX,DX	CX = CX and DX
AND ECX,EDI	ECX = ECX and EDI
AND CL,33H	CL = CL and 33H
AND DI,4FFFH	DI = DI and 4FFFH
AND ESI,34H	ESI = ESI and 34H
AND AX,[DI]	The word contents of the data segment memory location addressed by DI are ANDed with AX
AND ARRAY[SI],AL	The byte contents of the data segment memory location addressed by ARRAY plus SI are ANDed with AL
AND [EAX],CL	CL is ANDed with the byte contents of the data segment memory location addressed by ECX

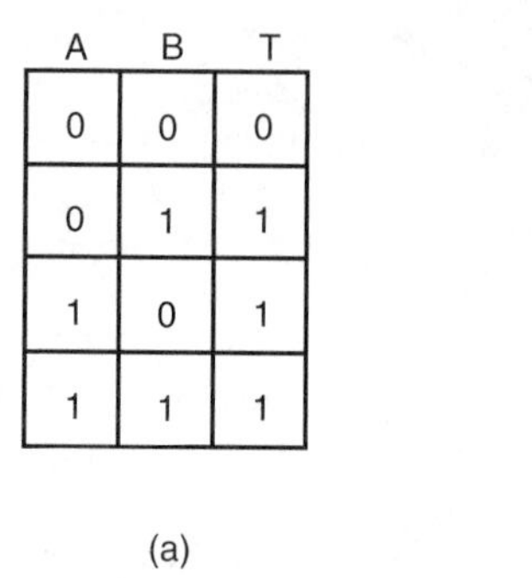

A	B	T
0	0	0
0	1	1
1	0	1
1	1	1

(a)

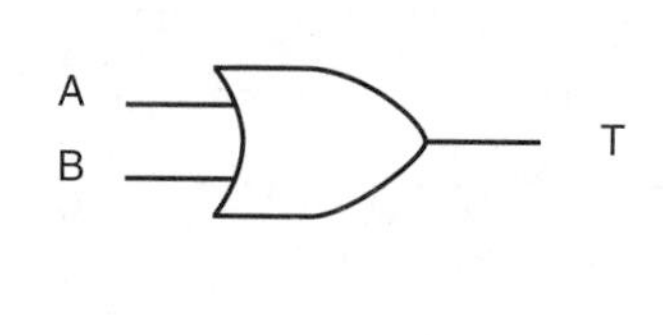

(b)

FIGURE 5–5 (a) The truth table for the OR operation and (b) the logic symbol of an OR gate.

```
  xxxx xxxx  Unknown number
+ 0000 1111  Mask
-----------
  xxxx 1111  Result
```

FIGURE 5–6 The operation of the OR function showing how bits of a number are set to one.

In embedded controller applications, the OR instruction can also replace discrete OR gates. This results in considerable savings because a quad, two-input OR gate (74HC32) costs about 40¢, while the OR instruction costs less than 1/100¢ to store in a read-only memory.

Figure 5–6 shows how the OR gate sets (1) any bit of a binary number. Here, an unknown number (XXXX XXXX) ORs with 0000 1111 to produce a result of XXXX 1111. The rightmost four bits set, whereas the leftmost four bits remain unchanged. The OR operation sets any bit; the AND operation clears any bit.

The OR instruction uses any of the addressing modes allowed to any other instruction except segment register addressing. Table 5–15 illustrates several example OR instructions with comments about their operation.

Suppose that two BCD numbers are multiplied and adjusted with the AAM instruction. The result appears in AX as a two-digit unpacked BCD number. Example 5–26 illustrates this multiplication and shows how to change the result into a two-digit ASCII-coded number using the OR instruction. Here, OR AX,3030H converts the 0305H found in AX to 3335H. The OR operation can be replaced with an ADD AX,3030H to obtain the same results.

TABLE 5–15 Example OR instructions.

Assembly Language	*Operation*
OR AH,BL	AL = AL or BL
OR SI,DX	SI = SI or DX
OR EAX,EBX	EAX = EAX or EBX
OR DH,0A3H	DH = DH or 0A3H
OR SP,990DH	SP = SP or 990DH
OR EBP,10	EBP = EBP or 10
OR DX,[BX]	DX is ORed with the word contents of data segment memory location addressed by BX
OR DATES[DI+2],AL	The byte contents of the data segment memory location addressed by DI plus 2 are ORed with AL

EXAMPLE 5–26

```
0000 B0 05          MOV  AL,5          ;load data
0002 B3 07          MOV  BL,7
0004 F6 E3          MUL  BL
0006 D4 0A          AAM                ;adjust
0008 0D 3030        OR   AX,3030H      ;convert to ASCII
```

Exclusive-OR

The **Exclusive-OR** instruction (XOR) differs from Inclusive-OR (OR). The difference is that a 1,1 condition of the OR function produces a 1; the 1,1 condition of the Exclusive-OR operation produces a 0. The Exclusive-OR operation excludes this condition; the Inclusive-OR includes it.

Figure 5–7 shows the truth table of the Exclusive-OR function. (Compare this with Figure 5–5 to appreciate the difference between these two OR functions.) If the inputs of the Exclusive-OR function are both 0 or both 1, the output is 0. If the inputs are different, the output is 1. Because of this, the Exclusive-OR is sometimes called a comparator.

The XOR instruction uses any addressing mode except segment register addressing. Table 5–16 lists several Exclusive-OR instructions and their operations.

As with the AND and OR functions, Exclusive-OR can replace discrete logic circuitry in embedded applications. The 74HC86 quad, two-input Exclusive-OR gate is replaced by one XOR instruction. The 74HC86 costs about 40¢, whereas the instruction costs less than 1/100¢ to store in the memory. Replacing just one 74HC86 saves a considerable amount of money, especially if many systems are built.

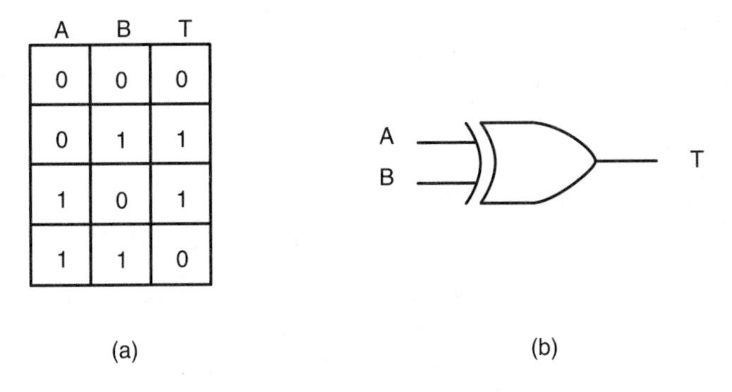

FIGURE 5–7 (a) The truth table for the Exclusive-OR operation and (b) the logic symbol of an Exclusive-OR gate.

TABLE 5–16 Example Exclusive-OR instructions.

Assembly Language	*Operation*
XOR CH,DL	CH = CH xor DL
XOR SI,BX	SI = SI xor BX
XOR EBX,EDI	EBX = EBX xor EDI
XOR AH,0EEH	AH = AH xor 0EEH
XOR DI,00DDH	DI = DI xor 00DDH
XOR ESI,100	ESI = ESI xor 100
XOR DX,[SI]	DX is Exclusive-ORed with the word contents of the data segment memory location addressed by SI
XOR DEAL[BP+2],AH	AH is Exclusive-ORed with the byte contents of the stack segment memory location addressed by BP plus 2

FIGURE 5–8 The operation of the Exclusive-OR function showing how bits of a number are inverted.

x x x x x x x x	Unknown number
⊕0 0 0 0 1 1 1 1	Mask
x x x x $\overline{x}$ $\overline{x}$ $\overline{x}$ $\overline{x}$	Result

The Exclusive-OR instruction is useful if some bits of a register or memory location must be inverted. This instruction allows part of a number to be inverted or complemented. Figure 5–8 shows how just part of an unknown quantity can be inverted by XOR. Notice that when a 1 Exclusive-ORs with X, the result is X. If a 0 Exclusive-ORs with X, the result is X.

Suppose that the leftmost 10 bits of the BX register must be inverted without changing the rightmost six bits. The XOR BX,0FFC0H instruction accomplishes this task. The AND instruction clears (0) bits, the OR instruction sets (1) bits, and now the Exclusive-OR instruction inverts bits. These three instructions allow a program to gain complete control over any bit stored in any register or memory location. This is ideal for control system applications in which equipment must be turned on (1), turned off (0), and toggled from on to off or off to on.

A common use for the Exclusive-OR instruction is to clear a register to zero. For example, the XOR CH,CH instruction clears register CH to 00H and requires two bytes of memory to store the instruction. Likewise, the MOV CH,00H instruction also clears CH to 00H, but requires three bytes of memory. Because of this saving, the XOR instruction is often used to clear a register in place of a move immediate.

Example 5–27 shows a short sequence of instructions that clears bits 0 and 1 of CX, sets bits 9 and 10 of CX, and inverts bit 12 of CX. The OR instruction is used to set bits, the AND instruction is used to clear bits, and the XOR instruction inverts bits.

EXAMPLE 5–27

```
0000 81 C9 0600          OR   CX,0600H          ;set bits 9 and 10
0004 83 E1 FC            AND  CX,0FFFCH         ;clear bits 0 and 1
0007 81 F1 1000          XOR  CX,1000H          ;invert bit 12
```

Test and Bit Test Instructions

The **TEST instruction** performs the AND operation. The difference is that the AND instruction changes the destination operand, whereas the TEST instruction does not. A TEST only affects the condition of the flag register, which indicates the result of the test. The TEST instruction uses the same addressing modes as the AND instruction. Table 5–17 lists some TEST instructions and their operations.

The TEST instruction functions in the same manner as a CMP instruction. The difference is that the TEST instruction normally tests a single bit (or occasionally multiple bits), whereas the CMP instruction tests the entire byte, word, or doubleword. The zero flag (Z) is a logic 1 (indicating a zero result) if the bit under test is a zero, and Z = 0 (indicating a nonzero result) if the bit under test is not zero.

TABLE 5–17 Example TEST instructions.

Assembly Language	*Operation*
TEST DL,DH	DL is ANDed with DH
TEST CX,BX	CX is ANDed with BX
TEST EDX,ECX	EDX is ANDed with ECX
TEST AH,4	AH is ANDed with 4
TEST EAX,256	EAX is ANDed with 256

Usually the TEST instruction is followed by either the JZ (**jump if zero**) or JNZ (**jump if not zero**) instruction. The destination operand is normally tested against immediate data. The value of immediate data is 1 to test the rightmost bit position, 2 to test the next bit, 4 for the next, and so on.

Example 5–28 lists a short program that tests the rightmost and leftmost bit positions of the AL register. Here, 1 selects the rightmost bit and 128 selects the leftmost bit. (Note: A 128 is an 80H.) The JNZ instruction follows each test to jump to different memory locations, depending on the outcome of the tests. The JNZ instruction jumps to the operand address (RIGHT or LEFT in the example) if the bit under test is not zero.

EXAMPLE 5–28

```
0000 A8 01              TEST AL,1          ;test right bit
0002 75 1C              JNZ  RIGHT         ;if set
0004 A8 80              TEST AL,128        ;test left bit
0006 75 38              JNZ  LEFT          ;if set
```

The 80386 through the Pentium 4 processors contain additional test instructions that test single bit positions. Table 5–18 lists the four different bit test instructions available to these microprocessors.

All four forms of the bit test instruction test the bit position in the destination operand selected by the source operand. For example, the BT AX,4 instruction tests bit position 4 in AX. The result of the test is located in the carry flag bit. If bit position 4 is a 1, carry is set; if bit position 4 is a 0, carry is cleared.

The remaining three bit test instructions also place the bit under test into the carry flag and change the bit under test afterward. The BTC AX,4 instruction complements bit position 4 after testing it, the BTR AX,4 instruction clears it (0) after the test, and the BTS AX,4 instruction sets it (1) after the test.

Example 5–29 repeats the sequence of instructions listed in Example 5–27. Here, the BTR instruction clears bits in CX, BTS sets bits in CX, and BTC inverts bits in CX.

EXAMPLE 5–29

```
0000 0F BA E9 09        BTS  CX,9          ;set bit 9
0004 0F BA E9 0A        BTS  CX,10         ;set bit 10
0008 0F BA F1 00        BTR  CX,0          ;clear bit 0
000C 0F BA F1 01        BTR  CX,1          ;clear bit 1
0010 0F BA F9 0C        BTC  CX,12         ;complement bit 12
```

NOT and NEG

Logical inversion, or the **one's complement** (NOT), and arithmetic sign inversion, or the **two's complement** (NEG), are the last two logic functions presented (except for shift and rotate in the

TABLE 5–18 Bit test instructions.

Assembly Language	*Operation*
BT	Tests a bit in the destination operand specified by the source operand
BTC	Tests and complements a bit in the destination operand specified by the source operand
BTR	Tests and resets a bit in the destination operand specified by the source operand
BTS	Tests and sets a bit in the destination operand specified by the source operand

TABLE 5–19 Example NOT and NEG instructions.

Assembly Language	*Operation*
NOT CH	CH is one's complemented
NEG CH	CH is two's complemented
NEG AX	AX is two's complemented
NOT EBX	EBX is one's complemented
NEG ECX	ECX is two's complemented
NOT TEMP	The contents of data segment memory location TEMP are one's complemented
NOT BYTE PTR[BX]	The byte contents of the data segment memory location addressed by BX are one's complemented

next section of the text). These are two of a few instructions that contain only one operand. Table 5–19 lists some variations of the NOT and NEG instructions. As with most other instructions, NOT and NEG can use any addressing mode except segment register addressing.

The NOT instruction inverts all bits of a byte, word, or doubleword. The NEG instruction two's complements a number, which means that the arithmetic sign of a signed number changes from positive to negative or from negative to positive. The NOT function is considered logical; the NEG function is considered an arithmetic operation.

5–5 SHIFT AND ROTATE

Shift and rotate instructions manipulate binary numbers at the binary bit level, as did the AND, OR, Exclusive-OR, and NOT instructions. Shifts and rotates find their most common applications in low-level software used to control I/O devices. The microprocessor contains a complete set of shift and rotate instructions that are used to shift or rotate any memory data or register.

Shift

Shift instructions position or move numbers to the left or right within a register or memory location. They also perform simple arithmetic such as multiplication by powers of 2^{+n} (left shift) and division by powers of 2^{-n} (right shift). The microprocessor's instruction set contains four different shift instructions: Two are logical shifts and two are arithmetic shifts. All four shift operations appear in Figure 5–9.

Notice in Figure 5–9 that there are two right shifts and two left shifts. The logical shifts move a 0 into the rightmost bit position for a logical left shift and a 0 into the leftmost bit position for a logical right shift. There are also two arithmetic shifts. The arithmetic shift left and logical left shift are identical. The arithmetic right shift and logical right shift are different because the arithmetic right shift copies the sign-bit through the number, whereas the logical right shift copies a 0 through the number.

Logical shift operations function with unsigned numbers, and arithmetic shifts function with signed numbers. Logical shifts multiply or divide unsigned data, and arithmetic shifts multiply or divide signed data. A shift left always multiplies by 2 for each bit position shifted, and a shift right always divides by 2 for each bit position shifted. Shifting a number two places multiplies or divides by 4.

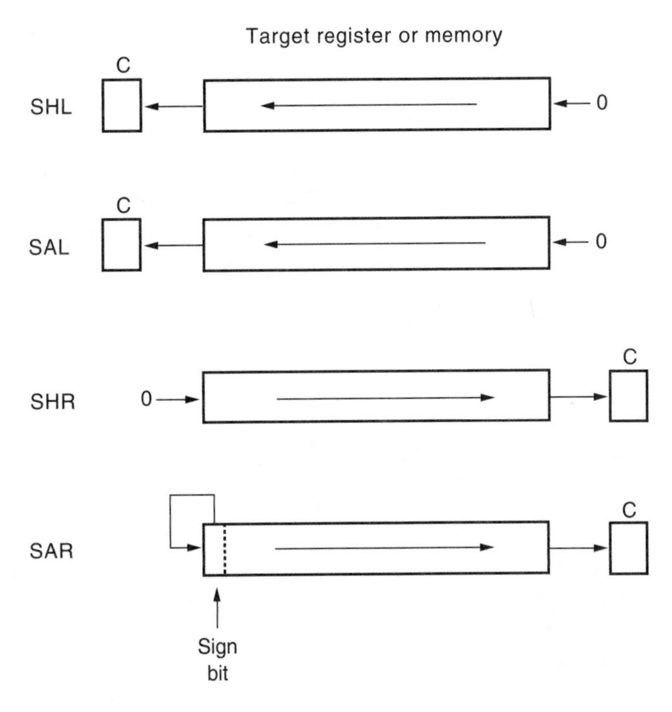

FIGURE 5–9 The shift instructions showing the operation and direction of the shift.

Table 5–20 illustrates some addressing modes allowed for the various shift instructions. There are two different forms of shifts that allow any register (except the segment register) or memory location to be shifted. One mode uses an immediate shift count, and the other uses register CL to hold the shift count. Note that CL must hold the shift count. When CL is the shift count, it does not change when the shift instruction executes. Note that the shift count is a modulo-32 count, which means that a shift count of 33 will shift the data one place (33/32 = remainder of 1).

Example 5–30 shows how to shift the DX register left 14 places in two different ways. The first method uses an immediate shift count of 14. The second method loads 14 into CL and then uses CL as the shift count. Both instructions shift the contents of the DX register logically to the left 14 binary bit positions or places.

EXAMPLE 5–30

```
0000 C1 E2 0E              SHL  DX,14

                 or

0003 B1 0E                 MOV  CL,14
0005 D3 E2                 SHL  DX,CL
```

TABLE 5–20 Example shift instructions.

Assembly Language	*Operation*
SHL AX,1	AX is logically shifted left 1 place
SHR BX,12	BX is logically shifted right 12 places
SHR ECX,10	ECX is logically shifted right 10 places
SAL DATA1,CL	The contents of data segment memory location DATA1 are arithmetically shifted left the number of spaces specified by CL
SAR SI,2	SI is arithmetically shifted right 2 places
SAR EDX,14	EDX is arithmetically shifted right 14 places

Suppose that the contents of AX must be multiplied by 10, as shown in Example 5–31. This can be done in two ways: by the MUL instruction or by shifts and additions. A number is doubled when it shifts left one place. When a number is doubled and then added to the number times 8, the result is 10 times the number. The number 10 decimal is 1010 in binary. A logic 1 appears in both the 2's and 8's positions. If 2 times the number is added to 8 times the number, the result is 10 times the number. Using this technique, a program can be written to multiply by any constant. This technique often executes faster than the multiply instruction found in earlier versions of the Intel microprocessor.

EXAMPLE 5–31

```
                         ;Multiply AX by 10 (1010)
                         ;
0000 D1 E0                      SHL  AX,1            ;AX times 2
0002 8B D8                      MOV  BX,AX
0004 C1 E0 02                   SHL  AX,2            ;AX times 8
0007 03 C3                      ADD  AX,BX           ;AX times 10
                         ;
                         ;Multiply AX by 18 (10010)
                         ;
0009 D1 E0                      SHL  AX,1            ;AX times 2
000B 8B D8                      MOV  BX,AX
000D C1 E0 03                   SHL  AX,3            ;AX times 16
0010 03 C3                      ADD  AX,BX           ;AX times 18
                         ;
                         ;Multiply AX by 5 (101)
                         ;
0012 8B D8                      MOV  BX,AX
0014 C1 E0 02                   SHL  AX,2            ;AX times 4
0017 03 C3                      ADD  AX,BX           ;AX times 5
```

Double-Precision Shifts (80386–Pentium 4 Only). The 80386 and above contain two double-precision shifts: SHLD (shift left) and SHRD (shift right). Each instruction contains three operands, instead of the two found with the other shift instructions. Both instructions function with two 16- or 32-bit registers, or with one 16- or 32-bit memory location and a register.

The SHRD AX,BX,12 instruction is an example of the double-precision shift right instruction. This instruction logically shifts AX right by 12 bit positions. The rightmost 12 bits of BX shift into the leftmost 12 bits of AX. The contents of BX remain unchanged by this instruction. The shift count can be an immediate count, as in this example, or it can be found in register CL, as with other shift instructions.

The SHLD EBX,ECX,16 instruction shifts EBX left. The leftmost 16 bits of ECX fill the rightmost 16 bits of EBX after the shift. As before, the contents of ECX, the second operand, remain unchanged. This instruction, as well as SHRD, affect the flag bits.

Rotate. Rotate instructions position binary data by rotating the information in a register or memory location, either from one end to another or through the carry flag. They are often used to shift or position numbers that are wider than 16 bits in the 8086–80286 microprocessors or wider than 32 bits in the 80386 through the Pentium 4. The four available rotate instructions appear in Figure 5–10.

Numbers rotate through a register or memory location, through the C flag (carry), or through a register or memory location only. With either type of rotate instruction, the programmer can select either a left or a right rotate. Addressing modes used with rotate are the same as those used with shifts. A rotate count can be immediate or located in register CL. Table 5–21 lists some of the possible rotate instructions. If CL is used for a rotate count, it does not change. As with shifts, the count in CL is a modulo-32 count.

Rotate instructions are often used to shift wide numbers to the left or right. The program listed in Example 5–32 shifts the 48-bit number in registers DX, BX, and AX left one

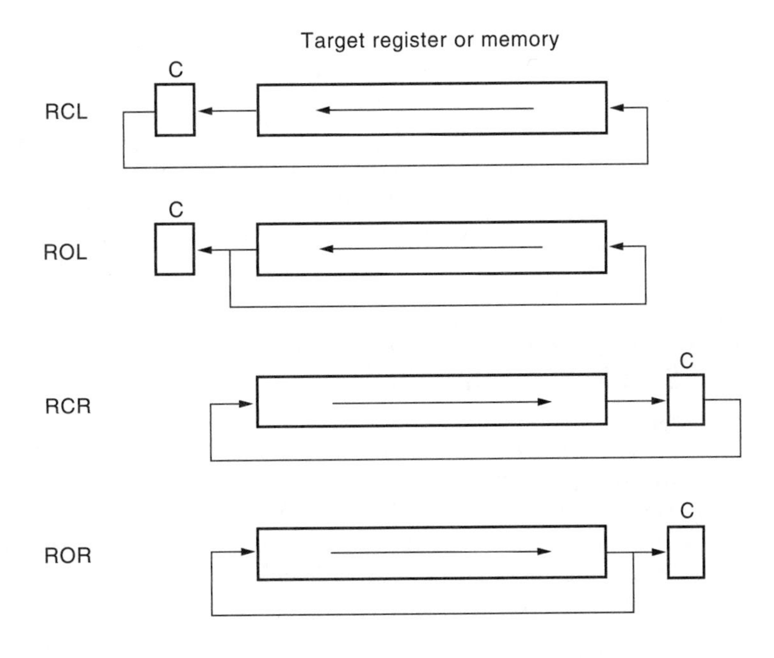

FIGURE 5–10 The rotate instructions showing the direction and operation of each rotate.

binary place. Notice that the least-significant 16 bits (AX) shift left first. This moves the leftmost bit of AX into the carry flag bit. Next, the rotate BX instruction rotates carry into BX, and its leftmost bit moves into carry. The last instruction rotates carry into DX, and the shift is complete.

EXAMPLE 5–32

```
0000 D1 E0          SHL  AX,1
0002 D1 D3          RCL  BX,1
0004 D1 D2          RCL  DX,1
```

Bit Scan Instructions

Although the bit scan instructions don't shift or rotate numbers, they do scan through a number searching for a 1-bit. Because this is accomplished within the microprocessor by shifting the number, bit scan instructions are included in this section of the text.

The bit scan instructions BSF (**bit scan forward**) and BSR (**bit scan reverse**) are available only in the 80386–Pentium 4 processors. Both forms scan through the source number, searching for the first 1-bit. The BSF instruction scans the number from the leftmost bit toward the right, and BSR scans the number from the rightmost bit toward the left. If a 1-bit is encountered, the

TABLE 5–21 Example rotate instructions.

Assembly Language	*Operation*
ROL SI,14	SI rotates left 14 places
RCL BL,6	BL rotates left through carry 6 places
ROL ECX,18	ECX rotates left 18 places
RCR AH,CL	AH rotates right through carry the number of places specified by CL
ROR WORD PTR[BP],2	The word contents of the stack segment memory location addressed by BP rotate right 2 places

zero flag is set and the bit position number of the 1-bit is placed into the destination operand. If no 1-bit is encountered (i.e., the number contains all zeros), the zero flag is cleared. Thus, the result is not-zero if no 1-bit is encountered.

For example, if EAX = 60000000H and the BSF EBX,EAX instruction executes, the number is scanned from the leftmost bit toward the right. The first 1-bit encountered is at bit position 30, which is placed into EBX and the zero flag bit is set. If the same value for EAX is used for the BSR instruction, the EBX register is loaded with 29 and the zero flag bit is set.

5–6 STRING COMPARISONS

As illustrated in Chapter 4, the string instructions are very powerful because they allow the programmer to manipulate large blocks of data with relative ease. Block data manipulation occurs with the string instructions MOVS, LODS, STOS, INS, and OUTS. This section discusses additional string instructions that allow a section of memory to be tested against a constant or against another section of memory. To accomplish these tasks, use the SCAS (**string scan**) or CMPS (**string compare**) instructions.

SCAS

The SCAS (string scan instruction) compares the AL register with a byte block of memory, the AX register with a word block of memory, or the EAX register (80386–Pentium 4) with a doubleword block of memory. The SCAS instruction subtracts memory from AL, AX, or EAX without affecting either the register or the memory location. The opcode used for byte comparison is SCASB, the opcode used for word comparison is SCASW, and the opcode used for doubleword comparison is SCASD. In all cases, the contents of the extra segment memory location addressed by DI are compared with AL, AX, or EAX. Recall that this default segment (ES) cannot be changed with a segment override prefix.

Like the other string instructions, SCAS instructions use the direction flag (D) to select either the auto-increment or auto-decrement operation for DI. They also repeat if prefixed by a conditional repeat prefix.

Suppose that a section of memory is 100 bytes long and begins at location BLOCK. This section of memory must be tested to see whether any location contains 00H. The program in Example 5–33 shows how to search this part of memory for 00H using the SCASB instruction. In this example, the SCASB instruction has an REPNE (**repeat while not equal**) prefix. The REPNE prefix causes the SCASB instruction to repeat until either the CX register reaches 0, or until an equal condition exists as the outcome of the SCASB instruction's comparison. Another conditional repeat prefix is REPE (**repeat while equal**). With either repeat prefix, the contents of CX decrement without affecting the flag bits. The SCASB instruction and the comparison it makes change the flags.

EXAMPLE 5–33

```
0000 BF 0011 R        MOV   DI,OFFSET BLOCK      ;address data
0003 FC               CLD                        ;auto-increment
0004 B9 0064          MOV   CX,100               ;load counter
0007 32 C0            XOR   AL,AL                ;clear AL
0009 F2/AE            REPNE SCASB
```

Suppose that you must develop a program that skips ASCII-coded spaces in a memory array. (This task appears in the procedure listed in Example 5–34.) This procedure assumes that the DI register already addresses the ASCII-coded character string and that the length of the string is 256 bytes or fewer. Because this program is to skip spaces (20H), the REPE prefix is used with a SCASB instruction. The SCASB instruction repeats the comparison, searching for a 20H, as long as an equal condition exists.

EXAMPLE 5–34

```
0000 FC                    CLD               ;auto-increment
0001 B9 0100               MOV  CX,256       ;load counter
0004 B0 20                 MOV  AL,20H       ;get space
0006 F3/AE                 REPE SCASB
```

CMPS

The CMPS (compare strings instruction) always compares two sections of memory data as bytes (CMPSB), words (CMPSW), or doublewords (CMPSD). Note that only the 80386 through Pentium 4 can use doublewords. The contents of the data segment memory location addressed by SI are compared with the contents of the extra segment memory location addressed by DI. The CMPS instruction increments or decrements both SI and DI. The CMPS instruction is normally used with either the REPE or REPNE prefix. Alternates to these prefixes are REPZ (**repeat while zero**) and REPNZ (**repeat while not zero**), but usually the REPE or REPNE prefixes are used in programming.

Example 5–35 illustrates a short procedure that compares two sections of memory searching for a match. The CMPSB instruction is prefixed with REPE. This causes the search to continue as long as an equal condition exists. When the CX register becomes 0 or an unequal condition exists, the CMPSB instruction stops execution. After the CMPSB instruction ends, the CX register is 0 or the flags indicate an equal condition when the two strings match. If CX is not 0 or the flags indicate a not-equal condition, the strings do not match.

EXAMPLE 5–35

```
0000 BE 0075 R             MOV  SI,OFFSET LINE    ;address LINE
0003 BF 007F R             MOV  DI,OFFSET TABLE   ;address TABLE
0006 FC                    CLD                    ;auto-increment
0007 B9 000A               MOV  CX,10             ;load counter
000A F3/A6                 REPE CMPSB             ;search
```

5–7 SUMMARY

1. Addition (ADD) can be 8, 16, or 32 bits. The ADD instruction allows any addressing mode except segment register addressing. Most flags (C, A, S, Z, P, and O) change when the ADD instruction executes. A different type of addition, add-with-carry (ADC), adds two operands and the contents of the carry flag (C). The 80486 through the Pentium 4 processors have an additional instruction (XADD) that combines an addition with an exchange.
2. The increment instruction (INC) adds 1 to the byte, word, or doubleword contents of a register or memory location. The INC instruction affects the same flag bits as ADD except the carry flag. The BYTE PTR, WORD PTR, and DWORD PTR directives appear with the INC instruction when the contents of a memory location are addressed by a pointer.
3. Subtraction (SUB) is a byte, word, or doubleword and is performed on a register or a memory location. The only form of addressing not allowed by the SUB instruction is segment register addressing. The subtract instruction affects the same flags as ADD and subtracts carry if the SBB form is used.

4. The decrement (DEC) instruction subtracts 1 from the contents of a register or a memory location. The only addressing modes not allowed with DEC are immediate or segment register addressing. The DEC instruction does not affect the carry flag and is often used with BYTE PTR, WORD PTR, or DWORD PTR.
5. The comparison (CMP) instruction is a special form of subtraction that does not store the difference; instead, the flags change to reflect the difference. Comparison is used to compare an entire byte or word located in any register (except segment) or memory location. An additional comparison instruction (CMPXCHG), which is a combination of comparison and exchange instructions, is found in the 80486–Pentium 4 processors. In the Pentium–Pentium 4 processors, the CMPXCHG8B instruction compares and exchanges quadword data.
6. Multiplication is byte, word, or doubleword, and it can be signed (IMUL) or unsigned (MUL). The 8-bit multiplication always multiplies register AL by an operand with the product found in AX. The 16-bit multiplication always multiplies register AX by an operand with the product found in DX–AX. The 32-bit multiplication always multiplies register EAX by an operand with the product found in EDX–EAX. A special IMUL immediate instruction exists on the 80186–Pentium 4 processors that contains three operands. For example, the IMUL BX,CX,3 instruction multiplies CX by 3 and leaves the product in BX.
7. Division is byte, word, or doubleword, and it can be signed (IDIV) or unsigned (DIV). For an 8-bit division, the AX register divides by the operand, after which the quotient appears in AL and the remainder appears in AH. In the 16-bit division, the DX–AX register divides by the operand, after which the AX register contains the quotient and DX contains the remainder. In the 32-bit division, the EDX–EAX register is divided by the operand, after which the EAX register contains the quotient and the EDX register contains the remainder. Note that the remainder after a signed division always assumes the sign of the dividend.
8. BCD data add or subtract in packed form by adjusting the result of the addition with DAA or the subtraction with DAS. ASCII data are added, subtracted, multiplied, or divided when the operations are adjusted with AAA, AAS, AAM, and AAD.
9. The AAM instruction has an interesting added feature that allows it to convert a binary number into unpacked BCD. This instruction converts a binary number between 00H–63H into unpacked BCD in AX. The AAM instruction divides AX by 10 and leaves the remainder in AL and quotient in AH.
10. The AND, OR, and Exclusive-OR instructions perform logic functions on a byte, word, or doubleword stored in a register or memory location. All flags change with these instructions, with carry (C) and overflow (O) cleared.
11. The TEST instruction performs the AND operation, but the logical product is lost. This instruction changes the flag bits to indicate the outcome of the test.
12. The NOT and NEG instructions perform logical inversion and arithmetic inversion. The NOT instruction one's complements an operand, and the NEG instruction two's complements an operand.
13. There are eight different shift and rotate instructions. Each of these instructions shifts or rotates a byte, word, or doubleword register or memory data. These instructions have two operands: the first is the location of the data shifted or rotated, and the second is an immediate shift or rotate count or CL. If the second operand is CL, the CL register holds the shift or rotate count. The 80386 through the Pentium 4 processors have two additional double-precision shifts (SHRD and SHLD).
14. The scan string (SCAS) instruction compares AL, AX, or EAX with the contents of the extra segment memory location addressed by DI.
15. The string compare (CMPS) instruction compares the byte, word, or doubleword contents of two sections of memory. One section is addressed by DI in the extra segment, and the other is addressed by SI in the data segment.

16. The SCAS and CMPS instructions repeat with the REPE or REPNE prefixes. The REPE prefix repeats the string instruction while an equal condition exists, and the REPNE repeats the string instruction while a not-equal condition exists.

5–8 QUESTIONS AND PROBLEMS

1. Select an ADD instruction that will:
   (a) add BX to AX
   (b) add 12H to AL
   (c) add EDI and EBP
   (d) add 22H to CX
   (e) add the data addressed by SI to AL
   (f) add CX to the data stored at memory location FROG
2. What is wrong with the ADD ECX,AX instruction?
3. Is it possible to add CX to DS with the ADD instruction?
4. If AX = 1001H and DX = 20FFH, list the sum and the contents of each flag register bit (C, A, S, Z, and O) after the ADD AX,DX instruction executes.
5. Develop a short sequence of instructions that adds AL, BL, CL, DL, and AH. Save the sum in the DH register.
6. Develop a short sequence of instructions that adds AX, BX, CX, DX, and SP. Save the sum in the DI register.
7. Develop a short sequence of instructions that adds ECX, EDX, and ESI. Save the sum in the EDI register.
8. Select an instruction that adds BX to DX, and also adds the contents of the carry flag (C) to the result.
9. Choose an instruction that adds 1 to the contents of the SP register.
10. What is wrong with the INC [BX] instruction?
11. Select a SUB instruction that will:
    (a) subtract BX from CX
    (b) subtract 0EEH from DH
    (c) subtract DI from SI
    (d) subtract 3322H from EBP
    (e) subtract the data address by SI from CH
    (f) subtract the data stored 10 words after the location addressed by SI from DX
    (g) subtract AL from memory location FROG
12. If DL = 0F3H and BH = 72H, list the difference after BH is subtracted from DL and show the contents of the flag register bits.
13. Write a short sequence of instructions that subtracts the numbers in DI, SI, and BP from the AX register. Store the difference in register BX.
14. Choose an instruction that subtracts 1 from register EBX.
15. Explain what the SBB [DI–4],DX instruction accomplishes.
16. Explain the difference between the SUB and CMP instructions.
17. When two 8-bit numbers are multiplied, where is the product found?
18. When two 16-bit numbers are multiplied, what two registers hold the product? Show the registers that contain the most- and least-significant portions of the product.
19. When two numbers multiply, what happens to the O and C flag bits?
20. Where is the product stored for the MUL EDI instruction?
21. What is the difference between the IMUL and MUL instructions?
22. Write a sequence of instructions that cube the 8-bit number found in DL. Load DL with 5 initially, and make sure that your result is a 16-bit number.

23. Describe the operation of the IMUL BX,DX,100H instruction.
24. When 8-bit numbers are divided, in which register is the dividend found?
25. When 16-bit numbers are divided, in which register is the quotient found?
26. What errors are detected during a division?
27. Explain the difference between the IDIV and DIV instructions.
28. Where is the remainder found after an 8-bit division?
29. Write a short sequence of instructions that divides the number in BL by the number in CL and then multiplies the result by 2. The final answer must be a 16-bit number stored in the DX register.
30. Which instructions are used with BCD arithmetic operations?
31. Which instructions are used with ASCII arithmetic operations?
32. Explain how the AAM instruction converts from binary to BCD.
33. Develop a sequence of instructions that converts the unsigned number in AX (values of 0–65535) into a 5-digit BCD number stored in memory, beginning at the location addressed by the BX register in the data segment. Note that the most-significant character is stored first and no attempt is made to blank leading zeros.
34. Develop a sequence of instructions that adds the 8-digit BCD number in AX and BX to the 8-digit BCD number in CX and DX. (AX and CX are the most-significant registers. The result must be found in CX and DX after the addition.)
35. Select an AND instruction that will:
    (a) AND BX with DX and save the result in BX
    (b) AND 0EAH with DH
    (c) AND DI with BP and save the result in DI
    (d) AND 1122H with EAX
    (e) AND the data addressed by BP with CX and save the result in memory
    (f) AND the data stored in four words before the location addressed by SI with DX and save the result in DX
    (g) AND AL with memory location WHAT and save the result at location WHAT
36. Develop a short sequence of instructions that clears (0) the three leftmost bits of DH without changing the remainder of DH and stores the result in BH.
37. Select an OR instruction that will:
    (a) OR BL with AH and save the result in AH
    (b) OR 88H with ECX
    (c) OR DX with SI and save the result in SI
    (d) OR 1122H with BP
    (e) OR the data addressed by BX with CX and save the result in memory
    (f) OR the data stored 40 bytes after the location addressed by BP with AL and save the result in AL
    (g) OR AH with memory location WHEN and save the result in WHEN
38. Develop a short sequence of instructions that sets (1) the rightmost five bits of DI without changing the remaining bits of DI. Save the results in SI.
39. Select the XOR instruction that will:
    (a) XOR BH with AH and save the result in AH
    (b) XOR 99H with CL
    (c) XOR DX with DI and save the result in DX
    (d) XOR lA23H with ESP
    (e) XOR the data addressed by EBX with DX and save the result in memory
    (f) XOR the data stored 30 words after the location addressed by BP with DI and save the result in DI
    (g) XOR DI with memory location WELL and save the result in DI

40. Develop a sequence of instructions that sets (1) the rightmost four bits of AX; clears (0) the leftmost three bits of AX; and inverts bits 7, 8, and 9 of AX.
41. Describe the difference between the AND and TEST instructions.
42. Select an instruction that tests bit position 2 of register CH.
43. What is the difference between the NOT and the NEG instruction?
44. Select the correct instruction to perform each of the following tasks:
    (a) shift DI right three places, with zeros moved into the leftmost bit
    (b) move all bits in AL left one place, making sure that a 0 moves into the rightmost bit position
    (c) rotate all the bits of AL left three places
    (d) rotate carry right one place through EDX
    (e) move the DH register right one place, making sure that the sign of the result is the same as the sign of the original number
45. What does the SCASB instruction accomplish?
46. For string instructions, DI always addresses data in the __________ segment.
47. What is the purpose of the D flag bit?
48. Explain what the REPE prefix does when coupled with the SCASB instruction.
49. What condition or conditions will terminate the repeated string instruction REPNE SCASB?
50. Describe what the CMPSB instruction accomplishes.
51. Develop a sequence of instructions that scans through a 300H-byte section of memory called LIST, located in the data segment, searching for 66H.
52. What happens if AH = 02H and DL = 43H when the INT 21H instruction is executed?

CHAPTER 6

Program Control Instructions

INTRODUCTION

The program control instructions direct the flow of a program and allow the flow to change. A change in flow often occurs after a decision made with the CMP or TEST instruction is followed by a conditional jump instruction. This chapter explains the program control instructions, including the jumps, calls, returns, interrupts, and machine control instructions.

The chapter also presents the relational assembly language statements (.IF, .ELSE, .ELSEIF, .ENDIF, .WHILE, .ENDW, .REPEAT, and .UNTIL) that are available in version 6.xx and above of MASM or TASM, with version 5.xx set for MASM compatibility. These relational assembly language commands allow the programmer to develop control flow portions of the program with C/C++ language efficiency.

CHAPTER OBJECTIVES

Upon completion of this chapter, you will be able to:

1. Use both conditional and unconditional jump instructions to control the flow of a program
2. Use the relational assembly language statements .IF, .REPEAT, .WHILE, and so forth in programs
3. Use the call and return instructions to include procedures in the program structure
4. Explain the operation of the interrupts and interrupt control instructions
5. Use machine control instructions to modify the flag bits
6. Use ENTER and LEAVE to enter and leave programming structures

6–1 THE JUMP GROUP

The main program control instruction, **jump** (*JMP*), allows the programmer to skip sections of a program and branch to any part of the memory for the next instruction. A conditional jump instruction allows the programmer to make decisions based upon numerical tests. The results of numerical tests are held in the flag bits, which are then tested by conditional jump instructions. Another instruction similar to the conditional jump, the conditional set, is explained with the conditional jump instructions in this section.

In this section of the text, all jump instructions are illustrated with their uses in sample programs. Also revisited are the LOOP and conditional LOOP instructions, first presented in Chapter 3, because they are also forms of the jump instruction.

Unconditional Jump (JMP)

Three types of unconditional jump instructions (see Figure 6–1) are available to the microprocessor: short jump, near jump, and far jump. The **short jump** is a two-byte instruction that allows jumps or branches to memory locations within +127 and –128 bytes from the address following the jump. The three-byte **near jump** allows a branch or jump within ±32K bytes (or anywhere in the current code segment) from the instruction in the current code segment. Remember that segments are cyclic in nature, which means that one location above offset address FFFFH is offset address 0000H. For this reason, if you jump two bytes ahead in memory and the instruction pointer addresses offset address FFFFH, the flow continues at offset address 0001H. Thus, a displacement of ±32K bytes allows a jump to any location within the current code segment. Finally, the five-byte **far jump** allows a jump to any memory location within the real memory system. The short and near jumps are often called **intrasegment jumps**, and the far jumps are often called **intersegment jumps**.

In the 80386 through the Pentium 4 processors, the near jump is within ±2G if the machine is operated in the protected mode, with a code segment that is 4G bytes long. If operated in the real mode, the near jump is within ±32K bytes. In the protected mode, the 80386 and above use a 32-bit displacement that is not shown in Figure 6–1.

Short Jump. Short jumps are called **relative jumps** because they can be moved, along with their related software, to any location in the current code segment without a change. This is because the jump address is not stored with the opcode. Instead of a jump address, a **distance**, or displacement, follows the opcode. The short jump displacement is a distance represented by a one-byte signed number whose value ranges between +127 and –128. The short jump instruction appears in Figure 6–2. When the microprocessor executes a short jump, the displacement is sign-extended and added to the instruction pointer (IP/EIP) to generate the jump address within the current code segment. The short jump instruction branches to this new address for the next instruction in the program.

Example 6–1 shows how short jump instructions pass control from one part of the program to another. It also illustrates the use of a **label** (a symbolic name for a memory address) with the jump instruction. Notice how one jump (JMP SHORT NEXT) uses the SHORT directive to force a short jump, while the other does not. Most assembler programs choose the best form of the jump instruction so the second jump instruction (JMP START) also assembles as a short jump. If

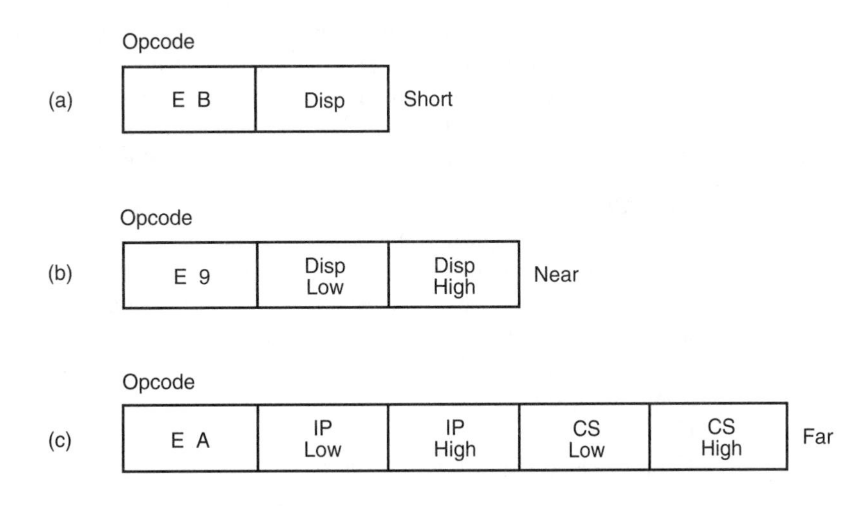

FIGURE 6–1 The three main forms of the JMP instruction. Note that Disp is either an 8- or 16-bit signed displacement or distance.

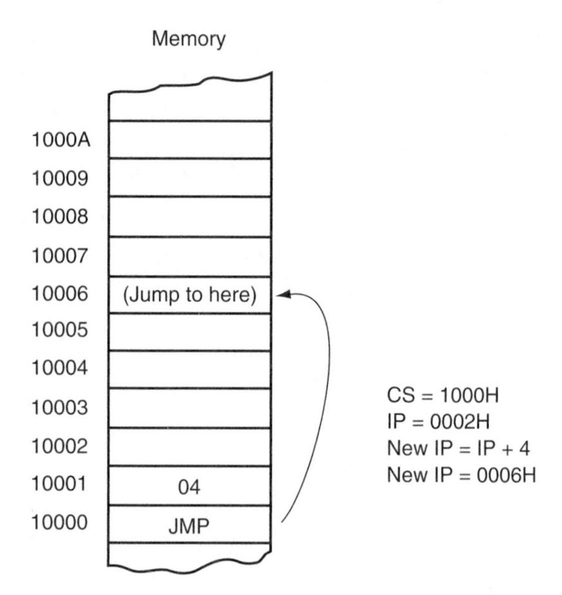

FIGURE 6–2 A short jump to four memory locations beyond the address of the next instruction.

the address of the next instruction (0009H) is added to the sign-extended displacement (0017H) of the first jump, the address of NEXT is at location 0017H + 0009H or 0020H.

EXAMPLE 6–1

```
0000 33 DB                  XOR   BX,BX
0002 B8 0001         START: MOV   AX,1
0005 03 C3                  ADD   AX,BX
0007 EB 17                  JMP   SHORT NEXT

                     <skipped memory locations>

0020 8B D8           NEXT:  MOV   BX,AX
0022 EB DE                  JMP   START
```

Whenever a jump instruction references an address, a label normally identifies the address. The JMP NEXT instruction is an example; it jumps to label NEXT for the next instruction. It is very rare to use an actual hexadecimal address with any jump instruction, but the assembler supports addressing in relation to the instruction pointer by using the $+a displacement. For example, the JMP $+2 instruction jumps over the next two memory locations (bytes) following the JMP instruction. The label NEXT must be followed by a colon (NEXT:) to allow an instruction to reference it for a jump. If a colon does not follow a label, you cannot jump to it. Note that the only time a colon is used after a label is when the label is used with a jump or call instruction. This is also true in Visual C++.

Near Jump. The near jump is similar to the short jump, except that the distance is farther. A near jump passes control to an instruction in the current code segment located within ±32K bytes from the near jump instruction. The distance is ±2G in the 80386 and above when operated in protected mode. The near jump is a three-byte instruction that contains an opcode followed by a signed 16-bit displacement. In the 80386 through the Pentium 4 processors, the displacement is 32 bits and the near jump is five bytes long. The signed displacement adds to the instruction pointer (IP) to generate the jump address. Because the signed displacement is in the range of ±32K, a near jump can jump to any memory location within the current real mode code segment. The protected mode code segment in the 80386 and above can be 4G bytes long, so the 32-bit

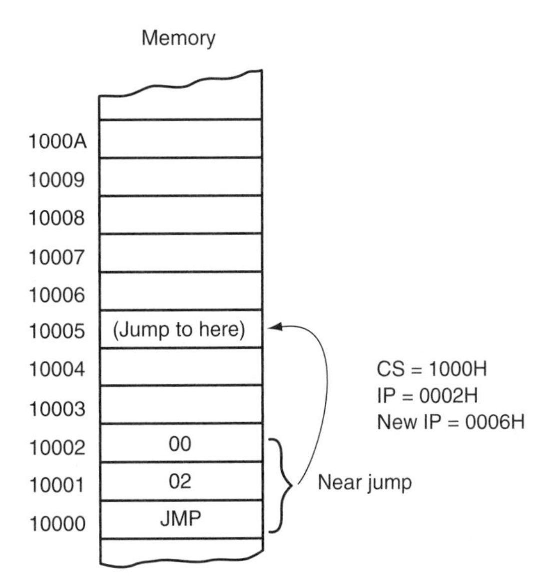

FIGURE 6–3 A near jump that adds the displacement (0002H) to the contents of IP.

displacement allows a near jump to any location within ±2G bytes. Figure 6–3 illustrates the operation of the real mode near jump instruction.

The near jump is also relocatable (as was the short jump) because it is also a relative jump. If the code segment moves to a new location in the memory, the distance between the jump instruction and the operand address remains the same. This allows a code segment to be relocated by simply moving it. This feature, along with the relocatable data segments, makes the Intel family of microprocessors ideal for use in a general-purpose computer system. Software can be written and loaded anywhere in the memory and function without modification because of the relative jumps and relocatable data segments.

Example 6–2 shows the same basic program that appeared in Example 6–1, except that the jump distance is greater. The first jump (JMP NEXT) passes control to the instruction at offset memory location 0200H within the code segment. Notice that the instruction assembles as E9 0200 R. The letter R denotes a **relocatable jump** address of 0200H. The relocatable address of 0200H is for the assembler program's internal use only. The actual machine language instruction assembles as E9 F6 01, which does not appear in the assembler listing. The actual displacement is 01F6H for this jump instruction. The assembler lists the jump address as 0200 R, so the address is easier to interpret as software is developed. If the linked execution file (.EXE) or command file (.COM) is displayed in hexadecimal code, the jump instruction appears as E9 F6 01.

EXAMPLE 6–2

```
0000 33DB                   XOR  BX,BX
0002 B8 0001        START:  MOV  AX,1
0005 03 C3                  ADD  AX,BX
0007 E9 0200 R              JMP  NEXT

                    <skipped memory locations>

0200 8B D8          NEXT:   MOV  BX,AX
0202 E9 0002 R              JMP  START
```

Far Jump. A far jump instruction (see Figure 6–4) obtains a new segment and offset address to accomplish the jump. Bytes 2 and 3 of this five-byte instruction contain the new offset address;

FIGURE 6–4 A far jump instruction replaces the contents of both CS and IP with four bytes following the opcode.

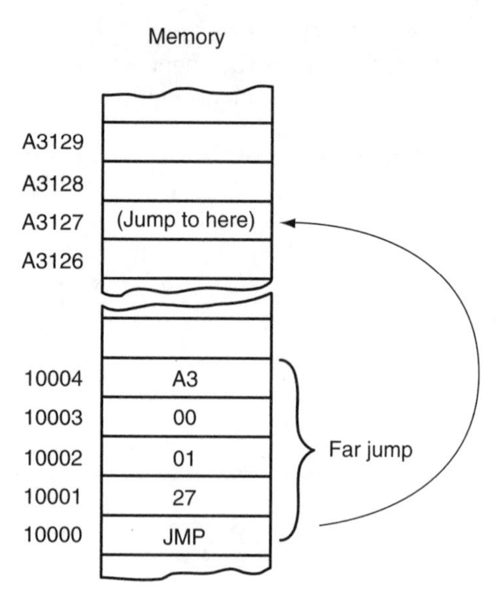

bytes 4 and 5 contain the new segment address. If the microprocessor (80286 through the Pentium 4) is operated in the protected mode, the segment address accesses a descriptor that contains the base address of the far jump segment. The offset address, which is either 16 or 32 bits, contains the offset address within the new code segment.

Example 6–3 lists a short program that uses a far jump instruction. The far jump instruction sometimes appears with the FAR PTR directive, as illustrated. Another way to obtain a far jump is to define a label as a **far label**. A label is far only if it is external to the current code segment or procedure. The JMP UP instruction in the example references a far label. The label UP is defined as a far label by the EXTRN UP:FAR directive. **External labels** appear in programs that contain more than one program file. Another way of defining a label as global is to use a double colon (LABEL::) following the label in place of the single colon. This is required inside procedure blocks that are defined as near if the label is accessed from outside the procedure block.

When the program files are joined, the linker inserts the address for the UP label into the JMP UP instruction. It also inserts the segment address in the JMP START instruction. The segment address in JMP FAR PTR START is listed as – – – – R for relocatable; the segment address in JMP UP is listed as – – – – E for external. In both cases, the – – – – is filled in by the linker when it links or joins the program files.

EXAMPLE 6–3

```
                                  EXTRN  UP:FAR

0000 33 DB                        XOR  BX,BX
0002 B8 0001              START:  ADD  AX,1
0005 E9 0200 R                    JMP  NEXT

                          <skipped memory locations>

0200 8B D8                NEXT:   MOV  BX,AX
0202 EA 0002 ---- R               JMP  FAR PTR START

0207 EA 0000 ---- R               JMP  UP
```

Jumps with Register Operands. The jump instruction can also use a 16- or 32-bit register as an operand. This automatically sets up the instruction as an **indirect jump**. The address of the jump is in the register specified by the jump instruction. Unlike the displacement associated with the near jump, the contents of the register are transferred directly into the instruction pointer. An indirect jump does not add to the instruction pointer, as with short and near jumps. The JMP AX instruction, for example, copies the contents of the AX register into the IP when the jump occurs. This allows a jump to any location within the current code segment. In the 80386 and above, a JMP EAX instruction also jumps to any location within the current code segment; the difference is that in protected mode the code segment can be 4G bytes long, so a 32-bit offset address is needed.

Example 6–4 shows how the JMP AX instruction accesses a jump table in the code segment. This DOS program reads a key from the keyboard and then modifies the ASCII code to 00H in AL for a '1', 01H for a '2', and 02H for a '3'. If a '1', '2', or '3' is typed, AH is cleared to 00H. Because the jump table contains 16-bit offset addresses, the contents of AX are doubled to 0, 2, or 4, so a 16-bit entry in the table can be accessed. Next, the offset address of the start of the jump table is loaded to SI, and AX is added to form the reference to the jump address. The MOV AX, [SI] instruction then fetches an address from the jump table, so the JMP AX instruction jumps to the addresses (ONE, TWO, or THREE) stored in the jump table.

EXAMPLE 6–4

```
                     ;Instructions that read 1, 2, or 3 from the keyboard.
                     ;The number is displayed as 1, 2, or 3 using a jump table
                     ;
                     .MODEL SMALL                   ;select SMALL model
0000                 .DATA                          ;start data segment
0000 0030 R          TABLE: DW   ONE                ;jump table
0002 0034 R                 DW   TWO
0004 0038 R                 DW   THREE
0000                 .CODE                          ;start code segment
                     .STARTUP                       ;start program
0017 B4 01           TOP:   MOV  AH,1               ;read key into AL
0019 CD 21                  INT  21H
001B 2C 31                  SUB  AL,31              ;convert to BCD
001D 72 F9                  JB   TOP                ;if key < 1
001F 32 02                  CMP  AL,2
0021 77 F4                  JA   TOP                ;if key > 3
0023 B4 00                  MOV  AH,0               ;double key code
0025 03 C0                  ADD  AX,AX
0027 BE 0000 R              MOV  SI,OFFSET TABLE    ;address TABLE
002A 03 F0                  ADD  SI,AX              ;form lookup address
002C 8B 04                  MOV  AX,[SI]            ;get ONE, TWO or THREE
002E FF E0                  JMP  AX                 ;jump to ONE, TWO or THREE
0030 B2 31           ONE:   MOV  DL,'1'             ;get ASCII 1
0032 EB 06                  JMP  BOT
0034 B2 32           TWO:   MOV  DL,'2'             ;get ASCII 2
0036 EB 02                  JMP  BOT
0038 B2 33           THREE: MOV  DL,'3'             ;get ASCII 3
003A B4 02           BOT:   MOV  AH,2               ;display number
003C CD 21                  INT  21H
                     .EXIT
                     END
```

Indirect Jumps Using an Index. The jump instruction may also use the [ ] form of addressing to directly access the jump table. The jump table can contain offset addresses for near indirect jumps, or segment and offset addresses for far indirect jumps. (This type of jump is also known as a **double-indirect** jump if the register jump is called an indirect jump.) The assembler assumes that the jump is near unless the FAR PTR directive indicates a far jump instruction. Here Example 6–5 repeats Example 6–4 by using the JMP TABLE [SI] instead of JMP AX. This reduces the length of the program.

EXAMPLE 6–5

```
                          ;Instructions that read 1, 2, or 3 from the keyboard.
                          ;The number is displayed as 1, 2, or 3 using a jump table
                          ;
                          .MODEL SMALL                ;select SMALL model
0000                      .DATA                       ;start data segment
0000 002D R               TABLE: DW   ONE             ;jump table
0002 0031 R                      DW   TWO
0004 0035 R                      DW   THREE
0000                      .CODE                       ;start code segment
                          .STARTUP                    ;start program
0017 B4 01                TOP:   MOV  AH,1            ;read key into AL
0019 CD 21                       INT  21H
001B 2C 31                       SUB  AL,31           ;convert to BCD
001D 72 F9                       JB   TOP             ;if key < 1
001F 32 02                       CMP  AL,2
0021 77 F4                       JA   TOP             ;if key > 3
0023 B4 00                       MOV  AH,0            ;double key code
0025 03 C0                       ADD  AX,AX
0027 B5 F0                       MOV  SI,AX           ;form lookup address
0029 FF A4 0000 R                JMP  TABLE[SI]       ;jump to ONE, TWO or THREE
002D B2 31                ONE:   MOV  DL,'1'          ;get ASCII 1
002F EB 06                       JMP  BOT
0031 B2 32                TWO:   MOV  DL,'2'          ;get ASCII 2
0033 EB 02                       JMP  BOT
0035 B2 33                THREE: MOV  DL,'3'          ;get ASCII 3
0037 B4 02                BOT:   MOV  AH,2            ;display number
0039 CD 21                       INT  21H
                          .EXIT
                          END
```

The mechanism used to access the jump table is identical with a normal memory reference. The JMP TABLE [S1] instruction points to a jump address stored at the code segment offset location addressed by S1. It jumps to the address stored in the memory at this location. Both the register and indirect indexed jump instructions usually address a 16-bit offset. This means that both types of jumps are near jumps. If a JMP FAR PTR [S1] or JMP TABLE [S1], with TABLE data defined with the DD directive, appears in a program, the microprocessor assumes that the jump table contains doubleword, 32-bit addresses (IP and CS).

Conditional Jumps and Conditional Sets

Conditional jump instructions are always short jumps in the 8086 through the 80286 microprocessors. This limits the range of the jump to within +127 bytes and –128 bytes from the location following the conditional jump. In the 80386 and above, conditional jumps are either short or near jumps. This allows these microprocessors to use a conditional jump to any location within the current code segment. Table 6–1 lists all the conditional jump instructions with their test conditions. Note that the Microsoft MASM version 6.x assembler automatically adjusts conditional jumps if the distance is too great.

The conditional jump instructions test the following flag bits: sign (S), zero (Z), carry (C), parity (P), and overflow (0). If the condition under test is true, a branch to the label associated with the jump instruction occurs. If the condition is false, the next sequential step in the program executes. For example, a JC will jump if the carry bit is set.

The operation of most conditional jump instructions is straightforward because they often test just one flag bit, although some test more than one. Relative magnitude comparisons require more complicated conditional jump instructions that test more than one flag bit.

Because both signed and unsigned numbers are used in programming, and because the order of these numbers is different, there are two sets of conditional jump instructions for

TABLE 6–1 Conditional jump instructions.

Assembly Language	*Tested Condition*	*Operation*
JA	Z = 0 and C = 0	Jump if above
JAE	C = 0	Jump if above or equal
JB	C = 1	Jump if below
JBE	Z = 1 or C = 1	Jump if below or equal
JC	C = 1	Jump if carry
JE or JZ	Z = 1	Jump if equal or jump if zero
JG	Z = 0 and S = 0	Jump if greater than
JGE	S = 0	Jump if greater than or equal
JL	S != O	Jump if less than
JLE	Z = 1 or S != O	Jump if less than or equal
JNC	C = 0	Jump if no carry
JNE or JNZ	Z = 0	Jump if not equal or jump if not zero
JNO	O = 0	Jump if no overflow
JNS	S = 0	Jump if no sign (positive)
JNP or JPO	P = 0	Jump if no parity or jump if parity odd
JO	O = 1	Jump if overflow
JP or JPE	P = 1	Jump if parity or jump if parity even
JS	S = 1	Jump if sign (negative)
JCXZ	CX = 0	Jump if CX is zero
JECXZ	ECX = 0	Jump if ECX equals zero

magnitude comparisons. Figure 6–5 shows the order of both signed and unsigned 8-bit numbers. The 16- and 32-bit numbers follow the same order as the 8-bit numbers, except that they are larger. Notice that an FFH (255) is above the 00H in the set of unsigned numbers, but an FFH (–1) is less than 00H for signed numbers. Therefore, an unsigned FFH is above 00H, but a signed FFH is less than 00H.

When signed numbers are compared, use the JG, JL, JGE, JLE, JE, and JNE instructions. The terms *greater than* and *less than* refer to signed numbers. When unsigned numbers are compared, use the JA, JB, JAB, JBE, JE, and JNE instructions. The terms *above* and *below* refer to unsigned numbers.

FIGURE 6–5 Signed and unsigned numbers follow different orders.

Unsigned numbers
255 FFH
254 FEH
132 84H
131 83H
130 82H
129 81H
128 80H
4 04H
3 03H
2 02H
1 01H
0 00H

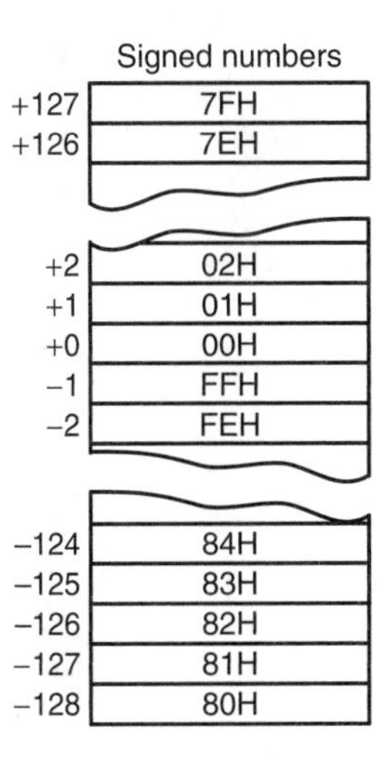

The remaining conditional jumps test individual flag bits, such as overflow and parity. Notice that JE has an alternative opcode, JZ. All instructions have alternates, but many aren't used in programming because they don't usually fit the condition under test. (The alternates appear in Appendix B with the instruction set listing.) For example, the JA instruction (jump if above) has the alternative JNBE (jump if not below or equal). A JA functions exactly like a JNBE, but a JNBE is awkward in many cases when compared to a JA.

The conditional jump instructions all test flag bits except for JCXZ (jump if CX = 0) and JECXZ (jump if ECX = 0). Instead of testing flag bits, JCXZ directly tests the contents of the CX register without affecting the flag bits, and JECXZ tests the contents of the ECX register. For the JCXZ instruction, if CX = 0, a jump occurs, and if CX != 0, no jump occurs. Likewise for the JECXZ instruction, if ECX = 0, a jump occurs; if CX != 0, no jump occurs.

A program that uses JCXZ appears in Example 6–6. Here, the SCASB instruction searches a table for 0AH. Following the search, a JCXZ instruction tests CX to see if the count has reached zero. If the count is zero, the 0AH is not found in the table. The carry flag is used in this example to pass the not found condition back to the calling program. Another method used to test to see if the data are found is the JNE instruction. If JNE replaces JCXZ, it performs the same function. After the SCASB instruction executes, the flags indicate a not-equal condition if the data were not found in the table.

EXAMPLE 6–6

```
                    ;Instructions that search a table of 100H bytes for 0AH
                    ;The offset address of TABLE is assumed to be in SI
                    ;
0017 B9 0064              MOV  CX,100          ;load counter
001A B0 0A                MOV  AL,0AH          ;load AL with 0AH
001C FC                   CLD                  ;auto-increment
001D F2/AE                REPNE SCASB          ;search for 0AH
001F F9                   STC                  ;set carry if found
0020 E3 01                JCXZ NOT_FOUND       ;if not found
0022                NOT_FOUND
```

The Conditional Set Instructions. In addition to the conditional jump instructions, the 80386 through the Pentium 4 processors also contain conditional set instructions. The conditions tested by conditional jumps are put to work with the conditional set instructions. The conditional set instructions set a byte to either 01H or clear a byte to 00H, depending on the outcome of the condition under test. Table 6–2 lists the available forms of the conditional set instructions.

These instructions are useful where a condition must be tested at a point much later in the program. For example, a byte can be set to indicate that the carry is cleared at some point in the program by using the SETNC MEM instruction. This instruction places 01H into memory location MEM if carry is cleared, and 00H into MEM if carry is set. The contents of MEM can be tested at a later point in the program to determine if carry is cleared at the point where the SETNC MEM instruction executed.

LOOP

The LOOP instruction is a combination of a decrement CX and the JNZ conditional jump. In the 8086 through the 80286 processors, LOOP decrements CX; if CX != 0, it jumps to the address indicated by the label. If CX becomes 0, the next sequential instruction executes. In the 80386 and above, LOOP decrements either CX or ECX, depending upon the instruction mode. If the 80386 and above operate in the 16-bit instruction mode, LOOP uses CX; if operated in the 32-bit instruction mode, LOOP uses ECX. This default is changed by the LOOPW (using CX) and LOOPD (using ECX) instructions in the 80386 through the Pentium 4.

TABLE 6–2 Conditional set instructions.

Assembly Language	*Tested Condition*	*Operation*
SETA	Z = 0 and C = 0	Set if above
SETAE	C = 0	Set if above or equal
SETB	C = 1	Set if below
SETBE	Z = 1 or C = 1	Set if below or equal
SETC	C = 1	Set if carry
SETE or SETZ	Z = 1	Set if equal or set if zero
SETG	Z = 0 and S = 0	Set if greater than
SETGE	S = 0	Set if greater than or equal
SETL	S != O	Set if less than
SETLE	Z = 1 or S != O	Set if less than or equal
SETNC	C = 0	Set if no carry
SETNE or SETNZ	Z = 0	Set if not equal or set if not zero
SETNO	O = 0	Set if no overflow
SETNS	S = 0	Set if no sign (positive)
SETNP or SETPO	P = 0	Set if no parity or set if parity odd
SETO	O = 1	Set if overflow
SETP or SETPE	P = 1	Set if parity or set if parity even
SETS	S = 1	Set if sign (negative)

Example 6–7 shows how data in one block of memory (BLOCK1) adds to data in a second block of memory (BLOCK2), using LOOP to control how many numbers add. The LODSW and STOSW instructions access the data in BLOCK1 and BLOCK2. The ADD AX,ES:[DI] instruction accesses the data in BLOCK2 located in the extra segment. The only reason that BLOCK2 is in the extra segment is that DI addresses extra segment data for the STOSW instruction. The .STARTUP directive only loads DS with the address of the data segment. In this example, the extra segment also addresses data in the data segment, so the contents of DS are copied to ES through the accumulator. Unfortunately, there is no direct move from segment register to segment register instruction.

EXAMPLE 6–7

```
                   ;A program that sums the contents of BLOCK1 and BLOCK2
                   ;and stores the results on top of the data in BLOCK2.
                   ;
                   .MODEL SMALL                  ;select SMALL model
0000               .DATA                         ;start data segment
0000 0064 [        BLOCK1 DW  100 DUP(?)         ;100 words for BLOCK1
           0000
               ]
00C8 0064 [        BLOCK2 DW  100 DUP(?)         ;100 words for BLOCK2
           0000
               ]
0000               .CODE                         ;start code segment
                   .STARTUP                      ;start program
0017 8C D8                MOV AX,DS              ;overlap DS and ES
0019 8E C0                MOV ES,AX
001B FC                   CLD                    ;select auto-increment
001C B9 0064              MOV CX,100             ;load counter
001F BE 0000 R            MOV SI,OFFSET BLOCK1   ;address BLOCK1
0022 BF 00C8 R            MOV DI,OFFSET BLOCK2   ;address BLOCK2
0025 AD            L1:    LODSW                  ;load AX with BLOCK1
0026 26:03 05             ADD AX,ES:[DI]         ;add BLOCK2
0029 AB                   STOSW                  ;save answer
002A E2 F9                LOOP L1                ;repeat 100 times
                   .EXIT
                   END
```

Conditional LOOPs. As with REP, the LOOP instruction also has conditional forms: LOOPE and LOOPNE. The LOOPE (**loop while equal**) instruction jumps if CX != 0 while an equal condition exists. It will exit the loop if the condition is not equal or if the CX register decrements to 0. The LOOPNE (**loop while not equal**) instruction jumps if CX != 0 while a not-equal condition exists. It will exit the loop if the condition is equal or if the CX register decrements to 0. In the 80386 through the Pentium 4 processors, the conditional LOOP instruction can use either CX or ECX as the counter. The LOOPEW/LOOPED or LOOPNEW/LOOPNED instructions override the instruction mode if needed.

As with the conditional repeat instructions, alternates exist for LOOPE and LOOPNE. The LOOPE instruction is the same as LOOPZ, and the LOOPNE instruction is the same as LOOPNZ. In most programs, only the LOOPE and LOOPNE apply.

6–2 CONTROLLING THE FLOW OF THE PROGRAM

It is much easier to use the assembly language statements .IF, .ELSE, .ELSEIF, and .ENDIF to control the flow of the program than it is to use the correct conditional jump statement. These statements always indicate a special assembly language command to MASM. Note that the control flow assembly language statements beginning with a period are only available to MASM version 6.xx, and not to earlier versions of the assembler such as 5.10. Other statements developed in this chapter include the .REPEAT-.UNTIL and .WHILE-.ENDW statements. These statements (the **dot commands**) do not function when using the Visual C++ inline assembler.

Example 6–8(a) shows how these statements are used to control the flow of a program by testing AL for the ASCII letters A through F. If the contents of AL are A through F, 7 is subtracted from AL.

Accomplishing the same task using the Visual C++ inline assembler is usually handled in C++ rather than in assembly language. Example 6–8(b) shows the same task using the inline assembler in visual C++ and conditional jumps in assembly language. It also shows how to use a label in an assembly block in Visual C++. This illustrates that it is more difficult to accomplish the same task without the dot commands. Never use uppercase for assembly language commands with the inline assembler because some of them are reserved by C++ and will cause problems.

EXAMPLE 6–8(a)

```
.IF AL >= 'A'  &&  AL <= 'F'
       SUB  AL,7
.ENDIF
SUB  AL,30H
```

EXAMPLE 6–8(b)

```
char temp;
_asm{
       mov  al,temp
       cmp  al,41h
       jb   Later
       cmp  al,46h
       ja   Later
       sub  al,7
Later:
       sub  al,30h
       mov  temp,al

}
```

TABLE 6–3 Relational operators used with the .IF statement in assembly language.

Operator	*Function*
==	Equal or the same as
!=	Not equal
>	Greater than
>=	Greater than or equal
<	Less than
<=	Less than or equal
&	Bit test
!	Logical inversion
&&	Logical AND
||	Logical OR
|	OR

In Example 6–8(a) notice how the && symbol represents the AND function in the .IF statement. There is no .if in Example 6–8(b) because the same operation was performed by using a few compare (CMP) instructions to accomplish the same task. See Table 6–3 for a complete list of relation operators used with the .IF statement. Note that many of these conditions (such as &&) are also used by many high-level languages such as C/C++.

Example 6–9 shows another example of the conditional .IF directive that converts all ASCII-coded letters to uppercase. First, the keyboard is read without echo using DOS INT 21H function 06H, and then the .IF statement converts the character into uppercase, if needed. In this example, the logical AND function (&&) is used to determine if the character is in lowercase. If it is lowercase, a 20H is subtracted, converting to uppercase. This program reads a key from the keyboard and converts it to uppercase before displaying it. Notice also how the program terminates when the control C key (ASCII = 03H) is typed. The .LISTALL directive causes all assembler-generated statements to be listed, including the label @Startup generated by the .STARTUP directive. The .EXIT directive also is expanded by .LISTALL to show the use of the DOS INT 21H function 4CH, which returns control to DOS.

EXAMPLE 6–9

```
                   ;A DOS program that reads the keyboard and converts all
                   ;lowercase data to uppercase before displaying it.
                   ;
                   ;This program is terminated with a control-C
                   ;
                   .MODEL TINY                      ;select tiny model
                   .LISTALL                         ;list all statements
0000               .CODE                            ;start code segment
                   .STARTUP                         ;start program
0100             * @Startup
0100 B4 06         MAIN1: MOV  AH,6                 ;read key without echo
0102 B2 FF                MOV  DL,0FFH
0104 CD 21                INT  21H
0106 74 F8                JE   MAIN1                ;if no key
0108 3C 03                CMP  AL,3                 ;test for control-C
010A 74 10                JE   MAIN2                ;if control-C

                          .IF  AL >= 'a' && AL <= 'z'

010C 3C 61       *               cmp  al,'a'
010E 72 06       *               jb   @C0001
0110 3C 7A       *               cmp  al,'z'
0112 77 02       *               ja   @C0001
```

```
0114 2C 20                         SUB   AL,20H

                           .ENDIF

0116              * @C0001:
0116 8A D0                   MOV   DL,AL              ;echo character to display
0118 CD 21                   INT   21H
011A EB E4                   JMP   MAIN1              ;repeat
011C                MAIN2:
                    .EXIT
011C B4 4C        *          MOV   AH,4CH
011E CD 21        *          INT   21H
                    END
```

In this program, a lowercase letter is converted to uppercase by the use of the .IF AL >= 'a' && AL <= 'z' statement. If AL contains a value that is greater than or equal to a lowercase a, and less than or equal to a lowercase z (a value of a–z), the statement between the .IF and .ENDIF executes. This statement (SUB AL,20H) subtracts 20H from the lowercase letter to change it to an uppercase letter. Notice how the assembler program implements the .IF statement (see lines that begin with *). The label @C0001 is an assembler-generated label used by the conditional jump statements placed in the program by the .IF statement.

Another example that uses the conditional .IF statement appears in Example 6–10. This program reads a key from the keyboard, and then converts it to hexadecimal code. This program is not listed in expanded form.

In this expanded, the .IF AL >='a' && AL<='f' statement causes the next instruction (SUB AL,57H) to execute if AL contains letters a through f, converting them to hexadecimal. If it is not between letters a and f, the next .ELSEIF statement tests it for the letters A through F. If it is the letters A through F, a 37H is subtracted from AL. If neither condition is true, a 30H is subtracted from AL before AL is stored at data segment memory location TEMP. The same conversion can be performed in a C++ function as illustrated in the program snippet of Example 6–10(b).

EXAMPLE 6–10(a)

```
                   ;A DOS program that reads key and stores its hexadecimal
                   ;value in memory location TEMP
                   ;
                   .MODEL SMALL                    ;select small model
0000               .DATA                           ;start data segment
0000 00            TEMP    DB      ?               ;define TEMP
0000               .CODE                           ;start code segment
                   .STARTUP                        ;start program
0017 B4 01                MOV  AH,1                ;read keyboard
0019 CD 21                INT  21H
                          .IF AL >= 'a' && AL <= 'f'
0023 2C 57                       SUB  AL,57H                 ;if lowercase
                          .ELSEIF .IF AL >= 'A' && AL <= 'F'
002F 2C 37                       SUB  AL,37H                 ;if uppercase
                          .ELSE
0033 2C 30                       SUB  AL,30H                 ;otherwise
                          .ENDIF
0035 A2 0000 R            MOV  TEMP,AL             ;save it in TEMP
                   .EXIT
                   END
```

EXAMPLE 6–10(b)

```
char Convert(char temp)
{
       if ( temp >= 'a' && temp <= 'f' )
              temp -= 0x57;
       else if ( temp >= 'A' && temp <= 'F' )
```

```
            temp -= 0x37;
        else
            temp -= 0x30;
        return temp;
}
```

WHILE Loops

As with most high-level languages, the assembler also provides the WHILE loop construct, available to MASM version 6.x. The .WHILE statement is used with a condition to begin the loop, and the .ENDW statement ends the loop.

Example 6–11 shows how the .WHILE statement is used to read data from the keyboard and store it into an array called BOP until the Enter key (0DH) is typed. This program assumes that BUF is stored in the extra segment because the STOSB instruction is used to store the keyboard data in memory. Note that the .WHILE loop portion of the program is shown in expanded form so that the statements inserted by the assembler (beginning with a *) can be studied. After the Enter key (0DH) is typed, the string is appended with a $ so it can be displayed with DOS INT 21H function number 9.

EXAMPLE 6–11

```
                         ;A DOS program that reads a character string from the
                         ;keyboard and then displays it again.
                         ;
                         .MODEL SMALL                      ;select small model
0000                     .DATA                             ;start data segment
0000 0D 0A               MES    DB  13,10                  ;return and line feed
0002 0100 [              BUF    DB 256 DUP(?)              ;character string buffer
           00
              ]
0000                     .CODE                             ;start code segment
                         .STARTUP                          ;start program
0017 8C D8                      MOV  AX,DX                 ;overlap DS with ES
0019 8C C0                      MOV  ES,AX
001B FC                         CLD                        ;select auto-increment
001C BF 0002 R                  MOV  DI,OFFSET BUF         ;address buffer

                                .WHILE AL != 0DH           ;loop while not enter
001F EB 05             *        jmp  @C0001
0021                   * @C0002:
0021 B4 01                      MOV  AH,1                  ;read key
0023 CD 21                      INT  21H
0025 AA                         STOSB                      ;store key code
                                .ENDW
0026                   * @C0001:
0026 3C 0D             *        cmp  al,0dh
0028 75 F7             *        jne  @C0002
002A C6 45 FF 24                MOV  BYTE PTR[DI-1]'&'
002E BA 0000 R                  MOV  DX,OFFSET MES
0031 B4 09                      MOV  AH,9
0033 CD 21                      INT  21H                   ;display MES
                         .EXIT
                         END
```

The program in Example 6–11 functions perfectly, as long as we arrive at the .WHILE statement with AL containing some other value except 0DH. This can be corrected by adding a MOV AL,0DH instruction before the .WHILE statement in Example 6–11. Although not shown in an example, the .BREAK and .CONTINUE statements are available for use with the while loop. The .BREAK statement is often followed by the .IF statement to select the break condition as in .BREAK .IF AL == 0DH. The .CONTINUE statement, which can be used to allow the DO–.WHILE loop to continue if a certain condition is met, can be used with

.BREAK. For example, .CONTINUE .IF AL == 15 allows the loop to continue if AL equals 15. Note that the .BREAK and .CONTINUE commands function in the same manner in a C++ program.

REPEAT-UNTIL Loops

Also available to the assembler is the REPEAT-UNTIL construct. A series of instructions is repeated until some condition occurs. The .REPEAT statement defines the start of the loop; the end is defined with the .UNTIL statement, which contains a condition. Note that .REPEAT and .UNTIL are available to version 6.x of MASM.

If Example 6–11 is again reworked by using the REPEAT-UNTIL construct, this appears to be the best solution. See Example 6–12 for the program that reads keys from the keyboard and stores keyboard data into extra segment array BUF until the Enter key is pressed. This program also fills the buffer with keyboard data until the Enter key (0DH) is typed. Once the Enter key is typed, the program displays the character string using DOS INT 21H function number 9, after appending the buffer data with the required dollar sign. Notice how the .UNTIL AL == 0DH statement generates code (statements beginning with *) to test for the Enter key.

EXAMPLE 6–12

```
                         ;A DOS program that reads a character string from the
                         ;keyboard and then displays it again.
                         ;
                         .MODEL SMALL                        ;select small model
0000                     .DATA                               ;start data segment
0000 0D 0A               MES    DB  13,10                    ;return and line feed
0002 0100[               BUF    DB 256 DUP(?)                ;character string buffer
         00
           ]
0000                     .CODE                               ;start code segment
                         .STARTUP                            ;start program
0017 8C D8                      MOV   AX,DX                  ;overlap DS with ES
0019 8C C0                      MOV   ES,AX
001B FC                         CLD                          ;select auto-increment
001C BF 0002 R                  MOV   DI,OFFSET BUF          ;address buffer

                                .REPEAT                      ;repeat until enter

001F                   * @C0001:
001F B4 01                      MOV   AH,1                   ;read key
0021 CD 21                      INT   21H
0023 AA                         STOSB                        ;store key code

                                .UNTIL AL == 0DH

0025 3C 0D             *        cmp   al,0dh
0027 75 F7             *        jne   @C0001
0028 C6 45 FF 24                MOV   BYTE PTR[DI-1]'&'
002C BA 0000 R                  MOV   DX,OFFSET MES
002E B4 09                      MOV   AH,9
0031 CD 21                      INT   21H                    ;display MES
                         .EXIT
                         END
```

There is also an .UNTILCXZ instruction available that uses the LOOP instruction to check CX for a repeat loop. The .UNTILCXZ instruction uses the CX register as a counter to repeat a loop a fixed number of times. Example 6–13 shows a sequence of instructions that uses the .UNTILCXZ instruction used to add the contents of byte-sized array ONE to byte-sized array

TWO. The sums are stored in array THREE. Note that each array contains 100 bytes of data, so the loop is repeated 100 times. This example assumes that array THREE is in the extra segment, and that arrays ONE and TWO are in the data segment. Notice how the LOOP instruction is inserted for the .UNTILCXZ.

EXAMPLE 6–13

```
012C B9 0064              MOV  CX,100              ;set count
012F BF 00C8 R            MOV  DI,OFFSET THREE     ;address arrays
0132 BE 0000 R            MOV  SI,OFFSET ONE
0135 BB 0064 R            MOV  BX,OFFSET TWO

                          .REPEAT

0138             * @C0001:
0138 AC                   LODSB
0139 02 07                ADD  AL,[BX]
013B AA                   STOSB
013C 43                   INC  BX

                          .UNTILCXZ

013D E2 F9       *        LOOP @C0001
```

6–3 PROCEDURES

The procedure (subroutine or **function**) is an important part of any computer system's architecture. A procedure is a group of instructions that usually performs one task. A procedure is a reusable section of the software that is stored in memory once, but used as often as necessary. This saves memory space and makes it easier to develop software. The only disadvantage of a procedure is that it takes the computer a small amount of time to link to the procedure and return from it. The CALL instruction links to the procedure, and the RET (**return**) instruction returns from the procedure.

The stack stores the return address whenever a procedure is called during the execution of a program. The CALL instruction pushes the address of the instruction following the CALL (**return address**) on the stack. The RET instruction removes an address from the stack so the program returns to the instruction following the CALL.

With the assembler, there are specific rules for storing procedures. A procedure begins with the PROC directive and ends with the ENDP directive. Each directive appears with the name of the procedure. This programming structure makes it easy to locate the procedure in a program listing. The PROC directive is followed by the type of procedure: NEAR or FAR. Example 6–14 shows how the assembler uses the definition of both a near (intrasegment) and far (intersegment) procedure. In MASM version 6.x, the NEAR or FAR type can be followed by the USES statement. The USES statement allows any number of registers to be automatically pushed to the stack and popped from the stack within the procedure. The USES statement is also illustrated in Example 6–14.

EXAMPLE 6–14

```
0000               SUMS    PROC NEAR
0000 03 C3                 ADD  AX,BX
0002 03 C1                 ADD  AX,CX
0004 03 C2                 ADD  AX,DX
0006 C3                    RET
0007               SUMS    ENDP
```

```
0007                    SUMS1   PROC FAR
0007 03 C3                      ADD  AX,BX
0009 03 C1                      ADD  AX,CX
000B 03 C2                      ADD  AX,DX
000D CB                         RET
000E                    SUMS1   ENDP

000E                    SUMS3   PROC NEAR  USE  BX CX DX
0011 03 C3                      ADD  AX,BX
0013 03 C1                      ADD  AX,CX
0015 03 C2                      ADD  AX,DX
                                RET
001B                    SUMS    ENDP
```

When these first two procedures are compared, the only difference is the opcode of the return instruction. The near return instruction uses opcode C3H and the far return uses opcode CBH. A near return removes a 16-bit number from the stack and places it into the instruction pointer to return from the procedure in the current code segment. A far return removes a 32-bit number from the stack and places it into both IP and CS to return from the procedure to any memory location.

Procedures that are to be used by all software (**global**) should be written as far procedures. Procedures that are used by a given task (**local**) are normally defined as near procedures. Most procedures are near procedures.

CALL

The CALL instruction transfers the flow of the program to the procedure. The CALL instruction differs from the jump instruction because a CALL saves a return address on the stack. The return address returns control to the instruction that immediately follows the CALL in a program when a RET instruction executes.

Near CALL. The near CALL instruction is three bytes long; the first byte contains the opcode, and the second and third bytes contain the displacement, or distance of ±32K in the 8086 through the 80286 processors. This is identical to the form of the near jump instruction. The 80386 and above use a 32-bit displacement, when operating in the protected mode, to allow a distance of ±2G bytes. When the near CALL executes, it first pushes the offset address of the next instruction on the stack. The offset address of the next instruction appears in the instruction pointer (IP or EIP). After saving this return address, it then adds the displacement from bytes 2 and 3 to the IP to transfer control to the procedure. There is no short CALL instruction. A variation on the opcode exists as CALLN, but this should be avoided in favor of using the PROC statement to define the CALL as near.

Why save the IP or EIP on the stack? The instruction pointer always points to the next instruction in the program. For the CALL instruction, the contents of IP/EIP are pushed onto the stack, so program control passes to the instruction following the CALL after a procedure ends. Figure 6–6 shows the return address (IP) stored on the stack and the call to the procedure.

Far CALL. The far CALL instruction is like a far jump because it can call a procedure stored in any memory location in the system. The far CALL is a five-byte instruction that contains an opcode followed by the next value for the IP and CS registers. Bytes 2 and 3 contain the new contents of the IP, and bytes 4 and 5 contain the new contents for CS.

The far CALL instruction places the contents of both IP and CS on the stack before jumping to the address indicated by bytes 2–5 of the instruction. This allows the far CALL to call a procedure located anywhere in the memory and return from that procedure.

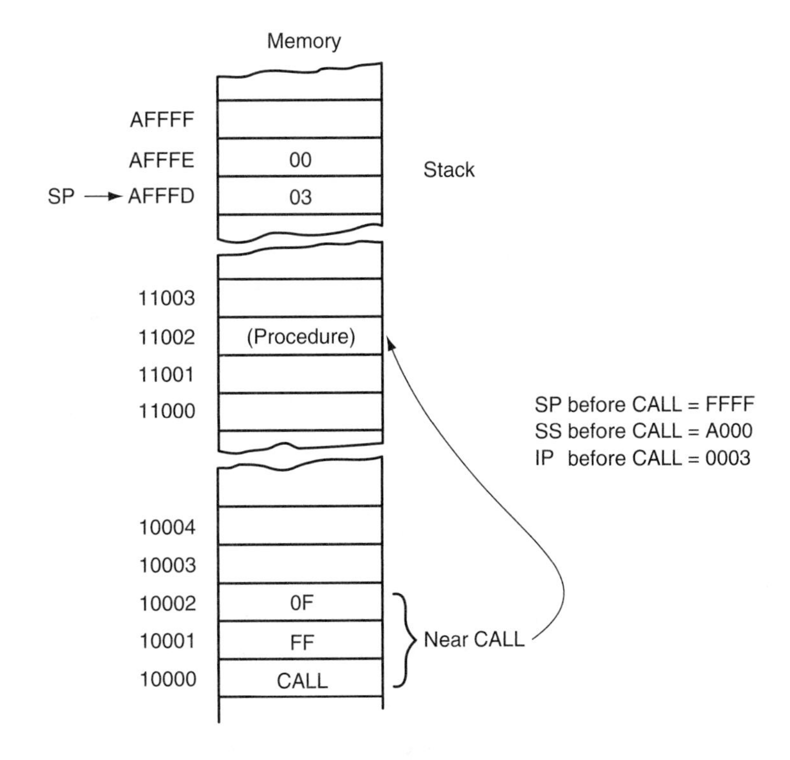

FIGURE 6–6 The effect of a near CALL on the stack and the instruction pointer.

Figure 6–7 shows how the far CALL instruction calls a far procedure. Here, the contents of IP and CS are pushed onto the stack. Next, the program branches to the procedure. A variant of the far call exists as CALLF, but this should be avoided in favor of defining the type of call instruction with the PROC statement.

CALLs with Register Operands. Like jump instructions, call instructions also may contain a register operand. An example is the CALL BX instruction, which pushes the contents of IP onto the stack. It then jumps to the offset address, located in register BX, in the current code segment. This type of CALL always uses a 16-bit offset address, stored in any 16-bit register except the segment registers.

Example 6–15 illustrates the use of the CALL register instruction to call a procedure that begins at offset address DISP. (This call could also call the procedure directly by using the CALL DISP instruction.) The OFFSET address DISP is placed into the BX register, and then the CALL BX instruction calls the procedure beginning at address DISP. This program displays "OK" on the monitor screen.

EXAMPLE 6–15

```
                    ;A DOS program that displays OK using the DISP procedure.
                    ;
                    .MODEL TINY                           ;select tiny model
0000                .CODE                                 ;start code segment
                    .STARTUP                              ;start program
0100 BB 0110 R             MOV  BX,OFFSET DISP   ;load BX with offset DISP
0103 B2 4F                 MOV  DL,'O'                    ;display O
0105 FF D3                 CALL BX
0107 B2 4B                 MOV  DL,'K'                    ;display K
0109 FF D3                 CALL BX
                    .EXIT
                    ;
                    ;Procedure that displays the ASCII character in DL
                    ;
```

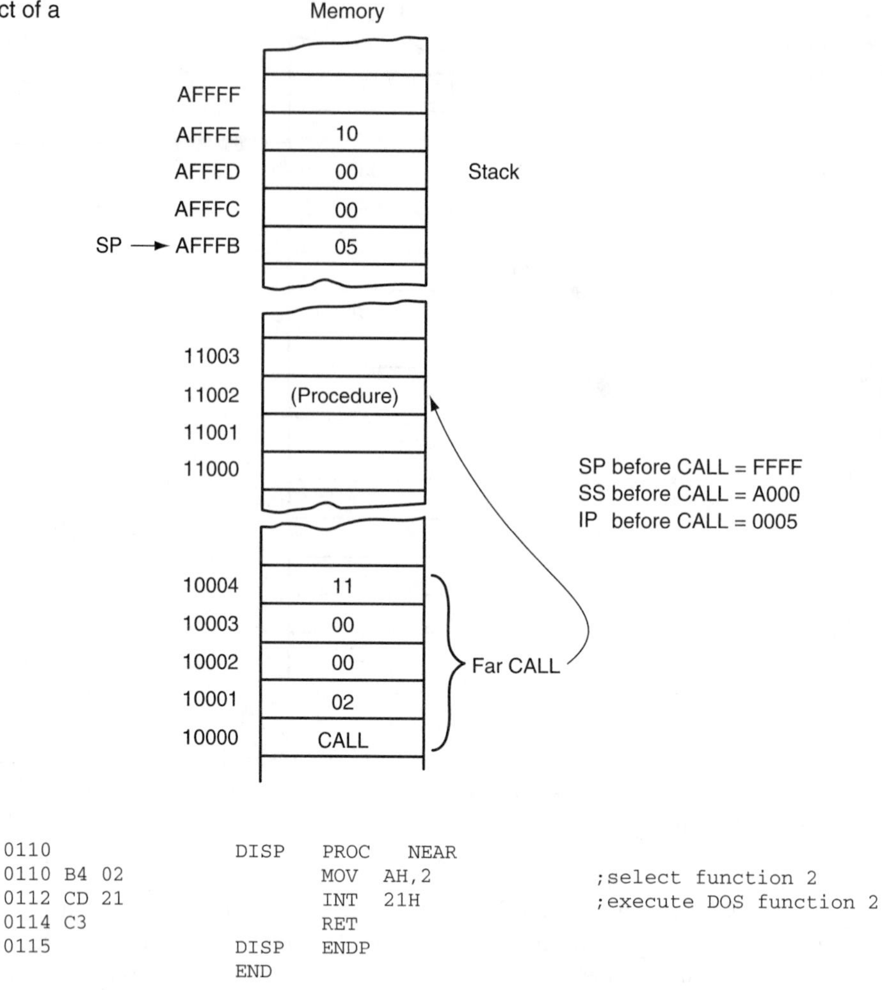

FIGURE 6–7 The effect of a far CALL instruction.

```
0110                 DISP   PROC   NEAR
0110 B4 02                  MOV  AH,2                ;select function 2
0112 CD 21                  INT  21H                 ;execute DOS function 2
0114 C3                     RET
0115                 DISP   ENDP
                     END
```

CALLs with Indirect Memory Addresses. A CALL with an indirect memory address is particularly useful whenever different subroutines need to be chosen in a program. This selection process is often keyed with a number that addresses a CALL address in a lookup table. This is essentially the same as the indirect jump that used a lookup table for a jump address earlier in this chapter.

Example 6–16 shows how to access a table of addresses using an indirect CALL instruction. This table illustrated in the example contains three separate subroutine addresses referenced by the numbers 0, 1, and 2. This example uses the scaled-index addressing mode to multiply the number in EBX by 2 so it properly accesses the correct entry in the lookup table.

EXAMPLE 6–16

```
              ;Instruction that calls procedure ZERO, ONE, or TWO
              ;depending on the value in EBX
              ;
       TABLE  DW    ZERO                    ;address of procedure ZERO
              DW    ONE                     ;address of procedure ONE
              DW    TWO                     ;address of procedure TWO

              CALL  TABLE[2*EBX]
```

The CALL instruction also can reference far pointers if the instruction appears as CALL FAR PTR [4*EBX] or as CALL TABLE [4*EBX], if the data in the table are defined as doubleword data with the DD directive. These instructions retrieve a 32-bit address (four bytes long) from the data segment memory location addressed by EBX and use it as the address of a far procedure.

RET

The return instruction (RET) removes a 16-bit number (**near return**) from the stack and places it into IP, or removes a 32-bit number (**far return**) and places it into IP and CS. The near and far return instructions are both defined in the procedure's PROC directive, which automatically selects the proper return instruction. With the 80386 through the Pentium 4 processors operating in the protected mode, the far return removes six bytes from the stack. The first four bytes contain the new value for EIP and the last two contain the new value for CS. In the 80386 and above, a protected mode near return removes four bytes from the stack and places them into EIP.

When IP/EIP or IP/EIP and CS are changed, the address of the next instruction is at a new memory location. This new location is the address of the instruction that immediately follows the most recent CALL to a procedure. Figure 6–8 shows how the CALL instruction links to a procedure and how the RET instruction returns in the 8086–Pentium 4 operating in the real mode.

There is one other form of the return instruction, which adds a number to the contents of the stack pointer (SP) after the return address is removed from the stack. A return that uses an immediate operand is ideal for use in a system that uses the C/C++ or PASCAL calling conventions. (This is true even though the C/C++ and PASCAL calling conventions require the caller to remove stack data for many functions.) These conventions push parameters on the stack before calling a procedure. If the parameters are to be discarded upon return, the return

FIGURE 6–8 The effect of a near return instruction on the stack and instruction pointer.

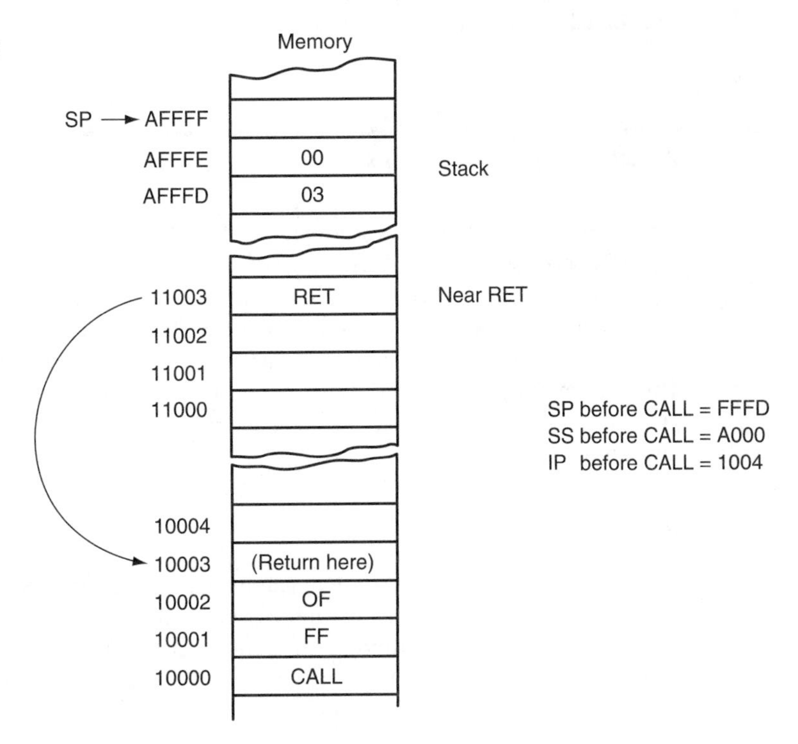

instruction contains a number that represents the number of bytes pushed to the stack as parameters.

Example 6–17 shows how this type of return erases the data placed on the stack by a few pushes. The RET 4 adds a 4 to SP after removing the return address from the stack. Because the PUSH AX and PUSH BX together place four bytes of data on the stack, this return effectively deletes AX and BX from the stack. This type of return rarely appears in assembly language programs, but it is used in high-level programs to clear stack data after a procedure. Notice how parameters are addressed on the stack by using the BP register, which by default addresses the stack segment. Parameter stacking is common in procedures written for C++ or PASCAL by using the C++ or PASCAL calling conventions.

EXAMPLE 6–17

```
0000 B8 001E              MOV   AX,30
0003 BB 0028              MOV   BX,40
0006 50                   PUSH AX                  ;stack parameter 1
0007 53                   PUSH BX                  ;stack parameter 2
0008 E8 0066              CALL ADDM                ;add stack parameters


0071                ADDM  PROC NEAR
0071 55                   PUSH BP                  ;save BP
0072 8B EC                MOV   BP,SP              ;address stack with BP
0074 8B 46 04             MOV   AX,[BP+4]          ;get parameter 1
0077 03 46 06             ADD   AX,[BP+6]          ;add parameter 2
007A 5D                   POP   BP                 ;restore BP
007B C2 0004              RET   4                  ;return, dump parameters
007E                ADDM  ENDP
```

As with the CALLN and CALLF instructions, there are also variants of the return instruction: RETN and RETF. As with the CALLN and CALLF instructions, these variants should also be avoided in favor of using the PROC statement to define the type of call and return.

6–4 INTRODUCTION TO INTERRUPTS

An interrupt is either a **hardware-generated CALL** (externally derived from a hardware signal) or a **software-generated CALL** (internally derived from the execution of an instruction or by some other internal event). At times, an internal interrupt is called an *exception*. Either type interrupts the program by calling an **interrupt service procedure** (ISP) or interrupt handler.

This section explains software interrupts, which are special types of CALL instructions. This section describes the three types of software interrupt instructions (INT, INTO, and INT 3), provides a map of the interrupt vectors, and explains the purpose of the special interrupt return instruction (IRET).

Interrupt Vectors

An **interrupt vector** is a four-byte number stored in the first 1024 bytes of the memory (00000H–003FFH) when the microprocessor operates in the real mode. In the protected mode, the vector table is replaced by an interrupt descriptor table that uses eight-byte descriptors to describe each of the interrupts. There are 256 different interrupt vectors, and each vector contains

TABLE 6–4 Interrupt vectors defined by Intel.

Number	*Address*	*Microprocessor*	*Function*
0	0H–3H	All	Divide error
1	4H–7H	All	Single-step
2	8–BH	All	NMI pin
3	CH–FH	All	Breakpoint
4	10H–13H	All	Interrupt on overflow
5	14H–17H	80186–Pentium 4	Bound instruction
6	18H–1BH	80186–Pentium 4	Invalid opcode
7	1CH–1FH	80186–Pentium 4	Coprocessor emulation
8	20H–23H	80386–Pentium 4	Double fault
9	24H–27H	80386	Coprocessor segment overrun
A	28H–2BH	80386–Pentium 4	Invalid task state segment
B	2CH–2FH	80386–Pentium 4	Segment not present
C	30H–33H	80386–Pentium 4	Stack fault
D	34H–37H	80386–Pentium 4	General protection fault (GPF)
E	38H–3BH	80386–Pentium 4	Page fault
F	3CH–3FH	—	Reserved
10	40H–43H	80286–Pentium 4	Floating-point error
11	44H–47H	80486SX	Alignment check interrupt
12	48H–4BH	Pentium–Pentium 4	Machine check exception
13–1F	4CH–7FH	—	Reserved
20–FF	80H–3FFH	—	User interrupts

the address of an interrupt service procedure. Table 6–4 lists the interrupt vectors with a brief description and the memory location of each vector for the real mode. Each vector contains a value for IP and CS that forms the address of the interrupt service procedure. The first two bytes contain the IP, and the last two bytes contain the CS.

Intel reserves the first 32 interrupt vectors for present and future microprocessor products. The remaining interrupt vectors (32–255) are available for the user. Some of the reserved vectors are for errors that occur during the execution of software, such as the divide error interrupt. Some vectors are reserved for the coprocessor. Still others occur for normal events in the system. In a personal computer, the reserved vectors are used for system functions, as detailed later in this section. Vectors 1–6, 7, 9, 16, and 17 function in the real mode and protected mode; the remaining vectors function only in the protected mode.

Interrupt Instructions

The microprocessor has three different interrupt instructions that are available to the programmer: INT, INTO, and INT 3. In the real mode, each of these instructions fetches a vector from the vector table, and then calls the procedure stored at the location addressed by the vector. In the protected mode, each of these instructions fetches an interrupt descriptor from the interrupt descriptor table. The descriptor specifies the address of the interrupt service procedure. The interrupt call is similar to a far CALL instruction because it places the return address (IP/EIP and CS) on the stack.

INTs. There are 256 different software interrupt instructions (INTs) available to the programmer. Each INT instruction has a numeric operand whose range is 0 to 255 (00H–FFH). For example, INT 100 uses interrupt vector 100, which appears at memory address 190H–193H. The address of the interrupt vector is determined by multiplying the interrupt type number times 4. For example, the INT 10H instruction calls the interrupt service procedure whose address is stored beginning at memory location 40H (10H × 4) in the real mode. In the protected mode, the interrupt descriptor is located by multiplying the type number by 8 instead of 4 because each descriptor is eight bytes long.

Each INT instruction is two bytes long. The first byte contains the opcode, and the second byte contains the vector type number. The only exception to this is INT 3, a one-byte special software interrupt used for breakpoints.

Whenever a software interrupt instruction executes, it (1) pushes the flags onto the stack, (2) clears the T and I flag bits, (3) pushes CS onto the stack, (4) fetches the new value for CS from the interrupt vector, (5) pushes IP/EIP onto the stack, (6) fetches the new value for IP/EIP from the vector, and (7) jumps to the new location addressed by CS and IP/EIP.

The INT instruction performs as a far CALL except that it not only pushes CS and IP onto the stack, but it also pushes the flags onto the stack. The INT instruction performs the operation of a PUSHF, followed by a far CALL instruction.

Notice that when the INT instruction executes, it clears the interrupt flag (I), which controls the external hardware interrupt input pin INTR (interrupt request). When I = 0, the microprocessor disables the INTR pin; when I = 1, the microprocessor enables the INTR pin.

Software interrupts are most commonly used to call system procedures because the address of the system function need not be known. The system procedures are common to all system and application software. The interrupts often control printers, video displays, and disk drives. Besides relieving the program from remembering the address of the system call, the INT instruction replaces a far CALL that otherwise would be used to call a system function. The INT instruction is two bytes long, whereas the far CALL is five bytes long. Each time that the INT instruction replaces a far CALL, it saves three bytes of memory in a program. This can amount to a sizable saving if the INT instruction appears often in a program, as it does for system calls.

IRET/IRETD. The interrupt return instruction (IRET) is used only with software or hardware interrupt service procedures. Unlike a simple return instruction (RET), the IRET instruction will (1) pop stack data back into the IP, (2) pop stack data back into CS, and (3) pop stack data back into the flag register. The IRET instruction accomplishes the same tasks as the POPF, followed by a far RET instruction.

Whenever an IRET instruction executes, it restores the contents of I and T from the stack. This is important because it preserves the state of these flag bits. If interrupts were enabled before an interrupt service procedure, they are automatically re-enabled by the IRET instruction because it restores the flag register.

In the 80386 through the Pentium 4 processors, the IRETD instruction is used to return from an interrupt service procedure that is called in the protected mode. It differs from the IRET because it pops a 32-bit instruction pointer (EIP) from the stack. The IRET is used in the real mode and the IRETD is used in the protected mode.

INT 3. An INT 3 instruction is a special software interrupt designed to function as a **breakpoint** to interrupt or break the flow of the software. The difference between it and the other software interrupts is that INT 3 is a one-byte instruction, while the others are two-byte instructions.

A breakpoint occurs for any software interrupt, but because INT 3 is one byte long, it is easier to use for this function. Breakpoints also help to debug faulty software.

INTO. Interrupt on overflow (INTO) is a conditional software interrupt that tests the overflow flag (O). If O = 0, the INTO instruction performs no operation; if O = 1 and an INTO instruction executes, an interrupt occurs via vector type number 4.

The INTO instruction appears in software that adds or subtracts signed binary numbers. With these operations, it is possible to have an overflow. Either the JO instruction or the INTO instruction detects the overflow condition.

An Interrupt Service Procedure. Suppose that, in a particular system, a procedure is required to add the contents of DI, SI, BP, and BX and then save the sum in AX. Because this is a common task in this system, occasionally it may be worthwhile to develop the task as a software interrupt. Although interrupts are usually reserved for system events, this example shows how an interrupt service procedure appears. Example 6–18 shows this software interrupt. The main difference between this procedure and a normal far procedure is that it ends with the IRET instruction instead of the RET instruction, and the contents of the flag register are saved on the stack during its execution. It is also important to save all registers that are changed by the procedure using USES.

EXAMPLE 6–18

```
0000                 INTS    PROC FAR USES AX
0000 03 C3                   ADD  AX,BX
0002 03 05                   ADD  AX,BP
0004 03 C7                   ADD  AX,DI
0006 03 C6                   ADD  AX,SI
0008 CF                      IRET
0009                 INTS    ENDP
```

Interrupt Control

Although this section does not explain hardware interrupts, two instructions are introduced that control the INTR pin. The **set interrupt flag** instruction (STI) places a 1 into the I flag bit, which enables the INTR pin. The **clear interrupt flag** instruction (CLI) places a 0 into the I flag bit, which disables the INTR pin. The STI instruction enables INTR and the CLI instruction disables INTR. In a software interrupt service procedure, hardware interrupts are enabled as one of the first steps. This is accomplished by the STI instruction. The reason interrupts are enabled early in an interrupt service procedure is that just about all of the I/O devices in the personal computer are interrupt-processed. If the interrupts are disabled for too long, severe system problems result.

Interrupts in the Personal Computer

The interrupts found in the personal computer differ somewhat from the ones presented in Table 6–4. The reason that they differ is that the original personal computers were 8086/8088-based systems. This meant that they only contained Intel-specified interrupts 0–4. This design has been carried forward so that newer systems are compatible with the early personal computers.

Access to the protected mode interrupt structure in use by Windows is accomplished through kernel functions Microsoft provides and cannot be directly addressed. Protected mode interrupts use an interrupt descriptor table, which is beyond the scope of the text at this point. Protected mode interrupts are discussed completely in later chapters.

Figure 6–9 illustrates the interrupts available in the author's computer. The interrupt assignments can be viewed in the control panel of Windows under Performance and Maintenance. Click on System and select Hardware and then Device Manager. Then click on View and select Device by Type and finally Interrupts.

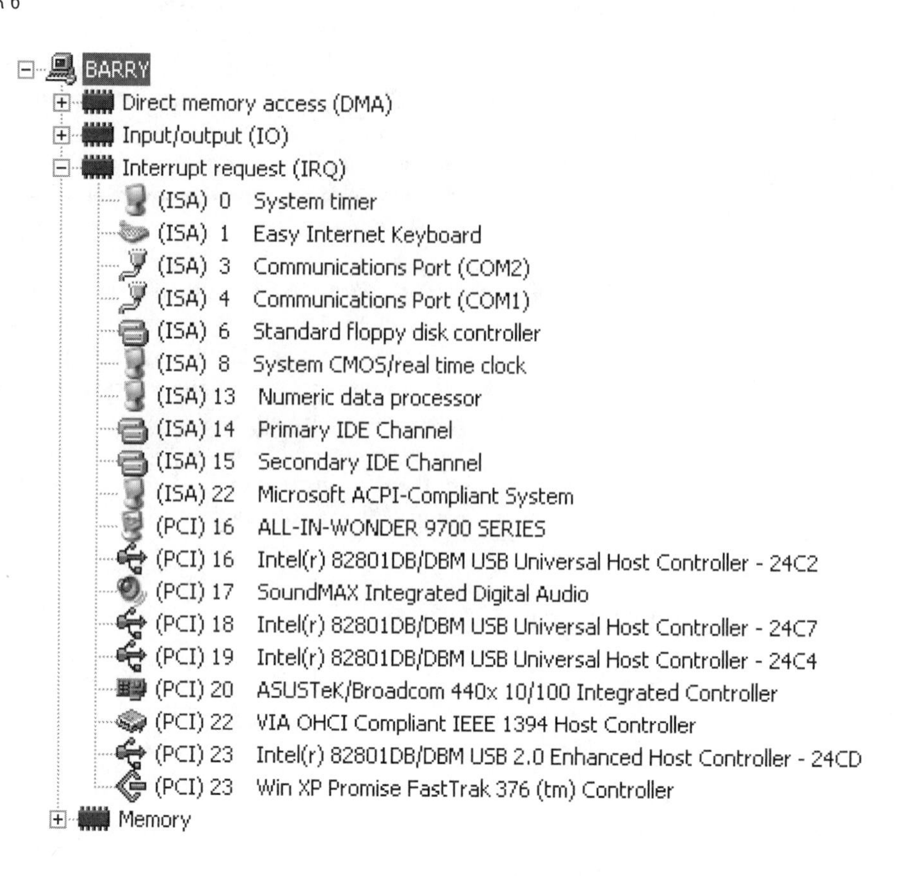

FIGURE 6–9 Interrupts in a typical personal computer.

6–5 MACHINE CONTROL AND MISCELLANEOUS INSTRUCTIONS

The last category of real mode instructions found in the microprocessor are the machine control and miscellaneous group. These instructions provide control of the carry bit, sample the BUSY/TEST pin, and perform various other functions. Because many of these instructions are used in hardware control, they need only be explained briefly at this point.

Controlling the Carry Flag Bit

The carry flag (C) propagates the carry or borrow in multiple-word/doubleword addition and subtraction. It also can indicate errors in assembly language procedures. Three instructions control the contents of the carry flag: STC (set carry), CLC (clear carry), and CMC (complement carry).

Because the carry flag is seldom used except with multiple-word addition and subtraction, it is available for other uses. The most common task for the carry flag is to indicate an error upon return from a procedure. Suppose that a procedure reads data from a disk memory file. This operation can be successful, or an error such as file-not-found can occur. Upon return from this procedure, if C = 1, an error has occurred; if C = 0, no error occurred. Most of the DOS and BIOS procedures use the carry flag to indicate error conditions. This flag is not available in Visual C/C++ for use with C++.

WAIT

The WAIT instruction monitors the hardware $\overline{BUSY}$ pin on the 80286 and 80386, and the $\overline{TEST}$ pin on the 8086/8088. The name of this pin was changed beginning with the 80286 microprocessor from $\overline{TEST}$ to $\overline{BUSY}$. If the WAIT instruction executes while the $\overline{BUSY}$ pin = 1, nothing happens and the next instruction executes. If the $\overline{BUSY}$ pin = 0 when the WAIT instruction executes, the microprocessor waits for the $\overline{BUSY}$ pin to return to a logic 1. This pin inputs a busy condition when at a logic 0 level.

The $\overline{BUSY}/\overline{TEST}$ pin of the microprocessor is usually connected to the $\overline{BUSY}$ pin of the 8087 through the 80387 numeric coprocessors. This connection allows the microprocessor to wait until the coprocessor finishes a task. Because the coprocessor is inside the 80486 through the Pentium 4, the $\overline{BUSY}$ pin is not present in these microprocessors.

HLT

The halt instruction (HLT) stops the execution of software. There are three ways to exit a halt: by an interrupt, by a hardware reset, or during a DMA operation. This instruction normally appears in a program to wait for an interrupt. It often synchronizes external hardware interrupts with the software system. Note that DOS and Windows both use interrupts extensively, so HLT will not halt the computer when operated under these operating systems.

NOP

When the microprocessor encounters a no operation instruction (NOP), it takes a short time to execute. In early years, before software development tools were available, a NOP, which performs absolutely no operation, was often used to pad software with space for future machine language instructions. If you are developing machine language programs, which are extremely rare, it is recommended that you place 10 or so NOPs in your program at 50-byte intervals. This is done in case you need to add instructions at some future point. A NOP may also find application in time delays to waste time. However, a NOP used for timing is not very accurate because of the cache and pipelines in modern microprocessors.

LOCK Prefix

The LOCK prefix appends an instruction and causes the $\overline{LOCK}$ pin to become a logic 0. The $\overline{LOCK}$ pin often disables external bus masters or other system components. The LOCK prefix causes the $\overline{LOCK}$ pin to activate just for the duration of a locked instruction. If more than one sequential instruction is locked, the $\overline{LOCK}$ pin remains a logic 0 for the duration of the sequence of locked instructions. The LOCK:MOV AL,[SI] instruction is an example of a locked instruction.

ESC

The escape (ESC) instruction passes instructions to the floating-point coprocessor from the microprocessor. Whenever an ESC instruction executes, the microprocessor provides the memory address, if required, but otherwise performs a NOP. Six bits of the ESC instruction provide the opcode to the coprocessor and begin executing a coprocessor instruction.

The ESC opcode never appears in a program as ESC and in itself is considered obsolete as an opcode. In its place are a set of coprocessor instructions (FLD, FST, FMUL, etc.) that assemble as ESC instructions for the coprocessor. More detail is provided in Chapter 13, which discusses the 8087–Pentium 4 numeric coprocessors.

BOUND

The BOUND instruction, first made available in the 80186 microprocessor, is a comparison instruction that may cause an interrupt (vector type number 5). This instruction compares the contents of any 16-bit or 32-bit register against the contents of two words or doublewords of memory: an upper and a lower boundary. If the value in the register compared with memory is not within the upper and lower boundary, a type 5 interrupt ensues. If it is within the boundary, the next instruction in the program executes.

For example, if the BOUND SI,DATA instruction executes, word-sized location DATA contains the lower boundary, and word-sized location DATA+2 bytes contains the upper boundary. If the number contained in SI is less than memory location DATA or greater than memory location DATA+2 bytes, a type 5 interrupt occurs. Note that when this interrupt occurs, the return address points to the BOUND instruction, not to the instruction following BOUND. This differs from a normal interrupt, where the return address points to the next instruction in the program.

ENTER and LEAVE

The ENTER and LEAVE instructions, first made available in the 80186 microprocessor, are used with stack frames, which are mechanisms used to pass parameters to a procedure through the stack memory. The stack frame also holds local memory variables for the procedure. Stack frames provide dynamic areas of memory for procedures in multiuser environments.

The ENTER instruction creates a stack frame by pushing BP onto the stack and then loading BP with the uppermost address of the stack frame. This allows stack frame variables to be accessed through the BP register. The ENTER instruction contains two operands. The first operand specifies the number of bytes to reserve for variables on the stack frame, and the second specifies the level of the procedure.

Suppose that an ENTER 8,0 instruction executes. This instruction reserves eight bytes of memory for the stack frame and the zero specifies level 0. Figure 6–10 shows the stack frame set up by this instruction. Note that this instruction stores BP onto the top of the stack. It then subtracts 8 from the stack pointer, leaving eight bytes of memory space for temporary data storage. The uppermost location of this eight-byte temporary storage area is addressed by BP. The LEAVE instruction reverses this process by reloading both SP and BP with their prior values. The ENTER and LEAVE instructions were used to call C++ functions in Windows 3.1, but since then, CALL has been used in modern versions of Windows for C++ functions.

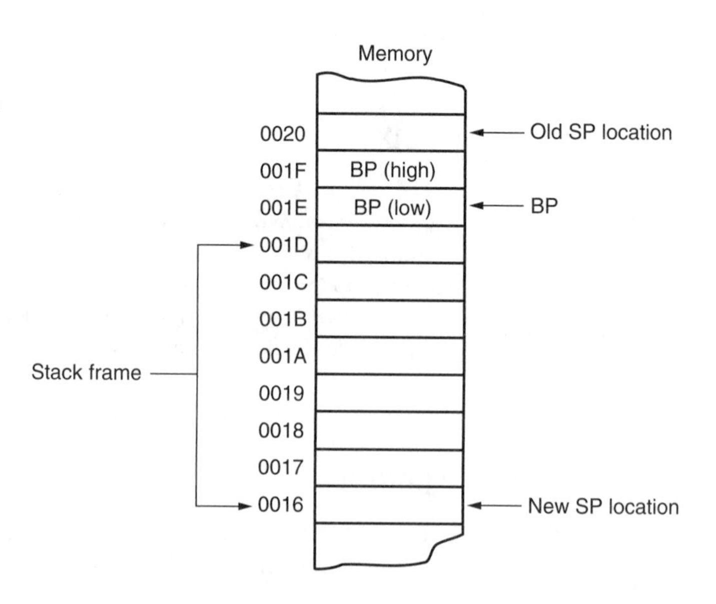

FIGURE 6–10 The stack frame created by the ENTER 8,0 instruction. Notice that BP is stored beginning at the top of the stack frame. This is followed by an 8-byte area called a stack frame.

6–6 SUMMARY

1. There are three types of unconditional jump instructions: short, near, and far. The short jump allows a branch to within +127 and –128 bytes. The near jump (using a displacement of ±32K) allows a jump to any location in the current code segment (intrasegment). The far jump allows a jump to any location in the memory system (intersegment). The near jump in an 80386 through a Pentium 4 is within ±2G bytes because these microprocessors can use a 32-bit signed displacement.
2. Whenever a label appears with a JMP instruction or conditional jump, the label, located in the label field, must be followed by a colon (LABEL:). For example, the JMP DOGGY instruction jumps to memory location DOGGY:.
3. The displacement that follows a short or near jump is the distance from the next instruction to the jump location.
4. Indirect jumps are available in two forms: (1) jump to the location stored in a register and (2) jump to the location stored in a memory word (near indirect) or doubleword (far indirect).
5. Conditional jumps are all short jumps that test one or more of the flag bits: C, Z, O, P, or S. If the condition is true, a jump occurs; if the condition is false, the next sequential instruction executes. Note that the 80386 and above allow a 16-bit signed displacement for the conditional jump instructions.
6. A special conditional jump instruction (LOOP) decrements CX and jumps to the label when CX is not 0. Other forms of loop include LOOPE, LOOPNE, LOOPZ, and LOOPNZ. The LOOPE instruction jumps if CX is not 0 and if an equal condition exists. In the 80386 through the Pentium 4, the LOOPD, LOOPED, and LOOPNED instructions also use the ECX register as a counter.
7. The 80386 through the Pentium 4 contain conditional set instructions that either set a byte to 01H or clear it to 00H. If the condition under test is true, the operand byte is set to 01H; if the condition under test is false, the operand byte is cleared to 00H.
8. The .IF and .ENDIF statements are useful in assembly language for making decisions. The instructions cause the assembler to generate conditional jump statements that modify the flow of the program.
9. The .WHILE and .ENDW statements allow an assembly language program to use the WHILE construction, and the .REPEAT and .UNTIL statements allow an assembly language program to use the REPEAT-UNTIL construct.
10. Procedures are groups of instructions that perform one task and are used from any point in a program. The CALL instruction links to a procedure and the RET instruction returns from a procedure. In assembly language, the PROC directive defines the name and type of procedure. The ENDP directive declares the end of the procedure.
11. The CALL instruction is a combination of a PUSH and a JMP instruction. When CALL executes, it pushes the return address on the stack and then jumps to the procedure. A near CALL places the contents of IP on the stack, and a far CALL places both IP and CS on the stack.
12. The RET instruction returns from a procedure by removing the return address from the stack and placing it into IP (near return), or IP and CS (far return).
13. Interrupts are either software instructions similar to CALL or hardware signals used to call procedures. This process interrupts the current program and calls a procedure. After the procedure, a special IRET instruction returns control to the interrupted software.
14. Real mode interrupt vectors are four bytes long and contain the address (IP and CS) of the interrupt service procedure. The microprocessor contains 256 interrupt vectors in the first 1K byte of memory. The first 32 are defined by Intel; the remaining 224 are user interrupts. In protected mode operation, the interrupt vector is eight bytes long, and the interrupt vector table may be relocated to any section of the memory system.

15. Whenever an interrupt is accepted by the microprocessor, the flags IP and CS are pushed on the stack. Besides pushing the flags, the T and I flag bits are cleared to disable both the trace function and the INTR pin. The final event that occurs for the interrupt is that the interrupt vector is fetched from the vector table and a jump to the interrupt service procedure occurs.
16. Software interrupt instructions (INT) often replace system calls. Software interrupts save three bytes of memory each time they replace CALL instructions.
17. A special return instruction (IRET) must be used to return from an interrupt service procedure. The IRET instruction not only removes IP and CS from the stack, it also removes the flags from the stack.
18. Interrupt on an overflow (INTO) is a conditional interrupt that calls an interrupt service procedure if the overflow flag (O) = 1.
19. The interrupt enable flag (I) controls the INTR pin connection on the microprocessor. If the STI instruction executes, it sets I to enable the INTR pin. If the CLI instruction executes, it clears I to disable the INTR pin.
20. The carry flag bit (C) is clear, set, and complemented by the CLC, STC, and CMC instructions.
21. The WAIT instruction tests the condition of the $\overline{BUSY}$ or $\overline{TEST}$ pin on the microprocessor. If $\overline{BUSY}$ or $\overline{TEST} = 1$, WAIT does not wait; but if $\overline{BUSY}$ or $\overline{TEST} = 0$, WAIT continues testing the $\overline{BUSY}$ or $\overline{TEST}$ pin until it becomes a logic 1. Note that the 8086/8088 contains the $\overline{TEST}$ pin, while the 80286–80386 contain the $\overline{BUSY}$ pin. The 80486 through the Pentium 4 do not contain a $\overline{BUSY}$ or $\overline{TEST}$ pin.
22. The LOCK prefix causes the $\overline{LOCK}$ pin to become a logic 0 for the duration of the locked instruction. The ESC instruction passes instruction to the numeric coprocessor.
23. The BOUND instruction compares the contents of any 16-bit register against the contents of two words of memory: an upper and a lower boundary. If the value in the register compared with memory is not within the upper and lower boundary, a type 5 interrupt ensues.
24. The ENTER and LEAVE instructions are used with stack frames. A stack frame is a mechanism used to pass parameters to a procedure through the stack memory. The stack frame also holds local memory variables for the procedure. The ENTER instruction creates the stack frame, and the LEAVE instruction removes the stack frame from the stack. The BP register addresses stack frame data.

6–7 QUESTIONS AND PROBLEMS

1. What is a short JMP?
2. Which type of JMP is used when jumping to any location within the current code segment?
3. Which JMP instruction allows the program to continue execution at any memory location in the system?
4. Which JMP instruction is five bytes long?
5. What is the range of a near jump in the 80386–Pentium 4 microprocessors?
6. Which type of JMP instruction (short, near, or far) assembles for the following:
   (a) if the distance is 0210H bytes
   (b) if the distance is 0020H bytes
   (c) if the distance is 10000H bytes
7. What can be said about a label that is followed by a colon?
8. The near jump modifies the program address by changing which register or registers?
9. The far jump modifies the program address by changing which register or registers?
10. Explain what the JMP AX instruction accomplishes. Identify it as a near or a far jump instruction.

11. Contrast the operation of a JMP DI with a JMP [DI].
12. Contrast the operation of a JMP [DI] with a JMP FAR PTR [DI].
13. List the five flag bits tested by the conditional jump instructions.
14. Describe how the JA instruction operates.
15. When will the JO instruction jump?
16. Which conditional jump instructions follow the comparison of signed numbers?
17. Which conditional jump instructions follow the comparison of unsigned numbers?
18. Which conditional jump instructions test both the Z and C flag bits?
19. When does the JCXZ instruction jump?
20. Which SET instruction is used to set AL if the flag bits indicate a zero condition?
21. The 8086 LOOP instruction decrements register ______ and tests it for a 0 to decide if a jump occurs.
22. The Pentium 4 LOOPD instruction decrements register ______ and tests it for a 0 to decide if a jump occurs.
23. Explain how the LOOPE instruction operates.
24. Develop a short sequence of instructions that stores 00H into 150H bytes of memory, beginning at extra segment memory location DATAZ. You must use the LOOP instruction to help perform this task.
25. Develop a sequence of instructions that searches through a block of 100H bytes of memory. This program must count all the unsigned numbers that are above 42H and all that are below 42H. Byte-sized data segment memory location UP must contain the count of numbers above 42H, and data segment location DOWN must contain the count of numbers below 42H.
26. Show the assembly language instructions generated by the following sequence:

```
.IF AL==3
    ADD AL,2
.ENDIF
```

27. What happens if the .WHILE 1 instruction is placed in a program?
28. Develop a short sequence of instructions that uses the REPEAT-UNTIL construct to copy the contents of byte-sized memory BLOCKA into byte-sized memory BLOCKB until 00H is moved.
29. What is the purpose of the .BREAK directive?
30. Using the WHILE construct, develop a sequence of instructions that add the byte-sized contents of BLOCKA to BLOCKB while the sum is not 12H.
31. What is a procedure?
32. Explain how the near and far CALL instructions function.
33. How does the near RET instruction function?
34. The last executable instruction in a procedure must be a(n) ______.
35. Which directive identifies the start of a procedure?
36. How is a procedure identified as near or far?
37. Explain what the RET 6 instruction accomplishes.
38. Write a near procedure that cubes the contents of the CX register. This procedure may not affect any register except CX.
39. Write a procedure that multiplies DI by SI and then divides the result by 100H. Make sure that the result is left in AX upon returning from the procedure. This procedure may not change any register except AX.
40. Write a procedure that sums EAX, EBX, ECX, and EDX. If a carry occurs, place a logic 1 in EDI. If no carry occurs, place a 0 in EDI. The sum should be found in EAX after the execution of your procedure.
41. What is an interrupt?
42. Which software instructions call an interrupt service procedure?
43. How many different interrupt types are available in the microprocessor?

44. What is the purpose of interrupt vector type number 0?
45. Illustrate the contents of an interrupt vector and explain the purpose of each part.
46. How does the IRET instruction differ from the RET instruction?
47. What is the IRETD instruction?
48. The INTO instruction only interrupts the program for what condition?
49. The interrupt vector for an INT 40H instruction is stored at which memory locations?
50. What instructions control the function of the INTR pin?
51. What instruction tests the $\overline{BUSY}$ pin?
52. When will the BOUND instruction interrupt a program?
53. An ENTER 16,0 instruction creates a stack frame that contains ______ bytes.
54. Which register moves to the stack when an ENTER instruction executes?
55. Which instruction passes opcodes to the numeric coprocessor?

CHAPTER 7

Using Assembly Language with C/C++

INTRODUCTION

Today, it is rare to develop a complete system using only assembly language. We often use C/C++ with some assembly language to develop a system. The assembly language portion usually solves tasks (difficult or inefficient to accomplish in C/C++) that often include control software for peripheral interfaces and driver programs that use interrupts. Another application of assembly language in C/C++ programs is for the MMX and SEC instructions that are part of the Pentium class microprocessors and are not supported in C/C++. Although C++ does have macros for these commands, they are more complicated to use than assembly language. This chapter develops the idea of mixing C/C++ and assembly language. Many applications in later chapters also illustrate the use of both assembly language and C/C++ to accomplish tasks for the microprocessor.

This text uses Microsoft Visual C/C++, but programs can often be adapted to any version of C/C++, as long as it is standard ANSI (American National Standards Institute) format C/C++. If you want, you can use C/C++ to enter and execute all the programming applications in this text. The 16-bit applications are written by using Microsoft Visual C/C++ version 1.52 or newer (available [CL.EXE] at no cost as a legacy application in the Microsoft Windows Driver Development Kit [DDK]); the 32-bit applications are written using Microsoft Visual C/C++ version 6 or newer and preferably Microsoft Visual C/C++ version .NET 2003. The examples in the text are written assuming that you have the latest version.

CHAPTER OBJECTIVES

Upon completion of this chapter, you will be able to:

1. Use assembly language in _asm blocks within C/C+
2. Learn the rules that apply to mixed language software development
3. Use common C/C++ data and structures with assembly language
4. Use both the 16-bit (DOS) interface and the 32-bit (Microsoft Windows) interface with assembly language code
5. Use assembly language objects with C/C++ programs

7–1 USING ASSEMBLY LANGUAGE WITH C++ FOR 16-BIT DOS APPLICATIONS

This section shows how to incorporate assembly language commands within a C/C++ program. This is important because the performance of a program often depends on the incorporation of assembly language sequences to speed its execution. As mentioned in the introduction to the chapter, assembly language is also used for I/O operations in embedded systems. This text assumes that you are using a version of the Microsoft C/C++ program, but any C/C++ program should function as shown if it supports inline assembly commands. The only change might be setting up the C/C++ package to function with assembly language. This section of the text assumes that you are building 16-bit applications for DOS. Make sure that your software can build 16-bit applications before attempting any of the programs in this section. If you build a 32-bit application and attempt to use the DOS INT 21H function, the program will crash because DOS calls are not directly allowed. In fact, they are inefficient to use in a 32-bit application.

To build a 16-bit DOS application, you will need the legacy 16-bit compiler usually found in the C:\WINDDK\2600.1106\bin\win_me\bin16 directory of the Windows DDK. (The **Windows Driver Development Kit** can be obtained for a small shipping charge from Microsoft Corporation.) The compiler is **CL.EXE** and the 16-bit linker program is LINK.EXE. Both are located in the directory or folder listed. Since the path in the computer that you are using probably points to the 32-bit linker program, it would be wise to work from this directory so the proper linker will be used when linking the object files generated by the compiler. Compilation and linking must be performed at the command line because there is no visual interface or editor provided with the compiler and linker. Programs are generated using either NotePad or DOS Edit.

Basic Rules and Simple Programs

Before assembly language code can be placed in a C/C++ program, some rules must be learned. Example 7–1 shows how to place assembly code inside an assembly language block within a short C/C++ program. Note that all the assembly code in this example is placed in the **_asm** block. Labels are used as illustrated by the label **big**: in this example. It is also extremely important to use lowercase characters for any inline assembly code. If you use uppercase you will find that some of the assembly language commands and registers are reserved or defined words in C/C++ language.

Example 7–1 uses no C/C++ commands except for the main procedure. Enter the program using either NotePad or Edit. The program reads one character from the console keyboard, and then filters it through assembly language so that only the numbers 0 through 9 are sent back to the video display. Although this programming example does not accomplish much, it does show how to set up and use some simple programming constructs in the C/C++ environment and also how to use the inline assembler.

EXAMPLE 7–1

```
//Accepts and displays one character of 1 through 9,
//all others are ignored.

void main(void)
{
	_asm
	{
		mov ah,8		;read key no echo
		int 21h
		cmp al,'0'		;filter key code
		jb  big
		cmp al,'9'
		ja  big
		mov dl,al		;echo 0 - 9
```

```
            mov ah,2
            int 21h
      big:
      }
}
```

The register AX was not saved in Example 7–1, but it was used by the program. It is very important to note that the AX, BX, CX, DX, and ES registers are never used by Microsoft C/C++. (The function of AX on a return from a procedure is explained later in this chapter.) These registers, which might be considered **scratchpad** registers, are available to use with assembly language. If you wish to use any of the other registers, make sure that you save them with a PUSH before they are used and restore them with a POP afterwards. If you fail to save the registers used by a program, the program may not function correctly and can crash the computer. If you are using the 80386 processor or above as a base for your program, you need not save EAX, EBX, ECX, EDX, and ES.

To compile the program, start the Command Prompt program located in the Start Menu under Accessories. Change the path to C:\WINDDK\2600.1106\bin\win_me\bin16 if that is where you have your Windows DDK. You will also need to go to the C:\WINDDK\2600.1106\lib\win_me directory and copy slibce.lib to the C:\WINDDK\2600.1106\bin\win_me\bin16 directory. Make sure you save the program in the same path and use the extension .c with the file name. If you use NotePad, make sure you select All Files under File Type when saving. To compile the program, type CL /G3 filename.c>. This will generate the .exe file (/G3 is the 80386) for the program. (See Table 7–1 for a list of the /G compiler switches.) Any errors that appear can be ignored by pressing the Enter key. These errors generate warnings that will not cause a problem when the program is executed. When the program is executed you will only see a number echoed back to the DOS screen.

Example 7–2 shows how to use variables from C with a short assembly language program. In this example, the char variable type (a byte in C) is used to save space for a few eight-bit bytes of data. The program itself performs the operation X + Y = Z , where X and Y are two one-digit numbers, and Z is the result. As you might imagine, you could use the inline assembly in C to learn assembly language and write many of the programs in this textbook. The semicolon adds comments to the listing in the _asm block, just as with the normal assembler.

EXAMPLE 7–2

```
void main(void)
{
      char a, b;
      _asm
      {
            mov  ah,1            ;read first digit
            int  21h
            mov  a,al
            mov  ah,1            ;read a + sign
```

TABLE 7–1 Compiler (16-bit) G options.

Compiler Switch	*Function*
/G1	Selects the 8088/8086
/G2	Selects the 80188/80186/80286
/G3	Selects the 80386
/G4	Selects the 80486
/G5	Selects the Pentium
/G6	Selects the Pentium Pro–Pentium 4

Note: The 32-bit C++ compiler does not recognize /G1 or /G2.

```
          int  21h
          cmp  al,'+'
          jne  end1           ;if not plus
          mov  ah,1
          int  21h            ;read second number
          mov  b,al
          mov  ah,2           ;display =
          mov  dl,'='
          int  21h
          mov  ah,0
          mov  al,a           ;generate sum
          add  al,b
          aaa                 ;ASCII adjust for addition
          add  ax,3030h
          cmp  ah,'0'
          je   down
          push ax             ;display 10's position
          mov  dl,ah
          mov  ah,2
          int  21h
          pop  ax
     down:
          mov  dl,al          ;display units position
          mov  ah,2
          int  21h
     end1:
     }
}
```

What Cannot Be Used from MASM Inside an _asm Block

Although MASM contains some nice features, such as conditional commands (.IF, .WHILE, .REPEAT, etc.), the inline assembler does not include the conditional commands from MASM, nor does it include the MACRO feature found in the assembler. Data allocation with the inline assembler is handled by C instead of by using DB, DW, DD, and so on. Just about all other features are supported by the inline assembler. These omissions from the inline assembler can cause some slight problems, as will be discussed in later sections of this chapter.

Using Character Strings

Example 7–3 illustrates a simple program that uses a character string defined with C and displays it so that each word is listed on a separate line. Notice the blend of both C statements and assembly language statements. The WHILE statement repeats the assembly language commands until the null (00H) is discovered at the end of the character string. If the null is not discovered, the assembly language instructions display a character from the string unless a space is located. For each space, the program displays a carriage return/line feed combination. This causes each word in the string to be displayed on a separate line.

EXAMPLE 7–3

```
//Example that displays showing one word per line

void main(void)
{
      char string1[]="This is my first test application using _asm.\n";
      int sc = -1;
      while (string1[sc++] !=0)
      {
            _asm
            {
                  push si
                  mov  si,sc                  ;get pointer
```

```
                mov  dl,string1[si]      ;get character
                cmp  dl,' '              ;if not space
                jne  next
                mov  ah,2                ;display new line
                mov  dl,10
                int  21h
                mov  dl,13
          next: mov  ah,2                ;display character
                int  21h
                pop  si
          }
     }
}
```

Suppose that you want to display more than one string in a program, but you still want to use assembly language to develop the software to display a string. Example 7–4 illustrates a program that creates a procedure displaying a character string. This procedure is called each time a string is displayed in the program. Note that this program displays one string on each line, unlike Example 7–3.

EXAMPLE 7–4

```
//A program illustrating an assembly language procedure that
//displays C language character strings

char string1[] = "This is my first test program using _asm.";
char string2[] = "This is the second line in this program.";
char string3[] = "This is the third.";

void main(void)
{
      Str  (string1);
      Str  (string2);
      Str  (string3);
}

Str  (char *string_adr)
{
      _asm
      {
            mov  bx,string_adr       ;get address of string
            mov  ah,2
      top:
            mov  dl,[bx]
            inc  bx
            cmp  al,0                ;if null
            je   bot
            int  21h                 ;display character
            jmp  top
      bot:
            mov  dl,13               ;display CR + LF
            int  21h
            mov  dl,10
            int  21h
      }
}
```

Using Data Structures

Data structures are an important part of most programs. This section shows how to interface a data structure created in C with an assembly language section that manipulates the data in the structure. Example 7–5 illustrates a short program that uses a data structure to store names, ages, and salaries. The program then displays each of the entries by using a few assembly language procedures. Although the string procedure displays a character string, shown in Example 7–4, no carriage return/line feed combination is displayed—instead, a space is displayed. The Crlf procedure displays a carriage return/line feed combination. The Numb procedure displays the integer.

EXAMPLE 7–5

```
//Program illustrating an assembly language procedure that
//displays the contents of a C data structure.

//A simple data structure

typedef struct records
{
       char first_name[16];
       char last_name[16];
       int age;
       int salary;
} RECORD;

//Fill some records

RECORD record[4] =
{ {"Bill","Boyd", 56, 23000},
  {"Page", "Turner", 32, 34000},
  {"Bull", "Dozer", 39, 22000},
  {"Hy", "Society", 48, 62000}
};

//Program

void main(void)
{
      int pnt = -1;
      while (pnt++ < 3)
      {
             Str(record[pnt].last_name);
             Str(record[pnt].first_name);
             Numb(record[pnt].age);
             Numb(record[pnt].salary);
             Crlf();
      }
}

Str (char *string_adr[])
{
      _asm
      {
             mov  bx,string_adr
             mov  ah,2
      top:
             mov  dl,[bx]
             inc  bx
             cmp  al,0
             je   bot
             int  21h
             jmp  top
      bot:
             mov  al,20h
             int  21h
      }
}
```

```
Crlf()
{
        _asm
        {
                mov  ah,2
                mov  dl,13
                int  21h
                mov  dl,10
                int  21h
        }
}

Numb (int temp)
{
        _asm
        {
                mov  ax,temp
                mov  bx,10
                push bx
        L1:
                mov  dx,0
                div  bx
                push dx
                cmp  ax,0
                jne  L1
        L2:
                pop  dx
                cmp  dl,bl
                je   L3
                mov  ah,2
                add  dl,30h
                int  21h
                jmp  L2
        L3:
                mov  dl,20h
                int  21h
        }
}
```

An Example of a Mixed-Language Program

To see how this technique can be applied to any program, Example 7–6 shows how the program can do some operations in assembly language and some in C. Here, the only assembly language portion of the program is the Dispn procedure that displays an integer and the Readnum procedure, which reads an integer. The program in Example 7–6 makes no attempt to detect or correct errors. Also, the program functions correctly only if the result is positive and less than 64K. Notice that this example uses assembly language to perform the I/O; the C portion performs all the other operations to form the shell of the program.

EXAMPLE 7–6

```
/*

A program that functions as a simple calculator to perform addition,
subtraction, multiplication, and division. The format is X <oper> Y =.

*/

int temp;

void main(void)
{
```

```
      int temp1, oper;
      while (1)
      {
            oper = Readnum();                  //get first number and operation
            temp1 = temp;
            if ( Readnum() == '=' )            //get second number
            {
                  switch (oper)
                  {
                        case '+':
                              temp += temp1;
                              break;
                        case '-':
                              temp = temp1 - temp;
                              break;
                        case '/':
                              temp = temp1 / temp;
                              break;
                        case '*':
                              temp *= temp1;
                              break;
                  }
                  Dispn(temp);                 //display result
            }
            else
                  Break;
            }
      }
}

int Readnum()
{
      int a;
      temp = 0;
      _asm
      {
      Readnum1:
            mov  ah,1
            int  21h
            cmp  al,30h
            jb   Readnum2
            cmp  al,39h
            ja   Readnum2
            sub  al,30h
            shl  temp,1
            mov  bx,temp
            shl  temp,2
            add  temp,bx
            add  byte ptr temp,al
            adc  byte ptr temp+1,0
            jmp  Readnum1
      Readnum2:
            Mov  ah,0
            mov  a,ax
      }
      return a;
}

Dispn (int DispnTemp)
{
      _asm
      {
            mov  ax,DispnTemp
            mov  bx,10
            push bx
      Dispn1:
            mov  dx,0
```

```
            div  bx
            push dx
            cmp  ax,0
            jne  Dispn1
        Dispn2:
            pop  dx
            cmp  dl,bl
            je   Dispn3
            add  dl,30h
            mov  ah,2
            int  21h
            jmp  Dispn2
        Dispn3:
            mov  dl,13
            int  21h
            mov  dl,10
            int  21h
        }
}
```

7–2 USING ASSEMBLY LANGUAGE WITH VISUAL C/C++ FOR 32-BIT APPLICATIONS

A major difference exists between 16-bit and 32-bit applications. The 32-bit applications are written using Microsoft Visual C/C++ for Windows and the 16-bit applications are written using Microsoft C++ for DOS. The main difference is that Visual C/C++ for Windows is more common today, but Visual C/C++ cannot easily call DOS functions such as INT 21H. It is suggested that embedded applications that do not require a visual interface be written in 16-bit C or C++, and applications that incorporate Windows or Windows CE (available for use on a ROM or Flash[1] device for embedded applications) use 32-bit Visual C/C++ for Windows.

A 32-bit application is written by using any of the 32-bit registers, and the memory space is essentially limited to 2G bytes for Windows. The only difference is that you may not use the DOS function calls; instead use the console getch() or getche() and putch C/C++ language functions available for use with DOS console applications. Embedded applications use direct assembly language instructions to access I/O devices in an embedded system. In the Visual interface, all I/O is handled by the Windows operating system framework.

An Example that Uses Console I/O to Access the Keyboard and Display

Example 7–7 illustrates a simple example that uses the console I/O commands to read and write data from the console. To enter this application (assuming Visual Studio **.NET 2003** is available), select a WIN32 console application in the new project option (see Figure 7–1). Notice that instead of using the customary stdio.h library, we use the conio.h library. This example program displays any number between 0 and 1000 in all number bases between base 2 and base 16. Notice that the main program is not called main as it was in earlier versions of C/C++, but is called _tmain in the current version of Visual C/C++ when used with a console application. The argc is the argument count passed to the _tmain procedure from the command line, and the argv[] is an array that contains the command line argument strings.

[1]Flash is a trademark of Intel Corporation.

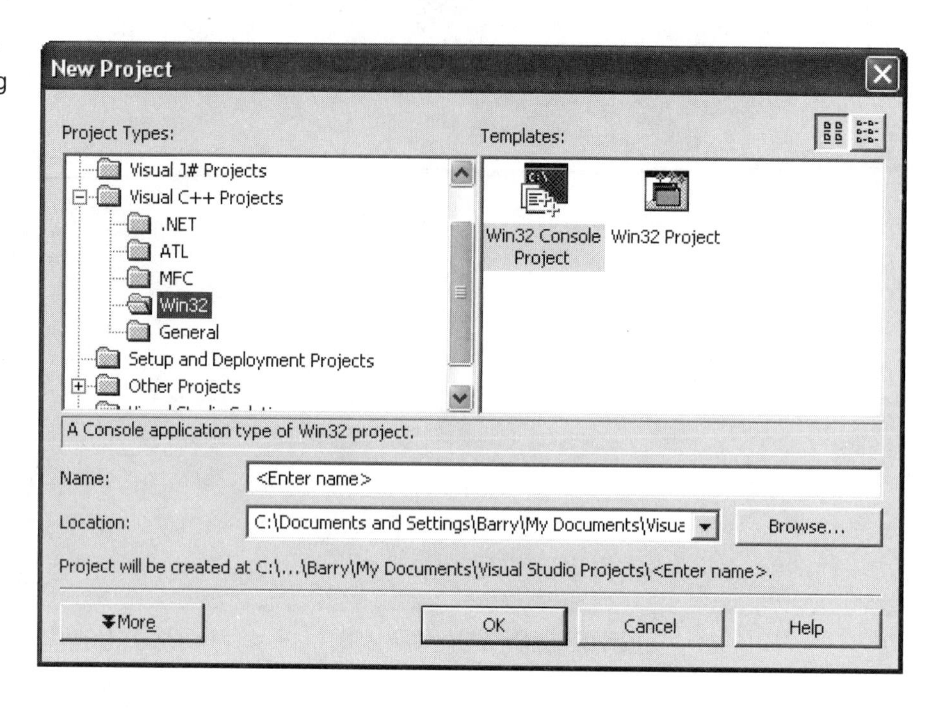

FIGURE 7–1 A screen shot of the New Project menu for writing Visual C/C++ console applications for Visual Studio .NET.

EXAMPLE 7–7

```
//Program that displays any number in all numbers bases
//between base 2 and base 16.

#include "stdafx.h"
#include <conio.h>

char *buffer = "Enter a number between 0 and 1000: ";
char *buffer1 = "Base :";
int a, b = 0;

int _tmain(int argc, _TCHAR* argv[])
{
      int i;
      _cputs(buffer);
      a = _getche();
      while ( a >= '0' && a <= '9' )
      {
            _asm sub a, 30h;
            b = b * 10 + a;
            a = _getche();
      }
      _putch(10);
      _putch(13);
      for (i = 2; i < 17; i++ )
      {
            _cputs(buffer1);
            disps(10,i );
            _putch(' ');
            disps(i, b);
            _putch(10);
            _putch(13);
      }
      getche();                          //wait for any key
      return 0;
}
```

```
void disps(int base, int data)
{
      int temp;
      _asm
      {
            mov   eax, data
            mov   ebx, base
            push ebx
      disps1:
            mov   edx,0
            div   ebx
            push edx
            cmp   eax,0
            jne   disps1
      disps2:
            pop   edx
            cmp   ebx,edx
            je    disps4
            add   dl,30h
            cmp   dl,39h
            jbe   disps3
            add   dl,7
      disps3:
            mov   temp,edx
      }
      _putch(temp);
      _asm jmp  disps2;
disps4:;
}
```

This example presents a nice mixture of assembly language and C/C++ language commands. The procedure disps (base,data) does most of the work for this program. It allows any integer (unsigned) to be displayed in any number base, which can be any value between base 2 and base 36. The upper limit occurs because we run out of letters for displaying number bases after the letter Z. If you need to convert to larger number bases, a new scheme for bases over 36 has to be developed. Perhaps the letters a through z could be used for bases 37 to 52. Example 7–7 displays the number that is entered in base 2 through base 16.

Directly Addressing I/O Ports

If a program is written that must access an actual port number, we can use console I/O commands such as the _inp(port) command to input byte data, and the _outp(port,byte_data) command to output byte data. When writing software for the personal computer, it is rare to directly address an I/O port, but when we write software for an embedded system, we often directly address an I/O port. An alternate to using the _inp and _out commands is assembly language, which is more efficient in most cases. Be aware that I/O ports may not be accessed in the Windows environment if you are using Windows NT, Windows 2000, or Windows XP. The only way that you can access the I/O ports in these systems is to develop a kernel driver. At this point in the text it would not be practical to develop such a driver. If you are using Windows 98 or even Windows 95 you can use inp and outp to access the I/O ports directly.

Developing a Visual C++ Application for Windows

This section of the text shows how to use Visual C++ to develop a dialog-based application for the Microsoft Foundation Classes library. Microsoft Foundation Classes (MFC) is a collection of classes that allow us to use the Windows interface without a great deal of difficulty. The easiest application to develop is the dialog application presented here. This basic application type can be used to program and test all of the software examples in this textbook in the Visual C++ programming environment.

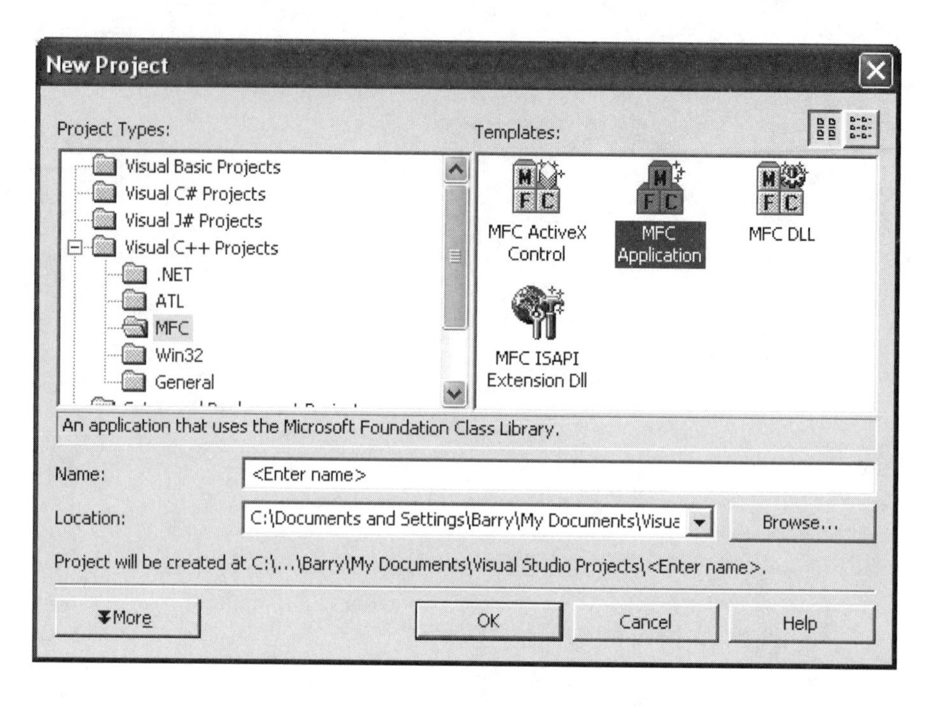

FIGURE 7–2 A screen shot of the New Project Wizard from Visual Studio.NET 2003.

To create a Visual C++ dialog-based application, start Visual Studio .NET 2003. Click on File and select New Project. Figure 7–2 illustrates what is displayed when you select the MFC application type under Visual C++ Projects. Enter a name for the project and select an appropriate path for the project, then click on OK.

Figure 7–3 illustrates the MFC Application Wizard for a project called First, which is the project name that was selected in the New Project wizard of Figure 7–2. The Application Wizard has many features, but for a dialog application all that needs to be done is to click on Application type and select Dialog-based and nothing else. Next, click on Finish.

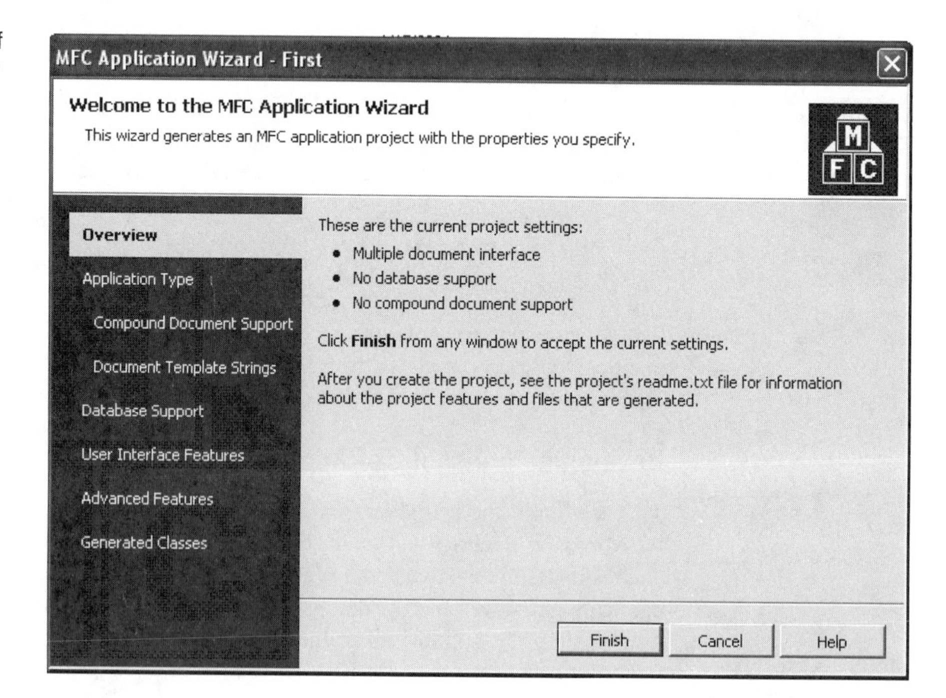

FIGURE 7–3 A screen shot of the MFC Application Wizard for project First.

FIGURE 7–4 A screen shot of the First dialog-based application.

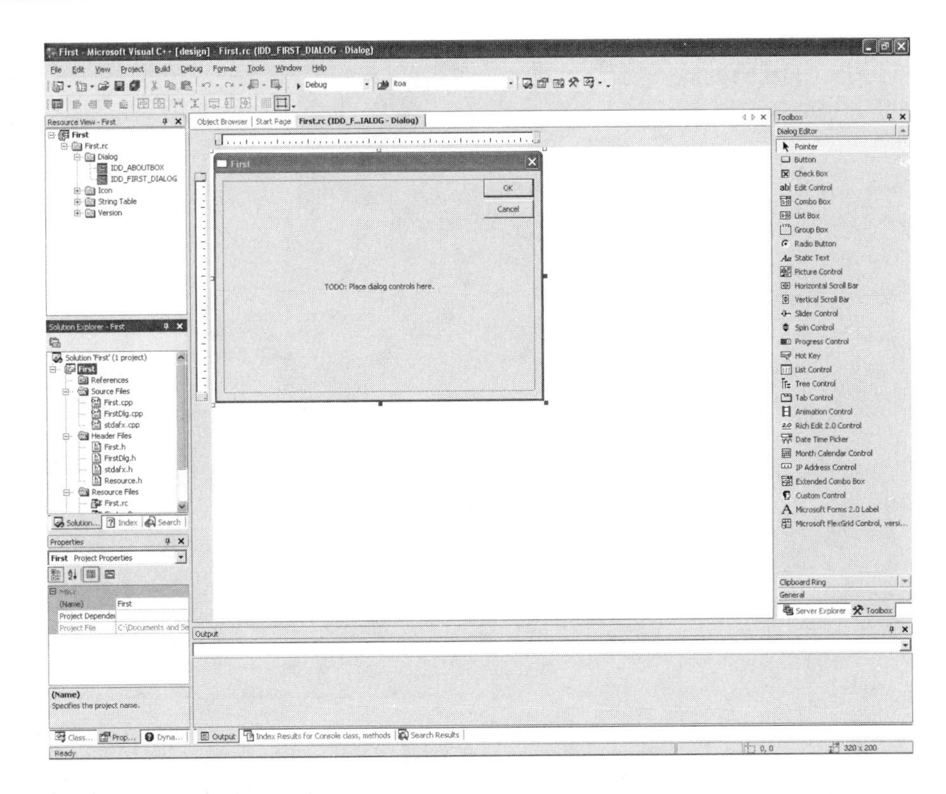

After a few moments the screen should appear as shown in Figure 7–4. In the middle section is the dialog box created by this application. To test the application, just find the blue arrow ▸ located below the Window menu at the top of the screen and click on it to compile, link, and execute the dialog application. (Answer yes to "Would you like to build the application?"). Click on the OK control to quit the application. You have just created and tested your very first Visual C++ dialog-based application.

Several items are shown in Figure 7–4 that are important for program creation and development. The right margin of the screen contains a toolbox that holds stock objects that can be placed on the application. The left margin contains three windows that display the Resource View, Solution Explorer, and the Properties window. The tabs located at the bottom of the Solution window allow the solution, index, or search windows to be displayed in this area. The tabs at the bottom of the Properties window allow the classes, properties, dynamic help, or output to be displayed in this window. Your screen may or may not look like the one illustrated in Figure 7–4 because it can be modified by selecting the Profile tab when starting Visual Studio .NET 2003.

To create a simple application for the dialog box, first click on the label "TODO: Place dialog controls here." Then click on the Delete key on the keyboard to erase it. You can also right-click on the label and select Delete to erase it. Windows is an events-driven system, so we need to create some object in our dialog application to initiate an event. This could be a button or almost any control selected from the toolbox. Click on the button control near the top of the toolbox, which selects the button. Now move the mouse pointer (do not drag the button) over to the dialog application in the middle of the screen and draw the button near the center (see Figure 7–5).

Once the button is placed on the screen an event handler needs to be added to the application to handle the act of pressing or clicking on the button. The event handlers are selected by going to the Properties window at the left margin of the screen and clicking on the lightning bolt. Locate the IDC_BUTTON1 object below the lightning bolt and click on the plus to expand the listing and display the events available to Button1. Locate BN_CLICKED, which is the event handler that responds to clicking on the button. You will see an arrow to the right of BN_CLICKED. Click on the

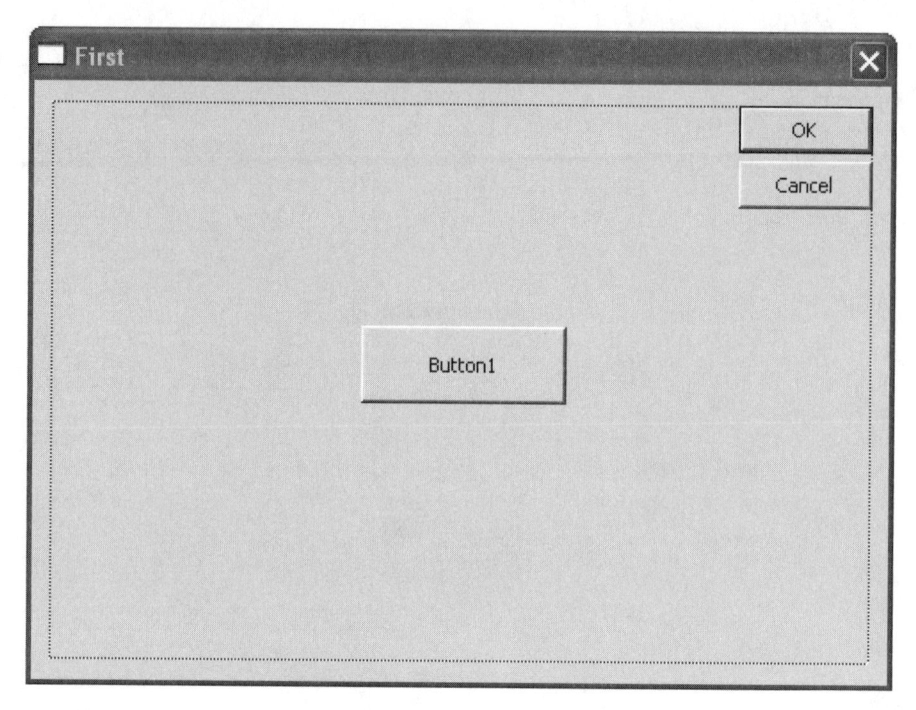

FIGURE 7–5 A screen shot of the button placed in the center of the dialog application.

arrow and select <add> OnBnClickedButtont1 to add a handler for the button-clicked event. When you do this the screen will display the OnBnClickedButtont1 procedure for the button-clicked event. This procedure is illustrated in Example 7–8. To test the button, change the software in Example 7–8 to the software in Example 7–9(a). Click on the blue arrow to compile, link, and execute the dialog application and click on Button1 when it is running. The label on Button1 will change to "WOW, Hello world!" if you made the button wide enough. You now have your first working application, but it does not use any assembly code. Example 7–9(a) used the SetDlgItemText(control ID, string) function to change the text displayed on IDC_BUTTON1. A variant that uses a character string object (CString) appears in Example 7–9(b).

EXAMPLE 7–8

```
void CFirstDlg::OnBnClickedButton1()
{
	// TODO: Add your control notification handler code here
}
```

EXAMPLE 7–9

```
//Version (a)

void CFirstDlg::OnBnClickedButton1()
{
	SetDlgItemText( IDC_BUTTON1, "WOW, Hello world!" );
}

//Version (b)

void CFirstDlg::OnBnClickedButton1()
{
	CString str1 = "WOW, Hello world!";
	SetDlgItemText( IDC_BUTTON1, str1 );
}
```

Now that a simple application has been written, we can modify it to illustrate a more complicated application as shown in Figure 7–6. The caption on the button has been changed to the

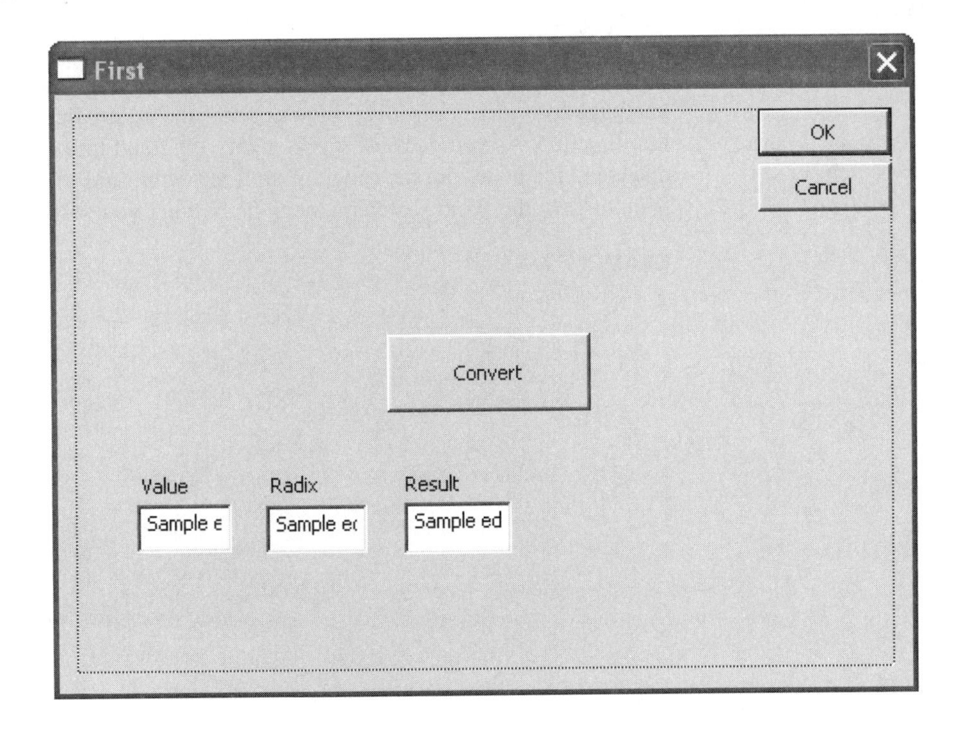

FIGURE 7–6 A screen shot of a complete application that displays any number in any number base (radix).

word Convert. To create this display, select the dialog box by going to the upper left margin (the Resource View) and clicking on the pluses next to First, then First.rc, and finally Dialog. Now double-click on IDD_FIRST_DIALOG to again display the dialog box in the middle of the screen. To change the caption on the Button1 object, click on the button and then go to the Properties window (lower left margin) and click on the properties button. Now scroll to the Caption entry in the list of properties that appears and change the caption to Convert. In Figure 7–6 notice that there are three static text controls and three edit controls below and to the left of the Convert button. You can find these two items near the top of the toolbox. Draw them on the screen in approximately the same places as in Figure 7–6. You will need to go to Properties for each static text control to change the captions as indicated.

Our goal in this example is to display any decimal number entered in the value box to any radix (number base) entered in the radix box. The result appears in the result box when the Convert button is clicked. To obtain the value from an edit control, use GetDlgItemInt(control ID). This function returns the integer value from the edit control. The difficult portion of this example is the conversion from base 10 to any number base. If we wrote the entire program in C++, it would appear as in Example 7–10. Note that only the OnBnClickedButton1 function is used in this example. The integer-to-ASCII function called itoa(int number, char string, int radix) converts an integer to any radix, and it also returns a string to the caller. This software assumes that the three edit boxes are (from left to right) IDC_EDIT1, IDC_EDIT2, and IDC_EDIT3.

EXAMPLE 7–10

```
void CFirstDlg::OnBnClickedButton1()
{
      char temp[10];
      SetDlgItemText(IDC_EDIT3, itoa( GetDlgItemInt( IDC_EDIT1 ), temp,
                     GetDlgItemInt( IDC_EDIT2 ) ) );
}
```

Since this is an assembly language text, we are not going to use the itoa function—and for good reason, because the function is quite large. To see just how large, you can put a breakpoint in the software to the left of the SetDlgItemText() function by left-clicking on the gray bar to the left of the line of code. A brown circle, the breakpoint, will appear. If you run the program, it will break (stop) at this point and enter the debugging mode so it can be viewed in assembly language form. Example 7–11 shows part of the code Microsoft inserts into a program for the itoa function. As you can see, this is a very long and time-consuming function. To display the disassembled code, run the program until it breaks, and then go to the Debug menu and select Windows. In the Windows menu, near the bottom, you will find the word "Disassembly."

EXAMPLE 7–11

```
static void __cdecl xtoa (
        unsigned long val,
        char *buf,
        unsigned radix,
        int is_neg
        )
{
102194E0  push       ebp
102194E1  mov        ebp,esp
102194E3  sub        esp,10h
       char *p;                  /* pointer to traverse string */
       char *firstdig;           /* pointer to first digit */
       char temp;                /* temp char */
       unsigned digval;          /* value of digit */
       p = buf;
102194E6  mov        eax,dword ptr [buf]
102194E9  mov        dword ptr [p],eax

       if (is_neg) {
102194EC  cmp        dword ptr [is_neg],0
102194F0  je         xtoa+29h (10219509h)
            /* negative, so output '-' and negate */
            *p++ = '-';
102194F2  mov        ecx,dword ptr [p]
102194F5  mov        byte ptr [ecx],2Dh
102194F8  mov        edx,dword ptr [p]
102194FB  add        edx,1
102194FE  mov        dword ptr [p],edx
            val = (unsigned long)(-(long)val);
10219501  mov        eax,dword ptr [val]
10219504  neg        eax
10219506  mov        dword ptr [val],eax
        }

        firstdig = p;            /* save pointer to first digit */
10219509  mov        ecx,dword ptr [p]
1021950C  mov        dword ptr [firstdig],ecx

        do {
           digval = (unsigned) (val % radix);
1021950F  mov        eax,dword ptr [val]
10219512  xor        edx,edx
10219514  div        eax,dword ptr [radix]
10219517  mov        dword ptr [digval],edx
           val /= radix;         /* get next digit */
1021951A  mov        eax,dword ptr [val]
1021951D  xor        edx,edx
1021951F  div        eax,dword ptr [radix]
10219522  mov        dword ptr [val],eax

            /* convert to ascii and store */
            if (digval > 9)
10219525  cmp        dword ptr [digval],9
10219529  jbe        xtoa+61h (10219541h)
                *p++ = (char) (digval - 10 + 'a');    /* a letter */
1021952B  mov        edx,dword ptr [digval]
1021952E  add        edx,57h
```

```
10219531  mov       eax,dword ptr [p]
10219534  mov       byte ptr [eax],dl
10219536  mov       ecx,dword ptr [p]
10219539  add       ecx,1
1021953C  mov       dword ptr [p],ecx
            else
1021953F  jmp       xtoa+75h (10219555h)
              *p++ = (char) (digval + '0');    /* a digit */
10219541  mov       edx,dword ptr [digval]
10219544  add       edx,30h
10219547  mov       eax,dword ptr [p]
1021954A  mov       byte ptr [eax],dl
1021954C  mov       ecx,dword ptr [p]
1021954F  add       ecx,1
10219552  mov       dword ptr [p],ecx
        } while (val > 0);
10219555  cmp       dword ptr [val],0
10219559  ja        xtoa+2Fh (1021950Fh)
       /* We now have the digit of the number in the buffer, but in reverse
          order. Thus we reverse them now. */

        *p-- = '\0';            /* terminate string; p points to last digit */
1021955B  mov       edx,dword ptr [p]
1021955E  mov       byte ptr [edx],0
10219561  mov       eax,dword ptr [p]
10219564  sub       eax,1
10219567  mov       dword ptr [p],eax

        do {
            temp = *p;
1021956A  mov       ecx,dword ptr [p]
1021956D  mov       dl,byte ptr [ecx]
1021956F  mov       byte ptr [temp],dl
            *p = *firstdig;
10219572  mov       eax,dword ptr [p]
10219575  mov       ecx,dword ptr [firstdig]
10219578  mov       dl,byte ptr [ecx]
1021957A  mov       byte ptr [eax],dl
            *firstdig = temp;  /* swap *p and *firstdig */
1021957C  mov       eax,dword ptr [firstdig]
1021957F  mov       cl,byte ptr [temp]
10219582  mov       byte ptr [eax],cl
            --p;
10219584  mov       edx,dword ptr [p]
10219587  sub       edx,1
1021958A  mov       dword ptr [p],edx
            ++firstdig;         /* advance to next two digits */
1021958D  mov       eax,dword ptr [firstdig]
10219590  add       eax,1
10219593  mov       dword ptr  [firstdig],eax
        } while (firstdig < p); /* repeat until halfway */
10219596  mov       ecx,dword ptr [firstdig]
10219599  cmp       ecx,dword ptr [p]
1021959C  jb        xtoa+8Ah (1021956Ah)
}
1021959E  mov       esp,ebp
102195A0  pop       ebp
102195A1  ret
```

As you can see, this is a substantial amount of code that can be significantly reduced if it is rewritten using the inline assembler. Example 7–12 depicts the assembly language version of the OnBnClickedButton1 function. This function is considerably shorter and executes many times faster than the function in Example 7–11. You may wish to change the OnBnClickedButton1 function and see if the dialog application functions correctly. This example points out the inefficiency of the code generated by a high-level language, which may not always be important, but many cases require tight and efficient code, and that can only be written in assembly language. My guess is that as a plateau is reached on processor speed, more things will be written in assembly language. In addition, the new instructions such as MMX and SEC are not available in high-level languages. They require a very good working knowledge of assembly code.

EXAMPLE 7–12

```
void CFirstDlg::OnBnClickedButton1()
{
       char temp[10];
       int value = GetDlgItemInt(IDC_EDIT1);
       int radix = GetDlgItemInt(IDC_EDIT2);
       _asm
       {
            lea  ebx,temp                  ;address string
            mov  ecx,radix                 ;get radix
            push ecx
            mov  eax,value                 ;test for negative
            test eax,80000000h
            jz   L1
            neg  eax
            mov  byte ptr[ebx], '-'
            inc  ebx
L1:
            mov  edx,0                     ;divide by radix
            div  ecx
            push edx                       ;save remainder
            cmp  eax,0
            jne  L1                        ;while eax != 0
L2:
            pop  edx                       ;convert to ASCII
            cmp  edx,ecx
            je   L4                        ;if done
            add  dl,30h
            cmp  dl,39h
            jbe  L3                        ;if number
            add  dl,7                      ;if letter
L3:
            mov  [ebx],dl                  ;save in string
            jmp  L2
L4:
            mov  byte ptr [ebx+1],0        ;save null to end string
       }
       SetDlgItemText(IDC_EDIT3, temp);
}
```

7–3 SEPARATE ASSEMBLY OBJECTS

As mentioned in the prior sections, the inline assembler is limited because it cannot use MACRO sequences and the conditional program flow directives presented in Chapter 6. In some cases, it is better to develop assembly language modules that are then linked with C++ for more flexibility. This is especially true if the application is being developed by a team of programmers. This section of the chapter details the use of different objects that are linked to form a program using both assembly language and C++. Information covered in this section applies to Microsoft Visual C+ for Windows.

Linking Assembly Language with Visual C++

Example 7–13 illustrates a flat model procedure that will be linked to a C++ program. We denote that the assembly module is a C++ module by using the letter C after the word "flat" in the model statement. The linkage specified by the letter C is the same for the C or C++ languages. The flat model allows assembly language software to be any length up to 2G bytes. Note that the .586

switch appears before the model statement, which causes the assembler to generate code that functions in the protected 32-bit mode. The Reverse procedure, shown in Example 7–13, accepts a character string from a C++ program, reverses its order, and returns to the C++ program. Notice how this program uses conditional program flow instructions, which are not available with the inline assembler described in prior sections of this chapter. The assembly language module can have any name and it can contain more than one procedure, as long as each procedure contains a PUBLIC statement defining the name of the procedure as public. Any parameters that are transferred between the C++ program and the assembly language program are indicated with the backslash following the name of the procedure. This names the parameter for the assembly language program (it can be a different name in C++) and indicates the size of the parameter. The only thing that is not different in the C++ calling program and the assembly program is the order of the parameters. In this example, the parameter is a pointer to a character string and the result is returned as a replacement for the original string.

EXAMPLE 7–13

```
;
;External function that reverses the order of a string of characters
;
.586                            ;select Pentium and 32-bit model
.model flat, C                  ;select flat model with C/C++ linkage
.stack 1024                     ;allocate stack space
.code                           ;start code segment

public Reverse                  ;define Reverse as a public function

Reverse proc uses esi, \        ;define procedure
arraychar:ptr                   ;define external pointer

      mov  esi,arraychar        ;address string
      mov  eax,0
      push eax                  ;indicate end of string

      .repeat                   ;push all the characters to the stack
             mov  al,[esi]
             push eax
             inc  esi
      .until byte ptr [esi] == 0

      mov esi,arraychar         ;address string start

      .while eax != 0           ;pop in reverse order
            pop eax
            mov [esi],al
            inc esi
      .endw
      Ret

Reverse     endp
End
```

Example 7–14 illustrates a C++ language program for DOS console applications that uses the Reverse assembly language procedure. The EXTERN statement is used to indicate that an external procedure called Reverse is to be used in the C++ program. The name of the procedure is case-sensitive, so make sure that it is spelled the same in both the assembly language module and the C++ language module. The EXTERN statement in Example 7–14 shows that the external assembly language procedure transfers a character string to the procedure and returns no data. If data are returned from the assembly language procedure, they are returned as a value in register EAX for bytes, words, or doublewords. If floating-point numbers are returned, they must be returned on the floating-point coprocessor stack. If a pointer is returned, it must be in EAX.

EXAMPLE 7–14

```
/* Program that reverses the order of a character string */

#include <stdio.h>
#include <conio.h>

extern "C" void Reverse(char *);

char chararray[17] = "So what is this?";

int main(int argc, char* argv[]
{
  printf ("%s \n", chararray);
  Reverse (char array);
  printf ("%s\n", chararray);
  getche();                                   //wait to see result
  return 0;
}
```

Once both the C++ program and the assembly language program are written, the Visual C++ development system must be set up to link the two together. For linking and assembly, we will assume that the assembly language module is called Reverse.txt (you cannot add an .asm extension file to the file list for inclusion in a project, so just use the .txt extension and add a .txt file) and the C++ language module is called Main.cpp. Both modules are stored in the C:\PROJECT\MINE directory or some other directory of your choosing. After the modules are placed in the same project workspace, the Programmer's WorkBench program is used to edit both assembly language and C++ language modules.

To set up the Visual C++ developer studio to compile, assemble, and link these files, follow these steps:

1. Start the developer studio and select New from the File menu.
   a. Choose New Project.
   b. When the Application Wizard appears, click on Visual C++ Projects.
   c. Select C++ Console Application, and name the project Mine.
   d. Then click on OK.
2. You will see the project in the Solution window at the left margin in the center. It will have a single file called Main.cpp, which is the C++ program file. Modify this to appear as in Example 7–14.
3. To add the assembly language module, right-click on the line Source Files and select Add from the menu. Choose Add New Item from the list. Scroll down the list of file types until you find Text Files and select it, then enter the file name as Reverse and click on Open. This creates the assembly module called Reverse.txt. You may enter the assembly code from Example 7–13 in this file.
4. Under the Source Files listing in the Solution Explorer, right-click on Reverse.txt and select Properties. Figure 7–7 shows what to enter in this wizard after you click on the Custom Build step. Make sure you enter the object file name (Reverse.obj) in the Outputs box and ml /c /Cx /coff Reverse.txt in the Command Line box. The Reverse assembly language file will assemble and be included in the project.
5. Assuming both Examples 7–13 and 7–14 have been entered and you have completed all steps, the program will function.

At last, you can execute the program. Click on the blue arrow. You should see two lines of ASCII text data displayed. The first line is in correct forward order and the second is in reverse order. Although this is a trivial application, it does illustrate how to create and link C++ language with assembly language.

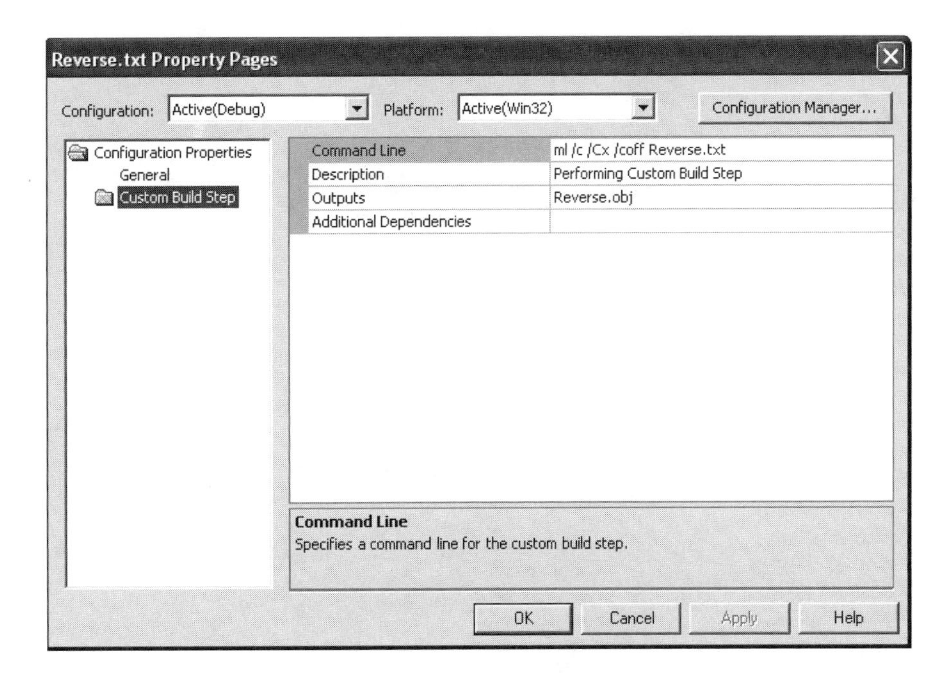

FIGURE 7–7 Using the assembler to assemble a module in Visual C++.

Now that we have a good understanding of interfacing assembly language with C++, we need a longer example that uses a few assembly language procedures with a C++ language program. Example 7–15 illustrates an assembly language package that includes a procedure (Scan) to test a character input against a lookup table and return a number that indicates the relative position in the table. A second procedure (Look) uses a number transferred to it and returns with a character string that represents Morse code. (The code is not important, but if you are interested, Table 7–2 lists Morse code.)

EXAMPLE 7–15

```
.586
.model flat, C
.data

table db  2,1,4,8,4,10,3,4     ;ABCD
      db  1,0,4,2,3,6,4,0      ;EFGH
      db  2,0,4,7,3,5,4,4      ;IJKL
      db  2,3,2,2,3,7,4,6      ;MNOP
      db  4,13,3,2,3,0,1,1     ;QRST
      db  3,1,4,1,3,3,4,9      ;UVWX
      db  4,11,4,12            ;YZ

.code
```

TABLE 7–2 Morse code.

A	._	J	___	S	...
B	_...	K	_._	T	_
C	_._.	L	._..	U	.._
D	_..	M	__	V	..._
E	.	N	_.	W	.__
F	.._.	O	___	X	_.._
G	__.	P	.__.	Y	_.__
H	....	Q	__._	Z	__..
I	..	R	._.		

```
Public Scan
Public Look

Scan proc uses ebx,\
char:dword

      mov  ebx,char
      .if  bl >= 'a' && bl <= 'z'
              sub  bl,20h
      .endif
      sub  bl,41h
      add  bl,bl
      add  ebx,offset table
      mov  ax,word ptr[ebx]
      ret

Scan  endp

Look  proc uses ebx ecx,\
numb:dword,\
pntr:ptr

      mov  ebx,pntr
      mov  eax,numb
      mov  ecx,0
      mov  cl,al
      .repeat
            shr  ah,1
            .if carry?
                   mov byte ptr[ebx],'-'
            .else
                   mov byte ptr[ebx],'.'
            .endif
                   inc ebx
            .untilcxz
            mov byte ptr[ebx],0
            ret

Look        endp
end
```

The lookup table in Example 7–15 contains two bytes for each character between A and Z. For example, the code for an A is a 2 followed by a 1. The 2 indicates how many dots or dashes are used to form the Morse-coded character and the 1 (01) is the code for the letter A (.—), where the 0 is a dot and the 1 is a dash. The Scan procedure accesses this lookup table to obtain the correct Morse code, which is returned in AX as a parameter to the C++ language call. The remaining assembly code is mundane.

Example 7–16 lists the C++ program, which calls the two procedures listed in Example 7–15. This software is simple to understand, so we do not explain it.

EXAMPLE 7–16

```
// Morse.cpp : Defines the entry point for the console application.

#include <iostream>
using namespace std;

extern "C" int Scan(int);
extern "C" void Look(int, char *);

int main(int argc, char* argv[])
{
      int a = 0;
      char chararray[] = "This, is the trick!\n";
      char chararray1[10];
```

```
        while ( chararray[a] != '\n' )
        {
              if ( chararray[a] < 'A' || chararray[a] > 'z' )
                    cout << chararray[a] << '\n';
              else
              {
                    Look ( Scan ( chararray[a] ), chararray1 );
                    cout << chararray[a] << " = " << chararray1 << '\n';

              }
              a++;
        }
        cout << "Type enter to quit!";
        cin.get();
        return 0;
}
```

Although the examples presented here are for console applications, the same method of instructing Visual Studio to assemble and link an assembly language module is also used for Visual applications for Windows. The main difference is that Windows applications do not use printf or cout. Chapter 8 explains how library files can be used with Visual C++ and also gives many more programming examples.

Adding New Assembly Language Instructions to C/C++ Programs

From time to time, as new microprocessors are introduced by Intel, new assembly language instructions are also introduced. These new instructions cannot be used in C++ unless you develop a macro for C++ to include them in the program. An example is the CPUID assembly language instruction. This will not function in an _asm block within C++ because the inline assembler does not recognize it. Another group of newer instructions includes the MMX and SEC instructions. These are also not recognized, but in order to illustrate how a new instruction is added that is not in the assembler, we show the technique. To use any new instructions, first look up the machine language code from Appendix B or from Intel's website at www.intel.com. For example, the machine code for the CPUID instruction is 0F A2. This two-byte instruction can be defined as a C++ macro, as illustrated in Example 7–17. To use the new macro in a C++ program, all we need to type is CPUID. The _emit macro stores the byte that follows it in the program.

EXAMPLE 7–17

```
#define CPUID _asm _emit 0x0f _asm _emit 0xa2
```

7–4 SUMMARY

1. The inline assembler is used to insert short, limited assembly language sequences into a C++ program. The main limitation of the inline assembler is that it cannot use macro sequences or conditional program flow instructions.
2. Two versions of the C++ language are available. One is designed for 16-bit DOS console applications and the other for 32-bit Windows applications. The type chosen for an application depends on the environment, but in most cases programmers today use Windows and the 32-bit Visual version.
3. The 16-bit assembly language applications use the DOS INT 21H commands to access devices in the system. The 32-bit assembly language applications cannot efficiently or easily access the DOS INT 21H function calls even though many are available.

4. The most flexible and often-used method of interfacing assembly language in a C++ program is through separate assembly language modules. The only difference is that these separate assembly language modules must be defined by using the C directive following the .model statement to define the module linkage as C/C++ compatible.
5. The PUBLIC statement is used in an assembly language module to indicate that the procedure name is public and available to use with another module. External parameters are defined in an assembly language module by using the name of the procedure in the PROC statement. Parameters are returned through the EAX register to the calling C/C++ procedure from the assembly language procedure.
6. Assembly language modules are declared external to the C++ program by using the extern directive. If the extern directive is followed by the letter C, the directive is used in a C/C++ language program.
7. When using Visual Studio, we can instruct it to assemble an assembly language module by clicking on Properties for the module and adding the assembler language program (ml /c /Cx /coff Filename.txt) and the output file as an object file (Filename.obj) in the Custom Build step for the module.
8. Assembly language modules can contain many procedures, but can never contain programs using the .startup directive.

7–5 QUESTIONS AND PROBLEMS

1. Does the inline assembler support assembly language macro sequences?
2. Can a byte be defined in the inline assembler by using the DB directive?
3. How are labels defined in the inline assembler?
4. Which registers can be used in assembly language (either inline or linked modules) without being saved?
5. What register is used to return integer data from assembly language to the C++ language caller?
6. What register is used to return floating-point data from assembler language to the C++ language caller?
7. Is it possible to use the .if statement in the inline assembler?
8. In Example 7–3, explain how the mov dl,string1[si] instruction accesses stringl data.
9. In Example 7–3, why was the SI register pushed and popped?
10. Notice in Example 7–5 that no C++ libraries (#include) are used. Do you think that compiled code for this program is smaller than for a program to accomplish the same task in the C++ language? Why?
11. What is the main difference between the 16-bit and 32-bit versions of C/C++ when using the inline assembler?
12. Can the INT 21H instruction, used to access DOS functions, be used in a program using the 32-bit version of the C/C++ compiler? Explain your answer.
13. What is the #include <conio.h> C/C++ library used for in a program?
14. Write a short C/C++ program that uses the _getche() function to read a key and the _putch() function to display the key. The program must end if an '@' is typed.
15. Would an embedded application that is not written for the PC ever use the conio.h library?
16. In Example 7–7, what is the purpose of the sequence of instructions _punch(10); followed by _punch(13);?
17. In Example 7–7, explain how a number is displayed in any number base.
18. Which is more flexible in its application, the inline assembler or assembly language modules that are linked to C++?

19. What is the purpose of a PUBLIC statement in an assembly code module?
20. How is an assembly code module prepared for use with the C++ language?
21. In a C++ language program, the extern void GetIt(int); statement indicates what about function GetIt?
22. How is a 16-bit word of data defined in C++?
23. What is a control in a C++ Visual program and where is it obtained?
24. What is an event in a C++ Visual program and what is an event handler?
25. In Example 7–13, what type of parameter is arraychar?
26. Can the edit screen of C++ Visual Studio be used to enter and edit an assembly language programming module?
27. How are external procedures that are written in assembly language indicated to a C++ program?
28. Show how the RDTSC instruction (opcode is 0F 31) could be added to a C++ program using the _emit macro.
29. In Example 7–15, explain what data type is used by Scan.
30. Write a short assembly language module to be used with C++ that rotates a number three places to the left. Call your procedure RotateLeft3 and assume the number is an 8-bit char (byte in assembly).
31. Repeat question 30, but write the same function in C++ without the assembler.
32. Write a short assembly language module that receives a parameter (byte-sized) and returns a byte-sized result to a caller. Your procedure must take this byte and convert it into an uppercase letter. If an uppercase letter or anything else appears, the byte should not be modified.
33. How is an MFC Visual C++ application executed from Visual Studio?
34. What are properties in a Visual C++ application?
35. What is an ActiveX control or object?
36. Show how a single assembly language instruction, such as inc ptr, is inserted into a Visual C++ program.

CHAPTER 8

Programming the Microprocessor

INTRODUCTION

This chapter develops programs and programming techniques using the MASM macro assembler program and the inline assembler program from Visual C++. The MASM assembler and the Visual C++ inline assembler were explained and demonstrated in prior chapters, but there are still more features to learn.

Some programming techniques explained in this chapter include macro sequences for MASM and stand-alone assembly language modules, keyboard and display manipulation, program modules, library files, using the mouse, and using timers. As an introduction to programming, the chapter provides a wealth of information. You will be able to easily develop programs for the personal computer by using MASM and the inline assembler as a springboard for Visual applications created for Windows.

CHAPTER OBJECTIVES

Upon completion of this chapter, you will be able to:

1. Use the MASM assembler and linker program to create programs that contain more than one module
2. Explain the use of EXTRN and PUBLIC as they apply to modular programming
3. Set up a library file that contains commonly used subroutines and learn how to use the DUMPBIN program
4. Write and use MACRO and ENDM to develop macro sequences used with linear programming in modules that link to C++ code
5. Show how both sequential and random access files are developed for use in a system
6. Develop programs using event handlers to perform keyboard and display tasks
7. Use conditional assembly language statements in programs
8. Use the mouse in program examples

8–1 MODULAR PROGRAMMING

Many programs are too large to be developed by one person. This means that programs are routinely developed by teams of programmers. The linker program is provided with Visual Studio so that programming modules can be linked together into a complete program. Linking is also available from the command prompt provided by Windows. This section of the text describes the linker, the linking task, library files, EXTRN, and PUBLIC as they apply to program modules and modular programming.

The Assembler and Linker

The **assembler program** converts a symbolic **source module** (file) into a hexadecimal **object file**. It is even a part of Visual Studio, located in the C:\Program Files\Microsoft Visual Studio .NET 2003\Vc7\bin folder. We have seen many examples of symbolic source files, written in assembly language, in prior chapters. Example 8–1 shows how the assembler dialog that appears as a source module named NEW.ASM is assembled. Note that this dialog is used with version 6.15 at the DOS command line. The version that comes with Visual C will not work for 16-bit DOS programs. If a 16-bit assembler and linker are needed, they can be obtained in the Windows Driver Development Kit (DDK). Whenever you create a source file, it should have the extension ASM, but as we learned in Chapter 7, that is not always possible. Source files are created by using NotePad or almost any other word processor or editor capable of generating an ASCII file.

EXAMPLE 8–1

```
C:\masm611\BIN>ml new.asm

Microsoft (R) Macro Assembler Version 6.11
Copyright (C) Microsoft Corp 1981-1993.  All rights reserved.

 Assembling: new.asm

Microsoft (R) Segmented Executable Linker Version 5.60.220 Sep 9 1994
Copyright (C) Microsoft Corp 1984-1993.  All rights reserved.

Object Modules [.obj]: new.obj
Run File [new.exe]: "new.exe"
List File [nul.map]: NUL
Libraries [.lib]:
Definitions File [nul.def]:
```

The assembler program (ML) requires the source file name following ML. In Example 8–1, the /Fl switch is used to create a listing file named NEW.LST. Although this is optional, it is recommended so the output of the assembler can be viewed for troubleshooting problems. The source listing file (.LST) contains the assembled version of the source file and its hexadecimal machine language equivalent. The cross-reference file (.CRF), which is not generated in this example, lists all labels and pertinent information required for cross-referencing. An object file is also generated by ML as an input to the linker program. In many cases we only need to generate an object file, which is accomplished by using the /c switch.

The **linker program**, which executes as the second part of ML, reads the object files that are created by the assembler program and links them together into a single execution file. An **execution file** is created with the file name extension EXE. Execution files are selected by typing the file name at the DOS prompt (C:\). An example execution file is FROG.EXE, which is executed by typing FROG at the command prompt.

If a file is short enough (less than 64K bytes long) it can be converted from an execution file to a **command file** (.COM). The command file is slightly different from the execution file in that the program must be originated at location 0100H before it can execute. This means that the program must be no larger than 64K–100H in length. The ML program generates a command file if the tiny model is used with a starting address of 100H. Command files are only used with DOS or if a true binary version (for an EPROM/FLASH burner) is needed. The main advantage of a command file is that it loads off the disk into the computer much more quickly than an execution file. It also requires less disk storage space than the equivalent execution file.

Example 8–2 shows the linker program protocol when it is used to link the files NEW, WHAT, and DONUT. The linker also links library files (LIBS) so procedures, located with LIBS, can be used with the linked execution file. To invoke the linker, type LINK at the command prompt, as illustrated in Example 8–2. Note that before files are linked, they must first be assembled and they must be error-free. ML not only links the files, but it also assembles them prior to linking.

EXAMPLE 8–2

```
C:\masm611\BIN>ml new.asm what.asm donut.asm
Microsoft (R) Macro Assembler Version 6.11
Copyright (C) Microsoft Corp 1981-1993.  All rights reserved.

 Assembling: new.asm
 Assembling: what.asm
 Assembling: donut.asm

Microsoft (R) Segmented Executable Linker Version 5.60.220 Sep 9 1994
Copyright (C) Microsoft Corp 1984-1993.  All rights reserved.

Object Modules [.obj]: new.obj+
Object Modules [.obj]: "what.obj"+
Object Modules [.obj]: "donut.obj"/t
Run File [new.com]: "new.com"
List File [nul.map]: NUL
Libraries [.lib]:
Definitions File [nul.def]:
```

In this example, after you type ML, the linker program asks for the "Object Modules," which are created by the assembler. We have three object modules, NEW, WHAT, and DONUT, in this example. If more than one object file exists, type the main program file first (NEW, in this example), followed by any other supporting modules.

Library files are entered after the file name and after the switch /LINK. In this example, library files were not entered. To use a library called NUMB.LIB while assembling a program called NEW.ASM, type ML NEW.ASM /LINK NUMB.LIB.

In the Windows environment you cannot link a program—you can only assemble a program. You must use Visual Studio to link the program files during the build. You can assemble a file or files and generate objects for use with Visual C++. Example 8–3 illustrates how a module is compiled, but not linked with ML. The /c switch (lowercase c) tells the assembler to compile and generate object files, /Cx preserves the case of all functions and variables, and /coff generates a **common object file format** output for the object files used in a 32-bit environment.

EXAMPLE 8–3

```
C:\Program Files\Microsoft Visual Studio .NET 2003\Vc7\bin>ml /c /Cx /coff new.asm
Microsoft (R) Macro Assembler Version 7.10.3077
Copyright (C) Microsoft Corporation.  All rights reserved.

 Assembling: new.asm
```

PUBLIC and EXTRN

The PUBLIC and EXTRN directives are very important in modular programming because they allow communications between modules. We use PUBLIC to declare that labels of code, data, or entire segments are available to other program modules. EXTRN (external) declares that labels are external to a module. Without these statements, modules could not be linked together to create a program. They might link, but one module would not be able to communicate to another.

The PUBLIC directive is placed in the opcode field of an assembly language statement to define a label as public, so that the label can be used (seen by) by other modules. The label declared as public can be a jump address, a data address, or an entire segment. Example 8–4 shows the PUBLIC statement used to define some labels and make them public to other modules in a program fragment. When segments are made public, they are combined with other public segments that contain data with the same segment name.

EXAMPLE 8–4

```
                        .model  flat,  c
                        .data

                                public Data1        ;declare Data1 and Data2 public
                                public Data2

                        Data1   db      100 dup(?)
0000 0064[
          00
            ]
0064 0064[              Data2   db      100 dup(?)
          00
            ]
                        .code
                        .startup

                                public Read         ;declare Read public

                        Read    proc    far
0006 B4 06                      mov ah,6
```

The EXTRN statement appears in both data and code segments to define labels as external to the segment. If data are defined as external, their sizes must be defined as BYTE, WORD, or DWORD. If a jump or call address is external, it must be defined as NEAR or FAR. Example 8–5 shows how the external statement is used to indicate that several labels are external to the program listed. Notice in this example that any external address or data is defined with the letter E in the hexadecimal assembled listing. It is assumed that Example 8–4 and Example 8–5 are linked together.

EXAMPLE 8–5

```
                        .model flat, c
                        .data
                                extrn Data1:byte
                                extrn Data2:byte
                                extrn Data3:word
                                extrn Data4:dword
                        .code
                                extrn Read:far
                        .startup
0005 Bf 0000 E                  mov     dx,offset Data1
0008 B9 000A                    mov     cx,10
000B                    Start:
000B 9A 0000 ---- E             call    Read
0010 AA                         stosb
0011 E2 F8                      loop    Start
                        .exit
                        End
```

Libraries

Library files are collections of common procedures that are collected in one place so they can be used by many different applications. These procedures are assembled and compiled into a library file by the LIB program that accompanies the MASM assembler program. You may have noticed when setting up Visual C++ to build the assembly language modules in Chapter 7 that many library files were in the link list used by Visual C++. The library file (FILENAME.LIB) is invoked when a program is linked with the linker program.

Why bother with library files? A library file is a good place to store a collection of related procedures. When the library file is linked with a program, only the procedures required by that program are removed from the library file and added to the program. If any amount of assembly language programming is to be accomplished efficiently, a good set of library files is essential to save many hours of recoding common functions.

Creating a Library File. A library file is created with the LIB command, which executes the LIB.EXE program that is supplied with Visual Studio. A library file is a collection of assembled .OBJ files that contains procedures or tasks written in assembly language or any other language. Example 8–6 shows two separate functions (UpperCase and LowerCase) included in a module that is written for Windows, which will be used to structure a library file. Please notice that the name of the procedure must be declared PUBLIC in a library file and does not necessarily need to match the file name, although it does in this example. A variable is transferred to each file, so the EXTRN statement also appears in each procedure to gain access to an external variable. Example 8–7 shows the C++ protocols that are required to use the functions in this library file in a C++ program, provided the library is linked to the program.

EXAMPLE 8–6

```
.586
.model flat,c
.code
       public UpperCase
       public LowerCase
UpperCase proc ,\
       Data1:byte
       mov   al,Data1
       .if al >= 'a' && al <= 'z'
              sub al,20h
       .endif
       ret
UpperCase endp

LowerCase proc ,\
       Data2:byte
       mov   al,Data2
       .if al >= 'A' && al <= 'Z'
              add al,20h
       .endif
       ret
LowerCase endp
End
```

EXAMPLE 8–7

```
extern "C" char UpperCase(char);
extern "C" char LowerCase(char);
```

The LIB program begins with the copyright message from Microsoft, followed by the prompt *Library name.* The library name chosen is case for the CASE.LIB file. Because this is a new file, the library program must be prompted with the object file name. You must first assemble CASE.ASM with ML. The actual LIB command is listed in Example 8–8. Notice that the LIB program is invoked with the object name following it on the command line.

EXAMPLE 8–8

```
C:\Program Files\Microsoft Visual Studio .NET 2003\Vc7\bin>lib case.obj
Microsoft (R) Library Manager Version 7.10.3077
Copyright (C) Microsoft Corporation.  All rights reserved.
```

A utility program called DUMPBIN.EXE is provided to display the contents of the library or any other file. Example 8–9 shows the outcome of a binary dump using the /all switch to show the library module CASE.LIB and all its components. Near the top of this listing are the public names for _UpperCase and _LowerCase. The Raw Data #1 section contains the actual hexadecimal-coded instructions for the two procedures.

EXAMPLE 8–9

```
C:\Program Files\Microsoft Visual Studio .NET 2003\Vc7\bin>dumpbin /all case.lib

Microsoft (R) COFF/PE Dumper Version 7.10.3077
Copyright (C) Microsoft Corporation.  All rights reserved.

Dump of file case.lib

File Type: LIBRARY

Archive member name at 8: /
401D4A83 time/date Sun Feb 01 13:50:43 2004
         uid
         gid
       0 mode
      22 size
correct header end

    2 public symbols

       C8 _LowerCase
       C8 _UpperCase

Archive member name at 66: /
401D4A83 time/date Sun Feb 01 13:50:43 2004
         uid
         gid
       0 mode
      26 size
correct header end

    1 offsets

        1       C8

    2 public symbols

        1 _LowerCase
        1 _UpperCase

Archive member name at C8: case.obj/
401D43A6 time/date Sun Feb 01 13:21:26 2004
         uid
         gid
  100666 mode
     228 size
correct header end
```

```
FILE HEADER VALUES
             14C machine (x86)
               3 number of sections
        401D43A6 time date stamp Sun Feb 01 13:21:26 2004
             124 file pointer to symbol table
               D number of symbols
               0 size of optional header
               0 characteristics

SECTION HEADER #1
   .text name
       0 physical address
       0 virtual address
      24 size of raw data
      8C file pointer to raw data (0000008C to 000000AF)
       0 file pointer to relocation table
       0 file pointer to line numbers
       0 number of relocations
       0 number of line numbers
60500020 flags
         Code
         16 byte align
         Execute Read

RAW DATA #1
  00000000: 55 8B EC 8A 45 08 3C 61 72 06 3C 7A 77 02 2C 20  U.ì.E.<ar.<zw.,
  00000010: C9 C3 55 8B EC 8A 45 08 3C 41 72 06 3C 5A 77 02  ÉAU.ì.E.<Ar.<Zw.
  00000020: 04 20 C9 C3                                      . ÉA

SECTION HEADER #2
   .data name
      24 physical address
       0 virtual address
       0 size of raw data
       0 file pointer to raw data
       0 file pointer to relocation table
       0 file pointer to line numbers
       0 number of relocations
       0 number of line numbers
C0500040 flags
         Initialized Data
         16 byte align
         Read Write

SECTION HEADER #3
.debug$S name
      24 physical address
       0 virtual address
      74 size of raw data
      B0 file pointer to raw data (000000B0 to 00000123)
       0 file pointer to relocation table
       0 file pointer to line numbers
       0 number of relocations
       0 number of line numbers
42100040 flags
         Initialized Data
         Discardable
         1 byte align
         Read Only

RAW DATA #3
  00000000: 04 00 00 00 F1 00 00 00 00 00 00 00 30 00 01 11  ....ñ.......0...
  00000010: 00 00 00 00 43 3A 5C 50 52 4F 47 52 41 7E 31 5C  ....C:\PROGRA~1\
  00000020: 4D 49 43 52 4F 53 7E 31 2E 4E 45 54 5C 56 63 37  MICROS~1.NET\Vc7
  00000030: 5C 62 69 6E 5C 63 61 73 65 2E 6F 62 6A 00 34 00  \bin\case.obj.4.
  00000040: 16 11 03 02 00 00 05 00 00 00 00 00 00 00 07 00  ................
  00000050: 0A 00 05 0C 4D 69 63 72 6F 73 6F 66 74 20 28 52  ....Microsoft (R
```

```
  00000060: 29 20 4D 61 63 72 6F 20 41 73 73 65 6D 62 6C 65  ) Macro Assemble
  00000070: 72 00 00 00                                      r...

COFF SYMBOL TABLE
000 00000000 DEBUG   notype       Filename     | .file
    C:\PROGRA~1\MICROS~1.NET\Vc7\bin\case.asm
004 000F0C05 ABS     notype       Static       | @comp.id
005 00000000 SECT1   notype       Static       | .text
    Section length   24, #relocs    0, #linenums    0, checksum         0
007 00000000 SECT2   notype       Static       | .data
    Section length    0, #relocs    0, #linenums    0, checksum         0
009 00000000 SECT3   notype       Static       | .debug$S
    Section length   74, #relocs    0, #linenums    0, checksum         0
00B 00000000 SECT1   notype ()    External     | _UpperCase
00C 00000012 SECT1   notype ()    External     | _LowerCase

String Table Size = 0x1A bytes

  Summary

           0 .data
          74 .debug$S
          24 .text
```

Once the library file is linked to your program file, only the library procedures actually used by your program are placed in the execution file. Don't forget to use the extern "C" statement in the C++ program to use a function from a library file.

Macros

A **macro** is a group of instructions that perform one task, just as a procedure performs one task. The difference is that a procedure is accessed via a CALL instruction, whereas a macro, and all the instructions defined in the macro, is inserted in the program at the point of usage. Creating a macro is very similar to creating a new opcode, which is actually a sequence of instructions, in this case, that can be used in the program. You type the name of the macro and any parameters associated with it, and the assembler then inserts them into the program. Macro sequences execute faster than procedures because there is no CALL or RET instruction to execute. The instructions of the macro are placed in your program by the assembler at the point where they are invoked. Be aware that macros will not function using the inline assembler; they only function in external assembly language modules.

The MACRO and ENDM directives delineate a macro sequence. The first statement of a macro is the MACRO instruction, which contains the name of the macro and any parameters associated with it. An example is MOVE MACRO A,B, which defines the macro name as MOVE. This new pseudo opcode uses two parameters: A and B. The last statement of a macro is the ENDM instruction, which is placed on a line by itself. Never place a label in front of the ENDM statement or the macro will not assemble.

Example 8–10 shows how a macro is created and used in a program. The first six lines of code define the macro. This macro moves the word-sized contents of memory location B into word-sized memory location A. After the macro is defined in the example, it is used twice. The macro is expanded by the assembler in this example, so that you can see how it assembles to generate the moves. Any hexadecimal machine language statement followed by a number (1, in this example) is a macro expansion statement. The expansion statements are not typed in the source program; they are generated by the assembler (if .LISTALL is included in the program) to show that the assembler has inserted them into the program. Notice that the comment in the macro is preceded with ;; instead of ; as is customary. Macro sequences must always be defined before they are used in a program, so they generally appear at the top of the code segment.

EXAMPLE 8–10

```
                         MOVE   MACRO A,B
                                PUSH AX
                                MOV  AX,B
                                MOV  A,AX
                                POP  AX
                                ENDM

                                MOVE VAR1,VAR2    ;;move VAR2 into VAR1

0000 50           1             PUSH AX
0001 A1 0002 R    1             MOV  AX,VAR2
0004 A3 0000 R    1             MOV  VAR1,AX
0007 58           1             POP  AX

                                MOVE VAR3,VAR4    ;;move VAR4 into VAR3
0008 50           1             PUSH AX
0009 A1 0006 R    1             MOV  AX,VAR4
000C A3 0004 R    1             MOV  VAR3,AX
000F 58           1             POP  AX
```

Local Variables in a Macro. Sometimes, macros contain local variables. A **local variable** is one that appears in the macro, but is not available outside the macro. To define a local variable, we use the LOCAL directive. Example 8–11 shows how a local variable, used as a jump address, appears in a macro definition. If this jump address is not defined as local, the assembler will flag it with errors on the second and subsequent attempts to use the macro.

EXAMPLE 8–11

```
                             FILL    MACRO WHERE, HOW_MANY    ;;fill memory
                                     LOCAL FILL1
                                     PUSH  SI
                                     PUSH  CX
                                     MOV   SI,OFFSET WHERE
                                     MOV   CX,HOW_MANY
                                     MOV   AL,0
                             FILL1:  MOV   [SI],AL
                                     INC   SI
                                     LOOP  FILL1
                                     POP   CX
                                     POP   SI
                                     ENDM

                                     FILL MES1,5

                         1           LOCAL FILL1
0014  56                 1           PUSH  SI
0015  51                 1           PUSH  CX
0016  BE 0000 R          1           MOV   SI,OFFSET MES1
0019  B9 0005            1           MOV   CX,5
001C  B0 00              1           MOV   AL,0
0029  88 04              1   ??0000: MOV   [SI],AL
002B  46                 1           INC   SI
002C  E2 FB              1           LOOP  ??0000
002E  59                 1           POP   CX
002F  5E                 1           POP   SI

                                     FILL  MES2,10

                         1           LOCAL FILL1
0030  56                 1           PUSH  SI
0031  51                 1           PUSH  CX
0032  BE 0014 R          1           MOV   SI,OFFSET MES2
0035  B9 000A            1           MOV   CX,10
0038  B0 00              1           MOV   AL,0
```

```
003A  88 04          1    ??0001:MOV   [SI],AL
003C  46             1           INC   SI
003D  E2 FB          1           LOOP  ??0001
003F  59             1           POP    CX
0040  5E             1           POP    SI
                          .EXIT
```

Example 8–11 shows a FILL macro that stores any number (parameter HOW_MANY) of 00H into the memory location addressed by parameter WHERE. Notice how the address FILL1 is treated when the macros are expanded. The assembler uses labels that start with ?? to designate them as assembler-generated labels.

The LOCAL directive must always be used on the line immediately following the MACRO statement or an error occurs. The LOCAL statement may have up to 35 labels, all separated with commas.

Placing MACRO Definitions in Their Own Module. Macro definitions can be placed in the program file as shown, or they can be placed in their own macro module. A file can be created that contains only macros to be included with other program files. We use the INCLUDE directive to indicate that a program file will include a module that contains external macro definitions. Although this is not a library file, for all practical purposes it functions as a library of macro sequences.

When macro sequences are placed in a file (often with the extension INC or MAC), they do not contain PUBLIC statements as does a library. If a file called MACRO.MAC contains macro sequences, the INCLUDE statement is placed in the program file as INCLUDE C:\ASSM\MACRO.MAC. Notice that the macro file is on drive C, subdirectory ASSM in this example. The INCLUDE statement includes these macros, just as if you had typed them into the file. No EXTRN statement is needed to access the macro statements that have been included. Programs may contain both macro include files and library files.

8–2 USING THE KEYBOARD AND VIDEO DISPLAY

Today, there are few programs that don't use the keyboard and video display. This section of the text explains how to use the keyboard and video display connected to the IBM PC or compatible computer running under Windows.

Reading the Keyboard

The keyboard of the personal computer is read by many different objects available to Visual C++. Data read from the keyboard are either in ASCII-coded or in extended ASCII-coded form. They are then either stored in 8-bit ASCII form or in 16-bit Unicode form. As mentioned in an earlier chapter, Unicode contains ASCII code in codes 0000H–00FFH. The remaining codes are used for foreign language character sets. We do not use cin or getch to read keys in Visual C++ as we do in a DOS C++ console application; instead we use objects that accomplish the same task.

The ASCII-coded data appear as outlined in Table 1–8 in Section 1–4. The extended character set of Table 1–9 applies to printed or displayed data only, and not to keyboard data. Notice that the ASCII codes in Table 1–8 correspond to most of the keys on the keyboard. Also available through the keyboard are extended ASCII-coded keyboard data. Table 8–1 lists most of the extended ASCII codes obtained with various keys and key combinations. Notice that most keys on the keyboard have alternative key codes. Each function key has four sets of codes selected by

TABLE 8–1 The keyboard scanning and extended ASCII codes as returned from the keyboard.

		Extended ASCII code with....			
Key	*Scan Code*	*Nothing*	*Shift*	*Control*	*Alternate*
Esc	01				01
1	02				78
2	03			03	79
3	04				7A
4	05				7B
5	06				7C
6	07				7D
7	08				7E
8	09				7F
9	0A				80
0	0B				81
-	0C				82
+	0D				83
Bksp	0E				0E
Tab	0F		0F	94	A5
Q	10				10
W	11				11
E	12				12
R	13				13
T	14				14
Y	15				15
U	16				16
I	17				17
O	18				18
P	19				19
[	1A				1A
]	1B				1B
Enter	1C				1C
Enter	1C				A6
Lctrl	1D				
Rctrl	1D				
A	1E				1E
S	1F				1F
D	20				20
F	21				21
G	22				22
H	23				23
J	24				24
K	25				25
L	26				26
;	27				27
'	28				28
`	29				29
Lshft	2A				
\	2B				

(continued on next page)

TABLE 8–1 *(continued)*

		Extended ASCII code with....			
Key	*Scan Code*	*Nothing*	*Shift*	*Control*	*Alternate*
Z	2C				2C
X	2D				2D
C	2E				2E
V	2F				2F
B	30				30
N	31				31
M	32				32
,	33				33
.	34				34
/	35				35
Gray/	35			95	A4
Rshft	36				
PrtSc	E0 2A E0 37				
L alt	38				
R alt	38				
Space	39				
Caps	3A				
F1	3B	3B	54	5E	68
F2	3C	3C	55	5F	69
F3	3D	3D	56	60	6A
F4	3E	3E	57	61	6B
F5	3F	3F	58	62	6C
F6	40	40	59	63	6D
F7	41	41	5A	64	6E
F8	42	42	5B	65	6F
F9	43	43	5C	66	70
F10	44	44	5D	67	71
F11	57	85	87	89	8B
F12	58	86	88	8A	8C
Num	45				
Scroll	46				
Home	E0 47	47	47	77	97
Up	48	48	48	8D	98
Pgup	E0 49	49	49	84	99
Gray–	4A				
Left	4B	4B	4B	73	9B
Center	4C				
Right	4D	4D	4D	74	9D
Gray +	4E				
End	E0 4F	4F	4F	75	9F
Down	E0 50	50	50	91	A0
Pgdn	E0 51	51	51	76	A1
Ins	E0 52	52	52	92	A2
Del	E0 53	53	53	93	A3
Pause	E0 10 45				

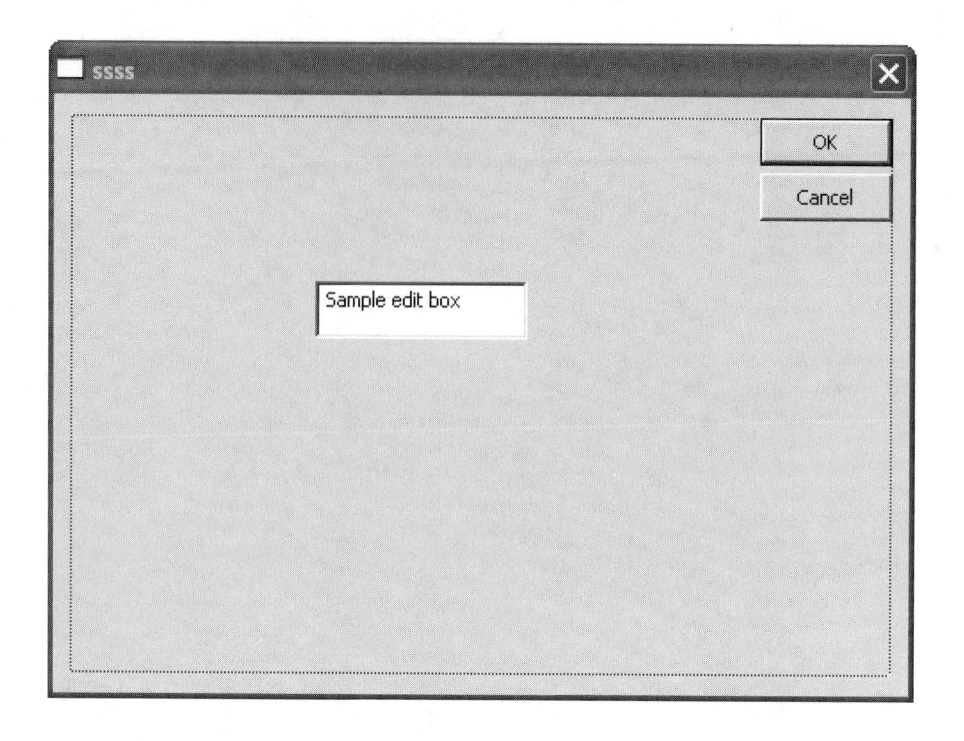

FIGURE 8–1 Keyboard use: example dialog application.

the function key alone, the Shift-function key combination, the alternate-function key combination, and the Control-function key combination.

Creating a Visual C++ application that contains a simple edit box (edit object) gives a better understanding of reading a key in Windows. Figure 8–1 shows such an application written as a dialog-based application. Recall how to create an MFC (Microsoft Foundation Classes) dialog-based application:

1. Start Visual Studio.
2. Click on File, then New, and finally Project to open a new project.
3. Select an MFC application under Visual C++ and give it a name, then click OK.
4. Under Application Type select Dialog-Based and then click on Finish.

Once the new dialog-based application is created, select the edit control from the toolbox and draw it on the screen of the dialog box, as illustrated in Figure 8–1. Next, delete the TOTO: message. Once the edit control is on the screen, right-click on it and select Add Variable. Give the edit box the variable control name Edit1.

Setting Focus. The first thing that should be added to the application is a set focus to the edit control. When focus is set the cursor moves to the object, in this case the edit box. Focus is set to a control by using Edit1.SetFocus(), which in our case is because the edit control is named Edit1. This statement is placed in the OnInitDialog function in the dialog class. In the OnInitDialog function, place the statement where the program lists the TODO: comment. Also, change the return on the OnInitDialog function to return false instead of return true. This is required to set focus. The application will now set focus to the edit control. This means that the blinking cursor appears inside the edit control.

When the application is executed and keys are typed into the edit control, the program reads the keyboard and displays each character as it is typed. In some cases this may be undesirable and may require some filtering. One such case is if the program requires that the user enter only hexadecimal data. In order to intercept keystrokes as they are typed, the function

PreTranslateMessage is used. The **PreTranslateMessage** function intercepts all Windows messages *before* Windows acts upon them. To insert this function into the dialog application, left-click on the edit control, then select Properties from the menu at the bottom of the screen.

To illustrate filtering, this application uses the PreTranslateMessage function to look at the keyboard character that is typed before the program sees the keystrokes so they can be modified. Here the program only allows the numbers 0 through 9 and the letters A through F to be typed from the keyboard. If a lowercase letter is typed, it is converted into uppercase. To accomplish this, use the WM_CHAR Windows message as illustrated in Example 8–12. In this example, C++ is used to filter the keyboard entry into the edit control.

EXAMPLE 8–12

```
BOOL CssssDlg::PreTranslateMessage(MSG* pMsg)
{
      if ( pMsg->message == WM_CHAR )
      {
            unsigned int key = pMsg->wParam;        //get key
            if ( key >= 'a' && key <= 'f' )         //make uppercase
                  key -= 32;
            if ( (key < '0' || key > 'F') || (key > '9' && key < 'A') )
                  return true;                      //ignore these
            pMsg->wParam = key;                     //replace key
      }
      return CDialog::PreTranslateMessage(pMsg);
}
```

The message structure (pMsg) associated with the PreTranslateMessage function contains information about the Windows message. All Windows messages are assigned names beginning with **WM_**. In this case the message intercepted by the function is WM_CHAR. The WM_CHAR message contains the ASCII character typed on the keyboard in the wParam of the pMsg structure. The outermost if statement looks for the WM_CHAR message, which only occurs when a key is typed. The next if statement tests the wParam for the lowercase letters a, b, c d, e, and f. If any of these keys are typed, 32 is subtracted from the wParam to convert to uppercase. The bias between uppercase and lowercase letters is a 32 decimal or 20H. That is, an A is 41H and an a is 61H, a difference of 20H.

If a lowercase letter a–f is not typed, the numbers 0–9 are next checked and if no number is present, a return true statement is executed. If a return true occurs from the PreTranlateMessage function, the keystroke is effectively erased or ignored. A return false causes Windows to issue a warning beep and disposes of the keystroke. In this case if a number is typed or the letters a through f or A through F, the keystroke is passed to the Windows framework by the normal return at the end of the PreTranslateMessage function. This filters the keystrokes so only A–F or 0–9 appears in the edit box.

Example 8–12 is repeated using the inline assembler to accomplish the same task in Example 8–13. In this example C++ seems to require less typing than assembly language, but it is important to be able to visualize both forms.

EXAMPLE 8–13

```
BOOL CssssDlg::PreTranslateMessage(MSG* pMsg)
{
      if ( pMsg->message == WM_CHAR )
      {
            unsigned int key = pMsg->wParam;
            _asm
            {
                  mov eax,key
                  cmp eax, 'a'
                  jb  PreTrans1
                  cmp eax, 'f'
                  ja  PreTrans1
```

```
                    sub eax,32          ;if a - f
                    mov key,eax
PreTrans1:
                    cmp eax, '0'
                    jb  PreTrans2
                    cmp eax, '9'
                    jbe PreTrans3
                    cmp eax, 'A'
                    jb  PreTrans2
                    cmp eax, 'F'
                    jbe PreTrans3
PreTrans2:
              }
              return true;
PreTrans3:
              pMsg->wParam = key;
        }
        return CDialog::PreTranslateMessage(pMsg);
}
```

If this code is added to the application and executed, the only keys that will appear in the edit control are 0–9 and A–F. Any amount of filtering can be done in a likewise manner in the Visual C++ environment. The properties of the edit control include uppercase, which could have been activated to shorten the filtering task, but here software accomplished this.

Using the Video Display

As with the keyboard, in Visual C++ objects are used to display information. The edit control can be used either to read data or display data as can most objects. Modify the application presented in Figure 8–1 so it contains an additional edit control as shown in Figure 8–2. Notice that a few static labels have been added to identify the contents of the edit controls. In this new application the keyboard data is still read into edit control Edit1, but when the Enter key is typed, a decimal version of the data entered into Edit1 appears in Edit2—the second edit control. Make sure that the second control is named Edit2 to maintain compatibility with the software presented here.

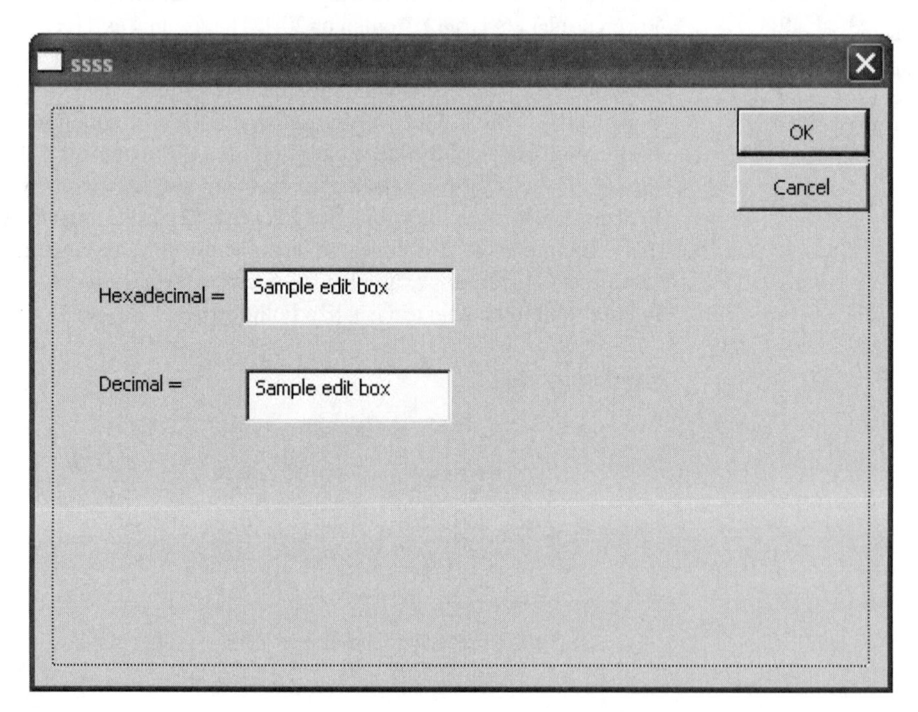

FIGURE 8–2 Converting from hexadecimal to decimal.

To cause the program to react to the Enter key, ASCII code 13 (0DH or 0x0d), modify the PreTranslateMessage function of Example 8–12 as shown in Example 8–14 (see italicized section). Notice how the Enter key is detected. Once the Enter key is detected, Edit1 can be converted to decimal for display in Edit2.

EXAMPLE 8–14

```
BOOL CssssDlg::PreTranslateMessage(MSG* pMsg)
{
        if ( pMsg->message == WM_KEYDOWN && pMsg->wParam == 13 )  //is it enter?
        {
              //software to convert and display decimal data goes here
        }
        else if ( pMsg->message == WM_CHAR )
        {
              unsigned int key = pMsg->wParam;
              if ( key >= 'a' && key <= 'f' )
                     key -= 32;
              if ( key < '0' || key > 'F' || key > '9' && key < 'A' )
                     return true;
              pMsg->wParam = key;
        }
        return CDialog::PreTranslateMessage(pMsg);
}
```

The main problem here is that data entered into an edit control can be accessed as a string or as a decimal number, but not a hexadecimal number. To obtain the string from an edit control, use the GetDlgItemText member function. In this example program in Example 8–15, this appears as GetDlgItemText(IDC_EDIT1, temp), where temp is a CString variable that receives the character string from the edit control. Once this is available, the string can be converted into binary so it can subsequently be converted to decimal for display in Edit2.

EXAMPLE 8–15

```
BOOL CssssDlg::PreTranslateMessage(MSG* pMsg)
{
        if ( pMsg->message == WM_KEYDOWN && pMsg->wParam == 13 ) //is it enter?
        {
               CString temp;
               int temp1 = 0;
               char temp2;
               GetDlgItemText(IDC_EDIT1, temp);
               for ( int a = 0; a < temp.GetLength(); a++)
               {
                             temp2 = temp.GetAt(a);
                             _asm
                             {
                                   mov  al,temp2   ;convert ASCII digit to binary
                                   sub  al,30h
                                   cmp  al,9
                                   jbe  L1
                                   sub  al,7
L1:
                                   shl  temp1,4
                                   or  byte ptr temp1,al
                             }
               }
               SetDlgItemInt(IDC_EDIT2,temp1);
               return true;
        }
        else if ( pMsg->message == WM_CHAR )
        {
        unsigned int key = pMsg->wParam;
```

```
                if ( key >= 'a' && key <= 'f' )
                        key -= 32;
                if ( key < '0' || key > 'F' || key > '9' && key < 'A' )
                        return true;
                pMsg->wParam = key;
        }
        return CDialog::PreTranslateMessage(pMsg);
}
```

Example 8–15 shows the completed application. When the Enter key is pressed, the program uses GetDlgItemText(IDC_EDIT1, temp) to obtain the character string from Edit1 and store it in CString temp. The next portion of the program (the for loop) scans through the character string one character at a time, converting each character to binary as it assembles the entire binary version of the string into integer temp1. Each digit is obtained from the character string by using the CString member function GetAt(int position). The ASCII character thus obtained is then stored into char size variable temp2 and assembly language is used to accomplish the conversion.

When converting from ASCII to binary, subtract 30h. This converts ASCII numbers (0–9) 30H through 39H to the binary numbers 0 through 9. It does not convert 41H through 46H (A through F) to binary because the result is 11H through 16H and not 0AH through 0FH. To adjust the values obtained for the letters, use an if statement to detect 11H through 16H and then subtract an additional 7 to convert 11H through 16H to 0Ah to 0FH. Once the ASCII digit is in binary form, the integer at temp1 is shifted left four binary places and the binary version of the ASCII digit is ORed to it to accumulate the converted digits in temp1.

Once the hexadecimal number from Edit1 is converted to binary in variable temp1, it is displayed in Edit2 using the SetDlgItemInt function. As before, a return true informs the Windows interface that the Enter key has been pressed. See Example 8–16 for the strictly C++ language version of this program in assembly language form as viewed in Debug in Visual Studio for comparison to the assembly code that is used in Example 8–15. Only the for loop is shown for comparison to the assembly language version of Example 8–15.

EXAMPLE 8–16

```
        //the for loop before disassembly

        for ( int a = 0; a < temp.GetLength(); a++)
        {
                temp1 <<= 4;
                if ( temp.GetAt(a) > '9' )
                        temp1 |= temp.GetAt(a) - 0x37;
                else
                        temp1 |= temp.GetAt(a) - 0x30;
        }



        //the for loop after disassembly

        for ( int a = 0; a < temp.GetLength(); a++)
00413936   mov          dword ptr [a],0
0041393D   jmp          CssssDlg::PreTranslateMessage+98h (413948h)
0041393F   mov          eax,dword ptr [a]
00413942   add          eax,1
00413945   mov          dword ptr [a],eax
00413948   mov          esi,esp
0041394A   lea          ecx,[temp]
0041394D   call         dword ptr [__imp_ATL::CSimpleStringT<char,1>::GetLength
(42BADCh)]
00413953   cmp          esi,esp
00413955   call         @ILT+2440(__RTC_CheckEsp) (41198Dh)
```

```
0041395A  cmp        dword ptr [a],eax
0041395D  jge        CssssDlg::PreTranslateMessage+121h (4139D1h)
          {
               temp1 <<= 4;
0041395F  mov        eax,dword ptr [temp1]
00413962  shl        eax,4
00413965  mov        dword ptr [temp1],eax
               if ( temp.GetAt(a) > '9' )
00413968  mov        esi,esp
0041396A  mov        eax,dword ptr [a]
0041396D  push       eax
0041396E  lea        ecx,[temp]
00413971  call       dword ptr [__imp_ATL::CSimpleStringT<char,1>::GetAt
(42BAD8h)]
00413977  cmp        esi,esp
00413979  call       @ILT+2440(__RTC_CheckEsp) (41198Dh)
0041397E  movsx      ecx,al
00413981  cmp        ecx,39h
00413984  jle        CssssDlg::PreTranslateMessage+0FAh (4139AAh)
                    temp1 |= ( temp.GetAt(a) - 0x37 );
00413986  mov        esi,esp
00413988  mov        eax,dword ptr [a]
0041398B  push       eax
0041398C  lea        ecx,[temp]
0041398F  call       dword ptr [__imp_ATL::CSimpleStringT<char,1>::GetAt
(42BAD8h)]
00413995  cmp        esi,esp
00413997  call       @ILT+2440(__RTC_CheckEsp) (41198Dh)
0041399C  movsx      ecx,al
0041399F  sub        ecx,37h
004139A2  or         ecx,dword ptr [temp1]
004139A5  mov        dword ptr [temp1],ecx
               else
004139A8  jmp        CssssDlg::PreTranslateMessage+11Ch (4139CCh)
                    temp1 |= ( temp.GetAt(a) - 0x30 );
004139AA  mov        esi,esp
004139AC  mov        eax,dword ptr [a]
004139AF  push       eax
004139B0  lea        ecx,[temp]
004139B3  call       dword ptr [__imp_ATL::CSimpleStringT<char,1>::GetAt
(42BAD8h)]
004139B9  cmp        esi,esp
004139BB  call       @ILT+2440(__RTC_CheckEsp) (41198Dh)
004139C0  movsx      ecx,al
004139C3  sub        ecx,30h
004139C6  or         ecx,dword ptr [temp1]
004139C9  mov        dword ptr [temp1],ecx
          }
004139CC  jmp        CssssDlg::PreTranslateMessage+8Fh (41393Fh)
```

Using Active X Controls in a Program

Active X controls are objects that are used with any visual programming language. They may be written in C++, BASIC, or any language. To select and use active X controls in a program, right-click on the dialog box and select Insert active X control. Realize that active X controls are owned by various companies and using them in a program may require paying a royalty to the owner of the control. As long as you do not sell a product you can use these controls without a fee. Active X controls usually have many more features than the stock controls that are provided by Visual Studio.

An application that uses an active X control appears in Figure 8–3. Even though this is a fairly simple application it does illustrate the use of an active X control to display a dynamic label. Only a static label (cannot be changed) is available as a stock control in Visual C++ and at times a dynamic label (can be changed) is needed. The member functions used most often in an active X control (assuming Visual Studio .NET) are put_ to mutate a control feature or get_ to

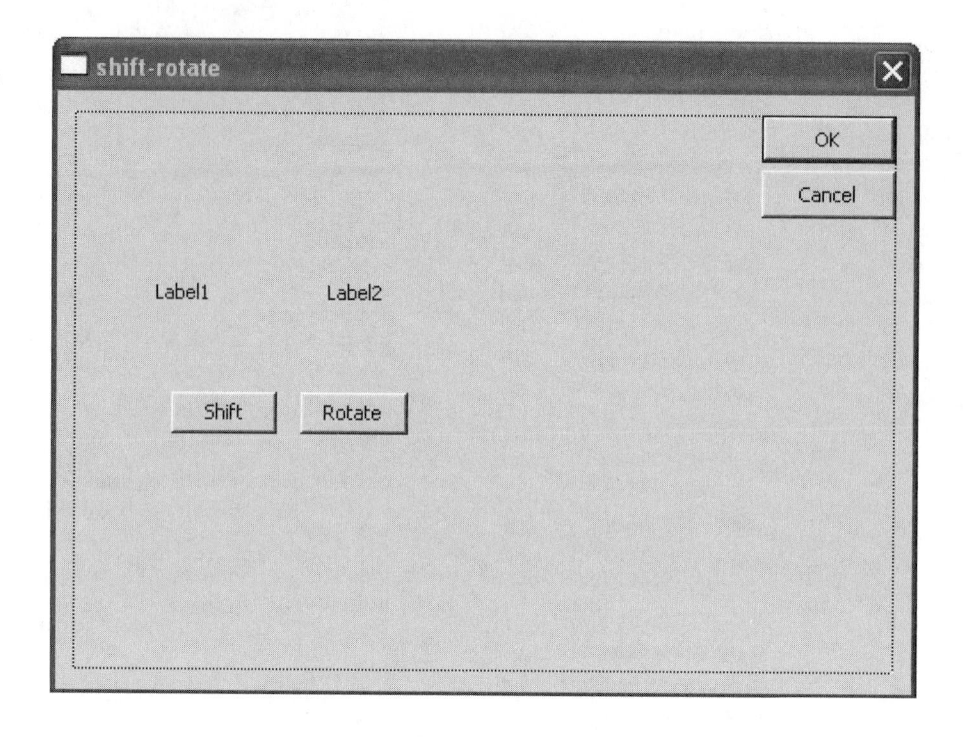

FIGURE 8–3 An application that illustrates active X controls as well as stock controls.

access a control feature. Figure 8–3 uses two stock command controls (recaptioned Shift and Rotate) and two active X Microsoft Forms 2.0 Labels. The Microsoft Forms 2.0 Labels are actually a part of Microsoft Word, so you must also have Microsoft Word installed to use these active X controls.

Once your dialog box appears as in Figure 8–3, add names for the two active X Forms 2.0 Labels: Label1 and Label2. Now add handlers for the two command button click functions (BN_CLICKED). (Recall that to do this go to Properties and click Button, then click on the lightning bolt to access the member handlers for the controls.) Now click on the icon just to the right of the lightning bolt and scroll down to WM_TIMER. Add the handler for the timer, which will be used in this programming example. Add two variables using the class view (click on the word Class to select the class view). Now right-click on the class that ends with Dlg—this is the dialog class where the handlers that were installed appear. Select Add and Add Variable and add a Boolean variable called Shift and a char variable called data1. These are class public variables visible from anywhere in the Dlg class.

Example 8–17 illustrates the software added to the application in the three handlers required to implement the application for Figure 8–3. The software for the two command button click handlers is short and easy to understand. The first statement in each places a caption on Label1. If the shift button is pressed, Shift Left = is displayed, and if the rotate button is pressed, Rotate Left = is displayed on Label1. The second statement merely makes Boolean variable shift true or false depending on the button pressed. In both command button handlers, data1 starts at 1 and 00000001 is displayed on Label2. Finally, a timer is started for the control button using the SetTimer function. The first variable in the timer is a number chosen for the timer, the second number is the timer cycle time in milliseconds, and the third is a zero for almost all Visual C++ applications. Every 500 milliseconds the timer goes off and calls the OnTimer function in the example. OnTimer therefore executes once per half second until the program is closed. This allows an event to occur every half second. In this example we will shift or rotate 00000001 and display it each half second.

EXAMPLE 8–17

```
void CshiftrotateDlg::OnBnClickedButton1()
{
	Label1.put_Caption("Shift Left = ");
	shift = true;		//select shift
	data1 = 1;
	Label2.put_Caption("00000001");
	SetTimer(1,500,0);
}

void CshiftrotateDlg::OnBnClickedButton2()
{
	Label1.put_Caption("Rotate Left = ");
	shift = false;		//select rotate
	data1 = 1;
	Label2.put_Caption("00000001");
	SetTimer(1,500,0);
}

void CshiftrotateDlg::OnTimer(UINT nIDEvent)
{
	if ( nIDEvent == 1 )
	{
		CString temp = "";
		char temp1 = data1;
		if ( shift )
			_asm shl temp1,1;	//shift
		else
			_asm rol temp1,1;	//rotate
		data1 = temp1;
		for (int a = 128; a > 0; a>>=1)
		{
			if ( ( temp1 & a ) == a )
				temp += "1";
			else
				temp += "0";
		}
		Label2.put_Caption(temp);
	}
	CDialog::OnTimer(nIDEvent);
}
```

The OnTimer function first checks to see if Timer 1 caused the OnTimer event. This is required in systems that use more than one timer. Here it would be optional because only one timer exists. Two local variables are used in this example; one is a character string (temp) and the other is a char called temp1. If shift is true, temp1 is shifted and if shift is false, temp1 is rotated. Recall that the command buttons set shift to either true or false to select the type of operation. Notice that a local variable is used here for the data that is shifted. The reason is that you cannot easily access a class variable from assembly language, so before data1 is used it is made local as temp1 and likewise after temp1 is stored back into data1.

The strangest portion of the OnTimer function is developing the numeric character string for display. The for construct initializes integer a as 128 (10000000) and continues while a is greater than zero. The loop expression of a for loop is often a++ or a––, but in this case it is a>>1. The a>>1 shifts the 10000000 right one place each time the loop executes so it changes to 01000000, then 00100000, and so on. The body of the for construct tests temp1 for a 1 in each bit position as a varies so it reconstructs the binary number in character string temp. The temp string is finally displayed as a caption for Label2 as the OnTimer function finishes.

Can the rotate instruction be accomplished in the C++ language? Yes, but not with a single instruction. Example 8–18 shows the if (shift) construct in C++. Notice how much more painful the rotate is in C++, and look at the code generated by C++. You may have been wondering why the assembly code generated by C++ appears to be very inefficient. The code is written to function correctly in all situations, so it is often far more elaborate than required for many applications. Because of this, the program can be made more efficient by fine tuning it.

EXAMPLE 8–18

```
                if ( shift )
00419E2C  mov         eax,dword ptr [this]
00419E2F  movzx       ecx,byte ptr [eax+270h]
00419E36  test        ecx,ecx
00419E38  je          CshiftrotateDlg::OnTimer+84h (419E44h)
                    temp1 <<= 1;
00419E3A  mov         al,byte ptr [temp1]
00419E3D  shl         al,1
00419E3F  mov         byte ptr [temp1],al
                else
00419E42  jmp         CshiftrotateDlg::OnTimer+0A4h (419E64h)
                    if ( ( temp1 & 0x80 ) == 0x80 )
00419E44  movsx       eax,byte ptr [temp1]
00419E48  and         eax,80h
00419E4D  je          CshiftrotateDlg::OnTimer+9Ch (419E5Ch)
                         temp1 = ( temp1 << 1 ) + 1;
00419E4F  movsx       eax,byte ptr [temp1]
00419E53  lea         ecx,[eax+eax+1]
00419E57  mov         byte ptr [temp1],cl
                    else
00419E5A  jmp         CshiftrotateDlg::OnTimer+0A4h (419E64h)
                         temp1 <<= 1;
00419E5C  mov         al,byte ptr [temp1]
00419E5F  shl         al,1
00419E61  mov         byte ptr [temp1],al
```

The Mouse

The mouse pointing device, as well as a trackball, is accessed from within the framework of Visual C++. Like many other devices controlled by Windows, the mouse can have message handlers installed in an application so that mouse movement and other functions can be used in a program. As we saw in prior examples, message handlers (event handlers) are installed in an application in the Properties section by clicking on the icon to the right of the lightning bolt. The mouse message handlers are listed in Table 8–2.

TABLE 8–2 Mouse messages.

Message	*Handler*	*Parameter 1*	*Parameter 2*	*Parameter 3*
WM_MOUSEMOVE	OnMouseMove	UINT nFlags	CPoint point	
WM_MOUSEWHEEL	OnMouseWheel	UINT nFlags	short zDelta	CPoint point
WM_LBUTTONDOWN	OnLButtonDown	UINT nFlags	CPoint point	
WM_RBUTTONDOWN	OnRButtonDown	UINT nFlags	CPoint point	
WM_LBUTTONUP	OnLButtonUp	UINT nFlags	CPoint point	
WM_RBUTTONUP	OnRButtonUp	UINT nFlags	CPoint point	
WM_LBUTTONDBLCLK	OnLButtonDblClk	UINT nFlags	CPoint point	
WM_RBUTTONDBLCLK	OnRButtonDblClk	UINT nFlags	CPoint point	

Notes: zDelta = the distance that the wheel is rotated. nFlags are: MK_CONTROL = Ctrl key down, MK_LBUTTON = left button down, MK_MBUTTON = middle button down, MK_RBUTTON = right button down, and MK_SHIFT = Shift key down. For a shift middle button combination use MK_SHIFT | MK_MBUTTON.

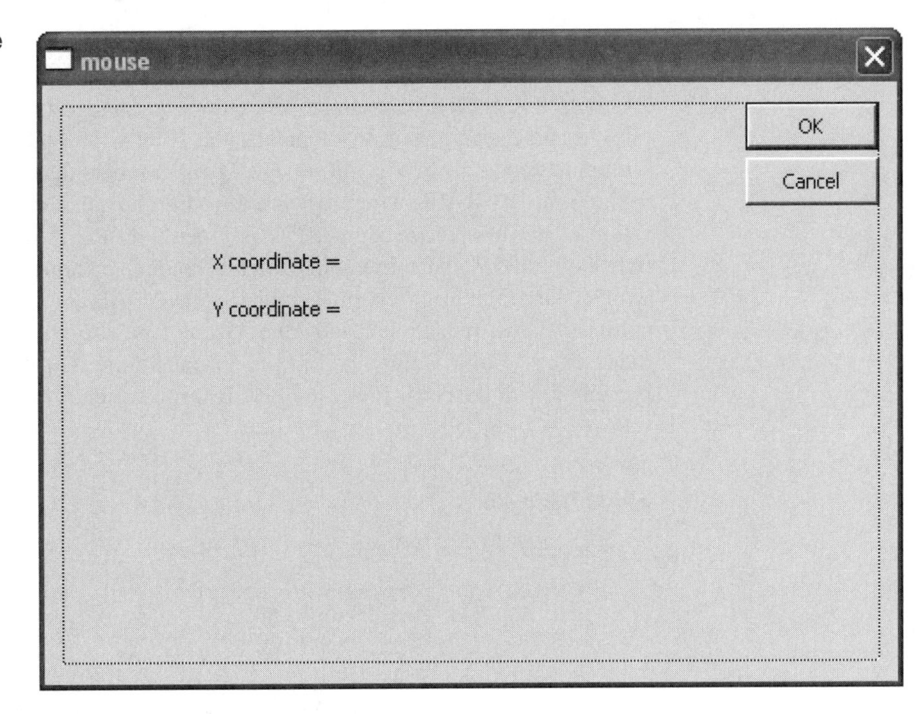

FIGURE 8–4 Display the mouse coordinates.

For an illustration of how to use the mouse, refer to the example application in Figure 8–4. This example shows the mouse point coordinates as the pointer is moved within the dialog application. Although the software listing (see Example 8–19) does not use any assembly language, it does illustrate how to obtain and display the position of the mouse pointer.

EXAMPLE 8–19

```
void CmouseDlg::OnMouseMove(UINT nFlags, CPoint point)
{
      char temp[10];
      CString coordinate;
      itoa(point.x, temp, 10);
      coordinate = "X coordinate = ";
      Label1.put_Caption(coordinate + temp);    //display x point
      itoa(point.y, temp, 10);
      coordinate = "Y coordinate = ";
      Label2.put_Caption(coordinate + temp);    //display y point
      CDialog::OnMouseMove(nFlags, point);
}
```

Example 8–19 illustrates the only part of the application that is modified to display the mouse coordinates. The OnMouseMove function is installed when the WM_MOUSEMOVE handler is installed in the program. This application uses two active X Microsoft Forms 2.0 Labels to display the mouse coordinates. These two objects are named Label1 and Label2. The OnMouseMove function returns the position of the mouse pointer in the CPoint data structure as members x and y. This example uses the itoa (integer to ASCII) function to convert the integer returned as the mouse point x or y into an ASCII character string. Three parameters are associated with this function. The first parameter is the integer to be converted by the function; the second parameter is the character string to receive the ASCII coded data; and the third parameter is the number base of the conversion (in this case, decimal). This example does not use the nFlags.

Install a mouse handler for the left mouse button (WM_LBUTTONDOWN) and one for the right mouse button (WM_RBUTTONDOWN). Modify your application as illustrated in Example 8–20. This causes the left button to change the color of the labels to red when clicked and the right button to change the color of the labels to blue. The color codes used with most functions in Visual C++ are 24-bit numbers that represent 16 million colors and are in the form BBGGRR, where BB is a two-digit hexadecimal number that represents the saturation of the blue video signal, GG is green, and RR is red. This means that color number 0xff0000 is 255 (the maximum) for blue video, 0 (none) for green, and 0 (none) for red, which causes the brightest blue to be displayed. All colors are made from the three primary colors of light: red, green, and blue. A color palette with the color codes can be found in index entry "Color Reference" in the Visual Studio help files. Note that the color numbers presented with the color tiles are in RGB form, which is backwards from the form needed for C++.

EXAMPLE 8–20

```
void CmouseDlg::OnRButtonDown(UINT nFlags, CPoint point)
{
        Label1.put_ForeColor(0xff0000);
        Label2.put_ForeColor(0xff0000);
        CDialog::OnRButtonDown(nFlags, point);
}

void CmouseDlg::OnLButtonDown(UINT nFlags, CPoint point)
{
        Label1.put_ForeColor(0x0000ff);
        Label2.put_ForeColor(0x0000ff);
        CDialog::OnLButtonDown(nFlags, point);
}
```

8–3 DATA CONVERSIONS

In computer systems, data are seldom in the correct form. One main task of the system is to convert data from one form to another. This section of the chapter describes conversions between binary and ASCII data. Binary data are removed from a register or memory and converted to ASCII for the video display. In many cases, ASCII data are converted to binary as they are typed on the keyboard. We also explain converting between ASCII and hexadecimal data.

Converting from Binary to ASCII

Conversion from binary to ASCII is accomplished in three ways: (1) by the AAM instruction if the number is less than 100, (2) by a series of decimal divisions (divide by 10), or (3) by using the itoa C++ function. Techniques 1 and 2 are presented in this section.

The AAM instruction converts the value in AX into a two-digit unpacked BCD number in AX. If the number in AX is 0062H (98 decimal) before AAM executes, AX contains 0908H after AAM executes. This is not ASCII code, but it is converted to ASCII code by adding 3030H to AX. Example 8–21 illustrates a program that uses the procedure which processes the binary value in AL (0–99) and displays it on the video screen as a decimal number. The procedure blanks a leading zero, which occurs for the numbers 0–9, with an ASCII space code. This example program displays the number 74 (testdata) on the video screen. To implement this program, create a dialog-based application in Visual C++ and place a single label called Label1 (do not forget to add this variable) on the dialog screen. The number 74 will appear if the code in Example 8–21 is placed at the TODO: location in OnIntiDialog function in the Dialog class (the class that ends with the letters Dlg).

EXAMPLE 8–21

```
        // TODO: Add extra initialization here

        char testdata = 74;
        char temp[3];
        _asm
        {
              mov al,testdata              ;get test data
              mov ah,0                     ;clear AH
              aam                          ;convert to BCD
              add    ah,20h
              cmp al,20h                   ;test for leading zero
              je D1                        ;if leading zero
              add ah,10h                   ;convert to ASCII
D1:
              mov temp[0],ah               ;save character
              add al,30h                   ;convert to ASCII
              mov temp[1],al
              mov temp[2],0                ;end string woth null
        }
        Label1.put_Caption(temp);          //display test number
```

The reason that AAM converts any number between 0 and 99 to a two-digit unpacked BCD number is because it divides AX by 10. The result is left in AX so AH contains the quotient and AL the remainder. This same scheme of dividing by 10 can be expanded to convert any whole number of any number system (if the divide-by number is changed) from binary to an ASCII-coded character string that can be displayed on the video screen. For example, if AX is divided by 8 instead of 10, the number is displayed in octal.

The algorithm (called **Horner's algorithm**) for converting from binary to decimal ASCII code is:

1. Divide by 10, then save the remainder on the stack as a significant BCD digit.
2. Repeat step 1 until the quotient is 0.
3. Retrieve each remainder and add 30H to convert to ASCII before displaying or printing.

Example 8–22 shows how the unsigned 32-bit content of EAX is converted to ASCII and displayed on the video screen. Here, we divide EAX by 10 and save the remainder on the stack after each division for later conversion to ASCII. After all the digits have been converted, the result is displayed on the video screen by removing the remainders from the stack and converting them to ASCII code. This program also blanks any leading zeros that occur. As mentioned, any number base can be used by changing the nbase variable in this example. Again, to implement this example create a dialog application with a single active X label called Label1. Place the software of the example in the OnInitDialog function.

If the number base is greater than 10, letters are used to represent characters beyond 9. To modify the software to function with number bases beyond 10, place cmp dl, 39h after the add dl, 30h statement and then jbe L2A followed by add dl, 7. Then prefix the mov temp[ebx], dl statement with L2A:.

EXAMPLE 8–22

```
        int testdata = 2349;
        int nBase = 10;
        char temp[20];
        _asm
        {
```

```
                mov   ebx,0             ;initialize pointer
                push  nBase             ;push base
                mov   eax, testdata     ;get test data
L1:
                mov   edx,0             ;clear edx
                div   nBase             ;divide by base
                push  edx               ;save remainder
                cmp   eax,0
                jnz   L1                ;repeat until 0
L2:
                pop   edx               ;get remainder
                cmp   edx,nBase
                je    L3                ;if finished
                add   dl,30h            ;convert to ASCII
                mov   temp[ebx],dl      ;save digit
                inc   ebx               ;point to next
                jmp   l2                ;repeat until done
L3:
                mov   temp[ebx],0       ;save null in string
        }
        Label1.put_Caption(temp);
```

Converting from ASCII to Binary

Conversions from ASCII to binary usually start with keyboard entry. If a single key is typed, the conversion occurs when 30H is subtracted from the number. If more than one key is typed, conversion from ASCII to binary still requires 30H to be subtracted, but there is one additional step. After subtracting 30H, the number is added to the result after the prior result is first multiplied by 10.

The algorithm for converting from ASCII to binary is:

1. Begin with a binary result of 0.
2. Subtract 30H from the character to convert it to BCD.
3. Multiply the result by 10, and then add the new BCD digit.
4. Repeat steps 2 and 3 for each character of the number.

Example 8–23 illustrates a program that implements this algorithm. Here, the binary number is displayed from variable temp in an edit control using the SetDlgItemInt function. Each time this program executes it reads a number from the char variable numb and converts it to binary for display on the edit control called Edit1.

EXAMPLE 8–23

```
        char numb[] = "2356";
        int temp = 0;
        _asm
        {
                mov  ebx,0              ;intialize pointer
                mov  ecx,ebx
B1:
                mov  cl,numb[ebx]       ;get digit
                inc  ebx                ;address next digit
                cmp  cl,0               ;if null found
                je   B2
                sub  cl,30h             ;convert from ASCII to BCD
                mov  eax,10             ;x10
                mul  temp
                add  eax,ecx            ;add digit
                mov  temp,eax           ;save result
                jmp  B1
B2:
        }
        SetDlgItemInt(IDC_EDIT1, temp);
```

Displaying and Reading Hexadecimal Data

Hexadecimal data are easier to read from the keyboard and display than decimal data. These types of data are not used at the application level, but at the system level. System-level data are often hexadecimal, and must either be displayed in hexadecimal form or read from the keyboard as hexadecimal data.

Reading Hexadecimal Data. Hexadecimal data appear as 0 to 9 and A to F. The ASCII codes obtained from the keyboard for hexadecimal data are 30H to 39H for the numbers 0 through 9, and 41H to 46H (A–F) or 61H to 66H (a–f) for the letters. To be useful, a program that reads hexadecimal data must be able to accept both lowercase and uppercase letters as well as numbers.

Example 8–24 shows two functions: One (Conv) converts the contents of an unsigned char from ASCII code to a single hexadecimal digit, and the other (Readh) converts a CString with up to eight hexadecimal digits into a numeric value that is returned as a 32-bit unsigned integer. This example illustrates a balanced mixture of C++ and assembly language to perform the conversion.

EXAMPLE 8–24

```
//function Readh converts a CString from hexadecimal to an
//unsigned integer.

unsigned int CasciiDlg::Readh(CString temp)
{
      unsigned int numb = 0;
      for ( int a = 0; a < temp.GetLength(); a++ )
      {
            numb <<= 4;
            numb += Conv(temp.GetAt(a));
      }
      return numb;
}

//function Conv converts a digit into binary

unsigned char CasciiDlg::Conv(unsigned char temp)
{
      _asm
      {
             cmp  temp,'9'
             jbe  Conv2              ;if 0 - 9
             cmp  temp,'a'
             jb   Conv1              ;if A - F
             sub  temp,20h           ;to uppercase
Conv1:
             sub  temp,7
Conv2:
             sub  temp,30h
      }
      return temp;
}
```

Displaying Hexadecimal Data. To display hexadecimal data, a number must be divided into two-, four-, or eight-bit sections that are converted into hexadecimal digits. Conversion is accomplished by adding 30H to the numbers 0 to 9 or 37H to the letters A to F for each section.

A function (Disph), which returns a CString of the contents of the unsigned integer parameter passed to the function, converts the unsigned integer into a two-, four-, or eight-digit character string as selected by parameter size. The function is listed in Example 8–25. Disph(number, 2) converts an unsigned integer number into a two-digit hexadecimal CString, where Disph(number, 4) converts it to a four-digit hexadecimal string and Disph(number, 8) converts it to an eight-digit hexadecimal character string.

EXAMPLE 8–25

```
CString CasciiDlg::Disph(unsigned int number, unsigned int size)
{
       CString temp;
       number <<= ( 8 - size ) * 4;       //adjust position
       for ( int a = 0; a < size; a++ )
       {
              char temp1;
              _asm
              {
                     rol number, 4;
                     mov al,byte ptr number
                     and al,0fh            ;make 0 - f
                     add al,30h            ;convert to ASCII
                     cmp al,39h
                     jbe Disph1
                     add al,7
Disph1:
                     mov temp1,al
              }
              temp +=  temp1;             //add digit to string
       }
       return temp;
}
```

Using Lookup Tables for Data Conversions

Lookup tables are often used to convert data from one form to another. A lookup table is formed in the memory as a list of data that is referenced by a procedure to perform conversions. In many lookup tables, the **XLAT instruction** is often used to look up data in a table, provided that the table contains eight-bit-wide data and its length is less than or equal to 256 bytes.

Converting from BCD to Seven-Segment Code. One simple application that uses a lookup table is BCD to seven-segment code conversion. Example 8–26 illustrates a lookup table that contains the seven-segment codes for the numbers 0 to 9. These codes are used with the seven-segment display pictured in Figure 8–5. This seven-segment display uses active high (logic 1) inputs to light a segment. The lookup table code (array temp1) is arranged so that the a segment is in bit position 0 and the g segment is in bit position 6. Bit position 7 is 0 in this example, but it can be used for displaying a decimal point, if required.

FIGURE 8–5 The seven-segment display.

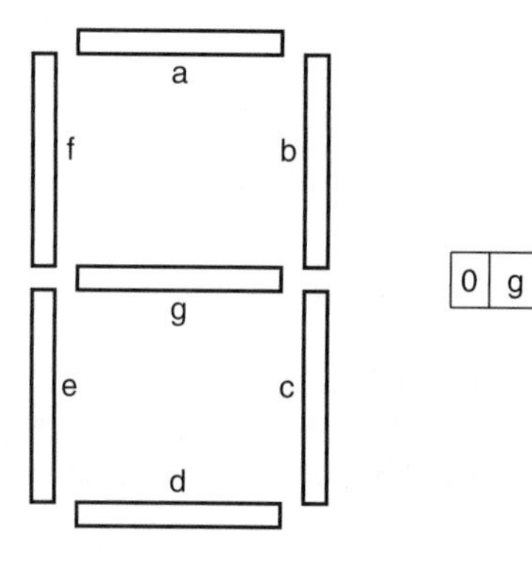

Control byte

0	g	f	e	d	c	b	a

EXAMPLE 8–26

```
unsigned char CasciiDlg::LookUp(unsigned char temp)
{
	char temp1[] = {0x3f, 6, 0x5b, 0x4f, 0x66, 0x6d, 0x7d, 7, 0x7f, 0x6f};
	_asm
	{
		lea  ebx,temp1
		mov  al,temp
		xlat
		mov  temp,al
	}
	return temp;
}
```

The LookUp function, which performs the conversion, contains only a few instructions and assumes that the temp parameter contains the BCD digit (0–9) to be converted to seven-segment code that is returned as an unsigned char. The first instruction addresses the lookup table by loading its address into EBX, and the others perform the conversion and return the seven-segment code as an unsigned char. Here the temp1 array is indexed by the BCD passed to the function in temp.

Using a Lookup Table to Access ASCII Data. Some programming techniques require that numeric codes be converted to ASCII character strings. For example, suppose that you need to display the days of the week for a calendar program. Because the number of ASCII characters in each day is different, some type of lookup table must be used to reference the ASCII-coded days of the week.

The program in Example 8–27 shows a table, formed as an array, which references ASCII-coded character strings. Each character string contains an ASCII-coded day of the week. The table contains references to each day of the week. The function that accesses the day of the week uses the day parameter, with the numbers 0 to 6 to refer to Sunday through Saturday. If day contains a 2 when this function is invoked, the word "Tuesday" is displayed on the video screen. Please note that this function does not use any assembly code, since we are merely accessing an element in an array using the day of the week as an index. It is shown so additional uses for arrays can be presented, as they may have application in programs used with embedded microprocessors.

EXAMPLE 8–27

```
CString CasciiDlg::GetDay(unsigned char day)
{
	CString temp[] =
	{
		"Sunday",
		"Monday",
		"Tuesday"
		"Wednesday",
		"Thursday",
		"Friday",
		"Saturday",
	};
	return temp[day];
}
```

An Example Program Using a Lookup Table

Figure 8–6 shows the screen of a dialog application called Display that displays a seven-segment-style character on the screen for each numeric key typed on the keyboard. As we learned in prior examples, the keyboard can be intercepted in a Visual C++ program using the PreTranslateMessage function, which is exactly what the program does to obtain the key from the keyboard. Next the

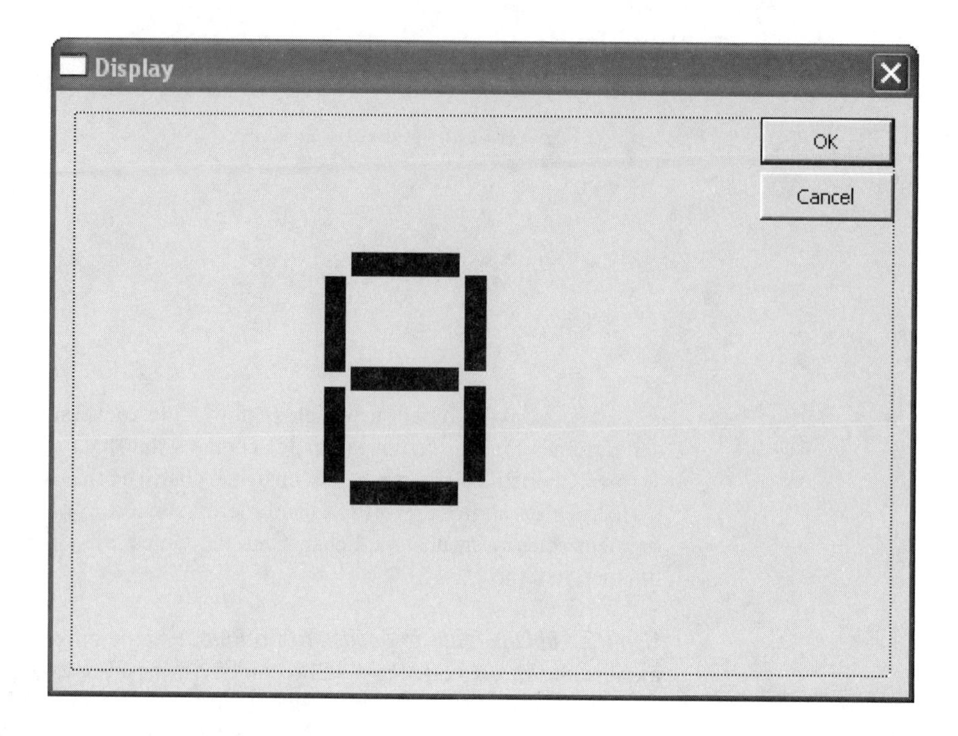

FIGURE 8–6 An example that displays a single seven-segment-style digit.

code typed is filtered so only 0–9 are accepted, and then a lookup table is used to access the seven-segment code for display.

The display digit is drawn using the active X Microsoft Forms 2.0 label objects. The horizontal bars are drawn using dimensions 40 × 8 and the vertical bars are drawn using dimensions 8 × 32. The dimensions of an object appear in the extreme lower right corner of the resource screen in Visual Studio. Make sure that you add the labels in the same order as the display; that is, add label a first, followed by label b, and so on, just as in the seven-segment display of Figure 8–5. Use Label1 through Label7 for the variable names of the labels and don't forget to add them as controls to the application.

Add the function listed in Example 8–28 called Clear to your program. This is used to clear the digit from the screen when the program first executes and also before a new digit is displayed. Notice that the background color of each label is changed to 0x8000000f. This is the color of the background of a dialog box in Windows, even though it is identified as the shading on the face of command buttons. Table 8–3 lists all the system color constants for Windows. Make sure that you place a Clear() in the TODO: area of the OnInitDialog procedure in the dialog class. This causes the display to be cleared when the program is executed at this point.

EXAMPLE 8–28

```
void CDisplayDlg::Clear(void)
{
        Label1.put_BackColor(0x8000000f);
        Label2.put_BackColor(0x8000000f);
        Label3.put_BackColor(0x8000000f);
        Label4.put_BackColor(0x8000000f);
        Label5.put_BackColor(0x8000000f);
        Label6.put_BackColor(0x8000000f);
        Label7.put_BackColor(0x8000000f);
}
```

Once a key is typed, the PreTranslateMessage function (see Example 8–29) filters the keystroke and converts the keystroke into seven-segment code using the lookup table. After the

TABLE 8–3 Windows system color constants.

Value	*Description*
0x80000000	Scroll bar color
0x80000001	Desktop color
0x80000002	Color of the title bar for the active window
0x80000003	Color of the title bar for the inactive window
0x80000004	Menu background color
0x80000005	Window background color
0x80000006	Window frame color
0x80000007	Color of text on menus
0x80000008	Color of text in windows
0x80000009	Color of text in caption, size box, and scroll arrow
0x8000000A	Border color of active window
0x8000000B	Border color of inactive window
0x8000000C	Background color of multiple-document interface (MDI) applications
0x8000000D	Background color of items selected in a control
0x8000000E	Text color of items selected in a control
0x8000000F	Color of shading on the face of command buttons
0x80000010	Color of shading on the edge of command buttons
0x80000011	Grayed (disabled) text
0x80000012	Text color on pushbuttons
0x80000013	Color of text in an inactive caption
0x80000014	Highlight color for 3-D display elements
0x80000015	Darkest shadow color for 3-D display elements
0x80000016	Second lightest 3-D color
0x80000017	Color of text in ToolTips
0x80000018	Background color of ToolTips

conversion to seven-segment code the ShowDigit function is called to show the digit on the screen. The ShowDigit function tests each bit of the seven-segment code and changes the background color of the proper segment into black.

EXAMPLE 8–29

```
BOOL CDisplayDlg::PreTranslateMessage(MSG* pMsg)
{
	char lookup[] = {0x3f, 6, 0x5b, 0x4f, 0x66, 0x6d, 0x7d, 7, 0x7f, 0x6f};
	char temp;
	if ( pMsg->message == WM_KEYDOWN )
	{
		if ( pMsg->wParam >= '0' && pMsg->wParam <= '9' )
		{
			temp = pMsg->wParam - 0x30;
			_asm			//lookup 7-segment code
			{
				lea  ebx,lookup
				mov  al,temp
				xlat
				mov  temp,al
			}
			ShowDigit(temp);	//display the digit
		}
		return true;			//finished with keystroke
	}
	return CDialog::PreTranslateMessage(pMsg);
}
```

```
void CDisplayDlg::ShowDigit(char code)
{
        Clear();
        if ( ( code & 1 ) == 1 )          //test a segment
                Label1.put_BackColor(0);
        if ( ( code & 2 ) == 2 )          //test b segment
                Label2.put_BackColor(0);
        if ( ( code & 4 ) == 4 )          //test c segment
                Label3.put_BackColor(0);
        if ( ( code & 8 ) == 8 )          //test d segment
                Label4.put_BackColor(0);
        if ( ( code & 16 ) == 16 )        //test e segment
                Label5.put_BackColor(0);
        if ( ( code & 32 ) == 32 )        //test f segment
                Label6.put_BackColor(0);
        if ( ( code & 64 ) == 64 )        //test g segment
                Label7.put_BackColor(0);
}
```

8–4 DISK FILES

Data are found stored on the disk in the form of files. The disk itself is organized in four main parts: the boot sector, the **file allocation table** (FAT), the root directory, and the data storage areas. The Windows NTFS (**New Technology File System**) contains a boot sector and a **master file table** (MFT). The first sector on the disk is the boot sector, which is used to load the disk operating system (DOS) from the disk into the memory when power is applied to the computer.

The FAT (or MFT) is where the operating system stores the names of files/subdirectories and their locations on the disk. All references to any disk file are handled through the FAT (or MFT). All other subdirectories and files are referenced through the **root directory** in the FAT system. The NTFS system does not have a root directory even though the file system may still appear to have a root directory. The disk files are all considered sequential access files, meaning that they are accessed a byte at a time, from the beginning of the file toward the end. Both the NTFS file system and the FAT file system are in use, with the hard disk drive on most modern Windows systems using NTFS and the floppy disk, CD-ROM, and DVD using the FAT system.

Disk Organization

Figure 8–7 illustrates the organization of sectors and tracks on the surface of the disk. This organization applies to both floppy and hard disk memory systems. The outer track is always track 0, and the inner track is 39 (double density) or 79 (high density) on floppy disks. The inner track on a hard disk is determined by the disk size, and could be 10,000 or higher for very large hard disks.

Figure 8–8 shows the organization of data on a disk. The length of the FAT is determined by the size of the disk. In the NTFS system, the length of the MFT is determined by the number of files stored on the disk. Likewise, the length of the root directory in a FAT volume is determined by the number of files and subdirectories located within it. The boot sector is always a single 512-byte-long sector located in the outer track at sector 0, the first sector.

The boot sector contains a **bootstrap loader** program that is read into RAM when the system is powered. The bootstrap loader then executes and loads the operating system into RAM. Next, the bootstrap loader passes control to the operating system program, allowing the computer to be under the control of and execute Windows, in most cases. This same sequence of events also occurs if the Linux operating system is found on the disk.

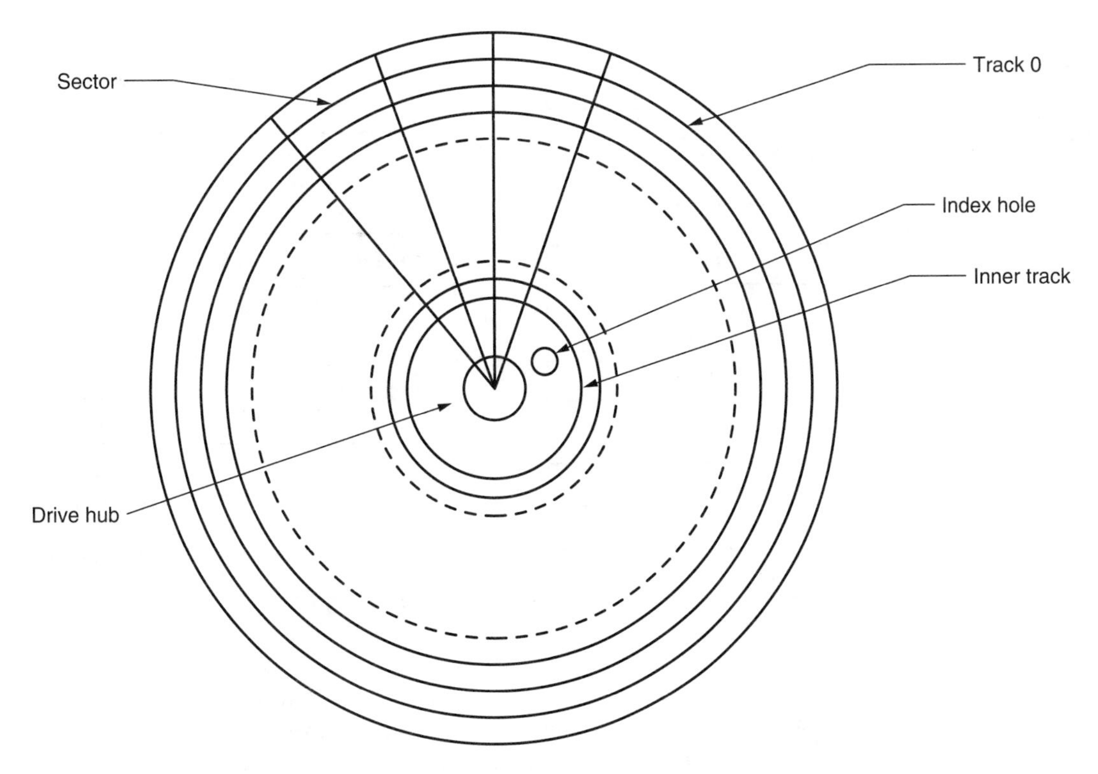

FIGURE 8–7 Structure of the disk.

The FAT indicates which sectors are free, which are corrupted (unusable), and which contain data. The FAT table is referenced each time that the operating system writes data to the disk so that it can find a free sector. Each free cluster is indicated by 0000H in the FAT and each occupied sector is indicated by the cluster number. A **cluster** can be anything from one sector to any number of sectors in length. Many hard disk memory systems use four sectors per cluster, which means that the smallest file is 512 bytes × 4, or 2048 bytes long. In a system that uses NTFS, the cluster size is usually 4K bytes, which is eight sectors long.

Figure 8–9 shows the format of each directory entry in the root, or in any other directory or subdirectory. Each entry contains the name, extension, attribute, time, date, location, and length.

FIGURE 8–8 Main data storage areas on a disk.

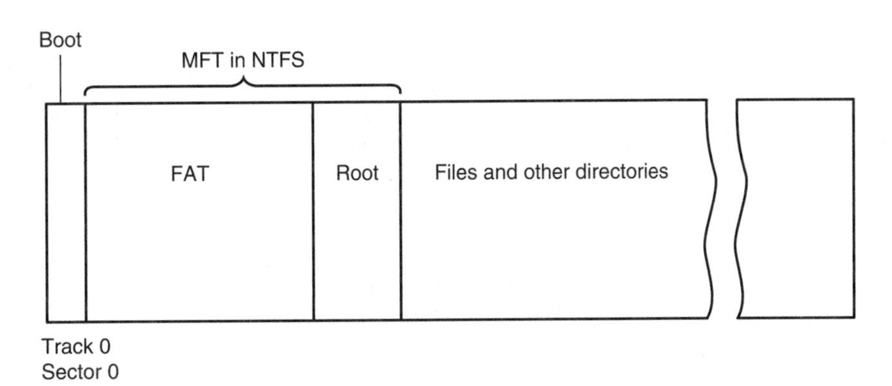

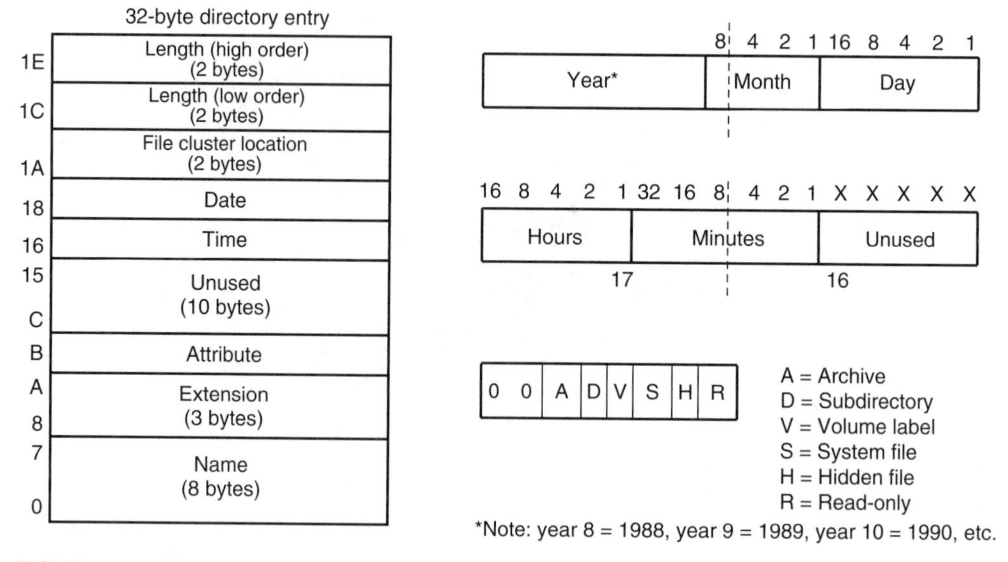

FIGURE 8–9 Format of any FAT directory or subdirectory entry.

The length of the file is stored as a 32-bit number. This means that a file can have a maximum length of 4G bytes. The location is the starting cluster number.

The Windows NTFS uses a much larger directory entry or record (1024 bytes) than that of the FAT system (32 bytes). The MFT record contains the file name, file date, attribute, and data. The data can be the entire contents of the file, or a pointer to where the data is stored on the disk, called a file run. Generally files that are smaller than about 1500 bytes fit into the MFT record. Longer files fit into a file run or file runs. A **file run** is a series of contiguous clusters that store the file data. Figure 8–10 illustrates an MFT record in the Windows NTFS file system. The information attribute contains the create date, last modification date, create time, last modification time, and file attributes such as read-only, archive, and so forth. The security attribute stores all security information for the file for limiting access to the file in the Windows system. The header stores information about the record type, size, name (optional), and whether it is resident or not.

File Names

Files and programs are stored on a disk and referenced both by a file name and an extension to the file name. With the DOS operating system, the *file name* may only be from one to eight characters long. The file name can contain just about any ASCII character, except for spaces or the “\ . / [ ] * , : < > I ; ? = characters. In addition to the file name, the file can have an optional one- to three-digit *extension* to the file name. Note that the name of a file and its extension are always separated by a period. If Windows 95 through Windows XP is in use, the file name can be of any length (up to 255 characters) and can even contain spaces. This is an improvement over the eight-character file name limitation of DOS. Also note that a Windows file can have more than one extension.

Header	Information Attribute	File Name Attribute	Data	Security Attribute

FIGURE 8–10 A record in the Master File Table in the NTFS system.

Directory and Subdirectory Names. The DOS file management system arranges the data and programs on a disk into directories and subdirectories. In Windows directories and subdirectories are called file folders. The rules that apply to file names also apply to file folder names. The disk is structured so that it contains a root directory when first formatted. The root directory or folder for a hard disk used as drive C is C:\. Any other folder is placed in the root directory. For example, C:\DATA is folder DATA in the root directory. Each folder placed in the root directory can also have subdirectories or subfolders. Examples are the subfolders C:\DATA\AREA1 and C:\DATA\AREA2, in which the folder DATA contains two subfolders: AREA1 and AREA2. Subfolders can also have additional subfolders. For example, C:\DATA\AREA2\LIST depicts folder DATA, subfolder AREA, which contains a subfolder called LIST.

Sequential Access Files

All DOS and Windows files are sequential files. A sequential file is stored and accessed from the beginning of the file toward the end, with the first byte and all bytes between it and the last accessed to read the last byte. Fortunately, files are read and written in C++ using the CFile object, which makes their access and manipulation easy. This section of the text describes how to create, read, write, delete, and rename a sequential access file.

File Creation. Before a file can be used, it must exist on the disk. A file is created by the CFile object using mode::Create as an attribute that directs CFile to create a file. A file is created with the Open member function of CFile as illustrated in Example 8–30. Here the name of the file that is created by the program is stored in a CString called FileName. Next the CFile object is initialized as File. Finally the if statement opens and creates the file. If the return from File.Open is false, Windows failed to create the file, and if it returns true, the file is opened successfully. In this example, if the file fails to open because the disk is full or the folder is not found a Windows message box displays *Cannot open file* followed by the file name, and then an exit from the program occurs when OK is clicked in the message box. To try this example, create a dialog application and place the code in the OnInitDialog function after the TODO: statement. Choose a folder name that does not exist (the name test should probably work) and run the application. You should see the error message. If you change the FileName so it does not include the folder, you will not get the error message.

EXAMPLE 8–30

```
CString FileName = "C:\\test\\Test.txt";
CFile File;

if ( !File.Open( FileName, CFile::modeWrite | CFile::modeCreate ) )
{
      MessageBox("Cannot open file:" + FileName, "Information", MB_OK);
      exit(0);      //if file cannot be opened quit program
}

//test.txt now exists in folder test with a length of 0 bytes
```

Writing to a File. Once a file exists it can be written to. In fact, it would be highly unusual to create a file without writing something to it. Data are written to a file one byte at a time. There is no provision for using CFile for writing any other type of data to a file. Data are always written starting at the very first byte in a file. Example 8–31 lists a program that creates a file in the root directory called Test1.txt and stores the letter A in each of its 256 bytes. If you execute this code and look at Test1.txt with NotePad you will see a file filled with 256 letter As. Note that the file should be closed when finished by using Close().

EXAMPLE 8–31

```
CString FileName = "C:\\Test1.txt";
CFile File;
unsigned char Buffer[256];

if ( !File.Open( FileName, CFile::modeWrite | CFile::modeCreate ) )
{
      AfxMessageBox("Cannot open file: " + FileName);
      exit(0);
}

for ( int a = 0; a < 256; a++ )              //fill the Buffer
      Buffer[a] = 'A';

File.Write( Buffer, sizeof( Buffer ) );      //write the buffer
File.Close();                                //close file
```

Suppose that a 32-bit integer must be written to a file. Since only bytes can be written, a method must be used to convert the four bytes of the integer into a form that can be written to a file. In C++ LOBYTE is used to access the low-order 8 bits of a short integer and HIBYTE accesses the high-order 8 bits of a short integer. Likewise, LOWORD accesses the low-order 16 bits of a 32-bit integer and HIWORD accesses the high-order 16 bits of a 32-bit integer. These four functions are used in C++ to access the separate bytes of a 32-bit integer. For example, HIWORD(LOWBYTE(number) accesses bits 23 through 16 of number. Assembly language can also accomplish the same task in fewer bytes, as listed in Example 8–32. If you look at the assembly code for each method, you see that the assembly language method is much shorter and much faster. If speed and size are important, then the assembly code is by far the best choice.

EXAMPLE 8–32

```
int number = 0x20000;
unsigned char Buf[4];

//C++ conversion

Buf[0] = LOWORD(LOBYTE(number));
Buf[1] = LOWORD(HIBYTE(number));
Buf[2] = HIWORD(LOBYTE(number));
Buf[3] = HIWORD(HIBYTE(number));

//Assembly language conversion

_asm
{
       mov  eax,number
       mov  Buf[0],al
       mov  Buf[1],ah
       bswap eax            ;little endian to big endian
       mov  Buf[2],ah
       mov  Buf[3],al
}
```

Reading File Data. File data are read from the beginning of the file toward the end using the Read member of CFile. Example 8–33 shows an example that reads the file written in Example 8–31 into a buffer. The Read function returns the number of bytes actually read from the file, but not used in this example. Notice that the example uses the CFileException class to report any error that might occur when the file is opened. The error message from Windows appears in Unicode variable Cause using the GetErrorMessage member of the CFileExcption class. The wait cursor (hourglass) is also displayed in this example to show that Windows is busy opening and reading the file. This occurs so quickly in this example that it probably will not be visible, but if a very large file is opened and read it will be visible.

EXAMPLE 8–33

```
CString FileName = "C:\\Test1a.txt";
CFile File;
CFileException ex;

unsigned char Buffer[256];
unsigned int FileByteCount;

CWaitCursor wait;                //display hour glass

if ( !File.Open( FileName, CFile::modeRead, &ex ) )
{
      TCHAR Cause[255];          //if file error
      CString Str;
      ex.GetErrorMessage(Cause, 255);
      Str = _T("File could not be opened. Error = ");
      Str += Cause;
      AfxMessageBox(Str);        //display error message
}
else                             //no file error
{
      FileByteCount = File.Read( Buffer, sizeof( Buffer ) );
      File.Close();
}


wait.Restore();                  //return to normal cursor
```

An Example Binary Dump Program. One tool not available with Windows is a program that displays the contents of a file in hexadecimal code. Although this may not be used by most programmers, it is used whenever software is developed so that the actual contents of a file can be viewed in hexadecimal format. Start a dialog application in Windows MFC and call it HexDump. Place an active X Microsoft Forms 2.0 Label and an active X Microsoft Rich Textbox Control onto the form as illustrated in Figure 8–11. Name the label Label1 and the Rich textbox Rich1. Under Properties

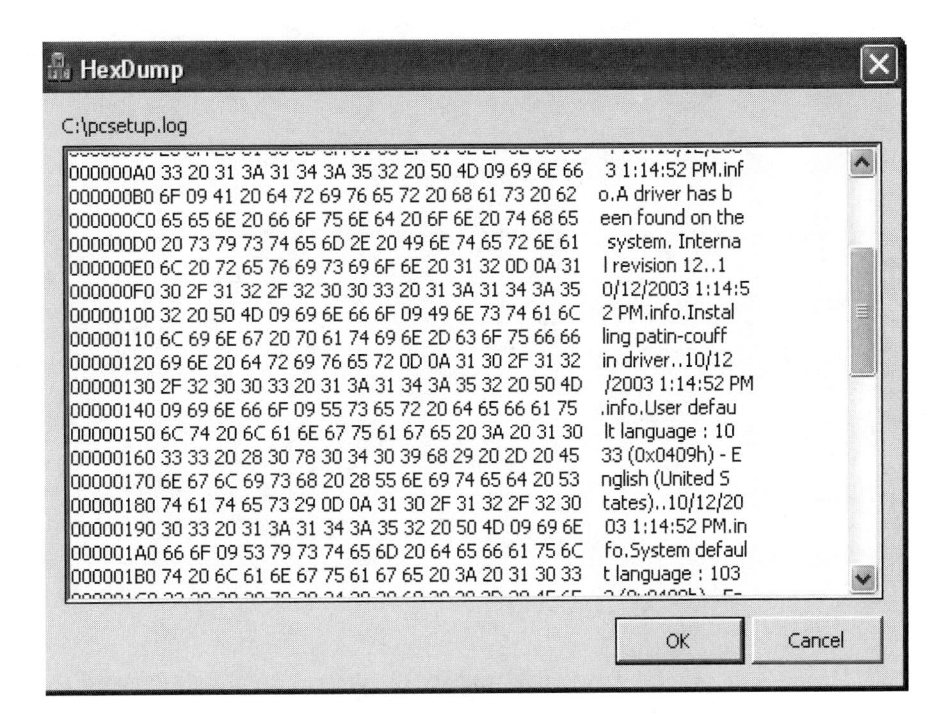

FIGURE 8–11 A screen shot of the HexDump program.

for the Rich Textbox Control make sure you change Locked to true and Scroll bars to Vertical. If you display a very large file you will want to be able to scroll down through the code.

This program uses the function (Disph) shown in Example 8–25 to display the address as an eight-digit hexadecimal address and also to display the contents of the address in hexadecimal form as a two-digit number. Add Disph to the HexDumpDlg class so its return value is a CString and it contains two integer parameters: one for the number and one for the number of digits, as shown in Example 8–25.

The program itself is placed in the OnIntiDialog function just below the TODO: statement. Example 8–34 shows the program to perform a hexadecimal dump. Note that to change the file requires that you change the name of the file in the program. This can be modified by using an edit box to enter the file name, but it was not done in this example for the sake of brevity. In this program 16 bytes are read at a time and formatted for display. This process continues until no bytes remain in the file and the FileByteCount is returned as a zero. The ASCII data that are displayed at the end of the hexadecimal listing are filtered so that any ASCII character under 32 (a space) are displayed as a period. This is important or control characters such as line feed, backspace, and the like will destroy the screen formatting of the ASCII text, and that is undesirable.

EXAMPLE 8–34

```
CString FileName = "C:\\Test1.txt";     //change to display a different file
CFile File;
CFileException ex;
unsigned char Buffer[16];
unsigned int FileByteCount;
unsigned int Address = 0;

Label1.put_Caption(FileName);
CWaitCursor wait;
if ( !File.Open( FileName, CFile::modeRead, &ex ) )
{
      TCHAR Cause[255];
      CString Str;
      ex.GetErrorMessage(Cause, 255);
      Str = _T("File could not be opened. Error = ");
      Str += Cause;
      AfxMessageBox(Str);
}
else
{
      while ( ( FileByteCount = File.Read( Buffer, sizeof( Buffer ) ) )!= 0)
      {
             CString ascii, line;
             line = Disph(Address, 8);
             for (int a = 0; a < FileByteCount; a++ )
             {
                   line += " " + Disph(Buffer[a], 2);
                   if ( Buffer[a] >= 32 )
                          ascii += Buffer[a];
                   "else"
                          ascii += ".";
             }
             Address += 16;
             Rich1.put_Text(Rich1.get_Text() + line + "    " + ascii + "\n");
      }
      File.Close();
}

wait.Restore();
```

The File Pointer and Seek. When a file is opened, written, or read, the file pointer addresses the current location in the sequential file. When a file is opened, the file pointer always addresses the first byte of the file. If a file is 1024 bytes long, and a read function reads 1023 bytes, the file pointer addresses the last byte of the file, but not the end of the file.

The **file pointer** is a 32-bit number that addresses any byte in a file. The CFile Seek member function is used to modify the file pointer. A file pointer can be moved from the start of the file (CFile::being), from the current location (CFile::current), or from the end of the file (CFile::end). In practice, all three directions of the move are used to access different parts of the file. The distance moved by the file pointer (in bytes) is specified as the first parameter in a Seek function.

Suppose that a file exists on the disk and that you must append the file with 256 bytes of new information. When the file is opened, the file pointer addresses the first byte of the file. If you attempt to write without moving the file pointer to the end of the file, the new data will *overwrite* the first 256 bytes of the file. Example 8–35 shows a a sequence of instructions that opens a file, moves the file pointer to the end of the file, writes 256 bytes of data, and then closes the file. This appends the file with 256 new bytes of data from area Buffer.

EXAMPLE 8–35

```
CString FileName = "C:\\test1.txt";
CFile File;
CFileException ex;
unsigned char Buffer[256];

CWaitCursor wait;
if ( !File.Open( FileName, CFile::modeWrite, &ex ) )
{
      TCHAR Cause[255];
      CString Str;
      ex.GetErrorMessage(Cause, 255);
      Str = _T("File could not be opened. Error = ");
      Str += Cause;
      AfxMessageBox(Str);
}
else
{
      File.Seek(0, CFile::end);       //go to end of file
      File.Write(Buffer, 256);        //append file
      File.Close();
}

wait.Restore();
```

One of the more difficult file maneuvers is inserting new data in the middle of the file. Figure 8–12 shows how this is accomplished by creating a second file. Notice that the part of the

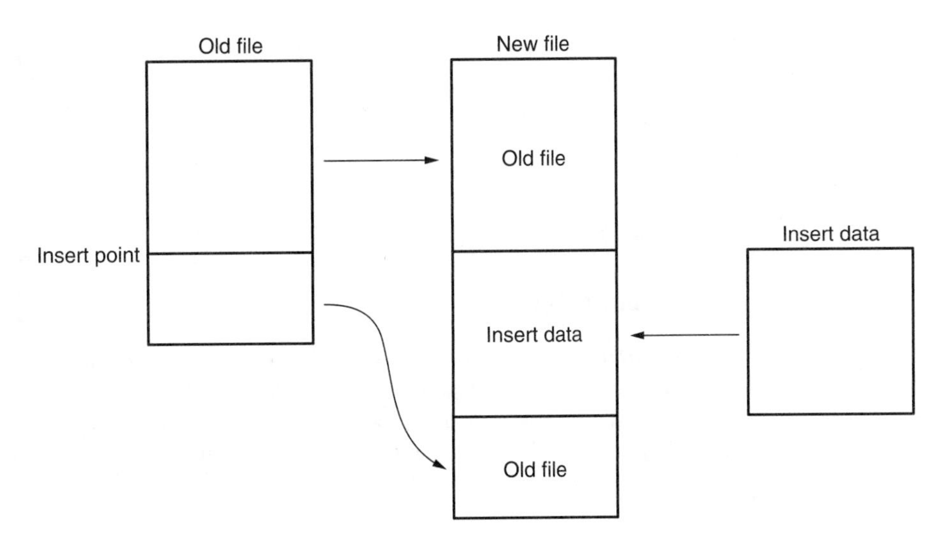

FIGURE 8–12 Inserting new data within an old file.

file before the insertion point is copied into the new file. This is followed by the new information before the remainder of the file is appended after the insertion in the new file. Once the new file is complete, the old file is deleted and the new file is renamed with the old file name.

Example 8–36 shows a program that inserts new data into an old file. This program copies the DATA.NEW file into the DATA.OLD file at a point after the first 256 bytes of the DATA.OLD file. The new data from Buffer2 are added to the file and this is followed by the remainder of the old file. New CFile member functions are used to delete the old file and rename the new file with the old file name.

EXAMPLE 8–36

```
CString FileName1 = "C:\\Data.old";
CString FileName2 = "C:\\Data.new";
CFile File1;
CFile File2;
CFileException ex;
unsigned char Buffer1[256];
unsigned char Buffer2[6];
unsigned int length;

CWaitCursor wait;
if ( !File1.Open( FileName1, CFile::modeRead, &ex ) )
{
        TCHAR Cause[255];
        CString Str;
        ex.GetErrorMessage(Cause, 255);
        Str = _T("File could not be opened. Error = ");
        Str += Cause;
        AfxMessageBox(Str);
}
else
{
        if ( !File2.Open( FileName2, CFile::modeCreate | CFile::modeWrite, &ex ))
        {
                TCHAR Cause[255];
                CString Str;
                ex.GetErrorMessage(Cause, 255);
                Str = _T("File could not be opened. Error = ");
                Str += Cause;
                AfxMessageBox(Str);
        }
        else
        {
                File1.Read(Buffer1, 256);
                File2.Write(Buffer1, 256);
                File2.Write(Buffer2, sizeof Buffer2);
                while ( ( length = File1.Read(Buffer1, 256) ) != 0 )
                        File2.Write(Buffer1, length);
                File1.Close();
                File2.Close();
                CFile::Remove( FileName1 );   //delete old file
                CFile::Rename( FileName2, FileName1 ); //rename new file
        }
}
wait.Restore();
```

Random Access Files

Random access files are developed through software using sequential access files. A random access file is addressed by a record number rather than by going through the file searching for data. The Seek function becomes very important when random access files are created. Random access files are much easier to use for large volumes of data, which are often called databases.

Creating a Random Access File. Planning is paramount when creating a random access file system. Suppose that a random access file is required for storing the names of customers. Each customer record requires 32 bytes for the last name, 32 bytes for the first name, and one byte for the middle initial. Each customer record contains two street address lines of 64 bytes each, a city line of 32 bytes, two bytes for the state code, and nine bytes for the Zip code. The basic customer information alone requires 236 bytes; additional information expands the record to 512 bytes. Because the business is growing, provisions are made for 5000 customers. This means that the total random access file is 2,560,000 bytes long.

Example 8–37 illustrates a short program that creates a file called CUST.FIL and inserts 5000 blank records of 512 bytes each. A blank record contains 00H in each byte. This appears to be a large file, but it fits on the smallest of hard disks.

EXAMPLE 8–37

```
CString FileName = "C:\\Cust.fil";
CFile File;

CFileException ex;
unsigned char Buffer[512];

for ( int a = 0; a < 512; a++ )         //fill buffer
      Buffer[a] = 0;

CWaitCursor wait;
if ( !File.Open( FileName, CFile::modeCreate | CFile::modeWrite, &ex ) )
{
      TCHAR Cause[255];
      CString Str;
      ex.GetErrorMessage(Cause, 255);
      Str = _T("File could not be opened. Error = ");
      Str += Cause;
      AfxMessageBox(Str);
}
else
{
      for ( int a = 0; a < 5000; a++ )  //create 5000 records
             File.Write(Buffer, 512);
      File.Close();
}
wait.Restore();
```

Reading and Writing a Record. Whenever a record must be read, the record number is found by using a Seek. Example 8–38 lists a function that is used to Seek a record. This function assumes that a file has been opened as CustomerFile and that the CUST.FIL remains open at all times.

Notice how the record number is multiplied by 512 to obtain a count to move the file pointer using a Seek. In each case, the file pointer is moved from the start of the file to the desired record.

EXAMPLE 8–38

```
void CCusDatabaseDlg::FindRecord(unsigned int RecordNumber)
{
      File.Seek( RecordNumber * 512, CFile::begin );
}
```

Other functions (listed in Example 8–39) are needed to manage the customer database. These include WriteRecord, ReadRecord, FindLastNameRecord, FindBlankRecord, and so on.

Some of these are listed in the example as well as the data structure that contains the information for each record.

EXAMPLE 8–39

```
struct BASE {                   //data base record
        char FirstName[32];
        char Mi[1];
        char LastName[32];
        char Street1[64];
        char Street2[64];
        char City[32];
        char State[2];
        char ZipCode[9];
        char Other[276];
}cus[2];                        //setup 2 record

void CCusDatabaseDlg::WriteRecord(unsigned int RecordNumber)
{
        FindRecord(RecordNumber);
        File.Write(cus[0].FirstName, 32);
        File.Write(cus[0].Mi, 1);
        File.Write(cus[0].LastName, 32);
        File.Write(cus[0].Street1, 64);
        File.Write(cus[0].Street2, 64);
        File.Write(cus[0].City, 32);
        File.Write(cus[0].State, 2);
        File.Write(cus[0].ZipCode, 9);
}

void CCusDatabaseDlg::ReadRecord(unsigned int RecordNumber)
{
        FindRecord(RecordNumber);
        File.Read(cus[0].FirstName, 32);
        File.Read(cus[0].Mi, 1);
        File.Read(cus[0].LastName, 32);
        File.Read(cus[0].Street1, 64);
        File.Read(cus[0].Street2, 64);
        File.Read(cus[0].City, 32);
        File.Read(cus[0].State, 2);
        File.Read(cus[0].ZipCode, 9);
}

unsigned int CCusDatabaseDlg::FindFirstName(void) //using cus[1].FirstName
{
        for ( int a = 0; a < 5000; a++ )
        {
              ReadRecord(a);
              if ( !strcmp(cus[0].FirstName, cus[1].FirstName ) )
                    return a;      //if found return record number
        }
        return 5000;               //if not found return 5001
}

unsigned int CCusDatabaseDlg::FindBlankRecord(void)
{
        for ( int a = 0; a < 5000; a++ )
        {
              ReadRecord(a);
              if ( strlen(cus[0].LastName) == 0 )
                    return a;
        }
        return 0;
}
```

8–5 EXAMPLE PROGRAMS

Now that many of the basic programming building blocks have been discussed, we present some example application programs. Although these example programs may seem trivial, they show some additional programming techniques and illustrate programming styles for the microprocessor.

Time/Date Display Program

Although this program does not use assembly language, it does demonstrate how to obtain the date and time from the Windows API and how to format it for display. It also illustrates how to use a timer in Visual C++. Example 8–40 illustrates a program that uses a timer, set to interrupt the program once per second, to display the time and date. The time and date are obtained by using the COleTimeDate::GetCurrentTime() member function to read the computer time and date into a variable called TimeDate. The Format member of COleTimeDate is used to format the TimeDate variable. Create a dialog application called TimeDate and place two active X Microsoft Forms 2.0 Labels on it as shown in Figure 8–13. Name the uppermost label LabelTime and the other label LabelDate.

Table 8–4 lists the Format options for the COleTimeDate variable so other styles can be used to display a time or date. This object contains the correct date and time from January 1, 100 through December 31, 9999. Make sure that this object is used for all dates and times so we will not have another year 2000 problem, at least not until the year 10,000.

EXAMPLE 8–40

```
//Place these 2 lines after TODO: in the OnInitDialog function

TimeDate();
SetTimer( 1, 1000, 0 );
```

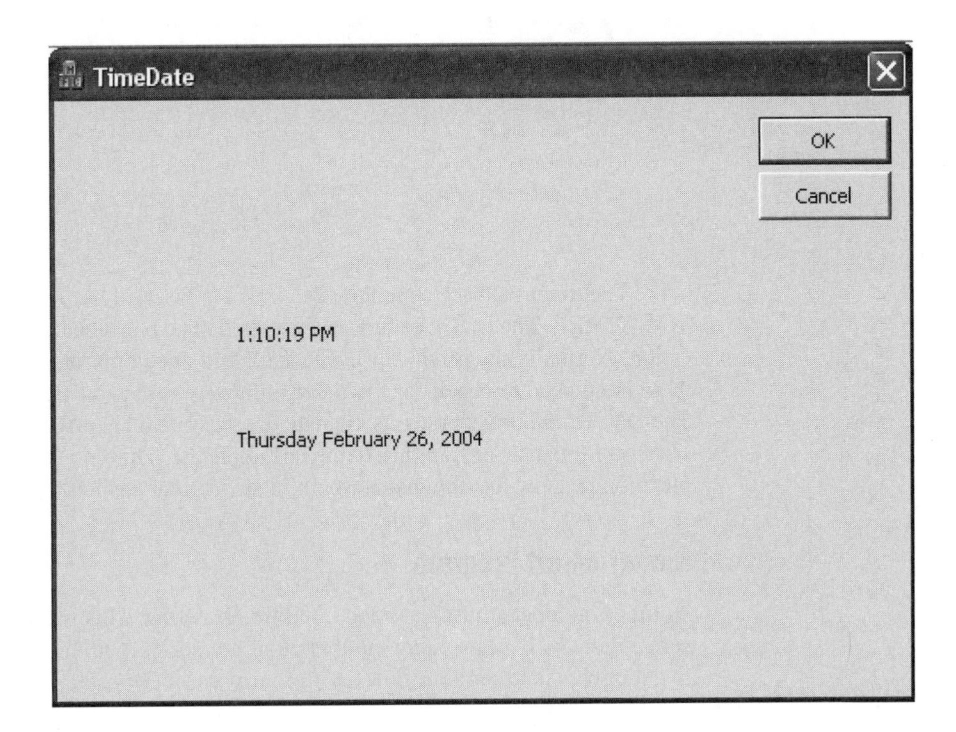

FIGURE 8–13 Displaying the time and date.

TABLE 8–4 Format command for Date and Time.

Code	*Function*
%a	Abbreviated weekday name
%A	Full weekday name
%b	Abbreviated month name
%B	Full month name
%c	Date and time for appropriate locale
%d	Day of month (01–31)
%H	24-hour format (00–23)
%I	12-hour format (01–12)
%j	Day of the year (001–366)
%m	Month (01–12)
%M	Minute (00–59)
%p	Current locale AM/PM
%S	Second (00–59)
%U, %W	Week (00–53) (%U = Sunday is first day) (%W = Monday is first day)
%w	Weekday number (0–6; Sunday is 0)
%x	Locale date
%X	Locale time
%y	Year (00–99)
%Y	Year (100–9999)
%z, %Z	Time zone
%%	Percent sign
#	Remove leading zero

```
//Functions for the TimeDate application

void CTimeDateDlg::TimeDate(void)
{
      COleDateTime DateTime = COleDateTime::GetCurrentTime();
      LabelTime.put_Caption( DateTime.Format( "%#I:%M:%S %p" ) );
      LabelDate.put_Caption( DateTime.Format( "%A %B %#d, %Y" ) );
}

void CTimeDateDlg::OnTimer(UINT nIDEvent)
{
      TimeDate();
      CDialog::OnTimer(nIDEvent);
}
```

The timer callback function (OnTimer) is inserted in the Properties section by clicking on WM_TIMER. The OnTimer function is called once per second (1000 milliseconds) in this application. Normally the nIDEvent is checked, but since only one timer is used in this application, there is no need to check for the timer number (in this case 1, the first parameter in SetTimer). The DateTime.Format converts the date or time into a character string that is displayed, with the selected format string, on LabelTime or LabelDate. This code can be easily inserted into any application to show the time/date or both in any format desired.

Numeric Sort Program

At times, numbers must be sorted into numeric order. This is often accomplished with a bubble sort. Figure 8–14 shows five numbers that are sorted with a bubble sort. Notice that the set of five numbers is tested four times with four passes. For each pass, two consecutive numbers are

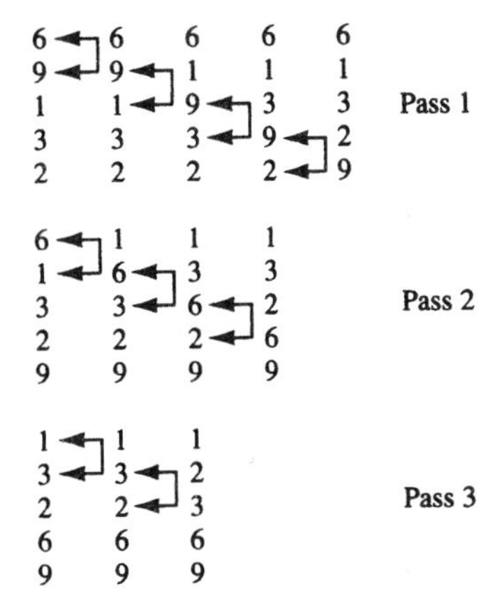

FIGURE 8–14 A bubble sort showing data as they are sorted. Note: Sorting five numbers may require four passes.

compared and sometimes exchanged. Also notice that during the first pass there are four comparisons, during the second three, and so forth.

Example 8–41 illustrates a program that accepts 10 numbers from the keyboard (32-bit integers). After these 32-bit numbers are accepted and stored in memory section Numbers, they are sorted by using the bubble-sorting technique. This bubble sort uses a swap flag to determine whether any numbers were exchanged in a pass. If no numbers were exchanged, the numbers are in order and the sort terminates. This early termination normally increases the efficiency of the sort because numbers are rarely completely out of order.

Once the numbers are sorted, they are displayed in ascending order. To specify an array, go to the header for the Dlg class and manually insert the unsigned int array Numbers[10]. The contents of the public section of the header are also illustrated in Example 8–41 with the program. Figure 8–15 shows how the application appears after it is executed.

FIGURE 8–15 The bubble sort application.

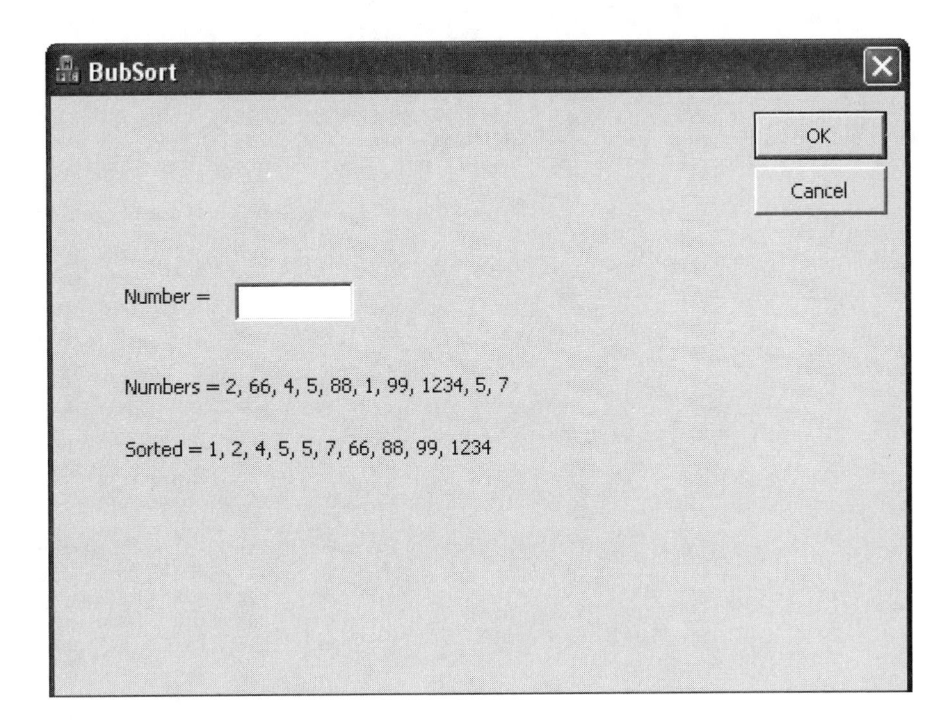

EXAMPLE 8–41

```
//Part of the header
// Implementation
protected:
       HICON m_hIcon;

       // Generated message map functions
       virtual BOOL OnInitDialog();
       afx_msg void OnSysCommand(UINT nID, LPARAM lParam);
       afx_msg void OnPaint();
       afx_msg HCURSOR OnQueryDragIcon();
       DECLARE_MESSAGE_MAP()
public:
       CEdit Edit1;
       CLabel1 Label1;
       CLabel1 Label2;
       int Count;
       UINT Numbers[10];    //added manually to header
       virtual BOOL PreTranslateMessage(MSG* pMsg);
};


//the OnInitDialog after TODO:

       Edit1.SetFocus();
       Label1.put_Caption("Numbers = ");
       Label2.put_Caption("Sorted = ");

       return false;  // return TRUE unless you set the focus to a control
}

//The PreTranslateMessage function (must be inserted in properties)

BOOL CBubSortDlg::PreTranslateMessage(MSG* pMsg)
{
       if ( Count != 10 && pMsg->message == WM_KEYDOWN && pMsg->wParam == 13 )
       {
             CString temp;
             char flag;
             char temp1[10];
             UINT *pNumbers = &Numbers[0];            //get array pointer
             Numbers[Count] = GetDlgItemInt(IDC_EDIT1);
             GetDlgItemText(IDC_EDIT1, temp);
             SetDlgItemText(IDC_EDIT1, "");
             Label1.put_Caption(Label1.get_Caption() + temp);
             if ( Count != 9 )
                    Label1.put_Caption(Label1.get_Caption() + ", ");
             Count++;
             if ( Count == 10 )
             {
                    _asm{
                           mov  ecx,9           ;9 for 10 numbers
L1:
                           mov  flag,0          ;clear flag
                           mov  edx,0
L2:
                           mov  ebx,pNumbers
                           mov  eax,[ebx+edx*4]
                           cmp  eax,[ebx+edx*4+4]
                           jbe  L3
                           push eax             ;swap
                           mov  eax,[ebx+edx*4+4]
                           mov  [ebx+edx*4], eax
                           pop  dword ptr [ebx+edx*4+4]
                           mov  flag,1          ;set flag
L3:
                           inc  edx
                           cmp  edx,ecx
```

```
                        jne  L2
                        cmp  flag,0
                        jz   L4:          ;if no swaps
                        loop L1
L4:
                }
                for ( int a = 0; a < 9; a++ )
                        Label2.put_Caption(Label2.get_Caption() +
                        itoa(Numbers[a], temp1, 10) + ", ");
                Label2.put_Caption(Label2.get_Caption() + itoa(Numbers[9],
                        temp1, 10));
            }
            return true;
        }
        return CDialog::PreTranslateMessage(pMsg);
}
```

Data Encryption

Data encryption seems to be in vogue at this time because of the security aspect of many systems. To illustrate simple data encryption for a character string, suppose that each character in a string is Exclusive-ORed with a number called an encryption key. This certainly changes the code of the character, but to make it a bit more random, suppose that the encryption key is changed after each character is encrypted. In this way patterns are much harder to detect in the encrypted message, making it harder to decipher.

To illustrate this simple scheme, Figure 8–16 shows a screen shot of the program to test the scheme using an edit control to accept a character string and an active X label to display the encrypted message. This example was generated using an initial encryption key of 0x45. If the initial value is changed, the encrypted message will change.

Example 8–42 lists the program used to generate the message in its encrypted form in Label1, the active X label. The first section is the initialization dialog used to start the program and set focus to Edit1, the edit control for entering the character string to be encrypted. The PreTranslateMessage handler, which must be inserted into the program, encrypts the string. Notice how the program uses assembly language to Exclusive-OR each character of the string with the EncryptionKey and then

FIGURE 8–16 Screen shot of the encryption program.

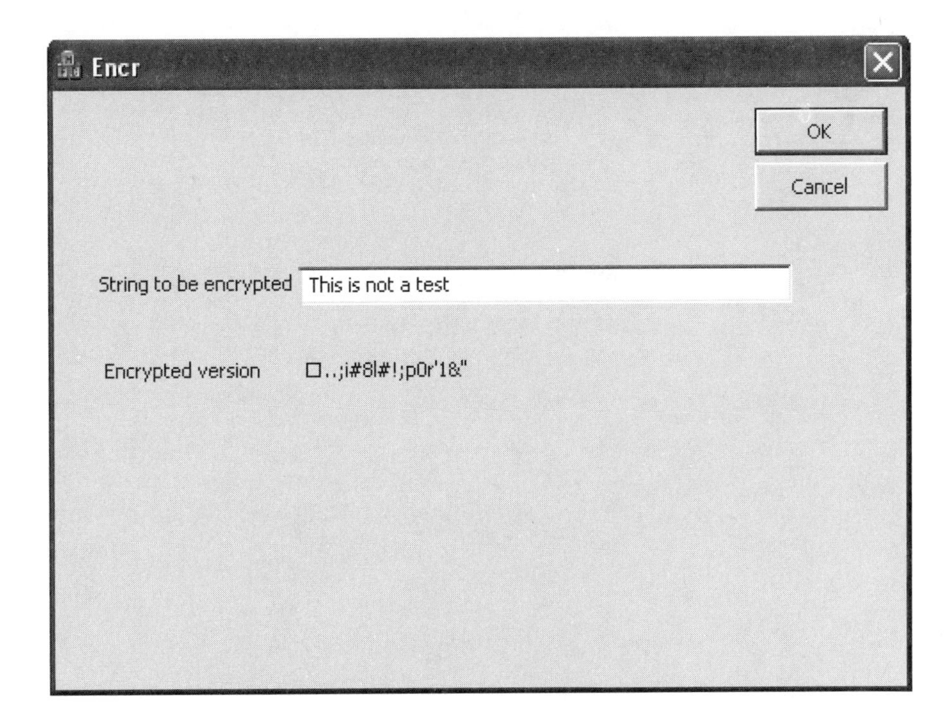

how the EncryptionKey is modified for the next character. This can always be modified to make it even more difficult to decipher. For example, suppose that the key is incremented on every other character and that is alternated with inverting the key, as shown in Example 8–43. Almost any combination of operations can be used to modify the key between passes to make it very difficult to decode. In practice we use a 128-bit key and the technique for modification is different, but nonetheless, this is basically how encryption is performed. Because Example 8–42 uses an 8-bit key, the encrypted message could be cracked by trying all 256 (2^8) possible keys, but if a 128-bit key is used, it requires far many more attempts (2^{128}) to crack—an almost impossible number of attempts.

EXAMPLE 8–42

```
//code added to the OnIntiDialog function in the Dlg class

       // TODO: Add extra initialization here

       Edit1.SetFocus();

       return false; // return TRUE unless you set the focus to a control
}

//PreTranslateMessage function

BOOL CEncrDlg::PreTranslateMessage(MSG* pMsg)
{
       char EncryptionKey = 0x45;       //Random Key
       CString temp;
       if ( pMsg->message == WM_KEYDOWN && pMsg->wParam == 13 )
       {
              GetDlgItemText(IDC_EDIT1, temp);
              for ( int a = 0; a < temp.GetLength(); a++ )
              {
                     char code = temp.GetAt(a);
                     _asm
                     {
                            mov  al,code
                            xor  al,EncryptionKey
                            mov  code,al
                            inc  EncryptionKey
                     }
                     temp.SetAt(a, code);
                     Label1.put_Caption(temp);
              }
              return true;
       }
       return CDialog::PreTranslateMessage(pMsg);
}
```

EXAMPLE 8–43

```
//just the assembly language part of the program

              _asm
                     {
                            mov  al,code
                            xor  al,EncryptionKey
                            mov  code,al
                            mov  eax,a
                            shr  eax,1
                            jc   L1
                            inc  EncryptionKey
                            jmp  L2
       L1:                  not  EncryptionKey
       L2:
                     }
```

8–6 SUMMARY

1. The assembler program (ML.EXE) assembles modules that contain PUBLIC variables and segments, plus EXTRN (external) variables. The linker program (LINK.EXE) links modules and library files to create a run-time program executed from the DOS command line. The run-time program usually has the extension EXE, but might contain the extension COM.
2. The MACRO and ENDM directives create a new opcode for use in programs. These macros are similar to procedures, except that there is no call or return. In place of them, the assembler inserts the code of the macro sequence into a program each time it is invoked. Macros can include variables that pass information and data to the macro sequence.
3. Setting focus to an object is accomplished by using the SetFocus() member variable found with most objects.
4. Data may be placed in an object using the SetDlgItemInt and SetDlgItemText functions.
5. The mouse driver is accessed from Windows by installing handlers for various Windows messages such as WM_MOUSEMOVE, etc.
6. Conversion from binary to BCD is accomplished with the AAM instruction for numbers that are less than 100 or by repeated division by 10 for larger numbers. Once the number is converted to BCD, 30H is added to convert each digit to ASCII code for the video display.
7. When converting from an ASCII number to BCD, 30H is subtracted from each digit. To obtain the binary equivalent, multiply by 10 and then add each new digit.
8. Lookup tables are used for code conversion with the XLAT instruction if the code is an 8-bit code. If the code is wider than eight bits, a short procedure that accesses a lookup table provides the conversion. Lookup tables are also used to hold addresses so that different parts of a program or different procedures can be selected.
9. Conditional assembly language statements allow portions of a program to be assembled if a condition is met. These statements are useful for tailoring software to an application.
10. The disk memory system contains tracks that hold information stored in sectors. Many disk systems store 512 bytes of information per sector. Data on the disk are organized in a boot sector, file allocation table, root directory, and data storage area. The boot sector loads the DOS system from the disk into the computer memory system. The FAT or MFT indicates which sectors are present and whether they contain data. The root directory contains files names and subdirectories through which all disk files are accessed. The data storage area contains all subdirectories and data files.
11. Files are manipulated with the CFile object in Visual C++. To read a disk file, the file must be opened, read, and then closed. To write to a disk file, it must be opened, written, and then closed. When a file is opened, the file pointer addresses the first byte of the file. To access data at other locations, the file pointer is moved using a Seek before data are read or written.
12. A sequential access file is a file that is accessed sequentially from the beginning to the end. A random access file is a file that is accessed at any point. Although all disk files are sequential, they can be treated as random access files by using software.

8–7 QUESTIONS AND PROBLEMS

1. The assembler converts a source file to a(n) _________ file.
2. What files are generated from the source file TEST.ASM if it is processed by ML.EXE?
3. The linker program links object files and _________ files to create an execution file.
4. What does the PUBLIC directive indicate when placed in a program module?
5. What does the EXTRN directive indicate when placed in a program module?

6. What directive appears with labels defined as external?
7. Describe how a library file works when it is linked to other object files by the linker program.
8. What assembler language directives delineate a macro sequence?
9. What is a macro sequence?
10. How are parameters transferred to a macro sequence?
11. Develop a macro called ADD32 that adds the 32-bit contents of DX-CX to the 32-bit contents of BX-AX.
12. How is the LOCAL directive used within a macro sequence?
13. Develop a macro called ADDLIST PARA1,PARA2 that adds the contents of PARA1 to PARA2. Each of these parameters represents an area of memory. The number of bytes added are indicated by register CX before the macro is invoked.
14. Develop a macro that sums a list of byte-sized data invoked by the macro ADDM LIST,LENGTH. The label LIST is the starting address of the data block and LENGTH is the number of data added. The result must be a 16-bit sum found in AX at the end of the macro sequence.
15. What is the purpose of the INCLUDE directive?
16. Modify the function in Example 8–12 so that it only filters the numbers 0 through 9 and ignores all other characters.
17. Modify the function in Example 8–12 so that it generates a random 8-bit number in class variable char Random. (Hint: To accomplish this, increment Random each time that the PreTranslateMessage function is called, no matter what Windows message is issued.)
18. Modify the software you developed in Question 17 so that it generates a random number between 9 and 62.
19. Modify the function listed in Example 8–15 so that the hexadecimal numbers use lowercase letters a through f instead of the uppercase letters.
20. Modify Example 8–17 so it will shift/rotate right or left. This is accomplished by adding a command button to select the direction.
21. What handlers are used to access the mouse in the Visual C++ programming environment?
22. How is the right mouse button detected in a program?
23. How is a double-click detected with the mouse?
24. Develop software that detects when both the right and left mouse buttons are pressed simultaneously.
25. What color is displayed for the value 0x00ffff?
26. What is displayed when color number 0x80000010 is used to set a color?
27. When a number is converted from binary to BCD, the _________ instruction accomplishes the conversion, provided the number is less than 100 decimal.
28. How is a large number (over 100 decimal) converted from binary to BCD?
29. How could a binary number be displayed as an octal number?
30. A BCD digit is converted to ASCII code by adding a(n) _________.
31. An ASCII-coded number is converted to BCD by subtracting _________.
32. Develop a function that reads an ASCII number IDC_EDIT1 using the GetDlgItemText() function from the keyboard and returns it as an unsigned int. The number in the edit box is an octal number that is converted to binary by the function.
33. Explain how a three-digit ASCII-coded number is converted to binary.
34. Develop a function that converts all lowercase ASCII-coded letters into uppercase ASCII-coded letters. Your procedure may not change any other character except the letters a–z and must return the converted character as a char.
35. Develop a lookup table that converts hexadecimal data 00H–0FH into the ASCII-coded characters that represent the hexadecimal digits. Make sure to show the lookup table and any software required for the conversion. (Hint: Create a function to perform the conversion.)

36. Explain the purpose of a boot sector, FAT, and root directory in the FAT system.
37. Explain the purpose of the MFT in the NTFS file system.
38. The surface of a disk is divided into tracks that are further subdivided into _________.
39. What is a bootstrap loader and where is it found?
40. What is a cluster?
41. The NTFS file system often uses clusters of _________ bytes in length.
42. What is the maximum length of a file?
43. What code is used to store the name of a file when long file names are in use?
44. DOS file names are at most _________ characters in length.
45. How many characters normally appear in an extension?
46. How many characters may appear in a long file name?
47. Develop a program that opens a file called TEST.LST, reads 512 bytes from the file into memory area Array, and closes the file.
48. Show how to rename file TEST.LST to TEST.LIS.
49. What is the purpose of the CFile Remove member function?
50. What is an active X control?
51. Write a program that reads any decimal number between 0 and 2G and displays the 16-bit binary version on the video display.
52. Write a program that displays the binary powers of 2 (in decimal) on the video screen for the powers 0 through 7. Your display shows 2^n = value for each power of 2.
53. Using the technique discussed in Question 17, develop a program that displays random numbers between 1 and 47 (or whatever) for your state's lottery.
54. Modify the program in Example 8–29 so it also displays the letters A, b, C, d, E, and F for a hexadecimal seven-segment display.
55. Modify Example 8–42 to encrypt the message using an algorithm of your own design.
56. Develop a Decryption function (for a CString) to accompany the encryption of Question 55.

CHAPTER 9

8086/8088 Hardware Specifications

INTRODUCTION

This chapter describes the pin functions of both the 8086 and 8088 microprocessors and provides details on the following hardware topics: clock generation, bus buffering, bus latching, timing, wait states, and minimum mode operation versus maximum mode operation. These simple microprocessors are explained first, because of their simple structures, as an introduction to the Intel microprocessor family.

Before it is possible to connect or interface anything to the microprocessor, it is necessary to understand the pin functions and timing. Thus, the information in this chapter is essential to a complete understanding of memory and I/O interfacing, which we cover in the later chapters of the text.

CHAPTER OBJECTIVES

Upon completion of this chapter, you will be able to:

1. Describe the function of each 8086 and 8088 pin
2. Understand the microprocessor's DC characteristics and indicate its fan-out to common logic families
3. Use the clock generator chip (8284A) to provide the clock for the microprocessor
4. Connect buffers and latches to the buses
5. Interpret the timing diagrams
6. Describe wait states and connect the circuitry required to cause various numbers of waits
7. Explain the difference between minimum and maximum mode operation

9–1 PIN-OUTS AND THE PIN FUNCTIONS

In this section, we explain the function and (in certain instances) the multiple functions of each of the microprocessor's pins. In addition, we discuss the DC characteristics to provide a basis for understanding the later sections on buffering and latching.

The Pin-Out

Figure 9–1 illustrates the pin-outs of the 8086 and 8088 microprocessors. As a close comparison reveals, there is virtually no difference between these two microprocessors—both are packaged in 40-pin **dual in-line** packages (DIPs).

As mentioned in Chapter 1, the 8086 is a 16-bit microprocessor with a 16-bit data bus and the 8088 is a 16-bit microprocessor with an 8-bit data bus. (As the pin-outs show, the 8086 has pin connections AD_0–AD_{15}, and the 8088 has pin connections AD_0–AD_7.) Data bus width is therefore the only major difference between these microprocessors. This allows the 8086 to transfer 16-bit data more efficiently.

There is, however, a minor difference in one of the control signals. The 8086 has an $M/\overline{IO}$ pin, and the 8088 has an $IO/\overline{M}$ pin. The only other hardware difference appears on Pin 34 of both integrated circuits: on the 8088, it is an SS0 pin, while on the 8086, it is a $\overline{BHE}$/S7 pin.

Power Supply Requirements

Both the 8086 and 8088 microprocessors require +5.0 V with a supply voltage tolerance of ±10%. The 8086 uses a maximum supply current of 360 mA, and the 8088 draws a maximum of 340 mA. Both microprocessors operate in ambient temperatures of between 32° F and 180° F. This range is not wide enough to be used outdoors in the winter or even in the summer, but extended temperature-range versions of the 8086 and 8088 microprocessors are available. There is also a CMOS version, which requires a very low supply current and has an extended temperature range. The 80C88 and 80C86 are CMOS versions that require only 10 mA of power supply current and function in temperature extremes of –40° F through +225° F.

DC Characteristics

It is impossible to connect anything to the pins of the microprocessor without knowing the input current requirement for an input pin and the output current drive capability for an output pin. This knowledge allows the hardware designer to select the proper interface components for use with the microprocessor without the fear of damaging anything.

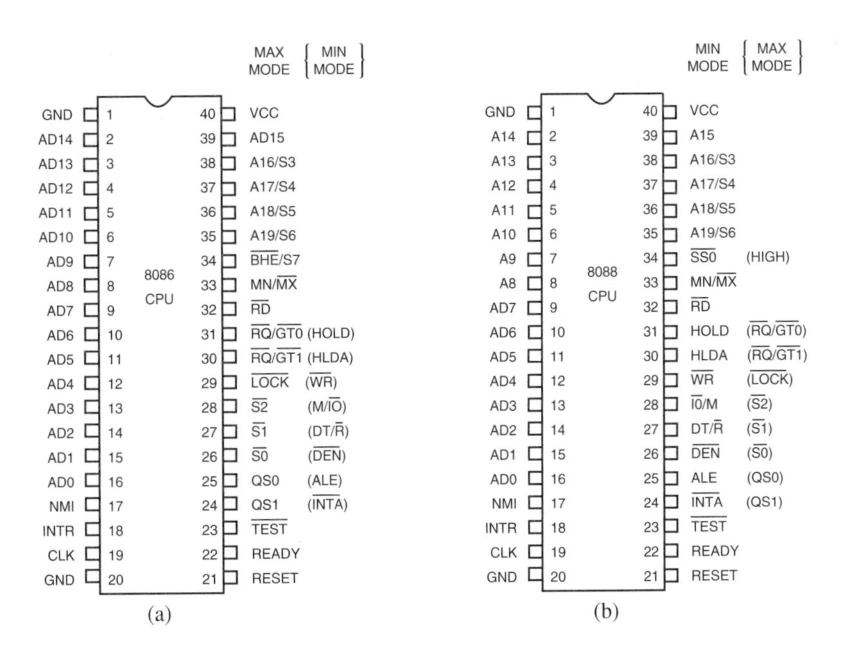

FIGURE 9–1 (a) The pin-out of the 8086 microprocessor; (b) the pin-out of the 8088 microprocessor.

TABLE 9–1 Input characteristics of the 8086 and 8088 microprocessors.

Logic Level	Voltage	Current
0	0.8 V maximum	±10 μA maximum
1	2.0 V minimum	±10 μA maximum

Input Characteristics. The input characteristics of these microprocessors are compatible with all of the standard logic components available today. Table 9–1 depicts the input voltage levels and the input current requirements for any input pin on either microprocessor. The input current levels are very small because the inputs are the gate connections of MOSFETs and represent only leakage currents.

Output Characteristics. Table 9–2 illustrates the output characteristics of all the output pins of these microprocessors. The logic 1 voltage level of the 8086/8088 is compatible with that of most standard logic families, but the logic 0 level is not. Standard logic circuits have a maximum logic 0 voltage of 0.4 V, and the 8086/8088 has a maximum of 0.45 V. Thus, there is a difference of 0.05 V.

This difference reduces the noise immunity from a standard level of 400 mV (0.8 V – 0.45 V) to 350 mV. (The **noise immunity** is the difference between the logic 0 output voltage and the logic 0 input voltage levels.) The reduction in noise immunity may result in problems with long wire connections or too many loads. It is therefore recommended that no more than 10 loads of any type or combination be connected to an output pin without buffering. If this loading factor is exceeded, noise will begin to take its toll in timing problems.

Table 9–3 lists some of the more common logic families and the recommended fan-out from the 8086/8088. The best choice of component types for the connection to an 8086/8088 output pin is an LS, 74ALS, or 74HC logic component. Note that some of the fan-out currents calculate to more than 10 unit loads. It is therefore recommended that if a fan-out of more than 10 unit loads is required, the system should be buffered.

Pin Connections

AD_7–AD_0 The 8088 **address/data bus** lines are the multiplexed address data bus of the 8088 and contain the rightmost eight bits of the memory address or I/O port number whenever ALE is active (logic 1) or data whenever ALE is inactive (logic 0). These pins are at their high-impedance state during a hold acknowledge.

A_{15}–A_8 The 8088 **address bus** provides the upper-half memory address bits that are present throughout a bus cycle. These address connections go to their high-impedance state during a hold acknowledge.

AD_{15}–AD_8 The 8086 **address/data bus** lines compose the upper multiplexed address/data bus on the 8086. These lines contain address bits A_{15}–A_8 whenever ALE is a logic 1, and data bus connections D_{15}–D_8 when ALE is a logic 0. These pins enter a high-impedance state when a hold acknowledge occurs.

A_{19}/S_6–A_{16}/S_3 The **address/status bus** bits are multiplexed to provide address signals A_{19}–A_{16} and also status bits S_6–S_3. These pins also attain a high-impedance state during the hold acknowledge.

Status bit S_6 is always a logic 0, bit S_5 indicates the condition of the IF flag bit, and S_4 and S_3 show which segment is accessed during the current

TABLE 9–2 Output characteristics of the 8086 and 8088 microprocessors.

Logic Level	Voltage	Current
0	0.45 V maximum	2.0 μA maximum
1	2.4 V minimum	−400 μA maximum

TABLE 9–3 Recommended fan-out from any 8086/8088 pin connection.

Family	*Sink Current*	*Source Current*	*Fan-Out*
TTL (74)	–1.6 mA	40 μA	1
TTL (74LS)	–0.4 mA	20 μA	5
TTL (74S)	–2.0 mA	50 μA	1
TTL (74ALS)	–0.1 mA	20 μA	10
TTL (74AS)	–0.5 mA	25 μA	10
TTL (74F)	–0.5 mA	25 μA	10
CMOS (74HC)	–10 μA	10 μA	10
CMOS (CD)	–10 μA	10 μA	10
NMOS	–10 μA	10 μA	10

bus cycle. See Table 9–4 for the truth table of S_4 and S_3. These two status bits could be used to address four separate 1M byte memory banks by decoding them as A_{21} and A_{20}.

$\overline{\text{RD}}$ Whenever the **read signal** is a logic 0, the data bus is receptive to data from the memory or I/O devices connected to the system. This pin floats to its high-impedance state during a hold acknowledge.

READY The **READY** input is controlled to insert wait states into the timing of the microprocessor. If the READY pin is placed at a logic 0 level, the microprocessor enters into wait states and remains idle. If the READY pin is placed at a logic 1 level, it has no effect on the operation of the microprocessor.

INTR **Interrupt request** is used to request a hardware interrupt. If INTR is held high when IF = 1, the 8086/8088 enters an interrupt acknowledge cycle (INTA becomes active) after the current instruction has completed execution.

$\overline{\text{TEST}}$ The **Test** pin is an input that is tested by the WAIT instruction. If $\overline{\text{TEST}}$ is a logic 0, the WAIT instruction functions as a NOP and if $\overline{\text{TEST}}$ is a logic 1, the WAIT instruction waits for $\overline{\text{TEST}}$ to become a logic 0. The $\overline{\text{TEST}}$ pin is most often connected to the 8087 numeric coprocessor.

NMI The **non-maskable interrupt** input is similar to INTR except that the NMI interrupt does not check to see whether the IF flag bit is a logic 1. If NMI is activated, this interrupt input uses interrupt vector 2.

RESET The **reset** input causes the microprocessor to reset itself if this pin is held high for a minimum of four clocking periods. Whenever the 8086 or 8088 is reset, it begins executing instructions at memory location FFFF0H and disables future interrupts by clearing the IF flag bit.

CLK The **clock** pin provides the basic timing signal to the microprocessor. The clock signal must have a duty cycle of 33% (high for one third of the clocking period and low for two thirds) to provide proper internal timing for the 8086/8088.

VCC This **power supply** input provides a +5.0 V, ±10% signal to the microprocessor.

TABLE 9–4 Function of status bits S_3 and S_4.

S_4	S_3	*Function*
0	0	Extra segment
0	1	Stack segment
1	0	Code or no segment
1	1	Data segment

GND The **ground** connection is the return for the power supply. Note that the 8086/8088 microprocessors have two pins labeled GND–both must be connected to ground for proper operation.

MN/$\overline{\text{MX}}$ The **minimum/maximum** mode pin selects either minimum mode or maximum mode operation for the microprocessor. If minimum mode is selected, the MN/$\overline{\text{MX}}$ pin must be connected directly to +5.0 V.

$\overline{\text{BHE}}/S_7$ The **bus high enable** pin is used in the 8086 to enable the most-significant data bus bits (D_{15}–D_8) during a read or a write operation. The state of S_7 is always a logic 1.

Minimum Mode Pins. Minimum mode operation of the 8086/8088 is obtained by connecting the MN/$\overline{\text{MX}}$ pin directly to +5.0 V. Do not connect this pin to +5.0 V through a pull-up register, or it will not function correctly.

IO/$\overline{\text{M}}$ or M/$\overline{\text{IO}}$ The **IO/$\overline{\text{M}}$** (8088) or the **M/$\overline{\text{IO}}$** (8086) pin selects memory or I/O. This pin indicates that the microprocessor address bus contains either a memory address or an I/O port address. This pin is at its high-impedance state during a hold acknowledge.

$\overline{\text{WR}}$ The **write line** is a strobe that indicates that the 8086/8088 is outputting data to a memory or I/O device. During the time that the **$\overline{\text{WR}}$** is a logic 0, the data bus contains valid data for memory or I/O. This pin floats to a high-impedance during a hold acknowledge.

$\overline{\text{INTA}}$ The **interrupt acknowledge** signal is a response to the INTR input pin. The $\overline{\text{INTA}}$ pin is normally used to gate the interrupt vector number onto the data bus in response to an interrupt request.

ALE **Address latch enable** shows that the 8086/8088 address/data bus contains address information. This address can be a memory address or an I/O port number. Note that the ALE signal does not float during a hold acknowledge.

DT/$\overline{\text{R}}$ The **data transmit/receive** signal shows that the microprocessor data bus is transmitting (DT/$\overline{\text{R}}$ = 1) or receiving (DT/$\overline{\text{R}}$ = 0) data. This signal is used to enable external data bus buffers.

$\overline{\text{DEN}}$ **Data bus enable** activates external data bus buffers.

HOLD The **hold input** requests a direct memory access (DMA). If the HOLD signal is a logic 1, the microprocessor stops executing software and places its address, data, and control bus at the high-impedance state. If the HOLD pin is a logic 0, the microprocessor executes software normally.

HLDA **Hold acknowledge** indicates that the 8086/8088 has entered the hold state.

$\overline{\text{SS0}}$ The **$\overline{\text{SS0}}$** status line is equivalent to the S_0 pin in maximum mode operation of the microprocessor. This signal is combined with IO/$\overline{\text{M}}$ and DT/$\overline{\text{R}}$ to decode the function of the current bus cycle (see Table 9–5).

Maximum Mode Pins. In order to achieve maximum mode for use with external coprocessors, connect the **MN/$\overline{\text{MX}}$** pin to ground.

$\overline{\text{S2}}$, $\overline{\text{S1}}$, and $\overline{\text{S0}}$ The **status bits** indicate the function of the current bus cycle. These signals are normally decoded by the 8288 bus controller described later in this chapter. Table 9–6 shows the function of these three status bits in the maximum mode.

$\overline{\text{RQ}}/\overline{\text{GT1}}$ and $\overline{\text{RQ}}/\overline{\text{GT0}}$ The **request/grant** pins request direct memory accesses (DMA) during maximum mode operation. These lines are bidirectional and are used to both request and grant a DMA operation.

TABLE 9–5 Bus cycle status (8088) using $\overline{SS0}$.

$IO/\overline{M}$	$DT/\overline{R}$	$\overline{SS0}$	Function
0	0	0	Interrupt acknowledge
0	0	1	Memory read
0	1	0	Memory write
0	1	1	Halt
1	0	0	Opcode fetch
1	0	1	I/O read
1	1	0	I/O write
1	1	1	Passive

TABLE 9–6 Bus control function generated by the bus controller (8288).

$\overline{S2}$	$\overline{S1}$	$\overline{S0}$	Function
0	0	0	Interrupt acknowledge
0	0	1	I/O read
0	1	0	I/O write
0	1	1	Halt
1	0	0	Opcode fetch
1	0	1	Memory read
1	1	0	Memory write
1	1	1	Passive

TABLE 9–7 Queue status bits.

QS_1	QS_0	Function
0	0	Queue is idle
0	1	First byte of opcode
1	0	Queue is empty
1	1	Subsequent byte of opcode

$\overline{LOCK}$ The **lock** output is used to lock peripherals off the system. This pin is activated by using the LOCK: prefix on any instruction.

QS_1 and QS_0 The **queue status** bits show the status of the internal instruction queue. These pins are provided for access by the numeric coprocessor (8087). See Table 9–7 for the operation of the queue status bits.

9–2 CLOCK GENERATOR (8284A)

This section describes the clock generator (8284A) and the RESET signal, and introduces the READY signal for the 8086/8088 microprocessors. (The READY signal and its associated circuitry are treated in detail in Section 9–5.)

The 8284A Clock Generator

The 8284A is an ancillary component to the 8086/8088 microprocessors. Without the clock generator, many additional circuits are required to generate the clock (CLK) in an 8086/8088-based system. The 8284A provides the following basic functions or signals: clock generation, RESET synchronization, READY synchronization, and a TTL-level peripheral clock signal. Figure 9–2 illustrates the pin-out of the 8284A clock generator.

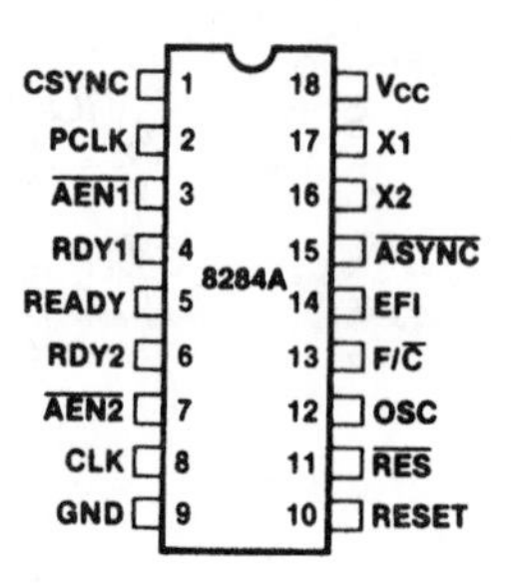

FIGURE 9–2 The pin-out of the 8284A clock generator.

Pin Functions. The 8284A is an 18-pin integrated circuit designed specifically for use with the 8086/8088 microprocessor. The following is a list of each pin and its function.

$\overline{AEN1}$ and $\overline{AEN2}$ The **address enable** pins are provided to qualify the bus ready signals, RDY1 and RDY2, respectively. Section 9–5 illustrates the use of these two pins, which are used to cause wait states, along with the RDY1 and RDY2 inputs. Wait states are generated by the READY pin of the 8086/8088 microprocessors, which is controlled by these two inputs.

RDY_1 and RDY_2 The **bus ready** inputs are provided, in conjunction with the $\overline{AEN1}$ and $\overline{AEN2}$ pins, to cause wait states in an 8086/8088-based system.

$\overline{ASYNC}$ The **ready synchronization** selection input selects either one or two stages of synchronization for the RDY_1 and RDY_2 inputs.

READY **Ready** is an output pin that connects to the 8086/8088 READY input. This signal is synchronized with the RDY_1 and RDY_2 inputs.

X_1 and X_2 The **crystal oscillator** pins connect to an external crystal used as the timing source for the clock generator and all its functions.

F/$\overline{C}$ The **frequency/crystal** select input chooses the clocking source for the 8284A. If this pin is held high, an external clock is provided to the EFI input pin; if it is held low, the internal crystal oscillator provides the timing signal. The external frequency input is used when the F/$\overline{C}$ pin is pulled high. EFI supplies the timing whenever the F/$\overline{C}$ pin is high.

CLK The **clock output** pin provides the CLK input signal to the 8086/8088 microprocessors and other components in the system. The CLK pin has an output signal that is one-third of the crystal or EFI input frequency, and has a 33% duty cycle, which is required by the 8086/8088.

PCLK The **peripheral clock** signal is one-sixth the crystal or EFI input frequency, and has a 50% duty cycle. The PCLK output provides a clock signal to the peripheral equipment in the system.

OSC The **oscillator output** is a TTL-level signal that is at the same frequency as the crystal or EFI input. The OSC output provides an EFI input to other 8284A clock generators in some multiple-processor systems.

$\overline{RES}$ The **reset input** is an active-low input to the 8284A. The $\overline{RES}$ pin is often connected to an RC network that provides power-on resetting.

RESET The **reset output** is connected to the 8086/8088 RESET input pin.

CSYNC The **clock synchronization** pin is used whenever the EFI input provides synchronization in systems with multiple processors. If the internal crystal oscillator is used, this pin must be grounded.

GND The **ground pin** connects to ground.

Vcc This **power supply** pin connects to +5.0 V with a tolerance of ±10%.

Operation of the 8284A

The 8284A is a relatively easy component to understand. Figure 9–3 illustrates the internal timing diagram of the 8284A clock generator.

Operation of the Clock Section. The top half of the logic diagram represents the clock and synchronization section of the 8284A clock generator. As the diagram shows, the crystal oscillator has two inputs: X_1 and X_2. If a crystal is attached to X_1 and X_2, the oscillator generates a square-wave signal at the same frequency as the crystal. The square-wave signal is fed to an AND gate and also to an inverting buffer that provides the OSC output signal. The OSC signal is sometimes used as an EFI input to other 8284A circuits in a system.

An inspection of the AND gate reveals that when $F/\overline{C}$ is a logic 0, the oscillator output is steered through to the divide-by-3 counter. If $F/\overline{C}$ is a logic 1, then EFI is steered through to the counter.

The output of the divide-by-3 counter generates the timing for ready synchronization, a signal for another counter (divide-by-2), and the CLK signal to the 8086/8088 microprocessors. The CLK signal is also buffered before it leaves the clock generator. Notice that the output of the first counter feeds the second. These two cascaded counters provide the divide-by-6 output at PCLK, the peripheral clock output.

Figure 9–4 shows how an 8284A is connected to the 8086/8088. Notice that $F/\overline{C}$ and CSYNC are grounded to select the crystal oscillator, and that a 15 MHz crystal provides the normal 5 MHz clock signal to the 8086/8088, as well as a 2.5 MHz peripheral clock signal.

Operation of the Reset Section. The reset section of the 8284A is very simple: It consists of a Schmitt trigger buffer and a single D-type flip-flop circuit. The D-type flip-flop ensures that the timing requirements of the 8086/8088 RESET input are met. This circuit applies the RESET signal to the microprocessor on the negative edge (1-to-0 transition) of each clock. The 8086/8088 microprocessors sample RESET at the positive edge (0-to-1 transition) of the clocks; therefore, this circuit meets the timing requirements of the 8086/8088.

Refer to Figure 9–4. Notice that an RC circuit provides a logic 0 to the $\overline{\textbf{RES}}$ input pin when power is first applied to the system. After a short time, the $\overline{\textbf{RES}}$ input becomes a logic 1 because the capacitor charges toward +5.0 V through the resistor. A pushbutton switch allows the microprocessor to be reset by the operator. Correct reset timing requires the RESET input to become a logic 1 no later than four clocks after system power is applied, and to be held high for at least 50 μs. The flip-flop makes certain that RESET goes high in four clocks, and the RC time constant ensures that it stays high for at least 50 μs.

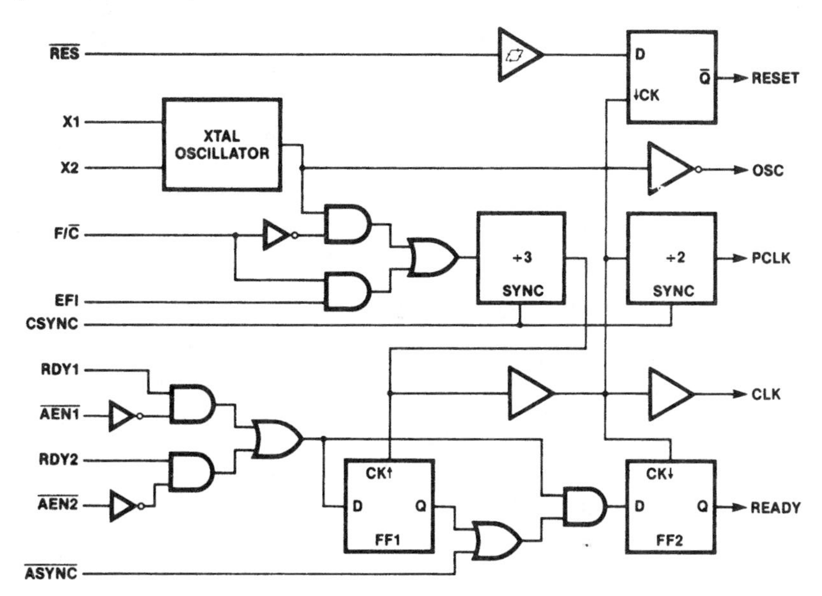

FIGURE 9–3 The internal block diagram of the 8284A clock generator.

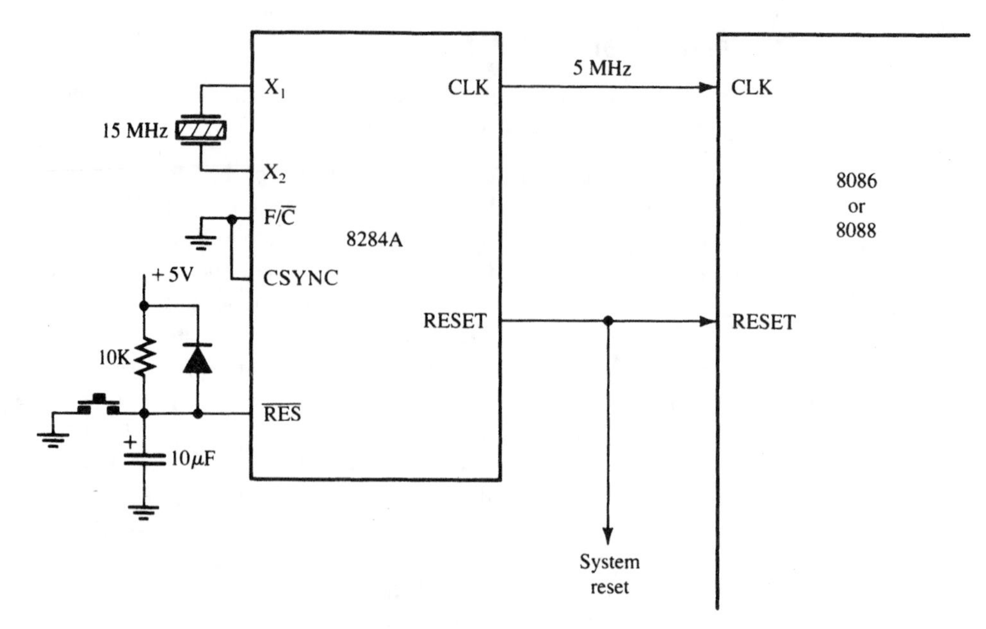

FIGURE 9–4 The clock generator (8284A) and the 8086 and 8088 microprocessors illustrating the connection for the clock and reset signals. A 15 MHz crystal provides the 5 MHz clock for the microprocessor.

9–3 BUS BUFFERING AND LATCHING

Before the 8086/8088 microprocessors can be used with memory or I/O interfaces, their multiplexed buses must be demultiplexed. This section provides the detail required to demultiplex the buses and illustrates how the buses are buffered for very large systems. (Because the maximum fan-out is 10, the system must be buffered if it contains more than 10 other components.)

Demultiplexing the Buses

The address/data bus on the 8086/8088 is multiplexed (shared) to reduce the number of pins required for the 8086/8088 microprocessor integrated circuit. Unfortunately, this burdens the hardware designer with the task of extracting or demultiplexing information from these multiplexed pins.

Why not leave the buses multiplexed? Memory and I/O require that the address remain valid and stable throughout a read or write cycle. If the buses are multiplexed, the address changes at the memory and I/O, which causes them to read or write data in the wrong locations.

All computer systems have three buses: (1) an address bus that provides the memory and I/O with the memory address or the I/O port number, (2) a data bus that transfers data between the microprocessor and the memory and I/O in the system, and (3) a control bus that provides control signals to the memory and I/O. These buses must be present in order to interface to memory and I/O.

Demultiplexing the 8088. Figure 9–5 illustrates the 8088 microprocessor and the components required to demultiplex its buses. In this case, two 74LS373 or 74LS573 transparent latches are used to demultiplex the address/data bus connections AD_7–AD_0 and the multiplexed address/status connections A_{19}/S_6–A_{16}/S_3.

These transparent latches, which are like wires whenever the address latch enable pin (ALE) becomes a logic 1, pass the inputs to the outputs. After a short time, ALE returns to its logic 0 condition, which causes the latches to remember the inputs at the time of the change to a logic 0. In this case, A_7–A_0 are stored in the bottom latch and A_{19}–A_{16} are stored in the top latch.

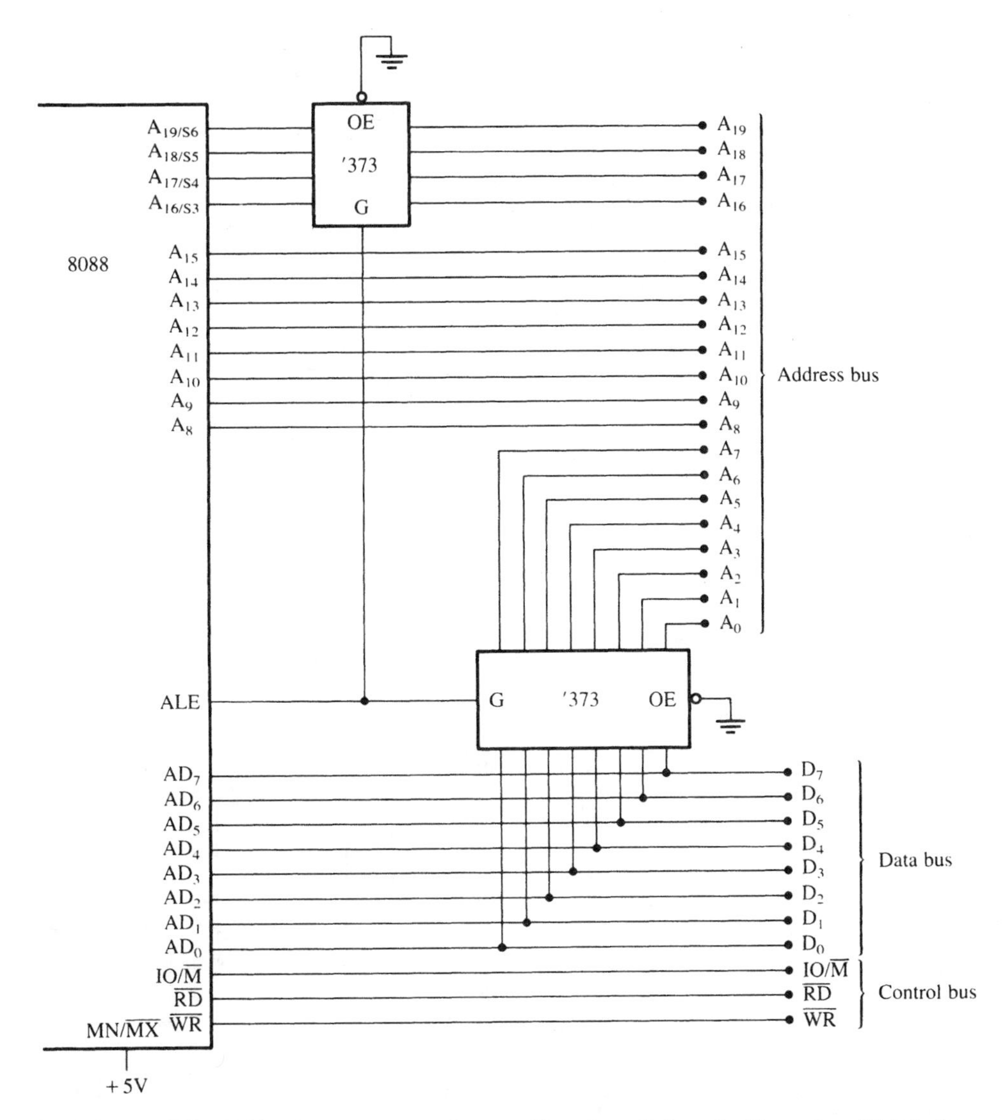

FIGURE 9–5 The 8088 microprocessor shown with a demultiplexed address bus. This is the model used to build many 8088-based systems.

This yields a separate address bus with connections A_{19}–A_0. These address connections allow the 8088 to address 1M byte of memory space. The fact that the data bus is separate allows it to be connected to any eight-bit peripheral device or memory component.

Demultiplexing the 8086. Like the 8088, the 8086 system requires separate address, data, and control buses. It differs primarily in the number of multiplexed pins. In the 8088, only AD_7–AD_0 and A_{19}/S_6–A_{16}/S_3 are multiplexed. In the 8086, the multiplexed pins include AD_{15}–AD_0, A_{19}/S_6–A_{16}/S_3, and $\overline{BHE}/S_7$. All of these signals must be demultiplexed.

Figure 9–6 illustrates a demultiplexed 8086 with all three buses: address (A_{19}–A_0 and $\overline{BHE}$), data (D_{15}–D_0), and control ($M/\overline{IO}$, $\overline{RD}$, and $\overline{WR}$).

This circuit shown in Figure 9–6 is almost identical to the one pictured in Figure 9–5, except that an additional 74LS373 latch has been added to demultiplex the address/data bus pins AD_{15}–AD_8 and a $\overline{BHE}/S_7$ input has been added to the top 74LS373 to select the high-order memory bank in the 16-bit memory system of the 8086. Here, the memory and I/O system see

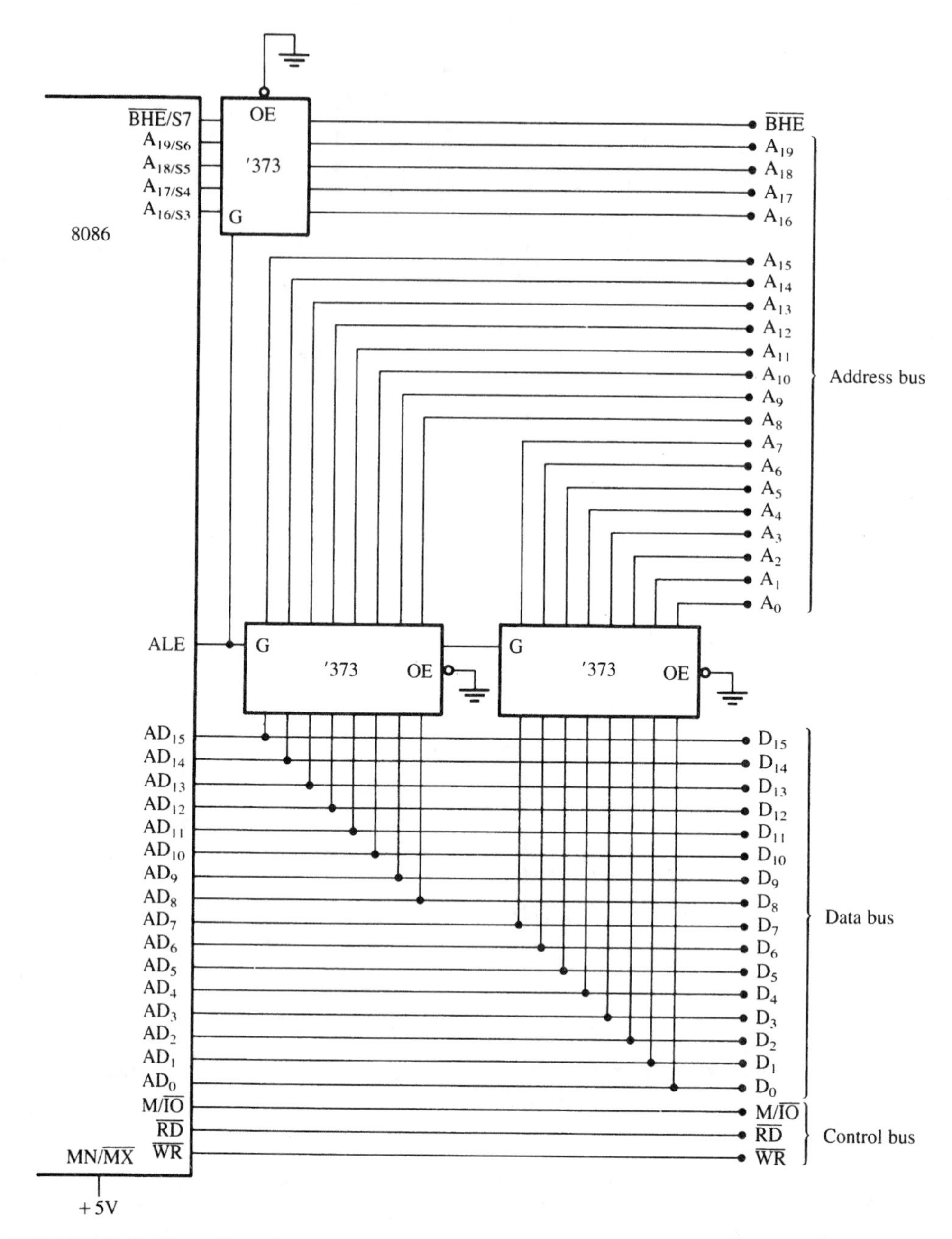

FIGURE 9–6 The 8086 microprocessor shown with a demultiplexed address bus. This is the model used to build many 8086-based systems.

the 8086 as a device with a 20-bit address bus (A_{19}–A_0), a 16-bit data bus (D_{15}–D_0), and a three-line control bus ($M/\overline{IO}$, $\overline{RD}$, and $\overline{WR}$).

The Buffered System

If more than 10 unit loads are attached to any bus pin, the entire 8086 or 8088 system must be buffered. The demultiplexed pins are already buffered by the 74LS373 or 74LS573 latches, which have been designed to drive the high-capacitance buses encountered in microcomputer systems. The buffer's output currents have been increased so that more TTL unit loads may be driven: A logic 0 output provides up to 32 mA of sink current, and a logic 1 output provides up to 5.2 mA of source current.

A fully buffered signal will introduce a timing delay to the system. This causes no difficulty unless memory or I/O devices are used, which function at near the maximum speed of the bus. Section 9–4 discusses this problem and the time delays involved in more detail.

The Fully Buffered 8088. Figure 9–7 depicts a fully buffered 8088 microprocessor. Notice that the remaining eight address pins, A_{15}–A_8, use a 74LS244 octal buffer; the eight data bus pins, D_7–D_0, use a 74LS245 octal bidirectional bus buffer; and the control bus signals, $M/\overline{IO}$, $\overline{RD}$, and

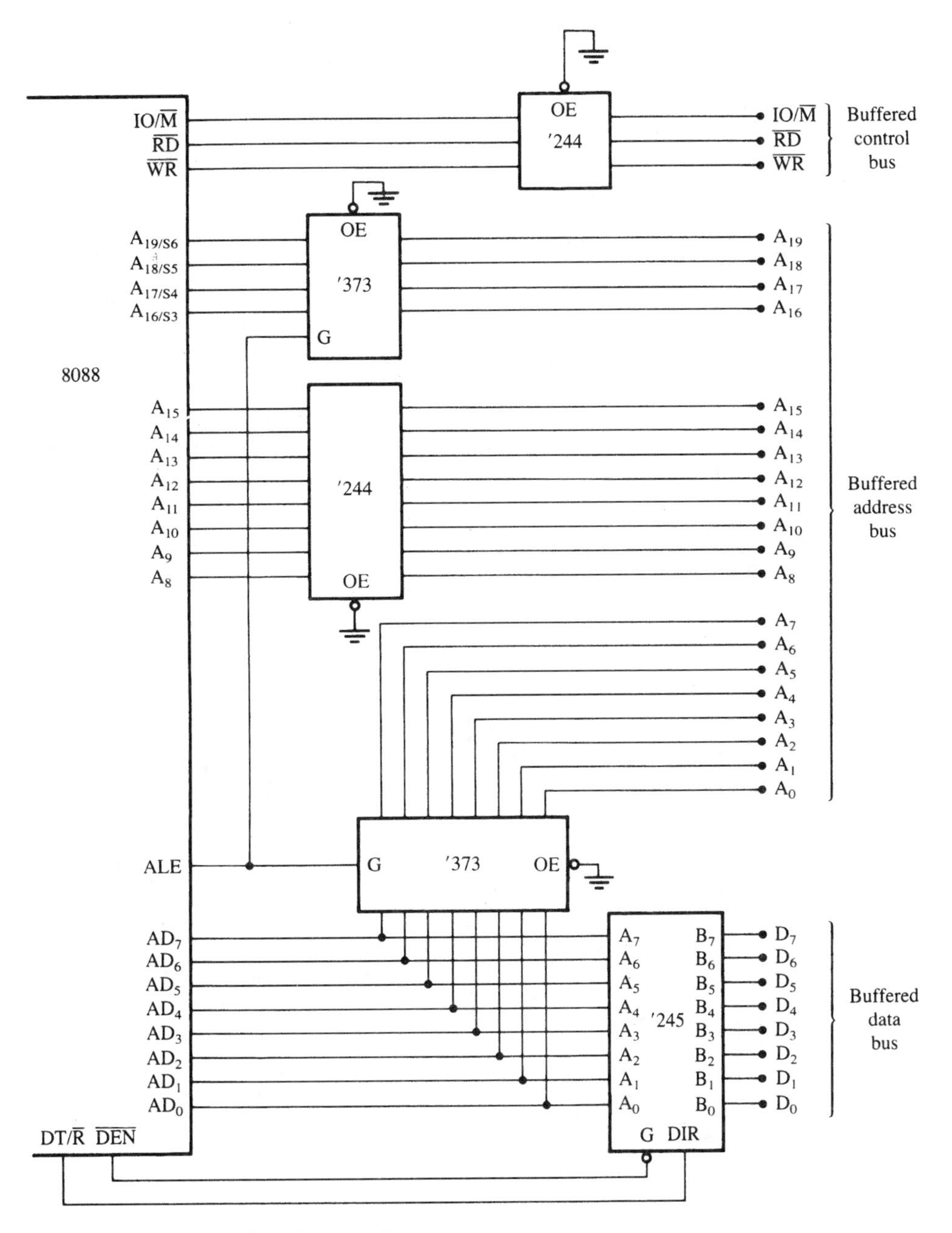

FIGURE 9–7 A fully buffered 8088 microprocessor.

$\overline{WR}$, use a 74LS244 buffer. A fully buffered 8088 system requires two 74LS244s, one 74LS245, and two 74LS373s. The direction of the 74LS245 is controlled by the DT/$\overline{R}$ signal and is enabled and disabled by the $\overline{DEN}$ signal.

The Fully Buffered 8086. Figure 9–8 illustrates a fully buffered 8086 microprocessor. Its address pins are already buffered by the 74LS373 address latches; its data bus employs two

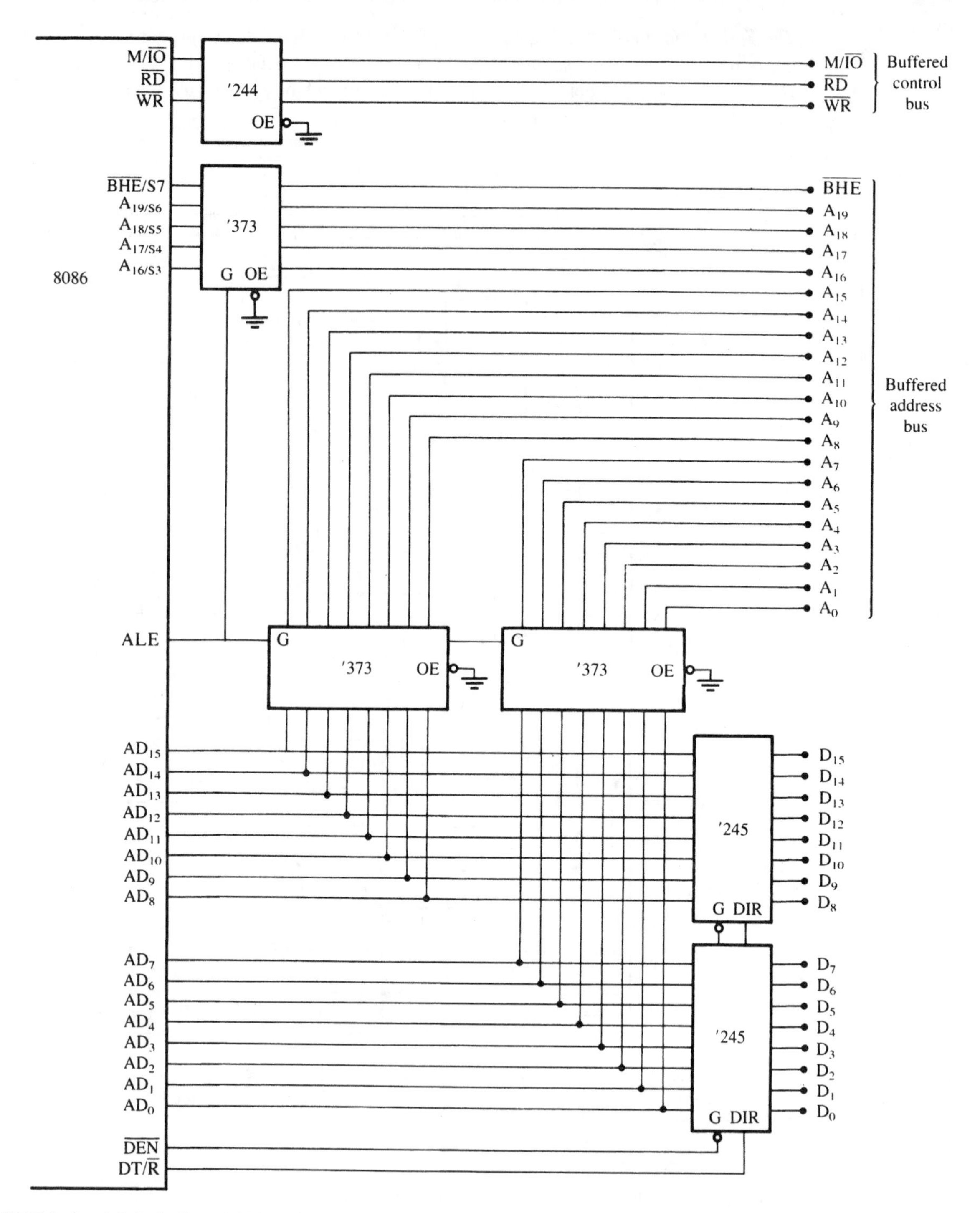

FIGURE 9–8 A fully buffered 8086 microprocessor.

74LS245 octal bidirectional bus buffers; and the control bus signals, M/$\overline{IO}$, $\overline{RD}$, and $\overline{WR}$ use a 74LS244 buffer. A fully buffered 8086 system requires one 74LS244, two 74LS245s, and three 74LS373s. The 8086 requires one more buffer than the 8088 because of the extra eight data bus connections, D_{15}–D_8. It also has a $\overline{BHE}$ signal that is buffered for memory-bank selection.

9–4 BUS TIMING

It is essential to understand system bus timing before choosing a memory or I/O device for interfacing to the 8086 or 8088 microprocessors. This section provides insight into the operation of the bus signals and the basic read and write timing of the 8086/8088. It is important to note that we discuss only the times that affect memory and I/O interfacing in this section.

Basic Bus Operation

The three buses of the 8086 and 8088—address, data, and control—function exactly the same way as those of any other microprocessor. If data are written to the memory (see the simplified timing for write in Figure 9–9), the microprocessor outputs the memory address on the address bus, outputs the data to be written into memory on the data bus, and issues a write ($\overline{WR}$) to memory and IO/$\overline{M}$ = 0 for the 8088 and M/$\overline{IO}$ = 1 for the 8086. If data are read from the memory (see the simplified timing for read in Figure 9–10), the microprocessor outputs the memory address on the address bus, issues a read memory signal ($\overline{RD}$), and accepts the data via the data bus.

Timing in General

The 8086/8088 microprocessors use the memory and I/O in periods called **bus cycles**. Each bus cycle equals four system-clocking periods (T states). Newer microprocessors divide the bus cycle into as few as two clocking periods. If the clock is operated at 5 MHz (the basic operating frequency for these two microprocessors), one 8086/8088 bus cycle is complete in 800 ns. This means that the microprocessor reads or writes data between itself and memory or I/O at a maximum rate of 1.25 million times a second. (Because of the internal queue, the 8086/8088 can execute 2.5 million instructions per second [MIPS] in bursts.) Other available versions of these microprocessors operate at much higher transfer rates due to higher clock frequencies.

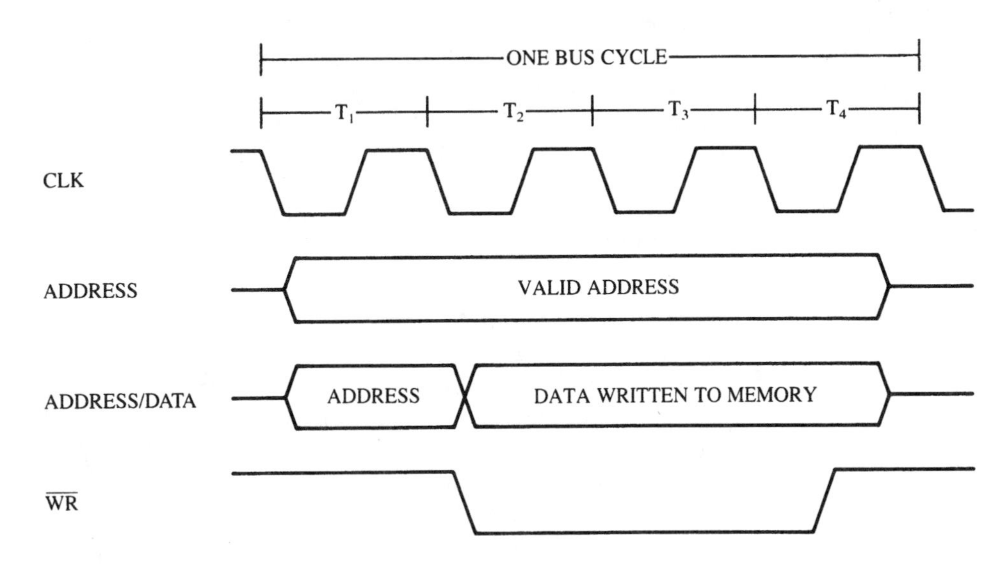

FIGURE 9–9 Simplified 8086/8088 write bus cycle.

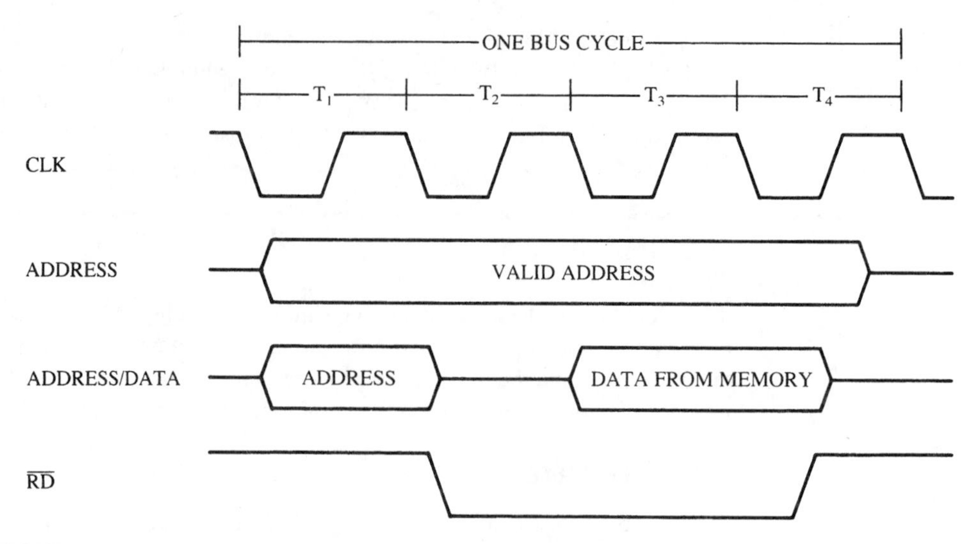

FIGURE 9–10 Simplified 8086/8088 read bus cycle.

During the first clocking period in a bus cycle, which is called T1, many things happen. The address of the memory or I/O location is sent out via the address bus and the address/data bus connections. (The address/data bus is multiplexed and sometimes contains memory-addressing information, sometimes data.) During T_1, control signals ALE, DT/$\overline{R}$, and IO/$\overline{M}$ (8088) or M/$\overline{IO}$ (8086) are also output. The IO/$\overline{M}$ or M/$\overline{IO}$ signal indicates whether the address bus contains a memory address or an I/O device (port) number.

During T_2, the 8086/8088 microprocessors issue the $\overline{RD}$ or $\overline{WR}$ signal, $\overline{DEN}$, and in the case of a write, the data to be written appear on the data bus. These events cause the memory or I/O device to begin to perform a read or a write. The $\overline{DEN}$ signal turns on the data bus buffers, if they are present in the system, so the memory or I/O can receive data to be written, or so the microprocessor can accept the data read from the memory or I/O for a read operation. If this happens to be a write bus cycle, the data are sent out to the memory or I/O through the data bus.

READY is sampled at the end of T_2, as illustrated in Figure 9–11. If READY is low at this time, T_3 becomes a wait state (T_W). (More detail is provided in Section 9–5.) This clocking period is provided to allow the memory time to access data. If the bus cycle happens to be a read bus cycle, the data bus is sampled at the end of T_3.

In T_4, all bus signals are deactivated in preparation for the next bus cycle. This is also the time when the 8086/8088 samples the data bus connections for data that are read from memory or I/O. In addition, at this point, the trailing edge of the $\overline{WR}$ signal transfers data to the memory or I/O, which activates and writes when the $\overline{WR}$ signal returns to a logic 1 level.

Read Timing

Figure 9–11 also depicts the read timing for the 8088 microprocessor. The 8086 read timing is identical except that the 8086 has 16 rather than eight data bus bits. A close look at this timing diagram should allow you to identify all the main events described for each T state.

The most important item contained in the read timing diagram is the amount of time allowed for the memory or I/O to read the data. Memory is chosen by its access time, which is the fixed amount of time that the microprocessor allows it to access data for the read operation. It is therefore extremely important that the memory chosen complies with the limitations of the system.

The microprocessor timing diagram does not provide a listing for memory access time. Instead, it is necessary to combine several times to arrive at the access time. To find memory access

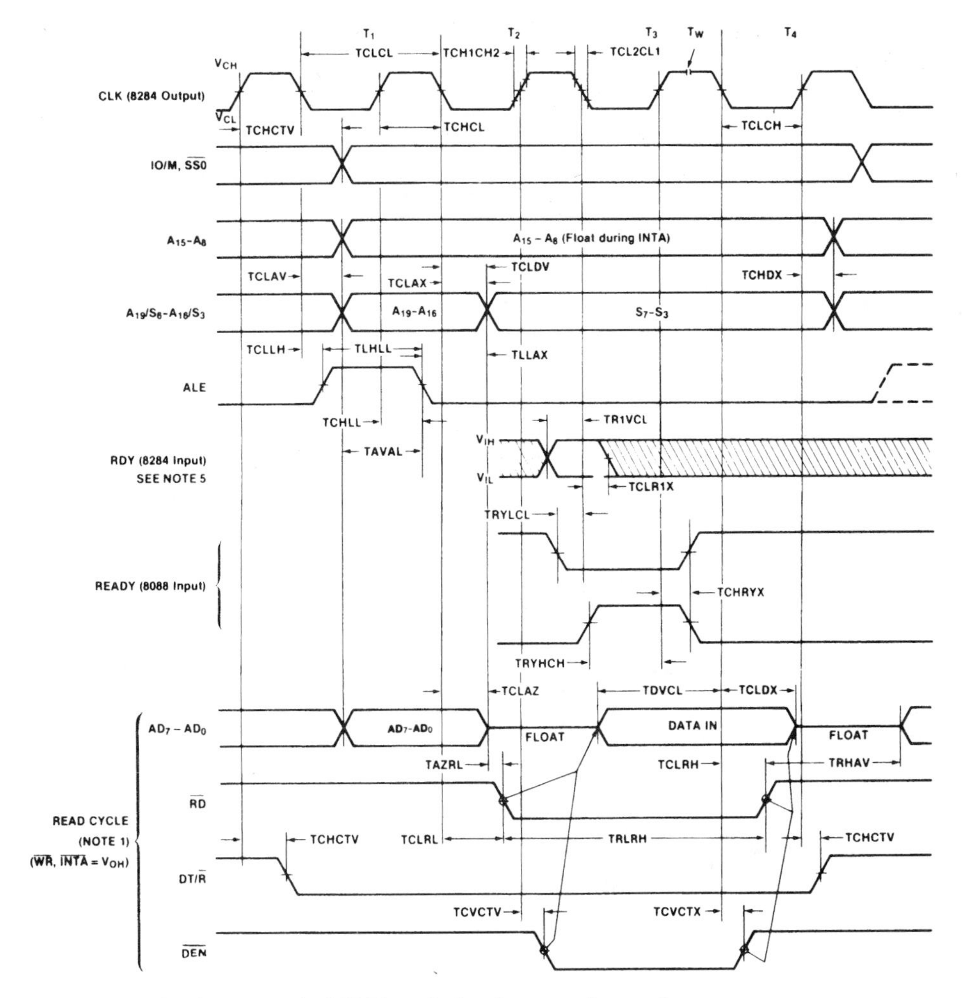

FIGURE 9–11 Minimum mode 8088 bus timing for a read operation.

time in this diagram, first locate the point in T_3 when data are sampled. If you examine the timing diagram closely, you will notice a line that extends from the end of T_3 down to the data bus. At the end of T_3, the microprocessor samples the data bus.

Memory access time starts when the address appears on the memory address bus and continues until the microprocessor samples the memory data at T_3. Approximately three T states elapse between these times. (See Figure 9–12 for the following times.) The address does not appear until T_{CLAV} time (110 ns if the clock is 5 MHz) after the start of T_1. This means that T_{CLAV} time must be subtracted from the three clocking states (600 ns) that separate the appearance of the address (T_1) and the sampling of the data (T_3). One other time must also be subtracted: the data setup time (T_{DVCL}), which occurs before T_3. Memory access time is thus three clocking states minus the sum of T_{CLAV} and T_{DVCL}. Because T_{DVCL} is 30 ns with a 5 MHz clock, the allowed memory access time is only 460 ns (access time = 600 ns – 110 ns – 30 ns).

The memory devices chosen for connection to the 8086/8088 operating at 5 MHz must be able to access data in less than 460 ns, because of the time delay introduced by the address decoders and buffers in the system. At least a 30- or 40-ns margin should exist for the operation of these circuits. Therefore, the memory speed should be no slower than about 420 ns to operate correctly with the 8086/8088 microprocessors.

FIGURE 9–12 8088 AC characteristics.

A.C. CHARACTERISTICS (8088: T_A = 0°C to 70°C, V_{CC} = 5V ±10%)*
(8088-2: T_A = 0°C to 70°C, V_{CC} = 5V ±5%)

MINIMUM COMPLEXITY SYSTEM TIMING REQUIREMENTS

		8088		8088-2			
Symbol	Parameter	Min.	Max.	Min.	Max.	Units	Test Conditions
TCLCL	CLK Cycle Period	200	500	125	500	ns	
TCLCH	CLK Low Time	118		68		ns	
TCHCL	CLK High Time	69		44		ns	
TCH1CH2	CLK Rise Time		10		10	ns	From 1.0V to 3.5V
TCL2CL1	CLK Fall Time		10		10	ns	From 3.5V to 1.0V
TDVCL	Data in Setup Time	30		20		ns	
TCLDX	Data in Hold Time	10		10		ns	
TR1VCL	RDY Setup Time into 8284 (See Notes 1, 2)	35		35		ns	
TCLR1X	RDY Hold Time into 8284 (See Notes 1, 2)	0		0		ns	
TRYHCH	READY Setup Time into 8088	118		68		ns	
TCHRYX	READY Hold Time into 8088	30		20		ns	
TRYLCL	READY Inactive to CLK (See Note 3)	−8		−8		ns	
THVCH	HOLD Setup Time	35		20		ns	
TINVCH	INTR, NMI, $\overline{TEST}$ Setup Time (See Note 2)	30		15		ns	
TILIH	Input Rise Time (Except CLK)		20		20	ns	From 0.8V to 2.0V
TIHIL	Input Fall Time (Except CLK)		12		12	ns	From 2.0V to 0.8V

A.C. CHARACTERISTICS (Continued)

TIMING RESPONSES

		8088		8088-2			
Symbol	Parameter	Min.	Max.	Min.	Max.	Units	Test Conditions
TCLAV	Address Valid Delay	10	110	10	60	ns	C_L = 20-100 pF for all 8088 Outputs in addition to internal loads
TCLAX	Address Hold Time	10		10		ns	
TCLAZ	Address Float Delay	TCLAX	80	TCLAX	50	ns	
TLHLL	ALE Width	TCLCH−20		TCLCH−10		ns	
TCLLH	ALE Active Delay		80		50	ns	
TCHLL	ALE Inactive Delay		85		55	ns	
TLLAX	Address Hold Time to ALE Inactive	TCHCL−10		TCHCL−10		ns	
TCLDV	Data Valid Delay	10	110	10	60	ns	
TCHDX	Data Hold Time	10		10		ns	
TWHDX	Data Hold Time After $\overline{WR}$	TCLCH−30		TCLCH−30		ns	
TCVCTV	Control Active Delay 1	10	110	10	70	ns	
TCHCTV	Control Active Delay 2	10	110	10	60	ns	
TCVCTX	Control Inactive Delay	10	110	10	70	ns	
TAZRL	Address Float to READ Active	0		0		ns	
TCLRL	$\overline{RD}$ Active Delay	10	165	10	100	ns	
TCLRH	$\overline{RD}$ Inactive Delay	10	150	10	80	ns	
TRHAV	$\overline{RD}$ Inactive to Next Address Active	TCLCL−45		TCLCL−40		ns	
TCLHAV	HLDA Valid Delay	10	160	10	100	ns	
TRLRH	$\overline{RD}$ Width	2TCLCL−75		2TCLCL−50		ns	
TWLWH	$\overline{WR}$ Width	2TCLCL−60		2TCLCL−40		ns	
TAVAL	Address Valid to ALE Low	TCLCH−60		TCLCH−40		ns	
TOLOH	Output Rise Time		20		20	ns	From 0.8V to 2.0V
TOHOL	Output Fall Time		12		12	ns	From 2.0V to 0.8V

The only other timing factor that may affect memory operation is the width of the $\overline{RD}$ strobe. On the timing diagram, the read strobe is given as T_{RLRH}. The time for this strobe is 325 ns (5 MHz clock rate), which is wide enough for almost all memory devices manufactured with an access time of 400 ns or less.

Write Timing

Figure 9–13 illustrates the write-timing diagram for the 8088 microprocessor. (Again, the 8086 is nearly identical, so it need not be presented here in a separate timing diagram.)

The main differences between read and write timing are minimal. The $\overline{RD}$ strobe is replaced by the $\overline{WR}$ strobe, the data bus contains information for the memory rather than information from the memory, and DT/$\overline{R}$ remains a logic 1 instead of a logic 0 throughout the bus cycle.

When interfacing some memory devices, timing may be especially critical between the point at which $\overline{WR}$ becomes a logic 1 and the time when the data are removed from the data bus. This is the case because, as you will recall, memory data are written at the trailing edge of the $\overline{WR}$ strobe. According to the timing diagram, this critical period is T_{WHDX} or 88 ns when the 8088 is operated with a 5 MHz clock. Hold time is often much less than this; it is, in fact, often 0 ns for memory devices. The width of the $\overline{WR}$ strobe is T_{WLWH} or 340 ns at a 5 MHz clock rate. This rate is compatible with most memory devices that have an access time of 400 ns or less.

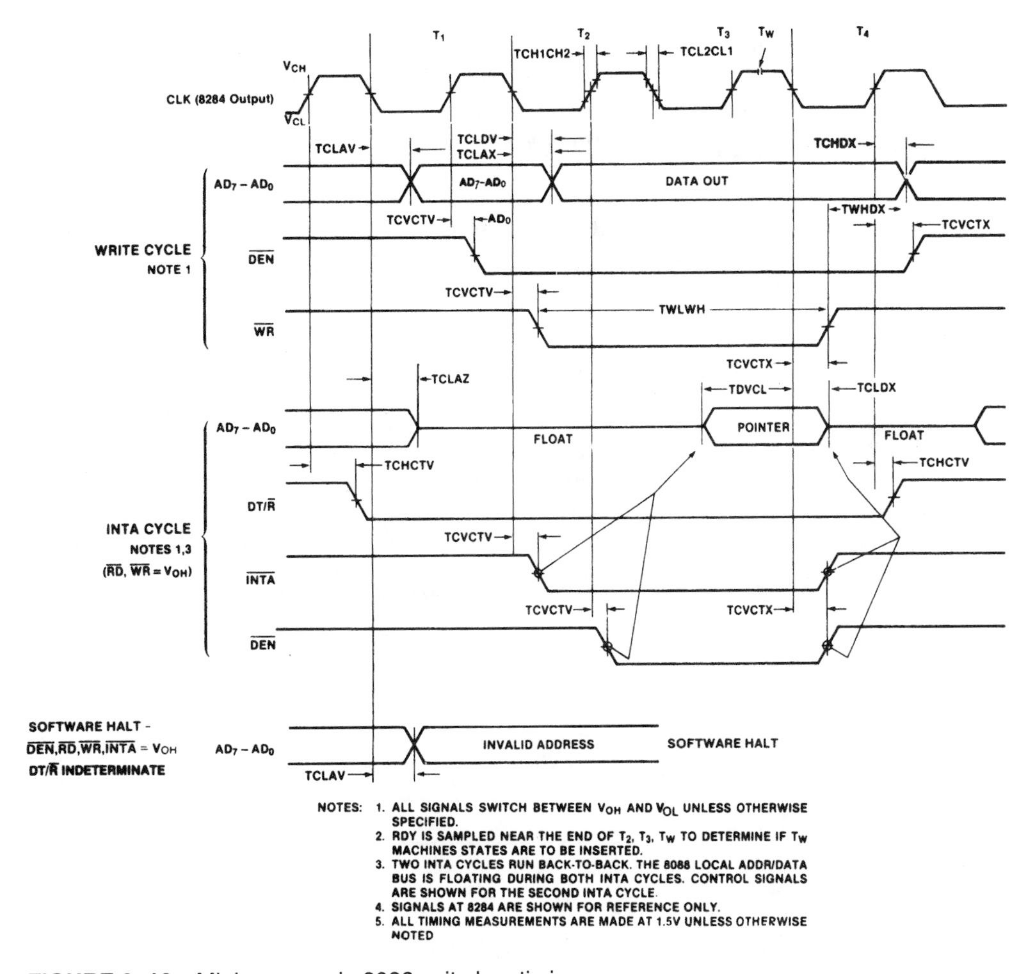

FIGURE 9–13 Minimum mode 8088 write bus timing.

9–5 READY AND THE WAIT STATE

As we mentioned earlier in this chapter, the READY input causes wait states for slower memory and I/O components. A wait state (T_W) is an extra clocking period inserted between T_2 and T_3 to lengthen the bus cycle. If one wait state is inserted, then the memory access time, normally 460 ns with a 5 MHz clock, is lengthened by one clocking period (200 ns) to 660 ns.

In this section, we discuss the READY synchronization circuitry inside the 8284A clock generator, show how to insert one or more wait states selectively into the bus cycle, and examine the READY input and the synchronization times it requires.

The READY Input

The READY input is sampled at the end of T_2 and again, if applicable, in the middle of T_W. If READY is a logic 0 at the end of T_2, T_3 is delayed and T_W is inserted between T_2 and T_3. READY is next sampled at the middle of T_W to determine whether the next state is T_W or T_3. It is tested for a logic 0 on the 1-to-0 transition of the clock at the end of T_2, and for a 1 on the 0-to-1 transition of the clock in the middle of T_W.

The READY input to the 8086/8088 has some stringent timing requirements. The timing diagram in Figure 9–14 shows READY causing one wait state (T_W), along with the required setup and hold times from the system clock. The timing requirement for this operation is met by the internal READY synchronization circuitry of the 8284A clock generator. When the 8284A is used for READY, the RDY (ready input to the 8284A) input occurs at the end of each T state.

RDY and the 8284A

RDY is the synchronized ready input to the 8284A clock generator. The timing diagram for this input is provided in Figure 9–15. Although it differs from the timing for the READY input to the

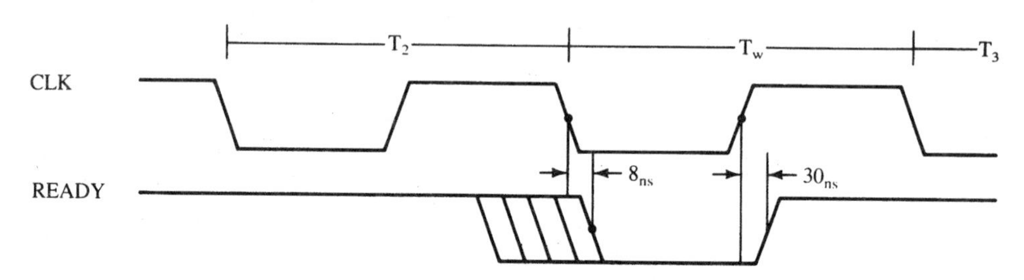

FIGURE 9–14 8086/8088 READY input timing.

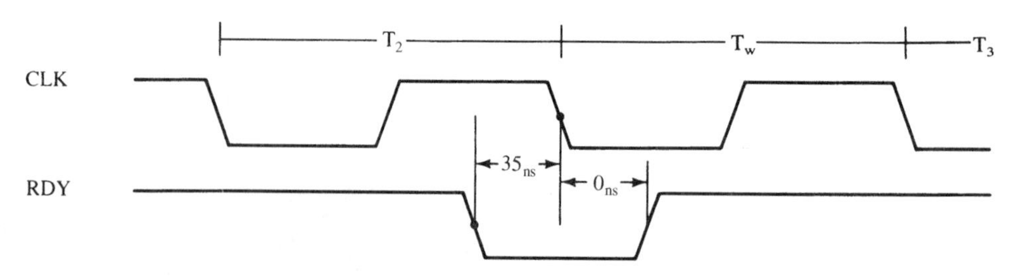

FIGURE 9–15 8284A RDY input timing.

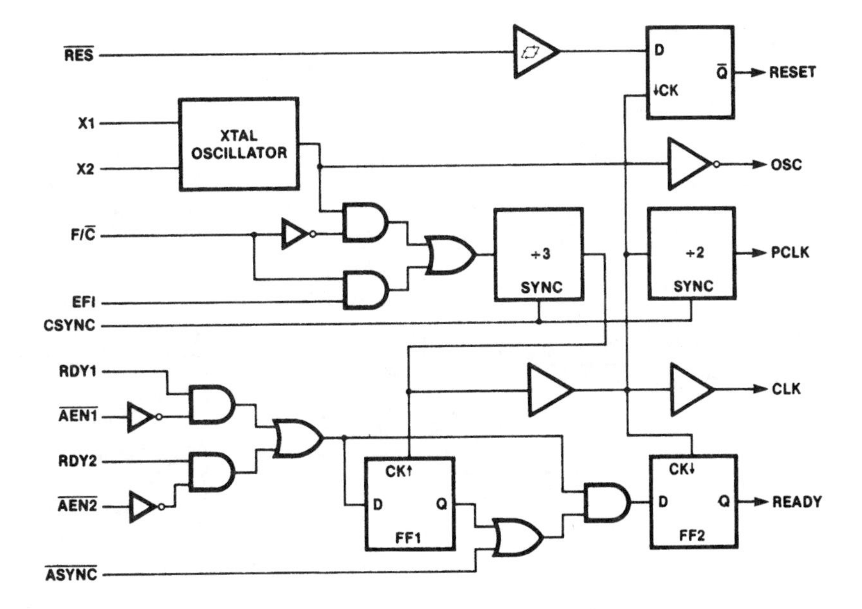

FIGURE 9–16 The internal block diagram of the 8284A clock generator. (Courtesy of Intel Corporation.)

8086/8088, the internal 8284A circuitry guarantees the accuracy of the READY synchronization provided to the 8086/8088 microprocessors.

Figure 9–16 again depicts the internal structure of the 8284A. The bottom half of this diagram is the READY synchronization circuitry. At the leftmost side, the RDY_1 and $\overline{AEN1}$ inputs are ANDed, as are the RDY_2 and $\overline{AEN2}$ inputs. The outputs of the AND gates are then ORed to generate the input to the one or two stages of synchronization. In order to obtain a logic 1 at the inputs to the flip-flops, RDY_1 ANDed with $\overline{AEN1}$ must be active or RDY_2 ANDed with $\overline{AEN2}$ must be active.

The ASYNC input selects one stage of synchronization when it is a logic 1 and two stages when it is a logic 0. If one stage is selected, then the RDY signal is kept from reaching the 8086/8088 READY pin until the next negative edge of the clock. If two stages are selected, the first positive edge of the clock captures RDY in the first flip-flop. The output of this flip-flop is fed to the second flip-flop, so on the next negative edge of the clock, the second flip-flop captures RDY.

Figure 9–17 illustrates a circuit used to introduce almost any number of wait states for the 8086/8088 microprocessors. Here, an eight-bit serial shift register (74LS164) shifts a logic 0 for one or more clock periods from one of its Q outputs through to the RDY_1 input of the 8284A. With appropriate strapping, this circuit can provide various numbers of wait states. Notice also how the shift register is cleared back to its starting point. The output of the register is forced high when the $\overline{RD}$, $\overline{WR}$, and $\overline{INTA}$ pins are all logic 1s. These three signals are high until state T_2, so the shift register shifts for the first time when the positive edge of the T_2 arrives. If one wait is desired, output Q_B is connected to the OR gate. If two waits are desired, output Q_C is connected, and so forth.

Notice in Figure 9–17 that this circuit does not always generate wait states. It is enabled from the memory only for memory devices that require the insertion of waits. If the selection signal from a memory device is a logic 0, the device is selected; then this circuit will generate a wait state.

Figure 9–18 illustrates the timing diagram for this shift register wait state generator when it is wired to insert one wait state. The timing diagram also illustrates the internal contents of the shift register's flip-flops to present a more detailed view of its operation. In this example, one wait state is generated.

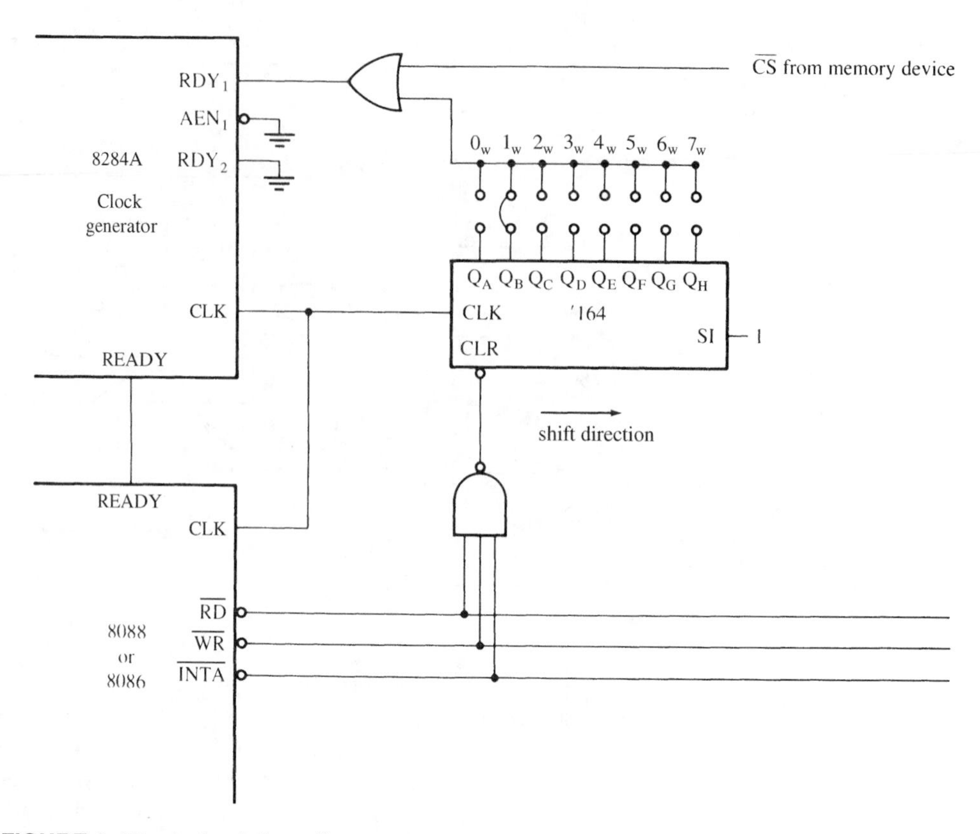

FIGURE 9–17 A circuit that will cause between 0 and 7 wait states.

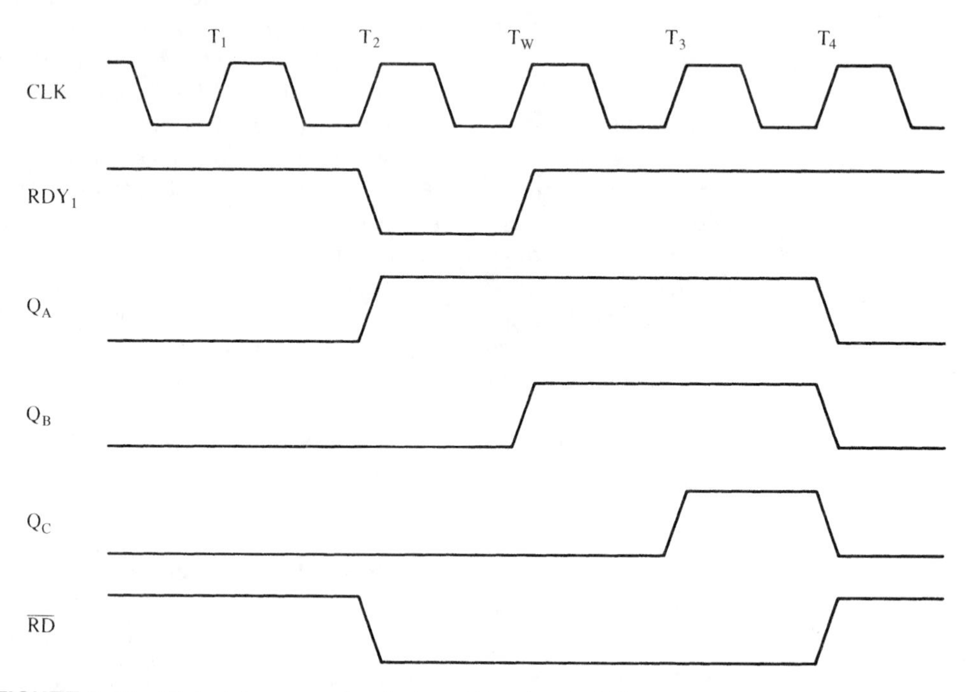

FIGURE 9–18 Wait state generation timing of the circuit of Figure 9–17.

9–6 MINIMUM MODE VERSUS MAXIMUM MODE

There are two available modes of operation for the 8086/8088 microprocessors: minimum mode and maximum mode. Minimum mode operation is obtained by connecting the mode selection pin **MN/$\overline{\text{MX}}$** to +5.0 V, and maximum mode is selected by grounding this pin. Both modes enable different control structures for the 8086/8088 microprocessors. The mode of operation provided by minimum mode is similar to that of the 8085A, the most recent Intel eight-bit microprocessor. The maximum mode is unique and designed to be used whenever a coprocessor exists in a system. Note that the maximum mode was dropped from the Intel family beginning with the 80286 microprocessor.

Minimum Mode Operation

Minimum mode operation is the least expensive way to operate the 8086/8088 microprocessors (see Figure 9–19 for the minimum mode 8088 system). It costs less because all the control signals for the memory and I/O are generated by the microprocessor. These control signals are identical to those of the Intel 8085A, an earlier eight-bit microprocessor. The minimum mode allows the 8085A eight-bit peripherals to be used with the 8086/8088 without any special considerations.

Maximum Mode Operation

Maximum mode operation differs from minimum mode in that some of the control signals must be externally generated. This requires the addition of an external bus controller—the 8288 bus controller (see Figure 9–20 for the maximum mode 8088 system). There are not enough pins on the 8086/8088 for bus control during maximum mode because new pins and new features have replaced some of them. Maximum mode is used only when the system contains external coprocessors such as the 8087 arithmetic coprocessor.

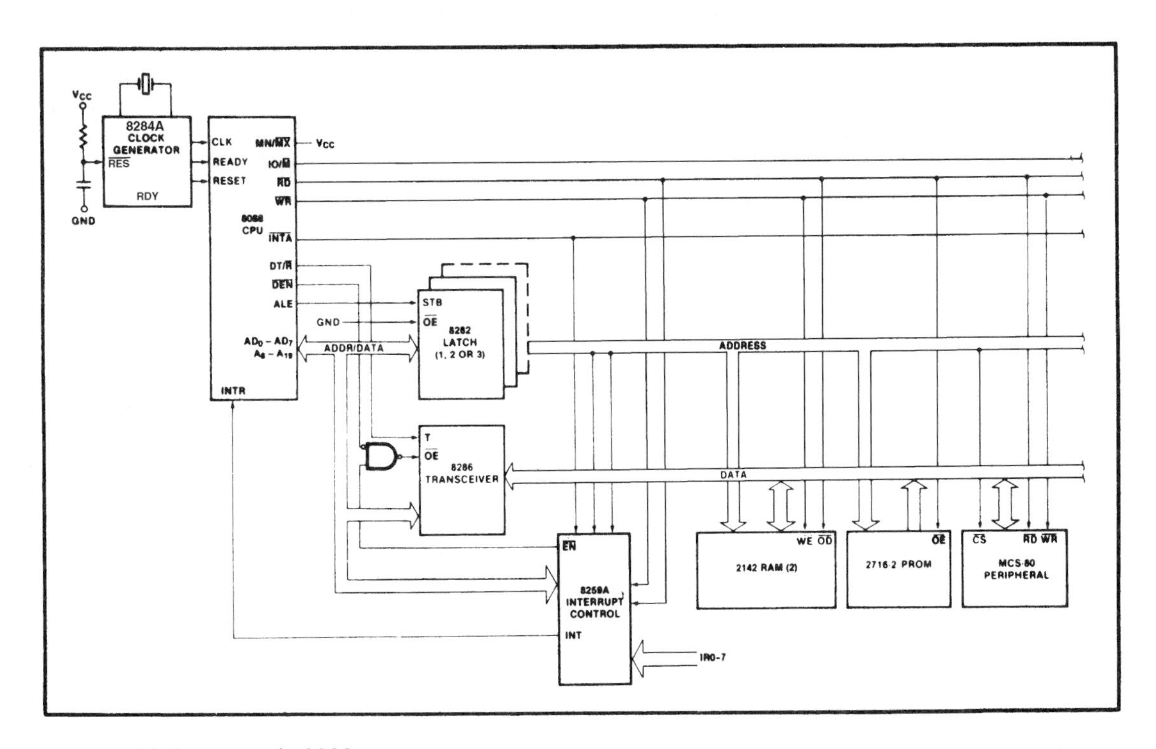

FIGURE 9–19 Minimum mode 8088 system.

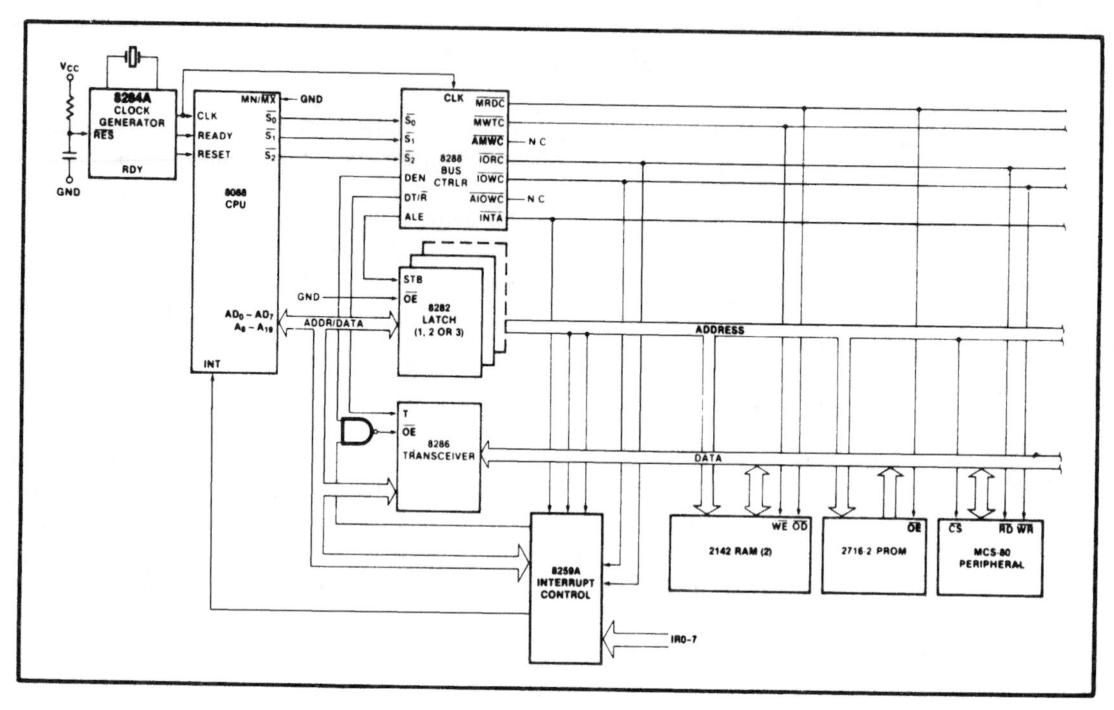

FIGURE 9–20 Maximum mode 8088 system.

The 8288 Bus Controller

An 8086/8088 system that is operated in maximum mode must have an 8288 bus controller to provide the signals eliminated from the 8086/8088 by the maximum mode operation. Figure 9–21 illustrates the block diagram and pin-out of the 8288 bus controller.

Notice that the control bus developed by the 8288 bus controller contains separate signals for I/O ($\overline{\text{IORC}}$ and $\overline{\text{IOWC}}$) and memory ($\overline{\text{MRDC}}$ and $\overline{\text{MWTC}}$). It also contains advanced memory ($\overline{\text{AMWC}}$) and I/O ($\overline{\text{AIOWC}}$) write strobes, and the $\overline{\text{INTA}}$ signal. These signals replace the minimum mode ALE, $\overline{\text{WR}}$, IO/$\overline{\text{M}}$, DT/$\overline{\text{R}}$, $\overline{\text{DEN}}$, and $\overline{\text{INTA}}$, which are lost when the 8086/8088 microprocessors are operated in the maximum mode.

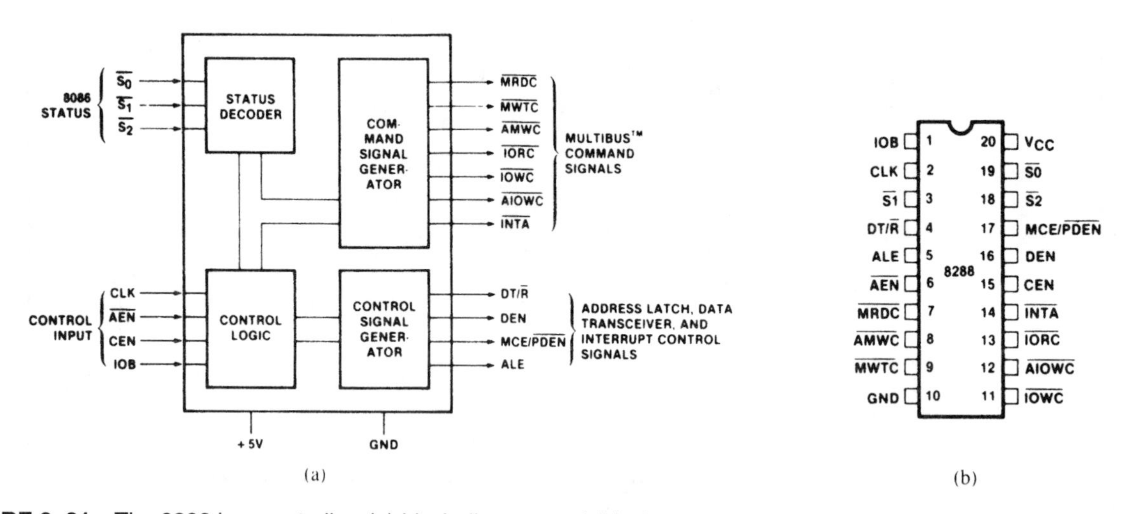

FIGURE 9–21 The 8288 bus controller; (a) block diagram and (b) pin-out.

Pin Functions

The following list provides a description of each pin of the 8288 bus controller.

S_2, S_1, and S_0 — **Status inputs** are connected to the status output pins on the 8086/8088 microprocessor. These three signals are decoded to generate the timing signals for the system.

CLK — The **clock** input provides internal timing and must be connected to the CLK output pin of the 8284A clock generator.

ALE — The **address latch enable** output is used to demultiplex the address/data bus.

DEN — The **data bus enable** pin controls the bidirectional data bus buffers in the system. Note that this is an active high output pin that is the opposite polarity from the $\overline{\text{DEN}}$ signal found on the microprocessor when operated in the minimum mode.

DT/$\overline{\text{R}}$ — The **data transmit/receive** signal is output by the 8288 to control the direction of the bidirectional data bus buffers.

$\overline{\text{AEN}}$ — The **address enable** input causes the 8288 to enable the memory control signals.

CEN — The **control enable** input enables the command output pins on the 8288.

IOB — The **I/O bus mode** input selects either the I/O bus mode or system bus mode operation.

$\overline{\text{AIOWC}}$ — The **advanced I/O write** is a command output used to provide I/O with an advanced I/O write control signal.

$\overline{\text{IORC}}$ — The **I/O read command** output provides I/O with its read control signal.

$\overline{\text{IOWC}}$ — The **I/O write command** output provides I/O with its main write signal.

$\overline{\text{AMWT}}$ — The **advanced memory write** control pin provides memory with an early or advanced write signal.

$\overline{\text{MWTC}}$ — The **memory write** control pin provides memory with its normal write control signal.

$\overline{\text{MRDC}}$ — The **memory read** control pin provides memory with a read control signal.

$\overline{\text{INTA}}$ — The **interrupt acknowledge** output acknowledges an interrupt request input applied to the INTR pin.

MCE/$\overline{\text{PDEN}}$ — The **master cascade/peripheral data** output selects cascade operation for an interrupt controller if IOB is grounded, and enables the I/O bus transceivers if IOB is tied high.

9–7 SUMMARY

1. The main differences between the 8086 and 8088 are (1) an eight-bit data bus on the 8088 and a 16-bit data bus on the 8086, (2) an $\overline{\text{SS0}}$ pin on the 8088 in place of $\overline{\text{BHE}}$/S7 on the 8086, and (3) an IO/$\overline{\text{M}}$ pin on the 8088 instead of an M/$\overline{\text{IO}}$ on the 8086.
2. Both the 8086 and 8088 require a single +5.0 V power supply with a tolerance of ±10%.
3. The 8086/8088 microprocessors are TTL-compatible if the noise immunity figure is derated to 350 mV from the customary 400 mV.
4. The 8086/8088 microprocessors can drive one 74XX, five 74LSXX, one 74SXX, 10 74ALSXX, and 10 74HCXX unit loads.

5. The 8284A clock generator provides the system clock (CLK), READY synchronization, and RESET synchronization.
6. The standard 5 MHz 8086/8088 operating frequency is obtained by attaching a 15 MHz crystal to the 8284A clock generator. The PCLK output contains a TTL-compatible signal at one-half the CLK frequency.
7. Whenever the 8086/8088 microprocessors are reset, they begin executing software at memory location FFFF0H (FFFF:0000) with the interrupt request pin disabled.
8. Because the 8086/8088 buses are multiplexed and most memory and I/O devices aren't, the system must be demultiplexed before interfacing with memory or I/O. Demultiplexing is accomplished by an eight-bit latch whose clock pulse is obtained from the ALE signal.
9. In a large system, the buses must be buffered because the 8086/8088 microprocessors are capable of driving only 10 unit loads, and large systems often have many more.
10. Bus timing is very important in the remaining chapters in the text. A bus cycle that consists of four clocking periods acts as the basic system timing. Each bus cycle is able to read or write data between the microprocessor and the memory or I/O system.
11. A bus cycle is broken into four states, or T periods: T_1 is used by the microprocessor to send the address to the memory or I/O and the ALE signal to the demultiplexers; T_2 is used to send data to memory for a write and to test the READY pin and activate control signals $\overline{RD}$ or $\overline{WR}$; T_3 allows the memory time to access data and allows data to be transferred between the microprocessor and the memory or I/O; and T_4 is where data are written.
12. The 8086/8088 microprocessors allow the memory and I/O 460 ns to access data when they are operated with a 5 MHz clock.
13. Wait states (T_w) stretch the bus cycle by one or more clocking periods to allow the memory and I/O additional access time. Wait states are inserted by controlling the READY input to the 8086/8088. READY is sampled at the end of T_2 and during T_w.
14. Minimum mode operation is similar to that of the Intel 8085A microprocessor. Maximum mode operation is new and specifically designed for the operation of the 8087 arithmetic coprocessor.
15. The 8288 bus controller must be used in the maximum mode to provide the control bus signals to the memory and I/O. This is because the maximum mode operation of the 8086/8088 removes some of the system's control signal lines in favor of control signals for the coprocessors. The 8288 reconstructs these removed control signals.

9–8 QUESTIONS AND PROBLEMS

1. List the differences between the 8086 and the 8088 microprocessors.
2. Is the 8086/8088 TTL-compatible? Explain your answer.
3. What is the fan-out from the 8086/8088 to the following devices:
   (a) 74XXX TTL
   (b) 74ALSXXX TTL
   (c) 74HCXXX CMOS
4. What information appears on the address/data bus of the 8088 while ALE is active?
5. What are the purposes of status bits S_3 and S_4?
6. What condition does a logic 0 on the 8086/8088 $\overline{RD}$ pin indicate?
7. Explain the operation of the $\overline{TEST}$ pin and the WAIT instruction.
8. Describe the signal that is applied to the CLK input pin of the 8086/8088 microprocessors.
9. What mode of operation is selected when MN/$\overline{MX}$ is grounded?
10. What does the $\overline{WR}$ strobe signal from the 8086/8088 indicate about the operation of the 8086/8088?

11. When does ALE float to its high-impedance state?
12. When DT/$\overline{R}$ is a logic 1, what condition does it indicate about the operation of the 8086/8088?
13. What happens when the HOLD input to the 8086/8088 is placed at its logic 1 level?
14. What three minimum mode 8086/8088 pins are decoded to discover whether the processor is halted?
15. Explain the operation of the $\overline{LOCK}$ pin.
16. What conditions do the QS_1 and QS_0 pins indicate about the 8086/8088?
17. What three housekeeping chores are provided by the 8284A clock generator?
18. By what factor does the 8284A clock generator divide the crystal oscillator's output frequency?
19. If the F/$\overline{C}$ pin is placed at a logic 1 level, the crystal oscillator is disabled. Where is the timing input signal attached to the 8284A under this condition?
20. The PCLK output of the 8284A is _______ MHz if the crystal oscillator is operating at 14 MHz.
21. The $\overline{RES}$ input to the 8284A is placed at a logic _______ level in order to reset the 8086/8088.
22. Which bus connections on the 8086 microprocessor are typically demultiplexed?
23. Which bus connections on the 8088 microprocessor are typically demultiplexed?
24. Which TTL-integrated circuit is often used to demultiplex the buses on the 8086/8088?
25. What is the purpose of the demultiplexed $\overline{BHE}$ signal on the 8086 microprocessor?
26. Why are buffers often required in an 8086/8088-based system?
27. What 8086/8088 signal is used to select the direction of the data flows through the 74LS245 bidirectional bus buffer?
28. A bus cycle is equal to clocking _______ periods.
29. If the CLK input to the 8086/8088 is 4 MHz, how long is one bus cycle?
30. What two 8086/8088 operations occur during a bus cycle?
31. How many MIPS is the 8086/8088 capable of obtaining when operated with a 10 MHz clock?
32. Briefly describe the purpose of each T state listed:
    (a) T_1
    (b) T_2
    (c) T_3
    (d) T_4
    (e) T_W
33. How much time is allowed for memory access when the 8086/8088 is operated with a 5 MHz clock?
34. How wide is $\overline{DEN}$ if the 8088 is operated with a 5 MHz clock?
35. If the READY pin is grounded, it will introduce _______ states into the bus cycle of the 8086/8088.
36. What does the $\overline{ASYNC}$ input to the 8284A accomplish?
37. What logic levels must be applied to $\overline{AEN1}$ and RDY_1 to obtain a logic 1 at the READY pin? (Assume that $\overline{AEN2}$ is at a logic 1 level.)
38. Contrast minimum and maximum mode 8086/8088 operation.
39. What main function is provided by the 8288 bus controller when used with 8086/8088 maximum mode operation?

CHAPTER 10

Memory Interface

INTRODUCTION

Whether simple or complex, every microprocessor-based system has a memory system. The Intel family of microprocessors is no different from any other in this respect. Almost all systems contain two main types of memory: read-only memory (ROM) and random access memory (RAM) or read/write memory. Read-only memory contains system software and permanent system data, while RAM contains temporary data and application software. This chapter explains how to interface both memory types to the Intel family of microprocessors. We demonstrate memory interface to an 8-, 16-, 32-, and 64-bit data bus by using various memory address sizes. This allows virtually any microprocessor to be interfaced to any memory system.

CHAPTER OBJECTIVES

Upon completion of this chapter, you will be able to:

1. Decode the memory address and use the outputs of the decoder to select various memory components
2. Use programmable logic devices (PLDs) to decode memory addresses
3. Explain how to interface both RAM and ROM to a microprocessor
4. Explain how error correction code (ECC) is used with memory
5. Interface memory to an 8-, 16-, 32-, and 64-bit data bus
6. Explain the operation of a dynamic RAM controller
7. Interface dynamic RAM to the microprocessor

10–1 MEMORY DEVICES

Before attempting to interface memory to the microprocessor, it is essential to completely understand the operation of memory components. In this section, we explain the functions of the four common types of memory: read-only memory (ROM), flash memory (EEPROM), static random access memory (SRAM), and dynamic random access memory (DRAM).

Memory Pin Connections

Pin connections common to all memory devices are the address inputs, data outputs or input/outputs, some type of selection input, and at least one control input used to select a read or write operation. See Figure 10–1 for ROM and RAM generic-memory devices.

Address Connections. All memory devices have address inputs that select a memory location within the memory device. Address inputs are almost always labeled from A_0, the least-significant address input, to A_n, where subscript n can be any value but is always labeled as one less than the total number of address pins. For example, a memory device with 10 address pins has its address pins labeled from A_0 to A_9. The number of address pins found on a memory device is determined by the number of memory locations found within it.

Today, the more common memory devices have between 1K (1024) to 512M (536,870,912) memory locations, with 1G and larger memory location devices on the horizon. A 1K memory device has 10 address pins (A_0–A_9); therefore, 10 address inputs are required to select any of its 1024 memory locations. It takes a 10-bit binary number (1024 different combinations) to select any single location on a 1024-location device. If a memory device has 11 address connections (A_0–A_{11}), it has 2048 (2K) internal memory locations. The number of memory locations can thus be extrapolated from the number of address pins. For example, a 4K memory device has 12 address connections, an 8K device has 13, and so forth. A device that contains 1M locations requires a 20-bit address (A_0–A_{19}).

The number 400H represents a 1K-byte section of the memory system. If a memory device is decoded to begin at memory address 10000H and it is a 1K device, its last location is at address 103FFH—one location less than 400H. Another important hexadecimal number to remember is 1000H, because 1000H is 4K. A memory device that contains a starting address of 14000H that is 4K bytes long ends at location 14FFFH—one location less than 1000H. A third number to remember is 64K, or 10000H. A memory that starts at location 30000H and ends at location 3FFFFH is a 64K-byte memory. Finally, because 1M of memory is common, a 1M memory contains 100000H memory locations.

Data Connections. All memory devices have a set of data outputs or input/outputs. The device illustrated in Figure 10–1 has a common set of input/output (I/O) connections. Today, many memory devices have bidirectional common I/O pins.

The data connections are the points at which data are entered for storage or extracted for reading. Data pins on memory devices are labeled DO through D_7 for an eight-bit-wide memory

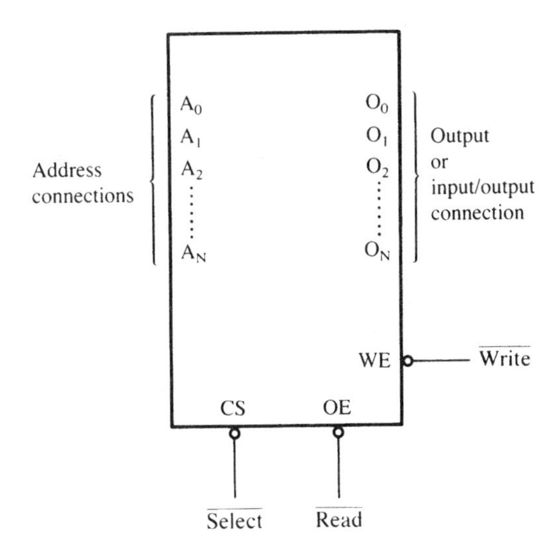

FIGURE 10–1 A pseudo-memory component illustrating the address, data, and control connections.

device. In this sample memory device there are eight I/O connections, meaning that the memory device stores eight bits of data in each of its memory locations. An eight-bit-wide memory device is often called a **byte-wide** memory. Although most devices are currently eight bits wide, some devices are 16 bits, four bits, or just one bit wide.

Catalog listings of memory devices often refer to memory locations times bits per location. For example, a memory device with 1K memory locations and eight bits in each location is often listed as a $1K \times 8$ by the manufacturer. A $16K \times 1$ is a memory device containing 16K one-bit memory locations. Memory devices are often classified according to total bit capacity. For example, a $1K \times 8$-bit memory device is sometimes listed as an 8K memory device, or a $64K \times 4$ memory is listed as a 256K device. These variations occur from one manufacturer to another.

Selection Connections. Each memory device has an input—sometimes more than one—that selects or enables the memory device. This type of input is most often called a **chip select** ($\overline{CS}$), **chip enable** ($\overline{CE}$), or simply **select** ($\overline{S}$) input. RAM memory generally has at least one $\overline{CS}$ or $\overline{S}$ input, and ROM has at least one $\overline{CE}$. If the $\overline{CE}$, $\overline{CS}$, or $\overline{S}$ input is active (a logic 0, in this case, because of the overbar), the memory device performs a read or write operation; if it is inactive (a logic 1, in this case), the memory device cannot do a read or a write because it is turned off or disabled. If more than one $\overline{CS}$ connection is present, all must be activated to read or write data.

Control Connections. All memory devices have some form of control input or inputs. A ROM usually has only one control input, while a RAM often has one or two control inputs.

The control input most often found on a ROM is the **output enable** ($\overline{OE}$) or **gate** ($\overline{G}$) connection, which allows data to flow out of the output data pins of the ROM. If $\overline{OE}$ and the selection input ($\overline{CE}$) are both active, the output is enabled; if $\overline{OE}$ is inactive, the output is disabled at its high-impedance state. The $\overline{OE}$ connection enables and disables a set of three-state buffers located within the memory device and must be active to read data.

A RAM memory device has either one or two control inputs. If there is only one control input, it is often called $R/\overline{W}$. This pin selects a read operation or a write operation only if the device is selected by the selection input ($\overline{CS}$). If the RAM has two control inputs, they are usually labeled $\overline{WE}$ (or $\overline{W}$), and $\overline{OE}$ (or $\overline{G}$). Here, $\overline{WE}$ (**write enable**) must be active to perform a memory write, and $\overline{OE}$ must be active to perform a memory read operation. When these two controls ($\overline{WE}$ and $\overline{OE}$) are present, they must never both be active at the same time. If both control inputs are inactive (logic 1s), data are neither written nor read, and the data connections are at their high-impedance state.

ROM Memory

The **read-only memory** (ROM) permanently stores programs and data that are resident to the system and must not change when the power supply is disconnected. The ROM is permanently programmed so that data are always present, even when power is disconnected. This type of memory is often called **nonvolatile memory**, because its contents *do not* change even if power is disconnected.

Today, the ROM is available in many forms. A device we call a ROM is purchased in mass quantities from a manufacturer and programmed during its fabrication at the factory. The EPROM (**erasable programmable read-only memory**), a type of ROM, is more commonly used when software must be changed often or when too few are in demand to make the ROM economical. For a ROM to be practical, we usually must purchase at least 10,000 devices to recoup the factory programming charge. An EPROM is programmed in the field on a device called an *EPROM programmer.* The EPROM is also erasable if exposed to high-intensity ultraviolet light for about 20 minutes or so, depending on the type of EPROM.

PROM memory devices are also available, although they are not as common today. The PROM (**programmable read-only memory**) is also programmed in the field by burning open tiny NI-chrome or silicon oxide fuses; but once it is programmed, it cannot be erased.

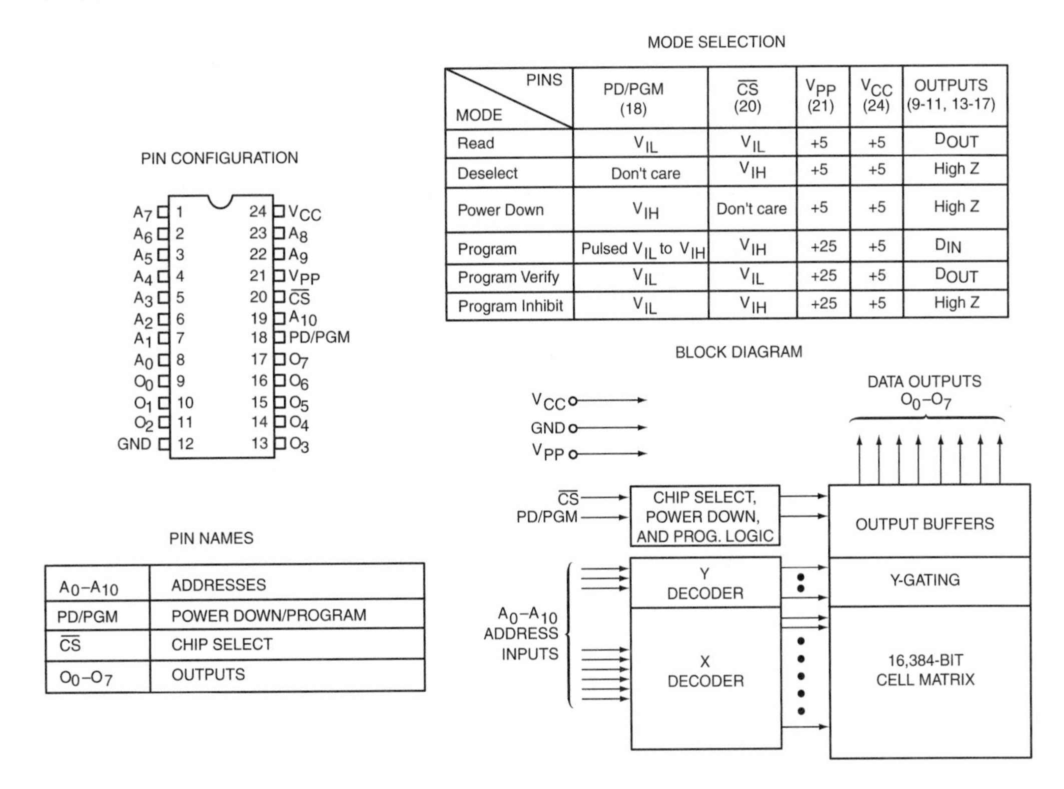

MODE SELECTION

MODE \ PINS	PD/PGM (18)	$\overline{CS}$ (20)	V_{PP} (21)	V_{CC} (24)	OUTPUTS (9-11, 13-17)
Read	V_{IL}	V_{IL}	+5	+5	D_{OUT}
Deselect	Don't care	V_{IH}	+5	+5	High Z
Power Down	V_{IH}	Don't care	+5	+5	High Z
Program	Pulsed V_{IL} to V_{IH}	V_{IH}	+25	+5	D_{IN}
Program Verify	V_{IL}	V_{IL}	+25	+5	D_{OUT}
Program Inhibit	V_{IL}	V_{IH}	+25	+5	High Z

PIN NAMES

A_0–A_{10}	ADDRESSES
PD/PGM	POWER DOWN/PROGRAM
$\overline{CS}$	CHIP SELECT
O_0–O_7	OUTPUTS

FIGURE 10–2 The pin-out of the 2716, 2K × 8 EPROM. (Courtesy of Intel Corporation.)

Still another, newer type of **read-mostly memory** (RMM) is called the **Flash memory**. The Flash memory[1] is also often called an EEPROM (**electrically erasable programmable ROM**), EAROM (**electrically alterable ROM**), or a NOVRAM (**nonvolatile RAM**). These memory devices are electrically erasable in the system, but they require more time to erase than a normal RAM. The Flash memory device is used to store setup information for systems such as the video card in the computer. It has all but replaced the EPROM in most computer systems for the BIOS memory. Some systems contain a password stored in the Flash memory device. Flash memory has its biggest impact in memory cards for digital cameras and memory in MP3 audio players.

Figure 10–2 illustrates the 2716 EPROM, which is representative of most common EPROMs. This device contains 11 address inputs and eight data outputs. The 2716 is a 2K × 8 read-only memory device. The 27XXX series of the EPROMs includes the following part numbers: 2704 (512 × 8), 2708 (1K × 8), 2716 (2K × 8), 2732 (4K × 8), 2764 (8K × 8), 27128 (16K × 8), 27256 (32K × 8), 27512 (64K × 8), and 271024 (128K × 8). Each of these parts contains address pins, eight data connections, one or more chip selection inputs ($\overline{CE}$), and an output enable pin ($\overline{OE}$).

Figure 10–3 illustrates the timing diagram for the 2716 EPROM. Data appear on the output connections only after a logic 0 is placed on both $\overline{CE}$ and $\overline{OE}$ pin connections. If $\overline{CE}$ and $\overline{OE}$ are not both logic 0s, the data output connections remain at their high-impedance or off states. Note that the V_{PP} pin must be placed at a logic 1 level for data to be read from the EPROM. In some cases, the V_{PP} pin is in the same position as the $\overline{WE}$ pin on the SRAM. This will allow a single socket to hold either an EPROM or an SRAM. An example is the 27256 EPROM and the 62256 SRAM, both 32K × 8 devices that have the same pin-out, except for V_{PP} on the EPROM and $\overline{WE}$ on the SRAM.

[1]Flash memory is a trademark of Intel Corporation.

A.C. Characteristics

$T_A = 0°C$ to $70°C$, V_{CC}[1] $= +5V \pm 5\%$, V_{PP}[2] $= V_{CC} \pm 0.6V$[3]

Symbol	Parameter	Limits Min.	Typ.[4]	Max.	Unit	Test Conditions
t_{ACC1}	Address to Output Delay		250	450	ns	PD/PGM = $\overline{CS}$ = V_{IL}
t_{ACC2}	PD/PGM to Output Delay		280	450	ns	$\overline{CS}$ = V_{IL}
t_{CO}	Chip Select to Output Delay			120	ns	PD/PGM = V_{IL}
t_{PF}	PD/PGM to Output Float	0		100	ns	$\overline{CS}$ = V_{IL}
t_{DF}	Chip Deselect to Output Float	0		100	ns	PD/PGM = V_{IL}
t_{OH}	Address to Output Hold	0			ns	PD/PGM = $\overline{CS}$ = V_{IL}

Capacitance[5] $T_A = 25°C$, f = 1 MHz

Symbol	Parameter	Typ.	Max.	Unit	Conditions
C_{IN}	Input Capacitance	4	6	pF	V_{IN} = 0V
C_{OUT}	Output Capacitance	8	12	pF	V_{OUT} = 0V

A.C. Test Conditions:

Output Load: 1 TTL gate and C_L = 100 pF
Input Rise and Fall Times: ≤20 ns
Input Pulse Levels: 0.8V to 2.2V
Timing Measurement Reference Level:
Inputs 1V and 2V
Outputs 0.8V and 2V

WAVEFORMS

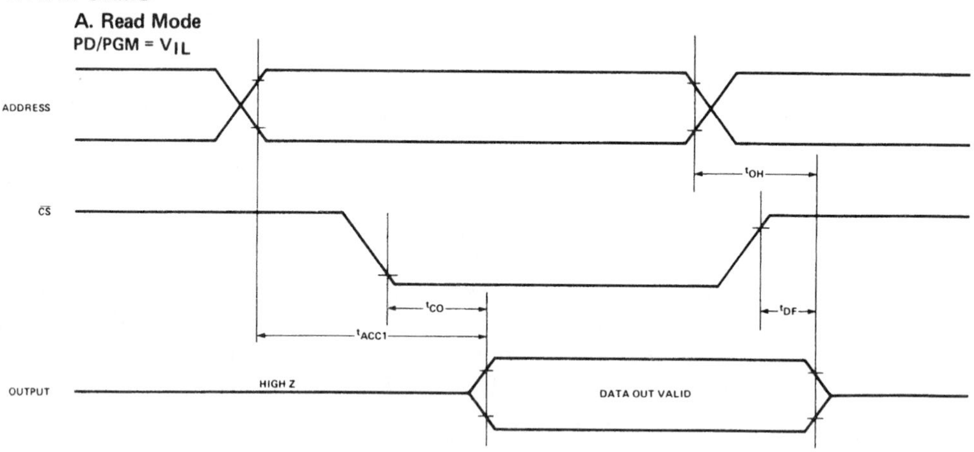

FIGURE 10–3 The timing diagram of AC characteristics of the 2716 EPROM. (Courtesy of Intel Corporation.)

One important piece of information provided by the timing diagram and data sheet is the memory access time—the time that it takes the memory to read information. As Figure 10–3 illustrates, memory access time (T_{ACC}) is measured from the appearance of the address at the address inputs until the appearance of the data at the output connections. This is based on the assumption that the $\overline{CE}$ input goes low at the same time that the address inputs become stable. Also, $\overline{OE}$ must be a logic 0 for the output connections to become active. The basic speed of this EPROM is 450 ns. (Recall that the 8086/8088 operated with a 5 MHz clock allowed the memory 460 ns to access data.) This type of memory component requires wait states to operate properly with the 8086/8088 microprocessors because of its rather long access time. If wait states are not desired, higher-speed versions of the EPROM are available at an additional cost. Today, EPROM memory is available with access times of as little as 100 ns. Obviously, wait states are required in modern microprocessors for any EPROM device.

Static RAM (SRAM) Devices

Static RAM memory devices retain data for as long as DC power is applied. Because no special action (except power) is required to retain stored data, these devices are called **static memory**. They are also called **volatile memory** because they will not retain data without power. The main

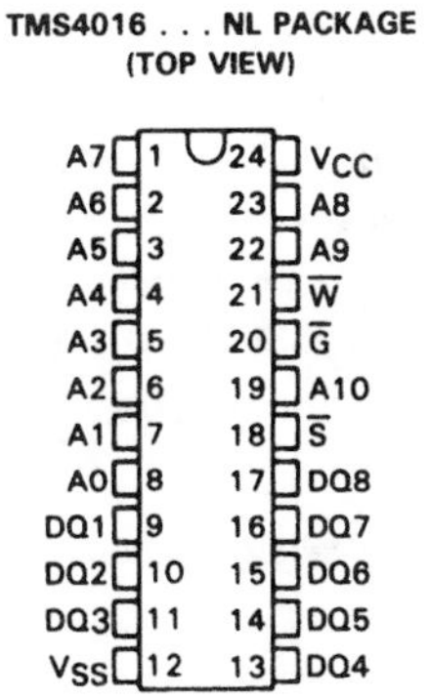

FIGURE 10–4 The pin-out of the TMS4016, 2K × 8 static RAM (SRAM). (Courtesy of Texas Instruments Incorporated.)

TMS4016 . . . NL PACKAGE
(TOP VIEW)

Pin	Name	Pin	Name
1	A7	24	V_{CC}
2	A6	23	A8
3	A5	22	A9
4	A4	21	$\overline{W}$
5	A3	20	$\overline{G}$
6	A2	19	A10
7	A1	18	$\overline{S}$
8	A0	17	DQ8
9	DQ1	16	DQ7
10	DQ2	15	DQ6
11	DQ3	14	DQ5
12	V_{SS}	13	DQ4

PIN NOMENCLATURE	
A0 – A10	Addresses
DQ1 – DQ8	Data In/Data Out
$\overline{G}$	Output Enable
$\overline{S}$	Chip Select
V_{CC}	+5-V Supply
V_{SS}	Ground
$\overline{W}$	Write Enable

difference between a ROM and a RAM is that a RAM is written under normal operation, whereas a ROM is programmed outside the computer and normally is only read. The SRAM, which stores temporary data, is used when the size of the read/write memory is relatively small. Today, a small memory is one that is less than 1M byte.

Figure 10–4 illustrates the 4016 SRAM, which is a 2K × 8 read/write memory. This device has 11 address inputs and eight data input/output connections. This device is representative of all SRAM devices, except for the number of address and data connections.

The control inputs of this RAM are slightly different from those presented earlier. The $\overline{OE}$ pin is labeled $\overline{G}$, the $\overline{CS}$ pin is $\overline{S}$, and the $\overline{WE}$ pin is $\overline{W}$. Despite the altered designations, the control pins function exactly the same as those outlined previously. Other manufacturers make this popular SRAM under the part numbers 2016 and 6116.

Figure 10–5 depicts the timing diagram for the 4016 SRAM. As the read cycle timing reveals, the access time is $t_{a(A)}$. On the slowest version of the 4016, this time is 250 ns, which is fast enough to connect directly to an 8088 or an 8086 operated at 5 MHz without wait states. Again, it is important to remember that the access time must be checked to determine the compatibility of memory components with the microprocessor.

Figure 10–6 illustrates the pin-out of the 62256, 32K × 8 static RAM. This device is packaged in a 28-pin integrated circuit and is available with access times of 120 ns or 150 ns. Other common SRAM devices are available in 8K × 8, 128K × 8, 256K × 8, 512K × 8, and 1M × 8 sizes, with access times of as little as 1.0 ns for SRAM used in computer cache memory systems.

Dynamic RAM (DRAM) Memory

About the largest static RAM available today is a 1M × 8. Dynamic RAM, on the other hand, is available in much larger sizes: up to 512M × 1. In all other respects, DRAM is essentially the same as SRAM, except that it retains data for only 2 or 4 ms on an integrated capacitor. After 2 or 4 ms, the contents of the DRAM must be completely rewritten (*refreshed*) because the capacitors, which store a logic 1 or logic 0, lose their charges.

Instead of requiring the almost impossible task of reading the contents of each memory location with a program and then rewriting them, the manufacturer has internally constructed the DRAM differently from the SRAM. In the DRAM, the entire contents of the memory are refreshed with 256 reads in a 2- or 4-ms interval. Refreshing also occurs during a write, a read, or during a special refresh cycle. Much more information about DRAM refreshing is provided in Section 10–6.

Another disadvantage of DRAM memory is that it requires so many address pins that the manufacturers have decided to multiplex the address inputs. Figure 10–7 illustrates a 64K × 4 DRAM, the TMS4464, which stores 256K bits of data. Notice that it contains only eight address inputs where it should contain 16—the number required to address 64K memory locations. The

electrical characteristics over recommended operating free-air temperature range (unless otherwise noted)

PARAMETER		TEST CONDITIONS		MIN	TYP†	MAX	UNIT
V_{OH}	High level voltage	$I_{OH} = -1$ mA,	$V_{CC} = 4.5$ V	2.4			V
V_{OL}	Low level voltage	$I_{OL} = 2.1$ mA,	$V_{CC} = 4.5$ V			0.4	V
I_I	Input current	$V_I = 0$ V to 5.5 V				10	μA
I_{OZ}	Off-state output current	$\overline{S}$ or $\overline{G}$ at 2 V or $\overline{W}$ at 0.8 V, $V_O = 0$ V to 5.5 V				10	μA
I_{CC}	Supply current from V_{CC}	$I_O = 0$ mA, $T_A = 0°C$ (worst case)	$V_{CC} = 5.5$ V,		40	70	mA
C_i	Input capacitance	$V_I = 0$ V,	f = 1 MHz			8	pF
C_O	Output capacitance	$V_O = 0$ V,	f = 1 MHz			12	pF

†All typical values are at $V_{CC} = 5$ V, $T_A = 25°C$.

timing requirements over recommended supply voltage range and operating free-air temperature range

PARAMETER		TMS4016-12		TMS4016-15		TMS4016-20		TMS4016-25		UNIT
		MIN	MAX	MIN	MAX	MIN	MAX	MIN	MAX	
$t_{c(rd)}$	Read cycle time	120		150		200		250		ns
$t_{c(wr)}$	Write cycle time	120		150		200		250		ns
$t_{w(W)}$	Write pulse width	60		80		100		120		ns
$t_{su(A)}$	Address setup time	20		20		20		20		ns
$t_{su(S)}$	Chip select setup time	60		80		100		120		ns
$t_{su(D)}$	Data setup time	50		60		80		100		ns
$t_{h(A)}$	Address hold time	0		0		0		0		ns
$t_{h(D)}$	Data hold time	5		10		10		10		ns

switching characteristics over recommended voltage range, $T_A = 0°C$ to $70°C$

PARAMETER		TMS4016-12		TMS4016-15		TMS4016-20		TMS4016-25		UNIT
		MIN	MAX	MIN	MAX	MIN	MAX	MIN	MAX	
$t_{a(A)}$	Access time from address		120		150		200		250	ns
$t_{a(S)}$	Access time from chip select low		60		75		100		120	ns
$t_{a(G)}$	Access time from output enable low		50		60		80		100	ns
$t_{v(A)}$	Output data valid after address change	10		15		15		15		ns
$t_{dis(S)}$	Output disable time after chip select high		40		50		60		80	ns
$t_{dis(G)}$	Output disable time after output enable high		40		50		60		80	ns
$t_{dis(W)}$	Output disable time after write enable low		50		60		60		80	ns
$t_{en(S)}$	Output enable time after chip select low	5		5		10		10		ns
$t_{en(G)}$	Output enable time after output enable low	5		5		10		10		ns
$t_{en(W)}$	Output enable time after write enable high	5		5		10		10		ns

NOTES: 3. $C_L = 100$pF for all measurements except $t_{dis(W)}$ and $t_{en(W)}$.
$C_L = 5$ pF for $t_{dis(W)}$ and $t_{en(W)}$.
4. t_{dis} and t_{en} parameters are sampled and not 100% tested.

FIGURE 10–5 (a) The AC characteristics of the TMS4016 SRAM. (b) The timing diagrams of the TMS4016 SRAM. (Courtesy of Texas Instruments Incorporated.)

timing waveform of read cycle (see note 5)

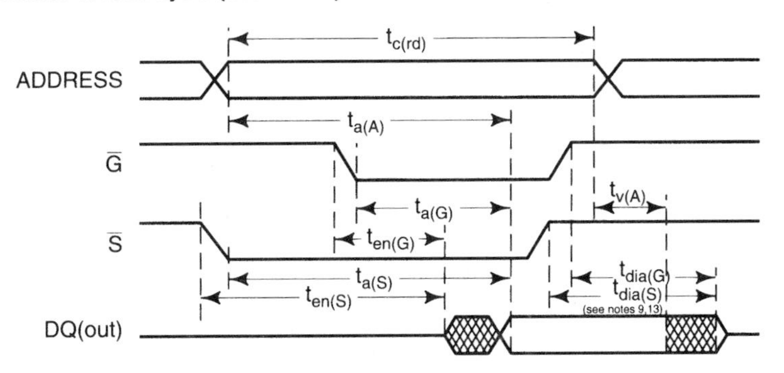

timing waveform of write cycle no. 1 (see note 6)

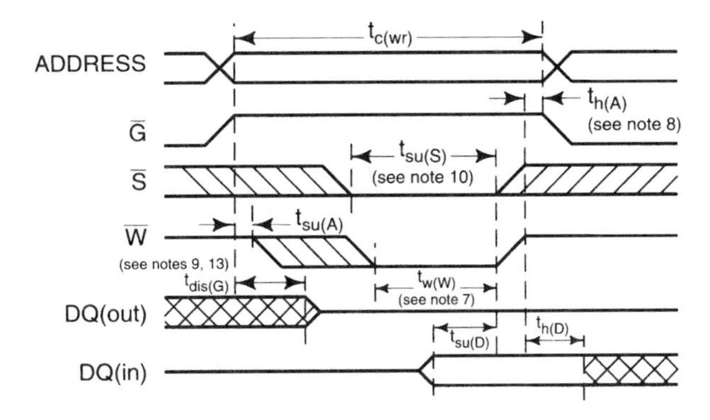

timing waveform of write cycle no. 2 (see notes 6 and 11)

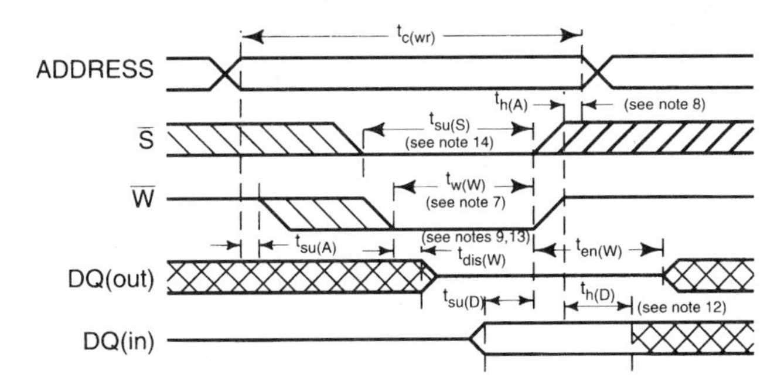

NOTES:
5. $\overline{W}$ is high Read Cycle.
6. $\overline{W}$ must be high during all address transitions.
7. A write occurs during the overlap of a low $\overline{S}$ and a low $\overline{W}$.
8. $t_{h(A)}$ is measured from the earlier of $\overline{S}$ or $\overline{W}$ going high to the end of the write cycle.
9. During this period, I/O pins are in the output state so that the input signals of opposite phase to the outputs must not be applied.
10. If the Slow transition occurs simultaneously with the $\overline{W}$ low transitions or after the $\overline{W}$ transition, output remains in a high impedance state.
11. G is continuously low ($G = V_{IL}$).
12. If $\overline{S}$ is low during this period, I/O pins are in the output state. Data input signals of opposite phase to the outputs must not be applied.
13. Transition is measured ± 200 mV from steady-state voltage.
14. If the $\overline{S}$ low transition occurs before the W low transition, then the data input signals of opposite phase to the outputs must not be applied for the duration of $t_{dis(W)}$ after the $\overline{W}$ low transition.

FIGURE 10–5 *(continued)*

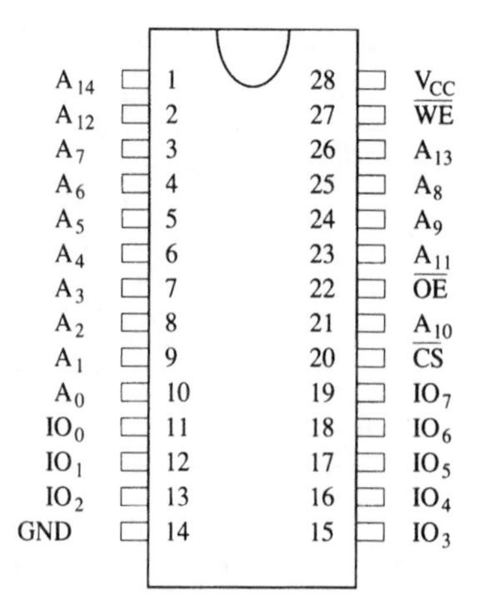

FIGURE 10–6 Pin diagram of the 62256, 32K × 8 static RAM.

PIN FUNCTION

A_0 - A_{14}	Addresses
IO_0 - IO_7	Data connections
$\overline{CS}$	Chip select
$\overline{OE}$	Output enable
$\overline{WE}$	Write enable
V_{CC}	+5V Supply
GND	Ground

TMS4464 . . . JL OR NL PACKAGE
(TOP VIEW)

Pin	Name	Pin	Name
1	$\overline{G}$	18	V_{SS}
2	DQ1	17	DQ4
3	DQ2	16	$\overline{CAS}$
4	$\overline{W}$	15	DQ3
5	$\overline{RAS}$	14	A0
6	A6	13	A1
7	A5	12	A2
8	A4	11	A3
9	V_{DD}	10	A7

(a)

PIN NOMENCLATURE	
A0-A7	Address Inputs
$\overline{CAS}$	Column Address Strobe
DQ1-DQ4	Data-In/Data-Out
$\overline{G}$	Output Enable
$\overline{RAS}$	Row Address Strobe
V_{DD}	+ 5-V Supply
V_{SS}	Ground
$\overline{W}$	Write Enable

(b)

FIGURE 10–7 The pin-out of the TMS4464, 64K × 4 dynamic RAM (DRAM). (Courtesy of Texas Instruments Incorporated.)

only way that 16 address bits can be forced into eight address pins is in two 8-bit increments. This operation requires two special pins: the **column address strobe** ($\overline{CAS}$) and **row address strobe** ($\overline{RAS}$). First, A_0–A_7 are placed on the address pins and strobed into an internal row latch by $\overline{RAS}$ as the row address. Next, the address bits A_8–A_{15} are placed on the same eight address inputs and strobed into an internal column latch by $\overline{CAS}$ as the column address (see Figure 10–8 for this timing). The 16-bit address held in these internal latches addresses the contents of one of the four-bit memory locations. Note that $\overline{CAS}$ also performs the function of the chip selection input to the DRAM.

Figure 10–9 illustrates a set of multiplexers used to strobe the column and row addresses into the eight address pins on a pair of TMS4464 DRAMs. Here, the $\overline{RAS}$ signal not only strobes

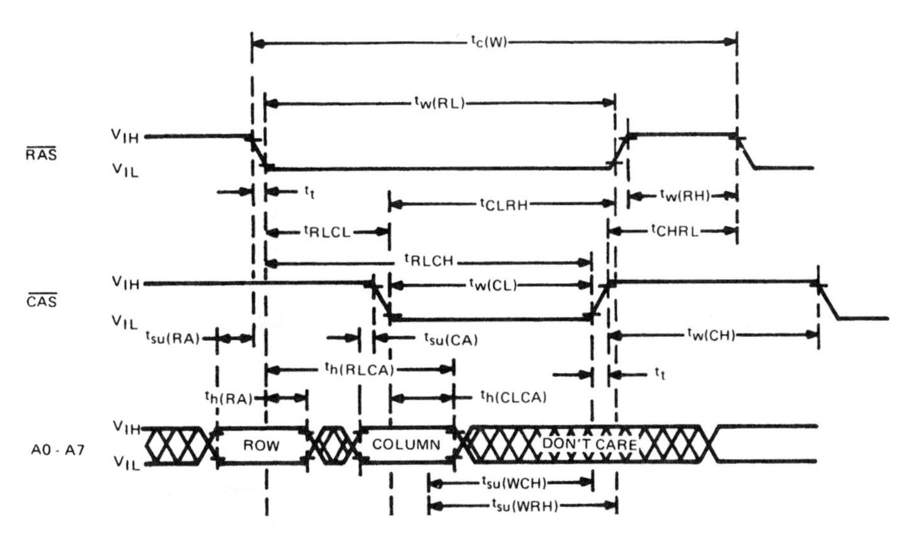

FIGURE 10–8 $\overline{RAS}$, $\overline{CAS}$, and address input timing for the TMS4464 DRAM. (Courtesy of Texas Instruments Incorporated.)

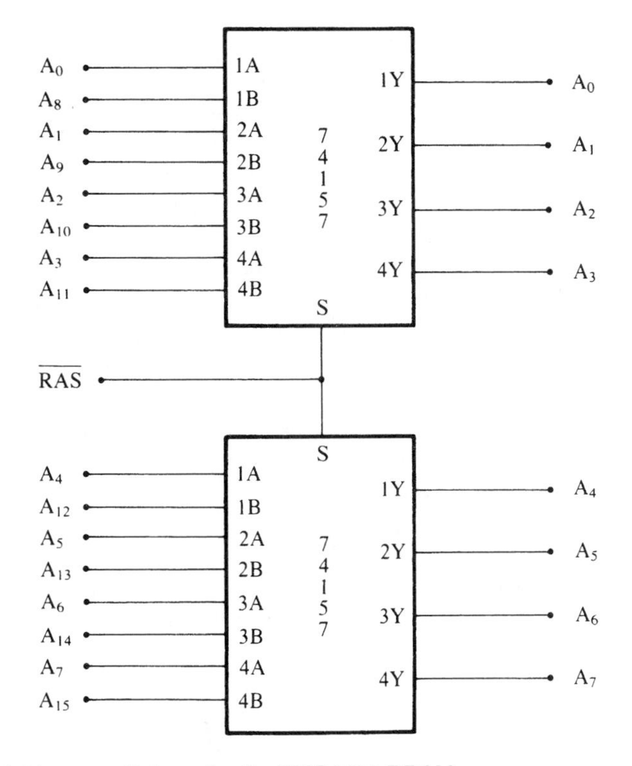

FIGURE 10–9 Address multiplexer for the TMS4464 DRAM.

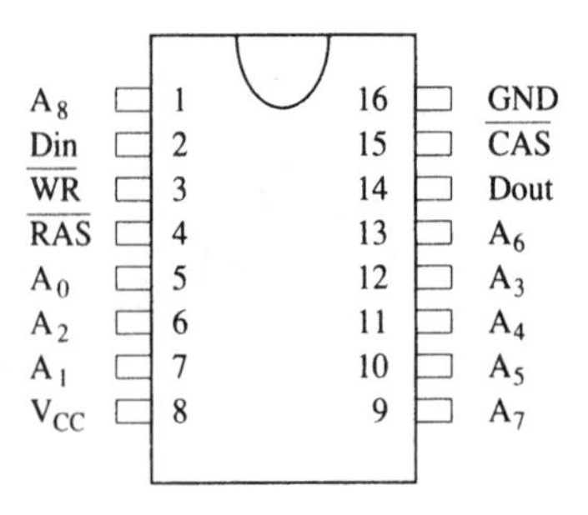

FIGURE 10–10 The 41256 dynamic RAM organized as a 256K × 1 memory device.

PIN FUNCTIONS

A_0 - A_8	Addresses
Din	Data in
Dout	Data out
$\overline{CAS}$	Column Address Strobe
$\overline{RAS}$	Row Address Strobe
$\overline{WR}$	Write enable
V_{CC}	+5V Supply
GND	Ground

the row address into the DRAMs, but it also selects which part of the address is applied to the address inputs. This is possible due to the long propagation-delay time of the multiplexers. When $\overline{RAS}$ is a logic 1, the B inputs are connected to the Y outputs of the multiplexers; when the $\overline{RAS}$ input goes to a logic 0, the A inputs connect to the Y outputs. Because the internal row address latch is edge-triggered, it captures the row address before the address at the inputs changes to the column address. More detail on DRAM and DRAM interfacing is provided in Section 10–6.

As with the SRAM, the R/$\overline{W}$ pin writes data to the DRAM when a logic 0, but there is no pin labeled $\overline{G}$ or enable. There also is no $\overline{S}$ (select) input to the DRAM. As mentioned, the $\overline{CAS}$ input selects the DRAM. If selected, the DRAM is written if R/$\overline{W}$ = 0 and read if R/$\overline{W}$ = 1.

Figure 10–10 shows the pin-out of the 41256 dynamic RAM. This device is organized as a 256K × 1 memory, requiring as little as 70 ns to access data.

More recently, larger DRAMs have become available that are organized as a 16M × 1, 256M × 1, and 512M × 1 memory. On the horizon is the 1G × 1 memory, which is in the planning stages. DRAM memory is often placed on small circuit boards called *SIMMs* (*Single In-Line Memory Modules*). Figure 10–11 shows the pin-outs of the two most common SIMMs. The 30-pin SIMM is organized most often as 1M × 8 or 1M × 9, and 4M × 8 or 4M × 9. Figure 10–11 illustrates a 4M × 9. The ninth bit is the parity bit. Also shown is a newer 72-pin SIMM. The 72-pin SIMMs are often organized as 1M × 32 or 1M × 36 (with parity). Other sizes are 2M × 32, 4M × 32, 8M × 32, and 16M × 32. These are also available with parity. Figure 10–11 illustrates a 4M × 36 SIMM, which has 16M bytes of memory.

Lately, many systems are using the Pentium–Pentium 4 microprocessors. These microprocessors have a 64-bit-wide data bus, which precludes the use of the 8-bit-wide SIMMs described here. Even the 72-pin SIMMs are cumbersome to use because they must be used in pairs to obtain a 64-bit-wide data connection. Today, the 64-bit-wide DIMMs (Dual In-Line Memory Modules) has become the standard in most systems. The memory on these modules is organized as 64 bits wide. The common sizes available are 16M bytes (2M × 64), 32M bytes (4M × 64), 64M bytes (8M × 64), 128M bytes (16M × 64), 256M bytes (32M × 64), and 512M bytes (64M × 64). The pin-out of the DIMM is illustrated in Figure 10–12. The DIMM module is available in DRAM, EDO, SDRAM, and DDR (double-data rate) forms, with or without an EPROM. The

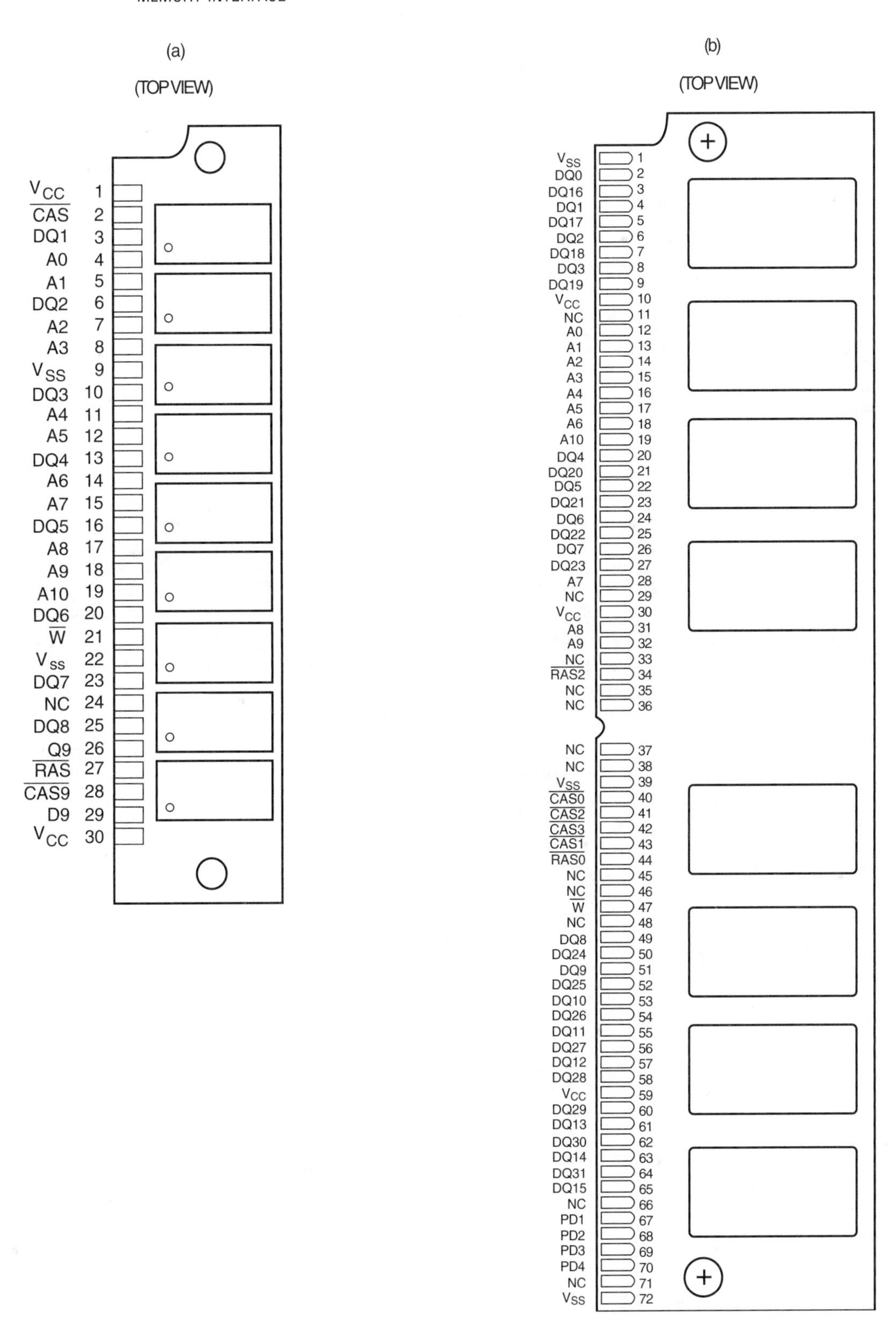

FIGURE 10–11 The pin-outs of the 30-pin and 72-pin SIMM. (a) A 30-pin SIMM organized as 4M × 9 and (b) a 72-pin SIMM organized as 4M × 36.

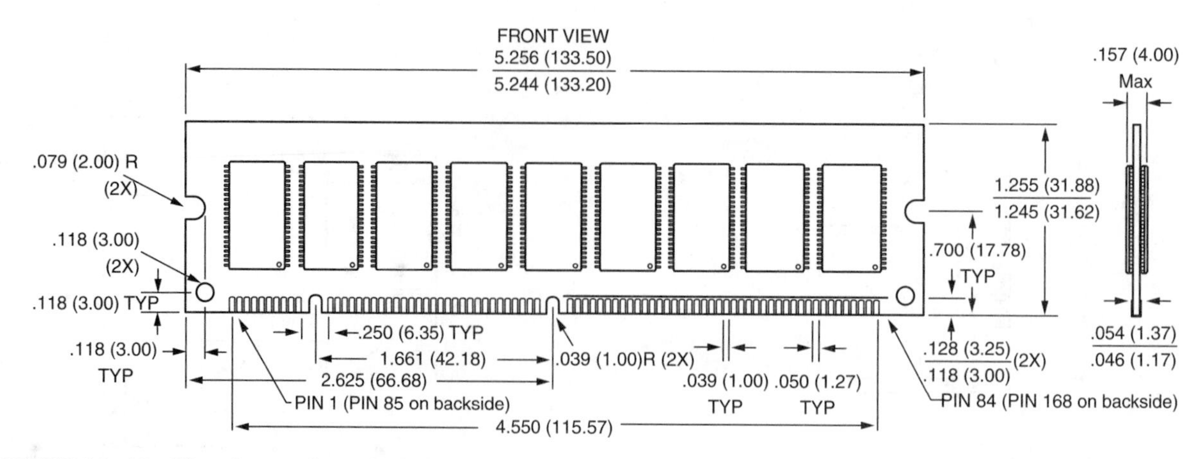

FIGURE 10–12 The pin-out of a 168-pin DIMM.

EPROM provides information to the system on the size and the speed of the memory device for plug-and-play applications.

A new addition to the memory market is the RIMM memory module from RAMBUS Corporation. Like the SDRAM, the RIMM contains 168 pins, but each pin is a two-level pin, bringing the total number of connections to 336. The fastest SDRAM currently available is the PC-4400 or 500 DDR, which operates at a rate of 4.4G bytes per second. By comparison, the 800 MHz RIMM operates at a rate of 3.2G bytes per second and is no longer supported in many systems. RDRAM had a fairly short life in the volatile computer market. The RIMM module is organized as a 32-bit-wide device. This means that to populate a Pentium 4 memory, RIMM memory is used in pairs. Intel claims that the Pentium 4 system using RIMM modules is 300% faster than a Pentium III using PC-100 memory. According to RAMBUS, the current 800 MHz RIMM has been increased to a speed of 1200 MHz, but still is not fast enough to garner much of a market share.

Currently the latest DRAM is the DDR (**double-data rate**) memory device. The DDR memory transfers data at each edge of the clock, making it operate at twice the speed of SDRAM. This does not affect the access time for the memory, so many wait states are still required to operate this type of memory, but it can be much faster than normal SDRAM memory and that includes RDRAM.

10–2 ADDRESS DECODING

In order to attach a memory device to the microprocessor, it is necessary to decode the address sent from the microprocessor. Decoding makes the memory function at a unique section or partition of the memory map. Without an address decoder, only one memory device can be connected to a microprocessor, which would make it virtually useless. In this section, we describe a few of the more common address-decoding techniques, as well as the decoders that are found in many systems.

Why Decode Memory?

When the 8088 microprocessor is compared to the 2716 EPROM, a difference in the number of address connections is apparent—the EPROM has 11 address connections and the microprocessor has 20. This means that the microprocessor sends out a 20-bit memory address whenever it reads or writes data. Because the EPROM has only 11 address pins, there is a mismatch that must be corrected. If only 11 of the 8088's address pins are connected to the memory, the 8088 will see only 2K bytes of memory instead of the 1M bytes that it "expects" the memory to contain. The decoder corrects the mismatch by decoding the address pins that do not connect to the memory component.

Simple NAND Gate Decoder

When the 2K × 8 EPROM is used, address connections A_{10}–A_0 of the 8088 are connected to address inputs A_{10}–A_0 of the EPROM. The remaining nine address pins (A_{19}–A_{11}) are connected to the inputs of a NAND gate decoder (see Figure 10–13). The decoder selects the EPROM from one of the 2K-byte sections of the 1M-byte memory system in the 8088 microprocessor.

In this circuit, a single NAND gate decodes the memory address. The output of the NAND gate is a logic 0 whenever the 8088 address pins attached to its inputs (A_{19}–A_{11}) are all logic 1s. The active low, logic 0 output of the NAND gate decoder is connected to the $\overline{CE}$ input pin that selects (enables) the EPROM. Recall that whenever $\overline{CE}$ is a logic 0, data will be read from the EPROM only if $\overline{OE}$ is also a logic 0. The $\overline{OE}$ pin is activated by the 8088 $\overline{RD}$ signal or the $\overline{MRDC}$ (**memory read control**) signal of other family members.

If the 20-bit binary address decoded by the NAND gate is written so that the leftmost nine bits are 1s and the rightmost 11 bits are don't cares (X), the actual address range of the EPROM can be determined. (A *don't care* is a logic 1 or a logic 0, whichever is appropriate.)

Example 10–1 illustrates how the address range for this EPROM is determined by writing down the externally decoded address bits (A_{19}–A_{11}) and the address bits decoded by the EPROM (A_{10}–A_0) as don't cares. We really do not care about the address pins on the EPROM because they are internally decoded. As the example illustrates, the don't cares are first written as 0s to locate the lowest address and then as 1s to find the highest address. Example 10–1 also shows these binary boundaries as hexadecimal addresses. Here, the 2K EPROM is decoded at memory address locations FF800H–FFFFFH. Notice that this is a 2K-byte section of the memory and is also located at the reset location for the 8086/8088 (FFFF0H), the most likely place for an EPROM in a system.

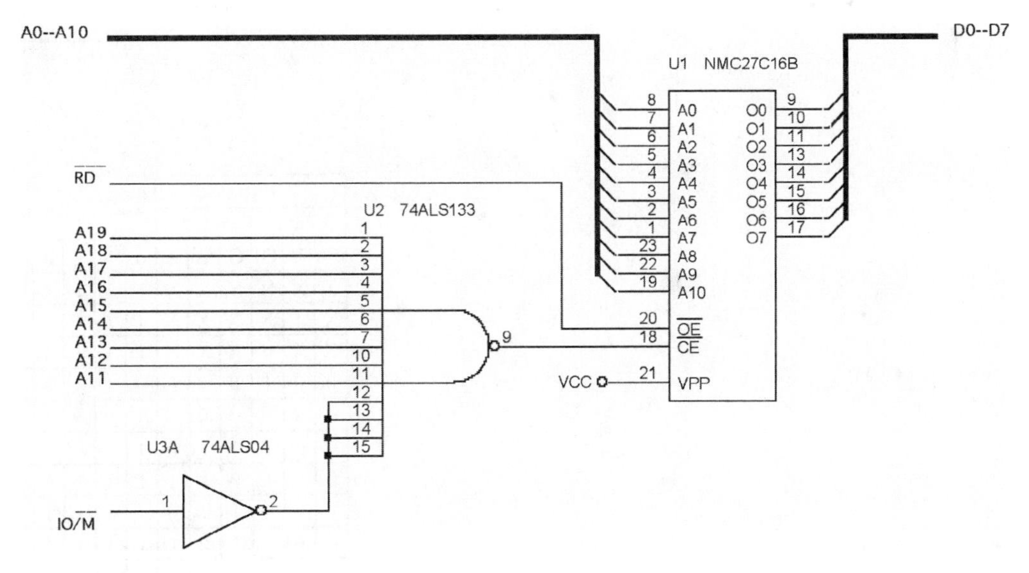

FIGURE 10–13 A simple NAND gate decoder that selects a 2716 EPROM for memory location FF800H–FFFFFH.

EXAMPLE 10–1

```
1111 1111 1XXX XXXX XXXX

          or

1111 1111 1000 0000 0000 = FF800H
          to
1111 1111 1111 1111 1111 = FFFFFH
```

Although this example serves to illustrate decoding, NAND gates are rarely used to decode memory because each memory device requires its own NAND gate decoder. Because of the excessive cost of the NAND gate decoder and the inverters that are often required, this option requires that an alternate be found.

The 3-to-8 Line Decoder (74LS138)

One of the more common integrated circuit decoders found in many microprocessor-based systems is the 74LS138 3-to-8 line decoder. Figure 10–14 illustrates this decoder and its truth table.

The truth table shows that only one of the eight outputs ever goes low at any time. For any of the decoder's outputs to go low, the three enable inputs ($\overline{G2A}$, $\overline{G2B}$, and G1) must all be active. To be active, the $\overline{G2A}$ and $\overline{G2B}$ inputs must both be low (logic 0), and G1 must be high (logic 1). Once the 74LS138 is enabled, the address inputs (C, B, and A) select which output pin goes low. Imagine eight EPROM $\overline{CE}$ inputs connected to the eight outputs of the decoder! This is

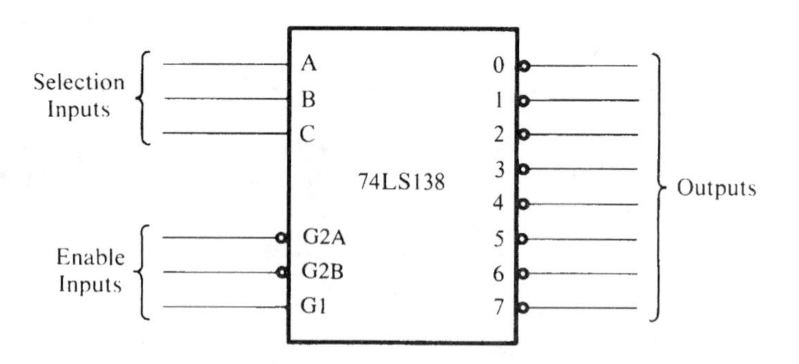

Inputs						Outputs							
Enable			Select										
$\overline{G2A}$	$\overline{G2B}$	G1	C	B	A	$\overline{0}$	$\overline{1}$	$\overline{2}$	$\overline{3}$	$\overline{4}$	$\overline{5}$	$\overline{6}$	$\overline{7}$
1	X	X	X	X	X	1	1	1	1	1	1	1	1
X	1	X	X	X	X	1	1	1	1	1	1	1	1
X	X	0	X	X	X	1	1	1	1	1	1	1	1
0	0	1	0	0	0	0	1	1	1	1	1	1	1
0	0	1	0	0	1	1	0	1	1	1	1	1	1
0	0	1	0	1	0	1	1	0	1	1	1	1	1
0	0	1	0	1	1	1	1	1	0	1	1	1	1
0	0	1	1	0	0	1	1	1	1	0	1	1	1
0	0	1	1	0	1	1	1	1	1	1	0	1	1
0	0	1	1	1	0	1	1	1	1	1	1	0	1
0	0	1	1	1	1	1	1	1	1	1	1	1	0

FIGURE 10–14 The 74LS138 3-to-8 line decoder and function table.

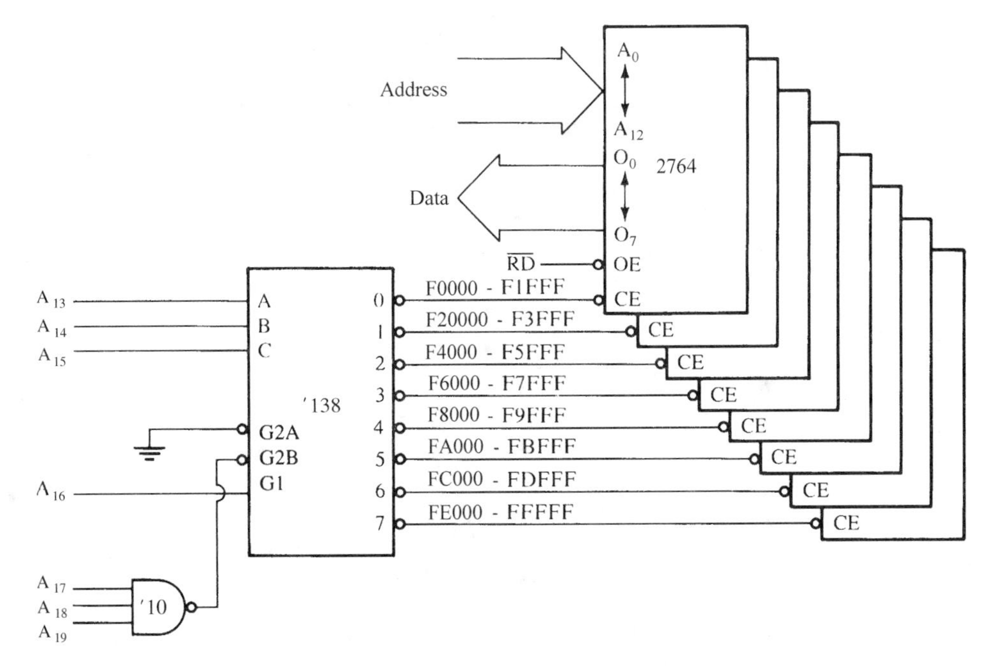

FIGURE 10–15 A circuit that uses eight 2764 EPROMs for a 64K × 8 section of memory in an 8088 microprocessor-based system. The addresses selected in this circuit are F0000H–FFFFFH.

a very powerful device because it selects eight different memory devices at the same time. Even today this device still finds wide application.

Sample Decoder Circuit. Notice that the outputs of the decoder, illustrated in Figure 10–15, are connected to eight different 2764 EPROM memory devices. Here, the decoder selects eight 8K-byte blocks of memory for a total memory capacity of 64K bytes. This figure also illustrates the address range of each memory device and the common connections to the memory devices. Notice that all of the address connections from the 8088 are connected to this circuit. Also, notice that the decoder's outputs are connected to the $\overline{CE}$ inputs of the EPROMs, and the $\overline{RD}$ signal from the 8088 is connected to the $\overline{OE}$ inputs of the EPROMs. This allows only the selected EPROM to be enabled and to send its data to the microprocessor through the data bus whenever $\overline{RD}$ becomes a logic 0.

In this circuit, a three-input NAND gate is connected to address bits A_{19}–A_{17}. When all three address inputs are high, the output of this NAND gate goes low and enables input $\overline{G2B}$ of the 74LS138. Input G1 is connected directly to A_{16}. In other words, in order to enable this decoder, the first four address connections (A_{19}–A_{16}) must all be high.

The address inputs C, B, and A connect to microprocessor address pins A_{15}–A_{13}. These three address inputs determine which output pin goes low and which EPROM is selected whenever the 8088 outputs a memory address within this range to the memory system.

Example 10–2 shows how the address range of the entire decoder is determined. Notice that the range is locations F0000H–FFFFFH. This is a 64K-byte span of the memory.

EXAMPLE 10–2

```
1111 XXXX XXXX XXXX XXXX

          or

1111 0000 0000 0000 0000 = FOOOOH
          to
1111 1111 1111 1111 1111 = FFFFFH
```

How is it possible to determine the address range of each memory device attached to the decoder's outputs? Again, the binary bit pattern is written down; this time the C, B, and A address inputs are not don't cares. Example 10–3 shows how output 0 of the decoder is made to go low to select the EPROM attached to that pin. Here, C, B, and A are shown as logic 0s.

EXAMPLE 10–3

```
     CBA
1111 000X XXXX XXXX XXXX

          or

1111 0000 0000 0000 0000 = F0000H
          to
1111 0001 1111 1111 1111 = F1FFFH
```

If the address range of the EPROM connected to output 1 of the decoder is required, it is determined in exactly the same way as that of output 0. The only difference is that now the C, B, and A inputs contain 001 instead of 000 (see Example 10–4). The remaining output address ranges are determined in the same manner by substituting the binary address of the output pin into C, B, and A.

EXAMPLE 10–4

```
     CBA
1111 001X XXXX XXXX XXXX

          or

1111 0010 0000 0000 0000 = F2000H
          to
1111 0011 1111 1111 1111 = F3FFFH
```

The Dual 2-to-4 Line Decoder (74LS139)

Another decoder that finds some application is the 74LS139 dual 2-to-4 line decoder. Figure 10–16 illustrates both the pin-out and the truth table for this decoder. The 74LS139 contains two separate 2-to-4 line decoders, each with its own address, enable, and output connections.

A more complicated decoder using the 74LS139 decoder appears in Figure 10–17. This circuit uses a 128K × 8 EPROM (271000) and a 128K × 8 SRAM (621000). The EPROM is decoded at memory locations E0000H–FFFFFH and the SRAM is decoded at addresses 00000H–1FFFFH. This is fairly typical of a small embedded system, where the EPROM is located at the top of the memory space and the SRAM at the bottom.

Output $\overline{Y0}$ of decoder U1A activates the SRAM whenever address bits A_{17} and A_{18} are both logic 0s if the IO/$\overline{M}$ signal is a logic 0 and address line A_{19} is a logic 0. This selects the SRAM for any address between 00000H and 1FFFFH. The second decoder (U1B) is slightly more complicated because the NAND gate (U4B) selects the decoder when IO/$\overline{M}$ is a logic 0 while A_{19} is a logic 1. This selects the EPROM for addresses E0000H through FFFFFH.

PLD Programmable Decoders

This section of the text explains the use of the programmable logic device, or PLD, as a decoder. There are three SPLD (**simple PLD**) devices that function in the same manner but have different names: PLA (**programmable logic array**), PAL (**programmable array logic**), and GAL (**gated array logic**). Although these devices have been in existence since the mid-1970s, they have only appeared in memory system and digital designs since the early 1990s. The PAL and the PLA are fuse-programmed, as is the PROM, and some PLD devices are erasable devices, as are EPROMs. In essence, all three devices are arrays of logic elements that are programmable.

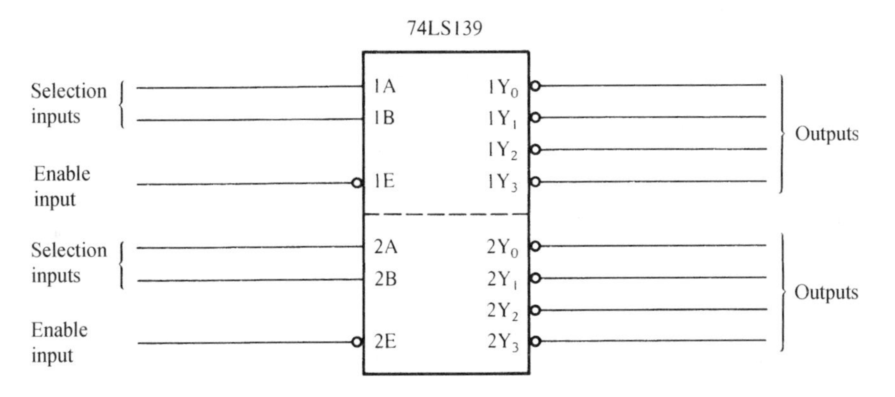

FIGURE 10–16 The pin-out and truth table of the 74LS139, dual 2-to-4 line decoder.

Inputs			Outputs			
$\overline{E}$	A	B	$\overline{Y_0}$	$\overline{Y_1}$	$\overline{Y_2}$	$\overline{Y_3}$
0	0	0	0	1	1	1
0	0	1	1	0	1	1
0	1	0	1	1	0	1
0	1	1	1	1	1	0
1	X	X	1	1	1	1

Other PLDs are also available, such as CPLDs (**complex programmable logic devices**), FPGAs (**field programmable gate arrays**), and FPICs (**field programmable interconnect**). These types of PLDs are much more complex than the SPLDs that are used more commonly in designing a complete system. If the concentration is on decoding addresses, the SPLD is used, and if the concentration is on a complete system, then the CPLD or FPIC is used to implement the design. These devices are also referred to as an ASIC (**application-specific integrated circuit**).

Combinatorial Programmable Logic Arrays. One of the two basic types of PALs is the combinatorial programmable logic array. This device is internally structured as a programmable array of combinational logic circuits. Figure 10–18 illustrates the internal structure of the PAL16L8 that is constructed with AND/OR gate logic. This device, which is representative of a PLD, has 10 fixed inputs, two fixed outputs, and six pins that are programmable as inputs or outputs. Each output signal is generated from a seven-input OR gate that has an AND gate attached to each input. The outputs of the OR gates pass through a three-state inverter that defines each output as an AND/NOR function. Initially, all of the fuses connect all of the vertical/horizontal connections illustrated in Figure 10–18. Programming is accomplished by blowing fuses to connect various inputs to the OR gate array. The wired-AND function is performed at each input connection, which allows a product term of up to 16 inputs. A logic expression using the PAL16L8 can have up to seven product terms with up to 16 inputs NORed together to generate the output expression. This device is ideal as a memory address decoder because of its structure. It is also ideal because the outputs are active low.

Fortunately, we don't have to choose the fuses by number for programming, as was customary when this device was first introduced. A PAL is programmed with a software package such as PALASM, the PAL assembler program. More recently, PLD design is accomplished using HDL (**hardware description language**) or VHDL (**verilog HDL**). The VHDL language and its syntax are currently the industry standard for programming PLD devices. Example 10–5 shows a program that decodes the same areas of memory as decoded in Figure 10–17. Note that this program was developed by using a text editor such as EDIT, available with Microsoft DOS version 7.1 with XP

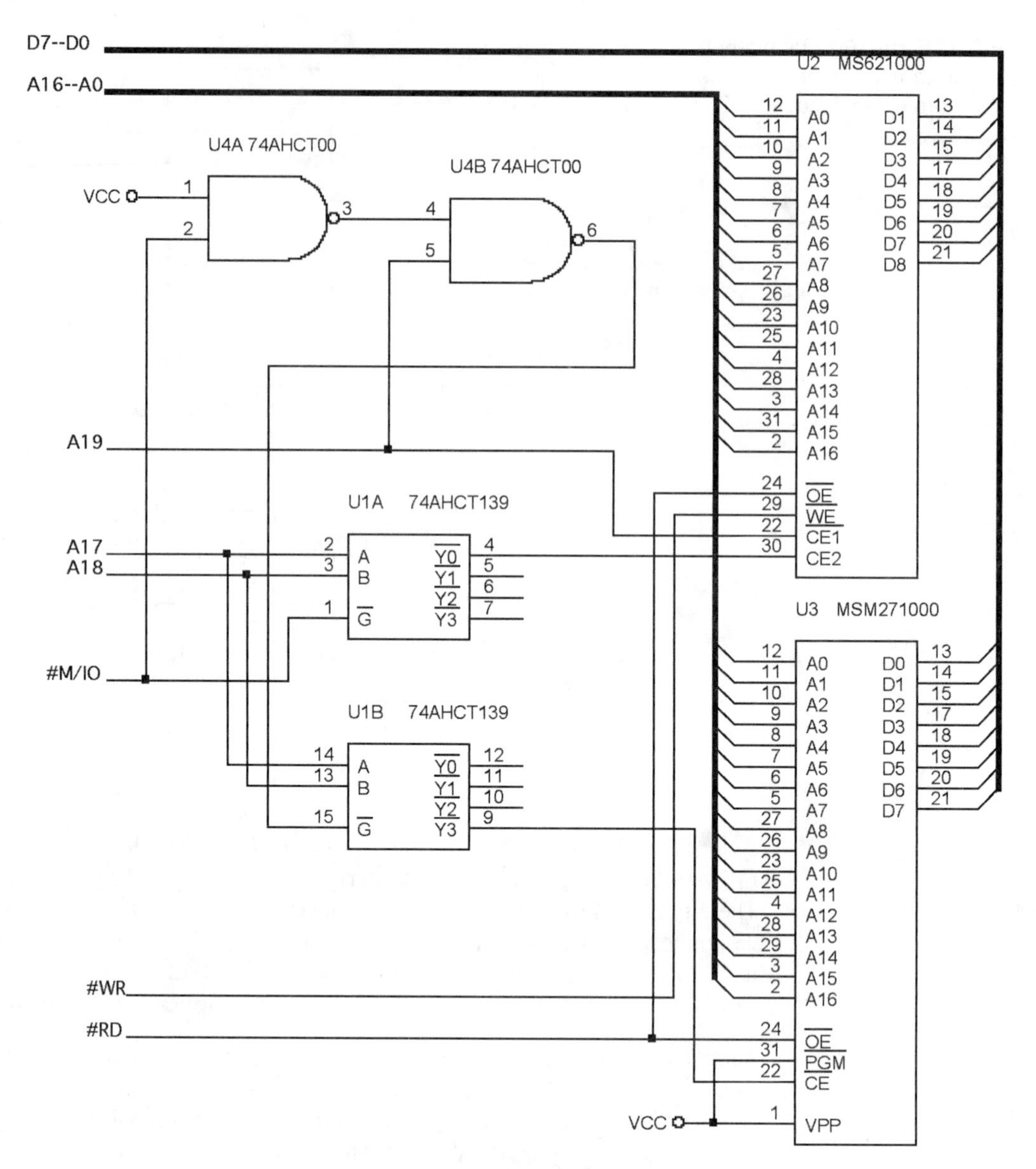

FIGURE 10–17 A sample memory system constructed with a 74HCT139.

or NotePad in Windows XP. The program can also be developed by using an editor that comes with any of the many programming packages for PLDs. Various editors attempt to ease the task of defining the pins, but we believe it is easier to use NotePad and the listing as shown.

EXAMPLE 10–5

```
-- VHDL code for the decoder of Figure 10-17

library ieee;
use ieee.std_logic_1164.all;

entity DECODER_10_17 is

port (
       A19, A18, A17, MIO: in STD_LOGIC;
```

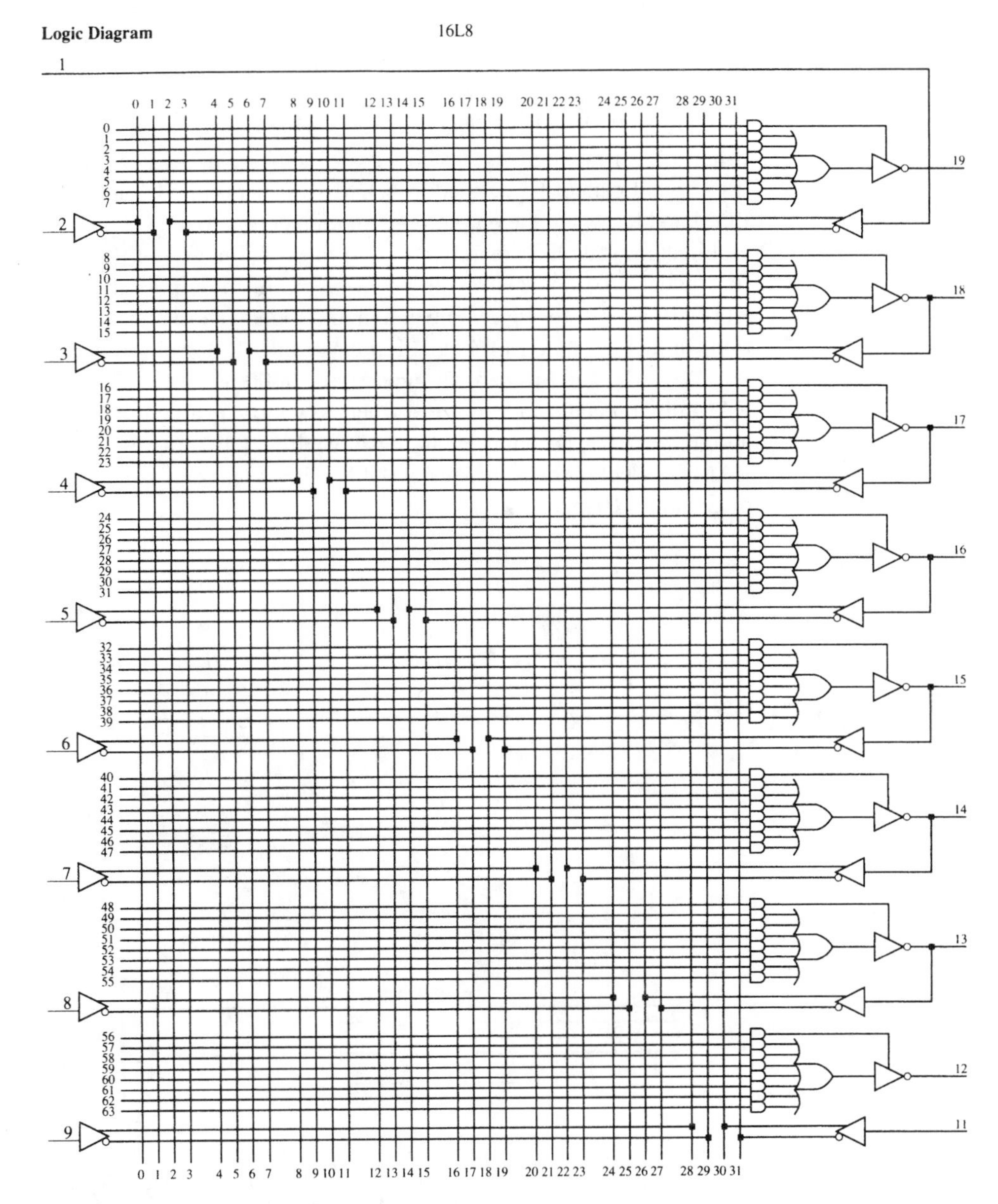

FIGURE 10–18 The PAL 16L8. (Copyright Advanced Micro Devices, Inc., 1988. Reprinted with permission of copyright owner. All rights reserved.)

```
        ROM, RAM, AX19: out STD_LOGIC
);

end;

architecture V1 of DECODER_10_17 is

begin

        ROM  <= A19 or A18 or A17 or MIO;
        RAM  <= A18 and A17 and (not MIO);
        AX19 <= not A19;

end V1;
```

Comments in VHDL programming begin with a pair of minus signs as illustrated in the first line of the VHDL code in Example 10–5. The library and use statements specify the standard IEEE library using standard logic. The entity statement names the VHDL module, in this case DECODER_10_17. The port statements define the in, out, and in-out pins used in the equations for the logic expression, which appears in the begin block. A_{19}, A_{18}, A_{17}, and MIO (this signal cannot be defined as $IO/\overline{M}$ so it was called MIO) are defined as input pins and ROM and RAM are the output pins for connection to the $\overline{CS}$ pins on the memory devices. The architecture statement merely refers to the version (V_1) of this design. Finally, the equations for the design are placed in the begin block. Each output pin has its own equation. The keyword not is used for logical inversion and the keyword and is used for the logical and operation. In this case the ROM equation causes the ROM pin to become a logic zero only when A_{19}, A_{18}, A_{17}, and MIO are all logic zeros (00000H–1FFFFH). The RAM equation causes the RAM pin to become a logic zero when A_{18} and A_{17} are all ones at the same time that MIO is a logic zero. A_{19} is connected to the active high CE2 pin after being inverted by the PLD. The RAM is selected for addresses 60000H–7FFFFH. See Figure 10–19 for the PLD realization of Example 10–5.

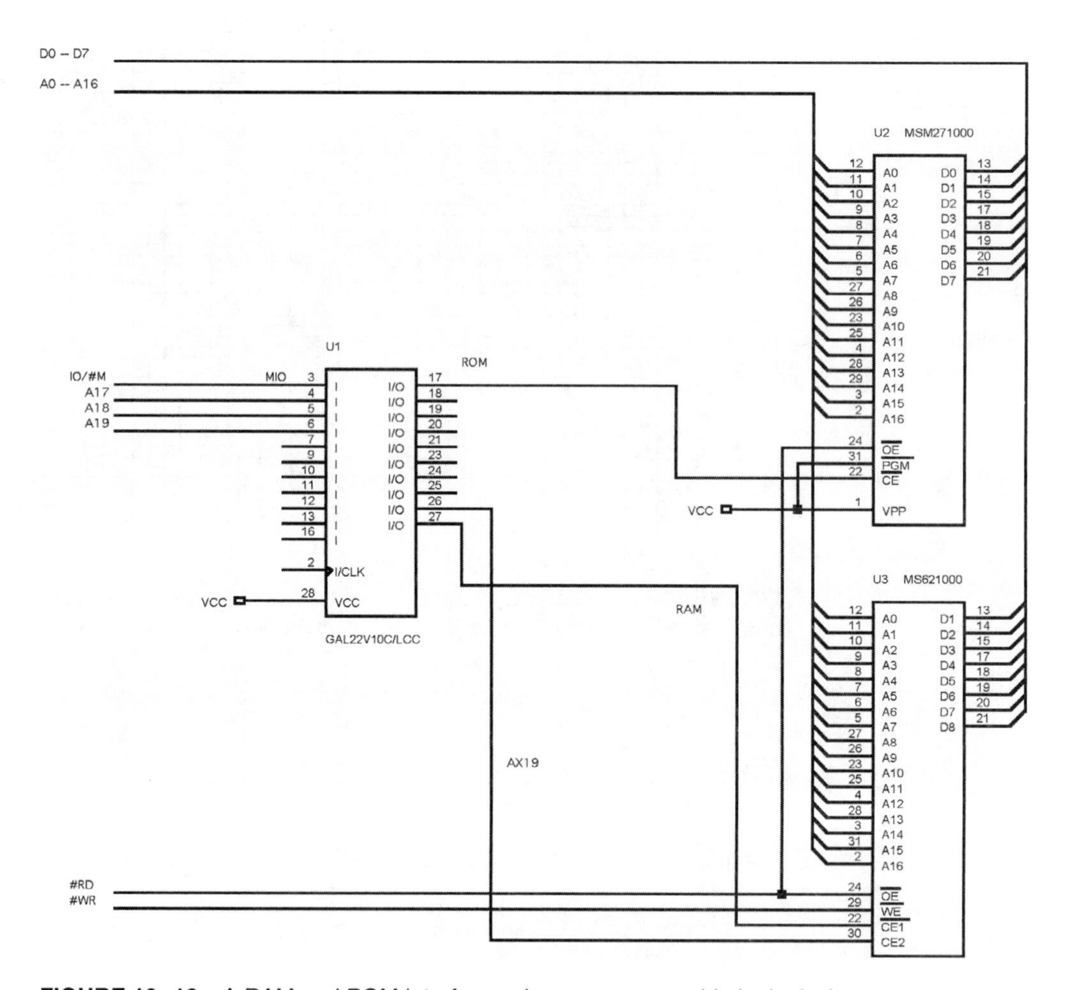

FIGURE 10–19 A RAM and ROM interface using a programmable logic device.

10–3 8088 AND 80188 (8-BIT) MEMORY INTERFACE

This text contains separate sections on memory interfacing for the 8088 and 80188 with their 8-bit data buses; the 8086, 80186, 80286, and 80386SX with their 16-bit data buses; the 80386DX and 80486 with their 32-bit data buses; and the Pentium–Pentium 4 with their 64-bit data buses. Separate sections are provided because the methods used to address the memory are slightly different in microprocessors that contain different data bus widths. Hardware engineers or technicians who wish to broaden their expertise in interfacing 16-bit, 32-bit, and 64-bit memory interface should cover all sections. This section is much more complete than the sections on the 16- , 32-, and 64-bit-wide memory interface, which cover material not explained in the 8088/80188 section.

In this section, we examine the memory interface to both RAM and ROM and explain the error-correction code (ECC), which still is currently available to memory system designers. Many home computer systems do not use ECC because of the cost, but business machines often do use it.

Basic 8088/80188 Memory Interface

The 8088 and 80188 microprocessors have an 8-bit data bus, which makes them ideal to connect to the common 8-bit memory devices available today. The 8-bit memory size makes the 8088, and especially the 80188, ideal as a simple controller. For the 8088/80188 to function correctly with the memory, however, the memory system must decode the address to select a memory component. It must also use the $\overline{RD}$, $\overline{WR}$, and $IO/\overline{M}$ control signals provided by the 8088/80188 to control the memory system.

The minimum mode configuration is used in this section and is essentially the same as the maximum mode system for memory interface. The main difference is that, in maximum mode, the $IO/\overline{M}$ signal is combined with $\overline{RD}$ to generate the $\overline{MRDC}$ signal, and $IO/\overline{M}$ is combined with $\overline{WR}$ to generate the $\overline{MWTC}$ signal. The maximum mode control signals are developed inside the 8288 bus controller. In minimum mode, the memory sees the 8088 or the 80188 as a device with 20 address connections (A_{19}–A_0), eight data bus connections (AD_7–AD_0), and the control signals $IO/\overline{M}$, $\overline{RD}$, and $\overline{WR}$.

Interfacing EPROM to the 8088. You will find this section very similar to Section 10–2 on decoders. The only difference is that, in this section, we discuss wait states and the use of the $IO/\overline{M}$ signal to enable the decoder.

Figure 10–20 illustrates an 8088/80188 microprocessor connected to three 27256 EPROMs, 32K × 8 memory devices. The 27256 has one more address input (A_{15}) than the 27128 and twice the memory. The 74HCT138 decoder in this illustration decodes three 32K × 8 blocks of memory for a total of 96K × 8 bits of the physical address space for the 8088/80188.

The decoder (74HCT138) is connected a little differently than might be expected because the slower version of this type of EPROM has a memory access time of 450 ns. Recall from Chapter 9 that when the 8088 is operated with a 5 MHz clock, it allows 460 ns for the memory to access data. Because of the decoder's added time delay (8 ns), it is impossible for this memory to function within 460 ns. In order to correct this problem, the output from the NAND gate can be used to generate a signal to enable the decoder and a signal for the wait state generator, covered in Chapter 9. (Note that the 80188 can internally insert from 0 to 15 wait states without any additional external hardware, so it does not require this NAND gate.) With a wait state inserted every time this section of the memory is accessed, the 8088 will allow 660 ns for the EPROM to access data. Recall that an extra wait state adds 200 ns (1 clock) to the access time. The 660 ns is ample time for a 450 ns memory component to access data, even with the delays introduced by the decoder and any buffers added to the data bus. The wait states are inserted in this system for memory locations C0000H–FFFFFH. If this creates a problem, a three-input OR gate can be added to the three outputs of the decoder to generate a wait signal only for the actual addresses for this system (E8000H–FFFFFH).

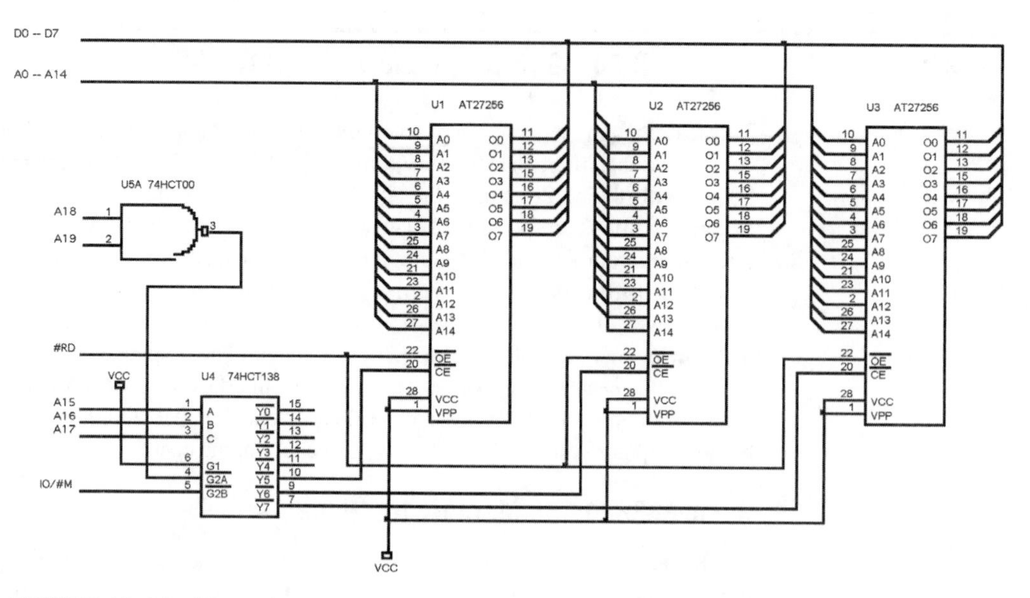

FIGURE 10–20 Three 27256 EPROMs interfaced to the 8088 microprocessor.

Notice that the decoder is selected for a memory address range that begins at location E8000H and continues through location FFFFFH—the upper 96K bytes of memory. This section of memory is an EPROM because FFFF0H is where the 8088 starts to execute instructions after a hardware reset. We often call location FFFF0H the **cold-start location**. The software stored in this section of memory would contain a JMP instruction at location FFFF0H that jumps to location E8000H so the remainder of the program can execute. In this circuit U_1 is decoded at addresses F8000H–FFFFFH, U_2 is decoded at F0000H–F7FFFH, and U_3 is decoded at E8000H–EFFFFH.

Interfacing RAM to the 8088. RAM is a little easier to interface than EPROM because most RAM memory components do not require wait states. An ideal section of the memory for the RAM is the very bottom, which contains vectors for interrupts. Interrupt vectors (discussed in more detail in Chapter 12) are often modified by software packages, so it is rather important to encode this section of the memory with RAM.

Figure 10–21 shows 16 62256, 32K × 8 static RAMs interfaced to the 8088, beginning at memory location 00000H. This circuit board uses two decoders to select the 16 different RAM memory components and a third to select the other decoders for the appropriate memory sections. Sixteen 32K RAMs fill memory from location 00000H through location 7FFFFH, for 512K bytes of memory.

The first decoder (U_4) in this circuit selects the other two decoders. An address beginning with 00 selects decoder U_3 and an address that begins with 01 selects decoder U_9. Notice that extra pins remain at the output of decoder U_4 for future expansion. These pins allow more 256K × 8 blocks of RAM for a total of 1M × 8, simply by adding the RAM and the additional secondary decoders.

Also notice from the circuit in Figure 10–21 that all the address inputs to this section of memory are buffered, as are the data bus connections and control signals $\overline{RD}$ and $\overline{WR}$. Buffering is important when many devices appear on a single board or in a single system. Suppose that three other boards like this are plugged into a system. Without the buffers on each board, the load on the system address, data, and control buses would be enough to prevent proper operation. (Excessive loading causes the logic 0 output to rise above the 0.8 V maximum allowed in a system.) Buffers are normally used if the memory will contain additions at some future date. If the memory will never grow, then buffers may not be needed.

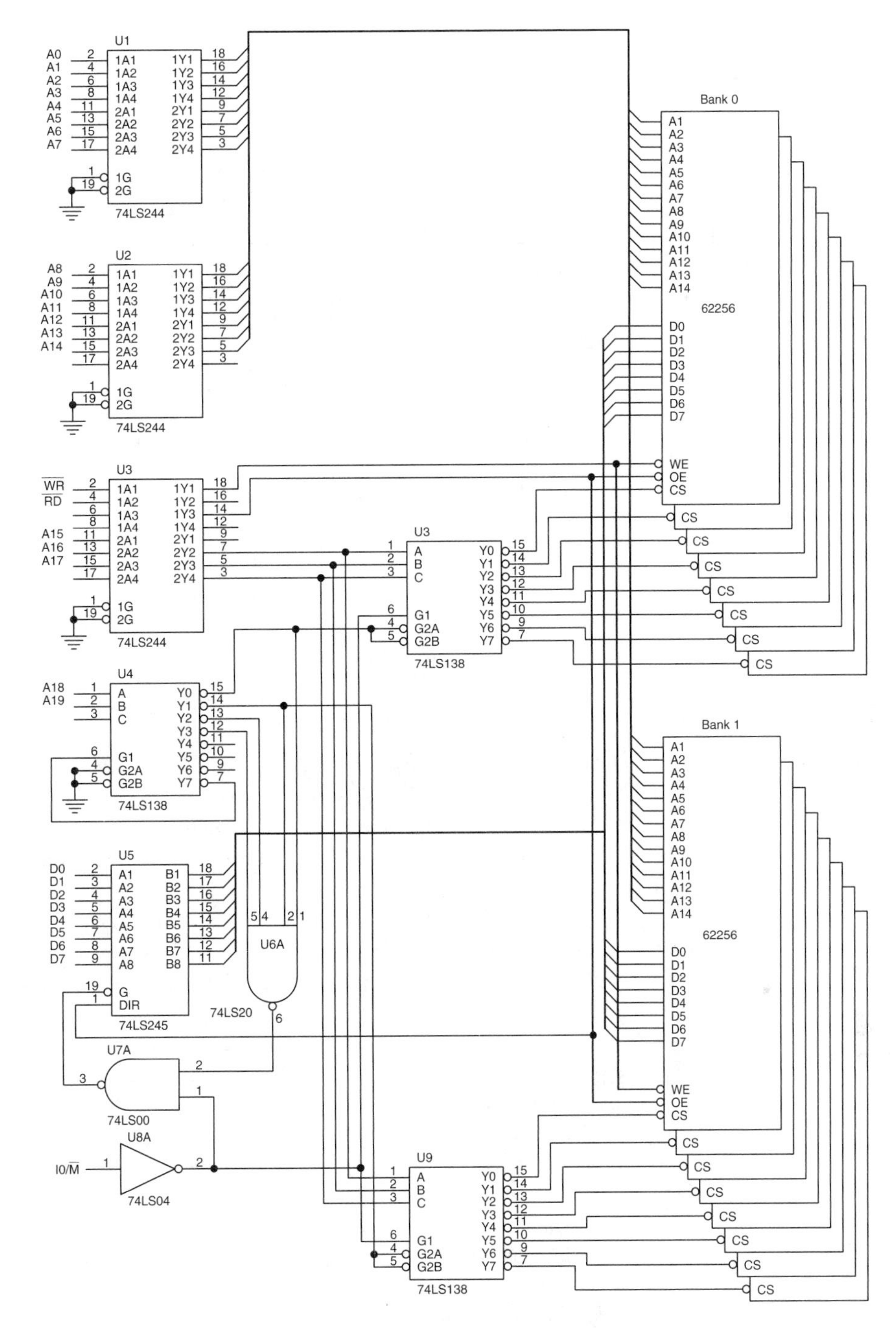

FIGURE 10–21 A 512K-byte static memory system using 16 62255 SRAMs.

Interfacing Flash Memory

Flash memory (EEPROM) is becoming commonplace for storing setup information on video cards, as well as for storing the system BIOS in the personal computer. It even finds application in MP3 audio players and USB pen drives. Flash memory is also found in many other applications to store information that is only changed occasionally.

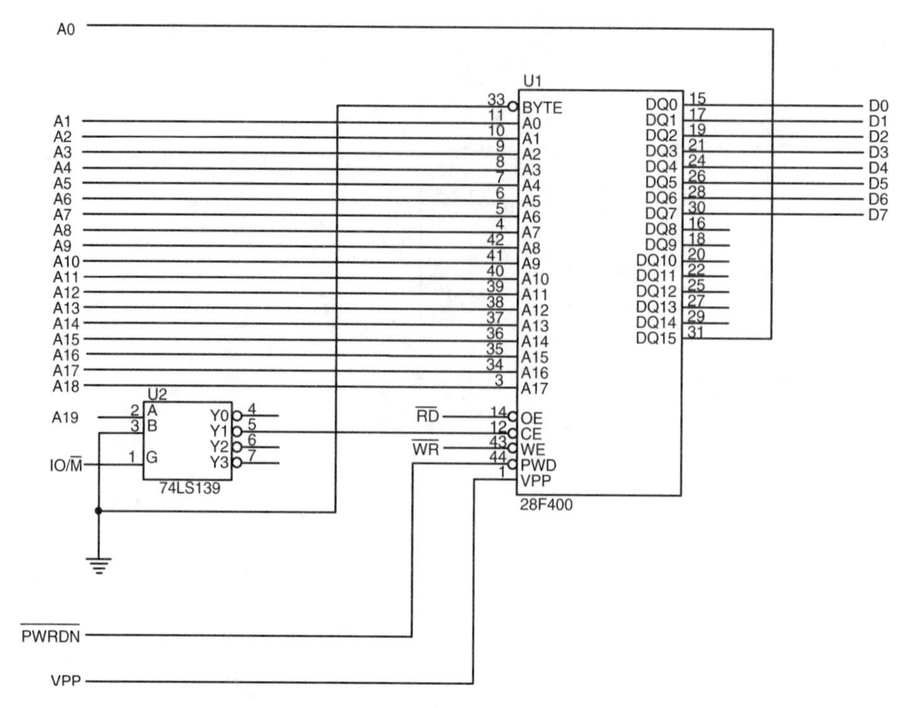

FIGURE 10–22 The 28F400 Flash memory device interfaced to the 8088 microprocessor.

The only difference between a Flash memory device and SRAM is that the Flash memory device requires a 12 V programming voltage to erase and write new data. The 12 V can be available either at the power supply or a 5-to-12 V converter designed for use with Flash memory can be obtained. The newest versions of Flash memory are erased with a 5.0 V or even a 3.3 V signal so that a converter is not needed.

Figure 10–22 illustrates the 28F400 Intel Flash memory device interfaced to the 8088 microprocessor. The 28F400 can be used as either a 512K × 8 memory device or as a 256K × 16 memory device. Because it is interfaced to the 8088, its configuration is 512K × 8. Notice that the control connections on this device are identical to that of an SRAM: $\overline{CE}$, $\overline{OE}$, and $\overline{WE}$. The only new pins are V_{PP}, which is connected to 12 V for erase and programming; $\overline{PWD}$, which selects the power-down mode when a logic 0 and is also used for programming; and $\overline{BYTE}$, which selects byte (0) or word (1) operation. Note that the pin DQ_{15} functions as the least-significant address input when operated in the byte mode. Another difference is the amount of time required to accomplish a write operation. The SRAM can perform a write operation in as little as 10 ns, but the Flash memory requires approximately 0.4 seconds to erase a byte. The topic of programming the Flash memory device is covered in Chapter 11, along with I/O devices. The Flash memory device has internal registers that are programmed by using I/O techniques not yet explained. This chapter concentrates on its interface to the microprocessor.

Notice in Figure 10–22 that the decoder chosen is the 74LS139 because only a simple decoder is needed for a Flash memory device this large. The decoder uses address connection A_{19} and $IO/\overline{M}$ as inputs. The A_{15} signal selects the Flash memory for locations 80000H through FFFFFH, and $IO/\overline{M}$ enables the decoder.

Error Correction

Error-correction schemes have been around for a long time, but integrated circuit manufacturers have only recently started to produce error-correcting circuits. One such circuit is the 74LS636, an 8-bit error correction and detection circuit that corrects any single-bit memory read error and flags any two-bit error, called SECDED (**single error correction/double error correction**).

This device is found in high-end computer systems because of the cost of implementing a system that uses error correction.

The newest computer systems are now using DDR memory with ECC (error-correction code). The scheme to correct the errors that might occur in these memory devices is identical to the scheme discussed in this text.

The 74LS636 corrects errors by storing five parity bits with each byte of memory data. This does increase the amount of memory required, but it also provides automatic error correction for single-bit errors. If more than two bits are in error, this circuit may not detect it. Fortunately, this is rare, and the extra effort required to correct more than a single-bit error is very expensive and not worth the effort. Whenever a memory component fails completely, its bits are all high or all low. In this case, the circuit flags the processor with a multiple-bit error indication.

Figure 10–23 depicts the pin-out of the 74LS636. Notice that it has eight data I/O pins, five check bit I/O pins, two control inputs (SO and SI), and two error outputs: single-error flag

pin assignments

J, N PACKAGES			
1	DEF	11	CB4
2	DB0	12	nc
3	DB1	13	CB3
4	DB2	14	CB2
5	DB3	15	CB1
6	DB4	16	CB0
7	DB5	17	S0
8	DB6	18	S1
9	DB7	19	SEF
10	GND	20	V_{CC}

(a)

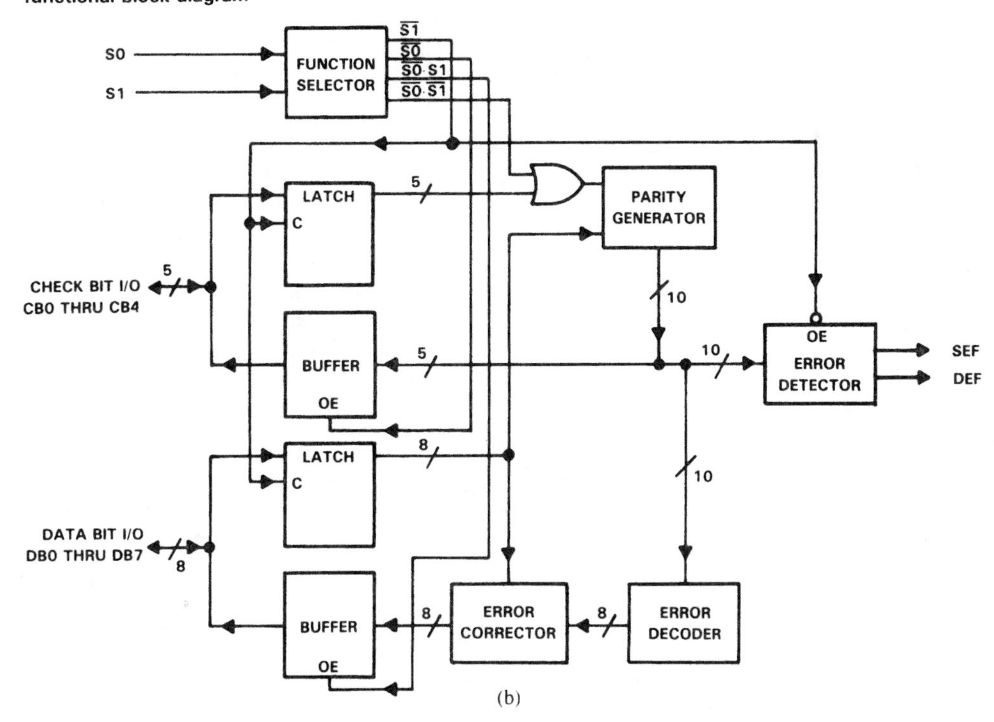

(b)

FIGURE 10–23 (a) The pin connections of the 74LS636. (b) The block diagram of the 74LS636. (Courtesy of Texas Instruments Incorporated.)

TABLE 10–1 Control bits S_0 and S_1 for the 74LS636.

S_1	S_0	Function	SEF	DEF
0	0	Write check word	0	0
0	1	Correct data word	*	*
1	0	Read data	0	0
1	1	Latch data	*	*

*These levels are determined by the error type.

(SEF) and double-error flag (DEF). The control inputs select the type of operation to be performed and are listed in the truth table of Table 10–1.

When a single error is detected, the 74LS636 goes through an error-correction cycle. It places 01 on S_0 and S_1 by causing a wait and then a read following error correction.

Figure 10–24 illustrates a circuit used to correct single-bit errors with the 74LS636 and to interrupt the processor through the NMI pin for double-bit errors. To simplify the illustration, we depict only one 2K × 8 RAM and a second 2K × 8 RAM to store the five-bit check code.

The connection of this memory component is different from that of the previous example. Notice that the S or $\overline{CS}$ pin is grounded, and data bus buffers control the flow to the system bus. This is necessary if the data are to be accessed from the memory before the $\overline{RD}$ strobe goes low.

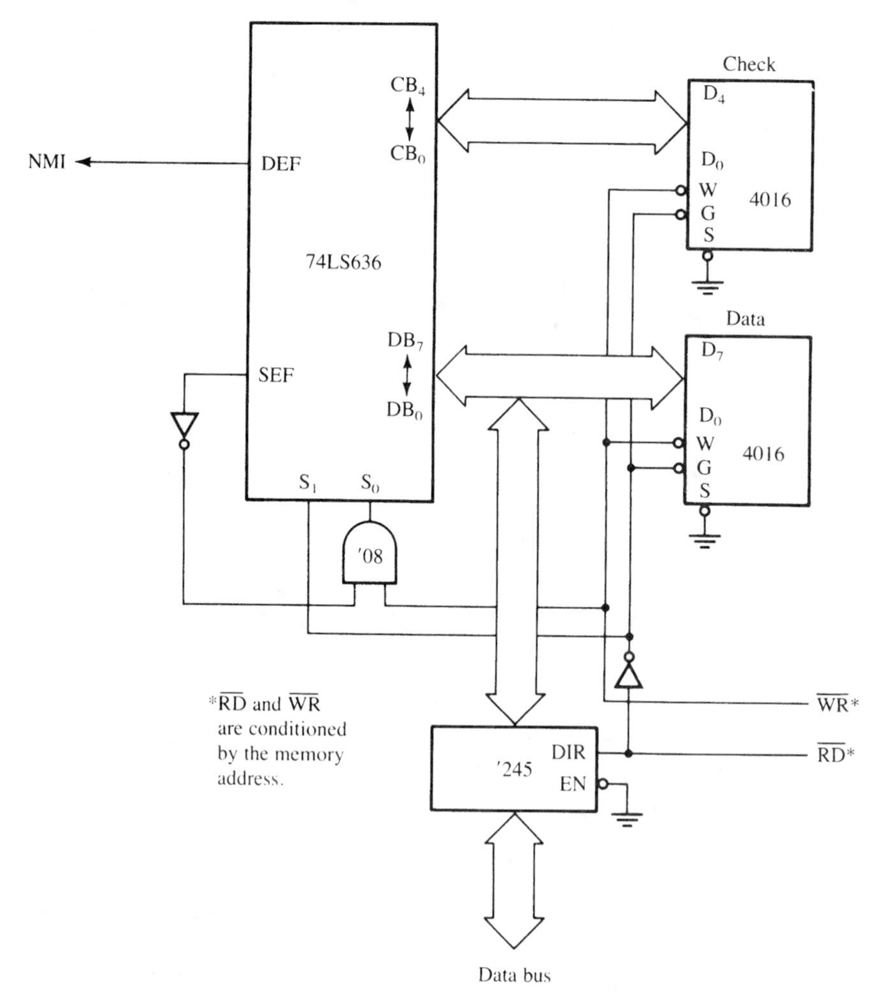

FIGURE 10–24 An error detection and correction circuit using the 74LS636.

On the next negative edge of the clock after an $\overline{RD}$ the 74LS636 checks the single-error flag (SEF) to determine whether an error has occurred. If it has, a correction cycle causes the single-error defect to be corrected. If a double error occurs, an interrupt request is generated by the double-error flag (DEF) output, which is connected to the NMI pin of the microprocessor.

Modern DDR error-correction memory (ECC) does not actually have logic circuitry on board that detects and corrects errors. Since the Pentium, the microprocessor incorporates the logic circuitry to detect/correct errors provided the memory can store the extra eight bits required for storing the ECC code. ECC memory is 72-bits wide using the eight additional bits to store the ECC code. If an error occurs, the microprocessor runs the correction cycle to correct the error. Some memory devices such as Samsung memory also perform an internal error check. The Samsung ECC uses three bytes to check every 256 bytes of memory, which is far more efficient. Additional information on the Samsung ECC algorithm is available at the Samsung Website.

10–4 8086, 80186, 80286, AND 80386SX (16-BIT) MEMORY INTERFACE

The 8086, 80186, 80286, and 80386SX microprocessors differ from the 8088/80188 in three ways: (1) The data bus is 16 bits wide instead of eight bits wide as on the 8088; (2) the $IO/\overline{M}$ pin of the 8088 is replaced with an $M/\overline{IO}$ pin; and (3) there is a new control signal called bus high enable ($\overline{BHE}$). The address bit A_0 or $\overline{BLE}$ is also used differently. (Because this section is based on information provided in Section 10–3, it is extremely important that you read the previous section first.) A few other differences exist between the 8086/80186 and the 80286/80386SX. The 80286/80386SX contains a 24-bit address bus (A_{23}–A_0) instead of the 20-bit address bus (A_{19}–A_0) of the 8086/80186. The 8086/80186 contain an $M/\overline{IO}$ signal while the 80286 system and 80386SX microprocessor contain control signals $\overline{MRDC}$ and $\overline{MWTC}$ instead of $\overline{RD}$ and $\overline{WR}$.

16-Bit Bus Control

The data bus of the 8086, 80186, 80286, and 80386SX is twice as wide as the bus for the 8088/80188. This wider data bus presents us with a unique set of problems. The 8086, 80186, 80286, and 80386SX must be able to write data to any 16-bit location—or any 8-bit location. This means that the 16-bit data bus must be divided into two separate sections (**banks**) that are eight bits wide so that the microprocessor can write to either half (8-bit) or both halves (16-bit). Figure 10–25 illustrates the two banks of the memory. One bank (**low bank**) holds all the even-numbered memory locations, and the other bank (**high bank**) holds all the odd-numbered memory locations.

The 8086, 80186, 80286, and 80386SX use the $\overline{BHE}$ signal (high bank) and the A_0 address bit or $\overline{BLE}$ (bus low enable) to select one or both banks of memory used for the data transfer. Table 10–2 depicts the logic levels on these two pins and the bank or banks selected.

Bank selection is accomplished in two ways: (1) A separate write signal is developed to select a write to each bank of the memory, or (2) separate decoders are used for each bank. As a careful comparison reveals, the first technique is by far the least costly approach to memory interface for the 8086, 80186, 80286, and 80386SX microprocessors. The second technique is only used in a system that must achieve the most efficient use of the power supply.

Separate Bank Decoders. The use of separate bank decoders is often the least effective way to decode memory addresses for the 8086, 80186, 80286, and 80386SX microprocessors. This method is sometimes used, but it is difficult to understand why in most cases. One reason may be to conserve energy, because only the bank or banks selected are enabled. This is not always the case, as with the separate bank read and write signals that are discussed later.

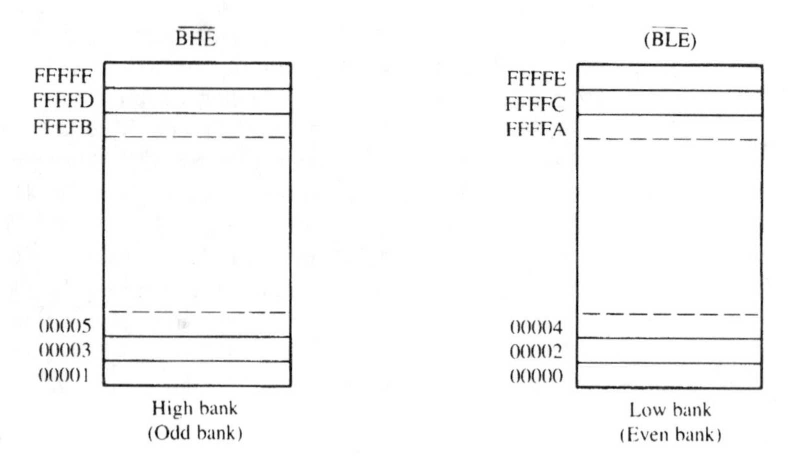

FIGURE 10–25 The high (odd) and low (even) 8-bit memory banks of the 8086/80286/80386SX microprocessors.

Figure 10–26 illustrates two 74LS138 decoders used to select 64K RAM memory components for the 80386SX microprocessor (24-bit address). Here, decoder U_2 has the $\overline{BLE}$ pin (A_0) attached to $\overline{G2A}$, and decoder U_3 has the $\overline{BHE}$ signal attached to its $\overline{G2A}$ input. Because the decoder will not activate until all of its enable inputs are active, decoder U_2 activates only for a 16-bit operation or an 8-bit operation from the low bank. Decoder U_3 activates for a 16-bit operation or an 8-bit operation to the high bank. These two decoders and the 16 64K-byte RAMs they control represent a 1M range of the 80386SX memory system. Decoder U_1 enables U_2 and U_3 for memory address range 000000H–0FFFFFH.

Notice in Figure 10–26 that the A_0 address pin does not connect to the memory because it does not exist on the 80386SX microprocessor. Also notice that address bus bit position A_1 is connected to memory address input A_0, A_2 is connected to A_1, and so forth. The reason is that A_0 from the 8086/80186 (or $\overline{BLE}$ from the 80286/80386SX) is already connected to decoder U_2 and does not need to be connected again to the memory. If A_0 or $\overline{BLE}$ were attached to the A_0 address pin of memory, every other memory location in each bank of memory would be used. This means that half of the memory would be wasted if A_0 or $\overline{BLE}$ were connected to A_0.

Separate Bank Write Strobes. The most effective way to handle bank selection is to develop a separate write strobe for each memory bank. This technique requires only one decoder to select a 16-bit wide memory, which often saves money and reduces the number of components in a system.

Why not also generate separate read strobes for each memory bank? This is usually unnecessary because the 8086, 80186, 80286, and 80386SX microprocessors read only the byte of data that they need at any given time from half of the data bus. If 16-bit sections of data are always presented to the data bus during a read, the microprocessor ignores the 8-bit section that it doesn't need, without any conflicts or special problems.

TABLE 10–2 Memory bank selection using $\overline{BHE}$ and $\overline{BLE}$ (A_0).

$\overline{BHE}$	$\overline{BLE}$	*Function*
0	0	Both banks enabled for a 16-bit transfer
0	1	High bank enabled for an 8-bit transfer
1	0	Low bank enabled for an 8-bit transfer
1	1	No bank enabled

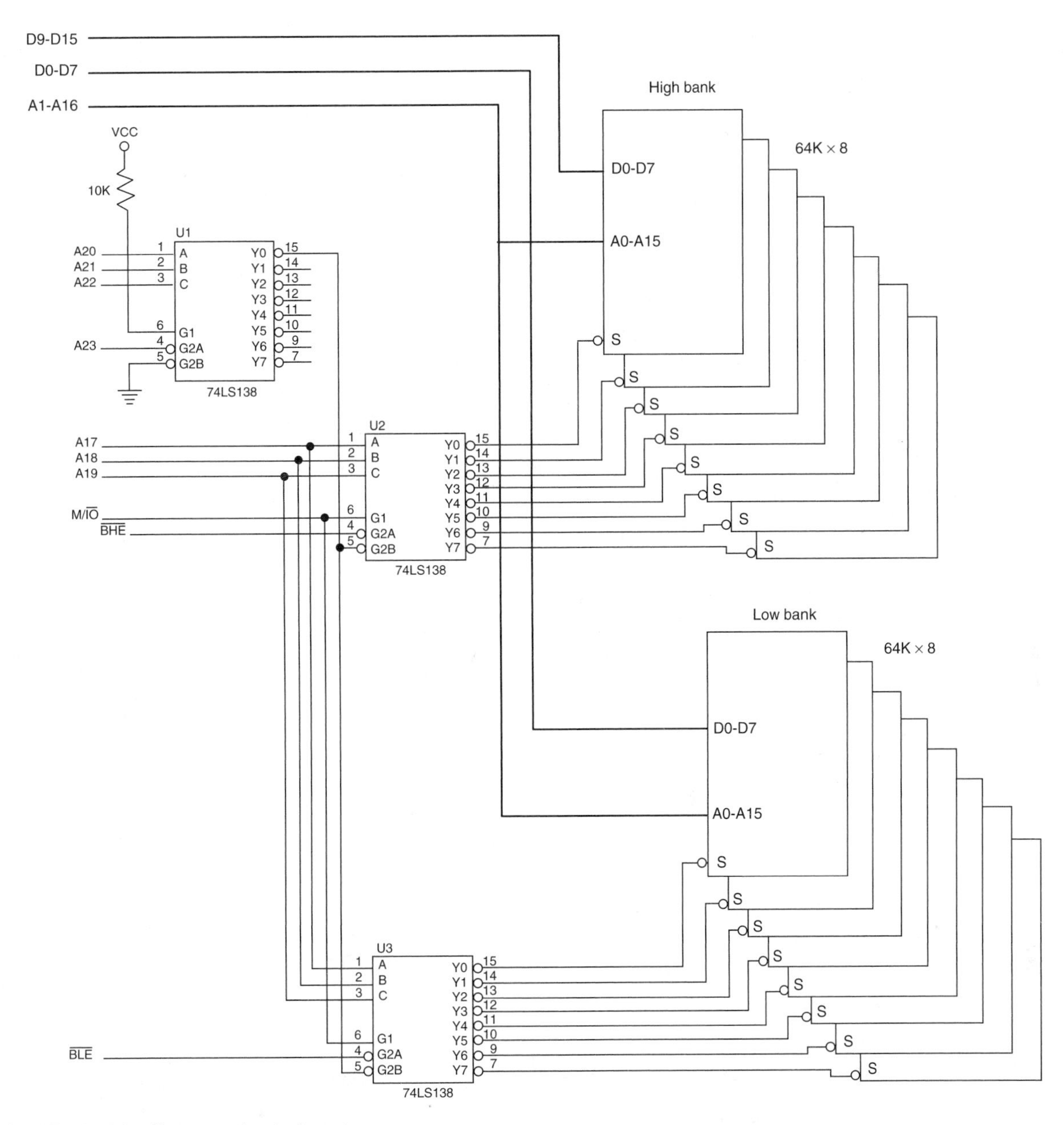

FIGURE 10–26 Separate bank decoders.

Figure 10–27 depicts the generation of separate 8086 write strobes for the memory. Here, a 74LS32 OR gate combines A_0 with $\overline{WR}$ for the low bank selection signal ($\overline{LWR}$), and $\overline{BHE}$ combines with $\overline{WR}$ for the high bank selection signal ($\overline{HWR}$). Write strobes for the 80286/80386SX are generated by using the $\overline{MWTC}$ signal instead of $\overline{WR}$.

A memory system that uses separate write strobes is constructed differently from either the 8-bit system (8088) or the system using separate memory banks. Memory in a system that uses separate write strobes is decoded as 16-bit-wide memory. For example, suppose that a memory system will contain 64K bytes of SRAM memory. This memory requires two 32K-byte memory devices (62256) so that a 16-bit-wide memory can be constructed. Because the memory is 16 bits

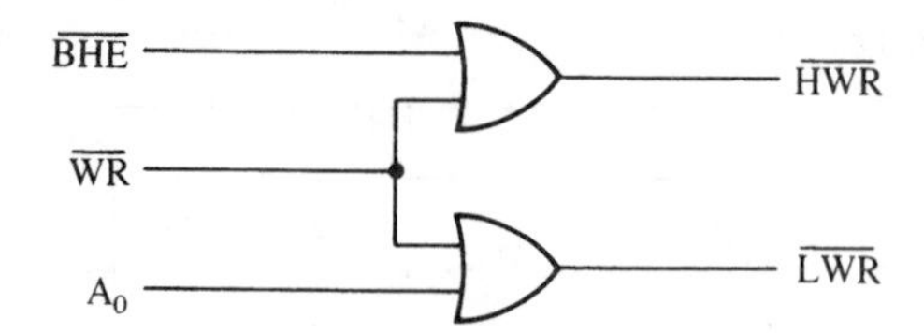

FIGURE 10–27 The memory bank write selection input signals: $\overline{HWR}$ (high bank write) and $\overline{LWR}$ (low bank write).

wide and another circuit generates the bank write signals, address bit A_0 becomes a don't care. In fact, A_0 is not even a pin on the 80386SX microprocessor.

Example 10–6 shows how a 16-bit-wide memory stored at locations 060000H–06FFFFH is decoded for the 80286 or 80386 microprocessor. Memory in this example is decoded, so bit A_0 is a don't care for the decoder. Bit positions A_1–A_{15} are connected to memory component address pins A_0–A_{14}. The decoder (GAL22V10) enables both memory devices by using address connection A_{23}–A_{15} to select memory whenever address 06XXXXH appears on the address bus.

EXAMPLE 10–6

```
0000 0110 0000 0000 0000 0000 = 060000H
          to
0000 0110 1111 1111 1111 1111 = 06FFFFH

0000 0110 XXXX XXXX xxxx XXXX = 06XXXXH
```

Figure 10–28 illustrates this simple circuit by using a GAL22V10 to both decode memory and generate the separate write strobe. The program for the GAL22V10 decoder is illustrated in

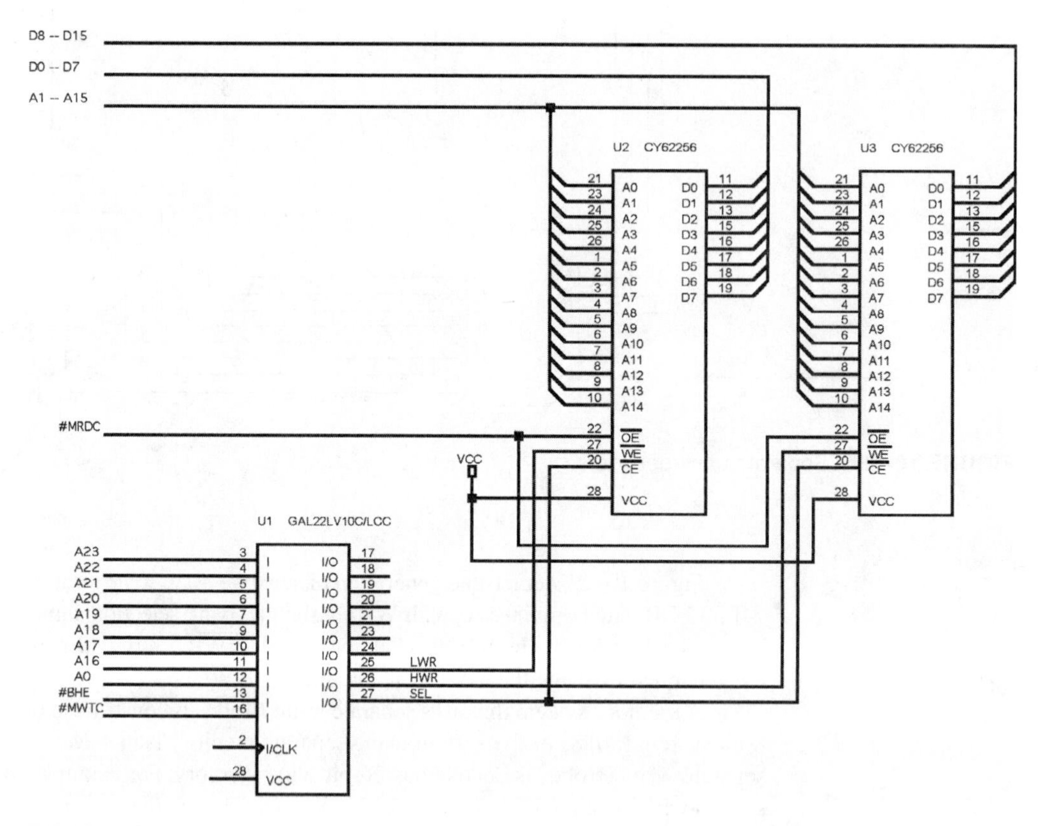

FIGURE 10–28 A 16-bit-wide memory interfaced at memory locations 06000H–06FFFH.

Example 10–7. Notice that not only is the memory selected, but both the lower and upper write strobes are also generated by the PLD.

EXAMPLE 10–7

```
-- VHDL code for the decoder of Figure 10-28

library ieee;
use ieee.std_logic_1164.all;

entity DECODER_10_28 is

port (
        A23, A22, A21, A20, A19, A18, A17, A16, A0, BHE, MWTC: in STD_LOGIC;
        SEL, LWR, HWR: out STD_LOGIC
);

end;

architecture V1 of DECODER_10_28 is

begin

        SEL <= A23 or A22 or A21 or A20 or A19 or (not A18) or (not A17) or A16;
        LWR <= A0 or MWTC;
        HWR <= BHE or MWTC;

end V1;
```

Figure 10–29 depicts a small memory system for the 8086 microprocessor that contains an EPROM section and a RAM section. Here, there are four 27128 EPROMs (16K × 8) that compose a 32K × 16-bit memory at locations F0000–FFFFFH and four 62256 (32K × 8) RAMs that compose an 64K × 16-bit memory at locations 00000H–1FFFFH. (Remember that even though the memory is 16 bits wide, it is still numbered in bytes.)

This circuit uses a 74LS139 dual 2-to-4 line decoder that selects EPROM with one half and RAM with the other half. It decodes memory that is 16 bits wide, not eight bits, as before. Notice that the $\overline{RD}$ strobe is connected to all the EPROM $\overline{OE}$ inputs and all the RAM $\overline{G}$ input pins. This is done because even if the 8086 is reading only eight bits of data, the application of the remaining eight bits to the data bus has no effect on the operation of the 8086.

The $\overline{LWR}$ and $\overline{HWR}$ strobes are connected to different banks of the RAM memory. Here, it does matter whether the microprocessor is doing a 16-bit or an 8-bit write. If the 8086 writes a 16-bit number to memory, both $\overline{LWR}$ and $\overline{HWR}$ go low and enable the $\overline{W}$ pins in both memory banks. But if the 8086 does an 8-bit write, only one of the write strobes goes low, writing to only one memory bank. Again, the only time the banks make a difference is for a memory write operation.

Notice that an EPROM decoder signal is sent to the 8086 wait state generator because EPROM memory usually requires a wait state. The signal comes from the NAND gate used to select the EPROM decoder section, so that if EPROM is selected, a wait state is requested.

Figure 10–30 illustrates a memory system connected to the 80386SX microprocessor by using a GAL22V10 as a decoder. This interface contains 256K bytes of EPROM in the form of four 27512 (64K × 8) EPROMs and 128K bytes of SRAM memory found in four 62256 (32K × 8) SRAMs.

Notice in Figure 10–30 that the PLD also generates the memory bank write signals $\overline{LWR}$ and $\overline{HWR}$. As can be gleaned from this circuit, the number of components required to interface memory has been reduced to just one, in most cases (the PLD). The program listing for the PLD is located in Example 10–8. The PLD decodes the 16-bit-wide memory addresses at locations 000000H–01FFFFH for the SRAM and locations FC0000H–FFFFFFH for the EPROM.

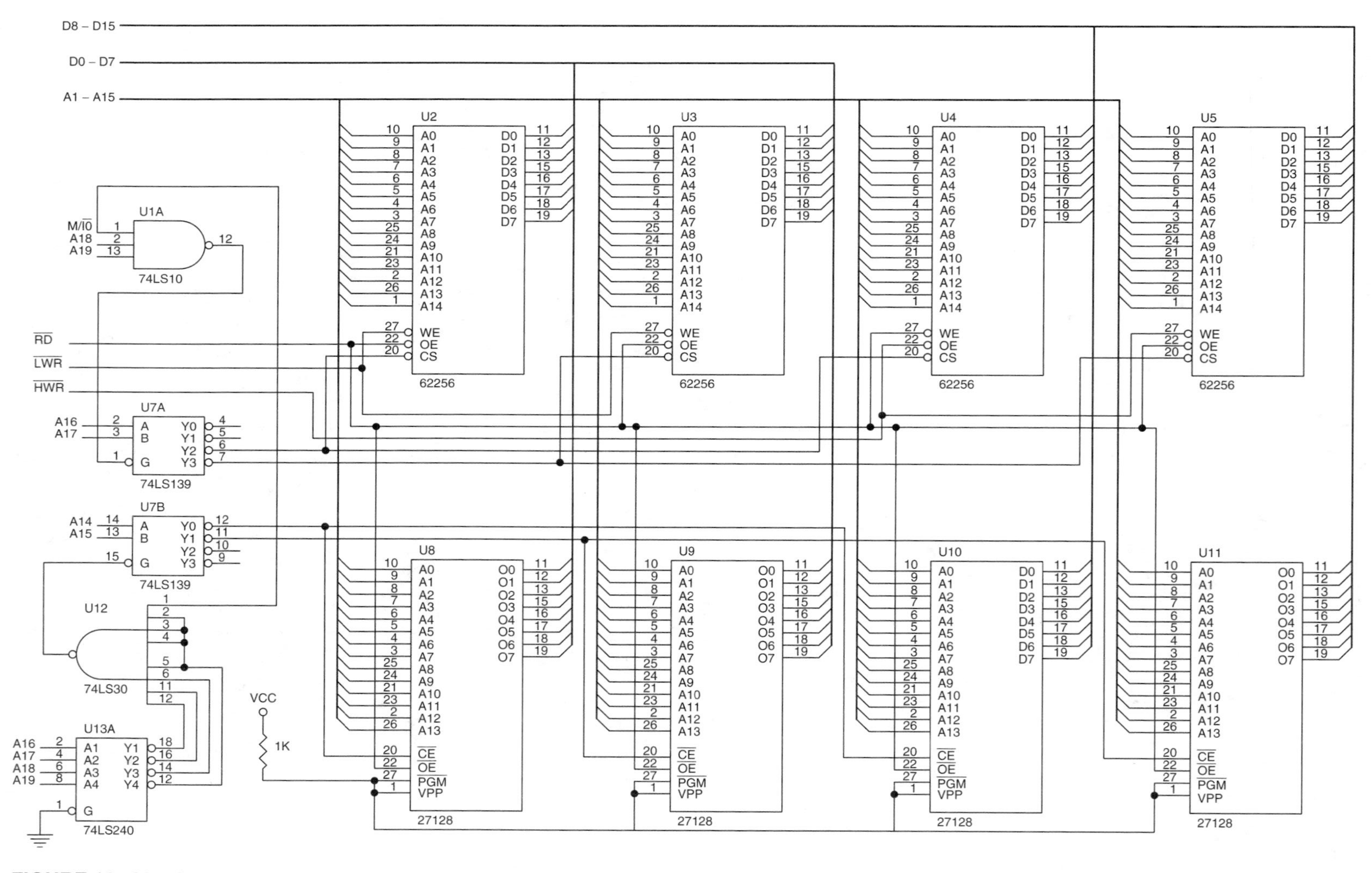

FIGURE 10–29 A memory system for the 8086 that contains a 64K-byte EPROM and a 128K-byte SRAM.

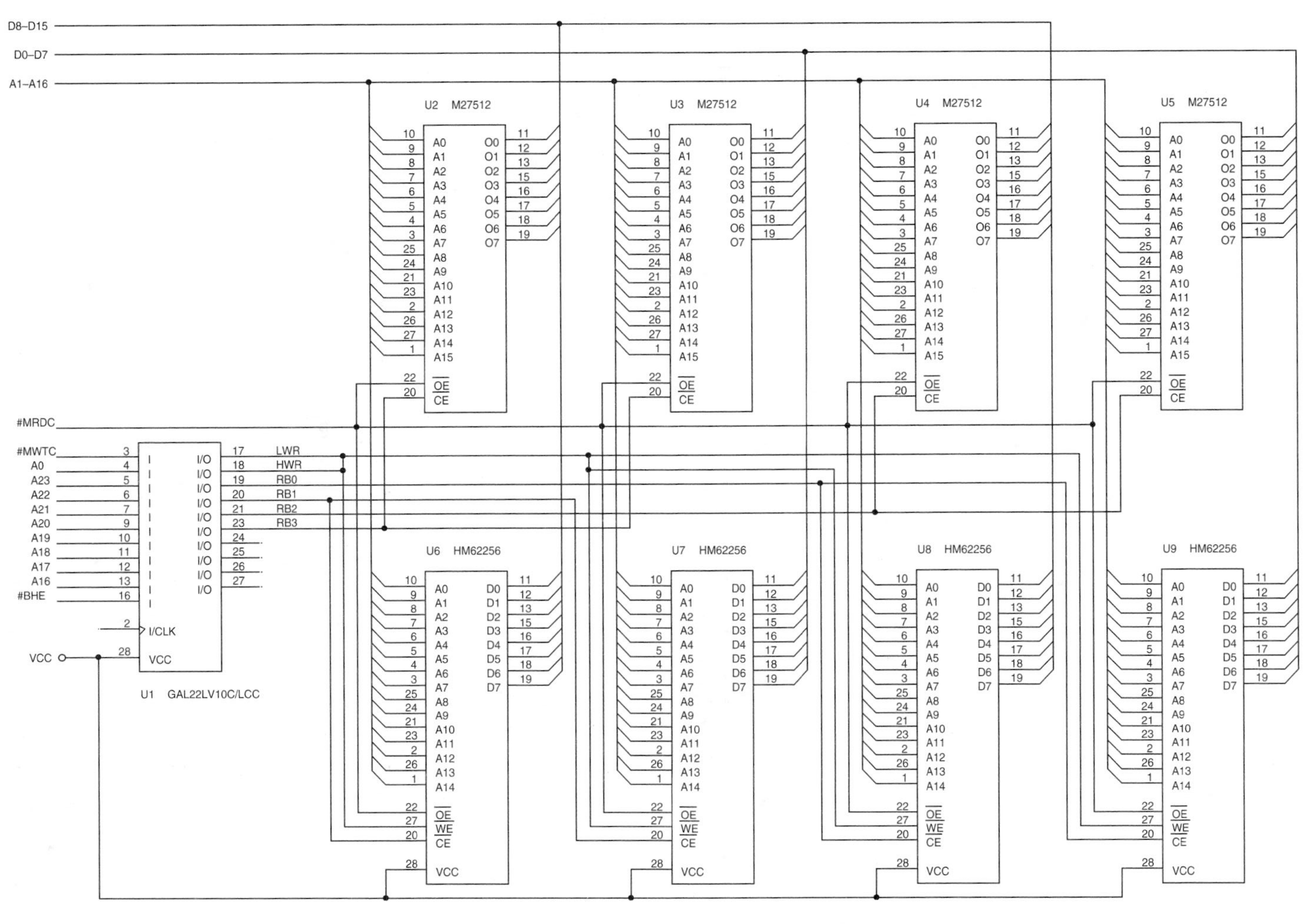

FIGURE 10–30 An 80386SX memory system containing 256K of EPROM and 128K of SRAM.

EXAMPLE 10–8

```
-- VHDL code for the decoder of Figure 10-30

library ieee;
use ieee.std_logic_1164.all;

entity DECODER_10_30 is

port (
        A23, A22, A21, A20, A19, A18, A17, A16, A0, BHE, MWTC: in STD_LOGIC;
        LWR, HWR, RB0, RB1, RB2, RB3: out STD_LOGIC
);

end;

architecture V1 of DECODER_10_30 is

begin

        LWR <= A0 or MWTC;
        HWR <= BHE or MWTC;
        RB0 <= A23 or A22 or A21 or A20 or A19 or A18 or A17 or A16;
        RB1 <= A23 or A22 or A21 or A20 or A19 or A18 or A17 or not(A16));
        RB2 <= not(A23 and A22 and A21 and A20 and A19 and A18 and A17);
        RB3 <= not(A23 and A22 and A21 and A20 and A19 and A18 and not(A17));

end V1;
```

10–5 80386DX AND 80486 (32-BIT) MEMORY INTERFACE

As with 8- and 16-bit memory systems, the microprocessor interfaces to memory through its data bus and control signals that select separate memory banks. The only difference with a 32-bit memory system is that the microprocessor has a 32-bit data bus and four banks of memory, instead of one or two. Another difference is that both the 80386DX and 80486 (both SX and DX) contain a 32-bit address bus that usually requires PLD decoders instead of integrated decoders because of the sizable number of address bits.

Memory Banks

The memory banks for both the 80386DX and 80486 microprocessors are illustrated in Figure 10–31. Notice that these large memory systems contain four 8-bit-wide banks that each contain up to 1G byte of memory. Bank selection is accomplished by the bank selection signals $\overline{BE3}$, $\overline{BE2}$, $\overline{BE1}$, and $\overline{BE0}$. If a 32-bit number is transferred, all four banks are selected; if a 16-bit number is transferred, two banks (usually $\overline{BE3}$ and $\overline{BE2}$ or $\overline{BE1}$ and $\overline{BE0}$) are selected; and if eight bits are transferred, a single bank is selected.

As with the 8086/80286/80386SX, the 80386DX and 80486 require separate write strobe signals for each memory bank. These separate write strobes are developed, as illustrated in Figure 10–32, by using a simple OR gate or other logic component.

32-Bit Memory Interface

As can be gathered from the prior discussion, a memory interface for the 80386DX or 80486 requires that we generate four bank write strobes and decode a 32-bit address. There are no integrated decoders, such as the 74LS138, that can easily accommodate a memory interface for the 80386DX or 80486 microprocessors. Note that address bits A_0 and A_1 are don't cares when 32-bit-wide memory is decoded. These address bits are used within the microprocessor to

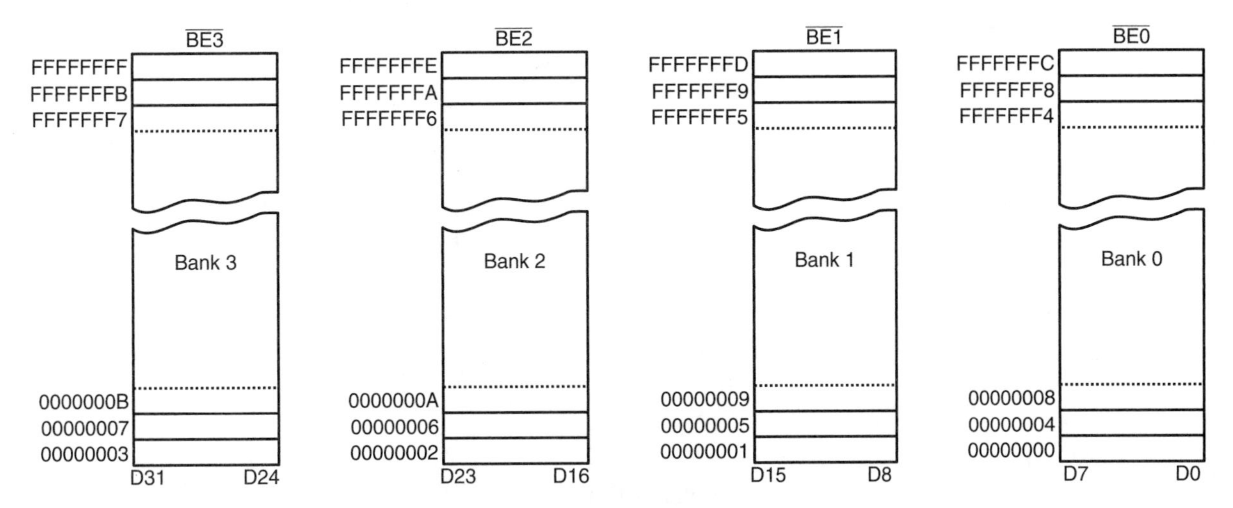

FIGURE 10–31 The memory organization for the 80386DX and 80486 microprocessors.

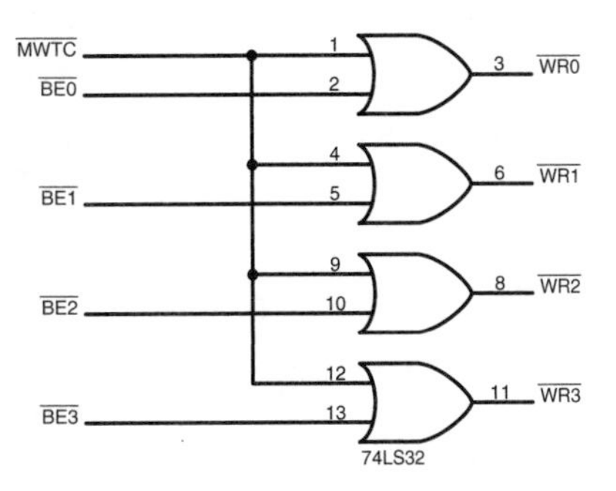

FIGURE 10–32 Bank write signals for the 80386DX and 80486 microprocessors.

generate the bank enable signals. Notice that the address bus connection A_2 connects to memory address pin A_0. This occurs because there is no A_0 or A_1 pin on the 80486 microprocessor.

Figure 10–33 shows a 512K × 8 SRAM memory system for the 80486 microprocessor. This interface uses eight 64K × 8 SRAM memory devices, a PLD, and an OR gate. The OR gate is required because of the number of address connections found on the microprocessor. This system places the SRAM memory at locations 02000000H–0203FFFFH. The program for the PLD device is found in Example 10–9.

EXAMPLE 10–9

```
library ieee;
use ieee.std_logic_1164.all;

entity DECODER_10_30 is

port (
       A30, A29, A28, A27, A26, A25, A24, A23, A22, A21, A19, BE0, BE1, BE2,
             BE3, MWTC: in STD_LOGIC;
       RB0, RB1, WR0, WR1, WR2, WR3: out STD_LOGIC
);

end;

architecture V1 of DECODER_10_30 is
```

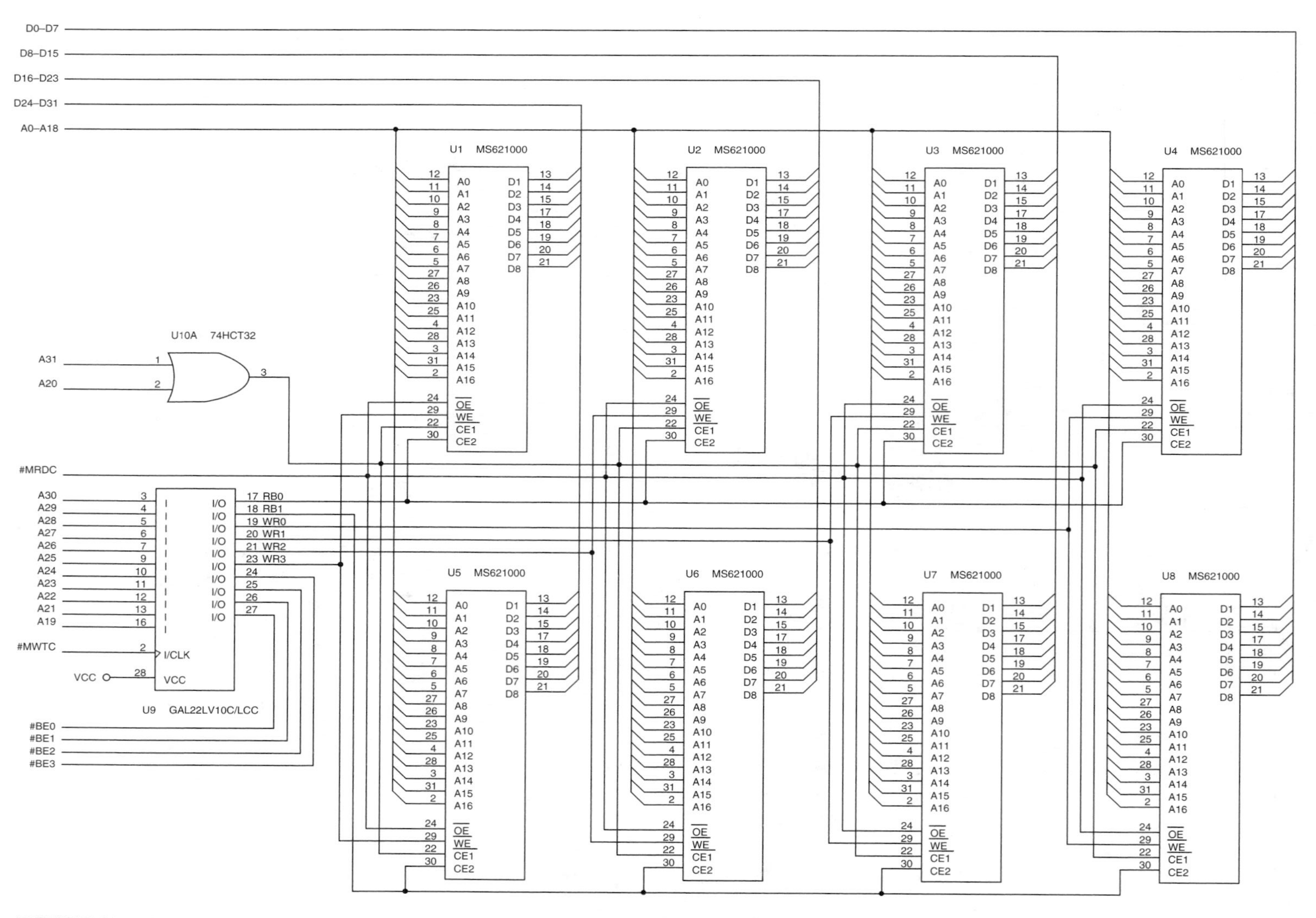

FIGURE 10–33 A small 512K-byte SRAM memory system for the 80486 microprocessor.

```
begin

        WR0 <= BE0 or MWTC;
        WR1 <= BE1 or MWTC;
        WR2 <= BE2 or MWTC;
        WR3 <= BE3 or MWTC;
        RB0 <= A30 or A29 or A28 or A27 or A26 or A25 or A24 or A23 or A22
                or A 21 or A19;
        RB1 <= A30 or A29 or A28 or A27 or A26 or A25 or A24 or A23 or A22
                or A 21 or not(A19);

end V1;
```

Although not mentioned in this section of the text, the 80386DX and 80486 microprocessors operate with very high clock rates that usually require wait states for memory access. Access time calculations for these microprocessors are discussed in Chapters 17 and 18. The interface provides a signal used with the wait state generator that is not illustrated in this section of the text. Other devices with these higher speed microprocessors are cache memory and interleaved memory systems, also presented in Chapter 17 with the 80386DX and 80486 microprocessors.

10–6 PENTIUM THROUGH PENTIUM 4 (64-BIT) MEMORY INTERFACE

The Pentium through Pentium 4 microprocessors (except for the P24T version of the Pentium) contain a 64-bit data bus, which requires either eight decoders (one per bank) or eight separate write signals. In most systems, separate write signals are used with this microprocessor when interfacing memory. Figure 10–34 illustrates the Pentium's memory organization and its eight memory banks. Notice that this is almost identical to the 80486, except that it contains eight banks instead of four.

As with earlier versions of the Intel microprocessor, this organization is required for upward memory compatibility. The separate write strobe signals are obtained by combining the bank enable signals with the $\overline{\text{MWTC}}$ signal, which is generated by combining the $\text{M}/\overline{\text{IO}}$ with $\text{W}/\overline{\text{R}}$. The circuit employed for bank write signals appears in Figure 10–35. As can be imagined, we often find a PLD used for bank write signal generation.

64-Bit Memory Interface

Figure 10–36 illustrates a small Pentium–Pentium 4 memory system. This system uses a PLD to decode the memory address. This system contains eight 27C4001 EPROM memory devices (512K × 8), interfaced to the Pentium–Pentium 4 at locations FFC00000H through FFFFFFFFH. This is a total memory size of 4M bytes organized so that each bank contains two memory components. Note that the Pentium Pro through the Pentium 4 can be configured with 36 address connections, allowing up to 64G of memory.

Memory decoding, as illustrated in Example 10–10, is similar to the earlier examples, except that with the Pentium–Pentium 4 the rightmost three address bits (A_2–A_0) are ignored. In this case, the decoder selects sections of memory that are 64 bits wide and contain 4M bytes of EPROM memory.

The A_0 address input of each memory device connects to the A_3 address output of the Pentium and above, and the A_1 address input of each memory device connects to the A_4 address output of the Pentium and above. This skewed address connection continues until the A_{18} address input to the memory is connected to the A_{22} address output of the Pentium. Address positions A_{22}–A_{31} are decoded by a PLD. The program for the PLD device is listed in Example 10–10 for memory locations FFC00000H–FFFFFFFFH.

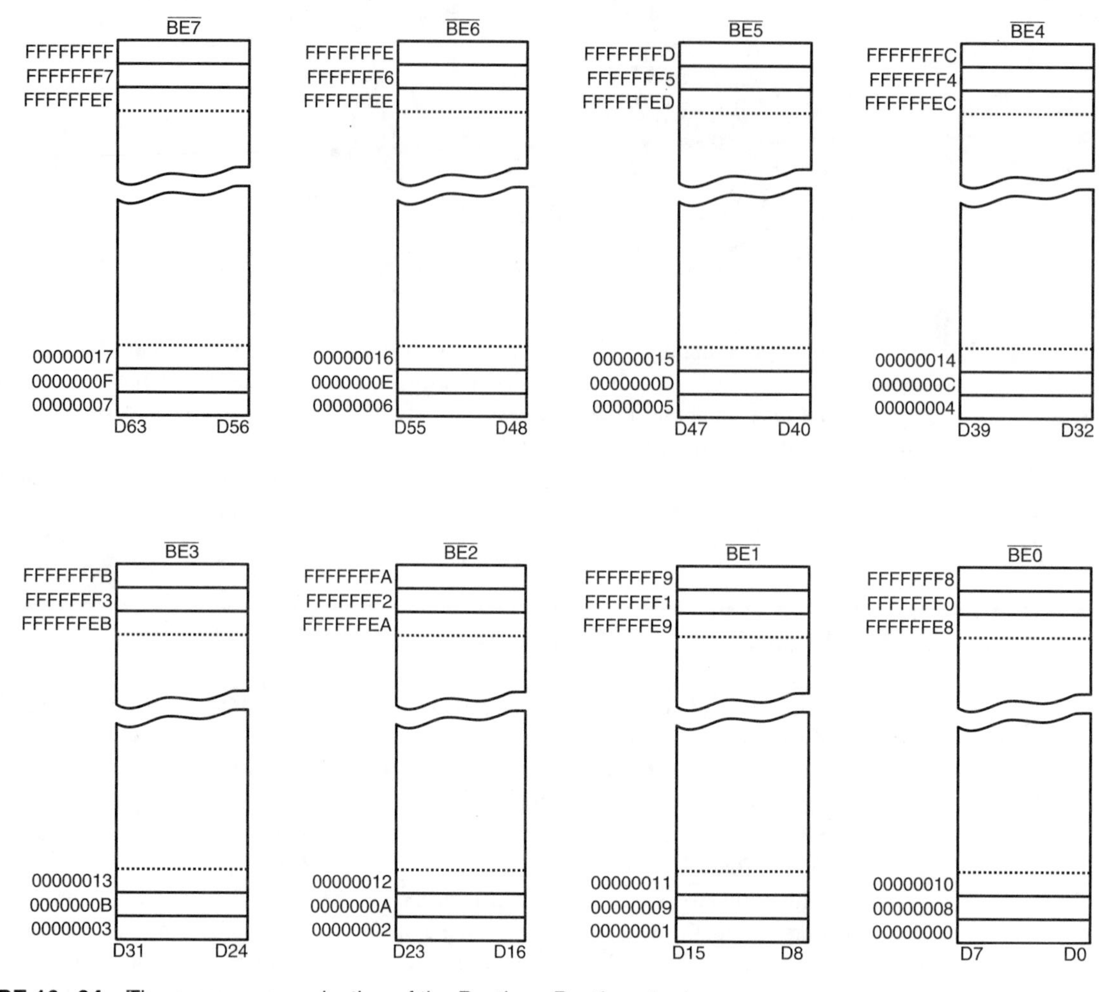

FIGURE 10–34 The memory organization of the Pentium–Pentium 4 microprocessors.

EXAMPLE 10–10

```
library ieee;
use ieee.std_logic_1164.all;

entity DECODER_10_36 is

port (
       A31, A30, A29, A28, A27, A26, A25, A24, A23, A22: in STD_LOGIC;
       SEL: out STD_LOGIC
);

end;

architecture V1 of DECODER_10_36 is

begin

       SEL <= not(A31 and A30 and A29 and A28 and A27 and A26 and A25 and A24
              and A23 and A22);

end V1;
```

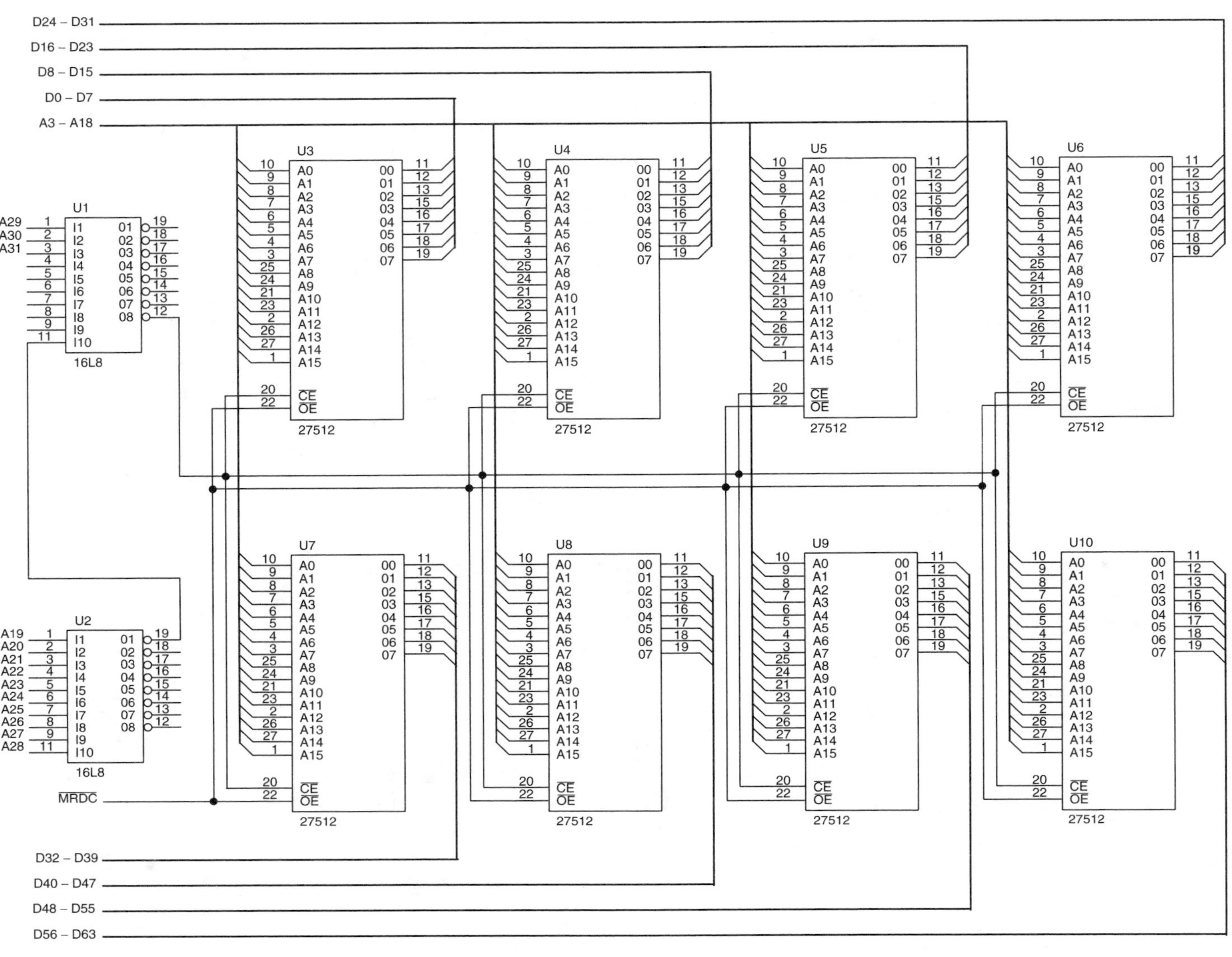

FIGURE 10–35 A small 512K-byte EPROM memory interfaced to the Pentium–Pentium 4 microprocessors.

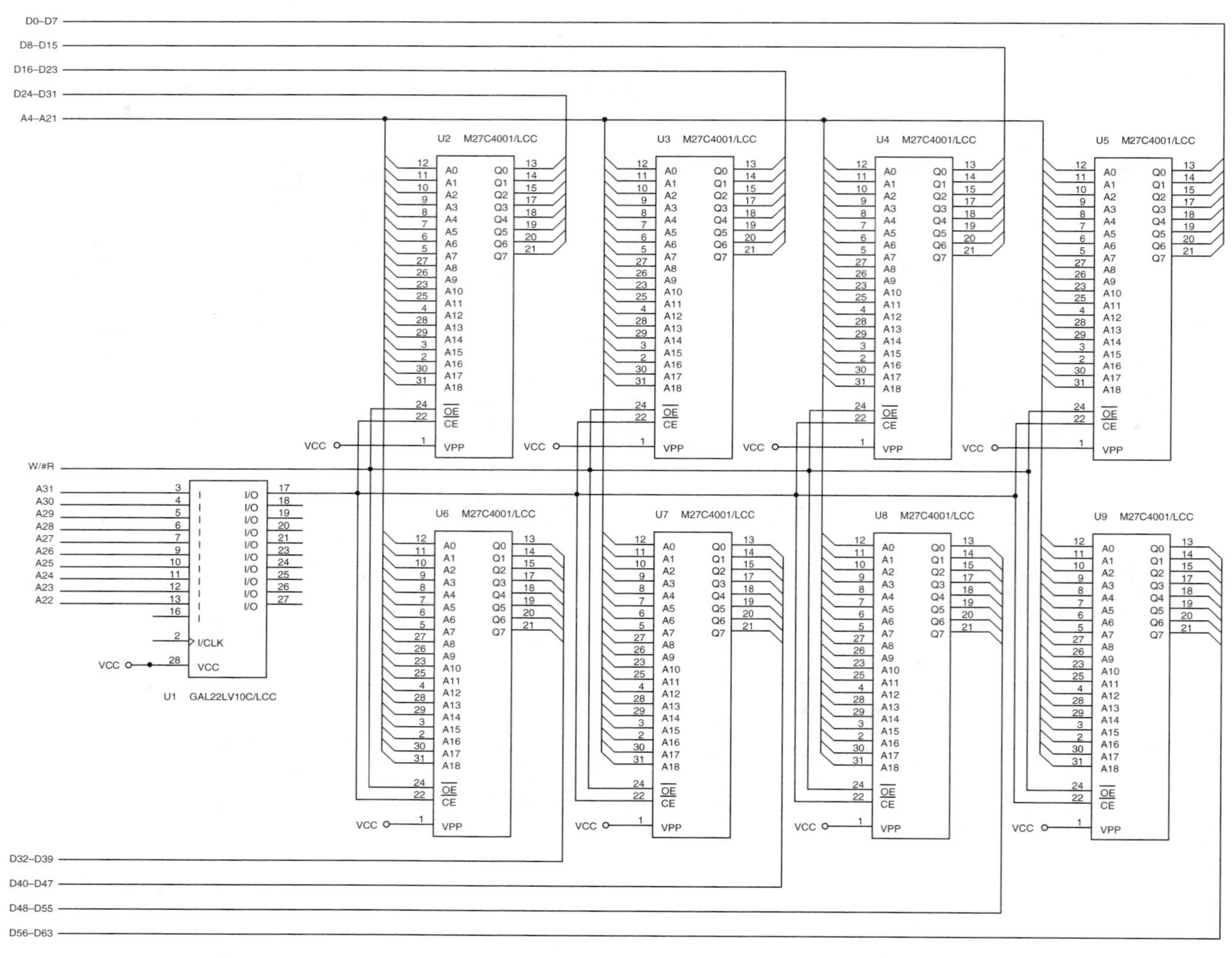

FIGURE 10–36 A small 4M-byte EPROM memory system for the Pentium–Pentium 4 microprocessors.

10–7 DYNAMIC RAM

Because RAM memory is often very large, it requires many SRAM devices at a great cost or just a few DRAMs (**dynamic RAMs**) at a much reduced cost. The DRAM memory, as briefly discussed in Section 10–1, is fairly complex because it requires address multiplexing and refreshing. Luckily, the integrated circuit manufacturers have provided a dynamic RAM controller that includes the address multiplexers and all the timing circuitry necessary for refreshing.

This section of the text covers the DRAM memory device in much more detail than in Section 10–1 and provides information on the use of a dynamic controller in a memory system.

DRAM Revisited

As mentioned in Section 10–1, a DRAM retains data for only 2–4 ms and requires the multiplexing of address inputs. Although address multiplexers were already covered in Section 10–1, the operation of the DRAM during refresh is explained in detail here.

As previously mentioned, a DRAM must be refreshed periodically because it stores data internally on capacitors that lose their charge in a short period of time. To refresh a DRAM, the contents of a section of the memory must periodically be read or written. Any read or write automatically refreshes an entire section of the DRAM. The number of bits that are refreshed depends on the size of the memory component and its internal organization.

Refresh cycles are accomplished by doing a read, a write, or a special refresh cycle that doesn't read or write data. The refresh cycle is internal to the DRAM and is often accomplished while other memory components in the system operate. This type of memory refresh is called *hidden refresh, transparent refresh*, or sometimes *cycle stealing*.

In order to accomplish a hidden refresh while other memory components are functioning, an $\overline{RAS}$-only cycle strobes a row address into the DRAM to select a row of bits to be refreshed. The $\overline{RAS}$ input also causes the selected row to be read out internally and rewritten into the selected bits. This recharges the internal capacitors that store the data. This type of refresh is hidden from the system because it occurs while the microprocessor is reading or writing to other sections of the memory.

The DRAM's internal organization contains a series of rows and columns. A 256K × 1 DRAM has 256 columns, each containing 256 bits, or rows organized into four sections of 64K bits each. Whenever a memory location is addressed, the column address selects a column (or internal memory word) of 1024 bits (one per section of the DRAM). Refer to Figure 10–37 for the internal structure of a 256K × 1 DRAM. Note that larger memory devices are structured similarly to the 256K × 1 device. The difference usually lies in either the size of each section or the number of sections in parallel.

Figure 10–38 illustrates the timing for an $\overline{RAS}$-only refresh cycle. The difference between the $\overline{RAS}$ and a read or write is that it applies only a refresh address, which is usually obtained from a 7- or 8-bit binary counter. The size of the counter is determined by the type of DRAM being refreshed. The refresh counter is incremented at the end of each refresh cycle so all the rows are refreshed in 2 or 4 ms, depending on the type of DRAM.

If there are 256 rows to be refreshed within 4 ms, as in a 256K × 1 DRAM, then the refresh cycle must be activated at least once every 15.6 μs in order to meet the refresh specification. For example, it takes the 8086/8088, running at a 5 MHz clock rate, 800 ns to do a read or a write. Because the DRAM must have a refresh cycle every 15.6 μs, for every 19 memory reads or writes, the memory system must run a refresh cycle or else memory data will be lost. This represents a loss of 5% of the computer's time, a small price to pay for the savings represented by using the dynamic RAM. In a modern system such as a 3.0 GHz Pentium 4, 15.6 μs is a great deal of

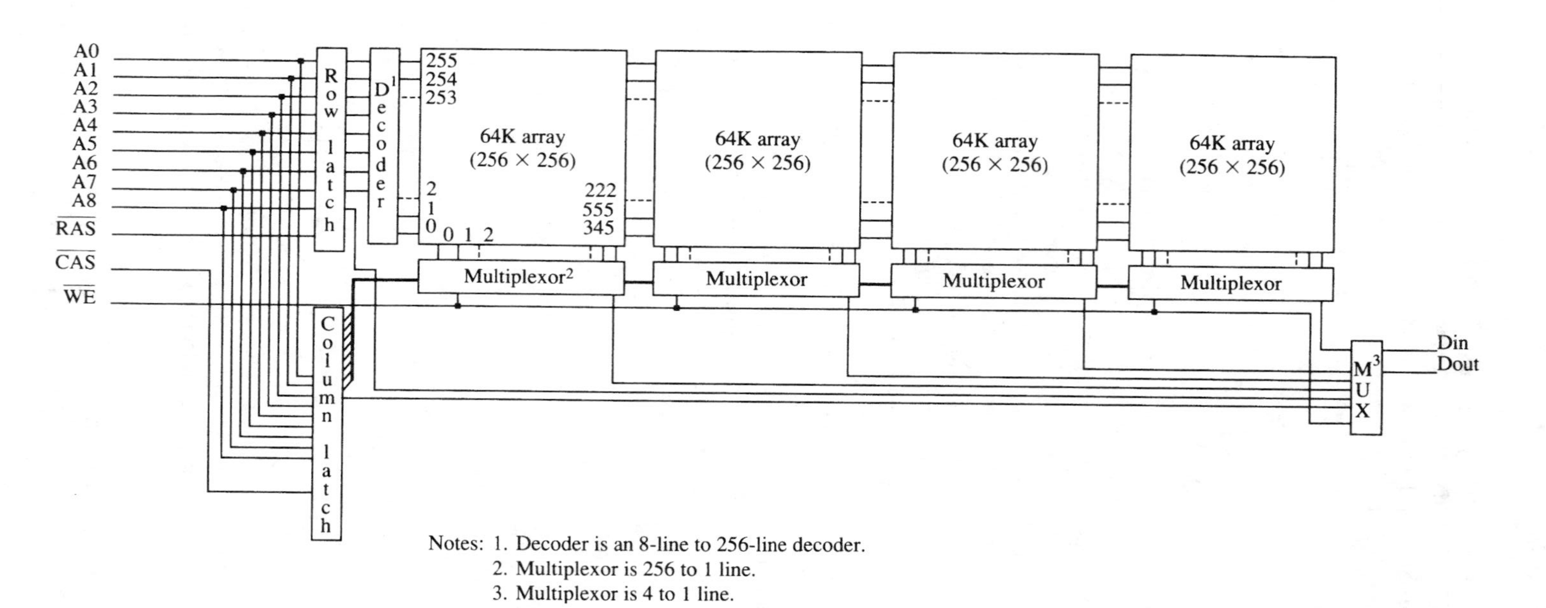

Notes: 1. Decoder is an 8-line to 256-line decoder.
2. Multiplexor is 256 to 1 line.
3. Multiplexor is 4 to 1 line.

FIGURE 10–37 The internal structure of a 256K × 1 DRAM. Note that each of the internal 256 words are 1024 bits wide.

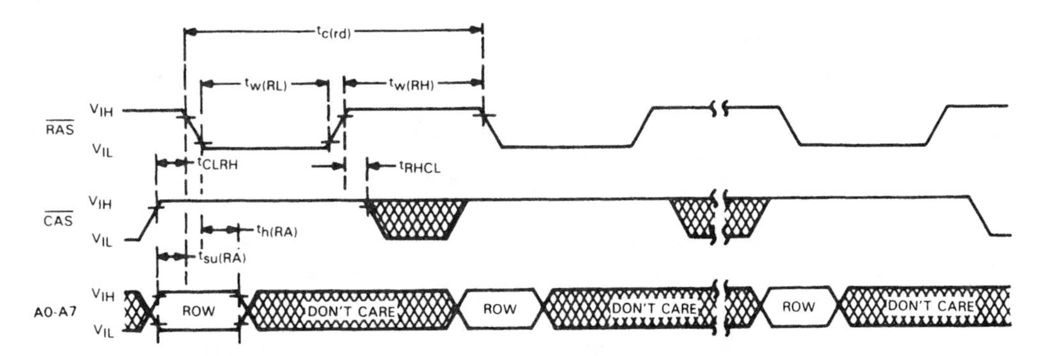

FIGURE 10–38 The timing diagram of the $\overline{RAS}$ refresh cycle for the TMS4464 DRAM. (Courtesy of Texas Instruments Corporation.)

time. Since the 3.0 GHz Pentium 4 executes an instruction in about one-third ns (many instructions execute in a single clock), it can execute about 46,000 instruction between refreshes. This means that in the new machines much less than 1% (≈0.002%) is required for a refresh.

EDO Memory

A slight modification to the structure of the DRAM changes the device into an EDO (**extended data output**) DRAM device. In the EDO memory, any memory access, including a refresh, stores the 256 bits selected by $\overline{RAS}$ into latches. These latches hold the next 256 bits of information, so in most programs, which are sequentially executed, the data are available without any wait states. This slight modification to the internal structure of the DRAM increases system performance by about 15 to 25%. Although EDO memory is no longer available, this technique is still employed in all modern DRAM.

SDRAM

Synchronous dynamic RAM (**SDRAM**) is used with most newer systems in one form or another because of its speed. Versions are available with access times of 10 ns for use with a 66 MHz system bus; 8 ns for use with a 100 MHz system bus; and 7 ns for the 133 MHz bus. At first, the access time may lead one to think that these devices operate without wait states, but that is not true. After all, DRAM access time is 60 ns and SDRAM access time is 10 ns. The 10 ns access time is misleading because it only applies to the second, third, and fourth 64-bit reads from the device. The first read requires the same number of waits as a standard DRAM.

When a burst transfer occurs to the SDRAM from the microprocessor, it takes three or four bus clocks before the first 64-bit number is read. Each subsequent number is read without wait states and in one bus cycle each. Because SDRAM bursts read four 64-bit numbers, and the second through the fourth require no waits and can be read in one bus cycle each, SDRAM outperforms standard DRAM or even EDO memory. This means that if it takes three bus cycles for the first number and three more for the next three, it takes a total of seven bus clocks to read four 64-bit numbers. If this is compared to DRAM, which takes three clocks per number or 12 clocks, you can see the increase in speed. Most estimates place SDRAM at about a 10% performance increase over EDO memory.

DDR

Double-data rate (DDR) memory is the latest improvement in a string of modifications to DRAM. The DDR memory transfers data at double the rate of SDRAM because it transfers data on each edge of the clock. Even though this seems as if it doubles the speed of the memory, it really does not. The main reason is that the access time problem still exists, with even the most advanced memory requiring an access time of 40 ns. If you think about a microprocessor running at many GHz, this is a very long time to wait for the memory. Hence the speed will not double as the name may suggest.

DRAM Controllers

In most systems, a DRAM controller-integrated circuit performs the task of address multiplexing and the generation of the DRAM control signals. Some newer embedded microprocessors, such as the 80186/80188, include the refresh circuitry as part of the microprocessor. Most modern computers contain the DRAM controller in the chip set so a stand-alone DRAM controller is not available. The DRAM controller in the chip set for the microprocessor times refresh cycles and inserts refresh cycles into the timing. The memory refresh is transparent to the microprocessor, since it really does not control refreshing.

10–8 SUMMARY

1. All memory devices have address inputs; data inputs and outputs, or just outputs; a pin for selection; and one or more pins that control the operation of the memory.
2. Address connections on a memory component are used to select one of the memory locations within the device. Ten address pins have 1024 combinations and therefore are able to address 1024 different memory locations.
3. Data connections on a memory are used to enter information to be stored in a memory location and also to retrieve information read from a memory location. Manufacturers list their memory as, for example, 4K × 4, which means that the device has 4K memory locations (4096) and that four bits are stored in each location.
4. Memory selection is accomplished via a chip selection pin ($\overline{CS}$) on many RAMs or a chip enable pin ($\overline{CE}$) on many EPROM or ROM memories.
5. Memory function is selected by an output enable pin ($\overline{OE}$) for reading data, which normally connects to the system read signal ($\overline{RD}$ or $\overline{MRDC}$). The write enable pin ($\overline{WE}$), for writing data, normally connects to the system write signal ($\overline{WR}$ or $\overline{MWTC}$).
6. An EPROM memory is programmed by an EPROM programmer and can be erased if exposed to ultraviolet light. Today, EPROMs are available in sizes from 1K × 8 all the way up to 512K × 8 and larger.
7. The Flash memory (EEPROM) is programmed in the system by using a 12 V or 5.0 V programming pulse.
8. Static RAM (SRAM) retains data for as long as the system power supply is attached. These memory types are available in sizes up to 128K × 8.
9. Dynamic RAM (DRAM) retains data for only a short period, usually 2–4 ms. This creates problems for the memory system designer because the DRAM must be refreshed periodically. DRAMs also have multiplexed address inputs that require an external multiplexer to provide each half of the address at the appropriate time.
10. Memory address decoders select an EPROM or RAM at a particular area of the memory. Commonly found address decoders include the 74LS138 3-to-8 line decoder, the 74LS139 2-to-4 line decoder, and programmed selection logic in the form of a PLD.
11. The PLD address decoder for microprocessors like the 8088 through the Pentium 4 reduce the number of integrated circuits required to complete a functioning memory system.

12. The 8088 minimum mode memory interface contains 20 address lines, eight data lines, and three control lines: $\overline{RD}$, $\overline{WR}$, and IO/$\overline{M}$. The 8088 memory functions correctly only when all these lines are used for memory interface.
13. The access speed of the EPROM must be compatible with the microprocessor to which it is interfaced. Many EPROMs available today have an access time of 450 ns, which is too slow for the 5 MHz 8088. In order to circumvent this problem, a wait state is inserted to increase memory access time to 660 ns.
14. Error-correction features are also available for memory systems, but these require the storage of many more bits. If an 8-bit number is stored with an error-correction circuit, it actually takes 13 bits of memory: five for an error checking code and eight for the data. Most error-correction integrated circuits are able to correct only a single-bit error.
15. The 8086/80286/80386SX memory interface has a 16-bit data bus and contains an M/$\overline{IO}$ control pin, whereas the 8088 has an 8-bit data bus and contains an IO/$\overline{M}$ pin. In addition to these changes, there is an extra control signal, bus high enable ($\overline{BHE}$).
16. The 8086/80386/80386SX memory is organized in two 8-bit banks: high bank and low bank. The high bank of memory is enabled by the $\overline{BHE}$ control signal and the low bank is enabled by the A_0 address signal or by the $\overline{BLE}$ control signal.
17. Two common schemes for selecting the banks in an 8086/80286/80386SX-based system include (1) a separate decoder for each bank and (2) separate $\overline{WR}$ control signals for each bank with a common decoder.
18. Memory interfaced to the 80386DX and 80486 is 32 bits wide, as selected by a 32-bit address bus. Because of the width of this memory, it is organized in four memory banks that are each eight bits wide. Bank selection signals are provided by the microprocessor as $\overline{BE3}$, $\overline{BE2}$, $\overline{BE1}$, and $\overline{BE0}$.
19. Memory interfaced to the Pentium–Pentium 4 is 64 bits wide, as selected by a 32-bit address bus. Because of the width of the memory, it is organized in eight banks that are each eight bits wide. Bank selection signals are provided by the microprocessor as $\overline{BE7}$–$\overline{BE0}$.
20. Dynamic RAM controllers are designed to control DRAM memory components. Many DRAM controllers today contain address multiplexers, refresh counters, and the circuitry required to do a periodic DRAM memory refresh.

10–9 QUESTIONS AND PROBLEMS

1. What types of connections are common to all memory devices?
2. List the number of words found in each memory device for the following numbers of address connections:
   (a) 8
   (b) 11
   (c) 12
   (d) 13
3. List the number of data items stored in each of the following memory devices and the number of bits in each datum:
   (a) 2K × 4
   (b) 1K × 1
   (c) 4K × 8
   (d) 16K × 1
   (e) 64K × 4
4. What is the purpose of the $\overline{CS}$ or $\overline{CE}$ pin on a memory component?
5. What is the purpose of the $\overline{OE}$ pin on a memory device?

6. What is the purpose of the $\overline{WE}$ pin on an SRAM?
7. How many bytes of storage do the following EPROM memory devices contain?
   (a) 2708
   (b) 2716
   (c) 2732
   (d) 2764
   (e) 27512
8. Why won't a 450 ns EPROM work directly with a 5 MHz 8088?
9. What can be stated about the amount of time needed to erase and write a location in a Flash memory device?
10. SRAM is an acronym for what type of device?
11. The 4016 memory has a $\overline{G}$ pin, an $\overline{S}$ pin, and a $\overline{W}$ pin. What are these pins used for in this RAM?
12. How much memory access time is required by the slowest 4016?
13. DRAM is an acronym for what type of device?
14. The 256M DIMM has 28 address inputs, yet it is a 256M DRAM. Explain how a 28-bit memory address is forced into 14 address inputs.
15. What are the purposes of the $\overline{CAS}$ and $\overline{RAS}$ inputs of a DRAM?
16. How much time is required to refresh the typical DRAM?
17. Why are memory address decoders important?
18. Modify the NAND gate decoder of Figure 10–13 to select the memory for address range DF800H–DFFFFH.
19. Modify the NAND gate decoder in Figure 10–13 to select the memory for address range 40000H–407FFH.
20. When the G1 input is high and both $\overline{G2A}$ and $\overline{G2B}$ are low, what happens to the outputs of the 74HCT138 3-to-8 line decoder?
21. Modify the circuit of Figure 10–15 to address memory range 70000H–7FFFFH.
22. Modify the circuit of Figure 10–15 to address memory range 40000H–4FFFFH.
23. Describe the 74LS139 decoder.
24. What is VHDL?
25. What are the five major keywords in VHDL for the five major logic functions (AND, OR, NAND, NOR, and invert)?
26. Equations are placed in what major block of a VHDL program?
27. Modify the circuit of Figure 10–19 by rewriting the PLD program to address memory at locations A0000H–BFFFFH for the ROM.
28. The $\overline{RD}$ and $\overline{WR}$ minimum mode control signals are replaced by what two control signals in the 8086 maximum mode?
29. Modify the circuit of Figure 10–20 to select memory at locations 60000H–77FFFH.
30. Modify the circuit of Figure 10–20 to select eight 27256 32K × 8 EPROMs at memory locations 40000H–7FFFFH.
31. Add another decoder to the circuit of Figure 10–21 so that an additional eight 62256 SRAMs are added at locations C0000H–FFFFFH.
32. The 74LS636 error-correction and detection circuit stores a check code with each byte of data. How many bits are stored for the check code?
33. What is the purpose of the SEF pin on the 74LS636?
34. The 74LS636 will correct ________ bits that are in error.
35. Outline the major difference between the buses of the 8086 and 8088 microprocessors.
36. What is the purpose of the $\overline{BHE}$ and A_0 pins on the 8086 microprocessor?
37. What is the $\overline{BLE}$ pin and what other pin has it replaced?
38. What two methods are used to select the memory in the 8086 microprocessor?
39. If $\overline{BHE}$ is a logic 0, then the ________ memory bank is selected.

40. If A_0 is a logic 0, then the ________ memory bank is selected.
41. Why don't separate bank read ($\overline{RD}$) strobes need to be developed when interfacing memory to the 8086?
42. Modify the circuit of Figure 10–28 so that the RAM is located at memory range 30000H–4FFFFH.
43. Develop a 16-bit-wide memory interface that contains SRAM memory at locations 200000H–21FFFFH for the 80386SX microprocessor.
44. Develop a 32-bit-wide memory interface that contains EPROM memory at locations FFFF0000H–FFFFFFFFH.
45. Develop a 64-bit-wide memory for the Pentium–Pentium 4 that contains EPROM at locations FFF00000H–FFFFFFFFH and SRAM at locations 00000000H–003FFFFFH.
46. On the Internet, search for the largest size EPROM you can find. List its size and manufacturer.
47. What is an $\overline{RAS}$-only cycle?
48. Can a DRAM refresh be done while other sections of the memory operate?
49. If a 1M $\times$ 1 DRAM requires 4 ms for a refresh and has 256 rows to be refreshed, no more than ________ of time must pass before another row is refreshed.
50. Scour the Internet to find the largest DRAM currently available.
51. Write a report on DDR memory. (Hint: Samsung makes them.)
52. Write a report that details RAMBUS RAM. Try to determine why this technology appears to have fallen by the wayside.

CHAPTER 11

Basic I/O Interface

INTRODUCTION

A microprocessor is great at solving problems, but if it can't communicate with the outside world, it is of little worth. This chapter outlines some of the basic methods of communications, both serial and parallel, between humans or machines and the microprocessor.

In this chapter, we first introduce the basic I/O interface and discuss decoding for I/O devices. Then, we provide detail on parallel and serial interfacing, both of which have a variety of applications. To study applications, we connect analog-to-digital and digital-to-analog converters, as well as both DC and stepper motors to the microprocessor.

CHAPTER OBJECTIVES

Upon completion of this chapter, you will be able to:

1. Explain the operation of the basic input and output interfaces
2. Decode 8-, 16-, and 32-bit I/O devices so that they can be used at any I/O port address
3. Define handshaking and explain how to use it with I/O devices
4. Interface and program the 82C55 programmable parallel interface
5. Interface LCD displays, LED displays, keyboards, ADC, DAC, and various other devices to the 82C55
6. Interface and program the 16550 serial communications interface adapter
7. Interface and program the 8254 programmable interval timer
8. Interface an analog-to-digital converter and a digital-to-analog converter to the microprocessor
9. Interface both DC and stepper motors to the microprocessor

11–1 INTRODUCTION TO I/O INTERFACE

In this section of the text I/O instructions (IN, INS, OUT, and OUTS) are explained and used in example applications. Also explained here is the concept of isolated (sometimes called direct or I/O mapped I/O) and memory-mapped I/O, the basic input and output interfaces, and handshaking. A working knowledge of these topics makes it easier to understand the connection and operation of the programmable interface components and I/O techniques presented in the remainder of this chapter and text.

The I/O Instructions

The instruction set contains one type of instruction that transfers information to an I/O device (OUT) and another to read information from an I/O device (IN). Instructions (INS and OUTS, found on all versions except the 8086/8088) are also provided to transfer strings of data between the memory and an I/O device. Table 11–1 lists all versions of each instruction found in the microprocessor's instruction set.

Instructions that transfer data between an I/O device and the microprocessor's accumulator (AL, AX, or EAX) are called **IN** and **OUT**. The I/O address is stored in register DX as a 16-bit I/O address or in the byte (p8) immediately following the opcode as an 8-bit I/O address. Intel calls the 8-bit form (p8) a **fixed address** because it is stored with the instruction, usually in a ROM. The 16-bit I/O address in DX is called a **variable address** because it is stored in a DX, and then used to address the I/O device. Other instructions that use DX to address I/O are the INS and OUTS instructions. I/O ports are eight bits in width so whenever a 16-bit port is accessed two consecutive 8-bit ports are actually addressed. A 32-bit I/O port is actually four 8-bit ports. For example, port 100H is accessed as a word, then 100H and 101H are actually accessed. Port 100H contains the least-significant part of the data and port 101H the most-significant part.

TABLE 11–1 Input/output instructions.

Instruction	*Data Width*	*Function*
IN AL, p8	8	A byte is input into AL from port p8
IN AX, p8	16	A word is input into AX from port p8
IN EAX, p8	32	A doubleword is input into EAX from port p8
IN AL, DX	8	A byte is input into AL from the port addressed by DX
IN AX, DX	16	A word is input into AX from the port addressed by DX
IN EAX, DX	32	A doubleword is input into EAX from the port addressed by DX
INSB	8	A byte is input from the port addressed by DI and stored into the extra segment memory location addressed by DI, then DI = DI ± 1
INSW	16	A word is input from the port addressed by DI and stored into the extra segment memory location addressed by DI, then DI = DI ± 2
INSD	32	A doubleword is input from the port addressed by DI and stored into the extra segment memory location addressed by DI, then DI = DI ± 4
OUT p8, AL	8	A byte is output from AL into port p8
OUT p8, AX	16	A word is output from AL into port p8
OUT p8, EAX	32	A doubleword is output from EAX into port p8
OUT DX, AL	8	A byte is output from AL into the port addressed by DX
OUT DX, AX	16	A word is output from AX into the port addressed by DX
OUT DX, EAX	32	A doubleword is output from EAX into the port addressed by DX
OUTSB	8	A byte is output from the data segment memory location addressed by SI into the port addressed by DX, then SI = SI ± 1
OUTSW	16	A word is output from the data segment memory location addressed by SI into the port addressed by DX, then SI = SI ± 2
OUTSD	32	A doubleword is output from the data segment memory location addressed by SI into the port addressed by DX, then SI = SI ± 4

Whenever data are transferred by using the IN or OUT instruction, the I/O address, often called a **port number** (or simply port), appears on the address bus. The external I/O interface decodes the port number in the same manner that it decodes a memory address. The 8-bit fixed port number (p8) appears on address bus connections A_7–A_0 with bits A_{15}–A_8 equal to 00000000_2. The address connections above A_{15} are undefined for an I/O instruction. The 16-bit variable port number (DX) appears on address connections A_{15}–A_0. This means that the first 256 I/O port addresses (00H–FFH) are accessed by both the fixed and variable I/O instructions, but any I/O address from 0100H to FFFFH is only accessed by the variable I/O address. In many dedicated systems, only the rightmost eight bits of the address are decoded, thus reducing the amount of circuitry required for decoding. In a PC computer, all 16 address bus bits are decoded with locations 0000H–03FFH, which are the I/O addresses used for I/O inside the PC on the ISA (**industry standard architecture**) bus.

The INS and OUTS instructions address an I/O device by using the DX register, but do not transfer data between the accumulator and the I/O device as do the IN and OUT instructions. Instead, these instructions transfer data between memory and the I/O device. The memory address is located by ES:DI for the INS instruction and by DS:SI for the OUTS instruction. As with other string instructions, the contents of the pointers are incremented or decremented, as dictated by the state of the direction flag (DF). Both INS and OUTS can be prefixed with the REP prefix, allowing more than one byte, word, or doubleword to be transferred between I/O and memory.

Isolated and Memory-Mapped I/O

There are two different methods of interfacing I/O to the microprocessor: **isolated I/O** and **memory-mapped I/O**. In the isolated I/O scheme, the IN, INS, OUT, and OUTS instructions transfer data between the microprocessor's accumulator or memory and the I/O device. In the memory-mapped I/O scheme, any instruction that references memory can accomplish the transfer. Both isolated and memory-mapped I/O are in use, so both are discussed in this text. The PC does not use memory-mapped I/O.

Isolated I/O. The most common I/O transfer technique used in the Intel microprocessor-based system is isolated I/O. The term *isolated* describes how the I/O locations are isolated from the memory system in a separate I/O address space. (Figure 11–1 illustrates both the isolated and memory-mapped address spaces for any Intel 80X86 or Pentium–Pentium 4 microprocessor.) The addresses for isolated I/O devices, called ports, are separate from the memory. Because the ports are separate, the user can expand the memory to its full size without using any of memory space for I/O devices. A disadvantage of isolated I/O is that the data transferred between I/O and the microprocessor must be accessed by the IN, INS, OUT, and OUTS instructions. Separate control signals for the I/O space are developed (using $M/\overline{IO}$ and $W/\overline{R}$), which indicate an I/O read ($\overline{IORC}$) or an I/O write ($\overline{IOWC}$) operation. These signals indicate that an I/O port address, which appears on the address bus, is used to select the I/O device. In the personal computer, isolated I/O ports are used for controlling peripheral devices. An 8-bit port address is used to access devices located on the system board, such as the timer and keyboard interface, while a 16-bit port is used to access serial and parallel ports as well as video and disk drive systems.

Memory-Mapped I/O. Unlike isolated I/O, memory-mapped I/O does not use the IN, INS, OUT, or OUTS instructions. Instead, it uses any instruction that transfers data between the microprocessor and memory. A memory-mapped I/O device is treated as a memory location in the memory map. The main advantage of memory-mapped I/O is that any memory transfer instruction can be used to access the I/O device. The main disadvantage is that a portion of the memory system is used as the I/O map. This reduces the amount of memory available to applications. Another advantage is that the $\overline{IORC}$ and $\overline{IOWC}$ signals have no function in a memory-mapped I/O system and may reduce the amount of circuitry required for decoding.

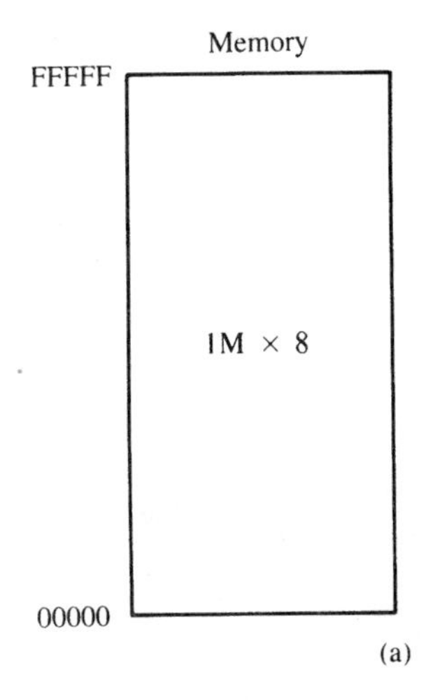

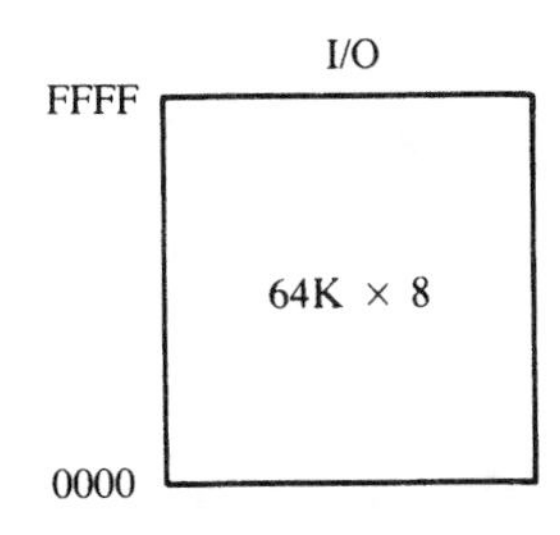

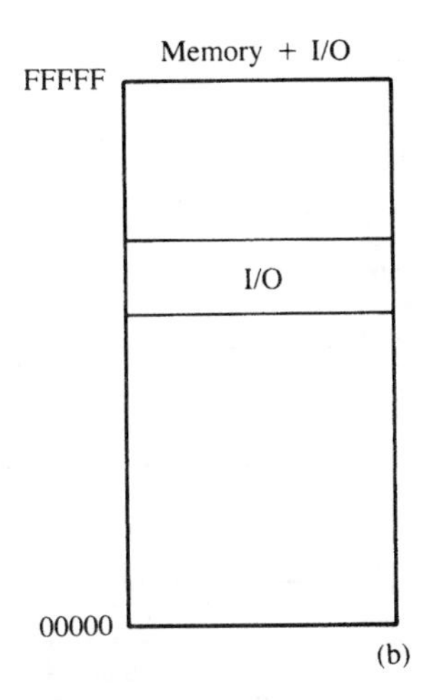

FIGURE 11–1 The memory and I/O maps for the 8086/8088 microprocessors. (a) Isolated I/O. (b) Memory-mapped I/O.

Personal Computer I/O Map

The personal computer uses part of the I/O map for dedicated functions. Figure 11–2 shows the I/O map for the PC. Note that I/O space between ports 0000H and 03FFH is normally reserved for the computer system and the ISA bus. The I/O ports located at 0400H–FFFFH are generally available for user applications, main-board functions, and the PCI bus. Note that the 80287 arithmetic coprocessor uses I/O addresses 00F8H–00FFH for communications. For this reason, Intel reserves I/O ports 00F0H–00FFH. The 80386–Pentium 4 use I/O ports 800000F8–800000FFH for communications to their coprocessors. The I/O ports located between 0000H and 00FFH are accessed via the fixed port I/O instructions; the ports located above 00FFH are accessed via the variable I/O port instructions.

Basic Input and Output Interfaces

The basic input device is a set of three-state buffers. The basic output device is a set of data latches. The term IN refers to moving data from the I/O device into the microprocessor, and the term OUT refers to moving data out of the microprocessor to the I/O device.

The Basic Input Interface. Three-state buffers are used to construct the 8-bit input port depicted in Figure 11–3. The external TTL data (simple toggle switches in this example) are connected to

FIGURE 11–2 The I/O map of a personal computer illustrating many of the fixed I/O areas.

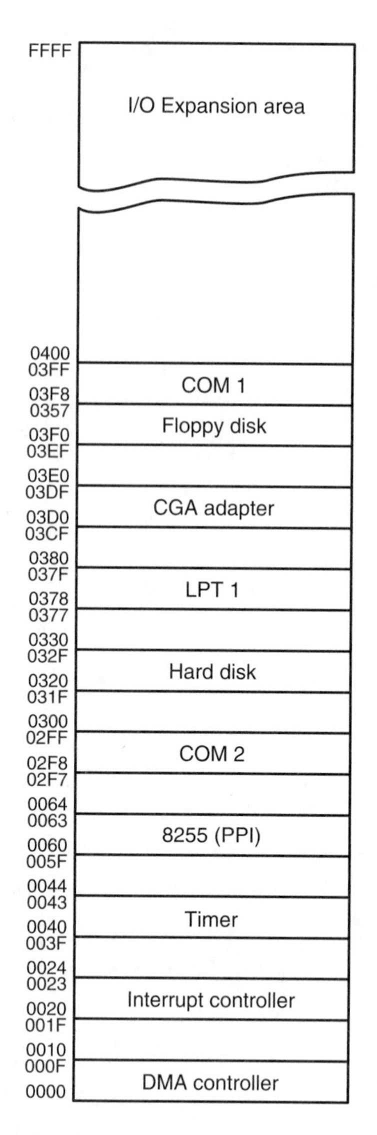

the inputs of the buffers. The outputs of the buffers connect to the data bus. The exact data bus connections depend on the version of the microprocessor. For example, the 8088 has data bus connections D_7–D_0, the 80386/80486 have connections D_{31}–D_0, and the Pentium–Pentium 4 have connections D_{63}–D_0. The circuit of Figure 11–3 allows the microprocessor to read the contents of the eight switches that connect to any 8-bit section of the data bus when the select signal $\overline{SEL}$ becomes a logic 0. Thus, whenever the IN instruction executes, the contents of the switches are copied into the AL register.

When the microprocessor executes an IN instruction, the I/O port address is decoded to generate the logic 0 on $\overline{SEL}$. A 0 placed on the output control inputs ($\overline{1G}$ and $\overline{2G}$) of the 74ALS244 buffer causes the data input connections (A) to be connected to the data output (Y) connections. If a logic 1 is placed on the output control inputs of the 74ALS244 buffer, the device enters the three-state high-impedance mode that effectively disconnects the switches from the data bus.

This basic input circuit is not optional and must appear any time that input data are interfaced to the microprocessor. Sometimes it appears as a discrete part of the circuit, as shown in Figure 11–3; many times it is built into a programmable I/O device.

Sixteen- or 32-bit data can also be interfaced to various versions of the microprocessor, but this is not nearly as common as using 8-bit data. To interface 16 bits of data, the circuit in

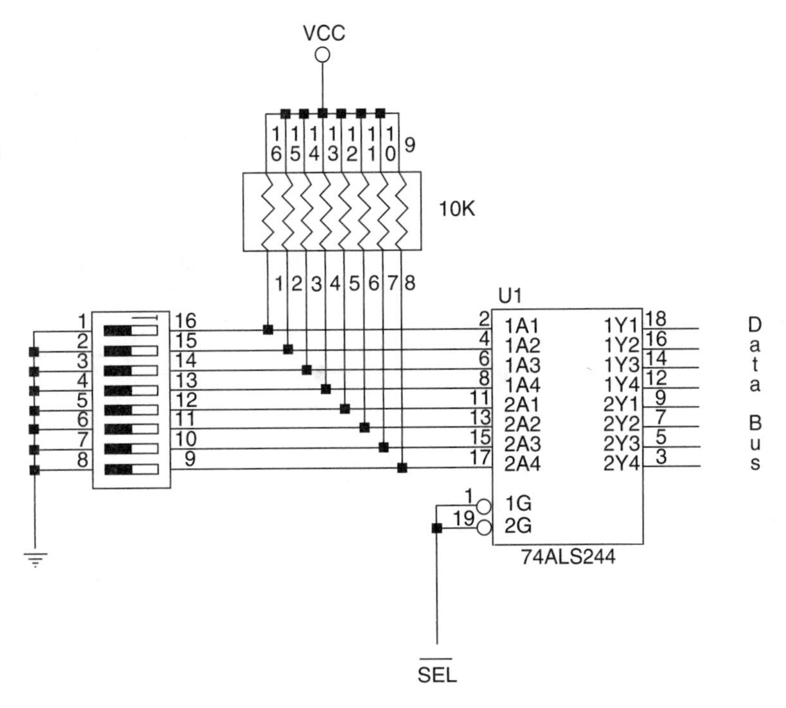

FIGURE 11–3 The basic input interface illustrating the connection of eight switches. Note that the 74ALS244 is a three-state buffer that controls the application of the switch data to the data bus.

Figure 11–3 is doubled to include two 74ALS244 buffers that connect 16 bits of input data to the 16-bit data bus. To interface 32 bits of data, the circuit is expanded by a factor of 4.

The Basic Output Interface. The basic output interface receives data from the microprocessor and usually must hold it for some external device. Its latches or flip-flops, like the buffers found in the input device, are often built into the I/O device.

Figure 11–4 shows how eight simple light-emitting diodes (LEDs) connect to the microprocessor through a set of eight data latches. The latch stores the number output by the microprocessor from the data bus so that the LEDs can be lit with any 8-bit binary number. Latches are needed to hold the data because when the microprocessor executes an OUT instruction, the data are only present on the data bus for less than 1.0 μs. Without a latch, the viewer would never see the LEDs illuminate.

When the OUT instruction executes, the data from AL, AX, or EAX are transferred to the latch via the data bus. Here, the D inputs of a 74ALS374 octal latch are connected to the data bus to capture the output data, and the Q outputs of the latch are attached to the LEDs. When a Q output becomes a logic 0, the LED lights. Each time that the OUT instruction executes, the $\overline{SEL}$ signal to the latch activates, capturing the data output to the latch from any 8-bit section of the data bus. The data are held until the next OUT instruction executes. Thus, whenever the output instruction is executed in this circuit, the data from the AL register appear on the LEDs.

Handshaking

Many I/O devices accept or release information at a much slower rate than the microprocessor. Another method of I/O control, called **handshaking** or **polling**, synchronizes the I/O device with the microprocessor. An example of a device that requires handshaking is a parallel printer that prints a few hundred characters per second (CPS). It is obvious that the microprocessor can send more than a few hundred CPS to the printer, so a way to slow the microprocessor down to match speeds with the printer must be developed.

Figure 11–5 illustrates the typical input and output connections found on a printer. Here, data are transferred through a series of data connections (D_7–D_0). BUSY indicates that the printer is busy. $\overline{STB}$ is a clock pulse used to send data to the printer for printing.

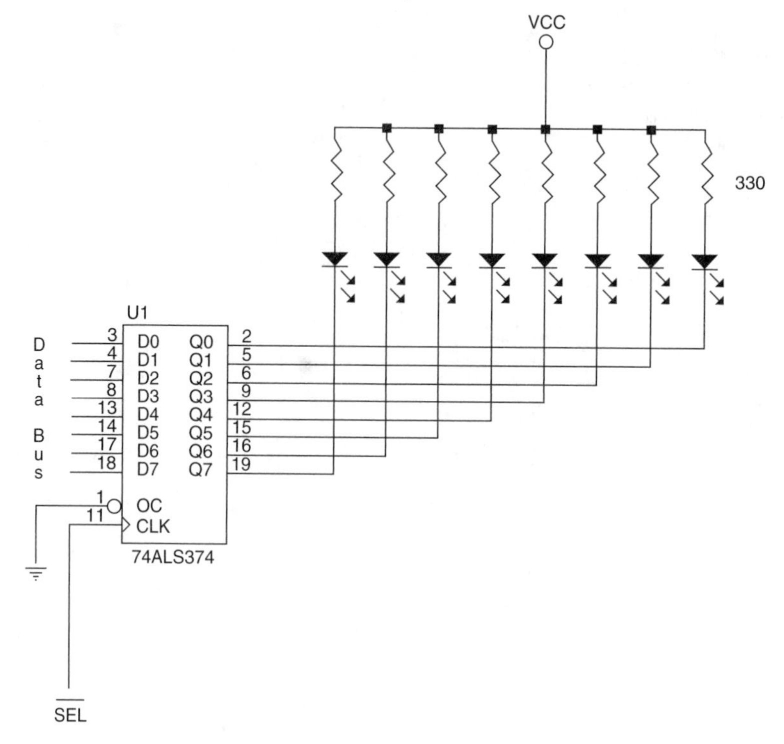

FIGURE 11–4 The basic output interface connected to a set of LED displays.

The ASCII data to be printed by the printer are placed on D_7–D_0, and a pulse is applied to the $\overline{STB}$ connection. The strobe signal sends or clocks the data into the printer so that they can be printed. As soon as the printer receives the data, it places a logic 1 on the BUSY pin, indicating that the printer is busy printing data. The microprocessor software polls or tests the BUSY pin to decide whether the printer is busy. If the printer is busy, the microprocessor waits; if it is not busy, the microprocessor sends the next ASCII character to the printer. Example 11–1 illustrates a simple procedure that tests the printer BUSY flag and then sends data to the printer if it is not busy. Here, the PRINT procedure prints the ASCII-coded contents of BL only if the BUSY flag is a logic 0, indicating that the printer is not busy. This procedure is called each time a character is to be printed.

EXAMPLE 11–1

```
;An assembly language procedure that prints the ASCII contents of BL.

PRINT PROC   NEAR

      .REPEAT                      ;test the busy flag
             IN AL,BUSY
             TEST AL,BUSY_BIT
      .UNTIL ZERO
      MOV AL,BL                    ;position data in AL
      OUT PRINTER,AL               ;print data
      RET

PRINT ENDP
```

Notes about Interfacing Circuitry

A part of interfacing requires some knowledge of electronics. This portion of the introduction to interfacing examines some of the many facets of electronic interfacing. Before a circuit or device

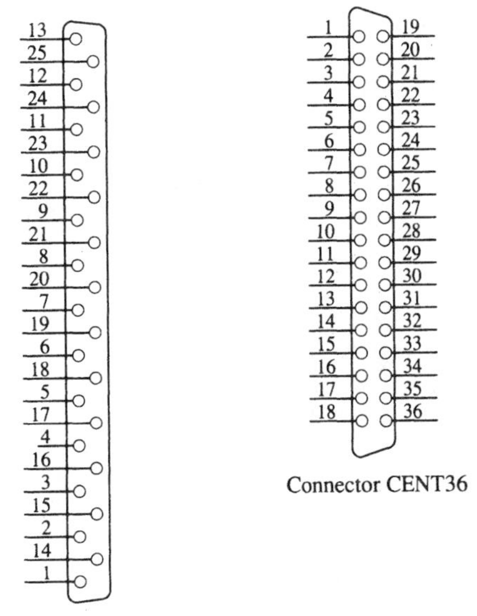

DB25 Pin number	CENT36 Pin number	Function	DB25 Pin number	CENT36 Pin number	Function
1	1	$\overline{\text{Data Strobe}}$	12	12	Paper empty
2	2	Data 0 (D0)	13	13	Select
3	3	Data 1 (D1)	14	14	Afd
4	4	Data 2 (D2)	15	32	$\overline{\text{Error}}$
5	5	Data 3 (D3)	16	—	$\overline{\text{RESET}}$
6	6	Data 4 (D4)	17	31	Select in
7	7	Data 5 (D5)	18—25	19—30	Ground
8	8	Data 6 (D6)	—	17	Frame ground
9	9	Data 7 (D7)	—	16	Ground
10	10	$\overline{\text{Ack}}$	—	33	Ground
11	11	Busy			

FIGURE 11–5 The DB25 connector found on computers and the Centronics 36-pin connector found on printers for the Centronics parallel printer interface.

can be interfaced to the microprocessor, the terminal characteristics of the microprocessor and its associated interfacing components must be known. (This subject was introduced at the start of Chapter 9.)

Input Devices. Input devices are already TTL and compatible, and therefore can be connected to the microprocessor and its interfacing components, or they are switch-based. Most switch-based devices are either open or connected. These are not TTL levels—TTL levels are a logic 0 (0.0 V–0.8 V) or a logic 1 (2.0 V–5.0 V).

For a switch-based device to be used as a TTL-compatible input device, some conditioning must be applied. Figure 11–6 shows a simple toggle switch that is properly connected to function as an input device. Notice that a pull-up resistor is used to ensure that when the switch is open, the output signal is a logic 1; when the switch is closed, it connects to ground, producing a valid logic 0 level. The value of the pull-up resistor is not critical—it merely assures that the signal is

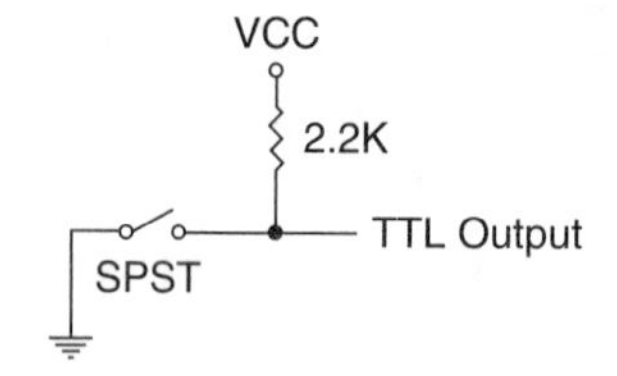

FIGURE 11–6 A single-pole, single-throw switch interfaced as a TTL device.

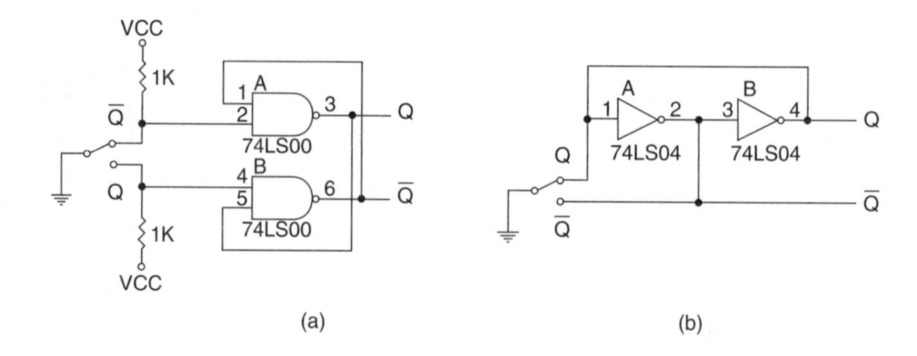

FIGURE 11–7 Debouncing switch contacts: (a) conventional debouncing and (b) practical debouncing.

at a logic 1 level. A standard range of values for pull-up resistors is usually anywhere between 1K Ω and 10K Ω.

Mechanical switch contacts physically bounce when they are closed, which can create a problem if a switch is used as a clocking signal for a digital circuit. To prevent problems with bounces, one of the two circuits depicted in Figure 11–7 can be constructed. The first circuit (a) is a classical textbook bounce eliminator; the second (b) is a more practical version of the same circuit. Because the first version costs more money to construct, in practice, the second would be used because it requires no pull-up resistors and only two inverters instead of two NAND gates.

You may notice that both circuits in Figure 11–7 are asynchronous flip-flops. The circuit of (b) functions in the following manner: Suppose that the switch is currently at position $\overline{Q}$. If it is moved toward Q but does not yet touch Q, the Q output of the circuit is a logic 0. The logic 0 state is remembered by the inverters. The output of inverter B connects to the input of inverter A. Because the output of inverter B is a logic 0, the output of inverter A is a logic 1. The logic 1 output of inverter A maintains the logic 0 output of inverter B. The flip-flop remains in this state until the moving switch-contact first touches the Q connection. As soon as the Q input from the switch becomes a logic 0, it changes the state of the flip-flop. If the contact bounces back away from the Q input, the flip-flop remembers and no change occurs, thus eliminating any bounce.

Output Devices. Output devices are far more diverse than input devices, but many are interfaced in a uniform manner. Before any output device can be interfaced, we must understand what the voltages and currents are from the microprocessor or a TTL interface component. The voltages are TTL-compatible from the microprocessor of the interfacing element. (Logic 0 = 0.0 V to 0.4 V; logic 1 = 2.4 V to 5.0 V.) The currents for a microprocessor and many microprocessor-interfacing components are less than for standard TTL components. (Logic 0 = 0.0 to 2.0 mA; logic 1 = 0.0 to 400 μA.)

Once the output currents are known, a device can be interfaced to one of the outputs. Figure 11–8 shows how to interface a simple LED to a microprocessor peripheral pin. Notice that a transistor driver is used in Figure 11–8(a) and a TTL inverter is used in Figure 11–8(b). The TTL inverter (standard version) provides up to 16 mA of current at a logic 0 level, which is more than enough to drive a standard LED. A standard LED requires 10 mA of forward bias current to light. In both circuits, we assume that the voltage drop across the LED is about 2.0 V. The

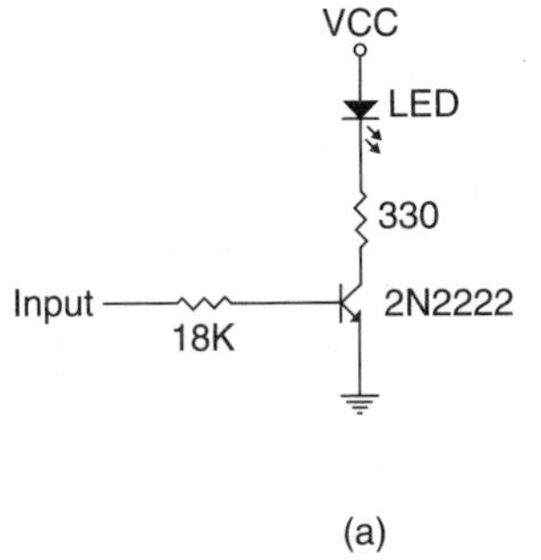

FIGURE 11–8 Interfacing an LED: (a) using a transistor and (b) using an inverter.

VCC
LED
330
Input
18K
2N2222
(a)

VCC
LED
330
A
1
2
Input
7404
(b)

data sheet for an LED states that the nominal drop is 1.65 V, but it is known from experience that the drop is anywhere between 1.5 V and 2.0 V. This means that the value of the current-limiting resistor is 3.0 V ÷ 10 mA or 300 Ω. Since 300 Ω is not a standard resistor value (the lowest cost), a 330 Ω resistor is chosen for this interface.

In the circuit of Figure 11–8(a), we elected to use a switching transistor in place of the TTL buffer. The 2N2222 is a good low-cost, general-purpose switching transistor that has a minimum gain of 100. In this circuit, the collector current is 10 mA, so the base current will be 1/100 of the collector current of 0.1 mA. To determine the value of the base current–limiting resistor, use the 0.1 mA base current and a voltage drop of 1.7 V across the base current–limiting resistor. The TTL input signal has a minimum value of 2.4 V and the drop across the emitter-base junction is 0.7 V. The difference is 1.7 V, which is the voltage drop across the resistor. The value of the resistor is 1.7 V ÷ 0.1 mA or 17K Ω. Because 17K Ω is not a standard value, an 18K Ω resistor is chosen.

Suppose that we need to interface a 12 V DC motor to the microprocessor and the motor current is 1A. Obviously, we cannot use a TTL inverter for two reasons: The 12 V signal would burn out the inverter and the amount of current far exceeds the 16 mA maximum current from the inverter. We cannot use a 2N2222 transistor either, because the maximum amount of current is 250 mA to 500 mA, depending on the package style chosen. The solution is to use a Darlington-pair.

Figure 11–9 illustrates a motor connected to the Darlington-pair. The Darlington-pair has a minimum current gain of 7000 and a maximum current of 4A. The value of the bias resistor is calculated exactly the same as the one used in the LED driver. The current through the resistor is 1.0 A ÷ 7000, or about 0.143 mA. The voltage drop across the resistor is 0.9 V because of the two diode drops (base/emitter junctions) instead of one. The value of the bias resistor is 0.9 V ÷ 0.143 mA or 6.29K Ω. The standard value of 6.2 K Ω is used in the circuit. The Darlington-pair must use a heat sink because of the amount of current going through it, and the diode must also be present to prevent the Darlington-pair from being destroyed by the inductive kickback from the motor. This circuit is also used to interface mechanical relays or just about any device that requires a large amount of current or a change in voltage.

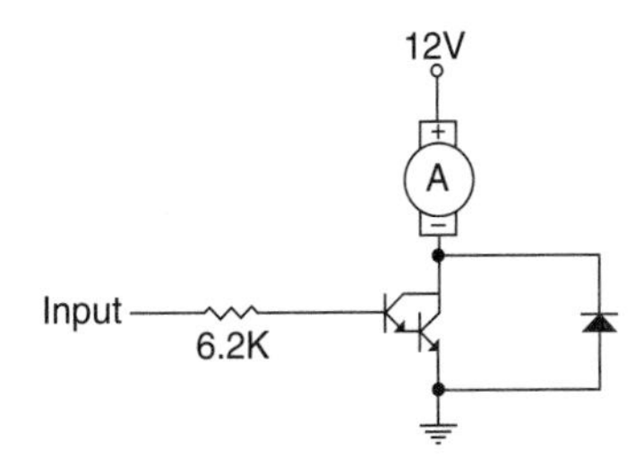

FIGURE 11–9 A DC motor interfaced to a system by using a Darlington-pair.

11–2 I/O PORT ADDRESS DECODING

I/O port address decoding is very similar to memory address decoding, especially for memory-mapped I/O devices. In fact, we do not discuss memory-mapped I/O decoding because it is treated the same as memory (except that the $\overline{IORC}$ and $\overline{IOWC}$ are not used because there is no IN or OUT instruction). The decision to use memory-mapped I/O is often determined by the size of the memory system and the placement of the I/O devices in the system.

The main difference between memory decoding and isolated I/O decoding is the number of address pins connected to the decoder. We decode A_{31}–A_0, A_{23}–A_0, or A_{19}–A_0 for memory, and A_{15}–A_0 for isolated I/O. Sometimes, if the I/O devices use only fixed I/O addressing, we decode only A_7–A_0. In the personal computer system, we always decode all 16 bits of the I/O port address. Another difference with isolated I/O is that $\overline{IORC}$ and $\overline{IOWC}$ activate I/O devices for a read or write operation. On earlier versions of the microprocessor, $IO/\overline{M} = 1$ and $\overline{RD}$ or $\overline{WR}$ are used to activate I/O devices. On the newest versions of the microprocessor, the $M/\overline{IO} = 0$ and $W/\overline{R}$ are combined and used to activate I/O devices.

Decoding 8-Bit I/O Port Addresses

As mentioned, the fixed I/O instruction uses an 8-bit I/O port address that appears on A_{15}–A_0 as 0000H–00FFH. If a system will never contain more than 256 I/O devices, we often decode only address connections A_7–A_0 for an 8-bit I/O port address. Thus, we ignore address connections A_{15}–A_8. Embedded systems often use 8-bit port addresses. Please note that the DX register can also address I/O ports 00H–FFH. If the address is decoded as an 8-bit address, we can never include I/O devices that use a 16-bit I/O address. The personal computer never uses or decodes an 8-bit address.

Figure 11–10 illustrates a 74ALS138 decoder that decodes 8-bit I/O ports F0H through F7H. (We assume that this system will only use I/O ports 00H–FFH for this decoder example.) This decoder is identical to a memory address decoder except we only connect address bits A_7–A_0 to the inputs of the decoder. Figure 11–11 shows the PLD version, using a GAL22V10 for this decoder. The PLD is a better decoder circuit because the number of integrated circuits has been reduced to one device. The VHDL program for the PLD appears in Example 11–2.

EXAMPLE 11–2

```
-- VHDL code for the decoder of Figure 11-11

library ieee;
use ieee.std_logic_1164.all;

entity DECODER_11_11 is

port (
        A7, A6, A5, A4, A3, A2, A1, A0: in STD_LOGIC;
        D0, D1, D2, D3, D4, D5, D6, D7: out STD_LOGIC
);

end;

architecture V1 of DECODER_11_11 is

begin

      D0 <= not( A7 and A6 and A5 and A4 and not A3 and not A2 and not A1 and
             not A0 );
      D1 <= not( A7 and A6 and A5 and A4 and not A3 and not A2 and not A1 and
             A0 );
      D2 <= not( A7 and A6 and A5 and A4 and not A3 and not A2 and A1 and
             not A0 );
```

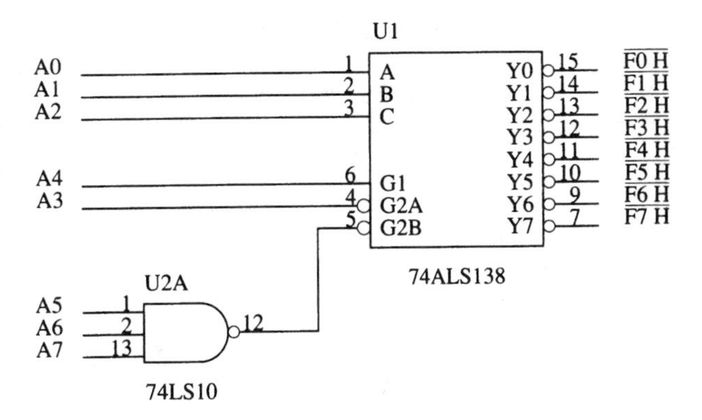

FIGURE 11–10 A port decoder that decodes 8-bit I/O ports. This decoder generates active low outputs for ports F0H–F7H.

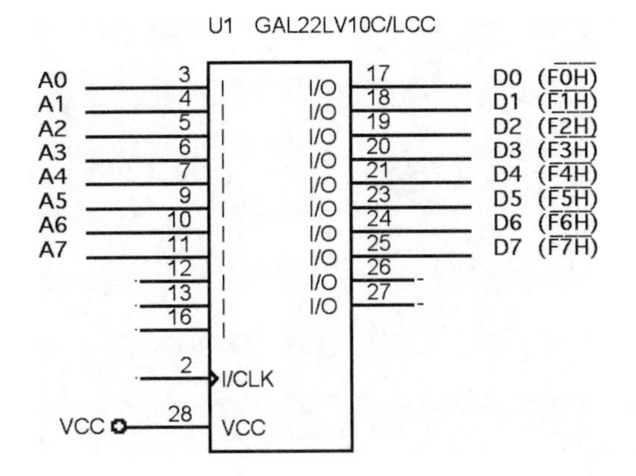

FIGURE 11–11 A PLD that generates part selection signals $\overline{\text{F0H}}$–$\overline{\text{F7H}}$.

```
        D3 <= not( A7 and A6 and A5 and A4 and not A3 and not A2 and A1 and
              A0 );
        D4 <= not( A7 and A6 and A5 and A4 and not A3 and A2 and not A1 and
              not A0 );
        D5 <= not( A7 and A6 and A5 and A4 and not A3 and A2 and not A1 and
              A0 );
        D6 <= not( A7 and A6 and A5 and A4 and not A3 and A2 and A1 and
              not A0 );
        D0 <= not( A7 and A6 and A5 and A4 and not A3 and A2 and A1 and
              A0 );

end V1;
```

Decoding 16-Bit I/O Port Addresses

Personal computer systems typically use 16-bit I/O addresses. It is relatively rare to find 16-bit port addresses in embedded systems. The main difference between decoding an 8-bit I/O address and a 16-bit I/O address is that eight additional address lines (A15–A8) must be decoded. Figure 11–12 illustrates a circuit that contains a PLD and a four-input NAND gate used to decode I/O ports EFF8H–EFFFH.

The NAND gate decodes part of the address (A15, A14, A13, and A11) because the PLD does not have enough address inputs. The output of the NAND gate connects to the Z input of the PLD and is decoded as part of the I/O port address. The PLD generates address strobes for I/O ports $\overline{\text{EFF8H}}$–$\overline{\text{EFFFH}}$. The program for the PLD is listed in Example 11–3.

EXAMPLE 11–3

```
-- VHDL code for the decoder of Figure 11-12

library ieee;
use ieee.std_logic_1164.all;
```

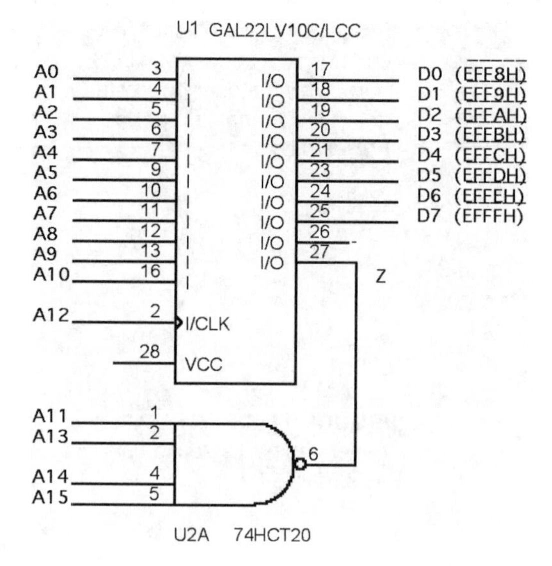

FIGURE 11–12 A PLD that decodes 16-bit I/O ports EFF8H through EFFFH.

```
entity DECODER_11_12 is

port (
        Z, A12, A10, A9, A8, A7, A6, A5, A4, A3, A2, A1, A0: in STD_LOGIC;
        D0, D1, D2, D3, D4, D5, D6, D7: out STD_LOGIC
);

end;

architecture V1 of DECODER_11_12 is

begin

        D0 <= not ( not Z and not A12 and A10 and A9 and A8 and A7 and A6 and A5
               and A4 and A3 and not A2 and not A1 and not A0 );
        D1 <= not ( not Z and not A12 and A10 and A9 and A8 and A7 and A6 and A5
               and A4 and A3 and not A2 and not A1 and A0 );
        D2 <= not ( not Z and not A12 and A10 and A9 and A8 and A7 and A6 and A5
               and A4 and A3 and not A2 and A1 and not A0 );
        D3 <= not ( not Z and not A12 and A10 and A9 and A8 and A7 and A6 and A5
               and A4 and A3 and not A2 and A1 and A0 );
        D4 <= not ( not Z and not A12 and A10 and A9 and A8 and A7 and A6 and A5
               and A4 and A3 and A2 and not A1 and not A0 );
        D5 <= not ( not Z and not A12 and A10 and A9 and A8 and A7 and A6 and A5
               and A4 and A3 and A2 and not A1 and A0 );
        D6 <= not ( not Z and not A12 and A10 and A9 and A8 and A7 and A6 and A5
               and A4 and A3 and A2 and A1 and not A0 );
        D7 <= not ( not Z and not A12 and A10 and A9 and A8 and A7 and A6 and A5
               and A4 and A3 and A2 and A1 and A0 );

end V1;
```

8- and 16-Bit-Wide I/O Ports

Now that I/O port addresses are understood and we learned that an I/O port address is probably simpler to decode than a memory address (because of the number of bits), interfacing between the microprocessor and 8- or 16-bit-wide I/O devices can be explained. Data transferred to an 8 bit I/O device exist in one of the I/O banks in a 16-bit microprocessor such as the 80386SX. There are 64K different 8-bit ports, but only 32K different 16-bit ports because a 16-bit port uses two 8-bit ports. The I/O system on such a microprocessor contains two 8-bit memory banks,

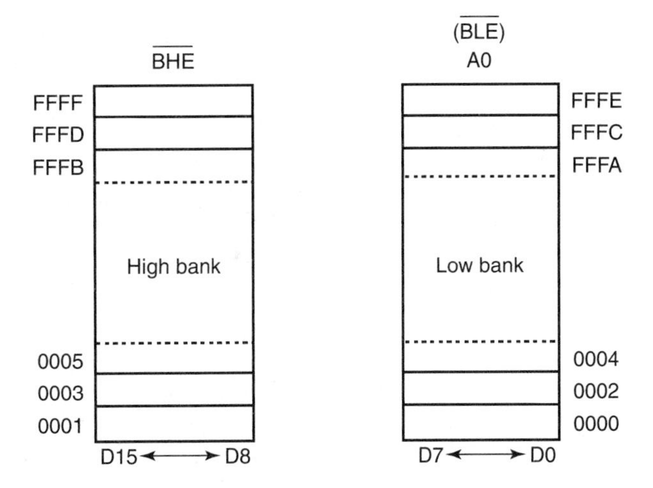

FIGURE 11–13 The I/O banks found in the 8086, 80186, 80286, and 80386SX.

just as memory does. This is illustrated in Figure 11–13, which shows the separate I/O banks for a 16-bit system such as the 80386SX.

Because two I/O banks exist, any 8-bit I/O write requires a separate write strobe to function correctly. I/O reads do not require separate read strobes. As with memory, the microprocessor reads only the byte it expects and ignores the other byte. The only time that a read can cause problems is when the I/O device responds incorrectly to a read operation. In the case of an I/O device that responds to a read from the wrong bank, we may need to include separate read signals. This case is discussed later in this chapter.

Figure 11–14 illustrates a system that contains two different 8-bit output devices, located at 8-bit I/O addresses 40H and 41H. Because these are 8-bit devices and because they appear in different I/O banks, separate I/O write signals are generated to clock a pair of latches that capture port data. Note that all I/O ports use 8-bit addresses. Thus, ports 40H and 41H can each be addressed as separate 8-bit ports, or together as one 16-bit port. The program for the PLD decoder used in Figure 11–14 is illustrated in Example 11–4.

EXAMPLE 11–4

```
-- VHDL code for the decoder of Figure 11-14

library ieee;
use ieee.std_logic_1164.all;

entity DECODER_11_14 is

port (
       BHE, IOWC, A7, A6, A5, A4, A3, A2, A1, A0: in STD_LOGIC;
       D0, D1: out STD_LOGIC
);

end;

architecture V1 of DECODER_11_14 is

begin

       D0 <= BHE or IOWC or A7 or not A6 or A5 or A4 or A3 or A2 or A1 or A0;
       D1 <= BHE or IOWC or A7 or not A6 or A5 or A4 or A3 or A2 or A1 or
             not A0;

end V1;
```

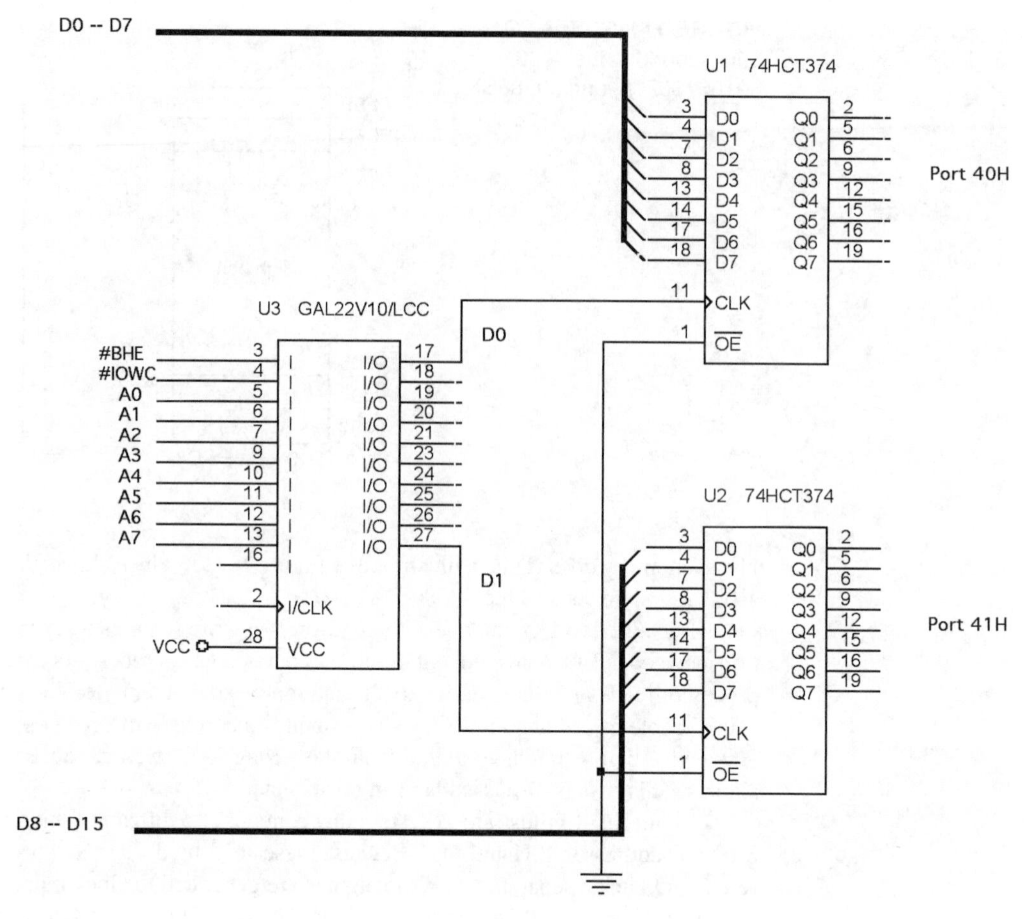

FIGURE 11–14 An I/O port decoder that selects ports 40H and 41H for output data.

When selecting 16-bit-wide I/O devices, the $\overline{\text{BLE}}$ (A_0) and $\overline{\text{BHE}}$ pins have no function because both I/O banks are selected together. Although 16-bit I/O devices are relatively rare, a few do exist for analog-to-digital and digit-to-analog converters, as well as for some video and disk interfaces.

Figure 11–15 illustrates a 16-bit input device connected to function at 8-bit I/O addresses 64H and 65H. Notice that the PLD decoder does not have a connection for address bits $\overline{\text{BLE}}$ (A_0) and $\overline{\text{BHE}}$ because these signals do not apply to 16-bit-wide I/O devices. The program for the PLD, illustrated in Example 11–5, shows how the enable signals are generated for the three-state buffers (74HCT244) used as input devices.

EXAMPLE 11–5

```
-- VHDL code for the decoder of Figure 11-15

library ieee;
use ieee.std_logic_1164.all;

entity DECODER_11_15 is

port (
       IORC, A7, A6, A5, A4, A3, A2, A1: in STD_LOGIC;
       D0: out STD_LOGIC
);
```

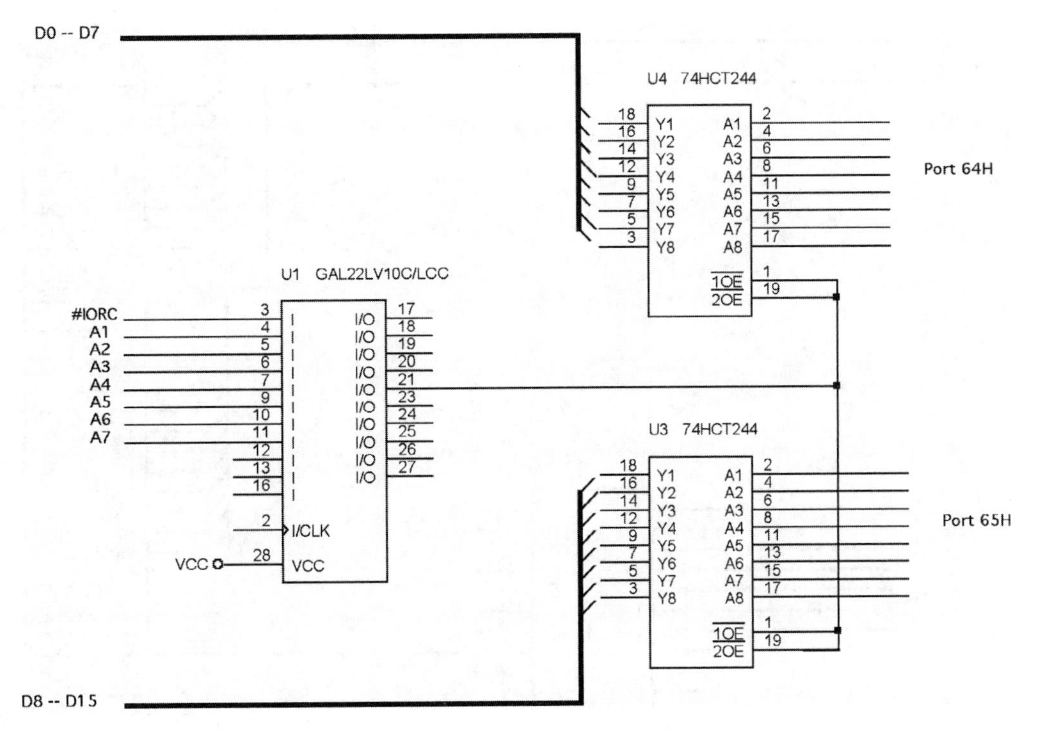

FIGURE 11–15 A 16-bit-wide port decoded at I/O addresses 64H and 65H.

```
end;

architecture V1 of DECODER_11_15 is

begin

     D0 <= IORC or A7 or not A6 or not A5 or A4 or A3 or not A2 or A1;

end V1;
```

32-Bit-Wide I/O Ports

Although 32-bit-wide I/O ports are not common, they may eventually become commonplace because of newer buses found in computer systems. The once-promising EISA system bus supports 32-bit I/O as well as the VESA local and current PCI bus, but not many I/O devices are 32 bits in width.

The circuit of Figure 11–16 illustrates a 32-bit input port for the 80386DX through the 80486DX microprocessors. As with prior interfaces, this circuit uses a single PLD to decode the I/O ports and four 74HCT244 buffers to connect the I/O data to the data bus. The I/O ports decoded by this interface are the 8-bit ports 70H–73H, as illustrated by the PLD program in Example 11–6. Again, we only decode an 8-bit I/O port address. When writing software to access this port it is crucial to use the address 70H for the 32-bit input as in the instruction IN EAX, 70H.

EXAMPLE 11–6

```
-- VHDL code for the decoder of Figure 11-16

library ieee;
use ieee.std_logic_1164.all;

entity DECODER_11_16 is

port (
```

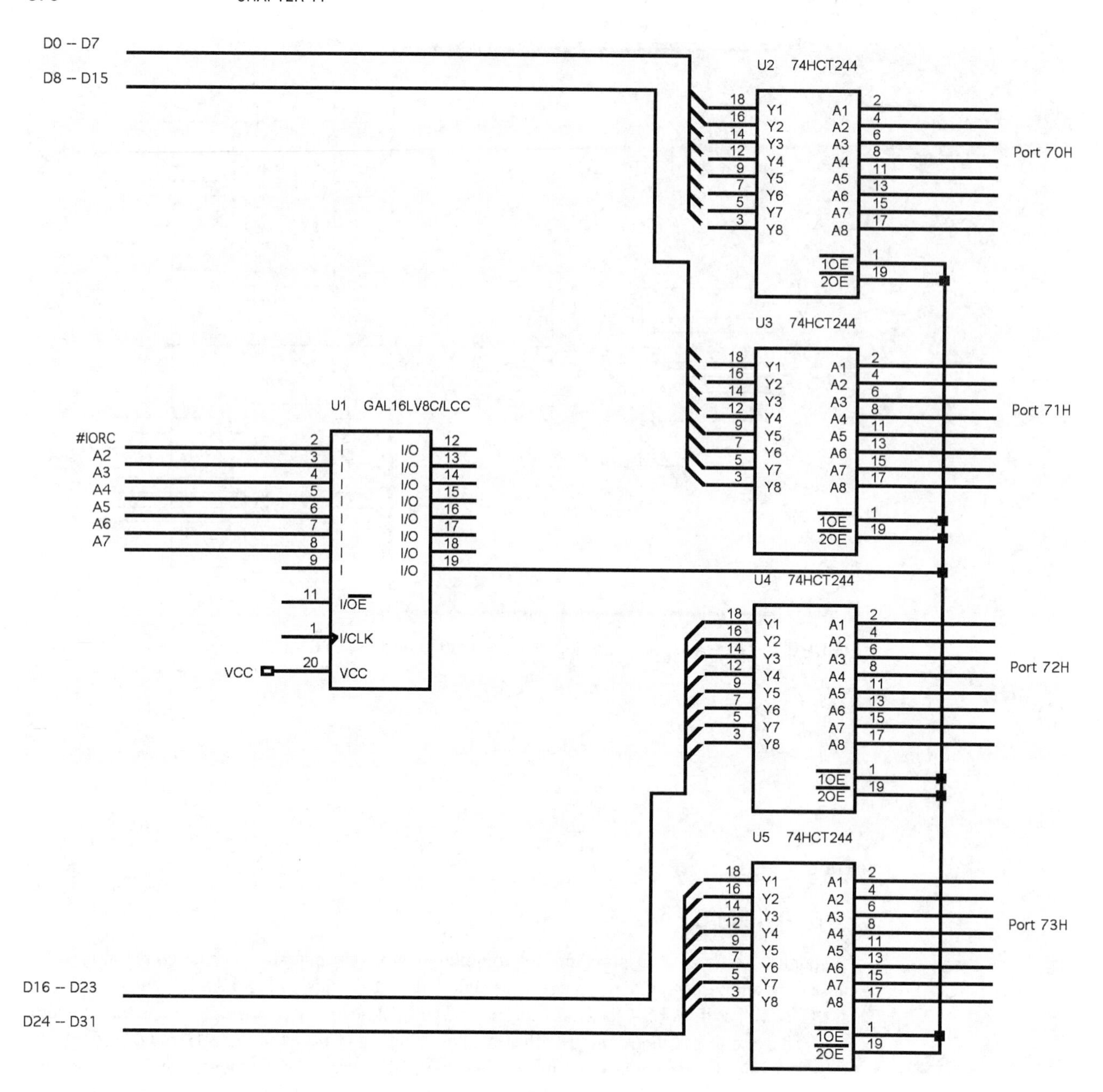

FIGURE 11–16 A 32-bit-wide port decoded at 70H through 73H for the 80486DX microprocessor.

```
        IORC, A7, A6, A5, A4, A3, A2: in STD_LOGIC;
        D0: out STD_LOGIC
);

end;

architecture V1 of DECODER_11_16 is

begin

        D0 <= IORC or A7 or not A6 or not A5 or not A4 or A3 or A2;

end V1;
```

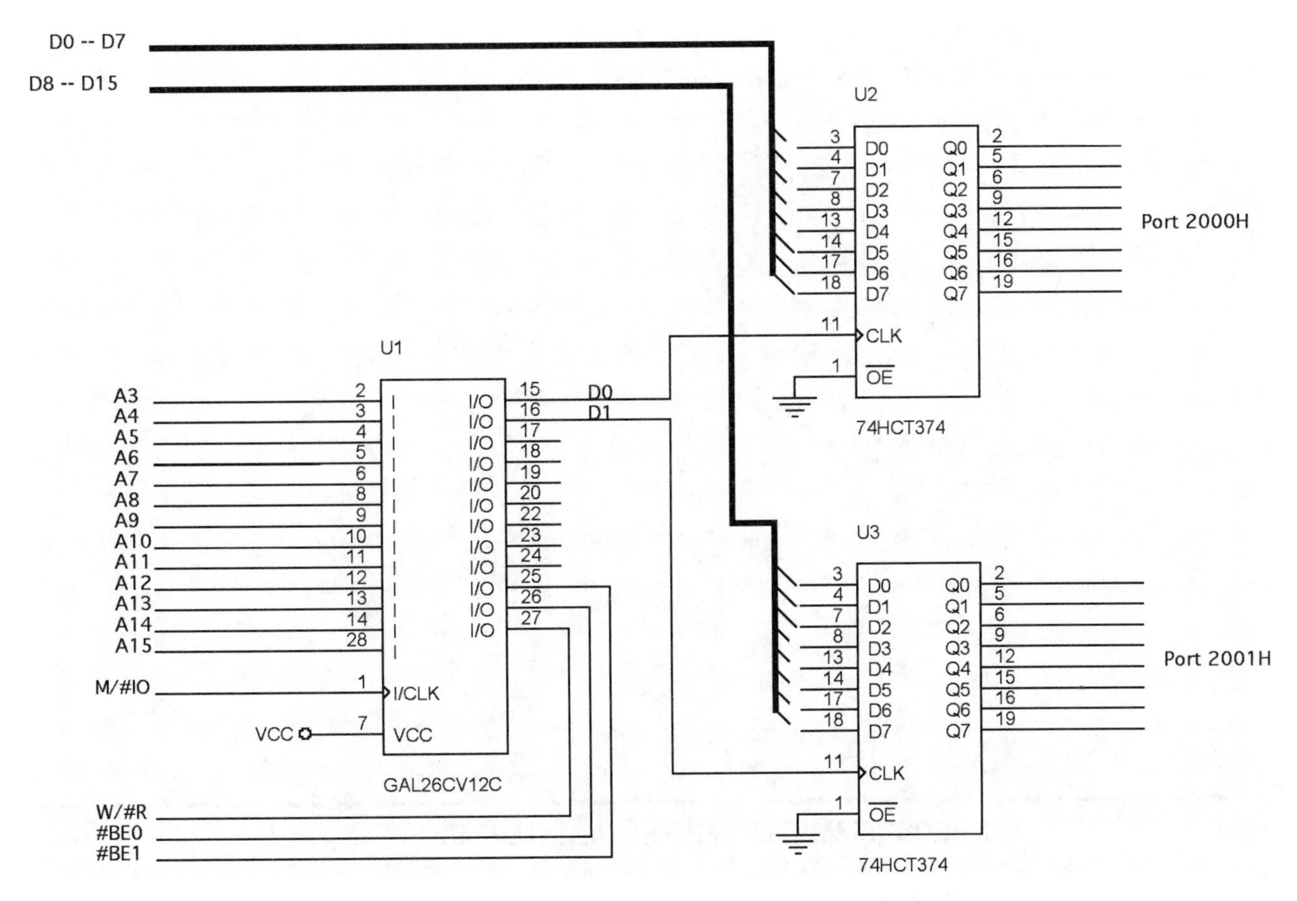

FIGURE 11–17 A Pentium 4 interfaced to a 16-bit-wide I/O port at port addresses 2000H and 2001H.

With the Pentium–Pentium 4 microprocessors and their 64-bit data buses, I/O ports appear in various banks, as determined by the I/O port address. For example, 8-bit I/O port 0034H appears in Pentium I/O bank 4, while the 16-bit I/O ports 0034H–0035H appear in Pentium banks 4 and 5. A 32-bit I/O access in the Pentium system can appear in any four consecutive I/O banks. For example, 32-bit I/O ports 0100H–0103H appear in banks 0–3. How is a 64-bit I/O device interfaced? The widest I/O transfers are 32 bits, and currently there are no 64-bit I/O instructions to support 64-bit transfers.

Suppose that we need to interface a simple 16-bit-wide output port at I/O port addresses 2000H and 2001H. The rightmost three bits of the lowest port address are 000 for port 2000H. This means that port 2000H is in memory bank 0. Likewise the rightmost three binary bits of I/O port 2001H are 001, which means that port 2001H is in bank 1. An interface is illustrated in Figure 11–17 and the PLD program is listed in Example 11–7.

The control signals $M/\overline{IO}$ and $W/\overline{R}$ must be combined to generate an I/O write signal for the latches and both $\overline{BE0}$ and $\overline{BE1}$ bank enable signals must be used to steer the write signal to the correct latch clock for address 2000H (bank 0) and 2001H (bank 1). The only problem that can arise in interfacing is when the I/O port spans across a 64-bit boundary, for example, a 16-bit-wide port located at 2007H and 2008H. In this case port 2007H uses bank 7 and 2008H uses bank 0, but the address that is decoded is different for each location: 0010 0000 0000 0XXX is decoded for 2007H and 0010 0000 0000 1XXX is decoded for 2008H. It is probably best to avoid situations such as this.

EXAMPLE 11–7

```
-- VHDL code for the decoder of Figure 11-17

library ieee;
use ieee.std_logic_1164.all;

entity DECODER_11_17 is

port (
       MIO, BE0, BE1, WR, A15, A14, A13, A12, A11, A10, A9, A8, A7, A6, A5, A4,
              A3: in STD_LOGIC;
       D0, D1: out STD_LOGIC
);

end;

architecture V1 of DECODER_11_17 is

begin

       D0 <= MIO or BE0 or not WR or A15 or A14 or not A13 or A12 or A11 or A10
              or A9 or A8 or A7 or A6 or A5 or A4 or A3;
       D1 <= MIO or BE1 or not WR or A15 or A14 or not A13 or A12 or A11 or A10
             or A9 or A8 or A7 or A6 or A5 or A4 or not A3;

end V1;
```

11–3 THE PROGRAMMABLE PERIPHERAL INTERFACE

The 82C55 **programmable peripheral interface** (PPI) is a very popular, low-cost interfacing component found in many applications. This is true even with all the programmable devices available for simple applications. The PPI, which has 24 pins for I/O that are programmable in groups of 12 pins, has groups that operate in three distinct modes of operation. The 82C55 can interface any TTL-compatible I/O device to the microprocessor. The 82C55 (CMOS version) requires the insertion wait states if operated with a microprocessor using higher than an 8 MHz clock. It also provides at least 2.5 mA of sink (logic 0) current at each output, with a maximum of 4.0 mA. Because I/O devices are inherently slow, wait states used during I/O transfers do not significantly affect the speed of the system. The 82C55 still finds application (compatible for programming, although it may not appear in the system as a discrete 82C55), even in the latest Pentium 4–based computer system. The modern computer uses a few 82C55s located inside the chip set for various features on the personal computer. The 82C55 is used for interface to the keyboard and the parallel printer port in many personal computers, but it is found as a function within an interfacing chip set. The chip set also controls the timer and reads data from the keyboard interface.

Basic Description of the 82C55

Figure 11–18 illustrates the pin-out diagram of the 82C55 in both the DIP format and the surface mount (flat pack). Its three I/O ports (labeled A, B, and C) are programmed as groups. Group A connections consist of port A (PA_7–PA_0) and the upper half of port C (PC_7–PC_4), and group B consists of port B (PB_7–PB_0) and the lower half of port C (PC_3–PC_0). The 82C55 is selected by its $\overline{CS}$ pin for programming and for reading or writing to a port. Register selection is accomplished through the A_1 and A_0 input pins that select an internal register for programming or operation. Table 11–2 shows the I/O port assignments used for programming and access to the I/O

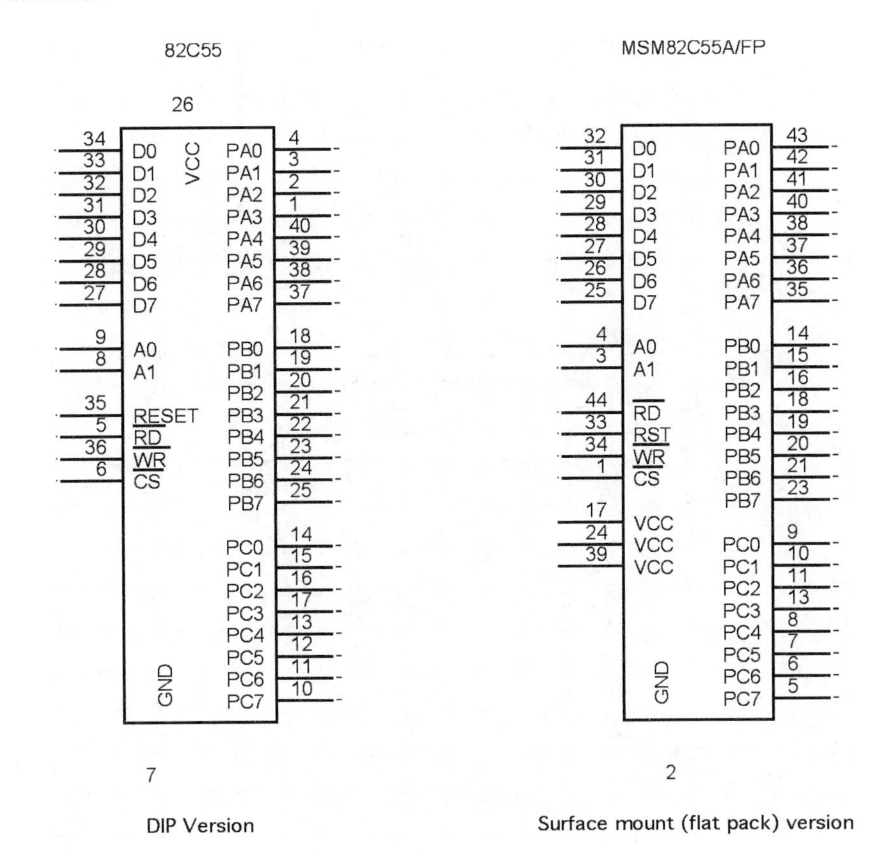

FIGURE 11–18 The pinout of the 82C55 peripheral interface adapter (PPI).

TABLE 11–2 I/O port assignments for the 82C55.

A_1	A_0	*Function*
0	0	Port A
0	1	Port B
1	0	Port C
1	1	Command register

ports. In the personal computer a pair of 82C55s, or their equivalents, are decoded at I/O ports 60H–63H and also at ports 378H–37BH.

The 82C55 is a fairly simple device to interface to the microprocessor and program. For the 82C55 to be read or written, the $\overline{CS}$ input must be a logic 0 and the correct I/O address must be applied to the A_1 and A_0 pins. The remaining port address pins are don't cares as far as the 82C55 is concerned, and are externally decoded to select the 82C55.

Figure 11–19 shows an 82C55 connected to the 80386SX so that it functions at 8-bit I/O port addresses C0H (port A), C2H (port B), C4H (port C), and C6H (command register). This interface uses the low bank of the 80386SX I/O map. Notice from this interface that all the 82C55 pins are direct connections to the 80386SX, except for the $\overline{CS}$ pin. The $\overline{CS}$ pin is decoded and selected by a 74ALS138 decoder.

The RESET input to the 82C55 initializes the device whenever the microprocessor is reset. A RESET input to the 82C55 causes all ports to be set up as simple input ports using mode 0 operation. Because the port pins are internally programmed as input pins after a RESET, damage is prevented when the power is first applied to the system. After a RESET, no other commands are needed to program the 82C55, as long as it is used as an input device for all three ports. Note that

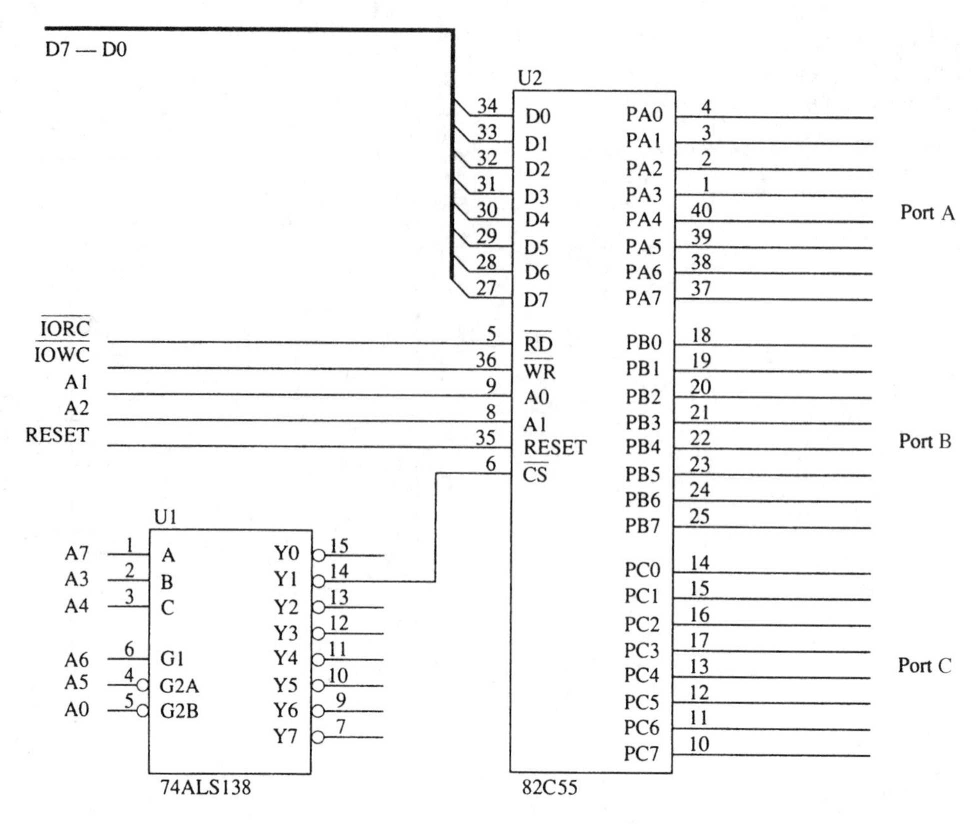

FIGURE 11–19 The 82C55 interfaced to the low bank of the 80386SX microprocessor.

an 82C55 is interfaced to the personal computer at port addresses 60H–63H for keyboard control, and also for controlling the speaker, timer, and other internal devices such as memory expansion. It is also used for the parallel printer port at I/O ports 378H–37BH.

Programming the 82C55

The 82C55 is programmed through the two internal command registers that are illustrated in Figure 11–20. Notice that bit position 7 selects either command byte A or command byte B. Command byte A programs the function of group A and B. Command byte B sets (1) or resets (0) bits of port C only if the 82C55 is programmed in mode 1 or 2.

Group B pins (port B and the lower part of port C) are programmed as either input or output pins. Group B operates in either mode 0 or mode 1. Mode 0 is the basic input/output mode that allows the pins of group B to be programmed as simple input and latched output connections. Mode 1 operation is the strobed operation for group B connections, where data are transferred through port B and handshaking signals are provided by port C.

Group A pins (port A and the upper part of port C) are programmed as either input or output pins. The difference is that group A can operate in modes 0, 1, and 2. Mode 2 operation is a bidirectional mode of operation for port A.

If a 0 is placed in bit position 7 of the command byte, command byte B is selected. This command allows any bit of port C to be set (1) or reset (0), if the 82C55 is operated in either mode 1 or 2. Otherwise, this command byte is not used for programming. The bit set/reset feature is often used in a control system to set or clear a control bit at port C. The bit set/reset function is glitch-free, which means that the other port C pins will not change during the bit set/reset command.

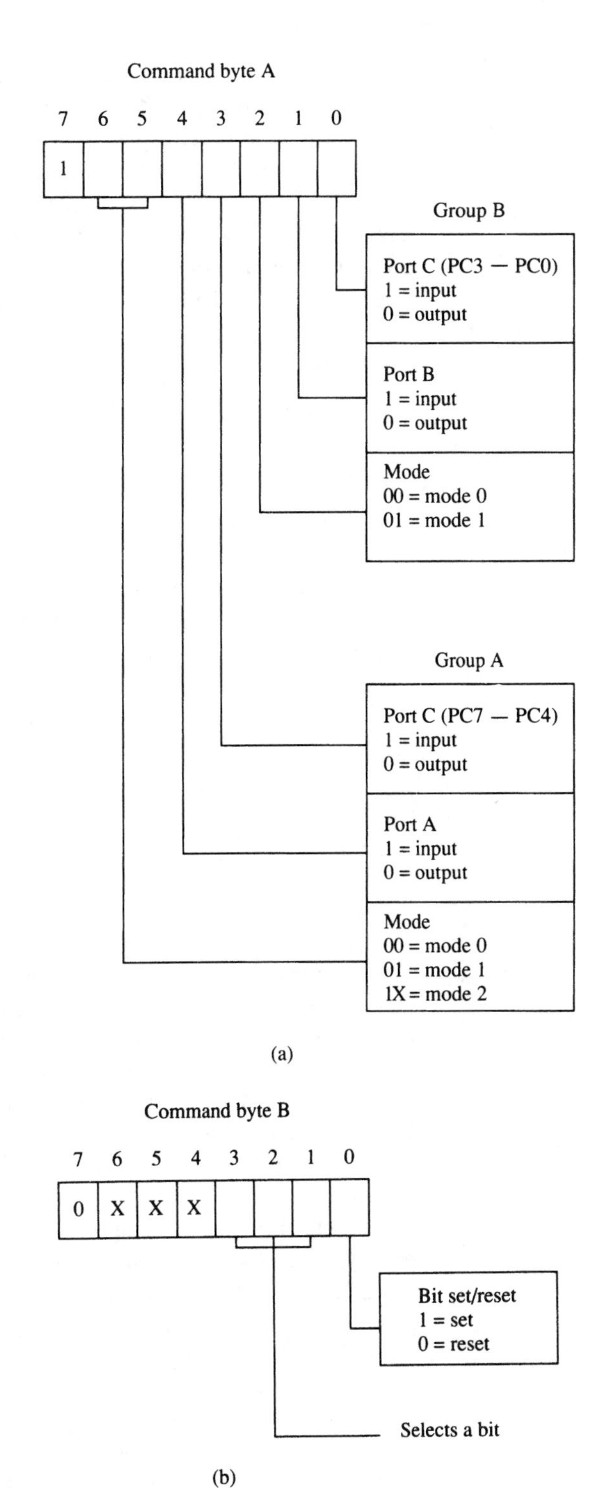

FIGURE 11–20 The command byte of the command register in the 82C55. (a) Programs ports A, B, and C. (b) Sets or resets the bit indicated in the select a bit field.

Mode 0 Operation

Mode 0 operation causes the 82C55 to function either as a buffered input device or as a latched output device. These are the same as the basic input and output circuits discussed in the first section of this chapter.

Figure 11–21 shows the 82C55 connected to a set of eight seven-segment LED displays. These are standard LEDs, but the interface can be modified with a change in resistor values for an organic LED (OLED) display or high-brightness LEDs. In this circuit, both ports A and B are programmed as (mode 0) simple latched output ports. Port A provides the segment data inputs to the display and port B provides a means of selecting one display position at a time for multiplexing the displays. The 82C55 is interfaced to an 8088 microprocessor through a PLD so that it functions at I/O port numbers 0700H–0703H. The program for the PLD is listed in Example 11–8. The PLD decodes the I/O address and develops the write strobe for the $\overline{WR}$ pin of the 82C55.

EXAMPLE 11–8

```
-- VHDL code for the decoder of Figure 11-21

library ieee;
use ieee.std_logic_1164.all;

entity DECODER_11_21 is

port (
        IOM, A15, A14, A13, A12, A11, A10, A9, A8, A7, A6, A5, A4, A3,
             A2: in STD_LOGIC;
        D0: out STD_LOGIC
);

end;

architecture V1 of DECODER_11_17 is

begin

        D0 <= not IOM or A15 or A14 or A13 or A12 or A11 or not A10
        or not A9 or not A8 or A7 or A6 or A5 or A4 or A3 or A2;

end V1;
```

The resistor values are chosen in Figure 11–21 so that the segment current is 80 mA. This current is required to produce an average current of 10 mA per segment as the displays are multiplexed. A six-digit display uses a segment current of 60 mA for an average of 10 mA per segment. In this type of display system, only one of the eight display positions is on at any given instant. The peak anode current in an eight-digit display is 560 mA (seven segments × 80 mA), but the average anode current is 80 mA. In a six-digit display, the peak current would be 420 mA (seven segments × 60 mA). Whenever displays are multiplexed, we increase the segment current from 10 mA (for a display that uses 10 mA per segment as the nominal current) to a value equal to the number of display positions times 10 mA. This means that a four-digit display uses 40 mA per segment, a five-digit display uses 50 mA, and so on.

In this display, the segment load resistor passes 80 mA of current and has a voltage of approximately 3.0 V across it. The LED (1.65 V nominally) and a few tenths are dropped across the anode switch and the segment switch, hence a voltage of 3.0 V appears across the segment load resistor. The value of the resistor is 3.0 V ÷ 180 mA = 37.5 Ω. The closest standard resistor value of 39 Ω is used in Figure 11–21 for the segment load.

The resistor in series with the base of the segment switch assumes that the minimum gain of the transistor is 100. The base current is therefore 80 mA ÷ 100 = 0.8 mA. The voltage across the base resistor is approximately 3.0 V (the minimum logic 1 voltage level of the 82C55), minus the drop across the emitter-base junction (0.7 V), or 2.3 V. The value of the base resistor is therefore 2.3 V ÷ 0.8 mA = 2.875 K Ω. The closest standard resistor value is 2.7 K Ω, but 2.2 K Ω is chosen for this circuit.

The anode switch has a single resistor on its base. The current through the resistor is 560 mA ÷ 100 = 5.6 mA because the minimum gain of the transistor is 100. This exceeds the

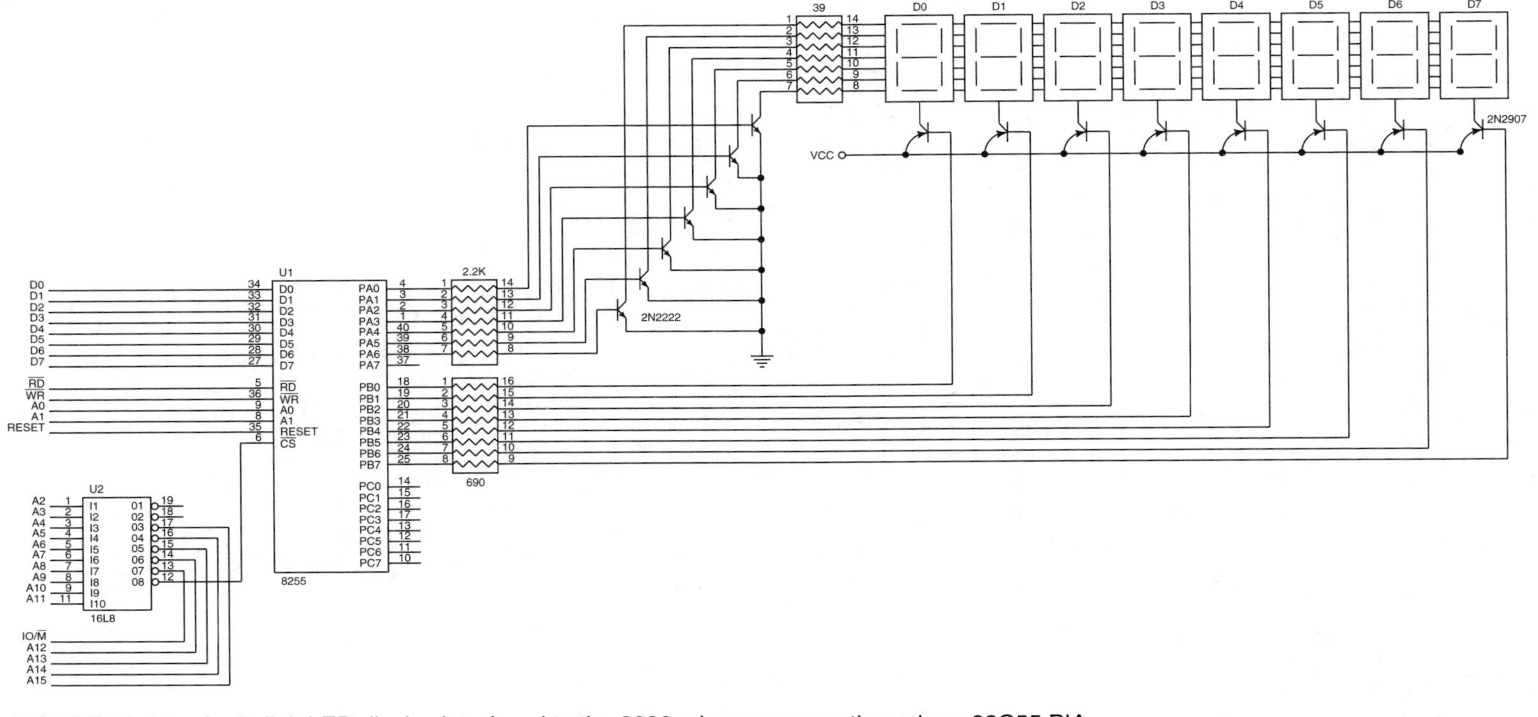

FIGURE 11–21 An 8-digit LED display interfaced to the 8088 microprocessor through an 82C55 PIA.

maximum current of 4.0 mA from the 82C55, but this is close enough so that it will work without problems. The maximum current assumes that you are using the port pin as a TIL input to another circuit. If the amount of current were over 8.0–10.0 mA, then appropriate circuitry (in the form of either a Darlington-pair or another transistor switch) would be required. Here, the voltage across the base resistor is 5.0 V, minus the drop across the emitter-base junction (0.7 V), minus the voltage at the port pin (0.4 V), for a logic 0 level. The value of the resistor is 3.9 V ÷ 5.66 mA = 68.9 Ω. The closest standard resistor value is 69 Ω, which is chosen for this example.

Before software to operate the display is examined, we must first program the 82C55. This is accomplished with the short sequence of instructions listed in Example 11–9. Here, ports A and B are both programmed as outputs.

EXAMPLE 11–9

```
;programming the 82C55 PIA

      MOV  AL,10000000B          ;command
      MOV  DX,703H               ;address port 703H
      OUT  DX,AL                 ;send command to port 703H
```

The procedure to multiplex the displays is listed in Example 11–10 in both assembly language and C++ with assembly language. For the display system to function correctly, we must call this procedure often. Notice that the procedure calls another procedure (DELAY) that causes a 1.0 ms time delay. The time delay is not illustrated in this example, but it is used to allow time for each display position to turn on. Manufacturers of LED displays recommend that the display flashes between 100 and 1500 times per second. Using a 1.0 ms time delay, each digit is displayed for 1.0 ms for a total display flash rate of 1000 Hz ÷ 8 or a flash rate of 125 Hz for all eight digits.

EXAMPLE 11–10

```
;An assembly language procedure that multiplexes the 8-digit display.
;This procedure must be called often for the display
;to appear correctly.

DISP   PROC  NEAR USES AX BX DX SI

       PUSHF
       MOV  BX,8                   ;load counter
       MOV  AH,7FH                 ;load selection pattern
       MOV  SI,OFFSET MEM-1        ;address display data
       MOV  DX,701H                ;address Port B

;display all 8 digits

       .REPEAT
              MOV  AL,AH           ;send selection pattern to Port B
              OUT  DX,AL
              DEC  DX
              MOV  AL,[BX+SI]      ;send data to Port A
              OUT  DX,AL
              CALL DELAY           ;wait 1.0 ms
              ROR  AH,1            ;adjust selection pattern
              INC  DX
              DEC  BX              ;decrement counter
       .UNTIL BX == 0

       POPF
       RET

DISP   ENDP

//A C++ function that multiplexes the 8-digit display
//uses char sized array MEM
```

```
void Disp()
{
        unsigned int *Mem = &MEM[0];        //point to array element 0
        for ( int a = 0; a < 8; a++ )
        {
              unsigned char b = 0xff ^ ( 1 << a );      //form select pattern
              _asm
              {
                     mov  al,b
                     mov  dx,701H
                     out  dx,al     ;send select pattern to Port B
                     mov  al,Mem[a]
                     dec  dx
                     out  dx,al     ;send data to Port A
              }
              Sleep(1);             ;wait 1.0 ms
        }
}
```

The display procedure (DISP) addresses an area of memory where the data, in seven-segment code, is stored for the eight display digits called MEM. The AH register is loaded with a code (7FH) that initially addresses the most-significant display position. Once this position is selected, the contents of memory location MEM +7 are addressed and sent to the most-significant digit. The selection code is then adjusted to select the next display digit. This process repeats eight times to display the contents of location MEM through MEM +7 on the eight display digits.

The time delay of 1.0 ms can be obtained by writing a procedure that uses the system clock frequency to determine how long each instruction requires to execute. The procedure listed in Example 11–11 causes a time delay of a duration determined by the number of times that the LOOP instruction executes. Here XXXX was used and will be filled in with a value after a few facts are discussed. The LOOP instruction requires a certain number of clocks to execute—how many can be discovered in Appendix B. Suppose that the interface is using the 80486 microprocessor running with a 20 MHz clock. Appendix B represents that the LOOP instruction requires 7/6 clocks. The first number is the number of clocks required when a jump to D1 occurs and the second number is when the jump does not occur. With a 20 MHz clock one clock requires $1 \div 20$ MHz = 50 ns. The LOOP instruction, in this case, requires 350 ns to execute in all but the very last iteration. To determine the count (XXXX) needed to accomplish a 1.0 ms time delay, divide 1.0 ms by 350 ns. In this case XXXX = 2,857 to accomplish a 1.0 ms time delay. If a larger count occurs a LOOPD instruction can be used with the ECX register. The time required to execute the MOV CX, XXXX, and RET instructions can usually be ignored.

Suppose a Pentium 4 with a 2.0 GHz clock is used for the delay. Here one clock is 0.5 ns and LOOP requires five clocks per iteration. This requires a count of 400,000, so LOOPD would be used with ECX.

EXAMPLE 11–11

```
;equation for the delay
;
;          Delay Time
;XXXX = -------------
;          time for LOOP
;

DELAY   PROC   NEAR USES CX

        MOV CX,XXXX
D1:
        LOOP D1
        RET

DELAY   ENDP
```

An LCD Display Interfaced to the 82C55

LCDs (**liquid crystal displays**) have replaced LED displays in many applications. The only disadvantage of the LCD display is that it is difficult to see in low-light situations in which the LED is still in limited use. If the price of the OLED becomes low enough, LCD displays will disappear.

Figure 11–22 illustrates the connection of the Optrex DMC-20481 LCD display interfaced to an 82C55. The DMC-20481 is a 4-line by 20-characters-per-line display that accepts ASCII code as input data. It also accepts commands that initialize it and control its application. As you can see in Figure 11–22, the LCD display has few connections. The data connections, which are attached to the 82C55 Port A, are used to input display data and to read information from the display.

There are four control pins on the display. The V_{EE} connection is used to adjust the contrast of the LED display and is normally connected to a 10 KQ potentiometer, as illustrated. The RS (register select) input selects data (RS = 1) or instructions (RS = 0). The E (enable) input must be a logic 1 for the DMC-20481 to read or write information and functions as a clock. Finally, the $R/\overline{W}$ pin selects a read or a write operation. Normally, the RS pin is placed at a 1 or 0, the $R/\overline{W}$ pin is set or cleared, data are placed on the data input pins, and then the E pin is pulsed to access the DMC-20481. This display also has two inputs (LEDA [anode] and LEDK [cathode]) for backlighting LED diodes, which are not shown in the illustration.

In order to program the DMC-20481 we must first initialize it. This applies to any display that uses the HD44780 (Hitachi) display driver integrated circuit. The entire line of small display panels from Optrex and most other manufacturers is programmed in the same manner. Initialization is accomplished via the following steps:

1. Wait at least 15 ms after V_{CC} rises to 5.0 V.
2. Output the function set command (30H), and wait at least 4.1 ms.
3. Output the function set command (30H) a second time, and wait at least 100 μs.
4. Output the function set command (30H) a third time, and wait at least 40 μs.
5. Output the function set command (38H) a fourth time, and wait at least 40 μs.
6. Output 08H to disable the display, and wait at least 40 μs.
7. Output 01H to home the cursor and clear the display, and wait at least 1.64 ms.
8. Output the enable display cursor off (0CH), and wait at least 40 μs.
9. Output 06H to select auto-increment, shift the cursor, and wait at least 40 μs.

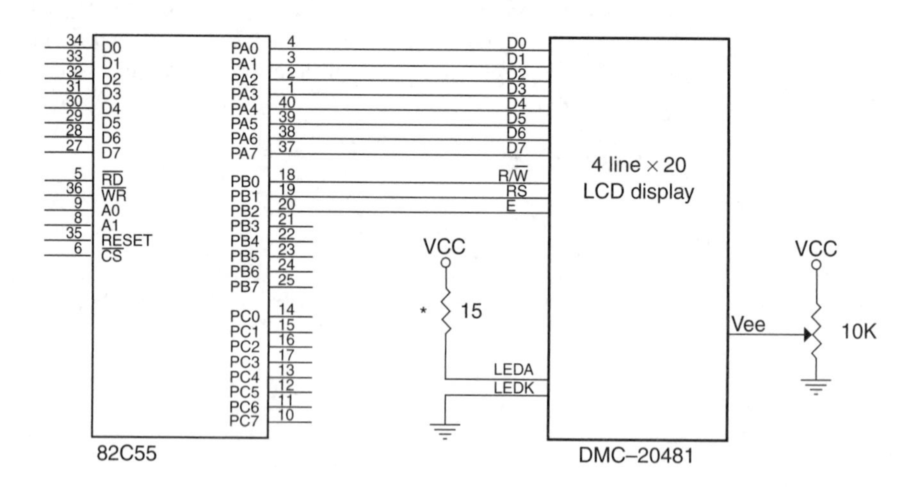

FIGURE 11–22 The DMC-20481 LCD display interfaced to the 82C55.

The software to accomplish the initialization of the LCD display is listed in Example 11–12. It is long, but the display controller requires the long initialization dialog. Note that the software for the three time delays is not included in the listing. If you are interfacing to a PC, you can use the RDTSC instruction as discussed in the Pentium chapter for the time delay. If you are developing the interface for another application, you must write separate time delays, which must provide the delay times indicated in the initialization dialog.

EXAMPLE 11–12

```
PORTA_ADDRESS    EQU 700H          ;set port addresses
PORTB_ADDRESS    EQU 701H
COMMAND_ADDRESS EQU 703H

;macro to send a command or data to the LCD display
;
SEND    MACRO   PORTA_DATA, PORTB_DATA, DELAY

        MOV   AL,PORTA_DATA         ;PORTA_DATA to Port A
        MOV   DX,PORTA_ADDRESS
        OUT   DX,AL
        MOV   AL,PORTB_DATA         ;PORTB_DATA to Port B
        MOV   DX,PORTB_ADDRESS
        OUT   DX,AL
        OR    AL,00000100B          ;Set E bit
        OUT   DX,AL                 ;send to Port B
        AND   AL,11111011B          ;Clear E bit
        NOP                         ;a small delay
        NOP
        OUT   DX,AL                 ;send to Port B
        MOV   BL,DELAY              ;BL = delay count
        CALL MS_DELAY               ;ms Time Delay
        ENDM

;Program to initialize the LCD display

START:
        MOV   AL,80H                ;Program the 82C55
        MOV   DX,COMMAND_ADDRESS
        OUT   DX,AL
        MOV   AL,0
        MOV   DX,PORTB_ADDRESS      ;Clear Port B
        SEND 30H, 2, 16             ;send 30H for 16 ms
        SEND 30H, 2, 5              ;send 30H for 5 ms
        SEND 30H, 2, 1              ;send 30H for 1 ms
        SEND 38H, 2, 1              ;send 38H for 1 ms
        SEND 8, 2, 1                ;send 8 for 1 ms
        SEND 1, 2, 2                ;send 1 for 2 ms
        SEND 0CH, 2, 1              ;send 0CH for 1 ms
        SEND 6, 2, 1                ;send 6 for 1 ms
```

The NOP instructions are added in the SEND macro to ensure that the E bit remains a logic 1 long enough to activate the LCD display. This process should work in most systems at most clock frequencies, but additional NOP instructions may be needed to lengthen this time in some cases. Also notice that equate statements are used to equate the port addresses to labels. This is done so that the software can be changed easily if the port numbers differ from those used in the program.

Before programming the display, the commands used in the initialization dialog must be explained. See Table 11–3 for a complete listing of the commands or instructions for the LCD display. Compare the commands sent to the LCD display in the initialization program to Table 11–3.

Once the LCD display is initialized, a few procedures are needed to display information and control the display. After initialization, time delays are no longer needed when sending data or many commands to the display. The clear display command still needs a time delay because

TABLE 11–3 Instructions for most LCD displays.

Instruction	*Code*	*Description*	*Time*
Clear display	0000 0001	Clears the display and homes the cursor	1.64 ms
Cursor home	0000 0010	Homes the cursor	1.64 ms
Entry mode set	0000 00AS	Sets cursor movement direction (A = 1, increment) and shift (S = 1, shift)	40 μs
Display on/off	0000 1DCB	Sets display on/off (D = 1, on) (C = 1, cursor on) (B = 1, cursor blink)	40 μs
Cursor/display shift	0001 SR00	Sets cursor movement and display shift (S = 1, shift display) (R = 1, right)	40 μs
Function set	001L NF00	Programs LCD circuit (L = 1, 8-bit interface) (N = 1, 2 lines) (F = 1, 5 × 10 characters) (F = 0, 5 × 7 characters)	40 μs
Set CGRAM address	01XX XXXX	Sets character generator RAM address	40 μs
Set DRAM address	10XX XXXX	Sets display RAM address	40 μs
Read busy flag	B000 0000	Reads busy flag (B = 1, busy)	0
Write data	Data	Writes data to the display RAM or the character generator RAM	40 μs
Read data	Data	Reads data from the display RAM or character generator RAM	40 μs

the busy flag is not used with that command. Instead of a time delay, the busy flag is tested to see whether the display has completed an operation. A procedure to test the busy flag appears in Example 11–13. The BUSY procedure tests the LCD display and only returns when the display has completed a prior instruction.

EXAMPLE 11–13

```
PORTA_ADDRESS    EQU 700H    ;set port addresses
PORTB_ADDRESS    EQU 701H
COMMAND_ADDRESS  EQU 703H

BUSY    PROC   NEAR USES DX AX

        PUSHF
        MOV  DX,COMMAND_ADDRESS
        MOV  AL,90H          ;program Port A as IN
        OUT  DX,AL
        .REPEAT
               MOV  AL,5     ;select read from LCD
               MOV  DX,PORTB_ADDRESS
               OUT  DX,AL    ;and pulse E
               NOP
               NOP
               MOV  AL,1
               OUT  DX,AL
```

```
                MOV  DX,PORTA_ADDRESS
                MOV  AL,DX     ;read busy command
                SHL  AL,1
         .UNTIL  !CARRY?       ;until not busy
         NOV  DX ,COMMAND_ADDRESS
         MOV  AL ,80H
         OUT  DX ,AL           ;program Port A as OUT
         POPF
         RET

BUSY     ENDP
```

Once the BUSY procedure is available, data can be sent to the display by writing another procedure called WRITE. The WRITE procedure uses BUSY to test before trying to write new data to the display. Example 11–14 shows the WRITE procedure, which transfers the ASCII character from the BL register to the current cursor position of the display. Note that the initialization dialog has sent the cursor for auto-increment, so if WRITE is called more than once, the characters written to the display will appear one next to the other, as they would on a video display.

EXAMPLE 11–14

```
WRITE  PROC   NEAR
       MOV   AL,BL                   ;BL to Port A
       MOV   DX,PORTA_ADDRESS
       OUT   DX,AL
       MOV   AL,0                    ;write ASCII
       MOV   DX,PORTB_ADDRESS
       OUT   DX,AL
       OR    AL,00000100B            ;Set E bit
       OUT   DX,AL                   ;send to Port B
       AND   AL,11111011B            ;Clear E bit
       NOP                           ;a small delay
       NOP
       OUT   DX,AL                   ;send to Port B
       CALL  BUSY                    ;wait for completion
       RET
WRITE  ENDP
```

The only other procedure that is needed for a basic display is the clear and home cursor procedure, called CLS, shown in Example 11–15. This procedure uses the SEND macro from the initialization software to send the clear command to the display. With CLS and the procedures presented thus far, you can display any message on the display, clear it, display another message, and basically operate the display. As mentioned earlier, the clear command requires a time delay (at least 1.64 ms) instead of a call to BUSY for proper operation.

EXAMPLE 11–15

```
CLS      PROC    NEAR
         SEND    1, 2, 2
         RET
CLS      ENDP
```

Additional procedures that could be developed might select a display RAM position. The display RAM address starts at 0 and progresses across the display until the last character address on the first line is location 19, location 20 is the first display position of the second line, and so forth. Once you can move the display address, you can change individual characters on the display and even read data from the display. These procedures are for you to develop if they are needed.

A word about the display RAM inside of the LCD display. The LCD contains 128 bytes of memory, addressed from 00H to 7FH. Not all of this memory is always used. For example, the one-line × 20-character display uses only the first 20 bytes of memory (00–13H.) The first line of any of these displays always starts at address 00H. The second line of any display powered by the HD44780 always begins at address 40H. For example, a two-line × 40-character display uses addresses 00H–27H to store ASCII-coded data from the first line. The second line is stored at addresses 40H–67H for this display. In the four-line displays, the first line is at 00H, the second is at 40H, the third is at 14H, and the last line is at 54H. The largest display device that uses the HD44780 is a two-line × 40-character display. The four-line by 40-character display uses an M50530 or a pair of HD44780s. Because information on these devices can be readily found on the Internet, they are not covered in the text.

A Stepper Motor Interlaced to the 82C55. Another device often interfaced to a computer system is the *stepper motor.* A stepper motor is a digital motor because it is moved in discrete steps as it traverses through 360°. A common stepper motor is geared to move perhaps 15° per step for an inexpensive stepper motor, to 1° per step for a more costly high-precision stepper motor. In all cases, these steps are gained through many magnetic poles and/or gearing. Notice that two coils are energized in Figure 11–23. If less power is required, one coil may be energized at a time, causing the motor to step at 45°, 135°, 225°, and 315°.

Figure 11–23 shows a four-coil stepper motor that uses an armature with a single pole. Notice that the stepper motor is shown four times with the armature (permanent magnetic) rotated to four discrete places. This is accomplished by energizing the coils, as shown. This is an illustration of full stepping. The stepper motor is driven by using NPN Darlington amplifier pairs to provide a large current to each coil.

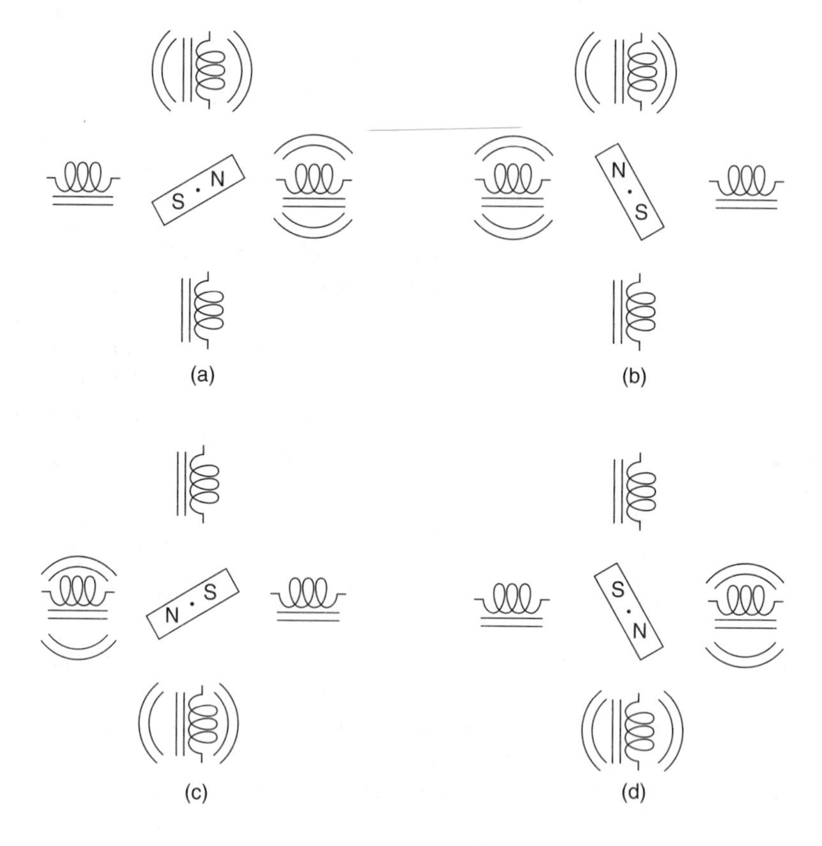

FIGURE 11–23 The stepper motor showing full-step operation. (a) 45° (b) 135° (c) 225° (d) 315°.

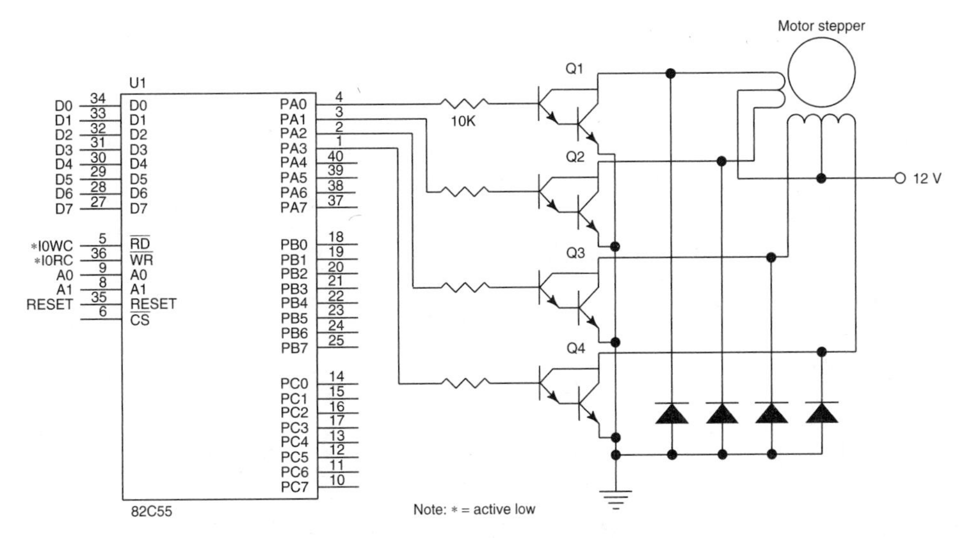

FIGURE 11–24 A stepper motor interfaced to the 82C55. This illustration does not show the decoder.

A circuit that can drive this stepper motor is illustrated in Figure 11–24, with the four coils shown in place. This circuit uses the 82C55 to provide it with the drive signals that are used to rotate the armature of the motor in either the right-hand or left-hand direction.

A simple procedure that drives the motor (assuming that port A is programmed in mode 0 as an output device) is listed in Example 11–16 in both assembly language and as a function in C++. This subroutine is called, with CX holding the number of steps and direction of the rotation. If CX is greater than 8000H, the motor spins in the right-hand direction; if CX is less than 8000H, it spins in the left-hand direction. For example, if the number of steps is 0003H, the motor moves in the left-hand direction three steps and if the number of steps is 8003H, it moves three steps in the right-hand direction. The leftmost bit of CX is removed and the remaining 15 bits contain the number of steps. Notice that the procedure uses a time delay (not illustrated) that causes a 1 ms time delay. This time delay is required to allow the stepper-motor armature time to move to its next position.

EXAMPLE 11–16

```
      PORT   EQU   40H

;An assembly language procedure that controls the stepper motor

STEP  PROC   NEAR USES CX AX

      MOV   AL,POS                           ;get position
      OR    CX,CX                            ;set flag bits
      IF   !ZERO?
            .IF !SIGN?                       ;if no sign
                  .REPEAT
                        ROL  AL,1            ;rotate step left
                        OUT  PORT,AL
                        CALL DELAY           ;wait 1 ms
                  .UNTILCXZ
            .ELSE
                  AND CX,7FFFH               ;make CX positive
                  .REPEAT
```

```
                        ROR  AL,1        ;rotate step right
                        OUT  PORT,AL
                        CALL DELAY       ;wait 1 ms
                  .UNTILCXZ
            .ENDIF
      .ENDIF
      MOV  POS,AL
      RET

STEP  ENDP

//A C++ function that controls the stepper motor

char Step(char Pos, short Step)
{
      char Direction = 0;
      if ( Step < 0 )
      {
            Direction = 1;
            Step =& 0x8000;
      }
      while ( Step )
      {
            if ( Direction )
                  if ( ( Pos & 1 ) == 1 )
                        Pos = ( Pos >> 1 ) | 0x80;
                  else
                        Pos >>= 1;
            else
                  if ( ( Pos & 0x80 ) == 0x80 )
                        Pos = ( Pos << 1 ) | 1;
                  else
                        Pos <<= 1;
            _asm
            {
                  mov al,Pos
                  out 40h, al
            }
      }
      return Pos;
}
```

The current position is stored in memory location POS, which must be initialized with 33H, 66H, 0EEH, or 99H. This allows a simple ROR (step right) or ROL (step left) instruction to rotate the binary bit pattern for the next step.

The C++ version has two parameters: Pos is the current position of the stepper motor and Step is the number of steps as described earlier. The new Pos is returned in the C++ version instead of being stored in a variable.

Stepper motors can also be operated in the half-step mode, which allows eight steps per sequence. This is accomplished by using the full-step sequence described with a half step obtained by energizing one coil interspersed between the full steps. Half-stepping allows the armature to be positioned at 0°, 90°, 180°, and 270°. The half-step position codes are 11H, 22H, 44H, and 88H. A complete sequence of eight steps would be: 11H, 33H, 22H, 66H, 44H, 0CCH, 88H, and 99H. This sequence could be either output from a lookup table or generated with software.

Key Matrix Interface. Keyboards come in a vast variety of sizes, from the standard 101-key QWERTY keyboards interfaced to the microprocessor to small specialized keyboards that may contain only four to 16 keys. This section of the text concentrates on the smaller keyboards that may be purchased preassembled or may be constructed from individual key switches.

Figure 11–25 illustrates a small key-matrix that contains 16 switches interfaced to ports A and B of an 82C55. In this example, the switches are formed into a 4×4 matrix, but any matrix

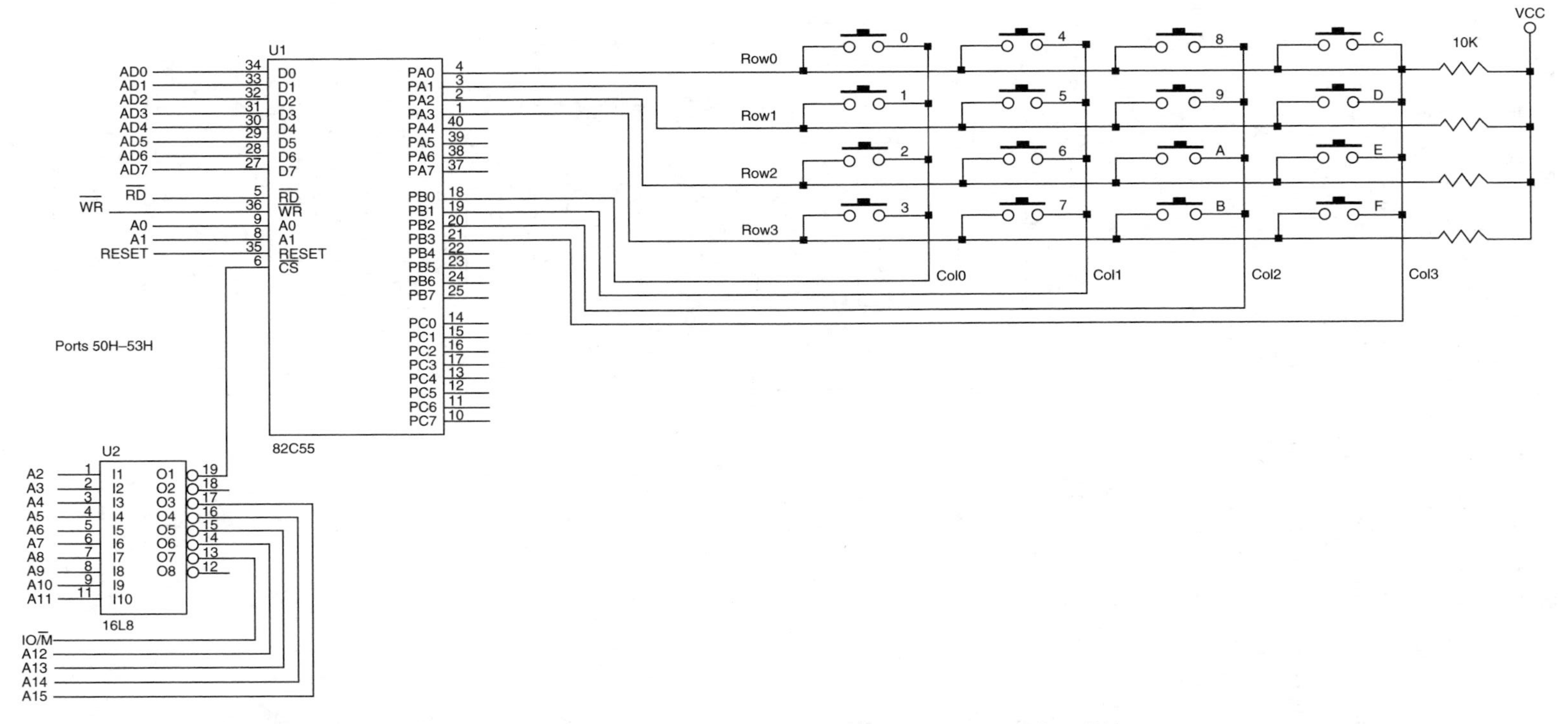

FIGURE 11–25 A 4 × 4 keyboard matrix connected to an 8088 microprocessor through the 82C55 PIA.

could be used such as a 2 × 8. Notice how the keys are organized into four rows (ROW_0–ROW_3) and four columns (COL_0–COL_3). Each row is connected to 5.0 V through a 10 KΩ pull-up resistor to ensure that the row is pulled high when no pushbutton switch is closed.

The 82C55 is decoded (the PLD program is not shown) at I/O ports 50H–53H for an 8088 microprocessor. Port A is programmed as an input port to read the rows and port B is programmed as an output port to select a column. For example, if 1110 is output to port B pins PB_3–PB_0, column 0 has a logic 1, so the four keys in column 0 are selected. Notice that with a logic 0 on PB0, the only switches that can place a logic 0 onto port A are switches 0–3. If switches 4–F are closed, the corresponding port A pins remain a logic 1. Likewise, if 1101 is output to port B, switches 4–7 are selected, and so forth.

A flowchart of the software required to read a key from the keyboard matrix and debounce the key is illustrated in Figure 11–26. Keys must be debounced, which is normally accomplished with a short time delay of 10 to 20 ms. The flowchart contains three main sections. The first waits for the

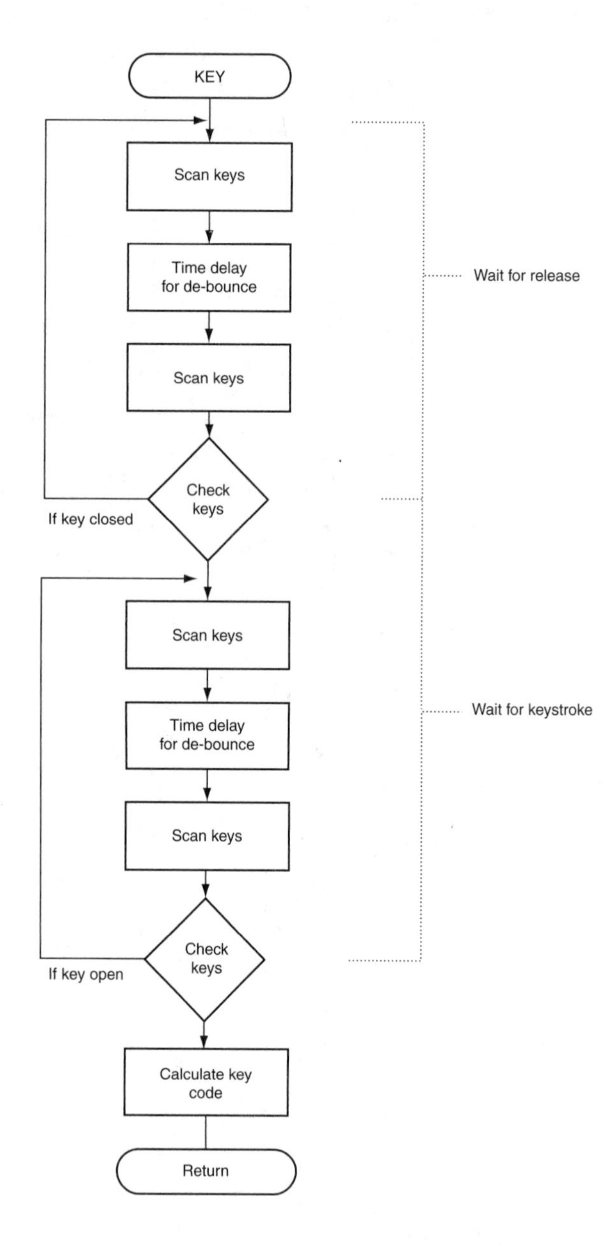

FIGURE 11–26 The flowchart of a keyboard-scanning procedure.

release of a key. This seems awkward, but software executes very quickly in a microprocessor and there is a possibility that the program will return to the top of this program before the key is released, so we must wait for a release first. Next, the flowchart shows that we wait for a keystroke. Once the keystroke is detected, the position of the key is calculated in the final part of the flowchart.

The software uses a procedure called SCAN to scan the keys and another called DELAY10 (not shown in this example) to waste 10 ms of time for debouncing. The main keyboard procedure is called KEY and it appears with the others in Example 11–17. Example 11–17 also lists a C++ function to accomplish a key read operation. Note that the KEY procedure is generic, so it can handle any keyboard configuration from a 1 × 1 matrix to an 8 × 8 matrix. Changing the two equates at the start of the program (ROWS and COLS) will change the configuration of the software for any size keyboard. Also note that the steps required to initialize the 82C55 so that port A is an input port and port B is an output port are not shown.

With certain keyboards that do not follow the way keys are scanned, a lookup table may be needed to convert the raw key codes returned by KEY into key codes that match the keys on the keyboard. The lookup software is placed just before returning from KEY. It is merely a MOV BX,OFFSET TABLE followed by the XLAT instruction.

EXAMPLE 11–17(a)

```
;assembly language version;

;KEY scans the keyboard and returns the key code in AL.

        COLS    EQU     4
        ROWS    EQU     4
        PORTA   EQU     50H
        PORTB   EQU     51H

KEY     PROC    NEAR USES CX BX

        MOV  BL,FFH                 ;compute row mask
        SHL  BL,ROWS

        MOV  AL,0
        OUT  PORTB,AL               ;place zeros on Port B

        .REPEAT                     ;wait for release
                .REPEAT
                        CALL SCAN
                .UNTIL  ZERO?
                CALL  DELAY10
                CALL  SCAN
        .UNTIL ZERO?
        .REPEAT                     ;wait for key
                .REPEAT
                        CALL SCAN
                .UNTIL !ZERO?
                CALL  DELAY10
                CALL  SCAN
        .UNTIL !ZERO?
        MOV  CX,00FEH
        .WHILE 1                    ;find column
                MOV  AL,CL
                OUT  PORTB,AL
                CALL SHORTDELAY     ;see text
                CALL SCAN
                .BREAK !ZERO?
                ADD  CH,COLS
                ROL  CL,1
        .ENDW
        .WHILE 1                    ;find row
                SHR  AL,1
```

```
            .BREAK .IF !CARRY?
            INC  CH
       .ENDW
       MOV  AL,CH                 ;get key code
       RET

KEY    ENDP

SCAN   PROC   NEAR

       IN   AL,PORTA              ;read rows
       OR   AL,BL
       CMP  AL,0FFH               ;test for no keys
       RET

SCAN   ENDP
```

EXAMPLE 11–17(b)

```
//C++ language version of keyboard scanning software

#define ROWS 4
#define COLS 4
#define PORTA 50h
#define PORTB 51h

char Key()
{
      char mask = 0xff << ROWS;
      _asm {
            mov  al,0                       ;select all columns
            out  PORTB,al
      }
      do {                                  //wait for release
            while ( Scan( mask ) );
            Delay();
      }while ( Scan( mask ) );
      Do {                                  //wait for key press
            while ( !Scan( mask ) );
            Delay();
      }while ( !Scan( mask ) );
      unsigned char select = 0xfe;
      char key = 0;
      _asm {
            mov  al,select
            out  PortB,al
      }
      ShortDelay();
      while( !Scan ( mask ) ){              //calculate key code
            _asm
            {
                     mov  al,select
                     rol  al,1
                     mov  select,al
                     out  PortB,al
            }
            ShortDelay();
            key += COLS;
      }
      _asm {
            in   al,PortA
            mov  select,al
      }
      while ( ( Select & 1 ) != 0 ) {
            Select <<= 1;
            key ++;
      }
```

```
        return key;
}

bool Scan(mask)
{
        bool flag;
        _asm
        {
                in   al,PORTA
                mov  flag,al
        }
        return ( flag | mask );
}
```

The SHORTDELAY procedure is needed because the computer changes port B at a very high rate of speed. The short time delay allows time for the data sent to port B to settle to their final state. In most cases, this is not needed if the scan rate (time between output instructions) of this part of the software does not exceed 30 KHz. If it does, the Federal Communications Commission (FCC) will not approve its application in any accepted system. Without FCC Type A or Type B certification the system cannot be sold.

Mode 1 Strobed Input

Mode 1 operation causes port A and/or port B to function as latching input devices. This allows external data to be stored into the port until the microprocessor is ready to retrieve it. Port C is also used in mode 1 operation—not for data, but for control or handshaking signals that help operate either or both port A and port B as strobed input ports. Figure 11–27 shows how both ports are structured for mode 1 strobed input operation and the timing diagram.

The strobed input port captures data from the port pins when the strobe ($\overline{STB}$) is activated. Note that the strobe captures the port data on the 0-to-1 transition. The $\overline{STB}$ signal causes data to be captured in the port, and it activates the IBF (**input buffer full**) and INTR (**interrupt request**) signals. Once the microprocessor, through software (IBF) or hardware (INTR), notices that data are strobed into the port, it executes an IN instruction to read the port $\overline{RD}$. The act of reading the port restores both IBF and INTR to their inactive states until the next datum is strobed into the port.

Signal Definitions for Mode 1 Strobed Input

$\overline{STB}$ The **strobe** input loads data into the port latch, which holds the information until it is input to the microprocessor via the IN instruction.

IBF **Input buffer full** is an output indicating that the input latch contains information.

INTR **Interrupt request** is an output that requests an interrupt. The INTR pin becomes a logic 1 when the $\overline{STB}$ input returns to a logic 1, and is cleared when the data are input from the port by the microprocessor.

INTE The **interrupt enable** signal is neither an input nor an output; it is an internal bit programmed via the port PC_4 (port A) or PC_2 (port B) bit position.

PC&, PC_6 The port C pins 7 and 6 are general-purpose I/O pins that are available for any purpose.

Strobed Input Example. An excellent example of a strobed input device is a keyboard. The keyboard encoder debounces the key switches and provides a strobe signal whenever a key is pressed and the data output contain the ASCII-coded key code. Figure 11–28 illustrates a keyboard connected to strobed input port A. Here $\overline{DAV}$ (data available) is activated for 1.0 μs each time that a key is typed on the keyboard. This causes data to be strobed into port A because $\overline{DAV}$

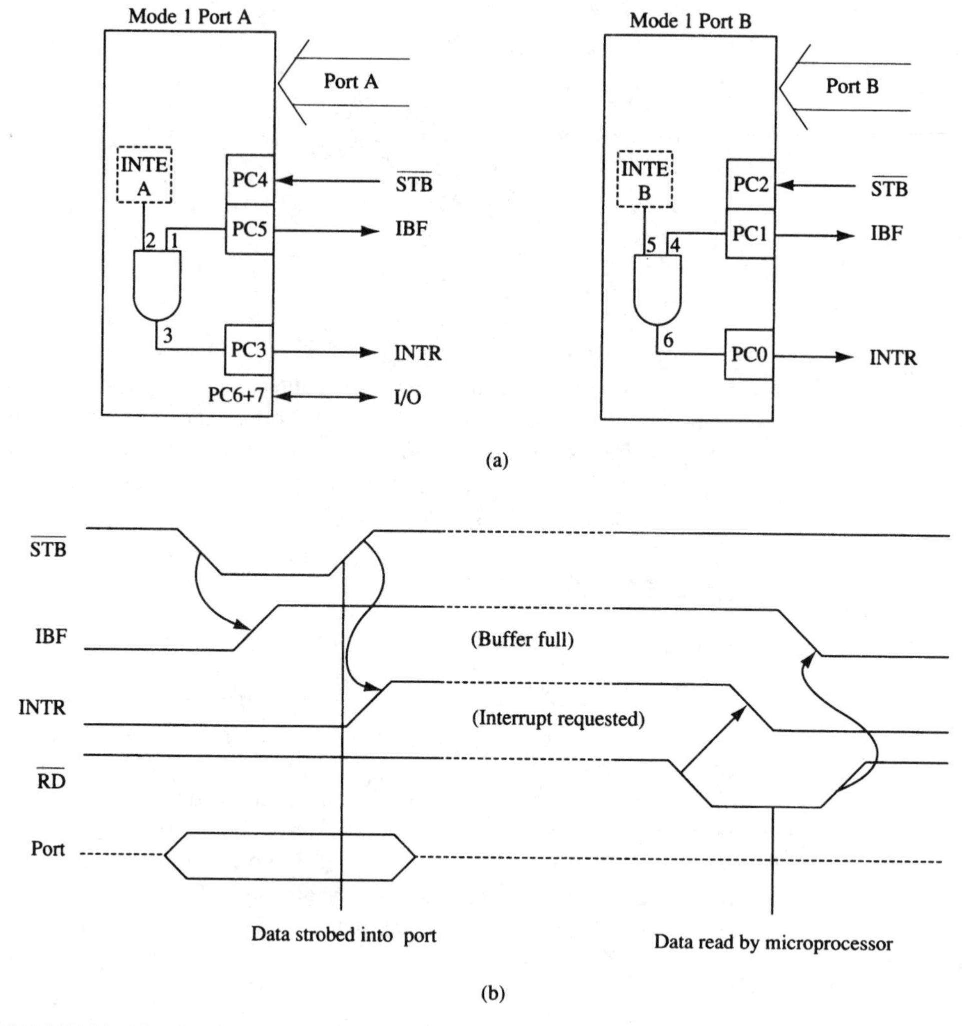

FIGURE 11–27 Strobed input operation (mode 1) of the 82C55. (a) Internal structure and (b) timing diagram.

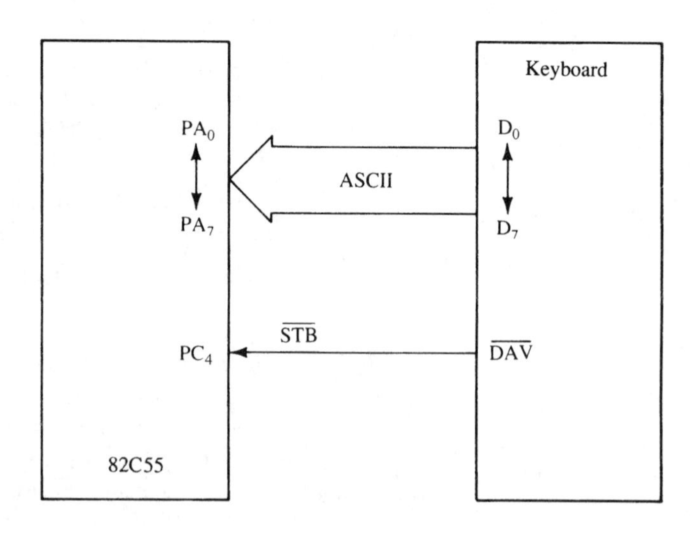

FIGURE 11–28 Using the 82C55 for strobed input operation of a keyboard.

is connected to the $\overline{STB}$ input of port A. Each time a key is typed, therefore, it is stored into port A of the 82C55. The $\overline{STB}$ input also activates the IBF signal, indicating that data are in port A.

Example 11–18 shows a procedure that reads data from the keyboard each time a key is typed. This procedure reads the key from port A and returns with the ASCII code in AL. To detect a key, port C is read and the IBF bit (bit position PC_5) is tested to see whether the buffer is full. If the buffer is empty (IBF = 0), then the procedure keeps testing this bit, waiting for a character to be typed on the keyboard.

EXAMPLE 11–18

```
;A procedure that reads the keyboard encoder and
;returns the ASCII key code in AL

      BIT5    EQU   20H
      PORTC   EQU   22H
      PORTA   EQU   20H

      READ    PROC  NEAR

              .REPEAT                         ;poll IBF bit
                     IN   AL,PORTC
                     TEST AL,BIT5
              .UNTIL !ZERO?
              IN   AL.PORTA                   ;get ASCII data
              RET

      READ    ENDP
```

Mode 1 Strobed Output

Figure 11–29 illustrates the internal configuration and timing diagram of the 82C55 when it is operated as a strobed output device under mode 1. Strobed output operation is similar to mode 0 output operation, except that control signals are included to provide handshaking.

Whenever data are written to a port programmed as a strobed output port, the $\overline{OBF}$ (**output buffer full**) signal becomes a logic 0 to indicate that data are present in the port latch. This signal indicates that data are available to an external I/O device that removes the data by strobing the $\overline{ACK}$ (acknowledge) input to the port. The $\overline{ACK}$ signal returns the $\overline{OBF}$ signal to a logic 1, indicating that the buffer is not full.

Signal Definitions for Mode 1 Strobed Output

$\overline{OBF}$ — **Output buffer full** is an output that goes low whenever data are output (OUT) to the port A or port B latch. This signal is set to a logic 1 whenever the $\overline{ACK}$ pulse returns from the external device.

$\overline{ACK}$ — The **acknowledge signal** causes the $\overline{OBF}$ pin to return to a logic 1 level. The $\overline{ACK}$ signal is a response from an external device, indicating that it has received the data from the 82C55 port.

INTR — **Interrupt request** is a signal that often interrupts the microprocessor when the external device receives the data via the $\overline{ACK}$ signal. This pin is qualified by the internal INTE (**interrupt enable**) bit.

INTE — **Interrupt enable** is neither an input nor an output; it is an internal bit programmed to enable or disable the INTR pin. The INTE A bit is programmed using the PC_6 bit and INTE B is programmed using the PC_2 bit.

PC_4, PC_5 — Port C pins PC_4 and PC_5 are general-purpose I/O pins. The bit set and reset command is used to set or reset these two pins.

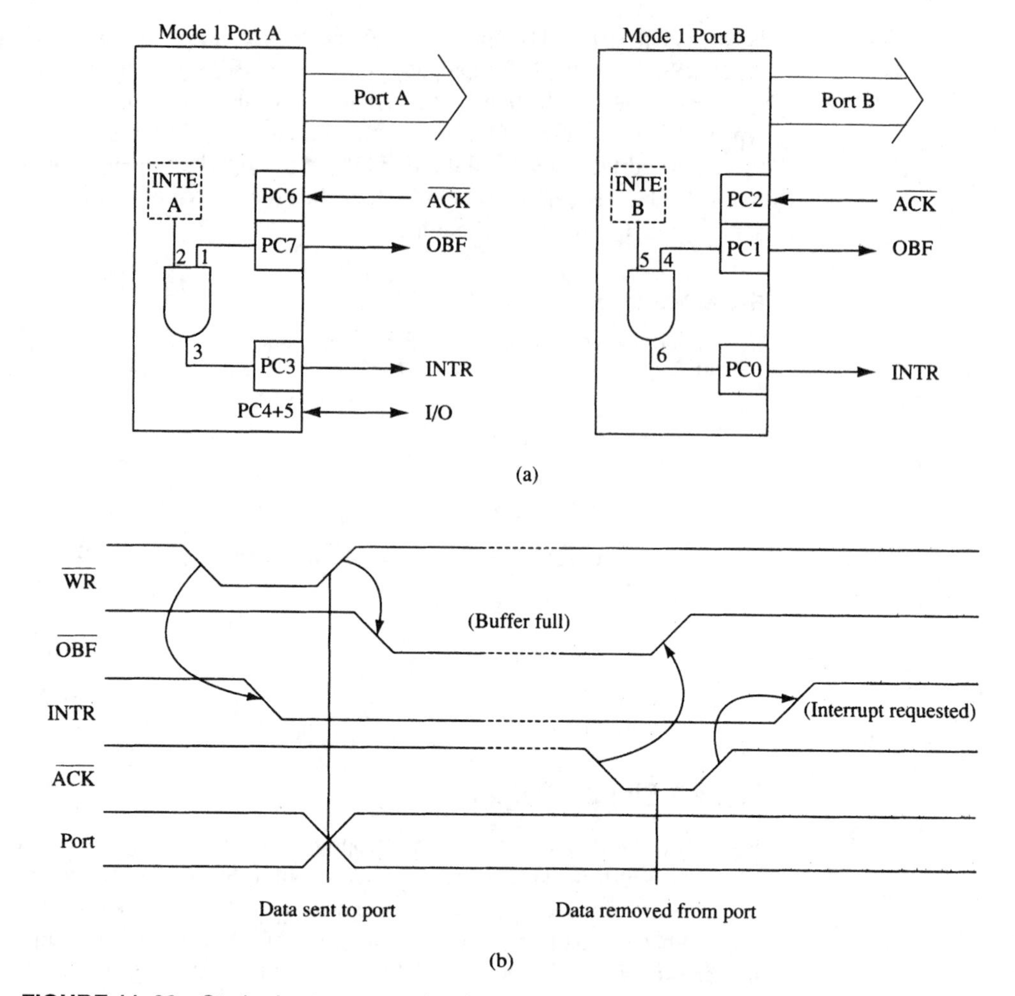

FIGURE 11–29 Strobed output operation (mode 1) of the 82C55. (a) Internal structure and (b) timing diagram.

Strobed Output Example. The printer interface discussed in Section 11–1 is used here to demonstrate how to achieve strobed output synchronization between the printer and the 82C55. Figure 11–30 illustrates port B connected to a parallel printer, with eight data inputs for receiving ASCII-coded data, a $\overline{DS}$ (**data strobe**) input to strobe data into the printer, and an $\overline{ACK}$ output to acknowledge the receipt of the ASCII character.

In this circuit, there is no signal to generate the $\overline{DS}$ signal to the printer, so PC_4 is used with software that generates the $\overline{DS}$ signal. The $\overline{ACK}$ signal that is returned from the printer acknowledges the receipt of the data and is connected to the $\overline{ACK}$ input of the 82C55.

Example 11–19 lists the software that sends the ASCII-coded character in AH to the printer. The procedure first tests $\overline{OBF}$ to decide whether the printer has removed the data from port B. If not, the procedure waits for the $\overline{ACK}$ signal to return from the printer. If $\overline{OBF}$ = 1, then the procedure sends the contents of AH to the printer through port B and also sends the $\overline{DS}$ signal.

EXAMPLE 11–19

```
;A procedure that transfers an ASCII character from AH to the printer
;connected to port B

BIT1    EQU   2
PORTC   EQU   63H
```

FIGURE 11–30 The 82C55 connected to a parallel printer interface that illustrates the strobed output mode of operation for the 82C55.

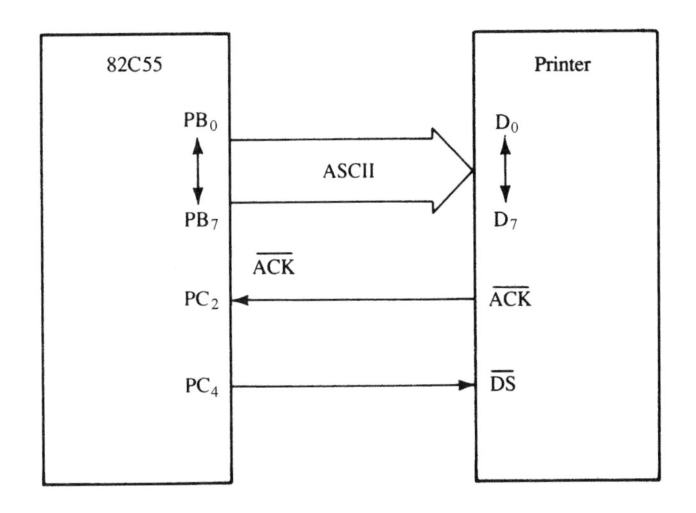


```
PORTB   EQU    61H
CMD     EQU    63H

PRINT   PROC   NEAR

        .REPEAT                      ;wait for printer ready
              IN    AL,PORTC
              TEST AL,BIT1
        .UNTIL !ZERO?
        MOV   AL,AH                  ;send ASCII
        OUT   PORTB,AL
        MOV   AL,8                   ;pulse data strobe
        OUT   CMD,AL
        MOV   AL,9
        OUT   CMD,AL
        RET

PRINT   ENDP
```

Mode 2 Bidirectional Operation

In mode 2, which is allowed with group A only, port A becomes bidirectional, allowing data to be transmitted and received over the same eight wires. Bidirectional bused data are useful when interfacing two computers. It is also used for the IEEE-488 parallel high-speed GPIB (**general-purpose instrumentation bus**) interface standard. Figure 11–31 shows the internal structure and timing diagram for mode 2 bidirectional operation.

Signal Definitions for Bidirectional Mode 2

INTR — **Interrupt request** is an output used to interrupt the microprocessor for both input and output conditions.

$\overline{\text{OBF}}$ — **Output buffer full** is an output indicating that the output buffer contains data for the bidirectional bus.

$\overline{\text{ACK}}$ — **Acknowledge** is an input that enables the three-state buffers so that data can appear on port A. If $\overline{\text{ACK}}$ is a logic 1, the output buffers of port A are at their high-impedance state.

$\overline{\text{STB}}$ — The **strobe** input loads the port A input latch with external data from the bidirectional port A bus.

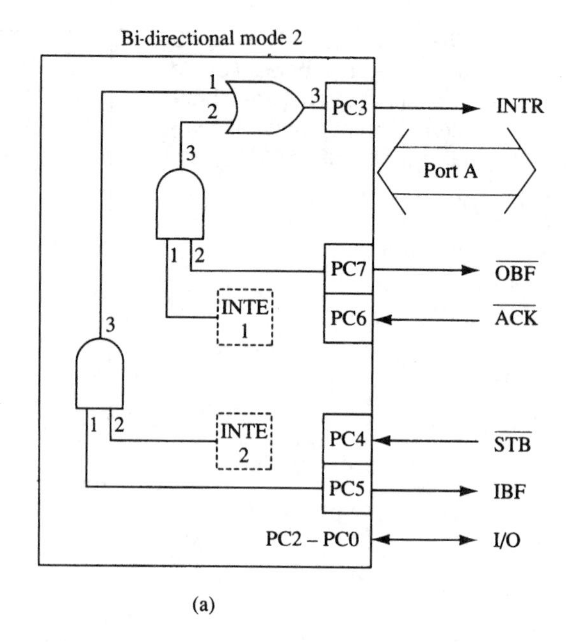

(a)

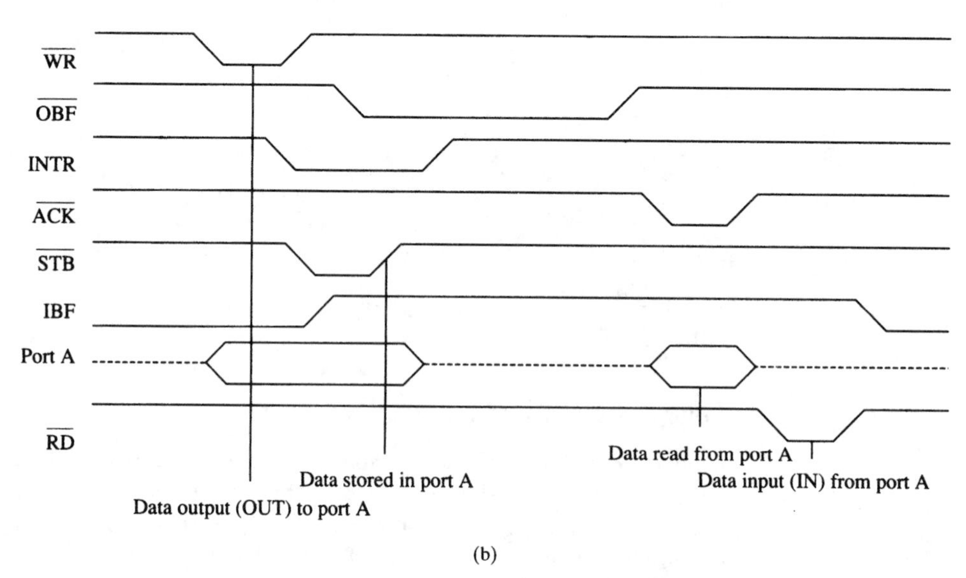

(b)

FIGURE 11–31 Mode 2 operation of the 82C55. (a) Internal structure and (b) timing diagram.

IBF — **Input buffer full** is an output used to signal that the input buffer contains data for the external bidirectional bus.

INTE — **Interrupt enable** are internal bits (INTE1 and INTE2) that enable the INTR pin. The state of the INTR pin is controlled through port C bits PC_6 (INTE1) and PC_4 (INTE2).

PC_0, PC_1, and PC_2 — These pins are general-purpose I/O pins in mode 2 controlled by the bit set and reset command.

The Bidirectional Bus. The bidirectional bus is used by referencing port A with the IN and OUT instructions. To transmit data through the bidirectional bus, the program first tests the $\overline{OBF}$ signal to determine whether the output buffer is empty. If it is, then data are sent to the output buffer via the OUT instruction. The external circuitry also monitors the $\overline{OBF}$ signal to decide whether the microprocessor has sent data to the bus. As soon as the output circuitry sees a logic 0 on $\overline{OBF}$, it sends back the $\overline{ACK}$ signal to remove it from the output buffer. The $\overline{ACK}$ signal sets the $\overline{OBF}$ bit and enables the three-state output buffers so that data may be read. Example 11–20 lists a procedure that transmits the contents of the AH register through bidirectional port A.

EXAMPLE 11–20

```
;A procedure transmits AH through the bidirectional bus

BIT7    EQU 80H
PORTC   EQU 62H
PORTA   EQU 60H

TRANS   PROC NEAR

        .REPEAT                         ;test OBF
             IN   AL,PORTC
             TEST AL,BIT7
        .UNTIL !ZERO?
        MOV  AL,AH                      ;send data
        OUT  PORTA,AL
        RET

TRANS   ENDP
```

To receive data through the bidirectional port A bus, the IBF bit is tested with software to decide whether data have been strobed into the port. If IBF = 1, then data are input using the IN instruction. The external interface sends data into the port by using the $\overline{STB}$ signal. When $\overline{STB}$ is activated, the IBF signal becomes a logic 1 and the data at port A are held inside the port in a latch. When the IN instruction executes, the IBF bit is cleared and the data in the port are moved into AL. Example 11–21 lists a procedure that reads data from the port.

EXAMPLE 11–21

```
;A procedure that reads data from the bidirectional bus into AL

BIT5  EQU  20H
PORTC EQU  62H
PORTA EQU  60H

READ  PROC NEAR

      .REPEAT                         ;test IBF
           IN   AL,PORTC
           TEST AL,BIT5
      .UNTIL !ZERO?
      IN   AL,PORTA
      RET

READ ENDP
```

The INTR (**interrupt request**) pin can be activated from both directions of data flow through the bus. If INTR is enabled by both INTE bits, then the output and input buffers both cause interrupt requests. This occurs when data are strobed into the buffer using $\overline{STB}$ or when data are written using OUT.

FIGURE 11–32 A summary of the port connections for the 82C55 PIA.

		Mode 0		Mode 1		Mode 2
Port A		IN	OUT	IN	OUT	I/O
Port B		IN	OUT	IN	OUT	Not used
Port C	0	IN	OUT	$INTR_B$	$INTR_B$	I/O
	1			IBF_B	$\overline{OBF}_B$	I/O
	2			$\overline{STB}_B$	$\overline{ACK}_B$	I/O
	3			$INTR_A$	$INTR_A$	INTR
	4			$\overline{STB}_A$	I/O	$\overline{STB}$
	5			IBF_A	I/O	IBF
	6			I/O	$\overline{ACK}_A$	$\overline{ACK}$
	7			I/O	$\overline{OBF}_A$	$\overline{OBF}$

82C55 Mode Summary

Figure 11–32 shows a graphical summary of the three modes of operation for the 82C55. Mode 0 provides simple I/O, mode 1 provides strobed I/O, and mode 2 provides bidirectional I/O. As mentioned, these modes are selected through the command register of the 82C55.

11–4 8254 PROGRAMMABLE INTERVAL TIMER

The 8254 programmable interval timer consists of three independent 16-bit programmable counters (**timers**). Each counter is capable of counting in binary or binary-coded decimal (BCD). The maximum allowable input frequency to any counter is 10 MHz. This device is useful wherever the microprocessor must control real-time events. Some examples of usage include a real-time clock, an events counter, and for motor speed and direction control.

This timer also appears in the personal computer decoded at ports 40H–43H to do the following:

1. Generate a basic timer interrupt that occurs at approximately 18.2 Hz.
2. Cause the DRAM memory system to be refreshed.
3. Provide a timing source to the internal speaker and other devices. The timer in the personal computer is an 8253 instead of an 8254.

8254 Functional Description

Figure 11–33 shows the pin-out of the 8254, which is a higher-speed version of the 8253, and a diagram of one of the three counters. Each timer contains a CLK input, a gate input, and an output (OUT) connection. The CLK input provides the basic operating frequency to the timer, the gate pin controls the timer in some modes, and the OUT pin is where we obtain the output of the timer.

The signals that connect to the microprocessor are the data bus pins (D_7–D_0), $\overline{RD}$, $\overline{WR}$, $\overline{CS}$, and address inputs A_1 and A_0. The address inputs are present to select any of the four internal registers used for programming, reading, or writing to a counter. The personal computer contains an 8253 timer or its equivalent, decoded at I/O ports 40H–43H. Timer zero is programmed to generate an 18.2 Hz signal that interrupts the microprocessor at interrupt vector 8 for a clock tick. The tick is often used to time programs and events in DOS. Timer 1 is programmed for 15 μs, which is used on the personal computer to request a DMA action used to refresh the dynamic RAM. Timer 2 is programmed to generate a tone from the personal computer speaker.

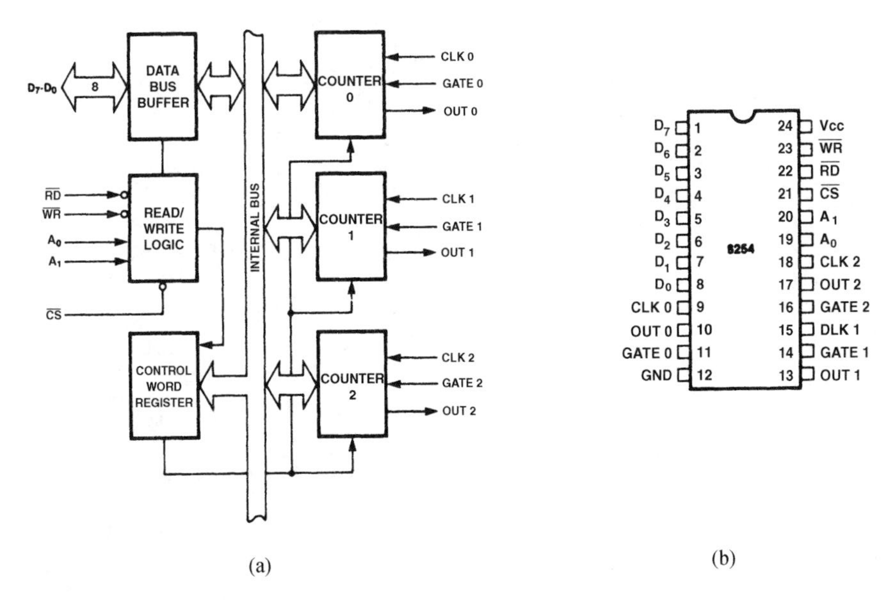

FIGURE 11–33 The 8254 programmable interval timer. (a) Internal structure and (b) pin-out. (Courtesy of Intel Corporation.)

Pin Definitions

A_0, A_1 The **address inputs** select one of four internal registers within the 8254. See Table 11–4 for the function of the A_1 and A_0 address bits.

CLK The **clock** input is the timing source for each of the internal counters. This input is often connected to the PCLK signal from the microprocessor system bus controller.

$\overline{CS}$ **Chip select** enables the 8254 for programming and for reading or writing a counter.

G The **gate input** controls the operation of the counter in some modes of operation.

GND **Ground** connects to the system ground bus.

OUT A **counter output** is where the waveform generated by the timer is available.

$\overline{RD}$ **Read** causes data to be read from the 8254 and often connects to the $\overline{IORC}$ signal.

V_{CC} **Power** connects to the +5.0 V power supply.

$\overline{WR}$ **Write** causes data to be written to the 8254 and often connects to the write strobe ($\overline{IOWC}$).

TABLE 11–4 Address selection inputs to the 8254.

A_1	A_0	*Function*
0	0	Counter 0
0	1	Counter 1
1	0	Counter 2
1	1	Control word

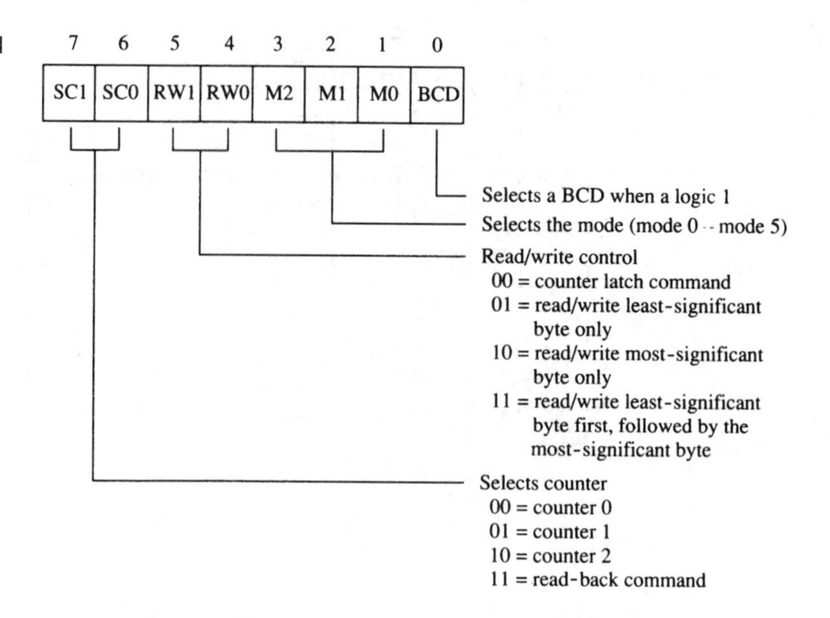

FIGURE 11–34 The control word for the 8254-2 timer.

Programming the 8254

Each counter is individually programmed by writing a control word, followed by the initial count. Figure 11–34 lists the program control word structure of the 8254. The **control word** allows the programmer to select the counter, mode of operation, and type of operation (read/write). The control word also selects either a binary or BCD count. Each counter may be programmed with a count of 1 to FFFFH. A count of 0 is equal to FFFFH+1 (65,536) or 10,000 in BCD. The minimum count of 1 applies to all modes of operation except modes 2 and 3, which have a minimum count of 2. Timer 0 is used in the personal computer with a divide-by count of 64K (FFFFH) to generate the 18.2 Hz (18.196 Hz) interrupt clock tick. Timer 0 has a clock input frequency of 4.77 MHz + 4 or 1.1925 MHz.

The control word uses the BCD bit to select a BCD count (BCD = 1) or a binary count (BCD = 0). The M_2, M_1, and M_0 bits select one of the six different modes of operation (000–101) for the counter. The RW_1 and RW_0 bits determine how the data are read from or written to the counter. The SC_1 and SC_0 bits select a counter or the special read-back mode of operation, discussed later in this section.

Each counter has a program control word used to select the way the counter operates. If two bytes are programmed into a counter, then the first byte (LSB) will stop the count, and the second byte (MSB) will start the counter with the new count. The order of programming is important for each counter, but programming of different counters may be interleaved for better control. For example, the control word may be sent to each counter before the counts for individual programming. Example 11–22 shows a few ways to program counters 1 and 2. The first method programs both control words, then the LSB of the count for each counter, which stops them from counting. Finally, the MSB portion of the count is programmed, starting both counters with the new count. The second example shows one counter programmed before the other.

EXAMPLE 11–22

```
PROGRAM CONTROL WORD 1 PROGRAM CONTROL WORD 2 PROGRAM LSB 1
PROGRAM LSB 2
PROGRAM MSB 1
PROGRAM MSB 2
;setup counter 1
;setup counter 2
```

```
;stop counter 1 and program LSB
;stop counter 2 and program LSB ;program MSB of counter 1 and start it
;program MSB of counter 2 and start it

     or

PROGRAM CONTROL WORD 1 PROGRAM LSB 1
PROGRAM MSB 1
PROGRAM CONTROL WORD 2 PROGRAM LSB 2
PROGRAM MSB 2

;setup counter 1
;stop counter 1 and program LSB ;program MSB of counter 1 and start it
;setup counter 2
;stop counter 2 and program LSB ;program MSB of counter 2 and start it
```

Modes of Operation. Six modes (mode 0–mode 5) of operation are available to each of the 8254 counters. Figure 11–35 shows how each of these modes functions with the CLK input, the gate (G) control signal, and the OUT signal. A description of each mode follows:

MODE 0 Allows the 8254 counter to be used as an events counter. In this mode, the output becomes a logic 0 when the control word is written and remains there until N plus the number of programmed counts. For example, if a count of 5 is programmed, the output will remain a logic 0 for 6 counts beginning with N. Note that the gate (G) input must be a logic 1 to allow the counter to count. If G becomes a logic 0 in the middle of the count, the counter will stop until G again becomes a logic 1.

MODE 1 Causes the counter to function as a retriggerable, monostable multivibrator (one-shot). In this mode the G input triggers the counter so that it develops a pulse at the OUT connection that becomes a logic 0 for the duration of the count. If the count is 10, then the OUT connection goes low for 10 clocking periods when triggered. If the G input occurs within the duration of the output pulse, the counter is again reloaded with the count and the OUT connection continues for the total length of the count.

MODE 2 Allows the counter to generate a series of continuous pulses that are one clock pulse wide. The separation between pulses is determined by the count. For example, for a count of 10, the output is a logic 1 for nine clock periods and low for one clock period. This cycle is repeated until the counter is programmed with a new count or until the G pin is placed at a logic 0 level. The G input must be a logic 1 for this mode to generate a continuous series of pulses.

MODE 3 Generates a continuous square wave at the OUT connection, provided that the G pin is a logic 1. If the count is even, the output is high for one half of the count and low for one half of the count. If the count is odd, the output is high for one clocking period longer than it is low. For example, if the counter is programmed for a count of 5, the output is high for three clocks and low for two clocks.

MODE 4 Allows the counter to produce a single pulse at the output. If the count is programmed as a 10, the output is high for 10 clocking periods and low for one clocking period. The cycle does not begin until the counter is loaded with its complete count. This mode operates as a software triggered one-shot. As with modes 2 and 3, this mode also uses the G input to enable the counter. The G input must be a logic 1 for the counter to operate for these three modes.

MODE 5 A hardware triggered one-shot that functions as mode 4, except that it is started by a trigger pulse on the G pin instead of by software. This mode is also similar to mode 1 because it is retriggerable.

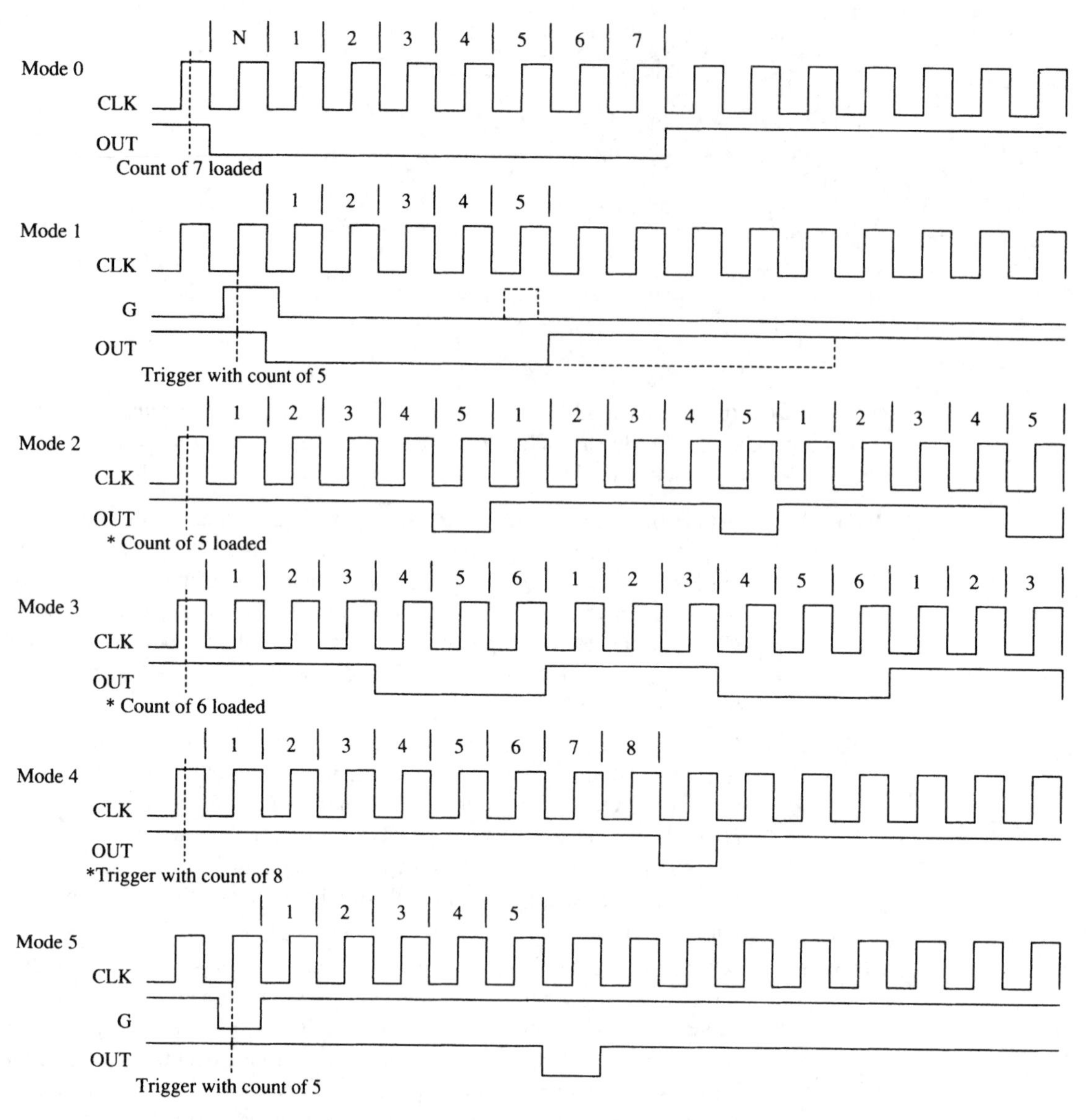

FIGURE 11–35 The six modes of operation for the 8254-2 programmable interval timer. *The G input stops the count when 0 in modes 2, 3, and 4.

Generating a Waveform with the 8254. Figure 11–36 shows an 8254 connected to function at I/O ports 0700H, 0702H, 0704H, and 0706H of an 80386SX microprocessor. The addresses are decoded by using a PLD that also generates a write strobe signal for the 8254, which is connected to the low-order data bus connections. The PLD also generates a wait signal for the microprocessor that causes two wait states when the 8254 is accessed. The wait state generator connected to the microprocessor actually controls the number of wait states inserted into the timing. The program for the PLD is not illustrated here because it is the same as many of the prior examples.

Example 11–23 lists the program that generates a 100 KHz square-wave at OUT0 and a 200 KHz continuous pulse at OUT1. Counter 0 uses mode 3 and counter 1 uses mode 2. The count programmed into counter 0 is 80 and the count for counter 1 is 40. These counts generate the desired output frequencies with an 8 MHz input clock.

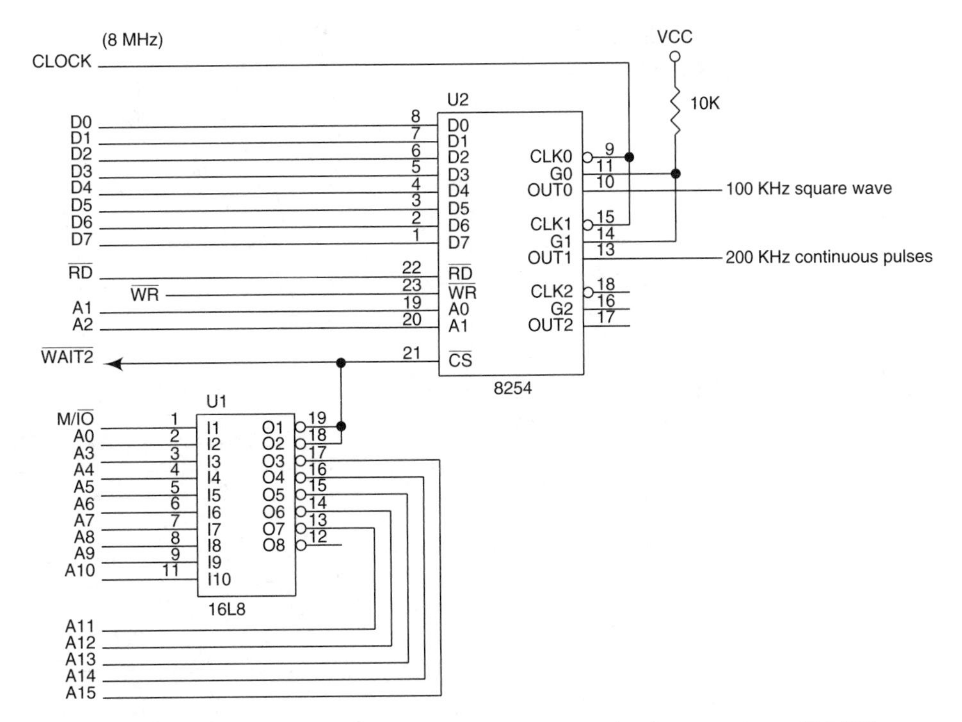

FIGURE 11–36 The 8254 interfaced to an 8 MHz 8086 so that it generates a 100 KHz square wave at OUT0 and a 200 KHz continuous pulse at OUT1.

EXAMPLE 11–23

```
;A procedure that programs the 8254 timer to function
;as illustrated in Figure 11-34

TIME    PROC    NEAR USES AX DX

        MOV  DX,706H          ;program counter 0 for mode 3
        MOV  AL,00110110B
        OUT  DX,AL
        MOV  AL,01110100B     ;program counter 1 for mode 2
        OUT  DX,AL

        MOV  DX,700H          ;program counter 0 with 80
        MOV  AL,80
        OUT  DX,AL
        MOV  AL,0
        OUT  DX,AL

        MOV  DX,702H          ;program counter 1 with 40
        MOV  AL,40
        OUT  DX,AL
        MOV  AL,0
        OUT  DX,AL

        RET

TIME    ENDP
```

Reading a Counter. Each counter has an internal latch that is read with the read counter port operation. These latches will normally follow the count. If the contents of the counter are needed, then the latch can remember the count by programming the counter latch control word (see Figure 11–37), which causes the contents of the counter to be held in a latch until it is read.

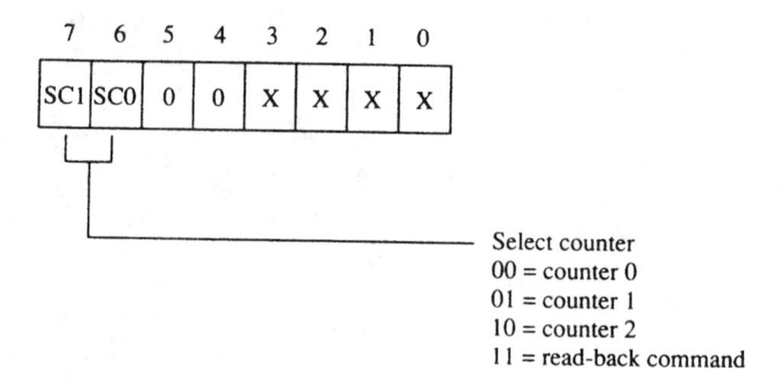

FIGURE 11–37 The 8254-2 counter latch control word.

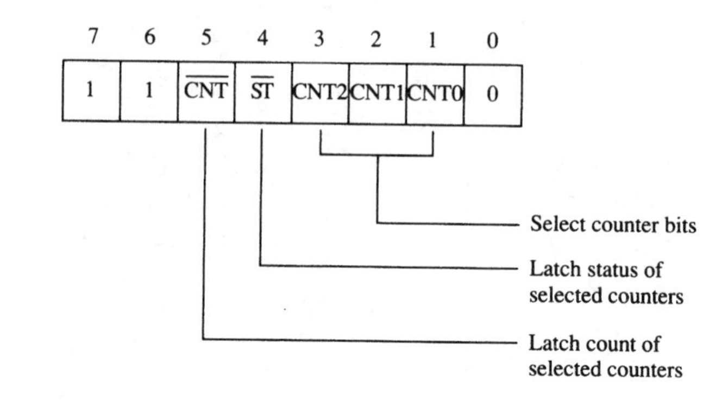

FIGURE 11–38 The 8254-2 read-back control word.

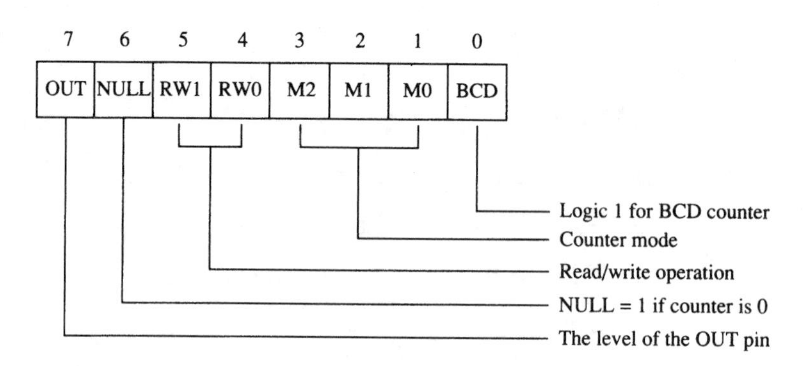

FIGURE 11–39 The 8254-2 status register.

Whenever a read from the latch or the counter is programmed, the latch tracks the contents of the counter.

When it is necessary for the contents of more than one counter to be read at the same time, we use the read-back control word, illustrated in Figure 11–38. With the read-back control word, the $\overline{\text{CNT}}$ bit is a logic 0 to cause the counters selected by CNT0, CNT1, and CNT2 to be latched. If the status register is to be latched, then the $\overline{\text{ST}}$ bit is placed at a logic 0. Figure 11–39 shows the status register, which shows the state of the output pin, whether the counter is at its null state (0), and how the counter is programmed.

DC Motor Speed and Direction Control

One application of the 8254 timer is as a motor speed controller for a DC motor. Figure 11–40 shows the schematic diagram of the motor and its associated driver circuitry. It also illustrates the interconnection of the 8254, a flip-flop, and the motor and its driver.

The operation of the motor driver circuitry is straightforward. If the Q output of the 74ALS112 is a logic 1, the base Q_2 is pulled up to +12 V through the base pull-up resistor, and

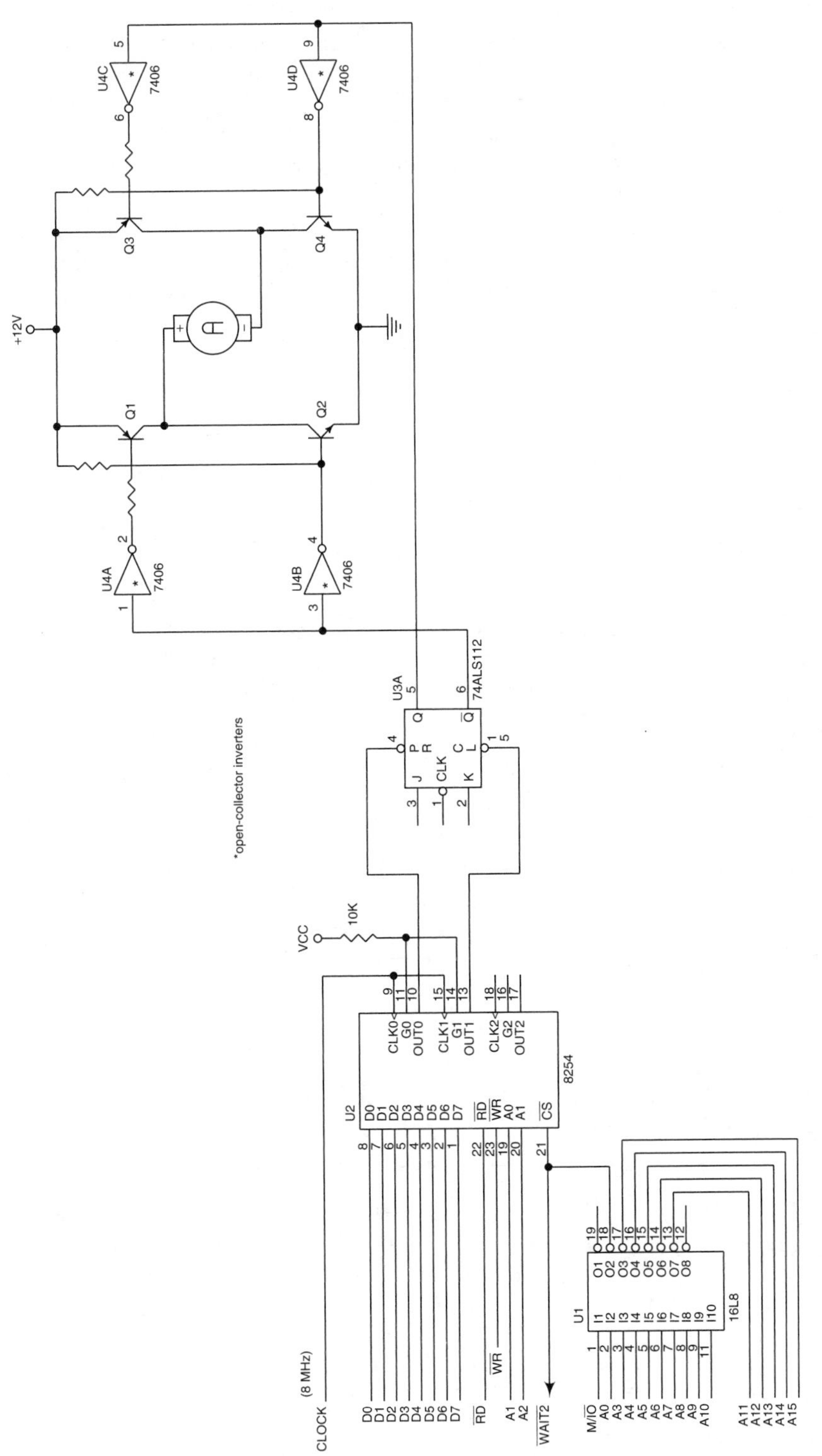

FIGURE 11–40 Motor speed and direction control using the 8254 timer.

the base of Q_2 is open circuited. This means that Q_1 is off and Q_2 is on, with ground applied to the positive lead of the motor. The bases of both Q_3 and Q_4 are pulled low to ground through the inverters. This causes Q_3 to conduct or turn on and Q_4 to turn off, applying ground to the negative lead of the motor. The logic 1 at the Q output of the flip-flop therefore connects +12 V to the positive lead of the motor and ground to the negative lead. This connection causes the motor to spin in its forward direction. If the state of the Q output of the flip-flop becomes a logic 0, then the conditions of the transistors are reversed and +12 V is attached to the negative lead of the motor, with ground attached to the positive lead. This causes the motor to spin in the reverse direction.

If the output of the flip-flop is alternated between a logic 1 and 0, the motor spins in either direction at various speeds. If the duty cycle of the Q output is 50%, the motor will not spin at all and exhibits some holding torque because current flows through it. Figure 11–41 shows some

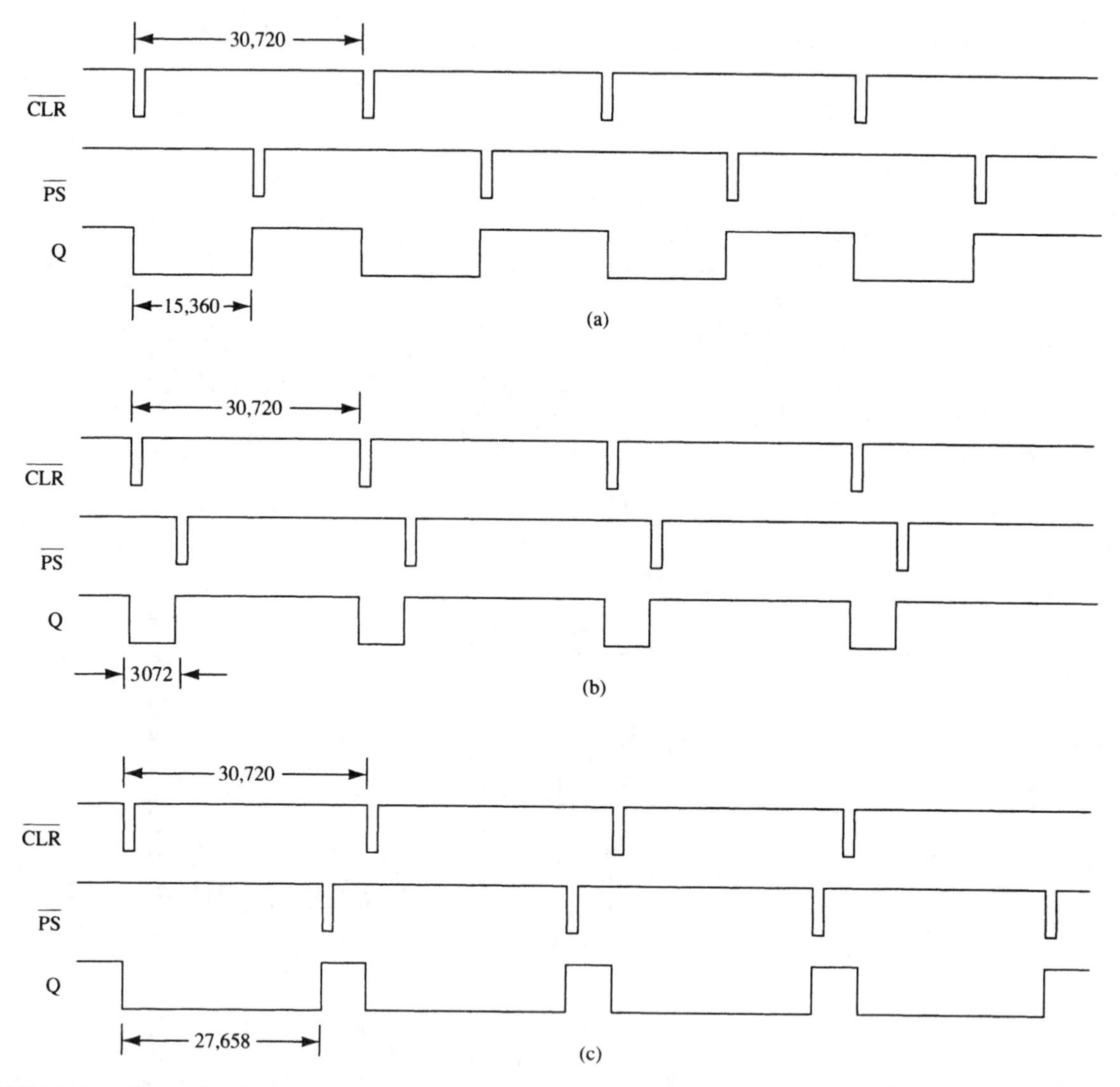

FIGURE 11–41 Timing for the motor speed and direction control circuit of Figure 11–45. (a) No rotation, (b) high-speed rotation in the reverse direction, and (c) high-speed rotation in the forward direction.

timing diagrams and their effects on the speed and direction of the motor. Notice how each counter generates pulses at different positions to vary the duty cycle at the Q output of the flip-flop. This output is also called *pulse width modulation*.

To generate these wave forms, counters 0 and 1 are both programmed to divide the input clock (PCLK) by 30,720. We change the duty cycle of Q by changing the point at which counter 1 is started in relationship to counter 0. This changes the direction and speed of the motor. But why divide the 8 MHz clock by 30,720? The divide rate of 30,720 is divisible by 256, so we can develop a short program that allows 256 different speeds. This also produces a basic operating frequency for the motor of about 260 Hz, which is low enough in frequency to power the motor. It is important to keep this operating frequency below 1000 Hz, but above 60 Hz.

Example 11–24 lists a procedure that controls the speed and direction of the motor. The speed is controlled by the value of AH when this procedure is called. Because we have an 8-bit number to represent speed, a 50% duty cycle, for a stopped motor, is a count of 128. By changing the value in AH when the procedure is called, we can adjust the motor speed. The speed of the motor will increase in either direction by changing the number in AH when this procedure is called. As the value in AH approaches 00H, the motor begins to increase its speed in the reverse direction. As the value of AH approaches FFH, the motor increases its speed in the forward direction.

EXAMPLE 11–24

```
;A procedure that controls the speed and direction of the motor
;in Figure 11-40.
;
;AH determines the speed and direction of the motor where
;AH is between 00H and FFH.

CNTR    EQU   706H
CNT0    EQU   700H
CNT1    EQU   702H
COUNT   EQU   30720

SPEED   PROC  NEAR USES BX DX AX

        MOV  BL,AH          ;calculate count
        MOV  AX,120
        MUL  BL
        MOV  BX,AX
        MOV  AX,COUNT
        SUB  AX,BX
        MOV  BX,AX

        MOV  DX,CNTR
        MOV  AL,00110100B   ;program control words
        OUT  DX,AL
        MOV  AL,01110100B
        OUT  DX,AL

        MOV  DX,CNT1        ;program counter 1
        MOV  AX,COUNT       ;to generate a clear
        OUT  DX,AL
        MOV  AL,AH
        OUT  DX,AL

        .REPEAT             ;wait for counter 1
               IN   AL,DX
               XCHG AL,AH
               IN   AL,DX
               XCHG AL,AH
        .UNTIL BX == AX
```

```
        MOV   DX,CNT0          ;program counter 0
        MOV   AX,COUNT         ;to generate a set
        OUT   DX,AL
        MOV   AL,AH
        OUT   DX,AL

        RET

SPEED   ENDP
```

The procedure adjusts the wave form at Q by first calculating the count at which counter 0 is to start in relationship to counter 1. This is accomplished by multiplying AH by 120 and then subtracting it from 30,720. This is required because the counters are down-counters that count from the programmed count to 0 before restarting. Next, counter 1 is programmed with a count of 30,720 and started so it generates the clear-wave form for the flip-flop. After counter 1 is started, it is read and compared with the calculated count. Once it reaches this count, counter 0 is started with a count of 30,720. From this point forward, both counters continue generating the clear and set wave forms until the procedure is again called to adjust the speed and direction of the motor.

11–5 16550 PROGRAMMABLE COMMUNICATIONS INTERFACE

The National Semiconductor Corporation's PC16550D is a programmable communications interface designed to connect to virtually any type of serial interface. The 16550 is a universal asynchronous receiver/transmitter (UART) that is fully compatible with the Intel microprocessors. The 16550 is capable of operating at 0–1.5 M baud. Baud rate is the number of bits transferred per second (bps), including start, stop, data, and parity. (Bps is bytes per second and bps is bits per second.) The 16550 also includes a programmable baud rate generator and separate FIFOs for input and output data to ease the load on the microprocessor. Each FIFO contains 16 bytes of storage. This is the most common communications interface found in modern microprocessor-based equipment, including the personal computer and many modems.

Asynchronous Serial Data

Asynchronous serial data are transmitted and received without a clock or timing signal. Figure 11–42 illustrates two frames of asynchronous serial data. Each frame contains a start bit, seven data bits, parity, and one stop bit. The figure shows a frame that contains one ASCII character and 10 bits. Most dial-up communications systems of the past, such as CompuServe, Prodigy, and America Online, used 10 bits for asynchronous serial data with even parity. Most Internet and bulletin board services also use 10 bits, but they normally do not use parity. Instead, eight data bits are transferred, replacing parity with a data bit. This makes byte transfers of non-ASCII data much easier to accomplish.

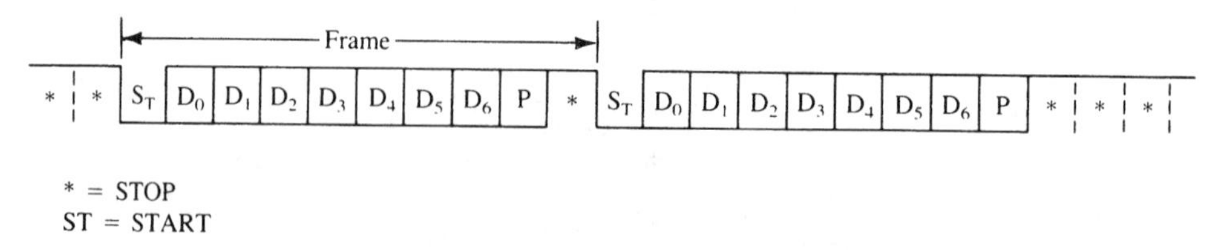

FIGURE 11–42 Asynchronous serial data.

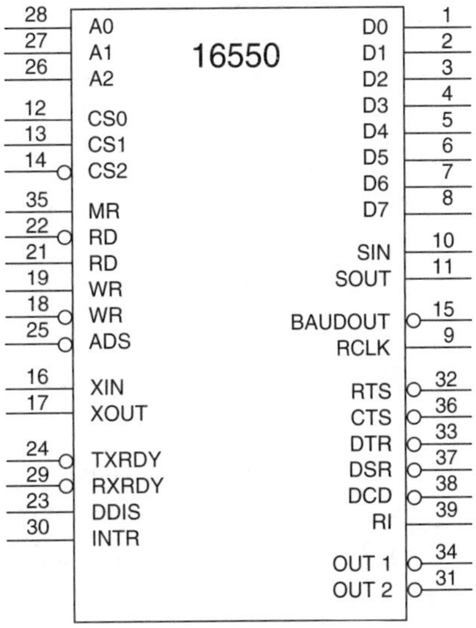

FIGURE 11–43 The pin-out of the 16550 UART.

16550 Functional Description

Figure 11–43 illustrates the pin-out of the 16550 UART. This device is available as a 40-pin DIP (**dual in-line package**) or as a 44-pin PLCC (**plastic leadless chip carrier**). Two completely separate sections are responsible for data communications: the receiver and the transmitter. Because each of these sections is independent, the 16550 is able to function in simplex, half-duplex, or full-duplex modes. One of the main features of the 16550 is its internal receiver and transmitter FIFO (first-in, first-out) memories. Because each is 16 bytes deep, the UART requires attention only from the microprocessor after receiving 16 bytes of data. It also holds 16 bytes before the microprocessor must wait for the transmitter. The FIFO makes this UART ideal when interfacing to high-speed systems because less time is required to service it.

An example **simplex** system is one in which the transmitter or receiver is used by itself, such as in an FM (**frequency modulation**) radio station. An example **half-duplex** system is a CB (**citizens band**) radio, on which we transmit and receive, but not both at the same time. The **full-duplex** system allows transmission and reception in both directions simultaneously. An example full-duplex system is the telephone.

The 16550 can control a **modem** (**modulator/demodulator**), which is a device that converts TTL levels of serial data into audio tones that can pass through the telephone system. Six pins on the 16650 are devoted to modem control: $\overline{\text{DSR}}$ (**data set ready**), $\overline{\text{DTR}}$ (**data terminal ready**), $\overline{\text{CTS}}$ (**clear-to-send**), $\overline{\text{RTS}}$ (**request-to-send**), $\overline{\text{RI}}$ (**ring indicator**), and $\overline{\text{DCD}}$ (**data carrier detect**). The modem is referred to as the **data set** and the 16550 is referred to as the **data terminal**.

16550 Pin Functions

A_0, A_1, A_2 — The **address inputs** are used to select an internal register for programming and also data transfer. See Table 11–5 for a list of each combination of the address inputs and the registers selected.

$\overline{\text{ADS}}$ — The **address strobe** input is used to latch the address lines and chip select lines. If not needed (as in the Intel system), connect this pin to ground. The $\overline{\text{ADS}}$ pin is designed for use with Motorola microprocessors.

$\overline{\text{BAUDOUT}}$ — The **Baud out** pin is where the clock signal generated by the Baud rate generator from the transmitter section is made available. It is most often

TABLE 11–5 The registers selected by A_0, A_1, and A_2.

A_2	A_1	A_0	*Function*
0	0	0	Receiver buffer (read) and transmitter holding (write)
0	0	1	Interrupt enable
0	1	0	Interrupt identification (read) and FIFO control (write)
0	1	1	Line control
1	0	0	Modem control
1	0	1	Line status
1	1	0	Modem status
1	1	1	Scratch

connected to the RCLK input to generate a receiver clock that is equal to the transmitter clock.

CS_0, CS_1, $\overline{CS2}$ The **chip select** inputs must all be active to enable the 16550 UART.

$\overline{CTS}$ The **clear-to-send** (if low) indicates that the modem or data set is ready to exchange information. This pin is often used in a half-duplex system to turn the line around.

D_0–D_7 The **data bus** pins are connected to the microprocessor data bus.

$\overline{DCD}$ The **data carrier detect** input is used by the modem to signal the 16550 that a carrier is present.

DDIS The **disable driver** output becomes a logic 0 to indicate that the microprocessor is reading data from the UART. DDIS can be used to change the direction of data flow through a buffer.

$\overline{DSR}$ **Data set ready** is an input to the 16550, indicating that the modem or data set is ready to operate.

$\overline{DTR}$ **Data terminal ready** is an output that indicates that the data terminal (16550) is ready to function.

INTR **Interrupt request** is an output to the microprocessor used to request an interrupt (INTR = 1) whenever the 16550 has a receiver error, it has received data, and if the transmitter is empty.

MR **Master reset** initializes the 16550 and should be connected to the system RESET signal.

$\overline{OUT1}$, $\overline{OUT2}$ **User-defined output pins** that can provide signals to a modem or any other device as needed in a system.

RCLK **Receiver clock** is the clock input to the receiver section of the UART. This input is always 16 times the desired receiver baud rate.

RD, $\overline{RD}$ **Read inputs** (either may be used) cause data to be read from the register specified by the address inputs to the UART.

$\overline{RI}$ The **ring indicator** input is placed at the logic 0 level by the modem to indicate that the telephone is ringing.

$\overline{RTS}$ **Request-to-send** is a signal to the modem indicating that the UART wishes to send data.

SIN, SOUT These are the **serial data pins**. SIN accepts serial data and SOUT transmits serial data.

$\overline{RXRDY}$ **Receiver ready** is a signal used to transfer received data via DMA techniques (see text).

$\overline{TXRDY}$ **Transmitter ready** is a signal used to transfer transmitter data via DMA techniques (see text).

WR, $\overline{WR}$ **Write** (either may be used) connects to the microprocessor write signal to transfer commands and data to the 16550.

XIN, XOUT These are the main **clock** connections. A crystal is connected across these pins to form a crystal oscillator, or XIN is connected to an external timing source.

Programming the 16550

Programming the 16550 is simple, although it may seem slightly more involved when compared to some of the other programmable interfaces described in this chapter. Programming is a two-part process that includes the initialization dialog and operational dialog.

In the personal computer, which uses the 16550 or its programming equivalent, the I/O port addresses are decoded at 3F8H through 3FFH for COM port 0 and 2F8H through 2FFH for COM port 2. Although the examples in this section of the chapter are not written specifically for the personal computer, they can be adapted by changing the port numbers to control the COM ports on the PC.

Initializing the 16550. Initialization dialog, which occurs after a hardware or software reset, consists of two parts: programming the line control register and the Baud rate generator. The line control register selects the number of data bits, number of stop bits, and parity (whether it's even or odd, or if parity is sent as a 1 or a 0). The Baud rate generator is programmed with a divisor that determines the Baud rate of the transmitter section.

Figure 11–44 illustrates the line control register. The line control register is programmed by outputting information to I/O port 011 (A_2, A_1, A_0). The rightmost two bits of the line control register select the number of transmitted data bits (5, 6, 7, or 8). The number of stop bits is selected by S in the line control register. If S = 0, one stop bit is used; if S = 1, 1.5 stop bits are used for five data bits, and two stop bits are used with six, seven, or eight data bits.

The next three bits are used together to send even or odd parity, to send no parity, or to send a 1 or a 0 in the parity bit position. To send even or odd parity, the ST (**stick**) bit must be placed at a logic 0 level, and parity enable must be a logic 1. The value of the parity bit then determines

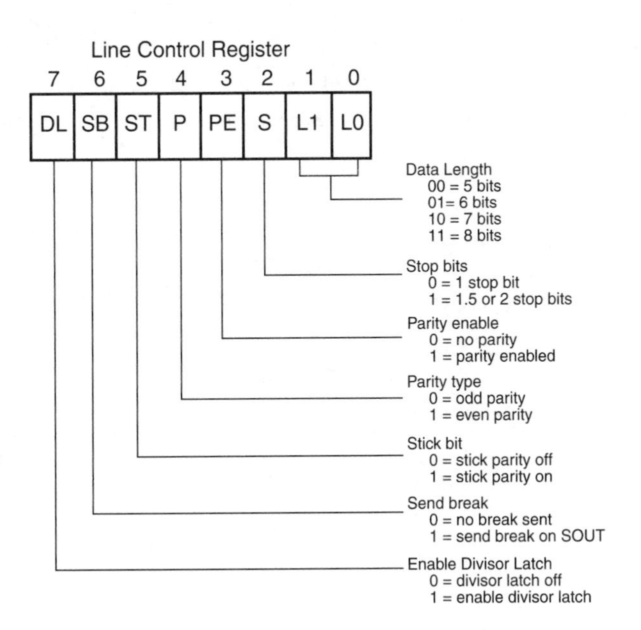

FIGURE 11–44 The contents of the 16550 line control register.

TABLE 11–6 The operation of the ST and parity bits.

ST	*P*	*PE*	*Function*
0	0	0	No parity
0	0	1	Odd parity
0	1	0	No parity
0	1	1	Even parity
1	0	0	Undefined
1	0	1	Send/receive 1
1	1	0	Undefined
1	1	1	Send/receive 0

even or odd parity. To send no parity (common in Internet connections), ST = 0 as well as the parity enable bit. This sends and receives data without parity. Finally, if a 1 or a 0 must be sent and received in the parity bit position for all data, ST = 1 with a 1 in parity enable. To send a 1 in the parity bit position, place a 0 in the parity bit; to send a 0, place a 1 in the parity bit. (See Table 11–6 for the operation of the parity and stick bits.)

The remaining bits in the line control register are used to send a break and to select programming for the Baud rate divisor. If bit position 6 of the line control register is a logic 1, a break is transmitted. As long as this bit is a 1, the break is sent from the SOUT pin. A break, by definition, is at least two frames of logic 0 data. The software in the system is responsible for timing the transmission of the break. To end the break, bit position 6 or the line control register is returned to a logic 0 level. The Baud rate divisor is only programmable when bit position 7 of the line control register is a logic 1.

Programming the Baud Rate. The Baud rate generator is programmed at I/O addresses 000 and 001 (A_2, A_1, A_0). Port 000 is used to hold the least-significant part of the 16-bit divisor and port 001 is used to hold the most-significant part. The value used for the divisor depends on the external clock or crystal frequency. Table 11–7 illustrates common Baud rates obtainable if an 18.432 MHz crystal is used as a timing source. It also shows the divisor values programmed into the Baud rate generator to obtain these Baud rates. The actual number programmed into the Baud rate generator causes it to produce a clock that is 16 times the desired baud rate. For example, if 240 is programmed into the Baud rate divisor, the Baud rate is (18.432 MHz ÷ 16) × 240 = 4800 Baud.

Sample Initialization. Suppose that an asynchronous system requires seven data bits, odd parity a Baud rate of 9600, and one stop bit. Example 11–25 lists a procedure that initializes the

TABLE 11–7 The divisor used with the Baud rate generator for an 18.432 MHz crystal illustrating common Baud rates.

Baud Rate	*Divisor Value*
110	10,473
300	3840
1200	920
2400	480
4800	240
9600	120
19,200	60
38,400	30
57,600	20
115,200	10

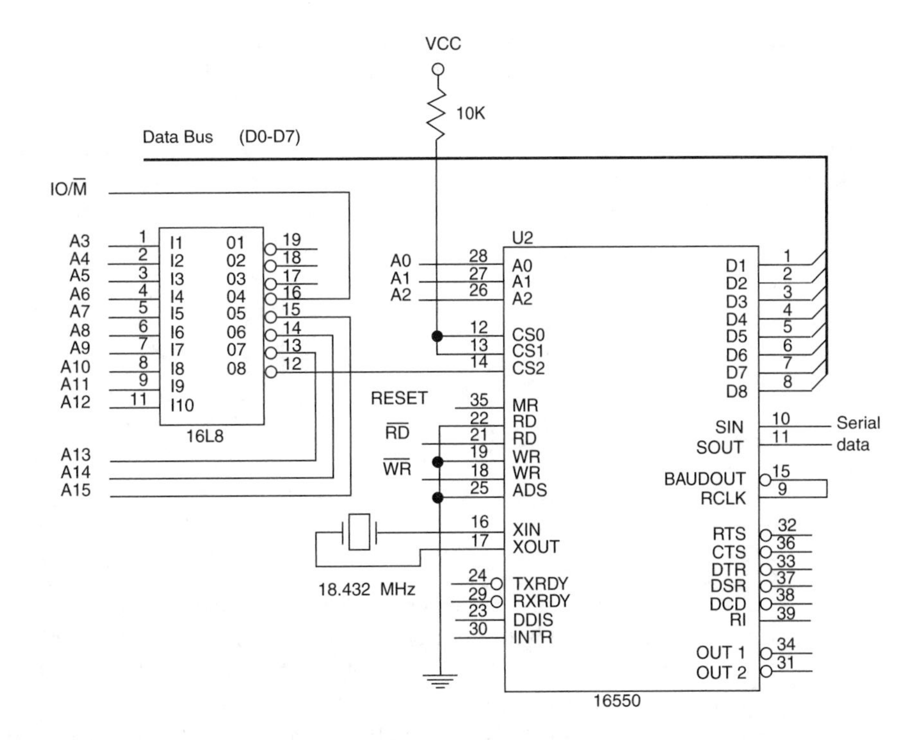

FIGURE 11–45 The 16550 interfaced to the 8088 microprocessor at ports 00F0H–00F7H.

16550 to function in this manner. Figure 11–45 shows the interface to the 8088 microprocessor, using a PLD to decode the 8-bit port addresses F0H through F7H. (The PLD program is not shown.) Here port F3H accesses the line control register and F0H and F1H access the Baud rate divisor registers. The last part of Example 11–25 is described with the function of the FIFO control register in the next few paragraphs.

EXAMPLE 11–25

```
;Initialization dialog for Figure 11-45
;Baud rate 9600, 7 data, odd parity, 1 stop

LINE   EQU   0F3H
LSB    EQU   0F0H
MSB    EQU   0F1H
FIFO   EQU   0F2H

INIT   PROC  NEAR

       MOV  AL,10001010B    ;enable Baud rate divisor
       OUT  LINE,AL

       MOV  AL,120          ;program Baud 9600
       OUT  LSB,AL
       MOV  AL,0
       OUT  MSB,AL

       MOV  AL,00001010B    ;program 7 data, odd
       OUT  LINE,AL         ;parity, 1 stop

       MOV  AL,00000111B    ;enable transmitter and
       OUT  FIFO,AL         ;receiver

       RET

INIT   ENDP
```

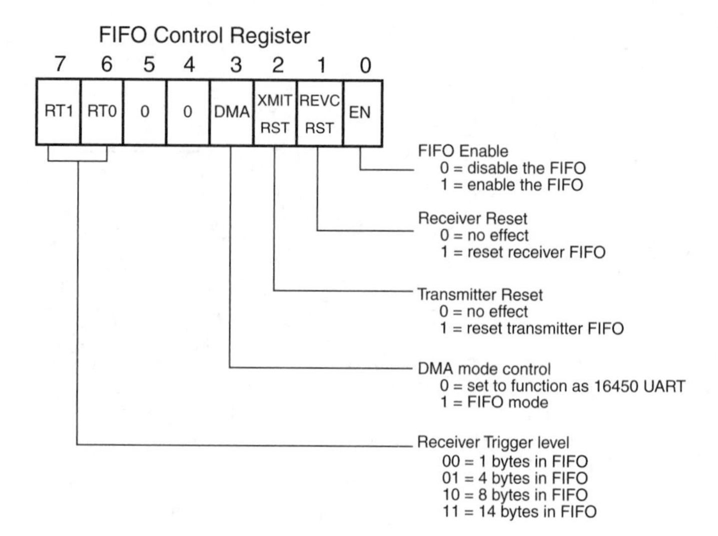

FIGURE 11–46 The FIFO control register of the 16550 UART.

After the line control register and Baud rate divisor are programmed into the 16550, it is still not ready to function. After programming the line control register and Baud rate, we still must program the FIFO control register, which is at port F2H in the circuit of Figure 11–45.

Figure 11–46 illustrates the FIFO control register for the 16550. This register enables the transmitter and receiver (bit 0 = 1), and clears the transmitter and receiver FIFOs. It also provides control for the 16550 interrupts, which are discussed in Chapter 12. Notice that the last section of Example 11–25 places a 7 into the FIFO control register. This enables the transmitter and receiver, and clears both FIFOs. The 16550 is now ready to operate, but without interrupts. Interrupts are automatically disabled when the MR (master reset) input is placed at a logic 1 by the system RESET signal.

Sending Serial Data. Before serial data can be sent or received through the 16550, we need to know the function of the line status register (see Figure 11–47). The line status register contains

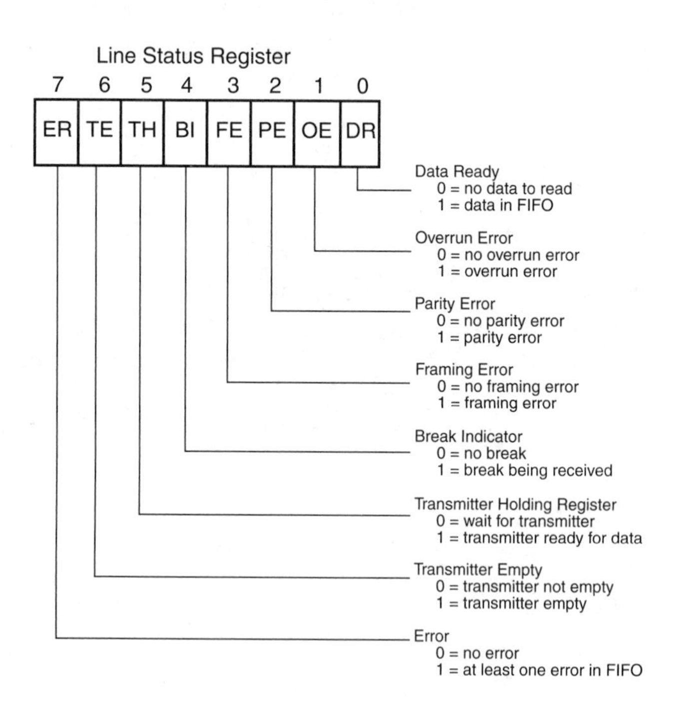

FIGURE 11–47 The contents of the line status register of the 16550 UART.

information about error conditions and the state of the transmitter and receiver. This register is tested before a byte is transmitted or can be received.

Suppose that a procedure (see Example 11–26) is written to transmit the contents of AH to the 16550 and out through its serial data pin (SOUT). The TH bit is polled by software to determine whether the transmitter is ready to receive data. This procedure uses the circuit of Figure 11–45.

EXAMPLE 11–26

```
;A procedure that transmits AH via the 16650 UART

LSTAT  EQU   0F5H
DATA   EQU   0F0H

SEND   PROC  NEAR USES AX

       .REPEAT                   ;test the TH bit
              IN   AL,LSTAT
              TEST AL,20H
       .UNTIL !ZERO?

       MOV  AL,AH                ;send data
       OUT  DATA,AL
       RET

SEND   ENDP
```

Receiving Serial Data. To read received information from the 16550, test the DR bit of the line status register. Example 11–27 lists a procedure that tests the DR bit to decide whether the 16550 has received any data. Upon the reception of data, the procedure tests for errors. If an error is detected, the procedure returns with AL equal to an ASCII '?'. If no error has occurred, then the procedure returns with AL equal to the received character.

EXAMPLE 11–27

```
;A procedure that receives data from the 16550 UART and
;returns it in AL.

LSTAT  EQU   0F5H
DATA   EQU   0F0H

REVC   PROC  NEAR

       .REPEAT
              IN   AL,LSTAT      ;test DR bit
              TEST AL,1
       .UNTIL !ZERO?

       TEST AL,0EH               ;test for any error
       .IF ZERO?                 ;no error
              IN   AL,DATA
       .ELSE                     ;any error
              MOV  AL,'?'
       .ENDIF
       RET

RECV   ENDP
```

UART Errors. The types of errors detected by the 16550 are parity error, framing error, and overrun error. A **parity error** indicates that the received data contain the wrong parity. A **framing error** indicates that the start and stop bits are not in their proper places. An **overrun error** indicates that data have overrun the internal receiver FIFO buffer. These errors should not occur during normal operation. If a parity error occurs, it indicates that noise was encountered during reception. A framing error occurs if the receiver is receiving data at an incorrect Baud rate. An overrun error occurs only if the software fails to read the data from the UART before the receiver FIFO is full. This example does not test the BI (break indicator bit) for a break condition. Note that a break is two consecutive frames of logic 0s on the SIN pin of the UART. The remaining registers, which are used for interrupt control and modem control, are developed in Chapter 12.

11–6 ANALOG-TO-DIGITAL (ADC) AND DIGITAL-TO-ANALOG (DAC) CONVERTERS

Analog-to digital (ADC) and digital-to-analog (DAC) converters are used to interface the microprocessor to the analog world. Many events that are monitored and controlled by the microprocessor are analog events. These can range from monitoring all forms of events, even speech, to controlling motors and like devices. In order to interface the microprocessor to these events, we must have an understanding of the interface and control of the ADC and DAC, which convert between analog and digital data.

The DAC0830 Digital-to-Analog Converter

A fairly common and low-cost digital-to-analog converter is the DAC0830 (a product of National Semiconductor Corporation). This device is an 8-bit converter that transforms an 8-bit binary number into an analog voltage. Other converters are available that convert from 10-, 12-, or 16-bit binary numbers into analog voltages. The number of voltage steps generated by the converter is equal to the number of binary input combinations. Therefore, an 8-bit converter generates 256 different voltage levels, a 10-bit converter generates 1024 levels, and so forth. The DAC0830 is a medium-speed converter that transforms a digital input to an analog output in approximately 1.0 μs.

Figure 11–48 illustrates the pin-out of the DAC0830. This device has a set of eight data bus connections for the application of the digital input code, and a pair of analog outputs labeled IOUT1 and IOUT2 that are designed as inputs to an external operational amplifier. Because this is an 8-bit converter, its output step voltage is defined as $-V_{REF}$ (reference voltage), divided by 255. For example, if the reference voltage is –5.0 V, its output step voltage is +.0196 V. Note that the output voltage is the opposite polarity of the reference voltage. If an input of $1001\ 0010_2$ is applied to the device, the output voltage will be the step voltage times $1001\ 0010_2$, or, in this case, +2.862 V. By changing the reference voltage to –5.1 V, the step voltage becomes +.02 V. The step voltage is also often called the **resolution** of the converter.

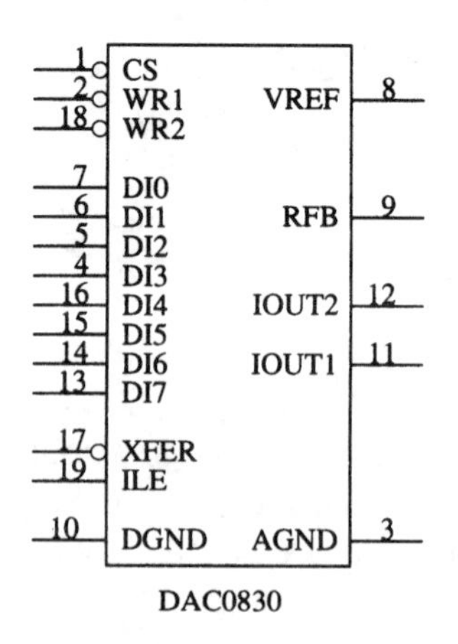

FIGURE 11–48 The pin-out of the DAC0830 digital-to-analog converter.

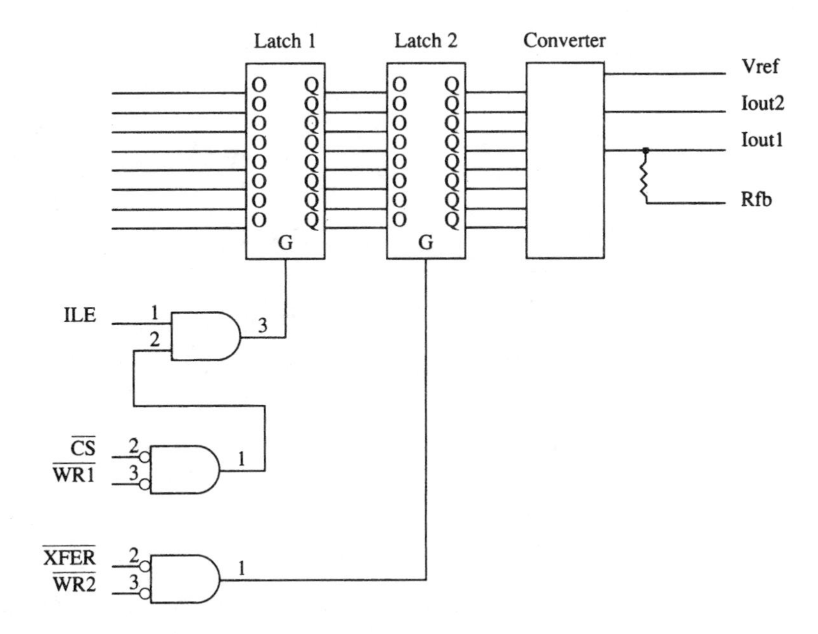

FIGURE 11–49 The internal structure of the DAC0830.

Internal Structure of the DAC0830. Figure 11–49 illustrates the internal structure of the DAC0830. Notice that this device contains two internal registers. The first is a holding register, and the second connects to the R–2R internal ladder converter. The two latches allow one byte to be held while another is converted. In many cases, we disable the first latch and only use the second for entering data into the converter. This is accomplished by connecting a logic 1 to ILE and a logic 0 to $\overline{\text{CS}}$ (chip select).

Both latches within the DAC0830 are transparent latches. That is, when the G input to the latch is a logic 1, data pass through the latch, but when the G input becomes a logic 0, data are latched or held. The converter has a reference input pin (V_{REF}) that establishes the full-scale output voltage. If –10 V is placed on V_{REF}, the full-scale (11111111_2) output voltage is +10 V. The output of the R–2R ladder within the converter appears at IOUT1 and IOUT2. These outputs are designed to be applied to an operational amplifier such as a 741 or similar device.

Connecting the DAC0830 to the Microprocessor. The DAC0830 is connected to the microprocessor, as illustrated in Figure 11–50. Here, a PLD is used to decode the DAC0830 at 8-bit I/O port address 20H. Whenever an OUT 20H,AL instruction is executed, the contents of data bus connections AD_0–AD_7 are passed to the converter within the DAC0830. The 741 operational amplifier, along with the –12 V zener reference voltage, causes the full-scale output voltage to equal +12 V. The output of the operational amplifier feeds a driver that powers a 12 V DC motor. This driver is a Darlington amplifier for large motors. This example shows the converter driving a motor, but other devices could be used as outputs.

The ADC080X Analog-to-Digital Converter

A common, low-cost ADC is the ADC080X, which belongs to a family of converters that are all identical, except for accuracy. This device is compatible with a wide range of microprocessors such as the Intel family. Although there are faster ADCs available and some have more resolution than eight bits, this device is ideal for many applications that do not require a high degree of accuracy. The ADC080X requires up to 100 μs to convert an analog input voltage into a digital output code.

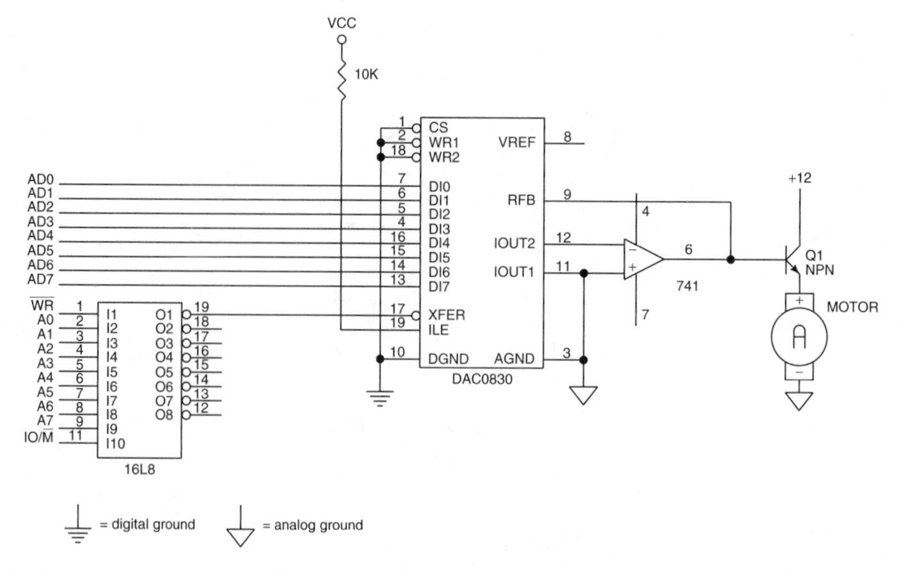

FIGURE 11–50 A DAC0830 interfaced to the 8086 microprocessor at 8-bit I/O location 20H.

Figure 11–51 shows the pin-out of the ADC0804 converter (a product of National Semiconductor Corporation). To operate the converter, the $\overline{WR}$ pin is pulsed with $\overline{CS}$ grounded to start the conversion process. Because this converter requires a considerable amount of time for the conversion, a pin labeled INTR signals the end of the conversion. Refer to Figure 11–52 for a timing diagram that shows the interaction of the control signals. As can be seen, we start the

FIGURE 11–51 The pin-out of the ADC0804 analog-to-digital converter.

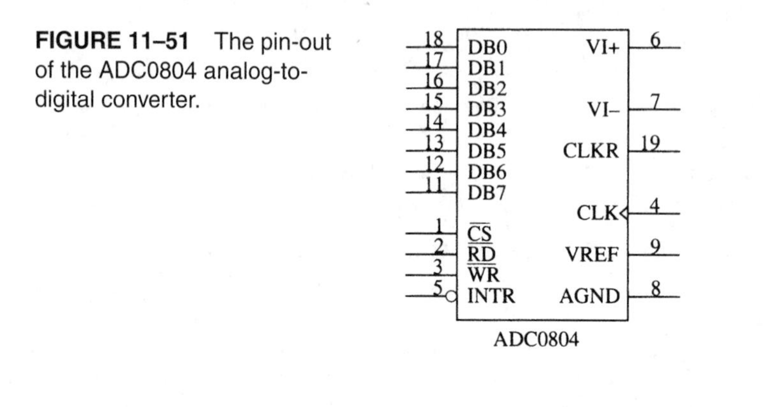

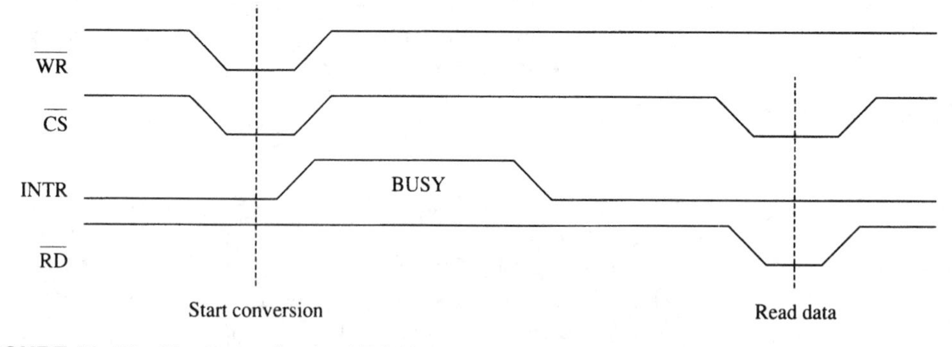

FIGURE 11–52 The timing for the ADC0804 analog-to-digital converter.

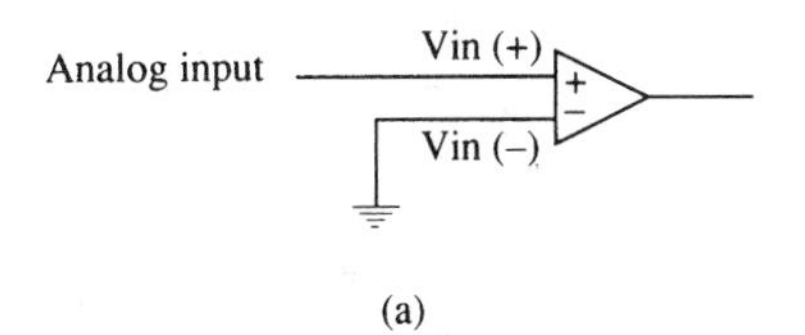

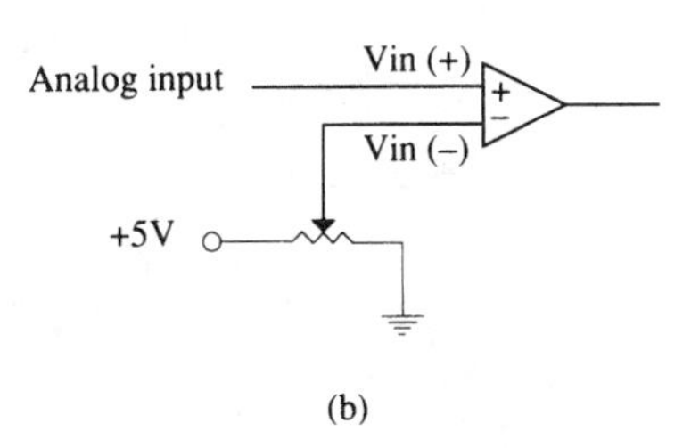

FIGURE 11–53 The analog inputs to the ADC0804 converter. (a) To sense a 0- to +5.0-V input. (b) To sense an input offset from ground.

converter with the $\overline{WR}$ pulse, we wait for INTR to return to a logic 0 level, and then we read the data from the converter. If a time delay is used that allows at least 100 μs of time, then we don't need to test the INTR pin. Another option is to connect the INTR pin to an interrupt input, so that when the conversion is complete, an interrupt occurs.

The Analog Input Signal. Before the ADC0804 can be connected to the microprocessor, its analog inputs must be understood. There are two analog inputs to the ADC0804: VIN(+) and VIN(–). These inputs are connected to an internal operational amplifier and are differential inputs, as shown in Figure 11–53. The differential inputs are summed by the operational amplifier to produce a signal for the internal analog-to-digital converter. Figure 11–53 shows a few ways to use these differential inputs. The first way (see Figure 11–53a) uses a single input that can vary between 0 V and +5.0 V. The second way (see Figure 11–53b) shows a variable voltage applied to the VIN(–) pin, so the zero reference for VIN(+) can be adjusted.

Generating the Clock Signal. The ADC0804 requires a clock source for operation. The clock can be an external clock applied to the CLK IN pin or it can be generated with an RC circuit. The permissible range of clock frequencies is between 100 KHz and 1460 KHz. It is desirable to use a frequency that is as close as possible to 1460 KHz, so conversion time is kept to a minimum.

If the clock is generated with an RC circuit, we use the CLK IN and CLK R pins connected to an RC circuit, as illustrated in Figure 11–54. When this connection is in use, the clock frequency is calculated by the following equation:

$$\text{Fclk} = \frac{1}{1.1\,\text{RC}}$$

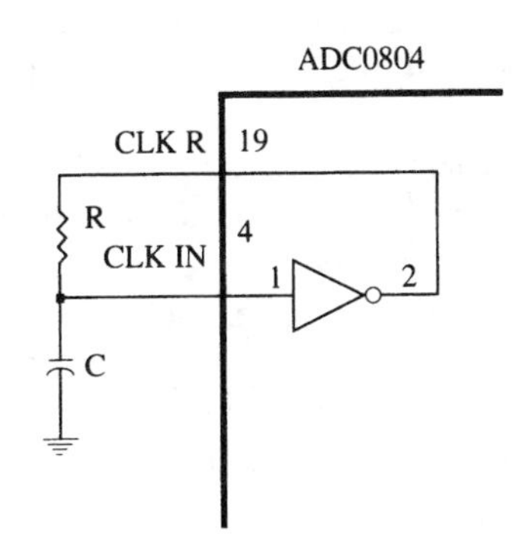

FIGURE 11–54 Connecting the RC circuit to the CLK IN and CLK R pins on the ADC0804.

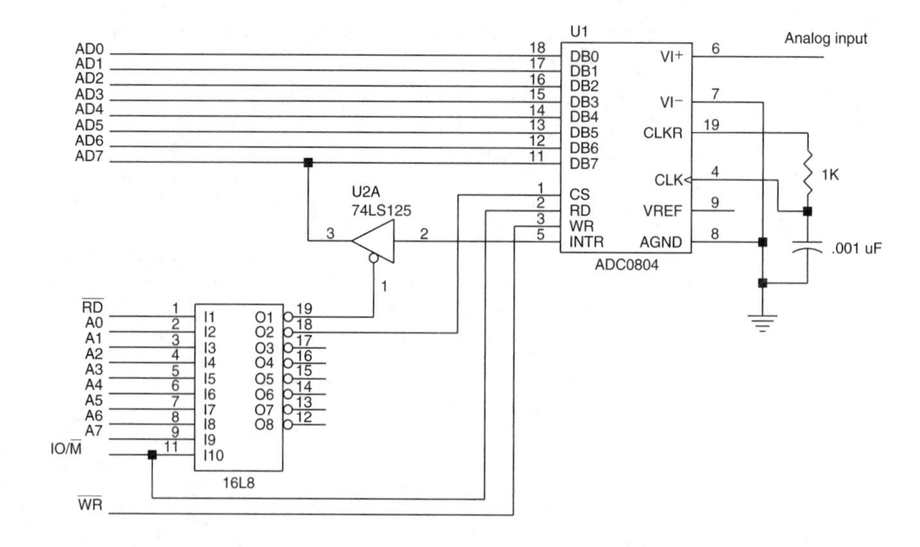

FIGURE 11–55 The ADC0804 interfaced to the microprocessor.

Connecting the ADC0804 to the Microprocessor. The ADC0804 is interfaced to the 8086 microprocessor, as illustrated in Figure 11–55. Note that the V_{REF} signal is not attached to anything, which is normal. Suppose that the ADC0804 is decoded at 8-bit I/O port address 40H for the data and port address 42H for the INTR signal, and a procedure is required to start and read the data from the ADC. This procedure is listed in Example 11–28. Notice that the INTR bit is polled and if it becomes a logic 0, the procedure ends with AL, containing the converted digital code.

EXAMPLE 11–28

```
ADC     PROC    NEAR

        OUT  40H,AL
        .REPEAT              ;test INTR
               IN   AL,42H
               TEST AL,80H
        .UNTIL ZERO?
        IN  AL,40H
        RET

ADC     ENDP
```

Using the ADC0804 and the DAC0830

This section of the text illustrates an example that uses both the ADC0804 and the DAC0830 to capture and replay audio signals or speech. In the past, we often used a speech synthesizer to generate speech, but the quality of the speech was poor. For human quality speech, we can use the ADC0804 to capture an audio signal and store it in memory for later playback through the DAC0830.

Figure 11–56 illustrates the circuitry required to connect the ADC0804 at I/O ports 0700H and 0702H. The DAC0830 is interfaced at I/O port 704H. These I/O ports are in the low bank of a 16-bit microprocessor such as the 8086 or 80386SX. The software used to run these converters appears in Example 11–29. This software reads a one-second burst of speech and then plays it back 10 times. One procedure, called READS, reads the speech and the other, called PLAYS, plays it back. The speech is sampled and stored in a section of memory called WORDS. The sample rate is chosen at 2048 samples per second, which renders acceptable-sounding speech.

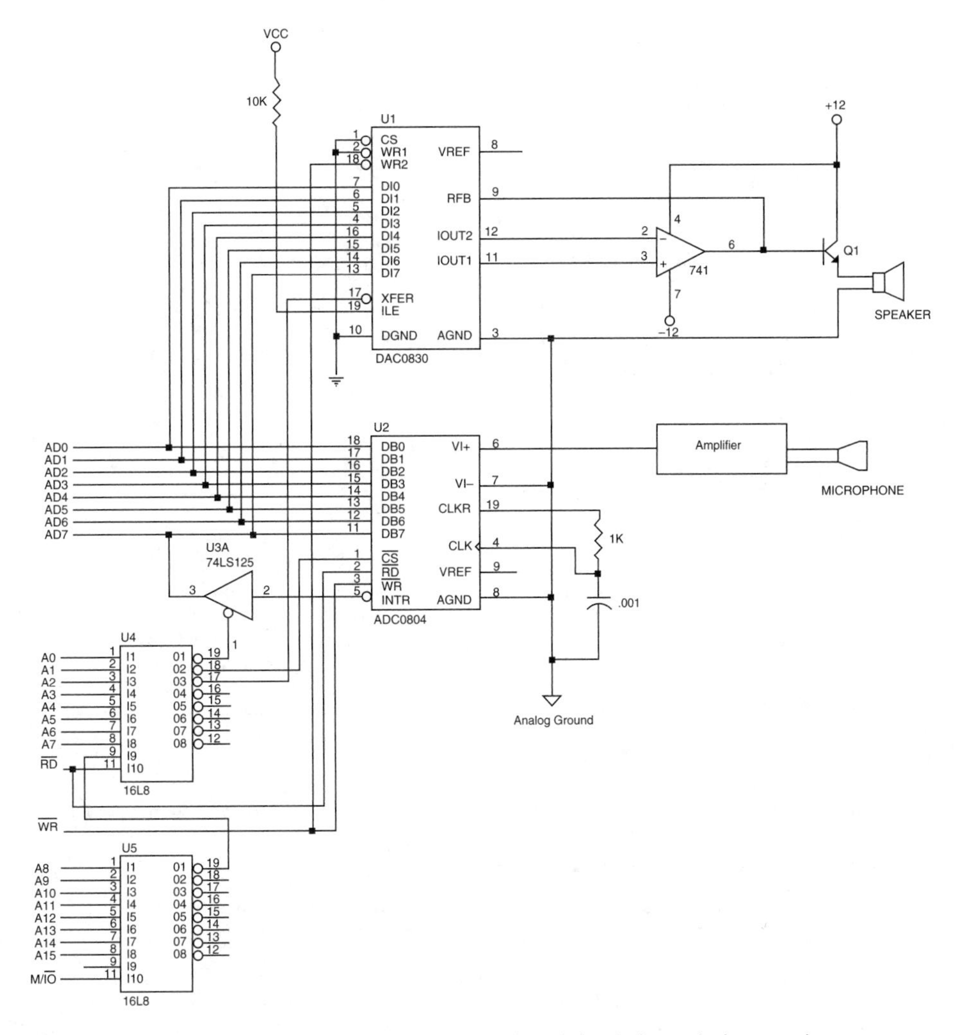

FIGURE 11–56 A circuit that stores speech and plays it back through the speaker.

EXAMPLE 11–29

```
;Software that records a second of speech and plays it back
;10 times.

;Assumes the clock frequency is 20 MHz on an 80386EX microprocessor

READS   PROC  NEAR USES ECX DX

        MOV  ECX,2048                 ;count = 2048
        MOV  DX,700H                  ;address port 700H
        .REPEAT
               OUT  DX,AL             ;start conversion
               ADD  DX,2              ;address status port
               .REPEAT                ;wait for converter
                      IN   AL,DX
                      TEST AL,80H
               .UNTIL ZERO?
               SUB  DX,2              ;address data port
               IN   AL,DX             ;get data
               MOV  WORDS[ECX-1]
               CALL DELAY             ;wait for 1/2048 sec
```

```
        .UNTILCXZ
        RET

READS   ENDP

PLAYS   PROC   NEAR USES DX ECX

        MOV   ECX,2048                  ;count = 2048
        MOV   DX,704H                   ;address DAC
        .REPEAT
               MOV   AL,WORDS[EAX-1]
               OUT   DX,AL              ;send byte to DAC
               CALL DELAY               ;wait for 1/2048 sec
        .UNTILCXZ
        RET
PLAYS   ENDP

DELAY   PROC    NEAR USES CX

        MOV   CX,888
        .REPEAT                         ;waste 1/2048 sec
        .UNTILCXZ
        RET

DELAY   ENDP
```

11–7 SUMMARY

1. The 8086–Pentium 4 microprocessors have two basic types of I/O instructions: IN and OUT. The IN instruction inputs data from an external I/O device into either the AL (8-bit) or AX (16-bit) register. The IN instruction is available as a fixed port instruction, a variable port instruction, or a string instruction (80286–Pentium 4) INSB or INSW. The OUT instruction outputs data from AL or AX to an external I/O device and is available as a fixed, variable, or string instruction OUTSB or OUTSW. The fixed port instruction uses an 8-bit I/O port address, while the variable and string I/O instructions use a 16-bit port number found in the DX register.
2. Isolated I/O, sometimes called direct I/O, uses a separate map for the I/O space, freeing the entire memory for use by the program. Isolated I/O uses the IN and OUT instructions to transfer data between the I/O device and the microprocessor. The control structure of the isolated I/O map uses $\overline{\text{IORC}}$ (I/O read control) and $\overline{\text{IOWC}}$ (I/O write control), plus the bank selection signals $\overline{\text{BHE}}$ and $\overline{\text{BLE}}$ (A0 on the 8086 and 80286), to effect the I/O transfer. The early 8086/8088 used the M/$\overline{\text{IO}}$ (IO/$\overline{\text{M}}$) signal with $\overline{\text{RD}}$ and $\overline{\text{WR}}$ to generate the I/O control signals.
3. Memory-mapped I/O uses a portion of the memory space for I/O transfers. This reduces the amount of memory available, but it negates the need to use the $\overline{\text{IORC}}$ and $\overline{\text{IOWC}}$ signals for I/O transfers. In addition, any instruction that addresses a memory location using any addressing mode can be used to transfer data between the microprocessor and the I/O device using memory-mapped I/O.
4. All input devices are buffered so that the I/O data are connected only to the data bus during the execution of the IN instruction. The buffer is either built into a programmable peripheral or located separately.
5. All output devices use a latch to capture output data during the execution of the OUT instruction. This is necessary because data appear on the data bus for less than 100 ns for an OUT instruction, and most output devices require the data for a longer time. In many cases, the latch is built into the peripheral.

6. Handshaking or polling is the act of two independent devices synchronizing with a few control lines. For example, the computer asks a printer if it is busy by inputting the BUSY signal from the printer. If it isn't busy, the computer outputs data to the printer and informs the printer that data are available with a data strobe ($\overline{DS}$) signal. This communication between the computer and the printer is a handshake or a poll.
7. Interfaces are required for most switch-based input devices and for most output devices that are not TTL-compatible.
8. The I/O port number appears on address bus connections A_7–A_0 for a fixed port I/O instruction and on A_{15}–A_0 for a variable port I/O instruction (note that A_{15}–A_8 contain zeros for an 8-bit port). In both cases, address bits above A_{15} are undefined.
9. Because the 8086/80286/80386SX microprocessors contain a 16-bit data bus and the I/O addresses reference byte-sized I/O locations, the I/O space is also organized in banks, as is the memory system. In order to interface an 8-bit I/O device to the 16-bit data bus, we often require separate write strobes (an upper and a lower) for I/O write operations. Likewise, the 80486 and Pentium–Pentium 4 also have I/O arranged in banks.
10. The I/O port decoder is much like the memory address decoder, except instead of decoding the entire address, the I/O port decoder decodes only a 16-bit address for variable port instructions and often an 8-bit port number for fixed I/O instructions.
11. The 82C55 is a programmable peripheral interface (PIA) with 24 I/O pins that are programmable in two groups of 12 pins each (group A and group B). The 82C55 operates in three modes: simple I/O (mode 0), strobed I/O (mode 1), and bidirectional I/O (mode 2). When the 82C55 is interfaced to the 8086 operating at 8 MHz, we insert two wait states because the speed of the microprocessor is faster than the 82C55 can handle.
12. The LCD display device requires a fair amount of software, but it displays ASCII-coded information.
13. The 8254 is a programmable interval timer containing three 16-bit counters that count in binary or binary-coded decimal (BCD). Each counter is independent and operates in six different modes: (1) events counter, (2) retriggerable, monostable multivibrator, (3) pulse generator, (4) square-wave generator, (5) software-triggered pulse generator, and (6) hardware-triggered pulse generator.
14. The 16550 is a programmable communications interface, capable of receiving and transmitting asynchronous serial data.
15. The DAC0830 is an 8-bit digital-to-analog converter that converts a digital signal to an analog voltage within 1.0 μs.
16. The ADC0804 is an 8-bit analog-to-digital converter that converts an analog signal into a digital signal within 100 μs.

11–8 QUESTIONS AND PROBLEMS

1. Explain which way the data flow for an IN and an OUT instruction.
2. Where is the I/O port number stored for a fixed I/O instruction?
3. Where is the I/O port number stored for a variable I/O instruction?
4. Where is the I/O port number stored for a string I/O instruction?
5. To which register are data input by the 16-bit IN instruction?
6. Describe the operation of the OUTSB instruction.
7. Describe the operation of the INSW instruction.
8. Contrast a memory-mapped I/O system with an isolated I/O system.
9. What is the basic input interface?
10. What is the basic output interface?

11. Explain the term *handshaking* as it applies to computer I/O systems.
12. An even-number I/O port address is found in the _________ I/O bank in the 8086 microprocessor.
13. In the Pentium 4, what bank contains I/O port number 000AH?
14. How many I/O banks are found in the Pentium 4 microprocessor?
15. Show the circuitry that generates the upper and lower I/O write strobes.
16. What is the purpose of a contact bounce eliminator?
17. Develop an interface to correctly drive a relay. The relay is 12 V and requires a coil current of 150 mA.
18. Develop a relay coil driver that can control a 5.0 V relay that requires 60 mA of coil current.
19. Develop an I/O port decoder, using a 74ALS138, which generates low-bank I/O strobes, for a 16-bit microprocessor, for the following 8-bit I/O port addresses: 10H, 12H, 14H, 16H, 18H, 1AH, 1CH, and 1EH.
20. Develop an I/O port decoder, using a 74ALS138, which generates high-bank I/O strobes, for a 16-bit microprocessor, for the following 8-bit I/O port addresses: 11H, 13H, 15H, 17H, 19H, 1BH, 1DH, and 1FH.
21. Develop an I/O port decoder, using a PLD, which generates 16-bit I/O strobes for the following 16-bit I/O port addresses: 1000H–1001H, 1002H–103H, 1004H–1005H, 1006H–1007H, 1008H–1009H, 100AH–100BH, 100CH–100DH, and 100EH–100FH.
22. Develop an I/O port decoder, using a PLD, which generates the following low-bank I/O strobes: 00A8H, 00B6H, and 00EEH.
23. Develop an I/O port decoder, using a PLD, which generates the following high-bank I/O strobes: 300DH, 300BH, 1005H, and 1007H.
24. Why are both $\overline{BHE}$ and $\overline{BLE}$ (A_0) ignored in a 16-bit port address decoder?
25. An 8-bit I/O device, located at I/O port address 0010H, is connected to which data bus connections in a Pentium 4?
26. An 8-bit I/O device, located at I/O port address 100DH, is connected to which data bus connections in a Pentium 4?
27. The 82C55 has how many programmable I/O pin connections?
28. List the pins that belong to group A and to group B in the 82C55.
29. Which two 82C55 pins accomplish internal I/O port address selection?
30. The $\overline{RD}$ connection on the 82C55 is attached to which 8086 system control bus connection?
31. Using a PLD, interface an 82C55 to the 8086 microprocessor so that it functions at I/O locations 0380H, 0382H, 0384H, and 0386H.
32. When the 82C55 is reset, its I/O ports are all initialized as _________.
33. What three modes of operation are available to the 82C55?
34. What is the purpose of the $\overline{STB}$ signal in strobed input operation of the 82C55?
35. Develop a time delay procedure for the 2.0 GHz Pentium 4 that waits for 80 μs.
36. Develop a time delay procedure for the 3.0 GHz Pentium 4 that waits for 12 ms.
37. Explain the operation of a simple four-coil stepper motor.
38. What sets the IBF pin in strobed input operation of the 82C55?
39. Write the software required to place a logic 1 on the PC7 pin of the 82C55 during strobed input operation.
40. How is the interrupt request pin (INTR) enabled in the strobed input mode of operation of the 82C55?
41. In strobed output operation of the 82C55, what is the purpose of the $\overline{ACK}$ signal?
42. What clears the $\overline{OBF}$ signal in strobed output operation of the 82C55?
43. Write the software required to decide whether PC4 is a logic 1 when the 82C55 is operated in the strobed output mode.
44. Which group of pins is used during bidirectional operation of the 82C55?
45. Which pins are general-purpose I/O pins during mode 2 operation of the 82C55?

46. Describe how the display is cleared in the LCD display.
47. How is a display position selected in the LCD display?
48. Write a short procedure that places an ASCII null string in display position 6 on the LCD display.
49. How is the busy flag tested in the LCD display?
50. What changes must be made to Figure 11–25 so that it functions with a keyboard matrix that contains three rows and five columns?
51. What time is usually used to debounce a keyboard?
52. Develop the interface to a three- by four-key telephone-style keypad. You will need to use a lookup table to convert to the proper key code.
53. The 8254 interval timer functions from DC to ________ Hz.
54. Each counter in the 8254 functions in how many different modes?
55. Interface an 8254 to function at I/O port addresses XX10H, XX12H, XX14H, and XX16H.
56. Write the software that programs counter 2 to generate an 80 KHz square wave if the CLK input to counter 2 is 8 MHz.
57. What number is programmed in an 8254 counter to count 300 events?
58. If a 16-bit count is programmed into the 8254, which byte of the count is programmed first?
59. Explain how the read-back control word functions in the 8254.
60. Program counter 1 of the 8254 so that it generates a continuous series of pulses that have a high time of 100 μs and a low time of 1 μs. Make sure to indicate the CLK frequency required for this task.
61. Why does a 50% duty cycle cause the motor to stand still in the motor speed and direction control circuit presented in this chapter?
62. What is asynchronous serial data?
63. What is Baud rate?
64. Program the 16550 for operation using six data bits, even parity, one stop bit, and a Baud rate of 19,200 using a 18.432 MHz clock. (Assume that the I/O ports are numbered 20H and 22H.)
65. If the 16550 is to generate a serial signal at a Baud rate of 2400 Baud and the Baud rate divisor is programmed for 16, what is the frequency of the signal?
66. Describe the following terms: *simplex*, *half-duplex*, and *full-duplex*.
67. How is the 16550 reset?
68. Write a procedure for the 16550 that transmits 16 bytes from a small buffer in the data segment address (DS is loaded externally) by SI (SI is loaded externally).
69. The DAC0830 converts an 8-bit digital input to an analog output in approximately________.
70. What is the step voltage at the output of the DAC0830 if the reference voltage is –2.55 V?
71. Interface a DAC0830 to the 8086 so that it operates at I/O port 400H.
72. Develop a program for the interface of Question 71 so the DAC0830 generates a triangular voltage wave-form. The frequency of this wave-form must be approximately 100 Hz.
73. The ADC080X requires approximately ________ to convert an analog voltage into a digital code.
74. The $\overline{\text{WR}}$ pin on the ADC080X is used for what purpose?
75. Interface an ADC080X at I/O port 0260H for data and 0270H to test the INTR pin.
76. Develop a program for the ADC080X in Question 75 so that it reads an input voltage once per 100 ms and stores the results in a memory array that is 100H bytes long.
77. Rewrite Example 11–29 using C++ with inline assembly code.

CHAPTER 12

Interrupts

INTRODUCTION

In this chapter, we expand our coverage of basic I/O and programmable peripheral interfaces by examining a technique called interrupt-processed I/O. An interrupt is a hardware-initiated procedure that interrupts whatever program is currently executing. This chapter provides examples and a detailed explanation of the interrupt structure of the entire Intel family of microprocessors.

CHAPTER OBJECTIVES

Upon completion of this chapter, you will be able to:

1. Explain the interrupt structure of the Intel family of microprocessors
2. Explain the operation of software interrupt instructions INT, INTO, INT 3, and BOUND
3. Explain how the interrupt enable flag bit (IF) modifies the interrupt structure
4. Describe the function of the trap interrupt flag bit (TF) and the operation of trap-generated tracing
5. Develop interrupt-service procedures that control lower-speed, external peripheral devices
6. Expand the interrupt structure of the microprocessor by using the 82S9A programmable interrupt controller and other techniques
7. Explain the purpose and operation of a real-time clock

12–1 BASIC INTERRUPT PROCESSING

In this section, we discuss the function of an interrupt in a microprocessor-based system, and the structure and features of interrupts available to the Intel family of microprocessors.

The Purpose of Interrupts

Interrupts are particularly useful when interfacing I/O devices that provide or require data at relatively low data transfer rates. In Chapter 11, for instance, we showed a keyboard example using strobed input operation of the 82C55. In that example, software polled the 82C55 and its IBF bit to decide whether data were available from the keyboard. If the person using the keyboard typed

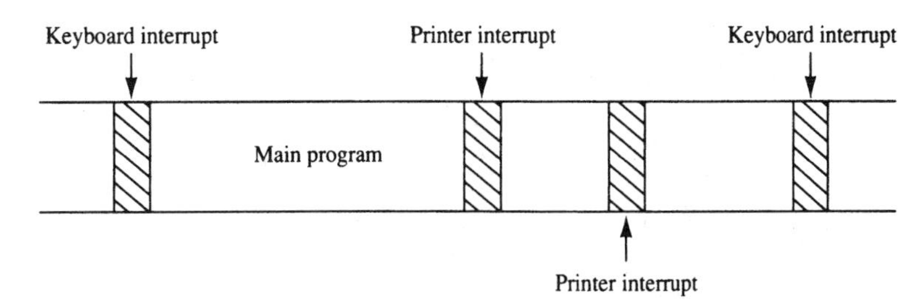

FIGURE 12–1 A time line that indicates interrupt usage in a typical system.

one character per second, the software for the 82C55 waited an entire second between each keystroke for the person to type another key. This process was such a tremendous waste of time that designers developed another process, *interrupt processing*, to handle this situation.

Unlike the polling technique, interrupt processing allows the microprocessor to execute other software while the keyboard operator is thinking about what key to type next. As soon as a key is pressed, the keyboard encoder debounces the switch and puts out one pulse that interrupts the microprocessor. The microprocessor executes other software until a key is actually pressed, when it reads the key and returns to the program that was interrupted. As a result, the microprocessor can print reports or complete any other task while the operator is typing a document and thinking about what to type next.

Figure 12–1 shows a time line that indicates a typist typing data on a keyboard, a printer removing data from the memory, and a program executing. The program is the main program that is interrupted for each keystroke and each character that is to print on the printer. Note that the keyboard interrupt service procedure, called by the keyboard interrupt, and the printer interrupt service procedure each take little time to execute.

Interrupts

The interrupts of the entire Intel family of microprocessors include two hardware pins that request interrupts (INTR and NMI), and one hardware pin ($\overline{\text{INTA}}$) that acknowledges the interrupt requested through INTR. In addition to the pins, the microprocessor also has software interrupts INT, INTO, INT 3, and BOUND. Two flag bits, IF (interrupt flag) and TF (trap flag), are also used with the interrupt structure and a special return instruction, IRET (or IRETD in the 80386, 80486, or Pentium–Pentium 4).

Interrupt Vectors. The interrupt vectors and vector table are crucial to an understanding of hardware and software interrupts. The **interrupt vector table** is located in the first 1024 bytes of memory at addresses 000000H–0003FFH. It contains 256 different four-byte interrupt vectors. An **interrupt vector** contains the address (segment and offset) of the interrupt service procedure.

Figure 12–2 illustrates the interrupt vector table for the microprocessor. The first five interrupt vectors are identical in all Intel microprocessor family members, from the 8086 to the Pentium. Other interrupt vectors exist for the 80286 that are upward-compatible to the 80386, 80486, and Pentium–Pentium 4, but not downward-compatible to the 8086 or 8088. Intel reserves the first 32 interrupt vectors for their use in various microprocessor family members. The last 224 vectors are available as user interrupt vectors. Each vector is four bytes long in the real mode and contains the **starting address** of the interrupt service procedure. The first two bytes of the vector contain the offset address and the last two bytes contain the segment address.

The following list describes the function of each dedicated interrupt in the microprocessor:

TYPE 0 The **divide error** whenever the result from a division overflows or an attempt is made to divide by zero.

TYPE 1 **Single-step** or **trap** occurs after the execution of each instruction if the trap (TF) flag bit is set. Upon accepting this interrupt, the TF bit is cleared so that the

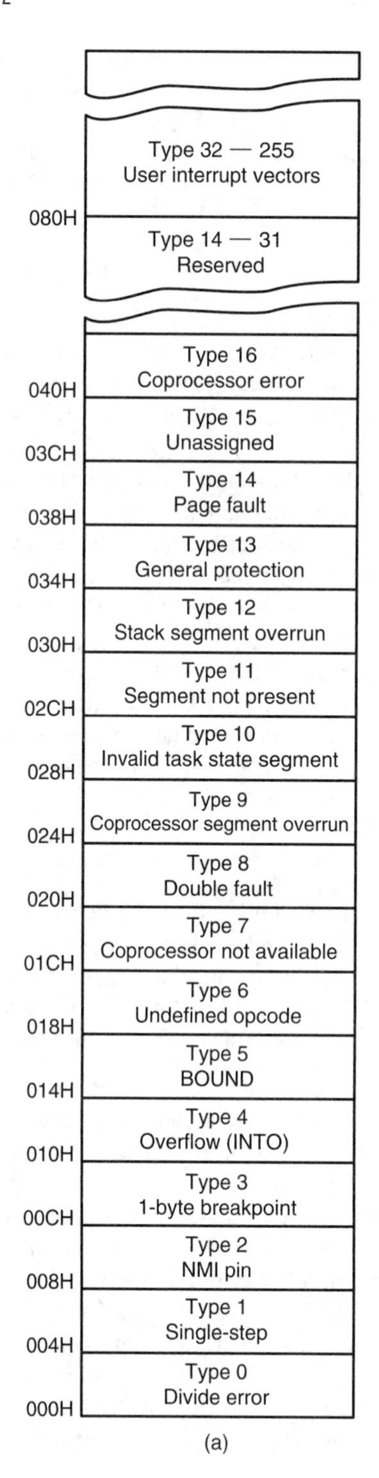

Any interrupt vector

3	Segment (high)
2	Segment (low)
1	Offset (high)
0	Offset (low)

(b)

FIGURE 12–2 (a) The interrupt vector table for the microprocessor and (b) the contents of an interrupt vector.

interrupt service procedure executes at full speed. (More detail is provided about this interrupt later in this section of the chapter.)

TYPE 2 The **non-maskable interrupt** occurs when a logic 1 is placed on the NMI input pin to the microprocessor. Non-maskable means that the input cannot be disabled.

TYPE 3 A special one-byte instruction (INT 3) that uses this vector to access its interrupt-service procedure. The INT 3 instruction is often used to store a **breakpoint** in a program for debugging.

TYPE 4 **Overflow** is a special vector used with the INTO instruction. The INTO instruction interrupts the program if an overflow condition exists, as reflected by the overflow flag (OF).

TYPE 5 The **BOUND** instruction compares a register with boundaries stored in the memory. If the contents of the register are greater than or equal to the first word in memory and less than or equal to the second word, no interrupt occurs because the contents of the register are within bounds. If the contents of the register are out of bounds, a type 5 interrupt ensues.

TYPE 6 An **invalid opcode** interrupt occurs whenever an undefined opcode is encountered in a program.

TYPE 7 The **coprocessor not available** interrupt occurs when a coprocessor is not found in the system, as dictated by the machine status word (MSW or CR0) coprocessor control bits. If an ESC or WAIT instruction executes and the coprocessor is not found, a type 7 exception or interrupt occurs.

TYPE 8 A **double fault** interrupt is activated whenever two separate interrupts occur during the same instruction.

TYPE 9 The **coprocessor segment overrun** occurs if the ESC instruction (coprocessor opcode) memory operand extends beyond offset address FFFFH in real mode.

TYPE 10 An **invalid task state segment** interrupt occurs in the protected mode if the TSS is invalid because the segment limit field is not 002BH or higher. In most cases, this is caused because the TSS is not initialized.

TYPE 11 The **segment not present** interrupt occurs when the protected mode P bit (P = 0) in a descriptor indicates that the segment is not present or not valid.

TYPE 12 A **stack segment overrun** occurs if the stack segment is not present (P = 0) in the protected mode or if the limit of the stack segment is exceeded.

TYPE 13 The **general protection fault** occurs for most protection violations in the 80286–Pentium 4 protected mode system. (These errors occur in Windows as general protection faults.) A list of these protection violations follows:

(a) Descriptor table limit exceeded
(b) Privilege rules violated
(c) Invalid descriptor segment type loaded
(d) Write to code segment that is protected
(e) Read from execute-only code segment
(f) Write to read-only data segment
(g) Segment limit exceeded
(h) CPL = IOPL when executing CTS, HLT, LGDT, LIDT, LLDT, LMSW, or LTR
(i) CPL > IOPL when executing CLI, IN, INS, LOCK, OUT, OUTS, and STI

TYPE 14 **Page fault** interrupts occur for any page fault memory or code access in the 80386, 80486, and Pentium–Pentium 4 microprocessors.

TYPE 16 **Coprocessor error** takes effect whenever a coprocessor error (ERROR = 0) occurs for the ESCape or WAIT instructions for the 80386, 80486, and Pentium–Pentium 4 microprocessors only.

TYPE 17 **Alignment checks** indicate that word and doubleword data are addressed at an odd memory location (or an incorrect location, in the case of a doubleword). This interrupt is active in the 80486 and Pentium–Pentium 4 microprocessors.

TYPE 18 A **machine check** activates a system memory management mode interrupt in the Pentium–Pentium 4 microprocessors.

Interrupt Instructions: BOUND, INTO, INT, INT 3, and IRET

Of the five software interrupt instructions available to the microprocessor, INT and INT 3 are very similar, BOUND and INTO are conditional, and IRET is a special interrupt return instruction.

The BOUND instruction, which has two operands, compares a register with two words of memory data. For example, if the instruction BOUND AX,DATA is executed, AX is compared with the contents of DATA and DATA+1 and also with DATA+2 and DATA+3. If AX is less than the contents of DATA and DATA+1, a type 5 interrupt occurs. If AX is greater than DATA+2 and DATA+3, a type 5 interrupt occurs. If AX is within the bounds of these two memory words, no interrupt occurs.

The INTO instruction checks the overflow flag (OF). If OF = 1, the INTO instruction calls the procedure whose address is stored in interrupt vector type number 4. If OF = 0, then the INTO instruction performs no operation and the next sequential instruction in the program executes.

The INT n instruction calls the interrupt service procedure that begins at the address represented in vector number n. For example, an INT 80H or INT 128 calls the interrupt service procedure whose address is stored in vector type number 80H (00200H–00203H). To determine the vector address, just multiply the vector type number (n) by 4, which gives the beginning address of the four-byte long interrupt vector. For example, INT 5 = 4×5 or 20 (14H). The vector for INT 5 begins at address 0014H and continues to 0017H. Each INT instruction is stored in two bytes of memory: The first byte contains the opcode, and the second byte contains the interrupt type number. The only exception to this is the INT 3 instruction, a one-byte instruction. The INT 3 instruction is often used as a breakpoint-interrupt because it is easy to insert a one-byte instruction into a program. Breakpoints are often used to debug faulty software.

The IRET instruction is a special return instruction used to return for both software and hardware interrupts. The IRET instruction is much like a far RET, because it retrieves the return address from the stack. It is unlike the near return because it also retrieves a copy of the flag register from the stack. An IRET instruction removes six bytes from the stack: two for the IP, two for the CS, and two for the flags.

In the 80386–Pentium 4, there is also an IRETD instruction because these microprocessors can push the EFLAG register (32 bits) on the stack, as well as the 32-bit EIP in the protected mode and 16-bit code segment register. If operated in the real mode, we use the IRET instruction with the 80386–Pentium 4 microprocessors.

The Operation of a Real Mode Interrupt

When the microprocessor completes executing the current instruction, it determines whether an interrupt is active by checking (1) instruction executions, (2) single-step, (3) NMI, (4) coprocessor segment overrun, (5) INTR, and (6) INT instruction in the order presented. If one or more of these interrupt conditions are present, the following sequence of events occurs:

1. The contents of the flag register are pushed onto the stack.
2. Both the interrupt (IF) and trap (TF) flags are cleared. This disables the INTR pin and the trap or single-step feature.
3. The contents of the code segment register (CS) are pushed onto the stack.
4. The contents of the instruction pointer (IP) are pushed onto the stack.
5. The interrupt vector contents are fetched, and then placed into both IP and CS so that the next instruction executes at the interrupt service procedure addressed by the vector.

Whenever an interrupt is accepted, the microprocessor stacks the contents of the flag register, CS and IP; clears both IF and TF; and jumps to the procedure addressed by the interrupt vector. After the flags are pushed onto the stack, IF and TF are cleared. These flags are returned to the state prior to the interrupt when the IRET instruction is encountered at the end of the interrupt service procedure. Therefore, if interrupts were enabled prior to the interrupt service procedure, they are automatically re-enabled by the IRET instruction at the end of the procedure.

The return address (in CS and IP) is pushed onto the stack during the interrupt. Sometimes the return address points to the next instruction in the program; sometimes it points to the instruction or point in the program where the interrupt occurred. Interrupt type numbers 0, 5, 6, 7, 8, 10, 11, 12, and 13 push a return address that points to the offending instruction, instead of to the next instruction in the program. This allows the interrupt service procedure to possibly retry the instruction in certain error cases.

Some of the protected mode interrupts (types 8, 10, 11, 12, and 13) place an error code on the stack following the return address. The error code identifies the selector that caused the interrupt. In cases where no selector is involved, the error code is a 0.

Operation of a Protected Mode Interrupt

In the protected mode, interrupts have exactly the same assignments as in the real mode, but the interrupt vector table is different. In place of interrupt vectors, protected mode uses a set of 256 interrupt descriptors that are stored in an interrupt descriptor table (IDT). The interrupt descriptor table is 256 × 8 (2K) bytes long, with each descriptor containing eight bytes. The interrupt descriptor table is located at any memory location in the system by the interrupt descriptor table address register (IDTR).

Each entry in the IDT contains the address of the interrupt service procedure in the form of a segment selector and a 32-bit offset address. It also contains the P bit (present) and DPL bits to describe the privilege level of the interrupt. Figure 12–3 shows the contents of the interrupt descriptor.

Real mode interrupt vectors can be converted into protected mode interrupts by copying the interrupt procedure addresses from the interrupt vector table and converting them to 32-bit offset addresses that are stored in the interrupt descriptors. A single selector and segment descriptor can be placed in the global descriptor table that identifies the first 1M byte of memory as the interrupt segment.

Other than the IDT and interrupt descriptors, the protected mode interrupt functions like the real mode interrupt. We return from both interrupts by using the IRET or IRETD instruction. The only difference is that in protected mode the microprocessor accesses the IDT instead of the interrupt vector table.

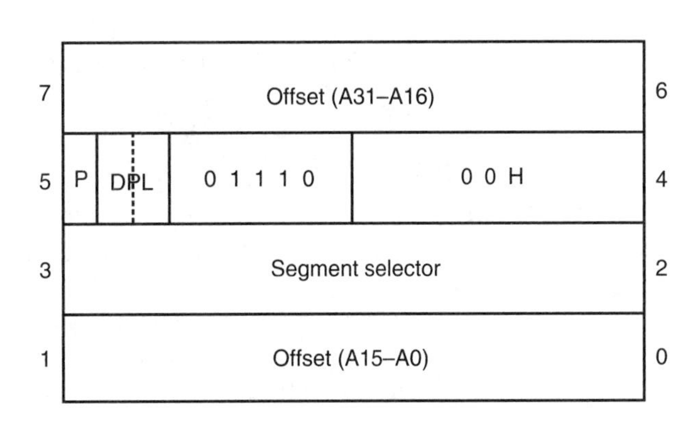

FIGURE 12–3 The protected mode interrupt descriptor.

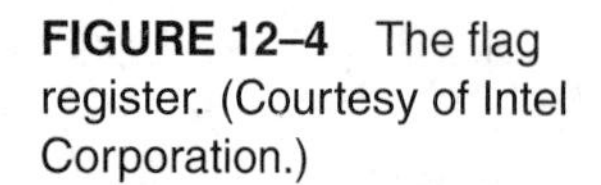
FIGURE 12–4 The flag register. (Courtesy of Intel Corporation.)

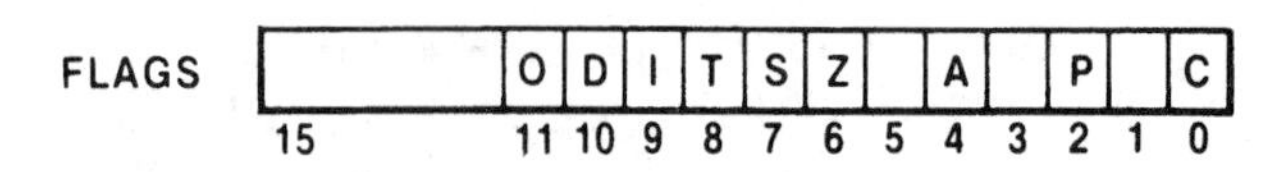

Interrupt Flag Bits

The interrupt flag (IF) and the trap flag (TF) are both cleared after the contents of the flag register are stacked during an interrupt. Figure 12–4 illustrates the contents of the flag register and the location of IF and TF. When the IF bit is set, it allows the INTR pin to cause an interrupt; when the IF bit is cleared, it prevents the INTR pin from causing an interrupt. When TF = 1, it causes a trap interrupt (type number 1) to occur after each instruction executes. This is why we often call trap a *single-step.* When TF = 0, normal program execution occurs. This flag bit allows debugging, as explained in Chapters 17–19, which detail the 80386–Pentium 4.

The interrupt flag is set and cleared by the STI and CLI instructions, respectively. There are no special instructions that set or clear the trap flag. Example 12–1 shows an interrupt service procedure that turns tracing on by setting the trap flag bit on the stack from inside the procedure. Example 12–2 shows an interrupt service procedure that turns tracing off by clearing the trap flag on the stack from within the procedure.

EXAMPLE 12–1

```
;A procedure that sets the TRAP flag bit to enable trapping

TRON  PROC    FAR USES AX BP

      MOV  BP,SP              ;get SP
      MOV  AX[BP+8]           ;retrieve flags from stack
      OR   AH,1               ;set trap flag
      MOV  [BP+8],AX
      IRET

TRON  ENDP
```

EXAMPLE 12–2

```
;A procedure that clears the TRAP flag to disable trapping

TROFF PROC    FAR USES AX BP

      MOV  BP,SP              ;get SP
      MOV  AX,[BP+8]          ;retrieve flags from stack
      AND  AH,0FEH            ;clear trap flag
      MOV  [BP+8],AX
      IRET

TROFF ENDP
```

In both examples, the flag register is retrieved from the stack by using the BP register, which, by default, addresses the stack segment. After the flags are retrieved, the TF bit is either set (TRON) or clears (TROFF) before returning from the interrupt service procedure. The IRET instruction restores the flag register with the new state of the trap flag.

Trace Procedure. Assuming that TRON is accessed by an INT 40H instruction and TROFF is accessed by an INT 41H instruction, Example 12–3 traces through a program immediately following the INT 40H instruction. The interrupt service procedure illustrated in Example 12–3 responds to interrupt type number 1 or a trap interrupt. Each time that a trap occurs—after each instruction executes following INT 40H—the TRACE procedure stores the contents of all the 32-bit microprocessor registers in an array called REGS. This provides a register trace of all the instructions between the INT 40H (TRON) and INT 41H (TROFF) if the contents of the registers stored in the array are saved.

EXAMPLE 12–3

```
REGS    DD    8 DUP(?)         ;space for registers

TRACE   PROC FAR USES EBX

        MOV   EBX,OFFSET REGS
        MOV   [EBX],EAX        ;save EAX
        POP   EAX
        PUSH  EAX
        MOV   [EBX+4],EAX      ;save EBX
        MOV   [EBX+8],ECX      ;save ECX
        MOV   [EBX+12],EDX     ;save EDX
        MOV   [EBX+16],ESP     ;save ESP
        MOV   [EBX+20],EBP     ;save EBP
        MOV   [EBX+24],ESI     ;save ESI
        MOV   [EBX+28],EDI     ;save EDI
        IRET

TRACE   ENDP
```

Storing an Interrupt Vector in the Vector Table

In order to install an interrupt vector—sometimes called a **hook**—the assembler must address absolute memory. Example 12–4 shows how a new vector is added to the interrupt vector table by using the assembler and a DOS function call. Here, the vector for INT 40H, for interrupt procedure NEW40, is installed in memory at real mode vector locations 100H–103H. The first thing accomplished by the procedure is that the old interrupt vector contents are saved in case we need to uninstall the vector. This step can be skipped if there is no need to uninstall the interrupt.

The function AX = 3100H for INT 21H, the DOS access function, installs the NEW40 procedure in memory until the computer is shut off. The number in DX is the length of the software in paragraphs (16-byte chunks). Refer to Appendix A for more detail about this DOS function.

EXAMPLE 12–4

```
.MODEL TINY
.CODE
.STARTUP
       JMP     START
OLD    DD      ?      ;space for old vector

NEW40  PROC    FAR

;
;Interrupt software for INT 40H
;

NEW40  ENDP

;start installation

START:
       MOV  AX,0                    ;address segment 0000H
       MOV  DS,AX
       MOV  AX,DS:[100H]            ;get INT 40H offset
       MOV  WORD PTR CS:OLD,AX      ;save it
       MOV  AX,DS:[102H]            ;get INT 40H segment
       MOV  WORD PTR CS:OLD+2,AX    ;save it
       MOV  DS:[100H],OFFSET NEW40  ;save offset
       MOV  DS:[102H],CS            ;save segment
```

```
        MOV  DX,OFFSET START
        SHR  DX,4
        INC  DX
        MOV  AX,3100H                  ;make NEW40 resident
        INT  21H

END
```

12–2 HARDWARE INTERRUPTS

The microprocessor has two hardware interrupt inputs: non-maskable interrupt (NMI) and interrupt request (INTR). Whenever the NMI input is activated, a type 2 interrupt occurs because NMI is internally decoded. The INTR input must be externally decoded to select a vector. Any interrupt vector can be chosen for the INTR pin, but we usually use an interrupt type number between 20H and FFH. Intel has reserved interrupts 00H through 1FH for internal and future expansion. The $\overline{INTA}$ signal is also an interrupt pin on the microprocessor, but it is an output that is used in response to the INTR input to apply a vector type number to the data bus connections D_7–D_0. Figure 12–5 shows the three user interrupt connections on the microprocessor.

The **non-maskable interrupt** (NMI) is an edge-triggered input that requests an interrupt on the positive edge (0-to-1 transition). After a positive edge, the NMI pin must remain a logic 1 until it is recognized by the microprocessor. Note that before the positive edge is recognized, the NMI pin must be a logic 0 for at least two clocking periods.

The NMI input is often used for parity errors and other major system faults, such as power failures. Power failures are easily detected by monitoring the AC power line and causing an NMI interrupt whenever AC power drops out. In response to this type of interrupt, the microprocessor stores all of the internal register in a battery-backed-up memory or an EEPROM. Figure 12–6 shows a power failure detection circuit that provides a logic 1 to the NMI input whenever AC power is interrupted.

In this circuit, an optical isolator provides isolation from the AC power line. The output of the isolator is shaped by a Schmitt-trigger inverter that provides a 60 Hz pulse to the trigger input of the 74LS122 retriggerable, monostable multivibrator. The values of R and C are chosen so that the 74LS122 has an active pulse width of 33 ms or 2 AC input periods. Because the 74LS122 is retriggerable, as long as AC power is applied, the Q output remains triggered at a logic 1 and $\overline{Q}$ remains a logic 0.

If the AC power fails, the 74LS122 no longer receives trigger pulses from the 74ALS14, which means that Q becomes a logic 0 and $\overline{Q}$ becomes a logic 1, interrupting the microprocessor through the NMI pin. The interrupt service procedure, not shown here, stores the contents of all internal registers and other data into a battery-backed-up memory. This system assumes that the system power supply has a large enough filter capacitor to provide energy for at least 75 ms after the AC power ceases.

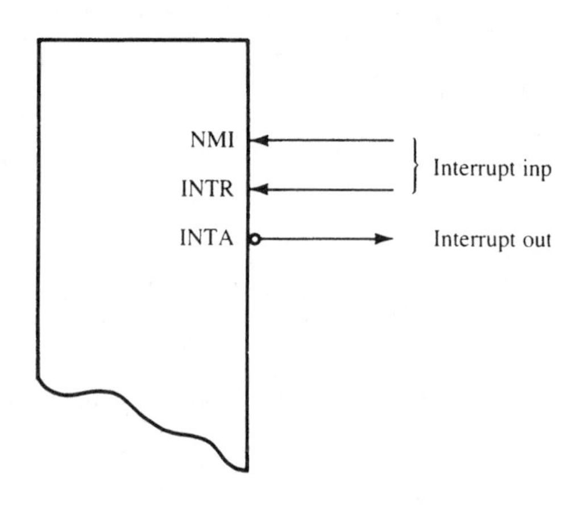

FIGURE 12–5 The interrupt pins on all versions of the Intel microprocessor.

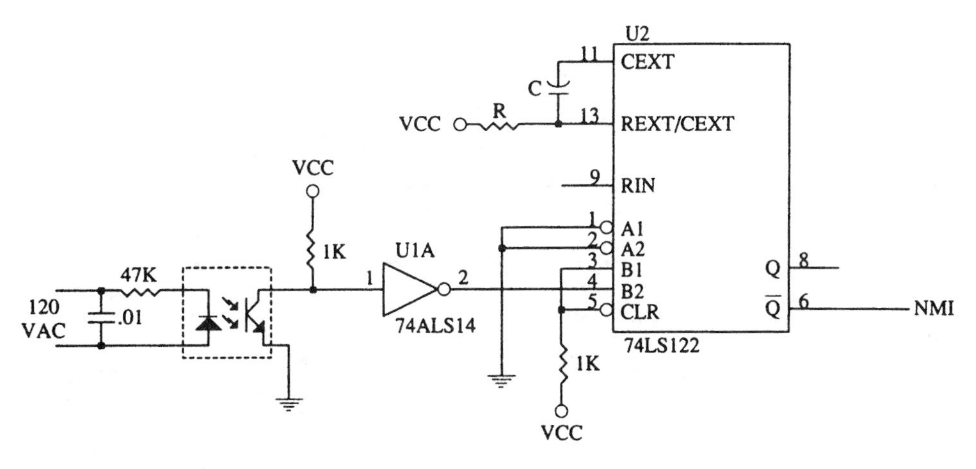

FIGURE 12–6 A power failure detection circuit.

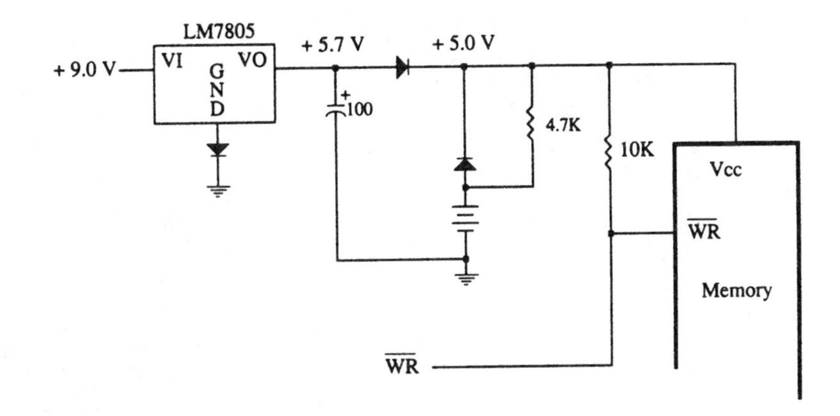

FIGURE 12–7 A battery-backed-up memory system using a NiCad, lithium, or gel cell.

Figure 12–7 shows a circuit that supplies power to a memory after the DC power fails. Here, diodes are used to switch supply voltages from the DC power supply to the battery. The diodes used are standard silicon diodes because the power supply to this memory circuit is elevated above +5.0 V to +5.7 V. The resistor is used to trickle-charge the battery, which is either NiCad, lithium, or a gel cell.

When DC power fails, the battery provides a reduced voltage to the V_{cc} connection on the memory device. Most memory devices will retain data with V_{cc} voltages as low as 1.5 V, so the battery voltage does not need to be +5.0 V. The $\overline{WR}$ pin is pulled to V_{cc} during a power outage, so no data will be written to the memory.

INTR and $\overline{INTA}$

The interrupt request input (INTR) is level-sensitive, which means that it must be held at a logic 1 level until it is recognized. The INTR pin is set by an external event and cleared inside the interrupt service procedure. This input is automatically disabled once it is accepted by the microprocessor and re-enabled by the IRET instruction at the end of the interrupt service procedure. The 80386–Pentium 4 use the IRETD instruction in the protected mode of operation.

The microprocessor responds to the INTR input by pulsing the $\overline{INTA}$ output in anticipation of receiving an interrupt vector type number on data bus connections D_7–D_0. Figure 12–8 shows the timing diagram for the INTR and $\overline{INTA}$ pins of the microprocessor. There are two $\overline{INTA}$ pulses generated by the system that are used to insert the vector type number on the data bus.

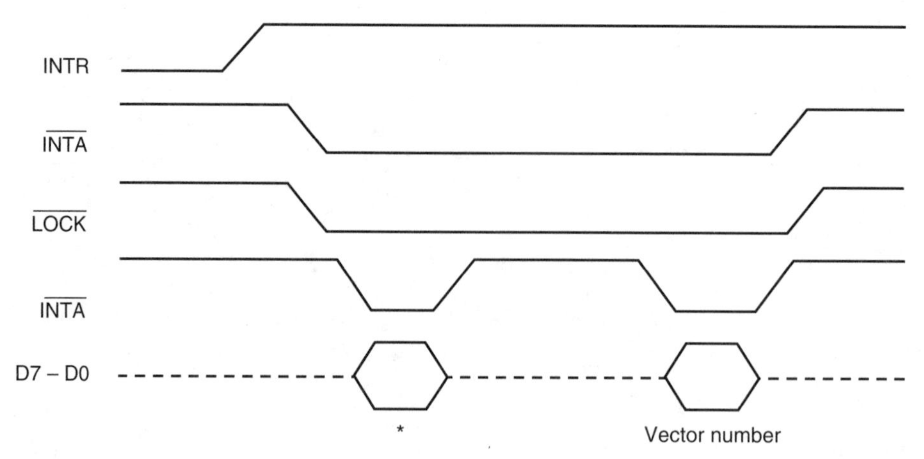

FIGURE 12–8 The timing of the INTR input and $\overline{\text{INTA}}$ output. *This portion of the data bus is ignored and usually contains the vector number.

FIGURE 12–9 A simple method for generating interrupt vector type number FFH in response to INTR.

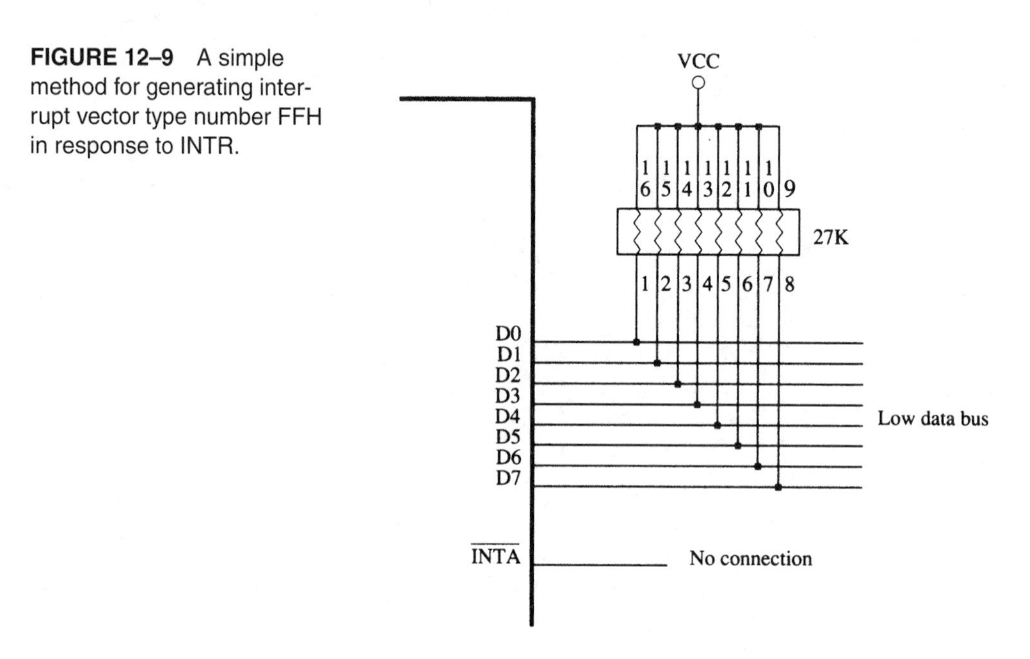

Figure 12–9 illustrates a simple circuit that applies interrupt vector type number FFH to the data bus in response to an INTR. Notice that the $\overline{\text{INTA}}$ pin is not connected in this circuit. Because resistors are used to pull the data bus connections (D_0–D_7) high, the microprocessor automatically sees vector type number FFH in response to the INTR input. This is the least expensive way to implement the INTR pin on the microprocessor.

Using a Three-State Buffer for INTA. Figure 12–10 shows how interrupt vector type number 80H is applied to the data bus (D_0–D_7) in response to an INTR. In response to the INTR, the microprocessor outputs the $\overline{\text{INTA}}$ that is used to enable a 74ALS244 three-state octal buffer. The octal buffer applies the interrupt vector type number to the data bus in response to the $\overline{\text{INTA}}$ pulse. The vector type number is easily changed with the DIP switches that are shown in this illustration.

Making the INTR Input Edge-Triggered. Often, we need an edge-triggered input instead of a level-sensitive input. The INTR input can be converted to an edge-triggered input by using a D-type flip-flop, as illustrated in Figure 12–11. Here, the clock input becomes an edge-triggered

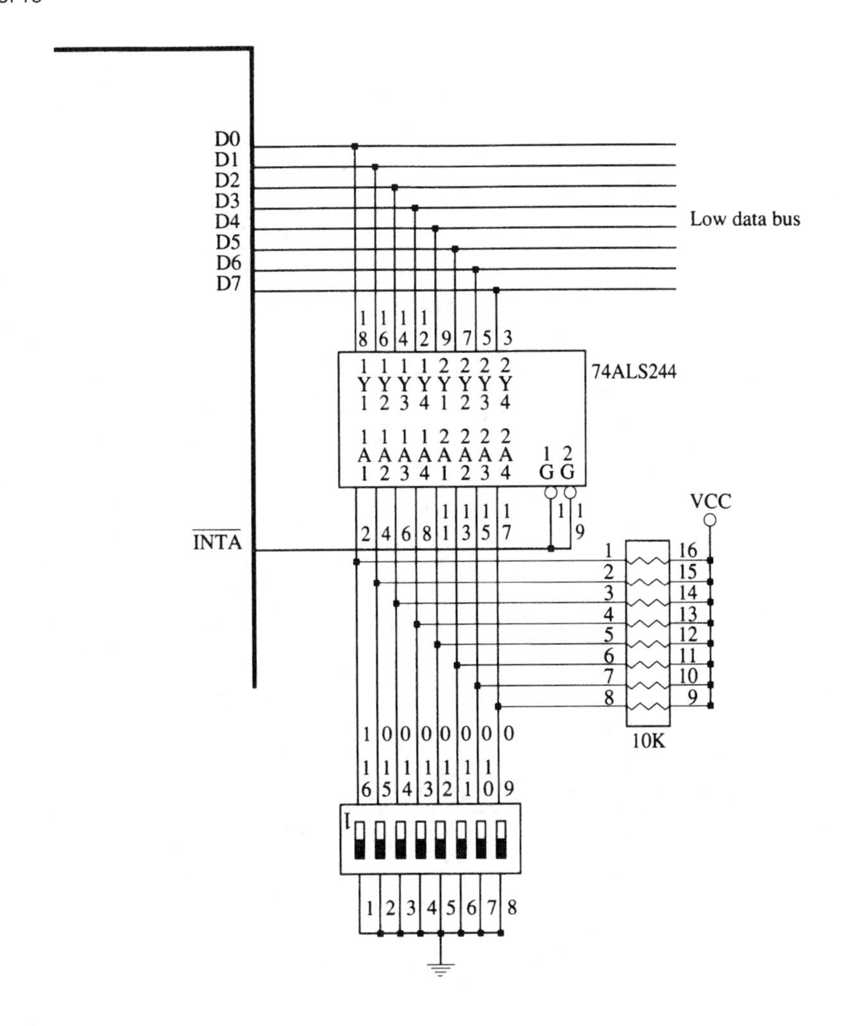

FIGURE 12–10 A circuit that applies any interrupt vector type number in response to $\overline{\text{INTA}}$. Here the circuit is applying type number 80H.

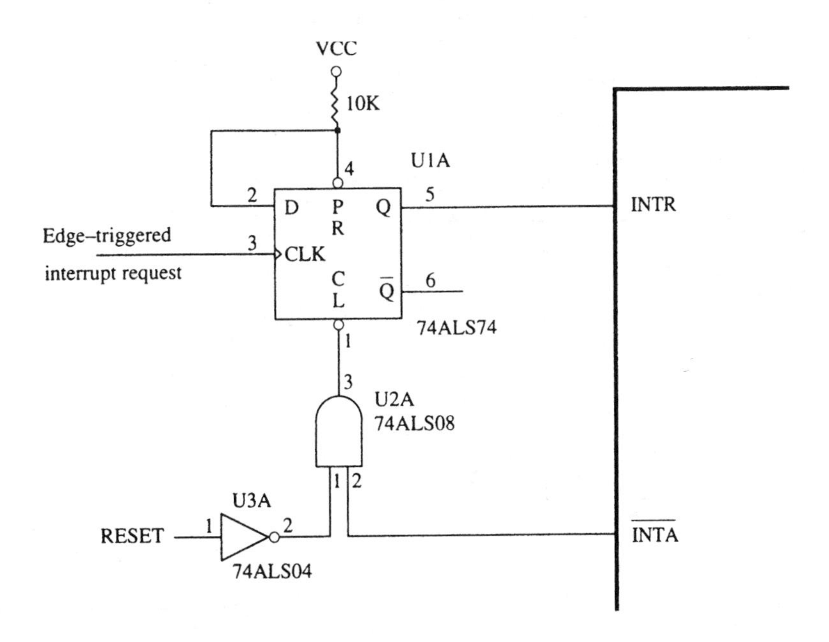

FIGURE 12–11 Converting INTR into an edge-triggered interrupt request input.

interrupt request input, and the clear input is used to clear the request when the $\overline{\text{INTA}}$ signal is output by the microprocessor. The RESET signal initially clears the flip-flop so that no interrupt is requested when the system is first powered.

The 82C55 Keyboard Interrupt

The keyboard example presented in Chapter 11 provides a simple example of the operation of the INTR input and an interrupt. Figure 12–12 illustrates the interconnection of the 82C55 with the microprocessor and the keyboard. It also shows how a 74ALS244 octal buffer is used to provide the microprocessor with interrupt vector type number 40H in response to the keyboard interrupt during the $\overline{\text{INTA}}$ pulse.

The 82C55 is decoded at 80386SX I/O port address 0500H, 0502H, 0504H, and 0506H by a PLD (the program is not illustrated). The 82C55 is operated in mode 1 (strobed input mode), so

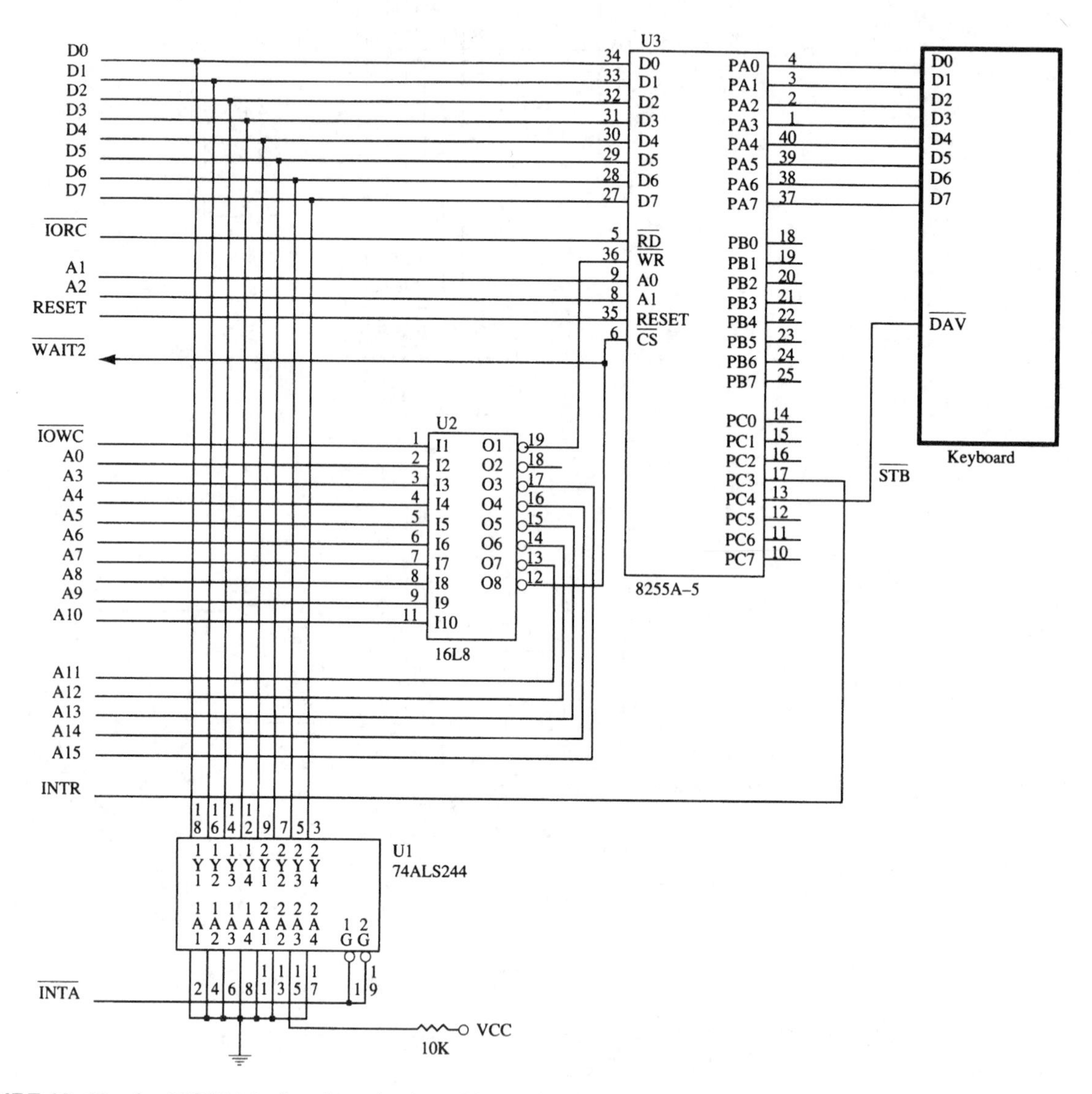

FIGURE 12–12 An 82C55 interfaced to a keyboard from the microprocessor system using interrupt vector 40H.

whenever a key is typed, the INTR output (PC3) becomes a logic 1 and requests an interrupt through the INTR pin on the microprocessor. The INTR pin remains high until the ASCII data are read from port A. In other words, every time a key is typed, the 82C55 requests a type 40H interrupt through the INTR pin. The $\overline{DAV}$ signal from the keyboard causes data to be latched into port A and causes INTR to become a logic 1.

Example 12–5 illustrates the interrupt service procedure for the keyboard. It is very important that all registers affected by an interrupt be saved before they are used. In the software required to initialize the 82C55 (not shown here), the FIFO is initialized so that both pointers are equal, the INTR request pin is enabled through the INTE bit inside the 82C55, and the mode of operation is programmed.

EXAMPLE 12–5

```
;An interrupt service procedure that reads a key from
;the keyboard depicted in Figure 12-12.

PORTA  EQU   500H
CNTR   EQU   506H

FIFO   DB    256 DUP(?)             ;queue

INP    DD    FIFO                   ;input pointer
OUTP   DD    FIFO                   ;output pointer

KEY    PROC  FAR USES EAX EBX EDX EDI

       MOV   EBX,CS:INP             ;get pointers
       MOV   EDI,CS:OUTP

       INC   BL
       .IF BX == DI                 ;if full
             MOV   AL,8
             MOV   DX,CNTR
             OUT   DX,AL            ;disable 82C55 interrupt
       .ELSE                        ;if not full
             DEC   BL
             MOV   DX,PORTA
             IN    AL,DX            ;read key code
             MOV   CS:[BX]          ;save in queue
             INC   BYTE PTR CS:INP
       .ENDIF
       IRET
KEY    ENDP
```

The procedure is short because the microprocessor already knows that keyboard data are available when the procedure is called. Data are input from the keyboard and then stored in the FIFO (first-in, first-out) buffer or queue. Most keyboard interfaces contain a FIFO that is at least 16 bytes in depth. The FIFO in this example is 256 bytes, which is more than adequate for a keyboard interface. Note how the INC BYTE PTR CX:INP is used to add 1 to the input pointer and also make sure that it always addresses data in the queue.

This procedure first checks to see whether the FIFO is full. A full condition is indicated when the input pointer (INP) is one byte below the output pointer (OUTP). If the FIFO is full, the interrupt is disabled with a bit set/reset command to the 82C55, and a return from the interrupt occurs. If the FIFO is not full, the data are input from port A, and the input pointer is incremented before a return occurs.

Example 12–6 shows the procedure that removes data from the FIFO. This procedure first determines whether the FIFO is empty by comparing the two pointers. If the pointers are equal, the FIFO is empty, and the software waits at the EMPTY loop where it continuously tests the pointers. The EMPTY loop is interrupted by the keyboard interrupt, which stores data into the FIFO so that it is no longer empty. This procedure returns with the character in register AH.

EXAMPLE 12–6

```
;A procedure that reads data from the queue of Example 12-5
;and returns it in AH;

READQ  PROC   FAR USES EBX EDI EDX

       .REPEAT
              MOV    EBX,CS:INP                 ;get pointers
              NOV    EDI,CS:OUTP
       .UNTIL EBX == EDI                        ;while empty

       MOV    AH,CS:[EDI]                       ;get data
       MOV    AL,9
       MOV    DX,CNTR
       OUT    DX,AL                             ;enable 52C55 interrupt
       INC    BYTE PTR CS:OUTP
       RET

READQ  ENDP
```

12–3 EXPANDING THE INTERRUPT STRUCTURE

This text covers three of the more common methods of expanding the interrupt structure of the microprocessor. In this section, we explain how, with software and some hardware modification of the circuit shown in Figure 12–10, it is possible to expand the INTR input so that it accepts seven interrupt inputs. We also explain how to "daisy-chain" interrupts by software polling. In the next section, we describe a third technique in which up to 63 interrupting inputs can be added by means of the 8259A programmable interrupt controller.

Using the 74ALS244 to Expand Interrupts

The modification shown in Figure 12–13 allows the circuit of Figure 12–10 to accommodate up to seven additional interrupt inputs. The only hardware change is the addition of an eight-input NAND gate, which provides the INTR signal to the microprocessor when any of the $\overline{\text{IR}}$ inputs becomes active.

Operation. If any of the $\overline{\text{IR}}$ inputs becomes a logic 0, then the output of the NAND gate goes to a logic 1 and requests an interrupt through the INTR input. The interrupt vector that is fetched during the $\overline{\text{INTA}}$ pulse depends on which interrupt request line becomes active. Table 12–1 shows the interrupt vectors used by a single interrupt request input.

If two or more interrupt request inputs are simultaneously active, a new interrupt vector is generated. For example, if $\overline{\text{IR1}}$ and $\overline{\text{IR0}}$ are both active, the interrupt vector generated is FCH (252). Priority is resolved at this location. If the $\overline{\text{IR0}}$ input is to have the higher priority, the vector address for $\overline{\text{IR0}}$ is stored at vector location FCH. The entire top half of the vector table and its 128 interrupt vectors must be used to accommodate all possible conditions of these seven interrupt request inputs. This seems wasteful, but in many dedicated applications it is a cost-effective approach to interrupt expansion.

Daisy-Chained Interrupt

Expansion by means of a daisy-chained interrupt is in many ways better than using the 74ALS244 because it requires only one interrupt vector. The task of determining priority is left to the interrupt service procedure. Setting priority for a daisy-chain does require additional software execution time, but in general this is a much better approach to expanding the interrupt structure of the microprocessor.

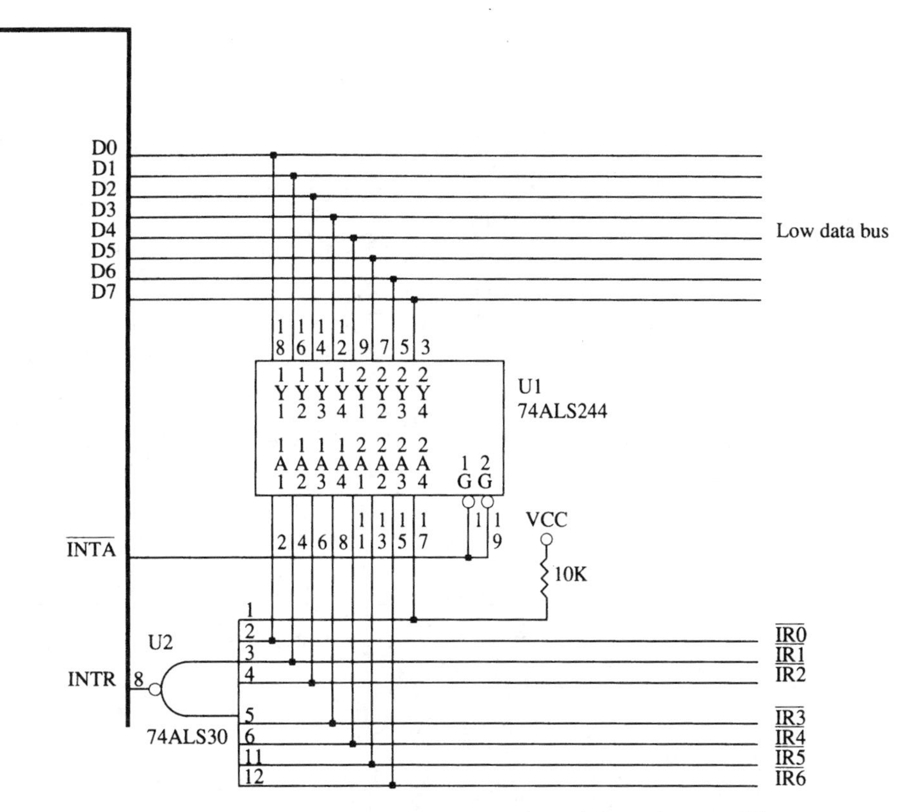

FIGURE 12–13 Expanding the INTR input from one to seven interrupt request lines.

TABLE 12–1 Single interrupt requests for Figure 12–13.

$\overline{IR6}$	$\overline{IR5}$	$\overline{IR4}$	$\overline{IR3}$	$\overline{IR2}$	$\overline{IR1}$	$\overline{IR0}$	*Vector*
1	1	1	1	1	1	0	FEH
1	1	1	1	1	0	1	FDH
1	1	1	1	0	1	1	FBH
1	1	1	0	1	1	1	F7H
1	1	0	1	1	1	1	EFH
1	0	1	1	1	1	1	DFH
0	1	1	1	1	1	1	BFH

Figure 12–14 illustrates a set of two 82C55 peripheral interfaces with their four INTR outputs daisy-chained and connected to the single INTR input of the microprocessor. If any interrupt output becomes a logic 1, so does the INTR input to the microprocessor, causing an interrupt.

When a daisy-chain is used to request an interrupt, it is better to pull the data bus connections (D_0–D_7) high by using pull-up resistors so interrupt vector FFH is used for the chain. Any interrupt vector can be used to respond to a daisy-chain. In the circuit, any of the four INTR outputs from the two 82C55s will cause the INTR pin on the microprocessor to go high, requesting an interrupt.

When the INTR pin does go high with a daisy-chain, the hardware gives no direct indication as to which 82C55 or which INTR output caused the interrupt. The task of locating which INTR output became active is up to the interrupt service procedure, which must poll the 82C55s to determine which output caused the interrupt.

Example 12–7 illustrates the interrupt service procedure that responds to the daisy-chain interrupt request. The procedure polls each 82C55 and each INTR output to decide which interrupt service procedure to utilize.

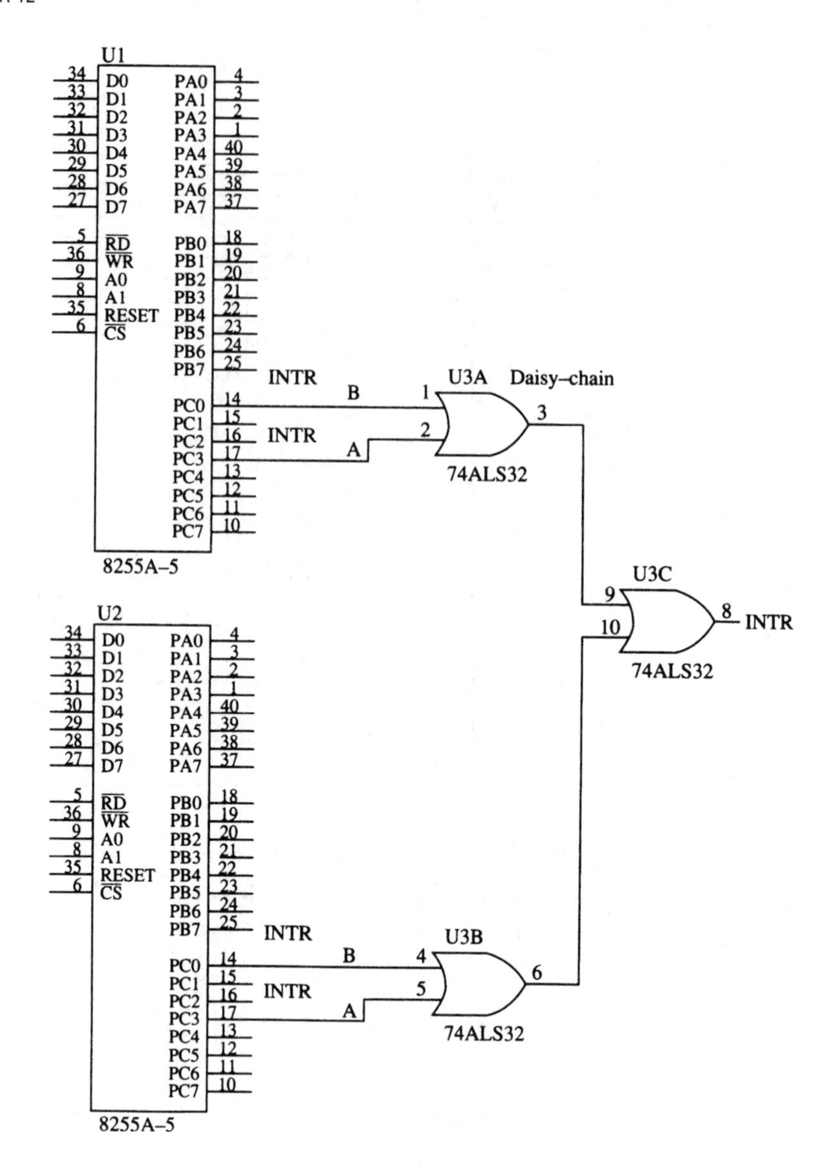

FIGURE 12–14 Two 82C55 PIAs connected to the INTR outputs are daisy-chained to produce an INTR signal.

EXAMPLE 12–7

```
;A procedure that services the daisy-chain interrupt
;of Figure 12-14.

C1      EQU   504H                 ;first 82C55
C2      EQU   604H                 ;second 82C55
MASK1   EQU   1                    ;INTRB
MASK2   EQU   8                    ;INTRA

POLL    PROC  FAR     USES EAX EDX

        MOV   DX,C1                ;address first 82C55
        IN    AL,DX
        TEST  AL,MASK1             ;test INTRB
        .IF !ZERO?

              ;LEVEL 1 interrupt software here

        .ENDIF
        TEST   AL,MASK2            ;test INTRA
```

```
        .IF !ZERO?

                ;LEVEL 2 interrupt software here

        .ENDIF
        MOV   DX,C2                   ;address second 82C55
        TEST  AL,MASK1                ;test INTRB
        .IF !ZERO?

                ;LEVEL 3 interrupt software here

        .ENDIF

                ;LEVEL 4 interrupt software here

POLL    ENDP
```

12–4 8259A PROGRAMMABLE INTERRUPT CONTROLLER

The 8259A programmable interrupt controller (PIC) adds eight vectored priority encoded interrupts to the microprocessor. This controller can be expanded, without additional hardware, to accept up to 64 interrupt requests. This expansion requires a master 8259A and eight 8259A slaves. A pair of these controllers still resides and is programmed as explained here in the latest chip sets from Intel and other manufacturers.

General Description of the 8259A

Figure 12–15 shows the pin-out of the 8259A. The 8259A is easy to connect to the microprocessor because all of its pins are direct connections except the $\overline{CS}$ pin, which must be decoded, and the $\overline{WR}$ pin, which must have an I/O bank write pulse. Following is a description of each pin on the 8259A:

D_0–D_7 The bidirectional **data connections** are normally connected to the data bus on the microprocessor.

IR_0–IR_7 **Interrupt request inputs** are used to request an interrupt and to connect to a slave in a system with multiple 8259As.

$\overline{WR}$ The **write input** connects to the write strobe signal ($\overline{IOWC}$) on the microprocessor.

$\overline{RD}$ The **read input** connects to the $\overline{IORC}$ signal.

INT The **interrupt output** connects to the INTR pin on the microprocessor from the master and is connected to a master IR pin on a slave.

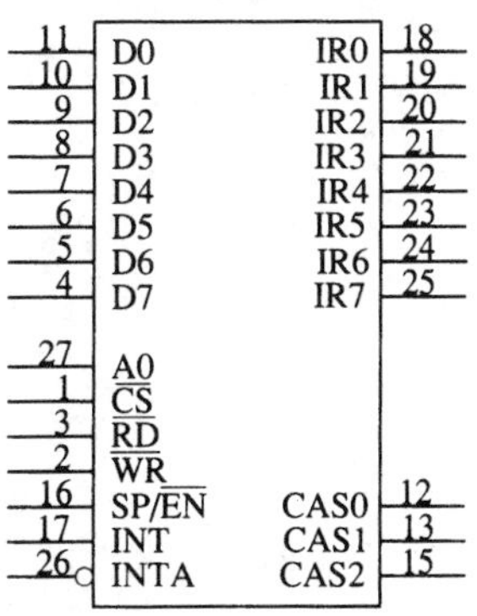

FIGURE 12–15 The pin-out of the 8259A programmable interrupt controller (PIC).

FIGURE 12–16 An 8259A interfaced to the 8086 microprocessor.

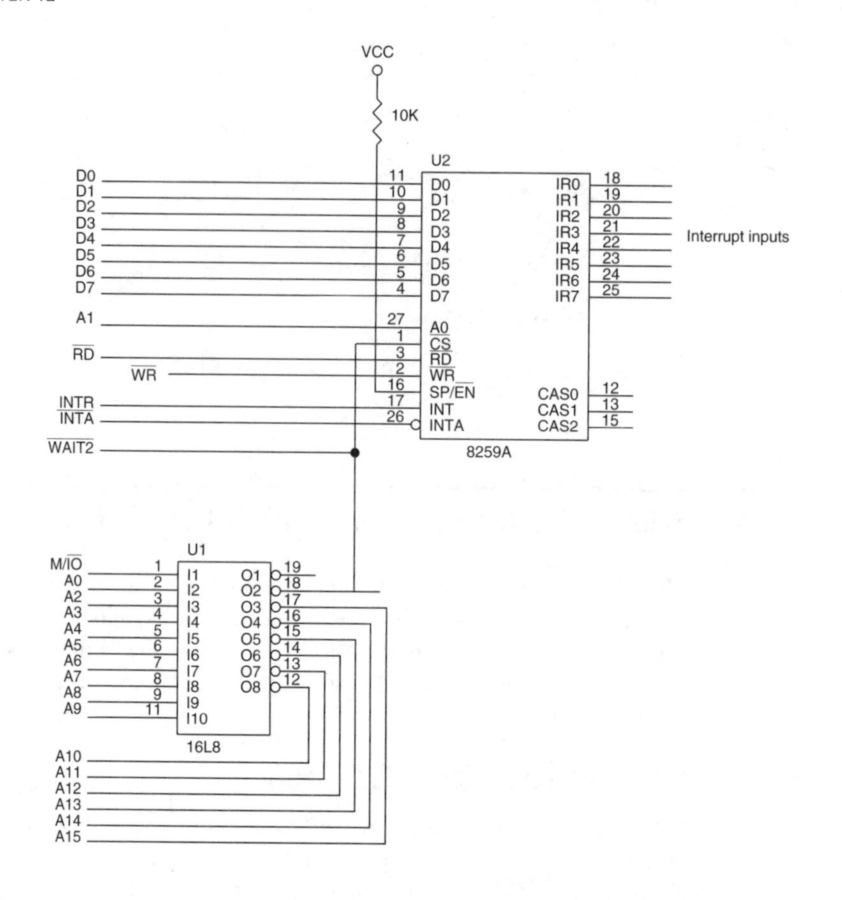

$\overline{\text{INTA}}$ — **Interrupt acknowledge** is an input that connects to the $\overline{\text{INTA}}$ signal on the system. In a system with a master and slaves, only the master $\overline{\text{INTA}}$ signal is connected.

A_0 — The **A_0 address input** selects different command words within the 8259A.

$\overline{\text{CS}}$ — **Chip select** enables the 8259A for programming and control.

SP/$\overline{\text{EN}}$ — **Slave program/enable buffer** is a dual-function pin. When the 8259A is in buffered mode, this is an output that controls the data bus transceivers in a large microprocessor-based system. When the 8259A is not in the buffered mode, this pin programs the device as a master (1) or a slave (0).

CAS_0–CAS_2 — The **cascade lines** are used as outputs from the master to the slaves for cascading multiple 8259As in a system.

Connecting a Single 8259A

Figure 12–16 shows a single 8259A connected to the microprocessor. Here the SP/$\overline{\text{EN}}$ pin is pulled high to indicate that it is a master. The 8259A is decoded at I/O ports 0400H and 0401H by the PLD (no program shown). Like other peripherals discussed in Chapter 11, the 8259A requires four wait states for it to function properly with a 16 MHz 80386SX and more for some other versions of the Intel microprocessor family.

Cascading Multiple 8259As

Figure 12–17 shows two 8259As connected to the microprocessor in a way that is often found in the ATX-style computer, which has two 8259As for interrupts. The XT- or PC-style

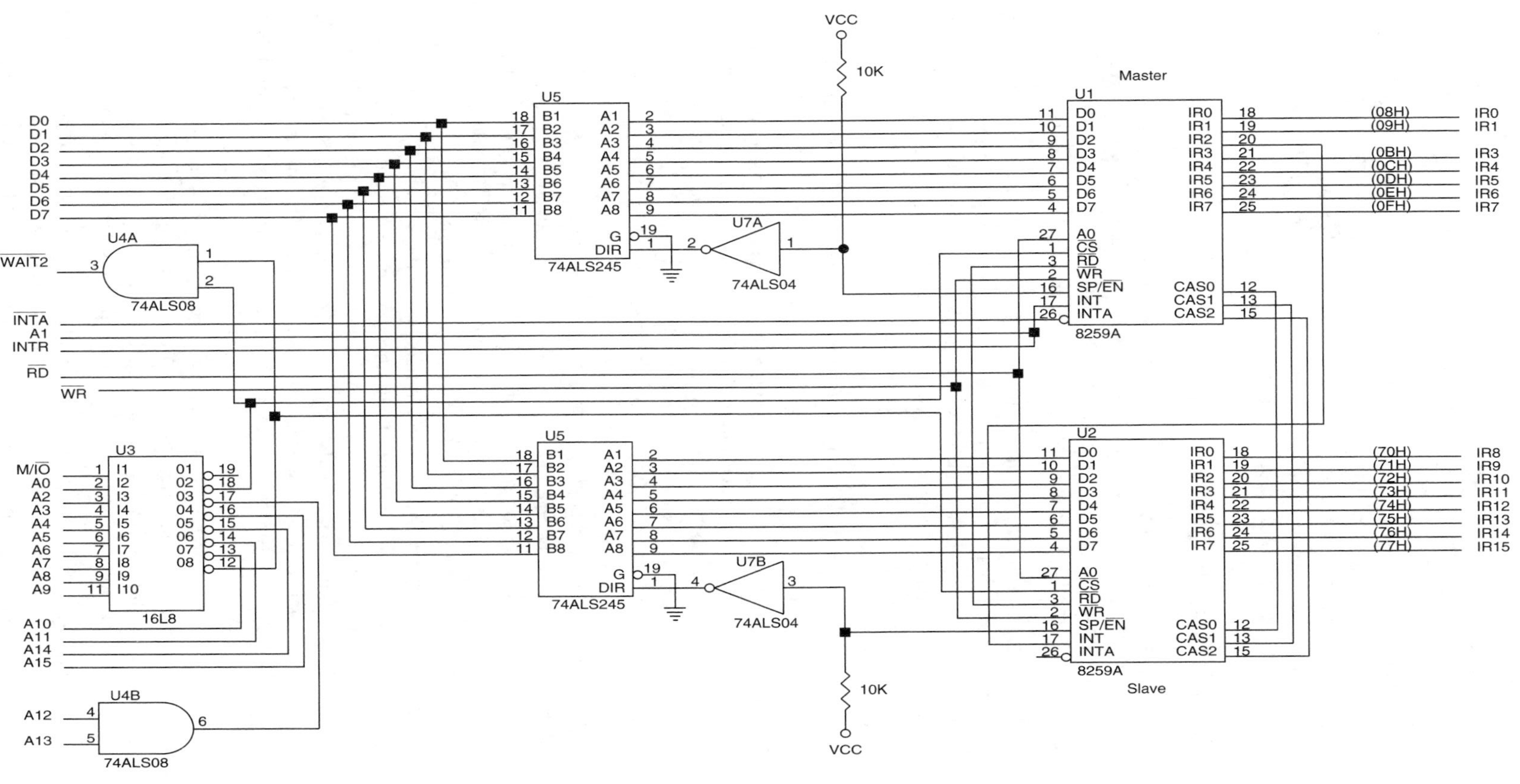

FIGURE 12–17 Two 8259As interfaced to the 8259A at I/O ports 0300H and 0302H for the master and 0304H and 0306H for the slave.

computers use a single 8259A controller at interrupt vectors 08H–0FH. The ATX-style computer uses interrupt vector 0AH as a cascade input from a second 8259A located at vectors 70H through 77H. Appendix A contains a table that lists the functions of all the interrupt vectors used.

This circuit uses vectors 08H–0FH and I/O ports 0300H and 0302H for U1, the master; and vectors 70H–77H and I/O ports 0304H and 0306H for U2, the slave. Notice that we also include data bus buffers to illustrate the use of the $SP/\overline{EN}$ pin on the 8259A. These buffers are used only in very large systems that have many devices connected to their data bus connections. In practice, we seldom find these buffers.

Programming the 8259A

The 8259A is programmed by initialization and operation command words. **Initialization command words** (ICWs) are programmed before the 8259A is able to function in the system and dictate the basic operation of the 8259A. **Operation command words** (OCWs) are programmed during the normal course of operation. The OCWs control the operation of the 8259A.

Initialization Command Words. There are four initialization command words (ICWs) for the 8259A that are selected when the A_0 pin is a logic one. When the 8259A is first powered up, it must be sent ICW_1, ICW_2, and ICW_4. If the 8259A is programmed in cascade mode by ICW_1, then we also must program ICW_3. So if a single 8259A is used in a system, ICW_1, ICW_2, and ICW_4 must be programmed. If cascade mode is used in a system, then all four ICWs must be programmed.

Refer to Figure 12–18 for the format of all four ICWs. The following is a description of each ICW:

ICW_1 Programs the basic operation of the 8259A. To program this ICW for 8086–Pentium 4 operation, place a logic 1 in bit IC_4. Bits AD_1, A_7, A_6, and A_5 are don't cares for microprocessor operation and only apply to the 8259A when used with an 8-bit 8085 microprocessor (not covered in this textbook). This ICW selects single or cascade operation by programming the SNGL bit. If cascade operation is selected, we must also program ICW_3. The LTIM bit determines whether the interrupt request inputs are positive edge-triggered or level-triggered.

ICW_2 Selects the vector number used with the interrupt request inputs. For example, if we decide to program the 8259A so it functions at vector locations 08H–0FH, we place 08H in this command word. Likewise, if we decide to program the 8259A for vectors 70H–77H, we place 70H in this ICW.

ICW_3 Only used when ICW_1 indicates that the system is operated in cascade mode. This ICW indicates where the slave is connected to the master. For example, in Figure 12–18 we connected a slave to IR_2. To program ICW_3 for this connection, in both master and slave, we place 04H in ICW_3. Suppose we have two slaves connected to a master using IR_0 and IR_1. The master is programmed with an ICW_3 of 03H; one slave is programmed with an ICW_3 of 01H and the other with an ICW_3 of 02H.

ICW_4 Programmed for use with the 8086–Pentium 4 microprocessors, but is not programmed in a system that functions with the 8085 microprocessor. The rightmost bit must be a logic 1 to select operation with the 8086–Pentium 4 microprocessors, and the remaining bits are programmed as follows:

SFNM—Selects the special fully-nested mode of operation for the 8259A if a logic I is placed in this bit. This allows the highest-priority interrupt request from a slave to be recognized by the master while it is processing another interrupt from a

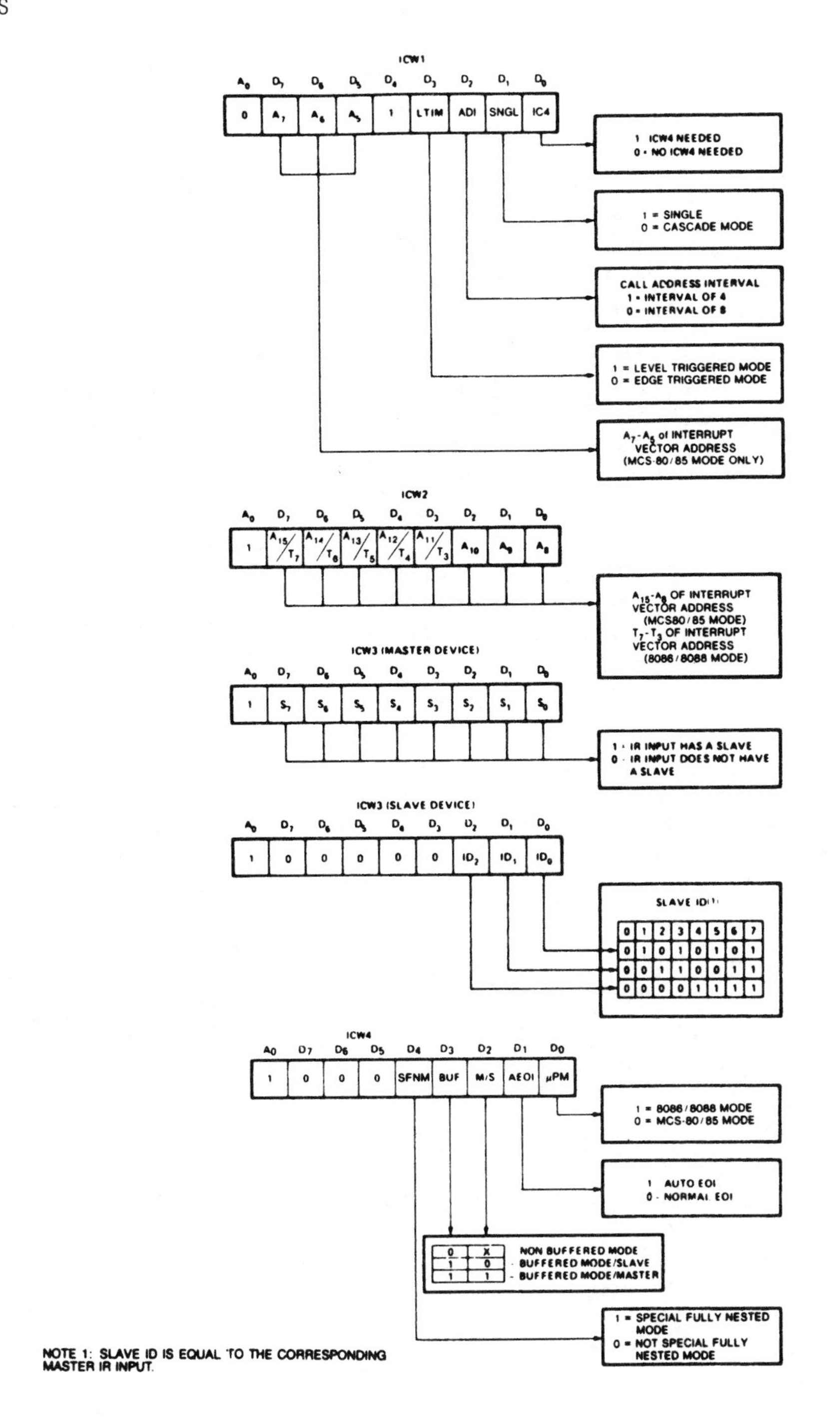

FIGURE 12–18 The 8259A initialization command words (ICWs). (Courtesy of Intel Corporation.)

slave. Normally, only one interrupt request is processed at a time and others are ignored until the process is complete.

BUF and M/S—Buffered and master slave are used together to select buffered operation or nonbuffered operation for the 8259A as a master or a slave.

AEOI—Selects automatic or normal end of interrupt (discussed more fully under operation command words). The EOI commands of OCW_2 are used only if the AEOI mode is not selected by ICW_4. If AEOI is selected, the interrupt automatically resets the interrupt request bit and does not modify priority. This is the preferred mode of operation for the 8259A and reduces the length of the interrupt service procedure.

Operation Command Words. The operation command words (OCWs) are used to direct the operation of the 8259A once it is programmed with the ICW. The OCWs are selected when the A_0 pin is at a logic 0 level, except for OCW_1, which is selected when A_0 is a logic 1. Figure 12–19 lists the binary bit patterns for all three operation command words of the 8259A. Following is a list describing the function of each OCW:

OCW_1 Used to set and read the interrupt mask register. When a mask bit is set, it will turn off (mask) the corresponding interrupt input. The mask register is read when OCW_1 is read. Because the state of the mask bits is unknown when the 8259A is first initialized, OCW_1 must be programmed after programming the ICW upon initialization.

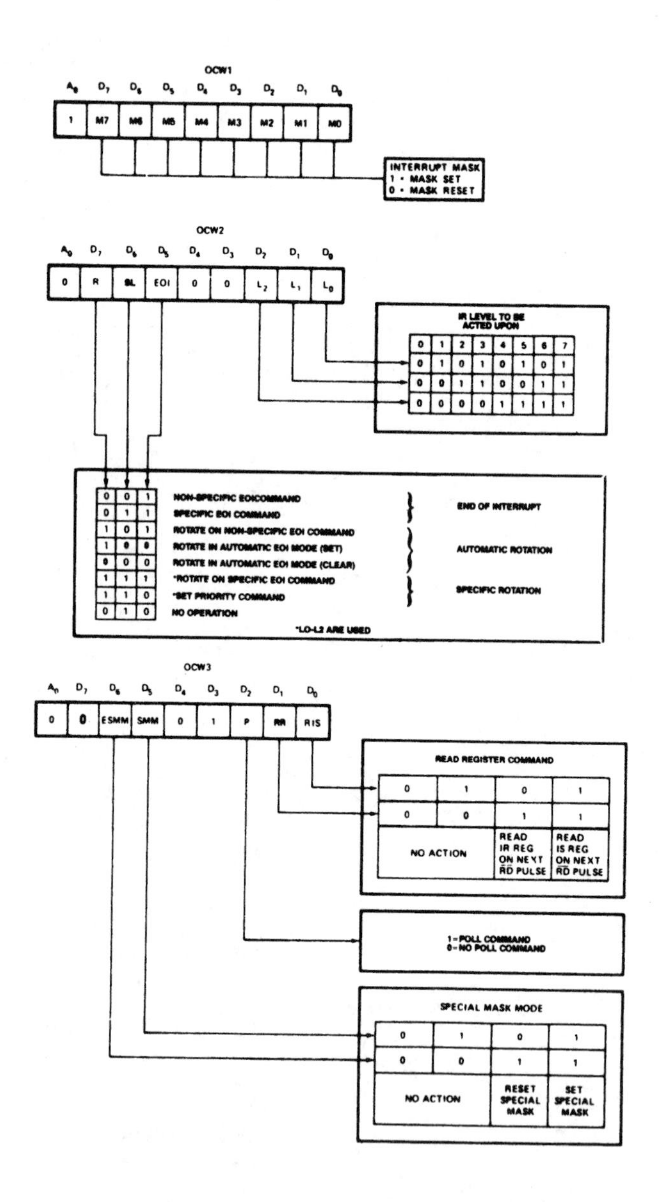

FIGURE 12–19 The 8259A operation command words (OCWs). (Courtesy of Intel Corporation.)

OCW_2 Programmed only when the AEOI mode is not selected for the 8259A. In this case, this OCW selects the way that the 8259A responds to an interrupt. The modes are listed as follows:

Nonspecific End-of-Interrupt—A command sent by the interrupt service procedure to signal the end of the interrupt. The 8259A automatically determines which interrupt level was active and resets the correct bit of the interrupt status register. Resetting the status bit allows the interrupt to take action again or a lower priority interrupt to take effect.

Specific End-of-Interrupt—A command that allows a specific interrupt request to be reset. The exact position is determined with bits L_2–L_0 of OCW_2.

Rotate-on-Nonspecific EOI—A command that functions exactly like the Nonspecific End-of-Interrupt command, except that it rotates interrupt priorities after resetting the interrupt status register bit. The level reset by this command becomes the lowest-priority interrupt. For example, if IR_4 was just serviced by this command, it becomes the lowest-priority interrupt input and IR_5 becomes the highest priority.

Rotate-on-Automatic EOI—A command that selects automatic EOI with rotating priority. This command must only be sent to the 8259A once if this mode is desired. If this mode must be turned off, use the clear command.

Rotate-on-Specific EOI—Functions as the specific EOI, except that it selects rotating priority.

Set priority—Allows the programmer to set the lowest priority interrupt input using the L_2–L_0 bits.

OCW_3 Selects the register to be read, the operation of the special mask register, and the poll command. If polling is selected, the P bit must be set and then output to the 8259A. The next read operation will read the poll word. The rightmost three bits of the poll word indicate the active interrupt request with the highest priority. The leftmost bit indicates whether there is an interrupt and must be checked to determine whether the rightmost three bits contain valid information.

Status Register. Three status registers are readable in the 8259A: interrupt request register (IRR), in-service register (ISR), and interrupt mask register (IMR). (See Figure 12–20 for all three status registers; they all have the same bit configuration.) The IRR is an 8-bit register that indicates which interrupt request inputs are active. The ISR is an 8-bit register that contains the level of the interrupt being serviced. The IMR is an 8-bit register that holds the interrupt mask bits and indicates which interrupts are masked off.

Both the IRR and ISR are read by programming OCW_3 and IMR is read through OCW_1. To read the IMR, $A_0 = 1$; to read either IRR or ISR, $A_0 = 0$. Bit positions D_0 and D_1 of OCW_3 select which register (IRR or ISR) is read when $A_0 = 0$.

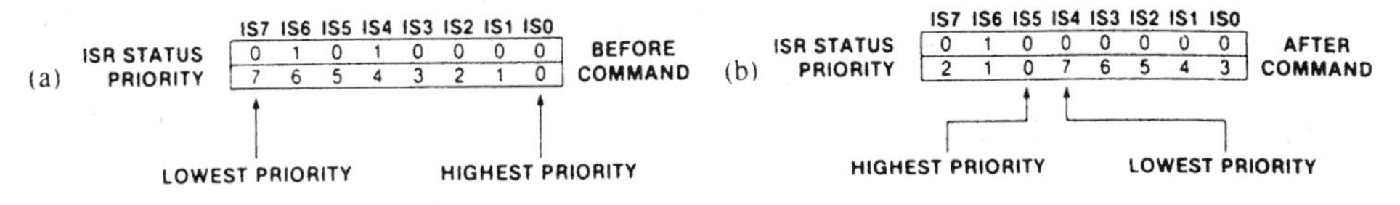

FIGURE 12–20 The 8259A in-service register (ISR). (a) Before IR_4 is accepted and (b) after IR_4 is accepted. (Courtesy of Intel Corporation.)

8259A Programming Example

Figure 12–21 illustrates the 8259A programmable interrupt controller connected to a 16550 programmable communications controller. In this circuit, the INTR pin from the 16550 is connected to the programmable interrupt controller's interrupt request input IR_0. An IR_0 occurs whenever (1) the transmitter is ready to send another character, (2) the receiver has received a character, (3) an error is detected while receiving data, and (4) a modem interrupt occurs. Notice that the 16550 is decoded at I/O ports 40H and 47H, and the 8259A is decoded at 8-bit I/O ports 48H and 49H. Both devices are interfaced to the data bus of an 8088 microprocessor.

Initialization Software. The first portion of the software for this system must program both the 16550 and the 8259A, and then enable the INTR pin on the 8088 so that interrupts can take effect. Example 12–8 lists the software required to program both devices and enable INTR. This software uses two memory FIFOs that hold data for the transmitter and for the receiver. Each memory FIFO is 16K bytes long and is addressed by a pair of pointers (input and output).

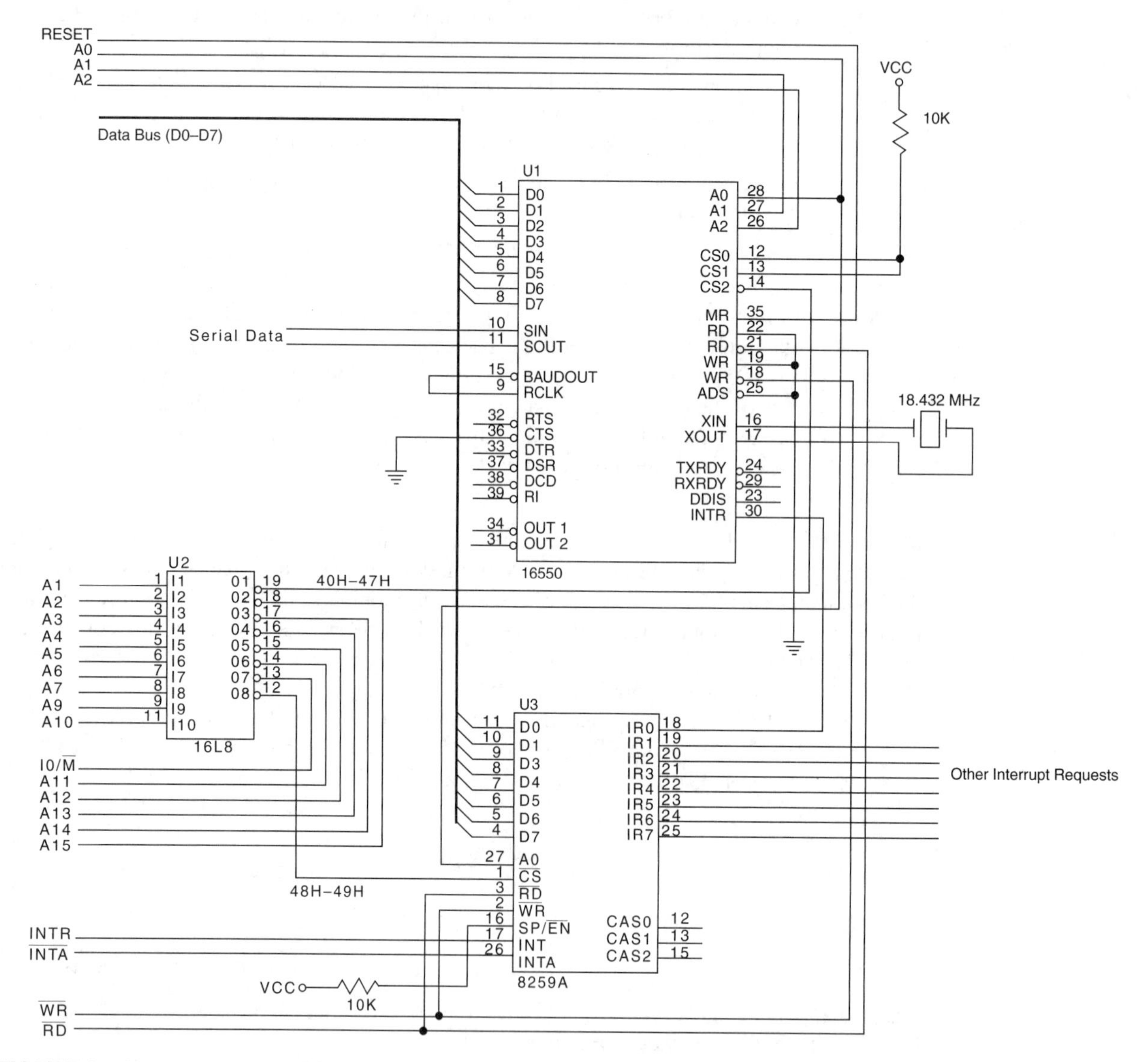

FIGURE 12–21 The 16550 UART interfaced to the 8088 microprocessor through the 8259A.

EXAMPLE 12–8

```
;Initialization software for the 16650 and 8259A
;of the circuit in Figure 12-21

PIC1     EQU  48H                  ;8259A control A0 = 0
PIC2     EQU  49H                  ;8259A control A0 = 1
ICW1     EQU  1BH                  ;8259A ICW1
ICW2     EQU  80H                  ;8259A ICW2
ICW4     EQU  3                    ;8259A ICW4
OCW1     EQU  0FEH                 ;8259A OCW1
LINE     EQU  43H                  ;16650 line register
LSB      EQU  40H                  ;16650 Baud divisor LSB
MSB      EQU  41H                  ;16650 Baud divisor MSB
FIFO     EQU  42H                  ;16650 FIFO register
ITR      EQU  41H                  ;16650 interrupt register

INIT PROC  NEAR

;
;setup 16650
;
       MOV  AL,10001010B           ;enable Baud rate divisor
       OUT  LINE,AL

       MOV  AL,120                 ;program Baud 9600
       OUT  LSB,AL
       MOV  AL,0
       OUT  MSB,AL

       MOV  AL,00001010B           ;program 7 data, odd
       OUT  LINE,AL                ;parity, 1 stop

       MOV  AL,00000111B           ;enable transmitter and
       OUT  FIFO,AL                ;receiver
;
;program 8259A
;
       MOV  AL,ICW1                ;program ICW1
       OUT  PIC1,AL
       MOV  AL,ICW2                ;program ICW2
       OUT  PIC2,AL
       MOV  AL,ICW4                ;program ICW4
       OUT  PIC2,AL
       MOV  AL,OCW1                ;program OCW1
       OUT  PIC2,AL
       STI                         ;enable INTR pin
;
;enable 16650 interrupts
;
       MOV  AL,5
       OUT  ITR,AL                 ;enable interrupts
       RET

INIT   ENDP
```

The first portion of the procedure (INIT) programs the 16550 UART for operation with seven data bits, odd parity, one stop bit, and a Baud rate clock of 9600. The FIFO control register also enables both the transmitter and receiver.

The second part of the procedure programs the 8259A, with its three ICWs and one OCW. The 8259A is set up so that it functions at interrupt vectors 80H–87H and operates with automatic EOI. The OCW enables the interrupt for the 16550 UART. The INTR pin of the microprocessor is also enabled by using the STI instruction.

The final part of the software enables the receiver and error interrupts of the 16550 UART through the interrupt control register. The transmitter interrupt is not enabled until data are available

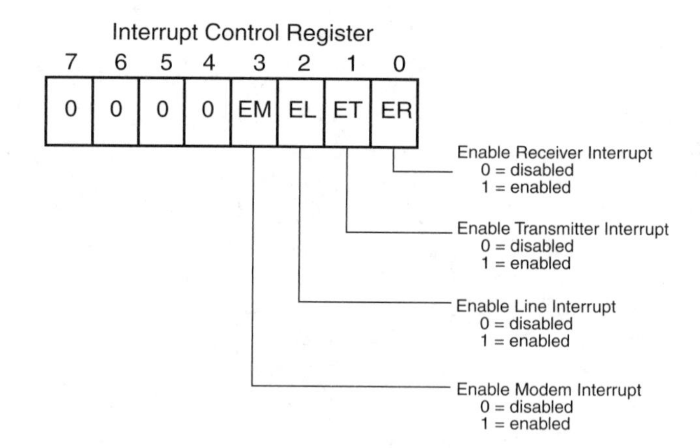

FIGURE 12–22 The 16550 interrupt control register.

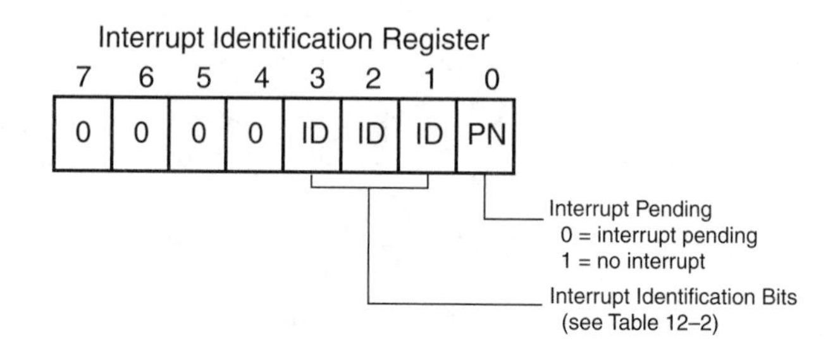

FIGURE 12–23 The 16550 interrupt identification register.

for transmission. See Figure 12–22 for the contents of the interrupt control register of the 16550 UART. Notice that the control register can enable or disable the receiver, transmitter, line status (error), and modem interrupts.

Handling the 16550 UART Interrupt Request. Because the 16550 generates only one interrupt request for various interrupts, the interrupt handler must poll the 16550 to determine what type of interrupt has occurred. This is accomplished by examining the interrupt identification register (see Figure 12–23). Note that the interrupt identification register (read-only) shares the same I/O port as the FIFO control register (write-only).

The interrupt identification register indicates whether an interrupt is pending, the type of interrupt, and whether the transmitter and receiver FIFO memories are enabled. See Table 12–2 for the contents of the interrupt control bits.

The interrupt service procedure must examine the contents of the interrupt identification register to determine what event caused the interrupt and pass control to the appropriate procedure for the event. Example 12–9 shows the first part of an interrupt handler that passes control to RECV for a receiver data interrupt, TRANS for a transmitter data interrupt, and ERR for a line status error interrupt. Note that the modem status is not tested in this example.

EXAMPLE 12–9

```
;Interrupt handler for the 16650 UART of Figure 12-21

INT80 PROC FAR USES AX BX DI SI

      IN   AL,42H                    ;read interrupt ID
      .IF AL == 6

               ;handle receiver error

      .ELSEIF AL == 2
               ;handle transmitter empty
```

TABLE 12–2 The interrupt control bits of the 16650.

Bit 3	*Bit 2*	*Bit 1*	*Bit 0*	*Priority*	*Type*	*Reset Control*
0	0	0	1	—	No interrupt	—
0	1	1	0	1	Receiver error (parity, framing, overrun, or break)	Reset by reading the register
0	1	0	0	2	Receiver data available	Reset by reading the data
1	1	0	0	2	Character time-out, nothing has been removed from the receiver FIFO for at least four character times	Reset by reading the data
0	0	1	0	3	Transmitter empty	Reset by writing the transmitter
0	0	0	0	4	Modem status	Reset by reading the modem status

```
                JMP     TRAN    ;example 12-13

        .ELSEIF AL == 4

                ;handle receiver ready

                JMP     RECV    ;example 12-11

        .ENDIF

        IRET

INT80   ENDP
```

Receiving data from the 16550 requires two procedures. One procedure reads the data register of the 16550 each time that the INTR pin requests an interrupt and stores it in the memory FIFO. The other procedure reads data from the memory FIFO from the main program.

Example 12–10 lists the procedure used to read data from the memory FIFO from the main program. This procedure assumes that the pointers (IIN and IOUT) are initialized in the initialization dialog for the system (not shown). The READ procedure returns with AL containing a character read from the memory FIFO. If the memory FIFO is empty, the procedure returns with the carry flag bit set to a logic 1. If AL contains a valid character, the carry flag bit is cleared upon return from READ.

Notice how the FIFO is reused by changing the address from the top of the FIFO to the bottom whenever it exceeds the start of the FIFO plus 16K. Notice that interrupts are enabled at the end of this procedure, in case they are disabled by a full memory FIFO condition by the RECV interrupt procedure.

EXAMPLE 12–10

```
;A procedure that reads one character from the FIFO
;and returns it on AL.  If the FIFO is empty the return
;occurs with carry = 1,

READC   PROC    NEAR USES BX DX

        MOV     DI,IOUT                 ;get pointer
        MOV     BX,IIN
        .IF     BX == DI                ;if empty
                STC                     ;set carry
```

```
        .ELSE                               ;if not empty
                MOV  AL,ES:[DI]             ;get data
                INC  DI                     ;increment pointer
                .IF DI == OFFSET FIFO+16*1024
                        MOV  DI,OFFSET FIFO
                .ENDIF
              MOV IOUT,DI
              CLC
        .ENDIF
        PUSHF                               ;enable receiver
        IN   AL,41H
        OR   AL,5
        OUT  41H,AL
        POPF
        RET

READC   ENDP
```

Example 12–11 lists the RECV interrupt service procedure that is called each time the 16550 receives a character for the microprocessor. In this example, the interrupt uses vector type number 80H, which must address the interrupt handler of Example 12–9. Each time that this interrupt occurs, the REVC procedure is accessed by the interrupt handler reading a character from the 16550. The RECV procedure stores the character in the memory FIFO. If the memory FIFO is full, the receiver interrupt is disabled by the interrupt control register within the 16550. This may result in lost data, but at least it will not cause the interrupt to overrun valid data already stored in the memory FIFO. Any error conditions detected by the 8251A store a ? (3FH) in the memory FIFO. Note that errors are detected by the ERR portion of the interrupt handler (not shown).

EXAMPLE 12–11

```
;RECV portion of the interrupt handler of Example 12-9

RECV:
      MOV   BX,IOUT                 ;get pointers
      MOV   DI,IIN
      MOV   SI,DI
      INC   SI
      .IF   SI == OFFSET FIFO+16*1024
             MOV SI,OFFSET FIFO
      .ENDIF
      .IF SI == BX                  ;if FIFO full
             IN    AL,41H           ;disable receiver
             AND   AL,0FAH
             OUT   41H,AL
       .ENDIF
       IN AL,40H                    ;read data
       STOSB
       MOV  IIN,SI
       MOV  AL,20H                  ;8259A EOI command
       OUT  49H,AL
       IRET
```

Transmitting Data to the 16550. Data are transmitted to the 16550 in much the same manner as they are received, except that the interrupt service procedure removes transmit data from a second 16K-byte memory FIFO.

Example 12–12 lists the procedure that fills the output FIFO. It is similar to the procedure listed in Example 12–10, except it determines whether the FIFO is full instead of empty.

EXAMPLE 12–12

```
;A procedure that places data into the memory FIFO for
;transmission by the transmitter interrupt. AL = the
;character transmitted.
```

```
SAVEC   PROC   NEAR USES BX DI SI

        MOV   SI,OIN                ;load pointers
        MOV   BX,OOUT
        MOV   DI,SI
        INC   SI
        .IF SI == OFFSET OFIFO+16*1024
               MOV   SI,OFFSET OFIFO
        .ENDIF
        .IF BX == SI                ;if OFIFO full
               STC
        .ELSE
               STOSB
               MOV   OIN,SI
               CLC
        .ENDIF
        PUSHF
        IN    AL,41H                ;enable transmitter
        OR    AL,1
        OUT   41H,AL
        RET

SAVEC   ENDP
```

Example 12–13 lists the interrupt service subroutine for the 16550 UART transmitter. This procedure is a continuation of the interrupt handler presented in Example 12–9 and is similar to the RECV procedure of Example 12–11, except that it determines whether the FIFO is empty rather than full. Note that we do not include an interrupt service procedure for the break interrupt or any errors.

EXAMPLE 12–13

```
;Interrupt service for the 16650 transmitter

TRAN:
        MOV BX,OIN                  ;load pointers
        MOV DI,OOUT
        .IF BX == DI                ;if empty
               IN    AL,41H
               AND   AL,0FDH        ;disable transmitter
               OUT   41H,AL
        .ELSE                       ;if not empty
               MOV   AL,ES:[DI]
               OUT   40H,AL         ;send data
               INC   DI
               .IF DI == OFFSET OFIFO+16*1024
                      MOV   DI,OFFSET OFIFO
               .ENDIF
               MOV   OFIFO,DI
        .ENDIF
        MOV  AL,20H                 ;send EOI to 8259A
        OUT  49H,AL
        IRET
```

The 16550 also contains a scratch register, which is a general-purpose register that can be used in any way deemed necessary by the programmer. Also contained in the 16550 are a modem control register and a modem status register. These registers allow the modem to cause interrupt and control the operation of the 16550 with a modem. See Figure 12–24 for the contents of both the modem status register and the modem control register.

The modem control register uses bit positions 0–3 to control various pins on the 16550. Bit position 4 enables the internal loop-back test for testing purposes. The modem status register allows the status of the modem pins to be tested; it also allows the modem pins to be checked for a change or, in the case of $\overline{RI}$, a trailing edge.

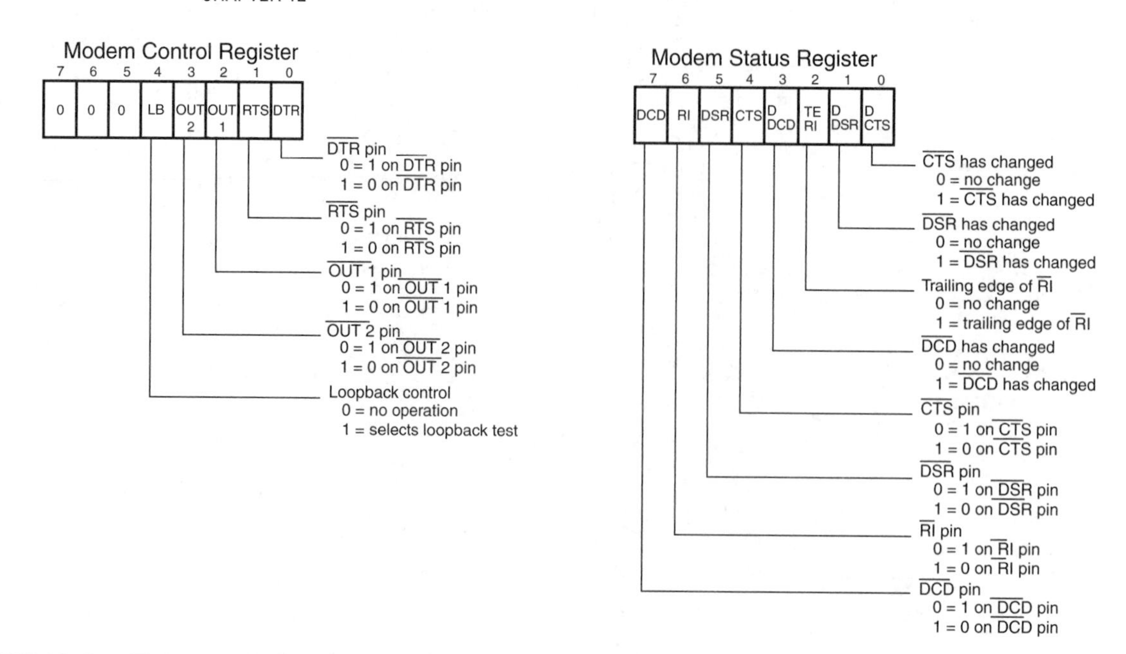

FIGURE 12–24 The 16550 modem control and modem status registers.

Figure 12–25 illustrates the 16550 UART, connected to an RS-232C interface that is often used to control a modem. Included in this interface are line driver and receiver circuits used to convert between TTL levels on the 16550 to RS-232C levels found on the interface. Note that RS-232C levels are usually +12 V for a logic 0 and –12 V for a logic 1 level.

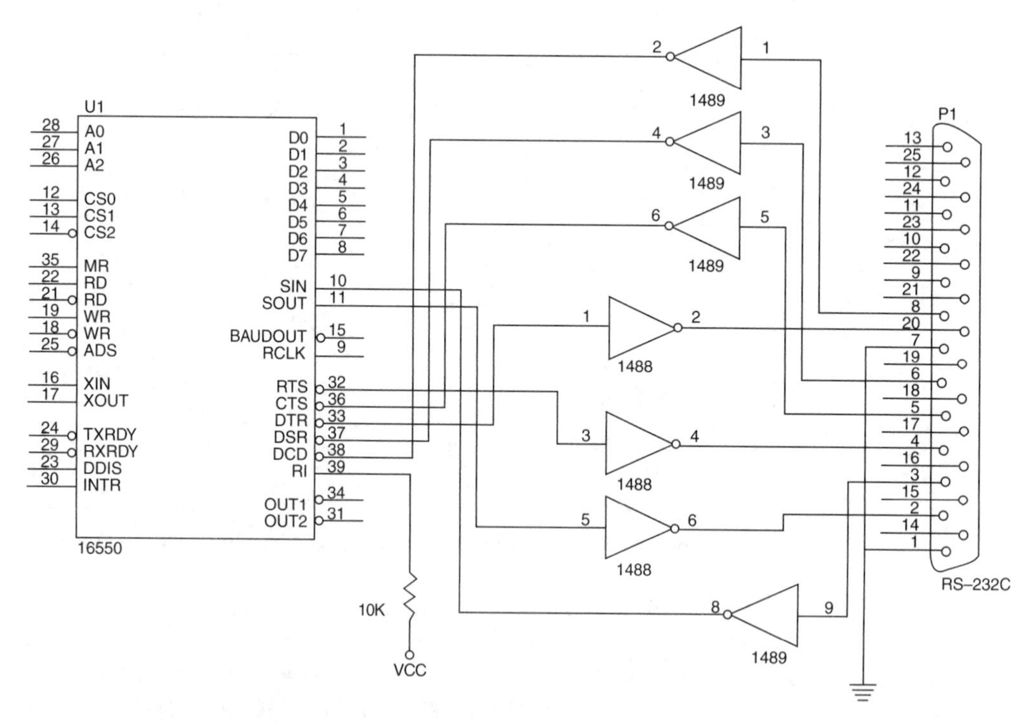

FIGURE 12–25 The 16550 interfaced to an RS-2332C using 1488 line drivers and 1489 line receivers.

In order to transmit or receive data through the modem, the $\overline{\text{DTR}}$ pin is activated (logic 0) and the UART then waits for the $\overline{\text{DSR}}$ pin to become a logic 0 from the modem, indicating that the modem is ready. Once this handshake is complete, the UART sends the modem a logic 0 on the $\overline{\text{RTS}}$ pin. When the modem is ready, it returns the $\overline{\text{CTS}}$ signal (logic 0) to the UART. Communications can now commence. The $\overline{\text{DCD}}$ signal from the modem is an indication that the modem has detected a carrier. This signal must also be tested before communications can begin.

12–5 INTERRUPT EXAMPLES

This section of the text presents a real-time clock and an interrupt-processed keyboard as examples of interrupt applications. A real-time (RTC) clock keeps time in real time—that is, in hours and minutes. It is also used for precision time delays. The example illustrated here keeps time in hours, minutes, seconds, and l/60 second, using four memory locations to hold the BCD time of day. The interrupt-processed keyboard uses a periodic interrupt to scan through the keys of the keyboard.

Real-Time Clock

Figure 12–26 illustrates a simple circuit that uses the 60 Hz AC power line to generate a periodic interrupt request signal for the NMI interrupt input pin. Although we are using a signal from the AC power line, which varies slightly in frequency from time to time, it is accurate over a period of time as mandated by the Federal Trade Commission (FTC).

The circuit uses a signal from the 120 V AC power line that is conditioned by a Schmitt trigger inverter before it is applied to the NMI interrupt input. Note that you must make certain that the power line ground is connected to the system ground in this schematic. The power line neutral (white wire) connection is the wide flat pin on the power line. The narrow flat pin is the hot (black wire) side or 120 V AC side of the line.

The software for the real-time clock contains an interrupt service procedure that is called 60 times per second and a procedure that updates the count located in four memory locations. Example 12–14 lists both procedures, along with the four bytes of memory used to hold the BCD time of day. The memory locations for the TIME are stored somewhere in the system memory at the segment address (SEGMENT) and at the offset address TIME, which is first loaded in the TIMEP procedure. The lookup table (LOOK) for the modulus or each counter is stored in the code segment with the procedure.

EXAMPLE 12–14

```
TIME    DB      ?              ;1/60 sec counter (÷60)
        DB      ?              ;second counter (÷60)
        DB      ?              ;minute counter (÷60)
        DB      ?              ;hour counter (÷24)

LOOK    DB      60H, 60H, 60H, 24H

TIMEP   PROC    FAR USES AX BX DS

        MOV  AX,SEGMENT               ;load segment address of TIME
        MOV  DS,AX
        MOV  BX,0                     ;initialize pointer
```

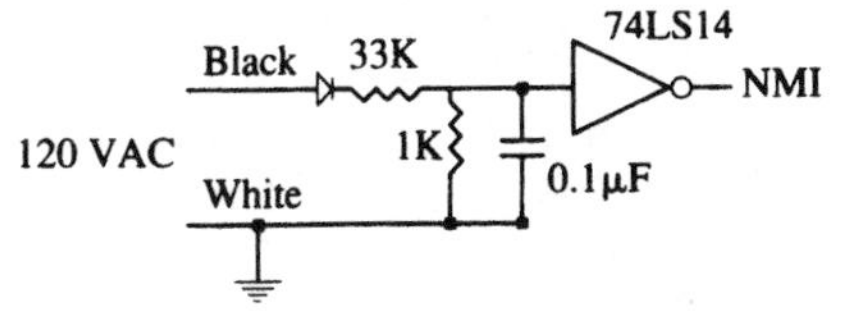

FIGURE 12–26 Converting the AC power line to a 60 Hz TTL signal for the NMI input.

```
        .REPEAT                         ;crank clock
              MOV  AL,DS:TIME[BX]
              ADD  AL,1                 ;increment count
              DAA                       ;adjust for BCD
              .IF AL == BYTE PTR CS:LOOK[BX]
                     MOV AL,0
              .ENDIF
               MOV  DS:TIME[BX],AL
               INC  BX
        .UNTIL AL != 0 || BX == 4
        IRET

TIMEP  ENDP
```

Another way to handle time is to use a single counter to store the time in memory and then determine the actual time with software. For example, the time can be stored in one single 32-bit counter (there are 5,184,000 1/60 sec in a day). In a counter such as this a count of 0 is 12:00:00:00 AM and a count of 5,183,999 is 11:59:59:59 PM. Example 12–15 shows the interrupt procedure for this type of RTC. This type of RTC requires the least of amount of time to execute.

EXAMPLE 12–15

```
TIME    DD     ?                       ;modulus 5,184,000 counter

TIMEP   PROC    FAR USES EAX

        MOV  AX,SEGMENT
        MOV  DS,AX

        INC  DS:TIME
        .IF DS:TIME == 5184000
               MOV  DWORD PTR DS:TIME,0
        .ENDIF
        .IRET

TIMEP  ENDP
```

Software to convert the count in the modulus 5,184,000 counter into hours, minutes, and seconds appears in Example 12–16. The procedure returns with the number of hours (0–23) in BL, number of minutes in BH, and number of seconds in AL. No attempt was made to retrieve the 1/60 second count.

EXAMPLE 12–16

```
;Time is returned as BL = hours, BH = minutes and AL = seconds

GETT  PROC   NEAR ECX EDX

        MOV  ECX,216000      ;divide by 216,000
        MOV  EAX,TIME
        SUB  EDX,EDX         ;clear EDX
        DIV  ECX             ;get hours
        MOV  BL,AL
        MOV  EAX,EDX
        MOV  ECX,3600        ;divide by 3600
        DIV  ECX             ;get minutes
        MOV  BH,AL
        SUB  EAX,EDX
        MOV  ECX,60          ;divide by 60
        DIV  ECX
        RET

GETT   ENDP
```

Suppose a time delay is needed. Time delays can be achieved using the RTC in Example 12–15 for any amount from 1/60 of a second to 24 hours. Example 12–17 shows a procedure that uses the RTC to perform time delays of the number of seconds passed to the procedure in the EAX register. This can be 1 second to an entire day's worth of seconds. It has an accuracy to within 1/60 second, the resolution of the RTC.

EXAMPLE 12–17

```
SEC     PROC     NEAR USES EAX EDX

        MOV    EDX,60
        MUL    EDX              ;get seconds as 1/60s count
        ADD    EAX,TIME         ;advance the TIME in EAX
        .IF    EAX >= 51840000
                 SUB EAX,5184000
        .ENDIF
        .REPEAT                 ;wait for TIME to catch up
        .UNTIL   EAX == TIME
        RET

SEC     ENDP
```

Interrupt-Processed Keyboard

The interrupt-processed keyboard scans through the keys on a keyboard through a periodic interrupt. Each time the interrupt occurs, the interrupt service procedure tests for a key or debounces the key. Once a valid key is detected, the interrupt service procedure stores the key code into a keyboard queue for later reading by the system. The basis for this system is a periodic interrupt that can be caused by a timer, RTC, or other device in the system. Note that most systems already have a periodic interrupt for the real-time clock. In this example, we assume the interrupt calls the interrupt service procedure every 10 ms or, if the RTC is used with a 60 Hz clock, every 16.7 ms.

Figure 12–27 shows the keyboard interfaced to an 82C55. It does not show the timer or other circuitry required to call the interrupt once in every 10 ms or 16.7 ms. (Not shown in the software is programming of the 82C55.) The 82C55 must be programmed so that port A is an input port, port B is an output port, and the initialization software must store 00H at port B. This interface uses memory that is stored in the code segment for a queue and a few bytes that keep track of the keyboard scanning. Example 12–18 lists the interrupt service procedure for the keyboard.

EXAMPLE 12–18

```
;Interrupt procedure for the keyboard of Figure 12-27

PORTA  EQU    1000H
PORTB  EQU    1001H

DBCNT  DB     0                      ;de-bounce counter
DBF    DB     0                      ;de-bounce flag
PNTR   DW     QUEUE                  ;input pointer to queue
OPNTR  DW     QUEUE                  ;output pointer to queue
QUEUE  DB     16 DUP(?)              ;16 byte queue

INTK   PROC   FAR USES AX BX DX

       MOV  DX,PORTA                 ;test for a key
       IN   AL,DX
       OR   AL,0F0H
       .IF AL != 0FFH                ;if key down
              INC  DBCNT             ;increment de-bounce count
              .IF DBCNT == 3         ;if key down for > 20 ms
                     DEC  DBCNT
```

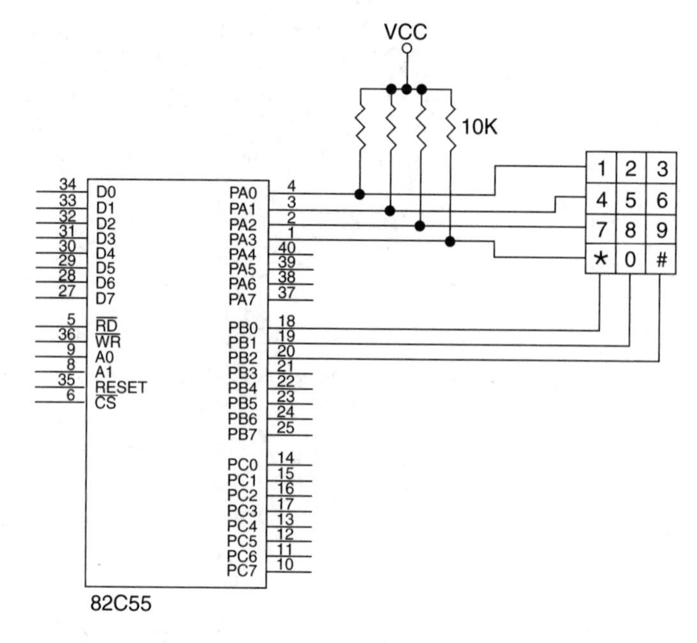

FIGURE 12–27 A telephone-style keypad interfaced to the 82C55.

```
                        .IF  DBF == 0
                             MOV  DBF,1
                             MOV  BX,00FEH
                             .WHILE 1        ;find key
                                    MOV  AL,BL
                                    MOV  DX,PORTB
                                    OUT  DX,AL
                                    ROL  BL,1
                                    MOV  DX,PORTA
                                    IN   AL,DX
                                    OR   AL,0F0H
                                    .BREAK .IF AL != 0
                                    ADD  BH,4
                             .ENDW
                             MOV  BL,AL
                             MOV  AL,0
                             MOV  DX,PORTB
                             OUT  DX,AL
                             DEC  BH
                             .REPEAT
                                    SHR  BL,1
                                    INC  BH
                             .UNTIL !CARRY?
                             MOV   AL,BH
                             MOV   BX,PNTR
                             MOV   [BX],AL        ;key code to queue
                             INC    BX
                             .IF BX == OFFSET QUEUE+16
                                    MOV  DX,OFFSET QUEUE
                             .ENDIF
                             MOV  PNTR,BX
                        .ENDIF
                  .ENDIF
          .ELSE                          ;if no key down
                  DEC  DBCNT             ;decrement de-bounce count
                  .IF SIGN?              ;if below zero
                        MOV  DBCNT,0
                        MOV  DBF,0
                  .ENDIF
          .ENDIF
          IRET

INTK      ENDP
```

The keyboard interrupt finds the key and stores the key code in the queue. The code stored in the queue is a raw code that does not indicate the key number. For example, the key code for the 1-key is 00H, the key code for the 4-key is 01H, and so on. There is no provision for a queue overflow in this software. It could be added, but in almost all cases it is difficult to out-type a 16-byte queue.

Example 12–19 illustrates a procedure that removes data from the keyboard queue. This procedure is not interrupt-driven and is called only when information from the keyboard is needed in a program. Example 12–20 shows the caller software for the key procedure.

EXAMPLE 12–19

```
LOOK    DB      1,4,7,10              ;lookup table
        DB      2,5,8,0
        DB      3,6,9,11

KEY     PROC    NEAR USES BX

        MOV  BX,OPNTR
        .IF BX == PNTR                ;if queue empty
               STC
        .ELSE
               MOV AL,[BX]            ;get queue data
               INC BX
               .IF BX == OFFSET QUEUE+16
                      MOV BX,OFFSET QUEUE
               .ENDIF
               MOV OPNTR,BX
               MOV BX,LOOK
               XLAT
               CLC
        .ENDIF
        RET

KEY     ENDP
```

EXAMPLE 12–20

```
.REPEAT
       CALL KEY
.UNTIL !CARRY?
```

12–6 SUMMARY

1. An interrupt is a hardware- or software-initiated call that interrupts the currently executing program at any point and calls a procedure. The procedure is called by the interrupt handler or an interrupt service procedure.
2. Interrupts are useful when an I/O device needs to be serviced only occasionally at low data transfer rates.
3. The microprocessor has five instructions that apply to interrupts: BOUND, INT, INT 3, INTO, and IRET. The INT and INT 3 instructions call procedures with addresses stored in the interrupt vector whose type is indicated by the instruction. The BOUND instruction is a conditional interrupt that uses interrupt vector type number 5. The INTO instruction is a conditional interrupt that interrupts a program only if the overflow flag is set. Finally, the IRET instruction is used to return from interrupt service procedures.
4. The microprocessor has three pins that apply to its hardware interrupt structure: INTR, NMI, and $\overline{\text{INTA}}$. The interrupt inputs are INTR and NMI, which are used to request interrupts, $\overline{\text{INTA}}$, an output used to acknowledge the INTR interrupt request.
5. Real mode interrupts are referenced through a vector table that occupies memory locations 0000H–03FFH. Each interrupt vector is four bytes long and contains the offset and segment

addresses of the interrupt service procedure. In protected mode, the interrupts reference the interrupt descriptor table (IDT) that contains 256 interrupt descriptors. Each interrupt descriptor contains a segment selector and a 32-bit offset address.
6. Two flag bits are used with the interrupt structure of the microprocessor: trap (TF) and interrupt enable (IF). The IF flag bit enables the INTR interrupt input, and the TF flag bit causes interrupts to occur after the execution of each instruction, as long as TF is active.
7. The first 32 interrupt vector locations are reserved for Intel's use, with many predefined in the microprocessor. The last 224 interrupt vectors are for the user's use and can perform any function desired.
8. Whenever an interrupt is detected, the following events occur: (1) the flags are pushed onto the stack, (2) the IF and TF flag bits are both cleared, (3) the IP and CS registers are both pushed onto the stack, and (4) the interrupt vector is fetched from the interrupt vector table and the interrupt service subroutine is accessed through the vector address.
9. Tracing or single-stepping is accomplished by setting the TF flag bit. This causes an interrupt to occur after the execution of each instruction for debugging.
10. The non-maskable interrupt input (NMI) calls the procedure whose address is stored at interrupt vector type number 2. This input is positive edge-triggered.
11. The INTR pin is not internally decoded, as is the NMI pin. Instead, $\overline{INTA}$ is used to apply the interrupt vector type number to data bus connections D_0–D_7 during the $\overline{INTA}$ pulse.
12. Methods of applying the interrupt vector type number to the data bus during $\overline{INTA}$ vary widely. One method uses resistors to apply interrupt type number FFH to the data bus, while another uses a three-state buffer to apply any vector type number.
13. The 8259A programmable interrupt controller (PIC) adds at least eight interrupt inputs to the microprocessor. If more interrupts are needed, this device can be cascaded to provide up to 64 interrupt inputs.
14. Programming the 8259A is a two-step process. First, a series of initialization command words (ICWs) is sent to the 8259A; then a series of operation command words (OCWs) is sent.
15. The 8259A contains three status registers: IMR (interrupt mask register), ISR (in-service register), and IRR (interrupt request register).
16. A real-time clock is used to keep time in real time. In most cases, time is stored in either binary or BCD form in several memory locations.

12–7 QUESTIONS AND PROBLEMS

1. What is interrupted by an interrupt?
2. Define the term *interrupt.*
3. What is called by an interrupt?
4. Why do interrupts free time for the microprocessor?
5. List the interrupt pins found on the microprocessor.
6. List the five interrupt instructions for the microprocessor.
7. What is an interrupt vector?
8. Where are the interrupt vectors located in the microprocessor's memory?
9. How many different interrupt vectors are found in the interrupt vector table?
10. Which interrupt vectors are reserved by Intel?
11. Explain how a type 0 interrupt occurs.
12. Where is the interrupt descriptor table located for protected mode operation?
13. Each protected mode interrupt descriptor contains what information?
14. Describe the differences between a protected and real mode interrupt.
15. Describe the operation of the BOUND instruction.

16. Describe the operation of the INTO instruction.
17. What memory locations contain the vector for an INT 44H instruction?
18. Explain the operation of the IRET instruction.
19. What is the purpose of interrupt vector type number 7?
20. List the events that occur when an interrupt becomes active.
21. Explain the purpose of the interrupt flag (IF).
22. Explain the purpose of the trap flag (TF).
23. How is IF cleared and set?
24. How is TF cleared and set?
25. The NMI interrupt input automatically vectors through which vector type number?
26. Does the $\overline{\text{INTA}}$ signal activate for the NMI pin?
27. The INTR input is ____________-sensitive.
28. The NMI input is ____________-sensitive.
29. When the $\overline{\text{INTA}}$ signal becomes a logic 0, it indicates that the microprocessor is waiting for an interrupt ____________ number to be placed on the data bus (D_0–D_7).
30. What is a FIFO?
31. Develop a circuit that places interrupt type number 86H on the data bus in response to the INTR input.
32. Develop a circuit that places interrupt type number CCH on the data bus in response to the INTR input.
33. Explain why pull-up resistors on D_0–D_7 cause the microprocessor to respond with interrupt vector type number FFH for the $\overline{\text{INTA}}$ pulse.
34. What is a daisy-chain?
35. Why must interrupting devices be polled in a daisy-chained interrupt system?
36. What is the 8259A?
37. How many 8259As are required to have 64 interrupt inputs?
38. What is the purpose of the IR_0–IR_7 pins on the 8259A?
39. When are the CAS_2–CAS_0 pins used on the 8259A?
40. Where is a slave INT pin connected on the master 8259A in a cascaded system?
41. What is an ICW?
42. What is an OCW?
43. How many ICWs are needed to program the 8259A when operated as a single master in a system?
44. Where is the vector type number stored in the 8259A?
45. Where is the sensitivity of the IR pins programmed in the 8259A?
46. What is the purpose of ICW_1?
47. What is a nonspecific EOI?
48. Explain priority rotation in the 8259A.
49. What is the purpose of IRR in the 8259A?
50. At which interrupt vectors is the master 8259A found in the personal computer?
51. At which interrupt vectors is the slave 8259A found in the personal computer?

CHAPTER 13

Direct Memory Access and DMA-Controlled I/O

INTRODUCTION

In previous chapters, we discussed basic and interrupt-processed I/O. Now we turn to the final form of I/O called direct memory access (DMA). The DMA I/O technique provides direct access to the memory while the microprocessor is temporarily disabled. This allows data to be transferred between memory and the I/O device at a rate that is limited only by the speed of the memory components in the system or the DMA controller. The DMA transfer speed can approach 33–150M-byte transfer rates with today's high-speed RAM memory components.

DMA transfers are used for many purposes, but more common are DRAM refresh, video displays for refreshing the screen, and disk memory system reads and writes. The DMA transfer is also used to do high-speed memory-to-memory transfers.

This chapter also explains the operation of disk memory systems and video systems that are often DMA-processed. Disk memory includes floppy, fixed, and optical disk storage. Video systems include digital and analog monitors.

CHAPTER OBJECTIVES

Upon completion of this chapter, you will be able to:

1. Describe a DMA transfer
2. Explain the operation of the HOLD and HLDA direct memory access control signals
3. Explain the function of the 8237 DMA controller when used for DMA transfers
4. Program the 8237 to accomplish DMA transfers
5. Describe the disk standards found in personal computer systems
6. Describe the various video interface standards that are found in the personal computer

13–1 BASIC DMA OPERATION

Two control signals are used to request and acknowledge a direct memory access (DMA) transfer in the microprocessor-based system. The HOLD pin is an input that is used to request a DMA action and the HLDA pin is an output that acknowledges the DMA action. Figure 13–1 shows the timing that is typically found on these two DMA control pins.

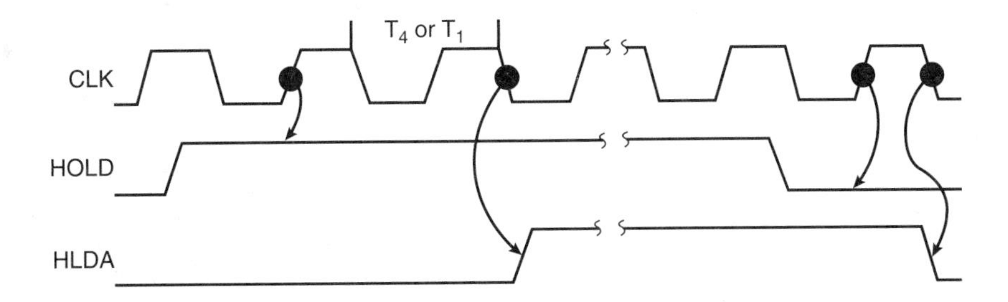

FIGURE 13–1 HOLD and HLDA timing for the microprocessor.

Whenever the HOLD input is placed at a logic 1 level, a DMA action (hold) is requested. The microprocessor responds, within a few clocks, by suspending the execution of the program and by placing its address, data, and control buses at their high-impedance states. The high-impedance state causes the microprocessor to appear as if it has been removed from its socket. This state allows external I/O devices or other microprocessors to gain access to the system buses so that memory can be accessed directly.

As the timing diagram indicates, HOLD is sampled in the middle of any clocking cycle. Thus, the hold can take effect any time during the operation of any instruction in the microprocessor's instruction set. As soon as the microprocessor recognizes the hold, it stops executing software and enters hold cycles. Note that the HOLD input has a higher priority than the INTR or NMI interrupt inputs. Interrupts take effect at the end of an instruction, whereas a HOLD takes effect in the middle of an instruction. The only microprocessor pin that has a higher priority than a HOLD is the RESET pin. Note that the HOLD input may not be active during a RESET or the reset is not guaranteed.

The HLDA signal becomes active to indicate that the microprocessor has indeed placed its buses at their high-impedance state, as can be seen in the timing diagram. Note that there are a few clock cycles between the time that HOLD changes and until HLDA changes. The HLDA output is a signal to the external requesting device that the microprocessor has relinquished control of its memory and I/O space. You could call the HOLD input a DMA request input and the HLDA output a DMA grant signal.

Basic DMA Definitions

Direct memory accesses normally occur between an I/O device and memory without the use of the microprocessor. A **DMA read** transfers data from the memory to the I/O device. A **DMA write** transfers data from an I/O device to memory. In both operations, the memory and I/O are controlled simultaneously, which is why the system contains separate memory and I/O control signals. This special control bus structure of the microprocessor allows DMA transfers. A DMA read causes both the $\overline{\text{MRDC}}$ and $\overline{\text{IOWC}}$ signals to activate simultaneously, transferring data from the memory to the I/O device. A DMA write causes both the $\overline{\text{MWTC}}$ and $\overline{\text{IORC}}$ signals to activate. These control bus signals are available to all microprocessors in the Intel family except the 8086/8088 system. The 8086/8088 require their generation with either a system controller or a circuit such as the one illustrated in Figure 13–2. The DMA controller provides the memory with its address and a signal from the controller ($\overline{\text{DACK}}$) selects the I/O device during the DMA transfer.

The data transfer speed is determined by the speed of the memory device or a DMA controller that often controls DMA transfers. If the memory speed is 50 ns, DMA transfers occur at rates of up to 1/50 ns or 20M bytes per second. If the DMA controller in a system functions at a maximum rate of 15 MHz and we still use 50 ns memory, the maximum transfer rate is 15 MHz because the DMA controller is slower than the memory. In many cases, the DMA controller slows the speed of the system when DMA transfers occur.

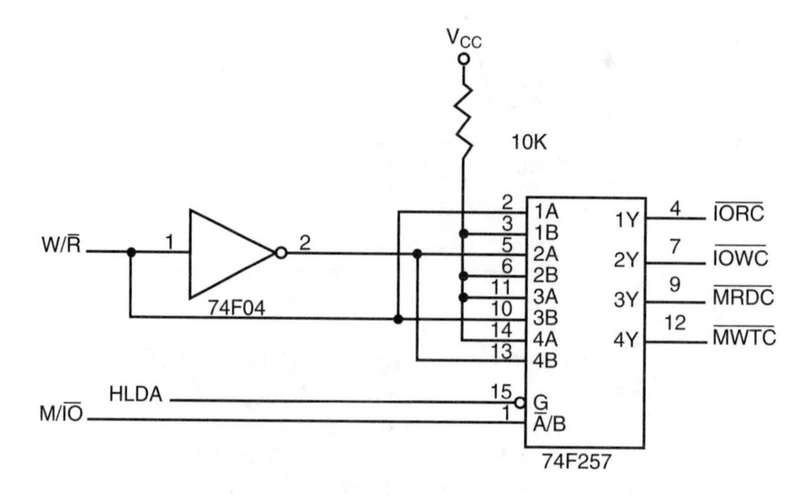

FIGURE 13–2 A circuit that generates system control signals in a DMA environment.

13–2 THE 8237 DMA CONTROLLER

The 8237 DMA controller supplies the memory and I/O with control signals and memory address information during the DMA transfer. The 8237 is actually a special-purpose microprocessor whose job is high-speed data transfer between memory and the I/O. Figure 13–3 shows the pin-out and block diagram of the 8237 programmable DMA controller. Although this device may not appear as a discrete component in modern microprocessor-based systems, it does appear within system controller chip sets found in most systems. Although not described because of its complexity, the chip set (82875P ISP or integrated system peripheral controller) and its integral set of two DMA controllers are programmed almost exactly (it does not support memory-to-memory transfers) like the 8237. The ISP also provides a pair of 8259A programmable interrupt controllers for the system.

The 8237 is a four-channel device that is compatible with the 8086/8088 microprocessors. The 8237 can be expanded to include any number of DMA channel inputs, although four channels seem to be adequate for many small systems. The 8237 is capable of DMA transfers at rates of up to 1.6M bytes per second. Each channel is capable of addressing a full 64K-byte section of memory and can transfer up to 64K bytes with a single programming.

Pin Definitions

CLK The **clock** input is connected to the system clock signal as long as that signal is 5 MHz or less. In the 8086/8088 system, the clock must be inverted for the proper operation of the 8237.

$\overline{CS}$ **Chip select** enables the 8237 for programming. The $\overline{CS}$ pin is normally connected to the output of a decoder. The decoder does not use the 8086/8088 control signal IO/$\overline{M}$(M/$\overline{IO}$) because it contains the new memory and I/O control signals ($\overline{MEMR}$, $\overline{MEMW}$, $\overline{IOR}$, and $\overline{IOW}$).

RESET The **reset** pin clears the command, status, request, and temporary registers. It also clears the first/last flip-flop and sets the mask register. This input primes the 8237 so it is disabled until programmed otherwise.

READY A logic 0 on the **ready** input causes the 8237 to enter wait states for slower memory components.

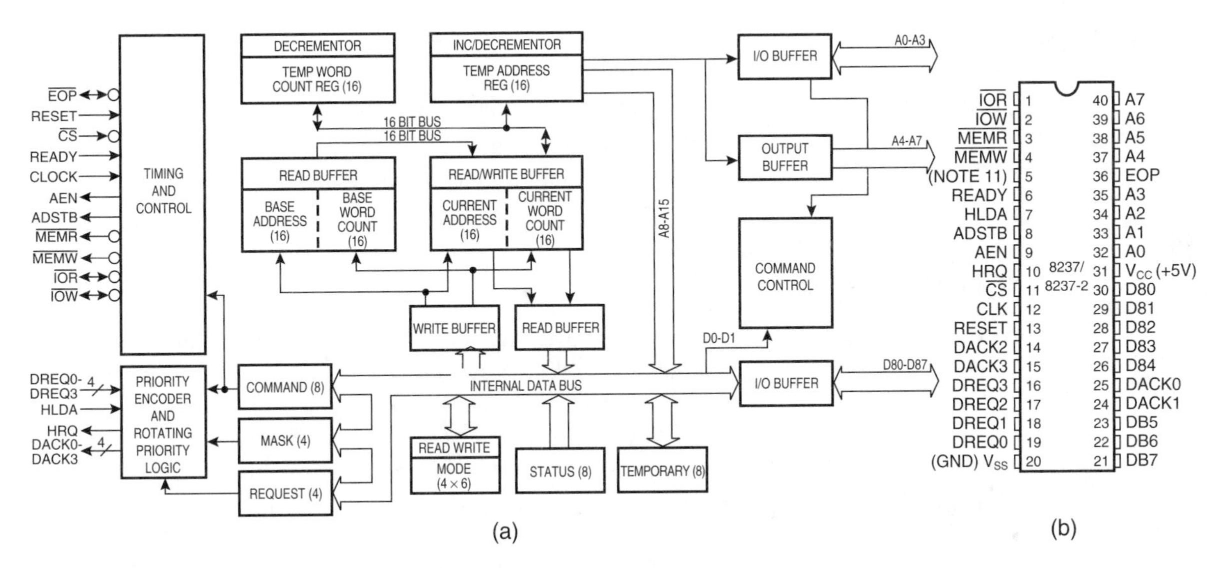

FIGURE 13–3 The 8237A-5 programmable DMA controller. (a) Block diagram and (b) pin-out. (Courtesy of Intel Corporation.)

HLDA	A **hold acknowledge** signals the 8237 that the microprocessor has relinquished control of the address, data, and control buses.
$DREQ_0$–$DREQ_3$	The **DMA request inputs** are used to request a DMA transfer for each of the four DMA channels. Because the polarity of these inputs is programmable, they are either active-high or active-low inputs.
DB_0–DB_7	The **data bus** pins are connected to the microprocessor data bus connections and are used during the programming of the DMA controller.
$\overline{IOR}$	**I/O read** is a bidirectional pin used during programming and during a DMA write cycle.
$\overline{IOW}$	**I/O write** is a bidirectional pin used during programming and during a DMA read cycle.
$\overline{EOP}$	**End-of-process** is a bidirectional signal that is used as an input to terminate a DMA process or as an output to signal the end of the DMA transfer. This input is often used to interrupt a DMA transfer at the end of a DMA cycle.
A_0–A_3	These **address pins** are outputs that select an internal register during programming and also provide part of the DMA transfer address during a DMA action.
HRQ	**Hold request** is an output that connects to the HOLD input of the microprocessor in order to request a DMA transfer.
$DACK_0$–$DACK_3$	**DMA channel acknowledge** outputs acknowledge a channel DMA request. These outputs are programmable as either active-high or active-low signals. The DACK outputs are often used to select the DMA-controlled I/O device during the DMA transfer.
AEN	The **address enable** signal enables the DMA address latch connected to the DB_7–DB_0 pins on the 8237. It is also used to disable any buffers in the system connected to the microprocessor.

ADSTB — **Address strobe** functions as ALE, except that it is used by the DMA controller to latch address bits A_{15}–A_8 during the DMA transfer.

$\overline{\text{MEMR}}$ — **Memory read** is an output that causes memory to read data during a DMA read cycle.

$\overline{\text{MEMW}}$ — **Memory write** is an output that causes memory to write data during a DMA write cycle.

Internal Registers

CAR — The **current address register** is used to hold the 16-bit memory address used for the DMA transfer. Each channel has its own current address register for this purpose. When a byte of data is transferred during a DMA operation, the CAR is either incremented or decremented, depending on how it is programmed.

CWCR — The **current word count register** programs a channel for the number of bytes (up to 64K) transferred during a DMA action. The number loaded into this register is one less than the number of bytes transferred. For example, if 10 is loaded into the CWCR, then 11 bytes are transferred during the DMA action.

BA and BWC — The **base address** (BA) and **base word count** (BWC) registers are used when auto-initialization is selected for a channel. In the auto-initialization mode, these registers are used to reload both the CAR and CWCR after the DMA action is completed. This allows the same count and address to be used to transfer data from the same memory area.

CR — The **command register** programs the operation of the 8237 DMA controller. Figure 13–4 depicts the function of the command register. The command register uses bit position 0 to select the memory-to-memory DMA transfer mode. Memory-to-memory DMA transfers use DMA channel 0 to hold the source address and DMA channel 1 to hold the

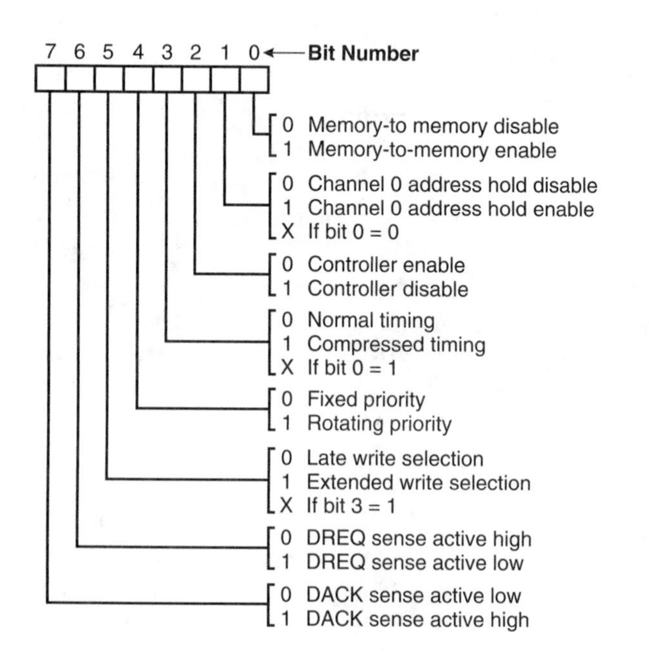

FIGURE 13–4 8237A-5 command register. (Courtesy of Intel Corporation.)

destination address. (This is similar to the operation of a MOVSB instruction.) A byte is read from the address accessed by channel 0 and saved within the 8237 in a temporary holding register. Next, the 8237 initiates a memory write cycle in which the contents of the temporary holding register are written into the address selected by DMA channel 1. The number of bytes transferred is determined by the channel 1 count register.

The channel 0 address hold enable bit (bit position 1) programs channel 0 for memory-to-memory transfers. For example, if you must fill an area of memory with data, channel 0 can be held at the same address while channel 1 changes for memory-to-memory transfer. This copies the contents of the address accessed by channel 0 into a block of memory accessed by channel 1.

The controller enable/disable bit (bit position 2) turns the entire controller on and off. The normal and compressed bit (bit position 3) determines whether a DMA cycle contains two (compressed) or four (normal) clocking periods. Bit position 5 is used in normal timing to extend the write pulse so it appears one clock earlier in the timing for I/O devices that require a wider write pulse.

Bit position 4 selects priority for the four DMA channel DREQ inputs. In the fixed priority scheme, channel 0 has the highest priority and channel 3 has the lowest. In the rotating priority scheme, the most recently serviced channel assumes the lowest priority. For example, if channel 2 just had access to a DMA transfer, it assumes the lowest priority and channel 3 assumes the highest priority position. Rotating priority is an attempt to give all channels equal priority.

The remaining two bits (bit positions 6 and 7) program the polarities of the DREQ inputs and the DACK outputs.

MR

The **mode register** programs the mode of operation for a channel. Note that each channel has its own mode register (see Figure 13–5), as selected by bit positions 1 and 0. The remaining bits of the mode register select the operation, auto-initialization, increment/decrement,

FIGURE 13–5 8237A-5 mode register. (Courtesy of Intel Corporation.)

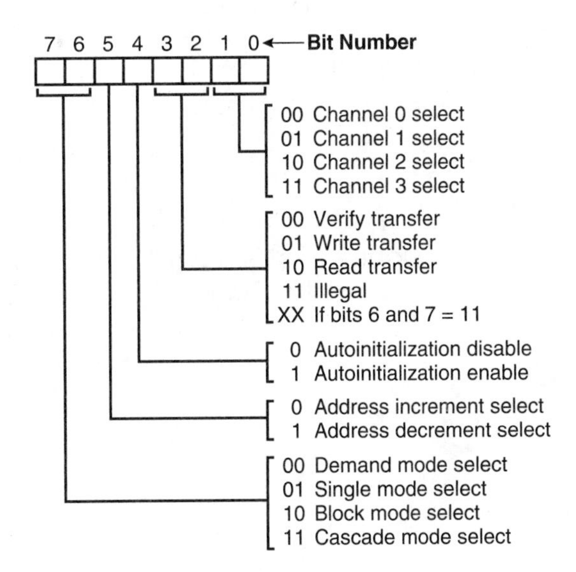

and mode for the channel. Verification operations generate the DMA addresses without generating the DMA memory and I/O control signals.

The modes of operation include demand mode, single mode, block mode, and cascade mode. Demand mode transfers data until an external EOP is input or until the DREQ input becomes inactive. Single mode releases the HOLD after each byte of data is transferred. If the DREQ pin is held active, the 8237 again requests a DMA transfer through the DRQ line to the microprocessor's HOLD input. Block mode automatically transfers the number of bytes indicated by the count register for the channel. DREQ need not be held active through the block mode transfer. Cascade mode is used when more than one 8237 is present in a system.

BR The **bus request register** is used to request a DMA transfer via software (see Figure 13–6). This is very useful in memory-to-memory transfers, where an external signal is not available to begin the DMA transfer.

MRSR The **mask register set/reset** sets or clears the channel mask, as illustrated in Figure 13–7. If the mask is set, the channel is disabled. Recall that the RESET signal sets all channel masks to disable them.

MSR The **mask register** (see Figure 13–8) clears or sets all of the masks with one command instead of individual channels, as with the MRSR.

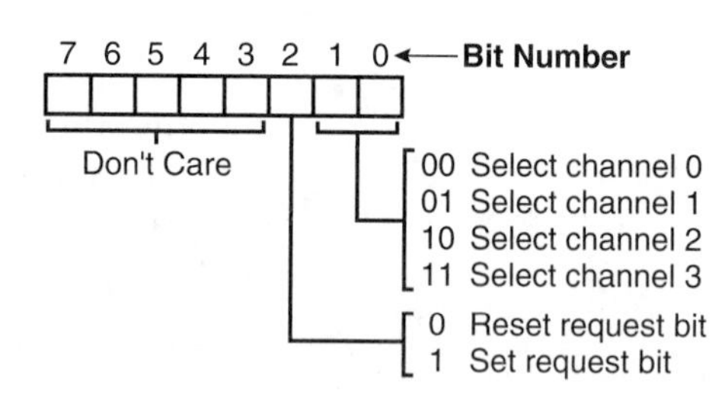

FIGURE 13–6 8237A-5 request register. (Courtesy of Intel Corporation.)

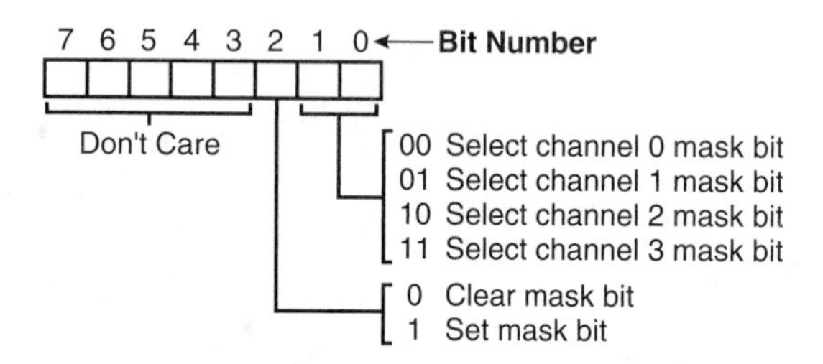

FIGURE 13–7 8237A-5 mask register set/reset mode. (Courtesy of Intel Corporation.)

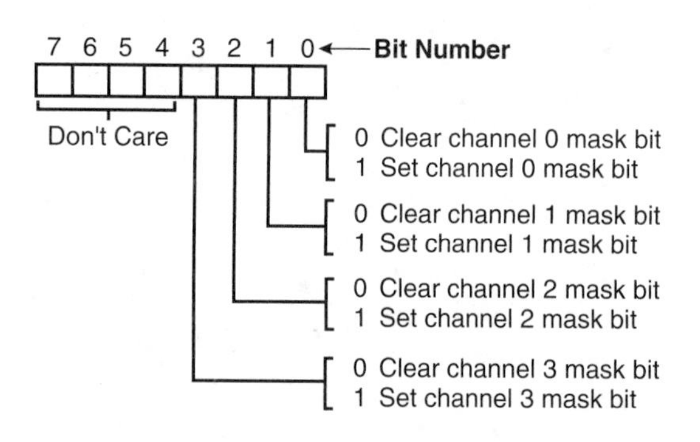

FIGURE 13–8 8237A-5 mask register. (Courtesy of Intel Corporation.)

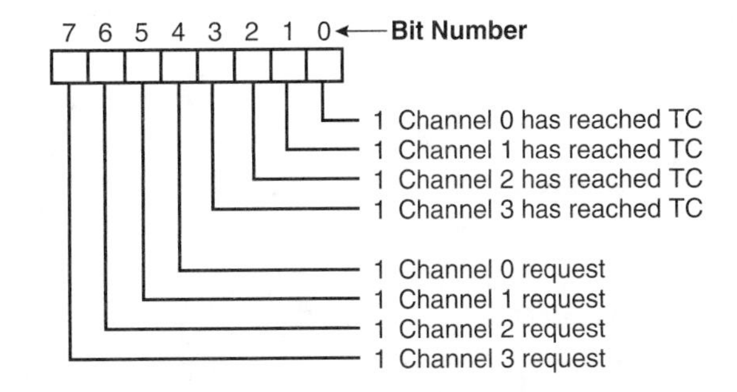

FIGURE 13–9 8237A-5 status register. (Courtesy of Intel Corporation.)

SR The **status register** shows the status of each DMA channel (see Figure 13–9). The TC bits indicate whether the channel has reached its terminal count (transferred all its bytes). Whenever the terminal count is reached, the DMA transfer is terminated for most modes of operation. The request bits indicate whether the DREQ input for a given channel is active.

Software Commands

Three software commands are used to control the operation of the 8237. These commands do not have a binary bit pattern, as do the various control registers within the 8237. A simple output to the correct port number enables the software command. Figure 13–10 shows the I/O port assignments that access all registers and the software commands.

The functions of the software commands are explained in the following list:

1. **Clear the first/last flip-flop**—Clears the first/last (F/L) flip-flop within the 8237. The F/L flip-flop selects which byte (low or high order) is read/written in the current address and current count registers. If F/L = 0, the low-order byte is selected; if F/L = 1, the high-order byte is selected. Any read or write to the address or count register automatically toggles the F/L flip-flop.
2. **Master clear**—Acts exactly the same as the RESET signal to the 8237. As with the RESET signal, this command disables all channels.
3. **Clear mask register**—Enables all four DMA channels.

FIGURE 13–10 8237A-5 command and control port assignments. (Courtesy of Intel Corporation.)

Signals						Operation
A3	A2	A1	A0	$\overline{IOR}$	$\overline{IOW}$	
1	0	0	0	0	1	Read Status Register
1	0	0	0	1	0	Write Command Register
1	0	0	1	0	1	Illegal
1	0	0	1	1	0	Write Request Register
1	0	1	0	0	1	Illegal
1	0	1	0	1	0	Write Single Mask Register Bit
1	0	1	1	0	1	Illegal
1	0	1	1	1	0	Write Mode Register
1	1	0	0	0	1	Illegal
1	1	0	0	1	0	Clear Byte Pointer Flip/Flop
1	1	0	1	0	1	Read Temporary Register
1	1	0	1	1	0	Master Clear
1	1	1	0	0	1	Illegal
1	1	1	0	1	0	Clear Mask Register
1	1	1	1	0	1	Illegal
1	1	1	1	1	0	Write All Mask Register Bits

Channel	Register	Operation	Signals							Internal Flip-Flop	Data Bus DB0-DB7
			$\overline{CS}$	$\overline{IOR}$	$\overline{IOW}$	A3	A2	A1	A0		
0	Base and Current Address	Write	0	1	0	0	0	0	0	0	A0-A7
			0	1	0	0	0	0	0	1	A8-A15
	Current Address	Read	0	0	1	0	0	0	0	0	A0-A7
			0	0	1	0	0	0	0	1	A8-A15
	Base and Current Word Count	Write	0	1	0	0	0	0	1	0	W0-W7
			0	1	0	0	0	0	1	1	W8-W15
	Current Word Count	Read	0	0	1	0	0	0	1	0	W0-W7
			0	0	1	0	0	0	1	1	W8-W15
1	Base and Current Address	Write	0	1	0	0	0	1	0	0	A0-A7
			0	1	0	0	0	1	0	1	A8-A15
	Current Address	Read	0	0	1	0	0	1	0	0	A0-A7
			0	0	1	0	0	1	0	1	A8-A15
	Base and Current Word Count	Write	0	1	0	0	0	1	1	0	W0-W7
			0	1	0	0	0	1	1	1	W8-W15
	Current Word Count	Read	0	0	1	0	0	1	1	0	W0-W7
			0	0	1	0	0	1	1	1	W8-W15
2	Base and Current Address	Write	0	1	0	0	1	0	0	0	A0-A7
			0	1	0	0	1	0	0	1	A8-A15
	Current Address	Read	0	0	1	0	1	0	0	0	A0-A7
			0	0	1	0	1	0	0	1	A8-A15
	Base and Current Word Count	Write	0	1	0	0	1	0	1	0	W0-W7
			0	1	0	0	1	0	1	1	W8-W15
	Current Word Count	Read	0	0	1	0	1	0	1	0	W0-W7
			0	0	1	0	1	0	1	1	W8-W15
3	Base and Current Address	Write	0	1	0	0	1	1	0	0	A0-A7
			0	1	0	0	1	1	0	1	A8-A15
	Current Address	Read	0	0	1	0	1	1	0	0	A0-A7
			0	0	1	0	1	1	0	1	A8-A15
	Base and Current Word Count	Write	0	1	0	0	1	1	1	0	W0-W7
			0	1	0	0	1	1	1	1	W8-W15
	Current Word Count	Read	0	0	1	0	1	1	1	0	W0-W7
			0	0	1	0	1	1	1	1	W8-W15

FIGURE 13–11 8237A-5 DMA channel I/O port addresses. (Courtesy of Intel Corporation.)

Programming the Address and Count Registers

Figure 13–11 illustrates the I/O port locations for programming the count and address registers for each channel. Notice that the state of the F/L flip-flop determines whether the LSB or MSB is programmed. If the state of the F/L flip-flop is unknown, the count and address could be programmed incorrectly. It is also important that the DMA channel be disabled before its address and count are programmed.

Four steps are required to program the 8237: (1) The F/L flip-flop is cleared using a clear F/L command; (2) the channel is disabled; (3) the LSB and then MSB of the address are programmed; and (4) the LSB and MSB of the count are programmed. Once these four operations are performed, the channel is programmed and ready to use. Additional programming is required to select the mode of operation before the channel is enabled and started.

The 8237 Connected to the 80X86 Microprocessor

Figure 13–12 shows an 80X86-based system that contains the 8237 DMA controller.

The address enable (AEN) output of the 8237 controls the output pins of the latches and the outputs of the 74LS257 (E). During normal 80X86 operation (AEN = 0), latches A and C and the multiplexer (E) provide address bus bits A_{19}–A_{16} and A_7–A_0. The multiplexer provides the system control signals as long as the 80X86 is in control of the system. During a DMA action (AEN = 1), latches A and C are disabled along with the multiplexer (E). Latches D and B now provide address bits A_{19}–A_{16} and A_{15}–A_8. Address bus bits A_7–A_0 are provided directly by the 8237 and contain a part of the DMA transfer address. The control signals $\overline{MEMR}$, $\overline{MEMW}$, $\overline{IOR}$, and $\overline{IOW}$ are provided by the DMA controller.

The address strobe output (ADSTB) of the 8237 clocks the address (A_{15}–A_8) into latch D during the DMA action so that the entire DMA transfer address becomes available on the address bus. Address bus bits A_{19}–A_{16} are provided by latch B, which must be programmed with these four address bits before the controller is enabled for the DMA transfer. The DMA operation of the 8237 is limited to a transfer of not more than 64K bytes within the same 64K-byte section of the memory.

The decoder (F) selects the 8237 for programming and the four-bit latch (B) for the uppermost four address bits. The latch in a PC is called the DMA page register (8 bits) that holds address bits A_{16}–A_{23} for a DMA transfer. A high page register also exists, but its address is chip-dependent. The port numbers for the DMA page registers are listed in Table 13–1 (these are for the Intel ISP). The decoder in this system enables the 8237 for I/O port addresses XX60H–XX7FH, and the I/O latch (B) for ports XX00H–XX1FH. Notice that the decoder output is combined with the $\overline{IOW}$ signal to generate an active-high clock for the latch (B).

During normal 80X86 operation, the DMA controller and integrated circuits B and D are disabled. During a DMA action, integrated circuits A, C, and E are disabled so that the 8237 can take control of the system through the address, data, and control buses.

In the personal computer, the two DMA controllers are programmed at I/O ports 0000H–000FH for DMA channels 0–3, and at ports 00C0H–00DFH for DMA channels 4–7. Note that the second controller is programmed at even addresses only, so the channel 4 base and current address is programmed at I/O port 00C0H and the channel 4 base and current count is programmed at port 00C2H. The page register, which holds address bits A_{23}–A_{16} of the DMA address, are located at I/O ports 0087H (CH-0), 0083H (CH-1), 0081H (CH-2), 0082H (CH-3), (no channel 4), 008BH (CH-5), 0089H (CH-6) and 008AH (CH-7). The page register functions as the address latch described with the examples in this text.

Memory-to-Memory Transfer with the 8237

The memory-to-memory transfer is much more powerful than even the automatically repeated MOVSB instruction. (Note: Most modern chip sets do not support the memory-to-memory feature.) Although the repeated MOVSB instruction tables show that the 8088 requires 4.2 μs per byte, the 8237 requires only 2.0 μs per byte, which is over twice as fast as a software data transfer. This is not true if an 80386, 80846, or Pentium through Pentium 4 is in use in the system.

Sample Memory-to-Memory DMA Transfer. Suppose that the contents of memory locations 10000H–13FFFH are to be transferred into memory locations 14000H–17FFFH. This is accomplished with a repeated string move instruction or, at a much faster rate, with the DMA controller.

Example 13–1 illustrates the software required to initialize the 8237 and program latch B in Figure 13–12 for this DMA transfer. This software is written for an embedded application. For it to function in the PC (if your chip set supports this feature), you must use the port addresses listed in Table 13–1 for the page registers.

EXAMPLE 13–1

```
;A procedure that transfers a block of data using the 8237A
;DMA controller in Figure 13-12.  This is a memory-to-memory
;transfer.
```

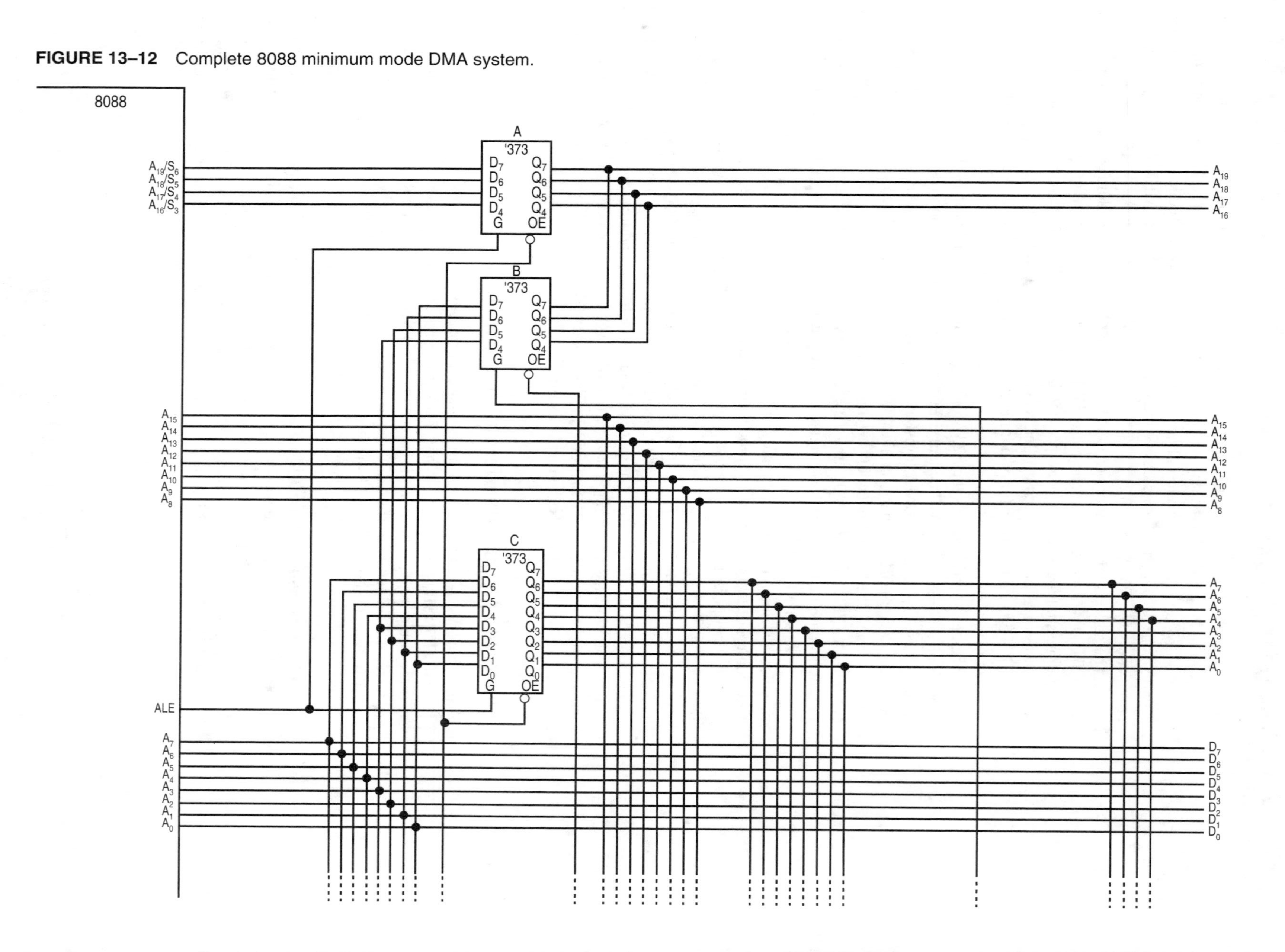

FIGURE 13–12 Complete 8088 minimum mode DMA system.

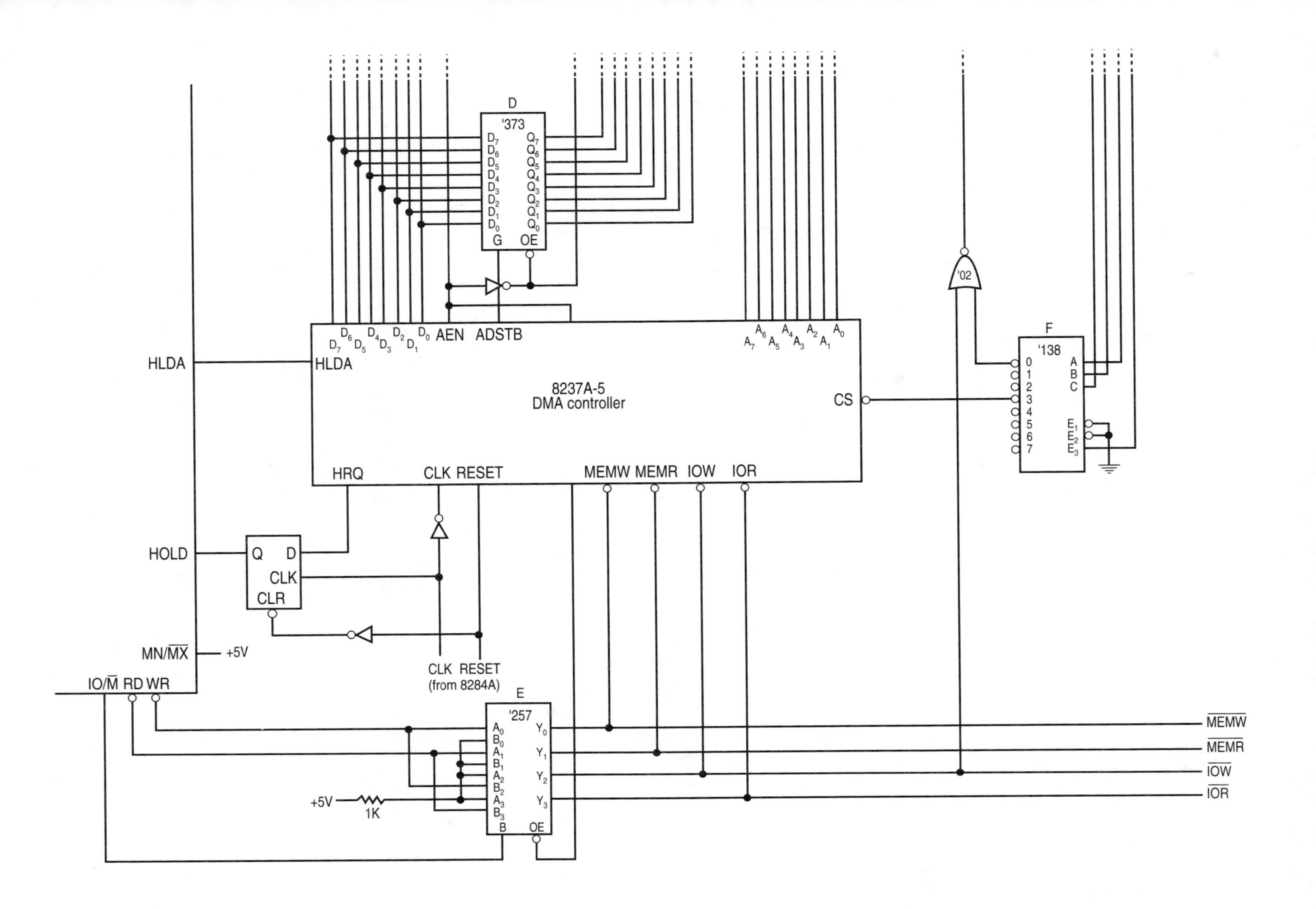

D
'373
G
OE
HLDA
HOLD
MN/$\overline{MX}$
+5V
IO/$\overline{M}$ RD WR
AEN
ADSTB
8237A-5
DMA controller
CS
HRQ
CLK RESET
MEMW MEMR IOW IOR
CLK
CLR
CLK RESET
(from 8284A)
E
'257
1K
'02
F
'138
$\overline{MEMW}$
$\overline{MEMR}$
$\overline{IOW}$
$\overline{IOR}$

TABLE 13–1 DMA page register ports.

Channel	*Port for A_{16}–A_{23}*	*Port for A_{24}–A_{31}*
0	87H	487H
1	83H	483H
2	81H	481H
3	82H	482H
4	8FH	48FH
5	8BH	48BH
6	89H	489H
7	8AH	486H

```
;Calling parameters:
;    SI = source address
;    DI = destination address
;    CX = count
;    ES = segment of source and destination

LATCHB EQU   10H
CLEARF EQU   7CH
CHOA   EQU   70H
CH1A   EQU   72H
CH1C   EQU   73H
MODE   EQU   7BH
CMMD   EQU   78H
MASKS  EQU   7FH
REQ    EQU   79H
STATUS EQU   78H

TRANS  PROC  NEAR USES AX

    MOV   AX,ES              ;program latch B
    MOV   AL,AH
    SHR   AL,4
    OUT   LATCHB,AL
    OUT   CLEARF,AL          ;clear F/L

    MOV   AX,ES              ;program source address
    SHL   AX,4
    ADD   AX,SI
    OUT   CH0A,AL
    MOV   AL,AH
    OUT   CH0A

    MOV   AX,ES              ;program destination address
    SHL   AX,4
    ADD   AX,DI
    OUT   CH1A,AL
    MOV   AL,AH
    OUT   CH1A,AL

    MOV   AX,CX              ;program count
    DEC   AX
    OUT   CH1C,AL
    MOV   AL,AH
    OUT   CH1C,AL

    MOV   AL,88H             ;program mode
    OUT   MODE,AL
    MOV   AL,85H
    OUT   MODE,AL

    MOV   AL,1               ;enable block transfer
    OUT   CMMD,AL
```

```
    MOV  AL,0EH             ;unmask channel 0
    OUT  MASKS,AL

    MOV  AL,4               ;start DMA
    OUT  REQ,AL

    .REPEAT                 ;wait for completion
          IN  AL,STATUS
    .UNTIL AL & 1
    RET

TRANS ENDP
```

Programming the DMA controller requires a few steps, as illustrated in Example 13–1. The leftmost digit of the five-digit address is sent to latch B. Next, the channels are programmed after the F/L flip-flop is cleared. Note that we use channel 0 as the source and channel 1 as the destination for a memory-to-memory transfer. The count is next programmed with a value that is one less than the number of bytes to be transferred. Next, the mode register of each channel is programmed, the command register selects a block move, channel 0 is enabled, and a software DMA request is initiated. Before return is made from the procedure, the status register is tested for a terminal count. Recall that the terminal count flag indicates that the DMA transfer is completed. The TC also disables the channel, preventing additional transfers.

Sample Memory Fill Using the 8237. In order to fill an area of memory with the same data, the channel 0 source register is programmed to point to the same address throughout the transfer. This is accomplished with the channel 0 hold mode. The controller copies the contents of this single memory location to an entire block of memory addressed by channel 1. This has many useful applications.

For example, suppose that a DOS video display must be cleared. This operation can be performed using the DMA controller with the channel 0 hold mode and a memory-to-memory transfer. If the video display contains 80 columns and 25 lines, it has 2000 display positions that must be set to 20H (an ASCII space) to clear the screen.

Example 13–2 shows a procedure that clears an area of memory addressed by ES:DI. The CX register transfers the number of bytes to be cleared to the CLEAR procedure. Notice that this procedure is nearly identical to Example 13–1, except that the command register is programmed so the channel 0 address is held. The source address is programmed as the same address as ES:DI, and then the destination is programmed as one location beyond ES:DI. Also note that this program is designed to function with the hardware in Figure 13–12 and will not function in the personal computer unless you have the same hardware.

EXAMPLE 13–2

```
;A procedure that clears the DOS mode video screen.= using the DMA
;controller as depicted in Figure 13-12.

;Calling sequence:
;     DI = offset address of area cleared
;     ES = segment address of area cleared
;     CX = number of bytes cleared

LATCHB EQU   10H
CLEARF EQU   7CH
CHOA   EQU   70H
CH1A   EQU   72H
CH1C   EQU   73H
MODE   EQU   7BH
CMMD   EQU   78H
MASKS  EQU   7FH
REQ    EQU   79H
```

```
STATUS EQU    78H
ZERO   EQU    0

CLEAR  PROC   NEAR USES AX

    MOV   AX,ES               ;program latch B
    MOV   AL,AH
    SHR   AL,4
    OUT   LATCHB,AL
    OUT   CLEARF,AL           ;clear F/L

    MOV   AL,ZERO             ;save zero in first byte
    MOV   ES:[DI],AL

    MOV   AX,ES               ;program source address
    SHL   AX,4
    ADD   AX,SI
    OUT   CH0A,AL
    MOV   AL,AH
    OUT   CH0A

    MOV   AX,ES               ;program destination address
    SHL   AX,4
    ADD   AX,DI
    OUT   CH1A,AL
    MOV   AL,AH
    OUT   CH1A,AL

    MOV   AX,CX               ;program count
    DEC   AX
    OUT   CH1C,AL
    MOV   AL,AH
    OUT   CH1C,AL

    MOV   AL,88H              ;program mode
    OUT   MODE,AL
    MOV   AL,85H
    OUT   MODE,AL

    MOV   AL,03H              ;enable block hold transfer
    OUT   CMMD,AL

    MOV   AL,0EH              ;enable channel 0
    OUT   MASKS,AL

    MOV   AL,4                ;start DMA
    OUT   REQ,AL

    .REPEAT
        IN   AL,STATUS
    .UNTIL AL & 1
    RET

CLEAR  ENDP
```

DMA-Processed Printer Interface

Figure 13–13 illustrates the hardware added to Figure 13–12 for a DMA-controlled printer interface. Little additional circuitry is added for this interface to a Centronics-type parallel printer. The latch is used to capture the data as it is sent to the printer during the DMA transfer. The write pulse passed through to the latch during the DMA action also generates the data strobe ($\overline{\text{DS}}$) signal to the printer through the single-shot. The $\overline{\text{ACK}}$ signal returns from the printer each time it is ready for additional data. In this circuit, $\overline{\text{ACK}}$ is used to request a DMA action through a flip-flop.

Notice that the I/O device is not selected by decoding the address on the address bus. During the DMA transfer, the address bus contains the memory address and cannot contain the

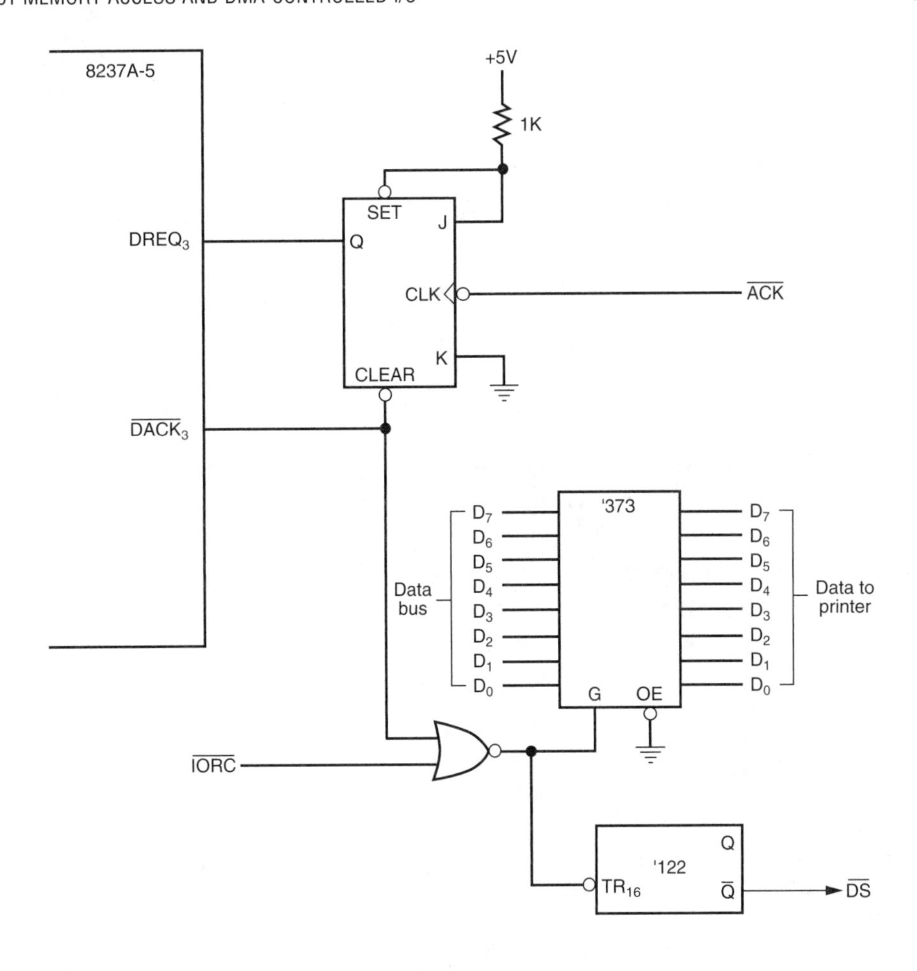

FIGURE 13–13 DMA-processed printer interface.

I/O port address. In place of the I/O port address, the $\overline{DACK3}$ output from the 8237 selects the latch by gating the write pulse through an OR gate.

Software that controls this interface is simple because only the address of the data and the number of characters to be printed are programmed. Once programmed, the channel is enabled, and the DMA action transfers a byte at a time to the printer interface each time that the interface receives the $\overline{ACK}$ single from the printer.

The procedure that prints data from the current data segment is illustrated in Example 13–3. This procedure programs the 8237, but doesn't actually print anything. Printing is accomplished by the DMA controller and the printer interface.

EXAMPLE 13–3

```
;A procedure that prints data via the printer interface in
;Figure 13-13

;Calling sequence:
;       BX = offset address of printer data
;       DS = segment address of printer data
;       CX = number of bytes to print

LATCHB EQU     10H
CLEARF EQU     7CH
CH3A   EQU     76H
CH1C   EQU     77H
MODE   EQU     7BH
```

```
CMMD    EQU    78H
MASKS   EQU    7FH
REQ     EQU    79H

PRINT   PROC   NEAR USES AX CX BX

        MOV    EAX,0
        MOV    AX,DS                 ;program latch B
        SHR    EAX,4
        PUSH   AX
        SHR    EAX,16
        OUT    LATCHB,AL

        POP    AX                    ;program address
        OUT    CH3A,AL
        MOV    AL,AH
        OUT    CH3A,AL

        MOV    AX,CX                 ;program count
        DEC    AX
        OUT    CH3C,AL
        MOV    AL,AH
        OUT    CH3C,AL

        MOV    AL,0BH                ;program mode
        OUT    MODE,AL

        MOV    AL,00H                ;enable block mode transfer
        OUT    CMMD,AL

        MOV    AL,7                  ;enable channel 3
        OUT    MASKS,AL
        RET

PRINT   ENDP
```

A secondary procedure is needed to determine whether the DMA action has been completed. Example 13–4 lists the secondary procedure that tests the DMA controller to see whether the DMA transfer is complete. The TESTP procedure is called before programming the DMA controller to see whether the prior transfer is complete.

EXAMPLE 13–4

```
;A procedure that tests for completion of the DMA action

STATUS   EQU    78H

TESTP    PROC   NEAR USES AX

         .REPEAT
                IN    AL,STATUS
         .UNTIL AL & 8
         RET

TESTP    ENDP
```

Printed data can be double-buffered by first loading buffer 1 with data to be printed. Next, the PRINT procedure is called to begin printing buffer 1. Because it takes very little time to program the DMA controller, a second buffer (buffer 2) can be filled with new printer data while the first buffer (buffer 1) is printed by the printer interface and DMA controller. This process is repeated until all data are printed.

13–3 SHARED-BUS OPERATION

Complex present-day computer systems have so many tasks to perform that some systems are using more than one microprocessor to accomplish the work. This is called a **multiprocessing** system. We also sometimes call this a **distributed** system. A system that performs more than one task is called a **multitasking** system. In systems that contain more than one microprocessor, some method of control must be developed and employed. In a distributed, multiprocessing, multitasking environment, each microprocessor accesses two buses: (1) the **local bus** and (2) the **remote** or **shared bus**.

This section of the text describes shared bus operation for the 8086 and 8088 microprocessors using the 8289 bus arbiter. The 80286 uses the 82289 bus arbiter and the 80386/80486 uses the 82389 bus arbiter. The Pentium–Pentium 4 directly support a multiuser environment, as described in Chapters 17, 18, and 19. These systems are much more complex and difficult to illustrate at this point in the text, but their terminology and operation is essentially the same as for the 8086/8088.

The local bus is connected to memory and I/O devices that are directly accessed by a single microprocessor without any special protocol or access rules. The remote (shared) bus contains memory and I/O that are accessed by any microprocessor in the system. Figure 13–14 illustrates this idea with a few microprocessors. Note that the personal computer is also configured in the same manner as the system in Figure 13–14. The bus master is the main microprocessor in the personal computer. What we call the local bus in the personal computer is the shared bus in this illustration. The ISA bus is operated as a slave to the personal computer's microprocessor as well as any other devices attached to the shared bus. The PCI bus can operate as a slave or a master.

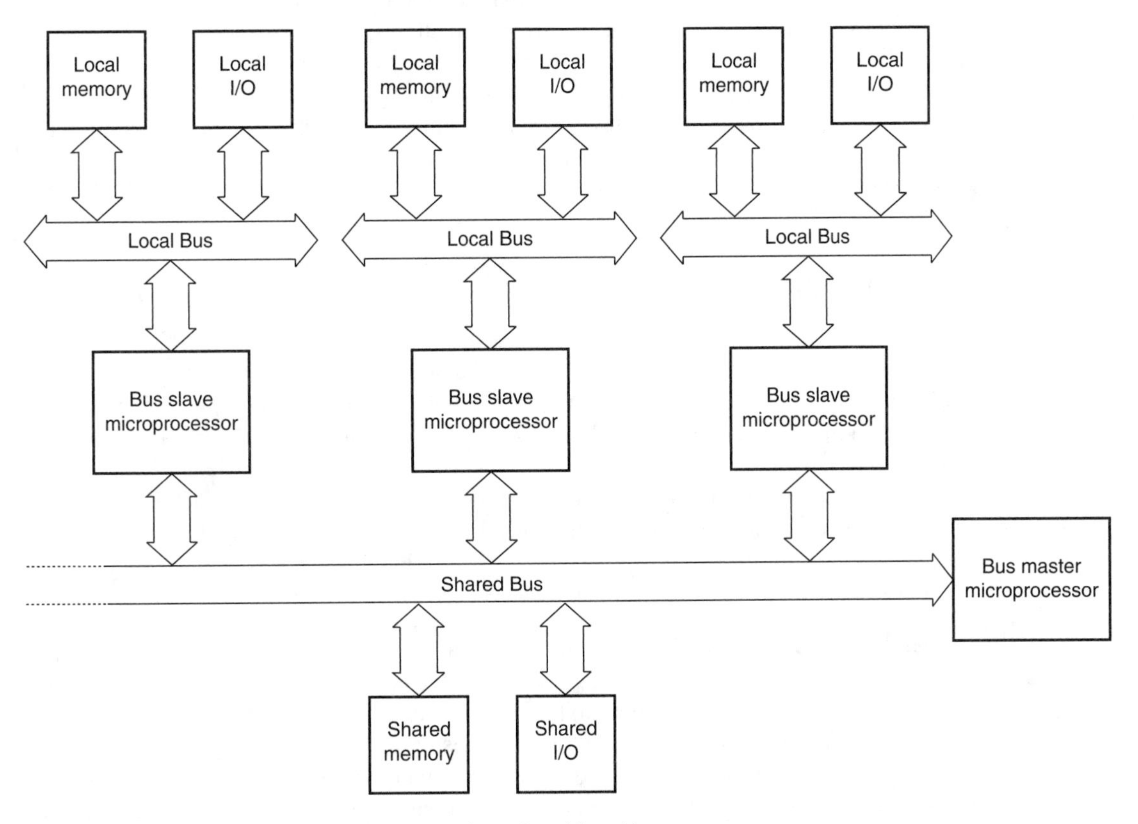

FIGURE 13–14 A block diagram illustrating the shared and local buses.

Types of Buses Defined

The local bus is the bus that is resident to the microprocessor. The local bus contains the resident or local memory and I/O. All microprocessors studied thus far in this text are considered to be local bus systems. The local memory and local I/O are accessed by the microprocessor that is directly connected to them.

A shared bus is one that is connected to all microprocessors in the system. The shared bus is used to exchange data between microprocessors in the system. A shared bus may contain memory and I/O devices that are accessed by all microprocessors in the system. Access to the shared bus is controlled by some form or arbiter that allows only a single microprocessor to access the system's shared bus space. As mentioned, the shared bus in the personal computer is what we often call the local bus in the personal computer because it is local to the microprocessor in the personal computer.

Figure 13–15 shows an 8088 microprocessor that is connected as a remote bus master. The term **bus master** applies to any device (microprocessor or otherwise) that can control a bus containing memory and I/O. The 8237 DMA controller presented earlier in the chapter is an example of a remote bus master. The DMA controller gained access to the system memory and I/O space to cause a data transfer. Likewise, a remote bus master gains access to the shared bus for the same purpose. The difference is that the remote bus master microprocessor can execute variable software, whereas the DMA controller can only transfer data.

Access to the shared bus is accomplished by using the HOLD pin on the microprocessor for the DMA controller. Access to the shared bus for the remote bus master is accomplished via a **bus arbiter**, which functions to resolve priority between bus masters and allows only one device at a time to access the shared bus.

Notice in Figure 13–15 that the 8088 microprocessor has an interface to both a local, resident bus and the shared bus. This configuration allows the 8088 to access local memory and I/O or, through the bus arbiter and buffers, the shared bus. The task assigned to the microprocessor might be data communications. It may, after collecting a block of data from the communications interface, pass those data on to the shared bus and shared memory so that other microprocessors attached to the system can access the data. This allows many microprocessors to share common data. In the same manner, multiple microprocessors can be assigned various tasks in the system, drastically improving throughput.

The Bus Arbiter

Before Figure 13–15 can be fully understood, the operation of the bus arbiter must be grasped. The 8289 bus arbiter controls the interface of a bus master to a shared bus. Although the 8289 is not the only bus arbiter, it is designed to function with the 8086/8088 microprocessors, so it is presented here. Each bus master or microprocessor requires an arbiter for the interface to the shared bus, which Intel calls the Multibus and IBM calls the Micro Channel.

The shared bus is used only to pass information from one microprocessor to another; otherwise, the bus masters function in their own local bus modes by using their own local programs, memory, and I/O space. Microprocessors connected in this kind of system are often called **parallel** or **distributed** processors because they can execute software and perform tasks in parallel.

8289 Architecture. Figure 13–16 illustrates the pin-out and block diagram of the 8289 bus arbiter. The left side of the block diagram depicts the connections to the microprocessor. The right side denotes the 8289 connection to the shared (remote) bus or Multibus.

The 8289 controls the shared bus by causing the READY input to the microprocessor to become a logic 0 (not ready) if access to the shared bus is denied. The **blocking** occurs whenever another microprocessor is accessing the shared bus. As a result, the microprocessor requesting access is blocked by the logic 0 applied to its READY input. When the READY pin is a logic 0, the microprocessor and its software wait until access to the shared bus is granted by the arbiter. In this manner, one microprocessor at a time gains access to the shared bus. No special instructions are required for bus arbitration with the 8289 bus arbiter because arbitration is accomplished strictly by the hardware.

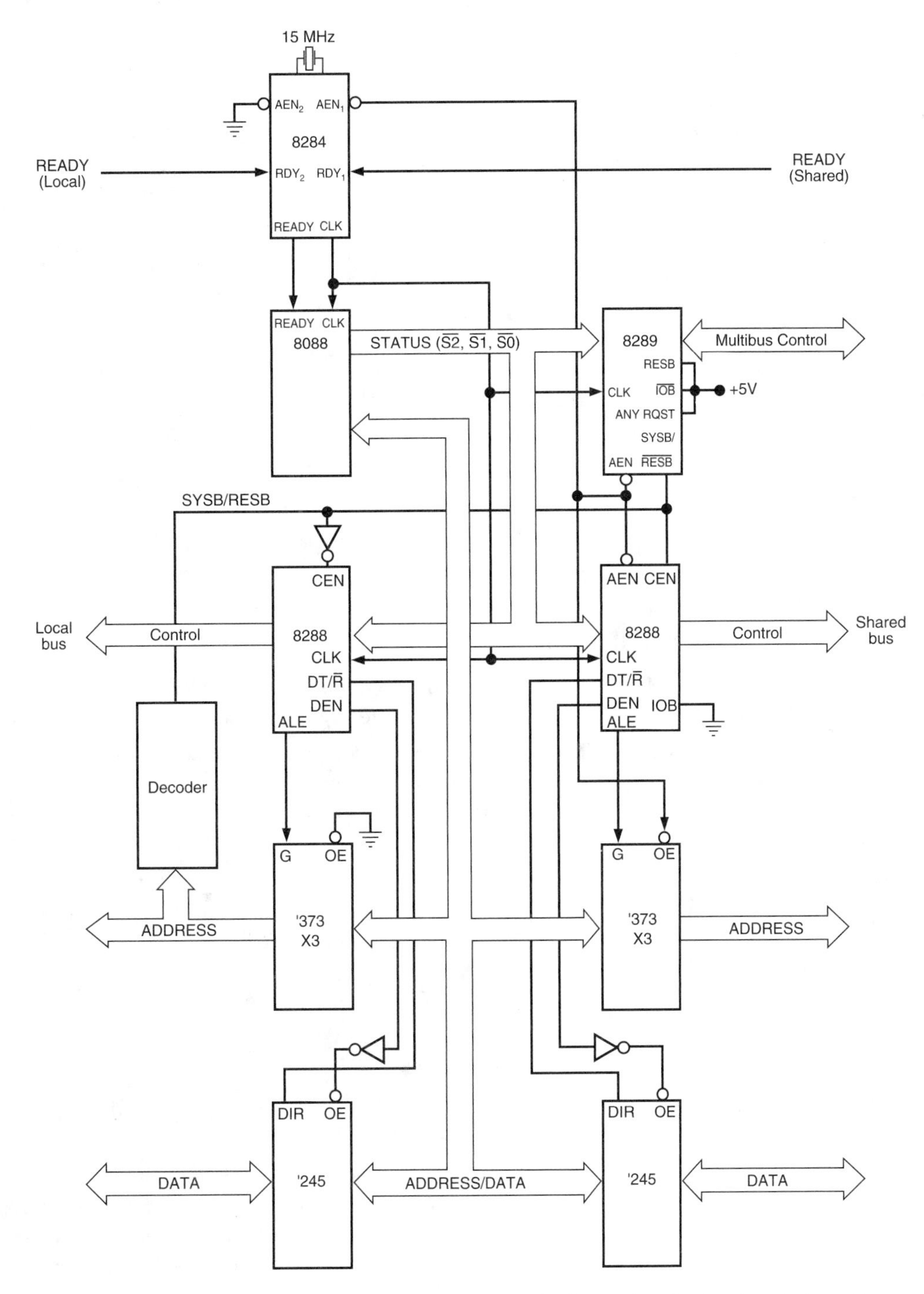

FIGURE 13–15 The 8088 operated in the remote mode, illustrating the local and shared bus connections.

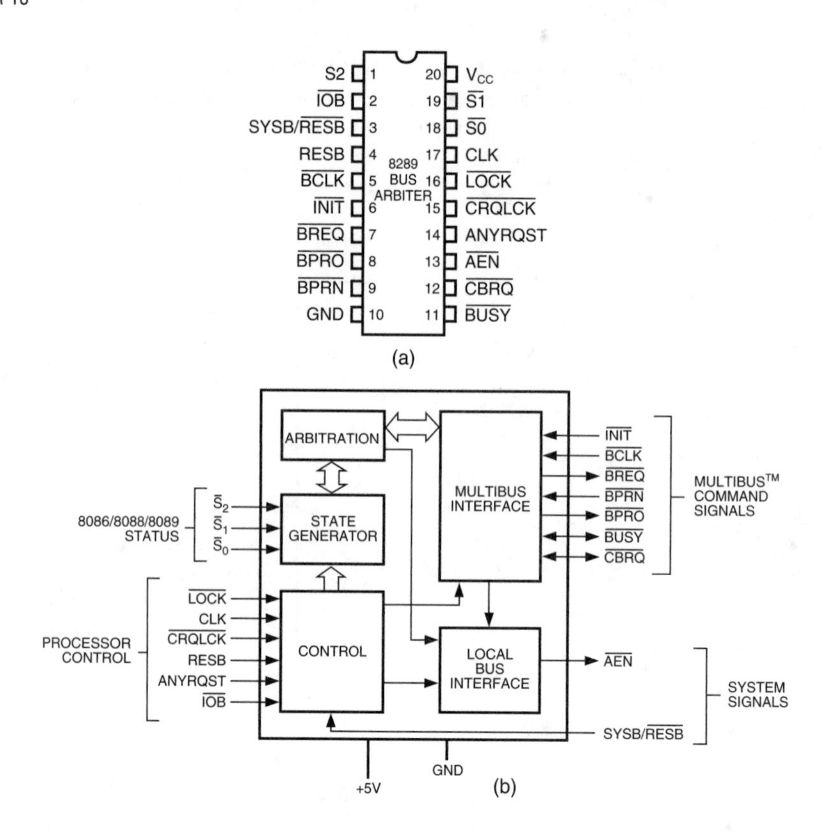

FIGURE 13–16 The 8289 pin-out and block diagram. (Courtesy of Intel Corporation.)

Pin Definitions

$\overline{\text{AEN}}$ The **address enable** output causes the bus drivers in a system to switch to their three-state, high-impedance state.

ANYRQST The **any request** input is a strapping option that prevents a lower-priority microprocessor from gaining access to the shared bus. If tied to a logic 0, normal arbitration occurs and a lower-priority microprocessor can gain access to the shared bus if $\overline{\text{CBRQ}}$ is also a logic O.

$\overline{\text{BCLK}}$ The **bus clock** input synchronizes all shared-bus masters.

$\overline{\text{BPRN}}$ The **bus priority input** allows the 8289 to acquire the shared bus on the next falling edge of the $\overline{\text{BCLK}}$ signal.

$\overline{\text{BPRO}}$ The **bus priority output** is a signal that is used to resolve priority in a system that contains multiple bus masters.

$\overline{\text{BREQ}}$ The **bus request** output is used to request access to the shared bus.

$\overline{\text{BUSY}}$ The **busy** input/output indicates, as an output, that an 8289 has acquired the shared bus. As an input, $\overline{\text{BUSY}}$ is used to detect that another 8289 has acquired the shared bus.

$\overline{\text{CBRQ}}$ The **common bus request** input/output is used when a lower-priority microprocessor is asking for the use of the shared bus. As an output, $\overline{\text{CBRQ}}$ becomes a logic 0 whenever the 8289 requests the shared bus and remains low until the 8289 obtains access to the shared bus.

CLK The **clock** input is generated by the 8284A clock generator and provides the internal timing source to the 8289.

$\overline{\text{CRQLCK}}$ The **common request lock** input prevents the 8289 from surrendering the shared bus to any of the 8289s in the system. This signal functions in conjunction with the $\overline{\text{CBRQ}}$ pin.

$\overline{\text{INIT}}$ The **initialization** input resets the 8289 and is normally connected to the system RESET signal.

$\overline{\text{IOB}}$ The **I/O bus** input selects whether the 8289 operates in a shared-bus system (if selected by RESB) with I/O ($\overline{\text{IOB}}$ = 0) or with memory and I/O ($\overline{\text{IOB}}$ = 1).

$\overline{\text{LOCK}}$ The **lock** input prevents the 8289 from allowing any other microprocessor from gaining access to the shared bus. An 8086/8088 instruction that contains a LOCK prefix will prevent other microprocessors from accessing the shared bus.

RESB The **resident-bus** input is a strapping connection that allows the 8289 to operate in systems that have either a shared-bus or resident-bus system. If RESB is a logic 1, the 8289 is configured as a shared-bus master. If RESB is a logic 0, the 8289 is configured as a local-bus master. When configured as a shared-bus master, access is requested through the SYSB/$\overline{\text{RESB}}$ input pin.

S_0, S_1, and S_2 The **status** inputs initiate shared-bus requests and surrenders. These pins connect to the 8288 system bus controller status pins.

SYSB/$\overline{\text{RESB}}$ The **system bus/resident bus** input selects the shared-bus system when placed at a logic 1 or the resident local bus when placed at a logic 0.

General 8289 Operation. As the pin descriptions demonstrate, the 8289 can be operated in three basic modes: (1) I/O peripheral-bus mode, (2) resident-bus mode, and (3) single-bus mode. See Table 13–2 for the connections required to operate the 8289 in these modes. In the I/O peripheral bus mode, all devices on the local bus are treated as I/O, including memory, and are accessed by all instructions. All memory references access the shared bus and all I/O access the resident-local bus. The resident-bus mode allows memory and I/O accesses on both the local and shared buses. Finally, the single-bus mode interfaces a microprocessor to a shared bus, but the microprocessor has no local memory or local I/O. In many systems, one microprocessor is set up as the shared-bus master (single-bus mode) to control the shared bus and become the shared-bus master. The shared-bus master controls the system through shared memory and I/O. Additional microprocessors are connected to the shared bus as resident- or I/O peripheral-bus masters. These additional bus masters usually perform independent tasks that are reported to the shared-bus master through the shared bus.

System Illustrating Single-Bus and Resident-Bus Connections. Single-bus operation interfaces a microprocessor to a shared bus that contains both I/O and memory resources that are shared by other microprocessors. Figure 13–17 illustrates three 8088 microprocessors, each

TABLE 13–2 8289 modes of operation.

Mode	*Pin Connections*
Single bus	$\overline{\text{IOB}}$ = 1 and RESB = 0
Resident bus	$\overline{\text{IOB}}$ = 1 and RESB = 1
I/O bus	$\overline{\text{IOB}}$ = 0 and RESB = 0
I/O bus and resident bus	$\overline{\text{IOB}}$ = 0 and RESB = 1

connected to a shared bus. Two of the three microprocessors operate in the resident-bus mode, while the third operates in the single-bus mode. Microprocessor A in Figure 13–17 operates in the single-bus mode and has no local bus. This microprocessor accesses only the shared memory and I/O space. Microprocessor A is often referred to as the system-bus master because it is responsible for coordinating the main memory and I/O tasks. The remaining two microprocessors (B and C) are connected in the resident-bus mode, which allows them access to both the shared bus and their own local buses. These resident-bus microprocessors are used to perform tasks that are independent from the system-bus master. In fact, the only time that the system-bus master is interrupted from performing its tasks is when one of the two resident-bus microprocessors needs to transfer data between itself and the shared bus. This connection allows all three microprocessors to perform tasks simultaneously, yet data can be shared between microprocessors when needed.

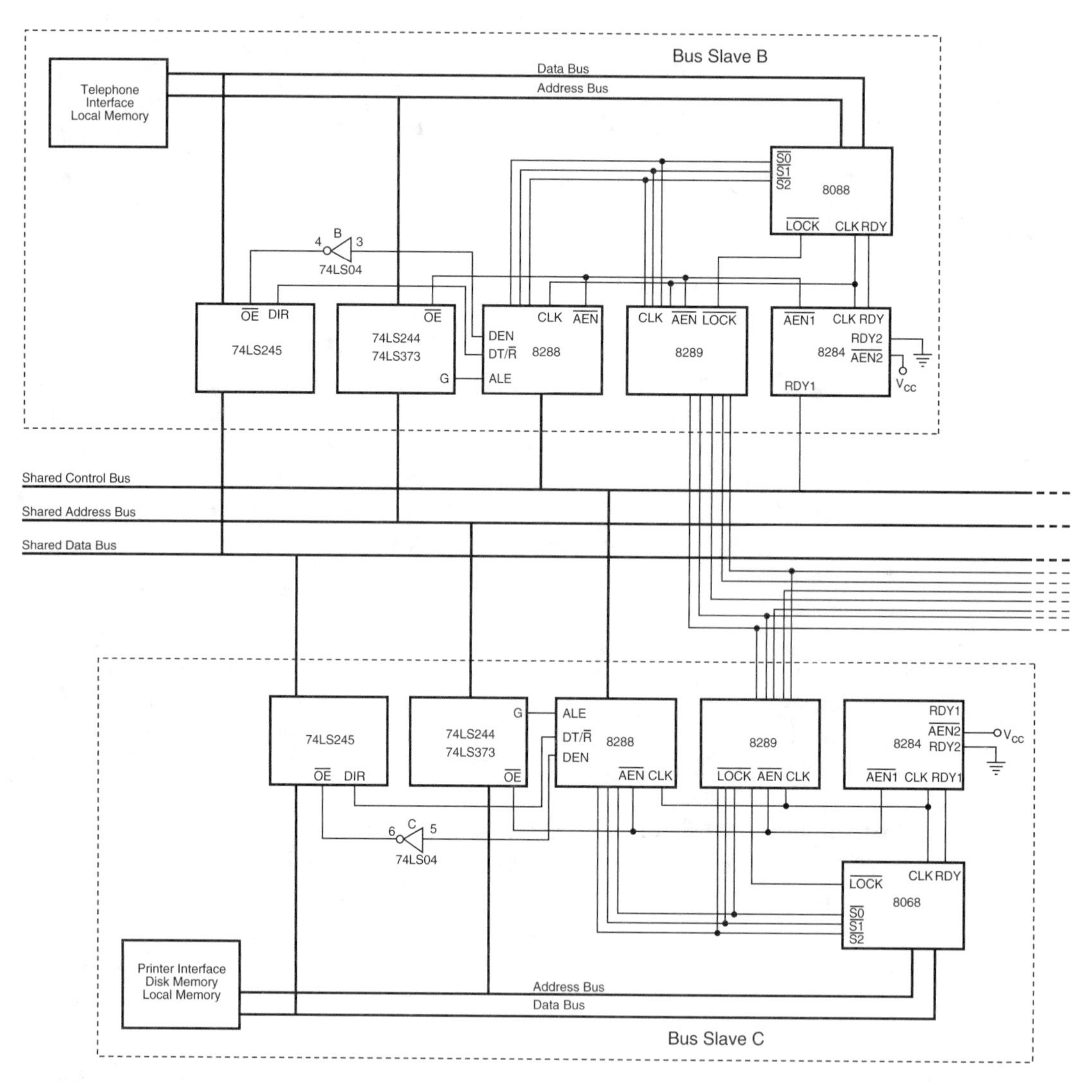

FIGURE 13–17 Three 8088 microprocessors that share a common bus system. Microprocessor A is the bus master in control of the shared memory and CRT terminal. Microprocessor B is a bus slave controlling its local telephone interface and memory. Microprocessor C is also a slave that controls a printer, disk memory system, and local memory.

In Figure 13–17, the bus master (A) allows the user to operate with a video terminal that allows the execution of programs and generally controls the system. Microprocessor B handles all telephone communications and passes this information to the shared memory in blocks. This means that microprocessor B waits for each character to be transmitted or received and controls the protocol used for the transfers. For example, suppose that a 1K-byte block of data is transmitted across the telephone interface at the rate of 100 characters per second. This means that the transfer requires 10 seconds. Rather than tie up the bus master for 10 seconds, microprocessor B patiently performs the data transfer from its own local memory and the local communications interface. This frees the bus master for other tasks. The only time microprocessor B interrupts the bus master is to transfer data between the shared memory and its local memory system. This data transfer between microprocessor B and the bus master requires only a few hundred microseconds.

Microprocessor C is used as a print spooler. Its only task is to print data on the printer. Whenever the bus master requires printed output, it transfers the task to microprocessor C. Microprocessor C then accesses the shared memory, captures the data to be printed, and stores them in its own local memory. Data are then printed from the local memory, freeing the bus master to perform other tasks. This allows the system to execute a program with the bus master, transfer data through the communications interface with microprocessor B, and print information on the printer with microprocessor C. These tasks all execute simultaneously. There is no limit to the number of microprocessors connected to a system or the number of tasks performed simultaneously using this technique. The only limit is that introduced by the system design and the designer's ingenuity. Lawrence Livermore Labs in California has a system that contains 4096 Pentium microprocessors.

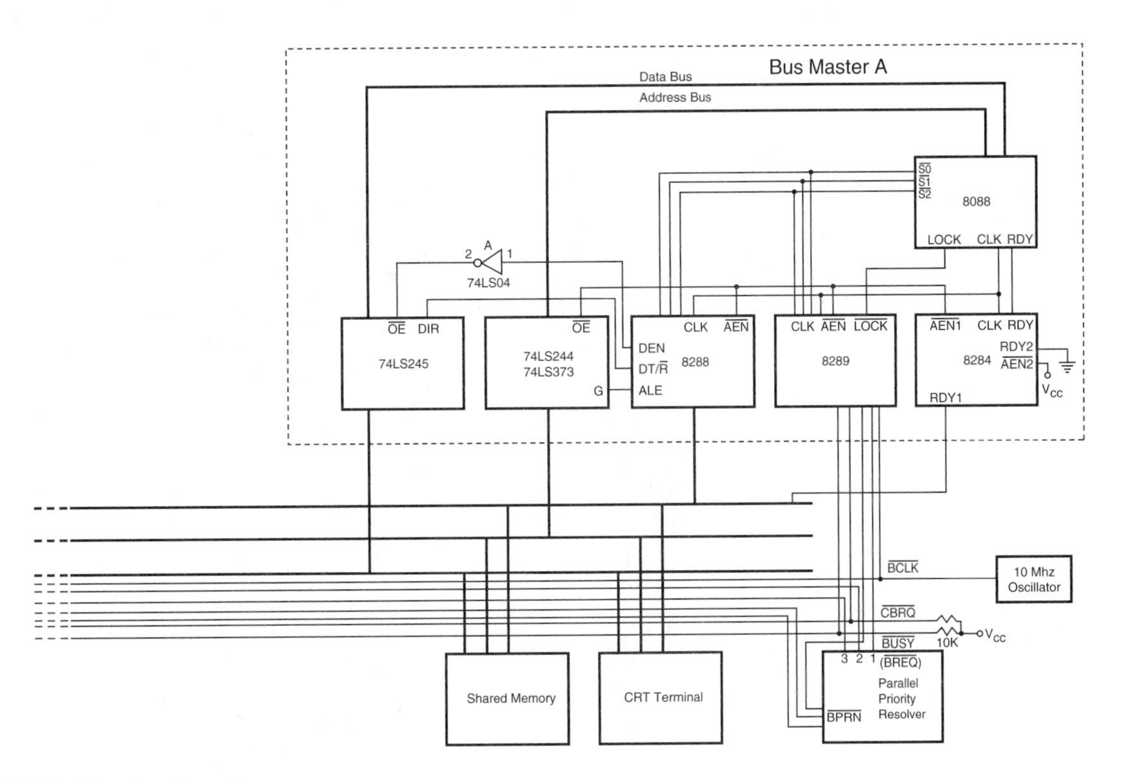

FIGURE 13–17 *(continued)*

13–4 DISK MEMORY SYSTEMS

Disk memory is used to store long-term data. Many types of disk storage systems are available today and they use magnetic media, except the optical disk memory that stores data on a plastic disk. Optical disk memory is either a **CD-ROM** (compact disk/read only memory) that is read, but never written, or a **WORM** (write once/read many) that is read most of the time, but can be written once by a laser beam. Also becoming available is optical disk memory that can be read and written many times, but there is still a limitation on the number of write operations allowed. The latest optical disk technology is called DVD (**digital-versatile disk**). This section of the chapter provides an introduction to disk memory systems so that they may be used with computer systems. It also provides details of their operation.

Floppy Disk Memory

Once the most common and the most basic form of disk memory was the floppy, or flexible disk. Today the floppy is beginning to vanish and may completely disappear shortly in favor of the USB pen drive. The floppy disk magnetic recording media have been made available in three sizes: 8" **standard**, 5¼" mini-floppy, and 3½" **micro-floppy**. Today, the 8" standard version and 5¼" mini-floppy have all but disappeared, giving way to the micro-floppy disks and more recently pen drives. The 8" disk is too large and difficult to handle and stockpile. To solve this problem, industry developed the 5¼" mini-floppy disk. Today, the micro-floppy disk has just about replaced the mini-floppy in newer systems because of its reduced size, ease of storage, and durability. Even so, systems are still marketed with the micro-floppy disk drives.

All disks and even the pen drives have several things in common. They are all organized so that data are stored in tracks. A **track** is a concentric ring of data that is stored on a surface of a disk. Figure 13–18 illustrates the surface of a 5¼" mini-floppy disk, showing a track that is divided into sectors. A **sector** is a common subdivision of a track that is designed to hold a reasonable amount

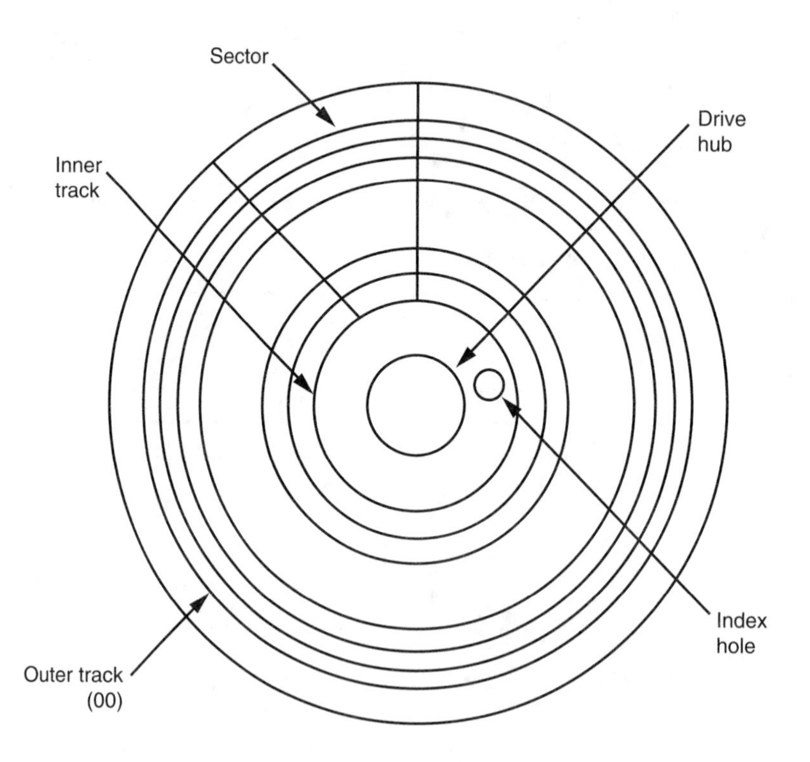

FIGURE 13–18 The format of a 5¼" mini-floppy disk.

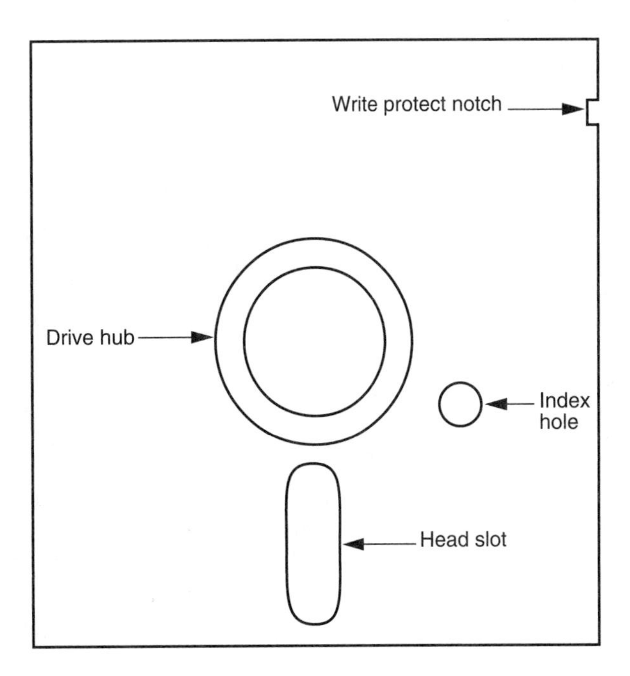

FIGURE 13–19 The 5¼" mini-floppy disk.

of data. In many systems, a sector holds either 512 or 1024 bytes of data. The size of a sector can vary from 128 bytes to the length of one entire track.

Notice in Figure 13–18 that there is a hole through the disk, labeled an **index hole**. The index hole is designed so that the electronic system that reads the disk can find the beginning of a track and its first sector (00). Tracks are numbered from track 00, the outermost track, in increasing value toward the center or innermost track. Sectors are often numbered from sector 00 on the outermost track to whatever value is required to reach the innermost track and its last sector.

The 5¼" Mini-Floppy Disk. Today, the 5¼" floppy is very difficult to find and is used only with older microcomputer systems. Figure 13–19 illustrates this mini-floppy disk. The floppy disk is rotated at 300 RPM inside its semi-rigid plastic jacket. The head mechanism in a floppy disk drive makes physical contact with the surface of the disk, which eventually causes wear and damage to the disk.

Most mini-floppy disks are double-sided. This means that data are written on both the top and bottom surfaces of the disk. A set of tracks called a **cylinder** consists of one top and one bottom track. Cylinder 00, for example, consists of the outermost top and bottom tracks.

Floppy disk data are stored in the double-density format, which uses a recording technique called MFM (**modified frequency modulation**) to store the information. Double-density, double-sided (**DSDD**) disks are normally organized with 40 tracks of data on each side of the disk. A double-density disk track is typically divided into nine sectors, with each sector containing 512 bytes of information. This means that the total capacity of a double-density, double-sided disk is 40 tracks per side × 2 sides × 9 sectors per track × 512 bytes per sector, or 368,640 (360K) bytes of information.

Also common are **high-density** (HD) mini-floppy disks. A high-density mini-floppy disk contains 80 tracks of information per side, with eight sectors per track. Each sector contains 1024 bytes of information. This gives the 5¼" high-density, mini-floppy disk a total capacity of 80 tracks per side × 2 sides × 15 sectors per track × 512 bytes per sector, or 1,228,800 (approximately 1.2M) bytes of information.

The magnetic recording technique used to store data on the surface of the disk is called **non-return to zero** (NRZ) recording. With NRZ recording, magnetic flux placed on the surface

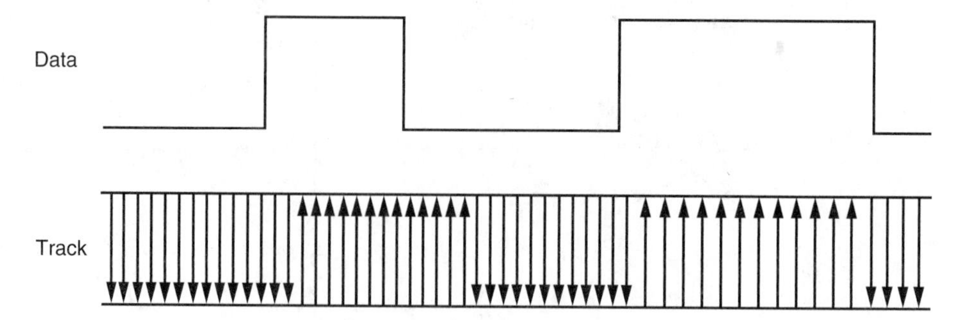

FIGURE 13–20 The non-return to zero (NRZ) recording technique.

of the disk never returns to zero. Figure 13–20 illustrates the information stored in a portion of a track. It also shows how the magnetic field encodes the data. Note that arrows are used in this illustration to show the polarity of the magnetic field stored on the surface of the disk.

The main reason that this form of magnetic encoding was chosen is that it automatically erases old information when new information is recorded. If another technique were used, a separate erase head would be required. The mechanical alignment of a separate erase head and a separate read/write head is virtually impossible. The magnetic flux density of the NRZ signal is so intense that it completely saturates (magnetizes) the surface of the disk, erasing all prior data. It also ensures that information will not be affected by noise because the amplitude of the magnetic field contains no information. The information is stored in the placement of the changes of the magnetic field.

Data are stored in the form of MFM (modified frequency modulation) in modern floppy disk systems. The MFM recording technique stores data in the form illustrated in Figure 13–21. Notice that each bit time is 2.0 μs wide on a double-density disk. This means that data are recorded at the rate of 500,000 bits per second. Each 2.0 μs bit time is divided into two parts: One part is designated to hold a clock pulse and the other holds a data pulse. If a clock pulse is present, it is 1.0 μs wide, as is a data pulse. Clock and data pulses are never present at the same time in one bit period. (Note that high-density disk drives halve these times so that a bit time is 1.0 μs and a clock or data pulse is 0.5 μs wide. This also doubles the transfer rate to 1 million bits per second—1 Mbps.)

If a data pulse is present, the bit time represents a logic 1. If no data or no clock is present, the bit time represents a logic 0. If a clock pulse is present with no data pulse, the bit time also represents a logic 0. The rules followed when data are stored using MFM are as follows:

1. A data pulse is always stored for a logic l.
2. No data and no clock is stored for the first logic 0 in a string of logic 0s.
3. The second and subsequent logic 0s in a row contain a clock pulse, but no data pulse.

The reason that a clock is inserted as the second and subsequent zero in a row is to maintain synchronization as data are read from the disk. The electronics used to recapture the data from the disk drive use a phase-locked loop to generate a clock and a data window. The phase-locked loop needs a clock or data to maintain synchronized operation.

FIGURE 13–21 Modified frequency modulation (MFM) used with disk memory.

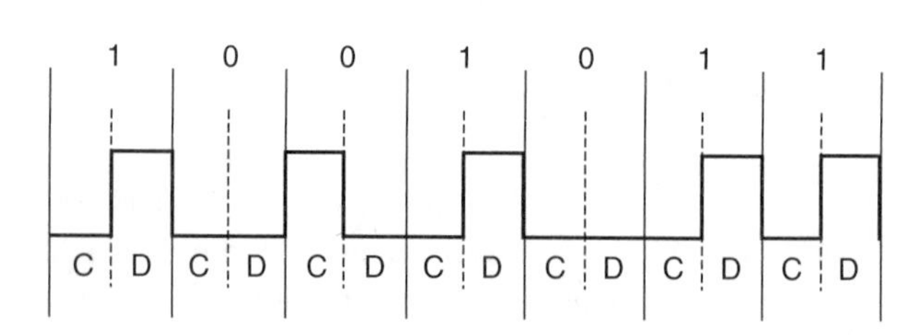

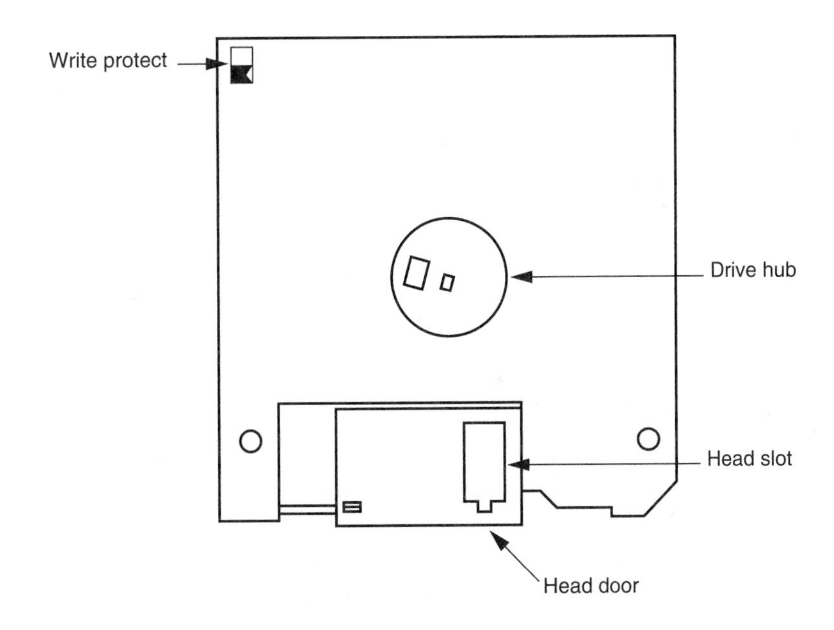

FIGURE 13–22 The 3½" micro-floppy disk.

The 3½" Micro-Floppy Disk. A popular disk size is the 3½" micro-floppy disk. Recently, this size of floppy disk has begun to be replaced by the USB pen drive as the dominant transportable media. The micro-floppy disk is a much improved version of the mini-floppy disk described earlier. Figure 13–22 illustrates the 3½" micro-floppy disk.

Disk designers noticed several shortcomings of the mini-floppy, which is a scaled-down version of the 8" standard floppy, soon after it was released. Probably one of the biggest problems with the mini-floppy is that it is packaged in a semi-rigid plastic cover that bends easily. The micro-floppy is packaged in a rigid plastic jacket that will not bend easily. This provides a much greater degree of protection to the disk inside the jacket.

Another problem with the mini-floppy is the head slot that continually exposes the surface of the disk to contaminants. This problem is also corrected on the micro-floppy because it is constructed with a spring-loaded sliding head door. The head door remains closed until the disk is inserted into the drive. Once inside the drive, the drive mechanism slides open the door, exposing the surface of the disk to the read/write heads. This provides a great deal of protection to the surface of the micro-floppy disk.

Yet another improvement is the sliding plastic write-protection mechanism on the micro-floppy disk. On the mini-floppy disk, a piece of tape was placed over a notch on the side of the jacket to prevent writing. This plastic tape easily became dislodged inside disk drives, causing problems. On the micro-floppy, an integrated plastic slide has replaced the tape write-protection mechanism. To write-protect (prevent writing) the micro-floppy disk, the plastic slide is moved to open the hole through the disk jacket. This allows light to strike a sensor that inhibits writing.

Still another improvement is the replacement of the index hole with a different drive mechanism. The drive mechanism on the mini-floppy allows the disk drive to grab the disk at any point. This requires an index hole so that the electronics can find the beginning of a track. The index hole is another trouble spot because it collects dirt and dust. The micro-floppy has a drive mechanism that is keyed so that it only fits one way inside the disk drive. The index hole is no longer required because of this keyed drive mechanism. Because of the sliding head mechanism and the fact that no index hole exists, the micro-floppy disk has no place to catch dust or dirt.

Two types of micro-floppy disks are widely available: the double-sided, double-density (DSDD) and the high-density (HD). The double-sided, double-density micro-floppy disk has 80 tracks per side, with each track containing nine sectors. Each sector contains 512 bytes of information. This allows 80 tracks per side × 2 sides × 9 sectors × 512 bytes per sector, or 737,280 (720K) bytes of data to be stored on a double-density, double-sided floppy disk.

The high-density, double-sided micro-floppy disk stores even more information. The high-density version has 80 tracks per side, but the number of sectors is doubled to 18 per track. This format still uses 512 bytes per sector, as did the double-density format. The total number of bytes on a high-density, double-sided micro-floppy disk is 80 tracks per side × 2 sides × 18 sectors per track × 512 bytes per sector, or 1,474,560 (1.44M) bytes of information.

Pen Drives

Pen drives, or as they are often called, Flash drives, are replacements for floppy disk drives that use Flash memory to store data. A driver, which is part of Windows (except for Windows 98), treats the pen drive as a floppy with tracks and sectors even though it really does not contain tracks and sectors. When a pen drive is connected to the USB bus, the operating system recognizes it and allows data to be transferred between it and the computer.

Newer pen drives use the USB 2.0 bus specification to transfer data at a much higher rate of speed than the older USB 1.1 specification. Transfer speeds for USB 1.1 are a read speed of 750 KBps and a write speed of 450 KBps. The USB 2.0 pen drives have a transfer speed of about 48 MBps. The pen drive is currently available in sizes up to 1G byte and have an erase cycle of up to 1,000,000 erases. The price is very reasonable when compared to the floppy disk.

Hard Disk Memory

Larger disk memory is available in the form of the **hard disk drive**. The hard disk drive is often called a **fixed disk** because it is not removable like the floppy disk. A hard disk is also often called a **rigid disk**. The term **Winchester drive** is also used to describe a hard disk drive, but less commonly today. Hard disk memory has a much larger capacity than floppy disk memory. Hard disk memory is available in sizes approaching 1 T (tera) byte of data. Common, low-cost (less than $1 per gigabyte) sizes are presently 20G bytes to 300G bytes.

There are several differences between the floppy disk and the hard disk memory. The hard disk memory uses a flying head to store and read data from the surface of the disk. A flying head, which is very small and light, does not touch the surface of the disk. It flies above the surface on a film of air that is carried with the surface of the disk as it spins. The hard disk typically spins at 3000 to 15,000 RPM, which is many times faster than the floppy disk. This higher rotational speed allows the head to fly (just as an airplane flies) just over the top of the surface of the disk. This is an important feature because there is no wear on the hard disk's surface, as there is with the floppy disk.

Problems can arise because of flying heads. One problem is a head crash. If the power is abruptly interrupted or the hard disk drive is jarred, the head can crash onto the disk surface, which can damage the disk surface or the head. To help prevent crashes, some drive manufacturers have included a system that automatically parks the head when power is interrupted. This type of disk drive has auto-parking heads. When the heads are parked, they are moved to a safe landing zone (unused track) when the power is disconnected. Some drives are not auto-parking; they usually require a program that parks the heads on the innermost track before power is disconnected. The innermost track is a safe landing area because it is the very last track filled by the disk drive. Parking is the responsibility of the operator in this type of disk drive.

Another difference between a floppy disk drive and a hard disk drive is the number of heads and disk surfaces. A floppy disk drive has two heads, one for the upper surface and one for the lower surface. The hard disk drive may have up to eight disk surfaces (four platters), with up to two heads per surface. Each time that a new cylinder is obtained by moving the head assembly, 16 new tracks are available under the heads. See Figure 13–23, which illustrates a hard disk system.

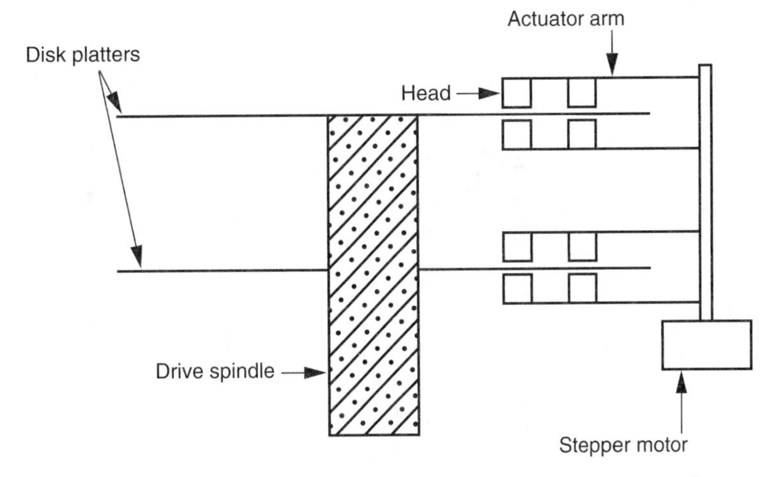

FIGURE 13–23 A hard disk drive that uses four heads per platter.

Heads are moved from track to track by using either a stepper motor or a voice coil. The stepper motor is slow and noisy, while the voice coil mechanism is quiet and quick. Moving the head assembly requires one step per cylinder in a system that uses a stepper motor to position the heads. In a system that uses a voice coil, the heads can be moved many cylinders with one sweeping motion. This makes the disk drive faster when seeking new cylinders.

Another advantage of the voice coil system is that a servo mechanism can monitor the amplitude of the signal as it comes from the read head and make slight adjustments in the position of the heads. This is not possible with a stepper motor, which relies strictly on mechanics to position the head. Stepper-motor-type head positioning mechanisms can often become misaligned with use, while the voice coil mechanism corrects for any misalignment.

Hard disk drives often store information in sectors that are 512 bytes long. Data are addressed in clusters of eight or more sectors, which contain 4096 bytes (or more) on most hard disk drives. Hard disk drives use either MFM or RLL to store information. MFM is described with floppy disk drives. Run-length limited (RLL) is described here.

A typical older MFM hard disk drive uses 18 sectors per track so that 18K bytes of data are stored per track. If a hard disk drive has a capacity of 40M bytes, it contains approximately 2280 tracks. If the disk drive has two heads, this means that it contains 1140 cylinders; if it contains four heads, then it has 570 cylinders. These specifications vary from disk drive to disk drive.

RLL Storage. **Run-length limited** (RLL) disk drives use a different method for encoding the data than MFM. The term RLL means that the run of zeros (zeros in a row) is limited. A common RLL encoding scheme in use today is RLL 2,7. This means that the run of zeros is always between two and seven. Table 13–3 illustrates the coding used with standard RLL.

TABLE 13–3 Standard RLL 2,7 encoding.

Input Data Stream	*RLL Output*
000	000100
10	0100
010	100100
0010	00100100
11	1000
011	001000
0011	00001000

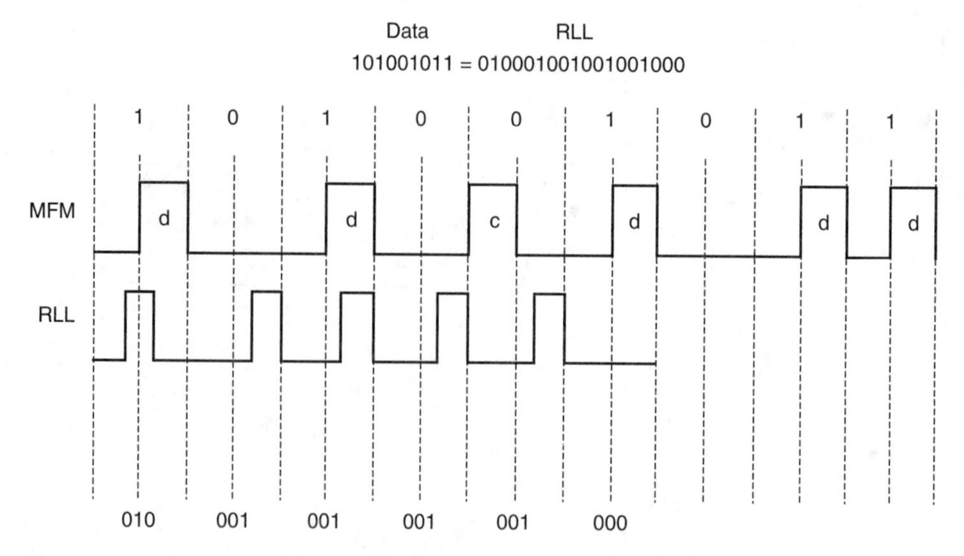

FIGURE 13–24 A comparison of MFM with RLL using data 101001011.

Data are first encoded by using Table 13–3 before being sent to the drive electronics for storage on the disk surface. Because of this encoding technique, it is possible to achieve a 50% increase in data storage on a disk drive when compared to MFM. The main difference is that the RLL drive often contains 27 tracks instead of the 18 found on the MFM drive. (Some RLL drives also use 35 sectors per track.)

In most cases, RLL encoding requires no change to the drive electronics or surface of the disk. The only difference is a slight decrease in the pulse width using RLL, which may require slightly finer oxide particles on the surface of the disk. Disk manufacturers test the surface of the disk and grade the disk drive as either an MFM-certified or an RLL-certified drive. Other than grading, there is no difference in the construction of the disk drive or the magnetic material that coats the surface of the disks.

Figure 13–24 shows a comparison of MFM data and RLL data. Notice that the amount of time (space) required to store RLL data is reduced when compared to MFM. Here 101001011 is coded in both MFM and RLL so that these two standards can be compared. Notice that the width of the RLL signal has been reduced so that three pulses fit in the same space as a clock and a data pulse for MFM. A 40M-byte MFM disk can hold 60M bytes of RLL-encoded data. Besides holding more information, the RLL drive can be written and read at a higher rate.

All hard disk drives use today RLL encoding. There are a number of disk drive interfaces in use today. The oldest is the ST-506 interface, which uses either MFM or RLL data. A disk system using this interface is also called either the MFM or RLL disk system. Newer standards are also in use today, which include ESDI, SCSI, and IDE. All of these newer standards use RLL, even though they normally do not call attention to it. The main difference is the interface between the computer and the disk drive. The IDE system is becoming the standard hard disk memory interface.

The **enhanced small disk interface** (ESDI) system, which has disappeared, is capable of transferring data between itself and the computer at rates approaching 10M bytes per second. An ST-506 interface can approach a transfer rate of 860K bytes per second.

The **small computer system interface** (SCSI) system is also in use because it allows up to seven different disk or other interfaces to be connected to the computer through the same interface controller. SCSI is found in some PC-type computers and also in the Apple Macintosh system. An improved version, SCSI-II, has started to appear in some systems. In the future, this interface may be replaced with IDE in most applications.

Today one of the most common systems is the **integrated drive electronics** (IDE) system, which incorporates the disk controller in the disk drive and attaches the disk drive to the host system through a small interface cable. This allows many disk drives to be connected to a system without worrying about bus conflicts or controller conflicts. IDE drives are found in newer IBM PS-2 systems and many clones. Even Apple computer systems are beginning to have IDE drives in place of the SCSI drives found in older Apple computers. The IDE interface is also capable of driving other I/O devices besides the hard disk. This interface also usually contains at least a 256K- to 8-Mbyte cache memory for disk data. The cache speeds disk transfers. Common access times for an IDE drive are often less than 8 ms, whereas the access time for a floppy-disk is about 200 ms.

Sometimes IDE is also called ATA, an acronym for **AT attachment**. The latest system is the serial ATA interface or SATA. This interface transfers serial data at rates of 150 MBps, which is faster than any IDE interface. It can do this because the logic 1 level is no longer 5.0 V. In the SATA interface, the logic 1 level is 0.5 V, which allows data to be transferred at higher rates because it takes less time for the signal to rise to 0.5 V than it takes to rise to 5.0 V. Speeds of this interface should eventually reach 600 MBps.

Optical Disk Memory

Optical disk memory (see Figure 13–25) is commonly available in two forms: the CD-ROM (compact disk/read only memory) and the WORM (write once/read many). The CD-ROM is the lowest cost optical disk, but it suffers from lack of speed. Access times for a CD-ROM are typically 300 ms or longer, about the same as a floppy disk. (Note that slower CD-ROM devices are on the market and should be avoided.) Hard disk magnetic memory can have access times as

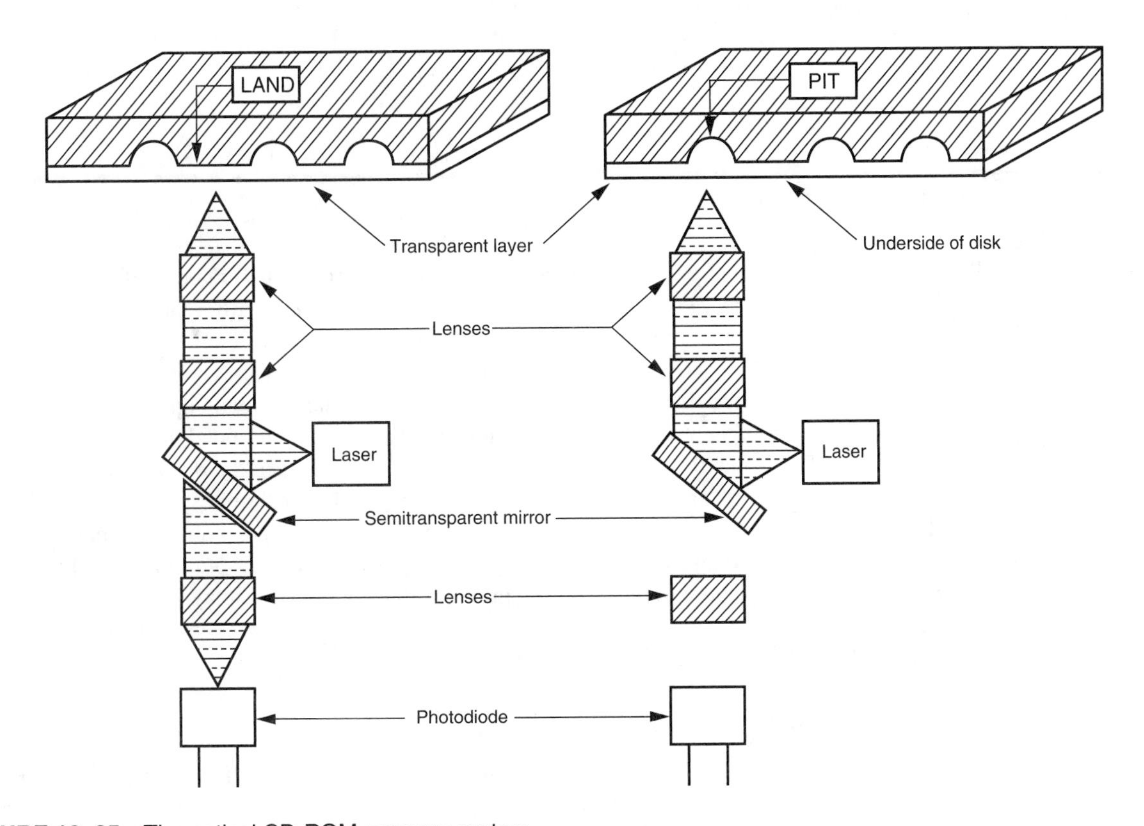

FIGURE 13–25 The optical CD-ROM memory system.

little as 11 ms. A CD-ROM stores 660M bytes of data, or a combination of data and musical passages. As systems develop and become more visually active, the use of the CD-ROM drive will become even more common.

The WORM drive sees far more commercial application than the CD-ROM. The problem is that its application is very specialized due to the nature of the WORM. Because data may be written only once, the main application is in the banking industry, insurance industry, and other massive data-storing organizations. The WORM is normally used to form an audit trail of transactions that are spooled onto the WORM and retrieved only during an audit. You might call the WORM an archiving device.

Many WORM and read/write optical disk memory systems are interfaced to the microprocessor by using the SCSI or ESDI interface standards used with hard disk memory. The difference is that the current optical disk drives are no faster than most floppy drives. Some CD-ROM drives are interfaced to the microprocessor through proprietary interfaces that are not compatible with other disk drives.

The main advantage of the optical disk is its durability. Because a solid state laser beam is used to read the data from the disk, and the focus point is below a protective plastic coating, the surface of the disk may contain small scratches and dirt particles and still be read correctly. This feature allows less care of the optical disk than a comparable floppy disk. About the only way to destroy data on an optical disk is to break it or deeply scar it.

The read/write CD-ROM drive is here and its cost is dropping rapidly. In the near future, we should start seeing the read/write CD-ROM replacing floppy disk drives. The main advantage is the vast storage available on the read/write CD-ROM. Soon, the format will change so that many G bytes of data will be available. The new versatile read/write CD-ROM, called a DVD, became available in late 1996 or early 1997. The DVD functions exactly like the CD-ROM except that the bit density is much higher. The CD-ROM stores 660M bytes of data, while the current-genre DVD stores 4.7G bytes or 9.4G bytes, depending on the current standard. Look for the DVD to eventually replace the CD-ROM format.

13–5 VIDEO DISPLAYS

Modem video displays are OEM (**original equipment manufacturer**) devices that are usually purchased and incorporated into a system. Today, there are many different types of video displays available in either color or monochrome versions.

Monochrome versions usually display information using amber, green, or paper-white displays. The paper-white displays were once extremely popular for many applications. The most common of these applications are desktop publishing and computer-aided drafting (CAD).

The color displays are more diverse and have all but replaced the black and white display. Color display systems are available that accept information as a composite video signal, much like your home television, as TTL voltage level signals (0 or 5 V), and as analog signals (0–0.7 V). Composite video displays are disappearing because the available resolution is too low. Today, many applications require high-resolution graphics that cannot be displayed on a composite display such as a home television receiver. Early composite video displays were found with the Commodore 64, Apple 2, and similar computer systems.

Video Signals

Figure 13–26 illustrates the signal sent to a composite video display. This signal is composed of several parts that are required for this type of display. The signals illustrated represent the signals sent to a color composite-video monitor. Notice that these signals include not only video, but

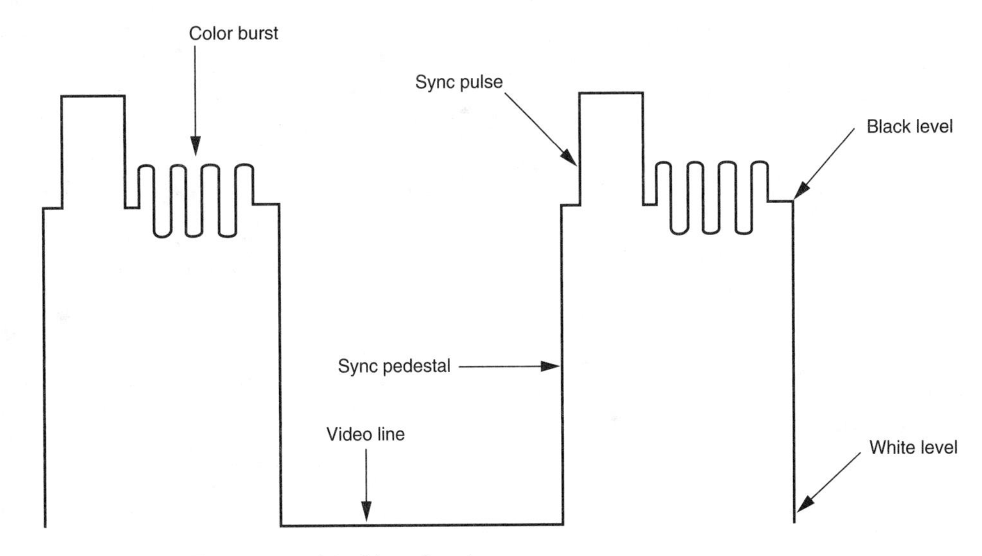

FIGURE 13–26 The composite video signal.

also include sync pulses, sync pedestals, and a color burst. Notice that no audio signal is illustrated because one often does not exist. Rather than include audio with the composite video signal, audio is developed in the computer and output from a speaker inside the computer cabinet. It can also be developed by a sound system and output in stereo to external speakers. The major disadvantages of the composite video display are the resolution and color limitations. Composite video signals were designed to emulate television video signals so that a home television receiver could function as a video monitor.

Most modern video systems use direct video signals that are generated with separate sync signals. In a direct video system, video information is passed to the monitor through a cable that uses separate lines for video and also synchronization pulses. Recall that these signals were combined in a composite video signal.

A monochrome (one color) monitor uses one wire for video, one for horizontal sync, and one for vertical sync. Often, these are the only signal wires found. A color video monitor uses three video signals. One signal represents red, another green, and the third blue. These monitors are often called RGB monitors for the video primary colors of light: red (R), green (G), and blue (B).

The TTL RGB Monitor

The RGB monitor is available as either an analog or TTL monitor. The RGB monitor uses TTL level signals (0 or 5 V) as video inputs and a fourth line called intensity to allow a change in intensity. The RGB video TTL display can display a total of 16 different colors. The TTL RGB monitor is used in the CGA (**color graphics adapter**) system found in older computer systems.

Table 13–4 lists these 16 colors and also the TTL signals present to generate them. Eight of the 16 colors are generated at high intensity and the other eight at low intensity. The three video colors are red, green, and blue. These are primary colors of light. The secondary colors are cyan, magenta, and yellow. Cyan is a combination of blue and green video signals, and is blue-green in color. Magenta is a combination of blue and red video signals, and is a purple color.

Yellow (high intensity) and brown (low intensity) are both a combination of red and green video signals. If additional colors are desired, TTL video is not normally used. A scheme was developed using low- and medium-color TTL video signals, which provided 32 colors, but it proved to have little application and never found widespread use in the field.

TABLE 13–4 The 16 colors found in a TTL display.

Intensity	*Red*	*Green*	*Blue*	*Color*
0	0	0	0	Black
0	0	0	1	Blue
0	0	1	0	Green
0	0	1	1	Cyan
0	1	0	0	Red
0	1	0	1	Magenta
0	1	1	0	Brown
0	1	1	1	White
1	0	0	0	Gray
1	0	0	1	Bright blue
1	0	1	0	Bright green
1	0	1	1	Bright cyan
1	1	0	0	Bright red
1	1	0	1	Bright magenta
1	1	1	0	Yellow
1	1	1	1	Bright white

Figure 13–27 illustrates the connector most often found on the TTL RGB monitor or a TTL monochrome monitor. The connector illustrated is a 9-pin connector. Two of the connections are used for ground, three for video, two for synchronization or retrace signals, and one for intensity. Notice that pin 7 is labeled normal video. This is the pin used on a monochrome monitor for the luminance or brightness signal. Monochrome TTL monitors use the same 9-pin connector as RGB TTL monitors.

The Analog RGB Monitor

In order to display more than 16 colors, an analog video display is required. These are often called analog RGB monitors. Analog RGB monitors still have three video input signals, but don't have the intensity input. Because the video signals are analog signals instead of two-level TTL signals, they are at any voltage level between 0.0 V and 0.7 V, which allows an infinite number of colors to be displayed. This is because an infinite number of voltage levels between the minimum and maximum could be generated. In practice, a finite number of levels are generated. This is usually either 256K, 16M, or 24M colors, depending on the standard.

Figure 13–28 illustrates the connector used for an analog RGB or analog monochrome monitor. Notice that the connector has 15 pins and supports both RGB and monochrome analog displays.

DB9

5 9 4 8 3 7 2 6 1

Pin	Function
1	Ground
2	Ground
3	Red video
4	Green video
5	Blue video
6	Intensity
7	Normal video
8	Horizontal retrace
9	Vertical retrace

FIGURE 13–27 The 9-pin connector found on a TTL monitor.

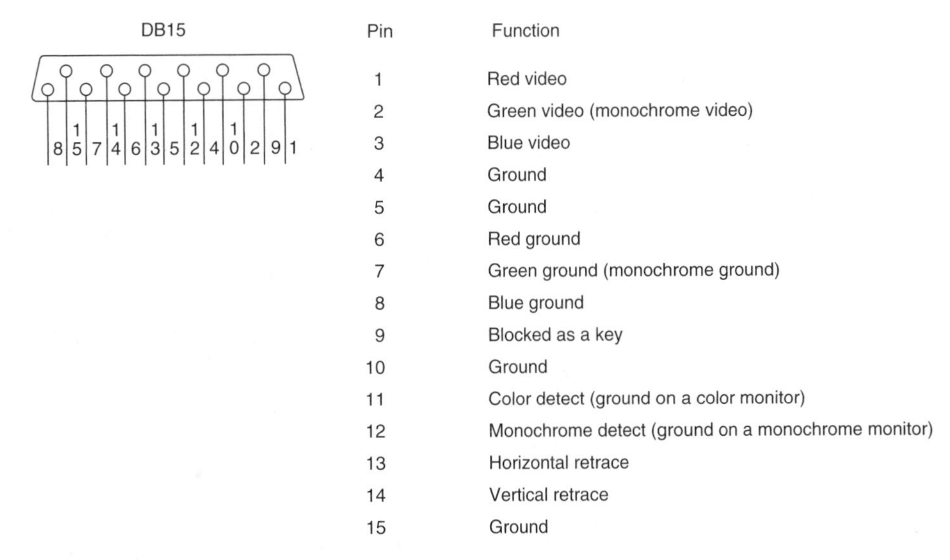

Pin	Function
1	Red video
2	Green video (monochrome video)
3	Blue video
4	Ground
5	Ground
6	Red ground
7	Green ground (monochrome ground)
8	Blue ground
9	Blocked as a key
10	Ground
11	Color detect (ground on a color monitor)
12	Monochrome detect (ground on a monochrome monitor)
13	Horizontal retrace
14	Vertical retrace
15	Ground

FIGURE 13–28 The 15-pin connector found on an analog monitor.

The way data are displayed on an analog RGB monitor depends upon the interface standard used with the monitor. Pin 9 is a key, which means that no hole exists on the female connector for this pin.

Another type of connector for the analog RGB monitor that is becoming common is called the DVI-D (digital visual interface) connector. The -D is for digital and is the most common interface of this type. Figure 13–29 illustrates the female connector found on newer monitors and video cards.

Most analog displays use a digital-to-analog converter (DAC) to generate each color video voltage. A common standard uses an eight-bit DAC for each video signal to generate 256 different voltage levels between 0 V and 0.7 V. There are 256 different red video levels, 256 different green video levels, and 256 different blue video levels. This allows 256 × 256 × 256, or 16,777,216 (16M) colors to be displayed.

Figure 13–30 illustrates the video generation circuit employed in many common video standards such as the short-lived EGA (**enhanced graphics adapter**) and VGA (**variable graphics array**), as used with an IBM PC. This circuit is used to generate VGA video. Notice

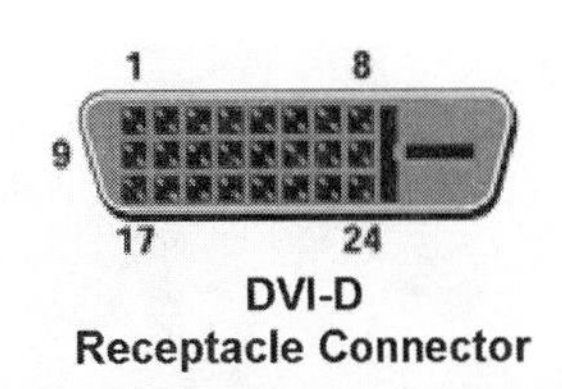

DIGITAL-ONLY CONNECTOR PIN ASSIGNMENTS					
Pin	Signal Assignment	Pin	Signal Assignment	Pin	Signal Assignment
1	Data2-	9	Data1-	17	Data0-
2	Data2+	10	Data1+	18	Data0+
3	Data2/4 Shield	11	Data1/3 Shield	19	Data0/5 Shield
4	Data4-	12	Data3-	20	Data5-
5	Data4+	13	Data3+	21	Data5+
6	DDC Clock	14	+5V Power	22	Clock Shield
7	DDC Data	15	Ground (for +5V)	23	Clock+
8	No Connect	16	Hot Plug Detect	24	Clock-

FIGURE 13–29 The DVI-D interface found on many newer monitors and video cards.

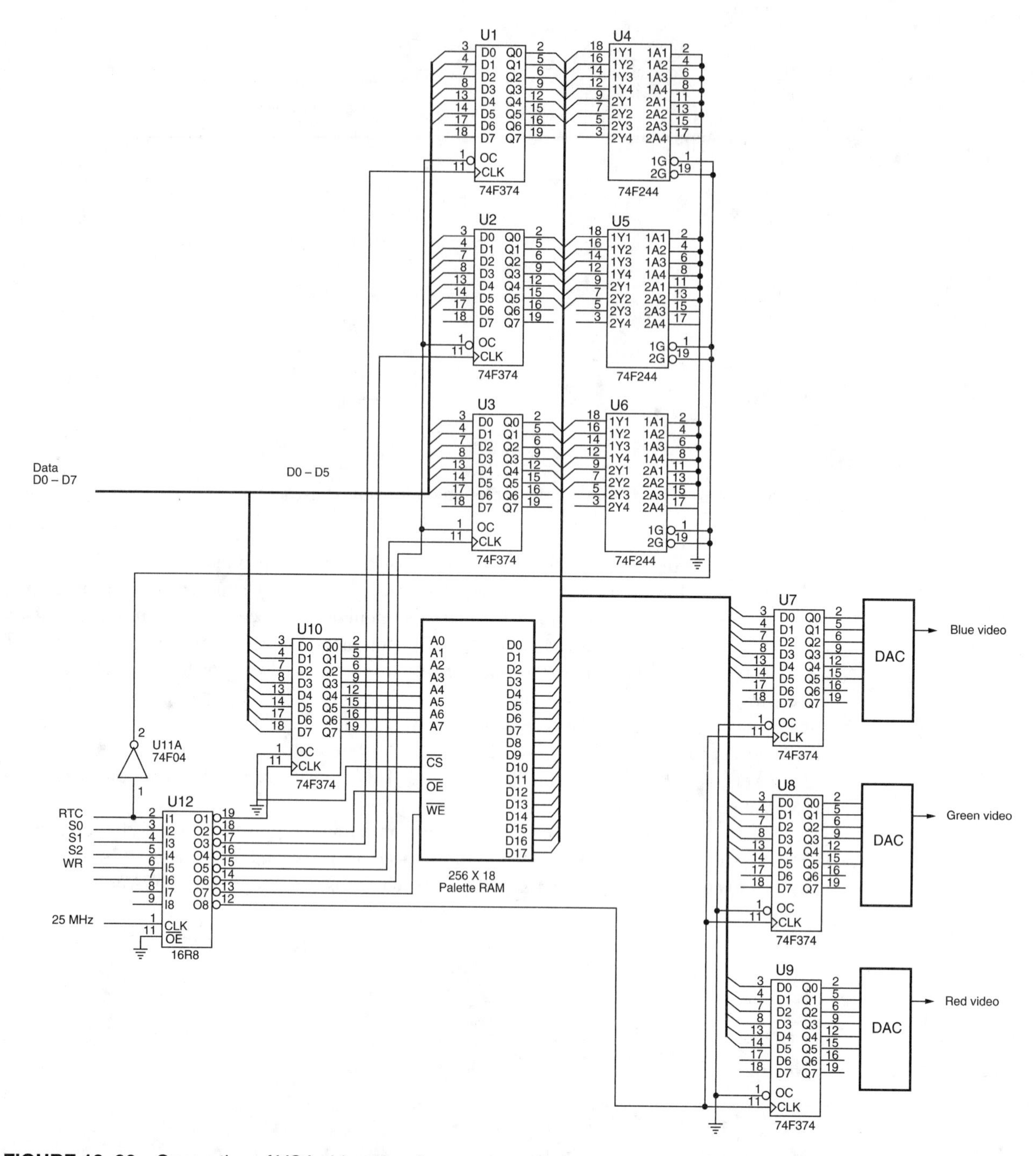

FIGURE 13–30 Generation of VGA video signals.

that each color is generated with an 18-bit digital code. Six of the 18 bits are used to generate each video color voltage when applied to the inputs of an eight-bit DAC.

A high-speed palette SRAM (access time of less than 40 ns) is used to store 256 different 18-bit codes that represent 256 different hues. This 18-bit code is applied to the digital-to-analog converters. The address input to the SRAM selects one of the 256 colors stored as 18-bit binary codes. This system allows 256 colors out of a possible 256K colors to be displayed at one time. In order to select any of the 256 colors, an 8-bit code that is stored in the computer's video display

RAM is used to specify the color of a picture element. If more colors are used in a system, the code must be wider. For example, a system that displays 1024 colors out of 256K colors requires a 10-bit code to address the SRAM that contains 1024 locations, each containing an 18-bit color code. Some newer systems use a larger palette SRAM to store up to 64K of different color codes.

Whenever a color is placed on the video display, provided that RTC is a logic 0, the system sends the 8-bit code that represents a color to the D_0–D_7 connections. The PLD then generates a clock pulse for U_{10}, which latches the color code. After 40 ns (one 25 MHz clock), the PLD generates a clock pulse for the DAC latches (U_7, U_8, and U_9). This amount of time is required for the palette SRAM to look up the 18-bit contents of the memory location selected by U_{10}. Once the color code (18-bit) is latched into U_7–U_9, the three DACs convert it to three video voltages for the monitor. This process is repeated for each 40 ns-wide picture element (pixel) that is displayed. The pixel is 40 ns wide because a 25 MHz clock is used in this system. Higher resolution is attainable if a higher clock frequency is used with the system.

If the color codes (18-bits) stored in the SRAM must be changed, this is always accomplished during retrace when RTC is a logic 1. This prevents any video noise from disrupting the image displayed on the monitor.

In order to change a color, the system uses the S_0, S_1, and S_2 inputs of the PLD to select U_1, U_2, U_3 and U_{10}. First, the address of the color to be changed is sent to latch U_{10}, which addresses a location in the palette SRAM. Next, each new video color is loaded into U_1, U_2, and U_3. Finally, the PLD generates a write pulse for the $\overline{WE}$ input to the SRAM to write the new color code into the palette SRAM.

Retrace occurs 70.1 times per second in the vertical direction and 31,500 times per second in the horizontal direction for a 640 × 480 display. During retrace, the video signal voltage sent to the display must be 0 V, which causes black to be displayed during the retrace. Retrace itself is used to move the electron beam to the upper left-hand corner for vertical retrace and to the left margin of the screen for horizontal retrace.

The circuit illustrated causes U_4–U_6 buffers to be enabled so that they apply 000000 each to the DAC latch for retrace. The DAC latches capture this code and generate 0 V for each video color signal to blank the screen. By definition, 0 V is considered to be the black level for video and 0.7 V is considered to be the full intensity on a video color signal.

The resolution of the display, for example, 640 × 480, determines the amount of memory required for the video interface card. If this resolution is used with a 256-color display (eight bits per pixel), then 640 × 480 bytes of memory (307,200) are required to store all of the pixels for the display. Higher-resolution displays are possible, but, as you can imagine, even more memory is required. A 640 × 480 display has 480 video raster lines and 640 pixels per line. A **raster line** is the horizontal line of video information that is displayed on the monitor. A pixel is the smallest subdivision of this horizontal line.

Figure 13–31 illustrates the video display, showing the video lines and retrace. The slant of each video line in this illustration is greatly exaggerated, as is the spacing between lines. This illustration shows retrace in both the vertical and horizontal directions. In the case of a VGA display, as described, the vertical retrace occurs exactly 70.1 times per second and the horizontal retrace occurs exactly 31,500 times per second.

In order to generate 640 pixels across one line, it takes 40 ns × 640, or 25.6 μs. A horizontal time of 31,500 Hz allows a horizontal line time of 1/31,500, or 31.746 μs. The difference between these two times is the retrace time allowed to the monitor. (The Apple Macintosh has a horizontal line time of 28.57 μs.)

Because the vertical retrace repetition rate is 70.1 Hz, the number of lines generated is determined by dividing the vertical time into the horizontal time. In the case of a VGA display (a 640 × 400 display), this is 449.358 lines. Only 400 of these lines are used to display information; the rest are lost during the retrace. Because 49.358 lines are lost during the retrace, the retrace time is 49.358 × 31.766 μs, or 1568 μs. It is during this relatively large amount of time that the color palette SRAM is changed or the display memory system is updated for a new video display.

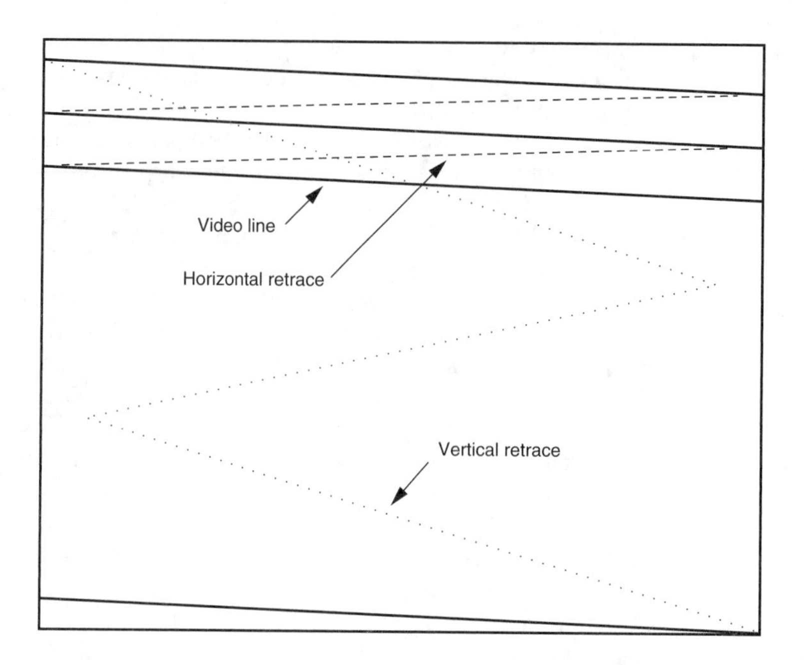

FIGURE 13–31 A video screen illustrating the raster lines and retrace.

In the Apple Macintosh computer (640 × 480), the number of lines generated is 525 lines. Of the total number of lines, 45 are lost during vertical retrace.

Other display resolutions are 800 × 600 and 1024 × 768. The 800 × 600 SVGA (super VGA) display is ideal for a 14" color monitor, while the 1024 × 768 EVGA or XVGA (extended VGA) is ideal for a 21" or 25" monitor used in CAD systems. These resolutions sound like just another set of numbers, but realize that an average home television receiver has a resolution of approximately 400 × 300. The high-resolution display available on computer systems is much clearer than that available as home television. A resolution of 1024 × 768 approaches that found in 35 mm film. The only disadvantage of the video display on a computer screen is the number of colors displayed at a time, but as time passes, this will surely improve. Additional colors allow the image to appear more realistically because of subtle shadings that are required for a true high-quality, lifelike image.

If a display system operates with a 60 Hz vertical time and a 15,600 Hz horizontal time, the number of lines generated is 15,600/60, or 260 lines. The number of usable lines in this system is most likely 240, where 20 are lost during vertical retrace. It is clear that the number of scanning lines is adjustable by changing the vertical and horizontal scanning rates. The vertical scanning rate must be greater than or equal to 50 Hz or flickering will occur. The vertical rate must not be higher than about 75 Hz or problems with the vertical deflection coil may occur. The electron beam in a monitor is positioned by an electrical magnetic field generated by coils in a yoke that surrounds the neck of the picture tube. Because the magnetic field is generated by coils, the frequency of the signal applied to the coil is limited.

The horizontal scanning rate is also limited by the physical design of the coils in the yoke. Because of this, it is normal to find the frequency applied to the horizontal coils within a narrow range. This is usually 30,000 Hz–37,000 Hz or 15,000 Hz–17,000 Hz. Some newer monitors are called multisync monitors because the deflection coil is taped so that it can be driven with different deflection frequencies. Sometimes, both the vertical and horizontal coils are both taped for different vertical and horizontal scanning rates.

High-resolution displays use either interlaced or noninterlaced scanning. The noninterlaced scanning system is used in all standards except the highest. In the interlaced system, the video image is displayed by drawing half the image first with all of the odd scanning lines, then the other half using the even scanning lines. Obviously, this system is more complex and is only more efficient because the scanning frequencies are reduced by 50% in an interlaced system. For example, a video system that uses 60 Hz for the vertical scanning frequency and 15,720 Hz for the horizontal frequency generates 262 (15,720/60) lines of video at the rate of 60 full frames per second. If the horizontal frequency is changed slightly to 15,750 Hz, 262.5 (15,750/60) lines are generated, so two full sweeps are required to draw one complete picture of 525 video lines. Notice how just a slight change in horizontal frequency doubles the number of raster lines.

13–6 SUMMARY

1. The HOLD input is used to request a DMA action, and the HLDA output signals that the hold is in effect. When a logic 1 is placed on the HOLD input, the microprocessor (1) stops executing the program; (2) places its address, data, and control buses at their high-impedance state; and (3) signals that the hold is in effect by placing a logic 1 on the HLDA pin.
2. A DMA read operation transfers data from a memory location to an external I/O device. A DMA write operation transfers data from an I/O device into the memory. Also available is a memory-to-memory transfer that allows data to be transferred between two memory locations by using DMA techniques.
3. The 8237 direct memory access (DMA) controller is a four-channel device that can be expanded to include an additional channel of DMA.
4. Disk memory comes in the form of floppy disk storage that is found as 3½" micro-floppy disks. Disks are found as double-sided, double-density (DSDD) or high-density (HD) storage devices. The DSDD 3½" disk stores 720K bytes of data and the HD 3½" disk stores 1.44M bytes of data.
5. Floppy disk memory data are stored using NRZ (non-return to zero) recording. This method saturates the disk with one polarity of magnetic energy for a logic 1 and the opposite polarity for a logic 0. In either case, the magnetic field never returns to 0. This technique eliminates the need for a separate erase head.
6. Data are recorded on disks by using either modified frequency modulation (MFM) or run-length limited (RLL) encoding schemes. The MFM scheme records a data pulse for a logic 1, no data or clock for the first logic 0 of a string of zeros, and a clock pulse for the second and subsequent logic 0 in a string of zeros. The RLL scheme encodes data so that 50% more information can be packed onto the same disk area. Most modern disk memory systems use the RLL encoding scheme.
7. Video monitors are either TTL or analog. The TTL monitor uses two discrete voltage levels of 0 V and 5.0 V. The analog monitor uses an infinite number of voltage levels between 0.0 V and 0.7 V. The analog monitor can display an infinite number of video levels, while the TTL monitor is limited to two video levels.
8. The color TTL monitor displays 16 different colors. This is accomplished through three video signals (red, green, and blue) and an intensity input. The analog color monitor can display an infinite number of colors through its three video inputs. In practice, the most common form of color analog display system (VGA) can display 16M different colors.
9. The video standards found today include VGA (640 × 480), SVGA (800 × 600), and EVGA or XVGA (1024 × 768). In all three cases, the video information can be 16M colors.

13–7 QUESTIONS AND PROBLEMS

1. Which microprocessor pins are used to request and acknowledge a DMA transfer?
2. Explain what happens whenever a logic 1 is placed on the HOLD input pin.
3. A DMA read transfers data from _________ to _________.

4. A DMA write transfers data from _________ to _________.
5. The DMA controller selects the memory location used for a DMA transfer through what bus signals?
6. The DMA controller selects the I/O device used during a DMA transfer by which pin?
7. What is a memory-to-memory DMA transfer?
8. Describe the effect on the microprocessor and DMA controller when the HOLD and HLDA pins are at their logic 1 levels.
9. Describe the effect on the microprocessor and DMA controller when the HOLD and HLDA pins are at their logic 0 levels.
10. The 8237 DMA controller is a(n) _________ channel DMA controller.
11. If the 8237 DMA controller is decoded at I/O ports 2000H–200FH, what ports are used to program channel 1?
12. Which 8237 DMA controller register is programmed to initialize the controller?
13. How many bytes can be transferred by the 8237 DMA controller?
14. Write a sequence of instructions that transfer data from memory locations 21000H–210FFH to 20000H–200FFH by using channel 2 of the 8237 DMA controller. You must initialize the 8237 and use the latch described in Section 12–1 to hold A_{19}–A_{16}.
15. Write a sequence of instructions that transfer data from memory to an external I/O device by using Channel 3 of the 8237. The memory area to be transferred is at locations 20000H–20FFFH.
16. What is a pen drive?
17. The 3½" disk is known as a(n) _________- floppy disk.
18. Data are recorded in concentric rings on the surface of a disk known as a(n) _________.
19. A track is divided into sections of data called _________.
20. On a double-sided disk, the upper and lower tracks together are called a(n) _________.
21. Why is NRZ recording used on a disk memory system?
22. Draw the timing diagram generated to write 1001010000 using MFM encoding.
23. Draw the timing diagram generated to write 1001010000 using RLL encoding.
24. What is a flying head?
25. Why must the heads on a hard disk be parked?
26. What is the difference between a voice coil head position mechanism and a stepper motor head positioning mechanism?
27. What is a WORM?
28. What is a CD-ROM?
29. How much data can be stored on a common DVD?
30. What is the difference between a TTL monitor and an analog monitor?
31. What are the three primary colors of light?
32. What are the three secondary colors of light?
33. What is a pixel?
34. A video display with a resolution of 800×600 contains _________ lines, with each line divided into _________ pixels.
35. Explain how a TTL RGB monitor can display 16 different colors.
36. What is a DVI connector?
37. Explain how an analog RGB monitor can display an infinite number of colors.
38. If an analog RGB video system uses 8-bit DACs, it can generate _________ different colors.
39. If a video system uses a vertical frequency of 60 Hz and a horizontal frequency of 32,400 Hz, how many raster lines are generated?

CHAPTER 14

The Arithmetic Coprocessor, MMX, and SIMD Technologies

INTRODUCTION

The Intel family of arithmetic coprocessors includes the 8087, 80287, 80387SX, 80387DX, and the 80487SX for use with the 80486SX microprocessor. The 80486DX–Pentium 4 microprocessors contain their own built-in arithmetic coprocessors. Be aware that some of the cloned 80486 microprocessors (from IBM and Cyrix) did not contain arithmetic coprocessors. The instruction sets and programming for all devices are almost identical; the main difference is that each coprocessor is designed to function with a different Intel microprocessor. This chapter provides detail on the entire family of arithmetic coprocessors. Because the coprocessor is a part of the 80486DX–Pentium 4, and because these microprocessors are commonplace, many programs now require or at least benefit from a coprocessor.

The family of coprocessors, which is labeled the 80X87, is able to multiply, divide, add, subtract, find the square root, and calculate the partial tangent, partial arctangent, and logarithms. Data types include 16-, 32-, and 64-bit signed integers; 18-digit BCD data; and 32-, 64-, and 80-bit floating-point numbers. The operations performed by the 80X87 generally execute many times faster than equivalent operations written with the most efficient programs that use the microprocessor's normal instruction set. With the improved Pentium coprocessor, operations execute about five times faster than those performed by the 80486 microprocessor with an equal clock frequency. Note that the Pentium can often execute a coprocessor instruction and two integer instructions simultaneously. The Pentium Pro through Pentium 4 coprocessors are similar in performance to the Pentium coprocessor, except that a few new instructions have been added: FCMOV and FCOMI.

The multimedia extensions (MMX) to the Pentium–Pentium 4 are instructions that share the arithmetic coprocessor register set. The MMX extension is a special internal processor designed to execute integer instructions at high speed for external multimedia devices. For this reason, the MMX instruction set and specifications have been placed in this chapter. The SIMD (single-instruction, multiple data) extensions, which are called SSE (streaming SIMD extensions), are similar to the MMX instructions, but function with floating-point numbers instead of integers and do not use the coprocessor register space as do MMX instructions.

CHAPTER OBJECTIVES

Upon completion of this chapter, you will be able to:

1. Convert between decimal data and signed integer, BCD, and floating-point data for use by the arithmetic coprocessor, MMX, and SSE technologies

2. Explain the operation of the 80X87 arithmetic coprocessor and the MMX and SSE units
3. Explain the operation and addressing modes of each arithmetic coprocessor, MMX, and SSE instruction
4. Develop programs that solve complex arithmetic problems using the arithmetic coprocessor, MMX, and SSE instructions

14–1 DATA FORMATS FOR THE ARITHMETIC COPROCESSOR

This section of the text presents the types of data used with all arithmetic coprocessor family members. (See Table 14–1 for a listing of all Intel microprocessors and their companion coprocessors.) These data types include signed integer, BCD, and floating-point. Each has a specific use in a system, and many systems require all three data types. Note that assembly language programming with the coprocessor is often limited to modifying the coding generated by a high-level language such as C/C++. In order to accomplish any such modification, the instruction set and some basic programming concepts are required, which are presented in this chapter.

Signed Integers

The signed integers used with the coprocessor are the same as those described in Chapter 1. When used with the arithmetic coprocessor, signed integers are 16- (word), 32- (doubleword integer), or 64-bits (quadword integer) wide. The long integer is new to the coprocessor and is not described in Chapter 1, but the principles are the same. Conversion between decimal and signed integer format is handled in exactly the same manner as for the signed integers described in Chapter 1. As you will recall, positive numbers are stored in true form with a leftmost sign-bit of 0, and negative numbers are stored in two's complement form with a leftmost sign-bit of 1.

The word integers range in value from –32,768 to +32,767, the doubleword integer range is $\pm 2 \times 10^9$, and the quadword integer range is $\pm 9 \times 10^{18}$. Integer data types are found in some applications that use the arithmetic coprocessor. See Figure 14–1, which shows these three forms of signed integer data.

Data are stored in memory using the same assembler directives described and used in earlier chapters. The DW directive defines words, DD defines doubleword integers, and DQ defines quadword integers. Example 14–1 shows how several different sizes of signed integers are defined for use by the assembler and arithmetic coprocessor.

EXAMPLE 14–1

```
0000 0002                DATA1  DW     2       ;16-bit integer
0002 FFDE                DATA2  DW     -34     ;16-bit integer
0004 000004D2            DATA3  DD     1234    ;32-bit integer
0008 FFFFFF9C            DATA4  DD     -100    ;32-bit integer
000C 0000000000005BA0    DATA5  DQ     23456   ;64-bit integer
0014 FFFFFFFFFFFFFF86    DATA6  DQ     -122    ;64-bit integer
```

TABLE 14–1 Microprocessor and coprocessor compatibility.

Microprocessor	*Coprocessor*
8086/8088	8087
80186/80188	80187
80286	80287
80386	80387
80486SX	80487SX
80486DX–Pentium 4	Built into microprocessor

FIGURE 14–1 Integer formats for the 80X87 family of arithmetic coprocessors: (a) word, (b) short, and (c) long.

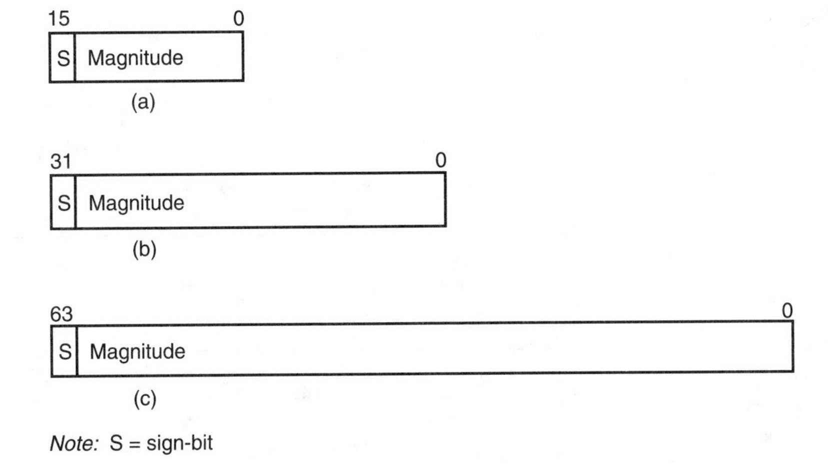

Binary-Coded Decimal (BCD)

The binary-coded decimal (BCD) form requires 80 bits of memory. Each number is stored as an 18-digit packed integer in nine bytes of memory as two digits per byte. The tenth byte contains only a sign-bit for the 18-digit signed BCD number. Figure 14–2 shows the format of the BCD number used with the arithmetic coprocessor. Note that both positive and negative numbers are stored in true form and never in ten's complement form. The DT directive stores BCD data in the memory as illustrated in Example 14–2. This form is rarely used because it is unique to the Intel coprocessor.

EXAMPLE 14–2

```
0000 00000000000000000200    DATA1  DT     200     ;define 10 byte
000A 80000000000000000010    DATA2  DT     -10     ;define 10 byte
0014 00000000000000010020    DATA3  DT     10020   ;define 10 byte
```

Floating-Point

Floating-point numbers are often called *real numbers* because they hold signed integers, fractions, and mixed numbers. A floating-point number has three parts: a sign-bit, a biased exponent, and a significand. Floating-point numbers are written in scientific binary notation. The Intel family of arithmetic coprocessors supports three types of floating-point numbers: single (32 bits), double (64 bits), and temporary (80 bits). See Figure 14–3 for the three forms of the floating-point number. Please note that the single form is also called a single-precision number and the double form is called a double-precision number. Sometimes the 80-bit temporary form is called an extended-precision number. The floating-point numbers and the operations performed by the arithmetic coprocessor conform to the IEEE-754 standard, as adopted by all major personal computer software producers. This includes Microsoft, which in 1995 stopped supporting the Microsoft floating-point format and also the ANSI floating-point standard that is popular in some mainframe computer systems.

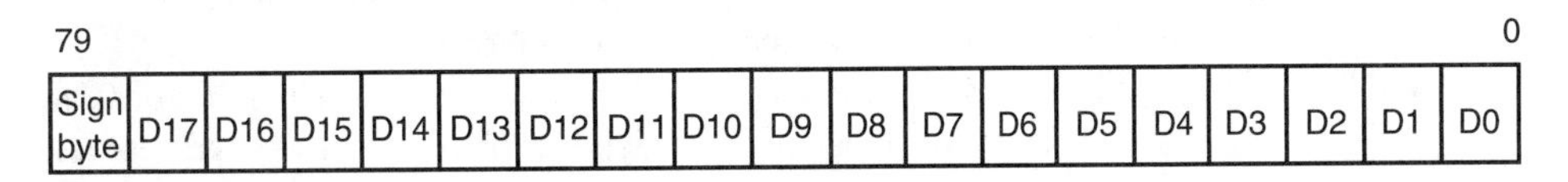

FIGURE 14–2 BCD data format for the 80X87 family of arithmetic coprocessors.

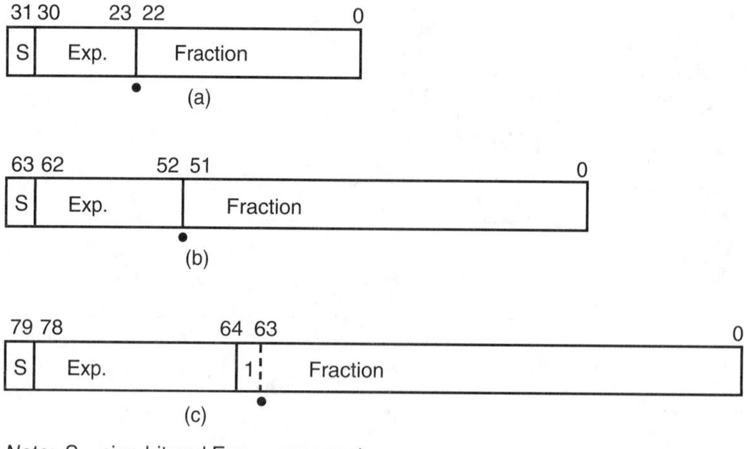

FIGURE 14–3 Floating-point (real) format for the 80X87 family of arithmetic coprocessors. (a) Short (single-precision) with a bias of 7FH, (b) long (double-precision) with a bias of 3FFH, and (c) temporary (extended-precision) with a bias of 3FFFH.

Converting to Floating-Point Form. Converting from decimal to the floating-point form is a simple task that is accomplished through the following steps:

1. Convert the decimal number to binary.
2. Normalize the binary number.
3. Calculate the biased exponent.
4. Store the number in the floating-point format.

These four steps are illustrated for the decimal number 100.25_{10} in Example 14–3. Here, the decimal number is converted to a single-precision (32-bit) floating-point number.

EXAMPLE 14–3

```
Step            Result

1               100.25 => 1100100.01

2               1100100.01 => 1.10010001 x 2^6

3               110 + 01111111 => 10000101

4               Sign => 0
                Exponent => 10000101
                Significand => 10010001000000000000000
```

In step 3 of Example 14–3, the biased exponent is the exponent, 2^6 or 110, plus a bias of 01111111 (7FH) or 10000101 (85H). All single-precision numbers use a bias of 7FH, double-precision numbers use a bias of 3FFH, and extended-precision numbers use a bias of 3FFFH.

In step 4 of Example 14–3, the information found in the prior steps is combined to form the floating-point number. The leftmost bit is the sign-bit of the number. In this case, it is a 0 because the number is $+100.25_{10}$. The biased exponent follows the sign-bit. The significand is a 23-bit number with an implied one-bit. Note that the significand of a number 1.XXXX is the XXXX portion. The 1. is an **implied one-bit** that is only stored in the extended temporary-precision form of the floating-point number as an explicit one-bit.

Some special rules apply to a few numbers. The number 0, for example, is stored as all zeros except for the sign-bit, which can be a logic 1 to represent a negative zero. The plus and minus infinity is stored as logic 1s in the exponent with a significand of all zeros and the sign-bit that represents plus or minus. A NAN (not-a-number) is an invalid floating-point result that has all ones in the exponent with a significand that is *not* all zeros.

Converting from Floating-Point Form. Conversion to a decimal number from a floating-point number is summarized in the following steps:

1. Separate the sign-bit, biased exponent, and significand.
2. Convert the biased exponent into a true exponent by subtracting the bias.
3. Write the number as a normalized binary number.
4. Convert it to a denormalized binary number.
5. Convert the denormalized binary number to decimal.

These five steps convert a single-precision floating-point number to decimal, as shown in Example 14–4. Notice how the sign-bit of 1 makes the decimal result negative. Also notice that the implied one-bit is added to the normalized binary result in step 3.

EXAMPLE 14–4

```
Step            Result

1               Sign => 1
                Exponent => 10000011
                Significand => 10010010000000000000000

2               100 = 10000011 - 01111111

3               1.1001001 x 2^4

4               11001.001

5               -25.125
```

Storing Floating-Point Data in Memory. Floating-point numbers are stored with the assembler using the DD directive for single-precision, DQ for double-precision, and DT for extended temporary-precision. Some examples of floating-point data storage are shown in Example 14–5. The author discovered that the Microsoft macro assembler contains an error that does not allow a plus sign to be used with positive floating-point numbers. A +92.45 must be defined as 92.45 for the assembler to function correctly. Microsoft has assured the author that this error has been corrected in version 6.11 of MASM if the REAL4, REAL8, or REAL10 directives are used in place of DD, DQ, and DT to specify floating-point data. The assembler provides access to the 8087 emulator if your system does not contain a microprocessor with a coprocessor. The emulator comes with all Microsoft high-level languages or as shareware programs such as EM87. Access the emulator by including the OPTION EMULATOR statement immediately following the .MODEL statement in a program. Be aware that the emulator does not emulate some of the coprocessor instructions. Do not use this option if your system contains a coprocessor. In all cases, you must include the .8087, .80187, .80287, .80387, .80487, .80587, or .80687 switch to enable the generation of coprocessor instructions.

EXAMPLE 14–5

```
0000 C377999A                DATA7  DD      -247.6      ;single-precision
0004 40000000                DATA8  DD      2.0         ;single precision
0008 486F4200                DATA9  REAL4   2,45E+5     ;single-precision
000C 4059100000000000        DATAA  DQ      100.25      ;double-precision
0014 3F543BF727136A40        DATAB  REAL8   0.001235    ;double-precision
001C 400487F34D6A161E4F76    DATAC  REAL10  33.9876     ;temporary-precision
```

14–2 THE 80X87 ARCHITECTURE

The 80X87 is designed to operate concurrently with the microprocessor. Note that the 80486DX–Pentium 4 microprocessors contain their own internal and fully compatible versions of the 80387. With other family members, the coprocessor is an external integrated circuit that parallels most of the connections on the microprocessor. The 80X87 executes 68 different instructions. The microprocessor executes all normal instructions and the 80X87 executes arithmetic coprocessor instructions. Both the microprocessor and coprocessor will execute their respective instructions simultaneously or concurrently. The numeric or arithmetic coprocessor is a special-purpose microprocessor that is especially designed to efficiently execute arithmetic and transcendental operations.

The microprocessor intercepts and executes the normal instruction set, and the coprocessor intercepts and executes only the coprocessor instructions. Recall that the coprocessor instructions are actually escape (ESC) instructions. These instructions are used by the microprocessor to generate a memory address for the coprocessor so that the coprocessor can execute a coprocessor instruction.

Internal Structure of the 80X87

Figure 14–4 shows the internal structure of the arithmetic coprocessor. Notice that this device is divided into two major sections: the control unit and the numeric execution unit.

The **control unit** interfaces the coprocessor to the microprocessor-system data bus. Both devices monitor the instruction stream. If the instruction is an ESCape (coprocessor) instruction, the coprocessor executes it; if not, the microprocessor executes it.

The **numeric execution unit** (NEU) is responsible for executing all coprocessor instructions. The NEU has an eight-register stack that holds operands for arithmetic instructions and the

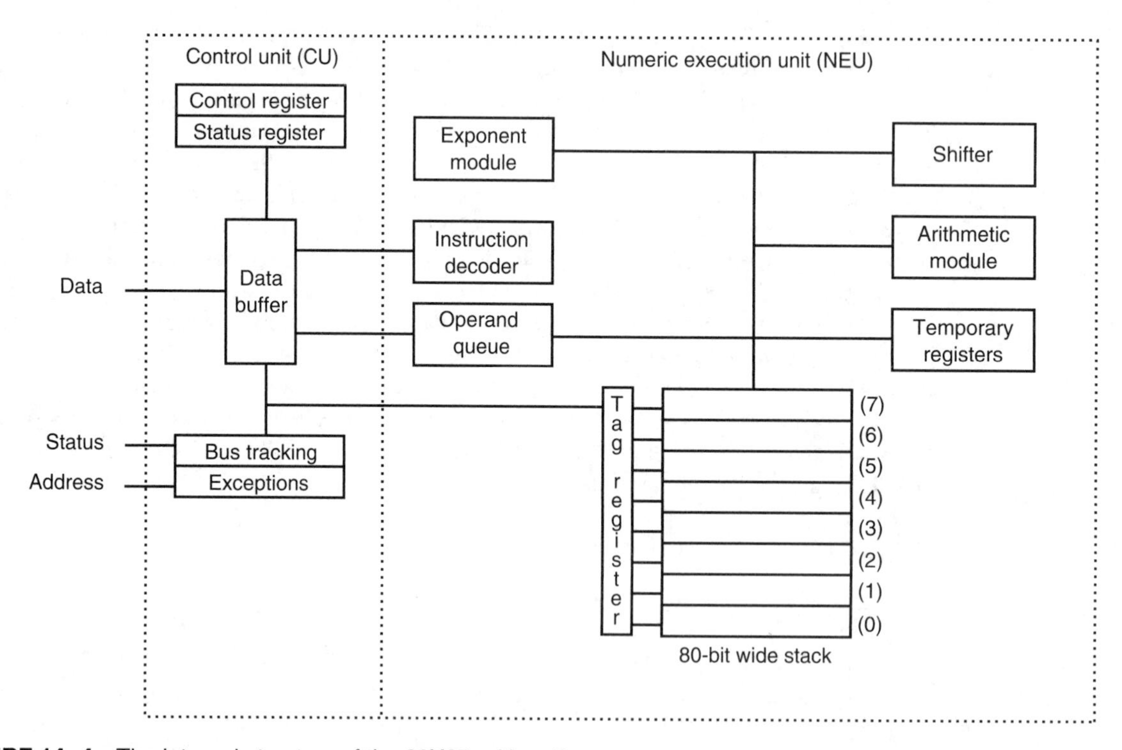

FIGURE 14–4 The internal structure of the 80X87 arithmetic coprocessor.

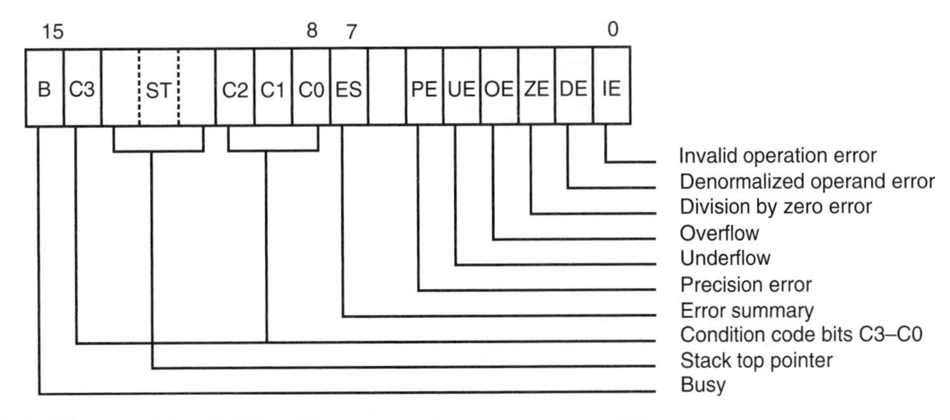

FIGURE 14–5 The 80X87 arithmetic coprocessor status register.

results of arithmetic instructions. Instructions either address data in specific stack data registers or use a push-and-pop mechanism to store and retrieve data on the top of the stack. Other registers in the NEU are status, control, tag, and exception pointers. A few instructions transfer data between the coprocessor and the AX register in the microprocessor. The FSTSW AX instruction is the only instruction available to the coprocessor that allows direct communications to the microprocessor through the AX register. Note that the 8087 does not contain the FSTSW AX instruction, but all newer coprocessors do contain it.

The stack within the coprocessor contains eight registers that are each 80 bits wide. These stack registers always contain an 80-bit extended-precision floating-point number. The only time that data appear as any other form is when they reside in the memory system. The coprocessor converts from signed integer, BCD, single-precision, or double-precision form as the data are moved between the memory and the coprocessor register stack.

Status Register. The status register (see Figure 14–5) reflects the overall operation of the coprocessor. The status register is accessed by executing the instruction (FSTSW), which stores the contents of the status register into a word of memory. The FSTSW AX instruction copies the status register directly into the microprocessor's AX register on the 80187 or above coprocessor. Once status is stored in memory or the AX register, the bit positions of the status register can be examined by normal software. The coprocessor/microprocessor communications are carried out through the I/O ports 00FAH–00FFH on the 80187 and 80287, and I/O ports 800000FAH–800000FFH on the 80386 through the Pentium 4. Never use these I/O ports for interfacing I/O devices to the microprocessor.

The newer coprocessors (80187 and above) use status bit position 6 (SF) to indicate a stack overflow or underflow error. Following is a list of the status bits, except for SF, and their applications:

B The **busy bit** indicates that the coprocessor is busy executing a task. Busy is tested by examining the status register or by using the FWAIT instruction. Newer coprocessors automatically synchronize with the microprocessor, so the busy flag need not be tested before performing additional coprocessor tasks.

C_0–C_3 The **condition code bits** indicate conditions about the coprocessor (see Table 14–2 for a complete listing of each combination of these bits and their functions). Note that these bits have different meanings for different instructions, as indicated in the table. The top of the stack is denoted as ST in this table.

TOP The **top-of-stack** (ST) bit indicates the current register addressed as the top-of-the-stack (ST). This is normally register ST(0).

TABLE 14–2 The coprocessor status register condition code bits.

Instruction	C_3	C_2	C_1	C_0	*Indication*
FTST, FCOM	0	0	X	0	ST > Operand
	0	0	X	1	ST < Operand
	1	0	X	1	ST = Operand
	1	1	X	1	ST is not comparable
FPREM	Q1	0	Q0	Q2	Rightmost 3 bits of quotient
	?	1	?	?	Incomplete
FXAM	0	0	0	0	+ un-normal
	0	0	0	1	+ NAN
	0	0	1	0	– un-normal
	0	0	1	1	– NAN
	0	1	0	0	+ normal
	0	1	0	1	+ ∞
	0	1	1	0	– normal
	0	1	1	1	– ∞
	1	0	0	0	+ 0
	1	0	0	1	Empty
	1	0	1	0	– 0
	1	0	1	1	Empty
	1	1	0	0	+ denormal
	1	1	0	1	Empty
	1	1	1	0	– denormal
	1	1	1	1	Empty

Notes: Un-normal = leading bits of the significand are zero; denormal = exponent is at its most negative value; normal = standard floating-point form; NAN (not-a-number) = an exponent of all ones and a significand not equal to zero; and the operand for TST is zero.

ES The **error summary** bit is set if any unmasked error bit (PE, UE, OE, ZE, DE, or IE) is set. In the 8087 coprocessor, the error summary also caused a coprocessor interrupt. Since the 80187, the coprocessor interrupt has been absent from the family.

PE The **precision error** indicates that the result or operands exceed the selected precision.

UE An **underflow error** indicates a nonzero result that is too small to represent with the current precision selected by the control word.

OE An **overflow error** indicates a result that is too large to be represented. If this error is masked, the coprocessor generates infinity for an overflow error.

ZE A **zero error** indicates the divisor was zero while the dividend is a noninfinity or nonzero number.

DE A **denormalized error** indicates that at least one of the operands is denormalized.

IE An **invalid error** indicates a stack overflow or underflow, indeterminate form ($0 \div 0$, $+\infty$, $-\infty$, etc.), or the use of a NAN as an operand. This flag indicates errors such as those produced by taking the square root of a negative number, and so on.

There are two ways to test the bits of the status register once they are moved into the AX register with the FSTSW AX instruction. One method uses the TEST instruction to test individual bits of the status register. The other uses the SAHF instruction to transfer the leftmost eight bits of the status register into the microprocessor's flag register. Both methods are illustrated in Example 14–6. This example uses the DIV instruction to divide the top of the stack by

TABLE 14–3 Coprocessor conditions tested with conditional jumps as illustrated in Example 14–6.

C_3	C_2	C_0	*Condition*	*Jump Instruction*
0	0	0	ST > Operand	JA (jump if ST above)
0	0	1	ST < Operand	JB (jump if ST below)
1	0	0	ST = Operand	JE (jump if ST equal)

the contents of DATA1 and the FSQRT instruction to find the square root of the top of the stack. The example also uses the FCOM instruction to compare the contents of the stack top with DATA1. Note that the conditional jump instructions are used with the SAHF instruction to test for the condition listed in Table 14–3. Although SAHF and conditional jumps cannot test all possible operating conditions of the coprocessor, they can help to reduce the complexity of certain tested conditions. Note that SAHF places C_0 into the carry flag, C_2 into the parity flag, and C_3 into the zero flag.

EXAMPLE 14–6

```
;testing for a divide by zero error

      FDIV    DATA1
      FSTSW   AX              ;copy status register into AX
      TEST    AX,4            ;test ZE bit
      JNZ     DIVIDE_ERROR

;testing for an invalid operation after a FSQRT

      FSQRT
      FSTSW   AX
      TEST    AX,1            ;test IE
      JNZ     FSQRT_ERROR

;testing with SAHF so conditional jumps can be used

      FCOM    DATA1
      FSTSW   AX
      SAHF                    ;copy coprocessor flags to flags
      JE      ST_EQUAL
      JB      ST_BELOW
      JA      ST_ABOVE
```

When the FXAM instruction and FSTSW AX are executed and followed by the SAHF instruction, the zero flag will contain C_3. The FXAM instruction could be used to test a divisor before a division for a zero value by using the JZ instruction following FXAM, FSTSW AX, and SAHF.

Control Register. The control register is pictured in Figure 14–6. The control register selects precision, rounding control, and infinity control. It also masks and unmasks the exception bits that correspond to the rightmost six bits of the status register. The FLDCW instruction is used to load a value into the control register.

Following is a description of each bit or grouping of bits found in the control register:

IC — **Infinity control** selects either affine or projective infinity. Affine allows positive and negative infinity; projective assumes infinity is unsigned.

RC — **Rounding control** determines the type of rounding, as defined in Figure 14–6.

PC — The **precision control** sets the precision of the result, as defined in Figure 14–6.

Exception masks — Determine whether the error indicated by the exception affects the error bit in the status register. If a logic 1 is placed in one of the exception control bits, the corresponding status register bit is masked off.

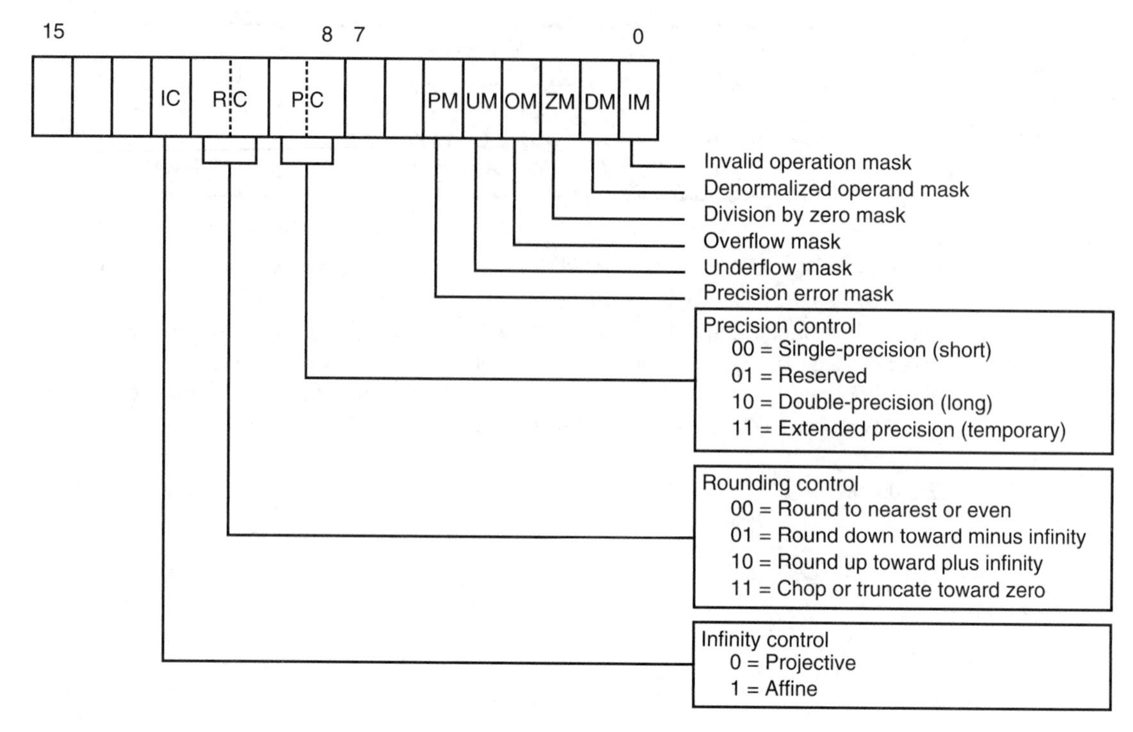

FIGURE 14–6 The 80X87 arithmetic coprocessor control register.

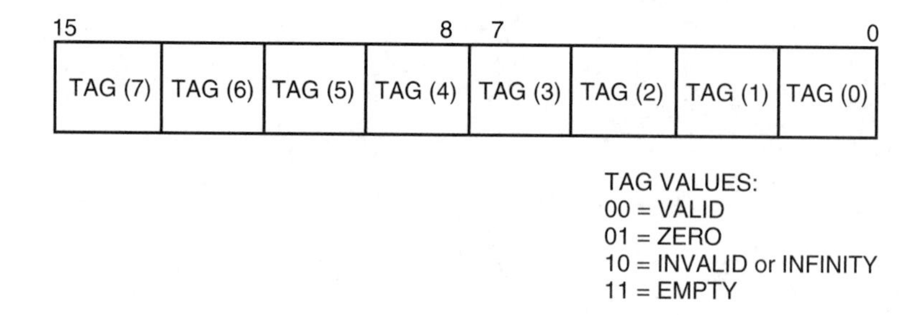

FIGURE 14–7 The 80X87 arithmetic coprocessor tag register.

Tag Register. The **tag register** indicates the contents of each location in the coprocessor stack. Figure 14–7 illustrates the tag register and the status indicated by each tag. The tag indicates whether a register is valid; zero; invalid or infinity; or empty. The only way that a program can view the tag register is by storing the coprocessor environment using the FSTENV, FSAVE, or FRSTOR instructions. Each of these instructions stores the tag register along with other coprocessor data.

14–3 INSTRUCTION SET

The arithmetic coprocessor executes over 68 different instructions. Whenever a coprocessor instruction references memory, the microprocessor automatically generates the memory address for the instruction. The coprocessor uses the data bus for data transfers during coprocessor instructions and the microprocessor uses it during normal instructions. Also note that the 80287 uses the Intel-reserved I/O ports 00F8H–00FFH for communications between the coprocessor and the microprocessor (even though the coprocessor only uses ports 00FCH–00FFH). These ports are used mainly for the FSTSW AX instruction. The 80387–Pentium 4 use I/O ports 800000F8H–800000FFH for these communications.

This section of the text describes the function of each instruction and lists its assembly language form. Because the coprocessor uses the microprocessor memory-addressing modes, not all forms of each instruction are illustrated. Each time that the assembler encounters a coprocessor mnemonic opcode, it converts it into a machine language ESC instruction. The ESC instruction represents an opcode to the coprocessor.

Data Transfer Instructions

There are three basic data transfers: floating-point, signed integer, and BCD. The only time that data ever appear in the signed integer or BCD form is in the memory. Inside the coprocessor, data are always stored as 80-bit extended-precision floating-point numbers.

Floating-Point Data Transfers. There are four traditional floating-point data transfer instructions in the coprocessor instruction set: FLD (load real), FST (store real), FSTP (store real and pop), and FXCH (exchange). A new instruction added to the Pentium Pro through Pentium 4 is called a conditional floating-point move instruction (described below). It uses the opcode FCMOV with a floating-point condition.

The FLD instruction loads floating-point memory data to the top of the internal stack, referred to as ST (stack top). This instruction stores the data on the top of the stack and then decrements the stack pointer by 1. Data loaded to the top of the stack are from any memory location or from another coprocessor register. For example, an FLD ST(2) instruction copies the contents of register 2 to the stack top, which is ST. The top of the stack is register 0 when the coprocessor is reset or initialized. Another example is the FLD DATA7 instruction, which copies the contents of memory location DATA 7 to the top of the stack. The size of the transfer is automatically determined by the assembler through the directives DD or REAL4 for single-precision, DQ or REAL 8 for double-precision, and DT or REAL10 for extended temporary-precision.

The FST instruction stores a copy of the top of the stack into the memory location or coprocessor register indicated by the operand. At the time of storage, the internal, extended temporary-precision floating-point number is rounded to the size of the floating-point number indicated by the control register.

The FSTP (floating-point store and pop) instruction stores a copy of the top of the stack into memory or any coprocessor register, and then pops the data from the top of the stack. You might think of FST as a copy instruction and FSTP as a removal instruction.

The FXCH instruction exchanges the register indicated by the operand with the top of the stack. For example, the FXCH ST(2) instruction exchanges the top of the stack with register 2.

Integer Data Transfer Instructions. The coprocessor supports three integer data transfer instructions: FILD (load integer), FIST (store integer), and FISTP (store integer and pop). These three instructions function as did FLD, FST, and FSTP, except that the data transferred are integer data. The coprocessor automatically converts the internal extended temporary-precision floating-point data to integer data. The size of the data is determined by the way that the label is defined with DW, DD, or DQ in the assembly language program.

BCD Data Transfer Instructions. Two instructions load or store BCD signed-integer data. The FBLD instruction loads the top of the stack with BCD memory data, and the FBSTP stores the top of the stack and does a pop.

The Pentium Pro through Pentium 4 FCMOV Instruction. The Pentium Pro–Pentium 4 microprocessors contain a new instruction called FCMOV, which also contains a condition. If the condition is true, the FCMOV instruction copies the source to the destination. The conditions tested

TABLE 14–4 The FCMOV instructions and conditions tested by them.

Instruction	*Condition*
FCMOVB	Move if below
FCMOVE	Move if equal
FCMOVBE	Move if below or equal
FCMOVU	Move if unordered
FCMOVNB	Move if not below
FCMOVNE	Move if not equal
FCMOVNBE	Move if not below or equal
FCMOVNU	Move if not unordered

by FCMOV and the opcodes used with FCMOV appear in Table 14–4. Notice that these conditions check for either an ordered or unordered condition. The testing for NAN and denormalized numbers is not checked with FCMOV.

Example 14–7 shows how the FCMOVB (move if below) instruction is used to copy the contents of ST(2) are the stack top (ST) if the contents of ST(2) are below ST. Notice that the FCOM instruction must be used to perform the compare and the contents of the status register must still be copied to the flags for this instruction to function. More about the FCMOV instruction appears with the FCOMI instruction, which is also new to the Pentium Pro through the Pentium 4 microprocessors.

EXAMPLE 14–7

```
FCOM    ST(2)          ;compare ST and ST(2)
FSTSW   AX             ;floating flags to AX
SAHF                   ;floating flags to flags
FCMOVB  ST(2)          ;copy ST(2) to ST if below

              OR

FCOMI   ST(2)
FCMOVB  ST(2)
```

Arithmetic Instructions

Arithmetic instructions for the coprocessor include addition, subtraction, multiplication, division, and calculating square roots. The arithmetic-related instructions are scaling, rounding, absolute value, and changing the sign.

Table 14–5 shows the basic addressing modes allowed for the arithmetic operations. Each addressing mode is shown with an example using the FADD (real addition) instruction. All arithmetic operations are floating-point, except some cases in which memory data are referenced as an operand.

TABLE 14–5 Coprocessor addressing modes.

Mode	*Form*	*Example*
Stack	ST(1),ST	FADD
Register	ST,ST(n)	FADD ST,ST(1)
	ST(n),ST	FADD ST(4),ST
Register with pop	ST(n),ST	FADDP ST(3),ST
Memory	Operand	FADD DATA3

Note: Stack address is fixed as ST(1),ST and includes a pop, so only the result remains at the top of the stack; and n = register number 0–7.

The classic stack form of addressing operand data (stack addressing) uses the top of the stack as the source operand and the next to the top of the stack as the destination operand. Afterward, a pop removes the source datum from the stack and only the result in the destination register remains at the top of the stack. To use this addressing mode, the instruction is placed in the program without any operands such as FADD or FSUB. The FADD instruction adds ST to ST(1) and stores the answer at the top of the stack; it also removes the original two data from the stack by popping. Note carefully that FSUB subtracts ST from ST(1) and leaves the difference at ST. Therefore, a reverse subtraction (FSUBR) subtracts ST(1) from ST and leaves the difference at ST. (Note that an error exists in Intel documentation, including the Pentium data book, which describes the operation of some reverse instructions.) Another use for reverse operations is for finding a reciprocal (1/X). This is accomplished, if X is at the top of the stack, by loading 1.0 to ST, followed by the FDIVR instruction. The FDIVR instruction divides ST(1) into ST or X into 1 and leaves the reciprocal (1/X) at ST.

The register-addressing mode uses ST for the top of the stack and ST(n) for another location, where n is the register number. With this form, one operand must be ST and the other is ST(n). Note that to double the top of the stack, the FADD ST,ST(0) instruction is used where ST(0) also addresses the top of the stack. One of the two operands in the register-addressing mode must be ST, while the other must be in the form ST(n), where n is a stack register 0–7. For many instructions, either ST or ST(n) can be the destination. It is fairly important that the top of the stack be ST(0). This is accomplished by resetting or initializing the coprocessor before using it in a program. Another example of register-addressing is FADD ST(1),ST where the contents of ST are added to ST(1) and the result is placed in ST(1).

The top of the stack is always used as the destination for the memory-addressing mode because the coprocessor is a stack-oriented machine. For example, the FADD DATA instruction adds the real number contents of memory location DATA to the top of the stack.

Arithmetic Operations. The letter P in an opcode specifies a register pop after the operation (FADDP compared to FADD). The letter R in an opcode (subtraction and division only) indicates reverse mode. The reverse mode is useful for memory data because memory data normally subtract from the top of the stack. A reversed subtract instruction subtracts the top of the stack from memory and stores the result in the top of the stack. For example, if the top of the stack contains a 10 and memory location DATA1 contains a 1, the FSUB DATA1 instruction results in a +9 on the stack top, and the FSUBR instruction results in a –9. Another example is FSUBR ST,ST(1), which will subtract ST from ST(1) and store the result on ST. A variant is FSUBR ST(1),ST, which will subtract ST(1) from ST and store the result on ST(1).

The letter I as a second letter in an opcode indicates that the memory operand is an integer. For example, the FADD DATA instruction is a floating-point addition, while the FIADD DATA is an integer addition that adds the integer at memory location DATA to the floating-point number at the top of the stack. The same rules apply to the FADD, FSUB, FMUL, and FDIV instructions.

Arithmetic-Related Operations. Other operations that are arithmetic in nature include FSQRT (square root), FSCALE (scale a number), FPREM/FPREM1 (find partial remainder), FRNDINT (round to integer), FXTRACT (extract exponent and significand), FABS (find absolute value), and FCHG (change sign). These instructions and the functions that they perform follow:

FSQRT Finds the **square root** of the top of the stack and leaves the resultant square root at the top of the stack. An invalid error occurs for the square root of a negative number. For this reason, the IE bit of the status register should be tested whenever an invalid result can occur. The IE bit can be tested by loading the status register to AX with the FSTSW AX instruction, followed by TEST AX,1 to test the IE status bit.

FSCALE Adds the contents of ST(1) (interpreted as an integer) to the exponent at the top of the stack. **FSCALE** multiplies or divides rapidly by powers of two. The value in ST(1) must be between 2^{-15} and 2^{+15}.

FPREM/FPREM1 Performs **modulo division** of ST by ST(1). The resultant remainder is found in the top of the stack and has the same sign as the original dividend. Note that a modulo division results in a remainder without a quotient. Note also that FPREM is supported for the 8086 and 80287, and FPREM1 should be used in newer coprocessors.

FRNDINT **Rounds** the top of the stack to an integer.

FXTRACT **Decomposes** the number at the top of the stack into two separate parts that represent the value of the unbiased exponent and the value of the significand. The extracted significand is found at the top of the stack and the unbiased exponent at ST(1). This instruction is often used to convert a floating-point number into a form that can be printed as a mixed number.

FABS **Changes the sign** of the top of the stack to positive.

FCHS **Changes the sign** from positive to negative or negative to positive.

Comparison Instructions

The comparison instructions all examine data at the top of the stack in relation to another element and return the result of the comparison in the status register condition code bits C_3–C_0. Comparisons that are allowed by the coprocessor are FCOM (floating-point compare), FCOMP (floating-point compare with a pop), FCOMPP (floating-point compare with two pops), FICOM (integer compare), FICOMP (integer compare and pop), FSTS (test), and FXAM (examine). New with the introduction of the Pentium Pro is the floating compare and move results to flags or FCOMI instruction. Following is a list of these instructions with a description of their functions:

FCOM **Compares** the floating-point data at the top of the stack with an operand, which may be any register or any memory operand. If the operand is not coded with the instruction, the next stack element ST(1) is compared with the stack top ST.

FCOMP/FCOMPP Both instructions perform as FCOM, but they also pop one or two data from the stack.

FICOM/FICOMP The top of the stack is compared with the integer stored at a memory operand. In addition to the compare, FICOMP also pops the top of the stack.

FTST **Tests** the contents of the top of the stack against a zero. The result of the comparison is coded in the status register condition code bits, as illustrated in Table 14–2 with the status register. Also, refer to Table 14–3 for a way of using SAHF and the conditional jump instruction with FTST.

FXAM **Examines** the stack top and modifies the condition code bits to indicate whether the contents are positive, negative, normalized, etc. Refer to the status register in Table 14–2.

FCOMI/FUCOMI New to the Pentium Pro through the Pentium 4, this instruction compares in exactly the same manner as the FCOM instruction, with one additional feature: It moves the floating-point flags into the flag register, just as the FNSTSW AX and SAHF instructions do in Example 14–8. Intel has combined the FCOM, FNSTSW AX, and

SAHF instructions to form FCOMI. Also available is the unordered compare or FUCOMI. Each is also available with a pop by appending the opcode with a P.

Transcendental Operations

The transcendental instructions include FPT AN (partial tangent), FPATAN (partial arctangent), FSIN (sine), FCOS (cosine), FSINCOS (sine and cosine), F2XM1 ($2^X - 1$), FYL2X (Y $\log_2$ X), and FYL2XP1 (Y $\log_2$ (X + 1)). A list of these operations follows with a description of each transcendental operation:

FPTAN Finds the **partial tangent** of $Y/X = \tan \theta$. The value of θ is at the top of the stack. It must be between 0 and n/4 radians for the 8087 and 80287, and must be less than 2^{63} for the 80387, 80486/7, and Pentium–Pentium 4 microprocessors. The result is a ratio found as ST = X and ST(1) = Y. If the value is outside of the allowable range, an invalid error occurs, as indicated by the status register IE bit. Also note that ST(7) must be empty for this instruction to function properly.

FPATAN Finds the **partial arctangent** as θ = ARCTAN X/Y. The value of X is at the top of the stack and Y is at ST(1). The values of X and Y must be as follows: $0 \leq Y < X < \infty$. The instruction pops the stack and leaves θ in radians at the top of the stack.

F2XM1 Finds the **function $2^X - 1$**. The value of X is taken from the top of the stack and the result is returned to the top of the stack. To obtain 2^X add one to the result at the top of the stack. The value of X must be in the range of –1 and +1. The F2XM1 instruction is used to derive the functions listed in Table 14–6. Note that the constants $\log_2 10$ and $\log_2 \varepsilon$ are built in as standard values for the coprocessor.

FSIN/FCOS Finds the **sine** or **cosine** of the argument located in ST expressed in radians ($360° = 2\pi$ radians), with the result found in ST. The values of ST must be less than 2^{63}.

FSINCOS Finds the **sine** and **cosine** of ST, expressed in radians, and leaves the results as ST = sine and ST(1) = cosine. As with FSIN or FCOS, the initial value of ST must be less than 2^{63}.

FYL2X Finds **Y $\log_2$ X**. The value X is taken from the stack top, and Y is taken from ST(1). The result is found at the top of the stack after a pop. The value of X must range between 0 and ∞, and the value of Y must be between $-\infty$ and $+\infty$. A logarithm with any positive base (b) is found by the equation $\text{LOG}_b X = (\text{LOG}_2 b)^{-1} \times \text{LOG}_2 X$.

FYL2P1 Finds **Y $\log_2$ (X + 1)**. The value of X is taken from the stack top and Y is taken from ST(1). The result is found at the top of the stack after a pop. The value of X must range between 0 and $1 - \sqrt{2}/2$ and the value of Y must be between $-\infty$ and $+\infty$.

TABLE 14–6 Exponential functions.

Function	*Equation*
10^Y	$2^Y \times \log_2 10$
ε^Y	$2^Y \times \log_2 \varepsilon$
X^Y	$2^Y \times \log_2 X$

TABLE 14–7 Constant operations.

Instruction	*Constant Pushed to ST*
FLDZ	+0.0
FLD1	+1.0
FLDPI	π
FLDL2T	$\log_2 10$
FLDL2E	$\log_2 \varepsilon$
FLDLG2	$\log_{10} 2$
FLDLN2	$\log \varepsilon\ 2$

Constant Operations

The coprocessor instruction set includes opcodes that return constants to the top of the stack. A list of these instructions appears in Table 14–7.

Coprocessor Control Instructions

The coprocessor has control instructions for initialization, exception handling, and task switching. The control instructions have two forms. For example, FINIT initializes the coprocessor, as does FNINIT. The difference is that FNINIT does not cause any wait states, while FINIT does cause waits. The microprocessor waits for the FINIT instruction by testing the BUSY pin on the coprocessor. All control instructions have these two forms. Following is a list of each control instruction with its function:

FINIT/FNINIT Performs a reset (**initialize**) operation on the arithmetic coprocessor (see Table 14–8 for the reset conditions). The coprocessor operates with a closure of projective (unsigned infinity), rounds to the nearest or even, and uses extended-precision when reset or initialized. It also sets register 0 as the top of the stack.

FSETPM Changes the addressing mode of the coprocessor to the **protected-addressing mode**. This mode is used when the microprocessor is also operated in the protected mode. As with the microprocessor, protected mode can only be exited by a hardware reset or, in the case of the 80386 through the Pentium 4, with a change to the control register.

TABLE 14–8 Coprocessor state after the FINIT instruction.

Field	*Value*	*Condition*
Infinity	0	Projective
Rounding	00	Round to nearest
Precision	11	Extended-precision
Error masks	11111	Error bits disabled
Busy	0	Not busy
C_0–C_3	????	Unknown
TOP	000	Register 000 or ST(0)
ES	0	No errors
Error bits	00000	No errors
All tags	11	Empty
Registers	ST(0)–ST(7)	Not changed

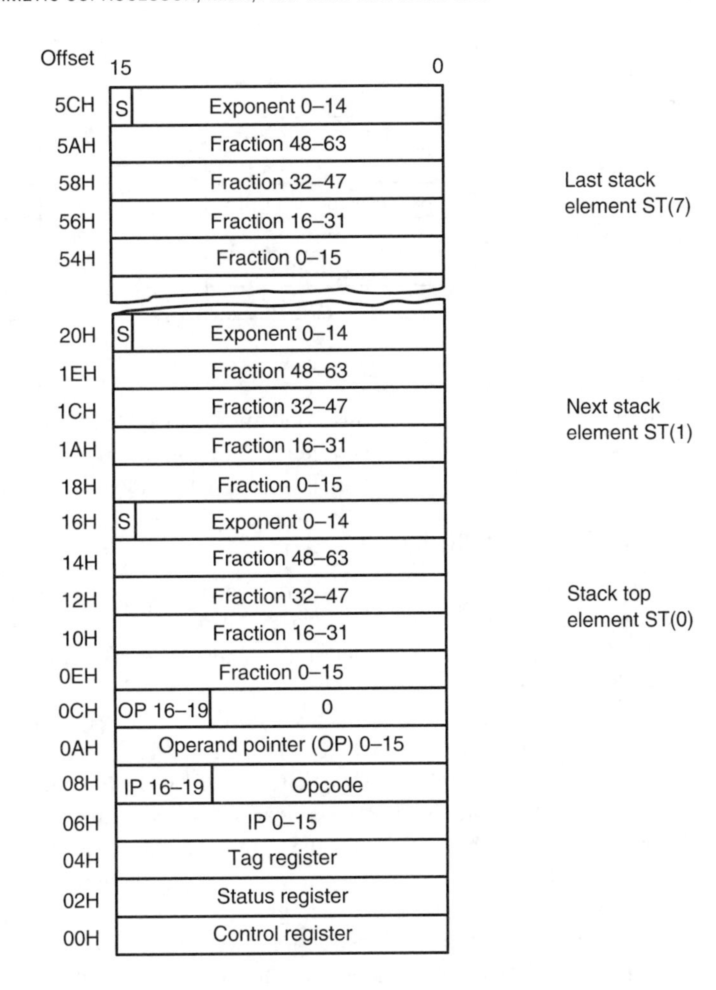

FIGURE 14–8 Memory format when the 80X87 registers are saved with the FSAVE instruction.

FLDCW	Loads the control register with the word addressed by the operand.
FSTCW	Stores the control register into the word-sized memory operand.
FSTSW AX	Copies the contents of the control register to the AX register. This instruction is not available to the 8087 coprocessor.
FCLEX	Clears the error flags in the status register and also the busy flag.
FSAVE	Writes the entire state of the machine to memory. Figure 14–8 shows the memory layout for this instruction.
FRSTOR	Restores the state of the machine from memory. This instruction is used to restore the information saved by FSAVE.
FSTENV	Stores the environment of the coprocessor, as shown in Figure 14–9.
FLDENV	Reloads the environment saved by FSTENV.
FINCSP	Increments the stack pointer.
FDECSP	Decrements the stack pointer.
FFREE	Frees a register by changing the destination register's tag to empty. It does not affect the contents of the register.
FNOP	Floating-point coprocessor NOP.

FIGURE 14–9 Memory format for the FSTENV instruction: (a) real mode and (b) protected mode.

(a)

Offset		
0CH	OP 16–19	0
0AH	Operand pointer 0–15	
08H	IP 16–19	Opcode
06H	Instruction pointer 0–15	
04H	Tag register	
02H	Status register	
00H	Control register	

(b)

Offset	
0CH	Operand selector
0AH	Operand offset
08H	CS selector
06H	IP offset
04H	Tag register
02H	Status register
00H	Control register

FWAIT Causes the microprocessor to wait for the coprocessor to finish an operation. FWAIT should be used before the microprocessor accesses memory data that are affected by the coprocessor.

Coprocessor Instructions

Although the microprocessor circuitry has not been discussed, the instruction sets of these coprocessors and their differences from the other versions of the coprocessor can be discussed. These newer coprocessors contain the same basic instructions provided by the earlier versions, with a few additional instructions.

The 80387, 80486, 80487SX, and Pentium through the Pentium 4 contain the following additional instructions: FCOS (cosine), FPREM1 (partial remainder), FSIN (sine), FSINCOS (sine and cosine), and FUCOM/FUCOMP/FUCOMPP (unordered compare). The sine and cosine instructions are the most significant addition to the instruction set. In the earlier versions of the coprocessor, the sine and cosine is calculated from the tangent. The Pentium Pro through the Pentium 4 contain two new floating-point instructions: FCMOV (a conditional move) and FCOMI (a compare and move to flags).

Table 14–9 lists the instruction sets for all versions of the coprocessor. It also lists the number of clocking periods required to execute each instruction. Execution times are listed for the 8087, 80287, 80387, 80486, 80487, and Pentium–Pentium 4. (The timings for the Pentium through the Pentium 4 are the same because the coprocessor is identical in each of these microprocessors.) To determine the execution time of an instruction, the clock time is multiplied times the listed execution time. The FADD instruction requires 70–143 clocks for the 80287. Suppose that an 8 MHz clock is used with the 80287. The clocking period is 1/8 MHz, or 125 ns. The FADD instruction requires between 8.75 μs and 17.875 μs to execute. Using a 33 MHz (33 ns) 80486DX2, this instruction requires between 0.264 μs and 0.66 μs to execute. On the Pentium the FADD instruction requires from 1–7 clocks, so if operated at 133 MHz (7.52 ns), the FADD requires between 0.00752 μs and 0.05264 μs. The Pentium Pro through the Pentium 4 are even faster than the Pentium. For a 2 GHz Pentium 4, which has a clock period of 0.5 ns, the FADD instruction takes between 0.5 ns and 3.5 ns.

Table 14–9 uses some shorthand notations to represent the displacement that may or may not be required for an instruction that uses a memory-addressing mode. It also uses the abbreviation *mmm* to represent a register/memory addressing mode and *rrr* to represent one of the floating-point coprocessor registers ST(0)–ST(7). The d (destination) bit that appears in some instruction opcodes defines the direction of the data flow, as in FADD ST,ST(2) or FADD ST(2),ST. The d bit is a logic 0 for flow toward ST, as in FADD ST,ST(2), where ST holds the sum after the addition; and a logic 1 for FADD ST(2),ST, where ST(2) holds the sum.

Also note that some instructions allow a choice of whether a wait is inserted. For example, the FSTSW AX instruction copies the status register into AX. The FNSTSW AX instruction also copies the status register to AX, but without a wait.

TABLE 14–9 The instruction set of the arithmetic coprocessor.

F2XM1 $2^{ST} - 1$

11011001 11110000

Example		Clocks
F2XM1	8087	310–630
	80287	310–630
	80387	211–476
	80486/7	140–279
	Pentium–Pentium 4	13–57

FABS Absolute value of ST

11011001 11100001

Example		Clocks
FABS	8087	10–17
	80287	10–17
	80387	22
	80486/7	3
	Pentium–Pentium 4	1

FADD/FADDP/FIADD Addition

11011000 oo000mmm disp	32-bit memory (FADD)
11011100 oo000mmm disp	64-bit memory (FADD)
11011d00 11000rrr	FADD ST,ST(rrr)
11011110 11000rrr	FADDP ST,ST(rrr)
11011110 oo000mmm disp	16-bit memory (FIADD)
11011010 oo000mmm disp	32-bit memory (FIADD)

Format	Examples		Clocks
FADD FADDP FIADD	FADD DATA FADD ST,ST(1) FADDP FIADD NUMBER FADD ST,ST(3) FADDP ST,ST(2) FADD ST(2),ST	8087	70–143
		80287	70–143
		80387	23–72
		80486/7	8–20
		Pentium–Pentium 4	1–7

FCLEX/FNCLEX Clear errors

11011011 11100010

Example	Clocks	
FCLEX FNCLEX	8087	2–8
	80287	2–8
	80387	11
	80486/7	7
	Pentium–Pentium 4	9

FCOM/FCOMP/FCOMPP/FICOM/FICOMP Compare

11011000 oo010mmm disp	32-bit memory (FCOM)
11011100 oo010mmm disp	64-bit memory (FCOM)
11011000 11010rrr	FCOM ST(rrr)
11011000 oo011mmm disp	32-bit memory (FCOMP)
11011100 oo011mmm disp	64-bit memory (FCOMP)
11011000 11011rrr	FCOMP ST(rrr)
11011110 11011001	FCOMPP
11011110 oo010mmm disp	16-bit memory (FICOM)
11011010 oo010mmm disp	32-bit memory (FICOM)
11011110 oo011mmm disp	16-bit memory (FICOMP)
11011010 oo011mmm disp	32-bit memory (FICOMP)

Format	Examples	Clocks	
FCOM FCOMP FCOMPP FICOM FICOMP	FCOM ST(2) FCOMP DATA FCOMPP FICOM NUMBER FICOMP DATA3	8087	40–93
		80287	40–93
		80387	24–63
		80486/7	15–20
		Pentium–Pentium 4	1–8

FCOMI/FUCOMI/COMIP/FUCOMIP Compare and Load Flags

11011011 11110rrr	FCOMI ST(rrr)
11011011 11101rrr	FUCOMI ST(rrr)
11011111 11110rrr	FCOMIP ST(rrr)
11011111 11101rrr	FUCOMIP ST(rrr)

Format	Examples	Clocks	
FCOM FUCOMI FCOMIP FUCOMIP	FCOMI ST(2) FUCOMI ST(4) FCOMIP ST(0) FUCOMIP ST(1)	8087	—
		80287	—
		80387	—
		80486/7	—
		Pentium–Pentium 4	—

FCMOVcc Conditional Move

11011010 11000rrr	FCMOVB ST(rrr)
11011010 11001rrr	FCMOVE ST(rrr)
11011010 11010rrr	FCMOVBE ST(rrr)
11011010 11011rrr	FCMOVU ST(rrr)
11011011 11000rrr	FCMOVNB ST(rrr)
11011011 11001rrr	FCMOVNE ST(rrr)
11011011 11010rrr	FCMOVENBE ST(rrr)
11011011 11011rrr	FCMOVNU ST(rrr)

Format	Examples		Clocks
FCMOVB	FCMOVB ST(2)	8087	—
FCMOVE	FCMOVE ST(3)	80287	—
		80387	—
		80486/7	—
		Pentium–Pentium 4	—

FCOS Cosine of ST

11011001 11111111

Example		Clocks
FCOS	8087	—
	80287	—
	80387	123–772
	80486/7	193–279
	Pentium–Pentium 4	18–124

FDECSTP Decrement stack pointer

11011001 11110110

Example		Clocks
FDECSTP	8087	6–12
	80287	6–12
	80387	22
	80486/7	3
	Pentium–Pentium 4	1

FDISI/FNDISI Disable interrupts

11011011 11100001

(Ignored on the 80287, 80387, 80486/7, Pentium–Pentium 4)

Example		Clocks
FDISI FNDISI	8087	2–8
	80287	—
	80387	—
	80486/7	—
	Pentium–Pentium 4	—

FDIV/FDIVP/FIDIV Division

11011000 oo110mmm disp	32-bit memory (FDIV)
11011100 oo100mmm disp	64-bit memory (FDIV)
11011d00 11111rrr	FDIV ST,ST(rrr)
11011110 11111rrr	FDIVP ST,ST(rrr)
11011110 oo110mmm disp	16-bit memory (FIDIV)
11011010 oo110mmm disp	32-bit memory (FIDIV)

Format	Examples		Clocks
FDIV FDIVP FIDIV	FDIV DATA FDIV ST,ST(3) FDIVP FIDIV NUMBER FDIV ST,ST(5) FDIVP ST,ST(2) FDIV ST(2),ST	8087	191–243
		80287	191–243
		80387	88–140
		80486/7	8–89
		Pentium–Pentium 4	39–42

FDIVR/FDIVRP/FIDIVR Division reversed

11011000 oo111mmm disp	32-bit memory (FDIVR)
11011100 oo111mmm disp	64-bit memory (FDIVR)
11011d00 11110rrr	FDIVR ST,ST(rrr)
11011110 11110rrr	FDIVRP ST,ST(rrr)
11011110 oo111mmm disp	16-bit memory (FIDIVR)
11011010 oo111mmm disp	32-bit memory (FIDIVR)

Format	Examples		Clocks
FDIVR FDIVRP FIDIVR	FDIVR DATA FDIVR ST,ST(3) FDIVRP FIDIVR NUMBER FDIVR ST,ST(5) FDIVRP ST,ST(2) FDIVR ST(2),ST	8087	191–243
		80287	191–243
		80387	88–140
		80486/7	8–89
		Pentium–Pentium 4	39–42

FENI/FNENI Disable interrupts

11011011 11100000

(Ignored on the 80287, 80387, 80486/7, Pentium–Pentium 4)

Example		Clocks
FENI FNENI	8087	2–8
	80287	—
	80387	—
	80486/7	—
	Pentium–Pentium 4	—

FFREE Free register

11011101 11000rrr

Format	Examples		Clocks
FFREE	FFREE FFREE ST(1) FFREE ST(2)	8087	9–16
		80287	9–16
		80387	18
		80486/7	3
		Pentium–Pentium 4	1

FINCSTP Increment stack pointer

11011001 11110111

Example		Clocks
FINCSTP	8087	6–12
	80287	6–12
	80387	21
	80486/7	3
	Pentium–Pentium 4	1

FINIT/FNINIT Initialize coprocessor

11011001 11110110

Example		Clocks
FINIT FNINIT	8087	2–8
	80287	2–8
	80387	33
	80486/7	17
	Pentium–Pentium 4	12–16

FLD/FILD/FBLD Load data to ST(0)

11011001 oo000mmm disp	32-bit memory (FLD)
11011101 oo000mmm disp	64-bit memory (FLD)
11011011 oo101mmm disp	80-bit memory (FLD)
11011111 oo000mmm disp	16-bit memory (FILD)
11011011 oo000mmm disp	32-bit memory (FILD)
11011111 oo101mmm disp	64-bit memory (FILD)
11011111 oo100mmm disp	80-bit memory (FBLD)

Format	Examples		Clocks
FLD FILD FBLD	FLD DATA FILD DATA1 FBLD DEC_DATA	8087	17–310
		80287	17–310
		80387	14–275
		80486/7	3–103
		Pentium–Pentium 4	1–3

FLD1 Load +1.0 to ST(0)

11011001 11101000

Example		Clocks
FLD1	8087	15–21
	80287	15–21
	80387	24
	80486/7	4
	Pentium–Pentium 4	2

FLDZ Load +0.0 to ST(0)

11011001 11101110

Example		Clocks
FLDZ	8087	11–17
	80287	11–17
	80387	20
	80486/7	4
	Pentium–Pentium 4	2

FLDPI Load π to ST(0)

11011001 11101011

Example		Clocks
FLDPI	8087	16–22
	80287	16–22
	80387	40
	80486/7	8
	Pentium–Pentium 4	3–5

FLDL2E Load $\log_2 e$ to ST(0)

11011001 11101010

Example		Clocks
FLDL2E	8087	15–21
	80287	15–21
	80387	40
	80486/7	8
	Pentium–Pentium 4	3–5

FLDL2T Load $\log_2 10$ to ST(0)

11011001 11101001

Example		Clocks
FLDL2T	8087	16–22
	80287	16–22
	80387	40
	80486/7	8
	Pentium–Pentium 4	3–5

FLDLG2 Load $\log_{10}2$ to ST(0)		
11011001 11101000		
Example		Clocks
FLDLG2	8087	18–24
	80287	18–24
	80387	41
	80486/7	8
	Pentium–Pentium 4	3–5

FLDLN2 Load $\log_e 2$ to ST(0)		
11011001 11101101		
Example		Clocks
FLDLN2	8087	17–23
	80287	17–23
	80387	41
	80486/7	8
	Pentium–Pentium 4	3–5

FLDCW Load control register			
11011001 oo101mmm disp			
Format	Examples		Clocks
FLDCW	FLDCW DATA FLDCW STATUS	8087	7–14
		80287	7–14
		80387	19
		80486/7	4
		Pentium–Pentium 4	7

FLDENV Load environment			
11011001 oo100mmm disp			
Format	Examples		Clocks
FLDENV	FLDENV ENVIRON FLDENV DATA	8087	35–45
		80287	25–45
		80387	71
		80486/7	34–44
		Pentium–Pentium 4	32–37

FMUL/FMULP/FIMUL Multiplication

11011000 oo001mmm disp	32-bit memory (FMUL)
11011100 oo001mmm disp	64-bit memory (FMUL)
11011d00 11001rrr	FMUL ST,ST(rrr)
11011110 11001rrr	FMULP ST,ST(rrr)
11011110 oo001mmm disp	16-bit memory (FIMUL)
11011010 oo001mmm disp	32-bit memory (FIMUL)

Format	Examples	Clocks	
FMUL FMULP FIMUL	FMUL DATA FMUL ST,ST(2) FMUL ST(2),ST FMULP FIMUL DATA3	8087	110–168
		80287	110–168
		80387	29–82
		80486/7	11–27
		Pentium–Pentium 4	1–7

FNOP No operation

11011001 11010000

Example	Clocks	
FNOP	8087	10–16
	80287	10–16
	80387	12
	80486/7	3
	Pentium–Pentium 4	1

FPATAN Partial arctangent of ST(0)

11011001 11110011

Example	Clocks	
FPATAN	8087	250–800
	80287	250–800
	80387	314–487
	80486/7	218–303
	Pentium–Pentium 4	17–173

FPREM Partial remainder

11011001 11111000

Example	Clocks	
FPREM	8087	15–190
	80287	15–190
	80387	74–155
	80486/7	70–138
	Pentium–Pentium 4	16–64

FPREM1 Partial remainder (IEEE)

11011001 11110101

Example	Clocks	
FPREM1	8087	—
	80287	—
	80387	95–185
	80486/7	72–167
	Pentium–Pentium 4	20–70

FPTAN Partial tangent of ST(0)

11011001 11110010

Example	Clocks	
FPTAN	8087	30–450
	80287	30–450
	80387	191–497
	80486/7	200–273
	Pentium–Pentium 4	17–173

FRNDINT Round ST(0) to an integer

11011001 11111100

Example	Clocks	
FRNDINT	8087	16–50
	80287	16–50
	80387	66–80
	80486/7	21–30
	Pentium–Pentium 4	9–20

FRSTOR Restore state

11011101 oo110mmm disp

Format	Examples	Clocks	
FRSTOR	FRSTOR DATA FRSTOR STATE FRSTOR MACHINE	8087	197–207
		80287	197–207
		80387	308
		80486/7	120–131
		Pentium–Pentium 4	70–95

FSAVE/FNSAVE Save machine state

11011101 oo110mmm disp

Format	Examples	Clocks	
FSAVE FNSAVE	FSAVE STATE FNSAVE STATUS FSAVE MACHINE	8087	197–207
		80287	197–207
		80387	375
		80486/7	143–154
		Pentium–Pentium 4	124–151

FSCALE Scale ST(0) by ST(1)

11011001 11111101

Example	Clocks	
FSCALE	8087	32–38
	80287	32–38
	80387	67–86
	80486/7	30–32
	Pentium–Pentium 4	20–31

FSETPM Set protected mode

11011011 11100100

Example	Clocks	
FSETPM	8087	—
	80287	2–18
	80387	12
	80486/7	—
	Pentium–Pentium 4	—

FSIN Sine of ST(0)

11011001 11111110

Example		Clocks
FSIN	8087	—
	80287	—
	80387	122–771
	80486/7	193–279
	Pentium–Pentium 4	16–126

FSINCOS Find sine and cosine of ST(0)

11011001 11111011

Example		Clocks
FSINCOS	8087	—
	80287	—
	80387	194–809
	80486/7	243–329
	Pentium–Pentium 4	17–137

FSQRT Square root of ST(0)

11011001 11111010

Example		Clocks
FSQRT	8087	180–186
	80287	180–186
	80387	122–129
	80486/7	83–87
	Pentium–Pentium 4	70

FST/FSTP/FIST/FISTP/FBSTP Store

11011001 oo010mmm disp	32-bit memory (FST)
11011101 oo010mmm disp	64-bit memory (FST)
11011101 11010rrr	FST ST(rrr)
11011011 oo011mmm disp	32-bit memory (FSTP)
11011101 oo011mmm disp	64-bit memory (FSTP)
11011011 oo111mmm disp	80-bit memory (FSTP)
11011101 11001rrr	FSTP ST(rrr)
11011111 oo010mmm disp	16-bit memory (FIST)
11011011 oo010mmm disp	32-bit memory (FIST)
11011111 oo011mmm disp	16-bit memory (FISTP)
11011011 oo011mmm disp	32-bit memory (FISTP)
11011111 oo111mmm disp	64-bit memory (FISTP)
11011111 oo110mmm disp	80-bit memory (FBSTP)

Format	Examples	Clocks	
FST	FST DATA	8087	15–540
FSTP	FST ST(3)	80287	15–540
FIST	FST	80387	11–534
FISTP	FSTP	80486/7	3–176
FBSTP	FIST DATA2	Pentium–Pentium 4	1–3
	FBSTP DATA6		
	FISTP DATA9		

FSTCW/FNSTCW Store control register

11011001 oo111mmm disp

Format	Examples	Clocks	
FSTCW	FSTCW CONTROL	8087	12–18
FNSTCW	FNSTCW STATUS	80287	12–18
	FSTCW MACHINE	80387	15
		80486/7	3
		Pentium–Pentium 4	2

FSTENV/FNSTENV Store environment

11011001 oo110mmm disp

Format	Examples	Clocks	
FSTENV	FSTENV CONTROL	8087	40–50
FNSTENV	FNSTENV STATUS	80287	40–50
	FSTENV MACHINE	80387	103–104
		80486/7	58–67
		Pentium–Pentium 4	48–50

FSTSW/FNSTSW Store status register

11011101 oo111mmm disp

Format	Examples	Clocks	
FSTSW FNSTSW	FSTSW CONTROL FNSTSW STATUS FSTSW MACHINE FSTSW AX	8087	12–18
		80287	12–18
		80387	15
		80486/7	3
		Pentium–Pentium 4	2–5

FSUB/FSUBP/FISUB Subtraction

11011000 oo100mmm disp	32-bit memory (FSUB)
11011100 oo100mmm disp	64-bit memory (FSUB)
11011d00 11101rrr	FSUB ST,ST(rrr)
11011110 11101rrr	FSUBP ST,ST(rrr)
11011110 oo100mmm disp	16-bit memory (FISUB)
11011010 oo100mmm disp	32-bit memory (FISUB)

Format	Examples	Clocks	
FSUB FSUBP FISUB	FSUB DATA FSUB ST,ST(2) FSUB ST(2),ST FSUBP FISUB DATA3	8087	70–143
		80287	70–143
		80387	29–82
		80486/7	8–35
		Pentium–Pentium 4	1–7

FSUBR/FSUBRP/FISUBR Reverse subtraction

11011000 oo101mmm disp	32-bit memory (FSUBR)
11011100 oo101mmm disp	64-bit memory (FSUBR)
11011d00 11100rrr	FSUBR ST,ST(rrr)
11011110 11100rrr	FSUBRP ST,ST(rrr)
11011110 oo101mmm disp	16-bit memory (FISUBR)
11011010 oo101mmm disp	32-bit memory (FISUBR)

Format	Examples	Clocks	
FSUBR FSUBRP FISUBR	FSUBR DATA FSUBR ST,ST(2) FSUBR ST(2),ST FSUBRP FISUBR DATA3	8087	70–143
		80287	70–143
		80387	29–82
		80486/7	8–35
		Pentium–Pentium 4	1–7

FTST Compare ST(0) with + 0.0

11011001 11100100

Example		Clocks
FTST	8087	38–48
	80287	38–48
	80387	28
	80486/7	4
	Pentium–Pentium 4	1–4

FUCOM/FUCOMP/FUCOMPP Unordered compare

11011101 11100rrr	FUCOM ST,ST(rrr)
11011101 11101rrr	FUCOMP ST,ST(rrr)
11011101 11101001	FUCOMPP

Format	Examples		Clocks
FUCOM FUCOMP FUCOMPP	FUCOM ST,ST(2) FUCOM FUCOMP ST,ST(3) FUCOMP FUCOMPP	8087	—
		80287	—
		80387	24–26
		80486/7	4–5
		Pentium–Pentium 4	1–4

FWAIT Wait

10011011

Example		Clocks
FWAIT	8087	4
	80287	3
	80387	6
	80486/7	1–3
	Pentium–Pentium 4	1–3

FXAM Examine ST(0)

11011001 11100101

Example		Clocks
FXAM	8087	12–23
	80287	12–23
	80387	30–38
	80486/7	8
	Pentium–Pentium 4	21

FXCH Exchange ST(0) with another register

11011001 11001rrr FXCH ST,ST(rrr)

Format	Examples	Clocks	
FXCH	FXCH ST,ST(1) FXCH FXCH ST,ST(4)	8087	10–15
		80287	10–15
		80387	18
		80486/7	4
		Pentium–Pentium 4	1

FXTRACT Extract components of ST(0)

11011001 11110100

Example	Clocks	
FXTRACT	8087	27–55
	80287	27–55
	80387	70–76
	80486/7	16–20
	Pentium–Pentium 4	13

FYL2X ST(1) x $\log_2$ ST(0)

11011001 11110001

Example	Clocks	
FYL2X	8087	900–1100
	80287	900–1100
	80387	120–538
	80486/7	196–329
	Pentium–Pentium 4	22–111

FXL2XP1 ST(1) x $\log_2$ [ST(0) + 1.0]

11011001 11111001

Example	Clocks	
FXL2XP1	8087	700–1000
	80287	700–1000
	80387	257–547
	80486/7	171–326
	Pentium–Pentium 4	22–103

Notes: d = direction, where d = 0 for ST as the destination, and d = 1 for ST as the source; rrr = floating-point register number; oo = mode; mmm = r/m field; and disp = displacement

14–4 PROGRAMMING WITH THE ARITHMETIC COPROCESSOR

This section of the chapter provides programming examples for the arithmetic coprocessor. Each example is chosen to illustrate a programming technique for the coprocessor.

Calculating the Area of a Circle

This first programming example illustrates a simple method of addressing the coprocessor stack. First, recall that the equation for calculating the area of a circle is $A = \pi R^2$. A program that performs this calculation is listed in Example 14–8. Note that this program takes test data from array RAD that contains five sample radii. The five areas are stored in a second array called AREA. No attempt is made in this program to use the data from the AREA array.

EXAMPLE 14–8

```
;A short procedure that finds the areas of 5 circles whose radii are stored
;in array RAD.

RAD     DD     2.34                     ;array of radii
        DD     5.66
        DD     9.33
        DD     234.5
        DD     23.4

AREA    DD     5 DUP(?)                 ;array for areas

FINDA   PROC   NEAR

        FLDPI                           ;load pi
        MOV ECX,0                       ;initialize pointer
        .REPEAT
               FLD    RAD[ECX*4]        ;get radius
               FMUL   ST,ST(0)          ;square radius
               FMUL   ST,ST(1)          ;multiply radius squared times pi
               FSTP   AREA[ECX*4]       ;store area
               INC    ECX               ;index next radius
        .UNTIL ECX = 5                  ;repeat 5 times
        FCOMP                           ;clear pi from coprocessor stack
        RET

FINDA   ENDP
```

Although this is a simple program, it does illustrate the operation of the stack. To provide a better understanding of the operation of the stack, Figure 14–10 shows the contents of the stack after each instruction of Example 14–8 executes. Note only one pass through the loop is illustrated because the program calculates five areas and each pass is identical.

The first instruction loads π to the top of the stack. Next, the contents of memory location RAD [ECX*4], one of the elements of the array, is loaded to the top of the stack. This pushes π

FIGURE 14–10 Operation of the coprocessor stack for one iteration of the loop in Example 14–8.

Instruction	ST(0)	ST(1)
FLDPI	π	
FLD RAD[ECX*4]	2.34	π
FMUL ST,ST(0)	5.4756	π
FMUL ST,ST(1)	17.202	π
FSTP AREA[ECX*4]	π	

to ST(1). Next, the FMUL ST,ST(0) instruction squares the radius on the top of the stack. The FMUL ST,ST(1) instruction forms the area. Finally, the top of the stack is stored in the AREA array and also pops the result from the stack in preparation for the next iteration.

Notice how care is taken to always remove all stack data. The last instruction before the RET pops π from the stack. This is important because if data remain on the stack at the end of the procedure, the stack top will no longer be register 0. This could cause problems because software assumes that the top of the stack is register 0. Another way of ensuring that the coprocessor is initialized is to place the FINIT (initialization) instruction at the start of the program.

Finding the Resonant Frequency

An equation commonly used in electronics is the formula for determining the resonant frequency of an LC circuit. The equation solved by the program illustrated in Example 14–9 is

$$Fr = \frac{1}{2\pi\sqrt{LC}}$$

This example uses L_1 for the inductance L, C_1 for the capacitor C, and RES for the resultant resonant frequency.

EXAMPLE 14–9

```
RES     DD      ?                       ;resonant frequency
L1      DD      0.0001                  ;1 mH inductor
C1      DD      47E-6                   ;47 µF capacitor

FR      PROC    NEAR

        FLD     L1                      ;get L
        FMUL    C1                      ;form LC
        FSQRT                           ;form square root of LC
        FLDPI                           ;get pi
        FADD    ST,ST(0)                ;form 2 pi
        FMUL                            ;form 2 pi square root LC
        FLD1
        FDIVR                           ;form reciprocal
        FSTP    RES
        RET

FR      ENDP
```

Notice the straightforward manner in which the program solves this equation. Very little extra data manipulation is required because of the stack inside the coprocessor. Notice how FDIVR, using classic stack addressing, is used to form the reciprocal. If you own a reverse Polish notation calculator, such as those produced by Hewlett-Packard, you are familiar with stack addressing. If not, using the coprocessor will increase your experience with this type of entry.

Finding the Roots Using the Quadratic Equation

This example illustrates how to find the roots of a polynomial expression ($ax^2 + bx + c = 0$) by using the quadratic equation. The quadratic equation is:

$$b \pm \frac{\sqrt{b^2 - 4ac}}{2a}$$

Example 14–10 illustrates a program that finds the roots (R_1 and R_2) for the quadratic equation. The constants are stored in memory locations A_1, B_1, and C_1. Note that no attempt is made to determine the roots if they are imaginary. This example tests for imaginary roots and exits to DOS with a zero in the roots (R_1 and R_2), if it finds them. In practice, imaginary roots could be solved for and stored in a separate set of result memory locations.

EXAMPLE 14–10

```
;A procedure that finds the roots of a polynomial equation using
;the quadratic equation. Imaginary roots are indicated if both
;R1 and R2 are returned as zero.

FOUR    DW    4              ;integer of 4
A1      DD    ?              ;value for a
B1      DD    ?              ;value for b
C1      DD    ?              ;value for c
R1      DD    ?              ;root 1
R2      DD    ?              ;root 2

ROOTS   PROC  NEAR

        FLDZ                 ;get 0.0
        FST   R1             ;clear roots
        FSTP  R2
        FLD   A1             ;from 2a
        FADD  ST,ST(0)
        FILD  FOUR           ;get 4
        FMUL  A1             ;from 4ac
        FMUL  C1
        FLD   B1             ;from b2
        FMUL  ST,ST(0)
        FSUBR                ;from b2 - 4ac
        FTST                 ;test result against zero
        SAHF
        .IF !ZERO?
              FSQRT                ;find square root of b2 - 4ac
              FSTSW AX             ;test for invalid error
              TEST  AX,1
              .IF !ZERO?
                    FCOMPP         ;clear stack
                    RET
              .ENDIF
        .ENDIF
        FLD   B1
        FSUB  ST,ST(1)
        FDIV  ST,ST(2)
        FSTP  R1             ;save root 1
        FLD   B1
        FADD
        FDIVR
        FSTP  R2             ;save root 2
        RET

ROOTS   ENDP
```

Using a Memory Array to Store Results

The next programming example illustrates the use of a memory array and the scaled-indexed addressing mode to access the array. Example 14–11 shows a program that calculates 100 values of inductive reactance. The equation for inductive reactance is $XL = 2\pi FL$. In this example, the frequency range is from 10 Hz to 1000 Hz for F and an inductance of 4 mH. Notice how the instruction FSTP XL[ECX*4 + 4] is used to store the reactance for each frequency, beginning with the last at 1000 Hz and ending with the first at 10 Hz. Also notice how the FCOMP instruction is used to clear the stack just before the RET instruction.

EXAMPLE 14–11

```
;A procedure that calculates the inductive reactance of L at a
;frequencies from 10 Hz to 1000 Hz in steps of 10 Hz and stores
;them in array called XL.

XL     DD     100 DUP(?)          ;array for XL
L      DD     4E-3                ;L = 4 mH
F      DW     10                  ;integer of 10 for F
```

```
XLS   PROC   NEAR

      MOV    ECX,100                    ;count = 100
      FLDPI                             ;get pi
      FADD   ST,ST(0)                   ;form 2 pi
      FMUL   L                          ;form 2 pi L
      .REPEAT
             FILD    F                  ;get F
             FMUL    ST,ST(1)           ;find XL
             FSTP    XL[ECX*4+4]        ;save result
             MOV     AX,F               ;add 10 to F
             ADD     AX,10
             MOV     F,AX
      .UNTILCXZ
      FCOMP                             ;clear stack
      RET

XLS   ENDP
```

Converting a Single-Precision Floating-Point Number to a String

This section of the text shows how to take the floating-point contents of a 32-bit single-precision floating-point number and store it as an ASCII character string. The procedure converts the floating-point number as a mixed number with an integer part and a fractional part, separated by a decimal point. In order to simplify the procedure, a limit is placed on the size of the mixed number so the integer portion is a 32-bit binary number (±2 G) and the fraction is a 24-bit binary number (1/16M). The procedure will not function properly for larger or smaller numbers.

Example 14–12 lists a procedure that converts the contents of memory location NUMB to a string stored in the STR array. The procedure first tests the sign of the number and stores a minus sign for a negative number. After storing a minus sign, if needed, the number is made positive by the FABS instruction. Next, it is divided into an integer and fractional part and stored at WHOLE and FRACT. Notice how the FRNDINT instruction is used to round (using the chop mode) the top of the stack to form the whole number part of NUMB. The whole number part is then subtracted from the original number to generate the fractional part. This is accomplished with the FSUB instruction that subtracts the contents of ST(1) from ST.

EXAMPLE 14–12

```
;A procedure that converts a floating-point number into an ASCII
;character string.

STR    DB     40 DUP(?)              ;storage for string
NUMB   DD     -2224.125              ;test number
WHOLE  DD     ?
FRACT  DD     ?
TEMP   DW     ?                      ;place for CW
TEN    DW     10                     ;integer of 10

FTOA   PROC   NEAR USES EBX ECX EDX

       MOV    ECX,0                  ;initialize pointer
       FSTCW  TEMP                   ;save current control word
       MOV    AX,TEMP                ;change rounding to chop
       PUSH   AX
       OR     AX,0C00H
       MOV    TEMP,AX
       FLDCW  TEMP
       FTST   NUMB                   ;test NUMB
       FSTSW  AX
       AND    AX,4500H               ;get C0, C2, and C3
       .IF AX == 100H                ;if negative
              MOV  STR[ECX],'-'
              INC  ECX
              FABS                   ;make positive
```

```
        .ENDIF
        FRNDINT                              ;round to integer
        FIST    WHOLE                        ;save whole number part
        FLD     NUMB                         ;form and store fraction
        FABS
        FSUBR
        FSTP    FRACT
        MOV     EAX,WHOLE                    ;convert whole part
        MOV     EBX,10
        PUSH    EBX
        .REPEAT
                MOV   EDX,0
                DIV   EBX
                ADD   DL,30H                 ;convert to ASCII
                PUSH  EDX
        .UNTIL  EAX == 0
        POP     EDX
        MOV     AH,3                         ;comma counter
        .WHILE  EDX !== 10                   ;whole part to ASCII
                POP   EBX
                DEC   AH
                .IF AH == 0 && EBX != 10
                         MOV STR[ECX],','
                         INC ECX
                         MOV AH,3
                .ENDIF
                MOV   STR[ECX],DL
                INC   ECX
                MOV   EDX,EBX
        .ENDW
        MOV     STR[ECX],'.'                 ;store decimal point
        INC     ECX
        POP     TEMP                         ;restore original CW
        FLDCW   TEMP
        FLD     FRACT                        ;convert fractional part
        .REPEAT
                FIMUL    TEN
                FIST     TEMP
                MOV      AX,TEMP
                ADD      AL,30H
                MOV      STR[ECX],AL
                INC      ECX
                FISUB    TEMP
                FXAM
                SAHF
        .UNTIL  ZERO?
        FCOMP                                ;clear stack
        MOV     STR[ECX],0                   ;store null
        RET

FTOA    ENDP
```

14–5 INTRODUCTION TO MMX TECHNOLOGY

The MMX[1] (**multimedia extensions**) technology adds 57 new instructions to the instruction set of the Pentium–Pentium 4 microprocessors. The MMX technology also introduces new general-purpose instructions. The new MMX instructions are designed for applications such as motion video, combined graphics with video, image processing, audio synthesis, speech synthesis and compression, telephony, video conferencing, 2D graphics, and 3D graphics. These instructions

[1]MMX is a registered trademark of Intel Corporation.

(new beginning with the Pentium) operate in parallel with other operations as the instructions for the arithmetic coprocessor.

Data Types

The MMX architecture introduces new packed data types. The data types are eight packed, consecutive 8-bit bytes; four packed, consecutive 16-bit words; and two packed, consecutive 32-bit doublewords. Bytes in this multibyte format have consecutive memory addresses and use the little endian form, like other Intel data. See Figure 14–11 for the format for these new data types.

The MMX technology registers have the same format as a 64-bit quantity in memory and have two data access modes: 64-bit access mode and 32-bit access mode. The 64-bit access mode is used for 64-bit memory and registers transfers for most instructions. The 32-bit access mode is used for 32-bit memory and also register transfers for most instructions. The 32-bit transfers occur between microprocessor registers, and the 64-bit transfers occur between floating-point coprocessor registers.

Figure 14–12 illustrates the internal register set of the MMX technology extension and how it uses the floating-point coprocessor register set. This technique is called **aliasing** because the floating-point registers are shared as the MMX registers. That is, the MMX registers (MM_0–MM_7) are the same as the floating-point registers. Note that the MMX register set is 64 bits wide and uses the rightmost 64-bits of the floating-point register set.

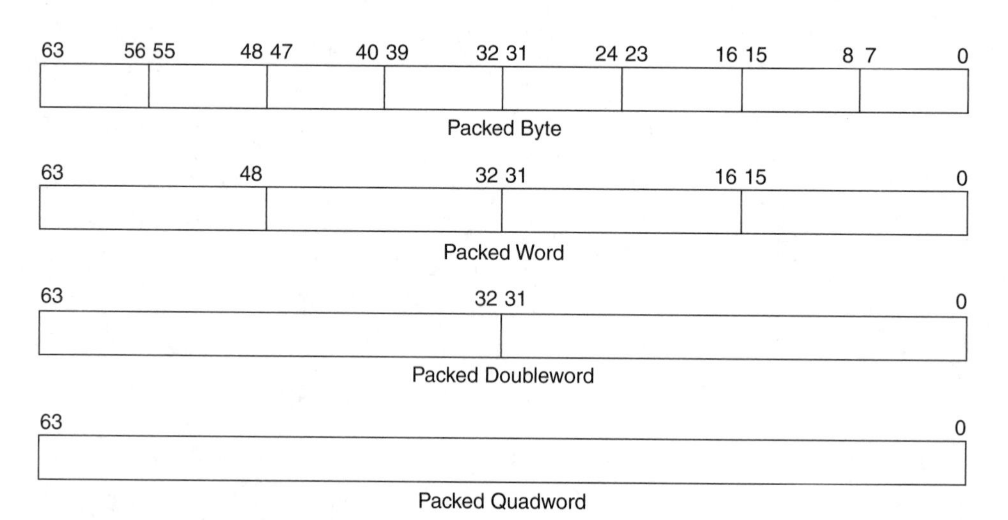

FIGURE 14–11 The structure of data stored in the MMX registers.

FIGURE 14–12 The structure of the MMX register set. Note that MM_0 and FF_0 through MM_7 and FP_7 interchange with each other.

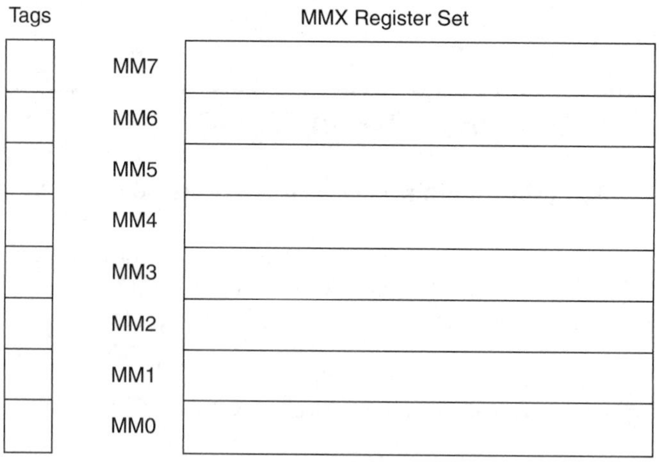

Tags

MMX Register Set

MM7
MM6
MM5
MM4
MM3
MM2
MM1
MM0

Instruction Set

The instruction for MMX technology includes arithmetic, comparison, conversion, logical, shift, and data transfer instructions. Although the instruction types are similar to the microprocessor's instruction set, the main difference is that the MMX instructions use the data types shown in Figure 14–11 instead of the normal data types used with the microprocessor.

Arithmetic Instructions. The set of arithmetic instructions includes addition, subtraction, multiplication, a special multiplication with an addition, and so on. Three additions exist. The PADD and PSUB instructions add or subtract packed signed or unsigned packed bytes, packed words, or packed doubleword data. The add instructions are appended with a B, W, or D to select the size, as in PADDB for a byte, PADDW for a word, and PADDD for a doubleword. The same is true for the PSUB instruction. The PMULHW and the PMULLW instructions perform multiplication on four pairs of 16-bit operands, producing 32-bit results. The PMULHW instruction multiplies the high-order 16-bits, and the PMULLW instruction multiplies the low-order 16-bits. The PMADDWD instruction multiplies and adds. After multiplying, the four 32-bit results are added to produce two 32-bit doubleword results.

The MMX instructions use operands just as the integer or floating-point instructions do. The difference is the register names (MM_0–MM_7). For example, the PADDB MM_1, MM_2 instruction adds the entire 64-bit contents of MM_2 to MM_1, byte by byte. The result is steered into MM_1. When each 8-bit section is added, any carries generated are dropped. For example, the byte A0H added to the byte 70H produces the byte sum of 10H. The true sum is 110H, but the carry is dropped. Note that the second operand or source can be a memory location containing the 64-bit packed source or an MMX register. You might say that this instruction performs the same function as eight separate byte-sized ADD instructions! If used in an application, this certainly speeds execution of the application. Like PADD, PSUB also does not carry or borrow. The difference is that if an overflow or underflow occurs, the difference becomes 7FH (+127) for an overflow and 80H (–128) for an underflow. Intel calls this **saturation**, because these values represent the largest and smallest signed bytes.

Comparison Instructions. There are two comparison instructions: PCMPEQ (equal) and PCMPGT (greater than). As with PADD and PSUB, there are three versions of each compare instruction: for example, PCMPEQUB (compares bytes), PCMPEQUW (compares words), and PCMPEQUD (compares doublewords). These instructions do not change the microprocessor flag bits; instead, the result is all ones for a true condition and all zeros for a false condition. For example, if the PCMPEQB MM_2, MM_3 instruction is executed and the least-significant bytes of MM_2 and MM_3 = 10H and 11H, respectively, the result found in the least-significant byte of MM_2 is 00H. This indicates that the least-significant bytes were not equal. If the least-significant byte contained an FFH, it indicates that the two bytes were equal.

Conversion Instructions. There are two basic conversion instructions: PACK and PUNPCK. PACK is available as PACKSS (signed saturation) and PACKUS (unsigned saturation). PUNPCK is available as PUNPCKH (unpack high data) and PUNPCKL (unpack low data). Similar to the prior instructions, these can be appended with B, W, or D for byte, word, and doubleword pack and unpack; but they must be used in combinations WB (word to byte) or DW (doubleword to word). For example, the PACKUSWB MM_3, MM_6 instruction packs the words from MM_6 into bytes in MM_3. If the unsigned word does not fit (too large) into a byte, the destination byte becomes an FFH. For signed saturation, we use the same values explained under addition.

Logic Instructions. The logic instructions are PAND (AND), PANDN (NAND), POR (OR), and PXOR (Exclusive-OR). These instructions do not have size extensions, and perform these bit-wise operations on all 64 bits of the data. For example, the POR MM_2, MM_3 instruction ORs all 64 bits of MM_3 with MM_2. The logical sum is placed in MM_2 after the OR operation.

Shift Instruction. This instruction contains logical shifts and an arithmetic shift right instruction. The logic shifts are PSLL (left) and PSRL (right). Variations are word (W), doubleword (D), and quadword (Q). For example, the PSLLQ $MM_{3,2}$ instruction shifts all 64 bits in MM_3 left two places. Another example is the PSLLD $MM_{3,2}$ instruction that shifts the two 32-bit doublewords in MM_3 left two places each.

The PSRA (arithmetic right shift) instruction functions in the same manner as the logical shifts, except that the sign-bit is preserved.

Data Transfer Instructions. There are two data transfer instructions: MOVED and MOVEQ. These instructions allow transfers between registers and between a register and memory. The MOVED instruction transfers 32 bits of data between an integer register or memory location and an MMX register. For example, the MOVED ECX, MM_2 instruction copies the rightmost 32 bits of MM_2 into ECX. There is no instruction to transfer the leftmost 32 bits of an MMX register. You could use a shift right before a MOVED to do the transfer.

The MOVEQ instruction copies all 64 bits of an MMX register between memory or another MMX register. The MOVEQ MM_2, MM_3 instruction transfers all 64 bits of MM_3 into MM_2.

EMMS Instruction. The EMMS (empty MMX-state) instruction sets (11) all the tags in the floating-point unit, so the floating-point registers are listed as empty. The EMMS instruction must be executed before the return instruction at the end of any MMX procedure, or a subsequent floating-point operation will cause a floating-point interrupt error, crashing Windows or any other application. If you plan to use floating-point instructions within an MMX procedure, you must use the EMMS instruction before executing the floating-point instruction. All other MMX instructions clear the tags, which indicate that all floating-point registers are in use.

Instruction Listing. Table 14–10 lists all the MMX instructions with the machine code so these instructions can be used with the assembler. At present, MASM does not support these new instructions unless you have upgraded to the latest version (6.15). The latest version can be found in the Windows Driver Development Kit (Windows DDK), which is available for a small shipping charge from Microsoft Corporation. It is also available in Visual Studio .NET 2003 (search for ML.EXE). Any MMX instruction can be used inside Visual C++ using the inline assembler.

Programming Example. Example 14–13 illustrates a simple programming example that uses the MMAX instructions to perform a task that takes eight times longer using normal microprocessor instructions. In this example an array of 1000 bytes of data (BLOCKA) is added to a second array of 1000 bytes (BLOCKB). The result is stored in a third array called BLOCKC. Example 14–13(a) lists a procedure that uses traditional assembly language to perform the addition and Example 14–13(b) shows the same process using MMX instructions.

EXAMPLE 14–13(a)

```
;Procedure that adds BLOCKA0 to BLOCKB and stores the sums in BLOCKC

BLOCKA DB      1000 DUP(?)
BLOCKB DB      1000 DUP(?)
BLOCKC DB      1000 DUP(?)

SUM     PROC    NEAR

         MOV   ECX,1000
         .REPEAT
                MOV        AL,BLOCKA[ECX-1]
                ADD        AL,BLOCKB[ECX-1]
                MOV        BLOCKC[ECX-1]
         .UNTILCXZ
         RET

SUM      ENDP
```

TABLE 14–10 The MMX instruction set extension.

EMMS Empty MMX state	
0000 1111 0111 1111	
Example	
EMMS	
MOVED Move doubleword	
0000 1111 0110 1110 11 xxx rrr	reg → xreg
Examples	
MOVED MM3, EDX MOVED MM4, EAX	
0000 1111 0111 1110 11 xxx rrr	xreg → reg
Examples	
MOVED EAX, MM3 MOVED EBP, MM7	
0000 1111 0110 1110 oo xxx mmm	mem → xreg
Examples	
MOVED MM3, DATA1 MOVED MM5, BIG_ONE	
0000 1111 0111 1110 oo xxx mmm	xreg → mem
Examples	
MOVED DATA2, MM3 MOVED SMALL_POTS, MM7	
MOVEQ Move quadword	
0000 1111 0110 1111 11 xxx1 xxx2	xreg2 → xreg1
Examples	
MOVEQ MM3, MM2 ;copies MM2 to MM3 MOVEQ MM7, MM3	
0000 1111 0111 1111 11 xxx1 xxx2	xreg1 → reg2
Examples	
MOVEQ MM3, MM2 ;copies MM3 to MM2 MOVEQ MM7, MM3	
0000 1111 0110 1111 oo xxx mmm	mem → xreg
Examples	
MOVEQ MM3, DATA1 MOVEQ MM5, DATA3	
0000 1111 0111 1111 oo xxx mmm	xreg → mem
Examples	
MOVEQ DATA2, MM0 MOVEQ SMALL_POTS, MM3	

PACKSSDW Pack signed doubleword to word	
0000 1111 0110 1011 11 xxx1 xxx2	xreg2 → xreg1
Examples	
PACKSSDW MM1, MM2 PACKSSDW MM7, MM3	
0000 1111 0111 1011 oo xxx mmm	mem → xreg
Examples	
PACKSSDW MM3, BUTTON PACKSSDW MM7, SOUND	

PACKSSWB Pack signed word to byte	
0000 1111 0110 0011 11 xxx1 xxx2	xreg2 → xreg1
Examples	
PACKSSWB MM1, MM2 PACKSSWB MM7, MM3	
0000 1111 0111 0011 oo xxx mmm	mem → xreg
Examples	
PACKSSWB MM3, BUTTON PACKSSWB MM7, SOUND	

PACKUSWB Pack unsigned word to byte	
0000 1111 0110 0111 11 xxx1 xxx2	xreg2 → xreg1
Examples	
PACKUSWB MM1, MM2 PACKUSWB MM7, MM3	
0000 1111 0111 0111 oo xxx mmm	mem → xreg
Examples	
PACKUSWB MM3, BUTTON PACKUSWB MM7, SOUND	

PADD Add with truncation	Byte, word, and doubleword
0000 1111 1111 11gg 11 xxx1 xxx2	xreg2 → xreg1
Examples	
PADDB MM1, MM2 PADDW MM7, MM3 PADDD MM3, MM4	
0000 1111 1111 11gg oo xxx mmm	mem → xreg
Examples	
PADDB MM3, BUTTON PADDW MM7, SOUND PADDD MM3, BUTTER	

PADDS Add with signed saturation	Byte and word
0000 1111 1110 11gg 11 xxx1 xxx2	xreg2 → xreg1
Examples	
PADDSB MM1, MM2 PADDSW MM7, MM3	
0000 1111 1110 11gg oo xxx mmm	mem → xreg
Examples	
PADDSB MM3, BUTTON PADDSW MM7, SOUND	

PADDUS Add with unsigned saturation	Byte and word
0000 1111 1101 11gg 11 xxx1 xxx2	xreg2 → xreg1
Examples	
PADDUSB MM1, MM2 PADDUSW MM7, MM3	
0000 1111 1101 11gg oo xxx mmm	mem → xreg
Examples	
PADDUSB MM3, BUTTON PADDUSW MM7, SOUND	

PAND And	
0000 1111 1101 1011 11 xxx1 xxx2	xreg2 → xreg1
Examples	
PAND MM1, MM2 PAND MM7, MM3	
0000 1111 1101 1011 oo xxx mmm	mem → xreg
Examples	
PAND MM3, BUTTON PAND MM7, SOUND	

PANDN Nand	
0000 1111 1101 1111 11 xxx1 xxx2	xreg2 → xreg1
Examples	
PANDN MM1, MM2 PANDN MM7, MM3	
0000 1111 1101 1111 oo xxx mmm	mem → xreg
Examples	
PANDN MM3, BUTTON PANDN MM7, SOUND	

PCMPEQU Compare for equality	Byte, word, and doubleword
0000 1111 0111 01gg 11 xxx1 xxx2	xreg2 → xreg1
Examples	
PCMPEQUB MM1, MM2 PCMPEQUW MM7, MM3 PCMPEQUD MM0, MM5	
0000 1111 0111 01gg oo xxx mmm	mem → xreg
Examples	
PCMPEQUB MM3, BUTTON PCMPEQUW MM7, SOUND PCMPEQUD MM0, FROG	

PCMPGT Compare for greater than	Byte, word, and doubleword
0000 1111 0110 01gg 11 xxx1 xxx2	xreg2 → xreg1
Examples	
PCMPGTB MM1, MM2 PCMPGTW MM7, MM3 PCMPGTD MM0, MM5	
0000 1111 0110 01gg oo xxx mmm	mem → xreg
Examples	
PCMPGTB MM3, BUTTON PCMPGTW MM7, SOUND PCMPGTD MM0, FROG	

PMADD Multiply and add	
0000 1111 1111 0101 11 xxx1 xxx2	xreg2 → xreg1
Examples	
PMADD MM1, MM2 PMADD MM7, MM3	
0000 1111 1111 0101 oo xxx mmm	mem → xreg
Examples	
PMADD MM3, BUTTON PMADD MM7, SOUND	

PMULH Multiplication–high	
0000 1111 1110 0101 11 xxx1 xxx2	xreg2 → xreg1
Examples	
PMULH MM1, MM2 PMULH MM7, MM3	
0000 1111 1110 0101 oo xxx mmm	mem → xreg
Examples	
PMULH MM3, BUTTON PMULH MM7, SOUND	

PMULL Multiplication–low	
0000 1111 1101 0101 11 xxx1 xxx2	xreg2 → xreg1
Examples	
PMULL MM1, MM2 PMULL MM7, MM3	
0000 1111 1101 0101 oo xxx mmm	mem → xreg
Examples	
PMULL MM3, BUTTON PMULL MM7, SOUND	

POR Or	
0000 1111 1110 1011 11 xxx1 xxx2	xreg2 → xreg1
Examples	
POR MM1, MM2 POR MM7, MM3	
0000 1111 1110 1011 oo xxx mmm	mem → xreg
Examples	
POR MM3, BUTTON POR MM7, SOUND	

PSLL Shift left	Word, doubleword, and quadword
0000 1111 1111 00gg 11 xxx1 xxx2	xreg2 → xreg1
Examples	
PSLLW MM1, MM2 PSLLD MM7, MM3 PSLLQ MM6, MM5	
0000 1111 1111 00gg oo xxx mmm	mem → xreg shift count in memory
Examples	
PSLLW MM3, BUTTON PSLLD MM7, SOUND PSLLQ MM2, COUNT1	
0000 1111 0111 00gg 11 110 mmm data8	xreg by count shift count is data8
Examples	
PSLLW MM3, 2 PSLLD MM0, 6 PSLLQ MM7, 1	

PSRA Shift arithmetic right	Word, doubleword, and quadword
0000 1111 1110 00gg 11 xxx1 xxx2	xreg2 → xreg1
Examples	
PSRAW MM1, MM2 PSRAD MM7, MM3 PSRAQ MM6, MM5	
0000 1111 1110 00gg oo xxx mmm	mem → xreg shift count in memory
Examples	
PSRAW MM3, BUTTON MM7, SOUND PSRAQ MM2, COUNT1	
0000 1111 0111 00gg 11 100 mmm data8	xreg by count shift count is data8
Examples	
PSRAW MM3, 2 PSRAD MM0, 6 PSRAQ MM7, 1	

PSRL Shift right	Word, doubleword, and quadword
0000 1111 1101 00gg 11 xxx1 xxx2	xreg2 → xreg1
Examples	
PSRLW MM1, MM2 PSRLD MM7, MM3 PSRLQ MM6, MM5	
0000 1111 1101 00gg oo xxx mmm	mem → xreg shift count in memory
Examples	
PSRLW MM3, BUTTON PSRLD MM7, SOUND PSRLQ MM2, COUNT1	
0000 1111 0111 00gg 11 010 mmm data8	xreg by count shift count is data8
Examples	
PSRLW MM3, 2 PSRLD MM0, 6 PSRLQ MM7, 1	

PSUB Subtract with truncation	Byte, word, and doubleword
0000 1111 1111 10gg 11 xxx1 xxx2	xreg2 → xreg1
Examples	
PSUBB MM1, MM2 PSUBW MM7, MM3 PSUBD MM3, MM4	
0000 1111 1111 10gg oo xxx mmm	mem → xreg
Examples	
PSUBB MM3, BUTTON PSUBW MM7, SOUND PSUBD MM3, BUTTER	

PSUBS Subtract with signed saturation	Byte, word, and doubleword
0000 1111 1110 10gg 11 xxx1 xxx2	xreg2 → xreg1
Examples	
PSUBSB MM1, MM2 PSUBSW MM7, MM3 PSUBSD MM3, MM4	
0000 1111 1110 10gg oo xxx mmm	mem → xreg
Examples	
PSUBSB MM3, BUTTON PSUBSW MM7, SOUND PSUBSD MM3, BUTTER	

PSUBUS Subtract with unsigned saturation	Byte, word, and doubleword
0000 1111 1101 10gg 11 xxx1 xxx2	xreg2 → xreg1
Examples	
PSUBUSB MM1, MM2 PSUBUSW MM7, MM3 PSUBUSD MM3, MM4	
0000 1111 1101 10gg oo xxx mmm	mem → xreg
Examples	
PSUBUSB MM3, BUTTON PSUBUSW MM7, SOUND PSUBUSD MM3, BUTTER	

PUNPCKH	Unpack high data to next larger	Byte, word, doubleword
0000 1111 0110 10gg 11 xxx1 xxx2		xreg2 → xreg1
Examples		
PUNPCKH MM1, MM2 PUNPCKH MM3, MM4		
0000 1111 0110 10gg oo xxx mmm		mem → xreg
Examples		
PUNPCKH MM7, WATER PUNPCKH MM2, DOGGY		

PUNPCKL	Unpack low data to next larger	Byte, word, doubleword
0000 1111 0110 00gg 11 xxx1 xxx2		xreg2 → xreg1
Examples		
PUNPCKL MM1, MM2 PUNPCKL MM3, MM4		
0000 1111 0110 00gg oo xxx mmm		mem → xreg
Examples		
PUNPCKL MM7, WATER PUNPCKL MM2, DOGGY		

PXOR	Bitwise Excluse-OR	Byte, word, doubleword
0000 1111 1110 1111 11 xxx1 xxx2		xreg2 → xreg1
Examples		
PXOR MM2, MM3 PXOR MM4, MM7 PXOR MM0, MM1		
0000 1111 1110 1111 oo xxx mmm		mem → xreg
Examples		
PXOR MM2, FROGS PXOR MM4, WALTER		

EXAMPLE 14–13(b)

```
;Procedure that adds BLOCKA0 to BLOCKB and stores the sums in BLOCKC

BLOCKA DB       1000 DUP(?)
BLOCKB DB       1000 DUP(?)
BLOCKC DB       1000 DUP(?)

SUMM    PROC    NEAR
        MOV     ECX,125
        .REPEAT
                MOVEQ  MM0,QWORD PTR BLOCKA[ECX-8]
                PADDB  MM0,QWORD PTR BLOCKB[ECX-8]
                MOVEQ  QWORD PTR BLOCKC[ECX-8],MM0
        .UNTILCXZ
        RET

SUMM    ENDP
```

If you closely compare the programs, notice that the MMX version executes the loop of three instructions 125 times, while the traditional software goes through its loop 1000 times. The MMX version will execute eight times faster. This occurs because eight bytes (QWORD) are added at a time.

14–6 INTRODUCTION TO SSE TECHNOLOGY

The latest type of instruction added to the instruction set of the Pentium 4 is SIMD (**single-instruction, multiple data**). As the name implies, a single instruction operates on multiple data in much the same way as do the MMX instructions, which are SIMD instructions that operate on multiple data. The MMX instruction set functions with integers; the SIMD instruction set functions with floating-point numbers as well as integers. The SIMD extension instructions first appeared in the Pentium III as SSE (**streaming SIMD extensions**) instructions. Later, SSE 2 instructions were added to the Pentium 4, and new to the Pentium 4 (the 90-nanometer E model) are SSE 3 instructions.

Recall that the MMX instructions shared registers with the arithmetic coprocessor. The SSE instructions use a new and separate register array to operate on data. Figure 14–13 illustrates an array of eight 128-bit-wide registers that function with the SSE instructions. These new registers are called **XMM registers** (XMM_0–XMM_7), which denotes extended multimedia registers. To accommodate this new 128-bit-wide data size, a new keyword is added called **OWORD**. An OWORD (**octalword**) designates a 128-bit variable, as in OWORD PTR for the SSE instruction set. A double quadword is also used at times to specify a 128-bit number.

Just as the MMX registers can contain multiple data types, so can the XMM registers of the SSE unit. Figure 14–14 illustrates the data types that can appear in any XMM register for various

127	0
XMM7	
XMM6	
XMM5	
XMM4	
XMM3	
XMM2	
XMM1	
XMM0	

FIGURE 14–13 The XMM registers used by the SSE instructions.

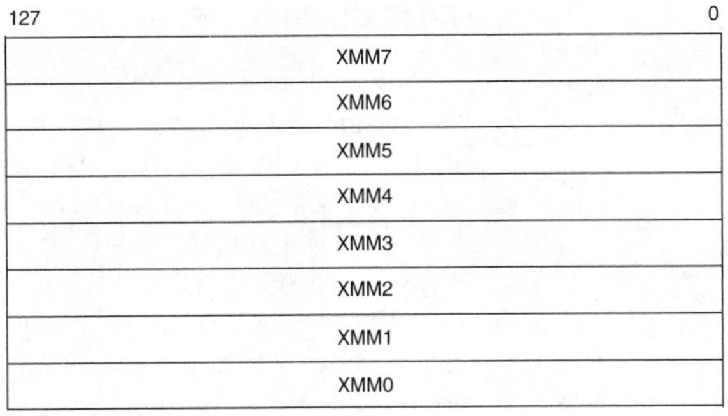

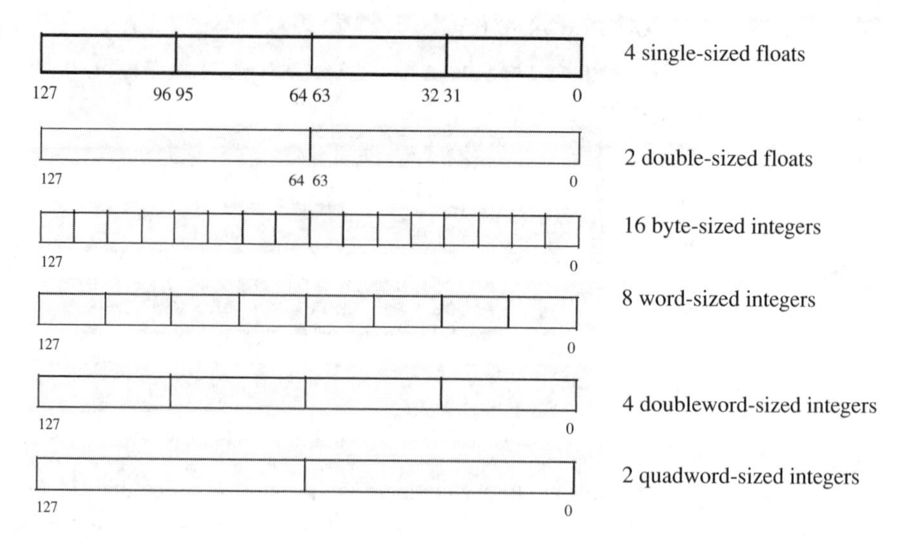

FIGURE 14–14 Data formats for the SSE 2 and SSE 3 instructions.

SSE instructions. An XMM register can hold four single-precision floating-point numbers or two double-precision floating-point numbers. XMM registers can also hold sixteen 8-bit integers, eight 16-bit integers, four 32-bit integers, or two 64-bit integers. This is a twofold increase in the capacity of the system when compared to the integers contained in MMX registers and hence a twofold increase in execution speeds of integer operations that use the XMM registers and SSE instructions. For new applications that are destined to execute on a Pentium 4 or newer microprocessor, the SSE instructions are used in place of the MMX instructions. Since not all machines are yet Pentium 4 class machines, there still is a need to include MMX technology instructions in a program for compatibility to these older systems.

Floating-Point Data

Floating-point data are operated upon as either packed or scalar and either single-precision or double-precision. The packed operation is performed on all sections at a time; the scalar form is only operated on the rightmost section of the register contents. Figure 14–15 shows both the packed and scalar operations on SSE data in XMM registers. The scalar form is comparable to the operation performed by the arithmetic coprocessor. Opcodes are appended with PS (packed single), SS (scalar single), PD (packed double), or SD (scaled double) to form the desired instruction. For example, the opcode for a multiply is MUL, but the opcode for a packed double is MULPD and MULSD for a scalar double multiplication. The single-precision multiplies are MULPS and MULSS. In other words, once the two-letter extension and its meaning are understood, it is relatively easy to master the new SSE instructions.

The Instruction Set

The SSE instructions have a few new types added to the instruction set. The floating-point unit does not have a reciprocal instruction, which is used quite often to solve complex equations. The reciprocal instruction ($\frac{1}{n}$) now appears in the SSE extensions as the RCP instruction, which generates reciprocals and is written as RCPPS, RCPSS, RCPPD, and RCPSD. There is also a reciprocal of a square root ($\frac{1}{\sqrt{n}}$) instruction, called RSQRT, which is written as RSQRTPS, RSQRTSS, RSQRTPD, and RSQRTSD.

The remainder of the instructions for the SSE unit are basically the same as for the microprocessor and MMX unit except for a few cases. The instruction table in Appendix B lists the instructions, but does not list the extensions (PS, SS, PD, and DS) to the instructions. Again note

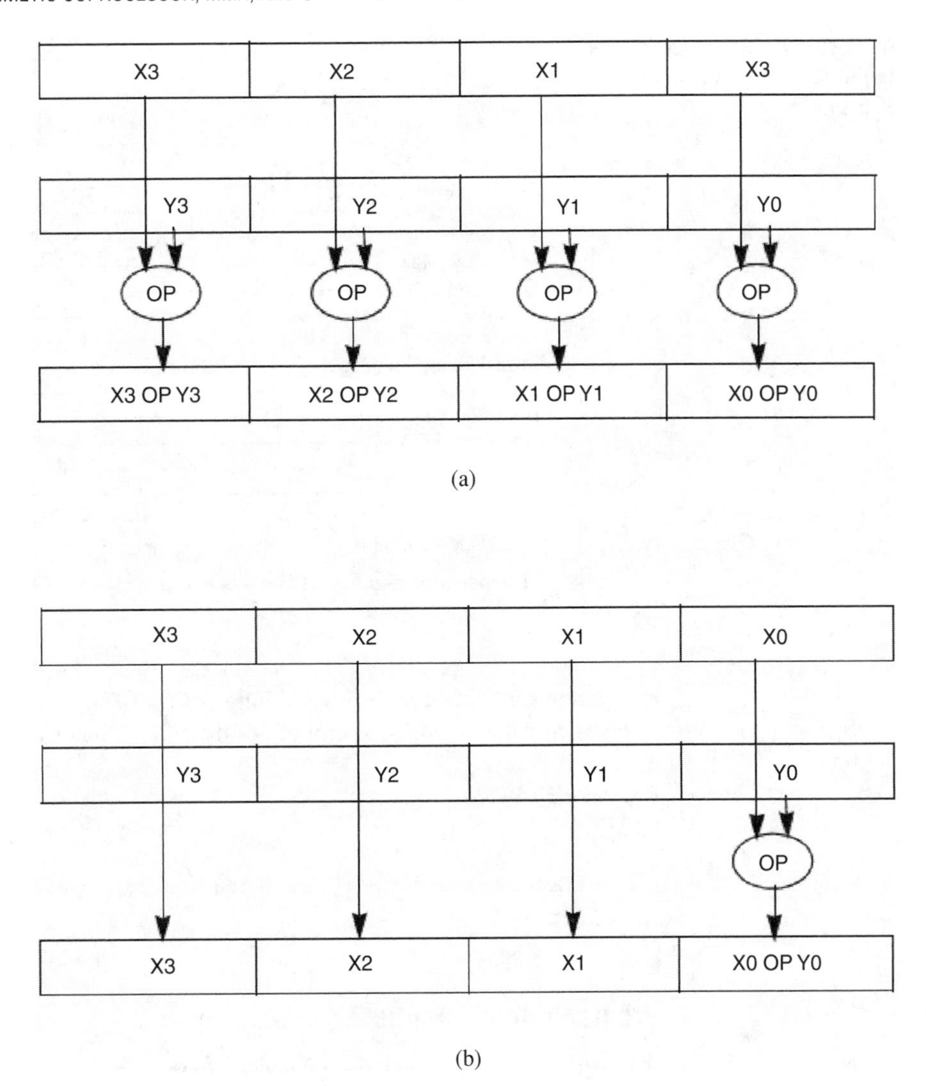

FIGURE 14–15 Packed (a) and scalar (b) operations for single-precision floating-point numbers.

that SSE 2 and SSE 3 contain double-precision operations and SSE does not. Instructions that start with the letter P operate on integer data that is byte, word, doubleword, or quadword sized. For example, the PADDB XMM_0, XMM_1 instruction adds the 16 byte-sized integers in the XMM_1 register to the 16 byte-sized integers in the XMM_0 register. PADDW adds 16-bit integers, PADDD adds doublewords, and PADDQ adds quadwords. The execution times are not provided by Intel so they do not appear in the appendix for these instructions.

The Control/Status Register

The SSE unit also contains a control/status register accessed as MXCSR. Figure 14–16 illustrates the MXCSR for the SSE unit. Notice that this register is very similar to the control/status register of the arithmetic coprocessor presented earlier in this chapter. This register sets the precision and rounding modes for the coprocessor as does the control register for the arithmetic coprocessor, and it provides information about the operation of the SSE unit.

The SSE control/status register is loaded from memory using the LDMXCSR and FXRSTOR instructions or stored into the memory using the STMXCSR and FXSAVE instructions. Suppose the rounding control (see Figure 14–6 for the state of the rounding control bits) needs to be

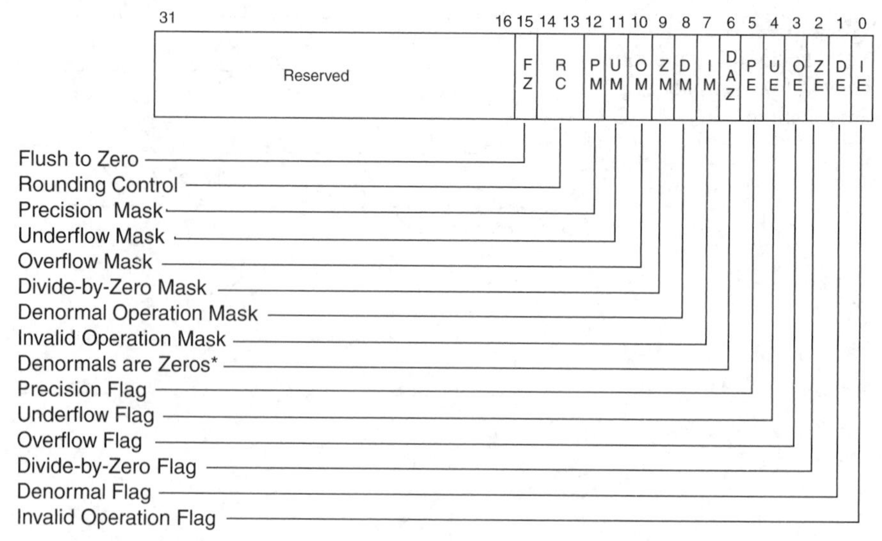

FIGURE 14–16 The MXCSR (control/status) register of the SSE unit.

changed to round toward positive infinity (RC = 10). Example 14–14 shows the software that changes only the rounding control bits of the control/status register.

EXAMPLE 14–14

```
;change the rounding control to 10.

        STMXCSR   CONTROL        ;save the control/status register
        BTS  CONTROL,14          ;set bit 14
        BTR  CONTROL,13          ;clear bit 13
        LDMXCSR CONTROL          ;reload control/status register
```

Programming Examples

A few programming examples are needed to show how to use the SSE unit. As mentioned, the SSE unit allows floating-point and integer operations on multiple data. Suppose that the capacitive reactance is needed for a circuit that contains a 1.0 μF capacitor at various frequencies from 100 Hz to 10,000 Hz in 100 Hz steps. The equation used to calculate capacitive reactance is:

$$XC = \frac{1}{2\pi FC}$$

Example 14–15 illustrates a procedure that generates the 100 outcomes for this equation using the SSE unit and single-precision floating-point data. The program listed in Example 14–15(a) uses the SSE unit to perform four calculations per iteration, while the program in Example 14–15(b) uses the floating-point coprocessor to calculate XC one at a time. Example 14–15(c) is yet another example in C++. Examine the loop to see that the first example goes through the loop 25 times and the second goes through the loop 100 times. Each time the loop executes in Example 14–15(a) it executes seven instructions ($25 \times 7 = 175$), which takes 175 instruction times. Example 14–15(b) executes eight instructions per iteration of its loop ($100 \times 8 = 800$), which requires 800 instruction times. By using this parallelism, the SSE unit allows the calculations to be accomplished in much less time than any other method. The C++ version in Example 14–15(c) uses the directive `__declspec(align(16))` before each variable to make certain that they are aligned properly in the memory. If these are missing, the program will not function because the SSE memory variables must be aligned on at least quadword boundaries (16). This final version executes about $4\frac{1}{2}$ times faster than Example 14–15(b).

EXAMPLE 14–15(a)

```
;using the SSE unit

XC      DD      100 DUP(?)
CAP     DD      1.0E-6, 1.0E-6, 1.0E-6, 1.0E-6
F       DD      100.0, 200.0, 300.0, 400.0
INCR    DD      400.0, 400.0, 400.0, 400.0
PI      DD      4 DUP(?)

FXC     PROC    NEAR

        MOV     ECX,0
        FLDPI                                           ;get π
        FADD    ST,ST(0)                                ;double π
        FST     PI                                      ;store four 2π
        FST     PI+4
        FST     PI+8
        FSTP    PI+12
        MOVUPS  XMM0,OWORD PTR PI                       ;get four 2πs
        MOVUPS  XMM1,OWORD PTR INCR                     ;get increment
        .REPEAT
             MOVUPS     XMM2,OWORD PTR F                ;load frequencies
             MULPS      XMM2,XMM0                       ;generate 2πFs
             MULPS      XMM2,CAPS                       ;2πFC
             RCPPS      XMM3,XMM2                       ;find reciprocal
             MOVUPS     OWORD PTR XC[ECX],XMM3          ;save four XCs
             ADD        ECX,16                          ;move pointer
             ADDPS      OWORD PTR F,XMM1                ;increment Fs
        .UNTIL ECX == 100
        RET

FXC     ENDP
```

EXAMPLE 14–15(b)

```
;using the coprocessor

XC      DD      100 DUP(?)
CAP     DD      1.0E-6
F       DD      0
INCR    DD      100.0

FXC1    PROC    NEAR

        FLDPI                           ;get π
        FADD ST,ST(0)                   ;form 2π
        FMUL ST,CAP                     ;form 2πC
        MOV  ECX,0
        .REPEAT
               FLD   F                  ;get frequency
               FADD  INCR               ;add increment
               FST   F                  ;save it for next time
               FMUL  ST,ST(1)           ;2πFC
               FLD1                     ;form reciprocal
               FDIVR
               FSTP  XC[ECX*4]          ;save XC
               INC   ECX
        .UNTIL ECX == 100
        FCOMP                           ;clear coprocessor stack
        RET

FXC1    ENDP
```

EXAMPLE 14–15(c)

```
void FindXC()
{
      //floating-point example using C++ with the inline assembler

      __declspec(align(16)) float f[4] = {-300,-200,-100,0};
      __declspec(align(16)) float pi[4];
      __declspec(align(16)) float caps[4] = {1.0E-6, 1.0E-6, 1.0E-6, 1.0E-6};
      __declspec(align(16)) float incr[4] = {400, 400, 400, 400};
      __declspec(align(16)) float Xc[100];
      _asm
      {
            Fldpi                                   ;form 2π
            fadd     st,st(0)
            fst      pi
            fst      pi+4
            fst      pi+8
            fstp     pi+12
            movaps   xmm0,oword ptr pi
            movaps   xmm1,oword ptr incr
            movaps   xmm3,oword ptr f
            mulps    xmm0,oword ptr caps           ;2πC
            mov      ecx,0
LOOP1:
            movaps   xmm2,xmm3
            addps    xmm2,xmm1
            movaps   xmm3,xmm2
            mulps    xmm2,xmm0
            rcpps    xmm2,xmm2                     ;reciprocal
            movaps   oword ptr Xc[ecx],xmm2
            add      ecx,16
            cmp      ecx,400
            jnz      LOOP1
      }
}
```

The first example in this section, Example 14–15, used floating-point numbers to perform multiple calculations, but the SSE unit can also operate on integers. The example illustrated in Example 14–16 uses integer operation to add BlockA to BlockB and store the sum in BlockC. Each block contains 4000 eight-bit numbers. Example 14–16(a) lists an assembly language procedure that forms the sums using the standard integer unit of the microprocessor, which requires 4000 iterations to accomplish.

EXAMPLE 14–16(a)

```
;A procedure that forms 4000 eight-bit sums

SUMS   PROC   NEAR

       MOV    ECX,0
       .REPEAT
              MOV  AL,BLOCKA[ECX]
              ADD  AL,BLOCKB[ECX]
              MOV  BLOCKC,[ECX]
              INC  ECX
       .UNTIL ECX == 4000
       RET

SUMS   EMDP
```

EXAMPLE 14–16(b)

```
;A procedure that uses SSE to form 4000 eight-bit sums

SUMS1  PROC   NEAR
```

```
        MOV    ECX,0
        .REPEAT
               MOVDQA XMM0,OWORD PTR BLOCKA[ECX]
               PADDB  XMM0,OWORD PTR BLOCKB[ECX]
               MOVDQA OWORD PTR BLOCKC[ECX]
               ADD    ECX,16
        .UNTIL ECX == 4000
        RET

SUMS1   ENDP
```

Both example programs generate 4000 sums, but the second example using the SSE unit does it by passing through its loop 250 times, while the first example requires 4000 passes. Hence the second example functions 16 times faster because of the SSE unit. Notice how the PADDB, an instruction presented with the MMX unit, is used with the SSE unit. The SSE unit uses the same commands as the MMX except the registers are different. The MMX unit uses 64-bit-wide MM registers and the SSE unit uses 128-bit-wide XMM registers.

Optimization

The compiler in Visual C++ does have optimization for the SSE unit, but it does not optimize the examples presented in this chapter. It will attempt to optimize a single equation in a statement if the SSE unit can be utilized for the equation. It does not look at a program for blocks of operations that can be optimized as in the examples presented here. Until a compiler and extensions are developed so parallel operations such as these can be included, programs that require high speeds will need hand-coded assembly language for optimization. This is especially true of the SSE unit.

14–7 SUMMARY

1. The arithmetic coprocessor functions in parallel with the microprocessor. This means that the microprocessor and coprocessor can execute their respective instructions simultaneously.
2. The data types manipulated by the coprocessor include signed integer, floating-point, and binary-coded decimal (BCD).
3. Three forms of integers are used with the coprocessor: word (16 bits), short (32 bits), and long (64 bits). Each integer contains a signed number in true magnitude for positive numbers and two's complement form for negative numbers.
4. A BCD number is stored as an 18-digit number in 10 bytes of memory. The most-significant byte contains the sign-bit, and the remaining nine bytes contain an 18-digit packed BCD number.
5. The coprocessor supports three types of floating-point numbers: single-precision (32 bits), double-precision (64 bits), and temporary extended-precision (80 bits). A floating-point number has three parts: the sign, biased exponent, and significand. In the coprocessor, the exponent is biased with a constant and the integer bit of the normalized number is not stored in the significand, except in the temporary extended-precision form.
6. Decimal numbers are converted to floating-point numbers by (a) converting the number to binary, (b) normalizing the binary number, (c) adding the bias to the exponent, and (d) storing the number in floating-point form.
7. Floating-point numbers are converted to decimal by (a) subtracting the bias from the exponent, (b) un-normalizing the number, and (c) converting it to decimal.
8. The 80287 uses I/O space for the execution of some of its instructions. This space is invisible to the program and is used internally by the 80286/80287 system. These 16-bit I/O

addresses (00F8H–00FFH) must not be used for I/O data transfers in a system that contains an 80287. The 80387, 80486/7, and Pentium through Pentium 4 use I/O addresses 800000F8H–800000FFH.

9. The coprocessor contains a status register that indicates busy, the conditions that follow a compare or test, the location of the top of the stack, and the state of the error bits. The FSTSW AX instruction, followed by SAHF, is often used with conditional jump instructions to test for some coprocessor conditions.
10. The control register of the coprocessor contains control bits that select infinity, rounding, precision, and error masks.
11. The following directives are often used with the coprocessor for storing data: DW (define word), DD (define doubleword), DQ (define quadword), and DT (define 10 bytes).
12. The coprocessor uses a stack to transfer data between itself and the memory system. Generally, data are loaded to the top of the stack or removed from the top of the stack for storage.
13. All internal coprocessor data are always in the 80-bit extended-precision form. The only time that data are in any other form is when they are stored or loaded from the memory.
14. The coprocessor addressing modes include the classic stack mode, register, register with a pop, and memory. Stack addressing is implied. The data at ST become the source, at ST(1) the destination, and the result is found in ST after a pop.
15. The coprocessor's arithmetic operations include addition, subtraction, multiplication, division, and square root calculation.
16. There are transcendental functions in the coprocessor's instruction set. These functions find the partial tangent or arctangent, 2X – 1, Y $\log_2$ X, and Y $\log_2$ (X + 1). The 80387, 80486/7, and Pentium–Pentium 4 also include sine and cosine functions.
17. Constants are stored inside the coprocessor that provide +0.0, +1.0, π, $\log_2 10$, $\log_2 \varepsilon$, $\log_2 2$, and $\log_\varepsilon 2$.
18. The 80387 functions with the 80386 microprocessor and the 80487SX functions with the 80486SX microprocessor, but the 80486DX and Pentium–Pentium 4 contain their own internal arithmetic coprocessor. The instructions performed by the earlier versions are available on these coprocessors. In addition to these instructions, the 80387, 80486/7, and Pentium–Pentium 4 also can find the sine and cosine.
19. The Pentium Pro through Pentium 4 contain two new floating-point instructions: FCMOV and FCOMI. The FCMOV instruction is a conditional move and the FCOMI performs the same task as FCOM, but it also places the floating-point flags in the system flag register.
20. The MMX extension uses the arithmetic coprocessor registers for MM_0–MM_7. Therefore, it is important that coprocessor software and MMX software do not try to use them at the same time.
21. The instructions for the MMX extensions perform arithmetic and logic operations on bytes (eight at a time), words (four at a time), doublewords (two at a time), and quadwords. The operations performed are addition, subtraction, multiplication, division, AND, OR, Exclusive-OR, and NAND.
22. Both the MMX unit and the SSE unit employ SIMD techniques to perform parallel operations on multiple data with a single instruction. The SSE unit performs operations on integers and floating-point numbers. The registers in the SSE unit are 128 bits in width and can hold (SSE 2 or newer) 16 bytes at a time or four single-precision floating-point numbers. The SSE unit contains registers XMM_0–XMM_7.
23. New applications written for the Pentium 4 should contain SSE instructions in place of MMX instructions.
24. The OWORD pointer has been added to address 128-bit-wide numbers, which are referred to as octal words or double quadwords.

14–8 QUESTIONS AND PROBLEMS

1. List the three types of data that are loaded or stored in memory by the coprocessor.
2. List the three integer data types, the range of the integers stored in them, and the number of bits allotted to each.
3. Explain how a BCD number is stored in memory by the coprocessor.
4. List the three types of floating-point numbers used with the coprocessor and the number of binary bits assigned to each.
5. Convert the following decimal numbers into single-precision floating-point numbers:
   (a) 28.75
   (b) 624
   (c) −0.615
   (d) +0.0
   (e) −1000.5
6. Convert the following single-precision floating-point numbers into decimal:
   (a) 11000000 11110000 00000000 00000000
   (b) 00111111 00010000 00000000 00000000
   (c) 01000011 10011001 00000000 00000000
   (d) 01000000 00000000 00000000 00000000
   (e) 01000001 00100000. 00000000 00000000
   (f) 00000000 00000000 00000000 00000000
7. Explain what the coprocessor does when a normal microprocessor instruction executes.
8. Explain what the microprocessor does when a coprocessor instruction executes.
9. What is the purpose of the C_3–C_0 bits in the status register?
10. What operation is accomplished with the FSTSW AX instruction?
11. What is the purpose of the IE bit in the status register?
12. How can SAHF and a conditional jump instruction be used to determine whether the top of the stack (ST) is equal to register ST(2)?
13. How is the rounding mode selected in the 80X87?
14. What coprocessor instruction uses the microprocessor's AX register?
15. What I/O ports are reserved for coprocessor use with the 80287?
16. How are data stored inside the coprocessor?
17. What is a NAN?
18. Whenever the coprocessor is reset, the top of the stack register is register number __________.
19. What does the term *chop* mean in rounding the control bits of the control register?
20. What is the difference between affine and projective infinity control?
21. What microprocessor instruction forms the opcodes for the coprocessor?
22. The FINIT instruction selects __________-precision for all coprocessor operations.
23. Using assembler pseudo-opcodes, form statements that accomplish the following:
   (a) Store 23.44 into the double-precision floating-point memory location FROG.
   (b) Store −123 into the 32-bit signed integer location $DATA_3$.
   (c) Store −23.8 into the single-precision floating-point memory location DATAL.
   (d) Reserve double-precision memory location $DATA_2$.
24. Describe how the FST DATA instruction functions. Assume that DATA is defined as a 64-bit memory location.
25. What does the FILD DATA instruction accomplish?
26. Form an instruction that adds the contents of register 3 to the top of the stack.
27. Describe the operation of the FADD instruction.
28. Choose an instruction that subtracts the contents of register 2 from the top of the stack and stores the result in register 2.

29. What is the function of the FBSTP DATA instruction?
30. What is the difference between a forward and a reverse division?
31. What is the purpose of the Pentium Pro FCOMI instruction?
32. What does a Pentium Pro FCMOVB instruction accomplish?
33. What must occur before executing any FCMOV instruction?
34. Develop a procedure that finds the reciprocal of the single-precision floating-point number. The number is passed to the procedure in EAX and must be returned as a reciprocal in EAX.
35. What is the difference between the FTST instruction and FXAM?
36. Explain what the F2XM1 instruction calculates.
37. Which coprocessor status register bit should be tested after the FSQRT instruction executes?
38. Which coprocessor instruction pushes π onto the top of the stack?
39. Which coprocessor instruction places 1.0 at the top of the stack?
40. What will FFREE ST(2) accomplish when executed?
41. Which instruction stores the environment?
42. What does the FSAVE instruction save?
43. Develop a procedure that finds the area of a rectangle ($A = L \times W$). Memory locations for this procedure are single-precision floating-point locations A, L, and W.
44. Write a procedure that finds the capacitive reactance ($XC = \frac{1}{2\pi FC}$). Memory locations for this procedure are single-precision floating-point locations XC, F, and $C1$ for C.
45. Develop a procedure that generates a table of square roots for the integers 2 through 10. The results must be stored as single-precision floating-point numbers in an array called ROOTS.
46. When is the FWAIT instruction used in a program?
47. What is the difference between the FSTSW and FNSTSW instructions?
48. Given the series/parallel circuit and equation illustrated in Figure 14–17, develop a program using single-precision values for R_1, R_2, R_3, and R_4 that finds the total resistance and stores the result at single-precision location RT.
49. Develop a procedure that finds the cosine of a single-precision floating-point number. The angle, in degrees, is passed to the procedure in EAX and the cosine is returned in EAX. Recall that FCOS finds the cosine of an angle expressed in radians.
50. Given two arrays of double-precision floating-point data ($ARRAY_1$ and $ARRAY_2$) that each contain 100 elements, develop a procedure that finds the product of $ARRAY_1$ times $ARRAY_2$ and then stores the double-precision floating-point result in a third array at ($ARRAY_3$).
51. Develop a procedure that takes the single-precision contents of register EBX times π and stores the result in register EBX as a single-precision floating-point number. You must use memory to accomplish this task.
52. Write a procedure that raises a single-precision floating-point number X to the power Y. Parameters are passed to the procedure with EAX = X and EBX = Y. The result is passed back to the calling sequence in ECX.

FIGURE 14–17 The series/parallel circuit (Question 48).

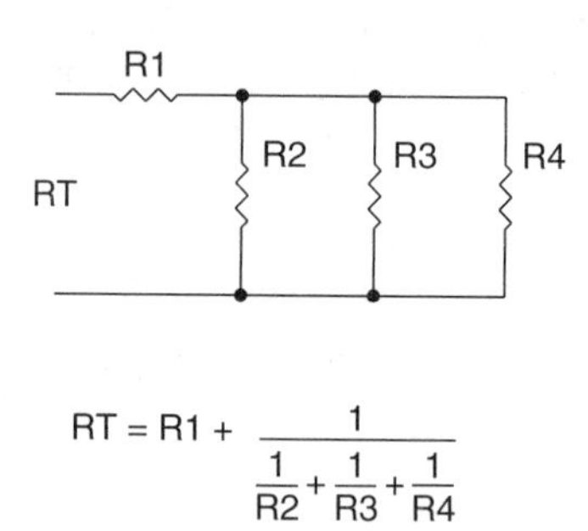

$$RT = R1 + \frac{1}{\frac{1}{R2} + \frac{1}{R3} + \frac{1}{R4}}$$

53. Given that the LOG_{10} X = $(LOG_2\ 10)^{-1} \times LOG_2$ X, write a procedure called LOG_{10} that finds the LOG_{10} of the value (X) at the stack top. Return the LOG_{10} at the stack top at the end the procedure.
54. Use the procedure developed in Question 53 to solve the equation

$$\text{Gain in decibels} = 20 \log_{10} \frac{V_{out}}{V_{in}}$$

The program should take arrays of single-precision values for V_{out} and V_{in} and store the decibel gains in a third array called DBG. These are 100 values V_{out} and V_{in}.
55. What is the MMX extension to the Pentium–Pentium 4 microprocessors?
56. What is the purpose of the EMMS instruction?
57. Where are the MM_0–MM_7 registers found in the microprocessor?
58. What is signed saturation?
59. What is unsigned saturation?
60. How could all of the MMX registers be stored in the memory with one instruction?
61. Write a short program that uses MMX instruction to multiply the word-size numbers in two arrays and stores the 32-bit results in a third array. The source arrays are 256 words long.
62. What are SIMD instructions?
63. What are SSE instructions?
64. The XMM registers are __________ bits wide.
65. A single XMM register can hold __________ single-precision floating-point numbers.
66. A single XMM register can hold __________ byte-sized integers.
67. What is an OWORD?
68. Can floating-point instructions for the arithmetic coprocessor execute at the same time as SSE instructions?
69. Develop a C++ function (using inline assembly code) that computes (using scalar SSE instructions and floating-point instructions) and returns a single-precision number that represents the resonant frequency from parameters (L and C) passed to it to solve the following equation:

$$Fr = \frac{1}{2\pi\sqrt{LC}}$$

CHAPTER 15

Bus Interface

INTRODUCTION

Many applications require knowledge of the bus systems located within the personal computer. At times, main boards from personal computers are used as core systems in industrial applications. These systems often require custom interfaces that are attached to one of the buses on the main board. This chapter presents the ISA (industry standard architecture) bus, the PCI (peripheral component interconnect) and PCI Express buses, the USB (universal serial bus), and the AGP (advanced graphics port). Also provided are some simple interfaces to many of these bus systems as design guides.

Although it is likely that they will not be on personal computers of the future, the parallel port and serial communications ports are discussed. These were the first I/O ports on the personal computer and they have stood the test of time, but the universal serial bus seems to have all but replaced them.

CHAPTER OBJECTIVES

Upon completion of this chapter, you will be able to:

1. Detail the pin connections and signal bus connections on the parallel and serial ports as well as on ISA, AGP, PCI, and PCI Express buses
2. Develop simple interfaces that connect to the parallel and serial ports and the ISA and PCI buses
3. Program interfaces located on boards that connect to the ISA and PCI buses
4. Describe the operation of the USB and develop some short programs that transfer data
5. Explain how the AGP increases the efficiency of the graphics subsystem

15–1 THE ISA BUS

The ISA, or **industry standard architecture**, bus has been around since the very start of the IBM-compatible personal computer system (circa 1982). In fact, any card from the very first personal computer will plug into and function in any of the modern Pentium 4–based computers provided it has an ISA slot. This is all made possible by the ISA bus interface found in some of these machines, which is still compatible with the early personal computers. The ISA bus has all but disappeared on the home PC, but is still found in many industrial applications and presented here for this reason.

Evolution of the ISA Bus

The ISA bus has changed from its early days. Over the years, the ISA bus has evolved from its original 8-bit standard to the 16-bit standard found in some systems today. The last computer systems that contained the ISA bus en masse was the Pentium III. When the Pentium 4 started to appear the ISA bus started to disappear. Along the way, there was even a 32-bit version called the EISA bus (**extended ISA**), but that seems to have all but disappeared. What remains today in some personal computers is an ISA slot (**connection**) on the main board that can accept either an 8-bit ISA card or a 16-bit ISA printed circuit card. The 32-bit printed circuit cards are the PCI bus or, in some older 80486-based machines, the VESA cards. The ISA bus has mostly vanished in home computers, but it is available as a special order for most main boards. The ISA bus is still found in many industrial applications, but its days now seem limited.

The 8-Bit ISA Bus Output Interface

Figure 15–1 illustrates the 8-bit ISA connector found on the main board of all personal computer systems (again, this may be combined with a 16-bit connector). The ISA bus connector contains the entire demultiplexed address bus (A_{19}–A_0) for the 1M-byte 8088 system, the 8-bit data bus (D_7–D_0), and the four control signals $\overline{MEMR}$, $\overline{MEMW}$, $\overline{IOR}$, and $\overline{IOW}$ for controlling I/O and any memory that might be placed on the printed circuit card. Memory is seldom added to any ISA bus card today because the ISA card only operates at an 8 MHz rate.

FIGURE 15–1 The 8-bit ISA bus.

Back of Computer

Pin #	Solder Side	Component Side
1	GND	$\overline{\text{IO CHK}}$
2	RESET	D7
3	+5V	D6
4	IRQ9	D5
5	−5V	D4
6	DRQ2	D3
7	−12V	D2
8	OWS	D1
9	+12V	D0
10	GND	IO RDY
11	$\overline{\text{MEMW}}$	AEN
12	$\overline{\text{MEMR}}$	A19
13	$\overline{\text{IOW}}$	A18
14	$\overline{\text{IOR}}$	A17
15	$\overline{\text{DACK3}}$	A16
16	DRQ3	A15
17	$\overline{\text{DACK1}}$	A14
18	DRQ1	A13
19	$\overline{\text{DACK0}}$	A12
20	CLOCK	A11
21	IRQ7	A10
22	IRQ6	A9
23	IRQ5	A8
24	IRQ4	A7
25	IRQ3	A6
26	$\overline{\text{DACK2}}$	A5
27	T/C	A4
28	ALE	A3
29	+5V	A2
30	OSC	A1
31	GND	A0

There might be an EPROM or Flash memory used for setup information on some ISA cards, but never any RAM.

Other signals, which are useful for I/O interface, are the **interrupt request lines** IRQ_2–IRQ_7. Note that IRQ_2 is redirected to IRQ_9 on modern systems and is so labeled on the connector in Figure 15–1. The DMA channels 0–3 control signals are also present on the connector. The **DMA request inputs** are labeled DRQ_1–DRQ_3 and the **DMA acknowledge outputs** are labeled $\overline{DACK0}$–$\overline{DACK3}$. Notice that the DRQ_0 input pin is missing because the early personal computers used it and the $DACK_0$ output as a refresh signal to refresh any DRAM that might be located on the ISA card. Today, this output pin contains a 15.2 μs clock signal that was used for refreshing DRAM. The remaining pins are for power and RESET.

Suppose that a series of four 8-bit latches must be interfaced to the personal computer for 32 bits of parallel data. This is accomplished by purchasing an ISA interface card (part number 4713-1) from a company like Vector Electronics or other companies. In addition to the edge connector for the ISA bus, the card also contains room at the back for interface connectors. A 37-pin subminiature D-type connector can be placed on the back of the card to transfer the 32 bits of data to the external source.

Figure 15–2 shows a simple interface for the ISA bus, which provides 32 bits of parallel TTL data. This example system illustrates some important points about any system interface. First, it is extremely important that the loading to the ISA bus be kept to one low-power (LS)

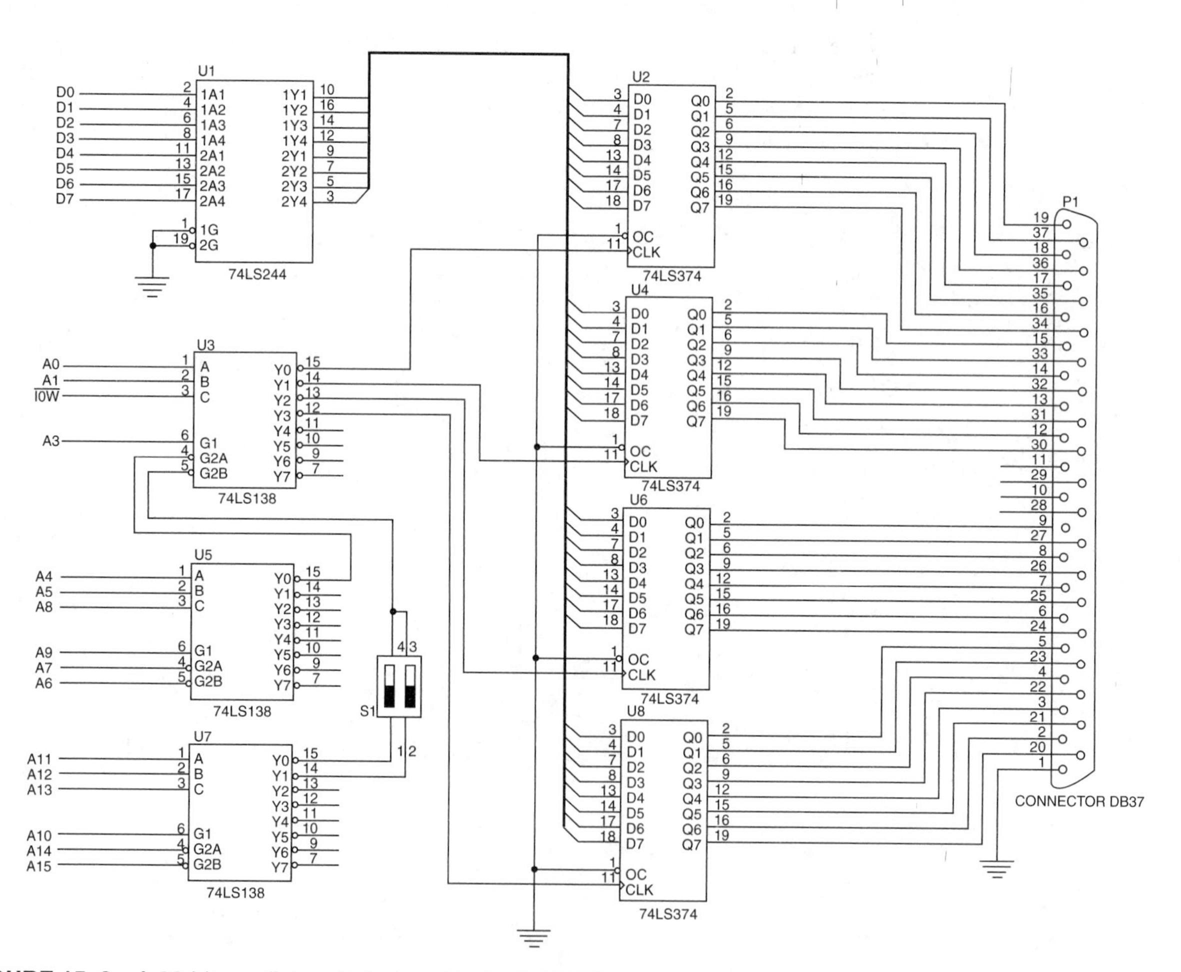

FIGURE 15–2 A 32-bit parallel port interfaced to the 8-bit ISA bus.

TABLE 15–1 The I/O port assignments of Figure 15–2.

DIP Switch	*Latch* U_2	*Latch* U_4	*Latch* U_6	*Latch* U_8
1–4 On	0608H or 060CH	0609H or 060DH	060AH or 060EH	060BH or 060FH
2–3 On	0E08H or 0E0CH	0E09H or 0E0DH	0E0AH or 0E0EH	0E0BH or 0E0FH

TTL load. In this circuit, a 74LS244 buffer is used to reduce the loading on the data bus. If the 74LS244 were not there, this system would present the data bus with four unit loads. If all bus cards were to present heavy loads, the system would not operate properly (or perhaps not at all).

Output from the ISA card is provided in this circuit by a 37-pin connector labeled P_1. The output pins from the circuit connect to P_1, and a ground wire is attached. You must provide ground to the outside world, or else the TTL data on the parallel ports are useless. If needed, the output control pins ($\overline{OC}$) on each of the 74LS374 latch chips can also be removed from ground and connected to the four remaining pins on P_1. This allows an external circuit to control the outputs from the latches.

A small DIP switch is placed on two of the outputs of D_7, so the address can be changed if an address conflict occurs with another card. This is unlikely, unless you plan to use two of these cards in the same system. Address connection A_2 is not decoded in this system so it becomes a don't care (x). See Table 15–1 for the addresses of each latch and each position of the S_1. Note that only one of the two switches may be on at a time and that each port has two possible addresses for each switch setting because A_2 is not connected.

In the personal computer, the ISA bus is designed to operate at I/O addresses 0000H through 03FFH. Depending on the version and manufacturer of the main board, ISA cards may or may not function above these locations. Some newer systems often allow ISA ports at locations above 03FFH, but older systems do not. The ports in this example may need to be changed for some systems. Some older cards only decode I/O addresses 0000H–03FFH and may have address conflicts if the port addresses above 03FFH conflict. The ports are decoded in this example by three 74LS138 decoders. It would be more efficient and cost-effective to decode the ports with a programmable logic device.

Figure 15–3 shows the circuit of Figure 15–2 reworked using a PLD to decode the addresses for the system. Notice that address bits A_{15}–A_4 are decoded by the PLD and the switch is connected to two of the PLD inputs. This change allows four different I/O port addresses for each latch, making the circuit more flexible. Table 15–2 shows the port number selected by switch 1–4 and switch 2–3. Example 15–1 shows the program for the PLD that causes the port assignments of Table 15–2.

EXAMPLE 15–1

```
-- VHDL code for the decoder of Figure 15-3

library ieee;
use ieee.std_logic_1164.all;

entity DECODER_15_3 is

port (
       IOW, A14, A13, A12, A11, A10, A9, A8, A7, A6
              A5, A4, A3, A2, A1, A0, S1, S2: in STD_LOGIC;
       U3, U4, U5, U6: out STD_LOGIC
);

end;

architecture V1 of DECODER_15_3 is
```

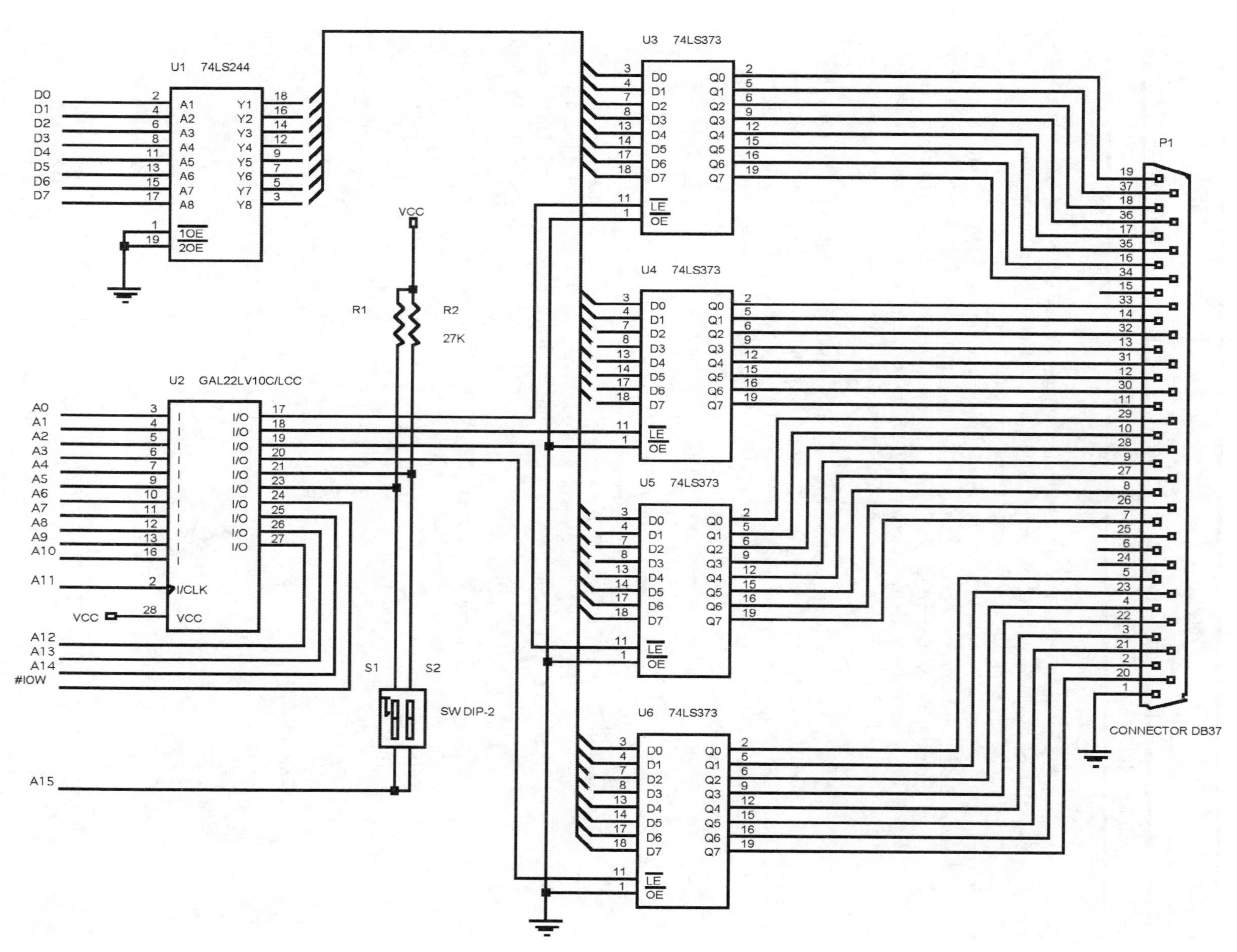

FIGURE 15–3 A 32-bit parallel interface for the ISA bus.

TABLE 15–2 Port assignments of Figure 15–3.

S_2	S_1	U_3	U_4	U_5	U_6
On	On	0300H	0301H	0302H	0303H
On	Off	0304H	0305H	0306H	0307H
Off	On	0308H	0309H	030AH	030BH
Off	Off	030CH	030DH	030EH	030FH

Note: On is a closed switch (0) and off is open (1).

```
begin

        U3 <= IOW or A14 or A13 or A12 or A11 or A10 or not A9 or not A8 or A7
              or A6 or A5 or A4 or A1 or A0 or (S2 or S1 or A3 or A2) and (S2 or
              not S1 or A3 or not A2) and (not S2 or S1 or not A3 or A2) and
              (not S2 or not S1 or not A3 or not A2);
        U4 <= IOW or A14 or A13 or A12 or A11 or A10 or not A9 or not A8 or A7
              or A6 or A5 or A4 or A1 or not A0 or (S2 or S1 or A3 or A2) and
              (S2 or not S1 or A3 or not A2) and (not S2 or S1 or not A3 or A2)
              and (not S2 or not S1 or not A3 or not A2);
        U5 <= IOW or A14 or A13 or A12 or A11 or A10 or not A9 or not A8 or A7
              or A6 or A5 or A4 or not A1 or A0 or (S2 or S1 or A3 or A2) and
              (S2 or not S1 or A3 or not A2) and (not S2 or S1 or not A3 or A2)
              and (not S2 or not S1 or not A3 or not A2);
        U6 <= IOW or A14 or A13 or A12 or A11 or A10 or not A9 or not A8 or A7
              or A6 or A5 or A4 or not A1 or not A0 or (S2 or S1 or A3 or A2)
              and (S2 or not S1 or A3 or not A2) and (not S2 or S1 or not A3 or
              A2) and (not S2 or not S1 or not A3 or not A2);

end V1;
```

Notice in Example 15–1 how the first term (U_3) generates a logic 0 on the output to the decoder only when both switches are in their off positions for I/O port 0300H. It also generates a clock for U_3 for I/O ports 304H, 308H, or 30CH depending on the switch settings. The second term (U_4) is active for ports 301H, 305H, 309H, or 30DH, depending on the switch settings. Again, refer to Table 15–2 for the complete set of port assignments for various switch settings. Since A_{15} is connected to the bottom of the switches, this circuit will also activate the latches for other I/O locations, because it is not decoded. I/O addresses 830XH will also generate clock signals to the latch because A_{15} is not decoded.

Example 15–2 shows two C++ functions that transfers an integer to the 32-bit port. Either of these functions sends data to the port; the first is more efficient, but the second may be more readable. (Example 15–2(c) shows Example 15–2(b) in disassembled form.) Two parameters are passed to the function: One is the data to be sent to the port, and the other is the base port address. The base address is 0300H, 0304H, 0308H, or 030CH and must match the switch settings of Figure 15–3.

EXAMPLE 15–2(a)

```
void OutPort(int address, int data)
{
        _asm
        {
                mov   edx,address
                mov   eax,data
                mov   ecx,4
OutPort1:
                out   dx,al              ;output 8-bits
                shr   eax,8              ;get next 8-bit section
```

```
            inc  dx                 ;address next port
            loop OutPort1           ;repeat 4 times
        }
}
```

EXAMPLE 15–2(b)

```
void OutPrt(int address, int data)
{
        for ( int a = address; a < address + 4; a++ )
        {
            _asm
            {
                    mov  edx,a
                    mov  eax,data
                    out  dx,al
            }
            data >>= 8;             //get next 8-bit section
        }
}
```

EXAMPLE 15–2(c)

```
//Example 15-2(b) disassembled

            for ( int a = address; a < address + 4; a++ )
00413823  mov          eax,dword ptr [address]
00413826  mov          dword ptr [a],eax
00413829  jmp          CSSEDlg::OutPrt+54h (413834h)
0041382B  mov          eax,dword ptr [a]
0041382E  add          eax,1
00413831  mov          dword ptr [a],eax
00413834  mov          eax,dword ptr [address]
00413837  add          eax,4
0041383A  cmp          dword ptr [a],eax
0041383D  jge          CSSEDlg::OutPrt+71h (413851h)
        {
            _asm
            {
                    mov  edx,a
0041383F  mov            edx,dword ptr [a]
                    mov eax,data
00413842  mov            eax,dword ptr [data]
                    out dx,al
00413845  out            dx,al
            }
            data >>= 8;             //get next 8-bit section
00413846  mov          eax,dword   ptr [data]
00413849  sar          eax,8
0041384C  mov          dword ptr [data],eax
        }
     0041384F  jmp          CSSEDlg::OutPrt+4Bh (41382Bh)
```

The 8-Bit ISA Bus Input Interface

To illustrate the input interface to the ISA bus, a pair of ADC804 analog-to-digital converters are interfaced to the ISA bus in Figure 15–4. The connections to the converters are made through a nine-pin DB_9 connector. The task of decoding the I/O port addresses is more complex, because each converter needs a write pulse to start a conversion, a read pulse to read the digital data once they have been converted from the analog input data, and a pulse to enable the selection of the $\overline{INTR}$ output. Notice that the $\overline{INTR}$ output is connected to data bus bit position D_0. When $\overline{INTR}$ is input to the microprocessor, the rightmost bit of AL is tested to determine whether the converter is busy.

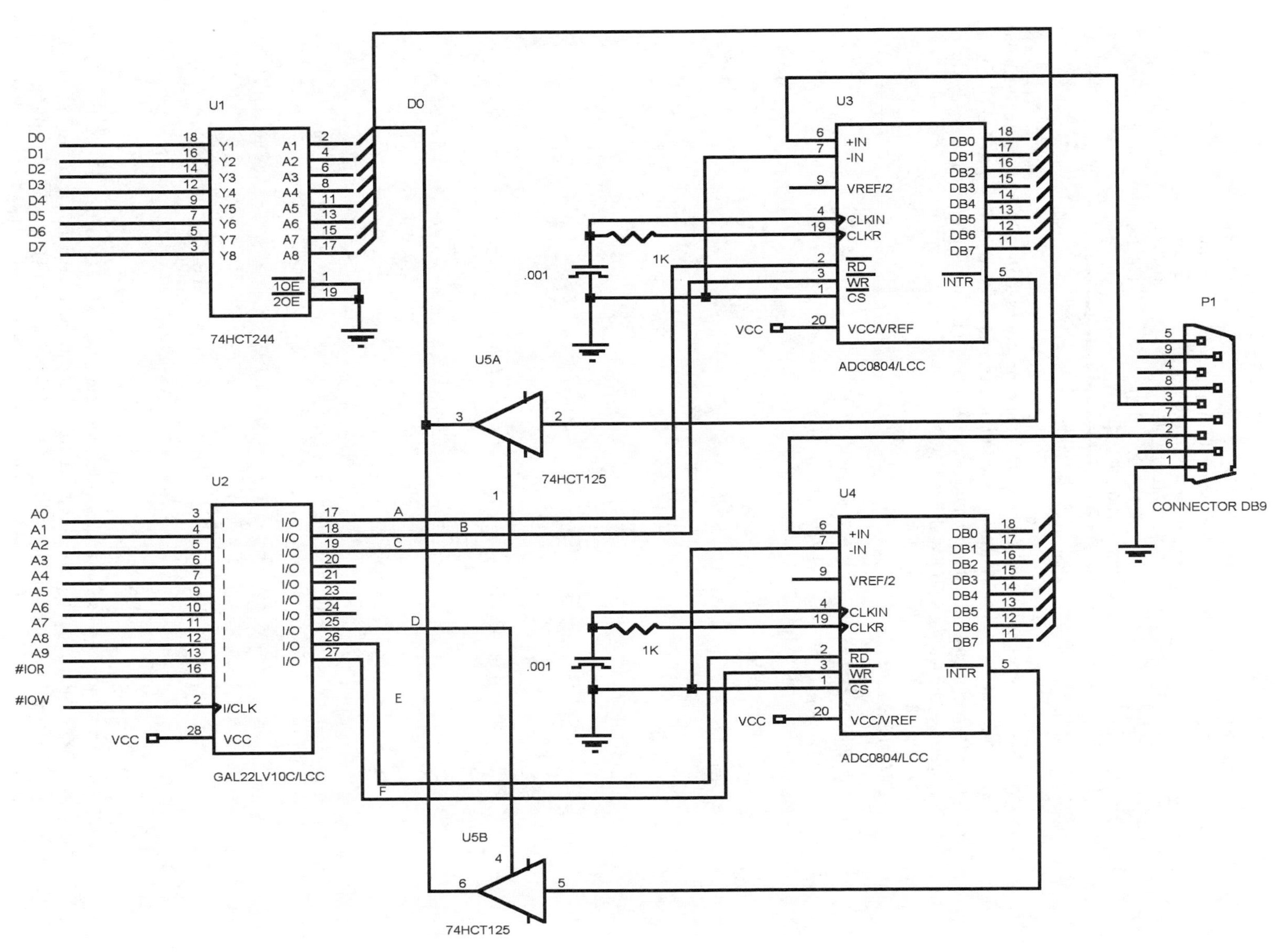

FIGURE 15–4 A pair of analog-to-digital converters interfaced to the ISA bus.

TABLE 15–3 Port assignments for Figure 15–4.

Device	*Port*
Start ADC (U_3)	0300H
Read ADC (U_3)	0300H
Read INTR (U_3)	0301H
Start ADC (U_4)	0302H
Read SDC (U_4)	0302H
Read INTR (U_4)	0303H

As before, great care is taken so that the connections to the ISA bus present one unit load to the system. Table 15–3 illustrates the I/O port assignment decoded by the PLD (see Example 15–3 for the program). In this example we assumed that the standard ISA bus is used, which only contains address connection A_0 through A_9.

EXAMPLE 15–3

```
-- VHDL code for the decoder of Figure 15-4

library ieee;
use ieee.std_logic_1164.all;

entity DECODER_15_4 is

port (
        IOW, IOR, A9, A8, A7, A6 A5, A4, A3, A2, A1, A0: in STD_LOGIC;
        A, B, C, D, E, F: out STD_LOGIC
);

end;

architecture V1 of DECODER_15_4 is

begin

        A <= not A9 or not A8 or A7 or A6 or A5 or A4 or A3 or A2 or A1 or A0 or
              IOR;

        B <= not A9 or not A8 or A7 or A6 or A5 or A4 or A3 or A2 or A1 or A0 or
              IOW;

        C <= not A9 or not A8 or A7 or A6 or A5 or A4 or A3 or A2 or A1 or not A0
              or IOR;

        D <= not A9 or not A8 or A7 or A6 or A5 or A4 or A3 or A2 or not A1 or
              not A0 or IOR;

        E <= not A9 or not A8 or A7 or A6 or A5 or A4 or A3 or A2 or not A1 or A0
              or IOR;

        F <= not A9 or not A8 or A7 or A6 or A5 or A4 or A3 or A2 or not A1 or A0
              or IOW;

end V1;
```

Example 15–4 lists a function that can read either ADC U_3 or U_4. The address is generated by passing either a 0 for U_3 or a 1 for U_4 to the address parameter of the function. The function starts the converter by writing to it, and then waits until the $\overline{\text{INTR}}$ pin returns to a logic 0, indicating that the conversion is complete before the data are read and returned by the function as a char.

EXAMPLE 15–4

```
char ADC(int address)
{
       char temp = 1;
       if ( address )
             address = 2;
       address += 0x300;
       _asm
       {                   ;start converter
             mov  edx,address
             out  dx,al
       }
       while ( temp )      //wait if busy
       {
             _asm
             {
                    mov edx,address
                    inc edx
                    in al,dx
                    mov temp,al
                    and al,1
             }
       }
       _asm
       {                   ;get data
             mov edx,address
             in  al,dx
             mov temp,al
       }
       return temp;
}
```

The 16-Bit ISA Bus

The only difference between the 8- and 16-bit ISA bus is that an additional connector is attached behind the 8-bit connector. A 16-bit ISA card contains two edge connectors: one plugs into the original 8-bit connector and the other plugs into the new 16-bit connector. Figure 15–5

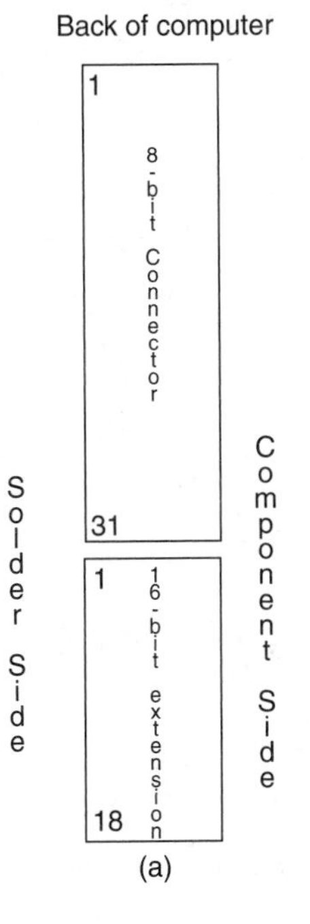

Back of computer

Pin#	Solder Side	Component Side
1	$\overline{MCS16}$	BHE
2	$\overline{IOCS16}$	A23
3	IRQ10	A22
4	IRQ11	A21
5	IRQ12	A20
6	IRQ15	A19
7	IRQ14	A18
8	$\overline{DACK0}$	A17
9	DRQ0	$\overline{MEMR}$
10	$\overline{DACK5}$	$\overline{MEMW}$
11	DRQ5	D8
12	$\overline{DACK6}$	D9
13	DRQ6	D10
14	$\overline{DACK7}$	D11
15	DRQ7	D12
16	+5V	D13
17	$\overline{MASTER}$	D14
18	GND	D15

(b)

FIGURE 15–5 The 16-bit ISA bus. (a) Both 8- and 16-bit connectors and (b) the pin-out of the 16-bit connector.

shows the pin-out of the additional connector and its placement in the computer in relation to the 8-bit connector. Unless additional memory is added on the ISA card, the extra address connections A_{23}–A_{20} do not serve any function for I/O operations. The added features that are most often used are the additional interrupt request inputs and the DMA request signals. In some systems, 16-bit I/O uses the additional eight data bus connections (D_8–D_{15}), but more often today the PCI bus is used for peripherals that are wider than eight bits. About the only recent interfaces found for the ISA bus are a few modems and sound cards.

15–2 THE PERIPHERAL COMPONENT INTERCONNECT (PCI) BUS

The PCI (**peripheral component interconnect**) bus is virtually the only bus found in the newest Pentium 4 systems and just about all the Pentium systems. In all of the newer systems, the ISA bus still exists by special order, but as an interface for older 8-bit and 16-bit interface cards. Many new systems contain only two ISA bus slots or no ISA slots. In time, the ISA bus may disappear, but it is still an important interface for many industrial applications. The PCI bus has replaced the VESA local bus. One reason is that the PCI bus has plug-and-play characteristics and the ability to function with a 64-bit data bus. A PCI interface contains a series of registers, located in a small memory device on the PCI interface, that contain information about the board. This same memory can provide plug-and-play characteristics to the ISA bus or any other bus. The information in these registers allows the computer to automatically configure the PCI card. This feature, called **plug-and-play (PnP)**, is probably the main reason that the PCI bus has become so popular in most systems.

Figure 15–6 shows the system structure for the PCI bus in a personal computer system. Notice that the microprocessor bus is separate and independent of the PCI bus. The microprocessor

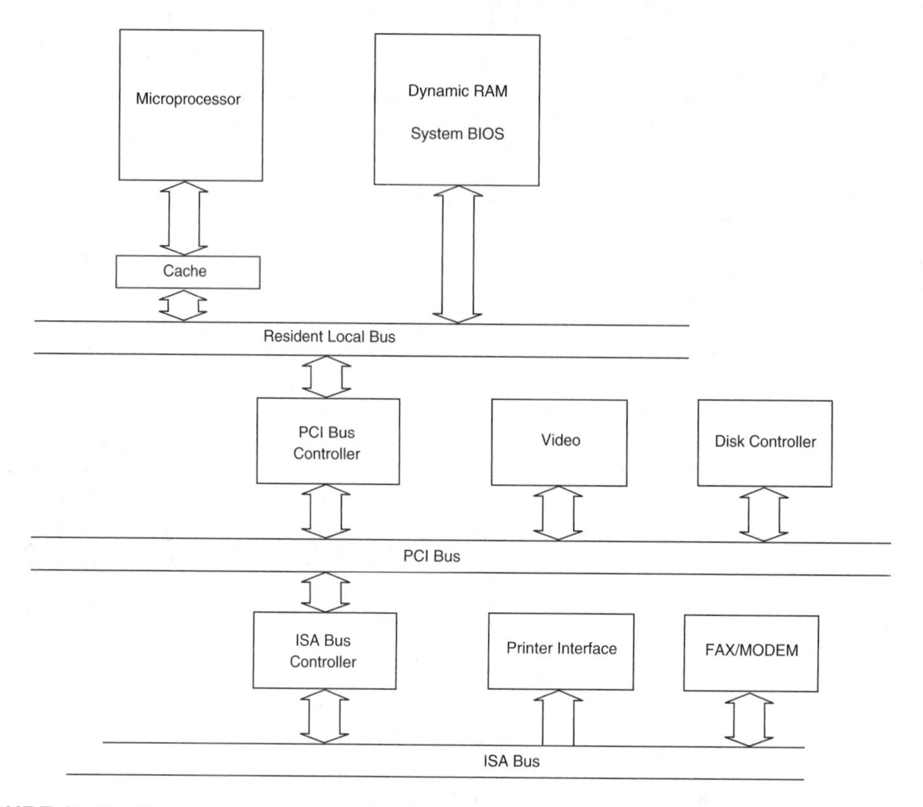

FIGURE 15–6 The system block diagram for the personal computer that contains a PCI bus.

connects to the PCI bus through an integrated circuit called a **PCI bridge**. This means that virtually any microprocessor can be interfaced to the PCI bus, as long as a PCI controller or bridge is designed for the system. In the future, all computer systems may use the same bus. Even the Apple Macintosh system is switching to the PCI bus. The resident local bus is often called a front side bus.

The PCI Bus Pin-Out

As with the other buses described in this chapter, the PCI bus contains all of the system control signals. Unlike the other buses, the PCI bus functions with either a 32-bit or a 64-bit data bus and a full 32-bit address bus. Another difference is that the address and data buses are multiplexed to reduce the size of the edge connector. These multiplexed pins are labeled AD_0–AD_{63} on the connector. The 32-bit card (which is found in most computers) has only connections 1 through 62, while the 64-bit card has all 94 connections. The 64-bit card can accommodate a 64-bit address if it is required at some point in the future. Figure 15–7 illustrates the PCI bus pin-out.

As with the other bus systems, the PCI bus is most often used for interfacing I/O components to the microprocessor. Memory could be interfaced, but it would operate only at a 33 MHz rate with the Pentium, which is half the speed of the 66 MHz resident local bus of the Pentium system. A more recent version of PCI (2.1-compliant) operates at 66 MHz and at 33 MHz for older interface cards. Pentium 4 systems use a 200 MHz system bus speed (although it is often listed as 800 MHz), but there is no planned modification to the PCI bus speed yet.

The PCI Address/Data Connections

The PCI address appears on AD_0–AD_{31} and it is multiplexed with data. In some systems, there is a 64-bit data bus that uses AD_{32}–AD_{63} for data transfer only. In the future, these pins can be used for extending the address to 64 bits. Figure 15–8 illustrates the timing diagram for the PCI bus, which shows the way that the address is multiplexed with data and also the control signals used for multiplexing.

During the first clocking period, the address of the memory or I/O location appears on the AD connections, and the command to a PCI peripheral appears on the $C/\overline{BE}$ pins. Table 15–4 illustrates the bus commands found on the PCI bus.

INTA Sequence During the interrupt acknowledge sequence, an interrupt controller (the controller that caused the interrupt) is addressed and interrogated for the interrupt vector. The byte-sized interrupt vector is returned during a byte read operation.

Special Cycle The special cycle is used to transfer data to all PCI components. During this cycle, the rightmost 16 bits of the data bus contain

TABLE 15–4 PC bus commands.

$C/\overline{BE3}$–$C/\overline{BE0}$	Command
0000	INTA sequence
0001	Special cycle
0010	I/O read cycle
0011	I/O write cycle
0100–1001	Reserved
1010	Configuration read
1011	Configuration write
1100	Memory multiple access
1101	Dual addressing cycle
1110	Line memory access
1111	Memory write with invalidation

FIGURE 15–7 The pin-out of the PCI bus.

Back of computer

Pin #	Solder Side	Component Side
1	−12V	$\overline{TRST}$
2	TCK	+12V
3	GND	TM5
4	TDO	TD1
5	+5V	+5V
6	+5V	$\overline{INTA}$
7	$\overline{INTB}$	$\overline{INTC}$
8	$\overline{INTD}$	+5V
9	$\overline{PRSNT 1}$	
10		+VI/O
11	$\overline{PRSNT 2}$	
12	KEY	KEY
13	KEY	KEY
14		
15	GND	$\overline{RST}$
16	CLK	VI/O
17	GND	$\overline{VNT}$
18	$\overline{REQ}$	GND
19	+V IO	
20	AD31	AD30
21	AD29	+3.3V
22	GND	AD28
23	AD27	AD26
24	AD25	GND
25	+3.3V	AD24
26	C/$\overline{BE3}$	IDSEL
27	AD23	+3.3V
28	GND	AD22
29	AD21	AD20
30	AD19	GND
31	+3.3V	AD18
32	AD17	AD16
33	C/$\overline{BE2}$	+3.3V
34	GND	$\overline{FRAME}$
35	$\overline{IRDY}$	GND
36	+3.3V	$\overline{TRDY}$
37	$\overline{DEVSEL}$	GND
38	GND	$\overline{STOP}$
39	$\overline{LOCK}$	+3.3V
40	$\overline{PERR}$	SDONE

Pin #	Solder Side	Component Side
41	+3.3V	$\overline{SBO}$
42	$\overline{SERR}$	GND
43	+3.3V	PAR
44	C/$\overline{BE1}$	AD15
45	AD14	+3.3V
46	GND	AD13
47	AD12	AD11
48	AD10	GND
49	GND	AD9
50	KEY	KEY
51	KEY	KEY
52	AD8	C/$\overline{BE0}$
53	AD7	+3.3V
54	+3.3V	AD6
55	AD5	AD4
56	AD3	GND
57	GND	AD2
58	AD1	AD0
59	+V IO	+V IO
60	$\overline{ACK64F}$	$\overline{REQ64}$
61	+5V	+5V
62	+5V	+5V
63		GND
64	GND	C/$\overline{BE7}$
65	C/$\overline{BE6}$	C/$\overline{BE5}$
66	C/$\overline{BE4}$	+V IO
67	GND	PAR64
68	AD63	AD62
69	AD61	GND
70	+V IO	AD60
71	AD59	AD58
72	AD57	GND
73	GND	AD56
74	AD55	AD54
75	AD53	+V IO
76	GND	AD52
77	AD51	AD50
78	AD49	GND
79	+V IO	AD48
80	AD47	AD46
81	AD45	GND
82	GND	AD44
83	AD43	AD42
84	AD41	+VI/O
85	GND	AD40
86	AD39	AD38
87	AD37	GND
88	+VI/O	AD36
89	AD35	AD34
90	AD33	GND
91	GND	AD32
92		
93		GND
94	GND	

Notes: (1) pins 63–94 exist only on the 64-bit PCI card
(2) + VI/O is 3.3V on a 3.3V board and +5V on a 5V board
(3) blank pins are reserved

0000H, indicating a processor shutdown, 0001H for a processor halt, or 0002H for 80X86 specific code or data.

I/O Read Cycle Data are read from an I/O device using the I/O address that appears on AD_0–AD_{15}. Burst reads are not supported for I/O devices.

I/O Write Cycle As with I/O read, this cycle accesses an I/O device, but writes data.

Memory Read Cycle Data are read from a memory device located on the PCI bus.

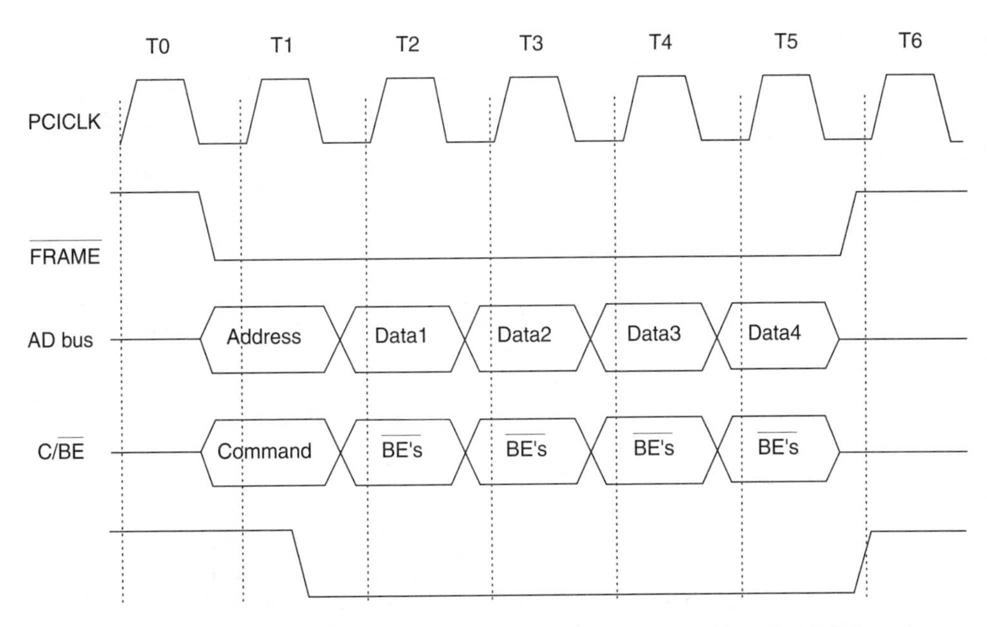

FIGURE 15–8 The basic burst mode timing for the PCI bus system. Note that this transfers either four 32-bit numbers (32-bit PCI) or four 64-bit numbers (64-bit PCI).

Memory Write Cycle	As with memory read, data are accessed in a device located on the PCI bus. The location is written.
Configuration Read	Configuration information is read from the PCI device using the configuration read cycle.
Configuration Write	The configuration write allows data to be written to the configuration area in a PCI device. Note that the address is specified by the configuration read.
Memory Multiple Access	This is similar to the memory read access, except that it is usually used to access many data instead of one.
Dual Addressing Cycle	Used for transferring address information to a 64-bit PCI device, which only contains a 32-bit data path.
Line Memory Addressing	Used to read more than two 32-bit numbers from the PCI bus.
Memory Write with Invalidation	This is the same as line memory access, but it is used with a write. This write bypasses the write-back function of the cache.

Configuration Space

The PCI interface contains a 256-byte configuration memory that allows the computer to interrogate the PCI interface. This feature allows the system to automatically configure itself for the PCI plug-board. Microsoft Corporation calls this plug-and-play (PnP). Figure 15–9 illustrates the configuration memory and its contents.

The first 64 bytes of the configuration memory contain the header that holds information about the PCI interface. The first 32-bit doubleword contains the unit ID code and the vendor ID code. The unit ID code is a 16-bit number (D_{31}–D_{16}) that is an FFFFH if the unit is not installed, and a number between 0000H and FFFEH that identifies the unit if it is installed. The class codes identify the class of the PCI interface. The class code is found in bits D_{31}–D_{16} of configuration

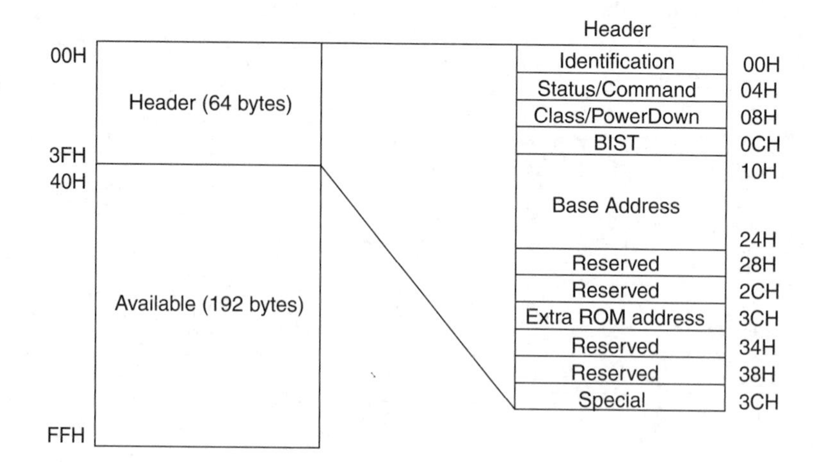

FIGURE 15–9 The contents of the configuration memory on a PCI expansion board.

memory at location 08H. Note that bits D_{15}–D_0 are defined by the manufacturer. The current class codes are listed in Table 15–5 and are assigned by the PCI SIG, which is the governing body for the PCI bus interface standard. The vendor ID (D_{15}–D_0) is also allocated by the PCI SIG.

The status word is loaded in bits D_{31}–D_{16} of configuration memory location 04H and the command is at bits D_{15}–D_0 of location 04H. Figure 15–10 illustrates the format of both the status and command registers.

TABLE 15–5 The class codes.

Class Code	*Function*
0000H	Older non-VGA device (not PnP)
0001H	Older VGA device (not PnP)
0100H	SCSI controller
0101H	IDE controller
0102H	Floppy disk controller
0103H	IPI controller
0180H	Other hard/floppy controller
0200H	Ethernet controller
0201H	Token ring controller
0202H	FDDI
0280H	Other network controller
0300H	VGA controller
0301H	XGA controller
0380H	Other video controller
0400H	Video multimedia
0480H	Other multimedia controller
0500H	RAM controller
0580H	Other memory bridge controller
0600H	Host bridge
0601H	ISA bridge
0602H	EISA bridge
0603H	MCA bridge
0604H	PCI–PCI bridge
0605H	PCMIA bridge
0680H	Other bridge
0700H–FFFEH	Reserved
FFFFH	Not installed

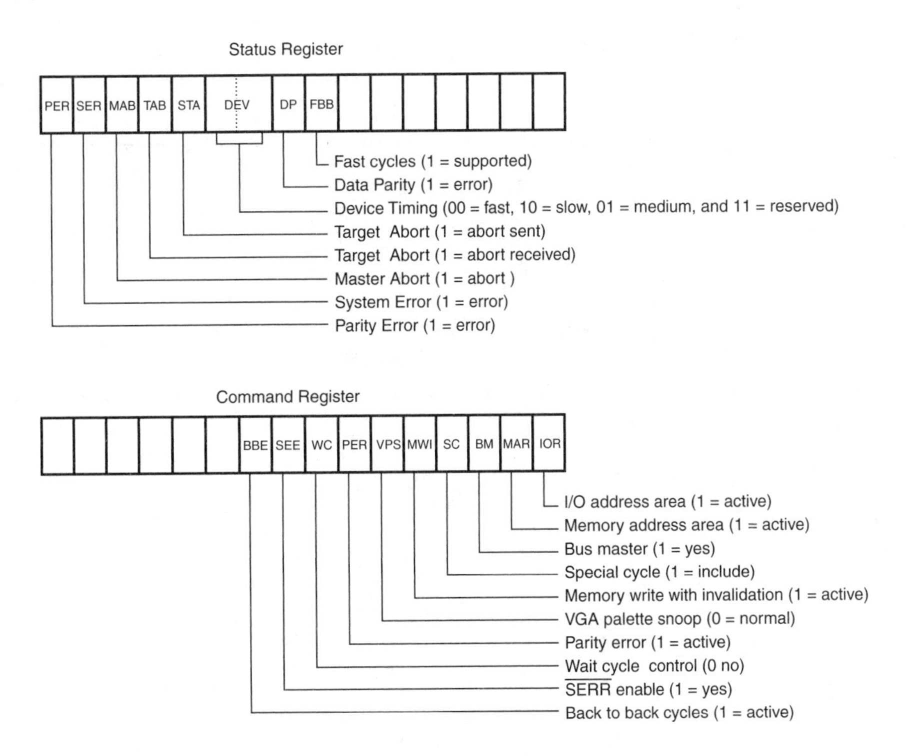

FIGURE 15–10 The contents of the status and control words in the configuration memory.

The base address space consists of a base address for the memory, a second for the I/O space, and a third for the expansion ROM. The first two doublewords of the base address space contain either the 32- or 64-bit base address for the memory present on the PCI interface. The next doubleword contains the base address of the I/O space. Note that even though the Intel microprocessors only use a 16-bit I/O address, there is room for expanding the I/O address to 32 bits. This allows systems that use the 680X0 family and PowerPC access to the PCI bus because they do have I/O space that is accessed via a 32-bit address. The 600X0 and PowerPC use memory-mapped I/O, discussed at the beginning of Chapter 11.

BIOS for PCI

Most modem Pentium–Pentium 4–based personal computers contain the PCI bus and an extension to the normal system BIOS that supports the PCI bus. These newer systems contain access to the PCI bus at interrupt vector 1AH. Table 15–6 lists the functions currently available through the DOS INT 1AH instruction with AH = 0B1H for the PCI bus.

Example 15–5 show how the BIOS is used to determine whether the PCI bus extension is available. Once the presence of the BIOS is established, the contents of the configuration memory can be read using the BIOS functions. Note that the BIOS does not support data transfers between the computer and the PCI interface. Data transfers are handled by drivers that are provided with the interface. These drivers control the flow of data between the microprocessor and the component found on the PCI interface.

EXAMPLE 15–5

```
;DOS program that determines whether PCI exists

.MODEL SMALL
.DATA
       MES1   DB      "PCI BUS IS PRESENT$"
       MES2   DB      "PCI BUS IS NOT FOUND$"
.CODE
```

TABLE 15–6 BIOS INT1AH functions for the PCI bus.

01H	BIOS Available?
Entry	AH = 0B1H AL = 01H
Exit	AH = 00H if PCI BIOS extension is available BX = version number EDX = ASCII string 'PCI' CARRY = 1 if no PCI extension present
02H	**PCI Unit Search**
Entry	AH = 0B1H AL = 02H CX = Unit DX = Manufacturer SI = index
Exit	AH = result code (see notes) BX = bus and unit number Carry = 1 for error
Notes	The result codes are: 00H = successful search 81H = function not supported 83H = invalid manufacturer ID code 86H = unit not found 87H = invalid register number
03H	**PCI Class Code Search**
Entry	AH = 0B1H AL = 03H ECX = class code SI = index
Exit	AH = result code (see notes for function 02H) BX = bus and unit number Carry = 1 for an error
06H	**Start Special Cycle**
Entry	AH = 0B1H AL = 06H BX = bus and unit number EDX = data
Exit	AH = result code (see notes for function 02H) Carry = 1 for error
Notes	The value passed in EDX is sent to the PCI bus during the address phase.
08H	**Configuration Byte-Sized Read**
Entry	AH = 0B1H AL = 08H BX = bus and unit number DI = register number
Exit	AH = result code (see notes for function 02H) CL = data from configuration register Carry = 1 for error

(continued on next page)

TABLE 15–6 *(continued)*

09H	Configuration Word-Sized Read
Entry	AH = 0B1H AL = 08H BX = bus and unit number DI = register number
Exit	AH = result code (see notes for function 02H) CX = data from configuration register Carry = 1 for error
0AH	Configuration Doubleword-Sized Read
Entry	AH = 0B1H AL = 08H BX = bus and unit number DI = register number
Exit	AH = result code (see notes for function 02H) ECX = data from configuration register Carry = 1 for error
0BH	Configuration Byte-Sized Write
Entry	AH = 0B1H AL = 08H BX = bus and unit number CL = data to be written to configuration register DI = register number
Exit	AH = result code (see notes for function 02H) Carry = 1 for error
0CH	Configuration Word-Sized Write
Entry	AH = 0B1H AL = 08H BX = bus and unit number CX = data to be written to configuration register DI = register number
Exit	AH = result code (see notes for function 02H) Carry = 1 for error
0DH	Configuration Doubleword-Sized Write
Entry	AH = 0B1H AL = 08H BX = bus and unit number ECX = data to be written to configuration register DI = register number
Exit	AH = result code (see notes for function 02H) Carry = 1 for error

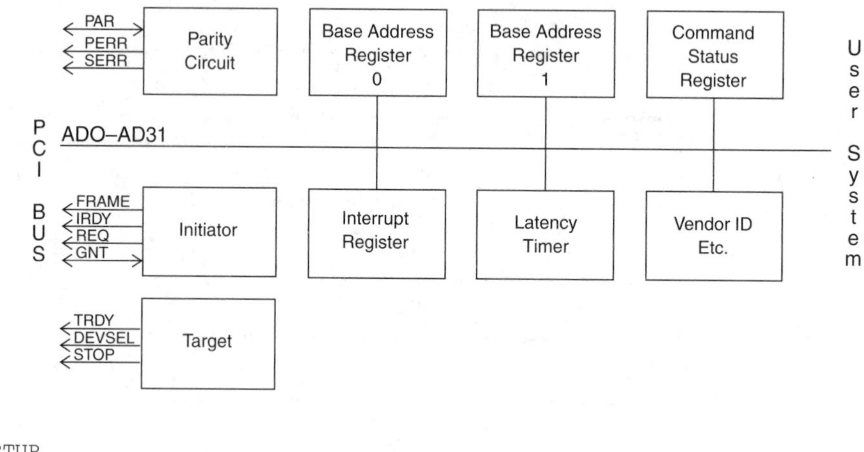

FIGURE 15–11 The block diagram of the PCI interface.

```
.STARTUP
       MOV  AH,0B1H          ;access PCI BIOS
       MOV  AL,1
       INT  1AH
       MOV  DX,OFFSET MES2
       .IF CARRY?            ;if PCI is present
              MOV  DX,OFFSET MES1
       .ENDIF
       MOV  AH,9             ;display MES1 or MES2
       INT  21H
       .EXIT
END
```

PCI Interface

The PCI interface is complex, and normally an integrated PCI bus controller is used for interfacing to the PCI bus. It requires memory (EPROM) to store vendor information and other information, as explained earlier in this section of the chapter. The basic structure of the PCI interface is illustrated in Figure 15–11. The contents of this block diagram illustrate the required components for a functioning PCI interface; it does not illustrate the interface itself. The registers, parity block, initiator, target, and vendor ID EPROM are required components of any PCI interface. If a PCI interface is constructed, a PCI controller is often used because of the complexity of this interface. The PCI controller provides the structures shown in Figure 15–11.

PCI Express Bus

The PCI Express, although not available yet in PCs, transfers data in serial at the rate of 2.5 GHz to legacy PCI applications, increasing the data link speed to 500 MBps to 16 GBps. The standard PCI bus delivers data at a speed of about 133 MBps, in comparison. The big improvement is on the motherboard, where the interconnections are in serial and at 2.5 GHz.

When available this might well replace current video cards on the AGP port with a higher speed version on the PCI Express bus (PCI-X). This serial technology allows main board manufacturers to use less space on the main board for interconnection and thus reduces costs.

15–3 THE PARALLEL PRINTER INTERFACE (LPT)

The parallel printer interface (LPT, which stands for line printer) is located on the rear of the personal computer, and as long as it is a part of the PC it can be used as an interface to the PC. The printer interface gives the user access to eight lines that can be programmed to receive or send parallel data.

TABLE 15–7 The signal descriptions for the connectors of Figure 15–12.

Signal	*Description*	*25-pin*	*36-pin*
#STR	Strobe to printer	1	1
D_0	Data bit 0	2	2
D_1	Data bit 1	3	3
D_2	Data bit 2	4	4
D_3	Data bit 3	5	5
D_4	Data bit 4	6	6
D_5	Data bit 5	7	7
D_6	Data bit 6	8	8
D_7	Data bit 7	9	9
#ACK	Acknowledge from printer	10	10
BUSY	Busy from printer	11	11
PAPER	Out of paper	12	12
ONLINE	Printer is online	13	13
#ALF	Low if printer issues LF after CR	14	14
#ERROR	Printer error	15	32
#RESET	Resets the printer	16	31
#SEL	Selects the printer	17	36
+5V	5V from printer	—	18
Protective Ground	Earth ground	—	17
Signal Ground	Signal ground	All other pins	All other pins

Note: # indicates an active low signal.

Port Details

The parallel port (LPT_1) is normally at I/O port addresses 378H, 379H, and 37AH from DOS or using a driver in Windows. The secondary (LPT2) port, if present, is located at I/O port addresses 278H, 279H, and 27AH. The following information applies to both ports, but LPT_1 port addresses are used throughout.

The Centronics Interface implemented by the parallel port uses two connectors, a 25-pin D-type on the back of the PC and a 36-pin Centronics on the back of the printer. The pin-outs of these connectors are listed in Table 15–7, and the connectors are shown in Figure 15–12.

The parallel port can work as both a receiver and a transmitter at its data pins (D_0–D_7). This allows devices other than printers, such as CD-ROMs, to be connected to and used by the PC through the parallel port. Anything that can receive and/or send data through an 8-bit interface can and often does connect to the parallel port (LPT_1) of a PC.

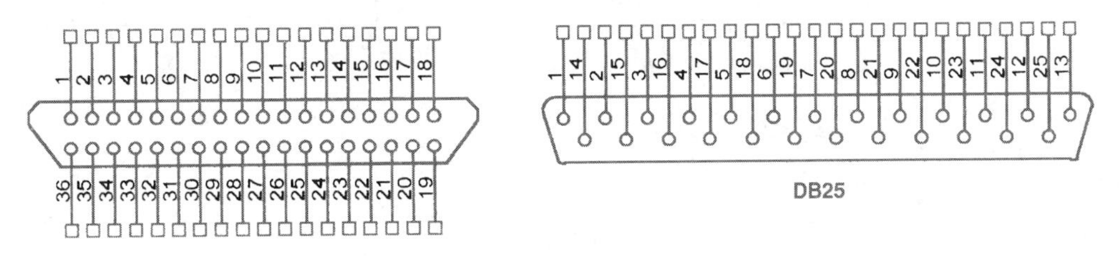

FIGURE 15–12 The connectors used for the parallel port.

FIGURE 15–13 Ports 378H, 379H, and 37AH as used by the parallel port.

Port 378H

The data port that connects to bits D_0–D_7 (pins 2–9)

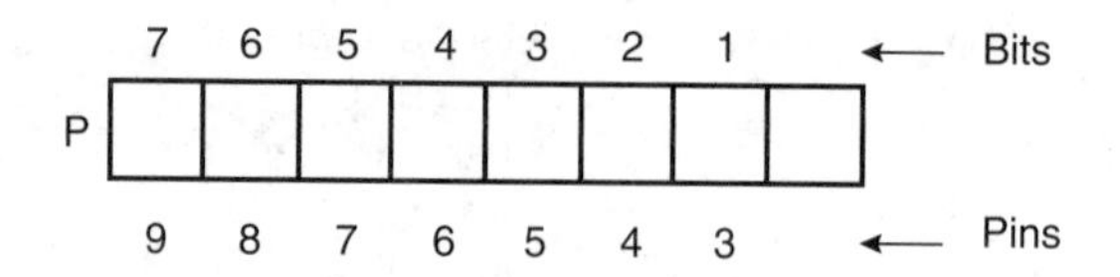

Port 379H

This is a read-only port that returns the information from the printer through signals such as BUSY, #ERROR, and so forth. (Careful! Some of the bits are inverted.)

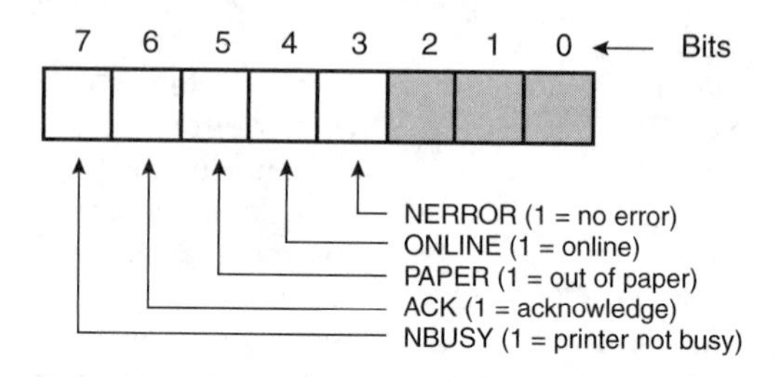

Port 37AH

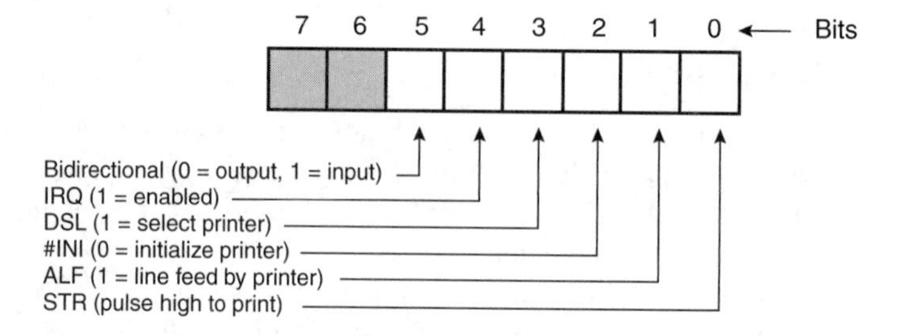

Figure 15–13 illustrates the contents of the data port (378H), the status register (379H), and an additional status port (37AH). Some of the status bits are true when they are a logic zero.

Using the Parallel Port Without ECP Support

For most systems since the PS/2 was released by IBM, you can basically follow the information presented in Figure 15–13 to use the parallel port without ECP. To read the port, it must first be initialized by sending 20H to register 37AH as illustrated in Example 15–6. As indicated in Figure 15–13, this sets the bidirectional bit that selects input operation for the parallel port. If the bit is cleared, output operation is selected.

EXAMPLE 15–6

```
MOV  AL,20H
MOV  DX,37AH
OUT  DX,AL
```

Once the parallel port is programmed as an input it is read as depicted in Example 15–7. Reading is accomplished by accessing the data port at address 378H.

EXAMPLE 15–7

```
MOV  DX,378H
IN   AL,DX
```

To write data to the parallel port, reprogram the command register at address 37A by writing 00H to program the bidirectional bit with a zero. Once the bidirectional bit is programmed, data are sent to the parallel port through the data port at address 378H. Example 15–8 illustrates how data are sent to the parallel port.

EXAMPLE 15–8

```
MOV  DX,378H
MOV  AL,WRITE_DATA
OUT  DX,AL
```

On older (80286-based) machines the bidirectional bit is missing from the interface. In order to read information from the parallel port, write 0FFH to the port (378H), then it can be read. These older systems do not have a register at location 37AH.

Accessing the printer port from Windows is difficult because a driver must be written to do so if Windows 2000 or Windows XP is in use. In Windows 98 or Windows ME, access to the port is accomplished as explained in this section.

There is a way to access the parallel port through Windows 2000 and Windows XP without writing a driver. A driver called UserPort (readily available on the Internet) opens up the protected I/O ports in Windows and allows direct access to the parallel port through assembly blocks in Visual C++ using port 378H. It also allows access to any I/O ports between 0000H and 03FFH. Another useful tool is available for a 30-day trial at www.jungo.com. The Jungo tool is a driver development tool, with many example drivers for most subsystems.

15–4 THE SERIAL COM PORTS

The serial communications ports are COM_1–COM_8, but most computers only have COM_1 and COM_2 installed. Some have a single communication port (COM_1). These ports are controlled and accessed in the DOS environment as described in Chapter 11 with the 16550 serial interface component and will not be discussed again. Instead, we will discuss the Windows API functions for operating the COM ports for the 16550 communications interface.

Communication Control

The serial ports are accessed through any version of Windows and Visual C++ by using a few system application interface (API) functions. An example of a short C++ function that accesses the serial ports is listed in Example 15–9. The function is called WriteComPort and it contains two parameters. The first parameter is the port, as in COM_1, COM_2, and so on, and the second parameter is the character to be sent through the port. A return true indicates that the character was sent and a return false indicates that a problem exists. To use the function to send the letter A through the COM_1 port call it with WriteComPort ("COM_1", "A"). This function is written to send only a single byte through the serial COM port, but it could be modified to send strings. To send 00H (no other number can be sent this way) through COM_2 use WriteComPort("COM_2", 0x00). Notice that the COM port is set to 9600 Baud, but this is easily changed by changing the CBR_9600 to another acceptable value. See Table 15–8 for the allowed Baud rates.

TABLE 15–8 Allowable Baud rates for the COM ports.

Keyword	*Speed in Bits per Second*
CBR_110	110
CBR_300	300
CBR_600	600
CBR_1200	1200
CBR_2400	2400
CBR_4800	4800
CBR_9600	9600
CBR_14400	14400
CBR_19200	19200
CBR_38400	38400
CBR_56000	56000
CBR_57600	57600
CBR_115200	115200
CBR_128000	128000
CBR_256000	256000

EXAMPLE 15–9

```
bool WriteComPort(CString PortSpecifier, CString data)
{
      DCB dcb;
      DWORD byteswritten;

      HANDLE hPort = CreateFile(PortSpecifier,
      GENERIC_WRITE,
      0,
      NULL,
      OPEN_EXISTING,
      0,
      NULL);

      if (!GetCommState(hPort,&dcb)){
            return false;
      }

      dcb.BaudRate = CBR_9600;          //9600 Baud
      dcb.ByteSize = 8;                 //8 data bits
      dcb.Parity = NOPARITY;            //no parity
      dcb.StopBits = ONESTOPBIT;        //1 stop

      if (!SetCommState(hPort,&dcb))
            return false;

      bool retVal = WriteFile(hPort,data,1,&byteswritten,NULL);
      CloseHandle(hPort);                 //close the handle
      return retVal;}
```

The CreateFile structure creates a handle to the COM ports that can be used to write data to the port. After getting and changing the state of the port to meet the Baud rate requirements, the WriteFile function sends data to the port. The parameters used with the WriteFile function are the file handle (hPort), the data to be written as a string, the number of bytes to write (1 in this example), and a place to store the number of bytes actually written to the port.

Receiving data through the COM port is a little more challenging because errors occur more frequently than with transmission. There are also many types of errors that can be detected that often should be reported to the user. Example 15–10 illustrates a C++ function that is used to read a character from the serial port called ReadByte. The ReadByte function returns either the character read from the port or an error code of 0×100 if the port could not be opened or 0×101 if the receiver detected an error. If data are not received, this function will hang because no timeouts were set.

EXAMPLE 15–10

```
int ReadByte(CString PortSpecifier)
{
       DCB dcb;
       int retVal;
       BYTE Byte;
       DWORD dwBytesTransferred;
       DWORD dwCommModemStatus;

       HANDLE hPort = CreateFile(PortSpecifier,
       GENERIC_READ,
       0,
       NULL,
       OPEN_EXISTING,
       0,
       NULL);

       if (!GetCommState(hPort,&dcb))
              return 0x100;

       dcb.BaudRate = CBR_9600;          //9600 Baud
       dcb.ByteSize = 8;                 //8 data bits
       dcb.Parity = NOPARITY;            //no parity
       dcb.StopBits = ONESTOPBIT;        //1 stop

       if (!SetCommState(hPort,&dcb))
              return 0x100;

       SetCommMask (hPort, EV_RXCHAR | EV_ERR);       //receive character event
       WaitCommEvent (hPort, &dwCommModemStatus, 0);  //wait for character

       if (dwCommModemStatus & EV_RXCHAR)
             ReadFile (hPort, &Byte, 1, &dwBytesTransferred, 0);  //read 1
       else if (dwCommModemStatus & EV_ERR)
              retVal = 0x101;
       retVal = Byte;
       CloseHandle(hPort);
       return retVal;
}
```

15–5 THE UNIVERSAL SERIAL BUS (USB)

The universal serial bus (USB) has solved a problem with the personal computer system. The current PCI sound cards use the internal PC power supply, which generates a tremendous amount of noise. Because the USB allows the sound card to have its own power supply, the noise associated with the PC power supply can be eliminated, allowing for high-fidelity sound without 60 Hz hum. Other benefits are ease of user connection and access to up to 127 different connections through a four-connection serial cable. This interface is ideal for keyboards, sound cards, simple video-retrieval devices, and modems. Data transfer speeds are 480 Mbps for full-speed USB 2.0 operation, 11 Mbps for USB 1.1 compliant transfers, and 1.5 Mbps for slow-speed operation.

Cable lengths are limited to five meters maximum for the full-speed interface and three meters maximum for the low-speed interface. The maximum power available through these cables is rated at 100 mA and maximum current at 5.0 V. If the amount of current exceeds 100 mA, Windows will display a yellow exclamation point next to the device, indicating an overload condition.

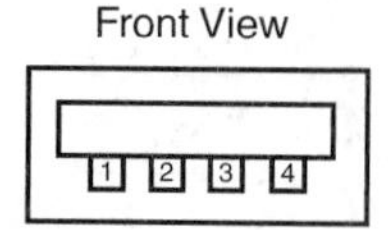

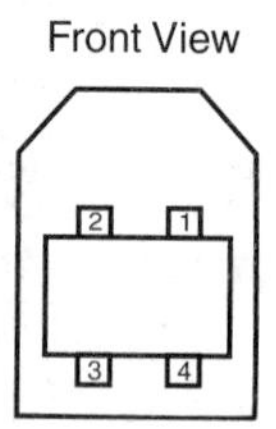

FIGURE 15–14 The front view of the two common types of USB connectors.

TABLE 15–9 USB pin configuration.

Pin Number	*Signal*
1	+ 5.0 V
2	– Data
3	+ Data
4	Ground

The Connector

Figure 15–14 illustrates the pin-out of the USB connector. There are two types of connectors specified and both are in use. In either case, there are four pins on each connector, which contain the signals indicated in Table 15–9. As mentioned, the +5.0 V and ground signals can be used to power devices connected to the bus as long as the amount of current does not exceed 100 mA per device. The data signals are biphase signals. When +data are at 5.0 V, –data are at zero volts and vice versa.

USB Data

The data signals are biphase signals that are generated using a circuit such as the one illustrated in Figure 15–15. The line receiver is also illustrated in Figure 15–15. Placed on the transmission pair is a noise-suppression circuit that is available from Texas Instruments (SN75240). Once the transceiver is in place, interfacing to the USB is complete. The 75773 integrated circuit from Texas Instruments functions as both the differential line driver and receiver for this schematic.

The next phase is learning how the signals interact on the USB. These signals allow data to be sent and received from the host computer system. The USB uses NRZI (non-return to zero, inverted) data encoding for transmitting packets. This encoding method does not change the signal level for the transmission of a logic 1, but the signal level is inverted for each change to a

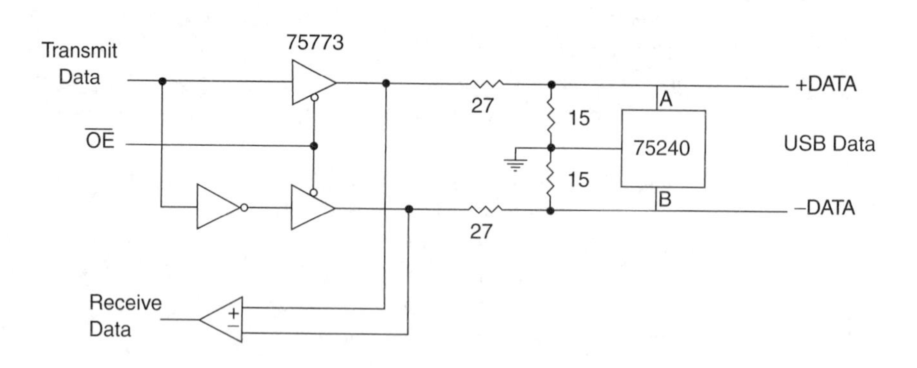

FIGURE 15–15 The interface to the USB using a pair of CMOS buffers.

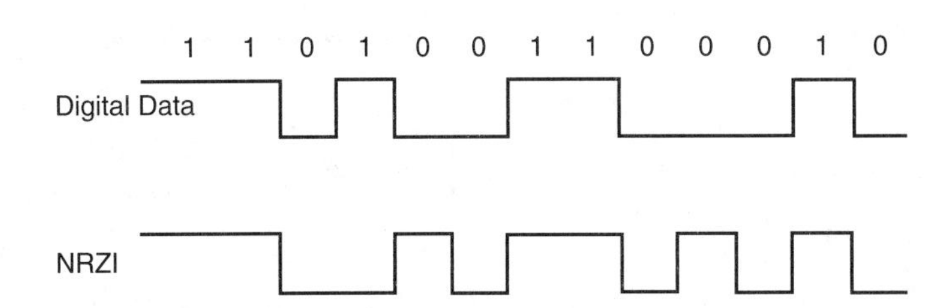

FIGURE 15–16 NRZI encoding used with the USB.

logic 0. Figure 15–16 illustrates a digital data stream and the USB signal produced using this encoding method.

The actual data transmitted includes sync bits using a method called *bit stuffing.* If a logic 1 is transmitted for more than six bits in a row, the bit stuffing technique adds an extra bit (logic 0) after six continuous 1s in a row. Because this lengthens the data stream, it is called bit stuffing. Figure 15–17 shows a bit-stuffed serial data stream and the algorithm used to create it from raw digital serial data. Bit stuffing ensures that the receiver can maintain synchronization for long strings of 1s. Data are always transmitted beginning with the least-significant bit first, followed by subsequent bits.

USB Commands

Now that the USB data format is understood, we will discuss the commands used to transfer data and select the receptor. To begin communications, the sync byte (80H) is transmitted first, followed by the packet identification byte (PID). The PID contains eight bits, but only the rightmost four bits contain the type of packet that follows, if any. The leftmost four bits of the PID are the

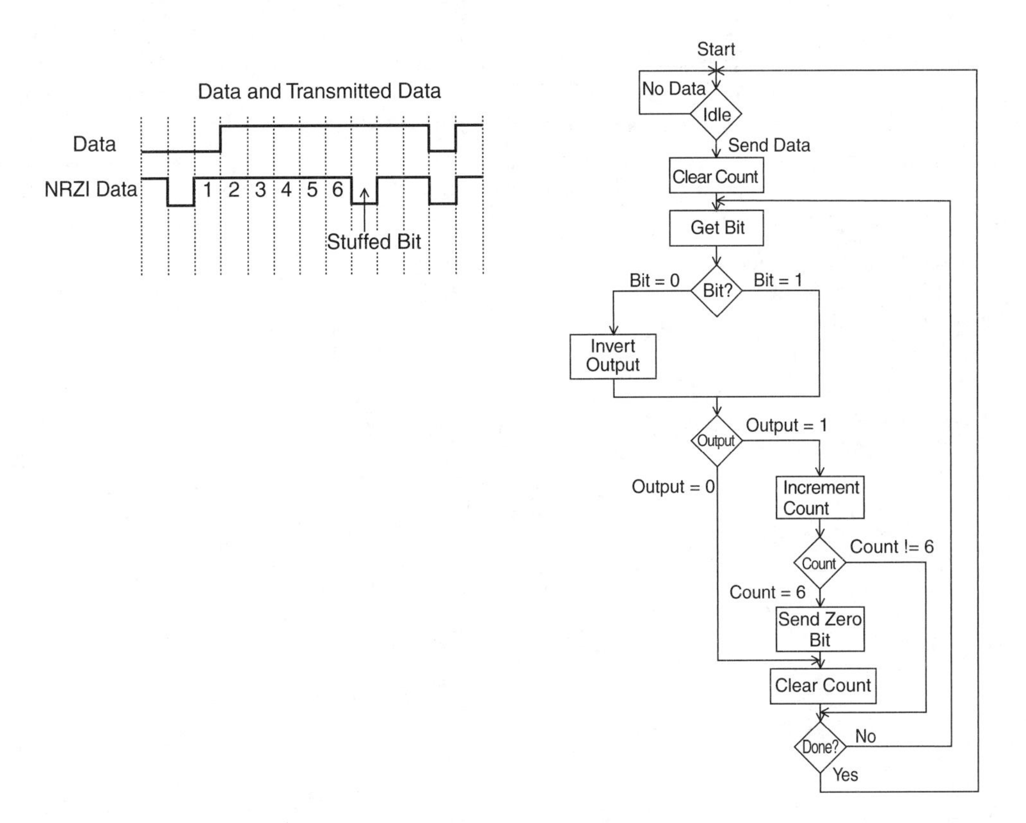

FIGURE 15–17 The data stream and the flowchart used to generate USB data.

TABLE 15–10 PID codes.

PID	*Name*	*Type*	*Description*
E_1	OUT	Token	Host → function transaction
D_2	ACK	Handshake	Receiver accepts packet
C_3	Data0	Data	Data packet (PID even)
A_5	SOF	Token	Start of frame
69	IN	Token	Function → host transaction
5A	NAK	Handshake	Receiver does not accept packet
4B	Data1	Data	Data packet (PID odd)
3C	PRE	Special	Host preamble
2D	Setup	Token	Setup command
1E	Stall	Token	Stalled

ones complementing the rightmost four bits. For example, if a command of 1000 is sent, the actual byte sent for the PID is 0111 1000. Table 15–10 shows the available four-bit PIDs and their eight-bit codes. Notice that PIDs are used as token indicators, as data indicators, and for handshaking.

Figure 15–18 lists the formats of the data, token, handshaking, and start-of-frame packets found on the USB. In the token packet, the ADDR (address field) contains the seven-bit address of the USB device. As mentioned earlier, there are up to 127 devices present on the USB at a time. The ENDP (endpoint) is a four-bit number used by the USB. Endpoint 0000 is used for initialization; other endpoint numbers are unique to each USB device.

There are two types of CRC (**cyclic redundancy checks**) used on the USB: One is a five-bit CRC and the other (used for data packets) is a 16-bit CRC. The five-bit CRC is generated with the $X_5 + X_2 + 1$ polynomial; the 16-bit CRC is generated with the $X_{16} + X_{15} + X_2 + 1$ polynomial. When constructing circuitry to generate or detect the CRC, the plus signs represent Exclusive-OR circuits. The CRC circuit or program is a serial checking mechanism. When using the five-bit CRC, a residual of 01100 is received for no error in all five bits of the CRC and the data bits. With the 16-bit CRC, the residual is 1000000000001101 for no error.

The USB uses the ACK and NAK tokens to coordinate the transfer of data packets between the host system and the USB device. Once a data packet is transferred from the host to the USB

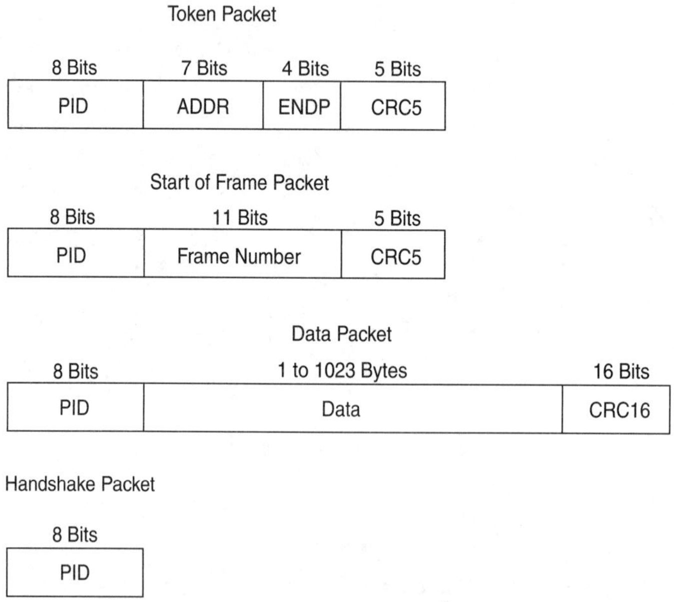

Token Packet

8 Bits	7 Bits	4 Bits	5 Bits
PID	ADDR	ENDP	CRC5

Start of Frame Packet

8 Bits	11 Bits	5 Bits
PID	Frame Number	CRC5

Data Packet

8 Bits	1 to 1023 Bytes	16 Bits
PID	Data	CRC16

Handshake Packet

8 Bits
PID

FIGURE 15–18 The types of packets and contents found on the USB.

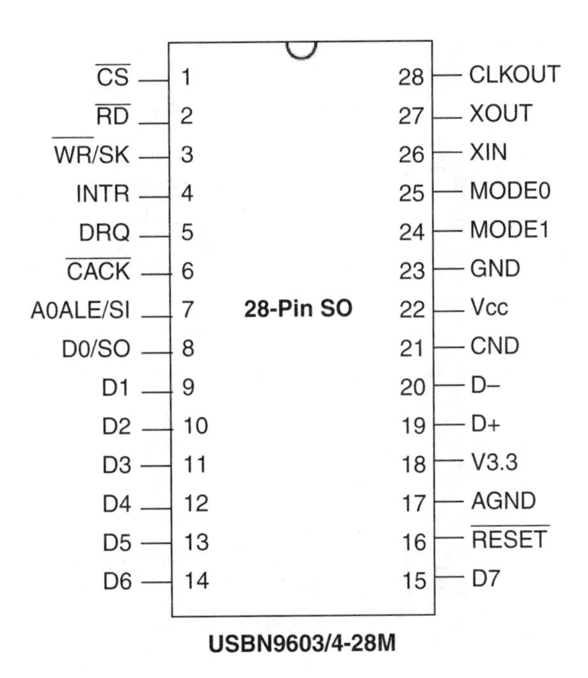

FIGURE 15–19 The USB bus node from National Semiconductor.

device, the USB device either transmits an ACK (acknowledge) or a NAK (not acknowledge) token back to the host. If the data and CRC are received correctly, the ACK is sent; if not, the NAK is sent. If the host receives a NAK token, it retransmits the data packet until the receiver finally receives it correctly. This method of data transfer is often called **stop and wait flow control**. The host must wait for the client to send an ACK or NAK before transferring additional data packets.

The USB Bus Node

National Semiconductor produces a USB bus interface that is fairly easy to interface to the microprocessor. Figure 15–19 illustrates the USBN9604 USB node. Connecting this device to a system using non-DMA access is accomplished by connecting the data bus to D_0–D_7, the control inputs $\overline{RD}$, $\overline{WR}$, and $\overline{CS}$, and a 24 MHz fundamental crystal across the X_{IN} and X_{OUT} pins. The USB bus connection is located on the D– and D+ pins. The simplest interface is achieved by connecting the two mode inputs to ground. This places the device into a nonmultiplexed parallel mode. In this mode the A_0 pin is used to select address (1) or data (0). Figure 15–20 shows this connection to the microprocessor decodes at I/O port addresses 0300H (data) and 0301H (address).

The USBN9604 is a USB bus transceiver that can receive and transmit USB data. This provides an interface point to the USB bus for a minimal cost of about two dollars.

Software for the USBN9604/3

The software presented here functions with the interface in Figure 15–20. Not provided is the driver software for the host system. Example 15–11 illustrates the code required to initialize the USB controller. The USBINT procedure sets the USB controller to use endpoint zero for data transfers.

EXAMPLE 15–11

```
SEND    MACRO  ADDR, DATA
        MOV  DX,301H
        MOV  AL,ADDR
        OUT  DX,AL
```

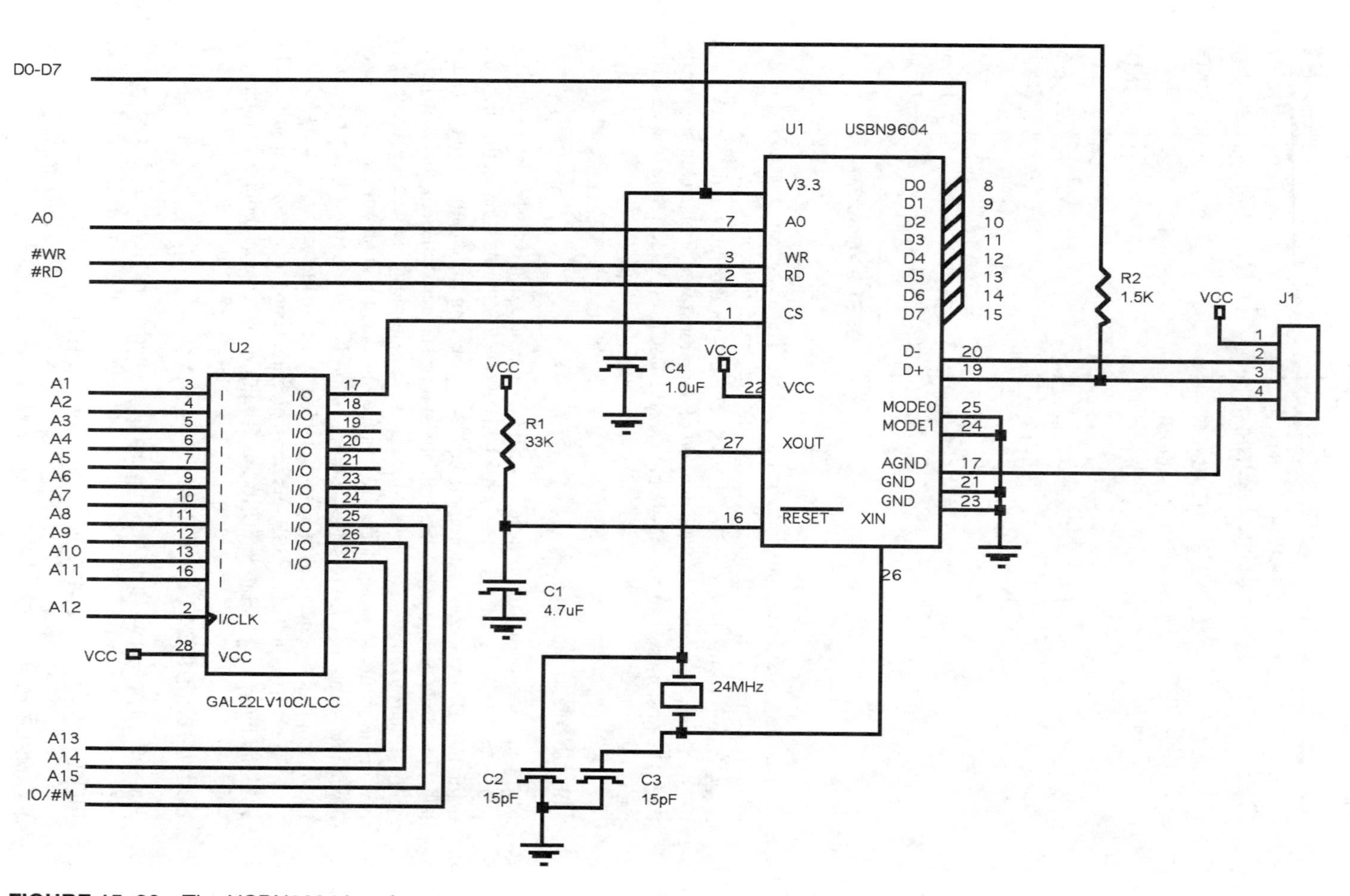

FIGURE 15–20 The USBN9604 interfaced to a microprocessor at I/O addresses 300H and 301H.

```
        MOV   DX,300H
        MOV   AL,DATA
        OUT   DX,AL
        ENDM

USBINT PROC NEAR

        SEND 0,5                    ;interrupts off, software reset USB
        SEND 0,4                    ;clear reset
        CALL DELAY1                 ;wait 1ms
        SEND 9,40H                  ;enable reset check
        SEND 0DH,3                  ;enable EP0 for receive data
        SEND 0BH,3                  ;enable EP0 for transmit data
        SEND 20H,0                  ;EP0 control to no default address
        SEND 4,80H                  ;set FAR to accept default address
        SEND 0,8CH                  ;USB is ready to send or receive data

USBINT ENDP
```

Once the USB controller is initialized, data can be sent or received to the host system through the USB. To accomplish data transmission, the procedure illustrated in Example 15–12 is called to send a one-byte packet using the TXD0 FIFO. This procedure uses the SEND macro listed in Example 15–11 to transfer the byte in BL through the USB to the host system.

EXAMPLE 15–12

```
TRANS   PROC    NEAR

        SEND 21H,BL                 ;send BL to FIFO
        SEND 23H,3                  ;send the byte

TRANS   ENDP
```

To receive data from the USB two functions are required. One tests to see if data are available and the other reads a byte from the USB and places it in the BL register. Both procedures are listed in Example 15–13. The STATUS procedure checks to see if data are in the receiver FIFO. If data are present, carry is set upon return and if no data are received, carry is cleared. The READS procedure retrieves a byte for the receiver FIFO and returns it in BL.

EXAMPLE 15–13

```
READ    MACRO   ADDR

        MOV   DX,301H
        MOV   AL,ADDR
        OUT   DX,AL
        MOV   DX,300H
        IN    AL,DX
        ENDM

STATUS PROC    NEAR

        SEND    6
        SEND    6
        SHL     AL,2
        RET

STATUS ENDP

READS   PROC    NEAR

        READ    25H
        RET

READS   ENDP
```

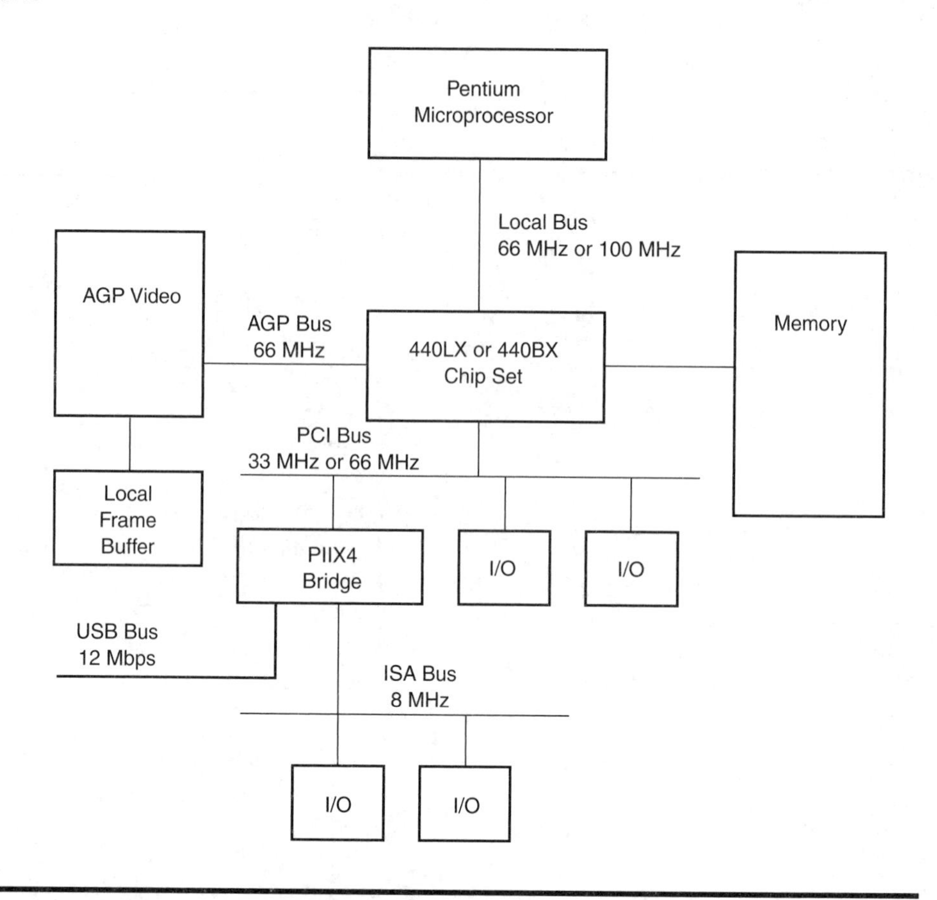

FIGURE 15–21 Structure of a modern computer, illustrating all the buses.

15–6 THE ACCELERATED GRAPHICS PORT (AGP)

The latest addition to most computer systems is the inclusion of the **accelerated graphics port** (AGP). The AGP operates at the bus clock frequency of the microprocessor. It is designed so that a transfer between the video card and the system memory can progress at a maximum speed. The AGP can transfer data at a maximum rate of 2G bytes per second. This port probably will never be used for any devices other than the video card, so we do not devote much space to it.

Figure 15–21 illustrates the interface of the AGP to a Pentium 4 system and the placement of other buses in the system. The main advantage of the AGP bus over the PCI bus is that the AGP can sustain transfers (using the 8X compliant system) at speeds up to 2G bytes per second. The 4X system transfers data at rates of over 1G byte per second. The PCI bus has a maximum transfer speed of about 133M bytes per second. The AGP is designed specifically to allow high-speed transfers between the video card frame buffer and the system memory through the chip set.

The future will probably bring the demise of the PCI bus and the incorporation of the USB into the chip set. It may even lead to the inclusion of the chip set into the microprocessor. At present, the system requires the 865 or 875 chip set and the ISA bridge, if the ISA is special ordered for the main board.

15–7 SUMMARY

1. The bus systems (ISA, PCI, and USB) allow I/O and memory systems to be interfaced to the personal computer.
2. The ISA bus is either eight or 16 bits, and supports either memory or I/O transfers at rates of 8 MHz.

3. The PCI (peripheral component interconnect) supports 32- or 64-bit transfers between the personal computer and memory or I/O at rates of 33 MHz. This bus also allows virtually any microprocessor to be interfaced to the PCI bus via the use of a bridge interface.
4. A plug-and-play (PnP) interface is one that contains a memory that holds configuration information for the system.
5. The parallel port called LPT_1 is used to transfer 8-bit data in parallel to printers and other devices.
6. The serial COM ports are used for serial data transfer. The Windows API is used in a Windows Visual C++ application to effect serial data transfer through the COM ports.
7. The universal serial bus (USB) has all but replaced the ISA bus in the most advanced systems. The USB has three data transfer rates: 1.5 Mbps, 12 Mbps, and 480 Mbps.
8. The USB uses the NRZI system to encode data, and uses bit stuffing for logic 1 transmission more than six bits long.
9. The accelerated graphics port (AGP) is a high-speed connection between the memory system and the video graphics card.

15–8 QUESTIONS AND PROBLEMS

1. The letters ISA are an acronym for what phrase?
2. The ISA bus system supports what size data transfers?
3. Is the ISA bus interface often used for memory expansion?
4. Develop an ISA bus interface that is decoded at addresses 310H–313H. This interface must contain an 8255 accessed via these port addresses. (Don't forget to buffer all inputs to the ISA bus card.)
5. Develop an ISA bus interface that decodes ports 0340H–0343H to control a single 8254 timer.
6. Develop a 32-bit PCI bus interface that adds a 27C256 EPROM at memory addresses FFFF0000H–FFFF7FFFH.
7. Given a 74LS244 buffer and a 74LS374 latch, develop an ISA bus interface that contains an 8-bit input port at I/O address 308H and an 8-bit output port at I/O address 30AH.
8. Create an ISA bus interface that allows four channels of analog output signals from 0 to 5.0 V each. These four channels must be decoded at I/O addresses 300H, 310H, 320H, and 330H. Also develop software that supports the four channels.
9. Redo Question 8, but instead of four output channels, use four ADCs to create four analog input channels at the same addresses.
10. Using an 8254 timer or timers, develop a darkroom timer on an ISA bus card. Your timer must generate a logic 0 for 1/100-second intervals from 1/100 second to five minutes. Use the system clock of 8 MHz as a timing source. The software you develop must allow the user to select the time from the keyboard. The output signal from the timer must be a logic 0 for the duration of the selected time and must be passed through an inverter to enable a solid-state relay that controls the photographic enlarger.
11. Interface a 16550 UART to the personal computer through the PCI bus interface. Develop software that transmits and receives data at Baud rates of 300, 1200, 9600, and 19,200. The UART must respond to I/O ports 1E3XH.
12. The ISA bus can transfer data that are _________ wide at the rate of 8 MHz.
13. Describe how the address can be captured from the PCI bus.
14. What is the purpose of the configuration memory found on the PCI bus interface?
15. Define the term "plug-and-play."
16. What is the purpose of the $C/\overline{BE}$ connection on the PCI bus system?
17. How is the BIOS tested for the PCI BIOS extension?

18. Develop a short program that interrogates the PCI bus, using the extension to the BIOS, and reads the 32-bit contents of configuration register 08H. For this problem, consider that the bus and unit numbers are 0000H.
19. What advantage does the PCI bus exhibit over the ISA bus?
20. How fast does the PCI Express bus transfer serial data?
21. The parallel port is decoded at which I/O addresses in a personal computer?
22. Can data be read from the parallel port?
23. The parallel port connecter found on the back of the computer has _________ pins.
24. Most computers contain at least one serial communication port. What is this port called?
25. Develop a C++ function that sends the letters ABC through the serial port and continues to do so until the letters ABC are returned through the serial port. Show all functions needed to accomplish this, including any initialization.
26. Modify Example 15–9 so it sends a character string of any length.
27. Search the Internet and detail, in a short report, variants as used in the Visual programming environment.
28. What data rates are available for use on the USB?
29. How are data encoded on the USB?
30. What is the maximum cable length for use with the USB?
31. Will the USB ever replace the ISA bus?
32. How many device addresses are available on the USB?
33. What is NRZI encoding?
34. What is a stuffed bit?
35. If the following raw data are sent on the USB, draw the waveform of the signal found on the USB: (110011000011001101 1010).
36. How long can a data packet be on the USB?
37. What is the purpose of the NAK and ACK tokens on the USB?
38. Describe the difference in data transfer rates on the PCI bus when compared with the AGP.
39. What is the transfer rate in a system using an 8X AGP video card?
40. On the Internet, locate a few video card manufacturers and find how much memory is available on AGP video cards. List the manufacturers and the amount of memory on the cards.
41. Using the Internet, write a report giving details of any USB controller.

CHAPTER 16

The 80186, 80188, and 80286 Microprocessors

INTRODUCTION

The Intel 80186/80188 and the 80286 are enhanced versions of the earlier versions of the 80X86 family of microprocessors. The 80186/80188 and 80286 are all 16-bit microprocessors that are upward-compatible from the 8086/8088. Even the hardware of these microprocessors is similar to the earlier versions. This chapter presents an overview of each microprocessor and points out the differences or enhancements of each version. The first part of the chapter describes the 80186/80188 microprocessors, and the last part shows the 80286 microprocessor.

New to recent editions is expanded coverage of the 80186/80188 family. Intel has added four new versions of each of these embedded controllers to its lineup of microprocessors. Each is a CMOS version and is designated with a two-letter suffix: XL, EA, EB, and EC. The 80C186XL and 80C188XL models are most similar to the earlier 80186/80188 models.

CHAPTER OBJECTIVES

Upon completion of this chapter, you will be able to:

1. Describe the hardware and software enhancements of the 80186/80188 and the 80286 microprocessors as compared to the 8086/8088
2. Detail the differences between the various versions of the 80186 and 80188 embedded controllers
3. Interface the 80186/80188 and the 80286 to memory and I/O
4. Develop software using the enhancements provided in these microprocessors
5. Describe the operation of the memory management unit (MMU) within the 80286 microprocessor
6. Define and detail the operation of a real-time operating system (RTOS)

16–1 80186/80188 ARCHITECTURE

The 80186 and 80188, like the 8086 and 8088, are nearly identical. The only difference between the 80186 and 80188 is the width of their data buses. The 80186 (like the 8086) contains a 16-bit data bus, while the 80188 (like the 8088) contains an 8-bit data bus. The internal register structure of the 80186/80188 is virtually identical to that of the 8086/8088. About the only difference is that the 80186/80188 contain additional reserved interrupt vectors and some very powerful built-in I/O features. The 80186 and 80188 are often called **embedded controllers** because of their application as a controller, not as a microprocessor-based computer.

Versions of the 80186/80188

As mentioned, the 80186 and 80188 are available in four different versions, which are all CMOS microprocessors. Table 16–1 lists each version and the major features provided. The 80C186XL and 80C188XL are the most basic versions of the 80186/80188; the 80C186EC and 80C188EC are the most advanced. This text details the 80C186XL/80C188XL, and then describes the additional features and enhancements provided in the other versions.

80186 Basic Block Diagram

Figure 16–1 provides the block diagram of the 80188 microprocessor that generically represents all versions except for the enhancements and additional features outlined in Table 16–1. Notice that this microprocessor has a great deal more internal circuitry than the 8088. The block diagrams of the 80186 and 80188 are identical except for the prefetch queue, which is four bytes in the 80188 and six bytes in the 80186. Like the 8088, the 80188 contains a bus interface unit (BIU) and an execution unit (ED).

In addition to the BIU and ED, the 80186/80188 family contains a clock generator, a programmable interrupt controller, programmable timers, a programmable DMA controller, and a programmable chip selection unit. These enhancements greatly increase the utility of the 80186/80188 and reduce the number of peripheral components required to implement a system. Many popular subsystems for the personal computer use the 80186/80188 microprocessors as

TABLE 16–1 The four versions of the 80186/80188 embedded controller.

Feature	*80C186XL 80C188XL*	*80C186EA 80C188EA*	*80C186EB 80C188EB*	*80C186EC 80C188EC*
80286-like instruction set	✔	✔	✔	✔
Power-save (green mode)	✔	✔		✔
Power down mode		✔	✔	✔
80C187 interface	✔	✔	✔	✔
ONCE mode	✔	✔	✔	✔
Interrupt controller	✔	✔	✔	✔ 8259-like
Timer unit	✔	✔	✔	✔
Chip selection unit	✔	✔	✔ enhanced	✔ enhanced
DMA controller	✔ 2-channel	✔ 2-channel		✔ 4-channel
Serial communications unit			✔	✔
Refresh controller	✔	✔	✔ enhanced	✔ enhanced
Watchdog timer				✔
I/O ports			✔ 16 bits	✔ 22 bits

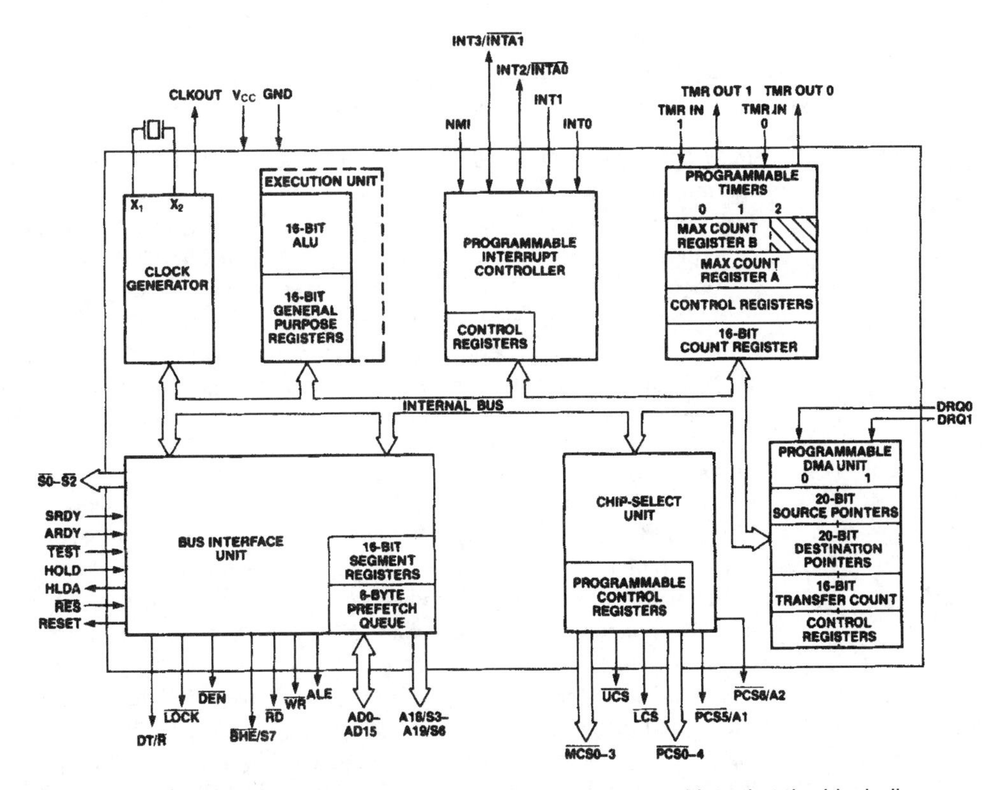

FIGURE 16–1 The block diagram of the 80186 microprocessor. Note that the block diagram of the 80188 is identical, except that $\overline{BHE}$/S7 is missing and AD_{15}–AD_8 are relabeled A_{15}–A_8. (Courtesy of Intel Corporation.)

caching disk controllers, local area network (LAN) controllers, and so forth. The 80186/80188 also finds application in the cellular telephone network as a switcher.

Software for the 80186/80188 is identical to that for the 80286 microprocessor, without the memory management instructions. This means that the 80286-like instructions for immediate multiplication, immediate shift counts, string I/O, PUSHA, POPA, BOUND, ENTER, and LEAVE all function on the 80186/80188 microprocessors.

80186/80188 Basic Features

In this segment of the text, we introduce the enhancements of the 80186/80188 microprocessors or embedded controllers that apply to all versions except where noted, but we do not provide exclusive coverage. More details on the operation of each enhancement and details of each advanced version are provided later in the chapter.

Clock Generator. The internal clock generator replaces the external 8284A clock generator used with the 8086/8088 microprocessors. This reduces the component count in a system.

The internal clock generator has three pin connections: X_1, X_2, and CLKOUT (or on some versions: CLKIN, OSCOUT, and CLKOUT). The X_1 (CLKIN) and X_2 (OSCOUT) pins are connected to a crystal that resonates at twice the operating frequency of the microprocessor. In the 8 MHz version of the 80186/80188, a 16 MHz crystal is attached to X_1 (CLKIN) and X_2 (OSCOUT). The 80186/80188 is available in 6 MHz, 8 MHz, 12 MHz, 16 MHz, or 25 MHz versions.

The CLKOUT pin provides a system clock signal that is one-half the crystal frequency, with a 50% duty cycle. The CLKOUT pin drives other devices in a system and provides a timing source to additional microprocessors in the system.

In addition to these external pins, the clock generator provides the internal timing for synchronizing the READY input pin, whereas in the 8086/8088 system, READY synchronization is provided by the 8284A clock generator.

Programmable Interrupt Controller. The programmable interrupt controller (PIC) arbitrates the internal and external interrupts and controls up to two external 8259A PICs. When an external 8259 is attached, the 80186/80188 microprocessors function as the master and the 8259 functions as the slave. The 80C186EC and 80C188EC models contain an 8259A-compatible interrupt controller in place of the one described here for the other versions (XL, EA, and EB).

If the PIC is operated without the external 8259, it has five interrupt inputs: INTO–INT3 and NMI. Note that the number of available interrupts depends on the version: The EB version has six interrupt inputs and the EC version has 16. This is an expansion from the two interrupt inputs available on the 8086/8088 microprocessors. In many systems, the five interrupt inputs are adequate.

Timers. The timer section contains three fully programmable 16-bit timers. Timers 0 and 1 generate waveforms for external use and are driven by either the master clock of the 80186/80188 or by an external clock. They are also used to count external events. The third timer, timer 2, is internal and clocked by the master clock. The output of timer 2 generates an interrupt after a specified number of clocks and can provide a clock to the other timers. Timer 2 can also be used as a watchdog timer because it can be programmed to interrupt the microprocessor after a certain length of time.

The 80C186EC and 80C188EC models have an additional timer called a *watchdog*. The watchdog timer is a 32-bit counter that is clocked internally by the CLKOUT signal (one-half the crystal frequency). Each time the counter hits zero, it reloads and generates a pulse on the WDTOUT pin that is four CLKOUT periods wide. This output can be used for any purpose: It can be wired to the reset input to cause a reset or to the NMI input to cause an interrupt. Note that if it is connected to the reset or NMI inputs, it is periodically reprogrammed so that it never counts down to zero. The purpose of a watchdog timer is to reset or interrupt the system if the software goes awry.

Programmable DMA Unit. The programmable DMA unit contains two DMA channels or four DMA channels in the 80C186EC/80C188EC models. Each channel can transfer data between memory locations, between memory and I/O, or between I/O devices. This DMA controller is similar to the 8237 DMA controller discussed in Chapter 13. The main difference is that the 8237 DMA controller has four DMA channels, as does the EC model.

Programmable Chip Selection Unit. The chip selection is a built-in programmable memory and I/O decoder. It has six output lines to select memory, seven lines to select I/O on the XL and EA models, and 10 lines that select either memory or I/O on the EB and EC models.

On the XL and EA models, the memory selection lines are divided into three groups that select memory for the major sections of the 80186/80188 memory map. The lower memory select signal enables memory for the interrupt vectors, the upper memory select signal enables memory for reset, and the middle memory select signals enable up to four middle memory devices. The boundary of the lower memory begins at location 00000H and the boundary of the upper memory ends at location FFFFFH. The sizes of the memory areas are programmable, and wait states (0–3 waits) can be automatically inserted with the selection of an area of memory.

On the XL and EA models, each programmable I/O selection signal addresses a 128-byte block of I/O space. The programmable I/O area starts at a base I/O address programmed by the user, and all seven 128-byte blocks are contiguous.

On the EB and EC models, there is an upper and lower memory chip selection pin and eight general-purpose memory or I/O chip selection pins. Another difference is that from 0 to 15 wait states can be programmed in these two versions of the 80186/80188 embedded controllers.

Power Save/Power Down Feature. The power save feature allows the system clock to be divided by 4, 8, or 16 to reduce power consumption. The power-saving feature is started by software and exited by a hardware event such as an interrupt. The power down feature stops the clock completely, but it is not available on the XL version. The power down mode is entered by execution of a HLT instruction and is exited by any interrupt.

Refresh Control Unit. The refresh control unit generates the refresh row address at the interval programmed. The refresh control unit does not multiplex the address for the DRAM—this is still the responsibility of the system designer. The refresh address is provided to the memory system at the end of the programmed refresh interval, along with the $\overline{\text{RFSH}}$ control signal. The memory system must run a refresh cycle during the active time of the $\overline{\text{RFSH}}$ control signal. More on memory and refreshing is provided in the section that explains the chip selection unit.

Pin-Out

Figure 16–2 illustrates the pin-out of the 80C186XL microprocessor. Note that the 80C186XL is packaged in either a 68-pin leadless chip carrier (LCC) or in a pin grid array (PGA). The LCC package and PGA packages are illustrated in Figure 16–3.

Pin Definitions. The following list defines each 80C186XL pin and notes any differences between the 80C186XL and 80C188XL microprocessors. The enhanced versions are described later in this chapter.

V_{CC} This is the system **power** supply connection for ±10%, +5.0 V.

V_{SS} This is the system **ground** connection.

X_1 and X_2 The **clock** pins are generally connected to a fundamental-mode parallel resonant crystal that operates an internal crystal oscillator. An external clock signal may be connected to the X_1 pin. The internal master clock operates at one-half the external crystal or clock input signal. Note that these pins are labeled CLKIN (X_1) and OSCOUT (X_2) on some versions of the 80186/80188.

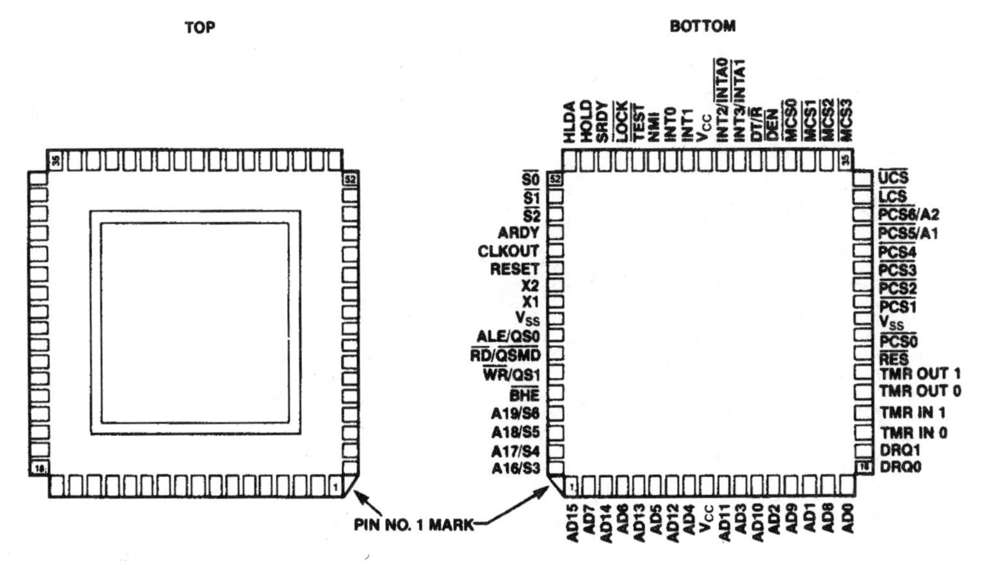

FIGURE 16–2 Pin-out of the 80186 microprocessor. (Courtesy of Intel Corporation.)

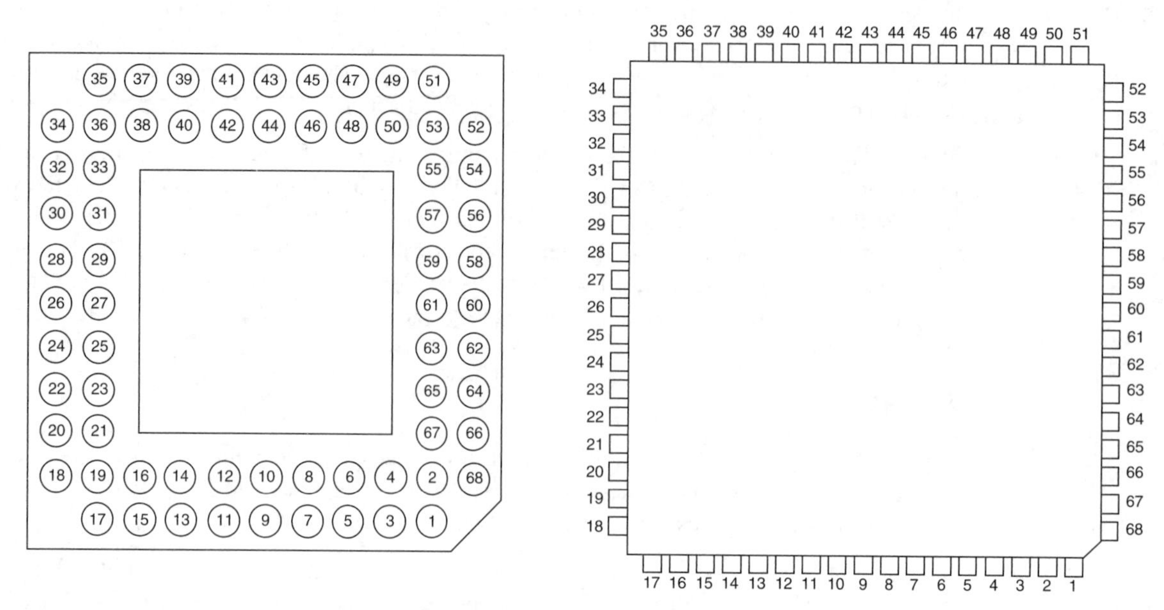

FIGURE 16–3 The bottom views of the PGA and LCC style versions of the 80C188XL microprocessor.

CLKOUT Clock out provides a **timing signal** to system peripherals at one-half the clock input frequency with a 50% duty cycle.

$\overline{\text{RES}}$ The **reset input** pin resets the 80186/80188. For a proper reset, the $\overline{\text{RES}}$ must be held low for at least 50 ms after power is applied. This pin is often connected to an RC circuit that generates a reset signal after power is applied. The reset location is identical to that of the 8086/8088 microprocessor—FFFF0H.

RESET The companion **reset output** pin (goes high for a reset) connects to system peripherals to initialize them whenever the $\overline{\text{RES}}$ input goes low.

$\overline{\text{TEST}}$ This test pin connects to the BUSY output of the 80187 numeric coprocessor. The $\overline{\text{TEST}}$ pin is interrogated with the FWAIT or WAIT instruction.

$T_{in}0$ and $T_{in}1$ These pins are used as **external clocking sources** to timers 0 and 1.

$T_{out}0$ and $T_{out}1$ These pins provide the **output signals** from timers 0 and 1, which can be programmed to provide square waves or pulses.

DRQ0 and DRQ1 These pins are active-high-level triggered **DMA request** lines for DMA channels 0 and 1.

NMI This is a **non-maskable interrupt** input. It is positive edge-triggered and always active. When NMI is activated, it uses interrupt vector 2.

INT_0, INT_1, $INT_2/\overline{INTA0}$, and $INT_3/\overline{INTA1}$ These are **maskable interrupt inputs**. They are active-high, and are programmed as either level or edge-triggered. These pins are configured as four interrupt inputs if no external 8259 is present, or as two interrupt inputs if the 8259A interrupt controller is present.

$A_{19}/\overline{ONCE}$, A_{18}, A_{17}, and A_{16} These are **multiplexed address/status** connections that provide the address (A_{19}–A_{16}) and status (S_6–S_3). Status bits found on address pins A_{18}–A_{16} have no system function and are used during manufacturing for testing. The A_{19} pin is an input for the $\overline{ONCE}$ function on a reset. If $\overline{ONCE}$ is held low on a reset, the microprocessor enters a testing mode.

AD_{15}–AD_0 These are **multiplexed address/data** bus connections. During T_1, the 80186 places A_{15}–A_0 on these pins; during T_2, T_3, and T_4, the 80186 uses these pins as the data bus for signals D_{15}–D_0. Note that the 80188 has pins AD_7–AD_0 and A_{15}–A_8.

$\overline{BHE}$ The **bus high enable** pin indicates (when a logic 0) that valid data are transferred through data bus connections D_{15}–D_8.

ALE **Address latch enable** is a output pin that contains ALE one-half clock cycle earlier than in the 8086. It is used to demultiplex the address/data and address/status buses. (Even though the status bits on A_{19}–A_{16} are not used in the system, they must still be demultiplexed.)

$\overline{WR}$ The **write** pin causes data to be written to memory or I/O.

$\overline{RD}$ The **read** pin causes data to be read from memory or I/O.

ARDY The **asynchronous READY** input informs the 80186/80188 that the memory or I/O is ready for the 80186/80188 to read or write data. If this pin is tied to +5.0 V, the microprocessor functions normally; if it is grounded, the microprocessor enters wait states.

SRDY The **synchronous READY** input is synchronized with the system clock to provide a relaxed timing for the ready input. Like ARDY, SRDY is tied to +5.0 V for no wait states.

$\overline{LOCK}$ The **lock** pin is an output controlled by the LOCK prefix. If an instruction is prefixed with LOCK, the $\overline{LOCK}$ pin becomes a logic 0 for the duration of the locked instruction.

S_2, S_1, and S_0 These are status bits that provide the system with the type of bus transfer in effect. See Table 16–2 for the states of the status bits. The **upper-memory chip select** pin selects memory on the upper portion of the memory map.

$\overline{UCS}$ The **upper-memory chip select** output is programmable to enable memory sizes of 1K–256K bytes ending at location FFFFFH. Note that this pin is programmed differently on the EB and EC versions and enables memory between 1K and 1M long.

TABLE 16–2 The S_2, S_1, and S_0 status bits.

S_2	S_1	S_0	Function
0	0	0	Interrupt acknowledge
0	0	1	I/O read
0	1	0	I/O write
0	1	1	Halt
1	0	0	Opcode fetch
1	0	1	Memory read
1	1	0	Memory write
1	1	1	Passive

$\overline{\textbf{LCS}}$	The **lower-memory chip select** pin enables memory beginning at location 00000H. This pin is programmed to select memory sizes from 1K to 256K bytes. Note that this pin functions differently for the EB and EC versions and enables memory between 1K and 1M bytes long.
$\overline{\textbf{MCS0}}$–$\overline{\textbf{MCS3}}$	The **middle-memory chip select** pins enable four middle memory devices. These pins are programmable to select an 8K–512K byte block of memory, containing four devices. Note that these pins are not present on the EB and EC versions.
$\overline{\textbf{PCS0}}$–$\overline{\textbf{PCS4}}$	These are five different **peripheral selection** lines. Note that the lines are not present on the EB and EC versions.
$\overline{\textbf{PCS5}}/\textbf{A}_1$ **and** $\overline{\textbf{PCS6}}/\textbf{A}_2$	These are programmed as peripheral selection lines or as internally latched address bits A_2 and A_1. These lines are not present on the EB and EC versions.
$\textbf{DT}/\overline{\textbf{R}}$	The **data transmit/receive** pin controls the direction of data bus buffers if attached to the system.
$\overline{\textbf{DEN}}$	The **data bus enable** pin enables the external data bus buffers.

DC Operating Characteristics

It is necessary to know the DC operating characteristics before attempting to interface or operate the microprocessor. The 80C186/801C88 microprocessors require between 42 mA and 63 mA of power-supply current. Each output pin provides 3.0 mA of logic 0 current and –2 mA of logic 1 current.

80186/80188 Timing

The timing diagram for the 80186 is provided in Figure 16–4. Timing for the 80188 is identical except for the multiplexed address connections, which are AD_7–AD_0 instead of AD_{15}–AD_0, and the $\overline{\text{BHE}}$, which does not exist on the 80188.

The basic timing for the 80186/80188 is composed of four clocking periods just as in the 8086/8088. A bus cycle for the 8 MHz version requires 500 ns, while the 16 MHz version requires 250 ns.

There are very few differences between the timing for the 80186/80188 and the 8086/8088. The most noticeable difference is that ALE appears one-half clock cycle earlier in the 80186/80188.

Memory Access Time. One of the more important points in any microprocessor's timing diagram is the memory access time. Access time calculations for the 80186/80188 are identical to that of the 8086/8088. Recall that the access time is the time allotted to the memory and I/O to provide data to the microprocessor after the microprocessor sends the memory or I/O its address.

A close examination of the timing diagram reveals that the address appears on the address bus T_{CLAV} time after the start of T_1. T_{CLAV} is listed as 44 ns for the 8 MHz version. (See Figure 16–5.) Data are sampled from the data bus at the end of T_3, but a setup time is required before the clock defined as T_{DVCL}. The times listed for T_{DVCL} are 20 ns for both versions of the microprocessor. Access time is therefore equal to three clocking periods minus both T_{CLAV} and T_{DVCL}. Access time for the 8 MHz microprocessor is 375 ns – 44 ns – 20 ns, or 311 ns. The access time for the 16 MHz version is calculated in the same manner, except that T_{CLAV} is 25 ns and T_{DVCL} is 15 ns.

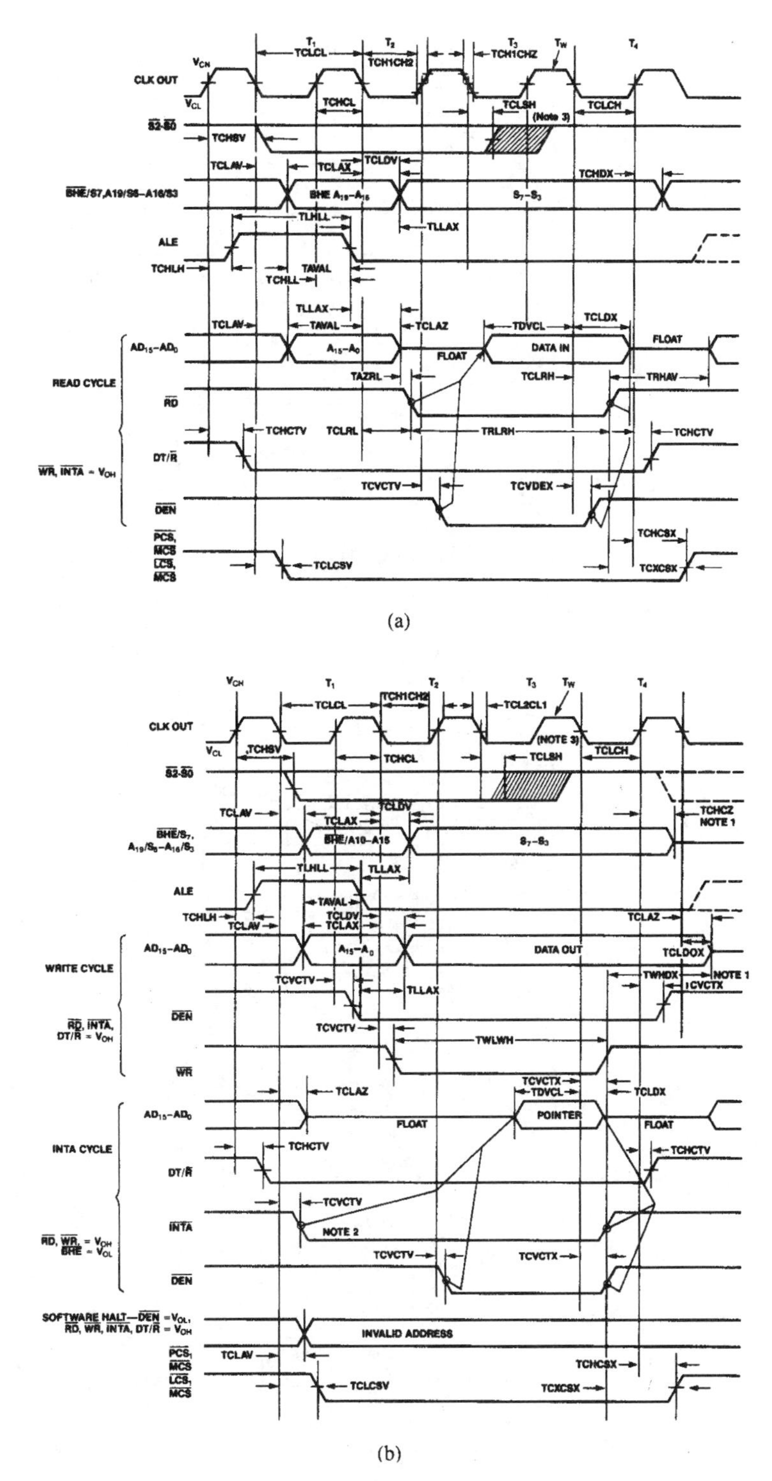

FIGURE 16–4 80186/80188 timing. (a) Read cycle timing and (b) write cycle timing. (Courtesy of Intel Corporation.)

80186 Master Interface Timing Responses

		80188 (8 MHz)		80188-6 (6 MHz)			
Symbol	**Parameters**	**Min.**	**Max.**	**Min.**	**Max.**	**Units**	**Test Conditions**
T_{CLAV}	Address Valid Delay	5	44	5	63	ns	C_L = 20-200 pF all outputs
T_{CLAX}	Address Hold	10		10		ns	
T_{CLAZ}	Address Float Delay	T_{CLAX}	35	T_{CLAX}	44	ns	
T_{CHCZ}	Command Lines Float Delay		45		56	ns	
T_{CHCV}	Command Lines Valid Delay (after float)		55		76	ns	
T_{LHLL}	ALE Width	$T_{CLCL-35}$		$T_{CLCL-35}$		ns	
T_{CHLH}	ALE Active Delay		35		44	ns	
T_{CHLL}	ALE Inactive Delay		35		44	ns	
T_{LLAX}	Address Hold to ALE Inactive	$T_{CHCL-25}$		$T_{CHCL-30}$		ns	
T_{CLDV}	Data Valid Delay	10	44	10	55	ns	
T_{CLDOX}	Data Hold Time	10		10		ns	
T_{WHDX}	Data Hold after WR	$T_{CLCL-40}$		$T_{CLCL-50}$		ns	
T_{CVCTV}	Control Active Delay 1	5	70	5	87	ns	
T_{CHCTV}	Control Active Delay 2	10	55	10	76	ns	
T_{CVCTX}	Control Inactive Delay	5	55	5	76	ns	
T_{CVDEX}	$\overline{DEN}$ Inactive Delay (Non-Write Cycle)		70		87	ns	
T_{AZRL}	Address Float to $\overline{RD}$ Active	0		0		ns	
T_{CLRL}	$\overline{RD}$ Active Delay	10	70	10	87	ns	
T_{CLRH}	$\overline{RD}$ Inactive Delay	10	55	10	76	ns	
T_{RHAV}	$\overline{RD}$ Inactive to Address Active	$T_{CLCL-40}$		$T_{CLCL-50}$		ns	
T_{CLHAV}	HLDA Valid Delay	10	50	10	67	ns	
T_{RLRH}	$\overline{RD}$ Width	$2T_{CLCL-50}$		$2T_{CLCL-50}$		ns	
T_{WLWH}	$\overline{WR}$ Width	$2T_{CLCL-40}$		$2T_{CLCL-40}$		ns	
T_{AVAL}	Address Valid to ALE Low	$T_{CLCH-25}$		$T_{CLCH-45}$		ns	
T_{CHSV}	Status Active Delay	10	55	10	76	ns	
T_{CLSH}	Status Inactive Delay	10	55	10	76	ns	
T_{CLTMV}	Timer Output Delay		60		75	ns	100 pF max
T_{CLRO}	Reset Delay		60		75	ns	
T_{CHQSV}	Queue Status Delay		35		44	ns	

80186 Chip-Select Timing Responses

Symbol	**Parameter**	**Min.**	**Max.**	**Min.**	**Max.**	**Units**	**Test Conditions**
T_{CLCSV}	Chip-Select Active Delay		66		80	ns	
T_{CXCSX}	Chip-Selct Hold from Command Inactive	35		35		ns	
T_{CHCSX}	Chip-Select Inactive Delay	5	35	5	47	ns	

Symbol	**Parameter**	**Min.**	**Max.**	**Units**	**Test Conditions**
TDVCL	Data in Setup (A/D)	20		ns	
TCLDX	Data in Hold (A/D)	10		ns	
TARYHCH	Asynchronous Ready (AREADY) active setup time*	20		ns	
TARYLCL	AREADY inactive setup time	35		ns	
TCHARYX	AREADY hold time	15		ns	
TSRYCL	Synchronous Ready (SREADY) transition setup time	35		ns	
TCLSRY	SREADY transition hold time	15		ns	
THVCL	HOLD Setup*	25		ns	
TINVCH	INTR, NMI, $\overline{TEST}$, TIMERIN, Setup*	25		ns	
TINVCL	DRQ0, DRQ1, Setup*	25		ns	

*To guarantee recognition at next clock.

FIGURE 16–5 80186 AC characteristics. (Courtesy of Intel Corporation.)

16–2 PROGRAMMING THE 80186/80188 ENHANCEMENTS

This section provides detail on the programming and operation of the 80186/80188 enhancements of all versions (XL, EA, EB, and EC). The next section details the use of the 80C188EB in a system that uses many of the enhancements discussed here. The only new feature not discussed here is the clock generator, which is described in the previous section on architecture.

Peripheral Control Block

All internal peripherals are controlled by a set of 16-bit-wide registers located in the peripheral block (PCB). The PCB (see Figure 16–6) is a set of 256 registers located in the I/O or memory space. Note that this set applies to the XL and EA versions. Later in this section, the EB and EC versions of the PCB are defined and described.

Whenever the 80186/80188 is reset, the peripheral control block is automatically located at the top of the I/O map (I/O addresses FF00H–FFFFH). In most cases, it stays in this area of I/O space, but the PCB may be relocated at any time to any other area of memory or I/O. Relocation is accomplished by changing the contents of the relocation register (see Figure 16–7) located at offset addresses FEH and FFH.

The relocation register is set to 20FFH when the 80186/80188 is reset. This locates the PCB at I/O addresses FF00H–FFFFH afterwards. To relocate the PCB, the user need only send a word OUT to I/O address FFFEH with a new bit pattern. For example, to relocate the PCB to

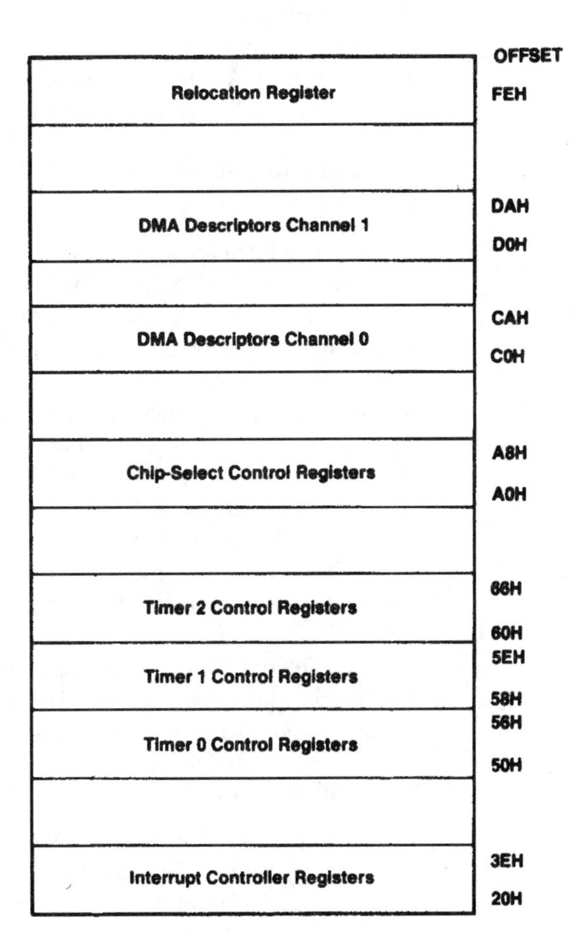

FIGURE 16–6 Peripheral control block (PCB) of the 80186/80188. (Courtesy of Intel Corporation.)

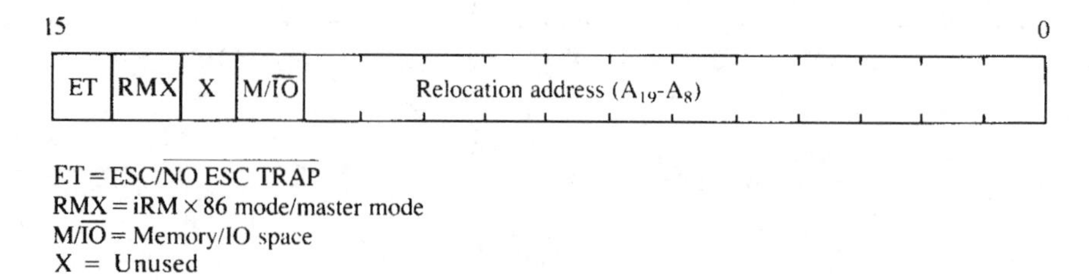

FIGURE 16–7 Peripheral control register.

memory locations 20000H–200FFH, send 1200H to I/O address FFFEH. Notice that M/$\overline{IO}$ is a logic 1 to select memory, and that 200H selects memory address 20000H as the base address of the PCB. Note that all accesses to the PCB must be word accesses because it is organized as 16-bit-wide registers. Example 16–1 shows the software required to relocate the PCB to memory locations 20000H–200FFH. Note that either an 8- or 16-bit output can be used to program the 80186; in the 80188, never use the OUT DX,AX instruction because it takes additional clocking periods to execute.

EXAMPLE 16–1

```
MOV   DX,0FFFEH                    ;address relocation register
MOV   AX,1200H                     ;new PCB location
OUT   DX,AL                        ;this can also be an OUT DX,AX
```

The EB and EC versions use a different address for programming the PCB location. Both versions have the PCB relocation register stored at offset XXA8H, instead of at offset XXFEH for the XL and EA versions. The bit pattern of these versions is the same as for the XL and EA versions, except that the RMX bit is missing.

Interrupts in the 80186/80188

The interrupts in the 80186/80188 are identical to the 8086/8088, except that additional interrupt vectors are defined for some of the internal devices. A complete listing of the reserved interrupt vectors appears in Table 16–3. The first five are identical to the 8086/8088.

The array BOUND instruction interrupt is requested if the boundary of an index register is outside the values set up in the memory. The unused opcode interrupt occurs whenever the 80186/80188 executes any undefined opcode. This is important if a program begins to run awry. Note that the unused opcode interrupt can be accessed by an instruction, but the assembler does not include it in the instruction set. On the Pentium Pro–Pentium 4 and some earlier Intel microprocessors, the 0F0BH or 0FB9H instruction will cause the program to call the procedure whose address is stored at the unused opcode interrupt vector.

The ESC opcode interrupt occurs if ESC opcodes D8H–DFH are executed. This occurs only if the ET (escape trap) bit of the relocation register is set. If an ESC interrupt occurs, the address stored on the stack by the interrupt points to the ESC instruction or to its segment override prefix, if one is used.

The internal hardware interrupts must be enabled by the I flag bit and must be unmasked to function. The I flag bit is set (enabled) with STI and cleared (disabled) with CLI. The remaining internally decoded interrupts are discussed with the timers and DMA controller, later in this section.

Interrupt Controller

The interrupt controller inside the 80186/80188 is a sophisticated device. It has many interrupt inputs that arrive from the five external interrupt inputs, the DMA controller, and the three timers. Figure 16–8 provides a block diagram of the interrupt structure of the 80186/80188 interrupt

TABLE 16–3 80186/80188 interrupt vectors.

Name	*Type*	*Address*	*Priority*
Divide error	0	00000–00003	1
Single-step	1	00004–00007	1A
NMI pin	2	00008–0000B	1
Breakpoint	3	0000C–0000F	1
Overflow	4	00010–00013	1
BOUND instruction	5	00014–00017	1
Unused opcode	6	00018–0001B	1
ESCape opcode	7	0001C–0001F	1
Timer 0	8	00020–00023	2A
Reserved	9	00024–00027	—
DMA 0	A	00028–0002B	4
DMA 1	B	0002C–0002F	5
INT_0	C	00030–00033	6
INT_1	D	00034–00037	7
INT_2	E	00038–0003B	8
INT_3	F	0003C–0003F	9
80187 coprocessor	10	00040–00043	1
Reserved	11	00044–00047	—
Timer 1	12	00048–0004B	2B
Timer 2	13	0004C–0004F	2C
Serial receiver	14	00050–00053	3A
Serial transmitter	15	00054–00057	3B

Note: Priority level 1 is the highest and 9 is the lowest. Some interrupts have the same priority. Only the EB and EC models contain the serial port.

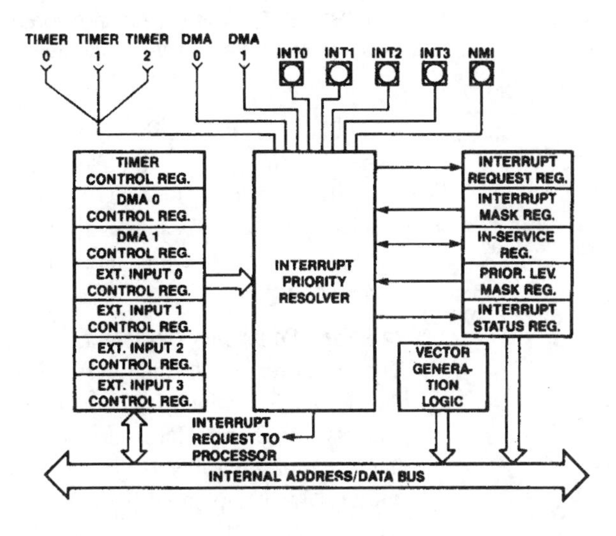

FIGURE 16–8 80186/80188 programmable interrupt controller. (Courtesy of Intel Corporation.)

controller. This controller appears in the XL, EA, and EB versions, but the EC version contains the exact equivalent to a pair of 8259As, as found in Chapter 12. In the EB version, the DMA inputs are replaced with inputs from the serial unit for receive and transmit.

The interrupt controller operates in two modes: master and slave mode. The mode is selected by a bit in the interrupt control register (EB and EC versions) called the *CAS bit.* If the CAS bit is a logic 1, the interrupt controller connects to external 8259A programmable interrupt controllers (see Figure 16–9); if CAS is a logic 0, the internal interrupt controller is selected. In many cases, there are enough interrupts within the 80186/80188, so the slave mode is not

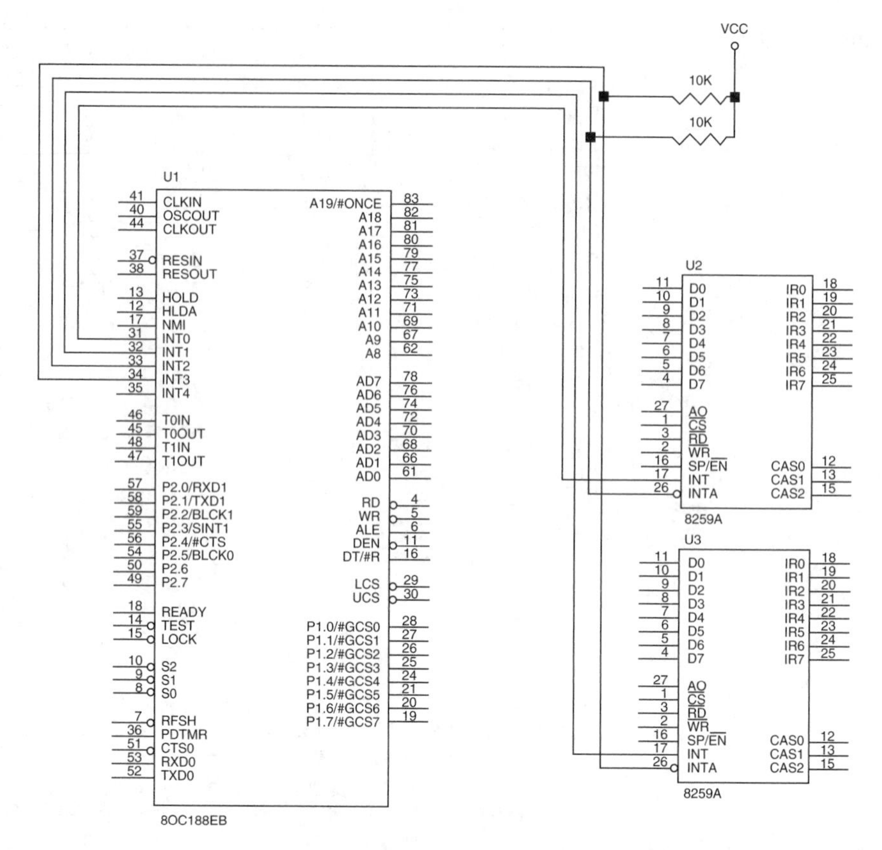

FIGURE 16–9 The interconnection between the 80C188EB and two 8259A programmable interrupt controllers. *Note:* Only the connections vital for this interface are shown.

normally used. In the XL and EA versions, the master and slave modes are selected in the peripheral control register at offset address FEH.

This portion of the text does not detail the programming of the interrupt controller. Instead, it is limited to a discussion of the internal structure of the interrupt controller. The programming and application of the interrupt controller is discussed in the sections that describe the timer and DMA controller.

Interrupt Controller Registers. Figure 16–10 illustrates the interrupt controller's registers. These registers are located in the peripheral control block beginning at offset address 22H. For the EC version, which is compatible with the 8259A, the interrupt controller ports are at offset addresses 00H and 02H for the master and ports 04H and 06H for the slave. In the EB version, the interrupt controller is programmed at offset address 02H. Note that the EB version has an additional interrupt input (INT4).

Slave Mode. When the interrupt controller operates in the slave mode, it uses up to two external 8259A programmable interrupt controllers for interrupt input expansion. Figure 16–9 shows how the external interrupt controllers connect to the 80186/80188 interrupt input pins for slave operation. Here, the INT_0 and INT_1 inputs are used as external connections to the interrupt request outputs of the 8259s, and $\overline{INTA0}$ (INT_2) and $\overline{INTA1}$ (INT_3) are used as interrupt acknowledge signals to the external controllers.

Interrupt Control Registers. There are interrupt control registers in both modes of operation, which each control a single interrupt source. Figure 16–11 depicts the binary bit pattern of each of these interrupt control registers. The mask bit enables (0) or disables (1) the interrupt input

XL and EA Versions		EB Version	
3EH	INT3 Control Register	1EH	INT3 Control Register
3CH	INT2 Control Register	1CH	INT2 Control Register
3AH	INT1 Control Register	1AH	INT1 Control Register
38H	INT0 Control Register	18H	INT0 Control Register
36H	DMA1 Control Register	16H	INT4 Control Register
34H	DMA0 Control Register	14H	Serial Control Register
32H	Timer Control Register	12H	Timer Control Register
30H	Interrupt Status	10H	Interrupt Status
2EH	Request	0EH	Request
2CH	In Service	0CH	In Service
2AH	PRIMSK	0AH	PRIMSK
28H	Interrupt Masks	08H	Interrupt Masks
26H	POLL Status	06H	POLL Status
24H	POLL	04H	POLL
22H	EOI	02H	EOI

FIGURE 16–10 The I/O offset port assignment for the interrupt control unit.

represented by the control word, and the priority bits set the priority level of the interrupt source. The highest priority level is 000, and the lowest is 111. The CAS bit is used to enable slave or cascade mode (0 enables slave mode), and the SFNM bit selects the **special fully nested mode**. The SFNM allows the priority structure of the 8259A to be maintained.

Interrupt Request Register. The interrupt request register contains an image of the interrupt sources in each mode of operation. Whenever an interrupt is requested, the corresponding interrupt request bit becomes a logic 1, even if the interrupt is masked. The request is cleared whenever the 80186/80188 acknowledges the interrupt. Figure 16–12 illustrates the binary bit pattern of the interrupt request register for both the master and slave modes.

Mask and Priority Mask Registers. The interrupt mask register has the same format as the interrupt register illustrated in Figure 16–12. If a source is masked (disabled), the corresponding bit of the interrupt mask register contains a logic 1; if enabled, it contains a logic 0. The interrupt mask register is read to determine which interrupt sources are masked and which are enabled. A source is masked by setting the source's mask bit in its interrupt control register.

The priority mask register, illustrated in Figure 16–13, shows the priority of the interrupt currently being serviced by the 80186/80188. The level of the interrupt is indicated by priority bits P_2–P_0. Internally, these bits prevent an interrupt by a lower priority source. These bits are automatically set to the next lower level at the end of an interrupt, as issued by the 80186/80188.

In-Service Register. The in-service register has the same binary bit pattern as the request register of Figure 16–12. The bit that corresponds to the interrupt source is set if the 80186/80188 is currently acknowledging the interrupt. The bit is reset at the end of an interrupt.

The Poll and Poll Status Registers. Both the interrupt poll and interrupt poll status registers share the same binary bit patterns as those illustrated in Figure 16–14. These registers have a bit

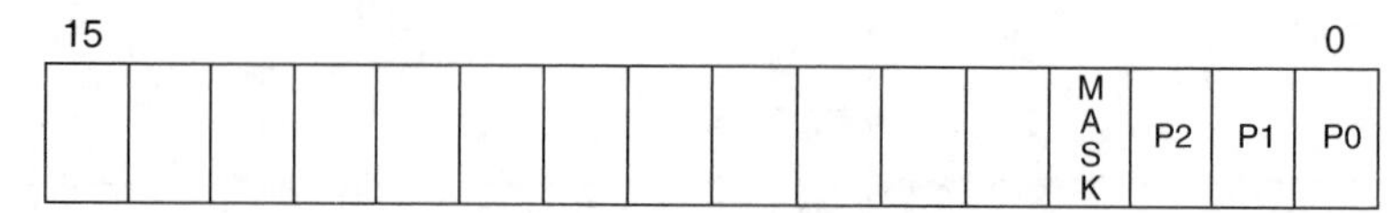

INT2, INT3, and INT4 Control Registers

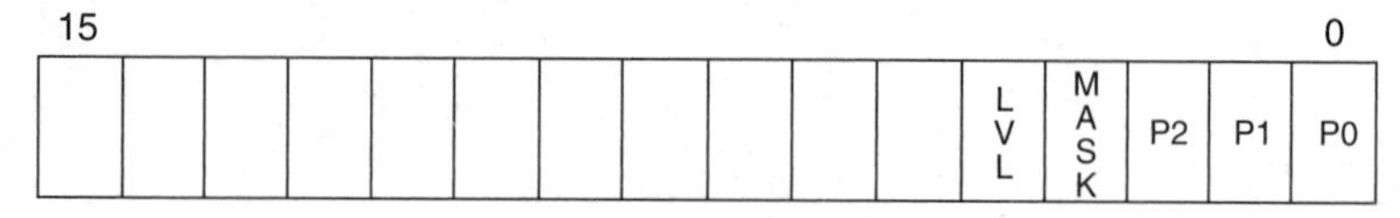

INT0 and INT1 Control Registers

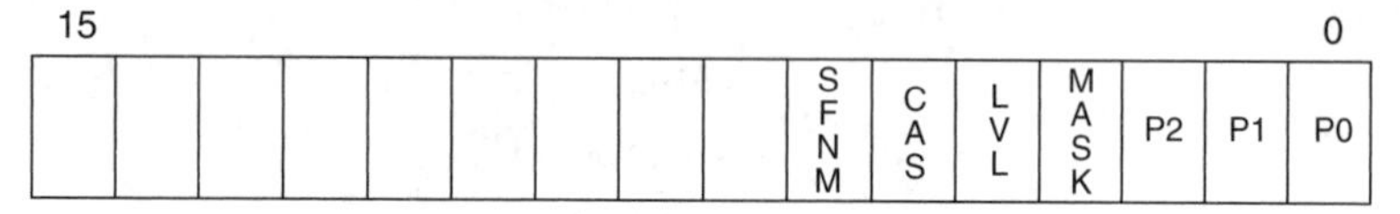

P2–P0 = Priority Level
Mask = 0 enables interrupt
LVL = 0 = edge and 1 = level triggering
CAS = 1 selects slave mode
SFNM = 1 selects special fully nested mode

FIGURE 16–11 The interrupt control registers.

FIGURE 16–12 The interrupt request register.

Interrupt Request Register (EB version)

15															0
								INT3	INT2	INT1	INT0	INT4	SER		TIM

Interrupt Request Register (XL and EA versions)

15															0
								INT3	INT2	INT1	INT0	DMA1	DMA0		TIM

FIGURE 16–13 The priority mask register.

Priority Mask Register

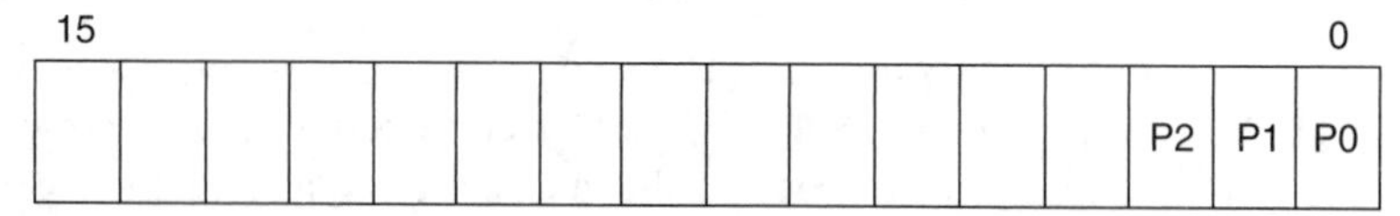

P2–P0 = Priority Level

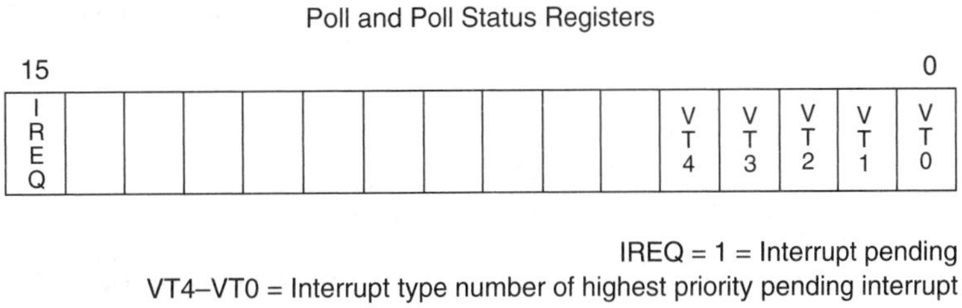

FIGURE 16–14 The poll and poll status registers.

Poll and Poll Status Registers

15															0
IREQ											VT4	VT3	VT2	VT1	VT0

IREQ = 1 = Interrupt pending
VT4–VT0 = Interrupt type number of highest priority pending interrupt

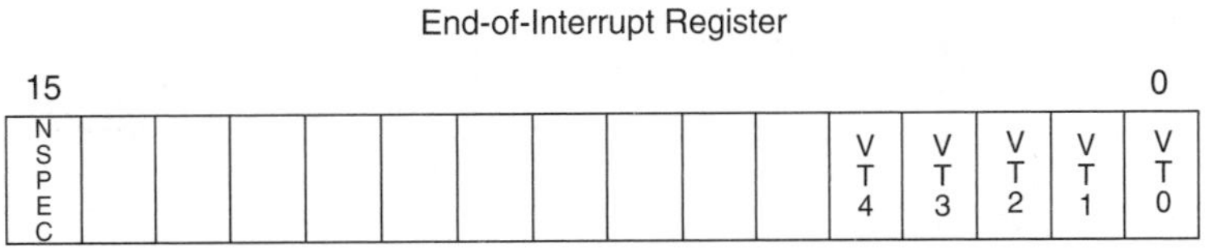

FIGURE 16–15 The end-of-interrupt (EOI) register.

End-of-Interrupt Register

15															0
NSPEC											VT4	VT3	VT2	VT1	VT0

(INT REQ) that indicates an interrupt is pending. This bit is set if an interrupt is received with sufficient priority, and cleared when an interrupt is acknowledged. The S bits indicate the interrupt vector type number of the highest priority pending interrupt.

The poll and poll status registers may appear to be identical because they contain the same information. However, they differ in function. When the interrupt poll register is read, the interrupt is acknowledged. When the interrupt poll status register is read, no acknowledge is sent. These registers are used only in the master mode, not in the slave mode.

End-of-interrupt Register. The end-of-interrupt (EOI) register causes the termination of an interrupt when written by a program. Figure 16–15 shows the contents of the EOI register for both the master and slave mode.

In the master mode, writing to the EOI register ends either a specific interrupt level (vector number) or whatever level is currently active (nonspecific). In the nonspecific mode, the NSPEC bit must be set before the EOI register is written to end a nonspecific interrupt. The nonspecific EOI clears the highest level interrupt bit in the in-service register. The specific EOI clears the selected bit in the in-service register, which informs the microprocessor that the interrupt has been serviced and another interrupt of the same type can be accepted. The nonspecific mode is used unless there is a special circumstance that requires a different order for interrupt acknowledges. If a specific EOI is required, the vector number is placed in the EOI command. For example, to clear the Timer 2 interrupt the EOI command is 13H (vector for timer 2).

In the slave mode, the level of the interrupt to be terminated is written to the EOI register. The slave mode does not allow a nonspecific EOI.

Interrupt Status Register. The format of the interrupt status register is depicted in Figure 16–16. In the master mode, T_2–T_0 indicates which timer (timer 0, timer 1, or timer 2) is causing an interrupt. This is necessary because all three timers have the same interrupt priority level. These bits are set when the timer requests an interrupt and are cleared when the interrupt is acknowledged. The DHLT (DMA halt) bit is only used in the master mode; when set, it stops a DMA action. Note that the interrupt status register is different for the EB version.

Interrupt Vector Register. The interrupt vector register is present only in the slave mode, and only in the XL and EA versions at offset address 20H. It is used to specify the most significant five bits of the interrupt vector type number. Figure 16–17 illustrates the format of this register.

Timers

The 80186/80188 microprocessors contain three fully programmable 16-bit timers and each is totally independent of the others. Two of the timers (timer 0 and timer 1) have input and output

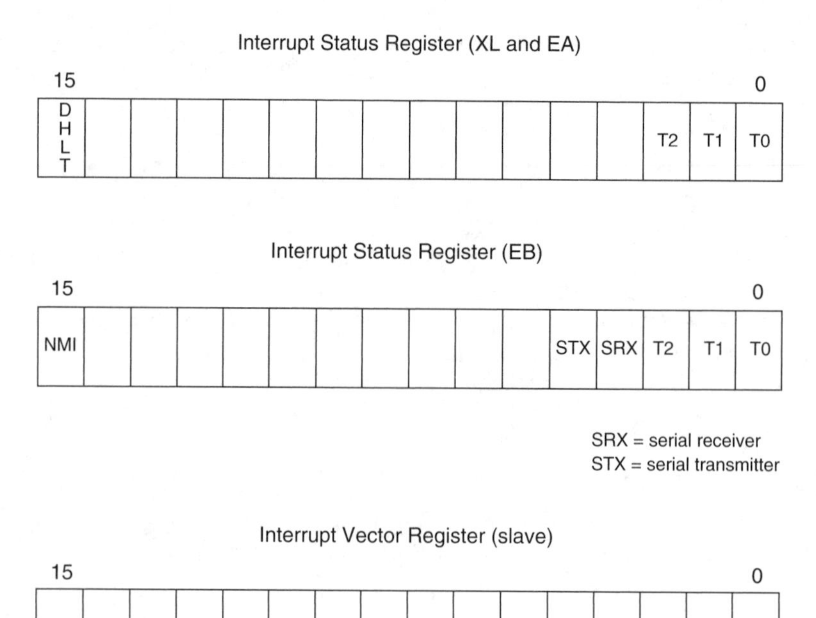

FIGURE 16–16 The interrupt status register.

FIGURE 16–17 The interrupt vector register.

pins that allow them to count external events or generate wave-forms. The third timer (timer 2) connects to the 80186/80188 clock. Timer 2 is used as a DMA request source, as a prescaler for other timers, or as a watchdog timer.

Figure 16–18 shows the internal structure of the timer unit. Notice that the timer unit contains one counting element that is responsible for updating all three counters. Each timer is actually a register that is rewritten from the counting element (a circuit that reads a value from a timer register and increments it before returning it). The counter element is also responsible for generating the outputs on the pins $T0_{OUT}$ and $T1_{OUT}$, reading the $T0_{IN}$ and $T1_{IN}$ pins, and causing a DMA request from the terminal count (TC) of timer 2 if timer 2 is programmed to request a DMA action.

Timer Register Operation. The timers are controlled by a block of registers in the peripheral control block (see Figure 16–19). Each timer has a count register, maximum-count register or

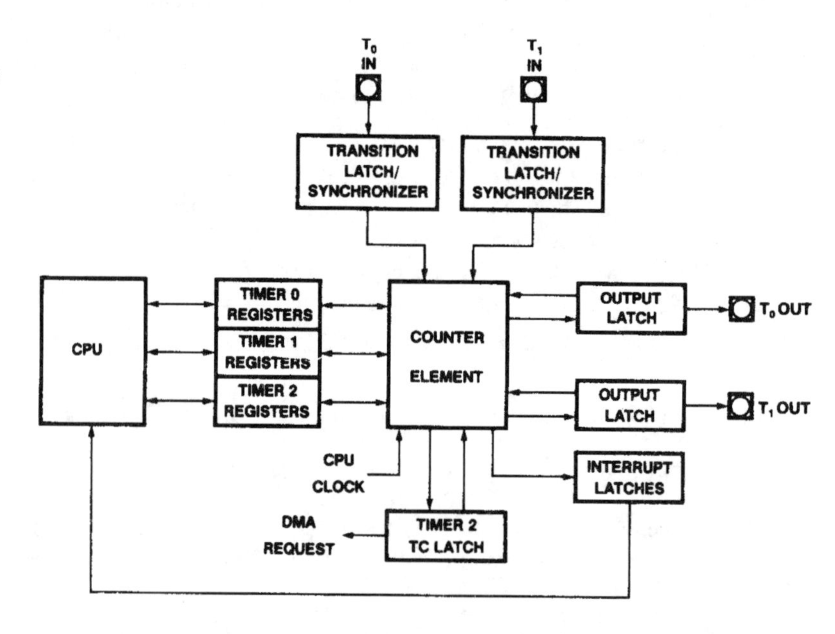

FIGURE 16–18 Internal structure of the 80186/80188 timers. (Courtesy of Intel Corporation.)

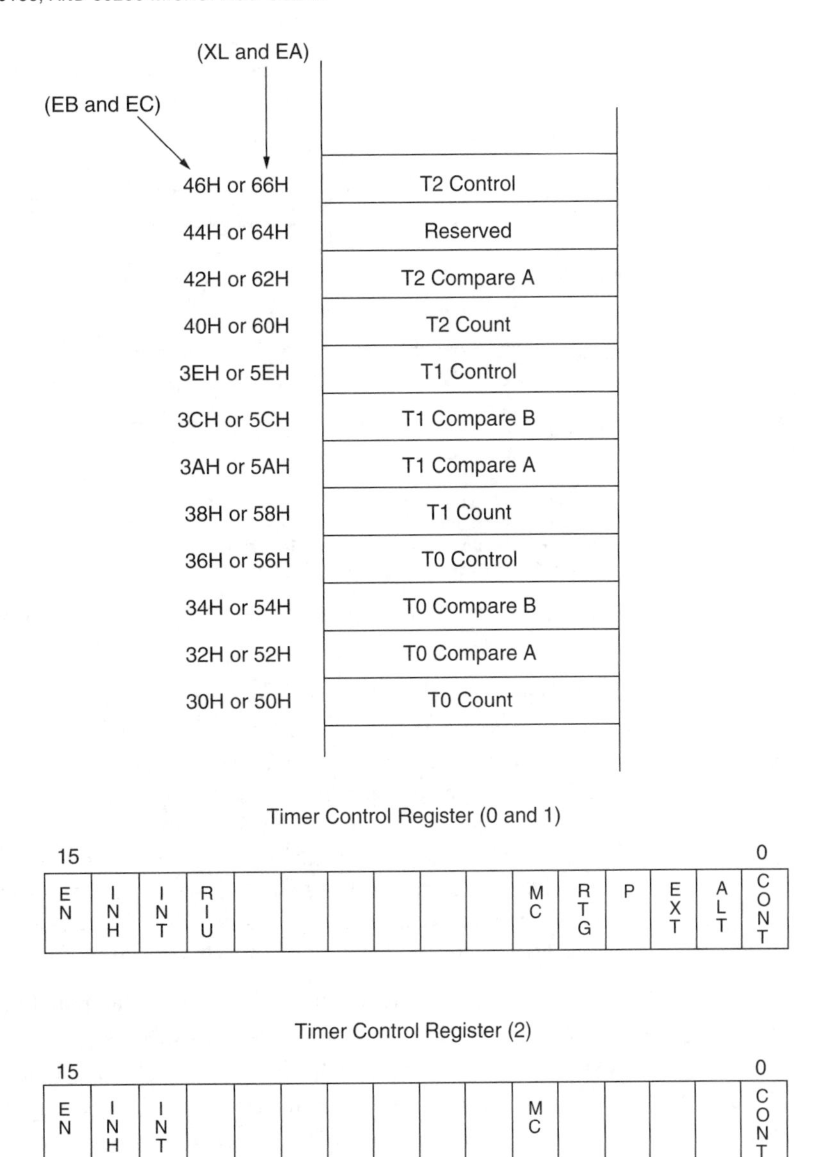

FIGURE 16–19 The offset locations and contents of the registers used to control the timers.

registers, and a control register. These registers may all be read or written at any time because the 80186/80188 microprocessors ensure that the contents never change during a read or write.

The timer count register contains a 16-bit number that is incremented whenever an input to the timer occurs. Timers 0 and 1 are incremented at the positive edge on an external input pin, every fourth 80186/80188 clock, or by the output of timer 2. Timer 2 is clocked on every fourth 80186/80188 clock pulse and has no other timing source. This means that in the 8 MHz version of the 80186/80188, timer 2 operates at 2 MHz, and the maximum counting frequency of timers 0 and 1 is 2 MHz. Figure 16–20 depicts these four clocking periods, which are not related to the bus timing.

Each timer has at least one maximum-count register, called a **compare** register (compare register A for timers 0 and 1), which is loaded with the maximum count of the count register to generate an output. Note that a timer is an up counter. Whenever the count register is equal to the maximum-count compare register, it is cleared to 0. With a maximum count of 0000H, the counter counts 65,536 times. For any other value, the timer counts the true value of the count. For example, if the maximum count is 0002H, the counter will count from 0 to 1 and then be cleared to 0—a modulus 2 counter has two states.

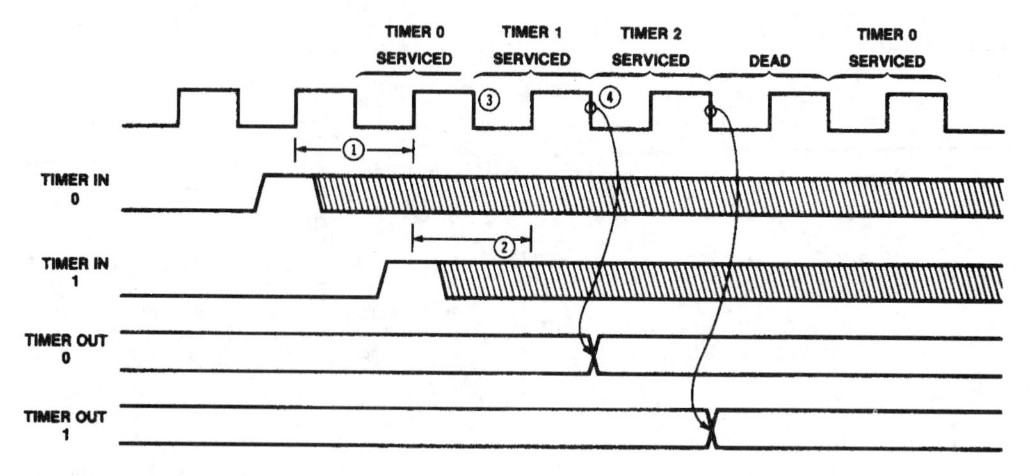

1. Timer in 0 resolution time
2. Timer in 1 resolution time
3. Modified count value written into 80186 timer 0 count register
4. Modified count value written into 80186 timer

FIGURE 16–20 Timing for the 80186/80188 timers. (Courtesy of Intel Corporation.)

Timers 0 and 1 each have a second maximum-count compare register (compare register B) that is selected by the control register for the timer. Either maximum-count compare register A or both maximum-count compare registers A and B are used with these timers, as programmed by the ALT bit in the control register for the timer. When both maximum-count compare registers are used, the timer counts up to the value in maximum-count compare register A, clears to 0, and then counts up to the count in maximum-count compare register B. This process is then repeated. Using both maximum-count registers allows the timer to count up to 131,072.

The control register (refer to Figure 16–19) of each timer is 16 bits wide and specifies the operation of the timer. A definition of each control bit follows:

EN The **enable** bit allows the timer to start counting. If EN is cleared, the timer does not count; if it is set, the timer counts.

INH The **inhibit** bit allows a write to the timer control register to affect the enable bit (EN). If INH is set, then the EN bit can be set or cleared to control the counting. If INH is cleared, EN is not affected by a write to the timer control register. This allows other features of the timer to be modified without enabling or disabling the timer.

INT The **interrupt** bit allows an interrupt to be generated by the timer. If INT is set, an interrupt will occur each time that the maximum count is reached in either maximum-count compare register. If this bit is cleared, no interrupt is generated. When the interrupt request is generated, it remains in force, even if the EN bit is cleared after the interrupt request.

RIU The **register in use** bit indicates which maximum-count compare register is currently in use by the timer. If RIU is a logic 0, then maximum-count compare register A is in use. This bit is a read-only bit, and writes do not affect it.

MC The **maximum count** bit indicates that the timer has reached its maximum count. This bit becomes a logic 1 when the timer reaches its maximum count and remains a logic 1 until the MC bit is cleared by writing a logic 0. This allows the maximum count to be detected by software.

RTG The **retrigger** bit is active only for external clocking (EXT = 0). The RTG bit is used only with timers 0 and 1 to select the operation of the timer input pins ($T0_{IN}$ and $T1_{IN}$).

If RTG is a logic 0, the external input will cause the timer to count if it is a logic 1; the timer will hold its count (stop counting) if it is a logic 0. If RTG is a logic 1, the external input pin clears the timer count to 0000H each time a positive-edge occurs.

P The **prescaler** bit selects the clocking source for timers 0 and 1. If EXT = 0 and P = 0, the source is one-fourth the system clock frequency. If P = 1, the source is timer 2.

EXT The **external** bit selects internal timing (EXT = 0) or external timing (EXT = 1). If EXT = 1, the timing source is applied to the $T0_{IN}$ or $T1_{IN}$ pins. In this mode, the timer increments after each positive-edge on the timer input pin. If EXT = 0, the clocking source is from one of the internal sources.

ALT The **alternate** bit selects single maximum-count mode (maximum-count compare register A) if a logic 0, or alternate maximum-count mode (maximum-count compare registers A and B) if a logic 1.

CONT The **continuous** bit selects continuous operation if a logic 1. In continuous operation, the counter automatically continues counting after it reaches its maximum count. If CONT is a logic 0, the timer will automatically stop counting and clear the EN bit. Note that whenever the 80186/80188 are reset, the timers are automatically disabled.

Timer Output Pin. Timers 0 and 1 have an output pin used to generate either square waves or pulses. To produce pulses, the timer is operated in single maximum-count mode (ALT = 0). In this mode, the output pin goes low for one clock period when the counter reaches its maximum count. By controlling the CONT bit in the control register, either a single pulse or continuous pulses can be generated.

To produce square waves or varying duty cycles, the alternate mode (ALT = 1) is selected. In this mode, the output pin is a logic 1 while maximum-count compare register A controls the timer; it is a logic 0 while maximum-count compare register B controls the timer. As with the single maximum-count mode, the timer can generate either a single square wave or continuous square waves. See Table 16–4 for the function of the ALT and CONT control bits.

Almost any duty cycle can be generated in the alternate mode. For example, suppose that a 10% duty cycle is required at a timer output pin. Maximum-count register A is loaded with a 10 and maximum-count register B is loaded with a 90 to produce an output that is a logic 1 for 10 clocks and a logic 0 for 90 clocks. This also divides the frequency of the timing source by a factor of 100.

Real-Time Clock Example. Many systems require the time of day. This is often called a **real-time clock** (RTC). A timer within the 80186/80188 can provide the timing source for software that maintains the time of day.

The hardware required for this application is not illustrated because all that is required is to connect the $T1_{IN}$ pin to +5.0 V through a pull-up resistor to enable timer 1. In the example, timers 1 and 2 are used to generate a one-second interrupt that provides the software with a timing source.

The software required to implement a real-time clock is listed in Examples 16–2 and 16–3. Example 16–2 illustrates the software required to initialize the timers. Example 16–3 shows an interrupt service procedure, which keeps time. There is another procedure in Example 16–3 that increments a BCD modulus counter. None of the software required to install the interrupt vector and set or display time of day is illustrated here.

TABLE 16–4 The ALT and CONT bits in the timer control register.

ALT	*CONT*	*Mode*
0	0	Single pulse
0	1	Continuous pulses
1	0	Single square wave
1	1	Continuous square waves

EXAMPLE 16–2

```
;software is written for the 80186/80188 EB version that
;initializes and starts both timer 1 and 2.

;address equates

T2_CA   EQU    0FF42H           ;timer 2 compare A register
T2_CON  EQU    0FF46H           ;timer 2 control register
T2_CNT  EQU    0FF40H           ;timer 2 count register
T1_CA   EQU    0FF3AH           ;timer 1 compare A register
T1_CON  EQU    0FF38H           ;timer 1 control register
T1_CNT  EQU    0FF3EH           ;timer 1 count register

        MOV   AX,20000          ;program timer 2 for 10 msec
        MOV   DX,T2_CA
        OUT   DX,AX

        MOV   AX,100            ;program timer 1 for 1 sec
        MOV   DX,T1_CA
        OUT   DX,AX

        MOV   AX,0              ;clear count registers
        MOV   DX,T2_CNT
        OUT   DX,AX
        MOV   DX,T1_CNT
        OUT   DX,AX

        MOV   AX,0C001H         ;enable timer 2 and start it
        MOV   DX,T2_CON
        OUT   DX,AX

        MOV   AX,0E009H         ;enable timer 1 with interrupt
        MOV   DX,T1_CON         ;and start it
        OUT   DX,AX
```

Timer 2 is programmed to divide by a factor of 20,000. This causes the clock (assuming a 2 MHz on the 8 MHz version of the 80186/80188) to be divided down to one pulse every 10 ms. The clock for timer 1 is derived internally from the timer 2 output. Timer 1 is programmed to divide the Timer 2 clock by 100 and generate a pulse once per second. The control register of timer 1 is programmed so that the one-second pulse internally generates an interrupt.

The interrupt service procedure is called once per second to keep time. The interrupt service procedure adds a one to the content of memory location SECONDS on each interrupt. Once every 60 seconds, the content of the next memory location (SECONDS + 1) is incremented. Finally, once per hour, the content of memory location SECONDS + 2 is incremented. The time is stored in these three consecutive memory locations in BCD, so the system software can easily access the time.

EXAMPLE 16–3

```
SECONDS      DB   ?
MINUTES      DB   ?
HOURS        DB   ?

INTRS   PROC   FAR  USES  DS  AX  SI

        MOV    AX,SEGMENT_ADDRESS
        MOV    DS,AX
        MOV    AH,60H                        ;load modulus 60
        MOV    SI,OFFSET SECONDS             ;address clock
        CALL   UPS                           ;increment seconds
        .IF ZERO?                            ;if seconds became 0
               CALL UPS                      ;increment minutes
               MOV  AH,24H                   ;load modulus 24
               .IF ZERO?                     ;if minutes became 0
                      CALL UPS               ;increment hours
```

```
                .ENDIF
        .ENDIF
        MOV   DX,0FF02H                          ;clear interrupt
        MOV   AX,8000H
        OUT   DX,AX

        RET

INTRS   ENDP

UPS     PROC  NEAR

        MOV   AL,[SI]
        ADD   AL,1                               ;increment counter
        DAA                                      ;make it BCD
        INC   SI
        .IF   AL == AH                           ;test for modulus
              MOV   AL,0
        .ENDIF
        MOV   [SI-1],AL
        RET

UPS     ENDP
```

DMA Controller

The DMA controller within the 80186/80188 has two fully independent DMA channels. Each has its own set of 20-bit address registers, so any memory or I/O location is accessible for a DMA transfer. In addition, each channel is programmable for auto-increment or auto-decrement to either source or destination registers. This controller is not available in the EB or EC versions. The EC version contains a modified four-channel DMA controller; the EB version contains no DMA controller. This text does not describe the DMA controller within the EC version.

Figure 16–21 illustrates the internal register structure of the DMA controller. These registers are located in the peripheral control block at offset addresses C0H–DFH.

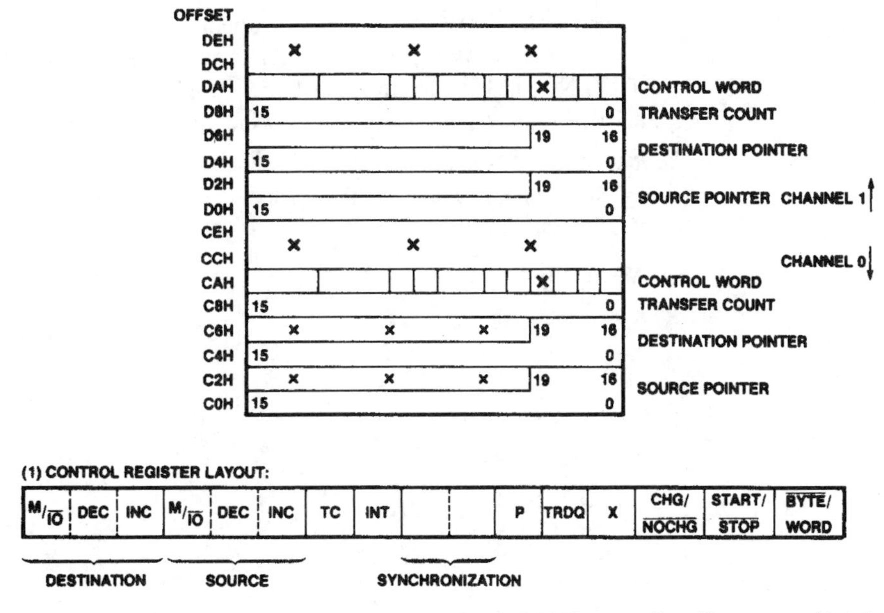

FIGURE 16–21 Register structure of the 80186/80188 DMA controller. (Courtesy of Intel Corporation.)

Notice that both DMA channel register sets are identical; each channel contains a control word, a source and destination pointer, and a transfer count. The transfer count is 16 bits wide and allows unattended DMA transfers of bytes (80188/80186) and words (80186 only). Each time that a byte or word is transferred, the count is decremented by one until it reaches 0000H—the terminal count.

The source and destination pointers are each 20 bits wide, so DMA transfers can occur to any memory location or I/O address without concern for segment and offset addresses. If the source or destination address is an I/O port, bits A_{19}–A_{16} must be 0000 or a malfunction may occur.

Channel Control Register. Each DMA channel contains its own channel control register (refer to Figure 16–21), which defines its operation. The leftmost six bits specify the operation of the source and destination registers. The $M/\overline{IO}$ bit indicates a memory or I/O location, DEC causes the pointer to be decremented, and INC causes the pointer to be incremented. If both the INC and DEC bits are 1, then the pointer is unchanged after each DMA transfer. Notice that memory-to-memory transfers are possible with this DMA controller.

The TC (terminal count) bit causes the DMA channel to stop transfers when the channel count register is decremented to 0000H. If this bit is a logic 1, the DMA controller continues to transfer data, even after the terminal count is reached.

The INT bit enables interrupts to the interrupt controller. If set, the INT bit causes an interrupt to be issued when the terminal count of the channel is reached.

The SYN bit selects the type of synchronization for the channel: 00 = no synchronization, 01 = source synchronization, and 10 = destination synchronization. When either unsynchronized or source synchronization is selected, data are transferred at the rate of 2M bytes per second. These two types of synchronization allow transfers to occur without interruption. If destination synchronization is selected, the transfer rate is slower (1.3M bytes per second), and the controller relinquishes control to the 80186/80188 after each DMA transfer.

The P bit selects the channel priority. If P = 1, the channel has the highest priority. If both channels have the same priority, the controller alternates transfers between channels.

The TRDQ bit enables DMA transfers from timer 2. If this bit is a logic 1, the DMA request originates from timer 2. This can prevent the DMA transfers from using all of the microprocessor's time for the transfer.

The $CHG/\overline{NOCHG}$ bit determines whether $START/\overline{STOP}$ changes for a write to the control register. The $START/\overline{STOP}$ bit starts or stops the DMA transfer. To start a DMA transfer, both $CHG/\overline{NOCHG}$ and $START/\overline{STOP}$ are placed at a logic 1 level.

The $BYTE/\overline{WORD}$ selects whether the transfer is byte- or word-sized.

Sample Memory-to-Memory Transfer. The built-in DMA controller is capable of performing memory-to-memory transfers. The procedure used to program the controller and start the transfer is listed in Example 16–4.

EXAMPLE 16–4

```
.MODEL SMALL
.186
.CODE
;
;Memory-to-Memory Transfer using DMA
;
;Calling Sequence for MOVES
;
;     DS:DI = source address
;     ED:DI = destination address
;     CX = Number of bytes
;
GETA   MACRO SEGA, OFFA, DMAA
       MOV   AX,SEGA          ;;get segment address
       SHL   AX,4             ;;shift left 4 places
```

```
        ADD   AX,OFFA          ;;add in offset
        MOV   DX,DMAA          ;;address DMA controller
        OUT   DX,AX            ;;program address
        PUSHF
        MOV   AX,SEGA
        SHR   AX,12
        POPF
        ADC   AX,0
        ADD   DX,2
        OUT   DX,AX
        ENDM

MOVES   PROC   NEAR

        GETA DS,SI,0FFC0H      ;program source address
        GETA ES,DI,0FFC4H      ;program destination address

        MOV   DX,0FFC8H        ;program count
        MOV   AX,CX
        OUT   DX,AX
        MOV   DX,0FFCAH        ;program DMA control
        MOV   AX,0B606H
        OUT   DX,AX            ;start transfer

        RET

MOVES   ENDP
```

The procedure in Example 16–4 transfers data from the data segment location addressed by SI into the extra segment location addressed by DI. The number of bytes transferred is held in register CX. This operation is identical to the REP MOVSB instruction, but execution occurs at a much higher speed through the use of the DMA controller.

Chip Selection Unit

The chip selection unit simplifies the interface of memory and I/O to the 80186/80188. This unit contains programmable chip selection logic. In small- and medium-sized systems, no external decoder is required to select memory and I/O. Large systems, however, may still require external decoders. There are two forms of the chip selection unit; one form found in the XL and EA versions differs from the unit found in the EB and EC versions.

Memory Chip Selects. Six pins (XL and EA versions) or 10 pins (EB and EC versions) are used to select different external memory components in a small- or medium-sized 80186/80188-based system. The $\overline{UCS}$ (**upper chip select**) pin enables the memory device located in the upper portion of the memory map that is most often populated with ROM. This programmable pin allows the size of the ROM to be specified and the number of wait states required. Note that the ending address of the ROM is FFFFFH. The $\overline{LCS}$ (**lower chip select**) pin selects the memory device (usually a RAM) that begins at memory location 00000H. As with the $\overline{UCS}$ pin, the memory size and number of wait states are programmable. The remaining four or eight chip select pins select middle memory devices. The four pins in the XL and EA version ($\overline{MCS3}$–$\overline{MCS0}$) are programmed for both the starting (base) address and memory size. Note that all devices must be of the same size. The eight pins ($\overline{GCS7}$–$\overline{GCS0}$) in the EB and EC versions are programmed by size and also by starting address, and can represent a memory device or an I/O device.

Peripheral Chip Selects. The 80186/80188 addresses up to seven external peripheral devices with pins $\overline{PCS6}$–$\overline{PCS0}$ (in the XL and EA versions). The $\overline{GCS}$ pins are used in the EB and EC versions to select up to eight memory or I/O devices. The base I/O address is programmed at any 1K-byte interval with port address block sizes of 128 bytes (64 bytes on the EB and EC versions).

TABLE 16–5 Wait state control bits R_2, R_1, and R_0 (XL and EA versions).

R_2	R_1	R_0	*Number of Waits*	*READY Required*
0	X	X	—	Yes
1	0	0	0	No
1	0	1	1	No
1	1	0	2	No
1	1	1	3	No

Programming the Chip Selection Unit for XL and EA Versions. The number of wait states in each section of the memory and the I/O are programmable. The 80186/80188 microprocessors have a built-in wait state generator that can introduce between 0 and 3 wait states (XL and EA version). Table 16–5 lists the logic levels required on bits R_2–R_0 in each programmable register to select various numbers of wait states. These three lines also select if an external READY signal is required to generate wait states. If READY is selected, the external READY signal is in parallel with the internal wait state generator. For example, if READY is a logic 0 for three clocking periods but the internal wait state generator is programmed to insert two wait states, three wait states are inserted.

Suppose that a 64K-byte EPROM is located at the top of the memory system and requires two wait states for proper operation. To select this device for this section of memory, the $\overline{\text{UCS}}$ pin is programmed for a memory range of F0000H–FFFFFH with two wait states. Figure 16–22 lists

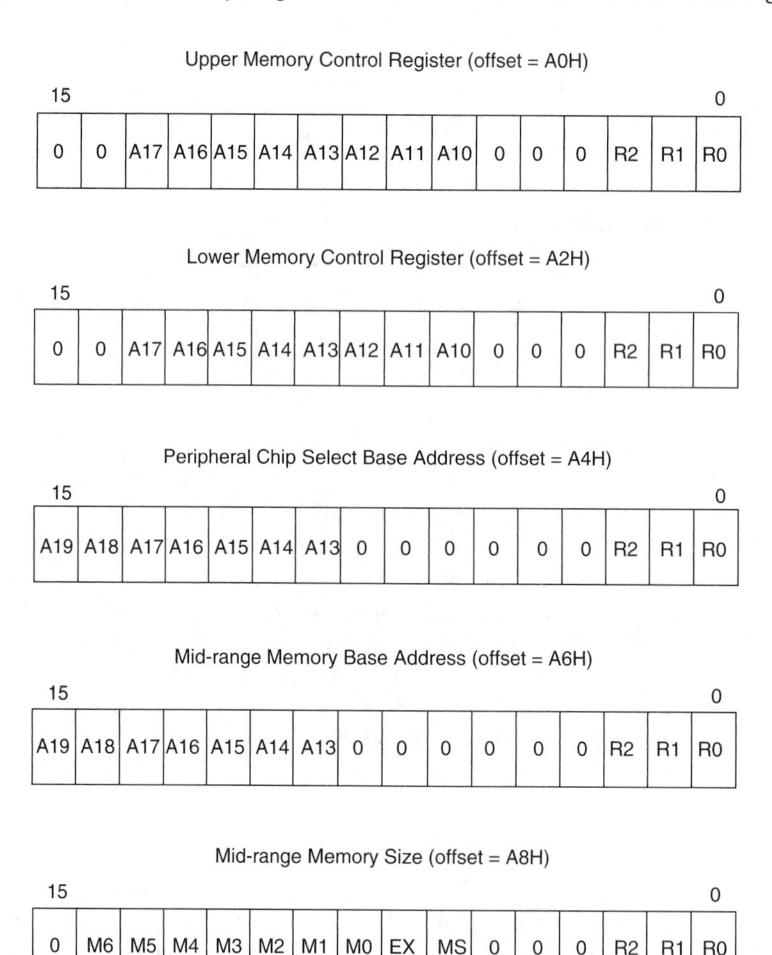

FIGURE 16–22 The chip selection registers for the XL and EA versions of the 80186/80188.

TABLE 16–6 Upper memory programming for the A_0H register (XL and EA versions).

Start Address	*Block Size*	*Value for No Waits, No READY*
FFC00H	1K	3FC4H
FF800H	2K	3F84H
FF000H	4K	3F04H
FE000H	8K	3E04H
FC000H	16K	3C04H
F8000H	32K	3804H
F0000H	64K	3004H
E0000H	128K	1004H
C0000H	256K	0004H

the control registers for all memory and I/O selections in the peripheral control block at offset addresses A_0–A_9H. Notice that the rightmost three bits of these control registers are from Table 16–5. The control register for the upper memory area is at location PCB offset address A_0H. This 16-bit register is programmed with the starting address of the memory area (F0000H, in this case) and the number of wait states. Please note that the upper two bits of the address must be programmed as 00, and that only address bits A_{17}–A_{10} are programmed into the control register. See Table 16–6 for examples illustrating the codes for various memory sizes. Because our example requires two wait states, the basic address is the same as in the table for a 64K device, except that the rightmost three bits are 110 instead of 100. The datum sent to the upper memory control register is 3006H.

Suppose that a 32K-byte SRAM that requires no waits and no READY input is located at the bottom of the memory system. To program the $\overline{\text{LCS}}$ pin to select this device, register A_2 is loaded in exactly the same manner as register A_0H. In this example, a 07FCH is sent to register A2H. Table 16–7 lists the programming values for the lower chip-selection output.

The central part of the memory is programmed via two registers: A_6H and A_8H. Register A_6H programs the beginning or base address of the middle memory select lines ($\overline{\text{MCS3}}$–$\overline{\text{MCS0}}$) and number of waits. Register A_8H defines the size of the block of memory and the individual memory device size (see Table 16–8). In addition to block size, the number of peripheral wait states are programmed as with other areas of memory. The EX (bit 7) and MS (bit 6) specify the peripheral selection lines, and will be discussed shortly.

For example, suppose that four 32K-byte SRAMs are added to the middle memory area, beginning at location 80000H and ending at location 9FFFFH with no wait states. To program the middle memory selection lines for this area of memory, we place the leftmost seven address bits in

TABLE 16–7 Lower memory programming for the A_2H register (XL and EA versions).

Ending Address	*Block Size*	*Value for No Waits, No READY*
003FFH	1K	0004H
007FFH	2K	0044H
00FFFH	4K	00C4H
01FFFH	8K	01C4H
03FFFH	16K	03C4H
07FFFH	32K	07C4H
0FFFFH	64K	0FC4H
1FFFFH	128K	1FC4H
3FFFFH	256K	3FC4H

TABLE 16–8 Middle memory programming for the A_8H register (XL and EA versions).

Block Size	*Chip Size*	*Value for No Waits, No READY, and EX = 0, MS = 1*
8K	2K	0144H
16K	4K	0344H
32K	8K	0744H
64K	16K	0F44H
128K	32K	1F44H
256K	64K	3F44H
512K	128K	7F44H

register A_6H, with bits 8–3 containing logic 0s, and the rightmost three bits containing the ready control bits. For this example, register A_6H is loaded with 8004H. Register A_8H is programmed with 1F44H, assuming that EX = 0 and MS = 1 and no wait states and no READY are required for the peripherals.

Register A4H programs the peripheral chip selection pins ($\overline{PCS6}$–$\overline{PCS0}$) along with the EX and MS bits of register A_8H. Register A_4H holds the beginning or base address of the peripheral selection lines. The peripherals may be placed in memory or in the I/O map. If they are placed in the I/O map, A_{19}–A_{16} of the port number must be 0000. Once the starting address is programmed on any 1K-byte I/O address boundary, the $\overline{PCS}$ pins are spaced at 128-byte intervals.

For example, if register A_4H is programmed with a 0204H, with no waits and no READY synchronization, the memory address begins at 02000H or the I/O port begins at 2000H. In this case, the I/O ports are: $\overline{PCS0}$ = 2000H, $\overline{PCS1}$ = 2080H, $\overline{PCS2}$ = 2100H, $\overline{PCS3}$ = 2180H, $\overline{PCS4}$ = 2200H, $\overline{PCS5}$ = 2280H, and $\overline{PCS6}$ = 2300H.

The MS bit of register A_8H selects memory mapping or I/O mapping for the peripheral select pins. If MS is a logic 0, then the $\overline{PCS}$ lines are decoded in the memory map; if it is a logic 1, then the $\overline{PCS}$ lines are in the I/O map.

The EX bit selects the function of the $\overline{PCS5}$ and $\overline{PCS6}$ pins. If EX = 1, these $\overline{PCS}$ pins select I/O devices; if EX = 0, these pins provide the system with latched address lines A_1 and A_2. The A_1 and A_2 pins are used by some I/O devices to select internal registers and are provided for this purpose.

Programming the Chip Selection Unit for EB and EC Versions. As mentioned earlier, the EB and EC versions have a different chip selection unit. These newer versions of the 80186/80188 contain an upper and lower memory chip selection pin as do earlier versions, but they do not contain middle selection and peripheral selection pins. In place of the middle and peripheral chip selection pins, the EB and EC versions contain eight general chip selection pins ($\overline{GCS7}$–$\overline{GCS0}$) that select either a memory device or an I/O device.

Programming is also different because each of the chip selection pins contains a starting address register and an ending address register. See Figure 16–23 for the offset address of each pin and the contents of the start and end registers.

Notice that programming for the EB and EC versions of the 80186/80188 is much easier than for the earlier XL and EA versions. For example, to program the $\overline{UCS}$ pin for an address that begins at location F0000H and ends at location FFFFFH (64K bytes), the starting address register (offset = A_4H) is programmed with F002H for a starting address of F0000H with two wait states. The ending address register (offset = A_6H) is programmed with 000EH for an ending address of FFFFFH for memory with no external ready synchronization. The other chip selection pins are programmed in a similar fashion.

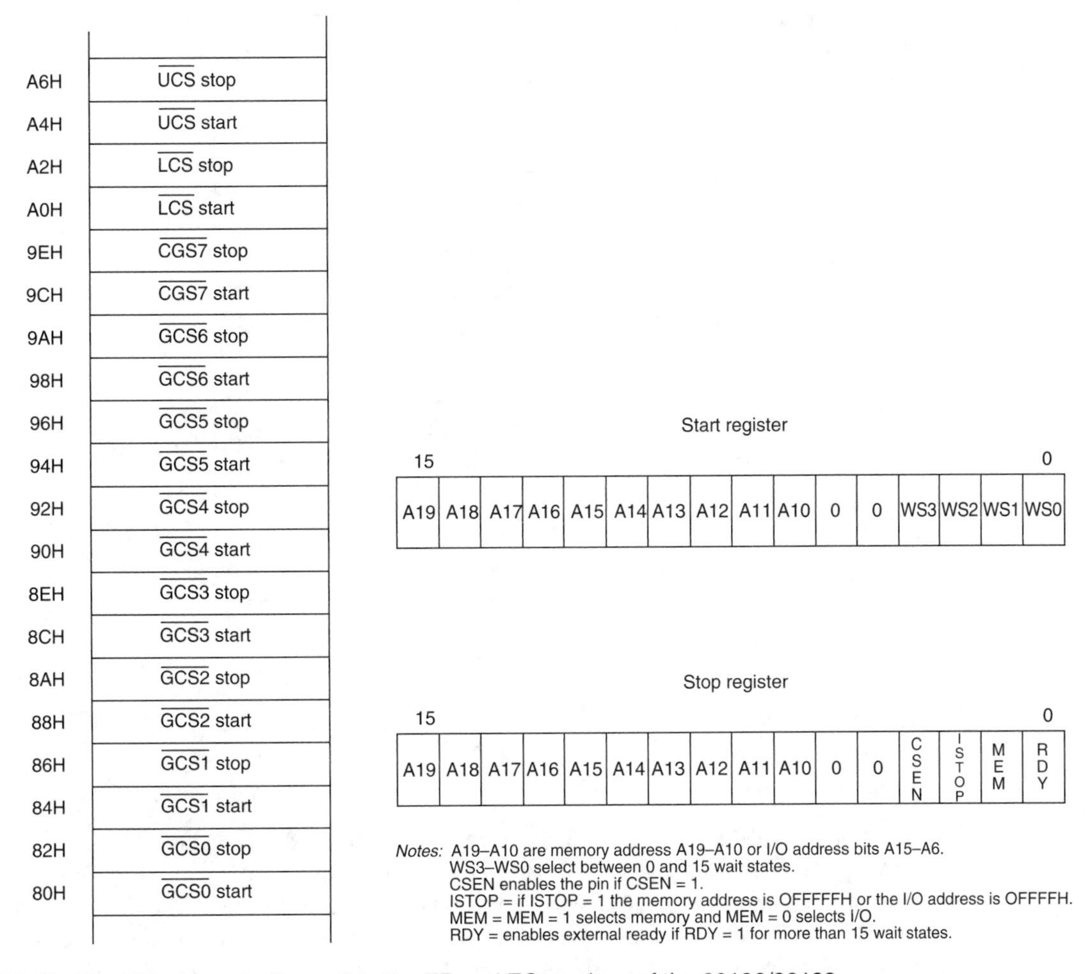

FIGURE 16–23 The chip selection unit in the EB and EC versions of the 80186/80188.

16–3 80C188EB EXAMPLE INTERFACE

Because the 80186/80188 microprocessors are designed as embedded controllers, this section of the text provides an example of such an application. The example illustrates simple memory and I/O attached to the 80C188EB microprocessor. It also lists the software required to program the 80C188EB and its internal registers after a system reset. Figure 16–24 illustrates the pin-out of the 80C188EB version of the 80188 microprocessor. Notice the differences between this version and the XL version presented earlier in the text.

The 80C188EB version contains some new features that were not present on earlier versions. These features include two I/O ports (P_1 and P_2) that are shared with other functions and two serial communications interfaces that are built into the processor. This version does not contain a DMA controller, as did the XL version.

The 80188EB can be interfaced with a small system designed to be used as a microprocessor trainer. The trainer illustrated in this text uses a 27256 EPROM for program storage, a 62256 SRAM for data storage, and an 8255 for a keyboard and LCD display interface. Figure 16–25 illustrates a small microprocessor trainer that is based on the 80C188EB microprocessor.

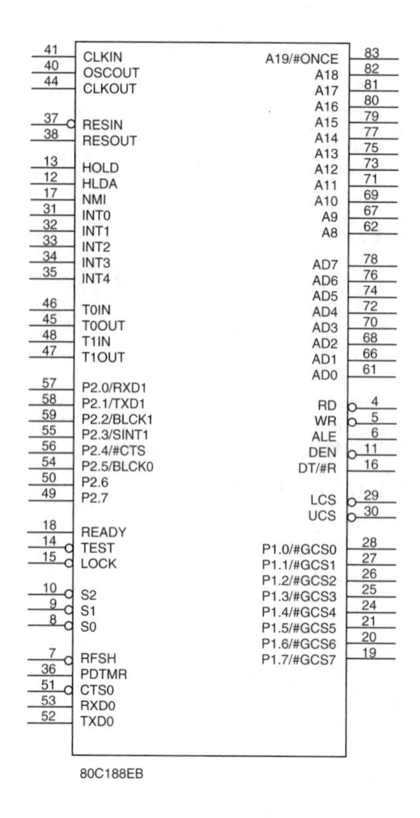

FIGURE 16–24 The pin-out of the 80C188EB version of the 80188 microprocessor.

Memory is selected by the $\overline{UCS}$ pin for the 27C256 EPROM and the $\overline{LCS}$ pin for the 62256 SRAM; the $\overline{GCS0}$ pin selects the 8255. The system software places the EPROM at memory addresses F8000H–FFFFFH; the SRAM at 00000H–07FFFH; and the 8255 at I/O ports 0000H–003FH (software uses ports 0, 1, 2, and 3). In this system, as is normally the case, we do not modify the address of the peripheral control block, which resides at I/O ports FF00H–FFFFH.

Example 16–5 lists the software required to initialize the 80C188EB microprocessor. This example completely programs the 80C188EB and also the entire system. The software is discussed in the next section of this chapter.

EXAMPLE 16–5

```
;Simple real-time operating test system for an 80188EB.

; INT 40H delays for BL milliseconds (range 1 to 99)
; INT 41H delays for BL seconds (range 1 to 59)
; Note* delay times must be written in hexadecimal!
; ie. 15 milliseconds is 15H.
; INT 42H displays character string on LCD
;       ES:BX addresses the NULL string
;       AL = where (80H line 1, C0H line2)
; INT 43H clears the LCD
; INT 44H reads a key from the keypad; AL = keycode

.MODEL TINY
.186            ;switch to the 80186/80188 instruction set
.CODE
.STARTUP

;program for the 80188EB microprocessor trainer.
;USE MASM 6.11
;command line = ML /AT FILENAME.ASM
```

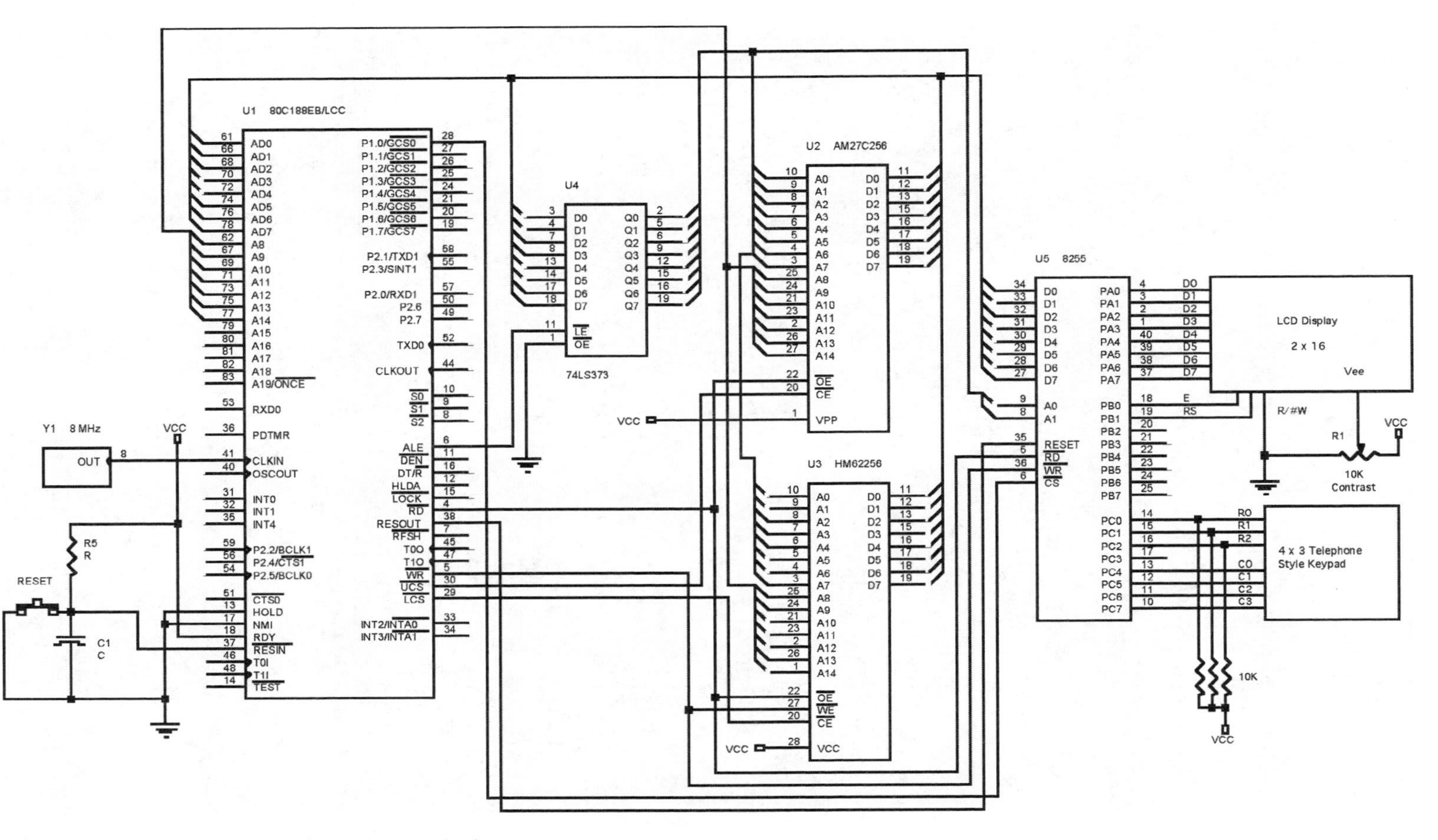

FIGURE 16–25 A sample 80C188EB-based microprocessor system.

```
;;;;;;;;;;;;;;;;;;;;;;;;;;;;;;;;;;;;;;;;;;;;;;;;;;;
;MACROs placed here
;;;;;;;;;;;;;;;;;;;;;;;;;;;;;;;;;;;;;;;;;;;;;;;;;;;

IO      MACRO    PORT,DATA
        MOV      DX,PORT
        MOV      AX,DATA
        OUT      DX,AL               ;;AL is more efficient
        ENDM

CS_IO   MACRO    PORT,START,STOP
        IO       PORT,START
        IO       PORT+2,STOP
        ENDM

SEND    MACRO    VALUE,COMMAND,DELAY
        MOV      AL,VALUE
        OUT      0,AL
        MOV      AL,COMMAND
        OUT      1,AL
        OR       AL,1
        OUT      1,AL
        AND      AL,2
        OUT      1,AL
        PUSH     BX
        MOV      BL,DELAY
        INT      40H
        POP      BX
        ENDM

BUT     MACRO
        IN   AL,2                     ;;test for key
        OR   AL,0F8H
        CMP  AL,0FFH
        ENDM

;;;;;;;;;;;;;;;;;;;;;;;;;;;;;;;;;;;;;;;;;;;;;;;;;;;;;;;;;;
;Initialization placed here
;;;;;;;;;;;;;;;;;;;;;;;;;;;;;;;;;;;;;;;;;;;;;;;;;;;;;;;;;;

        IO       0FFA6H,000EH        ;UCS Stop address
        CS_IO    0FFA0H,0,80AH       ;program LCS
        CS_IO    0FF80H,0,48H        ;program GCS0
        IO       0FF54H,1            ;Port 1 control
        IO       0FF5CH,0            ;Port 2 control
        IO       0FF58H,00FFH        ;port 2 direction
        IO       3,81H               ;program 8255
        MOV      AX,0                ;address segment 0000
        MOV      DS,AX               ;for DS, ES, and SS
        MOV      ES,AX
        MOV      SS,AX
        MOV      SP,8000H ;setup stack pointer (0000:8000)

        MOV      BX,OFFSET INTT-100H         ;install interrupt vectors
        .WHILE WORD PTR CS:[BX] != 0
               MOV       AX,CS:[BX]
               MOV       DI,CS:[BX+2]
               MOV       DS:[DI],AX
               MOV       DS:[DI+2],CS
               ADD       BX,4
        .ENDW

        MOV     BYTE PTR  DS:[40FH],0            ;don't display time
        IO      0FF40H,0             ;Timer 2 count
        IO      0FF42H,1000          ;Timer 2 compare
        IO      0FF46H,0E001H        ;Timer 2 control
        IO      0FF08H,00FCH         ;interrupt mask
        MOV     AL,0
```

```
        OUT     1,AL                    ;place E at 0 for LCD
        STI                             ;enable interrupts
        CALL    INIT                    ;initialize LCD

;;;;;;;;;;;;;;;;;;;;;;;;;;;;;;;;;;;;;;;;;;;;;;;;;;;;;;;;;;;;;;;;;;;;;;;;;;;;;;;;
;                                       SYSTEM SOFTWARE HERE
;;;;;;;;;;;;;;;;;;;;;;;;;;;;;;;;;;;;;;;;;;;;;;;;;;;;;;;;;;;;;;;;;;;;;;;;;;;;;;;;

;The following instructions are temporary to test the system and
;are replaced when a new system is created.

        MOV     BYTE PTR DS:[40FH],0FFH         ;set up to display time
        MOV     WORD PTR DS:[40CH],0            ;clear clock to 00:00:00 AM
        MOV     BYTE PTR DS:[40EH],0
        MOV     AX,CS                           ;line 1 message
        MOV     ES,AX
        MOV     AL,80H
        MOV     BX,OFFSET MES1 - 100H
        INT     42h

;System software placed here

        .WHILE 1        ;end of system loop
        .ENDW

;;;;;;;;;;;;;;;;;;;;;;;;;;;;;;;;;;;;;;;;;;;;;;;;;;;;;;;;;;;;;;;;;;;;;;;;;;;;;;;;
;procedures and data follow the system software
;;;;;;;;;;;;;;;;;;;;;;;;;;;;;;;;;;;;;;;;;;;;;;;;;;;;;;;;;;;;;;;;;;;;;;;;;;;;;;;;

MES1            DB      'The 80188 rules!',0

;Interrupt vector table

INTT    DW      TIM2-100H           ;interrupt procedure
        DW      13H * 4             ;vector address
        DW      DELAYM-100H
        DW      40H * 4
        DW      DELAYS-100H
        DW      41H * 4
        DW      STRING-100H
        DW      42H * 4
        DW      CLEAR-100H
        DW      43H * 4
        DW      KEY-100H
        DW      44H * 4
        DW      0                   ;end of table

;Interrupt service procedure for TIMER 2 interrupt. (Once per millisecond)

TIM2    PROC    FAR USES ES DS AX BX SI DX

        MOV     AX,0
        MOV     DS,AX
        MOV     ES,AX
        MOV     BX,409H                 ;address real-time clock - 1
        MOV     SI,OFFSET MODU-101H     ;address Modulus Table - 1
        .REPEAT
                INC     SI              ;point to modulus
                INC     BX              ;point to counter
                MOV     AL,[BX]         ;get counter
                ADD     AL,1            ;add 1
                DAA                     ;make it BCD
                .IF AL == BYTE PTR CS:[SI]  ;check modulus
                        MOV     AL,0
                .ENDIF
                MOV     [BX],AL   ;save new count
        .UNTIL !ZERO? || BX == 40FH
        IO      0FF02H,8000H            ;end of interrupt
        .IF     BYTE PTR DS:[40AH] == 0 && BYTE PTR DS:[40BH] == 0
                CALL    DISPLAY ;start Display Thread
```

```
        .ENDIF
        IRET

TIM2    ENDP

MODU    DB      0                               ;Mod 100
        DB      10H                             ;Mod 10
        DB      60H                             ;Mod 60
        DB      60H                             ;Mod 60
        DB      24H                             ;Mod 24

DISPLAY PROC NEAR       ;display time of day (once a second)

        .IF     BYTE PTR DS:[40FH] != 0             ;if display time is not zero
                STI                                 ;enable future interrupts
                MOV     BX,3F0H
                MOV     SI,40EH                     ;address clock
                MOV     AL,[SI]                     ;get hours
                .IF AL > 12H                        ; for AM / PM
                        SUB     AL,12H
                        DAS
                .ELSEIF AL == 0
                        MOV     AL,12H
                .ENDIF
                CALL    STORE
                MOV     BYTE PTR [BX],':'
                INC     BX
                MOV     AL,[SI]                     ;get minutes
                CALL    STORE
                MOV     BYTE PTR [BX],':'
                INC     BX
                MOV     AL,[SI]                     ;get seconds
                CALL    STORE
                MOV     DL,'A'                      ;for AM / PM
                .IF     BYTE PTR DS:[40EH] > 11H
                        MOV     DL,'P'
                .ENDIF
                MOV     BYTE PTR [BX],' '
                MOV     [BX+1],DL
                MOV     BYTE PTR [BX+2],'M'
                MOV     BYTE PTR [BX+3],0           ;end of string
                MOV     BX,3F0H                     ;display buffer
                MOV     AL,0C2H                     ;LCD starting position
                INT     42H
        .ENDIF
        RET

DISPLAY ENDP

STORE   PROC    NEAR

        PUSH    AX
        SHR     AL,4
        MOV     DL,AL
        POP     AX
        AND     AL,15
        MOV     DH,AL
        ADD     DX,3030H
        MOV     [BX],DX
        ADD     BX,2
        DEC     SI
        RET

STORE   ENDP

DELAYM  PROC    FAR USES DS BX AX

        STI                             ;enable future interrupts
        MOV     AX,0
        MOV     DS,AX
```

```
        MOV     AL,DS:[40AH]      ;get milli counter
        ADD     AL,BL             ;BL = no. of milliseconds
        DAA
        .REPEAT
        .UNTIL AL == DS:[40AH]
        IRET

DELAYM  ENDP

DELAYS  PROC    FAR USES DS BX AX

        STI                       ;enable future interrupts
        MOV      AX,0
        MOV      DS,AX
        MOV      AL,DS:[40CH]     ;get seconds
        ADD      AL,BL
        DAA
        .IF AL >= 60H
               SUB      AL,60H
               DAS
        .ENDIF
        .REPEAT
        .UNTIL AL == DS:[40CH]
        IRET

DELAYS  ENDP

INIT    PROC     NEAR

        MOV     BL,30H            ;wait 30 milliseconds
        INT     40H
        MOV     CX,4
        .REPEAT
                SEND      38H,0,6
        .UNTILCXZ
        SEND    8,0,2
        SEND    1,0,2
        SEND    12,0,2
        SEND    6,0,2
        RET

INIT_LCD ENDP

STRING  PROC    FAR USES BX AX   ;display string

        STI                                ;enable future interrupts
        SEND    AL,0,1                     ;send start position
        .REPEAT
                SEND     BYTE PTR ES:[BX],2,1
                INC      BX
        .UNTIL BYTE PTR ES:[BX] == 0
        IRET

STRING  ENDP

CLEAR   PROC    FAR USES AX BX   ;clear LCD

        STI                                ;enable future interrupts
        SEND    1,0,2
        IRET
CLEAR   ENDP

KEY     PROC     FAR USES BX        ;read key

        STI
        MOV  AL,0          ;clear C0 through C3
        OUT  2,AL
        .REPEAT            ;wait for key release
                .REPEAT
                         BUT
                .UNTIL ZERO?
```

```
                MOV   BL,12H          ;time delay
                INT   40H
                BUT
        .UNTIL ZERO?
        .REPEAT                 ;wait for key press
                .REPEAT
                          BUT
                .UNTIL   !ZERO?
                MOV   BL,  12H        ;time delay
                INT   40H
                BUT
        .UNTIL !ZERO?
        MOV  BX,0FDEFH
        .REPEAT
              MOV  AL,BL
              OUT  2,AL
              ADD  BH,3
              ROL  BL,1
              BUT
       .UNTIL !ZERO?
        .WHILE 1
              SHR  AL,1
              .BREAK .IF !CARRY?
              INC  BH
        .ENDW
        MOV  BX,OFFSET LOOK-100H
        XLAT CS:LOOK      ;specify code segment (EPROM)

        IRET

KEY     ENDP

LOOK    DB      3,2,1
        DB      6,5,4
        DB      9,8,7
        DB      10,0,11

;;;;;;;;;;;;;;;;;;;;;;;;;;;;;;;;;;;;;;;;;;;;;;;;;;;;;;;;;;;;;;;;;;;;;;;;;;;;;;;;
;ANY OTHER SUBROUTINES OR DATA THAT IS NEEDED SHOULD GO HERE
;;;;;;;;;;;;;;;;;;;;;;;;;;;;;;;;;;;;;;;;;;;;;;;;;;;;;;;;;;;;;;;;;;;;;;;;;;;;;;;;

        ORG     080F0H                  ;get to reset location
RESET:
        IO      0FFA4H,0F800H           ;UCS start address
        DB      0EAH                    ;JMP    F800:0000       (F8000H)
        DW      0000H,0F800H
END
```

16–4 REAL-TIME OPERATING SYSTEMS (RTOS)

This section of the text describes the real-time operating system (RTOS). Interrupts are used to develop the RTOS because they are used in embedded applications of the microprocessor. All systems, from the simplest embedded application to the most sophisticated system, must have an operating system.

What Is a Real-Time Operating System (RTOS)?

The RTOS is an operating system used in embedded applications that performs tasks in a predictable amount of time. Operating systems, like Windows, defer many tasks and do not guarantee their execution in a predictable time. The RTOS is much like any other operating system in that it contains the same basic sections. Figure 16–26 illustrates the basic structure of an operating system as it might be placed on an EPROM or Flash memory device.

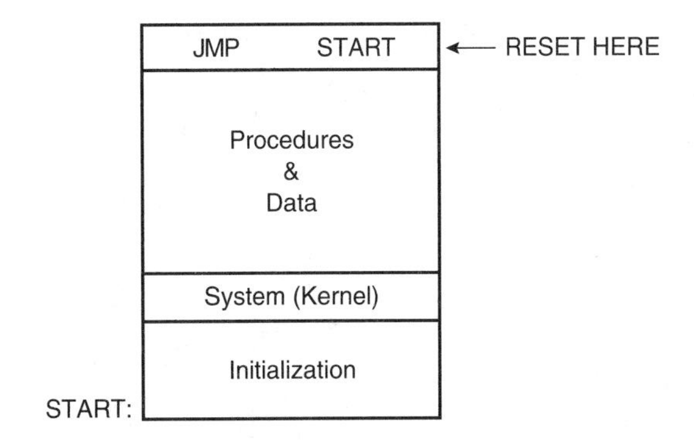

FIGURE 16–26 The structure of an RTOS operating system.

There are three components to all operating systems: (1) initialization, (2) the kernel, (3) data and procedures. If Example 16–5 (last section) is compared to Figure 16–26 all three sections will be seen. The initialization section is used to program all hardware components in the system, load drivers specific to a system, and program the contents of the microprocessor's registers. The kernel performs the basic system task, provides system calls or functions, and comprises the embedded system. The data and procedure section holds all procedures and any static data used by the operating system.

The RESET Section. The last part of the software in Example 16–5 shows the reset block of the RTOS. The ORG statement places the reset instructions at a location that is 16 bytes from the end of the memory device. In this case the EPROM is 32K bytes, which means it begins at 0000H and ends at 7FFFH. Recall that a 32K device has 15 address pins. The $\overline{\text{CS}}$ input selects the EPROM for locations F8000H through FFFFFH in the system. The ORG statement in the program places the origin of the reset section at location 80F0H because all tiny model (.COM) programs are assembled from offset address 100H even thought the first byte of the program is the first byte stored in the file. Because of this bias, all the addresses on the EPROM must be adjusted by 100H as is the ORG statement.

Only 16 bytes of memory exist for the reset instruction because the reset location is FFFF0H in the system. In this example there is only enough room to program the $\overline{\text{UCS}}$ starting address as F8000H before a jump to the start of the EPROM. Far jumps are not allowed in the tiny model, so it was forced by storing the actual hexadecimal opcode for a far jump (EAH).

Initialization Section. The initialization section of Example 16–5 begins in the reset block and continues at the start of the EPROM. If the initialization section is viewed, all of the programmable devices in the system are programmed and the segment registers are loaded. The initialization section also programs timer 2 so it causes an interrupt to the TIM_2 procedure each millisecond. The TIM_2 interrupt service procedure updates the clock once per second and is also the basis of precision time delays in the software.

The Kernel. The kernel in Example 16–5 is very short, because the system is incomplete and serves only as a test system. In this example all that the system does is display a sign-on message and display the time of day on the second line of the LCD. Once this is accomplished the system ends at an infinite WHILE loop. All system programs are infinite loops unless they crash.

An Example System

Figure 16–27 illustrates a simple embedded system based on the 80188EB embedded microprocessor. This schematic depicts the only the parts added to Figure 16–25 in order to read a temperature from the LM-70. This system contains a 2-line × 16 character-per-line LCD display that shows the time of day and the temperature. The system itself is stored on a small 32K × 8 EPROM. A 32K × 8 SRAM is included to act as a stack and store the time. A database holds the most recent temperatures and the times at which the temperatures were obtained.

The temperature sensor is located inside the LM_{70} digital temperature sensor manufactured by National Semiconductor Corporation for less than $1.00. The interface to the microprocessor is

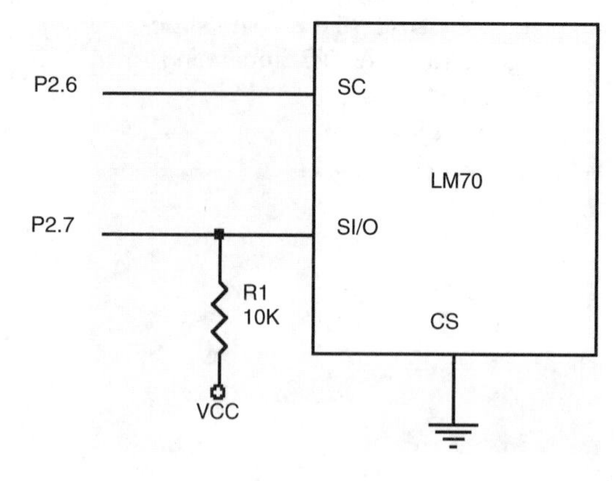

FIGURE 16–27 Additional circuitry for Figure 16–25 so it can read a temperature.

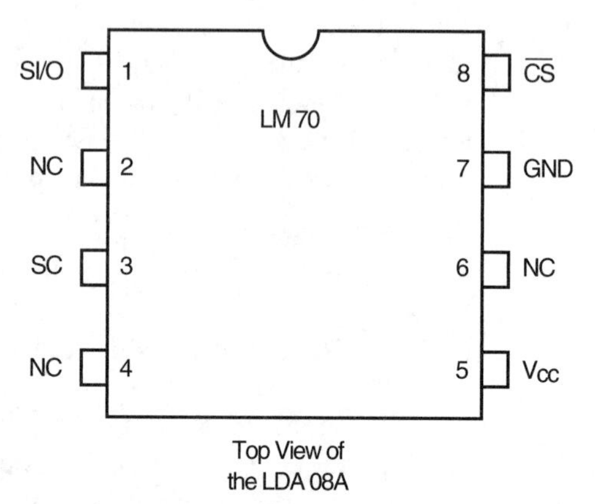

FIGURE 16–28 The LM_{70} temperature sensor.

in serial format, and the converter has a resolution of 10 bits plus a sign bit. Figure 16–28 illustrates the pin-out of the LM_{70} temperature sensor.

The LM_{70} transfers data to and from the microprocessor through the SIO pin, which is a bidirectional serial data pin. Information is clocked through the SIO pin by the SC (clock) pin. The LM_{70} contains three 16-bit registers: the configuration register, the temperature sensor register, and the identification register. The configuration register selects either the shutdown mode (XXFFH) or continuous conversion mode (XX00). The temperature register contains the signed temperature in the leftmost 11 bits of the 16-bit data word. If the temperature is negative it is in their respective complement forms. The identification register presents an 8100 when it is read.

When the temperature is read from the LM_{70}, it is read in Celsius and each step is equal to 0.25°C. For example, if the temperature register is 0000 1100 100X XXXX or a value of 100 decimal, the temperature is 25.0°C.

Example 16–6 illustrates the software added to the operating system listed in Example 16–5. The system samples the temperatures once per minute and stores them in a circular queue along with the day and the time in hour and minutes. The day is a number that starts at zero when the system is initialized. The size of the queue has been set to 16K bytes, so the most recent 4,096 measurements can be stored. In this example the keyboard is not used, but some of the system calls are used to display the temperature on line 1 of the display. The real-time clock is also interrogated to determine the start of each minute so a sample can be taken. The software in the listing replaces the software section in Example 16–5 where it states, ";System software placed here". This software replaces the infinite WHILE loop in the example.

The LM_{70} is initialized by outputting 16 bits of 0s to it and then read by reading all 16 bits of the temperature. The reading of the LM_{70} is accomplished in the software by the TEMP procedure and initialization is by the INITT procedure.

EXAMPLE 16–6

```
;System software that reads the temperature once per minute and
;logs it into a queue at 0500H - 44FFH.

       MOV   WORD PTR DS:[4FCH],500H    ;in queue pointer = 500H
       MOV   WORD PTR DS:[4FEH],500H    ;out queue pointer = 500H
       CALL  INITT          ;initialize LM70
       .while 1
              .IF DS:[40CH] == 0 && DS:[40BH] == 0 && DS:[40AH] == 0
                     CALL  TEMP             ;once per minute
                     CALL  ENQUE            ;queue temperature
                     CALL  DTEMP            ;display temperature
              .ENDIF
       .endw

INITT  PROC   NEAR          ;send 0000H to LM70 to reset it

       IO   0FF58H,003FH            ;p2.6 and p2.7 set to outputs
       MOV  CX,16                   ;bit count to 16
       MOV  DX,0FF5EH               ;address port 2 latch
       .REPEAT
                  MOV  AL,40H
                  OUT  DX,AL
                  MOV  AL,0
                  OUT  DX,AL
       .UNTILCXZ
       RET

INITT  ENDP

TEMP   PROC  NEAR           ;read temperature

       IO   0FF58K,00BFH            ;p2.7 as input
       MOV  CX,16
       MOV  BX,0
       .REPEAT                      ;get all 16-bits
              IO   0FF5EH,0C0H
              MOV DX,0FF5AH
              IN   AL,DX            ;read bit
              SHR AL,1
              RCR BX,1              ;into BX
              IO   0FF5EH,40H
       .UNTILCXZ
       MOV  AX,BX
       SAR  AX,6                    ;convert to integer temperature
       RET

TEMP   EMDP

ENQUE  PROC     NEAR USES AX        ;queue temperature, no check for full

       MOV  BX,DS:[4FCH]
       MOV  [BX],AX                 ;save temp
       ADD  BX,2
       MOV  AX,DS:[40DH]            ;get time HH:MM
       MOV  [BX],AX
       ADD  BX,2
       .IF BX == 4500H
              MOV BX,500H
       .ENDIF
       RET

ENQUE  ENDP

DTEMP  PROC

       MOV  BX,410H         ;address string buffer
       OR   AX,AX
```

```
        .IF SIGN?                   ;if negative
                MOV  BYTE PTR [BX],'-'
                NEG  AX
                INC  BX
        .ENDIF
        AAM                         ;convert to BCD
        .IF AH != 0
                ADD  AH,30H
                MOV  [BX],AH
                INC  BX
        .ENDIF
        ADD  AL,30H
        MOV  [BX],AL
        INC  BX
        MOV  BYTE PTR [BX],'°'
        MOV  BYTE PTR [BX+1],'C'
        MOV  AX,DS
        MOV  ES,AX
        MOV  AL,86H
        MOV  BX,410H
        INT  42H                    ;display temp on line 1

DTEMP   ENDP
```

A Threaded System

At times an operating system is needed that can process multiple threads. Multiple threads are handled by the kernel using a real-time clock interrupt. One method for scheduling processes in a small RTOS is to use a time slice to switch between various processes. The basic time slice can be any duration and is somewhat dependent on the execution speed of the microprocessor. For example, in a system using a 100 MHz clock, many instructions will execute in one or two clocks on a modern microprocessor. Assuming the machine executes one instruction every two clocks and a time slice of 1 ms is chosen, the machine can execute about 50,000 instructions per one time slice, which should be adequate for most systems. If a lower clock frequency is employed, then a time slice of 10 ms or even 100 ms is selected.

Each time slice is activated by a timer interrupt. The interrupt service procedure must look to the queue to determine if a task is available to execute, and if it is, it must start execution of the new task. If no new task is present, it must continue executing old tasks or enter an idle state and wait for a new task to be queued. The queue is circular and may contain any number of tasks for the system up to some finite limit. For example, it might be a small queue in a small system with 10 entries. The size is determined by the intended overall system needs and could be made larger or smaller.

Each scheduling queue entry must contain a pointer to the process (CS:IP) and the entire context state of the machine. Scheduling queue entries may also contain some form of a time-to-live entry in case of a deadlock, a priority entry, and an entry that can lengthen the slice activation time. In the following example, a priority entry or an entry to lengthen the amount of consecutive time slices allowed a program will not be used. The kernel will service processes strictly on a linear basis or in a round robin fashion as they come from the queue.

To implement a scheduler for the embedded system, procedures, or macros, are implemented to start a new application, kill an application when it completes, and pause an application if it needs time to access I/O. Each of these macros accesses a scheduling queue located in the memory system at an available address such as 0500H. The scheduling queue uses the data structure in Example 16–7 to make creating the queue fairly easy, and it will have room for 10 entries. This scheduling queue allows us to start up to 10 processes at a time.

EXAMPLE 16–7

```
PRESENT    DB   0       ;0 = not present
DUMMY1     DB   ?
RAX        DW   ?
```

```
RBX       DW   ?
RCX       DW   ?
RDX       DW   ?
RSP       DW   ?
RBP       DW   ?
RSI       DW   ?
RDI       DW   ?
RFLAG     DW   ?
RIP       DW   ?
RCS       DW   ?
RDS       DW   ?
RES       DW   ?
RSS       DW   ?
DUMMY2    DW   ?        ;padding to 32 bytes
```

The data structure of Example 16–7 is copied into memory 10 times to complete the queue structure during system initialization; hence, it contains no active process at initialization. We also need a queue pointer initialized to 500H. The queue pointer is stored at location 4FEH in this example. Example 16–8 provides one possible initialization. This stores the data structure in the RAM with 10 copies beginning at 500H. This software assumes that a system clock of 32 MHz operates timer 2 used as a prescaler to divide the clock input (system clock divided by 8) of 4 MHz by 40,000. This causes the output of timer 2 to be 1 KHz (1.0 ms). Timer 1 is programmed to divide the timer 2 clock signal by 10 to generate an interrupt every 10 ms.

EXAMPLE 16–8

```
;initialization of thread queue

        PUSH  DS
        MOV   AX,0
        MOV   DS,AX
        MOV   SI,500H
        MOV   BX,OFFSET PRESENT-100H     ;empty all queue entries
        MOV   CX,10
        .REPEAT
          MOV   BYTE PTR DS:[SI],0
          ADD   SI,32
        .UNTILCXZ
        MOV   DS:[4FEH],500H             ;set queue pointer
        POP   DS
        MOV   DX,0FF42H                  ;Timer 2 CMPA = 40000
        MOV   AX,40000
        OUT   DX,AL
        MOV   DX,0FF32H                  ;Timer 1 CMPA = 10
        MOV   AX,10
        OUT   DX,AL
        MOV   AX,0                       ;Clear Timer count registers
        MOV   DX,0FF30H
        OUT   DX,AL
        MOV   DX,0FF40H
        OUT   DX,AL
        MOV   DX,0FF46H
        MOV   AX,0C001H                  ;Start Timer 2
        OUT   DX,AL
        MOV   DX,0FF36H                  ;Start Timer 1
        MOV   AX,0E009H
        OUT   DX,AL
```

The NEW procedure (installed at INT 60H in Example 16–9) adds a process to the queue. It searches through the 10 entries until it finds a zero in the first byte (PRESENT), which indicates that the entry is empty. If it finds an empty entry, it places the starting address of the process into RCS and RIP and a 0200H into the RFLAG location. A 200H in RFLAG makes sure that the interrupt is enabled when the process begins, which prevents the system from crashing. The NEW procedure waits, if 10 processes are already scheduled, until a process ends. Each process is also assigned stack space in 256-byte sections beginning at offset address 7600H, so the lowest process has stack space 7500H–75FFH, the next has stack space 7600H–76FFH, and so on. The assignment of a stack area could be allocated by a memory manager algorithm.

EXAMPLE 16–9

```
INT60   PROC    FAR USES DS AX DX SI

        MOV   AX,0
        MOV   DS,AX                    ;address segment 0000
        STI                            ;interrupts on

        .REPEAT                        ;sticks here is full
              MOV    SI,500H
              HLT                      ;synchronize with RTC interrupt
              .WHILE BYTE PTR DS:[SI] != 0 && SI != 660H
                     ADD   SI,32
              .ENDW
        .UNTIL BYTE PTR DS:[SI] == 0

        MOV   BYTE PTR DS:[SI],0FFH ;activate process
        MOV   WORD PTR DS:[SI+18],200H      ;flags
        MOV   DS:[SI+20],BX            ;save IP
        MOV   DS:[SI+22,DX             ;save CS
        MOV   DS:[SI+28],SS            ;save SS
        MOV   AX,SI
        AND   AX,3FFH
        SHL   AX,3
        ADD   AX,7500H
        MOV   DS:[SI+10],AX            ;save SP
        IRET

INT60   ENDP
```

The final control procedure (KILL), located at interrupt vector 61H as illustrated in Example 16–10, kills an application by placing 00H into PRESENT of the queue data structure, which removes it from the scheduling queue.

EXAMPLE 16–10

```
INT61   PROC   FAR USES DS AX DX SI

        MOV  AX,0
        MOV  DS,AX
        MOV  SI,DS:[4FEH]             ;get queue pointer
        MOV  BYTE PTR DS:[SI],0       ;kill thread
        JMP  INT12A

INT61   ENDP
```

The PAUSE procedure is merely a call to the time slice procedure (INT 12H) that bails out of the process and returns control to the time slice procedure, prematurely ending the time slice for the process. This *early out* allows other processes to continue before returning to the current process.

The time slice interrupt service procedure for an 80188EB using a 10 ms time slice appears in Example 16–11. Because this is an interrupt service procedure, care has been taken to make it as efficient as possible. Example 16–11 illustrates the time slice procedure located at interrupt vector 12H for operation with timer 1 in the 80188EB microprocessor. Although not shown, this software assumes that timer 2 is used as a prescaler and timer 1 uses the signal from timer 2 to generate the 10 ms interrupt. The software also assumes that no other interrupt is in use in the system.

EXAMPLE 16–11

```
INT12  PROC   FAR USES DS AX DX SI

       MOV   AX,0
       MOV   DS,AX            ;address segment 0000
       MOV   SI,DS:[4FEH]     ;get queue pointer
       CALL  SAVES            ;save state
INT12A:                       ;from Kill
       ADD   SI,32            ;get next process
       .IF SI == 660H         ;make queue circular
             MOV  SI,500H
       .ENDIF
       .WHILE BYTE PTR DS:[SI] != 0FFH   ;find next process
              ADD  SI,32
              .IF SI == 660H
                     MOV  SI,500H
              .ENDIF
       .ENDW
       MOV   DX,0FF30H        ;get a complete 10ms time slice
       MOV   AX,0
       OUT   DX,AL
       MOV   DX,0FF02H        ;clear interrupt
       MOV   AX,8000H
       OUT   DX,AX
       JMP   LOADS            ;load state of next process

INT12  ENDP
```

The system startup (placed after system initialization) must be a single process inside an infinite wait loop. The startup is illustrated in Example 16–12.

EXAMPLE 16–12

```
SYSTEM_STARTUP:

;fork WAITS thread

      MOV   BX,OFFSET WAITS-100H   ;offset address of thread
      MOV   DX,0F800H              ;segment address of thread
      INT   60H                    ;start wait thread
      STI

;other system processes started here

WAITS:                       ;system idle loop
      .WHILE 1
             INT  12H        ;bail out
      .ENDW
```

Finally, the SAVES and LOADS procedures are used to load and save the machine context as a switch from one process to another occurs. These procedures are called by the time slice interrupt (INT 12H) and start a process (INT 60H) and are listed in Example 16–13.

EXAMPLE 16–13

```
SAVES   PROC   NEAR

        MOV  DS:[SI+4],BX          ;save BX
        MOV  DS:[SI+6],CX          ;save CX
        MOV  DS:[SI+10],SP         ;save SP
        MOV  DS:[SI+12],BP         ;save BP
        MOV  DS:[SI+16],DI         ;save DI
        MOV  DS:[SI+26],ES         ;save ES
        MOV  DS:[SI+28],SS         ;save SS
```

```
        MOV   BP,SP                     ;get SP
        MOV   AX,[SP+2]                 ;get SI
        MOV   DS:[SI+14],AX             ;save SI
        MOV   AX,[BP+4]                 ;get DX
        MOV   DS:[SI+8],AX              ;save DX
        MOV   AX,[BP+6]                 ;get AX
        MOV   DS:[SI+2],AX              ;save AX
        MOV   AX,[BP+8]                 ;get DS
        MOV   DS:[SI+24],AX             ;save DS
        MOV   AX,[BP+10]                ;get flags
        MOV   DS:[SI+18],AX             ;save flags
        MOV   AX,[BP+12]                ;get CS
        MOV   DS:[SI+22],AX             ;save CS
        MOV   AX,[BP+14]                ;get IP
        MOV   DS:[SI+20],AX             ;save IP
        RET

SAVES   ENDP

LOADS   PROC    FAR

        MOV   SS,DS:[SI+28]             ;get SS
        MOV   SP,DS:[SI+10]             ;get SP
        PUSH  WORD PTR DS:[SI+20]       ;PUSH IP
        PUSH  WORD PTR DS:[SI+22]       ;PUSH CS
        PUSH  WORD PTR DS:[SI+18]       ;PUSH Flags
        PUSH  WORD PTR DS:[SI+24]       ;PUSH DS
        PUSH  WORD PTR DS:[SI+2]        ;PUSH AX
        PUSH  WORD PTR DS:[SI+8]        ;PUSH DX
        PUSH  WORD PTR DS:[SI+14]       ;PUSH SI
        MOV   BX,DS:[SI+4]              ;get BX
        MOV   CX,DS:[SI+6]              ;get CX
        MOV   BP,DS:[SI+12]             ;get BP
        MOV   DI,DS:[SI+16]             ;get DI
        MOV   ES,DS:[SI+26]             ;get ES
        POP   SI
        POP   DX
        POP   AX
        POP   DS
        IRET

LOADS   ENDP
```

16–5 INTRODUCTION TO THE 80286

The 80286 microprocessor is an advanced version of the 8086 microprocessor that was designed for multiuser and multitasking environments. The 80286 addresses 16M bytes of physical memory and 1G byte of virtual memory by using its memory-management system. This section of the text introduces the 80286 microprocessor, which finds use in earlier AT-style personal computers that once pervaded the computer market and still find some applications. The 80286 is basically an 8086 that is optimized to execute instructions in fewer clocking periods than the 8086. The 80286 is also an enhanced version of the 8086 because it contains a memory manager. At this time, the 80286 no longer has a place in the personal computer system, but it does find applications in control systems as an embedded controller.

Hardware Features

Figure 16–29 shows the internal block diagram of the 80286 microprocessor. Note that like the 80186/80188, the 80286 does not incorporate internal peripherals; instead, it contains a memory-management unit (MMU) that is called the **address unit** in the block diagram.

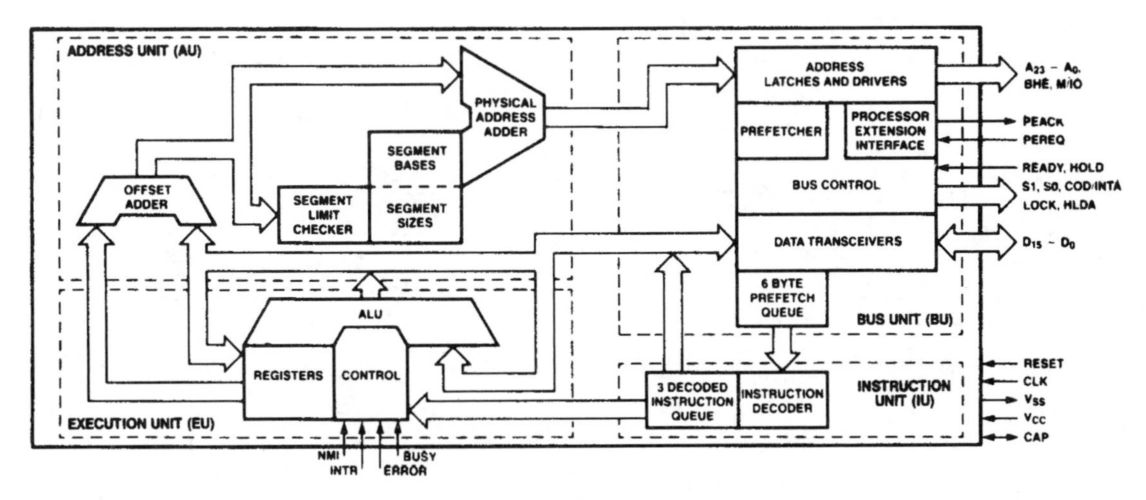

FIGURE 16–29 The block diagram of the 80286 microprocessor. (Courtesy of Intel Corporation.)

As a careful examination of the block diagram reveals, address pins A_{23}–A_0, $\overline{\text{BUSY}}$, CAP, $\overline{\text{ERROR}}$, $\overline{\text{PEREQ}}$, and $\overline{\text{PEACK}}$ are new or additional pins that do not appear on the 8086 microprocessor. The $\overline{\text{BUSY}}$, $\overline{\text{ERROR}}$, $\overline{\text{PEREQ}}$, and $\overline{\text{PEACK}}$ signals are used with the microprocessor extension or coprocessor, of which the 80287 is an example. (Note that the $\overline{\text{TEST}}$ pin is now referred to as the $\overline{\text{BUSY}}$ pin.) The address bus is now 24 bits wide to accommodate the 16M bytes of physical memory. The CAP pin is connected to a 0.047 μF, ±20% capacitor that acts as a 12 V filter and connects to ground. The pin-outs of the 8086 and 80286 are illustrated in Figure 16–30 for comparative purposes. Note that the 80286 does not contain a multiplexed address/data bus.

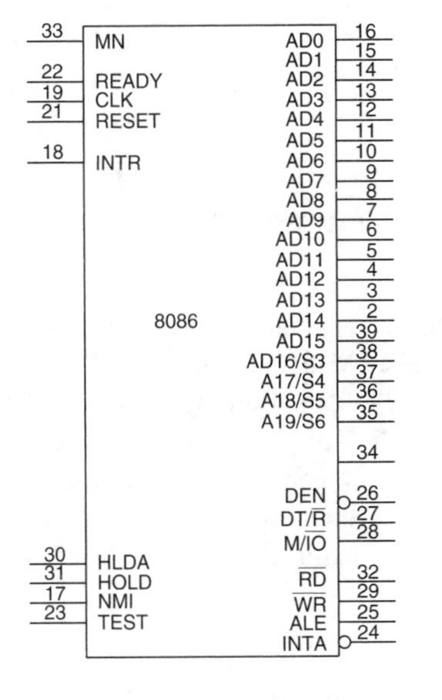

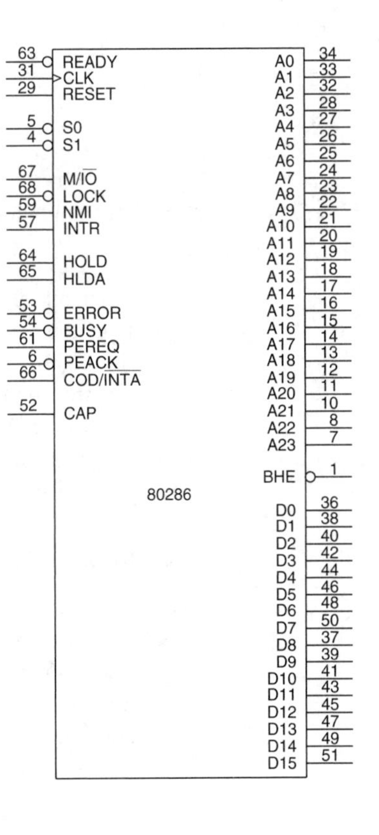

FIGURE 16–30 The 8086 and 80286 microprocessor pin-outs. Notice that the 80286 does not have a multiplexed address/data bus.

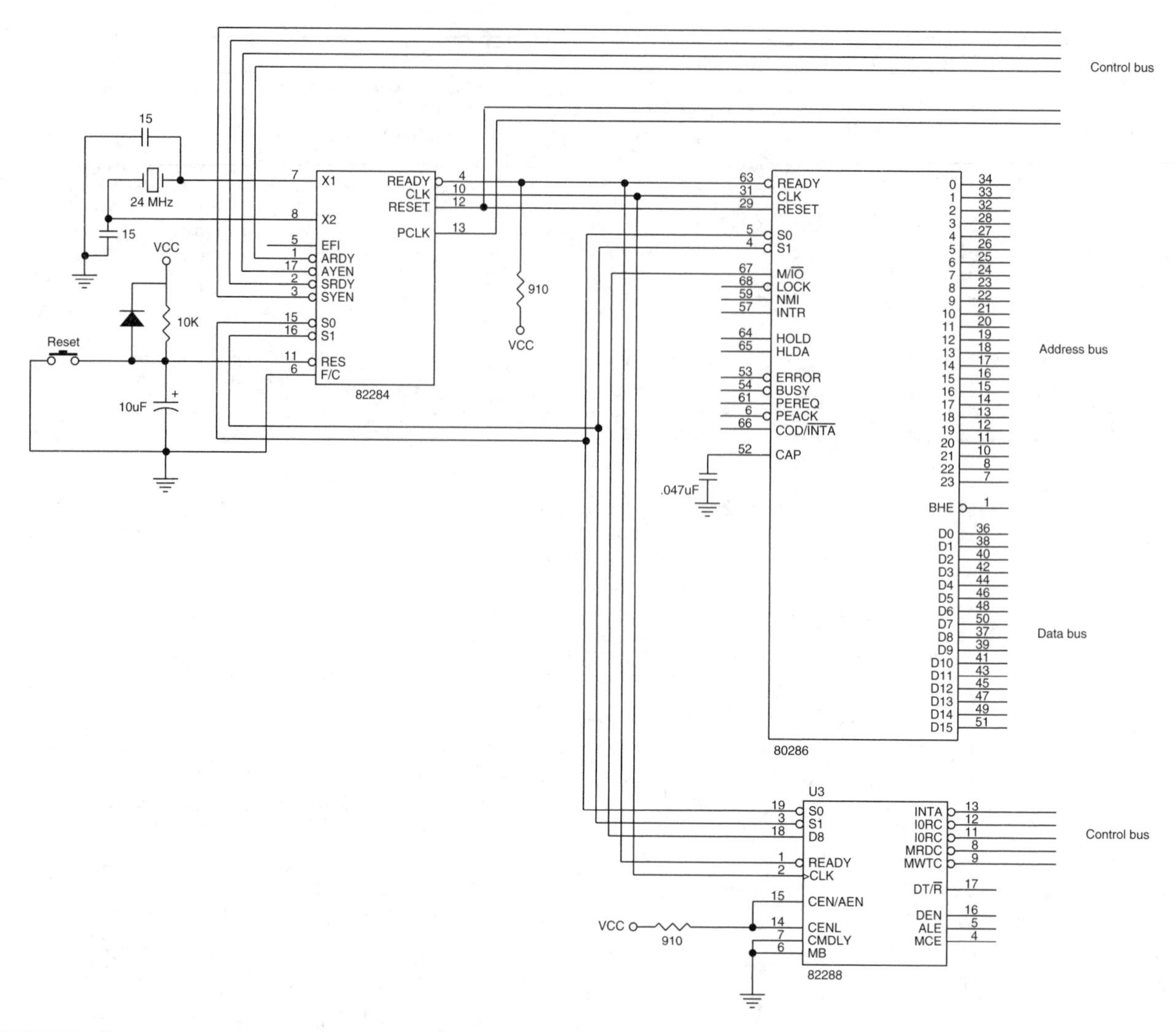

FIGURE 16–31 The interconnection of the 80286 microprocessor, 82284 clock generator, and 8288 system bus controller.

As mentioned in Chapter 1, the 80286 operates in both the real and protected modes. In the real mode, the 80286 addresses a 1M-byte memory address space and is virtually identical to the 8086. In the protected mode, the 80286 addresses a 16M-byte memory space.

Figure 16–31 illustrates the basic 80286 microprocessor-based system. Notice that the clock is provided by the 82284 clock generator (similar to the 8284A) and the system control signals are provided by the 82288 system bus controller (similar to the 8288). Also, note the absence of the latch circuits used to demultiplex the 8086 address/data bus.

Additional Instructions

The 80286 has even more instructions than its predecessors. These extra instructions control the virtual memory system through the memory manager of the 80286. Table 16–9 lists the additional 80286 instructions with a comment about the purpose of each instruction. These instructions are the only new instructions added to the 80286. Note that the 80286 contains the new instructions added to the 80186/80188 such as INS, OUTS, BOUND, ENTER, LEAVE, PUSHA, POP A, and the immediate multiplication and immediate shift and rotate counts.

TABLE 16–9 Additional 80286 instructions.

Instruction	*Purpose*
CLTS	Clear the task-switched bit
LDGT	Load global descriptor table register
SGDT	Store global descriptor table register
LIDT	Load interrupt descriptor table register
SIDT	Store interrupt descriptor table register
LLDT	Load local descriptor table register
SLDT	Store local descriptor table register
LMSW	Load machine status register
SMSW	Store machine status register
LAR	Load access rights
LSL	Load segment limit
SAR	Store access rights
ARPL	Adjust requested privilege level
VERR	Verify a read access
VERW	Verify a write access

Following are descriptions of instructions not explained in the memory-management section. The instructions described here are special and only used for the conditions indicated.

CLTS The **clear task-switched flag** (CLTS) instruction clears the TS (task-switched) flag bit to a logic 0. If the TS flag bit is a logic 1 and the 80287 numeric coprocessor is used by the task, an interrupt occurs (vector type 7). This allows the function of the coprocessor to be emulated with software. The CLTS instruction is used in a system and is considered a privileged instruction because it can be executed only in the protected mode at privilege level 0. There is no set TS flag instruction; this is accomplished by writing a logic 1 to bit position 3 (TS) of the machine status word (MSW) by using the LMSW instruction.

LAR The **load access rights** (LAR) instruction reads the segment descriptor and places a copy of the access rights byte into a 16-bit register. An example is the LAR AX,BX instruction that loads AX with the access rights byte from the descriptor selected by the selector value found in BX. This instruction is used to get the access rights so that they can be checked before a program uses the segment of memory described by the descriptor.

LSL The **load segment limit** (LSL) instruction loads a user-specified register with the segment limit. For example, the LSL AX,BX instruction loads AX with the limit of the segment described by the descriptor selected by the selector in BX. This instruction is used to test the limit of a segment.

ARPL The **adjust requested privilege level** (ARPL) instruction is used to test a selector so that the privilege level of the requested selector is not violated. An example is ARPL AX,CX: AX contains the requested privilege level and CX contains the selector value to be used to access a descriptor. If the requested privilege level is of a lower priority than the descriptor under test, the zero flag is set. This may require that a program adjust the requested privilege level or indicate a privilege violation.

VERR The **verify for read access** (VERR) instruction verifies that a segment can be read. Recall from Chapter 1 that a code segment can be read-protected. If the code segment can be read, the zero flag bit is set. The VERR AX instruction tests the descriptor selected by the AX register.

VERW The **verify for write access** (VERW) instruction is used to verify that a segment can be written. Recall from Chapter 1 that a data segment can be write-protected. If the data segment can be written, the zero flag bit is set.

The Virtual Memory Machine

A **virtual memory machine** is a machine that maps a larger memory space (1G byte for the 80286) into a much smaller physical memory space (16M bytes for the 80286), which allows a very large system to execute in smaller physical memory systems. This is accomplished by spooling the data and programs between the fixed disk memory system and the physical memory. Addressing a 1G-byte memory system is accomplished by the descriptors in the 80286 microprocessor. Each 80286 descriptor describes a 64K-byte memory segment and the 80286 allows 16K descriptors. This (64K × 16K) allows a maximum of 1G byte of memory to be described for the system.

As mentioned in Chapter 1, descriptors describe the memory segment in the protected mode. The 80286 has descriptors that define codes, data, stack segments, interrupts, procedures, and tasks. Descriptor accesses are performed by loading a segment register with a selector in the protected mode. The selector accesses a descriptor that describes an area of the memory. Additional details on descriptors and their applications are defined in Chapter 1, and also Chapters 17, 18, and 19. Please refer to these chapters for a detailed view of the protected mode memory management system.

16–6 SUMMARY

1. The 80186/80188 microprocessors contain the same basic instruction set as the 8086/8088 microprocessors, except that a few additional instructions are added. The 80186/80188 are thus enhanced versions of the 8086/8088 microprocessors. The new instructions include PUSHA, POPA, INS, OUTS, BOUND, ENTER, LEAVE, and immediate multiplication and shift/rotate counts.
2. Hardware enhancements to the 80186/80188 include a clock generator, a programmable interrupt controller, three programmable timers, a programmable DMA controller, a programmable chip selection logic unit, a watchdog timer, a dynamic RAM refresh logic circuit, and additional features on various versions.
3. The clock generator allows the 80186/80188 to operate from an external TTL-level clock source, or from a crystal attached to the X_1 (CLK_{IN}) and X_2 (OSCOUT) pins. The frequency of the crystal is twice the operating frequency of the microprocessor. The 80186/80188 microprocessors are available in speeds of 6–20 MHz.
4. The programmable interrupt controller arbitrates all internal and external interrupt requests. It is also capable of operating with two external 8259A interrupt controllers.
5. Three programmable timers are located within the 80186/80188. Each timer is a fully programmable, 16-bit counter used to generate wave-forms or count events. Two of the timers, timers 0 and 1, have external inputs and outputs. The third timer, timer 2, is clocked from the system clock and is used either to provide a clock for another timer or to request a DMA action.
6. The programmable DMA controller is a fully programmable, two-channel controller. DMA transfers are made between memory and I/O, I/O and I/O, or between memory locations. DMA requests occur from software, hardware, or the output of timer 2.
7. The programmable chip selection unit is an internal decoder that provides up to 13 output pins to select memory (6 pins) and I/O (7 pins). It also inserts 0–3 wait states, with or without external READY synchronization. On the EB and EC versions, the number of waits can be programmed from 0 to 15 with 10 chip selection pins.

8. The only difference between the timing of the 80186/80188 and the 8086/8088 is that ALE appears one-half clock pulse earlier. Otherwise, the timing is identical.
9. The 6 MHz version of the 80186/80188 allows 417 ns of access time for the memory; the 8 MHz version allows 309 ns of access time.
10. The intema180186/80188 peripherals are programmed via a peripheral control block (PCB), initialized at I/O ports FF00H–FFFFH. The PCB may be moved to any area of memory or I/O by changing the contents of the PCB relocation register at initial I/O locations FFFEH and FFFFH.
11. The 80286 is an 8086 that has been enhanced to include a memory-management unit (MMU). The 80286 is capable of addressing a 16M-byte physical memory space because of the management unit.
12. The 80286 contains the same instructions as the 80186/80188, except for a handful of additional instructions that control the memory-management unit.
13. Through the memory-management unit, the 80286 microprocessor addresses a virtual memory space of 1G byte, as specified by the 16K descriptors stored in two descriptor tables.

16–7 QUESTIONS AND PROBLEMS

1. List the differences between the 8086/8088 and the 80186/80188 microprocessors.
2. What hardware enhancements are added to the 80186/80188 that are not present in the 8086/8088?
3. The 80186/80188 is packaged in what types of integrated circuits?
4. If the 20 MHz crystal is connected to X_1 and X_2, what frequency signal is found at CLKOUT?
5. Describe the differences between the 80C188XL and the 80C188EB versions of the 80188 embedded controller.
6. The fan-out from any 80186/80188 pin is ________ for a logic 0.
7. How many clocking periods are found in an 80186/80188 bus cycle?
8. What is the main difference between the 8086/8088 and 80186/80188 timing?
9. What is the importance of memory access time?
10. How much memory access time is allowed by the 80186/80188 if operated with a 10 MHz clock?
11. Where is the peripheral control block located after the 80186/80188 are reset?
12. Write the software required to move the peripheral control block to memory locations 10000H–100FFH.
13. Which interrupt vector is used by the INT_0 pin on the 80186/80188 microprocessors?
14. How many interrupt vectors are available to the interrupt controller located within the 80186/80188 microprocessors?
15. Which two modes of operation are available to the interrupt controller?
16. What is the purpose of the interrupt control register?
17. Whenever an interrupt source is masked, the mask bit in the interrupt mask register is a logic ________.
18. What is the difference between the interrupt poll and interrupt poll status registers?
19. What is the purpose of the end-of-interrupt (EOI) register?
20. How many 16-bit timers are found within the 80186/80188?
21. Which timers have input and output pin connections?
22. Which timer connects to the system clock?
23. If two maximum-count compare registers are used with a timer, explain the operation of the timer.

24. What is the purpose of the INH timer control register bit?
25. What is the purpose of the P timer control register bit?
26. The timer control register bit ALT selects what type of operation for timers 0 and 1?
27. Explain how the timer output pins are used.
28. Develop a program that causes timer 1 to generate a continuous signal that is a logic 1 for 123 counts and a logic 0 for 23 counts.
29. Develop a program that causes timer 0 to generate a single pulse after 105 clock pulses on its input pin.
30. How many DMA channels are controlled by the DMA controller in the 80C186XL?
31. The DMA controller's source and destination registers are each _________- bits wide.
32. How is the DMA channel started with software?
33. The chip selection unit (XL and EA) has _________ pins to select memory devices.
34. The chip selection unit (XL and EA) has _________ pins to select peripheral devices.
35. The last location of the upper memory block, as selected by the $\overline{UCS}$ pin, is location _________.
36. The middle memory chip selection pins (XL and EA) are programmed for a(n) _________ size and a block size.
37. The lower memory area, as selected by $\overline{LCS}$, begins at address _________.
38. The internal wait state generator (EB and EC versions) is capable of inserting between zero and _________ wait states.
39. Program register A_8H (XL and EA) so that the mid-range memory block size is 128K bytes and a chip size of 32K.
40. What is the purpose of the EX bit in register A_8H?
41. Develop the software required to program the $\overline{GCS3}$ pin so that it selects memory from locations 20000H–2FFFFH and inserts two wait states.
42. Develop the software required to program the $\overline{GCS4}$ pin so that it selects an I/O device for ports 1000H–103FH and inserts one wait state.
43. The 80286 microprocessor addresses _________ bytes of physical memory.
44. When the memory manager is in use, the 80286 addresses _________ bytes of virtual memory.
45. The instruction set of the 80286 is identical to the _________, except for the memory-management instructions.
46. What is the purpose of the VERR instruction?
47. What is the purpose of the LSL instruction?
48. What is an RTOS?
49. How are multiple threads handled with the RTOS?
50. Search the Internet for at least two different RTOS and write a short report comparing them.

CHAPTER 17

The 80386 and 80486 Microprocessors

INTRODUCTION

The 80386 microprocessor is a full 32-bit version of the earlier 8086/80286 16-bit microprocessors, and represents a major advancement in the architecture—a switch from 16-bit architecture to 32-bit architecture. Along with this larger word size are many improvements and additional features. The 80386 microprocessor features multitasking, memory management, virtual memory (with or without paging), software protection, and a large memory system. All software written for the early 8086/8088 and the 80286 are upward-compatible to the 80386 microprocessor. The amount of memory addressable by the 80386 is increased from the 1M byte found in the 8086/8088 and the 16M bytes found in the 80286, to 4G bytes in the 80386. The 80386 can switch between protected mode and real mode without resetting the microprocessor. Switching from protected mode to real mode was a problem on the 80286 microprocessor because it required a hardware reset.

The 80486 microprocessor is an enhanced version of the 80386 microprocessor that executes many of its instructions in one clocking period. The 80486 microprocessor also contains an 8K-byte cache memory and an improved 80387 numeric coprocessor. (Note that the 80486DX4 contains a 16K-byte cache.) When the 80486 is operated at the same clock frequency as an 80386, it performs with an improvement in speed of about 50%. Chapter 18 details the Pentium and Pentium Pro. These microprocessors both contain a 16K cache memory and perform at better than twice the speed of the 80486 microprocessor. The Pentium and Pentium Pro also contain improved numeric coprocessors that operate five times faster than the 80486 numeric coprocessor. Chapter 19 deals with additional improvements in the Pentium II–Pentium 4 microprocessors.

CHAPTER OBJECTIVES

Upon completion of this chapter, you will be able to:

1. Contrast the 80386 and 80486 microprocessors with earlier Intel microprocessors
2. Describe the operation of the 80386 and 80486 memory management unit and paging unit
3. Switch between protected mode and real mode
4. Define the operation of additional 80386/80486 instructions and addressing modes
5. Explain the operation of a cache memory system
6. Detail the interrupt structure and direct memory access structure of the 80386/80486
7. Contrast the 80486 with the 80386 microprocessor
8. Explain the operation of the 80486 cache memory

17–1 INTRODUCTION TO THE 80386 MICROPROCESSOR

Before the 80386 or any other microprocessor can be used in a system, the function of each pin must be understood. This section of the chapter details the operation of each pin, along with the external memory system and I/O structures of the 80386.

Figure 17–1 illustrates the pin-out of the 80386DX microprocessor. The 80386DX is packaged in a 132-pin PGA (pin grid array). Two versions of the 80386 are commonly available. One is the 80386DX, which is illustrated and described in this chapter; the other is the 80386SX, which is a reduced bus version of the 80386. A new version of the 80386—the 80386EX—incorporates the AT bus system, dynamic RAM controller, programmable chip selection logic, 26 address pins, 16 data pins, and 24 I/O pins. Figure 17–2 illustrates the 80386EX embedded PC.

The 80386DX addresses 4G bytes of memory through its 32-bit data bus and 32-bit address. The 80386SX, more like the 80286, addresses 16M bytes of memory with its 24-bit address bus via its 16-bit data bus. The 80386SX was developed after the 80386DX for applications that didn't require the full 32-bit bus version. The 80386SX was found in many early personal computers that used the same basic motherboard design as the 80286. At the time that the 80386SX was popular, most applications, including Windows 3.11, required fewer than 16M bytes of memory, so the 80386SX is a popular and a less costly version of the 80386 microprocessor. Even though the 80486 has become a less expensive upgrade path for newer systems, the 80386 still can be used for many applications. For example, the 80386EX does not appear in computer systems, but it is becoming very popular in embedded applications.

As with earlier versions of the Intel family of microprocessors, the 80386 requires a single +5.0 V power supply for operation. The power supply current averages 550 mA for the 25 MHz version of the 80386, 500 mA for the 20 MHz version, and 450 mA for the 16 MHz version. Also available is a 33 MHz version that requires 600 mA of power supply current. The power supply current for the 80386EX is 320 mA when operated at 33 MHz. Note that during some modes of

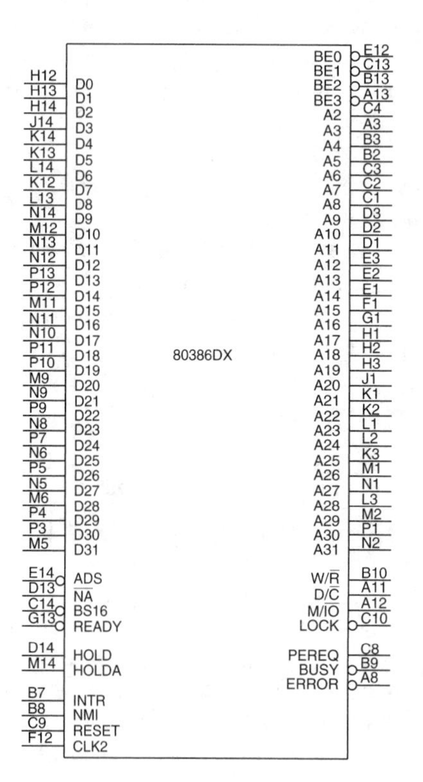

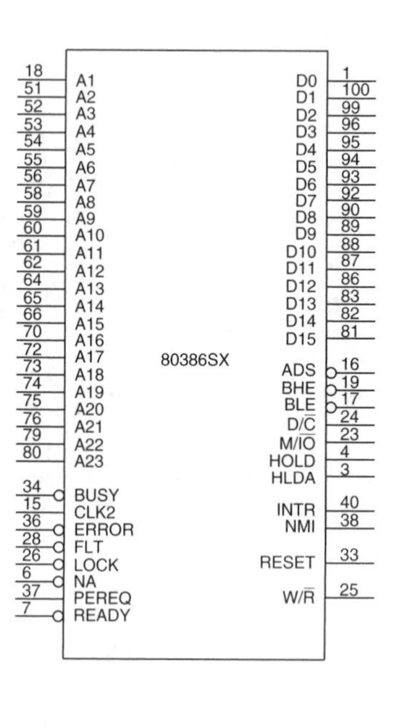

FIGURE 17–1 The pin-outs of the 80386DX and 80386SX microprocessors.

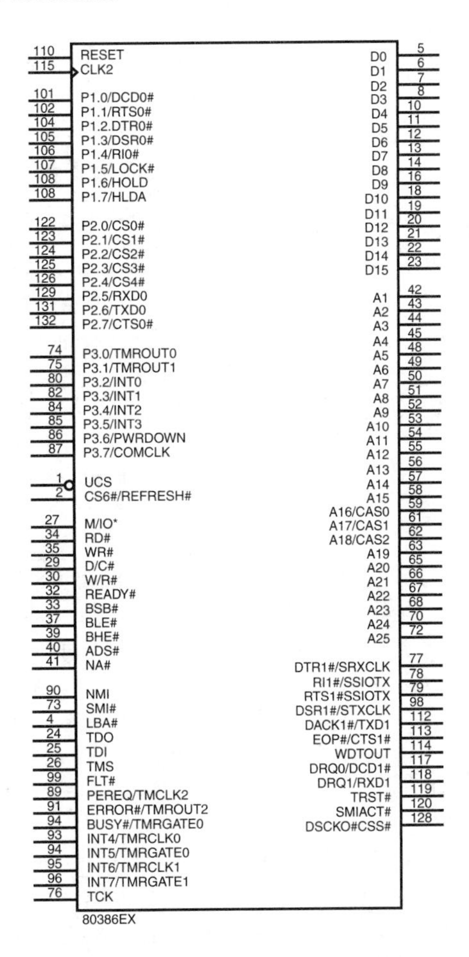

FIGURE 17–2 The 80386EX embedded PC.

normal operation, power supply current can surge to over 1.0 A. This means that the power supply and power distribution network must be capable of supplying these current surges. This device contains multiple V_{CC} and V_{SS} connections that must all be connected to +5.0 V and grounded for proper operation. Some of the pins are labeled N/C (no connection) and must not be connected. Additional versions of the 80386SX and 80386EX are available with a +3.3 V power supply. They are often found in portable notebook or laptop computers and are usually packaged in a surface mount device.

Each 80386 output pin is capable of providing 4.0 mA (address and data connections) or 5.0 mA (other connections). This represents an increase in drive current compared to the 2.0 mA available on earlier 8086, 8088, and 80286 output pins. The output current available on most 80386EX output pins is 8.0 mA. Each input pin represents a small load, requiring only ±10 μA of current. In some systems, except the smallest, these current levels require bus buffers.

The function of each 80386DX group of pins follows:

A_{31}–A_2 **Address bus** connections address any of the 1G × 32 (4G bytes) memory locations found in the 80386 memory system. Note that A_0 and A_1 are encoded in the bus enable ($\overline{BE3}$–$\overline{BE0}$) to select any or all of the four bytes in a 32-bit-wide memory location. Also note that because the 80386SX contains a 16-bit data bus in place of the 32-bit data bus found on the 80386DX, A_1 is present on the 80386SX, and the bank selection signals are replaced with $\overline{BHE}$ and $\overline{BLE}$. The $\overline{BHE}$ signal enables the upper data bus half; the $\overline{BLE}$ signal enables the lower data bus half.

D_{31}–D_0 — **Data bus** connections transfer data between the microprocessor and its memory and I/O system. Note that the 80386SX contains D_{15}–D_0.

$\overline{\text{BE3}}$–$\overline{\text{BE0}}$ — **Bank enable** signals select the access of a byte, word, or doubleword of data. These signals are generated internally by the microprocessor from address bits A_1 and A_0. On the 80386SX, these pins are replaced by $\overline{\text{BHE}}$, $\overline{\text{BLE}}$, and A_1.

M/$\overline{\text{IO}}$ — **Memory/IO** selects a memory device when a logic 1 or an I/O device when a logic 0. During the I/O operation, the address bus contains a 16-bit I/O address on address connections A_{15}–A_2.

W/$\overline{\text{R}}$ — **Write/$\overline{\text{Read}}$** indicates that the current bus cycle is a write when a logic 1 or a read when a logic 0.

$\overline{\text{ADS}}$ — The **address data strobe** becomes active whenever the 80386 has issued a valid memory or I/O address. This signal is combined with the W/$\overline{\text{R}}$ signal to generate the separate read and write signals present in the earlier 8086–80286 microprocessor-based systems.

RESET — **Reset** initializes the 80386, causing it to begin executing software at memory location FFFFFFF0H. The 80386 is reset to the real mode, and the leftmost 12 address connections remain logic 1s (FFFH) until a far jump or far call is executed. This allows compatibility with earlier microprocessors.

CLK_2 — **Clock times 2** is driven by a clock signal that is twice the operating frequency of the 80386. For example, to operate the 80386 at 16 MHz, apply a 32 MHz clock to this pin.

$\overline{\text{READY}}$ — **Ready** controls the number of wait states inserted into the timing to lengthen memory accesses.

$\overline{\text{LOCK}}$ — **Lock** becomes a logic 0 whenever an instruction is prefixed with the LOCK: prefix. This is used most often during DMA accesses.

D/$\overline{\text{C}}$ — **Data/control** indicates that the data bus contains data for or from memory or I/O when a logic 1. If D/$\overline{\text{C}}$ is a logic 0, the microprocessor is halted or executes an interrupt acknowledge.

$\overline{\text{BS16}}$ — **Bus size 16** selects either a 32-bit data bus ($\overline{\text{BS16}}$ = 1) or a 16-bit data bus ($\overline{\text{BS16}}$ = 0). In most cases, if an 80386DX is operated on a 16-bit data bus, we use the 80386SX that has a 16-bit data bus. On the 80386EX, the $\overline{\text{BS8}}$ pin selects an 8-bit data bus.

$\overline{\text{NA}}$ — **Next address** causes the 80386 to output the address of the next instruction or data in the current bus cycle. This pin is often used for pipelining the address.

HOLD — **Hold** requests a DMA action.

HLDA — **Hold acknowledge** indicates that the 80386 is currently in a hold condition.

$\overline{\text{PEREQ}}$ — The **coprocessor request** asks the 80386 to relinquish control and is a direct connection to the 80387 arithmetic coprocessor.

$\overline{\text{BUSY}}$ — **Busy** is an input used by the WAIT or FWAIT instruction that waits for the coprocessor to become available. This is also a direct connection to the 80387 from the 80386.

$\overline{\text{ERROR}}$ — **Error** indicates to the microprocessor that an error is detected by the coprocessor.

INTR — An **interrupt request** is used by external circuitry to request an interrupt.

NMI — A **non-maskable interrupt** requests a non-maskable interrupt as it did on the earlier versions of the microprocessor.

The Memory System

The physical memory system of the 80386DX is 4G bytes in size and is addressed as such. If virtual addressing is used, 64T bytes are mapped into the 4G bytes of physical space by the memory management unit and descriptors. (Note that virtual addressing allows a program to be larger than 4G bytes if a method of swapping with a large hard disk drive exists.) Figure 17–3 shows the organization of the 80386DX physical memory system.

The memory is divided into four 8-bit-wide memory banks, each containing up to 1G byte of memory. This 32-bit-wide memory organization allows bytes, words, or doublewords of memory data to be accessed directly. The 80386DX transfers up to a 32-bit-wide number in a single memory cycle, whereas the early 8088 requires four cycles to accomplish the same transfer, and the 80286 and 80386SX require two cycles. Today, the data width is important, especially with single-precision floating-point numbers that are 32 bits wide. High-level software normally uses floating-point numbers for data storage, so 32-bit memory locations speed the execution of high-level software when it is written to take advantage of this wider memory.

Each memory byte is numbered in hexadecimal as they were in prior versions of the family. The difference is that the 80386DX uses a 32-bit-wide memory address, with memory bytes numbered from location 00000000H to FFFFFFFFH.

The two memory banks in the 8086, 80286, and 80386SX system are accessed via $\overline{BLE}$ (A_0 on the 8086 and 80286) and $\overline{BHE}$. In the 80386DX, the memory banks are accessed via four bank enable signals, $\overline{BE3}$–$\overline{BE0}$. This arrangement allows a single byte to be accessed when one bank enable signal is activated by the microprocessor. It also allows a word to be addressed when two bank enable signals are activated. In most cases, a word is addressed in banks 0 and 1, or in banks 2 and 3. Memory location 00000000H is in bank 0, location 00000001H is in bank 1, location 00000002H is in bank 2, and location 00000003H is in bank 3. The 80386DX does not contain address connections A_0 and A_1 because these have been encoded as the bank enable signals. Likewise, the 80386SX does not contain the A_0 address pin because it is encoded in the $\overline{BLE}$ and $\overline{BHE}$ signals. The 80386EX addresses data either in two banks for a 16-bit-wide memory system if $\overline{BS8}$ = 1 or as an 8-bit system if $\overline{BS8}$ = 0.

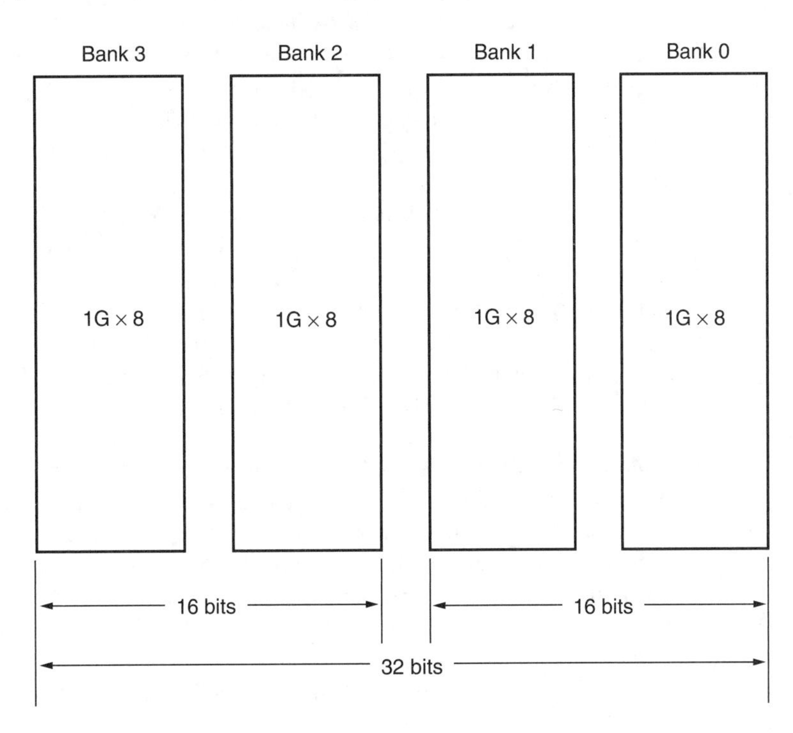

FIGURE 17–3 The memory system for the 80386 microprocessor. Notice that the memory is organized as four banks, each containing 1G byte. Memory is accessed as 8-, 16-, or 32-bit data.

Buffered System. Figure 17–4 shows the 80386DX connected to buffers that increase fan-out from its address, data, and control connections. This microprocessor is operated at 25 MHz using a 50 MHz clock input signal that is generated by an integrated oscillator module. Oscillator modules are often used to provide a clock in modern microprocessor-based equipment. The HLDA signal is used to enable all buffers in a system that uses direct memory access. Otherwise, the buffer enable pins are connected to ground in a non-DMA system.

Pipelines and Caches. The cache memory is a buffer that allows the 80386 to function more efficiently with lower DRAM speeds. A **pipeline** is a special way of handling memory accesses so the memory has additional time to access data. A_{16} MHz 80386 allows memory devices with access times of 50 ns or less to operate at full speed. Obviously, there are few DRAMs currently available with these access times. In fact, the fastest DRAMs currently in use have an access time of 40 ns or longer. This means that some technique must be found to interface these memory devices, which are slower than required by the microprocessor. Three techniques are available: interleaved memory, caching, and a pipeline.

The pipeline is the preferred means of interfacing memory because the 80386 microprocessor supports pipelined memory accesses. Pipelining allows memory an extra clocking period to access data. The extra clock extends the access time from 50 ns to 81 ns on an 80386 operating with a 16 MHz clock. The **pipe**, as it is often called, is set up by the microprocessor. When an instruction is fetched from memory, the microprocessor often has extra time before the next instruction is fetched. During this extra time, the address of the next instruction is sent out from the address bus ahead of time. This extra time (one clock period) is used to allow additional access time to slower memory components.

Not all memory references can take advantage of the pipe, which means that some memory cycles are not pipelined. These nonpipelined memory cycles request one wait state if the normal pipeline cycle requires no wait states. Overall, a pipe is a cost-saving feature that reduces the access time required by the memory system in low-speed systems.

Not all systems can take advantage of the pipe. Those systems typically operate at 20, 25, or 33 MHz. In these higher-speed systems, another technique must be used to increase the memory system speed. The **cache** memory system improves overall performance of the memory systems for data that are accessed more than once. Note that the 80486 contains an internal cache called a **level 1 cache** and the 80386 can only contain an external cache called a **level 2 cache**.

A **cache** is a high-speed memory (SRAM) system that is placed between the microprocessor and the DRAM memory system. Cache memory devices are usually static RAM memory components with access times of less than 10 ns. In many cases, we see level 2 cache memory systems with sizes between 32K and 1M byte. The size of the cache memory is determined more by the application than by the microprocessor. If a program is small and refers to little memory data, a small cache is beneficial. If a program is large and references large blocks of memory, the largest cache size possible is recommended. In many cases, a 64K-byte cache improves speed sufficiently, but the maximum benefit is often derived from a 256K-byte cache. It has been found that increasing the cache size much beyond 256K provides little benefit to the operating speed of the system that contains an 80386 microprocessor.

Interleaved Memory Systems. An **interleaved memory system** is another method of improving the speed of a system. Its only disadvantage is that it costs considerably more memory because of its structure. Interleaved memory systems are present in some systems, so memory access times can be lengthened without the need for wait states. In some systems, an interleaved memory may still require wait states, but may reduce their number. An interleaved memory system requires two or more complete sets of address buses and a controller that provides addresses for each bus. Systems that employ two complete buses are called a **two-way interleave**; systems that use four complete buses are called a **four-way interleave**.

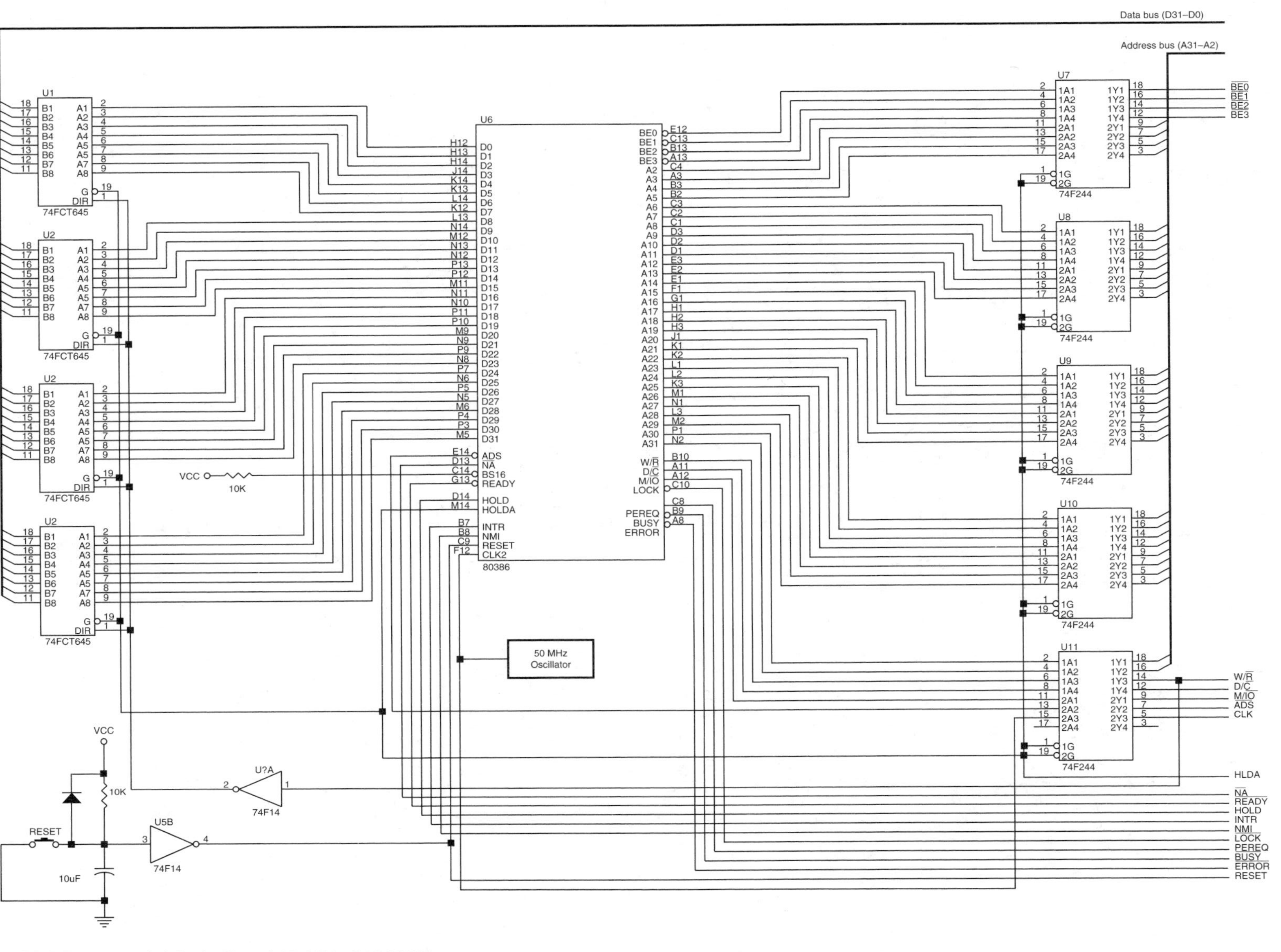

FIGURE 17–4 A fully buffered 25 MHz 80386DX.

An interleaved memory is divided into two or four parts. For example, if an interleaved memory system is developed for the 80386SX microprocessor, one section contains the 16-bit addresses 000000H–000001H, 000004H–000005H, and so on; the other section contains addresses 000002–000003, 000006H–000007H, and so forth. While the microprocessor accesses locations 000000H–000001H, the interleave control logic generates the address strobe signal for locations 000002H–000003H. This selects and accesses the word at location 000002H–000003H, while the microprocessor processes the word at location 000000H–000001H. This process alternates memory sections, thus increasing the performance of the memory system.

Interleaving increases the amount of access time provided to the memory because the address is generated to select the memory before the microprocessor accesses it. This is because the microprocessor pipelines memory addresses, sending the next address out before the data are read from the last address.

The problem with interleaving, although not major, is that the memory addresses must be accessed so that each section is alternately addressed. This does not always happen as a program executes. Under normal program execution, the microprocessor alternately addresses memory approximately 93% of the time. For the remaining 7%, the microprocessor addresses data in the same memory section, which means that in these 7% of the memory accesses, the memory system must cause wait states because of the reduced access time. The access time is reduced because the memory must wait until the previous data are transferred before it can obtain its address. This leaves the memory with less access time; therefore, a wait state is required for accesses in the same memory bank.

See Figure 17–5 for the timing diagram of the address as it appears at the microprocessor address pins. This timing diagram shows how the next address is output before the current data are accessed. It also shows how access time is increased by using interleaved memory addresses for each section of memory compared to a noninterleaved access, which requires a wait state.

Figure 17–6 pictures the interleave controller. Admittedly, this is a complex logic circuit, which needs some explanation. First, if the SEL input (used to select this section of the memory) is inactive (logic 0), then the $\overline{\text{WAIT}}$ signal is a logic 1. Also, both ALE0 and ALE1, used to strobe the address to the memory sections, are both logic 1s, causing the latches connected to them to become transparent.

As soon as the SEL input becomes a logic 1, this circuit begins to function. The A1 input is used to determine which latch (U2B or U5A) becomes a logic 0, selecting a section of the memory. Also the ALE pin that becomes a logic 0 is compared with the previous state of the ALE pins. If the same section of memory is accessed a second time, the $\overline{\text{WAIT}}$ signal becomes a logic 0, requesting a wait state.

Figure 17–7 illustrates an interleaved memory system that uses the circuit of Figure 17–6. Notice how the ALE_0 and ALE_1 signals are used to capture the address for either section of memory. The memory in each bank is 16 bits wide. If accesses to memory require 8-bit data, the system causes wait states in most cases. As a program executes, the 80386SX fetches instructions 16 bits at a time from normally sequential memory locations. Program execution uses interleaving in most cases. If a system is going to access mostly eight-bit data, it is doubtful that memory interleaving will reduce the number of wait states.

The access time allowed by an interleaved system, such as the one shown in Figure 17–7, is increased to 112 ns from 69 ns by using a 16 MHz system clock. (If a wait state is inserted, access time with a 16 MHz clock is 136 ns, which means that an interleaved system performs at about the same rate as a system with one wait state.) If the clock is increased to 20 MHz, the interleaved memory requires 89.6 ns, whereas standard, noninterleaved memory interfaces allow 48 ns for memory access. At this higher clock rate, 80 ns DRAMs function properly without wait states when the memory addresses are interleaved. If an access to the same section occurs, a wait state is inserted.

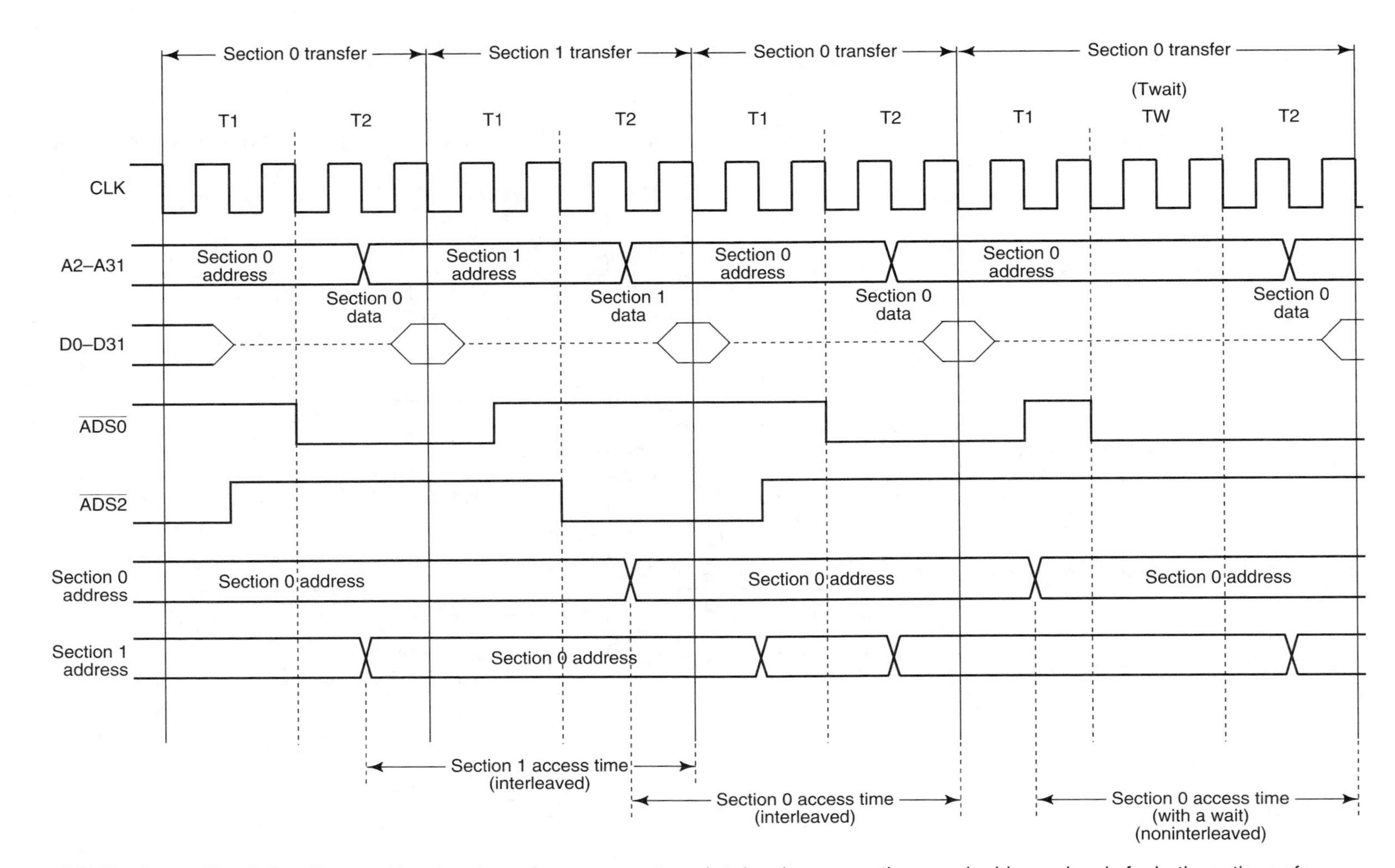

FIGURE 17–5 The timing diagram of an interleaved memory system showing the access times and address signals for both sections of memory.

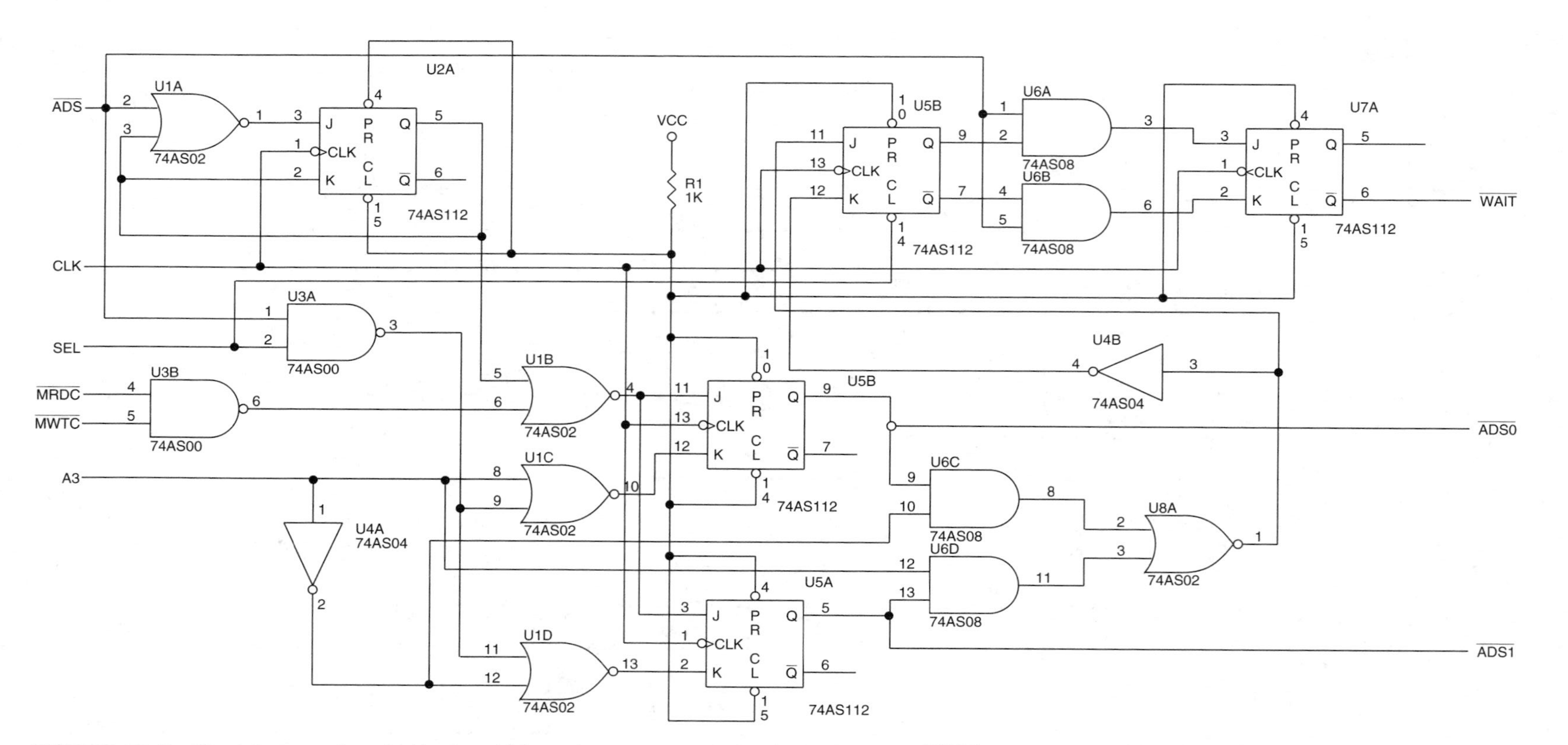

FIGURE 17–6 The interleaved control logic, which generates separate ADS signals and a $\overline{\text{WAIT}}$ signal used to control interleaved memory.

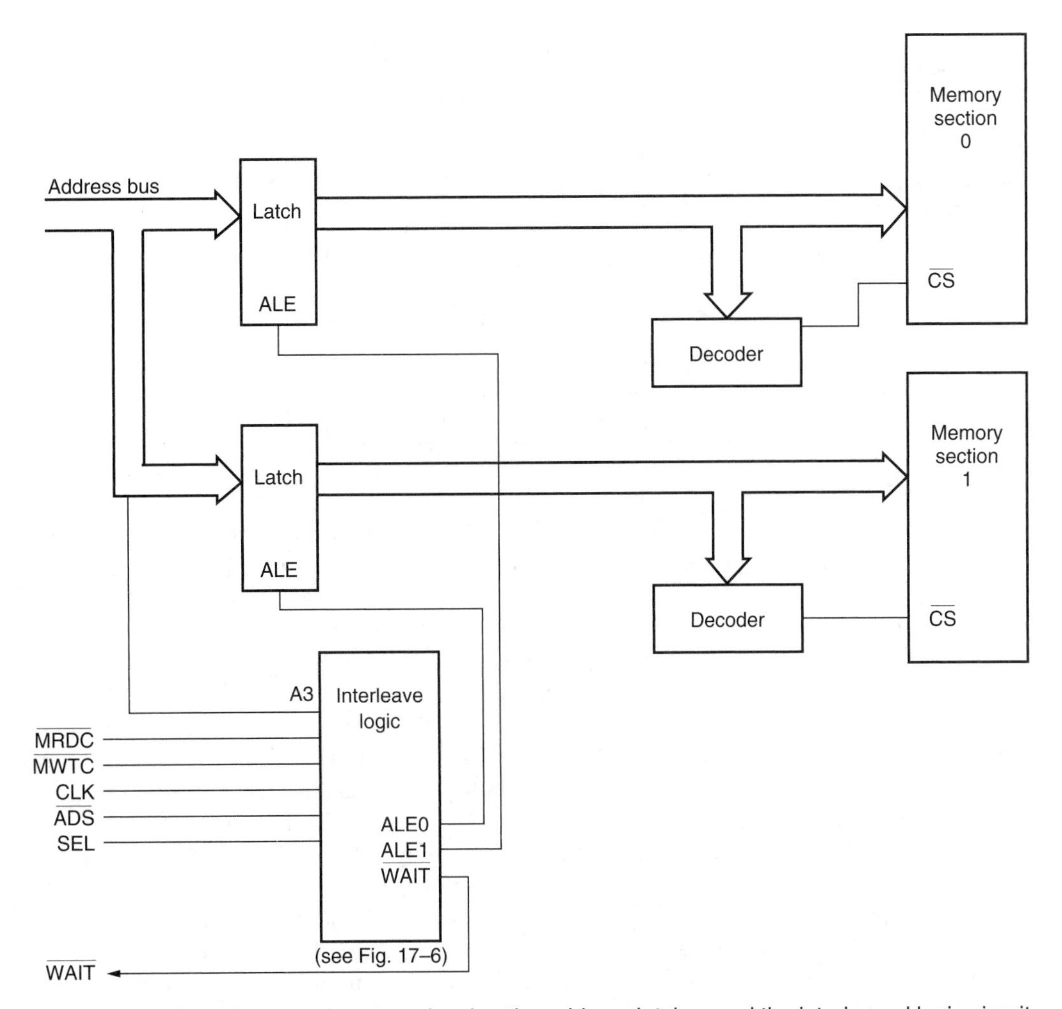

FIGURE 17–7 An interleaved memory system showing the address latches and the interleaved logic circuit.

The Input/Output System

The 80386 input/output system is the same as that found in any Intel 8086 family microprocessor-based system. There are 64K different bytes of I/O space available if isolated I/O is implemented. With isolated I/O, the IN and OUT instructions are used to transfer I/O data between the microprocessor and I/O devices. The I/O port address appears on address bus connections A_{15}–A_2, with $\overline{BE3}$–$\overline{BE0}$ used to select a byte, word, or doubleword of I/O data. If memory-mapped I/O is implemented, then the number of I/O locations can be any amount up to 4G bytes. With memory-mapped I/O, any instruction that transfers data between the microprocessor and memory system can be used for I/O transfers because the I/O device is treated as a memory device. Almost all 80386 systems use isolated I/O because of the I/O protection scheme provided by the 80386 in protected mode operation.

Figure 17–8 shows the I/O map for the 80386 microprocessor. Unlike the I/O map of earlier Intel microprocessors, which were 16 bits wide, the 80386 uses a full 32-bit-wide I/O system divided into four banks. This is identical to the memory system, which is also divided into four banks. Most I/O transfers are 8 bits wide because we often use ASCII code (a seven-bit code) for transferring alphanumeric data between the microprocessor and printers and keyboards. This may change if Unicode, a 16-bit alphanumeric code, becomes common and replaces ASCII code. Recently, I/O devices that are 16 and even 32 bits wide have appeared for systems such as disk

FIGURE 17–8 The isolated I/O map for the 80386 microprocessor. Here four banks of 8 bits each are used to address 64K different I/O locations. I/O is numbered from location 0000H to FFFFH.

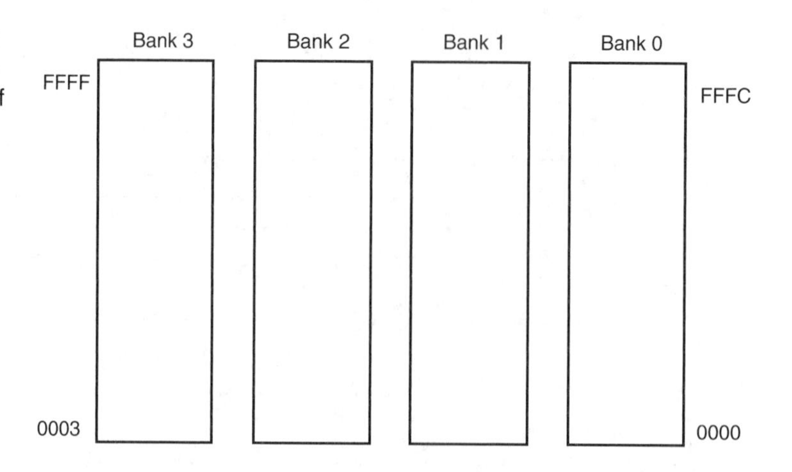

memory and video display interfaces. These wider I/O paths increase the data transfer rate between the microprocessor and the I/O device when compared to 8-bit transfers.

The I/O locations are numbered from 0000H to FFFFH. A portion of the I/O map is designated for the 80387 arithmetic coprocessor. Although the port numbers for the coprocessor are well above the normal I/O map, it is important that they be taken into account when decoding I/O space (overlaps). The coprocessor uses I/O locations 800000F8H–800000FFH for communications between the 80387 and 80386. The 80287 numeric coprocessor designed to use with the 80286 uses the I/O addresses 00F8H–00FFH for coprocessor communications. Because we often decode only address connections A_{15}–A_2 to select an I/O device, be aware that the coprocessor will activate devices 00F8H–00FFH unless address line A_{31} is also decoded. This should present no problem because you really should not be using I/O ports 00F8H–00FFH for any purpose.

The only new feature that was added to the 80386 with respect to I/O is the I/O privilege information added to the tail end of the TSS when the 80386 is operated in protected mode. As described in the section on memory management, an I/O location can be blocked or inhibited in the protected mode. If the blocked I/O location is addressed, an interrupt (type 13, general fault) is generated. This scheme is added so that I/O access can be prohibited in a multiuser environment. Blocking is an extension of the protected mode operation, as are privilege levels.

Memory and I/O Control Signals

The memory and I/O are controlled with separate signals. The $M/\overline{IO}$ signal indicates whether the data transfer is between the microprocessor and the memory ($M/\overline{IO} = 1$) or I/O ($M/\overline{IO} = 0$). In addition to $M/\overline{IO}$, the memory and I/O systems must read or write data. The $W/\overline{R}$ signal is a logic 0 for a read operation and a logic 1 for a write operation. The $\overline{ADS}$ signal is used to qualify the $M/\overline{IO}$ and $W/\overline{R}$ control signals. This is a slight deviation from earlier Intel microprocessors, which didn't use $\overline{ADS}$ for qualification.

See Figure 17–9 for a simple circuit that generates four control signals for the memory and I/O devices in the system. Notice that two control signals are developed for memory control ($\overline{MRDC}$ and $\overline{MWTC}$) and two for I/O control ($\overline{IORC}$ and $\overline{IOWC}$). These signals are consistent with the memory and I/O control signals generated for use in earlier versions of the Intel microprocessor.

Timing

Timing is important for understanding how to interface memory and I/O to the 80386 microprocessor. Figure 17–10 shows the timing diagram of a nonpipelined memory read cycle. Note that the timing is referenced to the CLK_2 input signal and that a bus cycle consists of four clocking periods.

Each bus cycle contains two clocking states with each state (T_1 and T_2) containing two clocking periods. Note in Figure 17–10 that the access time is listed as time number 3. The 16 MHz

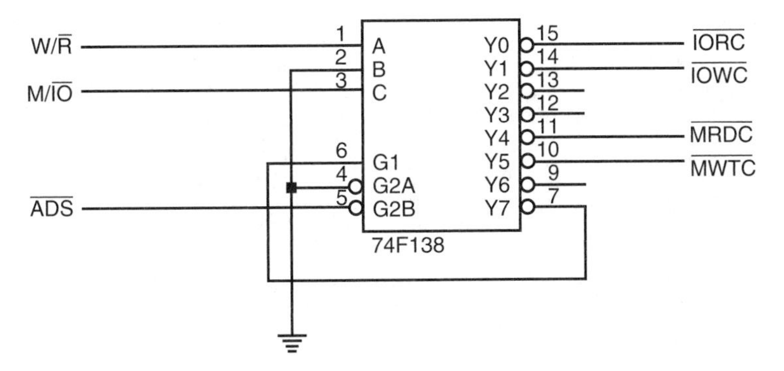

FIGURE 17–9 Generation of memory and I/O control signals for the 80386, 80486, and Pentium.

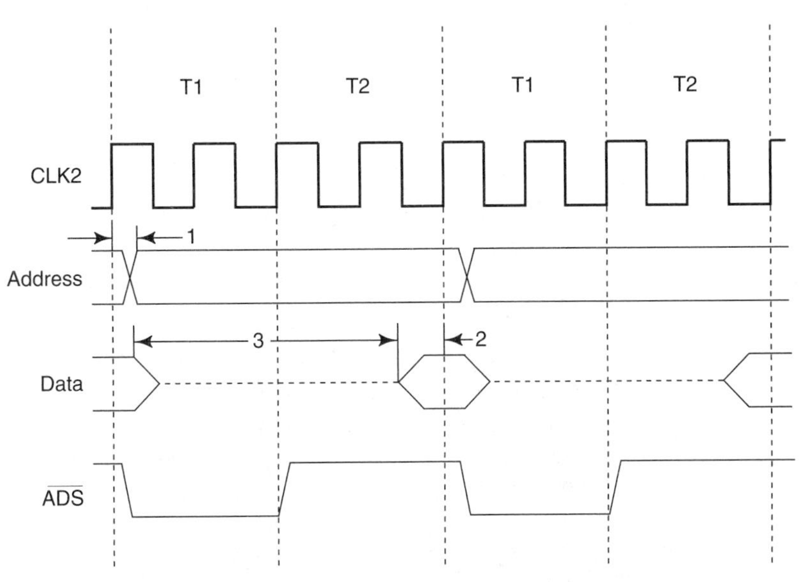

FIGURE 17–10 The nonpipelined read timing for the 80386 microprocessor.

	33 MHz	25 MHz	20 MHz	16 MHz
Time 1:	4–15 ns	4–21 ns	4–30 ns	4–36 ns
Time 2:	5 ns	7 ns	11 ns	11 ns
Time 3:	46 ns	52 ns	59 ns	78 ns

version allows memory an access time of 78 ns before wait states are inserted in this nonpipelined mode of operation. To select the nonpipelined mode, we place a logic 1 on the $\overline{\text{NA}}$ pin.

Figure 17–11 illustrates the read timing when the 80386 is operated in the pipelined mode. Notice that additional time is allowed to the memory for accessing data because the address is sent out early. Pipelined mode is selected by placing a logic 0 on the $\overline{\text{NA}}$ pin and by using address latches to capture the pipelined address. The clock pulse that is applied to the address latches comes from the $\overline{\text{ADS}}$ signal. Address latches must be used with a pipelined system, as well as with interleaved memory banks. The minimum number of interleaved banks of two and four have been successfully used in some applications.

Notice that the pipelined address appears one complete clocking state before it normally appears with nonpipelined addressing. In the 16 MHz version of the 80386, this allows an additional 62.5 ns for memory access. In a nonpipelined system, a memory access time of 78 ns is allowed to the memory system; in a pipelined system, 140.5 ns is allowed. The advantages of the pipelined system are that no wait states are required (in many, but not all bus cycles) and much lower-speed memory devices may be connected to the microprocessor. The disadvantage is that we need to interleave memory to use a pipe, which requires additional circuitry and occasional wait states.

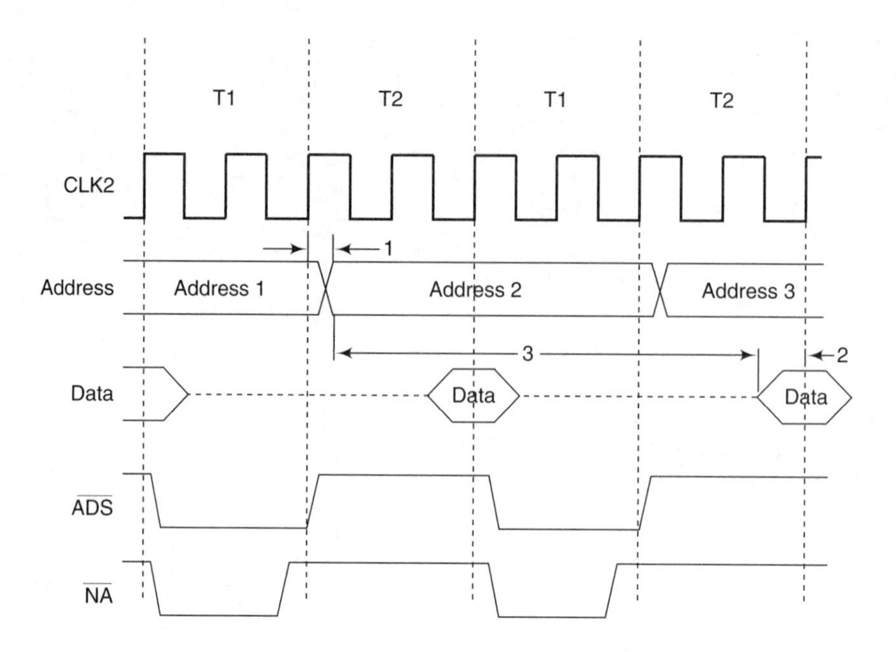

FIGURE 17–11 The pipelined read timing for the 80386 microprocessor.

Wait States

Wait states are needed if memory access times are long compared with the time allowed by the 80386 for memory access. In a nonpipelined 33 MHz system, memory access time is only 46 ns. Currently, only a few DRAM memories exist that have an access time of 46 ns. This means that often wait states must be introduced to access the DRAM (one wait for 60 ns DRAM) or an EPROM that has an access time of 100 ns (two waits). Note that this wait state is built into a motherboard and cannot be removed.

The $\overline{\text{READY}}$ input controls whether or not wait states are inserted into the timing. The $\overline{\text{READY}}$ input on the 80386 is a dynamic input that must be activated during each bus cycle. Figure 17–12 shows a few bus cycles with one normal (no wait) cycle and one that contains a single wait state. Notice how the $\overline{\text{READY}}$ is controlled to cause 0 or 1 wait.

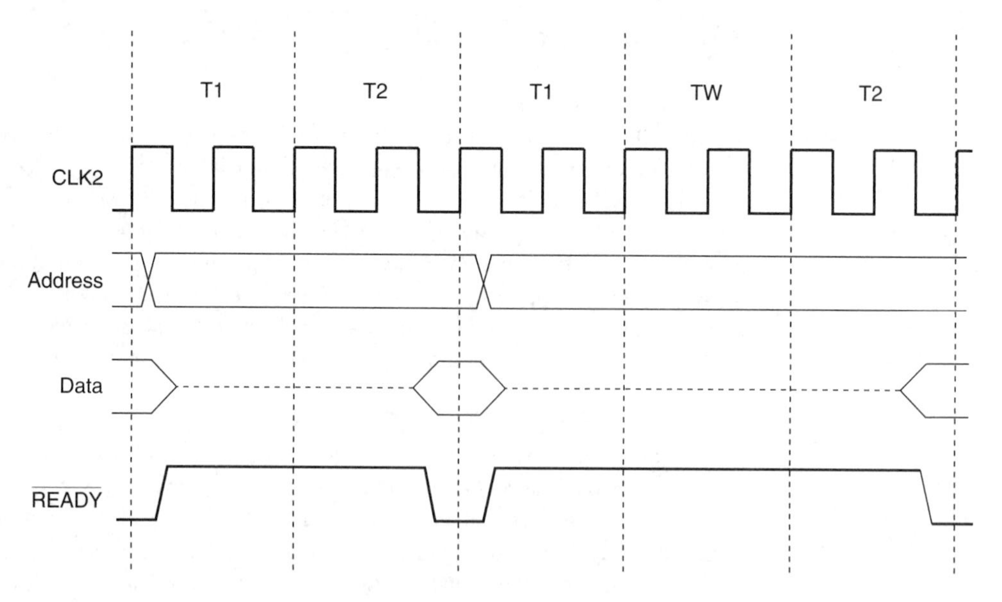

FIGURE 17–12 A nonpipelined 80386 with 0 and 1 wait states.

The $\overline{READY}$ signal is sampled at the end of a bus cycle to determine whether the clock cycle is T_2 or TW. If $\overline{READY} = 0$ at this time, it is the end of the bus cycle or T_2. If $\overline{READY}$ is 1 at the end of a clock cycle, the cycle is a TW and the microprocessor continues to test $\overline{READY}$, searching for a logic 0 and the end of the bus cycle.

In the nonpipelined system, whenever $\overline{ADS}$ becomes a logic 0, a wait state is inserted if $\overline{READY} = 1$. After $\overline{ADS}$ returns to a logic 1, the positive edges of the clock are counted to generate the $\overline{READY}$ signal. The $\overline{READY}$ signal becomes a logic 0 after the first clock to insert 0 wait states. If one wait state is inserted, the $\overline{READY}$ line must remain a logic 1 until at least two clocks have elapsed. If additional wait states are desired, then additional time must elapse before $\overline{READY}$ is cleared. This essentially allows any number of wait states to be inserted into the timing.

Figure 17–13 shows a circuit that inserts 0 through 3 wait states for various memory addresses. In the example, one wait state is produced for a DRAM access and two wait states for an

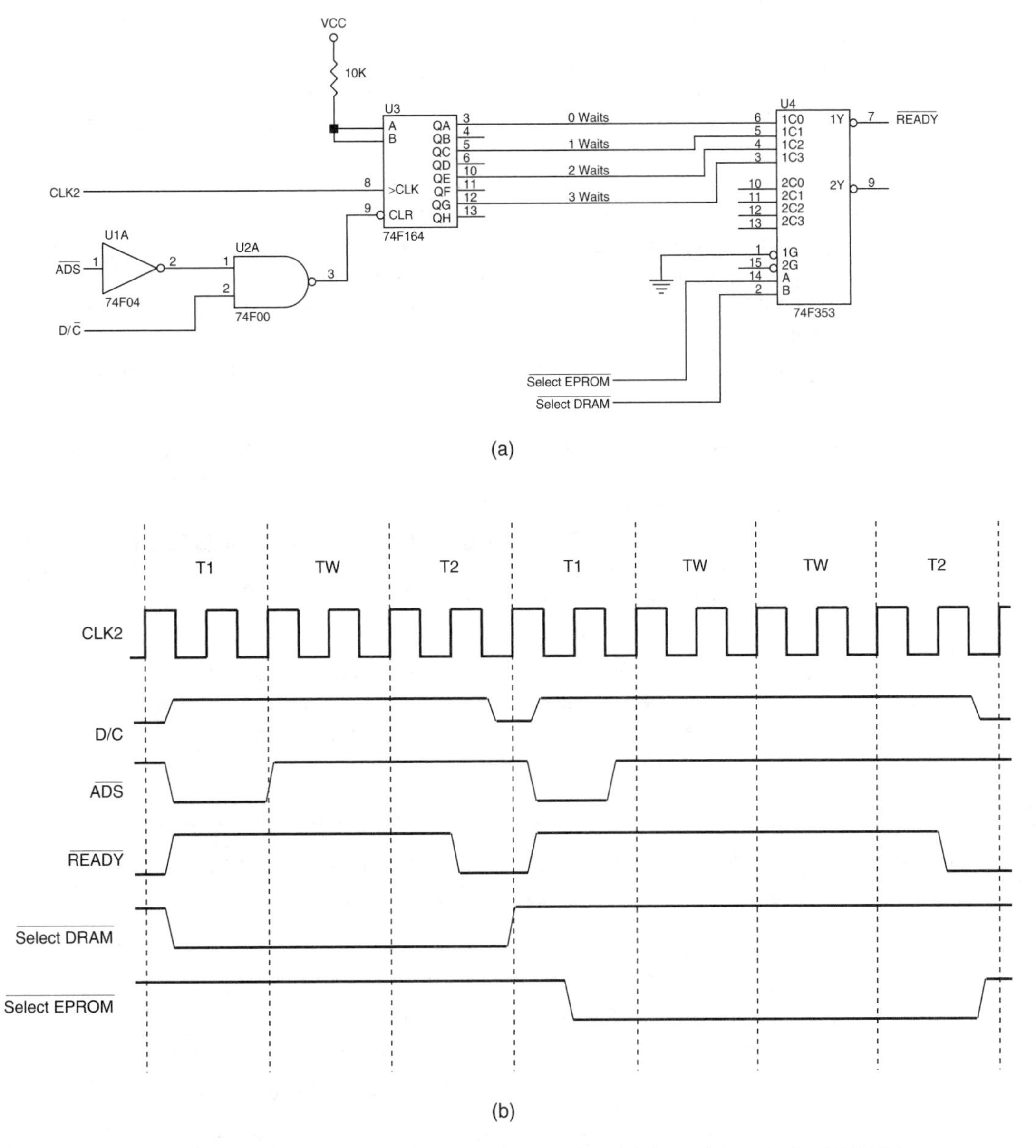

FIGURE 17–13 (a) Circuit and (b) timing that selects 1 wait state for DRAM and 2 waits for EPROM.

EPROM access. The 74F164 clears whenever $\overline{ADS}$ is low and $D/\overline{C}$ is high. It begins to shift after $\overline{ADS}$ returns to a logic 1 level. As it shifts, the 00000000 in the shift register begins to fill with logic 1s from the QA connection toward the QH connection. The four different outputs are connected to an inverting multiplexer that generates the active low $\overline{READY}$ signal.

17–2 SPECIAL 80386 REGISTERS

A new series of registers, not found in earlier Intel microprocessors, appears in the 80386 as control, debug, and test registers. Control registers CR_0–CR_3 control various features, DR_0–DR_7 facilitate debugging, and registers TR_6 and TR_7 are used to test paging and caching.

Control Registers

In addition to the EFLAGS and EIP as described earlier, there are other control registers found in the 80386. Control register 0 (CR_0) is identical to the MSW (machine status word) found in the 80286 microprocessor, except that it is 32 bits wide instead of 16 bits wide. Additional control registers are CR_1, CR_2, and CR_3.

Figure 17–14 illustrates the control register structure of the 80386. Control register CR_1 is not used in the 80386, but is reserved for future products. Control register CR_2 holds the linear page address of the last page accessed before a page fault interrupt. Finally, control register CR_3 holds the base address of the page directory. The rightmost 12 bits of the 32-bit page table address contain zeros and combine with the remainder of the register to locate the start of the 4K-long page table.

Register CR_0 contains a number of special control bits that are defined as follows in the 80386:

PG Selects **page table translation** of linear addresses into physical addresses when PG = 1. Page table translation allows any linear address to be assigned any physical memory location.

ET Selects the **80287 coprocessor** when ET = 0 or the **80387 coprocessor** when ET = l. This bit was installed because there was no 80387 available when the 80386 first appeared. In most systems, ET is set to indicate that an 80387 is present in the system.

TS Indicates that the 80386 has **switched tasks** (in protected mode, changing the contents of TR places a 1 in TS). If TS = 1, a numeric coprocessor instruction causes a type 7 (coprocessor not available) interrupt.

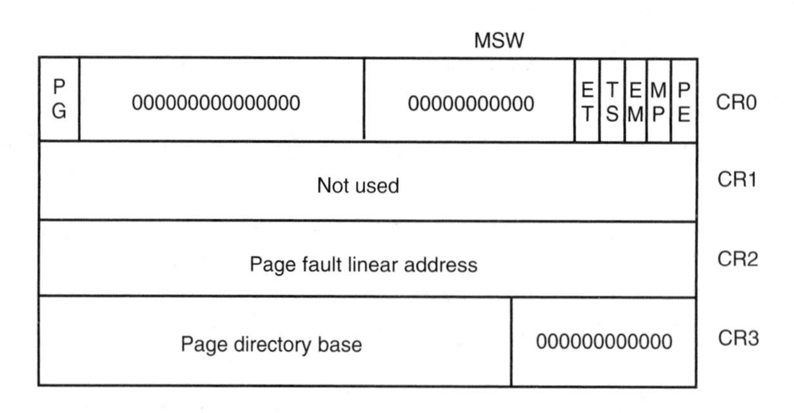

FIGURE 17–14 The control register structure of the 80386 microprocessor.

EM The **emulate** bit is set to cause a type 7 interrupt for each ESC instruction. (ESCape instructions are used to encode instructions for the 80387 coprocessor.) Once this feature was used to emulate interrupts with software, the function of the coprocessor. Emulation reduces the system cost, but it often requires at least 100 times longer to execute the emulated coprocessor instructions.

MP Is set to indicate that the arithmetic **coprocessor is present** in the system.

PE Is set to select the **protected mode** of operation for the 80386. It may also be cleared to reenter the real mode. This bit can only be set in the 80286. The 80286 could not return to real mode without a hardware reset, which precludes its use in most systems that use protected mode.

Debug and Test Registers

Figure 17–15 shows the sets of debug and test registers. The first four debug registers contain 32-bit linear breakpoint addresses. (A linear address is a 32-bit address generated by a microprocessor instruction that may or may not be the same as the physical address.) The breakpoint addresses, which may locate an instruction or datum, are constantly compared with the addresses generated by the program. If a match occurs, the 80386 will cause a type 1 interrupt (TRAP or debug interrupt) to occur, if directed by debug registers DR_6 and DR_7. This feature is a much-expanded version of the basic trapping or tracing allowed with the earlier Intel microprocessors through the type 1 interrupt. The breakpoint addresses are very useful in debugging faulty software. The control bits in DR_6 and DR_7 are defined as follows:

BT If set (1), the debug interrupt was caused by a task switch.

BS If set, the debug interrupt was caused by the TF bit in the flag register.

BD If set, the debug interrupt was caused by an attempt to read the debug register with the GD bit set. The GD bit protects access to the debug registers.

B_3–B_0 Indicate which of the four debug breakpoint addresses caused the debug interrupt.

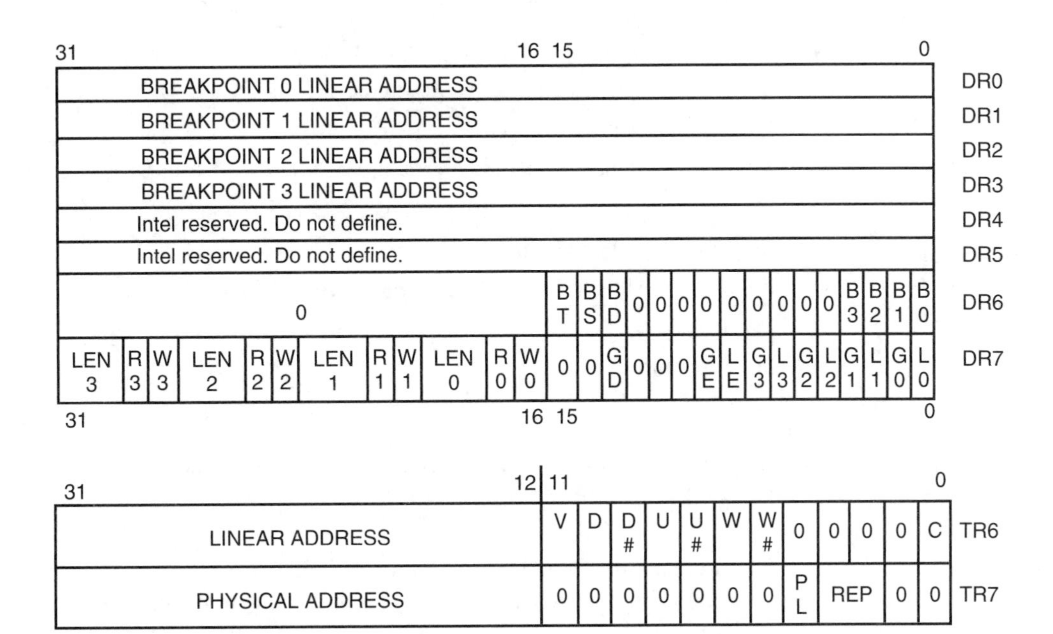

FIGURE 17–15 The debug and test registers of the 80386. (Courtesy of Intel Corporation.)

LEN Each of the four length fields pertains to each of the four breakpoint addresses stored in DR_0–DR_3. These bits further define the size of access at the breakpoint address as 00 (byte), 01 (word), or 11 (doubleword).

RW Each of the four read/write fields pertains to each of the four breakpoint addresses stored in DR_0–DR_3. The RW field selects the cause of action that enabled a breakpoint address as 00 (instruction access), 01 (data write), and 11 (data read and write).

GD If set, GD prevents any read or write of a debug register by generating the debug interrupt. This bit is automatically cleared during the debug interrupt so that the debug registers can be read or changed, if needed.

GE If set, selects a global breakpoint address for any of the four breakpoint address registers.

LE If set, selects a local breakpoint address for any of the four breakpoint address registers.

The test registers, TR_6 and TR_7, are used to test the **translation look-aside buffer** (TLB). The TLB is used with the paging unit within the 80386. The TLB, which holds the most commonly used page table address translations, reduces the number of memory reads required for looking up page translation addresses in the page translation tables. The TLB holds the most common 32 entries from the page table, and it is tested with the TR_6 and TR_7 test registers.

Test register TR_6 holds the tag field (linear address) of the TLB, and TR_7 holds the physical address of the TLB. To write a TLB entry, perform the following steps:

1. Write TR_7 for the desired physical address, PL, and REP values.
2. Write TR_6 with the linear address, making sure that C = 0.

To read a TLB entry:

1. Write TR_6 with the linear address, making sure that C = 1.
2. Read both TR_6 and TR_7. If the PL bit indicates a hit, then the desired values of TR_6 and TR_7 indicate the contents of the TLB.

The bits found in TR_6 and TR_7 indicate the following conditions:

V Shows that the entry in the TLB is valid.

D Indicates that the entry in the TLB is invalid or dirty.

U A bit for the TLB.

W Indicates that the area addressed by the TLB entry is writable.

C Selects a write (0) or immediate lookup (1) for the TLB.

PL Indicates a hit if a logic 1.

REP Selects which block of the TLB is written.

Refer to the section on memory management and the paging unit for more detail on the function of the TLB.

17–3 80386 MEMORY MANAGEMENT

The **memory-management** unit (MMU) within the 80386 is similar to the MMU inside the 80286, except that the 80386 contains a paging unit not found in the 80286. The MMU performs the task of converting linear addresses, as they appear as outputs from a program, into physical addresses that access a physical memory location located anywhere within the memory system. The 80386 uses the paging mechanism to allocate any physical address to any logical address. Therefore, even though the program is accessing memory location A0000H with an instruction, the actual physical address could be memory location 100000H, or any other location if paging is enabled. This feature allows virtually any software, written to operate at any memory location, to function in an 80386 because any linear location can become any physical location. Earlier Intel microprocessors did not have this flexibility. Paging is used with DOS to relocate 80386 and 80486 memory at addresses above FFFFFH and into spaces between ROMs at locations D0000–DFFFFH and other areas as they are available. The area between ROMs is often referred to as *upper memory*; the area above FFFFFH is referred to as *extended memory*.

Descriptors and Selectors

Before the memory paging unit is discussed, we examine the descriptor and selector for the 80386 microprocessor. The 80386 uses descriptors in much the same fashion as the 80286. In both microprocessors, a **descriptor** is a series of eight bytes that describe and locate a memory segment. A **selector** (segment register) is used to index a descriptor from a table of descriptors. The main difference between the 80286 and 80386 is that the latter has two additional selectors (FS and GS) and the most-significant two bytes of the descriptor are defined for the 80386. Another difference is that 80386 descriptors use a 32-bit base address and a 20-bit limit, instead of the 24-bit base address and 16-bit limit found on the 80286.

The 80286 addresses a 16M-byte memory space with its 24-bit base address and has a segment length limit of 64K bytes, due to the 16-bit limit. The 80386 addresses a 4G-byte memory space with its 32-bit base address and has a segment length limit of 1M byte or 4G bytes, due to a 20-bit limit that is used in two different ways. The 20-bit limit can access a segment with a length of 1M byte if the granularity bit (G) = 0. If G = 1, the 20-bit limit allows a segment length of 4G bytes.

The granularity bit is found in the 80386 descriptor. If G = 0, the number stored in the limit is interpreted directly as a limit, allowing it to contain any limit between 00000H and FFFFFH for a segment size up to 1M byte. If G = 1, the number stored in the limit is interpreted as 00000XXXH–FFFFFXXXH, where the XXX is any value between 000H and FFFH. This allows the limit of the segment to range between 0 bytes to 4G bytes in steps of 4K bytes. A limit of 00001 H indicates that the limit is 4K bytes when G = 1 and 1 byte when G = 0. An example is a segment that begins at physical address 10000000H. If the limit is 00001H and G = 0, this segment begins at 10000000H and ends at 10000001H. If G = 1 with the same limit (00001H), the segment begins at location 10000000H and ends at location 10001FFFH.

Figure 17–16 shows how the 80386 addresses a memory segment in the protected mode using a selector and a descriptor. Note that this is identical to the way that a segment is addressed by the 80286. The difference is the size of the segment accessed by the 80386. The selector uses

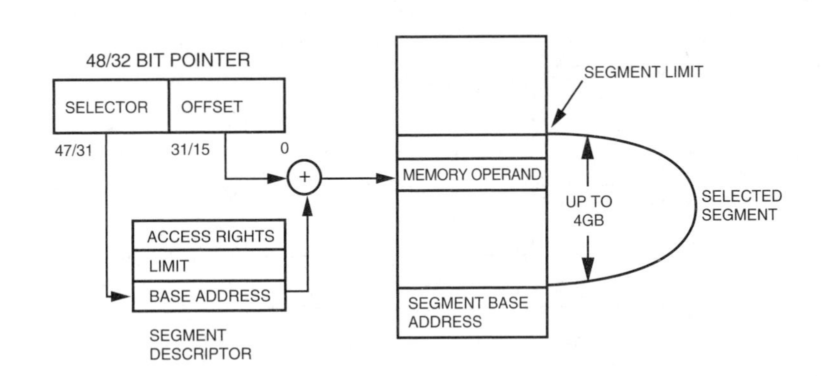

FIGURE 17–16 Protected mode addressing using a segment register as a selector. (Courtesy of Intel Corporation.)

its leftmost 13 bits to select a descriptor from a descriptor table. The TI bit indicates either the local (TI = 1) or global (TI = 0) descriptor table. The rightmost two bits of the selector define the requested privilege level of the access.

Because the selector uses a 13-bit code to access a descriptor, there are at most 8192 descriptors in each table—local or global. Because each segment (in an 80386) can be 4G bytes in length, 16,384 segments can be accessed at a time with the two descriptor tables. This allows the 80386 to access a virtual memory size of 64T bytes. Of course, only 4G bytes of memory actually exist in the memory system (1T byte = 1024G bytes). If a program requires more than 4G bytes of memory at a time, it can be swapped between the memory system and a disk drive or other form of large volume storage.

The 80386 uses descriptor tables for both global (GDT) and local (LDT) descriptors. A third descriptor table appears for interrupt (IDT) descriptors or gates. The first six bytes of the descriptor are the same as in the 80286, which allows 80286 software to be upward compatible with the 80386. (An 80286 descriptor used 00H for its most significant two bytes.) See Figure 17–17 for the 80286 and 80386 descriptor. The base address is 32 bits in the 80386, the limit is 20 bits, and a G bit selects the limit multiplier (1 or 4K times). The fields in the descriptor for the 80386 are defined as follows:

Limit (L_{19}–L_0) Defines the starting 32-bit address of the segment within the 4G-byte physical address space of the 80386 microprocessor. Defines the limit of the segment in units of bytes if the G bit = 0, or in units of 4K bytes if G = 1. This allows a segment to be of any length from 1 byte to 1M bytes if G = 0, and from 4K bytes to 4G bytes if G = 1. Recall that the limit indicates the last byte in a segment.

Access Rights Determines privilege level and other information about the segment. This byte varies with different types of descriptors and is elaborated with each descriptor type.

G The granularity bit selects a multiplier of 1 or 4K times for the limit field. If G = 0, the multiplier is 1; if G = 1, the multiplier is 4K.

D Selects the default instruction mode. If D = 0, the registers and memory pointers are 16 bits wide, as in the 80286; if D = 1, they are 32 bits wide, as in the 80386. This bit determines whether prefixes are required for 32-bit data and index registers. If D = 0, then a prefix is required to access 32-bit registers and to use 32-bit pointers. If D = 1, then a prefix is required to access 16-bit registers and 16-bit pointers. The USE_{16} and USE_{32} directives appended to the SEGMENT statement in assembly language control the setting of the D bit. In the real mode, it is always assumed that the registers are 16-bits wide, so any instruction that references a 32-bit register or pointer must be prefixed. The current version of DOS assumes D = 0, and most Windows programs assume D = 1.

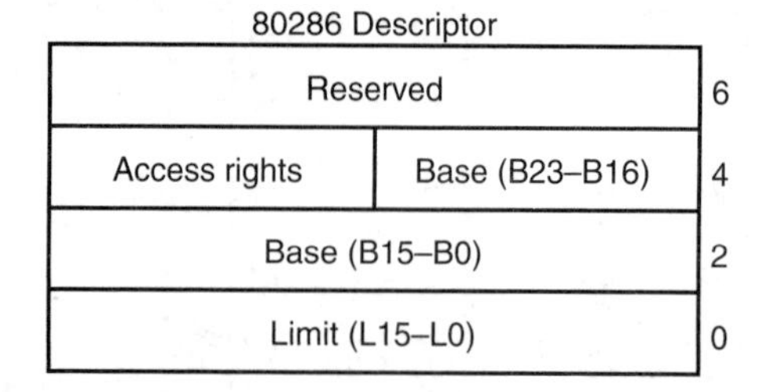

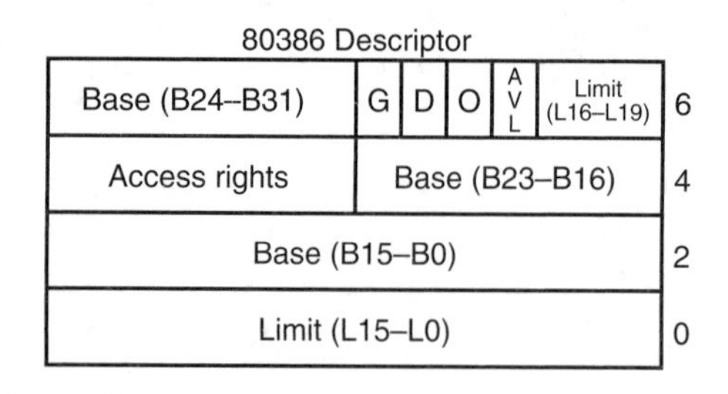

FIGURE 17–17 The descriptors for the 80286 and 80386 microprocessors.

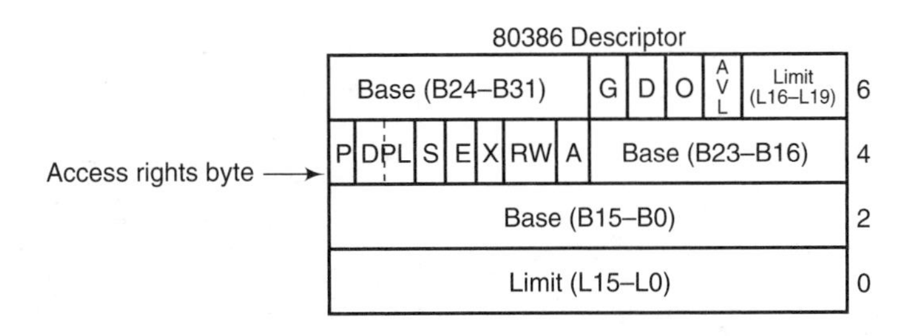

FIGURE 17–18 The format of the 80386 segment descriptor.

AVL This bit is available to the operating system to use in any way that it sees fit. It often indicates that the segment described by the descriptor is available.

Descriptors appear in two forms in the 80386 microprocessor: the segment descriptor and the system descriptor. The segment descriptor defines data, stack, and code segments; the system descriptor defines information about the system's tables, tasks, and gates.

Segment Descriptors. Figure 17–18 shows the segment descriptor. This descriptor fits the general form, as dictated in Figure 17–17, but the access rights bits are defined to indicate how the data, stack, or code segment described by the descriptor functions. Bit position 4 of the access rights byte determines whether the descriptor is a data or code segment descriptor (S = 1) or a system segment descriptor (S = 0). Note that the labels used for these bits may vary in different versions of Intel literature, but they perform the same tasks.

Following is a description of the access rights bits and their function in the segment descriptor:

P **Present** is a logic 1 to indicate that the segment is present. If P = 0 and the segment is accessed through the descriptor, a type 11 interrupt occurs. This interrupt indicates that a segment was accessed that is not present in the system.

DPL **Descriptor privilege level** sets the privilege level of the descriptor; 00 has the highest privilege and 11 has the lowest. This is used to protect access to segments. If a segment is accessed with a privilege level that is lower (higher in number) than the DPL, a privilege violation interrupt occurs. Privilege levels are used in multiuser systems to prevent access to an area of the system memory.

S **Segment** indicates a data or code segment descriptor (S = 1) or a system segment descriptor (S = 0).

E **Executable** selects a data (stack) segment (E = 0) or a code segment (E = 1). E also defines the function of the next two bits (X and RW).

X If E = 0, then X indicates the direction of **expansion** for the data segment. If X = 0, the segment expands upward, as in a data segment; if X = 1, the segment expands downward as in a stack segment. If E = 1, then X indicates whether the privilege level of the code segment is ignored (X = 0) or observed (X = 1).

RW If E = 0, then the **read/write** bit (RW) indicates that the data segment may be written (RW = 1) or not written (RW = 0). If E = 1, then RW indicates that the code segment may be read (RW = 1) or not read (RW = 0).

A **Accessed** is set each time that the microprocessor accesses the segment. It is sometimes used by the operating system to keep track of which segments have been accessed.

System Descriptor. The system descriptor is illustrated in Figure 17–19. There are 16 possible system descriptor types (see Table 17–1 for the different descriptor types), but not all are used in the 80386 microprocessor. Some of these types are defined for the 80286 so that the 80286 software is compatible with the 80386. Some of the types are new and unique to the 80386; some have yet to be defined and are reserved for future Intel products.

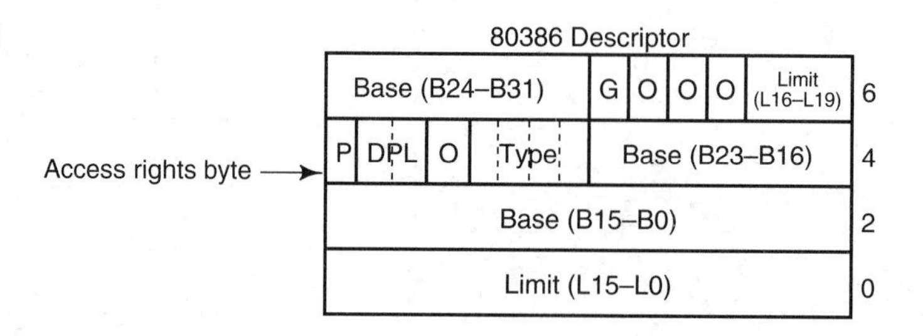

FIGURE 17–19 The general format of an 80386 system descriptor.

TABLE 17–1 System descriptor types.

Type	*Purpose*
0000	Invalid
0001	Available 80286 TSS
0010	LDT
0011	Busy 80286 TSS
0100	80286 task gate
0101	Task gate (80386 and above)
0110	80286 interrupt gate
0111	80286 trap gate
1000	Invalid
1001	Available 80386 and above TSS
1010	Reserved
1011	Busy 80386 and above TSS
1100	80386 and above CALL gate
1101	Reserved
1110	80386 and above interrupt gate
1111	80386 and above trap gate

Descriptor Tables

The descriptor tables define all the segments used in the 80386 when it operates in the protected mode. There are three types of descriptor tables: the global descriptor table (GDT), the local descriptor table (LDT), and the interrupt descriptor table (IDT). The registers used by the 80386 to address these three tables are called the global descriptor table register (GDTR), the local descriptor table register (LDTR), and the interrupt descriptor table register (IDTR). These registers are loaded with the LGDT, LLDT, and LIDT instructions, respectively.

The **descriptor table** is a variable-length array of data, with each entry holding an eight-byte-long descriptor. The local and global descriptor tables hold up to 8192 entries each, and the interrupt descriptor table holds up to 256 entries. A descriptor is indexed from either the local or global descriptor table by the selector that appears in a segment register. Figure 17–20 shows a segment register and the selector that it holds in the protected mode. The leftmost 13 bits index a descriptor; the TI bit selects either the local (TI = 1) or global (TI = 1) descriptor table; and the RPL bits indicate the requested privilege level.

Whenever a new selector is placed into one of the segment registers, the 80386 accesses one of the descriptor tables and automatically loads the descriptor into a program-invisible cache

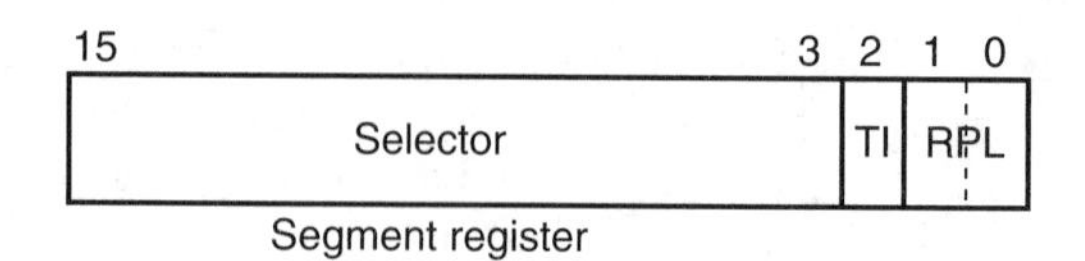

FIGURE 17–20 A segment register showing the selector, T_1 bit, and requested privilege level (RPL) bits.

portion of the segment register. As long as the selector remains the same in the segment register, no additional accesses are required to the descriptor table. The operation of fetching a new descriptor from the descriptor table is program-invisible because the microprocessor automatically accomplishes this each time that the segment register contents are changed in the protected mode.

Figure 17–21 shows how a sample global descriptor table (GDT), which is stored at memory address 00010000H, is accessed through the segment register and its selector. This table contains four entries. The first is a null (0) descriptor. Descriptor 0 must always be a null descriptor. The other entries address various segments in the 80386 protected mode memory system. In this illustration, the data segment register contains 0008H. This means that the selector is indexing descriptor location 1 in the global descriptor table (TI = 0), with a requested privilege level of 00. Descriptor 1 is located eight bytes above the base descriptor table address, beginning at location 00010008H. The descriptor located in this memory location accesses a base address of 00200000H and a limit of 100H. This means that this descriptor addresses memory locations 00200000H–00200100H. Because this is the DS (data segment) register, the data segment is located at these locations in the memory system. If data are accessed outside of these boundaries, an interrupt occurs.

The local descriptor table (LDT) is accessed in the same manner as the global descriptor table (GDT). The only difference in access is that the TI bit is cleared for a global access and set for a local access. Another difference exists if the local and global descriptor table registers are examined. The global descriptor table register (GDTR) contains the base address of the global descriptor table and the limit. The local descriptor table register (LDTR) contains only a selector,

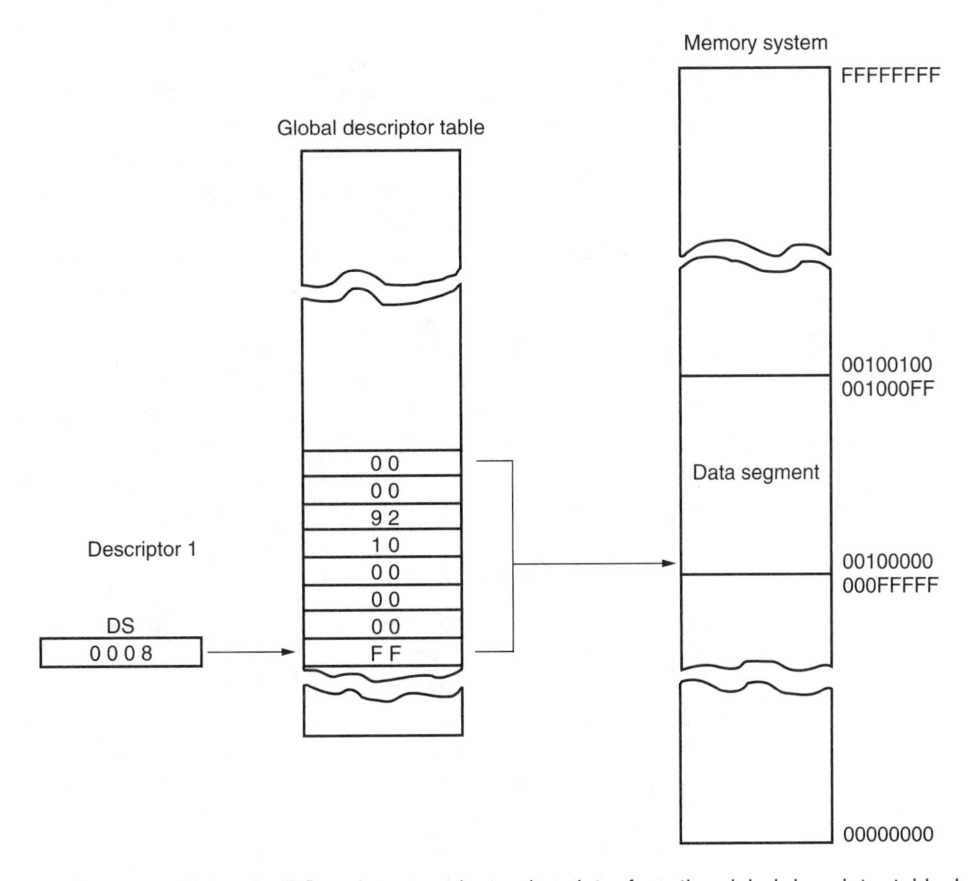

FIGURE 17–21 Using the DS register to select a descriptor from the global descriptor table. In this example, the DS register accesses memory locations 00100000H–001000FFH as a data segment.

FIGURE 17–22 The gate descriptor for the 80386 microprocessor.

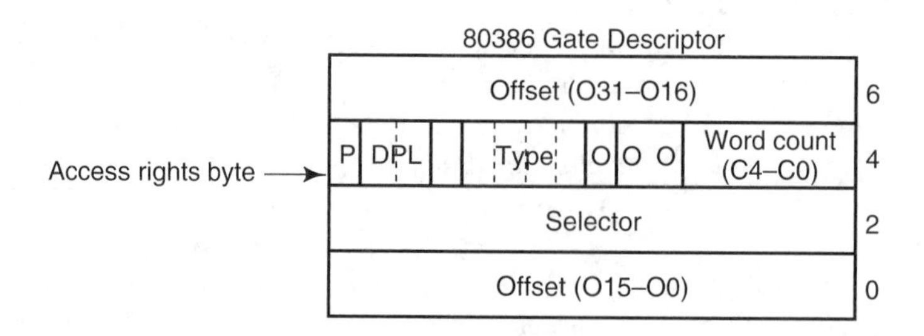

and it is 16 bits wide. The contents of the LDTR address a type 0010 system descriptor that contains the base address and limit of the LDT. This scheme allows one global table for all tasks, but allows many local tables, one or more for each task, if necessary. Global descriptors describe memory for the system, while local descriptors describe memory for applications or tasks.

Like the GDT, the interrupt descriptor table (IDT) is addressed by storing the base address and limit in the interrupt descriptor table register (IDTR). The main difference between the GDT and IDT is that the IDT contains only interrupt gates. The GDT and LDT contain segment and system descriptors, but never contain interrupt gates.

Figure 17–22 shows the gate descriptor, a special form of the system descriptor described earlier. (Refer to Table 17–1 for the different gate descriptor types.) Notice that the gate descriptor contains a 32-bit offset address, a word count, and a selector. The 32-bit offset address points to the location of the interrupt service procedure or other procedure. The word count indicates how many words are transferred from the caller's stack to the stack of the procedure accessed by a call gate. This feature of transferring data from the caller's stack is useful for implementing high-level languages such as C/C++. Note that the word count field is not used with an interrupt gate. The selector is used to indicate the location of the task state segment (TSS) in the GDT or LDT if it is a local procedure.

When a gate is accessed, the contents of the selector are loaded into the task register (TR), causing a task switch. The acceptance of the gate depends on the privilege and priority levels. A return instruction (RET) ends a call gate procedure and a return from interrupt instruction (IRET) ends an interrupt gate procedure. Tasks are usually accessed with a CALL or an INT instruction, where the call instruction addresses a call gate in the descriptor table and the interrupt addresses an interrupt descriptor.

The difference between real mode interrupts and protected mode interrupts is that the interrupt vector table is an IDT in the protected mode. The IDT still contains up to 256 interrupt levels, but each level is accessed through an interrupt gate instead of an interrupt vector. Thus, interrupt type number 2 (INT 2) is located at IDT descriptor number 2 at 16 locations above the base address of the IDT. This also means that the first IK byte of memory no longer contains interrupt vectors, as it did in the real mode. The IDT can be located at any location in the memory system.

The Task State Segment (TSS)

The **task state segment** (TSS) descriptor contains information about the location, size, and privilege level of the task state segment, just like any other descriptor. The difference is that the TSS described by the TSS descriptor does not contain data or code. It contains the state of the task and linkage so tasks can be nested (one task can call a second, which can call a third, and so forth). The TSS descriptor is addressed by the **task register** (TR). The contents of the TR are changed by the LTR instruction. Whenever the protected mode program executes a JMP or CALL instruction, the contents of TR are also changed. The LTR instruction is used to initially access a task during system initialization. After initialization, the CALL or JUMP instructions normally switch tasks. In most cases, we use the CALL instruction to initiate a new task.

The TSS is illustrated in Figure 17–23. As can be seen, the TSS is quite a formidable data structure, containing many different types of information. The first word of the TSS is labeled

back-link. This is the selector that is used, on a return (RET or IRET), to link back to the prior TSS by loading the back-link selector into the TR. The following word must contain a 0. The second through the seventh doublewords contain the ESP and ESS values for privilege levels 0–2. These are required in case the current task is interrupted so these privilege level (PL) stacks

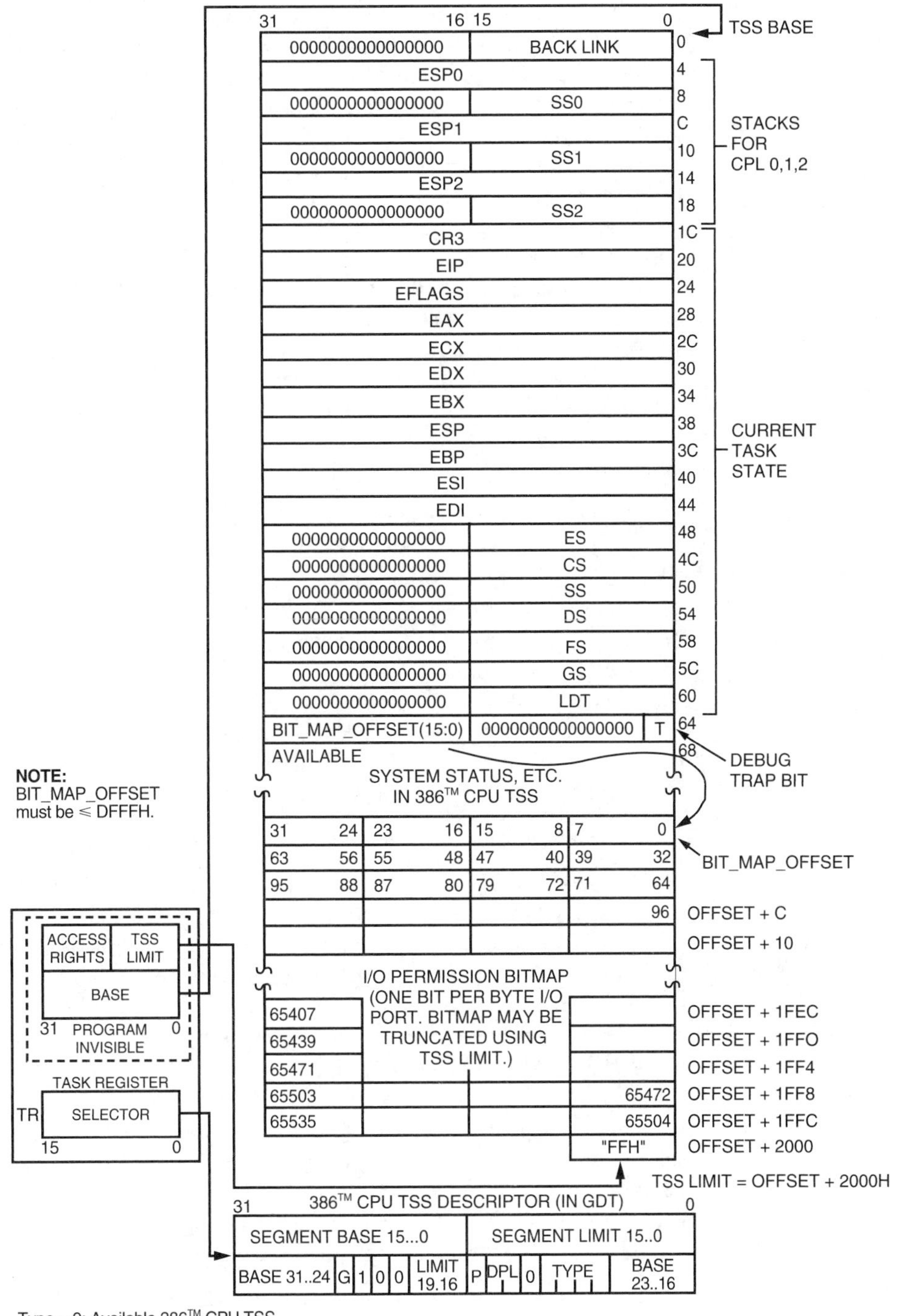

FIGURE 17–23 The task state segment (TSS) descriptor. (Courtesy of Intel Corporation.)

can be addressed. The eighth word (offset 1CH) contains the contents of CR_3, which stores the base address of the prior state's page directory register. This must be restored if paging is in effect. The contents of the next 17 doublewords are loaded into the registers indicated. Whenever a task is accessed, the entire state of the machine (all of the registers) is stored in these memory locations and then reloaded from the same locations in the new TSS. The last word (offset 66H) contains the I/O permission bit map base address.

The I/O permission bit map allows the TSS to block I/O operations to inhibited I/O port addresses via an I/O permission denial interrupt. The permission denial interrupt is type number 13, the general protection fault interrupt. The I/O permission bit map base address is the offset address from the start of the TSS. This allows the same permission map to be used by many TSSs.

Each I/O permission bit map is 64K bits long (8K bytes), beginning at the offset address indicated by the I/O permission bit map base address. The first byte of the I/O permission bit map contains I/O permission for I/O ports 0000H–0007H. The rightmost bit contains the permission for port number 0000H. The leftmost bit contains the permission for port number 0007H. This sequence continues for the very last port address (FFFFH) stored in the leftmost bit of the last byte of the I/O permission bit map. A logic 0 placed in an I/O permission bit map bit enables the I/O port address, while a logic 1 inhibits or blocks the I/O port address. At present, only Windows NT, Windows 2000, and Windows XP uses the I/O permission scheme to disable I/O ports dependent on the application or the user.

To review the operation of a task switch, which requires only 17 μs to execute on an 80386 microprocessor, we list the following steps:

1. The gate contains the address of the procedure or location jumped to by the task switch. It also contains the selector number of the TSS descriptor and the number of words transferred from the caller to the user stack area for parameter passing.
2. The selector is loaded into TR from the gate. (This step is accomplished by a CALL or JMP that refers to a valid TSS descriptor.)
3. The TR selects the TSS.
4. The current state is saved in the current TSS and the new TSS is accessed with the state of the new task (all the registers) loaded into the microprocessor. The current state is saved at the TSS selector currently found in the TR. Once the current state is saved, a new value (by the JMP or CALL) for the TSS selector is loaded into TR and the new state is loaded from the new TSS.

The return from a task is accomplished by the following steps:

1. The current state of the microprocessor is saved in the current TSS.
2. The back-link selector is loaded to the TR to access the prior TSS so that the prior state of the machine can be returned to and be restored to the microprocessor. The return for a called TSS is accomplished by the IRET instruction.

17–4 MOVING TO PROTECTED MODE

In order to change the operation of the 80386 from the real mode to the protected mode, several steps must be followed. Real mode operation is accessed after a hardware reset or by changing the PE bit to a logic 0 in CR_0. Protected mode is accessed by placing a logic 1 into the PE bit of CR_0; before this is done, however, some other things must be initialized. The following steps accomplish the switch from the real mode to the protected mode:

1. Initialize the interrupt descriptor table so that it contains valid interrupt gates for at least the first 32 interrupt type numbers. The IDT may (and often does) contain up to 256 eight-byte interrupt gates defining all 256 interrupt types.

2. Initialize the global descriptor table (GDT) so that it contains a null descriptor at descriptor 0 and valid descriptors for at least one code, one stack, and one data segment.
3. Switch to protected mode by setting the PE bit in CR_0.
4. Perform an intersegment (far) JMP to flush the internal instruction queue.
5. Load all the data selectors (segment registers) with their initial selector values.
6. The 80386 is now operating in the protected mode, using the segment descriptors that are defined in GDT and IDT.

Figure 17–24 shows the protected system memory map set up by following steps 1–6. The software for this task is listed in Example 17–1. This system contains one data segment descriptor and one code segment descriptor with each segment set to 4G bytes in length. This is the simplest protected mode system possible (called the **flat model**): loading all the segment registers, except code, with the same data segment descriptor from the GDT. The privilege level is initialized to 00, the highest level. This system is most often used when one user has access to the microprocessor and requires the entire memory space. This program is designed for use in a system that does not use DOS or does not shell from Windows to DOS. Later in this section, we show how to go to protected mode in a DOS environment. (Please note that the software in Example 17–1 is designed for a stand-alone system such as the 80386EX embedded microprocessor, and not for use in the PC.)

Example 17–1 does not store any interrupt vectors in the interrupt descriptor table, because none are used in the example. If interrupt vectors are used, then software must be included to load the addresses of the interrupt service procedures into the IDT. The software must be generated as two separate parts that are then connected together and burned on a ROM. The first part is written as shown in the real mode and the second part (see comment in listing) with the assembler set to generate protected mode code using the 32-bit flat model. This software will not function on a personal computer because it is written to function on an embedded system. The code must be converted to a binary file using EXE2BIN after it is assembled and before burning to a ROM.

FIGURE 17–24 The memory map for Example 17–1.

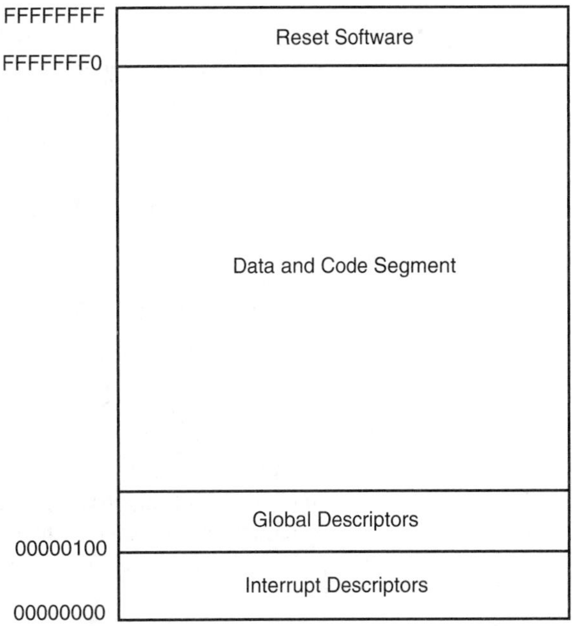

EXAMPLE 17–1

```
.MODEL SMALL
.386P
ADR     STRUC
        DW      ?               ;address structure
        DD      ?
ADR     ENDS

.DATA
IDT     DQ      32 DUP(?)       ;interrupt descriptor table
GDT     DQ      8               ;global descriptor table
DESC1   DW      0FFFFH          ;code segment descriptor
        DW      0
        DB      0
        DB      9EH
        DB      8FH
        DB      0
DESC2   DW      0FFFFH          ;data segment descriptor
        DW      0
        DB      0
        DB      92H
        DB      8FH
        DB      0
IDTR    ADR     <0FFH,IDT>      ;IDTR data
GDTR    ADR     <17H,GDT>       ;GDTR data
JADR    ADR     <8,PM>          ;Far JMP data
.CODE
.STARTUP
        MOV  AX,0               ;initialize DS
        MOV  DS,AX

        LIDT IDTR               ;initialize IDTR
        LGDT GDTR               ;initialize GDTR

        MOV  EAX,CR0            ;set PE
        OR   EAX,1
        MOV  CR0,EAX

        JMP  JADR               ;far jump to PM
PM::                            ;force a far label

;the software that follows must be developed separately
;with the assembler set to generate 32-bit protected mode code
;
; ie:     .MODEL FLAT

        MOV  AX,10H             ;load segment registers
        MOV  DS,AX              ;now in protected mode
        MOV  ES,AX
        MOV  SS,AX
        MOV  FS,AX
        MOV  GS,AX
        MOV  SP,0FFFF000H

;other initialization appears here

end
```

In more complex systems (very unlikely to appear in embedded systems), the steps required to initialize the system in the protected mode are more involved. For complex systems that are often multiuser systems, the registers are loaded by using the task state segment (TSS). The steps required to place the 80386 into protected mode operation for a more complex system using a task switch follow:

1. Initialize the interrupt descriptor table so that it refers to valid interrupt descriptors with at least 32 descriptors in the IDT.

2. Initialize the global descriptor table so that it contains a task state segment (TSS) descriptor, and the initial code and data segments required for the initial task.
3. Initialize the task register (TR) so that it points to a TSS.
4. Switch to protected mode by using an intersegment (far) jump to flush the internal instruction queue. This loads the TR with the current TSS selector and initial task.
5. The 80386 is now operating in the protected mode under control of the first task.

Example 17–2 illustrates the software required to initialize the system and switch to protected mode by using a task switch. The initial system task operates at the highest level of protection (00) and controls the entire operating environment for the 80386. In many cases, it is used to boot (load) software that allows many users to access the system in a multiuser environment. As in Example 17–1 this software will not function on a personal computer and is designed to function only on an embedded system.

EXAMPLE 17–2

```
.MODEL SMALL
.386P
.DATA
ADR     STRUC                       ;structure for 48 bit address
        DW      ?                   ;selector
        DD      ?                   ;offset
ADR     ENDS

DESC    STRUC                       ;structure of a descriptor
        DW      ?
        DW      ?
        DB      ?
        DB      ?
        DB      ?
        DB      ?
DESC    ENDS

TSS     STRUC                       ;structure of TSS
        DD      18 DUP(?)
        DD      18H                 ;ES
        DD      10H                 ;CS
        DD      4 DUP(18H)
        DD      28H                 ;LDT
        DD      IOBP                ;IO privilege map
TSS     ENDS

GDT     DESC    <>                          ;null
        DESC    <2067H,TS1,0,89H,90H,0>     ;TSS descriptor
        DESC    <-1,0,0,9AH,0CFH,0>         ;code segment
        DESC    <-1,0,0,92H,0CFH,0>         ;data segment
        DESC    <0,0,0,0,0,0>               ;LDT for TSS

LDT     DESC    <>                  ;null

IOBP    DB      2000H DUP(0)        ;all I/O on

IDT     DQ      32 DUP(?)           ;IDT

TS1     TSS     <>                  ;make TSS

IDTA    ADR     <0FFH,IDT>          ;IDTR
GDTA    ADR     <27H,GDT>           ;GDTR
JADR    ADR     <10H,PM>            ;jump address

.CODE
.STARTUP

        MOV  AX,0
        MOV  DS,AX
```

```
        LGDT GDTA
        LIDT IDTA

        MOV  EAX,CR0
        OR   EAX,1
        MOV  CR0,EAX

        MOV  AX,8
        LTR  AX
        JMP  JADR
PM:

        ;protected mode

END
```

Neither Example 17–1 nor 17–2 is written to function in the personal computer environment. The personal computer environment requires the use of either the VCPI (virtual control program interface) driver provided by the HIMEM.SYS driver in DOS or the DPMI (**DOS protected mode interface**) driver provided by Windows when shelling to DOS. Example 17–3 shows how to switch to protected mode using DPMI and then display the contents of any area of memory. This includes memory in the extended memory area or anywhere else. This DOS application allows the contents of any memory location to be displayed in hexadecimal format on the monitor, including locations above the first 1M byte of the memory system.

EXAMPLE 17–3

```
                         ;A program that displays the contents of any area of memory
                         ;including extended memory.
                         ;***command line syntax***
                         ;EDUMP XXXX,YYYY   where XXXX is the start address and YYYY is
                         ;the end address.
                         ;Note:  this program must be executed from WINDOWS.
                         ;
                                .MODEL SMALL
                                .386
                                .STACK 1024              ;stack area of 1,024 bytes
0000                            .DATA
0000  00000000            ENTRY DD   ?                   ;DPMI entry point
0004  00000000            EXIT  DD   ?                   ;DPMI exit point
0008  00000000            FIRST DD   ?                   ;first address
000C  00000000            LAST1 DD   ?                   ;last address
0010  0000                MSIZE DW   ?                   ;memory needed for DPMI
0012  0D 0A 0A 50 61      ERR1  DB   13,10,10,'Parameter error.$'
      72 61 6D 65 74
      65 72 20 65 72
      72 6F 72 2E 24
0026  0D 0A 0A 44 50      ERR2  DB   13,10,10,'DPMI not present.$'
      4D 49 20 6E 6F
      74 20 70 72 65
      73 65 6E 74 2E
      24
003B  0D 0A 0A 4E 6F      ERR3  DB   13,10,10,'Not enough real memory.$'
      74 20 65 6E 6F
      75 67 68 20 72
      65 61 6C 20 6D
      65 6D 6F 72 79
      2E 24
0056  0D 0A 0A 43 6F      ERR4  DB   13,10,10,'Could not move to protected mode.$'
      75 6C 64 20 6E
      6F 74 20 6D 6F
      76 65 20 74 6F
      20 70 72 6F 74
      65 63 74 65 64
      20 6D 6F 64 65
      2E 24
```

```
007B  0D 0A 0A 43 61 ERR5  DB   13,10,10,'Cannot allocate selector.$'
      6E 6E 6F 74 20
      61 6C 6C 6F 63
      61 74 65 20 73
      65 6C 65 63 74
      6F 72 2E 24
0098  0D 0A 0A 43 61 ERR6  DB   13,10,10,'Cannot use base address.$'
      6E 6E 6F 74 20
      75 73 65 20 62
      61 73 65 20 61
      64 64 72 65 73
      73 2E 24
00B4  0D 0A 0A 43 61 ERR7  DB   13,10,10,'Cannot allocate 64K to limit.$'
      6E 6E 6F 74 20
      61 6C 6C 6F 63
      61 74 65 20 36
      34 4B 20 74 6F
      20 6C 69 6D 69
      74 2E 24
00D5  0D 0A 24       CRLF  DB   13,10,'$'
00D8  50 72 65 73 73 MES1  DB   'Press any key...$'
      20 61 6E 79 20
      6B 65 79 2E 2E
      2E 24
                     ;
                     ;register array storage for DPMI function 0300H
                     ;
00E9  = 00E9         ARRAY EQU  THIS BYTE
00E9  00000000       REDI  DD   0                  ;EDI
00ED  00000000       RESI  DD   0                  ;ESI
00F1  00000000       REBP  DD   0                  ;EBP
00F5  00000000             DD   0                  ;reserved
00F9  00000000       REBX  DD   0                  ;EBX
00FD  00000000       REDX  DD   0                  ;EDX
0101  00000000       RECX  DD   0                  ;ECX
0105  00000000       REAX  DD   0                  ;EAX
0109  0000           RFLAG DW   0                  ;flags
010B  0000           RES   DW   0                  ;ES
010D  0000           RDS   DW   0                  ;DS
010F  0000           RFS   DW   0                  ;FS
0111  0000           RGS   DW   0                  ;GS
0113  0000           RIP   DW   0                  ;IP
0115  0000           RCS   DW   0                  ;CS
0117  0000           RSP   DW   0                  ;SP
0119  0000           RSS   DW   0                  ;SS
0000                       .CODE
                           .STARTUP
0010  8C C0                MOV  AX,ES
0012  8C DB                MOV  BX,DS              ;find size of program and data
0014  2B D8                SUB  BX,AX
0016  8B C4                MOV  AX,SP              ;find stack size
0018  C1 E8 04             SHR  AX,4
001B  40                   INC  AX
001C  03 D8                ADD  BX,AX              ;BX = length in paragraphs
001E  B4 4A                MOV  AH,4AH
0020  CD 21                INT  21H                ;modify memory allocation
0022  E8 00D1              CALL GETDA              ;get command line information
0025  73 0A                JNC  MAIN1              ;if parameters are good
0027  B4 09                MOV  AH,9               ;parameter error
0029  BA 0012 R            MOV  DX,OFFSET ERR1
002C  CD 21                INT  21H
002E  E9 00AA              JMP  MAINE              ;exit to DOS
0031                 MAIN1:
0031  E8 00AB              CALL ISDPMI             ;is DPMI loaded?
0034  72 0A                JC   MAIN2              ;if DPMI present
0036  B4 09                MOV  AH,9
0038  BA 0026 R            MOV  DX,OFFSET ERR2
```

```
003B  CD 21                 INT   21H               ;display DPMI not present
003D  E9 009B               JMP   MAINE             ;exit to DOS
0040                  MAIN2:
0040  B8 0000               MOV   AX,0              ;indicate 0 memory needed
0043  83 3E 0010 R 00       CMP   MSIZE,0
0048  74 F6                 JE    MAIN2             ;if DPMI needs no memory
004A  8B 1E 0010 R          MOV   BX,MSIZE          ;get amount
004E  B4 48                 MOV   AH,48H
0050  CD 21                 INT   21H               ;allocate memory for DPMI
0052  73 09                 JNC   MAIN3
0054  B4 09                 MOV   AH,9              ;if not enough real memory
0056  BA 003B R             MOV   DX,OFFSET ERR3
0059  CD 21                 INT   21H
005B  EB 7E                 JMP   MAINE             ;exit to DOS
005D                  MAIN3:
005D  8E C0                 MOV   ES,AX
005F  B8 0000               MOV   AX,0              ;16-bit application
0062  FF 1E 0000 R          CALL  DS:ENTRY          ;switch to protected mode
0066  73 09                 JNC   MAIN4
0068  B4 09                 MOV   AH,9              ;if switch failed
006A  BA 0056 R             MOV   DX,OFFSET ERR4
006D  CD 21                 INT   21H
006F  EB 6A                 JMP   MAINE             ;exit to DOS
                      ;
                      ;PROTECTED MODE
                      ;
0071                  MAIN4:
0071  B8 0000               MOV   AX,0000H          ;get local selector
0074  B9 0001               MOV   CX,1              ;only one is needed
0077  CD 31                 INT   31H
0079  72 48                 JC    MAIN7             ;if error
007B  8B D8                 MOV   BX,AX             ;save selector
007D  8E C0                 MOV   ES,AX             ;load ES with selector
007F  B8 0007               MOV   AX,0007H          ;set base address
0082  8B 0E 000A R          MOV   CX,WORD PTR FIRST+2
0086  8B 16 0008 R          MOV   DX,WORD PTR FIRST
008A  CD 31                 INT   31H
008C  72 3D                 JC    MAIN8             ;if error
008E  B8 0008               MOV   AX,0008H
0091  B9 0000               MOV   CX,0
0094  BA FFFF               MOV   DX,0FFFFH         ;set limit to 64K
0097  CD 31                 INT   31H
0099  72 38                 JC    MAIN9             ;if error
009B  B9 0018               MOV   CX,24             ;load line count
009E  BE 0000               MOV   SI,0              ;load offset
00A1                  MAIN5:
00A1  E8 00F4               CALL  DADDR             ;display address, if needed
00A4  E8 00CE               CALL  DDATA             ;display data
00A7  46                    INC   SI                ;point to next data
00A8  66| A1 0008 R         MOV   EAX,FIRST         ;test for end
00AC  66| 3B 06 000C R      CMP   EAX,LAST1
00B1  74 07                 JE    MAIN6             ;if done
00B3  66| FF 06 0008 R      INC   FIRST
00B8  EB E7                 JMP   MAIN5
00BA                  MAIN6:
00BA  B8 0001               MOV   AX,0001H          ;release descriptor
00BD  8C C3                 MOV   BX,ES
00BF  CD 31                 INT   31H
00C1  EB 18                 JMP   MAINE             ;exit to DOS
00C3                  MAIN7:
00C3  BA 007B R             MOV   DX,OFFSET ERR5
00C6  E8 0096               CALL  DISPS             ;display cannot allocate selector
00C9  EB 10                 JMP   MAINE             ;exit to DOS
00CB                  MAIN8:
00CB  BA 0098 R             MOV   DX,OFFSET ERR6
00CE  E8 008E               CALL  DISPS             ;display cannot use base address
00D1  EB E7                 JMP   MAIN6             ;release descriptor
```

```
00D3                  MAIN9:
00D3  BA 00B4 R              MOV  DX,OFFSET ERR7
00D6  E8 0086                CALL DISPS              ;display cannot allocate 64K limit
00D9  EB DF                  JMP  MAIN6              ;release descriptor
00DB                  MAINE:
                             .EXIT
                      ;
                      ;The ISDPMI procedure tests for the presence of DPMI.
                      ;***exit parameters***
                      ;carry = 1; if DPMI is present
                      ;carry = 0; if DPMI is not present
                      ;
00DF                  ISDPMI PROC NEAR

00DF  B8 1687                MOV  AX,1687H           ;get DPMI status
00E2  CD 2F                  INT  2FH                ;DOS multiplex
00E4  0B C0                  OR   AX,AX
00E6  75 0D                  JNZ  ISDPMI1            ;if no DPMI
00E8  89 36 0010 R           MOV  MSIZE,SI           ;save amount of memory needed
00EC  89 3E 0000 R           MOV  WORD PTR ENTRY,DI
00F0  8C 06 0002 R           MOV  WORD PTR ENTRY+2,ES
00F4  F9                     STC
00F5                  ISDPMI1:
00F5  C3                     RET

00F6                  ISDPMI ENDP
                      ;
                      ;The GETDA procedure retrieves the command line parameters
                      ;for memory display in hexadecimal.
                      ;FIRST = the first address from the command line
                      ;LAST1 = the last address from the command line
                      ;***return parameters***
                      ;carry = 1; if error
                      ;carry = 0; for no error
                      ;
00F6                  GETDA  PROC NEAR

00F6  1E                     PUSH DS
00F7  06                     PUSH ES
00F8  1F                     POP  DS
00F9  07                     POP  ES                 ;exchange ES with DS
00FA  BE 0081                MOV  SI,81H             ;address command line
00FD                  GETDA1:
00FD  AC                     LODSB                   ;skip spaces
00FE  3C 20                  CMP  AL,' '
0100  74 FB                  JE   GETDA1             ;if space
0102  3C 0D                  CMP  AL,13
0104  74 1E                  JE   GETDA3             ;if enter = error
0106  4E                     DEC  SI                 ;adjust SI
0107                  GETDA2:
0107  E8 0020                CALL GETNU              ;get first number
010A  3C 2C                  CMP  AL,','
010C  75 16                  JNE  GETDA3             ;if no comma = error
010E  66| 26: 89 16 0008 R MOV  ES:FIRST,EDX
0114  E8 0013                CALL GETNU              ;get second number
0117  3C 0D                  CMP  AL,13
0119  75 09                  JNE  GETDA3             ;if error
011B  66| 26: 89 16 000C R MOV  ES:LAST1,EDX
0121  F8                     CLC                     ;indicate no error
0122  EB 01                  JMP  GETDA4             ;return no error
0124                  GETDA3:
0124  F9                     STC                     ;indicate error
0125                  GETDA4:
0125  1E                     PUSH DS                 ;exchange ES with DS
0126  06                     PUSH ES
0127  1F                     POP  DS
0128  07                     POP  ES
0129  C3                     RET
```

```
012A                    GETDA ENDP
                        ;
                        ;The GETNU procedure extracts a number from the command line
                        ;and returns with it in EDX and last command line character in
                        ;AL as a delimiter.
                        ;
012A                    GETNU  PROC NEAR

012A  66| BA 00000000          MOV  EDX,0              ;clear result
0130                    GETNU1:
0130  AC                       LODSB                   ;get digit from command line
                               .IF  AL >= 'a' && AL <= 'z'
0139  2C 20                       SUB  AL,20H          ;make uppercase
                              .ENDIF
013B  2C 30                    SUB  AL,'0'             ;convert from ASCII
013D  72 12                    JB   GETNU2             ;if not a number
                               .IF  AL > 9             ;convert A-F from ASCII
0143  2C 07                        SUB  AL,7
                              .ENDIF
0145  3C 0F                    CMP  AL,0FH
0147  77 08                    JA   GETNU2             ;if not 0-F
0149  66| C1 E2 04             SHL  EDX,4
014D  02 D0                    ADD  DL,AL              ;add digit to EDX
014F  EB DF                    JMP  GETNU1             ;get next digit
0151                    GETNU2:
0151  8A 44 FF                 MOV  AL,[SI-1]          ;get delimiter
0154  C3                       RET

0155                    GETNU  ENDP
                        ;
                        ;The DISPC procedure displays the ASCII character found
                        ;in register AL.
                        ;***uses***
                        ;INT21H
                        ;
0155                    DISPC  PROC NEAR

0155  52                       PUSH DX
0156  8A D0                    MOV  DL,AL
0158  B4 06                    MOV  AH,6
015A  E8 0084                  CALL INT21H             ;do real INT 21H
015D  5A                       POP  DX
015E  C3                       RET

015F                    DISPC ENDP
                        ;
                        ;The DISPS procedure displays a character string from
                        ;protected mode addressed by DS:EDX.
                        ;***uses***
                        ;DISPC
                        ;
015F                    DISPS  PROC NEAR

015F  66| 81 E2 0000FFFF       AND  EDX,0FFFFH
0166  67& 8A 02                MOV  AL,[EDX]           ;get character
0169  3C 24                    CMP  AL,'$'             ;test for end
016B  74 07                    JE   DISP1              ;if end
016D  66| 42                   INC  EDX                ;address next character
016F  E8 FFE3                  CALL DISPC              ;display character
0172  EB EB                    JMP  DISPS              ;repeat until $
0174                    DISP1:
0174  C3                       RET

0175                    DISPS  ENDP
                        ;
                        ;The DDATA procedure displays a byte of data at the location
                        ;addressed by ES:SI. The byte is followed by one space.
                        ;***uses***
                        ;DIP and DISPC
```

```
                        ;
0175                    DDATA PROC  NEAR

0175  26: 8A 04               MOV  AL,ES:[SI]         ;get byte
0178  C0 E8 04                SHR  AL,4
017B  E8 000C                 CALL DIP                ;display first digit
017E  26: 8A 04               MOV  AL,ES:[SI]         ;get byte
0181  E8 0006                 CALL DIP                ;display second digit
0184  B0 20                   MOV  AL,' '             ;display space
0186  E8 FFCC                 CALL DISPC
0189  C3                      RET

018A                    DDATA ENDP
                        ;
                        ;The DIP procedure displays the right nibble found in AL as a
                        ;hexadecimal digit.
                        ;***uses***
                        ;DISPC
                        ;
018A                    DIP   PROC NEAR

018A  24 0F                   AND  AL,0FH             ;get right nibble
018C  04 30                   ADD  AL,30H             ;convert to ASCII
                              .IF  AL > 39H           ;if A-F
0192  04 07                       ADD  AL,7
                              .ENDIF
0194  E8 FFBE                 CALL DISPC              ;display digit
0197  C3                      RET

0198                    DIP   ENDP
                        ;
                        ;The DADDR procedure displays the hexadecimal address found
                        ;in DS:FIRST if it is a paragraph boundary.
                        ;***uses***
                        ;DIP, DISPS, DISPC, and INT21H
                        ;
0198                    DADDR PROC NEAR

0198  66| A1 0008 R           MOV  EAX,FIRST          ;get address
019C  A8 0F                   TEST AL,0FH             ;test for XXXXXXX0
019E  75 40                   JNZ  DADDR4             ;if not, don't display address
01A0  BA 00D5 R               MOV  DX,OFFSET CRLF
01A3  E8 FFB9                 CALL DISPS              ;display CR and LF
01A6  49                      DEC  CX                 ;decrement line count
01A7  75 18                   JNZ  DADDR2             ;if not end of page
01A9  BA 00D8 R               MOV  DX,OFFSET MES1     ;if end of page
01AC  E8 FFB0                 CALL DISPS              ;display press any key
01AF                    DADDR1:
01AF  B4 06                   MOV  AH,6               ;get any key, no echo
01B1  B2 FF                   MOV  DL,0FFH
01B3  E8 002B                 CALL INT21H             ;do real INT 21H
01B6  74 F7                   JZ   DADDR1             ;if nothing typed
01B8  BA 00D5 R               MOV  DX,OFFSET CRLF
01BB  E8 FFA1                 CALL DISPS              ;display CRLF
01BE  B9 0018                 MOV  CX,24              ;reset line count
01C1                    DADDR2:
01C1  51                      PUSH CX                 ;save line count
01C2  B9 0008                 MOV  CX,8               ;load digit count
01C5  66| 8B 16 0008 R        MOV  EDX,FIRST          ;get address
01CA                    DADDR3:
01CA  66| C1 C2 04            ROL  EDX,4
01CE  8A C2                   MOV  AL,DL
01D0  E8 FFB7                 CALL DIP                ;display digit
01D3  E2 F5                   LOOP DADDR3             ;repeat 8 times
01D5  59                      POP  CX                 ;retrieve line count
01D6  B0 3A                   MOV  AL,':'
01D8  E8 FF7A                 CALL DISPC              ;display colon
01DB  B0 20                   MOV  AL,' '
```

```
01DD  E8 FF75                CALL DISPC            ;display space
01E0                 DADDR4:
01E0  C3                     RET

01E1                 DADDR   ENDP
                     ;
                     ;The INT21H procedure gains access to the real mode DOS
                     ;INT 21H instruction with the parameters intact.
                     ;
01E1                 INT21H PROC NEAR

01E1  66| A3 0105 R          MOV  REAX,EAX         ;save registers
01E5  66| 89 1E 00F9 R       MOV  REBX,EBX
01EA  66| 89 0E 0101 R       MOV  RECX,ECX
01EF  66| 89 16 00FD R       MOV  REDX,EDX
01F4  66| 89 36 00ED R       MOV  RESI,ESI
01F9  66| 89 3E 00E9 R       MOV  REDI,EDI
01FE  66| 89 2E 00F1 R       MOV  REBP,EBP
0203  9C                     PUSHF
0204  58                     POP  AX
0205  A3 0109 R              MOV  RFLAG,AX
0208  06                     PUSH ES               ;do DOS interrupt
0209  B8 0300                MOV  AX,0300H
020C  BB 0021                MOV  BX,21H
020F  B9 0000                MOV  CX,0
0212  1E                     PUSH DS
0213  07                     POP  ES
0214  BF 00E9 R              MOV  DI,OFFSET ARRAY
0217  CD 31                  INT  31H
0219  07                     POP  ES
021A  A1 0109 R              MOV  AX,RFLAG         ;restore registers
021D  50                     PUSH AX
021E  9D                     POPF
021F  66| 8B 3E 00E9 R       MOV  EDI,REDI
0224  66| 8B 36 00ED R       MOV  ESI,RESI
0229  66| 8B 2E 00F1 R       MOV  EBP,REBP
022E  66| A1 0105 R          MOV  EAX,REAX
0232  66| 8B 1E 00F9 R       MOV  EBX,REBX
0237  66| 8B 0E 0101 R       MOV  ECX,RECX
023C  66| 8B 16 00FD R       MOV  EDX,REDX
0241  C3                     RET

0242                 INT21H ENDP
                             END
```

You might notice that the DOS INT 21 H function call must be treated differently when operating in the protected mode. The procedure that calls a DOS INT 21H is at the end of Example 17–3. Because this is extremely long and time consuming, we have tended to move away from using the DOS interrupts from a Windows application. The best way to develop software for Windows is through the use of C/C++ with the inclusion of assembly language procedures for arduous tasks.

17–5 VIRTUAL 8086 MODE

One special mode of operation not discussed thus far is the virtual 8086 mode. This special mode is designed so that multiple 8086 real-mode software applications can execute at one time. The PC operates in this mode for DOS applications using the DOS emulator cmd.exe (the command prompt). Figure 17–25 illustrates two 8086 applications mapped into the 80386 using the virtual mode. The operating system allows multiple applications to execute, usually done through a technique called **time-slicing**. The operating system allocates a set amount of time to each task.

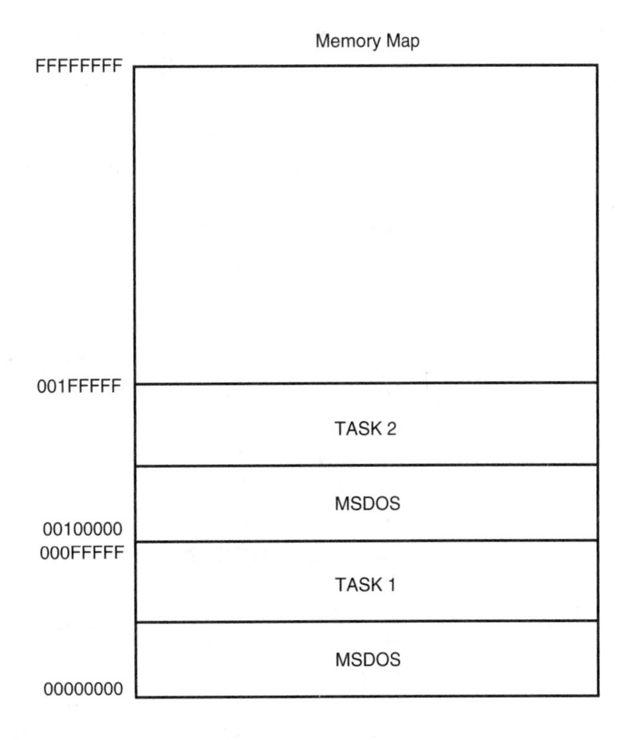

FIGURE 17–25 Two tasks resident in an 80386 operated in the virtual 8086 mode.

For example, if three tasks are executing, the operating system can allocate 1 ms to each task. This means that after each millisecond, a task switch occurs to the next task. In this manner, all tasks receive a portion of the microprocessor's execution time, resulting in a system that appears to execute more than one task at a time. The task times can be adjusted to give any task any percentage of the microprocessor's execution time.

A system that can use this technique is a print spooler. The print spooler can function in one DOS partition and be accessed 10% of the time. This allows the system to print using the print spooler, but it doesn't detract from the system because it uses only 10% of the system time.

The main difference between 80386 protected mode operation and the virtual 8086 mode is the way the segment registers are interpreted by the microprocessor. In the virtual 8086 mode, the segment registers are used as they are in the real mode: as a segment address and an offset address capable of accessing a 1M-byte memory space from locations 00000H–FFFFFH. Access to many virtual 8086 mode systems is made possible by the paging unit that is explained in the next section. Through paging, the program still accesses memory below the 1M-byte boundary, yet the microprocessor can access a physical memory space at any location in the 4G-byte range of the memory system.

Virtual 8086 mode is entered by changing the VM bit in the EFLAG register to a logic 1. This mode is entered via an IRET instruction if the privilege level is 00. This bit cannot be set in any other manner. An attempt to access a memory address above the 1M-byte boundary will cause a type 13 interrupt to occur.

The virtual 8086 mode can be used to share one microprocessor with many users by partitioning the memory so that each user has its own DOS partition. User 1 can be allocated memory locations 00100000H–01FFFFFH, user 2 can be allocated locations 0020000H–01FFFFFFH, and so forth. The system software located at memory locations 00000000H–000FFFFFH can then share the microprocessor between users by switching from one to another to execute software. In this manner, one microprocessor is shared by many users.

17–6 THE MEMORY PAGING MECHANISM

The paging mechanism allows any linear (logical) address, as it is generated by a program, to be placed into any physical memory page, as generated by the paging mechanism. A **linear memory page** is a page that is addressed with a selector and an offset in either the real or protected mode. A **physical memory page** is a page that exists at some actual physical memory location. For example, linear memory location 20000H could be mapped into physical memory location 30000H, or any other location, with the paging unit. This means that an instruction that accesses location 20000H actually accesses location 30000H.

Each 80386 memory page is 4K bytes long. Paging allows the system software to be placed at any physical address with the paging mechanism. Three components are used in page address translation: the page directory, the page table, and the actual physical memory page. Note that EEM386.EXE, the extended memory manager, uses the paging mechanism to simulate expanded memory in extended memory and to generate upper memory blocks between system ROMs.

The Page Directory

The page directory contains the location of up to 1024 page translation tables. Each page translation table translates a logic address into a physical address. The page directory is stored in the memory and accessed by the page descriptor address register (CR_3) (see Figure 17–14). Control register CR_3 holds the base address of the page directory, which starts at any 4K-byte boundary in the memory system. The MOV CR_3,reg instruction is used to initialize CR_3 for paging. In a virtual 8086 mode system, each 8086 DOS partition would have its own page directory.

The page directory contains up to 1024 entries, which are each four bytes long. The page directory itself occupies one 4K-byte memory page. Each entry in the page directory (see Figure 17–26) translates the leftmost 10 bits of the memory address. This 10-bit portion of the linear address is used to locate different page tables for different page table entries. The page table address (A_{32}–A_{12}), stored in a page directory entry, accesses a 4K-byte-long page translation table. To completely translate any linear address into any physical address requires 1024 page tables that are each 4K bytes long, plus the page table directory, which is also 4K bytes long. This translation scheme requires up to 4M plus 4K bytes of memory for a full address translation. Only the largest operating systems support this size address translation. Many commonly found operating systems translate only the first 16M bytes of the memory system if paging is enabled. This includes programs such as Windows. This translation requires four entries in the page directory (16 bytes) and four complete page tables (16K bytes).

The page table directory entry control bits, as illustrated in Figure 17–26, each perform the following functions:

D — **Dirty** is undefined for page table directory entries by the 80386 microprocessor and is provided for use by the operating system.

A — **Accessed** is set to a logic 1 whenever the microprocessor accesses the page directory entry.

R/W and U/S — **Read/write** and **user/supervisor** are both used in the protection scheme, as listed in Table 17–2. Both bits combine to develop paging priority level protection for level 3, the lowest user level.

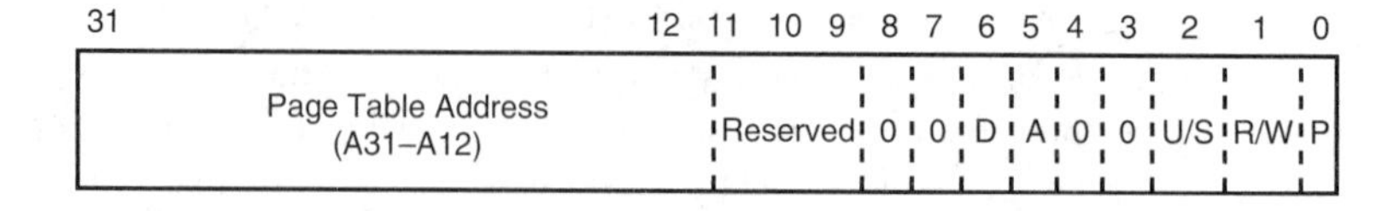

FIGURE 17–26 The page table directory entry.

TABLE 17–2 Protection for level 3 using U/S and R/W.

U/S	*R/W*	*Access Level 3*
0	0	None
0	1	None
1	0	Read-only
1	1	Write-only

P — **Present**, if a logic 1, indicates that the entry can be used in address translation. If P = 0, the entry cannot be used for translation. A not present entry can be used for other purposes, such as indicating that the page is currently stored on the disk. If P = 0, the remaining bits of the entry can be used to indicate the location of the page on the disk memory system.

The Page Table

The page table contains 1024 physical page addresses, accessed to translate a linear address into a physical address. Each page table translates a 4M section of the linear memory into 4M of physical memory. The format for the page table entry is the same as for the page directory entry (refer to Figure 17–26). The main difference is that the page directory entry contains the physical address of a page table, while the page table entry contains the physical address of a 4K-byte physical page of memory. The other difference is the D (dirty bit), which has no function in the page directory entry, but indicates that a page has been written to in a page table entry.

Figure 17–27 illustrates the paging mechanism in the 80386 microprocessor. Here, the linear address 00C03FFCH, as generated by a program, is converted to physical address XXXXXFFCH, as translated by the paging mechanism. (Note: XXXXX is any 4K-byte physical page address.) The paging mechanism functions in the following manner:

1. The 4K-byte long page directory is stored as the physical address located by CR_3. This address is often called the **root address**. One page directory exists in a system at a time. In the 8086 virtual mode, each task has its own page directory, allowing different areas of physical memory to be assigned to different 8086 virtual tasks.
2. The upper 10 bits of the linear address (bits 31–22), as determined by the descriptors described earlier in this chapter or by a real address, are applied to the paging mechanism to select an entry in the page directory. This maps the page directory entry to the leftmost 10 bits of the linear address.
3. The page table is addressed by the entry stored in the page directory. This allows up to 4K page tables in a fully populated and translated system.
4. An entry in the page table is addressed by the next 10 bits of the linear address (bits 21–12).
5. The page table entry contains the actual physical address of the 4K-byte memory page.
6. The rightmost 12 bits of the linear address (bits 11–0) select a location in the memory page.

The paging mechanism allows the physical memory to be assigned to any linear address through the paging mechanism. For example, suppose that linear address 20000000H is selected by a program, but this memory location does not exist in the physical memory system. The 4K byte linear page is referenced as locations 20000000H–20000FFFH by the program. Because this section of physical memory does not exist, the operating system might assign an existing physical memory page such as 12000000H–12000FFFH to this linear address range.

In the address translation process, the leftmost 10 bits of the linear address select page directory entry 200H located at offset address 800H in the page directory. This page directory entry contains the address of the page table for linear addresses 20000000H–203FFFFFH. Linear address

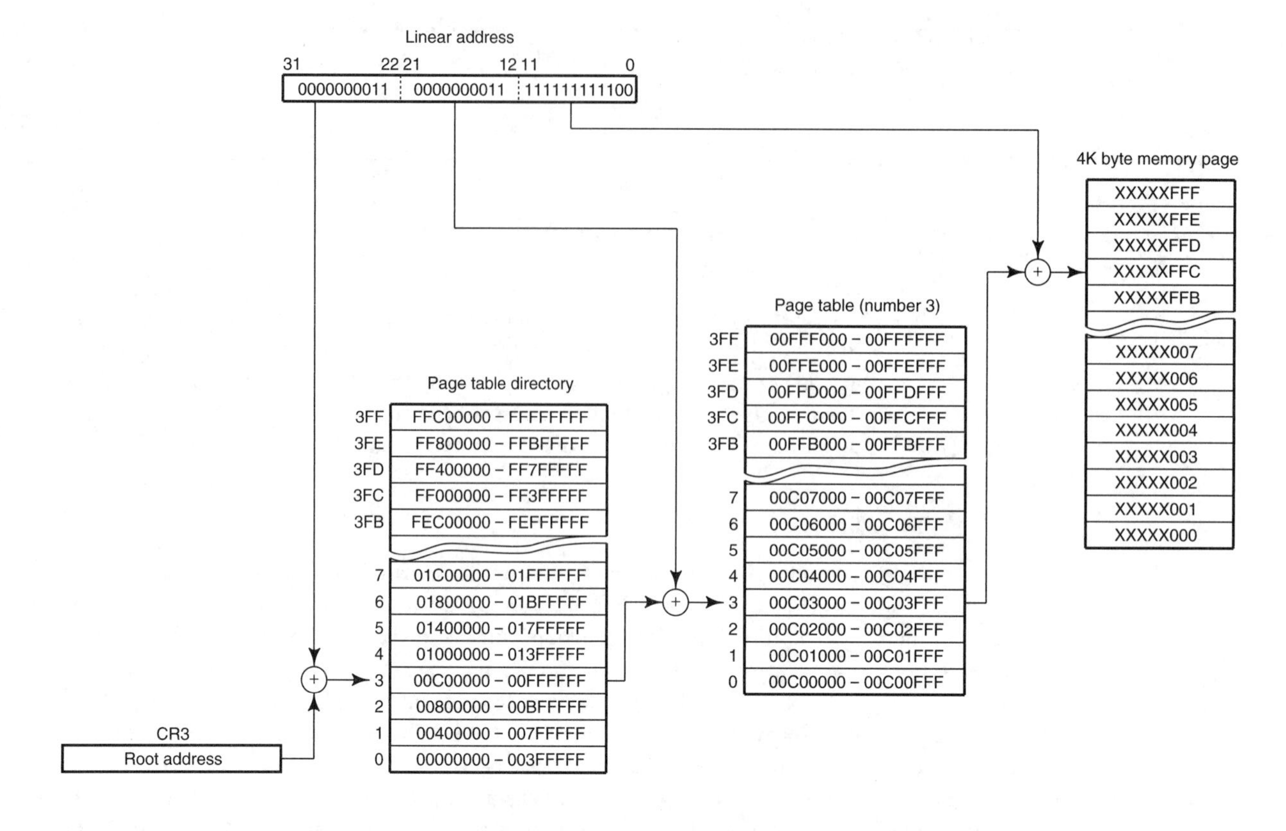

FIGURE 17–27 The translation of linear address 00C03FFCH to physical memory address XXXXXFFCH. The value of XXXXX is determined by the page table entry (not shown here).

bits (21–12) select an entry in this page table that corresponds to a 4K-byte memory page. For linear addresses 20000000H–20000FFFH, the first entry (entry 0) in the page table is selected. This first entry contains the physical address of the actual memory page, or 12000000H–12000FFFH in this example.

Take, for example, a typical DOS-based computer system. The memory map for the system appears in Figure 17–28. Note from the map that there are unused areas of memory, which can be paged to a different location, giving a DOS real mode application program more memory. The normal DOS memory system begins at location 00000H and extends to location 9FFFFH, which is 640K bytes of memory. Above location 9FFFFH, we find sections devoted to video cards, disk cards, and the system BIOS ROM. In this example, an area of memory just above 9FFFFH is unused (A0000–AFFFFH). This section of the memory could be used by DOS, so that the total application-memory area is 704K instead of 640K. Be careful when using A0000H–AFFFFH for additional RAM because the video card uses this area for bit-mapped graphics in mode 12H and 13H.

This section of memory can be used by mapping it into extended memory at locations 1002000H–11FFFFH. Software to accomplish this translation and initialize the page table directory, and page tables required to set up memory, are illustrated in Example 17–4. Note that this procedure

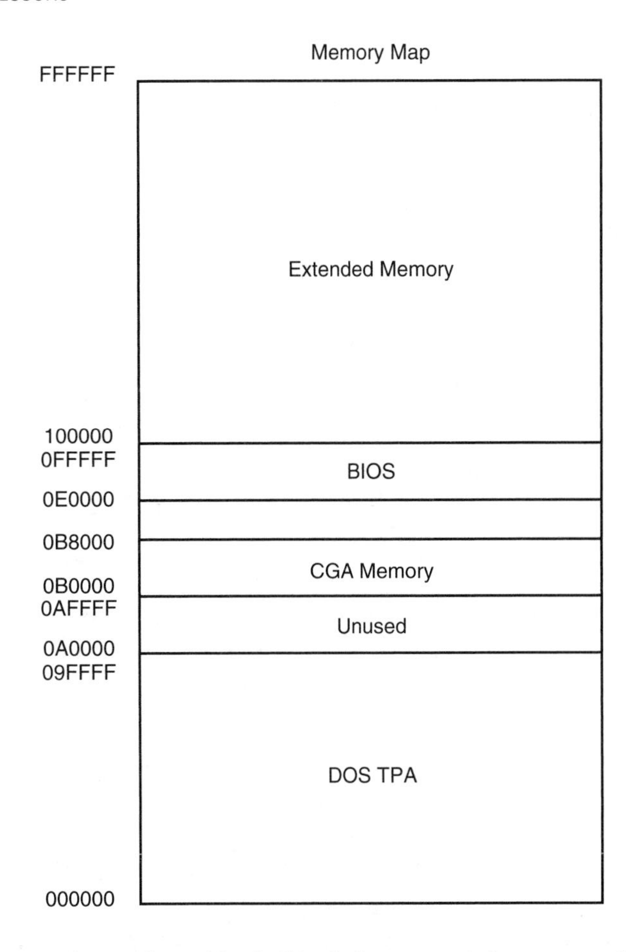

FIGURE 17–28 Memory map for an AT-style clone.

initializes the page table directory, a page table, and loads CR_3. It does not switch to protected mode and it does enable paging. Note that paging functions in real mode memory operation.

EXAMPLE 17–4

```
.MODEL SMALL
.386P
.DATA

;page directory

PDIR   DD      4

;page table 0

TAB0   DD      1024 dup(?)

.CODE
.STARTUP
       MOV  EAX,0
       MOV  AX,CS
       SHL  EAX,4
       ADD  EAX,OFFSET TAB0
       AND  EAX,0FFFFF000H
       ADD  EAX,7
       MOV  PDIR,EAX              ;address page directory
       MOV  ECX,256
       MOV  EDI,OFFSET TAB0
```

```
        MOV   AX,DS
        MOV   ES,AX
        MOV   EAX,7
        .REPEAT                          ;remap 00000H-9FFFFH
              STOSD                      ;to 00000H-9FFFFH
              ADD EAX,4096
        .UNTILCXZ
        MOV   EAX,102007H
        MOV   ECX,16
        .REPEAT                          ;remap A0000H-AFFFFH
              STOSD                      ;to 102000H-11FFFFH
              ADD  EAX,4096
        .UNTILCXZ
        MOV   EAX,0
        MOV   AX,DS
        SHL   EAX,4
        ADD   EAX,OFFSET PDIR            ;load CR3
        MOV   CR3,EAX

;additional software to remap other areas of memory

        END
```

17–7 INTRODUCTION TO THE 80486 MICROPROCESSOR

The 80486 microprocessor is a highly integrated device, containing well over 1.2 million transistors. Located within this device circuit are a memory-management unit (MMU), a complete numeric coprocessor that is compatible with the 80387, a high-speed level 1 cache memory that contains 8K bytes of space, and a full 32-bit microprocessor that is upward-compatible with the 80386 microprocessor. The 80486 is currently available as a 25 MHz, 33 MHz, 50 MHz, 66 MHz, or 100 MHz device. Note that the 66 MHz version is double-clocked and the 100 MHz version is triple-clocked. In 1990, Intel demonstrated a 100 MHz version (not double-clocked) of the 80486 for *Computer Design* magazine, but it has yet to be released. Advanced Micro Devices (AMD) has produced a 40 MHz version that is also available in an 80 MHz (double-clocked) and a 120 MHz (triple-clocked) form. The 80486 is available as an 80486DX or an 80486SX. The only difference between these devices is that the 80486SX does not contain the numeric coprocessor, which reduces its price. The 80487SX numeric coprocessor is available as a separate component for the 80486SX microprocessor.

This section details the differences between the 80486 and 80386 microprocessors. These differences are few, as shall be seen. The most notable differences apply to the cache memory system and parity generator.

Pin-Out of the 80486DX and 80486SX Microprocessors

Figure 17–29 illustrates the pin-out of the 80486DX microprocessor, a 168-pin PGA. The 80486SX, also packaged in a 168-pin PGA, is not illustrated because only a few differences exist. Note that pin B_{15} is NMI on the 80486DX and pin A_{15} is NMI on the 80486SX. The only other differences are that pin A_{15} is $\overline{IGNNE}$ (ignore numeric error) on the 80486DX (not present on the 80486SX); pin C_{14} is $\overline{FERR}$ (floating-point error) on the 80486DX; and pins B_{15} and C_{14} on the 80486SX are not connected.

When connecting the 80486 microprocessor, all V_{CC} and V_{SS} pins must be connected to the power supply for proper operation. The power supply must be capable of supplying 5.0 V ± 10%, with up to 1.2 A of surge current for the 33 MHz version. The average supply current is 650 mA for the 33 MHz version. Intel has also produced a 3.3 V version that requires an average of 500 mA at a triple-clock speed of 100 MHz. Logic 0 outputs allow up to 4.0 mA of current, and logic 1 outputs allow up to 1.0 mA. If larger currents are required, as they often are, then the 80486

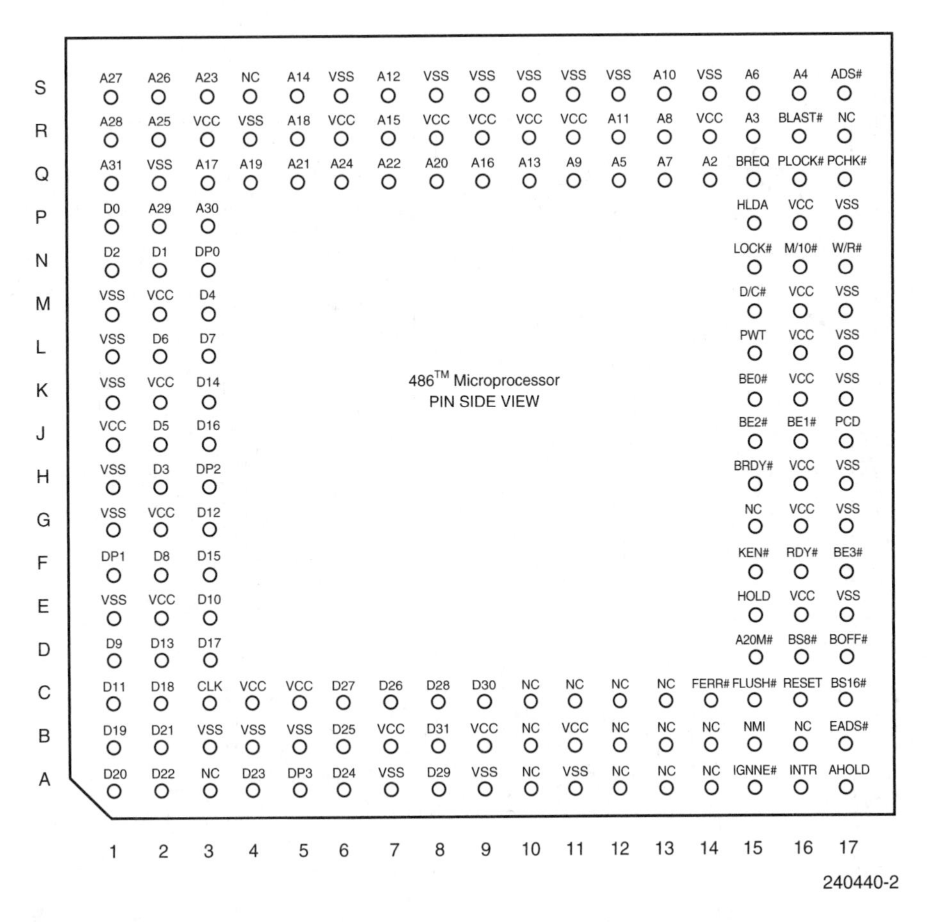

FIGURE 17–29 The pin-out of the 80486. (Courtesy of Intel Corporation.)

must be buffered. Figure 17–30 shows a buffered 80486DX system. In the circuit shown, only the address, data, and parity signals are buffered.

Pin Definitions

A_{31}–A_2 — **Address outputs** A_{31}–A_2 provide the memory and I/O with the address during normal operation; during a cache line invalidation, A_{31}–A_4 are used to drive the microprocessor.

$\overline{A20M}$ — **Address bit 20 mask** causes the 80486 to wrap its address around from location 000FFFFFH to 00000000H, as does the 8086 microprocessor. This provides a memory system that functions like the 1M-byte real memory system in the 8086 microprocessor.

$\overline{ADS}$ — **Address data strobe** becomes a logic 0 to indicate that the address bus contains a valid memory address.

AHOLD — **Address hold** input causes the microprocessor to place its address bus connections at their high-impedance state, with the remainder of the buses staying active. It is often used by another bus master to gain access for a cache invalidation cycle.

$\overline{BE3}$–$\overline{BE0}$ — **Byte enable** outputs select a bank of the memory system when information is transferred between the microprocessors and its memory and I/O space.

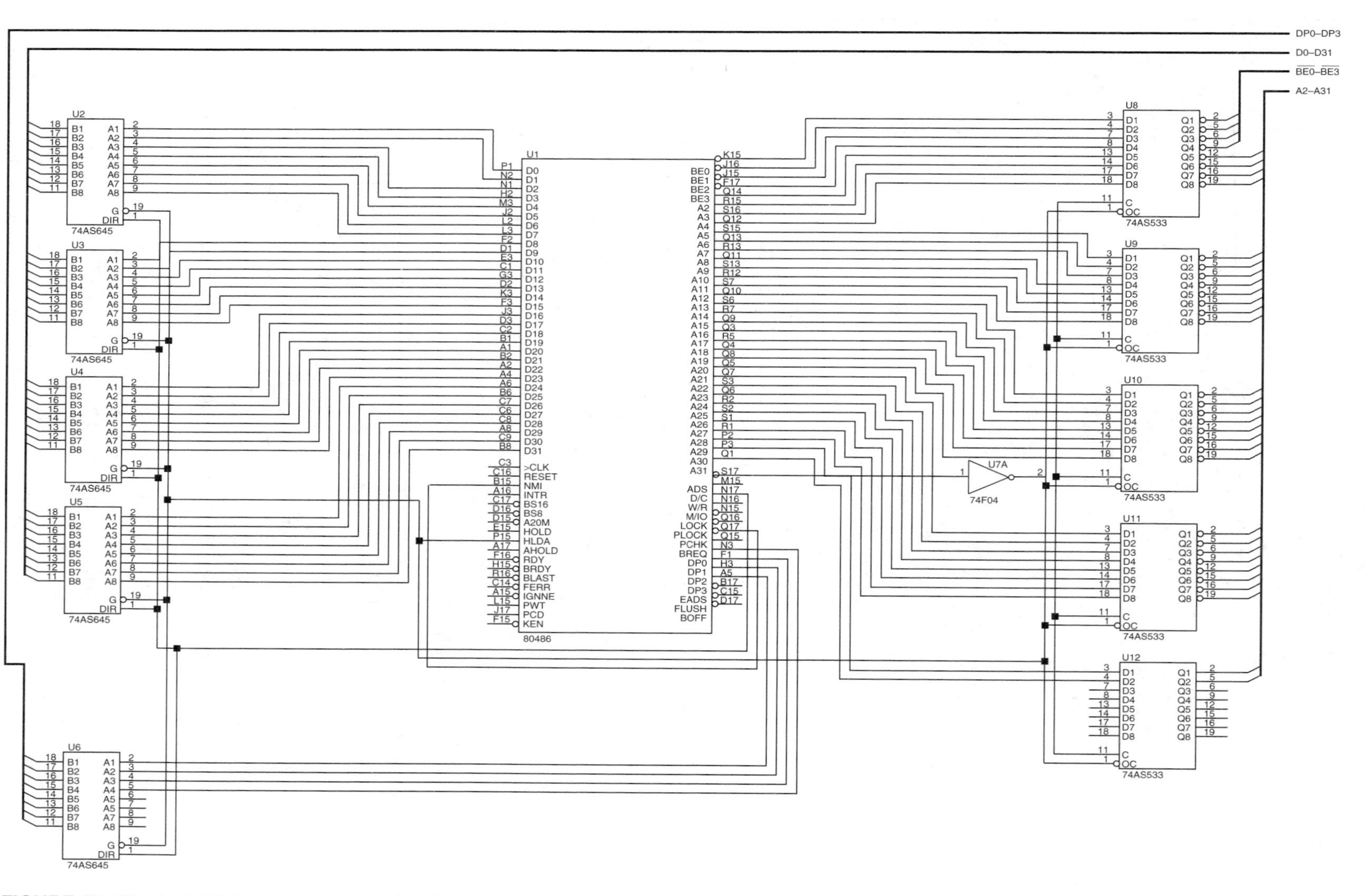

FIGURE 17–30 An 80486 microprocessor showing the buffered address, data, and parity buses.

The $\overline{BE3}$ signal enables D_{31}–D_{24}, $\overline{BE2}$ enables D_{23}–D_{16}, $\overline{BE1}$ enables D_{15}–D_8, and $\overline{BE0}$ enables D_7–D_0.

$\overline{BLAST}$ The **burst last** output shows that the burst bus cycle is complete on the next activation of the $\overline{BRDY}$ signal.

$\overline{BOFF}$ The **back-off** input causes the microprocessor to place its buses at their high-impedance state during the next clock cycle. The microprocessor remains in the bus hold state until the $\overline{BOFF}$ pin is placed at a logic 1 level.

$\overline{BRDY}$ The **burst ready** input is used to signal the microprocessor that a burst cycle is complete.

BREQ The **bus request** output indicates that the 80486 has generated an internal bus request.

$\overline{BS8}$ The **bus size 8** input causes the 80486 to structure itself with an 8-bit data bus to access byte-wide memory and I/O components.

$\overline{BS16}$ The **bus size 16** input causes the 80486 to structure itself with a 16-bit data bus to access word-wide memory and I/O components.

CLK The **clock** input provides the 80486 with its basic timing signal. The clock input is a TTL-compatible input that is 25 MHz to operate the 80486 at 25 MHz.

D_{31}–D_0 The **data bus** transfers data between the microprocessor and its memory and I/O system. Data bus connections D_7–D_0 are also used to accept the interrupt vector type number during an interrupt acknowledge cycle.

$D/\overline{C}$ The **data/control** output indicates whether the current operation is a data transfer or control cycle. See Table 17–3 for the function of $D/\overline{C}$, $M/\overline{IO}$, and $W/\overline{R}$.

DP_3–DP_0 **Data parity** I/O provides even parity for a write operation and check parity for a read operation. If a parity error is detected during a read, the $\overline{PCHK}$ output becomes a logic 0 to indicate a parity error. If parity is not used in a system, these lines must be pulled high to +5.0 V or to 3.3 V in a system that uses a 3.3 V supply.

$\overline{EADS}$ The **external address strobe** input is used with AHOLD to signal that an external address is used to perform a cache-invalidation cycle.

$\overline{FERR}$ The floating-point error output indicates that the floating-point coprocessor has detected an error condition. It is used to maintain compatibility with DOS software.

$\overline{FLUSH}$ The **cache flush** input forces the microprocessor to erase the contents of its 8K-byte internal cache.

HLDA The **hold acknowledge** output indicates that the HOLD input is active and that the microprocessor has placed its buses at their high-impedance state.

TABLE 17–3 Bus cycle identification.

$m/\overline{IO}$	$D/\overline{C}$	$W/\overline{R}$	*Bus Cycle Type*
0	0	0	Interrupt acknowledge
0	0	1	Halt/special
0	1	0	I/O read
0	1	1	I/O write
1	0	0	Opcode fetch
1	0	1	Reserved
1	1	0	Memory read
1	1	1	Memory write

HOLD The **hold** input requests a DMA action. It causes the address, data, and control buses to be placed at their high-impedance state and also, once recognized, causes HLDA to become a logic 0.

$\overline{\text{IGNNE}}$ The **ignore numeric error** input causes the coprocessor to ignore floating-point errors and to continue processing data. This signal does not affect the state of the $\overline{\text{FERR}}$ pin.

INTR The **interrupt request** input requests a maskable interrupt, as it does in all other family members.

$\overline{\text{KEN}}$ The **cache enable** input causes the current bus to be stored in the internal cache.

$\overline{\text{LOCK}}$ The **lock** output becomes a logic 0 for any instruction that is prefixed with the lock prefix.

M/$\overline{\text{IO}}$ **Memory/IO** defines whether the address bus contains a memory address or an I/O port number. It is also combined with the W/$\overline{\text{R}}$ signal to generate memory and I/O read and write control signals.

NMI The **non-maskable interrupt** input requests a type 2 interrupt.

PCD The **page cache disable** output reflects the state of the PCD attribute bit in the page table entry or the page directory entry.

$\overline{\text{PCHK}}$ The **parity check** output indicates that a parity error was detected during a read operation on the DP_3–DP_0 pins.

$\overline{\text{PLOCK}}$ The **pseudo-lock** output indicates that the current operation requires more than one bus cycle to perform. This signal becomes a logic 0 for arithmetic coprocessor operations that access 64- or 80-bit memory data.

PWT The **page write through** output indicates the state of the PWT attribute bit in the page table entry or the page directory entry.

$\overline{\text{RDY}}$ The **ready** input indicates that a non-burst bus cycle is complete. The $\overline{\text{RDY}}$ signal must be returned, or the microprocessor places wait states into its timing until $\overline{\text{RDY}}$ is asserted.

RESET The **reset** input initializes the 80486, as it does in other family members. Table 17–4 shows the effect of the RESET input on the 80486 microprocessor.

W/$\overline{\text{R}}$ **Write/read** signals that the current bus cycle is either a read or a write.

TABLE 17–4 State of the microprocessor after a RESET.

Register	*Initial Value with Self-Test*	*Initial Value without Self-Test*
EAX	00000000H	?
EDX	00000400H + ID*	00000400H + ID*
EFLAGS	00000002H	00000002H
EIP	0000FFF0H	0000FFF0H
ES	0000H	0000H
CS	F000H	F000H
DS	0000H	0000H
SS	0000H	0000H
GS	0000H	0000H
FS	0000H	0000H
IDTR	Base = 0, limit = 3FFH	Base = 0, limit = 3FFH
CR_0	60000010H	60000010H
DR_7	00000000H	00000000H

*Revision ID number is supplied by Intel for revisions to the microprocessor.

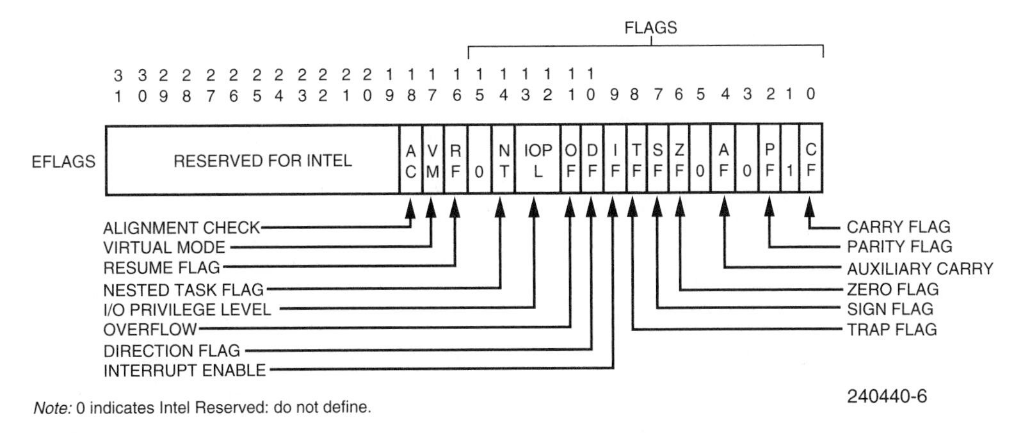

FIGURE 17–31 The EFLAG register of the 80486. (Courtesy of Intel Corporation.)

Basic 80486 Architecture

The architecture of the 80486DX is almost identical to that of the 80386. Added to the 80386 architecture inside the 80486DX is a math coprocessor and an 8K-byte level 1 cache memory. The 80486SX is almost identical to an 80386 with an 8K-byte cache, but no numeric coprocessor.

The extended flag register (EFLAG) is illustrated in Figure 17–31. As with other family members, the rightmost flag bits perform the same functions for compatibility. The only new flag bit is the AC (**alignment check**), used to indicate that the microprocessor has accessed a word at an odd address or a doubleword stored at a non-doubleword boundary. Efficient software and execution require that data be stored at word or doubleword boundaries.

80486 Memory System

The memory system for the 80486 is identical to the 80386 microprocessor. The 80486 contains 4G bytes of memory, beginning at location 00000000H and ending at location FFFFFFFFH. The major change to the memory system is internal to the 80486 in the form of an 8K-byte cache memory, which speeds the execution of instructions and the acquisition of data. Another addition is the parity checker/generator built into the 80486 microprocessor.

Parity Checker/Generator. Parity is often used to determine if data are correctly read from a memory location. To facilitate this, Intel has incorporated an internal parity generator/detector. Parity is generated by the 80486 during each write cycle. Parity is generated as even parity, and a parity bit is provided for each byte of memory. The parity check bits appear on pins DP_0–DP_3, which are also parity inputs as well as outputs. These are typically stored in memory during each write cycle and read from memory during each read cycle.

On a read, the microprocessor checks parity and generates a parity check error, if it occurs, on the $\overline{PCHK}$ pin. A parity error causes no change in processing unless the user applies the $\overline{PCHK}$ signal to an interrupt input. Interrupts are often used to signal a parity error in DOS-based computer systems. Figure 17–32 shows the organization of the 80486 memory system that includes parity storage. Note that this is the same as for the 80386, except for the parity bit storage. If parity is not used, Intel recommends that the DP_0–DP_3 pins be pulled up to +5.0 V.

Cache Memory. The cache memory system caches (stores) data used by a program and also the instructions of the program. The cache is organized as a four-way set associative cache, with each location (line) containing 16 bytes or four doublewords of data. The cache operates as a write-through cache. Note that the cache changes only if a miss occurs. This means that data written to a

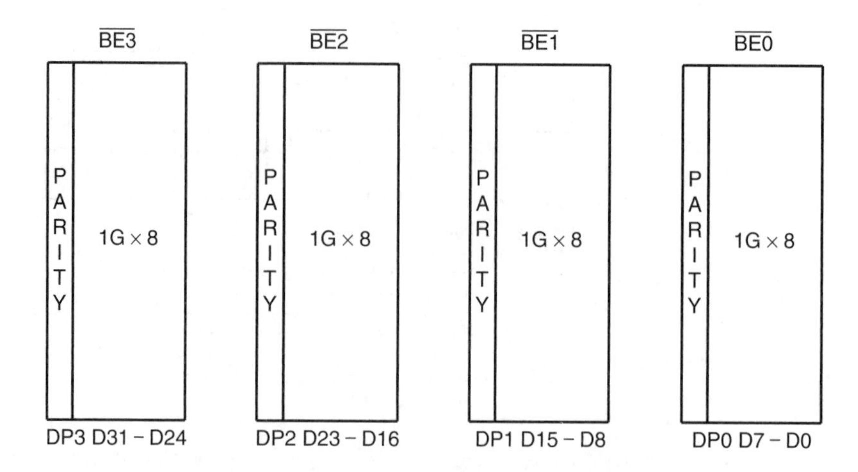

FIGURE 17–32 The organization of the 80486 memory, showing parity.

memory location not already cached are not written to the cache. In many cases, much of the active portion of a program is found completely inside the cache memory. This causes execution to occur at the rate of one clock cycle for many of the instructions that are commonly used in a program. About the only way that these efficient instructions are slowed is when the microprocessor must fill a line in the cache. Data are also stored in the cache, but it has less of an impact on the execution speed of a program because data are not referenced repeatedly as many portions of a program are.

Control register 0 (CR_0) is used to control the cache with two new control bits not present in the 80386 microprocessor. (See Figure 17–33 for CR_0 in the 80486 microprocessor.) The CD (cache disable) and NW (noncache write-through) bits are new to the 80486 and are used to control the 8K-byte cache. If the CD bit is a logic 1, all cache operations are inhibited. This setting is used only for debugging software and normally remains cleared. The NW bit is used to inhibit cache write-through operations. As with CD, cache write-through is inhibited only for testing. For normal program operation, CD = 0 and NW = 0.

Because the cache is new to the 80486 microprocessor and the cache is filled by using burst cycles not present on the 80386, some detail is required to understand bus-filling cycles. When a bus line is filled, the 80486 must acquire four 32-bit numbers from the memory system to fill a line in the cache. Filling is accomplished with a burst cycle. The burst cycle is a special memory in which four 32-bit numbers are fetched from the memory system in five clocking periods. This assumes that the speed of the memory is sufficient and that no wait states are required. If the clock frequency of the 80486 is 33 MHz, we can fill a cache line in 167 ns, which is very efficient considering that a normal, nonburst 32-bit memory read operation requires two clocking periods.

Memory Read Timing. Figure 17–34 illustrates the read timing for the 80486 for a nonburst memory operation. Note that two clocking periods are used to transfer data. Clocking period T_1 provides the memory address and control signals, and clocking period T_2 is where the data are transferred between the memory and the microprocessor. Note that the $\overline{RDY}$ must become a logic 0 to cause data to be transferred and to terminate the bus cycle. Access time for a nonburst access is determined by taking two clocking periods, minus the time required for the address to appear on the address bus connection, minus a setup time for the data bus connections. For the 20 MHz version of the 80486, two clocking periods require 100 ns minus 28 ns for address setup time and 6 ns for data setup time. This yields a nonburst access time of 100 ns – 34 ns, or 76 ns.

31							24	23							16	15							8	7							0
PG	CE	WT											AM		WP											NE		TS	EM	MP	PE

FIGURE 17–33 Control register zero (CR0) for the 80486 microprocessor.

FIGURE 17–34 The non-burst read timing for the 80486 microprocessor.

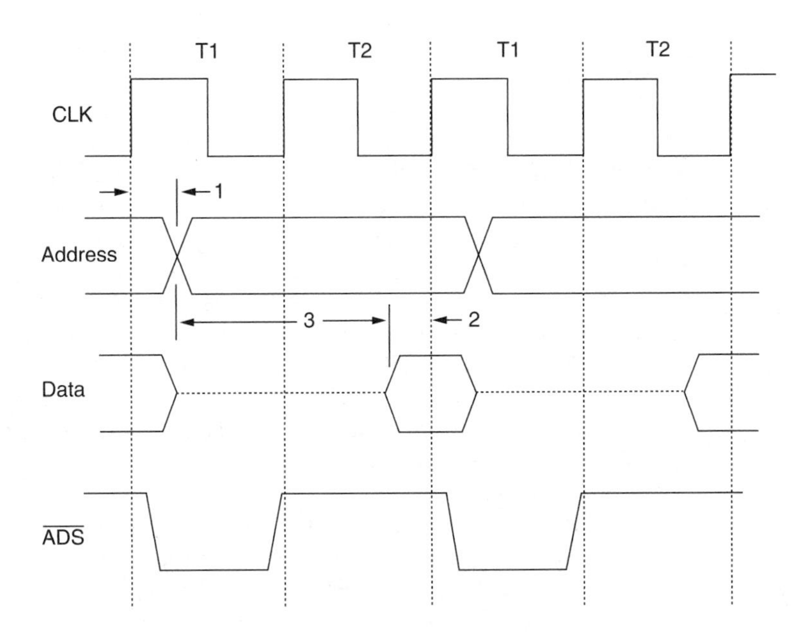

Of course, if decoder time and delay times are included, the access time allowed the memory is even less for no wait-state operation. If a higher frequency version of the 80486 is used in a system, memory access time is still less.

The 80486 33 MHz, 66 MHz, and 100 MHz processors all access bus data at a 33 MHz rate. In other words, the microprocessor may operate at 100 MHz, but the system bus operates at 33 MHz. Notice that the nonburst access timing for the 33 MHz system bus allows 60 ns – 24 ns = 36 ns. It is obvious that wait states are required for operation with standard DRAM memory devices.

Figure 17–35 illustrates the timing diagram for filling a cache line with four 32-bit numbers using a burst. Note that the addresses (A_{31}–A_4) appear during T_1 and remain constant throughout the burst cycle. Also, note that A_2 and A_3 change during each T_1 after the first to address four consecutive 32-bit numbers in the memory system. As mentioned, cache fills using

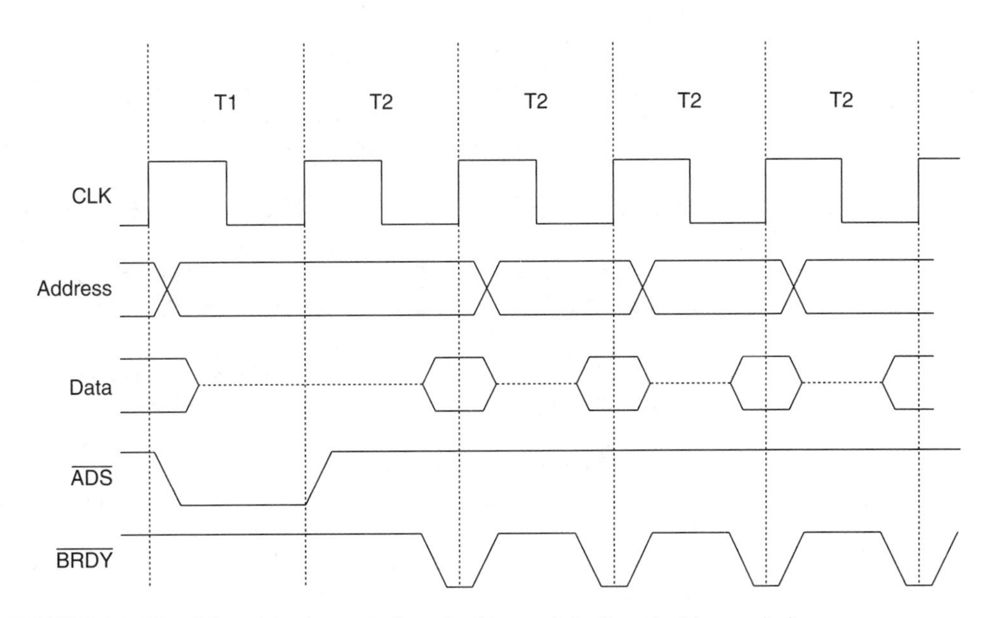

FIGURE 17–35 A burst cycle reads four doublewords in five clocking periods.

bursts require only five clocking periods (one T_1 and four T_2s) to fill a cache line with four doublewords of data. Access time using a 20 MHz version of the 80486 for the second and subsequent doublewords is 50 ns – 28 ns – 5 ns, or 17 ns, assuming no delays in the system. To use burst mode transfers, we need high-speed memory. Because DRAM memory access times are 40 ns at best, we are forced to use SRAM for burst cycle transfers. The 33 MHz system allows an access time of 30 ns – 19 ns – 5 ns, or 6 ns for the second and subsequent bytes. If an external counter is used in place of address bits A_2 and A_3, the 19 ns can be eliminated and the access time becomes 30 ns – 5 ns, or 25 ns, which is enough time for even the slowest SRAM connected to the system as a cache. This circuit is often called a *synchronous burst mode cache* if a SRAM cache is used with the system. Note that the $\overline{BRDY}$ pin acknowledges a burst transfer rather than the $\overline{RDY}$ pin, which acknowledges a normal memory transfer.

The PWT controls how the cache functions for a write operation of the external cache memory; it does not control writing to the internal cache. The logic level of this bit is found on the PWT pin of the 80486 microprocessor. Externally, it can be used to dictate the write-though policy of the external cache.

The PCD bit controls the on-chip cache. If the PCD = 0, the on-chip cache is enabled for the current page of memory. Note that 80386 page table entries place a logic 0 in the PCD bit position, enabling caching. If PCD = 1, the on-chip cache is disabled. Caching is disabled, regardless of the condition of $\overline{KEN}$, CD, and NW.

17–8 SUMMARY

1. The 80386 microprocessor is an enhanced version of the 80286 microprocessor and includes a memory-management unit that is enhanced to provide memory paging. The 80386 also includes 32-bit extended registers, and a 32-bit address and data bus. A scaled-down version of the 80386DX with a 16-bit data bus and 24-bit address bus is available as the 80386SX microprocessor. The 80386EX is a complete AT-style personal computer on a chip.
2. The 80386 has a physical memory size of 40 bytes that can be addressed as a virtual memory with up to 64T bytes. The 80386 memory is 32 bits wide, and it is addressed as bytes, words, or doublewords.
3. When the 80386 is operated in the pipelined mode, it sends the address of the next instruction or memory data to the memory system prior to completing the execution of the current instruction. This allows the memory system to begin fetching the next instruction or data before the current is completed. This increases access time, thus reducing the speed of the memory.
4. A cache memory system allows data that are frequently read to be accessed in less time because they are stored in high-speed semiconductor memory. If data are written to memory, they are also written to the cache, so the most current data are always present in the cache.
5. The I/O structure of the 80386 is almost identical to the 80286, except that I/O can be inhibited when the 80386 is operated in the protected mode through the I/O bit protection map stored with the TSS.
6. In the 80386 microprocessor, interrupts have been expanded to include additional predefined interrupts in the interrupt vector table. These additional interrupts are used with the memory-management system.
7. The 80386 memory manager is similar to the 80286, except that the physical addresses generated by the MMU are 32 bits wide instead of 24 bits wide. The 80386 MMU is also capable of paging.
8. The 80386 is operated in the real mode (8086 mode) when it is reset. The real mode allows the microprocessor to address data in the first 1M byte of memory. In the protected mode, the 80386 addresses any location in its 4G bytes of physical address space.

9. A descriptor is a series of eight bytes that specify how a code or data segment is used by the 80386. The descriptor is selected by a selector that is stored in one of the segment registers. Descriptors are used only in the protected mode.
10. Memory management is accomplished through a series of descriptors, stored in descriptor tables. To facilitate memory management, the 80386 uses three descriptor tables: the global descriptor table (GDT), the local descriptor table (LDT), and the interrupt descriptor table (IDT). The GDT and LDT each hold up to 8192 descriptors; the IDT holds up to 256 descriptors. The GDT and LDT describe code and data segments as well as tasks. The IDT describes the 256 different interrupt levels through interrupt gate descriptors.
11. The TSS (task state segment) contains information about the current task and the previous task. Appended to the end of the TSS is an I/O bit protection map that inhibits selected I/O port addresses.
12. The memory paging mechanism allows any 4K-byte physical memory page to be mapped to any 4K-byte linear memory page. For example, memory location 00A00000H can be assigned memory location A0000000H through the paging mechanism. A page directory and page tables are used to assign any physical address to any linear address. The paging mechanism can be used in the protected mode or the virtual mode.
13. The 80486 microprocessor is an improved version of the 80386 microprocessor that contains an 8K-byte cache and an 80387 arithmetic coprocessor; it executes many instructions in one clocking period.
14. The 80486 microprocessor executes a few new instructions that control the internal cache memory and allow addition (XADD) and comparison (CMPXCHG) with an exchange and a byte swap (BSW AP) operation. Other than these few additional instructions, the 80486 is 100% upward-compatible with the 80386 and 80387.
15. A new feature found in the 80486 is the BIST (built-in self-test) that tests the microprocessor, coprocessor, and cache at reset time. If the 80486 passes the test, EAX contains a zero.

17–9 QUESTIONS AND PROBLEMS

1. The 80386 microprocessor addresses __________ bytes of memory in the protected mode.
2. The 80386 microprocessor addresses __________ bytes of virtual memory through the memory-management unit.
3. Describe the differences between the 80386DX and the 80386SX.
4. Draw the memory map of the 80386 when operated in the (a) protected mode; (b) real mode.
5. How much current is available on various 80386 output pin connections? Compare these currents with the currents available at the output pin connection of an 8086 microprocessor.
6. Describe the 80386 memory system, and explain the purpose and operation of the bank selection signals.
7. Explain the action of a hardware reset on the address bus connections of the 80386.
8. Explain how pipelining lengthens the access time for many memory references in the 80386 microprocessor-based system.
9. Briefly describe how the cache memory system functions.
10. I/O ports in the 80386 start at I/O address __________ and extend to I/O address __________.
11. What I/O ports communicate data between the 80386 and its companion 80387 coprocessor?
12. Compare and contrast the memory and I/O connections found on the 80386 with those found in earlier microprocessors.
13. If the 80386 operates at 20 MHz, what clocking frequency is applied to the CLK_2 pin?

14. What is the purpose of the $\overline{BS16}$ pin on the 80386 microprocessor?
15. Define the purpose of each of the control registers (CR_0, CR_1, CR_2, and CR_3) found within the 80386.
16. Define the purpose of each 80386 debug register.
17. The debug registers cause which level of interrupt?
18. What are the test registers?
19. Select an instruction that copies control register 0 into EAX.
20. Describe the purpose of PE in CR_0.
21. Form an instruction that accesses data in the FS segment at the location indirectly addressed by the DI register. The instruction should store the contents of EAX into this memory location.
22. What is scaled index addressing?
23. Is the following instruction legal? MOV AX,[EBX+ECX]
24. Explain how the following instructions calculate the memory address:
    (a) ADD [EBX+8*ECX],AL
    (b) MOV DATA[EAX+EBX],CX
    (c) SUB EAX,DATA
    (d) MOV ECX,[EBX]
25. What is the purpose of interrupt type number 7?
26. Which interrupt vector type number is activated for a protection privilege violation?
27. What is a double interrupt fault?
28. If an interrupt occurs in the protected mode, what defines the interrupt vectors?
29. What is a descriptor?
30. What is a selector?
31. How does the selector choose the local descriptor table?
32. What register is used to address the global descriptor table?
33. How many global descriptors can be stored in the GDT?
34. Explain how the 80386 can address a virtual memory space of 64T bytes when the physical memory contains only 4G bytes of memory.
35. What is the difference between a segment descriptor and a system descriptor?
36. What is the task state segment (TSS)?
37. How is the TSS addressed?
38. Describe how the 80386 switches from the real mode to the protected mode.
39. Describe how the 80386 switches from the protected mode to the real mode.
40. What is virtual 8086 mode operation of the 80386 microprocessor?
41. How is the paging directory located by the 80386?
42. How many bytes are found in a page of memory?
43. Explain how linear memory address D0000000H can be assigned to physical memory address C0000000H with the paging unit of the 80386.
44. What are the differences between an 80386 and 80486 microprocessor?
45. What is the purpose of the $\overline{FLUSH}$ input pin on the 80486 microprocessor?
46. Compare the register set of the 80386 with the 80486 microprocessor.
47. What differences exist in the flags of the 80486 when compared to the 80386 microprocessor?
48. Which pins are used for parity checking on the 80486 microprocessor?
49. The 80486 microprocessor uses __________ parity.
50. The cache inside the 80486 microprocessor is __________ - K bytes.
51. A cache line is filled by reading __________ bytes from the memory system.
52. What is an 80486 burst?
53. Define the term *cache-write through.*
54. What is a BIST?

CHAPTER 18

The Pentium and Pentium Pro Microprocessors

INTRODUCTION

The Pentium microprocessor signals an improvement to the architecture found in the 80486 microprocessor. The changes include an improved cache structure, a wider data bus width, a faster numeric coprocessor, a dual integer processor, and branch prediction logic. The cache has been reorganized to form two caches that are each 8K bytes in size, one for caching data, and the other for instructions. The data bus width has been increased from 32 bits to 64 bits. The numeric coprocessor operates about five times faster than the 80486 numeric coprocessor. A dual integer processor often allows two instructions per clock. Finally, the branch prediction logic allows programs that branch to execute more efficiently. Notice that these changes are internal to the Pentium, which makes software upward-compatible from earlier Intel 80X86 microprocessors. A later improvement to the Pentium was the addition of the MMX instructions.

The Pentium Pro is a still faster version of the Pentium. It contains a modified internal architecture that can schedule up to five instructions for execution and an even faster floating-point unit. The Pentium Pro also contains a 256K-byte or 512K-byte level 2 cache in addition to the 16K-byte (8K for data and 8K for instruction) level 1 cache. The Pentium Pro includes error correction circuitry (ECC) to correct a one-bit error and indicate a two-bit error. Also added are four additional address lines, giving the Pentium Pro access to an astounding 64G bytes of directly addressable memory space.

CHAPTER OBJECTIVES

Upon completion of this chapter, you will be able to:

1. Contrast the Pentium and Pentium Pro with the 80386 and 80486 microprocessors
2. Describe the organization and interface of the 64-bit-wide Pentium memory system and its variations
3. Contrast the changes in the memory-management unit and paging unit when compared to the 80386 and 80486 microprocessors
4. Detail the new instructions found with the Pentium microprocessor
5. Explain how the superscaler dual integer units improve performance of the Pentium microprocessor
6. Describe the operation of the branch prediction logic
7. Detail the improvements in the Pentium Pro when compared with the Pentium
8. Explain how the dynamic execution architecture of the Pentium Pro functions

18–1 INTRODUCTION TO THE PENTIUM MICROPROCESSOR

Before the Pentium or any other microprocessor can be used in a system, the function of each pin must be understood. This section of the chapter details the operation of each pin, along with the external memory system and I/O structures of the Pentium microprocessor.

Figure 18–1 illustrates the pin-out of the Pentium microprocessor, which is packaged in a huge 237-pin PGA (pin grid array). The Pentium was made available in two versions: the full-blown Pentium and the P24T version called the Pentium OverDrive. The P24T version contains a 32-bit data bus, compatible for insertion into 80486 machines, which contains the P24T socket.

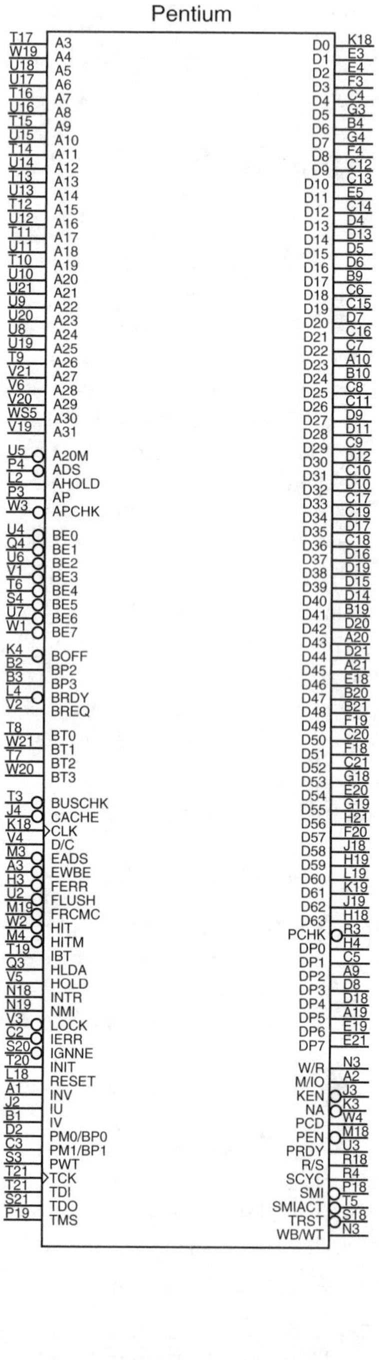

FIGURE 18–1 The pin-out of the Pentium microprocessor.

Pentium

T17 A3
W19 A4
U18 A5
U17 A6
T16 A7
U16 A8
T15 A9
U15 A10
T14 A11
U14 A12
T13 A13
U13 A14
T12 A15
U12 A16
T11 A17
U11 A18
T10 A19
U10 A20
U21 A21
U9 A22
U20 A23
U8 A24
U19 A25
T9 A26
V21 A27
V6 A28
V20 A29
WS5 A30
V19 A31
U5 A20M
P4 ADS
L2 AHOLD
P3 AP
W3 APCHK
U4 BE0
Q4 BE1
U6 BE2
V1 BE3
T6 BE4
S4 BE5
U7 BE6
W1 BE7
K4 BOFF
B2 BP2
B3 BP3
L4 BRDY
V2 BREQ
T8 BT0
W21 BT1
T7 BT2
W20 BT3
T3 BUSCHK
J4 CACHE
K18 CLK
V4 D/C
M3 EADS
A3 EWBE
H3 FERR
U2 FLUSH
M19 FRCMC
W2 HIT
M4 HITM
T19 IBT
Q3 HLDA
V5 HOLD
N18 INTR
N19 NMI
V3 LOCK
C2 IERR
S20 IGNNE
T20 INIT
L18 RESET
A1 INV
J2 IU
B1 IV
D2 PM0/BP0
C3 PM1/BP1
S3 PWT
T21 TCK
T21 TDI
S21 TDO
P19 TMS

D0 K18
D1 E3
D2 E4
D3 F3
D4 C4
D5 G3
D6 B4
D7 G4
D8 F4
D9 C12
D10 C13
D11 E5
D12 C14
D13 D4
D14 D13
D15 D5
D16 D6
D17 B9
D18 C6
D19 C15
D20 D7
D21 C16
D22 C7
D23 A10
D24 B10
D25 C8
D26 C11
D27 D9
D28 D11
D29 C9
D30 D12
D31 C10
D32 D10
D33 C17
D34 C19
D35 D17
D36 C18
D37 D16
D38 D19
D39 D15
D40 D14
D41 B19
D42 D20
D43 A20
D44 D21
D45 A21
D46 E18
D47 B20
D48 B21
D49 F19
D50 C20
D51 F18
D52 C21
D53 G18
D54 E20
D55 G19
D56 H21
D57 F20
D58 J18
D59 H19
D60 L19
D61 K19
D62 J19
D63 H18
PCHK R3
DP0 H4
DP1 C5
DP2 A9
DP3 D8
DP4 D18
DP5 A19
DP6 E19
DP7 E21
W/R N3
M/IO A2
KEN J3
NA K3
PCD W4
PEN M18
PRDY U3
R/S R18
SCYC R4
SMI P18
SMIACT T5
TRST S18
WB/WT N3

The P24T version also comes with a fan built into the unit. The most notable difference in the pin-out of the Pentium, when compared to earlier 80486 microprocessors, is that there are 64 data bus connections instead of 32, which require a larger physical footprint.

As with earlier versions of the Intel family of microprocessors, the early versions of the Pentium require a single +5.0 V power supply for operation. The power supply current averages 3.3 A for the 66 MHz version of the Pentium, and 2.91 A for the 60 MHz version. Because these currents are significant, so are the power dissipations of these microprocessors: 13 W for the 66 MHz version and 11.9 W for the 60 MHz version. The current versions of the Pentium, 90 MHz and above, use a 3.3 V power supply with reduced current consumption. At present, a good heat sink with considerable airflow is required to keep the Pentium cool. The Pentium contains multiple V_{CC} and V_{SS} connections that must all be connected to +5.0 V or +3.3 V and ground for proper operation. Some of the pins are labeled N/C (no connection) and must not be connected. The latest versions of the Pentium have been improved to reduce the power dissipation. For example, the 233 MHz Pentium requires 3.4 A or current, which is only slightly more than the 3.3 A required by the early 66 MHz version.

Each Pentium output pin is capable of providing 4.0 mA of current at a logic 0 level and 2.0 mA at a logic 1 level. This represents an increase in drive current, compared to the 2.0 mA available on earlier 8086, 8088, and 80286 output pins. Each input pin represents a small load requiring only 15 μA of current. In some systems, except the smallest, these current levels require bus buffers.

The function of each Pentium group of pins follows:

$\overline{\text{A20}}$ The **address A20 mask** is an input that is asserted in the real mode to signal the Pentium to perform address wraparound, as in the 8086 microprocessor, for use of the HIMEM.SYS driver.

A_{31}–A_3 **Address bus** connections address any of the 512K × 64 memory locations found in the Pentium memory system. Note that A_0, A_1, and A_2 are encoded in the bus enable ($\overline{\text{BE7}}$–$\overline{\text{BE0}}$), described elsewhere, to select any or all of the eight bytes in a 64-bit-wide memory location.

$\overline{\text{ADS}}$ The **address data strobe** becomes active whenever the Pentium has issued a valid memory or I/O address. This signal is combined with the W/$\overline{\text{R}}$ and M/$\overline{\text{IO}}$ signals to generate the separate read and write signals present in the earlier 8086–80286 microprocessor-based systems.

AHOLD **Address hold** is an input that causes the Pentium to hold the address and AP signals for the next clock.

$\overline{\text{APCHK}}$ **Address parity** provides even parity for the memory address on all Pentium-initiated memory and I/O transfers. The AP pin must also be driven with even parity information on all inquire cycles in the same clocking period as the $\overline{\text{EADS}}$ signal. Address parity check becomes a logic 0 whenever the Pentium detects an address parity error.

$\overline{\text{BE7}}$–$\overline{\text{BE0}}$ **Bank enable** signals select the access of a byte, word, doubleword, or quadword of data. These signals are generated internally by the microprocessor from address bits A_0, A_1, and A_2.

$\overline{\text{BOFF}}$ The **back-off** input aborts all outstanding bus cycles and floats the Pentium buses until $\overline{\text{BOFF}}$ is negated. After $\overline{\text{BOFF}}$ is negated, the Pentium restarts all aborted bus cycles in their entirety.

BP_3–BP_0 The **breakpoint** pins BP_3–BP_0 indicate a breakpoint match when the debug registers are programmed to monitor for matches.

PM_1–PM_0 The **performance monitoring** pins PM_1 and PM_0 indicate the settings of the performance monitoring bits in the debug mode control register.

$\overline{\text{BRDY}}$ The **burst ready** input signals the Pentium that the external system has applied or extracted data from the data bus connections. This signal is used to insert wait states into the Pentium timing.

BREQ The **bus request** output indicates that the Pentium has generated a bus request.

BT_3–BT_0 The **branch trace** outputs provide bits 2–0 of the branch target linear address and the default operand size on BT_3. These outputs become valid during a branch trace special message cycle.

$\overline{\text{BUSCHK}}$ The **bus check** input allows the system to signal the Pentium that the bus transfer has been unsuccessful.

$\overline{\text{CACHE}}$ The **cache output** indicates that the current Pentium cycle can cache data.

CLK The **clock** is driven by a clock signal that is at the operating frequency of the Pentium. For example, to operate the Pentium at 66 MHz, apply a 66 MHz clock to this pin.

D_{63}–D_0 **Data bus** connections transfer byte, word, doubleword, and quadword data between the microprocessor and its memory and I/O system.

$D/\overline{C}$ **Data/control** indicates that the data bus contains data for or from memory or I/O when a logic 1. If $D/\overline{C}$ is a logic 0, the microprocessor is either halted or executing an interrupt acknowledge.

DP_7–DP_0 **Data parity** is generated by the Pentium and detects its eight memory banks through these connections.

$\overline{\text{EADS}}$ The **external address strobe** input signals that the address bus contains an address for an inquire cycle.

$\overline{\text{EWBE}}$ The **external write buffer empty** input indicates that a write cycle is pending in the external system.

$\overline{\text{FERR}}$ The **floating-point error** is comparable to the $\overline{\text{ERROR}}$ line in the 80386 and shows that the internal coprocessor has erred.

$\overline{\text{FLUSH}}$ The **flush cache** input causes the cache to flush all write-back lines and invalidate its internal caches. If the $\overline{\text{FLUSH}}$ input is a logic 0 during a reset operation, the Pentium enters its test mode.

$\overline{\text{FRCMC}}$ The **functional redundancy check** is sampled during a reset to configure the Pentium in the master (1) or checker mode (0).

$\overline{\text{HIT}}$ **Hit** shows that the internal cache contains valid data in the inquire mode.

$\overline{\text{HITM}}$ **Hit modified** shows that the inquire cycle found a modified cache line. This output is used to inhibit other master units from accessing data until the cache line is written to memory.

HOLD **Hold** requests a DMA action.

HLDA **Hold acknowledge** indicates that the Pentium is currently in a hold condition.

IBT **Instruction branch taken** indicates that the Pentium has taken an instruction branch.

$\overline{\text{IERR}}$ The **internal error** output shows that the Pentium has detected an internal parity error or functional redundancy error.

$\overline{\text{IGNNE}}$ The **ignore numeric error** input causes the Pentium to ignore a numeric coprocessor error.

INIT The **initialization** input performs a reset without initializing the caches, write-back buffers, and floating-point registers. This may not be used to reset the microprocessor in lieu of RESET after power-up.

INTR The **interrupt request** is used by external circuitry to request an interrupt.

INV The **invalidation** input determines the cache line state after an inquiry.

IU The **U-Pipe instruction complete** output shows that the instruction in the U-pipe is complete.

IV The **V-Pipe instruction complete** output shows that the instruction in the V-pipe is complete.

$\overline{\text{KEN}}$ The **cache enable** input enables internal caching.

$\overline{\text{LOCK}}$ **LOCK** becomes a logic 0 whenever an instruction is prefixed with the LOCK: prefix. This is most often used during DMA accesses.

M/$\overline{\text{IO}}$ **Memory/IO** selects a memory device when a logic 1 or an I/O device when a logic 0. During the I/O operation, the address bus contains a 16-bit I/O address on address connections A_{15}–A_3.

$\overline{\text{NA}}$ **Next address** indicates that the external memory system is ready to accept a new bus cycle.

NMI The **non-maskable interrupt** requests a non-maskable interrupt, just as on the earlier versions of the microprocessor.

PCD The **page cache disable** output shows that the internal page caching is disabled by reflecting the state of the CR_3 PCD bit.

$\overline{\text{PCHK}}$ The **parity check** output signals a parity check error for data read from memory or I/O.

$\overline{\text{PEN}}$ The **parity enable** input enables the machine check interrupt or exception.

PRDY The **probe ready** output indicates that the probe mode has been entered for debugging.

PWT The **page write-through** output shows the state of the PWT bit in CR_3. This pin is provided for use with the Intel Debugging Port and causes an interrupt.

RESET **Reset** initializes the Pentium, causing it to begin executing software at memory location FFFFFFF0H. The Pentium is reset to the real mode and the leftmost 12 address connections remain logic 1s (FFFH) until a far jump or far call is executed. This allows compatibility with earlier microprocessors. See Table 18–1 for the state of the Pentium after a hardware reset.

SCYC The **split cycle** output signals a misaligned LOCKed bus cycle.

$\overline{\text{SMI}}$ The **system management interrupt** input causes the Pentium to enter the system management mode of operation.

$\overline{\text{SMIACT}}$ The **system management interrupt active** output shows that the Pentium is operating in the system management mode.

TCK The **testability clock** input selects the clocking function in accordance with the IEEE 1149.1 Boundary Scan interface.

TDI The **test data input** is used to test data clocked into the Pentium with the TCK signal.

TDO The **test data output** is used to gather test data and instructions shifted out of the Pentium with TCK.

TMS The **test mode select** input controls the operation of the Pentium in test mode. The test reset input allows the test mode to be reset.

W/$\overline{\text{R}}$ **Write/read** indicates that the current bus cycle is a write when a logic 1 or a read when a logic 0.

WB/$\overline{\text{WT}}$ **Write-back/write-through** selects the operation for the Pentium data cache.

TABLE 18–1 State of the Pentium after a RESET.

Register	*RESET Value*	*RESET + BIST Value*
EAX	0	0 (if test passes)
EDX	0500XXXXH	0500XXXXH
EBX, ECX, ESP, EBP, ESI, and EDI	0	0
EFLAGS	2	2
EIP	0000FFF0H	0000FFF0H
CS	F000H	F000H
DS, ES, FS, GS, and SS	0	0
GDTR and TSS	0	0
CR_0	60000010H	60000010H
CR_2, CR_3, and CR_4	0	0
DR_0–DR_3	0	0
DR_6	FFFF0FF0H	FFFF0FF0H
DR_7	00000400H	00000400H

The Memory System

The memory system for the Pentium microprocessor is 4G bytes in size, just as in the 80386DX and 80486 microprocessors. The difference lies in the width of the memory data bus. The Pentium uses a 64-bit data bus to address memory organized in eight banks that each contain 512M bytes of data. See Figure 18–2 for the organization of the Pentium physical memory system.

The Pentium memory system is divided into eight banks that each store a byte of data with a parity bit. The Pentium, like the 80486, employs internal parity generation and checking logic for the memory system's data bus information. (Note that most Pentium systems do not use parity checks, but it is available.) The 64-bit-wide memory is important to double-precision floating-point data. Recall that a double-precision floating-point number is 64 bits wide. Because of the change to a 64-bit-wide data bus, the Pentium is able to retrieve floating-point data with one read cycle, instead of two as in the 80486. This causes the Pentium to function at a higher throughput than an 80486. As with earlier 32-bit Intel microprocessors, the memory system is numbered in bytes, from byte 00000000H to byte FFFFFFFFH.

Memory selection is accomplished with the bank enable signals ($\overline{BE7}$–$\overline{BE0}$). These separate memory banks allow the Pentium to access any single byte, word, doubleword, or quadword with one memory transfer cycle. As with earlier memory selection logic, eight separate write strobes are generated for writing to the memory system.

A new feature added to the Pentium is its capability to check and generate parity for the address bus (A_{31}–A_5) during certain operations. The AP pin provides the system with parity

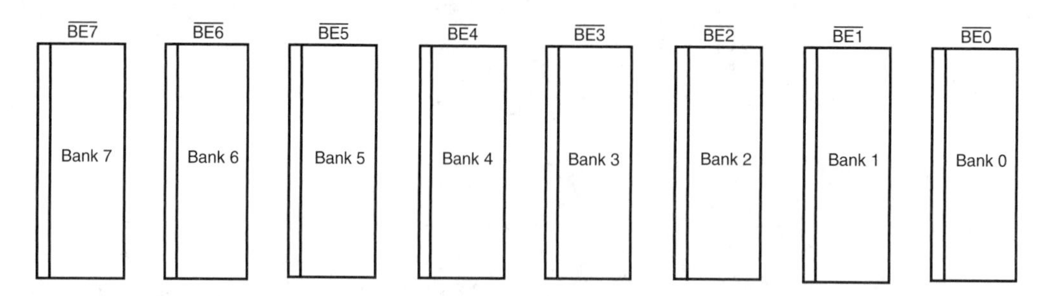

FIGURE 18–2 The 8-byte-wide memory banks of the Pentium microprocessor.

information and the $\overline{APCHK}$ indicates a bad parity check for the address bus. The Pentium takes no action when an address parity error is detected. The error must be assessed by the system and the system must take appropriate action (an interrupt), if so desired.

Input/Output System

The input/output system of the Pentium is completely compatible with earlier Intel microprocessors. The I/O port number appears on address lines A_{15}–A_3 with the bank enable signals used to select the actual memory banks used for the I/O transfer.

Beginning with the 80386 microprocessor, I/O privilege information is added to the TSS segment when the Pentium is operated in the protected mode. Recall that this allows I/O ports to be selectively inhibited. If the blocked I/O location is accessed, the Pentium generates a type 13 interrupt to signal an I/O privilege violation.

System Timing

As with any microprocessor, the system timing signals must be understood in order to interface the microprocessor. This portion of the text details the operation of the Pentium through its timing diagrams and shows how to determine memory access times.

The basic Pentium nonpipelined memory cycle consists of two clocking periods: T_1 and T_2. See Figure 18–3 for the basic nonpipelined read cycle. Notice from the timing diagram that the 66 MHz Pentium is capable of 33 million memory transfers per second. This assumes that the memory can operate at that speed.

Also notice from the timing diagram that the $W/\overline{R}$ signal becomes valid if $\overline{ADS}$ is a logic 0 at the positive edge of the clock (end of T_1). This clock must be used to qualify the cycle as a read or a write.

During T_1, the microprocessor issues the $\overline{ADS}$, $W/\overline{R}$, address, and $M/\overline{IO}$ signals. In order to qualify the $W/\overline{R}$ signal and generate appropriate $\overline{MRDC}$ and $\overline{MWTC}$ signals, we use a flip-flop

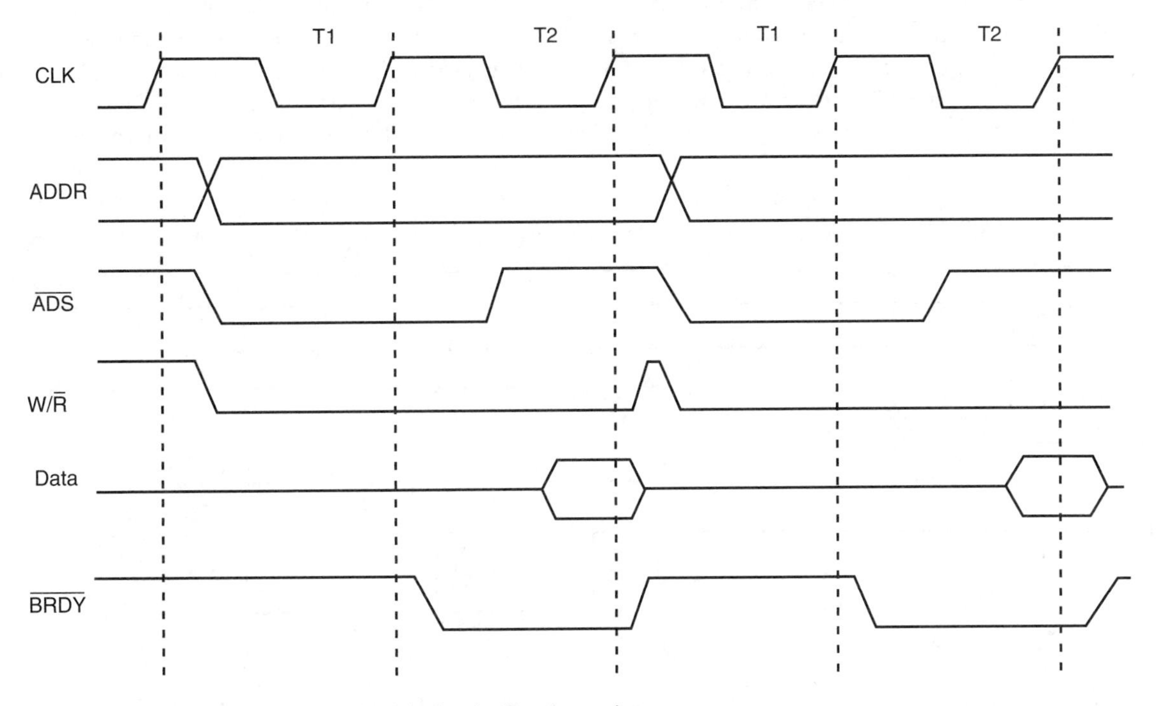

FIGURE 18–3 The nonpipelined read cycle for the Pentium microprocessor.

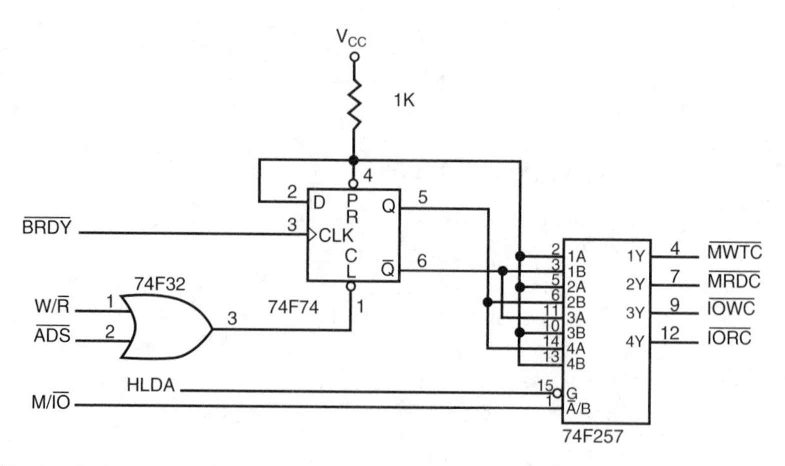

FIGURE 18–4 A circuit that generates the memory and I/O control signals.

to generate the W/$\overline{R}$ signal. Then a two-line-to-one-line multiplexer generates the memory and I/O control signals. See Figure 18–4 for a circuit that generates the memory and I/O control signals for the Pentium microprocessor.

During T_2, the data bus is sampled in synchronization with the end of T_2 at the positive transition of the clock pulse. The setup time before the clock is given as 3.8 ns, and the hold time after the clock is given as 2.0 ns. This means that the data window around the clock is 5.8 ns. The address appears on the 8.0 ns maximum after the start of T_1. This means that the Pentium microprocessor operating at 66 MHz allows 30.3 ns (two clocking periods), minus the address delay time of 8.0 ns and minus the data setup time of 3.8 ns. Memory access time without any wait states is 30.3 – 8.0 – 3.8, or 18.5 ns. This is enough time to allow access to a SRAM, but not to any DRAM without inserting wait states into the timing. The SRAM is normally found in the form of an external level 2 cache.

Wait states are inserted into the timing by controlling the $\overline{\text{BRDY}}$ input to the Pentium. The $\overline{\text{BRDY}}$ signal must become a logic 0 by the end of T_2 or additional T_2 states are inserted into the timing. See Figure 18–5 for a read cycle timing diagram that contains wait states for slower memory. The effect of inserting wait states into the timing is to lengthen the timing, allowing additional time to the memory to access data. In the timing shown, the access time has been lengthened so that

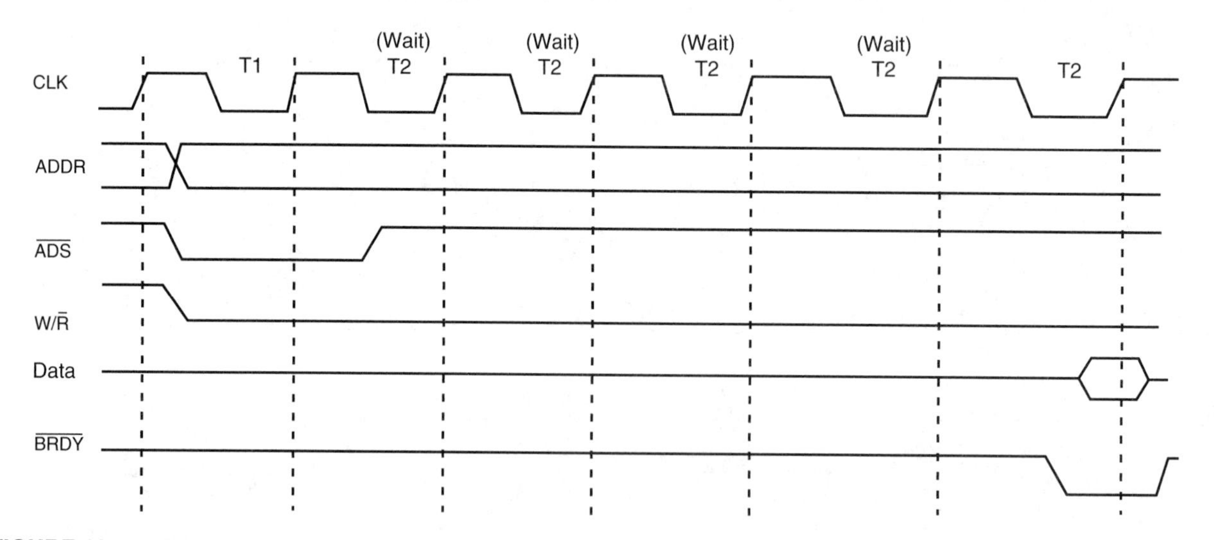

FIGURE 18–5 The Pentium timing diagram with four wait states inserted for an access time of 79.5 ns.

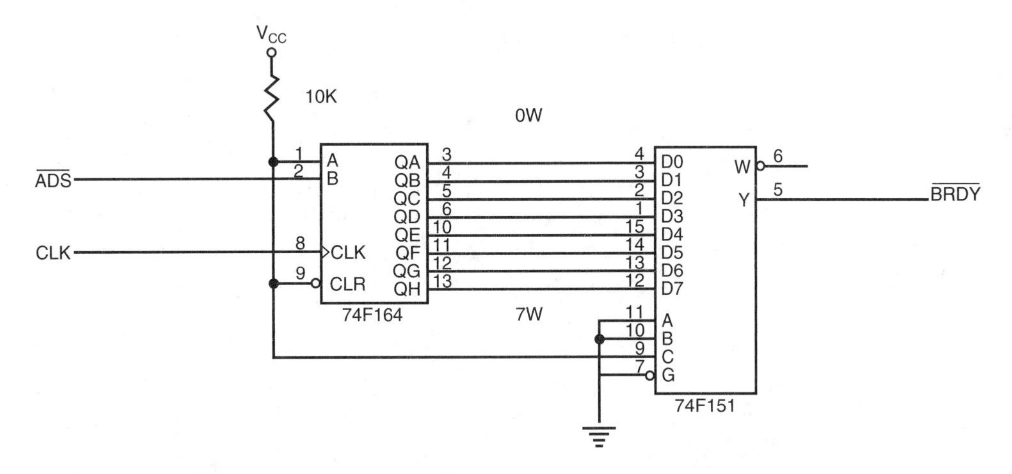

FIGURE 18–6 A circuit that generates wait states by delaying ADS. This circuit is wired to generate four wait states.

standard 60 ns DRAM can be used in a system. Note that this requires the insertion of four wait states of 15.2 ns (one clocking period) each to lengthen the access time to 79.5 ns. This is enough time for the DRAM and any decoder in the system to function.

The $\overline{\text{BRDY}}$ signal is a synchronous signal generated by using the system clock. Figure 18–6 illustrates a circuit that can be used to generate $\overline{\text{BRDY}}$ for inserting any number of wait states into the Pentium timing diagram. You may recall a similar circuit inserting wait states into the timing diagram of the 80386 microprocessor. The $\overline{\text{ADS}}$ signal is delayed between 0 and 7 clocking periods by the 74F161 shift register to generate the $\overline{\text{BRDY}}$ signal. The exact number of wait states is selected by the 74F151 eight-line-to-one-line multiplexer. In this example, the multiplexer selects the four-wait output from the shift register.

A more efficient method of reading memory data is via the burst cycle. The burst cycle in the Pentium transfers four 64-bit numbers per burst cycle in five clocking periods. A burst without wait states requires that the memory system transfers data every 15.2 ns. If a level 2 cache is in place, this speed is no problem as long as the data are read from the cache. If the cache does not contain the data, then wait states must be inserted, which will reduce the data throughput. See Figure 18–7 for the Pentium burst cycle transfer without wait states. As before, wait states can be inserted to allow more time to the memory system for accesses.

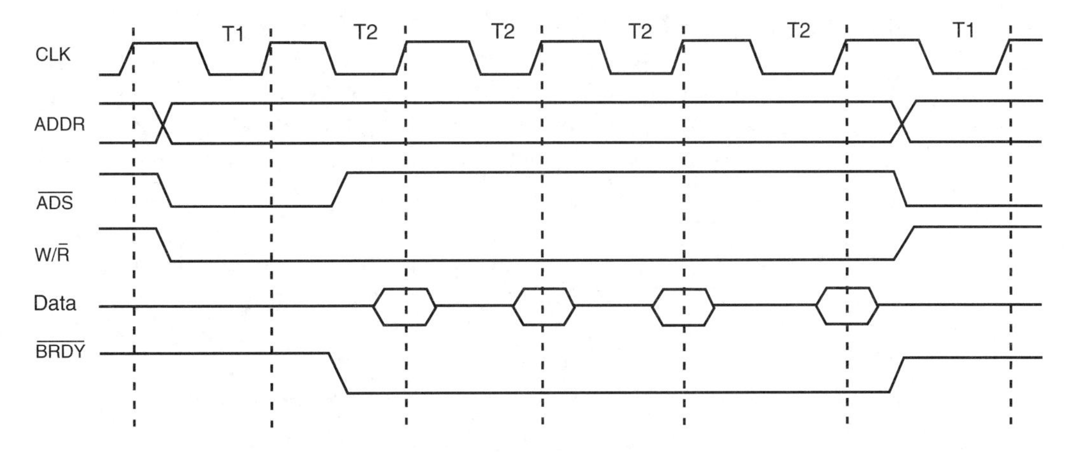

FIGURE 18–7 The Pentium burst cycle operation that transfers four 64-bit data between the microprocessor and memory.

Branch Prediction Logic

The Pentium microprocessor uses a branch prediction logic to reduce the time required for a branch caused by internal delays. These delays are minimized because when a branch instruction (short or near only) is encountered, the microprocessor begins prefetch instruction at the branch address. The instructions are loaded into the instruction cache, so when the branch occurs, the instructions are present and allow the branch to execute in one clocking period. If for any reason the branch prediction logic errs, the branch requires an extra three clocking periods to execute. In most cases, the branch prediction is correct and no delay ensues.

Cache Structure

The cache in the Pentium has been changed from the one found in the 80486 microprocessor. The Pentium contains two 8K-byte cache memories instead of one as in the 80486. There is an 8K-byte data cache and an 8K-byte instruction cache. The instruction cache stores only instructions, while the data cache stores data used by instructions.

In the 80486 with its unified cache, a program that was data-intensive quickly filled the cache, allowing little room for instructions. This slowed the execution speed of the 80486 microprocessor. In the Pentium, this cannot occur because of the separate instruction cache.

Superscalar Architecture

The Pentium microprocessor is organized with three execution units. One executes floating-point instructions, and the other two (U-pipe and V-pipe) execute integer instructions. This means that it is possible to execute three instructions simultaneously. For example, the FADD ST,ST(2) instruction, MOV EAX,10H instruction, and MOV EBX,12H instruction can all execute simultaneously because none of these instructions depend on each other. The FADD ST,ST(2) instruction is executed by the coprocessor; the MOV EAX,10H is executed by the U-pipe; and the MOV EBX,12H instruction is executed by the V-pipe. Because the floating-point unit is also used for MMX instructions, if available, the Pentium can execute two integers and one MMX instruction simultaneously.

Software should be written to take advantage of this feature by looking at the instructions in a program, and then modifying them when cases are discovered in which dependent instructions can be separated by nondependent instructions. These changes can result in up to a 40% execution speed improvement in some software. Make sure that any new compiler or other application package takes advantage of this new superscalar feature of the Pentium.

18–2 SPECIAL PENTIUM REGISTERS

The Pentium is essentially the same microprocessor as the 80386 and 80486, except that some additional features and changes to the control register set have occurred. This section highlights the differences between the 80386 control register structure and the flag register.

Control Registers

Figure 18–8 shows the control register structure for the Pentium microprocessor. Note that a new control register, CR_4, has been added to the control register array.

This section of the text only explains the new Pentium components in the control registers. See Figure 17–14 for a description and illustration of the 80386 control registers. Following is a description of the new control bits and new control register CR_4:

CD **Cache disable** controls the internal cache. If CD = 1, the cache will not fill with new data for cache misses, but it will continue to function for cache hits. If CD = 0, misses will cause the cache to fill with new data.

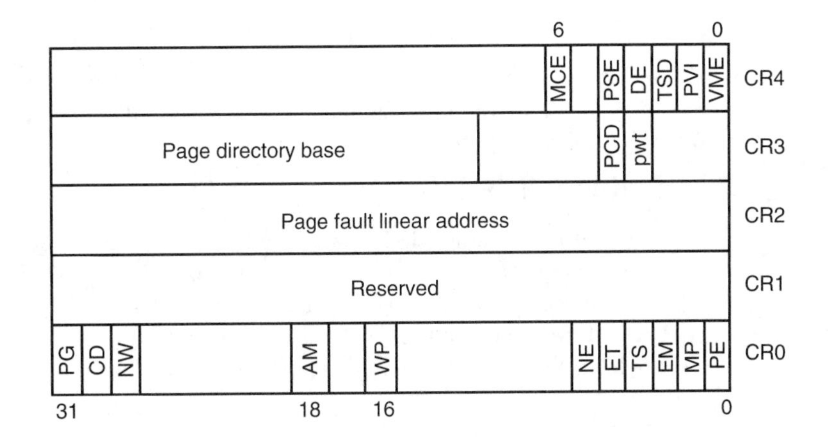

FIGURE 18–8 The structure of the Pentium control registers.

NW **Not write-through** selects the mode of operation for the data cache. If NW = 1, the data cache is inhibited from cache write-through.

AM **Alignment mask** enables alignment checking when set. Note that alignment checking only occurs for protected mode operation when the user is at privilege level 3.

WP **Write protect** protects user-level pages against supervisor-level write operations. When WP = 1, the supervisor can write to user-level segments.

NE **Numeric error** enables standard numeric coprocessor error detection. If NE = 1, the $\overline{\text{FERR}}$ pin becomes active for a numeric coprocessor error. If NE = 0, any coprocessor error is ignored.

VME **Virtual mode extension** enables support for the virtual interrupt flag in protected mode. If VME = 0, virtual interrupt support is disabled.

PVI **Protected mode virtual interrupt** enables support for the virtual interrupt flag in protected mode.

TSD **Time stamp disable** controls the RDTSC instruction.

DE **Debugging extension** enables I/O breakpoint debugging extensions when set.

PSE **Page size extension** enables 4M-byte memory pages when set.

MCE **Machine check enable** enables the machine checking interrupt.

The Pentium contains new features that are controlled by CR_4 and a few bits in CR_0. These new features are explained in later sections of the text.

EFLAG Register

The extended flag (EFLAG) register has been changed in the Pentium microprocessor. Figure 18–9 pictures the contents of the EFLAG register. Note that four new flag bits have been added to this register to control or indicate conditions about some of the new features in the Pentium. Following is a list of the four new flags and the function of each:

ID The **identification flag** is used to test for the CPUID instruction. If a program can set and clear the ID flag, the processor supports the CPUID instruction.

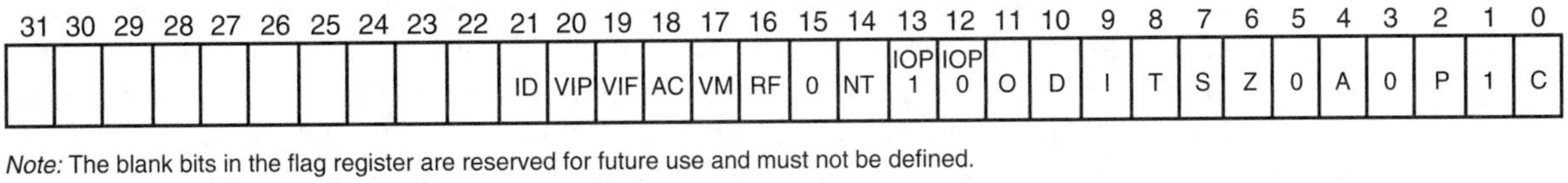

31	30	29	28	27	26	25	24	23	22	21	20	19	18	17	16	15	14	13	12	11	10	9	8	7	6	5	4	3	2	1	0
										ID	VIP	VIF	AC	VM	RF	0	NT	IOP 1	IOP 0	O	D	I	T	S	Z	0	A	0	P	1	C

Note: The blank bits in the flag register are reserved for future use and must not be defined.

FIGURE 18–9 The structure of the Pentium EFLAG register.

VIP **Virtual interrupt pending** indicates that a virtual interrupt is pending.

VIF **Virtual interrupt** is the image of the interrupt flag IF used with VIP.

AC **Alignment check** indicates the state of the AM bit in control register 0.

Built-In Self-Test (BIST)

The built-in self-test (BIST) is accessed on power-up by placing a logic 1 on INIT while the RESET pin changes from 1 to 0. The BIST tests 70% of the internal structure of the Pentium in approximately 150 μs. Upon completion of the BIST, the Pentium reports the outcome in register EAX. If EAX = 0, the BIST has passed and the Pentium is ready for operation. If EAX contains any other value, the Pentium has malfunctioned and is faulty.

18–3 PENTIUM MEMORY MANAGEMENT

The memory-management unit within the Pentium is upward-compatible with the 80386 and 80486 microprocessors. Many of the features of these earlier microprocessors are basically unchanged in the Pentium. The main change is in the paging unit and a new system memory-management mode.

Paging Unit

The paging mechanism functions with 4K-byte memory pages or with a new extension available to the Pentium with 4M-byte memory pages. As detailed in Chapters 1 and 17, the size of the paging table structure can become large in a system that contains a large memory. Recall that to fully repage 4G bytes of memory, the microprocessor requires slightly over 4M bytes of memory just for the page tables. In the Pentium, with the new 4M-byte paging feature, this is dramatically reduced to just a single page directory and no page tables. The new 4M-byte page sizes are selected by the PSE bit in control register 0.

The main difference between 4K paging and 4M paging is that in the 4M paging scheme there is no page table entry in the linear address. See Figure 18–10 for the 4M paging system in the Pentium microprocessor. Pay close attention to the way the linear address is used with this scheme. Notice that the leftmost 10 bits of the linear address select an entry in the page directory (just as with 4K pages). Unlike 4K pages, there are no page tables; instead, the page directory addresses a 4M-byte memory page.

Memory-Management Mode

The system memory-management mode (SMM) is on the same level as protected mode, real mode, and virtual mode, but it is provided to function as a manager. The SMM is not intended to be used as an application or a system-level feature. It is intended for high-level system functions such as power management and security, which most Pentiums use during operation.

Access to the SMM is accomplished via a new external hardware interrupt applied to the $\overline{\text{SMI}}$ pin on the Pentium. When the SMM interrupt is activated, the processor begins executing system-level software in an area of memory called the system management RAM, or SMMRAM, called the SMM state dump record. The $\overline{\text{SMI}}$ interrupt disables all other interrupts that are normally handled by user applications and the operating system. A return from the SMM interrupt is accomplished with a new instruction called RSM. RSM returns from the memory-management mode interrupt and returns to the interrupted program at the point of the interruption.

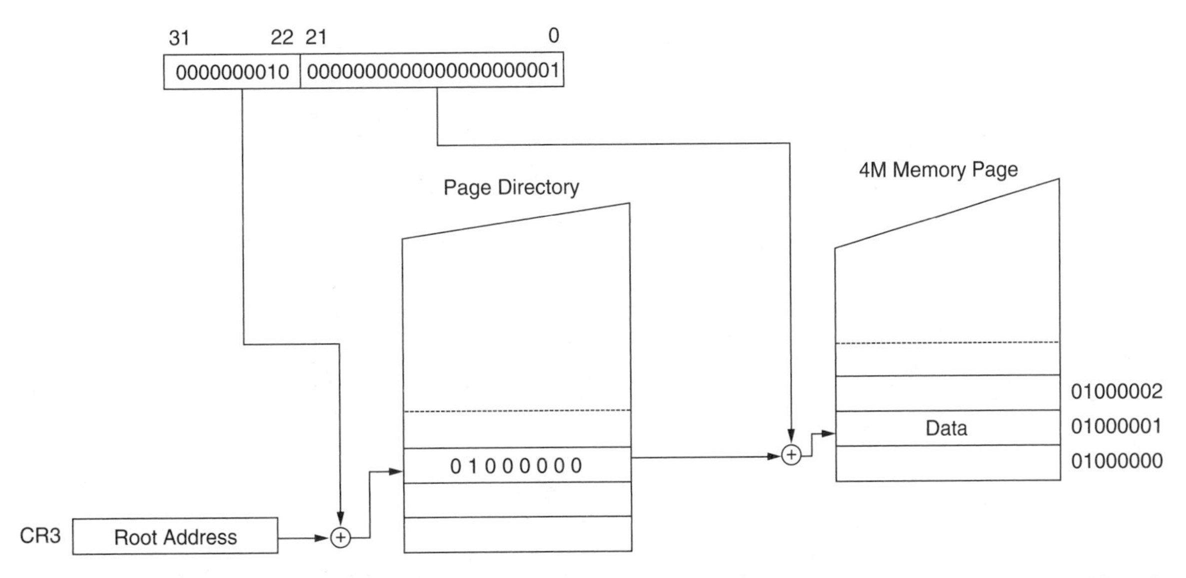

FIGURE 18–10 The linear address 00200001H repaged to memory location 01000002H in 4M-byte pages. Note that there are no page tables.

The SMM interrupt calls the software, initially stored at memory location 38000H, using CS = 3000H and EIP = 8000H. This initial state can be changed using a jump to any location within the first 1M byte of the memory. An environment similar to real mode memory addressing is entered by the management mode interrupt, but it is different because, instead of being able to address the first 1M of memory, SMM mode allows the Pentium to treat the memory system as a flat, 4G-byte system.

In addition to executing software that begins at location 38000H, the SMM interrupt also stores the state of the Pentium in what is called a **dump record**. The dump record is stored at memory locations 3FFA8H through 3FFFFH, with an area at locations 3FE00H through 3FEF7H that is reserved by Intel. The dump record allows a Pentium-based system to enter a sleep mode and reactivate at the point of program interruption. This requires that the SMMRAM be powered during the sleep period. Many laptop computers have a separate battery to power the SMMRAM for many hours during sleep mode. Table 18–2 lists the contents of the dump record.

The Halt auto restart and I/O trap restarts are used when the SMM mode is exited by the RSM instruction. These data allow the RSM instruction to return to the halt-state or return to the interrupt I/O instruction. If neither a halt nor an I/O operation is in effect upon entering the SMM mode, the RSM instruction reloads the state of the machine from the state dump and returns to the point of interruption.

The SMM mode can be used by the system before the normal operating system is placed in the memory and executed. It can also be used periodically to manage the system, provided that normal software doesn't exist at locations 38000H–3FFFFH. If the system relocates the SMRAM before booting the normal operating system, it becomes available for use in addition to the normal system.

The base address of the SMM mode SMRAM is changed by modifying the value in the state dump base address register (locations 3FEF8H through 3F3FBH) after the first memory-management mode interrupt. When the first RSM instruction is executed, returning control back to the interrupted system, the new value from these locations changes the base address of the SMM interrupt for all future uses. For example, if the state dump base address is changed to 000E8000H, all subsequent SMM interrupts use locations E8000H–EFFFFH for the Pentium state dump. These locations are compatible with DOS and Windows.

TABLE 18–2 Pentium SMM state dump record.

Offset Address	*Register*
FFFCH	CR_0
FFF8H	CR_3
FFF4H	EFLAGS
FFF0H	EIP
FFECH	EDI
FFE8H	ESI
FFE4H	EBP
FFE0H	ESP
FFDCH	EBX
FFD8H	EDX
FFD4H	ECX
FFD0H	EAX
FFCCH	DR_6
FFC8H	DR_7
FFC4H	TR
FFC0H	LDTR
FFBCH	GS
FFB8H	FS
FFB4H	DS
FFB0H	SS
FFACH	CS
FFA8H	ES
FF04H–FFA7H	Reserved
FF02H	Halt auto start
FF00H	I/O trap restart
FEFCH	SMM revision identifier
FED8H	State dump base
FE00H–FEF7H	Reserved

Note: The offset addresses are initially located at base address 00003000H.

18–4 NEW PENTIUM INSTRUCTIONS

The Pentium contains only one new instruction that functions with normal system software; the remainder of the new instructions are added to control the memory-management mode feature and serializing instructions. Table 18–3 lists the new instructions added to the Pentium instruction set.

The CMPXCHG8B instruction is an extension of the CMPXCHG instruction added to the 80486 instruction set. The CMPXCHG8B instruction compares the 64-bit number stored in EDX and EAX with the contents of a 64-bit memory location or register pair. For example, the CMPXCHG8B $DATA_2$ instruction compares the eight bytes stored in memory location $DATA_2$ with the 64-bit number in EDX and EAX. If $DATA_2$ equals EDX:EAX, the 64-bit number stored in ECX:EBX is stored in memory location $DATA_2$. If they are not equal, the contents of $DATA_2$ are stored into EDX:EAX. Note that the zero flag bit indicates that the contents of EDX:EAX were equal or not equal to $DATA_2$.

The CPUID instruction reads the CPU identification code and other information from the Pentium. Table 18–4 shows different information returned from the CPUID instruction for various input values for EAX. To use the CPUID instruction, first load EAX with the input value and then execute CPUID. The information is returned in the registers indicated in the table.

TABLE 18–3 New Pentium instructions.

Instruction	*Function*
CMPXCHG8B	Compare and exchange eight bytes
CPUID	Return CPU identification code
RDTSC	Read time-stamp counter
RDMSR	Read model-specific register
WRMSR	Write model-specific register
RSM	Return from system management interrupt

TABLE 18–4 CPUID instruction information.

Input Value (EAX)	*Result after CPUID*
0	EAX = 1 for all microprocessors
	EBX–EDX–ECX = Vendor information
1	EAX (bits 3–0) = Stepping ID
	EAX (bits 7–4) = Model
	EAX (bits 11–8) = Family
	EAX (bits 13–12) = Type
	EAX (bits 31–14) = Reserved
	EDX (bit 0) = CPU contains FPU
	EDX (bit 1) = Enhanced 8086 virtual mode supported
	EDX (bit 2) = I/O breakpoints supported
	EDX (bit 3) = Page size extensions supported
	EDX (bit 4) = Time-stamp counter supported
	EDX (bit 5) = Pentium-style MSR supported
	EDX (bit 6) = Reserved
	EDX (bit 7) = Machine check exception supported
	EDX (bit 8) = CMPXCHG8B supported
	EDX (bit 9) = 3.3 V microprocessor
	EDX (bit 10–31) = Reserved

If a 0 is placed in EAX before executing the CPUID instruction, the microprocessor returns the vendor identification in EBX, EDX, and EBX. For example, the Intel Pentium returns "GenuineIntel" in ASCII code with the "Genu" in the EBX, "ineI" in EDX, and "ntel" in ECX . The EDX register returns information if EAX is loaded with a 1 before executing the CPUID instruction.

Example 18–1 illustrates a short program that reads the vendor information with the CPUID instruction. This software was placed into the TODO: section of the OnInitDialog function of a simple dialog application. It then displays it on the video screen in an ActiveX label as illustrated in Figure 18–11. The CPUID instruction functions in both the real and protected mode and can be used in any Windows application.

EXAMPLE 18–1

```
CString temp;
int a, b, c;
_asm
{
        mov  eax,0
        cpuid
        mov  a,ebx
        mov  b,edx
        mov  c,ecx
}
for (int d = 0; d < 4; d++ )
```

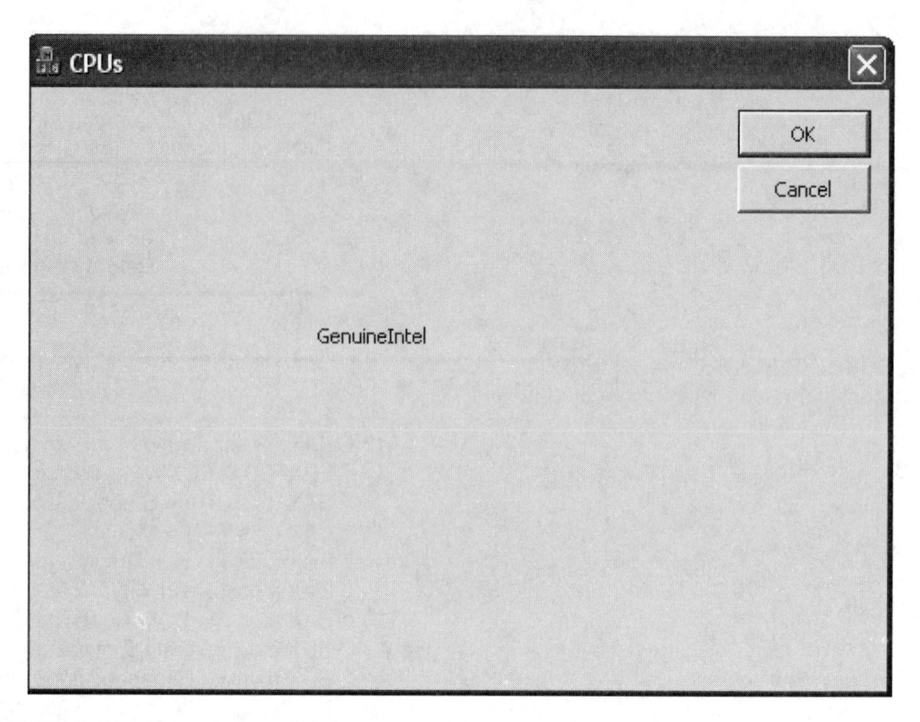

FIGURE 18–11 Screen shot of the program of Example 18–1 using the CPUID instruction.

```
{
        temp += (char)a;
        a >>= 8;
}
for (d = 0; d < 4; d++ )
{
        temp += (char)b;
        b >>= 8;
}
for (d = 0; d < 4; d++ )
{
        temp += (char)c;
        c >>= 8;
}
Label1.put_Caption(temp);
```

The RDTSC instruction reads the time-stamp counter into EDX:EAX. The time-stamp counter counts CPU clocks from the time the microprocessor is reset, where the time stamp counter is initialized to an unknown count. Because this is a 64-bit count, a 1 GHz microprocessor can accumulate a count of over 580 years before the time-stamp counter rolls over. This instruction functions only in real mode or privilege level 0 in protected mode.

Example 18–2 shows a class written for Windows that provides member functions for accurate time delays and also member functions to measure software execution times. This class is added by right-clicking on the project name and inserting an MFC generic class named TimeD. It contains three member functions called Start, Stop, and Delay.

The Start() function is used to start a measurement and Stop() is used to end a time measurement. The Stop() function returns a double floating-point value that is the amount of time in microseconds between Start() and Stop().

The Delay function causes a precision time delay based on the time-stamp counter. The parameter transferred to the Delay function is in milliseconds. This means that a Delay(1000) causes exactly 1000 ms of delay.

When TimeD is initialized in a program, it reads the microprocessor frequency in MHz from the Windows registry file using the RegQueryValueEx function after opening it with the RegOpenKeyEx function. The microprocessor clock frequency is returned in the MicroFrequency class variable.

EXAMPLE 18–2

```
#include "StdAfx.h"
#include ".\timed.h"

int MicroFrequency;                    //frequency in MHz
_int64 Count;

TimeD::TimeD(void)
{
        HKEY hKey;
        DWORD dataSize;
                        // Get the processor frequency

        if ( RegOpenKeyEx (HKEY_LOCAL_MACHINE,
                "Hardware\\Description\\System\\CentralProcessor\\0",
                0, KEY_QUERY_VALUE, &hKey) == ERROR_SUCCESS )
        {
                RegQueryValueEx (hKey, _T("~MHz"), NULL, NULL,
                        (LPBYTE)&MicroFrequency, &dataSize);
                RegCloseKey (hKey);
        }
}

TimeD::~TimeD(void)
{
}

void TimeD::Start(void)
{
        _asm
        {
                rdtsc                   ;get and store TSC
                mov  dword ptr Count,eax
                mov  dword ptr Count+4,edx
        }
}

double TimeD::Stop(void)
{
        _asm
        {
                rdtsc
                sub  eax,dword ptr Count
                mov  dword ptr Count,eax
                sbb  edx,dword ptr Count+4
                mov  dword ptr Count+4,edx
        }
    return (double)Count/MicroFrequency;
}
void TimeD::Delay(__int64 milliseconds)
{

        milliseconds *= 1000;               //convert to microseconds
        milliseconds *= MicroFrequency;     //convert to raw count
        _asm {
                mov   ebx, dword ptr milliseconds       ;64-bit delay in ms
                mov   ecx, dword ptr milliseconds+4
                rdtsc                       ;get count
                add   ebx, eax
                adc   ecx, edx              ;advance count by delay
```

```
Delay_LOOP1:                                    ;wait for count to catch up

              rdtsc
              cmp    edx, ecx
              jb     Delay_LOOP1
              cmp    eax, ebx
              jb     Delay_LOOP1
       }
}
```

If need an additional Delay could be added to the class to cause delays in microseconds, but a restriction should be made so it is no less than about 2 or 3 microseconds, because of the time that it takes to add the time to the count from the time-stamp counter.

Example 18–3 shows a sample dialog application that uses Delay() to wait for a second after clicking the button before changing the foreground color of an ActiveX Label. What does not appear in the example is that at the beginning of the dialog class an #inlcude "TimeD.h" statement appears. The software itself is in the TODO: section of the OnInitDialog function.

EXAMPLE 18–3

```
void CRDTSCDlg::OnBnClickedButton1()
{
       TimeD timer;
       timer.Delay(1000);
       Label1.put_ForeColor(0xff0000);
}
```

The RDMSR and WRMSR instructions allow the model-specific registers to be read or written. The model-specific registers are unique to the Pentium and are used to trace, check performance, test, and check for machine errors. Both instructions use ECX to convey the register number to the microprocessor and use EDX:EAX for the 64-bit-wide read or write. Note that the register addresses are 0H–13H. See Table 18–5 for a list of the Pentium model-specific registers

TABLE 18–5 The Pentium model-specific registers.

Address (ECX)	*Size*	*Function*
00H	64 bits	Machine check exception address
01H	5 bits	Machine check exception type
02H	14 bits	TR_1 parity reversal test register
03H	—	—
04H	4 bits	TR_2 instruction cache end bits
05H	32 bits	TR_3 cache data
06H	32 bits	TR_4 cache tag
07H	15 bits	TR_4 cache control
08H	32 bits	TR_6 TLB command
09H	32 bits	TR_7 TLB data
0AH	—	—
0BH	32 bits	TR_9 BTB tag
0CH	32 bits	TR_{10} BTB target
0DH	12 bits	TR_{11} BTB control
0EH	10 bits	TR_{12} new feature control
0FH	—	—
10H	64 bits	Time-stamp counter (can be written)
11H	26 bits	Events counter selection and control
12H	40 bits	Events counter 0
13H	40 bits	Events counter 1

and their contents. As with the RDTSC instruction, these model-specific registers operate in the real or privilege level 0 of protected mode.

Never use an undefined value in ECX before using the RDMSR or WRMSR instructions. If ECX = 0 before the read or write machine-specific register instruction, the value returned, EDX:EAX, is the machine check exception address. (EDX:EAX is where all data reside when written or read from the model-specific registers.) If ECX = 1, the value is the machine check exception type; if ECX = 0EH, test register 12 (TR_{12}) is accessed. Note that these are internal registers designed for in-house testing. The contents of these registers are proprietary to Intel and should not be used during normal programming.

18–5 INTRODUCTION TO THE PENTIUM PRO MICROPROCESSOR

Before this or any other microprocessor can be used in a system, the function of each pin must be understood. This section of the chapter details the operation of each pin, along with the external memory system and I/O structures of the Pentium Pro microprocessor. Figure 18–12 illustrates the pin-out of the Pentium Pro microprocessor, which is packaged in an immense 387-pin PGA (pin grid array). Currently, the Pentium Pro is available in two versions. One version contains a 256K level 2 cache; the other contains a 512K level 2 cache. The most notable difference in the pin-out of the Pentium Pro, when compared to the Pentium, is that there are provisions for a 36-bit address bus, which allows access to 64G bytes of memory. This is meant for future use because no system today contains anywhere near that amount of memory.

As with most recent versions of the Pentium microprocessor, the Pentium Pro requires a single +3.3 V or +2.7 V power supply for operation. The power supply current is a maximum of 9.9 A for the 150 MHz version of the Pentium Pro, which also has a maximum power dissipation of 26.7 W. At present, a good heat sink with considerable airflow is required to keep the Pentium Pro cool. As with the Pentium, the Pentium Pro contains multiple V_{CC} and V_{SS} connections that must all be connected for proper operation. The Pentium Pro contains $V_{CC}P$ pins (primary V_{CC}) that connect to +3.1 V, $V_{CC}S$ (secondary V_{CC}) pins that connect to +3.3 V, and $V_{CC}5$ (standard V_{CC}) pins that connect to +5.0 V. Some pins are labeled N/C (no connection) and must not be connected.

Each Pentium Pro output pin is capable of providing an ample 48.0 mA of current at a logic 0 level. This represents a considerable increase in drive current, compared to the 2.0 mA available on earlier microprocessor output pins. Each input pin represents a small load, requiring only 15 μA of current. Because of the 48.0 mA of drive current available on each output, only an extremely large system requires bus buffers.

Internal Structure of the Pentium Pro

The Pentium Pro is structured differently than earlier microprocessors. Early microprocessors contained an execution unit and a bus interface unit with a small cache buffering the execution unit for the bus interface unit. This structure was modified in later microprocessors, but the modifications were just additional stages within the microprocessors. The Pentium architecture is also a modification, but more significant than earlier microprocessors. Figure 18–13 shows a block diagram of the internal structure of the Pentium Pro microprocessor.

The system buses, which communicate to the memory and I/O, connect to an internal level 2 cache that is often on the main board in most other microprocessor systems. The level 2 cache in the Pentium Pro is either 256K bytes or 512K bytes. The integration of the level 2 cache speeds processing and reduces the number of components in a system.

The bus interface unit (BIU) controls the access to the system buses through the level 2 cache, as it does in most other microprocessors. Again, the difference is that the level 2 cache is

FIGURE 18–12 The pin-out of the Pentium Pro microprocessor.

PENTIUM PRO

Pin	Signal
S5	A3
S3	A4
Q5	A5
S1	A6
Q3	A7
Q7	A8
Q1	A9
Q9	A10
N5	A11
N1	A12
N7	A13
N3	A14
L5	A15
L3	A16
N9	A17
L7	A18
J1	A19
J5	A20
J3	A21
G1	A22
J7	A23
G3	A24
L9	A25
G7	A26
G5	A27
J9	A28
E1	A29
E3	A30
G9	A31
E5	A32
E7	A33
E9	A34
C1	A35
AE3	ADS
AE9	AERR
U1	AP0
S9	AP1
A19	BCLK
C5	BERR
AC43	BINIT
U7	BNR
AE43	BP2
AC39	BP3
AC41	BPM0
AA39	BPM1
U5	BPR1
AC5	BR0
W3	BR1
AA1	BR2
U9	BR3
B2	CPUPRES
Y5	DEFER
AC45	DEP0
Y43	DEP1
W39	DEP2
AC47	DEP3
W41	DEP4
AA47	DEP5
Y45	DEP6
U39	DEP7
AA3	DRDY
C17	FERR
A15	FLUSH
C9	FRCERR
AC3	HIT
AA7	HITM
C3	IERR
A9	IGNNE
C11	INIT
AG43	LINT0/INTR
AG41	LINT1/NMI
AA9	LOCK
Y41	RESET
AC47	RP
AC9	RS0
AE5	RS1
AE7	RS2
U3	RSP
W1	SMI
Y9	TRDY
AG3	UP
A11	A20M

Signal	Pin
D0	C25
D1	A27
D2	C27
D3	A29
D4	C29
D5	A31
D6	C31
D7	C33
D8	A33
D9	A35
D10	A39
D11	A41
D12	C35
D13	A43
D14	A37
D15	C37
D16	A45
D17	C39
D18	C43
D19	C45
D20	C41
D21	C47
D22	E39
D23	E41
D24	E45
D25	E43
D26	E47
D27	G39
D28	G45
D29	G41
D30	G43
D31	G47
D32	J39
D33	J45
D34	J47
D35	J41
D36	L45
D37	L39
D38	J43
D39	L47
D40	L41
D41	N47
D42	N45
D43	L43
D44	N39
D45	N41
D46	Q47
D47	N43
D48	Q45
D49	Q43
D50	S47
D51	Q39
D52	Q41
D53	S45
D54	S43
D55	U47
D56	U45
D57	S41
D58	W47
D59	S39
D60	U43
D61	W45
D62	Y47
D63	W43
PICCLK	AA43
PICD0	AA41
PICD1	AE41
PLL1	C19
PLL2	C23
PRDY	Y39
PREQ	AA45
PWRGOOD	AG7
REQ0	W9
REQ1	W7
REQ2	Y3
REQ3	Y1
REQ4	W5
STPCLK	A3
TCK	A5
TDI	A13
TDO	C13
THERTRIP	A17
TMS	C15
TRST	A7

integrated. The BIU generates the memory address and control signals, and passes and fetches data or instructions to either a level 1 data cache or a level 1 instruction cache. Each of these are 8K bytes in size at present and may be made larger in future versions of the microprocessor. Earlier versions of the Intel microprocessor contained a unified cache that held both instructions and data. The implementation of separate caches improves performance because data-intensive programs no longer fill the cache with data.

The instruction cache is connected to the instruction fetch and decode unit (IFDU). Although not shown, the IFDU contains three separate instruction decoders that decode three instructions

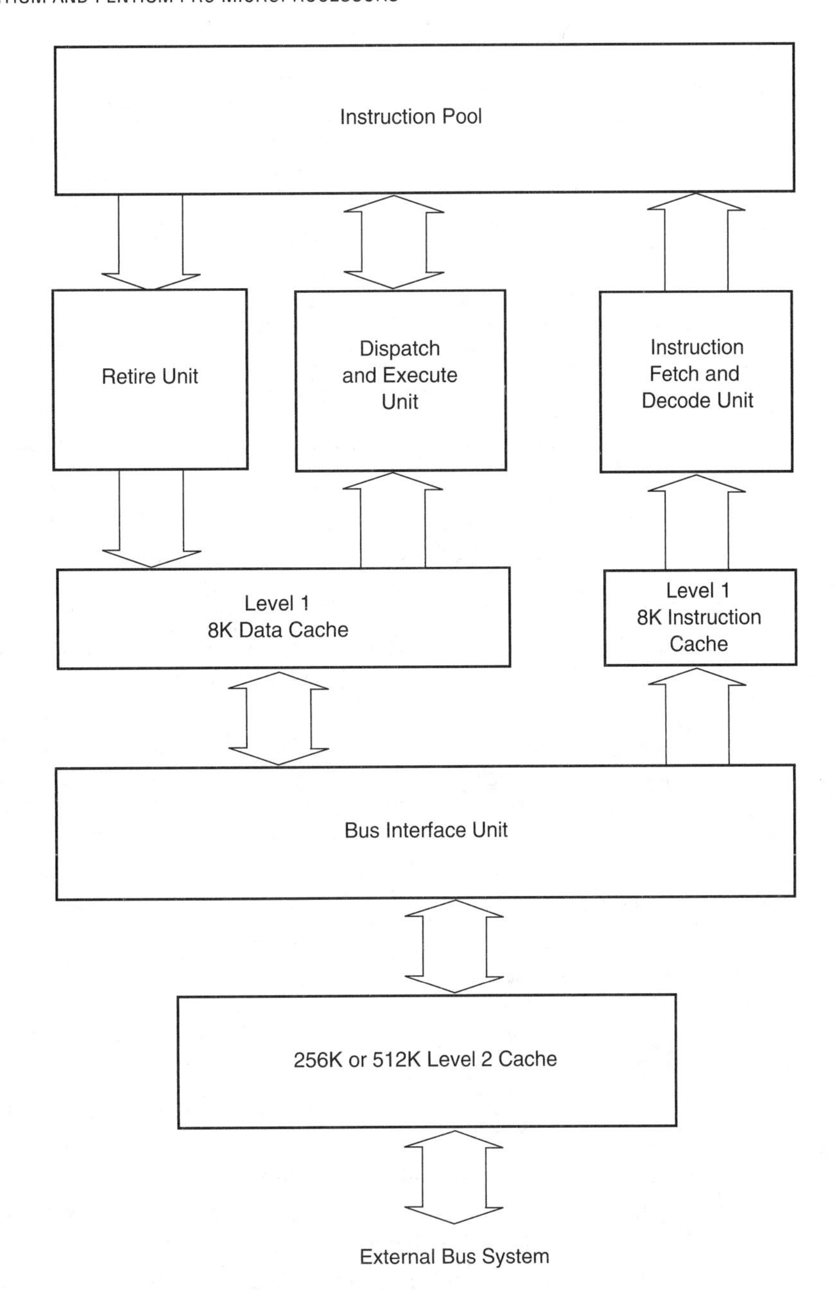

FIGURE 18–13 The internal structure of the Pentium Pro microprocessor.

simultaneously. Once decoded, the outputs of the three decoders are passed to the instruction pool, where they remain until the dispatch and execution unit or retire unit obtains them. Also included within the IFDU is a branch prediction logic section that looks ahead in code sequences that contain conditional jump instructions. If a conditional jump is located, the branch prediction logic tries to determine the next instruction in the flow of a program.

Once decoded instructions are passed to the instruction pool, they are held for processing. The instruction pool is a content-addressable memory, but Intel never states its size in the literature.

The dispatch and execute unit (DEU) retrieves decoded instructions from the instruction pool when they are complete, and then executes them. The internal structure of the DEU is illustrated in Figure 18–14. Notice that the DEU contains three instruction execution units: two for processing integer instructions and one for floating-point instructions. This means that the

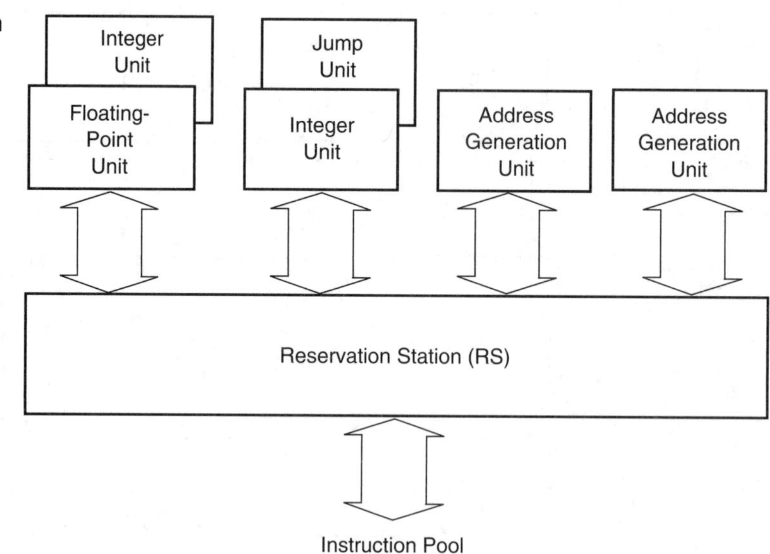

FIGURE 18–14 The Pentium Pro dispatch and execution unit (DEU).

Pentium Pro can process two integer instructions and one floating-point instruction simultaneously. The Pentium also contains three execution units, but the architecture is different because the Pentium does not contain a jump execution unit or address generation units, as does the Pentium Pro. The reservation station (RS) can schedule up to five events for execution and process four simultaneously. Note that there are two station components connected to one of the address generation units that do not appear in the illustration of Figure 18–14.

The last internal structure of the Pentium Pro is the retire unit (RU). The RU checks the instruction pool and removes decoded instructions that have been executed. The RU can remove three decoded instructions per clock pulse.

Pin Connections

The number of pins on the Pentium Pro has increased from the 237 pins on the Pentium to 387 pins on the Pentium Pro. Following is a description of each pin or grouping of pins:

$\overline{\textbf{A20M}}$ — The **address A20 mask** is an input that is asserted in the real mode to signal the Pentium Pro to perform address wraparound, as in the 8086 microprocessor, for use of the HIMEM.SYS driver.

$\overline{\textbf{A35}}$–$\overline{\textbf{A3}}$ — **Address bus** connections address any of the 8G × 64 memory locations found in the Pentium Pro memory system.

$\overline{\textbf{ADS}}$ — The **address data strobe** becomes active whenever the Pentium Pro has issued a valid memory or I/O address.

$\overline{\textbf{AP1}}$, $\overline{\textbf{AP0}}$ — **Address parity** provides even parity for the memory address on all Pentium Pro–initiated memory and I/O transfers. The $\overline{\text{AP0}}$ output provides parity for address connections A_{23}–A_3, and the $\overline{\text{AP1}}$ output provides parity for address connections A_{35}–A_{24}.

$\overline{\textbf{ASZ1}}$, $\overline{\textbf{ASZ0}}$ — **Address size** inputs are driven to select the size of the memory access. Table 18–6 illustrates the size of the memory access for the binary bit patterns on these two inputs to the Pentium Pro.

BCLK — The **bus clock** input determines the operating frequency of the Pentium Pro microprocessor. For example, if BCLK is 66 MHz, various internal clocking speeds are selected by the logic levels applied to the pins in Table 18–7. A BLCK frequency of 66 MHz runs the system bus at 66 MHz.

TABLE 18–6 Memory size dictated by the ASZ pins.

$\overline{ASZ1}$	$\overline{ASZ0}$	*Memory Size*
0	0	0–4G
0	1	4G–64G
1	X	Reserved

TABLE 18–7 The BCLK signal and its effect on the Pentium clock speed.

LINT1/NMI	*LINT0/INTR*	$\overline{IGNNE}$	$\overline{A20M}$	*Ratio*	*BCLK = 50 MHz*	*BCLK = 66 MHz*
0	0	0	0	2	100 MHz	133 MHz
0	0	0	1	4	200 MHz	266 MHz
0	0	1	0	3	150 MHz	200 MHz
0	0	1	1	5	250 MHz	333 MHz
0	1	0	0	5/2	125 MHz	166 MHz
0	1	0	1	9/2	225 MHz	300 MHz
0	1	1	0	7/2	175 MHz	233 MHz
0	1	1	1	11/2	275 MHz	366 MHz
1	1	1	1	2	100 MHz	133 MHz

$\overline{BERR}$ The **bus error** input/output either signals a bus error along or is asserted by an external device to cause a machine check interrupt or a non-maskable interrupt.

$\overline{BINIT}$ **Bus initialization** is active on power-up to initialize the bus system.

$\overline{BNR}$ **Block next request** is used to halt the system in a multiple microprocessor system.

$\overline{BP3}$, $\overline{BP2}$ The **breakpoint status** outputs indicate the status of the Pentium Pro breakpoints.

$\overline{BPM1}$, $\overline{BPM0}$ The **breakpoint monitor** outputs indicate the status of the breakpoints and programmable counters.

$\overline{BPRI}$ The **priority agent bus request** is an input that causes the microprocessor to cease bus requests.

$\overline{BR3}$–$\overline{BR0}$ The **bus request** inputs allow up to four Pentium Pro microprocessors to coexist on the same bus system.

$\overline{BREQ3}$–$\overline{BREQ0}$ **Bus request signals** are used for multiple microprocessors on the same system bus.

$\overline{D63}$–$\overline{D0}$ **Data bus** connections transfer byte, word, doubleword, and quadword data between the microprocessor and its memory and I/O system.

$\overline{DBSY}$ **Data bus busy** is asserted to indicate that the data bus is busy transferring data.

$\overline{DEFER}$ The **defer** input is asserted during the snoop phase to indicate that the transaction cannot be guaranteed in-order completion.

$\overline{DEN}$ The **defer enable** signal is driven to the bus on the second phase of a request phase.

$\overline{DEP7}$–$\overline{DEP0}$ **Data bus ECC protection** signals provide error-correction codes for correcting a single-bit error and detecting a double-bit error.

$\overline{FERR}$ The **floating-point error**, comparable to the ERROR line in the 80386, shows that the internal coprocessor has erred.

$\overline{\text{FLUSH}}$ The **flush** cache input causes the cache to flush all write-back lines and invalidate its internal caches. If the $\overline{\text{FLUSH}}$ input is a logic 0 during a reset operation, the Pentium Pro enters its test mode.

$\overline{\text{FRCERR}}$ **Functional redundancy check error** is used if two Pentium Pro microprocessors are configured in a pair.

$\overline{\text{HIT}}$ **Hit** shows that the internal cache contains valid data in the inquire mode.

$\overline{\text{HITM}}$ **Hit modified** shows that the inquire cycle found a modified cache line. This output is used to inhibit other master units from accessing data until the cache line is written to memory.

$\overline{\text{IERR}}$ **Internal error** output shows that the Pentium Pro has detected an internal parity error or functional redundancy error.

$\overline{\text{IGNNE}}$ The **ignore numeric error** input causes the Pentium Pro to ignore a numeric coprocessor error.

INIT The **initialization** input performs a reset without initializing the caches, write-back buffers, and floating-point registers. This input may not be used to reset the microprocessor in lieu of RESET after power-up.

INTR The **interrupt request** is used by external circuitry to request an interrupt.

$\overline{\text{LEN1}}$, $\overline{\text{LEN0}}$ Length signals (bit 0 and 1) indicate the size of the data transfer, as illustrated in Table 18–8.

$\overline{\text{LINT1}}$, $\overline{\text{LINT0}}$ The **local interrupt** inputs function as NMI and INTR, and also set the clock divider frequency on reset.

$\overline{\text{LOCK}}$ $\overline{\text{LOCK}}$ becomes a logic 0 whenever an instruction is prefixed with the LOCK: prefix. This is most often used during DMA accesses.

NMI The **non-maskable interrupt** requests a non-maskable interrupt, as it did on the earlier versions of the microprocessor.

PICCLK The **clock** signal input is used for synchronous data transfers.

PICD The **processor interface serial data** is used to transfer bidirectional serial messages between Pentium Pro microprocessors.

PWRGOOD **Power good** is an input that is placed at a logic 1 level when the power supply and clock have stabilized.

$\overline{\text{REQ4}}$–$\overline{\text{REQ0}}$ **Request signals** (bits 0–4) define the type of data-transfer operation, as illustrated in Tables 18–9 and 18–10.

$\overline{\text{RESET}}$ **Reset** initializes the Pentium Pro, causing it to begin executing software at memory location FFFFFFF0H. The Pentium Pro is reset to the real mode and the leftmost 12 address connections remain logic 1s (FFFH) until a far jump or far call is executed. This allows compatibility with earlier microprocessors.

$\overline{\text{RP}}$ **Request parity** provides a means of requesting the Pentium Pro to check parity.

TABLE 18–8 The $\overline{\text{LEN}}$ bits and data size.

$\overline{\text{LEN1}}$	*$\overline{\text{LEN0}}$*	*Data Transfer Size*
0	0	0–8 bytes
0	1	16 bytes
1	0	32 bytes
1	1	Reserved

TABLE 18–9 Function of the request inputs on the first clock pulse.

$\overline{REQ4}$	$\overline{REQ3}$	$\overline{REQ2}$	$\overline{REQ1}$	$\overline{REQ0}$	*Function*
0	0	0	0	0	Deferred reply
0	0	0	0	1	Reserved
0	1	0	0	0	Case 1*
0	1	0	0	1	Case 2*
1	0	0	0	0	I/O read
1	0	0	0	1	I/O write
X	X	0	1	0	Memory read
X	X	0	1	1	Memory write
X	X	1	0	0	Memory code read
X	X	1	1	0	Memory data read
X	X	1	X	1	Memory write

*See Table 18–10 for the second clock pulse for Case 1 and Case 2.

TABLE 18–10 Function of the request inputs for Case 1 and Case 2.

Case	$\overline{REQ4}$	$\overline{REQ3}$	$\overline{REQ2}$	$\overline{REQ1}$	$\overline{REQ0}$	*Function*
1	X	X	X	0	0	Interrupt acknowledge
1	X	X	X	0	1	Special transaction
1	X	X	X	1	X	Reserved
2	X	X	X	0	0	Branch trace message
2	X	X	X	0	1	Reserved
2	X	X	X	1	X	Reserved

$\overline{\text{RS2}}$–$\overline{\text{RS0}}$ The **response status** inputs cause the Pentium Pro to perform the functions listed in Table 18–11.

$\overline{\text{RSP}}$ The **response parity** input applies a parity error signal from an external parity checker.

$\overline{\text{SMI}}$ The **system management interrupt** input causes the Pentium Pro to enter the system management mode of operation.

$\overline{\text{SMMEM}}$ The **system memory-management mode** signal becomes a logic 0 whenever the Pentium Pro is executing in the system memory-management mode interrupt and address space.

$\overline{\text{SPCLK}}$ The **split lock** signal is placed at a logic 0 level to indicate that the transfer will contain four locked transactions.

TABLE 18–11 Operation of the response status inputs.

$\overline{RS2}$	$\overline{RS1}$	$\overline{RS0}$	*Function*	$\overline{HITM}$	$\overline{DEFER}$
0	0	0	Idle state	X	X
0	0	1	Retry	0	1
0	1	0	Defer	0	1
0	1	1	Reserved	0	1
1	0	0	Hard failure	X	X
1	0	1	Normal, no data	0	0
1	1	0	Implicit write-back	1	X
1	1	1	Normal, with data	0	0

$\overline{\text{SPCLK}}$ **Stop clock** causes the Pentium Pro to enter the power-down state when placed at a logic 0 level.

TCK The **testability clock** input selects the clocking function in accordance with the IEEE 1149.1 Boundary Scan interface.

TDI The **test data input** is used to test data clocked into the Pentium Pro with the TCK signal.

TDO The **test data output** is used to gather test data and instructions shifted out of the Pentium with TCK.

TMS The **test mode select** input controls the operation of the Pentium Pro in test mode.

$\overline{\text{TRDY}}$ The **target ready** input is asserted when the target is ready for a data transfer operation.

The Memory System

The memory system for the Pentium Pro microprocessor is 4G bytes in size, just as in the 80386DX–Pentium microprocessors, but access to an area between 4G and 64G is made possible by additional address signals A_{32}–A_{35}. The Pentium Pro uses a 64-bit data bus to address memory organized in eight banks that each contain 8G bytes of data. Note that the additional memory is enabled with bit position 5 of CR4 and is accessible only when 2M paging is enabled. Note also that 2M paging is new to the Pentium Pro to allow memory above 4G to be accessed. More information is presented on Pentium Pro paging later in this chapter. Refer to Figure 18–15 for the organization of the Pentium Pro physical memory system.

The Pentium Pro memory system is divided into eight banks that each store a byte of data with a parity bit. Most Pentium and Pentium Pro microprocessor-based systems forgo the use of the parity bit. The Pentium Pro, like the 80486 and Pentium, employs internal parity generation and checking logic for the memory system data bus information. The 64-bit-wide memory is important to double-precision floating-point data. Recall that a double-precision floating-point number is 64 bits wide. As with earlier Intel microprocessors, the memory system is numbered in bytes from byte 000000000H to byte FFFFFFFFFH. This nine-digit hexadecimal address is employed in a system that addresses 64G of memory.

Memory selection is accomplished with the bank enable signals ($\overline{\text{BE7}}$–$\overline{\text{BE0}}$). In the Pentium Pro microprocessor, the bank enable signals are presented on the address bus (A_{15}–A_8) during the second clock cycle of a memory or I/O access. These must be extracted from the

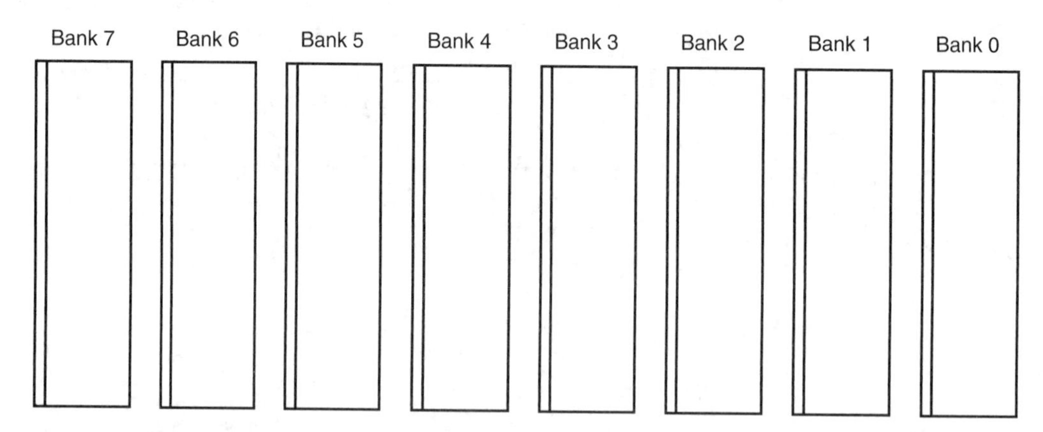

FIGURE 18–15 The eight memory banks in the Pentium Pro system. Note that each bank is 8 bits wide and 8G long if 36-bit addressing is enabled.

address bus to access memory banks. The separate memory banks allow the Pentium Pro to access any single byte, word, doubleword, or quadword with one memory transfer cycle. As with earlier memory selection logic, we often generate eight separate write strobes for writing to the memory system. The memory write information is provided on the request lines from the microprocessor during the second clock phase of a memory or I/O access.

A new feature added to the Pentium and Pentium Pro is the capability to check and generate parity for the address bus during certain operations. The $\overline{AP}$ pin (Pentium) or pins (Pentium Pro) provide the system with parity information, and the $\overline{APCHK}$ (Pentium) or $\overline{AP}$ pins (Pentium Pro) indicate a bad parity check for the address bus. The Pentium Pro takes no action when an address-parity error is detected. The error must be assessed by the system, and the system must take appropriate action (an interrupt) if so desired.

New to the Pentium Pro is a built-in error-correction circuit (ECC) that allows the correction of a one-bit error and the detection of a two-bit error. To accomplish the detection and correction of errors, the memory system must have room for an extra 8-bit number that is stored with each 64-bit number. The extra eight bits are used to store an error-correction code that allows the Pentium Pro to automatically correct any single-bit error. A 1M × 64 is a 64M SDRAM without ECC, and a 1M × 72 is an SDRAM with EEC support. The ECC code is much more reliable than the old parity scheme, which is rarely used in modern systems. The only drawback of the ECC scheme is the additional cost of SDRAM that is 72 bits wide.

Input/Output System

The input/output system of the Pentium Pro is completely compatible with earlier Intel microprocessors. The I/O port number appears on address lines A_{15}–A_3 with the bank enable signals used to select the actual memory banks used for the I/O transfer.

System Timing

As with any microprocessor, the system timing signals must be understood in order to interface the microprocessor. This portion of the text details the operation of the Pentium Pro through its timing diagrams and shows how to determine memory access times.

The basic Pentium Pro memory cycle consists of two sections: the address phase and the data phase. During the address phase, the Pentium Pro sends the address (T_1) to the memory and I/O system, and also the control signals (T_2). The control signals include the ATTR lines (A_{31}–A_{24}), the DID lines (A_{23}–A_{16}), the bank enable signals (A_{15}–A_8), and the EXF lines (A_7–A_3). See Figure 18–16 for the basic timing cycle. The type of memory cycle appears on the request pins. During the data phase, four 64-bit-wide numbers are fetched or written to the memory. This operation is most

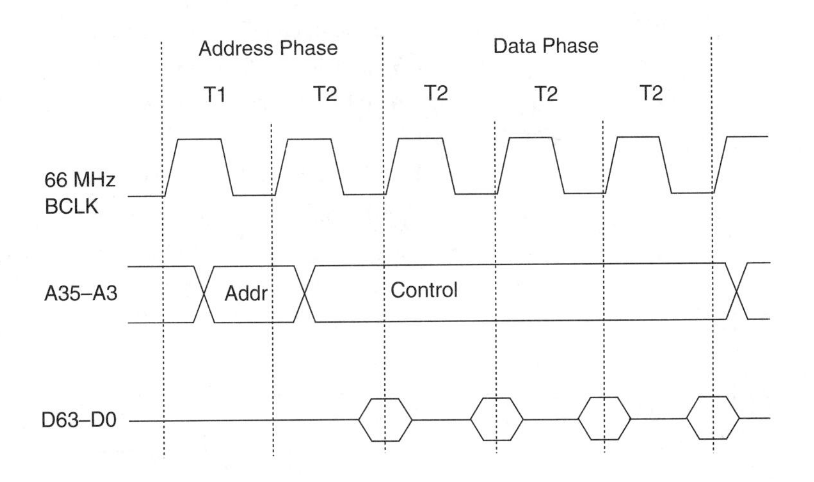

FIGURE 18–16 The basic Pentium Pro timing.

common because data from the main memory are transferred between the internal 256K or 512K write-back cache and the memory system. Operations that write a byte, word, or doubleword, such as I/O transfers, use the bank selection signals and have only one clock in the data transfer phase. Notice from the timing diagram that the 66 MHz Pentium Pro is capable of 33 million memory transfers per second. (This assumes that the memory can operate at that speed.)

The setup time before the clock is given as 5.0 ns and the hold time after the clock is given as 1.5 ns. This means that the data window around the clock is 6.5 ns. The address appears on the 8.0 ns maximum after the start of T_1. This means that the Pentium Pro microprocessor operating at 66 MHz allows 30 ns (two clocking periods), minus the address delay time of 8.0 ns and also minus the data setup time of 5.0 ns. Memory access time without any wait states is 30 – 8.0 – 5.0, or 17.0 ns. This is enough time to allow access to a SRAM, but not to any DRAM without inserting wait states into the timing.

Wait states are inserted into the timing by controlling the $\overline{TRDY}$ input to the Pentium Pro. The $\overline{TRDY}$ signal must become a logic 0 by the end of T_2; otherwise, additional T_2 states are inserted into the timing. Note that 60 ns DRAM requires the insertion of four wait states of 15 ns (one clocking period) each to lengthen the access time to 77 ns. This is enough time for the DRAM and any decoder in the system to function. Because many EPROM or Flash memory devices require an access time of 100 ns, EPROM or Flash requires the addition of seven wait states to lengthen the access time to 122 ns.

18–6 SPECIAL PENTIUM PRO FEATURES

The Pentium Pro is essentially the same microprocessor as the 80386, 80486, and Pentium, except that some additional features and changes to the control register set have occurred. This section highlights the differences between the 80386 control register structure and the Pentium Pro control register.

Control Register 4

Figure 18–17 shows control register 4 of the Pentium Pro microprocessor. Notice that CR_4 has two new control bits that are added to the control register array.

This section of the text explains only the two new Pentium Pro components in control register 4. (Refer to Figure 18–8 for a description and illustration of the Pentium control registers.) Following is a description of the Pentium CR_4 bits and the new Pentium Pro control bits in control register CR_4:

VME **Virtual mode extension** enables support for the virtual interrupt flag in protected mode. If VME = 0, virtual interrupt support is disabled.

PVI **Protected mode virtual interrupt** enables support for the virtual interrupt flag in protected mode.

TSD **Time stamp disable** controls the RDTSC instruction.

DE **Debugging extension** enables I/O breakpoint debugging extensions when set.

PSE **Page size extension** enables 4M-byte memory pages when set in the Pentium, or 2M-byte pages when set in the Pentium Pro whenever PSE is also set.

31	7	6	5	4	3	2	1	0
	PGE	MCE	PAE	PSE	DE	TSD	PVI	VME

FIGURE 18–17 The new control register 4 (CR_4) in the Pentium Pro microprocessor.

PAE **Page address extension** enables address lines A_{35}–A_{32} whenever a special new addressing mode, controlled by PGE, is enabled for the Pentium Pro.

MCE **Machine check** enable enables the machine checking interrupt.

PGE **Page extension** controls the new, larger 64G addressing mode whenever it is set along with PAE and PSE.

18–7 SUMMARY

1. The Pentium microprocessor is almost identical to the earlier 80386 and 80486 microprocessors. The main difference is that the Pentium has been modified internally to contain a dual cache (instruction and data) and a dual integer unit. The Pentium also operates at a higher clock speed of 66 MHz.
2. The 66 MHz Pentium requires 3.3 A of current, and the 60 MHz version requires 2.91 A. The power supply must be a +5.0 V supply with a regulation of ±5%. Newer versions of the Pentium require a 3.3 V or 2.7 V power supply.
3. The data bus on the Pentium is 64 bits wide and contains eight byte-wide memory banks selected with bank enable signals ($\overline{BE7}$–$\overline{BE0}$).
4. Memory access time, without wait states, is only about 18 ns in the 66 MHz Pentium. In many cases, this short access time requires wait states that are introduced by controlling the $\overline{BRDY}$ input to the Pentium.
5. The superscalar structure of the Pentium contains three independent processing units: a floating-point processor and two integer processing units labeled U and V by Intel.
6. The cache structure of the Pentium is modified to include two caches. One 8K × 8 cache is designed as an instruction cache; the other 8K × 8 cache is a data cache. The data cache can be operated as either a write-through or a write-back cache.
7. A new mode of operation called the system memory-management (SMM) mode has been added to the Pentium. The SMM mode is accessed via the system memory-management interrupt applied to the $\overline{SMI}$ input pin. In response to $\overline{SMI}$, the Pentium begins executing software at memory location 38000H.
8. New instructions include the CMPXCHG8B, RSM, RDMSR, WRMSR, and CPUID. The CMPXCHG8B instruction is similar to the 80486 CMPXCHG instruction. The RSM instruction returns from the system memory-management interrupt. The RDMSR and WRMSR instructions read or write to the machine-specific registers. The CPUID instruction reads the CPU identification code from the Pentium.
9. The built-in self-test (BIST) allows the Pentium to be tested when power is first applied to the system. A normal power-up reset activates the RESET input to the Pentium. A BIST power-up reset activates INIT and then deactivates the RESET pin. EAX is equal to a 00000000H in the BIST passes.
10. A new proprietary Intel modification to the paging unit allows 4M-byte memory pages instead of the 4K-byte pages. This is accomplished by using the page directory to address 1024 page tables that each contains 4M of memory.
11. The Pentium Pro is an enhanced version of the Pentium microprocessor that contains not only the level 1 caches found inside the Pentium, but also the level 2 cache of 256K or 512K found on most main boards.
12. The Pentium Pro operates by using the same 66 MHz bus speed as the Pentium and the 80486. It uses an internal clock generator to multiply the bus speed by various factors to obtain higher internal execution speeds.
13. The only significant software difference between the Pentium Pro and earlier microprocessors is the addition of the FCMOV and CMOV instructions.

14. The only hardware difference between the Pentium Pro and earlier microprocessors is the addition of 2M paging and four extra address lines that allow access to a memory address space of 64G bytes.
15. Error correction code has been added to the Pentium Pro, which corrects any single-bit error and detects any two-bit error.

18–8 QUESTIONS AND PROBLEMS

1. How much memory is accessible to the Pentium microprocessor?
2. How much memory is accessible to the Pentium Pro microprocessor?
3. The memory data bus width is __________ in the Pentium.
4. What is the purpose of the DP_0–DP_7 pins on the Pentium?
5. If the Pentium operates at 66 MHz, what frequency clock signal is applied to the CLK pin?
6. What is the purpose of the $\overline{BRDY}$ pin on the Pentium?
7. What is the purpose of the AP pin on the Pentium?
8. How much memory access time is allowed by the Pentium, without wait states, when it is operated at 66 MHz?
9. What Pentium pin is used to insert wait states into the timing?
10. A wait state is an extra __________ clocking period.
11. Explain how two integer units allow the Pentium to execute two nondependent instructions simultaneously.
12. How many caches are found in the Pentium and what are their sizes?
13. How wide is the Pentium memory data sample window for a memory read operation?
14. Can the Pentium execute three instructions simultaneously?
15. What is the purpose of the $\overline{SMI}$ pin?
16. What is the system memory-management mode of operation for the Pentium?
17. How is the system memory-management mode exited?
18. Where does the Pentium begin to execute software for an $\overline{SMI}$ interrupt input?
19. How can the system memory-management unit dump address be modified?
20. Explain the operation of the CMPXCHG8B instruction.
21. What information is returned in register EAX after the CPUID instruction executes with an initial value of 0 in EAX?
22. What new flag bits are added to the Pentium microprocessor?
23. What new control register is added to the Pentium microprocessor?
24. Describe how the Pentium accesses 4M pages.
25. Explain how the time-stamp counter functions and how it can be used to time events.
26. Contrast the Pentium with the Pentium Pro microprocessor.
27. Where are the bank enable signals found in the Pentium Pro microprocessor?
28. How many address lines are found in the Pentium Pro system?
29. What changes have been made to CR_4 in the Pentium Pro and for what purpose?
30. Compare access times in the Pentium system with the Pentium Pro system.
31. What is ECC?
32. What type of SDRAM must be purchased to use ECC?

CHAPTER 19

The Pentium II, Pentium III, and Pentium 4 Microprocessors

INTRODUCTION

The Pentium II, Pentium III, and Pentium 4 microprocessors may well signal the end of the evolution of 32-bit architecture with the advent of the Itanium[1] and Itanium II microprocessors from Intel. The Itanium is a 64-bit architecture microprocessor. The Pentium II, Pentium III, and Pentium 4 architectures are extensions of the Pentium Pro architecture, with some differences. The most notable difference is that the internal cache from the Pentium Pro architecture has been moved out of the microprocessor in the Pentium II. Another major change is that the Pentium II is not available in integrated circuit form. Instead, the Pentium II is found on a small plug-in circuit board called a cartridge along with the level 2 cache chip. Various versions of the Pentium II are available. The Celeron[2] is a version of the Pentium II that does not contain the level 2 cache on the Pentium II circuit board. The Xeon[3] is an enhanced version of the Pentium II that contains up to a 2M-byte cache on the circuit board.

Similar to the Pentium II, early Pentium III microprocessors were packaged in a cartridge instead of an integrated circuit. More recent versions, such as the Coppermine, are again packaged in an integrated circuit (370 pins). The Pentium III Coppermine, like the Pentium Pro, contains an internal cache. The Pentium 4 is packaged in a larger integrated circuit, with 423 or 478 pins. The Pentium 4 also uses physically smaller transistors, which makes it much smaller and faster than the Pentium III. To date Intel has released versions of the Pentium 4 that operate at frequencies over 3 GHz with a limit of possibly 10 GHz at some future time. Also available to the Pentium 4 are the extreme model (P4E) with a 1M-byte cache and the extreme edition model (P4EE) with a 2M-byte cache. The P4E and P4EE versions are now available in the new 90 nm (0.09 micron) form as compared to earlier P_4 microprocessors that use the 0.13 micron form.

CHAPTER OBJECTIVES

Upon completion of this chapter, you will be able to:

1. Detail the differences between the Pentium II, Pentium III, and Pentium 4 and prior Intel microprocessors

---

[1]Itanium is a registered trademark of Intel Corporation.

[2]Celeron is a registered trademark of Intel Corporation.

[3]Xeon is a registered trademark of Intel Corporation.

2. Explain how the architectures of the Pentium II, Pentium III, and Pentium 4 improve system speed
3. Explain how the basic architecture of the computer system has changed by using the Pentium II, Pentium III, and Pentium 4 microprocessors
4. Detail the changes to the CPUID instruction and model-specific registers
5. Describe the operation of the SYSENTER and SYSEXIT instructions
6. Describe the operation of the FXSAVE and FXRSTOR instructions

19–1 INTRODUCTION TO THE PENTIUM II MICROPROCESSOR

Before the Pentium II or any other microprocessor can be used in a system, the function of each pin must be understood. This section of the chapter details the operation of each pin, along with the external memory system and I/O structures of the Pentium II microprocessor.

Figure 19–1 illustrates the basic outline of the Pentium II microprocessor slot 1 connector and the signals used to interface to the chip set. Figure 19–2 shows a simplified diagram of the components on the cartridge, and the placement of the Pentium II cartridge and bus components in the typical Pentium II system. There are 242 pins on the slot 1 connector for the microprocessor. (This is a reduction from the number of pins found on the Pentium and the Pentium II microprocessors.) The Pentium II is packaged on a printed circuit board instead of the integrated circuits of the past Intel microprocessors. The level 1 cache is 32K bytes as it was in the Pentium Pro, but the level 2 cache is no longer inside the integrated circuit. Intel changed the architecture so that a level 2 cache could be placed very close to the microprocessor. This change makes the microprocessor less expensive and still allows the level 2 cache to operate efficiently. The Pentium level 2 cache operates at one-half the microprocessor clock frequency, instead of the 66 MHz of the Pentium microprocessor. A 400 MHz Pentium II has a cache speed of 200 MHz. The Pentium II is available in three versions. The first is the full-blown Pentium II, which is the Pentium II for the slot 1 connector. The second is the Celeron, which is like the Pentium II except that the slot 1 circuit board does not contain a level 2 cache; the level 2 cache in the Celeron system is located on the main board and operates at 66 MHz. The most recent version is the Xeon, which, because it uses a level 2 cache of 512K, 1M, or 2M, represents a significant speed improvement over the Pentium II. The Xeon's level 2 cache operates at the clock frequency of the microprocessor. A 400 MHz Xeon has a level 2 cache speed of 400 MHz, which is twice the speed of the regular Pentium II.

The early versions of the Pentium II require a 5.0 V, 3.3 V, and variable voltage power supply for operation. The main variable power supply voltages vary from 3.5 V to as low as 1.8 V at the microprocessor. This is known as the core microprocessor voltage. The power-supply current averages 14.2 A to 8.4 A, depending on the operating frequency and voltage of the Pentium II. Because these currents are significant, so is the power dissipation of these microprocessors. At present, a good heat sink with considerable airflow is required to keep the Pentium II cool. Luckily, the heat sink and fan are built into the Pentium II cartridge. The latest versions of the Pentium II have been improved to reduce the power dissipation.

Each Pentium II cartridge output pin is capable of providing at least 36 mA of current at a logic 0 level on the signal connections. Some of the output control signals provide only 14 mA of current. Another change to the Pentium II is that the outputs are open-drain and require an external pull-up resister for proper operation.

The function of each Pentium II group of pins follows:

$\overline{\text{A20}}$ — **Address A_{20} mask** is an input that is asserted in the real mode to signal the Pentium II to perform address wraparound, as in the 8086 microprocessor, for use of the HIMEM.SYS driver.

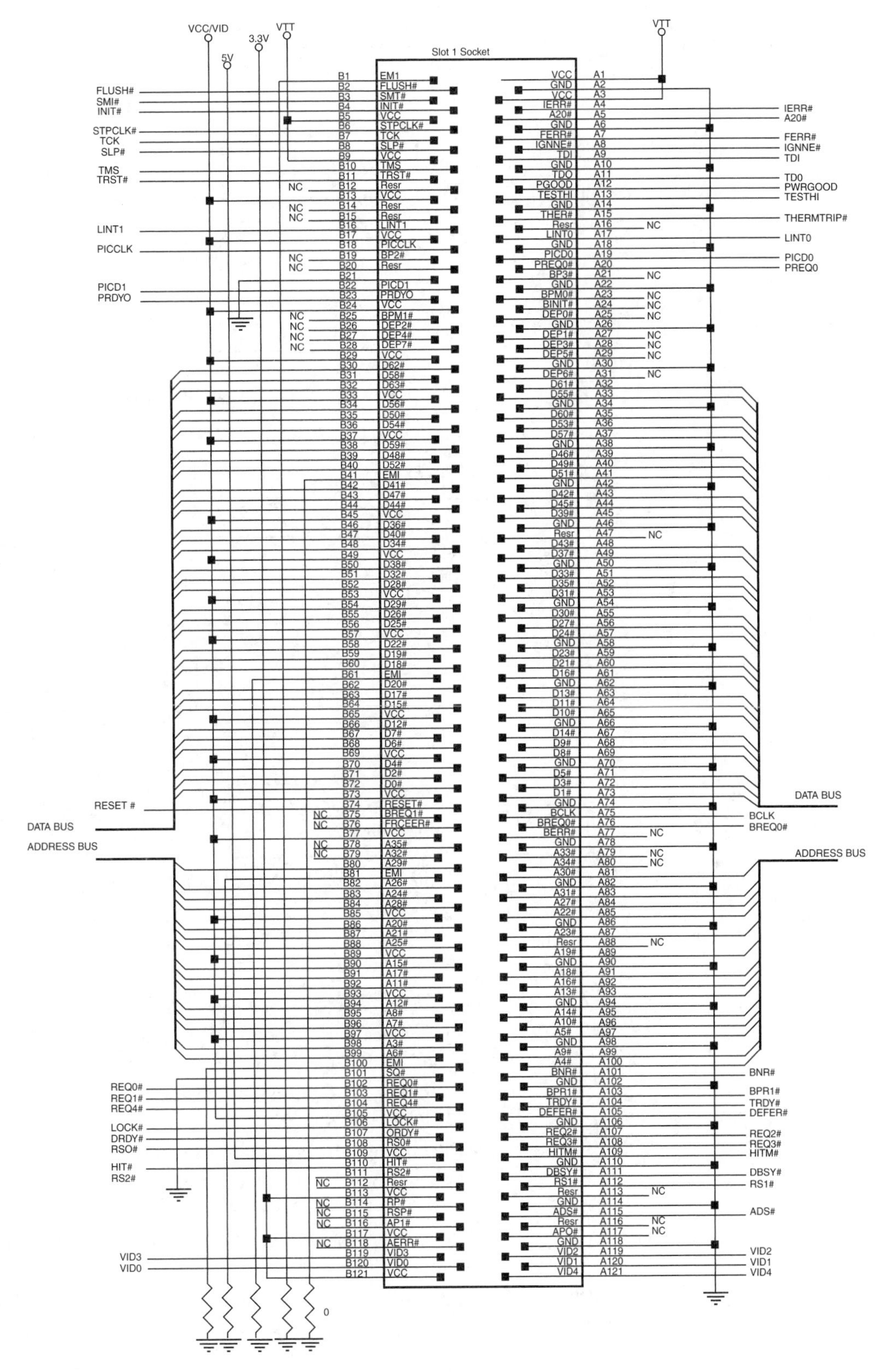

FIGURE 19–1 The pin-out of the slot 1 connector showing the connections to the system.

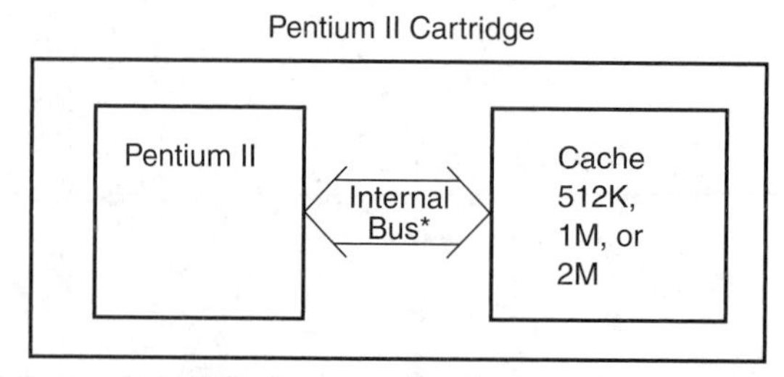

* The bus speed is 1/2 Pentium speed or the same as the Pentium speed in the Xeon.

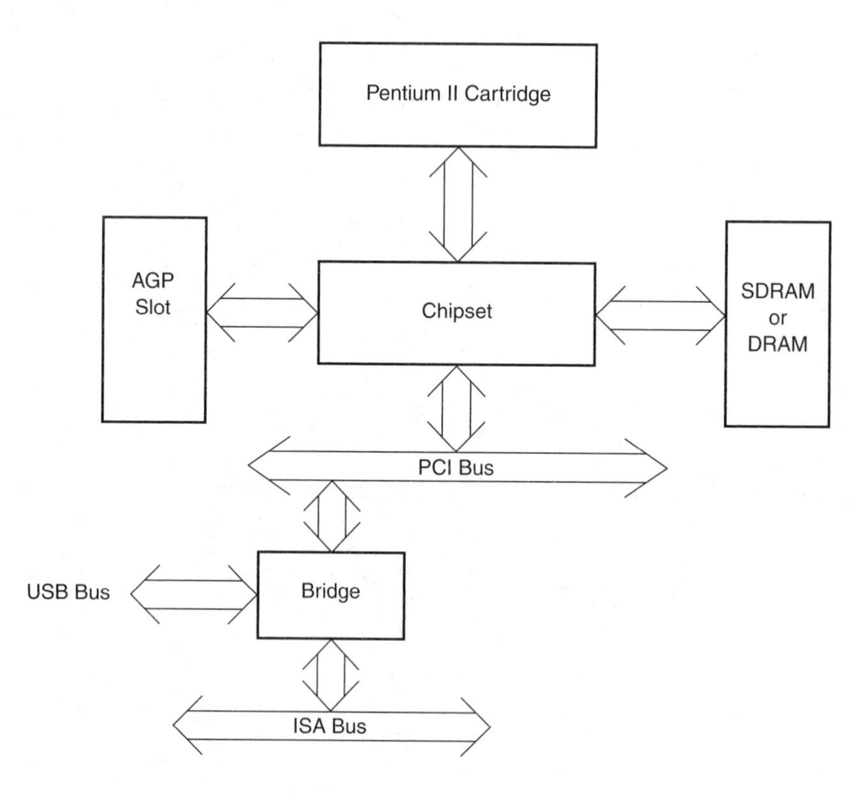

FIGURE 19–2 The structure of the Pentium II cartridge and the structure of the Pentium II system.

$\overline{A35}$–$\overline{A3}$ — **Address buses**, which are active low connections, address any of the memory locations found in the Pentium II memory system. Note that A_0, A_1, and A_2 are encoded in the bus enable ($\overline{BE7}$– $\overline{BE0}$), which are generated by the chip set, to select any or all of the eight bytes in a 64-bit-wide memory location.

$\overline{ADS}$ — **Address data strobe** is an input that is activated to indicate to the Pentium II that the system is ready to perform a memory or I/O operation. This signal causes the microprocessor to provide the address to the system.

$\overline{AERR}$ — **Address error** is an input used to cause the Pentium II to check for an address parity error if it is activated.

$\overline{AP1}$, $\overline{AP0}$ — **Address parity** inputs indicate an address parity error.

$\overline{BCLK}$ — **Bus clock** is an input that sets the bus clock frequency. This is either 66 MHz or 100 MHz in the Pentium II.

BERR — **Bus error** is asserted to indicate that an error has occurred on the bus system.

BINIT — **Bus initialization** is a logic 0 during system reset or initialization. It is an input to indicate that a bus error has occurred and the system needs to be reinitialized.

BNR — **Bus not ready** is an input used to insert wait states into the timing for the Pentium II. Placing a logic 0 on this pin causes the Pentium II to enter stall states or wait states.

BP3, BP2, PM1/BP1, and PM0/BP0 — The **breakpoint** pins BP3– BP0 indicate a breakpoint match when the debug registers are programmed to monitor for matches. The **performance monitoring** pins PM1 and PM0 indicate the settings of the performance monitoring bits in the debug mode control register.

BPRI — The **bus priority request input** is used to request the system bus from the Pentium II.

BR1 and BR0 — **Bus requests** indicate that the Pentium II has generated a bus request. During initialization, the BR0 pin must be activated.

BSEL — **Bus select** is currently not used by the Pentium II and must be connected to ground for proper operation.

D63–D0 — **Data bus** connections transfer byte, word, doubleword, and quadword data between the microprocessor and its memory and I/O system.

DEFER — The **defer** signal indicates that the external system cannot complete the bus cycle.

EP7–EP0 — **Data EEC** pins are used in the error-correction scheme of the Pentium II and normally connect to an extra 8-bit memory section. This means that ECC memory modules are 72 bits wide instead of 64 bits wide.

DRDY — **Data ready** is activated to indicate that the system is presenting valid data to the Pentium II.

EMI — **Electromagnetic interference** must be grounded to prevent the Pentium II from generating or receiving noise.

FERR — **Floating-point error**, comparable to the ERROR line in the 80386, shows that the internal coprocessor has erred.

FLUSH — The **flush cache** input causes the cache to flush all write-back lines and invalidate its internal caches. If the FLUSH input is a logic 0 during a reset operation, the Pentium enters its test mode.

FRCERR — **Functional redundancy check** is sampled during a reset to configure the Pentium II in the master (1) or checker (0) mode.

HIT — **Hit** shows that the internal cache contains valid data in the inquire mode.

HITM — **Hit modified** shows that the inquire cycle found a modified cache line. This output is used to inhibit other master units from accessing data until the cache line is written to memory.

IERR — The **internal error** output shows that the Pentium II has detected an internal error or functional redundancy error.

IGNNE — The **ignore numeric error** input causes the Pentium II to ignore a numeric coprocessor error.

$\overline{\text{INIT}}$	The **initialization** input performs a reset without initializing the caches, write-back buffers, and floating-point registers. This input may not be used to reset the microprocessor in lieu of RESET after power-up.
INTR	**Interrupt request** is used by external circuitry to request an interrupt.
$LINT_1$, $LINT_0$	**Local APIC interrupt** signals must connect the appropriate pins of all APIC bus agents. When the APIC is disabled, the $LINT_0$ signal becomes INTR, a maskable interrupt request signal; $LINT_1$ becomes NMI, a non-maskable interrupt.
$\overline{\text{LOCK}}$	$\overline{\text{LOCK}}$ becomes a logic 0 whenever an instruction is prefixed with the LOCK: prefix. This is most often used during DMA accesses.
NMI	**Non-maskable interrupt** requests a non-maskable interrupt as it did on the earlier versions of the microprocessor.
PICCLK	This **clock** signal must be $\frac{1}{4}$ the frequency of $\overline{\text{BCLK}}$.
$PICD_1$, $PICD_0$	Used for serial messages between the Pentium II and APIC.
PRDY	The **probe ready** output indicates that the probe mode has been entered for debugging.
$\overline{\text{PREQ}}$	The **probe request** is used to request debugging.
PWRGOOD	The **power good** input that indicates that the system power supply is operational.
$\overline{\text{REQ4}}$–$\overline{\text{REQ0}}$	**Request signals** communicate commands between bus controllers and the Pentium II.
$\overline{\text{RESET}}$	**Reset** initializes the Pentium II, causing it to begin executing software at memory locations FFFFFFF0H or 000FFFF0H. The A_{35}–A_{32} address bits are set as logic 0s during the reset operation. The Pentium II is reset to the real mode and the leftmost 12 address connections remain logic 1s (FFFH) until a far jump or far call is executed. This allows compatibility with earlier microprocessors. See Table 19–1 for the state of the Pentium II after a hardware reset.

TABLE 19–1 State of the Pentium II after a reset.

Register	*Reset*	*Reset + BIST*
EAX	0	0 (if test passes)
EDX	0500XXXXH	0500XXXXH
EBX, ECX, ESP, EBP, ESI, and EDI	0	0
EFLAGS	2	2
EIP	0000FFF0H	0000FFF0H
CS	F000H	F000H
DS, ES, FS, GS, and SS	0	0
GDTR and TSS	0	0
CR_0	60000010H	60000010H
CR_2, CR_3, and CR_4	0	0
DR_3–DR_0	0	0
DR_6	FFFF0FF0H	FFFF0FF0H
DR_7	00000400H	00000400H

Notes: BIST = built-in self-test, XXXX = Pentium II version number.

$\overline{\text{RP}}$ — **Request parity** is used to request parity.

$\overline{\text{RS2}}$–$\overline{\text{RS0}}$ — **Request status** inputs are used to request the current status of the Pentium II.

$\overline{\text{RSP}}$ — The **response parity** input is activated to request parity.

$\overline{\text{SLOTOCC}}$ — The **slot occupied** output is a logic 0 if slot zero contains either a Pentium II or a dummy terminator.

$\overline{\text{SLP}}$ — **Sleep** is an input that, when inserted in the stop-grant state, causes the Pentium II to enter the sleep state.

$\overline{\text{SMI}}$ — The **system management interrupt** input causes the Pentium II to enter the system management mode of operation.

$\overline{\text{STPCLK}}$ — The **stop clock** input causes the Pentium II to enter the low-power stop-grant state.

TCK — The **testability clock** input selects the clocking function in accordance with the IEEE 1149.1 Boundary Scan interface.

TDI — The **test data input** is used to test data clocked into the Pentium II with the TCK signal.

TDO — The **test data output** is used to gather test data and instructions shifted out of the Pentium II with TCK.

TESTHI — **Test high** is an input that must be connected to +2.5 V through a 1K–10K Ω resister for proper Pentium II operation.

$\overline{\text{THERMTRIP}}$ — **Thermal sensor trip** is an output that becomes a zero when the temperature of the Pentium II exceeds 130°C.

TMS — The **test mode select** input controls the operation of the Pentium in test mode.

$\overline{\text{TRDY}}$ — **Target ready** is an input that is used to cause the Pentium II to perform a write-back operation.

$\overline{\text{VID4}}$–$\overline{\text{VID0}}$ — **Voltage data** output pins are either open or grounded signals that indicate what supply voltage is currently required by the Pentium II. The power supply must apply the request voltage to the Pentium II, as listed in Table 19–2.

The Memory System

The memory system for the Pentium II microprocessor is 64G bytes in size, just like the Pentium Pro microprocessor. Both microprocessors address a memory system that is 64 bits wide with an address bus that is 36 bits wide. Most systems use SDRAM operating at 66 MHz or 100 MHz for the Pentium II. The SDRAM for the 66 MHz system has an access time of 10 ns and the SDRAM for the 100 MHz system has an access time of 8 ns. The memory system, which connects to the chip set, is not illustrated in this chapter. Refer to earlier chapters to see the organization of a 64-bit-wide memory system without ECC.

The Pentium II memory system is divided into eight or nine banks that each store a byte of data. If the ninth byte is present, it stores an error-checking code (ECC). The Pentium II, like the 80486–Pentium Pro, employs internal parity generation and checking logic for the memory system's data bus information. (Note that most Pentium II systems do not use parity checks, but it is available.) If parity checks are employed, each memory bank contains a ninth bit. The 64-bit-wide memory is important to double-precision floating-point data. Recall that a double-precision floating-point number is 64 bits wide. As with the Pentium Pro, the memory system is numbered in bytes from byte 000000000H to byte FFFFFFFFFH. Please note that none of the current chip sets support more than 1G bytes of system memory, so the additional address connections are for

TABLE 19–2 Power supply voltages for the Pentium II as requested by the $\overline{VID}$ pins.

$\overline{VID4}$	$\overline{VID3}$	$\overline{VID2}$	$\overline{VID1}$	$\overline{VID0}$	V_{CC}
0	0	0	0	0	2.05 V
0	0	0	0	1	2.00 V
0	0	0	1	0	1.95 V
0	0	0	1	1	1.90 V
0	0	1	0	0	1.85 V
0	0	1	0	1	1.80 V
0	0	1	1	0	—
0	0	1	1	1	—
0	1	0	0	0	—
0	1	0	0	1	—
0	1	0	1	0	—
0	1	0	1	1	—
0	1	1	0	0	—
0	1	1	0	1	—
0	1	1	1	0	—
0	1	1	1	1	—
1	0	0	0	0	3.5 V
1	0	0	0	1	3.4 V
1	0	0	1	0	3.3 V
1	0	0	1	1	3.2 V
1	0	1	0	0	3.1 V
1	0	1	0	1	3.0 V
1	0	1	1	0	2.9 V
1	0	1	1	1	2.8 V
1	1	0	0	0	2.7 V
1	1	0	0	1	2.6 V
1	1	0	1	0	2.5 V
1	1	0	1	1	2.4 V
1	1	1	0	0	2.3 V
1	1	1	0	1	2.2 V
1	1	1	1	0	2.1 V
1	1	1	1	1	—

future expansion. Figure 19–3 illustrates the basic memory map of the Pentium II system, using the AGP for the video card.

The memory map for the Pentium II system is similar to the map illustrated in earlier chapters, except that an area of the memory is used for the AGP area. The AGP area allows the video card and Windows to access the video information in a linear address space. This is unlike the 128K-byte window in the DOS area for a standard VGA video card. The benefit is much faster video updates because the video card does not need to page through the 128K-byte DOS video memory.

Transfers between the Pentium II and the memory system are controlled by the 440 LX or 440 BX chip set. Data transfers between the Pentium II and the chip set are eight bytes wide. The chip set communicates to the microprocessor through the five $\overline{REQ}$ signals, as listed in Table 19–3. In essence, the chip set controls the Pentium II, which is a departure from the traditional method of connecting a microprocessor to the system directly to the memory.

The Pentium II only connects directly to the cache, which is on the Pentium II cartridge. As mentioned, the Pentium II cache operates at one-half the clock frequency of the microprocessor. Therefore, a 400 MHz Pentium II cache operates at 200 MHz. The Pentium II Xeon

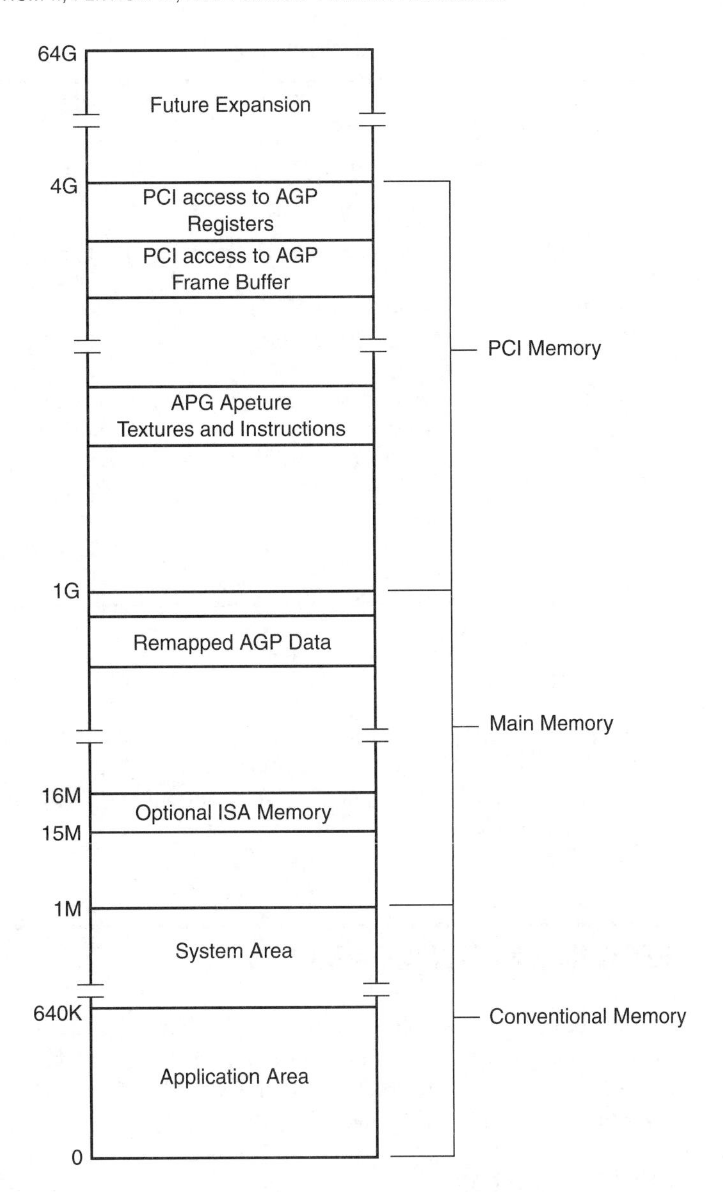

FIGURE 19–3 The memory map of a Pentium II–based computer system.

cache operates at the same frequency as the microprocessor, which means that the Xeon, with its 512K, 1M, or 2M cache, outperforms the standard Pentium II.

Input/Output System

The input/output system of the Pentium II is completely compatible with earlier Intel microprocessors. The I/O port number appears on address lines A_{15}–A_3 with the bank-enable signals used to select the actual memory banks used for the I/O transfer. Transfers are controlled by the chip set, which is a departure from the standard microprocessor architecture before the Pentium II.

Beginning with the 80386 microprocessor, I/O privilege information is added to the TSS segment when the Pentium II is operated in the protected mode. Recall that this allows I/O ports to be selectively inhibited. If the blocked I/O location is accessed, the Pentium II generates a type 13 interrupt to signal an I/O privilege violation.

TABLE 19–3 The $\overline{REQ}$ signals to the Pentium II.

$\overline{REQ4}$–$\overline{REQ0}$	Name	Comment
00000	Deferred reply	Deferred replies are issued for previously deferred transactions
00001	Reserved	—
00010	Memory read & invalidate	Memory read from DRAM or PCI write to DRAM from PCI
00011	Reserved	—
00100	Memory code read	Memory read cycle
00101	Memory write-back	Memory write-back cycle
00110	Memory data read	Memory read cycle
00111	Memory write	Normal memory write cycle
01000	Interrupt acknowledge or special cycle	Interrupt acknowledge for PCI bus
01001	Reserved	—
10000	I/O read	I/O read operation
10001	I/O write	I/O write operation

System Timing

As with any microprocessor, the system timing signals must be understood in order to interface the microprocessor, or so it was at one time. Because the Pentium II is designed to be controlled by the chip set, the timing signals between the microprocessor and the chip set have become proprietary to Intel.

19–2 PENTIUM II SOFTWARE CHANGES

The Pentium II microprocessor core is a Pentium Pro. This means that the Pentium II and the Pentium Pro are essentially the same device for software. This section of the text lists the changes to the CPUID instruction and the SYSENTER, SYSEXIT, FXSAVE, and FXRSTORE instructions (the only modifications to the software).

CPUID Instruction

Table 19–4 lists the values passed between the Pentium II and the CPUID instruction. These are changed from earlier versions of the Pentium microprocessor.

The version information returned after executing the CPUID instruction with a logic 0 in EAX is returned in EAX. The family ID is returned in bits 8 to 11; the model ID is returned in bits 4 to 7. The stepping ID is returned in bits 0 to 3. For the Pentium II, the model number is 6 and the family ID is a 3. The stepping number refers to an update number—the higher the stepping number, the newer the version.

The features are indicated in the EDX register after executing the CPUID instruction with a zero in EAX. Only two new features are returned in EDX for the Pentium II. Bit position 11 indicates whether the microprocessor supports the two new fast call instructions, SYSENTER and SYSEXIT. Bit position 23 indicates whether the microprocessor supports the MMX instruction set introduced in Chapter 14. The remaining bits are identical to earlier versions of the microprocessor and are not described. Bit 16 indicates whether the microprocessor supports the page attribute table or PAT. Bit 17 indicates whether the microprocessor supports the page size extension

TABLE 19–4 CPUID instruction for the Pentium II.

Input EAX	*Output Register*	*Contents*
0	EAX	Maximum allowed input to EAX for CPUID
0	EBX	"uneG"
0	ECX	"Inei"
0	EDX	"letn"
1	EAX	Version number
1	EDX	Feature information
2	EAX	Cache data
2	EBX	Cache data
2	ECX	Cache data
2	EDX	Cache data

found with the Pentium Pro and Pentium II microprocessors. The page size extension allows memory above 4G through 64G to be addressed. Finally, bit 24 indicates whether the fast floating-point save (FXSAVE) and restore (FXRSTOR) instructions are implemented.

SYSENTER and SYSEXIT Instructions

The SYSENTER and SYSEXIT instructions use the fast call facility introduced in the Pentium II microprocessor. Please note that these instructions function only in ring 0 (privilege level 0) in protected mode. Windows operates in ring 0, but does not allow applications access to ring 0. These new instructions are meant for operating system software because they will not function at any other privilege level.

The SYSENTER instruction uses some of the model-specific registers to store CS, EIP, and ESP to execute a fast call to a procedure defined by the model-specific register. The fast call is different from a regular call because it does not push the return address onto the stack as a regular call. Table 19–5 illustrates the model-specific register used with SYSENTER and SYSEXIT. Note that the model-specific registers are read with the RDMSR instruction and written with the WRMSR instruction.

To use the RDMSR or WRMSR instructions, place the register number in the ECX register. If the WRMSR is used, place the new data for the register in EDS:EAX. For the SYSENTER instruction, you need use only the EAX register, but place a zero into EDX. If the RDMSR register instruction is used in a program the data are returned in the EDX:EAX registers.

To use the SYSENTER instruction, you must first load the model-specific registers with the address of the system entrance point into the SYSENTER_CS, SYSENTER_ESP, and SYSENTER_EIP registers. This would normally be the entrance address and stack area of an operating system such as Windows 2000 or Windows XP. Note that this instruction is meant as a system instruction to access code or software in ring 0. The stack segment register is loaded with the value placed into SYSENTER_CS plus 8. In other words, the selector pair addressed by SYSENTER_CS selector value is loaded into CS and SS. The value of the stack offset is loaded into SYSENTER_ESP.

TABLE 19–5 The model-specific registers used with SYSENTER and SYSEXIT.

Name	*Number*	*Function*
SYSENTER_CS	174H	SYSENTER target code segment
SYSENTER_ESP	175H	SYSENTER target stack pointer
SYSENTER_EIP	176H	SYSENTER target instruction pointer

TABLE 19–6 Selectors addressed by the SYSENTER_CS value.

SYSENTER_CS (MSR 174H)	*Function*
SYSENTER_CS value	SYSENTER code segment selector
SYSENTER_CS value + 8	SYSENTER stack segment selector
SYSENTER_CS value + 16	SYSEXIT code segment selector
SYSENTER_CS value + 24	SYSEXIT stack segment selector

The SYSEXIT instruction loads CS and SS with the selector pair addressed by SYSENTER_CS plus 16 and 24. Table 19–6 illustrates the selectors from the global selector table, as addressed by SYSENTER_CS. In addition to the code and stack segment selector and the memory segments that they represent, the SYSEXIT instruction passes the value in EDX to the EIP register and the value in ECX to the ESP register. The SYSEXIT instruction returns control back to application ring 3. As mentioned, these instructions appear to have been designed for quick entrance and return from the Windows or Windows NT operating systems on the personal computer.

To use SYSENTER and SYSEXIT, the SYSENTER instruction must pass the return address to the system. This is accomplished by loading the EDX register with the return offset and by placing the segment address into the global descriptor table at location SYSENTER_CS+16. The stack segment is transferred by loading the stack segment selector into SYSENTER_CS+24 and the ESP into the ECX.

FXSAVE and FXRSTOR Instructions

The last two new instructions added to the Pentium II microprocessor are the FXSAVE and FXRSTOR instructions, which are almost identical to the FSAVE and FRSTOR instructions detailed in Chapter 14. The main difference is that the FXSAVE instruction is designed to properly store the state of the MMX machine, while the FSAVE properly stores the state of the floating-point coprocessor. The FSAVE instruction stores the entire tag field, whereas the FXSAVE instruction only stores the valid bits of the tag field. The valid tag field is used to reconstruct the restore tag field when the FXRSTOR instruction executes. This means that if the MMX state of the machine is saved, use the FXSAVE instruction; if the floating-point state of the machine is saved, use the FSAVE instruction. For new applications, it is recommended that the FXSAVE and FXRSTOR instructions should be used to save the MMX state and floating-point state of the machine. Do not use the FSAVE and FRSTOR instructions in new applications.

19–3 THE PENTIUM III

The Pentium III microprocessor is an improved version of the Pentium II microprocessor. Even though it is newer than the Pentium II, it is still based on the Pentium Pro architecture.

There are two versions of the Pentium III. One version is available with a nonblocking 512K-byte cache and packaged in the slot 1 cartridge, and the other version is available with a 256K-byte advanced transfer cache and packaged in an integrated circuit. The slot 1–version cache runs at half the processor speed, and the integrated-cache version runs at the processor clock frequency. As shown in most benchmarks of cache performance, increasing the cache size from 256K bytes to 512K bytes only improves performance by a few percent.

Chip Sets

The chip set for the Pentium III is different from that of the Pentium II. The Pentium III uses an Intel 810, 815, or 820 chip set. The 815 is most commonly found in newer systems that use the Pentium III. A few other vendors' chip sets are available, but problems with drivers for new peripherals, such as the video cards, have been reported. An 840 chip set also was developed for the Pentium III, but Intel did not make it available.

Bus

The Coppermine version of the Pentium III increases the bus speed to either 100 MHz or 133 MHz. The faster version allows transfers between the microprocessor and the memory at higher speeds. The last release version of the Pentium III was a 1 GHz microprocessor with a 133 MHz bus.

Suppose that you have a 1-GHz microprocessor that uses a 133-MHz memory bus. You might think that the memory bus speed could be faster to improve performance, and we agree. However, the connections between the microprocessor and the memory preclude using a higher speed for the memory. If we decided to use a 200-MHz bus speed, we must recognize that a wavelength at 200 MHz is 300,000,000/200,000,000 or 3/2 meter. An antenna is 1/4 of a wavelength. At 200 MHz, an antenna is 14.8 inches. We do not want to radiate energy at 200 MHz, so we need to keep the printed circuit board connections shorter than 1/4-wavelength. In practice, we would keep the connections to no more than 1/10 of 1/4-wavelength. This means that the connections in a 200 MHz system should be no longer than 1.48 inches. This size presents the main board manufacturer with a problem when placing the sockets for a 200 MHz memory system. A 200 MHz bus system may be the limit for the technology. If the bus is tuned, there may be a way to go higher in frequency; only time will determine if it is possible. At present all that can be done is a play on words in advertisements such as 800M bytes per second to rate a bus. (Since 64 bits [8 bytes] are transferred at a time 800M bytes per second is really 100 MHz.)

Will it be possible to exceed the 200 MHz memory system? Yes, if we develop a new technology for interconnecting the microprocessor, chip set, and memory. At present the memory functions in bursts of four 64-bit numbers each time we read the main memory. This burst of 32 bytes is read into the cache. The main memory requires three wait states at 100 MHz to access the first 64-bit number and then zero wait states for each of the three remaining 64-bit-wide numbers for a total of seven 100 MHz bus clocks. This means we are reading data at 70 ns/32 = 2.1875 ns per byte, which is a bus speed of 457M bytes per second. This is slower than the clock on a 1 GHz microprocessor, but because most programs are cyclic and the instructions are stored in an internal cache, we can and often do approach the operating frequency of the microprocessor.

Pin-Out

Figure 19–4 shows the pin-out of the socket 370 version of the Pentium III microprocessor. This integrated circuit is packaged in a 370-pin, pin grid array (PGA) socket. It is designed to function with one of the chip sets available from Intel. In addition to the full version of the Pentium III, the Celeron, which uses a 66 MHz memory bus speed, is available. The Pentium III Xeon, also manufactured by Intel, allows larger cache sizes for server applications.

19–4 THE PENTIUM 4

The most recent version of the Pentium Pro architecture microprocessor is the Pentium 4 microprocessor from Intel. So far the Pentium II, Pentium III, and Pentium 4 are all versions of the Pentium Pro architecture. The Pentium 4 was released initially in November 2000 with a speed of 1.3 GHz. It is currently available in speeds up to 3.4 GHz. Two packages are available for this integrated microprocessor, the 423-pin PGA and the 478-pin FC-PGA2. Both versions of the original issue of the Pentium 4 used the 0.18 micron technology for fabrication. The most recent versions use either the 0.13 micron technology or the 90 nm (0.09 micron) technology. Intel is currently developing a 65 nm technology for future products. As with earlier versions of the Pentium III, the Pentium 4 uses a 100-MHz memory bus speed, but because it is quad pumped,

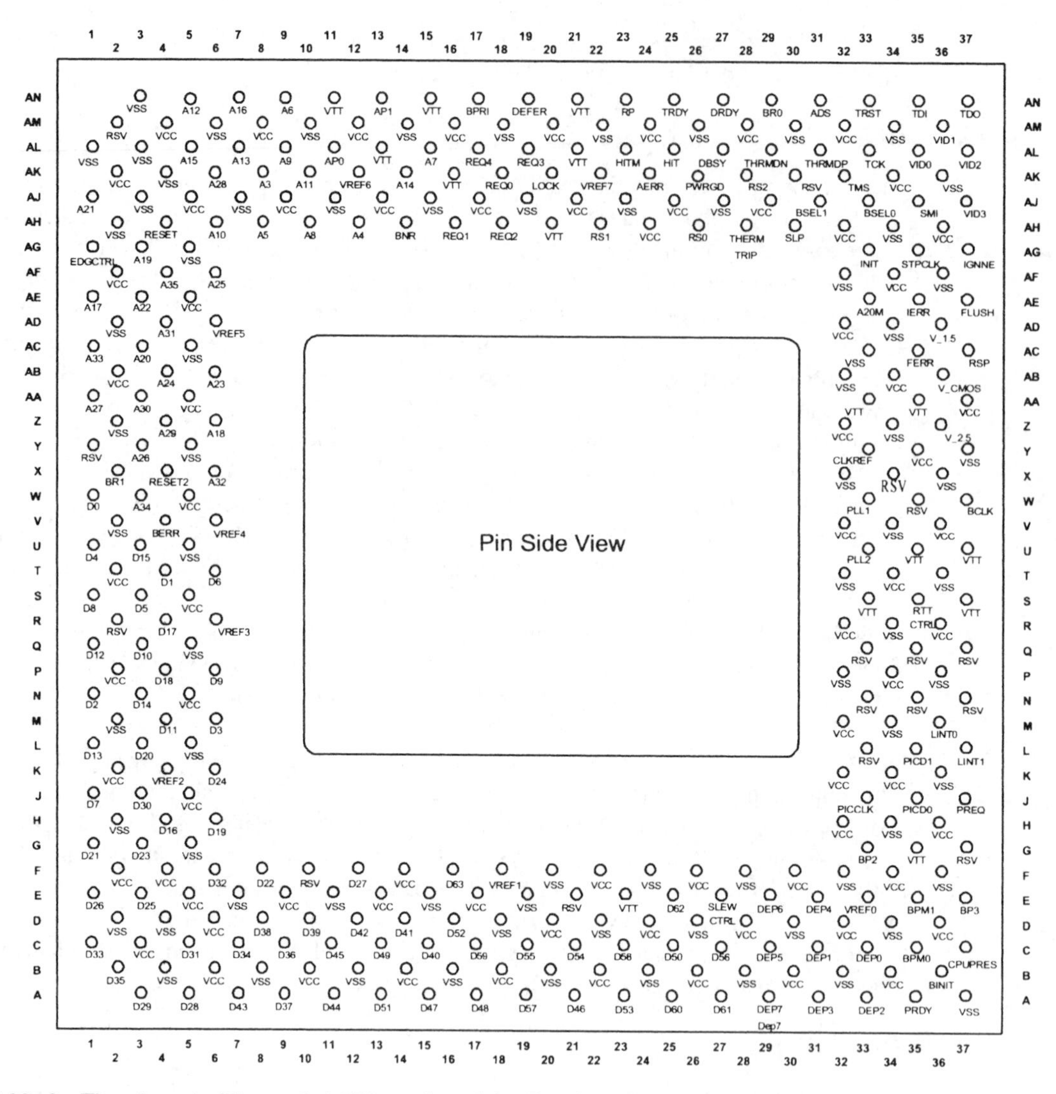

FIGURE 19–4 The pin-out of the socket 370 version of the Pentium III microprocessor. (Courtesy of Intel Corporation.)

the bus speed can approach 400 MHz. More recent versions use the 133 MHz bus listed as 533 MHz because of quad pumping or 200 MHz listed as 800 MHz. Figure 19–5 illustrates the pin-out of the 423-pin PGA of the Pentium 4 microprocessor.

Memory Interface

The memory interface to the Pentium 4 typically uses the Intel 850, 865, or 875 chip set. The 850 chip set provides a dual-pipe memory bus to the microprocessor with each pipe interfaced to a 32-bit-wide section of the memory. The two pipes function together to comprise the 64-bit-wide data path to the microprocessor. Because of the dual pipe arrangement, the memory must be populated with pairs of RDRAM memory devices operating at either 600 MHz or 800 MHz. According to Intel this arrangement provides a 300% increase in speed over a memory populated with PC-100 memory.

Intel has abandoned RDRAM in favor of DDR (double data rate) memory in the 865 and 875 chip sets. Apparently the claim of a 300% increase in RDRAM speed failed to prove factual.

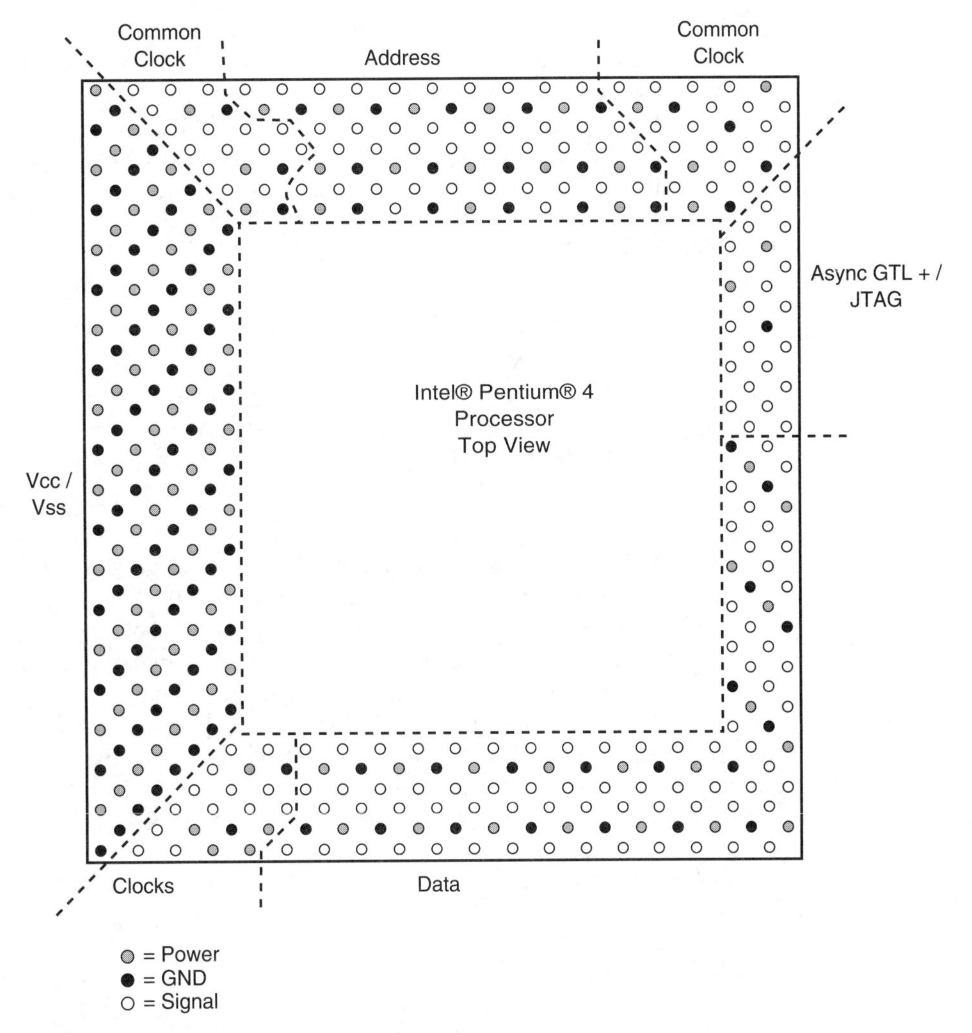

FIGURE 19–5 The pin-out of the Pentium 4, 423 PGA. (Courtesy of Intel Corporation.)

In addition to the inclusion of support for DDR, memory support for the serial ATA disk interface has also been added.

Register Set

The Pentium 4 register set is nearly identical to all other versions of the Pentium except that the MMX registers are separate entities from the floating-point registers. In addition, eight 128-bit-wide XMM registers are added for use with the SIMD (single-instruction, multiple data) instructions as explained in Chapter 14 and the extended 128-bit packed doubled floating-point numbers.

You might think of the XMM registers as double-wide MMX registers that can hold a pair of 64-bit double-precision floating-point numbers or four single-precision floating-point numbers. Likewise they can also hold 16-byte-wide numbers as the MMX registers hold eight-byte-wide numbers. The XMM registers are double-width MMX registers.

If the new patch for MASM 6.15 is downloaded from Microsoft, programs can be assembled using both the MMX and XMM instructions. The ML.EXE program is also found in Microsoft Visual Studio.NET 2003. To assemble programs that include MMX instructions, use the .MMX switch. For programs that include the SIMD instructions, use the .XMM switch.

Example 19–1 illustrates a very simple program that uses the MMX instructions to add two eight-byte-wide numbers together. Notice how the .MMX switch is used to select the MMX instruction set. The MOVQ instructions transfer numbers between memory and the MMX registers. The MMX registers are numbered from MM_0 to MM_7. You can also use the MMX and SIMD instructions in Microsoft Visual C++ using the inline assembler if you download the latest patch from Microsoft for Visual Studio version 6.0 or use a newer version of Visual Studio.

EXAMPLE 19–1

```
.MMX
.DATA
       DATA1  DQ     1FFH
       DATA2  DQ     101H
       DATA3  DQ     ?
.CODE
       MOVQ   MM0,DATA1
       MOVQ   MM1,DATA2
       PADDB  MM0,MM1
       MOVQ   DATA3,MM0
```

Similarly, the XMM software can be used in a program with the .XMM switch. Most modern programs use the XMM registers and the XMM instruction set to accomplish multimedia and other high-speed operations. Example 19–2 shows a short program that illustrates the use of a few XMM instructions. This program multiplies two sets of four single-precision floating-point numbers and stores the four products into the four doublewords at ANS. In order to enable access to octalwords (128-byte-wide numbers), we use the OWORD PTR directive. Also notice that the FLAT model is used with the C profile. The SIMD instructions only function in protected mode so the program uses the FLAT model format. This means that the .686 and .XMM switches are both placed before the model statement.

EXAMPLE 19–2

```
.686
.XMM
.MODEL FLAT,C
.DATA
       DATA1  DD  1.0        ;define four floats for DATA1
              DD  2.0
              DD  3.0
              DD  4.0
       DATA2  DD  6.3        ;define four floats for DATA2
              DD  4.6
              DD  4.5
              DD  -2.3
       ANS    DD  4 DUP(?)
.CODE
       MOVAPS XMM0,OWORD PTR DATA1
       MOVAPS XMM1,OWORD PTR DATA2
       MULPS  XMM0,XMM1
       MOVAPS OWORD PTR ANS,XMM0

;additional code here

END
```

Hyper-Threading Technology

The most recent innovation and new to the Pentium is called hyper-threading technology. This significant advancement combines two microprocessors into a single package. To understand this new technology, refer to Figure 19–6, which shows a traditional dual processor system and a hyper-threaded system.

FIGURE 19–6 Systems illustrating dual processors and a hyper-threaded processor.

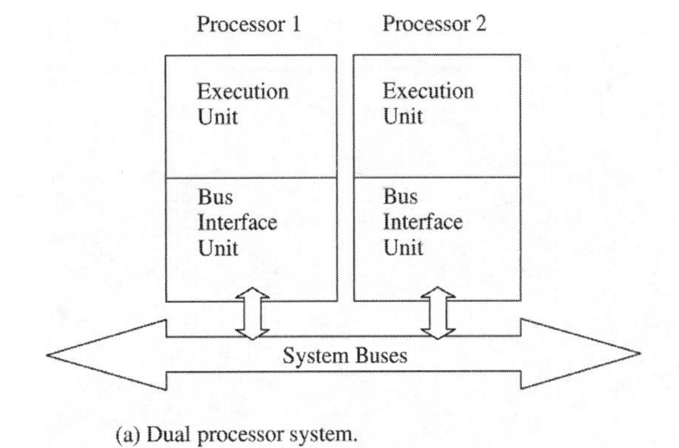

(a) Dual processor system.

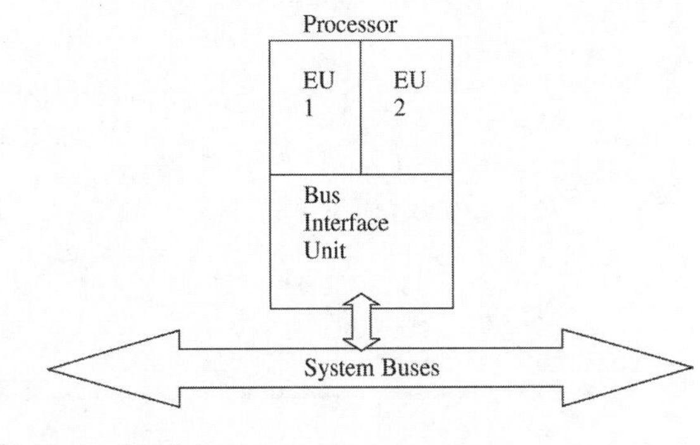

(b) A hyper-threaded system.

The hyper-threaded processor contains two execution units that each contain a complete set of the registers capable of running software independently or concurrently. These two separate machine contexts share a common bus interface unit. During machine operation each processor is capable of running a thread (process) independently, increasing the execution speed of an application that is written using multiple threads. The bus interface unit contains the level 2 and level 3 caches and the interface to the memory and I/O structure of the machine. When either microprocessor needs to access memory or I/O it must share the bus interface unit.

The bus interface unit is in use to access memory, but since memory is accessed in bursts that fill caches, it is often idle. Because of this a second processor can use this idle time to access memory while the first processor is busy executing instructions. Does the speed of the system double? Yes and no. Some threads can run independently of each other as long as they do not access the same area of memory. If each thread accesses the same area of memory, the machine can actually run slower with hyper-threaded technology. This does not occur very often, so in most cases the system performance increases with hyper-threading achieving nearly the same performance as a dual processor system.

Eventually most machines will use hyper-threading technology, which means that more attention should be given to developing software that is multithreaded. Each thread runs on a different processor in a system that has either dual processors or hyper-threaded processors, increasing performance. In the future the architecture may include even more processors to handle additional threads.

CPUID

As in earlier versions of the Pentium, the CPUID instruction accesses information that indicates the type of microprocessor as well as the features supported by the microprocessor. In the ever-evolving

TABLE 19–7 Pentium 4 CPUID instructions.

EAX Input Value	*Output Registers*	*Notes*
0	EAX = Maximum input value EBX = "uneG" ECX = "Iene" EDX = "Ietn"	"GenuineIntel" is returned in little endian format
1	EAX = Version information EBX = Feature information ECX = Extended feature information EDX = Feature information	Feature information
2	EAX, EBX, ECX, and EDX	Cache and TLB information
3	ECX and EDX	Serial number in the Pentium III only
4	EAX, EBX, ECX, and EDX	Deterministic cache parameters
5	EAX, EBX, ECX, and EDX	Monitor/Mwait information
80000000H	EAX	Extended function information
80000001H	EAX	Reserved
80000002H, 80000003H, and 80000004H	EAX, EBX, ECX, and EDX	Processor brand string
80000006H	ECX	Cache information

series of microprocessors it is important to be able to access this information so that efficient software can be written to operate on many different versions of the microprocessor.

Table 19–7 lists the latest features available to the CPUID instructions. To access these features, EAX is loaded with the input number listed in the table, then the CPUID instruction is executed. The CPUID instruction usually returns information in the EAX, EBX, ECX, and EDX registers in the real or protected mode. As can be gleaned from the table, additional features have been added to the CPUID instructions when compared to previous versions.

In Chapter 18 software was developed to read and display the data available after the CPUID instruction was invoked with EAX = 1. Here we deal with reading the processor brand string and prepare it for display it in a Visual C++ function. The brand string, if supported, contains the frequency at which the microprocessor is certified to operate and also the genuine Intel keyword. The BrandString function (see Example 19–3) returns a CString that contains the information stored in the CPUID members 80000002H–80000004H. This software requires a Pentium 4 system for proper operation as tested for in the BrandString function. The Convert function reads the contents of EAX, EBX, ECX, and EDX from the register specified as the parameter and converts them to a CString that is returned. The author's system shows that the brand string is

" Intel(R) Pentium(R) 4 CPU 3.06GHz"

EXAMPLE 19–3

```
CString CCPUIDDlg::BrandString(void)
{
        CString temp;
        int temp1;
        _asm
        {
                mov  eax,80000000h
```

```
            cpuid
            mov   temp1,eax
        }
        if ( temp1 >= 0x80000004 )      //if brand string present
        {
            temp += Convert(0x80000002);      //read register 80000002H
            temp += Convert(0x80000003);      //read register 80000003H
            temp += Convert(0x80000004);      //read register 80000004H
        }
        return temp;
}

CString CCPUIDDlg::Convert(int EAXvalue)
{
        CString temp = "                "; //must be 16 spaces
        int temp1, temp2, temp3, temp4;
        _asm
        {
            mov   eax,EAXvalue
            cpuid
            mov   temp1,eax
            mov   temp2,ebx
            mov   temp3,ecx
            mov   temp4,edx
        }
        for ( int a = 0; a <4; a++ )
        {
            temp.SetAt(a, temp1);
            temp.SetAt(a + 4, temp2);
            temp.SetAt(a + 8, temp3);
            temp.SetAt(a + 12, temp4);
            temp1 >>= 8;
            temp2 >>= 8;
            temp3 >>= 8;
            temp4 >> =8;
        }
        return temp;
}
```

The other information available about the system is returned in EAX, EBX, ECX, and EDX after executing CPUID after loading EAX with a 1. The EAX register contains the version information as the model, family, and stepping information, as illustrated in Figure 19–7. The EBX register contains information about the cache, such as the size of the cache line flushed by the CFLUSH instruction in bits 15–8 and the ID assigned to the local APIC on reset in bits

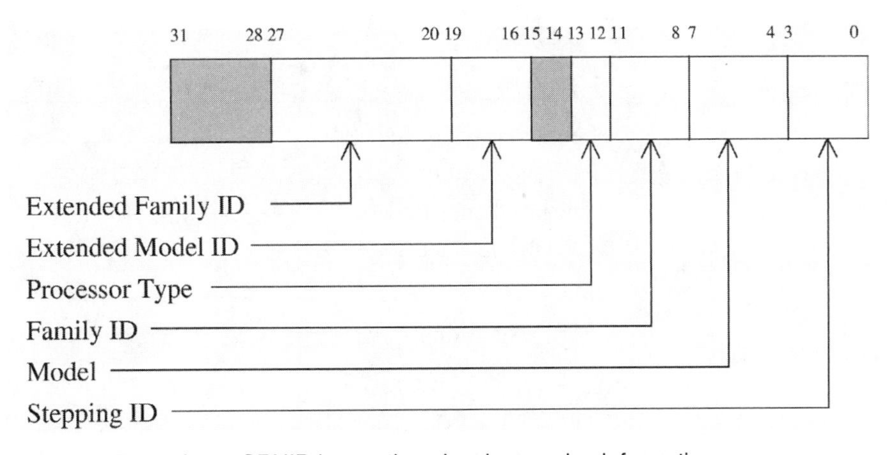

FIGURE 19–7 EAX after a CPUID instruction showing version information.

31–24. Bits 23–16 indicate how many internal processors are available for hyper-threading (two for the current Pentium 4 microprocessor). Example 19–4 shows a function that identifies the number of processors in a hyper-threaded CPU and returns it as a character string. If more than nine processors are eventually added to the microprocessor, then the software in Example 19–4 would need to be modified.

EXAMPLE 19–4

```
CString CCPUIDDlg::GetProcessorCount(void)
{
        CString temp = "This CPU has ";
        char temp1;
        _asm
        {
                mov  eax,1
                cpuid
                mov  temp1,31h
                bt   edx,28            ;check for hyper-threading
                jnc   GetPro1          ;if no hyper-threading, temp1 = 1
                bswap ebx
                add  bh,30h
                mov  temp1,bh
GetPro1:        }
        return temp + temp1 + " processors.";
}
```

Feature information for the microprocessor is returned in ECX and EDX as indicated in Figures 19–8 and 19–9. Each bit is a logic 1 if the feature is present. For example, if hyper-threading is needed in an application bit, position 28 is tested in EDX to see if hyper-threading is supported. This appears in Example 19–4 along with reading the number of processors found in a hyper-threaded microprocessor. The BT instruction tests the bit indicated and places it into the carry flag. If the bit under test is a 1, then the resultant carry is one and if the bit under test is a 0, the resultant carry is zero.

Model-Specific Registers

As with earlier versions of the Pentium, the Pentium 4 also contains model-specific registers that are read with the RDMSR instruction and written with the WRMSR instruction. The Pentium 4 has 1743 model-specific registers numbered from 0H to 6CFH. Intel does not provide information

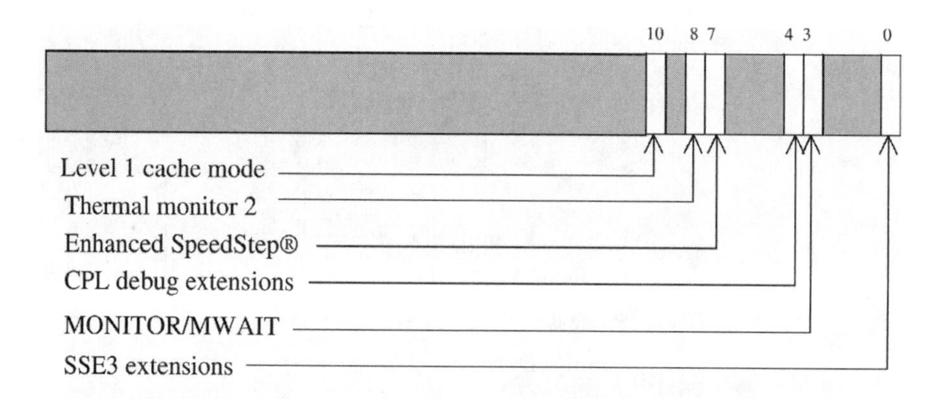

FIGURE 19–8 ECX after a CPUID instruction showing the version extensions.

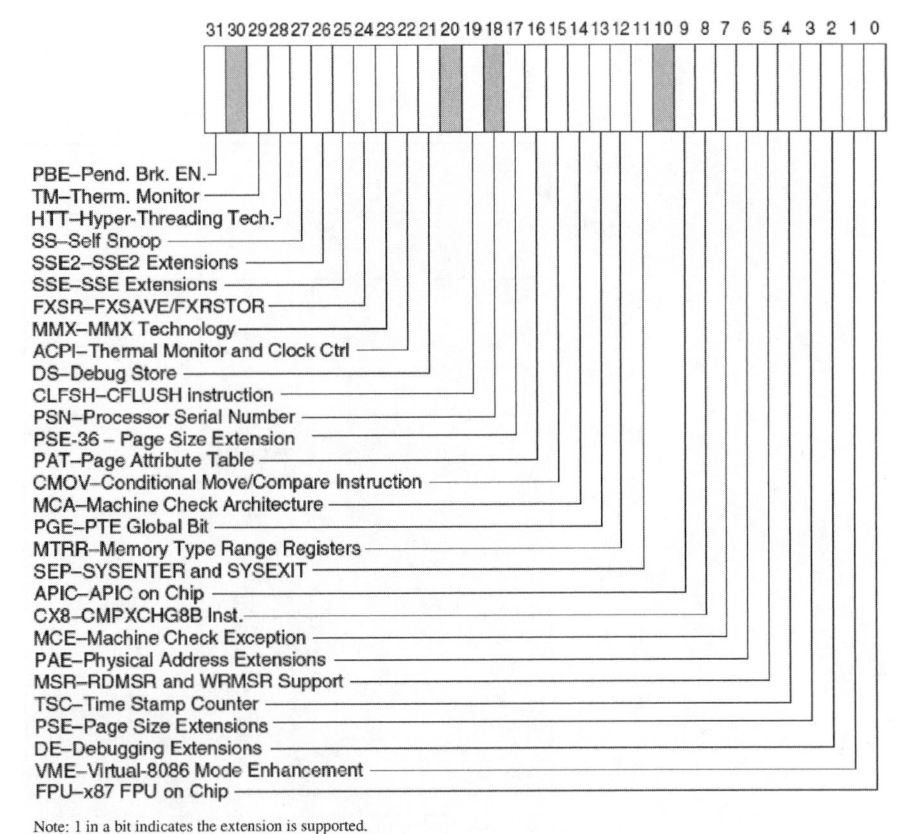

FIGURE 19–9 EDX after a CPUID instruction showing the version extensions.

on all of them. The registers not identified are either reserved by Intel or used for some undocumented feature or function.

Both the read and write model-specific register instructions function in the same manner. Register ECX is loaded with the register number to be accessed, and the data are transferred through the EDX:EAX register pair as a 64-bit number where EDX is the most-significant 32 bits and EAX is the least-significant bits. These registers must be accessed in either the real mode (DOS) or in ring 0 of the protected mode. These registers are normally accessed by the operating system and cannot be accessed in normal Visual C++ programming.

Performance-Monitoring Registers

Another feature in the Pentium 4 is a set of performance-monitoring registers (PMR) that, like the model-specific registers, can only be used in real mode or at ring 0 of protected mode. The only register that can be accessed via user software is the time-stamp counter, which is a performance-monitoring register. The remaining PMRs are accessed with the RDPMR. This instruction is similar to the RDMSR instruction in that it uses ECX to specify the register number and the result appears in EDX:EAX. There is no write instruction for the PMRs.

64-Bit Extension Technology

At the time of this writing, Intel has announced its 64-bit extension technology for the Intel 32-bit architecture family, but has yet to announce the release of a microprocessor that supports it. The instruction set and architecture is backwards compatible to the 8086, which means that

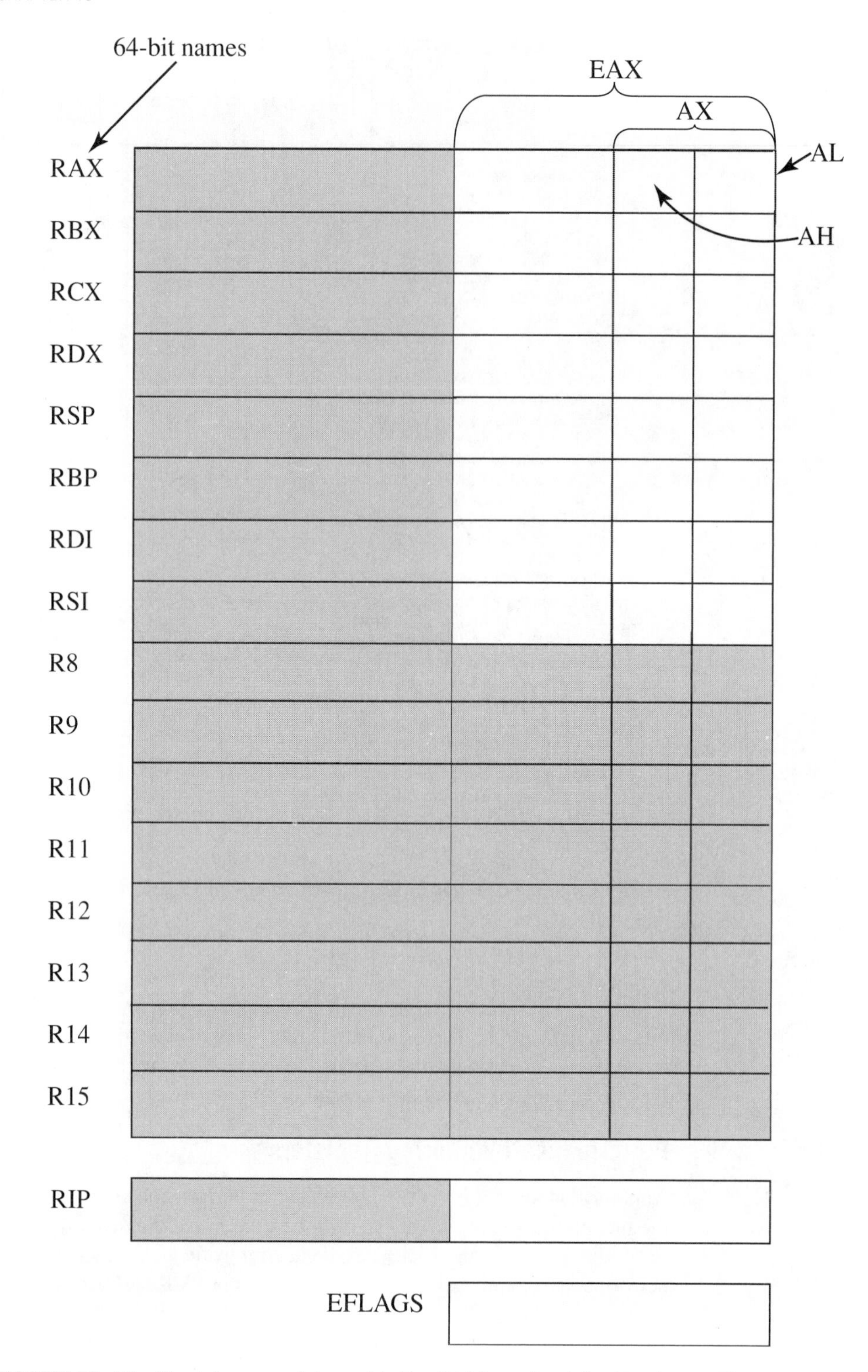

FIGURE 19–10 The integer register set in the 64-bit mode of the Pentium 4. The shaded areas are new to the Pentium 4 operated in the 64-bit mode.

the instructions and register set have remained compatible. What is changed is that the register set is stretched to 64 bits in width in place of the current 32-bit-wide registers. Refer to Figure 19–10 for the programming model of the Pentium 4 in 64-bit mode.

Notice that the register set now contains sixteen 64-bit-wide general-purpose registers, RAX, RBX, RCX, RDX, RSP, RBP, RDI, RSI, R_8–R_{15}. The instruction pointer is also stretched to a width of 64 bits, allowing the microprocessor to address memory using a 64-bit memory address. This allows the microprocessor to address as much memory as the specific implementation of the microprocessor has address pins.

The registers are addressed as 64-bit, 32-bit, 16-bit, or 8-bit registers. An example is R_8 (64 bits), R8D (32 bits), R8W (16 bits), and R8L (8 bits). There is no way to address the high byte (as in BH) for a numbered register; only the low byte of a numbered register can be addressed. Legacy addressing such as MOV AH,AL functions correctly, but addressing a legacy high-byte register and a numbered low-byte register is not allowed. In other words, MOV AH,R9L is not allowed, but MOV AL,R9L is allowed. If the MOV AH,R9L instruction is included in a program no error will occur; instead, the instruction will be changed to MOV BPL,R9L. AH, BH, CH, and DH are changed to the low-order 8 bits (the L is for low order) of BPL, SPL, DIL, and SIL respectively. Otherwise the legacy registers can be mixed with the new numbered registers R_8–R_{15} as in MOV R_{11}, RAX, MOV R11D, ECX, or MOV BX, R14W.

Another addition to the architecture is a set of additional SSE registers numbered XMM_8–XMM_{15}. These registers are accessed by the SSE, SSE_2, or SSE_3 instructions. Otherwise, the SSE unit has not been changed. The control and debug registers are expanded to 64 bits in width. And a new model-specific register is added to control the extended features at address C0000080H. Figure 19–11 depicts the extended feature control register.

SCE The **system CALL enable** bit is set to enable the SYSCALL and SYSRET instructions in the 64-bit mode.

LME The **mode enable** bit is set to allow the microprocessor to use the 64-bit extended mode.

LMA The **mode active** bit shows that the microprocessor is operating in the 64-bit extended mode.

The protected mode descriptor table registers are expanded in the extended 64-bit mode so that each descriptor table register, GDTR, LDTR, IDTR, and the task register (TR) hold a 64-bit base address instead of a 32-bit base address. The biggest change is that the base address and limits of the segment descriptors are ignored. The system uses a base address of 0000000000000000H for the code segment and the DS, ES, and SS segments are ignored.

Paging is also modified to include a paging unit that supports the translation of a 64-bit linear address into a 52-bit physical address. Intel states that in the first version of this 64-bit Pentium the linear address will be 48 bits and the physical address will be 40 bits. This means that there will be a 40-bit address to support 1T (tera) byte of physical memory translated from a linear address space of 256T bytes. The 52-bit address accesses 4P (peta) bytes of memory and a 64-bit linear address accesses 16E (exa) bytes of memory. The translation is accomplished with additional tables in the paging unit. In place of two tables (a page directory and a page table), the 64-bit extended paging unit uses four levels of page tables.

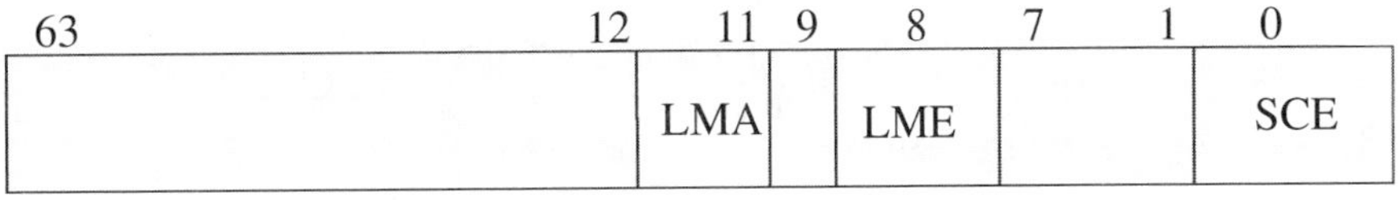

63–12	11	9	8	7–1	0
	LMA		LME		SCE

FIGURE 19–11 The contents of the extended feature model-specific register.

19–5 SUMMARY

1. The Pentium II differs from earlier microprocessors because instead of being offered as an integrated circuit, the Pentium II is available on a plug-in cartridge or printed circuit board.
2. The level 2 cache for the Pentium II is mounted inside the cartridge, except for the Celeron, which has no level 2 cache. The cache speed is one-half the Pentium II clock speed, except in the Xeon, where it is at the same speed as the Pentium II. All versions of the Pentium II contain an internal level 1 cache that stores 32K bytes of data.
3. The Pentium II is the first Intel microprocessor that is controlled from an external bus controller. Unlike earlier versions of the microprocessor, which issued read and write signals, the Pentium II is ordered to read or write information by an external bus controller.
4. The Pentium II operates at clock frequencies from 233 MHz to 450 MHz with bus speeds of 66 MHz or 100 MHz. The level 2 cache can be 512K, 1M, or 2M bytes in size. The Pentium II contains a 64-bit data bus and a 36-bit address bus that allow up to 64G bytes of memory to be accessed.
5. The new instructions added to the Pentium II are SYSENTER, SYSEXIT, FXSAVE, and FXRSTOR.
6. The SYSENTER and SYSEXIT commands are optimized to access the operating system in privilege level 0 from a privilege level 3 access. These instructions operate at a much higher speed than a task switch or even a call and return combination.
7. The FXSAVE and FXRSTOR instructions are optimized to properly store the state of both the MMX technology unit and the floating-point coprocessor.
8. The Pentium III microprocessor is an extension of the Pentium Pro architecture with the addition of the SIMD instruction set that uses the XMM registers.
9. The Pentium 4 microprocessor is an extension of the Pentium Pro architecture, which includes enhancements that allow it to operate at higher clock frequencies than previously possible because of the 0.13 micron and the latest 90 nm fabrication technologies.
10. The Pentium 4 microprocessor requires a modified ATX power supply and case to function properly in a system.
11. Version 6.15 of the MASM program and Visual Studio version 6 now support the new MMX and SIMD instructions using the .686 switch with the .MMX and .XMM switches.
12. The Pentium II, Pentium III, and Pentium 4 microprocessors are all variations of the Pentium Pro microprocessor.
13. Future Pentium 4 microprocessors will purportedly use the 64-bit extension to the 32-bit architecture. At the time of this writing no device or concrete information is available.

19–6 QUESTIONS AND PROBLEMS

1. What is the size of the level 1 cache in the Pentium II microprocessor?
2. What sizes are available for the level 2 cache in the Pentium II microprocessor? (List all versions.)
3. What is the difference between the level 2 cache on the Pentium-based system and the Pentium II–based system?
4. What is the difference between the level 2 cache in the Pentium Pro and the Pentium II?
5. The speed of the Pentium II Xeon level 2 cache is _________ times faster than the cache in the Pentium II (excluding the Celeron).
6. How much memory can be addressed by the Pentium II?
7. Is the Pentium II available in integrated circuit form?

8. How many pin connections are found on the Pentium II cartridge?
9. What is the purpose of the PICD control signals?
10. What happened to the read and write pins on the Pentium II?
11. At what bus speeds does the Pentium II operate?
12. How fast is the SDRAM connected to the Pentium II system for a 100 MHz bus speed version?
13. How wide is the Pentium II memory if ECC is employed?
14. What new model-specific registers (MSR) have been added to the Pentium II microprocessor?
15. What new CPUID identification information has been added to the Pentium II microprocessor?
16. How is a model-specific register addressed and what instruction is used to read it?
17. Write software that stores 12H into model-specific register 175H.
18. Write a short procedure that determines whether the microprocessor contains the SYSENTER and SYSEXIT instructions. Your procedure must return carry set if the instructions are present, and return carry cleared if not present.
19. How is the return address transferred to the system when using the SYSENTER instruction?
20. How is the return address retrieved when using the SYSEXIT instruction to return to the application?
21. The SYSENTER instruction transfers control to software at what privilege level?
22. The SYSEXIT instruction transfers control to software at what privilege level?
23. What is the difference between the FSAVE and the FXSAVE instructions?
24. The Pentium III is an extension of the ________ architecture.
25. What new instructions appear in the Pentium III microprocessor that do not appear in the Pentium Pro microprocessor?
26. What changes to the power supply does the Pentium 4 microprocessor require?
27. Write a short program that reads and displays the serial number of the Pentium III microprocessor on the video screen.
28. Develop a short C++ function that returns a bool value of true if the Pentium 4 supports hyper-threaded technology and false if it does not support it.
29. Develop a short C++ function that returns a bool value of true if the Pentium 4 supports SSE, SSE_2, and SSE_3 extensions.
30. Compare, in your own words, hyper-threading to dual processing. Postulate on the possibility of including additional processors beyond two.

APPENDIX A

The Assembler, Visual C++, and DOS

This appendix is provided so that the assembler can be used for program development in the DOS environment and also the Visual C++ environment. The DOS environment is essentially gone (unless Windows 98 is still in use), but it lives on through the emulation program called CMD.EXE in the accessory folder of Microsoft Windows. Some may shed a tear at the departure of DOS, but realize that the DOS environment was a vast headache to the many of us that spent years programming in it. It had only a 1M memory system and drivers were a problem, especially in recent years. Microsoft never really provided a decent protected mode DOS. DOS displayed text information well, but graphics were another story because of the DOS architecture of the video memory and the lack of drivers.

Windows solved many of the problems that plagued DOS and ushered in the GUI age, which is a great improvement over DOS text-based applications. Windows is just so much easier for a human to use and control. The author remembers the old days when he had to write batch files so his wife could use his computer. Now she is a real pro because of Windows. Windows is a tremendous system—bar none.

THE ASSEMBLER

Although the assembler program is not often used as a stand-alone programming medium, it still finds some application in developing modules that are linked to Visual C++ programs (see Chapter 7). The program itself is provided with Visual C++ in the `C:\Program Files\Microsoft Visual Studio .NET 2003\Vc7\bin` directory as ML.EXE. Also found in the same directory is the LIB.EXE (library) program for creating library collections and the LINK program used for linking object modules.

Example A–1 illustrates how to assemble a program written in assembly language. The example uses a file called WOW.TXT (it does not need the .ASM extension even though it is often used for assembly language modules). The file WOW.TXT is compiled for errors (/c = lowercase c) and generates a listing file (/Fl) called WOW.LST. If other switches are needed just type `ML/?` at the command prompt to display a list of the switches. The /coff (c object file format) switch might also be included to generate an object file that can be linked to a Visual C++ program as in `ML/c/coff WOW.TXT`.

EXAMPLE A–1

```
ML /c /FlWOW.LST WOW.TXT
```

TABLE A–1 Commonly used assembler models.

Model	*Description*
.TINY	All data and code must fit into a single 64K-byte memory segment. Tiny programs assemble as DOS.COM files and must use an origin at 0100H for the code.
.SMALL	A two-segment model with a single code segment and a single data segment. Small programs generate DOS.EXE files and have an origin of 0000H.
.FLAT	The flat model uses a single segment of up to 4G bytes in length. Flat programs are programs that will only function in Windows with an origin of 00000000H.

If the LINK program is used from Visual C++, you cannot generate a DOS-compatible execution file because it is a 32-bit linker. The 16-bit linker for DOS is not in the Visual Studio package. If DOS software must be developed, obtain the Windows Driver Development Kit (Windows DDK) from Microsoft. The DDK contains the 16-bit linker needed to develop DOS applications. The linker is located in the `C:\WINDDK\2600.1106\bin\win_me\bin16` folder of the DDK. In addition to the linker program a 16-bit version of the C++ language for DOS appears as CL.EXE. These are provided for legacy applications.

Example A–2 shows how to link a program generated by the assembler. This assumes that you are using the 16-bit DOS real mode linker program. The 32-bit linker is normally used from Visual C++ for Windows applications. Here the object file generated by Example A–1 is linked to produce an executable program called WOW.EXE or, if the tiny model is in effect, WOW.COM.

EXAMPLE A–2

```
LINK WOW.OBJ
```

Assembler Memory Models

Although the flat model is most often used with Visual C++, there are other memory models that are used with DOS applications and embedded program development. Table A–1 lists the most commonly used models for these applications. The origin is set by the .STARTUP directive in a DOS program and automatically in a flat program.

Table A–2 lists the default information for each of the models listed in Table A–1. If additional information on models is required, please visit the Microsoft Web site and search for assembler models.

SELECTED DOS FUNCTION CALLS

Not all DOS function calls are included because it is doubtful that they will all be used. The most recent version of DOS has function calls from function 00H to function 6CH. This text only lists the function calls that are used for simple applications. Many of the function calls were from DOS version 1.0 and have been obsolete for many years and others are used to access the disk system, which is accessed in Visual C++.

TABLE A–2 Defaults for the more common assembly language models.

Model	*Directives*	*Name*	*Align*	*Combine*	*Class*	*Group*
.TINY	.CODE	_TEXT	Word	PUBLIC	'CODE'	DGROUP
	.FARDATA	FAR_DATA	Para	Private	'FAR_DATA'	
	.FARDATA?	FAR_BSS	Para	Private	'FAR_BSS'	
	.DATA	_DATA	Word	PUBLIC	'DATA'	DGROUP
	.CONST	CONST	Word	PUBLIC	'CONST'	DGROUP
	.DATA?	_BSS	Word	PUBLIC	'BSS'	DGROUP
.SMALL	.CODE	_TEXT	Word	PUBLIC	'CODE'	
	.FARDATA	FAR_DATA	Para	Private	'FAR_DATA'	
	.FARDATA?	FAR_BSS	Para	Private	'FAR_BSS'	
	.DATA	_DATA	Word	PUBLIC	'DATA'	DGROUP
	.CONST	CONST	Word	PUBLIC	'CONST'	DGROUP
	.DATA?	_BSS	Word	PUBLIC	'BSS'	DGROUP
	.STACK	STACK	Para	STACK	'STACK'	DGROUP
.FLAT	.CODE	_TEXT	Dword	PUBLIC	'CODE'	
	.FARDATA	_DATA	Dword	PUBLIC	'DATA'	
	.FARDATA?	_BSS	Dword	PUBLIC	'FBSS'	
	.DATA	_DATA	Dword	PUBLIC	'DATA'	DGROUP
	.CONST	CONST	Dword	PUBLIC	'CONST'	DGROUP
	.DATA?	_BSS	Dword	PUBLIC	'BSS'	DGROUP
	.STACK	STACK	Dword	STACK	'STACK'	DGROUP

To use a DOS function call in a DOS program, place the function number in AH and other data that might be necessary in other registers, as indicated in Table A–3. Example A–3 shows an example of DOS function number 01H. This function reads the DOS keyboard and returns an ASCII character in AL. Once everything is loaded, execute the INT 21H instruction to perform the task.

EXAMPLE A–3

```
MOV  AH,01H          ;load DOS function number
INT  21H             ;access DOS

;returns with AL = ASCII key code
```

USING VISUAL C++

Many of the new examples in the text use Visual C++ .net 2003. Very few, if any, programs are written in assembly language. If assembly language is used, it normally appears in a C++ program to accomplish a special task or to increase the performance of a section of a program.

Not everyone is familiar with the C++ environment so this section has been added to act as a guide in setting up programs that use assembly language within Visual C++. The easiest application type for this is a Dialog-based application using the Microsoft Foundation Classes (MFC).

TABLE A–3 DOS function calls.

00H	TERMINATE A PROGRAM
Entry	AH = 00H CS = program segment prefix address
Exit	DOS is entered
01H	READ THE KEYBOARD
Entry	AH = 01H
Exit	AL = ASCII character
Notes	If AL = 00H, the function call must be invoked again to read an extended ASCII character. Refer to Chapter 8, Table 8–1 for a listing of the extended ASCII keyboard codes. This function call automatically echoes whatever is typed to the video screen.
02H	WRITE TO STANDARD OUTPUT DEVICE
Entry	AH = 02H DL = ASCII character to be displayed
Notes	This function call normally displays data on the video display.
03H	READ CHARACTER FROM COM1
Entry	AH = 03H
Exit	AL = ASCII character read from the communications port
Notes	This function call reads data from the serial communications port.
04H	WRITE TO COM1
Entry	AH = 04H DL = character to be sent out of COM1
Notes	This function transmits data through the serial communications port. The COM port assignment can be changed to use other COM ports with functions 03H and 04H by using the DOS MODE command to reassign COM1 to another COM port.

05H	WRITE TO LPT1
Entry	AH = 05H DL = ASCII character to be printed
Notes	Prints DL on the line printer attached to LPT1. Note that the line printer port can be changed with the DOS MODE command.

06H	DIRECT CONSOLE READ/WRITE
Entry	AH = 06H DL = 0FFH or DL = ASCII character
Exit	AL = ASCII character
Notes	If DL = 0FFH on entry, then this function reads the console. If DL = ASCII character, then this function displays the ASCII character on the console (CON) video screen. If a character is read from the console keyboard, the zero flag (ZF) indicates whether a character was typed. A zero condition indicates that no key was typed, and a not-zero condition indicates that AL contains the ASCII code of the key or a 00H. If AL = 00H, the function must again be invoked to read an extended ASCII character from the keyboard. Note that the key does not echo to the video screen.

07H	DIRECT CONSOLE INPUT WITHOUT ECHO
Entry	AH = 07H
Exit	AL = ASCII character
Notes	This functions exactly like function number 06H with DL = 0FFH, but it will not return from the function until the key is typed.

08H	READ STANDARD INPUT WITHOUT ECHO
Entry	AH = 08H
Exit	AL = ASCII character
Notes	Performs like function 07H, except that it reads the standard input device. The standard input device can be assigned as either the keyboard or the COM port. This function also responds to a control-break, where function 06H and 07H do not. A control-break causes INT 23H to execute. By default, this functions as does function 07H.

09H	DISPLAY A CHARACTER STRING
Entry	AH = 09H DS:DX = address of the character string
Notes	The character string must end with an ASCII $ (24H). The character string can be of any length and may contain control characters such as carriage return (0DH) and line feed (0AH).

0AH	BUFFERED KEYBOARD INPUT
Entry	AH = 0AH DS:DX = address of keyboard input buffer
Notes	The first byte of the buffer contains the size of the buffer (up to 255). The second byte is filled with the number of characters typed upon return. The third byte through the end of the buffer contains the character string typed, followed by a carriage return (0DH). This function continues to read the keyboard (displaying data as typed) until either the specified number of characters are typed or until the enter key is typed.

0BH	TEST STATUS OF THE STANDARD INPUT DEVICE
Entry	AH = 0BH
Exit	AL = status of the input device
Notes	This function tests the standard input device to determine if data are available. If AL = 00, no data are available. If AL = 0FFH, then data are available that must be input using function number 08H.

0CH	CLEAR KEYBOARD BUFFER AND INVOKE KEYBOARD FUNCTION
Entry	AH = 0CH AL = 01H, 06H, 07H, or 0AH
Exit	See exit for functions 01H, 06H, 07H, or 0AH
Notes	The keyboard buffer holds keystrokes while programs execute other tasks. This function empties or clears the buffer and then invokes the keyboard function located in register AL.

Create a Dialog Application

Start Visual Studio and the screen in Figure A–1 should appear. Click on New Project to start a new C++ project. The screen in Figure A–2 should appear. Click on the + next to Visual C++ Projects, then click on MFC, and finally click on MFC Application. At this point enter the project name and path where indicated. In this example we chose the project name MyNewProject. Once all of this has been accomplished, click on OK.

At this point you should see the screen illustrated in Figure A–3. This allows you to select options for the program through the Application Wizard. The only feature that needs to be changed at this point is the Application Type. Click on Application Type and you should see the screen in Figure A–4. The only thing that needs to be changed on this form is to select the dialog-based application. Once this is checked, click on Finish to generate the dialog application.

To summarize the process for creating a dialog-based application:

1. Start Visual Studio .net 2003.
2. Click on New Project.

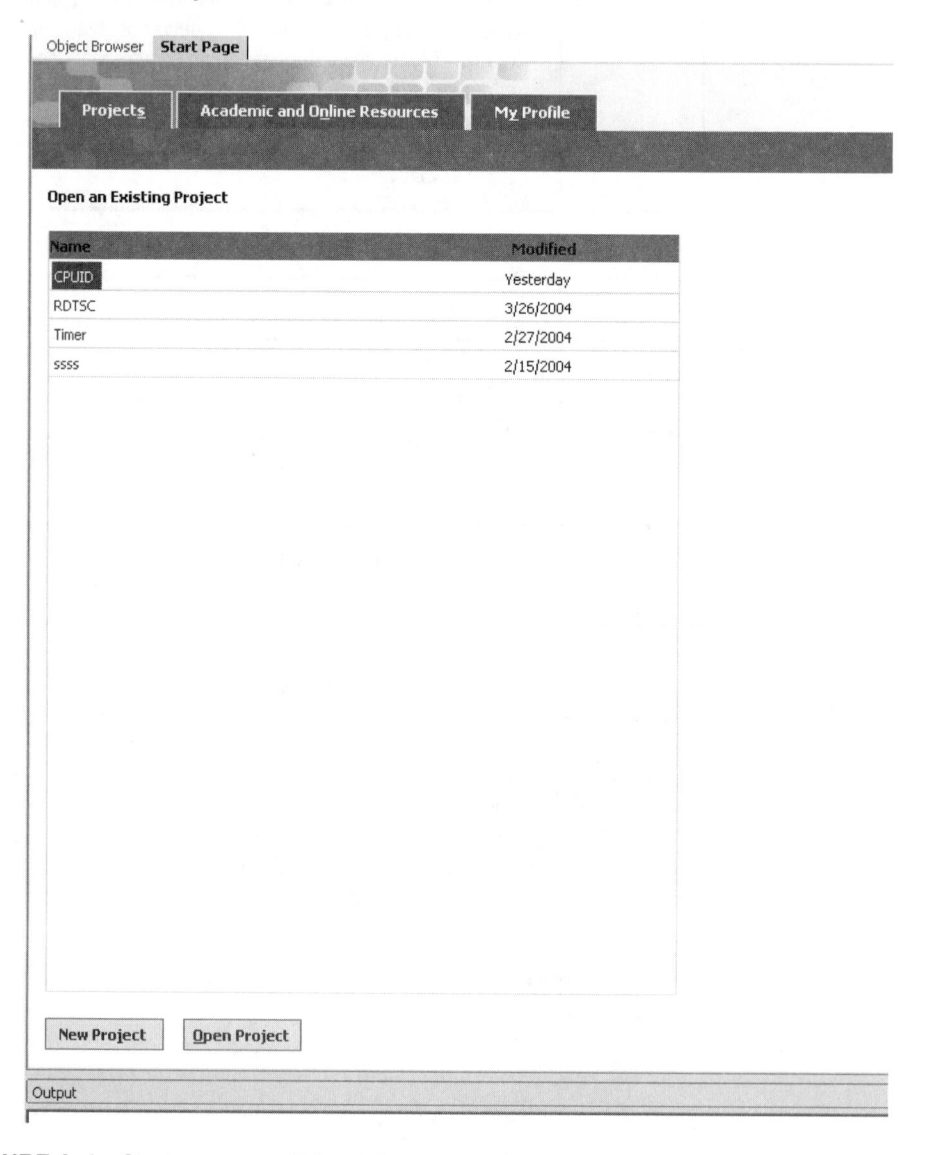

FIGURE A–1 Startup screen of Visual Studio .net 2003.

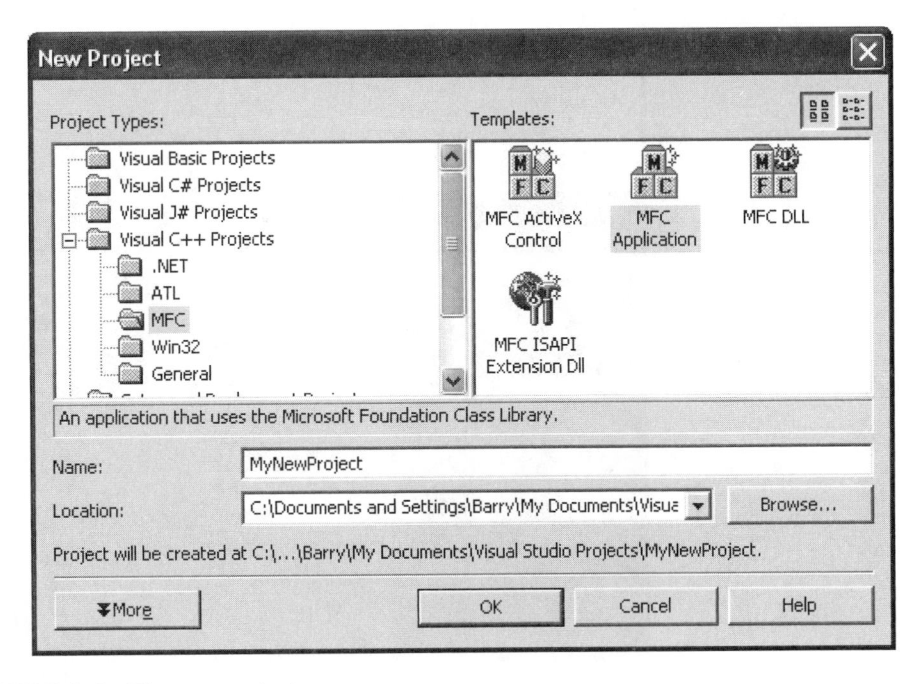

FIGURE A–2 The new project screen.

3. Click on MFC Application and enter a project name, then click OK.
4. Select Application Type and choose Dialog-Based.
5. Check Dialog-Based application and then click on Finish.

After accomplishing these five steps the screen illustrated in Figure A–5 will appear. This is the dialog-based application in the resource editor. At this point you are ready to start placing

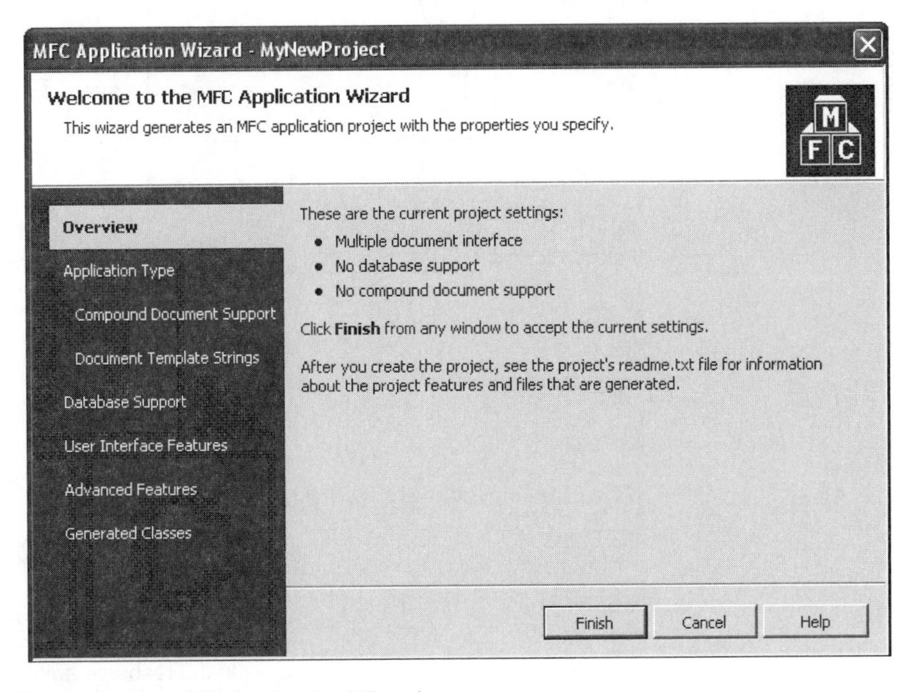

FIGURE A–3 The MFC Application Wizard.

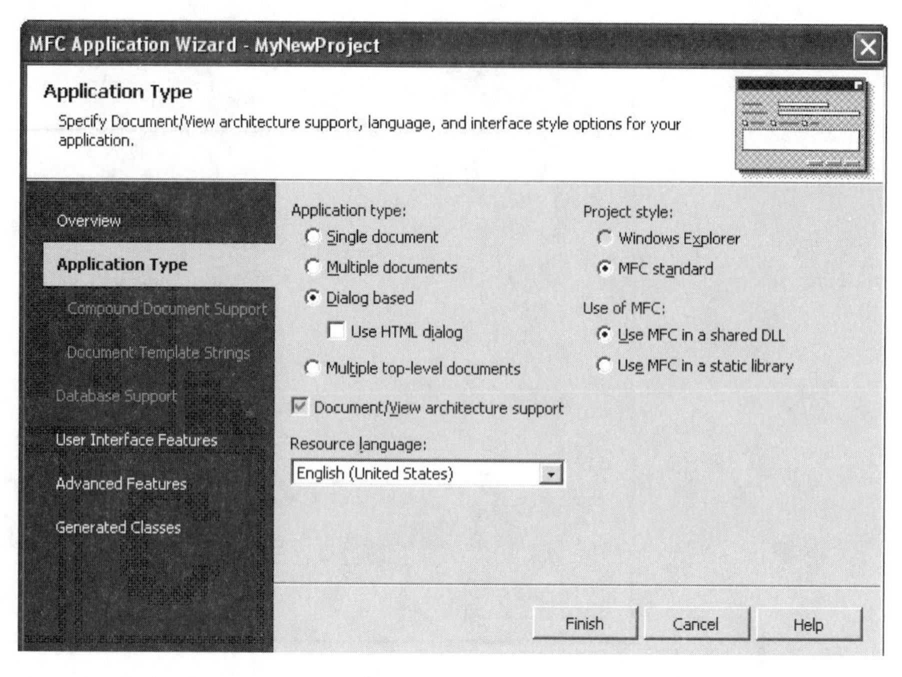

FIGURE A–4 Selecting the application type.

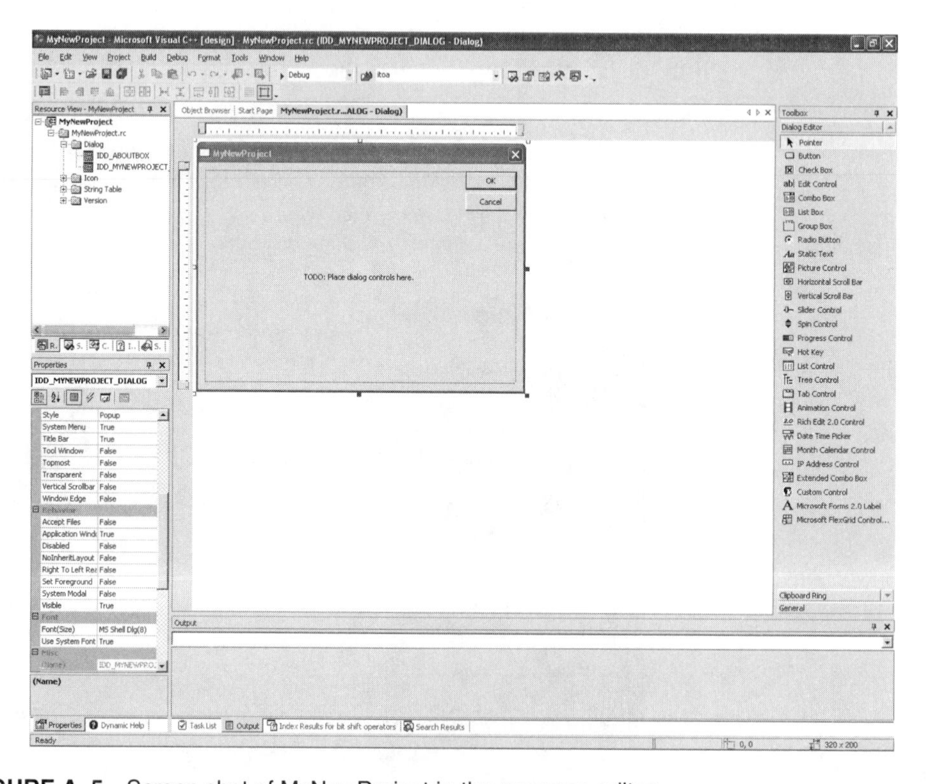

FIGURE A–5 Screen shot of MyNewProject in the resource editor.

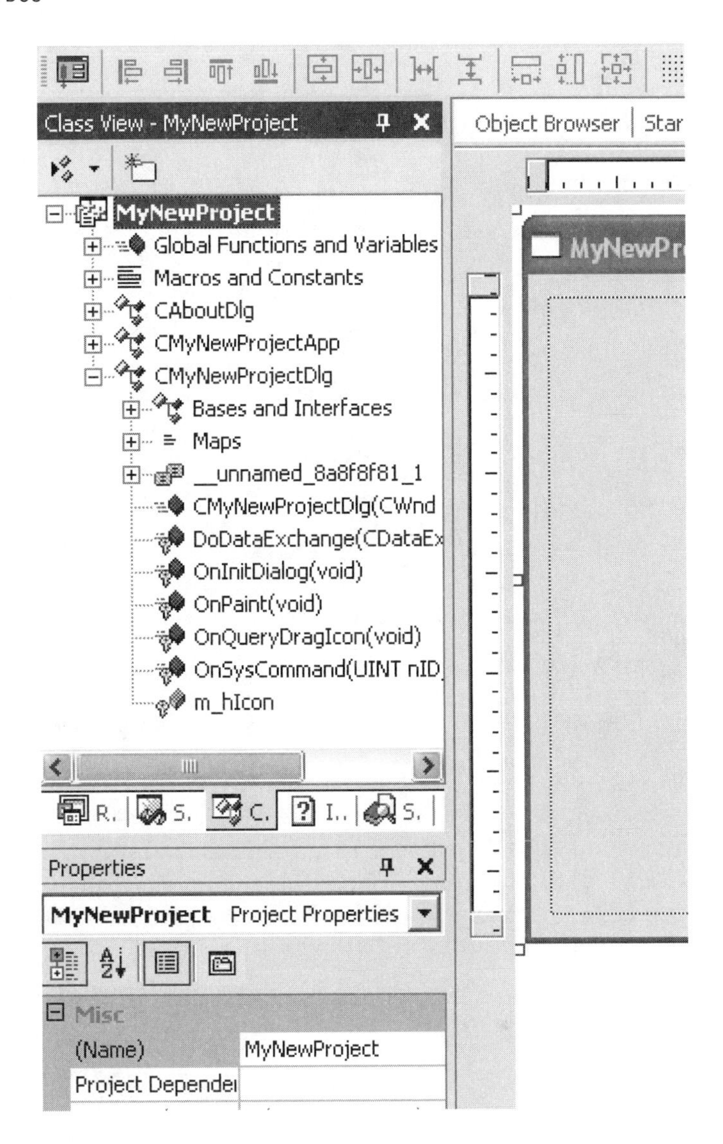

FIGURE A–6 The class view of MyNewProject.

objects onto the dialog form. The screen you see and the one displayed in the figure may vary somewhat, determined by the profile at the main Visual Studio page. You can change it by clicking in the My Profile tab in Figure A–1.

The window in the upper left corner of Figure A–5 has tab below it. These tabs are used to select different view of the program. Notice that the R tab is highlighted because the view is the Resource View. If you click the C tab the Class View will appear as in Figure A–6. In Figure A–6 the pluses next to MyNewProject and CMyNewProjectDlg were clicked on to expand the dlg (dialog) class. This is where the software is located for the dialog box as installed by Visual Studio for this application. Many of the programs in the text use the OnInitDialog function for setting up the dialog box.

Figure A–7 illustrates the software in the OnInitDialog function. To view this, double-click on OnIntiDialog in the class view. The MFC framework calls the OnInitDialog function before displaying the dialog box on the screen. Any software added for initialization is placed near the bottom of Figure A–7 where a TODO: comment appears. Once you arrive at this point you are ready to enter and execute any of the Visual C++ with assembly programs in the textbook.

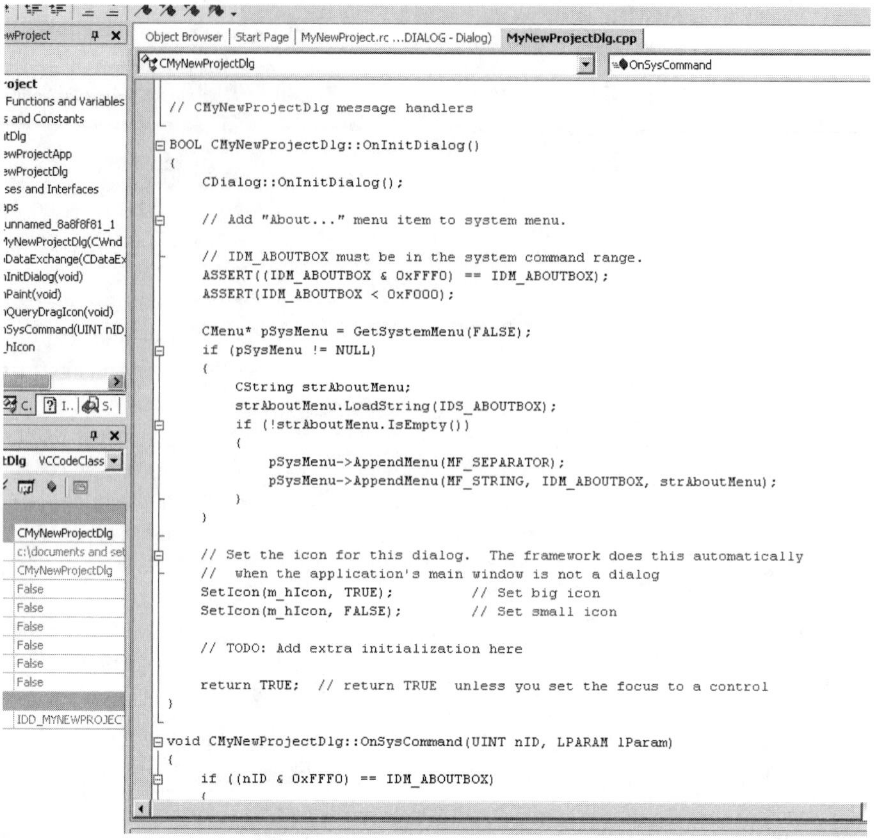

```
// CMyNewProjectDlg message handlers

BOOL CMyNewProjectDlg::OnInitDialog()
{
    CDialog::OnInitDialog();

    // Add "About..." menu item to system menu.

    // IDM_ABOUTBOX must be in the system command range.
    ASSERT((IDM_ABOUTBOX & 0xFFF0) == IDM_ABOUTBOX);
    ASSERT(IDM_ABOUTBOX < 0xF000);

    CMenu* pSysMenu = GetSystemMenu(FALSE);
    if (pSysMenu != NULL)
    {
        CString strAboutMenu;
        strAboutMenu.LoadString(IDS_ABOUTBOX);
        if (!strAboutMenu.IsEmpty())
        {
            pSysMenu->AppendMenu(MF_SEPARATOR);
            pSysMenu->AppendMenu(MF_STRING, IDM_ABOUTBOX, strAboutMenu);
        }
    }

    // Set the icon for this dialog.  The framework does this automatically
    //  when the application's main window is not a dialog
    SetIcon(m_hIcon, TRUE);         // Set big icon
    SetIcon(m_hIcon, FALSE);        // Set small icon

    // TODO: Add extra initialization here

    return TRUE;  // return TRUE  unless you set the focus to a control
}

void CMyNewProjectDlg::OnSysCommand(UINT nID, LPARAM lParam)
{
    if ((nID & 0xFFF0) == IDM_ABOUTBOX)
    {
```

FIGURE A–7 Contents of the OnInitDialog function.

APPENDIX B

Instruction Set Summary

The instruction set summary contains a complete alphabetical listing of the entire 8086–Pentium 4 instruction set. The coprocessor and MMX instructions are listed in Chapter 14 and are not repeated in this appendix. The SIMD instructions appear at the end of this appendix after the main instruction set summary.

Each instruction entry lists the mnemonic opcode plus a brief description of the purpose of the instruction. Also listed is the binary machine language coding of each instruction and any other data required to form the instruction, such as the displacement or immediate data. Listed to the right of each binary machine language version of the instruction are the flag bits and any change that might occur for the instruction. The flags are described in the following manner: A blank indicates no effect or change; a ? indicates a change with an unpredictable outcome; a * indicates a change with a predictable outcome; a 1 indicates the flag is set; and a 0 indicates that the flag is cleared. If the flag bits ODITSZAPC are not illustrated with an instruction, the instruction does not modify any of these flags.

Some information about the bit settings in binary machine language versions of the instructions is presented before the instruction listing. Table B–1 lists the modifier bits, coded as 00 in the instruction listing.

Table B–2 lists the memory-addressing modes available using a register field coding of mmm. This table applies to all versions of the microprocessor, as long as the operating mode is 16 bits.

Table B–3 lists the register selections provided by the rrr field in an instruction. This table includes the register selections for 8-, 16-, and 32-bit registers.

Table B–4 lists the segment register bit assignment (rrr) found with the MOV, PUSH, and POP instructions.

TABLE B–1 The modifier bits, coded as oo in the instruction listing.

oo	*Function*
00	If mmm = 110 a displacement follows the opcode; otherwise no displacement is used
01	An 8-bit signed displacement follows the opcode
10	A 16- or 32-bit signed displacement follows the opcode
11	mmm specifies a register instead of an addressing mode

TABLE B–2 The 16-bit register/memory (mmm) field description.

mmm	*16-bit*
000	DS:[BX+SI]
001	DS:[BX+DI]
010	SS:[BP+SI]
011	SS:[BP+DI]
100	DS:[SI]
101	DS:[DI]
110	SS:[BP]
111	DS:[BX]

TABLE B–3 The register (rrr) field.

rrr	*W = 0*	*W = 1 (16-bit)*	*W = 1 (32-bit)*
000	AL	AX	EAX
001	CL	CX	ECX
010	DL	DX	EDX
011	BL	BX	EBX
100	AH	SP	ESP
101	CH	BP	EBP
110	DH	SI	ESI
111	BH	DI	EDI

TABLE B–4 Register field assignments (rrr) for the segment registers.

rrr	*Segment Register*
000	ES
001	CS
010	SS
011	DS
100	FS
101	GS

When the 80386–Pentium 4 are used, some of the definitions provided in Table B–1 through B–3 change. See Tables B–5 and B–6 for these changes as they apply to the 80386–Pentium 4 microprocessors.

TABLE B–5 Index registers specified by rrr when the 80386–Pentium 4 are operated in 32-bit mode.

rrr	*Index Register*
000	DS:[EAX]
001	DS:[ECX]
010	DS:[EDX]
011	DS:[EBX]
100	(see Table B–6)
101	SS:[EBP]
110	DS:[ESI]
111	DS:[EDI]

TABLE B–6 Possible combinations of oo, mmm, and rrr for the 80386–Pentium 4 microprocessors using 32-bit addressing.

oo	*mmm*	*rrr (base in scaled-index byte)*	*Addressing Mode*
00	000	—	DS:[EAX]
00	001	—	DS:[ECX]
00	010	—	DS:[EDX]
00	011	—	DS:[EBX]
00	100	000	DS:[EAX+scaled index]
00	100	001	DS:[ECX+scaled index]
00	100	010	DS:[EDX+scaled index]
00	100	011	DS:[EBX+scaled index]
00	100	100	SS:[ESP+scaled index]
00	100	101	DS:[disp32+scaled index]
00	100	110	DS:[ESI+scaled index]
00	100	111	DS:[EDI+scaled index]
00	101	—	DS:disp32
00	110	—	DS:[ESI]
00	111	—	DS:[EDI]
01	000	—	DS:[EAX+disp8]
01	001	—	DS:[ECX+disp8]
01	010	—	DS:[EDX+disp8]
01	011	—	DS:[EBX+disp8]
01	100	000	DS:[EAX+scaled index+disp8]
01	100	001	DS:[ECX+scaled index+disp8]
01	100	010	DS:[EDX+scaled index+disp8]
01	100	011	DS:[EBX+scaled index+disp8]
01	100	100	SS:[ESP+scaled index+disp8]
01	100	101	SS:[EBP+scaled index+disp8]
01	100	110	DS:[ESI+scaled index+disp8]
01	100	111	DS:[EDI+scaled index+disp8]
01	101	—	SS:[EBP+disp8]
01	110	—	DS:[ESI+disp8]
01	111	—	DS:[EDI+disp8]
10	000	—	DS:[EAX+disp32]
10	001	—	DS:[ECX+disp32]
10	010	—	DS:[EDX+disp32]
10	011	—	DS:[EBX+disp32]
10	100	000	DS:[EAX+scaled index+disp32]
10	100	001	DS:[ECX+scaled index+disp32]
10	100	010	DS:[EDX+scaled index+disp32]
10	100	011	DS:[EBX+scaled index+disp32]
10	100	100	SS:[ESP+scaled index+disp32]
10	100	101	SS:[EBP+scaled index+disp32]
10	100	110	DS:[ESI+scaled index+disp32]
10	100	111	DS:[EDI+scaled index+disp32]
10	101	—	SS:[EBP+disp32]
10	110	—	DS:[ESI+disp32]
10	111	—	DS:[EDI+disp32]

Note: disp8 = 8-bit displacement and disp32 = 32-bit displacement.

In order to use the scaled-index addressing modes listed in Table B–6, code 00 and mmm in the second byte of the opcode. The scaled-index byte is usually the third byte and contains three fields. The leftmost two bits determine the scaling factor (00 = ×1, 01 = ×2, 10 = ×4, or 11 = ×8). The next three bits toward the right contain the scaled-index register number (this is obtained from Table B–5). The rightmost three bits are from the mmm field listed in Table B–6. For example, the MOV AL,[EBX+2*ECX] instruction has a scaled-index byte of 01 001 011, where 01 = X_2, 001 = ECX, and 011 = EBX.

Some instructions are prefixed to change the default segment or to override the instruction mode. Table B–7 lists the segment and instruction mode override prefixes with append at the beginning of an instruction if they are used to form the instruction. For example, the MOV AL,ES:[BX] instruction used the extra segment because of the override prefix ES:.

In the 8086 and 8088 microprocessors, the effective address calculation required additional clocks that are added to the times in the instruction set summary. These additional times are listed in Table B–8. No such times are added to the 80286–Pentium 4. Note that the instruction set summary does not include clock times for the Pentium Pro through the Pentium 4. Intel has not released these times and has decided that the RDTSC instruction can be used to have the microprocessor count the number of clocks required for a given application. Even though the timings do not appear for these new microprocessors, they are very similar to the Pentium, which can be used as a guide.

TABLE B–7 Override prefixes.

Prefix Byte	*Purpose*
26H	ES: segment override
2EH	CS: segment override
36H	SS: segment override
3EH	DS: segment override
64H	FS: segment override
65H	GS: segment override
66H	Memory operand instruction mode override
67H	Register operand instruction mode override

TABLE B–8 Effective address calculations for the 8086 and 8088 microprocessors.

Type	*Clocks*	*Example Instruction*
Base or index	5	MOV CL,[DI]
Displacement	3	MOV AL,DATA1
Base plus index	7	MOV AL,[BP+SI]
Displacement plus base or index	9	MOV DH,[DI+20H]
Base plus index plus displacement	11	MOV CL,[BX+DI+2]
Segment override	ea + 2	MOV AL,ED:[DI]

INSTRUCTION SET SUMMARY

AAA ASCII adjust AL after addition

00110111

O	D	I	T	S	Z	A	P	C
?				?	?	*	?	*

Example: AAA

Microprocessor	Clocks
8086	8
8088	8
80286	3
80386	4
80486	3
Pentium–Pentium 4	3

AAD ASCII adjust AX before division

11010101 00001010

O	D	I	T	S	Z	A	P	C
?				*	*	?	*	?

Example: AAD

Microprocessor	Clocks
8086	60
8088	60
80286	14
80386	19
80486	14
Pentium–Pentium 4	10

AAM ASCII adjust AX after multiplication

11010100 00001010

O	D	I	T	S	Z	A	P	C
?				*	*	?	*	?

Example: AAM

Microprocessor	Clocks
8086	83
8088	83
80286	16
80386	17
80486	15
Pentium–Pentium 4	18

AAS — ASCII adjust AL after subtraction

00111111

O	D	I	T	S	Z	A	P	C
?				?	?	*	?	*

Example	Microprocessor	Clocks
AAS	8086	8
	8088	8
	80286	3
	80386	4
	80486	3
	Pentium–Pentium 4	3

ADC — Addition with carry

000100dw oorrrmmm disp

O	D	I	T	S	Z	A	P	C
*				*	*	*	*	*

Format	Examples	Microprocessor	Clocks
ADC reg,reg	ADC AX,BX ADC AL,BL ADC EAX,EBX ADC CX,SI ADC ESI,EDI	8086	3
		8088	3
		80286	3
		80386	3
		80486	1
		Pentium–Pentium 4	1 or 3
ADC mem,reg	ADC DATAY,AL ADC LIST,SI ADC DATA2[DI],CL ADC [EAX],BL ADC [EBX+2*ECX],EDX	8086	16 + ea
		8088	24 + ea
		80286	7
		80386	7
		80486	3
		Pentium–Pentium 4	1 or 3

ADC reg,mem	ADC BL,DATA1 ADC SI,LIST1 ADC CL,DATA2[SI] ADC CX,[ESI] ADC ESI,[2*ECX]	8086	9 + ea
		8088	13 + ea
		80286	7
		80386	6
		80486	2
		Pentium–Pentium 4	1 or 2

100000sw oo010mmm disp data

Format	Examples	Microprocessor	Clocks
ADC reg,imm	ADC CX,3 ADC DI,1AH ADC DL,34H ADC EAX,12345 ADC CX,1234H	8086	4
		8088	4
		80286	3
		80386	2
		80486	1
		Pentium–Pentium 4	1 or 3
ADC mem,imm	ADC DATA4,33 ADC LIST,'A' ADC DATA3[DI],2 ADC BYTE PTR[EBX],3 ADC WORD PTR[DI],669H	8086	17 + ea
		8088	23 + ea
		80286	7
		80386	7
		80486	3
		Pentium–Pentium 4	1 or 3
ADC acc,imm	ADC AX,3 ADC AL,1AH ADC AH,34H ADC EAX,2 ADC AL,'Z'	8086	4
		8088	4
		80286	3
		80386	2
		80486	1
		Pentium–Pentium 4	1

ADD Addition

000000dw oorrrmmm disp

O	D	I	T	S	Z	A	P	C
*				*	*	*	*	*

Format	Examples	Microprocessor	Clocks

ADD reg,reg	ADD AX,BX
	ADD AL,BL
	ADD EAX,EBX
	ADD CX,SI
	ADD ESI,EDI
		8086	3

		8088	3
		80286	2
		80386	2
		80486	1
		Pentium–Pentium 4	1 or 3

ADD mem,reg	ADD DATAY,AL
	ADD LIST,SI
	ADD DATA6[DI],CL
	ADD [EAX],CL
	ADD [EDX+4*ECX],EBX
		8086	16 + ea

		8088	24 + ea
		80286	7
		80386	7
		80486	3
		Pentium–Pentium 4	1 or 3

ADD reg,mem	ADD BL,DATA2
	ADD SI,LIST3
	ADD CL,DATA2[DI]
	ADD CX,[EDI]
	ADD ESI,[ECX+200H]
		8086	9 + ea

		8088	13 + ea
		80286	7
		80386	6
		80486	2
		Pentium–Pentium 4	1 or 2

100000sw oo000mmm disp data

Format	Examples	Microprocessor	Clocks

ADD reg,imm	ADD CX,3
	ADD DI,1AH
	ADD DL,34H
	ADD EDX,1345H
	ADD CX,1834H
		8086	4

		8088	4
		80286	3
		80386	2
		80486	1
		Pentium–Pentium 4	1 or 3

Format	Examples	Microprocessor	Clocks

ADD mem,imm	ADD DATA4,33
	ADD LIST,'A'
	ADD DATA3[DI],2
	ADD BYTE PTR[EBX],3
	ADD WORD PTR[DI],669H
		8086	17 + ea

		8088	23 + ea
		80286	7
		80386	7
		80486	3
		Pentium–Pentium 4	1 or 3

ADD acc,imm	ADD AX,3
	ADD AL,1AH
	ADD AH,34H
	ADD EAX,2
	ADD AL,'Z'
		8086	4

		8088	4
		80286	3
		80386	2
		80486	1
		Pentium–Pentium 4	1

AND Logical AND

001000dw oorrrmmm disp

O	D	I	T	S	Z	A	P	C
0				*	*	?	*	0

Format	Examples	Microprocessor	Clocks
AND reg,reg	AND CX,BX AND DL,BL AND ECX,EBX AND BP,SI AND EDX,EDI	8086	3
		8088	3
		80286	2
		80386	2
		80486	1
		Pentium–Pentium 4	1 or 3
AND mem,reg	AND BIT,AL AND LIST,DI AND DATAZ[BX],CL AND [EAX],BL AND [ESI+4*ECX],EDX	8086	16 + ea
		8088	24 + ea
		80286	7
		80386	7
		80486	3
		Pentium–Pentium 4	1 or 3

AND reg,mem	AND BL,DATAW AND SI,LIST AND CL,DATAQ[SI] AND CX,[EAX] AND ESI,[ECX+43H]	8086	9 + ea
		8088	13 + ea
		80286	7
		80386	6
		80486	2
		Pentium–Pentium 4	1 or 2

100000sw oo100mmm disp data

Format	Examples	Microprocessor	Clocks
AND reg,imm	AND BP,1 AND DI,10H AND DL,34H AND EBP,1345H AND SP,1834H	8086	4
		8088	4
		80286	3
		80386	2
		80486	1
		Pentium–Pentium 4	1 or 3
AND mem,imm	AND DATA4,33 AND LIST,'A' AND DATA3[DI],2 AND BYTE PTR[EBX],3 AND DWORD PTR[DI],66H	8086	17 + ea
		8088	23 + ea
		80286	7
		80386	7
		80486	3
		Pentium–Pentium 4	1 or 3
AND acc,imm	AND AX,3 AND AL,1AH AND AH,34H AND EAX,2 AND AL,'r'	8086	4
		8088	4
		80286	3
		80386	2
		80486	1
		Pentium–Pentium 4	1

ARPL — Adjust requested privilege level

01100011 oorrrmmm disp

O	D	I	T	S	Z	A	P	C
					*			

Format	Examples	Microprocessor	Clocks
ARPL reg,reg	ARPL AX,BX ARPL BX,SI ARPL AX,DX ARPL BX,AX ARPL SI,DI	8086	—
		8088	—
		80286	10
		80386	20
		80486	9
		Pentium–Pentium 4	7
ARPL mem,reg	ARPL DATAY,AX ARPL LIST,DI ARPL DATA3[DI],CX ARPL [EBX],AX ARPL [EDX+4*ECX],BP	8086	—
		8088	—
		80286	11
		80386	21
		80486	9
		Pentium–Pentium 4	7

BOUND — Check array against boundary

01100010 oorrrmmm disp

Format	Examples	Microprocessor	Clocks
BOUND reg,mem	BOUND AX,BETS BOUND BP,LISTG BOUND CX,DATAX BOUND BX,[DI] BOUND SI,[BX+2]	8086	—
		8088	—
		80286	13
		80386	10
		80486	7
		Pentium–Pentium 4	8

BSF Bit scan forward

00001111 10111100 oorrrmmm disp

O	D	I	T	S	Z	A	P	C
?				?	*	?	?	?

Format	Examples	Microprocessor	Clocks
BSF reg,reg	BSF AX,BX BSF BX,SI BSF EAX,EDX BSF EBX,EAX BSF SI,DI	8086	—
		8088	—
		80286	—
		80386	10 + 3n
		80486	6–42
		Pentium–Pentium 4	6–42
BSF reg,mem	BSF AX,DATAY BSF SI,LIST BSF CX,DATA3[DI] BSF EAX,[EBX] BSF EBP,[EDX+4*ECX]	8086	—
		8088	—
		80286	—
		80386	10 + 3n
		80486	7–43
		Pentium–Pentium 4	6–43

BSR Bit scan reverse

00001111 10111101 oorrrmmm disp

O	D	I	T	S	Z	A	P	C
?				?	*	?	?	?

Format	Examples	Microprocessor	Clocks
BSR reg,reg	BSR AX,BX BSR BX,SI BSR EAX,EDX BSR EBX,EAX BSR SI,DI	8086	—
		8088	—
		80286	—
		80386	10 + 3n
		80486	6–103
		Pentium–Pentium 4	7–71

BSR reg,mem	BSR AX,DATAY BSR SI,LIST BSR CX,DATA3[DI] BSR EAX,[EBX] BSR EBP,[EDX+4*ECX]	8086	—
		8088	—
		80286	—
		80386	10 + 3n
		80486	7–104
		Pentium–Pentium 4	7–72

BSWAP Byte swap

00001111 11001rrr

Format	Examples	Microprocessor	Clocks
BSWAP reg32	BSWAP EAX BSWAP EBX BSWAP EDX BSWAP ECX BSWAP ESI	8086	—
		8088	—
		80286	—
		80386	—
		80486	1
		Pentium–Pentium 4	1

BT Bit test

00001111 10111010 oo100mmm disp data

O	D	I	T	S	Z	A	P	C
								*

Format	Examples	Microprocessor	Clocks
BT reg,imm8	BT AX,2 BT CX,4 BT BP,10H BT CX,8 BT BX,2	8086	—
		8088	—
		80286	—
		80386	3
		80486	3
		Pentium–Pentium 4	4

Format	Examples	Microprocessor	Clocks
BT mem,imm8	BT DATA1,2 BT LIST,2 BT DATA2[DI],3 BT [EAX],1 BT FROG,6	8086	—
		8088	—
		80286	—
		80386	6
		80486	3
		Pentium–Pentium 4	4

00001111 10100011 disp

Format	Examples	Microprocessor	Clocks
BT reg,reg	BT AX,CX BT CX,DX BT BP,AX BT SI,CX BT EAX,EBX	8086	—
		8088	—
		80286	—
		80386	3
		80486	3
		Pentium–Pentium 4	4 or 9
BT mem,reg	BT DATA4,AX BT LIST,BX BT DATA3[DI],CX BT [EBX],DX BT [DI],DI	8086	—
		8088	—
		80286	—
		80386	12
		80486	8
		Pentium–Pentium 4	4 or 9

BTC Bit test and complement

00001111 10111010 oo111mmm disp data

O	D	I	T	S	Z	A	P	C
								*

Format	Examples	Microprocessor	Clocks
BTC reg,imm8	BTC AX,2 BTC CX,4 BTC BP,10H BTC CX,8 BTC BX,2	8086	—
		8088	—
		80286	—
		80386	6
		80486	6
		Pentium–Pentium 4	7 or 8
BTC mem,imm8	BTC DATA1,2 BTC LIST,2 BTC DATA2[DI],3 BTC [EAX],1 BTC FROG,6	8086	—
		8088	—
		80286	—
		80386	7 or 8
		80486	8
		Pentium–Pentium 4	8

00001111 10111011 disp

Format	Examples	Microprocessor	Clocks
BTC reg,reg	BTC AX,CX BTC CX,DX BTC BP,AX BTC SI,CX BTC EAX,EBX	8086	—
		8088	—
		80286	—
		80386	6
		80486	6
		Pentium–Pentium 4	7 or 13
BTC mem,reg	BTC DATA4,AX BTC LIST,BX BTC DATA3[DI],CX BTC [EBX],DX BTC [DI],DI	8086	—
		8088	—
		80286	—
		80386	13
		80486	13
		Pentium–Pentium 4	7 or 13

BTR Bit test and reset

00001111 10111010 oo110mmm disp data

O	D	I	T	S	Z	A	P	C
								*

Format	Examples	Microprocessor	Clocks
BTR reg,imm8	BTR AX,2 BTR CX,4 BTR BP,10H BTR CX,8 BTR BX,2	8086	—
		8088	—
		80286	—
		80386	6
		80486	6
		Pentium–Pentium 4	7 or 8
BTR mem,imm8	BTR DATA1,2 BTR LIST,2 BTR DATA2[DI],3 BTR [EAX],1 BTR FROG,6	8086	—
		8088	—
		80286	—
		80386	8
		80486	8
		Pentium–Pentium 4	7 or 8

00001111 10110011 disp

Format	Examples	Microprocessor	Clocks
BTR reg,reg	BTR AX,CX BTR CX,DX BTR BP,AX BTR SI,CX BTR EAX,EBX	8086	—
		8088	—
		80286	—
		80386	6
		80486	6
		Pentium–Pentium 4	7 or 13
BTR mem,reg	BTR DATA4,AX BTR LIST,BX BTR DATA3[DI],CX BTR [EBX],DX BTR [DI],DI BTC [DI],DI	8086	—
		8088	—
		80286	—
		80386	13
		80486	13
		Pentium–Pentium 4	7 or 13

BTS Bit test and set

00001111 10111010 oo101mmm disp data

O	D	I	T	S	Z	A	P	C
								*

Format	Examples	Microprocessor	Clocks
BTS reg,imm8	BTS AX,2 BTS CX,4 BTS BP,10H BTS CX,8 BTS BX,2	8086	—
		8088	—
		80286	—
		80386	6
		80486	6
		Pentium–Pentium 4	7 or 8
BTS mem,imm8	BTS DATA1,2 BTS LIST,2 BTS DATA2[DI],3 BTS [EAX],1 BTS FROG,6	8086	—
		8088	—
		80286	—
		80386	8
		80486	8
		Pentium–Pentium 4	7 or 8

00001111 10101011 disp

Format	Examples	Microprocessor	Clocks
BTS reg,reg	BTS AX,CX BTS CX,DX BTS BP,AX BTS SI,CX BTS EAX,EBX	8086	—
		8088	—
		80286	—
		80386	6
		80486	6
		Pentium–Pentium 4	7 or 13
BTS mem,reg	BTS DATA4,AX BTS LIST,BX BTS DATA3[DI],CX BTS [EBX],DX BTS [DI],DI	8086	—
		8088	—
		80286	—
		80386	13
		80486	13
		Pentium–Pentium 4	7 or 13

CALL	Call procedure (subroutine)		
11101000 disp Format	Examples	Microprocessor	Clocks
CALL label (near)	CALL FOR_FUN CALL HOME CALL ET CALL WAITING CALL SOMEONE	8086	19
		8088	23
		80286	7
		80386	3
		80486	3
		Pentium–Pentium 4	1
10011010 disp Format	Examples	Microprocessor	Clocks
CALL label (far)	CALL FAR PTR DATES CALL WHAT CALL WHERE CALL FARCE CALL WHOM	8086	28
		8088	36
		80286	13
		80386	17
		80486	18
		Pentium–Pentium 4	4
11111111 oo010mmm Format	Examples	Microprocessor	Clocks
CALL reg (near)	CALL AX CALL BX CALL CX CALL DI CALL SI	8086	16
		8088	20
		80286	7
		80386	7
		80486	5
		Pentium–Pentium 4	2

Format	Examples	Microprocessor	Clocks
CALL mem (near)	CALL ADDRESS CALL NEAR PTR [DI] CALL DATA1 CALL FROG CALL ME_NOW	8086	21 + ea
		8088	29 + ea
		80286	11
		80386	10
		80486	5
		Pentium–Pentium 4	2

11111111 oo011mmm

Format	Examples	Microprocessor	Clocks
CALL mem (far)	CALL FAR_LIST[SI] CALL FROM_HERE CALL TO_THERE CALL SIXX CALL OCT	8086	16
		8088	20
		80286	7
		80386	7
		80486	5
		Pentium–Pentium 4	2

CBW Convert byte to word (AL ⇒ AX)

10011000

Example	Microprocessor	Clocks
CBW	8086	2
	8088	2
	80286	2
	80386	3
	80486	3
	Pentium–Pentium 4	3

CDQ Convert doubleword to quadword (EAX ⇒ EDX:EAX)

11010100 00001010

Example	Microprocessor	Clocks
CDQ	8086	—
	8088	—
	80286	—
	80386	2
	80486	2
	Pentium–Pentium 4	2

CLC Clear carry flag

11111000

O	D	I	T	S	Z	A	P	C
								0

Example	Microprocessor	Clocks
CLC	8086	2
	8088	2
	80286	2
	80386	2
	80486	2
	Pentium–Pentium 4	2

CLD Clear direction flag

11111100

O	D	I	T	S	Z	A	P	C
	0							

Example	Microprocessor	Clocks
CLD	8086	2
	8088	2
	80286	2
	80386	2
	80486	2
	Pentium–Pentium 4	2

CLI	Clear interrupt flag	
11111010	O D I T S Z A P C 0	
Example	Microprocessor	Clocks
CLI	8086	2
	8088	2
	80286	3
	80386	3
	80486	5
	Pentium–Pentium 4	7

CLTS	Clear task switched flag (CR0)	
00001111 00000110		
Example	Microprocessor	Clocks
CLTS	8086	—
	8088	—
	80286	2
	80386	5
	80486	7
	Pentium–Pentium 4	10

CMC	Complement carry flag	
10011000	O D I T S Z A P C *	
Example	Microprocessor	Clocks
CMC	8086	2
	8088	2
	80286	2
	80386	2
	80486	2
	Pentium–Pentium 4	2

CMOVcondition Conditional move

00001111 0100cccc oorrrmmm

Format	Examples	Microprocessor	Clocks
CMOVcc reg,mem	CMOVNZ AX,FROG CMOVC EAX,[EDI] CMOVNC BX,DATA1 CMOVP EBX,WAITING CMOVNE DI,[SI]	8086	—
		8088	—
		80286	—
		80386	—
		80486	—
		Pentium–Pentium 4	—

Condition Codes	Mnemonic	Flag	Description
0000	CMOVO	O = 1	Move if overflow
0001	CMOVNO	O = 0	Move if no overflow
0010	CMOVB	C = 1	Move if below
0011	CMOVAE	C = 0	Move if above or equal
0100	CMOVE	Z = 1	Move if equal/zero
0101	CMOVNE	Z = 0	Move if not equal/zero
0110	CMOVBE	C = 1 + Z = 1	Move if below or equal
0111	CMOVA	C = 0 • Z = 0	Move if above
1000	CMOVS	S = 1	Move if sign
1001	CMOVNS	S = 0	Move if no sign
1010	CMOVP	P = 1	Move if parity
1011	CMOVNP	P = 0	Move if no parity
1100	CMOVL	S • O	Move if less than
1101	CMOVGE	S = 0	Move if greater than or equal
1110	CMOVLE	Z = 1 + S • O	Move if less than or equal
1111	CMOVG	Z = 0 + S = O	Move if greater than

CMP Compare

001110dw oorrrmmm disp

O	D	I	T	S	Z	A	P	C
*				*	*	*	*	*

Format	Examples	Microprocessor	Clocks

CMP reg,reg	CMP AX,BX
	CMP AL,BL
	CMP EAX,EBX
	CMP CX,SI
	CMP ESI,EDI
		8086	3

		8088	3
		80286	2
		80386	2
		80486	1
		Pentium–Pentium 4	1 or 2

CMP mem,reg	CMP DATAY,AL
	CMP LIST,SI
	CMP DATA6[DI],CL
	CMP [EAX],CL
	CMP [EDX+4*ECX],EBX
		8086	9 + ea

		8088	13 + ea
		80286	7
		80386	5
		80486	2
		Pentium–Pentium 4	1 or 2

CMP reg,mem	CMP BL,DATA2
	CMP SI,LIST3
	CMP CL,DATA2[DI]
	CMP CX,[EDI]
	CMP ESI,[ECX+200H]
		8086	9 + ea

		8088	13 + ea
		80286	6
		80386	6
		80486	2
		Pentium–Pentium 4	1 or 2

100000sw oo111mmm disp data

Format	Examples	Microprocessor	Clocks

CMP reg,imm	CMP CX,3
	CMP DI,1AH
	CMP DL,34H
	CMP EDX,1345H
	CMP CX,1834H
		8086	4

		8088	4
		80286	3
		80386	2
		80486	1
		Pentium–Pentium 4	1 or 2

CMP mem,imm	CMP DATAS,3
	CMP BYTE PTR[EDI],1AH
	CMP DADDY,34H
	CMP LIST,'A'
	CMP TOAD,1834H
		8086	10 + ea

		8088	14 + ea
		80286	6
		80386	5
		80486	2
		Pentium–Pentium 4	1 or 2

0001111w data Format	Examples	Microprocessor	Clocks

CMP acc,imm	CMP AX,3
	CMP AL,1AH
	CMP AH,34H
	CMP EAX,1345H
	CMP AL,'Y'
		8086	4

		8088	4
		80286	3
		80386	2
		80486	1
		Pentium–Pentium 4	1

CMPS Compare strings

1010011w

O	D	I	T	S	Z	A	P	C
*				*	*	*	*	*

Format	Examples	Microprocessor	Clocks
CMPSB CMPSW CMPSD	CMPSB CMPSW CMPSD CMPSB DATA1,DATA2 REPE CMPSB REPNE CMPSW	8086	32
		8088	30
		80286	8
		80386	10
		80486	8
		Pentium–Pentium 4	5

CMPXCHG Compare and exchange

00001111 1011000w 11rrrrrr

O	D	I	T	S	Z	A	P	C
*				*	*	*	*	*

Format	Examples	Microprocessor	Clocks
CMPXCHG reg,reg	CMPXCHG EAX,EBX CMPXCHG ECX,EDX	8086	—
		8088	—
		80286	—
		80386	—
		80486	6
		Pentium–Pentium 4	6

0001111w data

Format	Examples	Microprocessor	Clocks
CMPXCHG mem,reg	CMPXCHG DATAD,EAX CMPXCHG DATA2,EDI	8086	—
		8088	—
		80286	—
		80386	—
		80486	7
		Pentium–Pentium 4	6

CMPXCHG8B Compare and exchange 8 bytes

00001111 11000111 oorrrmmm

O	D	I	T	S	Z	A	P	C
					*			

Format	Examples	Microprocessor	Clocks
CMPXCHG8B mem64	CMPXCHG8B DATA3	8086	—
		8088	—
		80286	—
		80386	—
		80486	—
		Pentium–Pentium 4	10

CPUID CPU identification code

00001111 10100010

Example	Microprocessor	Clocks
CPUID	8086	—
	8088	—
	80286	—
	80386	—
	80486	—
	Pentium–Pentium 4	14

CWD	Convert word to doubleword (AX ⇒ DX:AX)	
10011000 Example	Microprocessor	Clocks
CWD	8086	5
	8088	5
	80286	2
	80386	2
	80486	3
	Pentium–Pentium 4	2

CWDE	Convert word to extended doubleword (AX ⇒ EAX)	
10011000 Example	Microprocessor	Clocks
CWDE	8086	—
	8088	—
	80286	—
	80386	3
	80486	3
	Pentium–Pentium 4	3

DAA	Decimal adjust AL after addition	
00100111	O D I T S Z A P C ? * * * * *	
Example	Microprocessor	Clocks
DAA	8086	4
	8088	4
	80286	3
	80386	4
	80486	2
	Pentium–Pentium 4	3

DAS	Decimal adjust AL after subtraction		
00101111		O D I T S Z A P C ? * * * * *	
Example		Microprocessor	Clocks
DAS		8086	4
		8088	4
		80286	3
		80386	4
		80486	2
		Pentium–Pentium 4	3

DEC	Decrement		
1111111w oo001mmm disp		O D I T S Z A P C * * * * *	
Format	Examples	Microprocessor	Clocks
DEC reg8	DEC BL DEC BH DEC CL DEC DH DEC AH	8086	3
		8088	3
		80286	2
		80386	2
		80486	1
		Pentium–Pentium 4	1 or 3
DEC mem	DEC DATAY DEC LIST DEC DATA6[DI] DEC BYTE PTR [BX] DEC WORD PTR [EBX]	8086	15 + ea
		8088	23 + ea
		80286	7
		80386	6
		80486	3
		Pentium–Pentium 4	1 or 3

01001rrr Format	Examples	Microprocessor	Clocks
DEC reg16 DEC reg32	DEC CX DEC DI DEC EDX DEC ECX DEC BP	8086	3
		8088	3
		80286	2
		80386	2
		80486	1
		Pentium–Pentium 4	1

DIV Divide

1111011w oo110mmm disp

O	D	I	T	S	Z	A	P	C
?				?	?	?	?	?

Format	Examples	Microprocessor	Clocks
DIV reg	DIV BL DIV BH DIV ECX DIV DH DIV CX	8086	162
		8088	162
		80286	22
		80386	38
		80486	40
		Pentium–Pentium 4	17–41
DIV mem	DIV DATAY DIV LIST DIV DATA6[DI] DIV BYTE PTR [BX] DIV WORD PTR [EBX]	8086	168
		8088	176
		80286	25
		80386	41
		80486	40
		Pentium–Pentium 4	17–41

ENTER Create a stack frame

11001000 data

Format	Examples	Microprocessor	Clocks
ENTER imm,0	ENTER 4,0 ENTER 8,0 ENTER 100,0 ENTER 200,0 ENTER 1024,0	8086	—
		8088	—
		80286	11
		80386	10
		80486	14
		Pentium–Pentium 4	11
ENTER imm,1	ENTER 4,1 ENTER 10,1	8086	—
		8088	—
		80286	12
		80386	15
		80486	17
		Pentium–Pentium 4	15
ENTER imm,imm	ENTER 3,6 ENTER 100,3	8086	—
		8088	—
		80286	12
		80386	15
		80486	17
		Pentium–Pentium 4	15 + 2n

ESC Escape (obsolete–see coprocessor)

HLT	Halt	
11110100 Example	Microprocessor	Clocks
HLT	8086	2
	8088	2
	80286	2
	80386	5
	80486	4
	Pentium–Pentium 4	varies

IDIV	Integer (signed) division		
1111011w oo111mmm disp		O D I T S Z A P C ? ? ? ? ? ?	
Format	Examples	Microprocessor	Clocks
IDIV reg	IDIV BL IDIV BH IDIV ECX IDIV DH IDIV CX	8086	184
		8088	184
		80286	25
		80386	43
		80486	43
		Pentium–Pentium 4	22–46
IDIV mem	IDIV DATAY IDIV LIST IDIV DATA6[DI] IDIV BYTE PTR [BX] IDIV WORD PTR [EBX]	8086	190
		8088	194
		80286	28
		80386	46
		80486	44
		Pentium–Pentium 4	22–46

IMUL	Integer (signed) multiplication		
1111011w oo101mmm disp		O D I T S Z A P C * ? ? ? ? *	
Format	Examples	Microprocessor	Clocks
IMUL reg	IMUL BL IMUL CX IMUL ECX IMUL DH IMUL AL	8086	154
		8088	154
		80286	21
		80386	38
		80486	42
		Pentium–Pentium 4	10–11
IMUL mem	IMUL DATAY IMUL LIST IMUL DATA6[DI] IMUL BYTE PTR [BX] IMUL WORD PTR [EBX]	8086	160
		8088	164
		80286	24
		80386	41
		80486	42
		Pentium–Pentium 4	10–11
011010s1 oorrmmm disp data			
Format	Examples	Microprocessor	Clocks
IMUL reg,imm	IMUL CX,16 IMUL DI,100 IMUL EDX,20	8086	—
		8088	—
		80286	21
		80386	38
		80486	42
		Pentium–Pentium 4	10
IMUL reg,reg,imm	IMUL DX,AX,2 IMUL CX,DX,3 IMUL BX,AX,33	8086	—
		8088	—
		80286	21
		80386	38
		80486	42
		Pentium–Pentium 4	10

Format	Examples	Microprocessor	Clocks
IMUL reg,mem,imm	IMUL CX,DATAY,99	8086	—
		8088	—
		80286	24
		80386	38
		80486	42
		Pentium–Pentium 4	10

00001111 10101111 oorrmmm disp

Format	Examples	Microprocessor	Clocks
IMUL reg,reg	IMUL CX,DX IMUL DI,BX IMUL EDX,EBX	8086	—
		8088	—
		80286	—
		80386	38
		80486	42
		Pentium–Pentium 4	10
IMUL reg,mem	IMUL DX,DATAY IMUL CX,LIST IMUL ECX,DATA6[DI]	8086	—
		8088	—
		80286	—
		80386	41
		80486	42
		Pentium–Pentium 4	10

IN Input data from port

1110010w port#

Format	Examples	Microprocessor	Clocks
IN acc,pt	IN AL,12H IN AX,12H IN AL,0FFH IN AX,0A0H IN EAX,10H	8086	10
		8088	14
		80286	5
		80386	12
		80486	14
		Pentium–Pentium 4	7

1110110w Format	Examples	Microprocessor	Clocks
IN acc,DX	IN AL,DX IN AX,DX IN EAX,DX	8086	8
		8088	12
		80286	5
		80386	13
		80486	14
		Pentium–Pentium 4	7

INC Increment

1111111w oo000mmm disp

O	D	I	T	S	Z	A	P	C
*				*	*	*	*	

Format	Examples	Microprocessor	Clocks
INC reg8	INC BL INC BH INC AL INC AH INC DH	8086	3
		8088	3
		80286	2
		80386	2
		80486	1
		Pentium–Pentium 4	1 or 3
INC mem	INC DATA3 INC LIST INC COUNT INC BYTE PTR [DI] INC WORD PTR [ECX]	8086	15 + ea
		8088	23 + ea
		80286	7
		80386	6
		80486	3
		Pentium–Pentium 4	1 or 3
INC reg16 INC reg32	INC CX INC DX INC BP INC ECX INC ESP	8086	3
		8088	3
		80286	2
		80386	2
		80486	1
		Pentium–Pentium 4	1

INS Input string from port

0110110w

Format	Examples	Microprocessor	Clocks
INSB INSW INSD	INSB INSW INSD INS DATA2 REP INSB	8086	—
		8088	—
		80286	5
		80386	15
		80486	17
		Pentium–Pentium 4	9

INT Interrupt

11001101 type

Format	Examples	Microprocessor	Clocks
INT type	INT12H INT15H INT 21H INT 2FH INT 10H	8086	51
		8088	71
		80286	23
		80386	37
		80486	30
		Pentium–Pentium 4	16–82

INT 3 Interrupt 3

11001100

Example	Microprocessor	Clocks
INT 3	8086	52
	8088	72
	80286	23
	80386	33
	80486	26
	Pentium–Pentium 4	13–56

INTO Interrupt on overflow

11001110

Example	Microprocessor	Clocks
INTO	8086	53
	8088	73
	80286	24
	80386	35
	80486	28
	Pentium–Pentium 4	13–56

INVD Invalidate data cache

00001111 00001000

Example	Microprocessor	Clocks
INTVD	8086	—
	8088	—
	80286	—
	80386	—
	80486	4
	Pentium–Pentium 4	15

IRET/IRETD Return from interrupt

11001101 data

O	D	I	T	S	Z	A	P	C
*	*	*	*	*	*	*	*	*

Format	Examples	Microprocessor	Clocks
IRET IRETD	IRET IRETD IRET 100	8086	32
		8088	44
		80286	17
		80386	22
		80486	15
		Pentium–Pentium 4	8–27

Jcondition	Conditional jump		
0111cccc disp Format	Examples	Microprocessor	Clocks
Jcnd label (8-bit disp)	JA ABOVE JB BELOW JG GREATER JE EQUAL JZ ZERO	8086	16/4
		8088	16/4
		80286	7/3
		80386	7/3
		80486	3/1
		Pentium–Pentium 4	1
00001111 1000cccc disp Format	Examples	Microprocessor	Clocks
Jcnd label (16-bit disp)	JNE NOT_MORE JLE LESS_OR_SO	8086	—
		8088	—
		80286	—
		80386	7/3
		80486	3/1
		Pentium–Pentium 4	1

Condition Codes	Mnemonic	Flag	Description
0000	JO	O = 1	Jump if overflow
0001	JNO	O = 0	Jump if no overflow
0010	JB/NAE	C = 1	Jump if below
0011	JAE/JNB	C = 0	Jump if above or equal
0100	JE/JZ	Z = 1	Jump if equal/zero
0101	JNE/JNZ	Z = 0	Jump if not equal/zero
0110	JBE/JNA	C = 1 + Z = 1	Jump if below or equal
0111	JA/JNBE	C = 0 • Z = 0	Jump if above
1000	JS	S = 1	Jump if sign
1001	JNS	S = 0	Jump if no sign
1010	JP/JPE	P = 1	Jump if parity
1011	JNP/JPO	P = 0	Jump if no parity
1100	JL/JNGE	S • O	Jump if less than
1101	JGE/JNL	S = 0	Jump if greater than or equal
1110	JLE/JNG	Z = 1 + S • O	Jump if less than or equal
1111	JG/JNLE	Z = 0 + S = O	Jump if greater than

JCXZ/JECXZ Jump if CX (ECX) equals zero

11100011

Format	Examples	Microprocessor	Clocks
JCXZ label JECXZ label	JCXZ ABOVE JCXZ BELOW JECXZ GREATER JECXZ EQUAL JCXZ NEXT	8086	18/6
		8088	18/6
		80286	8/4
		80386	9/5
		80486	8/5
		Pentium–Pentium 4	6/5

JMP Jump

11101011 disp

Format	Examples	Microprocessor	Clocks
JMP label (short)	JMP SHORT UP JMP SHORT DOWN JMP SHORT OVER JMP SHORT CIRCUIT JMP SHORT JOKE	8086	15
		8088	15
		80286	7
		80386	7
		80486	3
		Pentium–Pentium 4	1

11101001 disp

Format	Examples	Microprocessor	Clocks
JMP label (near)	JMP VERS JMP FROG JMP UNDER JMP NEAR PTR OVER	8086	15
		8088	15
		80286	7
		80386	7
		80486	3
		Pentium–Pentium 4	1

11101010 disp Format	Examples	Microprocessor	Clocks
JMP label (far)	JMP NOT_MORE JMP UNDER JMP AGAIN JMP FAR PTR THERE	8086	15
		8088	15
		80286	11
		80386	12
		80486	17
		Pentium–Pentium 4	3

11111111 oo100mmm Format	Examples	Microprocessor	Clocks
JMP reg (near)	JMP AX JMP EAX JMP CX JMP DX	8086	11
		8088	11
		80286	7
		80386	7
		80486	3
		Pentium–Pentium 4	2
JMP mem (near)	JMP VERS JMP FROG JMP CS:UNDER JMP DATA1[DI+2]	8086	18 + ea
		8088	18 + ea
		80286	11
		80386	10
		80486	5
		Pentium–Pentium 4	4

11111111 oo101mmm Format	Examples	Microprocessor	Clocks
JMP mem (far)	JMP WAY_OFF JMP TABLE JMP UP JMP OUT_OF_HERE	8086	24 + ea
		8088	24 + ea
		80286	15
		80386	12
		80486	13
		Pentium–Pentium 4	4

LAHF Load AH from flags

10011111

Example	Microprocessor	Clocks
LAHF	8086	4
	8088	4
	80286	2
	80386	2
	80486	3
	Pentium–Pentium 4	2

LAR Load access rights byte

00001111 00000010 oorrrmmm disp

O	D	I	T	S	Z	A	P	C
					*			

Format	Examples	Microprocessor	Clocks
LAR reg,reg	LAR AX,BX LAR CX,DX LAR ECX,EDX	8086	—
		8088	—
		80286	14
		80386	15
		80486	11
		Pentium–Pentium 4	8
LAR reg,mem	LAR CX,DATA1 LAR AX,LIST3 LAR ECX,TOAD	8086	—
		8088	—
		80286	16
		80386	16
		80486	11
		Pentium–Pentium 4	8

LDS	Load far pointer to DS and register		
11000101 oorrrmmm Format	Examples	Microprocessor	Clocks

LDS reg,mem	LDS DI,DATA3
	LDS SI,LIST2
	LDS BX,ARRAY_PTR
	LDS CX,PNTR
		8086	16 + ea

		8088	24 + ea
		80286	7
		80386	7
		80486	6
		Pentium–Pentium 4	4

LEA	Load effective address		
10001101 oorrrmmm disp Format	Examples	Microprocessor	Clocks
LEA reg,mem	LEA DI,DATA3 LEA SI,LIST2 LEA BX,ARRAY_PTR LEA CX,PNTR	8086	2 + ea
		8088	2 + ea
		80286	3
		80386	2
		80486	2
		Pentium–Pentium 4	1

LEAVE	Leave high-level procedure	
11001001 Example	Microprocessor	Clocks
LEAVE	8086	—
	8088	—
	80286	5
	80386	4
	80486	5
	Pentium–Pentium 4	3

LES — Load far pointer to ES and register

11000100 oorrrmmm

Format	Examples	Microprocessor	Clocks
LES reg,mem	LES DI,DATA3 LES SI,LIST2 LES BX,ARRAY_PTR LES CX,PNTR	8086	16 + ea
		8088	24 + ea
		80286	7
		80386	7
		80486	6
		Pentium–Pentium 4	4

LFS — Load far pointer to FS and register

00001111 10110100 oorrrmmm disp

Format	Examples	Microprocessor	Clocks
LFS reg,mem	LFS DI,DATA3 LFS SI,LIST2 LFS BX,ARRAY_PTR LFS CX,PNTR	8086	—
		8088	—
		80286	—
		80386	7
		80486	6
		Pentium–Pentium 4	4

LGDT — Load global descriptor table

00001111 00000001 oo010mmm disp

Format	Examples	Microprocessor	Clocks
LGDT mem64	LGDT DESCRIP LGDT TABLED	8086	—
		8088	—
		80286	11
		80386	11
		80486	11
		Pentium–Pentium 4	6

LGS	Load far pointer to GS and register		
00001111 10110101 oorrrmmm disp			
Format	Examples	Microprocessor	Clocks
LGS reg,mem	LGS DI,DATA3 LGS SI,LIST2 LGS BX,ARRAY_PTR LGS CX,PNTR	8086	—
		8088	—
		80286	—
		80386	7
		80486	6
		Pentium–Pentium 4	4

LIDT	Load interrupt descriptor table		
00001111 00000001 oo011mmm disp			
Format	Examples	Microprocessor	Clocks
LIDT mem64	LIDT DATA3 LIDT LIST2	8086	—
		8088	—
		80286	12
		80386	11
		80486	11
		Pentium–Pentium 4	6

LLDT	Load local descriptor table		
00001111 00000000 oo010mmm disp			
Format	Examples	Microprocessor	Clocks
LLDT reg	LLDT BX LLDT DX LLDT CX	8086	—
		8088	—
		80286	17
		80386	20
		80486	11
		Pentium–Pentium 4	9

LLDT mem	LLDT DATA1 LLDT LIST3 LLDT TOAD	8086	—
		8088	—
		80286	19
		80386	24
		80486	11
		Pentium–Pentium 4	9

LMSW Load machine status word (80286 only)

00001111 00000001 oo110mmm disp

Format	Examples	Microprocessor	Clocks
LMSW reg	LMSW BX LMSW DX LMSW CX	8086	—
		8088	—
		80286	3
		80386	10
		80486	2
		Pentium–Pentium 4	8
LMSW mem	LMSW DATA1 LMSW LIST3 LMSW TOAD	8086	—
		8088	—
		80286	6
		80386	13
		80486	3
		Pentium–Pentium 4	8

LOCK	Lock the bus		
11110000 Format	Examples	Microprocessor	Clocks
LOCK:inst	LOCK:XCHG AX,BX LOCK:ADD AL,3	8086	2
		8088	3
		80286	0
		80386	0
		80486	1
		Pentium–Pentium 4	1

LODS	Load string operand		
1010110w Format	Examples	Microprocessor	Clocks
LODSB LODSW LODSD	LODSB LODSW LODSD LODS DATA3	8086	12
		8088	15
		80286	5
		80386	5
		80486	5
		Pentium–Pentium 4	2

LOOP/LOOPD	Loop until CX = 0 or ECX = 0		
11100010 disp Format	Examples	Microprocessor	Clocks
LOOP label LOOPD label	LOOP NEXT LOOP BACK LOOPD LOOPS	8086	17/5
		8088	17/5
		80286	8/4
		80386	11
		80486	7/6
		Pentium–Pentium 4	5/6

LOOPE/LOOPED Loop while equal

11100001 disp

Format	Examples	Microprocessor	Clocks
LOOPE label LOOPED label LOOPZ label LOOPZD label	LOOPE AGAIN LOOPED UNTIL LOOPZ ZORRO LOOPZD WOW	8086	18/6
		8088	18/6
		80286	8/4
		80386	11
		80486	9/6
		Pentium–Pentium 4	7/8

LOOPNE/LOOPNED Loop while not equal

11100000 disp

Format	Examples	Microprocessor	Clocks
LOOPNE label LOOPNED label LOOPNZ label LOOPNZD label	LOOPNE FORWARD LOOPNED UPS LOOPNZ TRY_AGAIN LOOPNZD WOO	8086	19/5
		8088	19/5
		80286	8/4
		80386	11
		80486	9/6
		Pentium–Pentium 4	7/8

LSL Load segment limit

00001111 00000011 oorrrmmm disp

O	D	I	T	S	Z	A	P	C
					*			

Format	Examples	Microprocessor	Clocks
LSL reg,reg	LSL AX,BX LSL CX,BX LSL EDX,EAX	8086	—
		8088	—
		80286	14
		80386	25
		80486	10
		Pentium–Pentium 4	8

Format	Examples	Microprocessor	Clocks
LSL reg,mem	LSL AX,LIMIT LSL EAX,NUM	8086	—
		8088	—
		80286	16
		80386	26
		80486	10
		Pentium–Pentium 4	8

LSS Load far pointer to SS and register

00001111 10110010 oorrrmmm disp

Format	Examples	Microprocessor	Clocks
LSS reg,mem	LSS DI,DATA1 LSS SP,STACK_TOP LSS CX,ARRAY	8086	—
		8088	—
		80286	—
		80386	7
		80486	6
		Pentium–Pentium 4	4

LTR Load task register

00001111 00000000 oo001mmm disp

Format	Examples	Microprocessor	Clocks
LTR reg	LTR AX LTR CX LTR DX	8086	—
		8088	—
		80286	17
		80386	23
		80486	20
		Pentium–Pentium 4	10

LTR mem16	LTR TASK LTR NUM	8086	—
		8088	—
		80286	19
		80386	27
		80486	20
		Pentium–Pentium 4	10

MOV Move data

100010dw oorrrmmm disp

Format	Examples	Microprocessor	Clocks

MOV reg,reg	MOV CL,CH
	MOV BH,CL
	MOV CX,DX
	MOV EAX,EBP
	MOV ESP,ESI
		8086	2

		8088	2
		80286	2
		80386	2
		80486	1
		Pentium–Pentium 4	1

MOV mem,reg	MOV DATA7,DL
	MOV NUMB,CX
	MOV TEMP,EBX
	MOV [ECX],BL
	MOV [DI],DH
		8086	9 + ea

		8088	13 + ea
		80286	3
		80386	2
		80486	1
		Pentium–Pentium 4	1

MOV reg,mem	MOV DL,DATA8
	MOV DX,NUMB
	MOV EBX,TEMP+3
	MOV CH,TEMP[EDI]
	MOV CL,DATA2
		8086	10 + ea

		8088	12 + ea
		80286	5
		80386	4
		80486	1
		Pentium–Pentium 4	1

1100011w oo000mmm disp data Format	Examples	Microprocessor	Clocks

MOV mem,imm	MOV DATAF,23H
	MOV LIST,12H
	MOV BYTE PTR [DI],2
	MOV NUMB,234H
	MOV DWORD PTR[ECX],1
		8086	10 + ea

		8088	14 + ea
		80286	3
		80386	2
		80486	1
		Pentium–Pentium 4	1

1011wrrr data Format	Examples	Microprocessor	Clocks

MOV reg,imm	MOV BX,22H
	MOV CX,12H
	MOV CL,2
	MOV ECX,123456H
	MOV DI,100
		8086	4

		8088	4
		80286	3
		80386	2
		80486	1
		Pentium–Pentium 4	1

101000dw disp Format	Examples	Microprocessor	Clocks

MOV mem,acc	MOV DATAF,AL
	MOV LIST,AX
	MOV NUMB,EAX
		8086	10

		8088	14
		80286	3
		80386	2
		80486	1
		Pentium–Pentium 4	1

MOV acc,mem	MOV AL,DATAE
	MOV AX,LIST
	MOV EAX,LUTE
		8086	10

		8088	14
		80286	5
		80386	4
		80486	1
		Pentium–Pentium 4	1

100011d0 oosssmmm disp Format	Examples	Microprocessor	Clocks

MOV seg,reg	MOV SS,AX
	MOV DS,DX
	MOV ES,CX
	MOV FS,BX
	MOV GS,AX
		8086	2

		8088	2
		80286	2
		80386	2
		80486	1
		Pentium–Pentium 4	1

MOV seg,mem	MOV SS,STACK_TOP
	MOV DS,DATAS
	MOV ES,TEMP1
		8086	8 + ea

		8088	12 + ea
		80286	2
		80386	2
		80486	1
		Pentium–Pentium 4	2 or 3

MOV reg,seg	MOV BX,DS
	MOV CX,FS
	MOV CX,ES
		8086	2

		8088	2
		80286	2
		80386	2
		80486	1
		Pentium–Pentium 4	1

MOV mem,seg	MOV DATA2,CS
	MOV TEMP,DS
	MOV NUMB1,SS
	MOV TEMP2,GS
		8086	9 + ea

		8088	13 + ea
		80286	3
		80386	2
		80486	1
		Pentium–Pentium 4	1

00001111 001000d0 11rrrmmm Format	Examples	Microprocessor	Clocks

MOV reg,cr	MOV EBX,CR0
	MOV ECX,CR2
	MOV EBX,CR3
		8086	—

		8088	—
		80286	—
		80386	6
		80486	4
		Pentium–Pentium 4	4

MOV cr,reg	MOV CR0,EAX
	MOV CR1,EBX
	MOV CR3,EDX
		8086	—

		8088	—
		80286	—
		80386	10
		80486	4
		Pentium–Pentium 4	12–46

00001111 001000d1 11rrrmmm Format	Examples	Microprocessor	Clocks

MOV reg,dr	MOV EBX,DR6
	MOV ECX,DR7
	MOV EBX,DR1
		8086	—

		8088	—
		80286	—
		80386	22
		80486	10
		Pentium–Pentium 4	11

MOV dr,reg	MOV DR0,EAX
	MOV DR1,EBX
	MOV DR3,EDX
		8086	—

		8088	—
		80286	—
		80386	22
		80486	11
		Pentium–Pentium 4	11

00001111 001001d0 11rrrmmm Format	Examples	Microprocessor	Clocks
MOV reg,tr	MOV EBX,TR6 MOV ECX,TR7	8086	—
		8088	—
		80286	—
		80386	12
		80486	4
		Pentium–Pentium 4	11
MOV tr,reg	MOV TR6,EAX MOV TR7,EBX	8086	—
		8088	—
		80286	—
		80386	12
		80486	6
		Pentium–Pentium 4	11

MOVS Move string data

1010010w Format	Examples	Microprocessor	Clocks
MOVSB MOVSW MOVSD	MOVSB MOVSW MOVSD MOVS DATA1,DATA2	8086	18
		8088	26
		80286	5
		80386	7
		80486	7
		Pentium–Pentium 4	4

MOVSX Move with sign extend

00001111 1011111w oorrrmmm disp

Format	Examples	Microprocessor	Clocks
MOVSX reg,reg	MOVSX BX,AL MOVSX EAX,DX	8086	—
		8088	—
		80286	—
		80386	3
		80486	3
		Pentium–Pentium 4	3
MOVSX reg,mem	MOVSX AX,DATA34 MOVSX EAX,NUMB	8086	—
		8088	—
		80286	—
		80386	6
		80486	3
		Pentium–Pentium 4	3

MOVZX Move with zero extend

00001111 1011011w oorrrmmm disp

Format	Examples	Microprocessor	Clocks
MOVZX reg,reg	MOVZX BX,AL MOVZX EAX,DX	8086	—
		8088	—
		80286	—
		80386	3
		80486	3
		Pentium–Pentium 4	3
MOVZX reg,mem	MOVZX AX,DATA34 MOVZX EAX,NUMB	8086	—
		8088	—
		80286	—
		80386	6
		80486	3
		Pentium–Pentium 4	3

MUL Multiply

1111011w oo100mmm disp

O	D	I	T	S	Z	A	P	C
*				?	?	?	?	*

Format	Examples	Microprocessor	Clocks
MUL reg	MUL BL MUL CX MUL EDX	8086	118
		8088	143
		80286	21
		80386	38
		80486	42
		Pentium–Pentium 4	10 or 11
MUL mem	MUL DATA9 MUL WORD PTR [ESI]	8086	139
		8088	143
		80286	24
		80386	41
		80486	42
		Pentium–Pentium 4	11

NEG Negate

1111011w oo011mmm disp

O	D	I	T	S	Z	A	P	C
*				*	*	*	*	*

Format	Examples	Microprocessor	Clocks
NEG reg	NEG BL NEG CX NEG EDI	8086	3
		8088	3
		80286	2
		80386	2
		80486	1
		Pentium–Pentium 4	1 or 3

Format	Examples	Microprocessor	Clocks
NEG mem	NEG DATA9 NEG WORD PTR [ESI]	8086	16 + ea
		8088	24 + ea
		80286	7
		80386	6
		80486	3
		Pentium–Pentium 4	1 or 3

NOP No operation

10010000

Example	Microprocessor	Clocks
NOP	8086	3
	8088	3
	80286	3
	80386	3
	80486	3
	Pentium–Pentium 4	1

NOT One's complement

1111011w oo010mmm disp

Format	Examples	Microprocessor	Clocks
NOT reg	NOT BL NOT CX NOT EDI	8086	3
		8088	3
		80286	2
		80386	2
		80486	1
		Pentium–Pentium 4	1 or 3
NOT mem	NOT DATA9 NOT WORD PTR [ESI]	8086	16 + ea
		8088	24 + ea
		80286	7
		80386	6
		80486	3
		Pentium–Pentium 4	1 or 3

OR Inclusive-OR

000010dw oorrrmmm disp

O	D	I	T	S	Z	A	P	C
0				*	*	?	*	0

Format	Examples	Microprocessor	Clocks
OR reg,reg	OR AX,BX OR AL,BL OR EAX,EBX OR CX,SI OR ESI,EDI	8086	3
		8088	3
		80286	2
		80386	2
		80486	1
		Pentium–Pentium 4	1 or 2
OR mem,reg	OR DATAY,AL OR LIST,SI OR DATA2[DI],CL OR [EAX],BL OR [EBX+2*ECX],EDX	8086	16 + ea
		8088	24 + ea
		80286	7
		80386	7
		80486	3
		Pentium–Pentium 4	1 or 3
OR reg,mem	OR BL,DATA1 OR SI,LIST1 OR CL,DATA2[SI] OR CX,[ESI] OR ESI,[2*ECX]	8086	9 + ea
		8088	13 + ea
		80286	7
		80386	6
		80486	2
		Pentium–Pentium 4	1 or 3

100000sw oo001mmm disp data

Format	Examples	Microprocessor	Clocks
OR reg,imm	OR CX,3 OR DI,1AH OR DL,34H OR EDX,1345H OR CX,1834H	8086	4
		8088	4
		80286	3
		80386	2
		80486	1
		Pentium–Pentium 4	1 or 3

OR mem,imm	OR DATAS,3 OR BYTE PTR[EDI],1AH OR DADDY,34H OR LIST,'A' OR TOAD,1834H	8086	17 + ea
		8088	25 + ea
		80286	7
		80386	7
		80486	3
		Pentium–Pentium 4	1 or 3

0000110w data

Format	Examples	Microprocessor	Clocks
OR acc,imm	OR AX,3 OR AL,1AH OR AH,34H OR EAX,1345H OR AL,'Y'	8086	4
		8088	4
		80286	3
		80386	2
		80486	1
		Pentium–Pentium 4	1

OUT Output data to port

1110011w port#

Format	Examples	Microprocessor	Clocks
OUT pt,acc	OUT 12H,AL OUT 12H,AX OUT 0FFH,AL OUT 0A0H,AX OUT 10H,EAX	8086	10
		8088	14
		80286	3
		80386	10
		80486	10
		Pentium–Pentium 4	12–26

1110111w

Format	Examples	Microprocessor	Clocks
OUT DX,acc	OUT DX,AL OUT DX,AX OUT DX,EAX	8086	8
		8088	12
		80286	3
		80386	11
		80486	10
		Pentium–Pentium 4	12–26

OUTS Output string to port

0110111w

Format	Examples	Microprocessor	Clocks
OUTSB OUTSW OUTSD	OUTSB OUTSW OUTSD OUTS DATA2 REP OUTSB	8086	—
		8088	—
		80286	5
		80386	14
		80486	10
		Pentium–Pentium 4	13–27

POP Pop data from stack

01011rrr

Format	Examples	Microprocessor	Clocks
POP reg	POP CX POP AX POP EDI	8086	8
		8088	12
		80286	5
		80386	4
		80486	1
		Pentium–Pentium 4	1

10001111 oo000mmm disp

Format	Examples	Microprocessor	Clocks
POP mem	POP DATA1 POP LISTS POP NUMBS	8086	17 + ea
		8088	25 + ea
		80286	5
		80386	5
		80486	4
		Pentium–Pentium 4	3

00sss111 Format	Examples	Microprocessor	Clocks
POP seg	POP DS POP ES POP SS	8086	8
		8088	12
		80286	5
		80386	7
		80486	3
		Pentium–Pentium 4	3

00001111 10sss001 Format	Examples	Microprocessor	Clocks
POP seg	POP FS POP GS	8086	—
		8088	—
		80286	—
		80386	7
		80486	3
		Pentium–Pentium 4	3

POPA/POPAD Pop all registers from stack

01100001 Example	Microprocessor	Clocks
POPA POPAD	8086	—
	8088	—
	80286	19
	80386	24
	80486	9
	Pentium–Pentium 4	5

POPF/POPFD — Pop flags from stack

10010000

O	D	I	T	S	Z	A	P	C
*	*	*	*	*	*	*	*	*

Example	Microprocessor	Clocks
POPF POPFD	8086	8
	8088	12
	80286	5
	80386	5
	80486	6
	Pentium–Pentium 4	4 or 6

PUSH — Push data onto stack

01010rrr

Format	Examples	Microprocessor	Clocks
PUSH reg	PUSH CX PUSH AX PUSH EDI	8086	11
		8088	15
		80286	3
		80386	2
		80486	1
		Pentium–Pentium 4	1

11111111 oo110mmm disp

Format	Examples	Microprocessor	Clocks
PUSH mem	PUSH DATA1 PUSH LISTS PUSH NUMBS	8086	16 + ea
		8088	24 + ea
		80286	5
		80386	5
		80486	4
		Pentium–Pentium 4	1 or 2

00ss110 Format	Examples	Microprocessor	Clocks
PUSH seg	PUSH ES PUSH CS PUSH DS	8086	10
		8088	14
		80286	3
		80386	2
		80486	3
		Pentium–Pentium 4	1

00001111 10sss000 Format	Examples	Microprocessor	Clocks
PUSH seg	PUSH FS PUSH GS	8086	—
		8088	—
		80286	—
		80386	2
		80486	3
		Pentium–Pentium 4	1

011010s0 data Format	Examples	Microprocessor	Clocks
PUSH imm	PUSH 2000H PUSH 53220 PUSHW 10H PUSH ',' PUSHD 100000H	8086	—
		8088	—
		80286	3
		80386	2
		80486	1
		Pentium–Pentium 4	1

PUSHA/PUSHAD Push all registers onto stack

01100000

Example	Microprocessor	Clocks
PUSHA PUSHAD	8086	—
	8088	—
	80286	17
	80386	18
	80486	11
	Pentium–Pentium 4	5

PUSHF/PUSHFD Push flags onto stack

10011100

Example	Microprocessor	Clocks
PUSHF PUSHFD	8086	10
	8088	14
	80286	3
	80386	4
	80486	3
	Pentium–Pentium 4	3 or 4

RCL/RCR/ROL/ROR Rotate

1101000w ooTTTmmm disp

O	D	I	T	S	Z	A	P	C
*								*

TTT = 000 = ROL, TTT = 001 = ROR, TTT = 010 = RCL, and TTT = 011 = RCR

Format	Examples	Microprocessor	Clocks
ROL reg,1 ROR reg,1	ROL CL,1 ROL DX,1 ROR CH,1 ROR SI,1	8086	2
		8088	2
		80286	2
		80386	3
		80486	3
		Pentium–Pentium 4	1 or 3

Format	Examples	Microprocessor	Clocks
RCL reg,1 RCR reg,1	RCL CL,1 RCL SI,1 RCR AH,1 RCR EBX,1	8086	2
		8088	2
		80286	2
		80386	9
		80486	3
		Pentium–Pentium 4	1 or 3
ROL mem,1 ROR mem,1	ROL DATAY,1 ROL LIST,1 ROR DATA2[DI],1 ROR BYTE PTR [EAX],1	8086	15 + ea
		8088	23 + ea
		80286	7
		80386	7
		80486	4
		Pentium–Pentium 4	1 or 3
RCL mem,1 RCR mem,1	RCL DATA1,1 RCL LIST,1 RCR DATA2[SI],1 RCR WORD PTR [ESI],1	8086	15 + ea
		8088	23 + ea
		80286	7
		80386	10
		80486	4
		Pentium–Pentium 4	1 or 3

1101001w ooTTTmmm disp

Format	Examples	Microprocessor	Clocks
ROL reg,CL ROR reg,CL	ROL CH,CL ROL DX,CL ROR AL,CL ROR ESI,CL	8086	8 + 4n
		8088	8 + 4n
		80286	5 + n
		80386	3
		80486	3
		Pentium–Pentium 4	4

Format	Examples	Microprocessor	Clocks
RCL reg,CL RCR reg,CL	RCL CH,CL RCL SI,CL RCR AH,CL RCR EBX,CL	8086	8 + 4n
		8088	8 + 4n
		80286	5 + n
		80386	9
		80486	3
		Pentium–Pentium 4	7–27
ROL mem,CL ROR mem,CL	ROL DATAY,CL ROL LIST,CL ROR DATA2[DI],CL ROR BYTE PTR [EAX],CL	8086	20 + 4n
		8088	28 + 4n
		80286	8 + n
		80386	7
		80486	4
		Pentium–Pentium 4	4
RCL mem,CL RCR mem,CL	RCL DATA1,CL RCL LIST,CL RCR DATA2[SI],CL RCR WORD PTR [ESI],CL	8086	20 + 4n
		8088	28 + 4n
		80286	8 + n
		80386	10
		80486	9
		Pentium–Pentium 4	9–26

1100000w ooTTTmmm disp data

Format	Examples	Microprocessor	Clocks
ROL reg,imm ROR reg,imm	ROL CH,4 ROL DX,5 ROR AL,2 ROR ESI,14	8086	—
		8088	—
		80286	5 + n
		80386	3
		80486	2
		Pentium–Pentium 4	1 or 3

RCL reg,imm RCR reg,imm	RCL CL,2 RCL SI,12 RCR AH,5 RCR EBX,18	8086	—
		8088	—
		80286	5 + n
		80386	9
		80486	8
		Pentium–Pentium 4	8–27
ROL mem,imm ROR mem,imm	ROL DATAY,4 ROL LIST,3 ROR DATA2[DI],7 ROR BYTE PTR [EAX],11	8086	—
		8088	—
		80286	8 + n
		80386	7
		80486	4
		Pentium–Pentium 4	1 or 3
RCL mem,imm RCR mem,imm	RCL DATA1,5 RCL LIST,3 RCR DATA2[SI],9 RCR WORD PTR [ESI],8	8086	—
		8088	—
		80286	8 + n
		80386	10
		80486	9
		Pentium–Pentium 4	8–27

RDMSR Read model specific register

00001111 00110010

Example	Microprocessor	Clocks
RDMSR	8086	—
	8088	—
	80286	—
	80386	—
	80486	—
	Pentium–Pentium 4	20–24

REP Repeat prefix

11110011 1010010w Format	Examples	Microprocessor	Clocks
REP MOVS	REP MOVSB REP MOVSW REP MOVSD REP MOVS DATA1,DATA2	8086	9 + 17n
		8088	9 + 25n
		80286	5 + 4n
		80386	8 + 4n
		80486	12 + 3n
		Pentium–Pentium 4	13 + n

11110011 1010101w Format	Examples	Microprocessor	Clocks
REP STOS	REP STOSB REP STOSW REP STOSD REP STOS ARRAY	8086	9 + 10n
		8088	9 + 14n
		80286	4 + 3n
		80386	5 + 5n
		80486	7 + 4n
		Pentium–Pentium 4	9 + n

11110011 0110110w Format	Examples	Microprocessor	Clocks
REP INS	REP INSB REP INSW REP INSD REP INS ARRAY	8086	—
		8088	—
		80286	5 + 4n
		80386	12 + 5n
		80486	17 + 5n
		Pentium–Pentium 4	25 + 3n

11110011 0110111w Format	Examples	Microprocessor	Clocks
REP OUTS	REP OUTSB REP OUTSW REP OUTSD REP OUTS ARRAY	8086	—
		8088	—
		80286	5 + 4n
		80386	12 + 5n
		80486	17 + 5n
		Pentium–Pentium 4	25 + 4n

REPE/REPNE Repeat conditional

11110011 1010011w Format	Examples	Microprocessor	Clocks
REPE CMPS	REPE CMPSB REPE CMPSW REPE CMPSD REPE CMPS DATA1,DATA2	8086	9 + 22n
		8088	9 + 30n
		80286	5 + 9n
		80386	5 + 9n
		80486	7 + 7n
		Pentium–Pentium 4	9 + 4n

11110011 1010111w Format	Examples	Microprocessor	Clocks
REPE SCAS	REPE SCASB REPE SCASW REPE SCASD REPE SCAS ARRAY	8086	9 + 15n
		8088	9 + 19n
		80286	5 + 8n
		80386	5 + 8n
		80486	7 + 5n
		Pentium–Pentium 4	9 + 4n

11110010 1010011w

Format	Examples	Microprocessor	Clocks
REPNE CMPS	REPNE CMPSB REPNE CMPSW REPNE CMPSD REPNE CMPS ARRAY,LIST	8086	9 + 22n
		8088	9 + 30n
		80286	5 + 9n
		80386	5 + 9n
		80486	7 + 7n
		Pentium–Pentium 4	8 + 4n

11110010 101011w

Format	Examples	Microprocessor	Clocks
REPNE SCAS	REPNE SCASB REPNE SCASW REPNE SCASD REPNE SCAS ARRAY	8086	9 + 15n
		8088	9 + 19N
		80286	5 + 8n
		80386	5 + 8n
		80486	7 + 5n
		Pentium–Pentium 4	9 + 4n

RET Return from procedure

11000011

Example	Microprocessor	Clocks
RET (near)	8086	16
	8088	20
	80286	11
	80386	10
	80486	5
	Pentium–Pentium 4	2

11000010 data Format	Examples	Microprocessor	Clocks
RET imm (near)	RET 4 RET 100H	8086	20
		8088	24
		80286	11
		80386	10
		80486	5
		Pentium–Pentium 4	3

11001011 Example	Microprocessor	Clocks
RET (far)	8086	26
	8088	34
	80286	15
	80386	18
	80486	13
	Pentium–Pentium 4	4–23

11001010 data Format	Examples	Microprocessor	Clocks
RET imm (far)	RET 4 RET 100H	8086	25
		8088	33
		80286	11
		80386	10
		80486	5
		Pentium–Pentium 4	4–23

RSM Resume from system management mode

00001111 10101010

O	D	I	T	S	Z	A	P	C
*	*	*	*	*	*	*	*	*

Example	Microprocessor	Clocks
RSM	8086	—
	8088	—
	80286	—
	80386	—
	80486	—
	Pentium–Pentium 4	83

SAHF Store AH into flags

10011110

O	D	I	T	S	Z	A	P	C
				*	*	*	*	*

Example	Microprocessor	Clocks
SAHF	8086	4
	8088	4
	80286	2
	80386	3
	80486	2
	Pentium–Pentium 4	2

SAL/SAR/SHL/SHR Shift

1101000w ooTTTmmm disp

O	D	I	T	S	Z	A	P	C
*				*	*	?	*	*

TTT = 100 = SHL/SAL , TTT = 101 = SHR, and TTT = 111 = SAR

Format	Examples	Microprocessor	Clocks
SAL reg,1	SAL CL,1	8086	2
SHL reg,1	SHL DX,1	8088	2
SHR reg,1	SAR CH,1	80286	2
SAR reg,1	SHR SI,1	80386	3
		80486	3
		Pentium–Pentium 4	1 or 3

SAL mem,1 SHL mem,1 SHR mem,1 SAR mem,1	SAL DATA1,1 SHL BYTE PTR [DI],1 SAR NUMB,1 SHR WORD PTR[EDI],1	8086	15 + ea
		8088	23 + ea
		80286	7
		80386	7
		80486	4
		Pentium–Pentium 4	1 or 3

1101001w ooTTTmmm disp

Format	Examples	Microprocessor	Clocks
SAL reg,CL SHL reg,CL SAR reg,CL SHR reg,CL	SAL CH,CL SHL DX,CL SAR AL,CL SHR ESI,CL	8086	8 + 4n
		8088	8 + 4n
		80286	5 + n
		80386	3
		80486	3
		Pentium–Pentium 4	4
SAL mem,CL SHL mem,CL SAR mem,CL SHR mem,CL	SAL DATAU,CL SHL BYTE PTR [ESI],CL SAR NUMB,CL SHR TEMP,CL	8086	20 + 4n
		8088	28 + 4n
		80286	8 + n
		80386	7
		80486	4
		Pentium–Pentium 4	4

1100000w ooTTTmmm disp data

Format	Examples	Microprocessor	Clocks
SAL reg,imm SHL reg,imm SAR reg,imm SHR reg,imm	SAL CH,4 SHL DX,10 SAR AL,2 SHR ESI,23	8086	—
		8088	—
		80286	5 + n
		80386	3
		80486	2
		Pentium–Pentium 4	1 or 3

Format	Examples	Microprocessor	Clocks
SAL mem,imm SHL mem,imm SAR mem,imm SHR mem,imm	SAL DATAU,3 SHL BYTE PTR [ESI],15 SAR NUMB,3 SHR TEMP,5	8086	—
		8088	—
		80286	8 + n
		80386	7
		80486	4
		Pentium–Pentium 4	1 or 3

SBB Subtract with borrow

000110dw oorrrmmm disp

O	D	I	T	S	Z	A	P	C
*				*	*	*	*	*

Format	Examples	Microprocessor	Clocks
SBB reg,reg	SBB CL,DL SBB AX,DX SBB CH,CL SBB EAX,EBX SBB ESI,EDI	8086	3
		8088	3
		80286	2
		80386	2
		80486	1
		Pentium–Pentium 4	1 or 2
SBB mem,reg	SBB DATAJ,CL SBB BYTES,CX SBB NUMBS,ECX SBB [EAX],CX	8086	16 + ea
		8088	24 + ea
		80286	7
		80386	6
		80486	3
		Pentium–Pentium 4	1 or 3
SBB reg,mem	SBB CL,DATAL SBB CX,BYTES SBB ECX,NUMBS SBB DX,[EBX+EDI]	8086	9 + ea
		8088	13 + ea
		80286	7
		80386	7
		80486	2
		Pentium–Pentium 4	1 or 2

100000sw oo011mmm disp data Format	Examples	Microprocessor	Clocks
SBB reg,imm	SBB CX,3 SBB DI,1AH SBB DL,34H SBB EDX,1345H SBB CX,1834H	8086	4
		8088	4
		80286	3
		80386	2
		80486	1
		Pentium–Pentium 4	1 or 3
SBB mem,imm	SBB DATAS,3 SBB BYTE PTR[EDI],1AH SBB DADDY,34H SBB LIST,'A' SBB TOAD,1834H	8086	17 + ea
		8088	25 + ea
		80286	7
		80386	7
		80486	3
		Pentium–Pentium 4	1 or 3

0001110w data Format	Examples	Microprocessor	Clocks
SBB acc,imm	SBB AX,3 SBB AL,1AH SBB AH,34H SBB EAX,1345H SBB AL,'Y'	8086	4
		8088	4
		80286	3
		80386	2
		80486	1
		Pentium–Pentium 4	1

SCAS Scan string

1010111w

O	D	I	T	S	Z	A	P	C
*				*	*	*	*	*

Format	Examples	Microprocessor	Clocks
SCASB SCASW SCASD	SCASB SCASW SCASD SCAS DATAF REP SCASB	8086	15
		8088	19
		80286	7
		80386	7
		80486	6
		Pentium–Pentium 4	4

SETcondition Conditional set

00001111 1001cccc oo000mmm

Format	Examples	Microprocessor	Clocks
SETcnd reg8	SETA BL SETB CH SETG DL SETE BH SETZ AL	8086	—
		8088	—
		80286	—
		80386	4
		80486	3
		Pentium–Pentium 4	1 or 2
SETcnd mem8	SETE DATAK SETAE LESS_OR_SO	8086	—
		8088	—
		80286	—
		80386	5
		80486	3
		Pentium–Pentium 4	1 or 2

Condition Codes	Mnemonic	Flag	Description
0000	SETO	O = 1	Set if overflow
0001	SETNO	O = 0	Set if no overflow
0010	SETB/SETAE	C = 1	Set if below
0011	SETAE/SETNB	C = 0	Set if above or equal
0100	SETE/SETZ	Z = 1	Set if equal/zero
0101	SETNE/SETNZ	Z = 0	Set if not equal/zero
0110	SETBE/SETNA	C = 1 + Z = 1	Set if below or equal
0111	SETA/SETNBE	C = 0 • Z = 0	Set if above
1000	SETS	S = 1	Set if sign
1001	SETNS	S = 0	Set if no sign
1010	SETP/SETPE	P = 1	Set if parity
1011	SETNP/SETPO	P = 0	Set if no parity
1100	SETL/SETNGE	S • O	Set if less than
1101	SETGE/SETNL	S = 0	Set if greater than or equal
1110	SETLE/SETNG	Z = 1 + S • O	Set if less than or equal
1111	SETG/SETNLE	Z = 0 + S = O	Set if greater than

SGDT/SIDT/SLDT Store descriptor table registers

00001111 00000001 oo000mmm disp

Format	Examples	Microprocessor	Clocks
SGDT mem	SGDT MEMORY SGDT GLOBAL	8086	—
		8088	—
		80286	11
		80386	9
		80486	10
		Pentium–Pentium 4	4

00001111 00000001 oo001mmm disp

Format	Examples	Microprocessor	Clocks
SIDT mem	SIDT DATAS SIDT INTERRUPT	8086	—
		8088	—
		80286	12
		80386	9
		80486	10
		Pentium–Pentium 4	4

00001111 00000000 oo000mmm disp

Format	Examples	Microprocessor	Clocks
SLDT reg	SLDT CX SLDT DX	8086	—
		8088	—
		80286	2
		80386	2
		80486	2
		Pentium–Pentium 4	2
SLDT mem	SLDT NUMBS SLDT LOCALS	8086	—
		8088	—
		80286	3
		80386	2
		80486	3
		Pentium–Pentium 4	2

SHLD/SHRD Double precision shift

00001111 10100100 oorrrmmm disp data

O	D	I	T	S	Z	A	P	C
?				*	*	?	*	*

Format	Examples	Microprocessor	Clocks
SHLD reg,reg,imm	SHLD AX,CX,10 SHLD DX,BX,8 SHLD CX,DX,2	8086	—
		8088	—
		80286	—
		80386	3
		80486	2
		Pentium–Pentium 4	4
SHLD mem,reg,imm	SHLD DATAQ,CX,8	8086	—
		8088	—
		80286	—
		80386	7
		80486	3
		Pentium–Pentium 4	4

00001111 10101100 oorrrmmm disp data

Format	Examples	Microprocessor	Clocks
SHRD reg,reg,imm	SHRD CX,DX,2	8086	—
		8088	—
		80286	—
		80386	3
		80486	2
		Pentium–Pentium 4	4
SHRD mem,reg,imm	SHRD DATAZ,DX,4	8086	—
		8088	—
		80286	—
		80386	7
		80486	2
		Pentium–Pentium 4	4

00001111 10100101 oorrrmmm disp Format	Examples	Microprocessor	Clocks
SHLD reg,reg,CL	SHLD BX,DX,CL	8086	—
		8088	—
		80286	—
		80386	3
		80486	3
		Pentium–Pentium 4	4 or 5
SHLD mem,reg,CL	SHLD DATAZ,DX,CL	8086	—
		8088	—
		80286	—
		80386	7
		80486	3
		Pentium–Pentium 4	4 or 5

00001111 10101101 oorrrmmm disp Format	Examples	Microprocessor	Clocks
SHRD reg,reg,CL	SHRD AX,DX,CL	8086	—
		8088	—
		80286	—
		80386	3
		80486	3
		Pentium–Pentium 4	4 or 5
SHRD mem,reg,CL	SHRD DATAZ,DX,CL	8086	—
		8088	—
		80286	—
		80386	7
		80486	3
		Pentium–Pentium 4	4 or 5

SMSW	Store machine status word (80286)		
00001111 00000001 oo100mmm disp			
Format	Examples	Microprocessor	Clocks
SMSW reg	SMSW AX SMSW DX SMSW BP	8086	—
		8088	—
		80286	2
		80386	10
		80486	2
		Pentium–Pentium 4	4
SMSW mem	SMSW DATAQ	8086	—
		8088	—
		80286	3
		80386	3
		80486	3
		Pentium–Pentium 4	4

STC	Set carry flag	
11111001	O D I T S Z A P C 1	
Example	Microprocessor	Clocks
STC	8086	2
	8088	2
	80286	2
	80386	2
	80486	2
	Pentium–Pentium 4	2

STD Set direction flag

11111101

O	D	I	T	S	Z	A	P	C
	1							

Example	Microprocessor	Clocks
STD	8086	2
	8088	2
	80286	2
	80386	2
	80486	2
	Pentium–Pentium 4	2

STI Set interrupt flag

11111011

O	D	I	T	S	Z	A	P	C
		1						

Example	Microprocessor	Clocks
STI	8086	2
	8088	2
	80286	2
	80386	3
	80486	5
	Pentium–Pentium 4	7

STOS Store string data

1010101w

Format	Examples	Microprocessor	Clocks
STOSB STOSW STOSD	STOSB STOSW STOSD STOS DATA_LIST REP STOSB	8086	11
		8088	15
		80286	3
		80386	40
		80486	5
		Pentium–Pentium 4	3

STR	Store task register		
00001111 00000000 oo001mmm disp			
Format	Examples	Microprocessor	Clocks
STR reg	STR AX STR DX STR BP	8086	—
		8088	—
		80286	2
		80386	2
		80486	2
		Pentium–Pentium 4	2
STR mem	STR DATA3	8086	—
		8088	—
		80286	2
		80386	2
		80486	2
		Pentium–Pentium 4	2

SUB	Subtract		
000101dw oorrrmmm disp		O D I T S Z A P C * * * * * *	
Format	Examples	Microprocessor	Clocks

SUB reg,reg	SUB CL,DL
	SUB AX,DX
	SUB CH,CL
	SUB EAX,EBX
	SUB ESI,EDI
		8086	3

		8088	3
		80286	2
		80386	2
		80486	1
		Pentium–Pentium 4	1 or 2

Format	Examples	Microprocessor	Clocks

SUB mem,reg	SUB DATAJ,CL
	SUB BYTES,CX
	SUB NUMBS,ECX
	SUB [EAX],CX
		8086	16 + ea

		8088	24 + ea
		80286	7
		80386	6
		80486	3
		Pentium–Pentium 4	1 or 3

SUB reg,mem	SUB CL,DATAL
	SUB CX,BYTES
	SUB ECX,NUMBS
	SUB DX,[EBX+EDI]
		8086	9 + ea

		8088	13 + ea
		80286	7
		80386	7
		80486	2
		Pentium–Pentium 4	1 or 2

100000sw oo101mmm disp data

Format	Examples	Microprocessor	Clocks

SUB reg,imm	SUB CX,3
	SUB DI,1AH
	SUB DL,34H
	SUB EDX,1345H
	SUB CX,1834H
		8086	4

		8088	4
		80286	3
		80386	2
		80486	1
		Pentium–Pentium 4	1 or 3

SUB mem,imm	SUB DATAS,3
	SUB BYTE PTR[EDI],1AH
	SUB DADDY,34H
	SUB LIST,'A'
	SUB TOAD,1834H
		8086	17 + ea

		8088	25 + ea
		80286	7
		80386	7
		80486	3
		Pentium–Pentium 4	1 or 3

0010110w data Format	Examples	Microprocessor	Clocks

SUB acc,imm	SUB AL,3
	SUB AX,1AH
	SUB EAX,34H
		8086	4

		8088	4
		80286	3
		80386	2
		80486	1
		Pentium–Pentium 4	1

TEST Test operands (logical compare)

1000001w oorrrmmm disp Format	Examples	O D I T S Z A P C 0 * * ? * 0 Microprocessor	Clocks
TEST reg,reg	TEST CL,DL TEST BX,DX TEST DH,CL TEST EBP,EBX TEST EAX,EDI	8086	5
		8088	5
		80286	2
		80386	2
		80486	1
		Pentium–Pentium 4	1 or 2
TEST mem,reg reg,mem	TEST DATAJ,CL TEST BYTES,CX TEST NUMBS,ECX TEST [EAX],CX TEST CL,POPS	8086	9 + ea
		8088	13 + ea
		80286	6
		80386	5
		80486	2
		Pentium–Pentium 4	1 or 2

1111011sw oo000mmm disp data Format	Examples	Microprocessor	Clocks
TEST reg,imm	TEST BX,3 TEST DI,1AH TEST DH,44H TEST EDX,1AB345H TEST SI,1834H	8086	4
		8088	4
		80286	3
		80386	2
		80486	1
		Pentium–Pentium 4	1 or 2
TEST mem,imm	TEST DATAS,3 TEST BYTE PTR[EDI],1AH TEST DADDY,34H TEST LIST,'A' TEST TOAD,1834H	8086	11 + ea
		8088	11 + ea
		80286	6
		80386	5
		80486	2
		Pentium–Pentium 4	1 or 2

1010100w data Format	Examples	Microprocessor	Clocks
TEST acc,imm	TEST AL,3 TEST AX,1AH TEST EAX,34H	8086	4
		8088	4
		80286	3
		80386	2
		80486	1
		Pentium–Pentium 4	1

VERR/VERW Verify read/write

00001111 00000000 oo100mmm disp

O	D	I	T	S	Z	A	P	C
					*			

Format	Examples	Microprocessor	Clocks
VERR reg	VERR CX VERR DX VERR DI	8086	—
		8088	—
		80286	14
		80386	10
		80486	11
		Pentium–Pentium 4	7
VERR mem	VERR DATAJ VERR TESTB	8086	—
		8088	—
		80286	16
		80386	11
		80486	11
		Pentium–Pentium 4	7

00001111 00000000 oo101mmm disp

Format	Examples	Microprocessor	Clocks
VERW reg	VERW CX VERW DX VERW DI	8086	—
		8088	—
		80286	14
		80386	15
		80486	11
		Pentium–Pentium 4	7
VERW mem	VERW DATAJ VERW TESTB	8086	—
		8088	—
		80286	16
		80386	16
		80486	11
		Pentium–Pentium 4	7

WAIT Wait for coprocessor

10011011

Example	Microprocessor	Clocks
WAIT FWAIT	8086	4
	8088	4
	80286	3
	80386	6
	80486	6
	Pentium–Pentium 4	1

WBINVD Write-back cache invalidate data cache

00001111 00001001

Example	Microprocessor	Clocks
WBINVD	8086	—
	8088	—
	80286	—
	80386	—
	80486	5
	Pentium–Pentium 4	2000+

WRMSR Write to model specific register

00001111 00110000

Example	Microprocessor	Clocks
WRMSR	8086	—
	8088	—
	80286	—
	80386	—
	80486	—
	Pentium–Pentium 4	30–45

XADD Exchange and add

00001111 1100000w 11rrrrrr

O	D	I	T	S	Z	A	P	C
*				*	*	*	*	*

Format	Examples	Microprocessor	Clocks
XADD reg,reg	XADD EBX,ECX XADD EDX,EAX XADD EDI,EBP	8086	—
		8088	—
		80286	—
		80386	—
		80486	3
		Pentium–Pentium 4	3 or 4

00001111 1100000w oorrrmmm disp

Format	Examples	Microprocessor	Clocks
XADD mem,reg	XADD DATA5,ECX XADD [EBX],EAX XADD [ECX+4],EBP	8086	—
		8088	—
		80286	—
		80386	—
		80486	4
		Pentium–Pentium 4	3 or 4

XCHG Exchange

1000011w oorrrmmm

Format	Examples	Microprocessor	Clocks
XCHG reg,reg	XCHG CL,DL XCHG BX,DX XCHG DH,CL XCHG EBP,EBX XCHG EAX,EDI	8086	4
		8088	4
		80286	3
		80386	3
		80486	3
		Pentium–Pentium 4	3

Format	Examples	Microprocessor	Clocks
XCHG mem,reg reg,mem	XCHG DATAJ,CL XCHG BYTES,CX XCHG NUMBS,ECX XCHG [EAX],CX XCHG CL,POPS	8086	17 + ea
		8088	25 + ea
		80286	5
		80386	5
		80486	5
		Pentium–Pentium 4	3

10010reg

Format	Examples	Microprocessor	Clocks
XCHG acc,reg reg,acc	XCHG BX,AX XCHG AX,DI XCHG DH,AL XCHG EDX,EAX XCHG SI,AX	8086	3
		8088	3
		80286	3
		80386	3
		80486	3
		Pentium–Pentium 4	2

XLAT Translate

11010111

Example	Microprocessor	Clocks
XLAT	8086	11
	8088	11
	80286	5
	80386	3
	80486	4
	Pentium–Pentium 4	4

XOR Exclusive-OR

000110dw oorrrmmm disp

O	D	I	T	S	Z	A	P	C
0				*	*	?	*	0

Format	Examples	Microprocessor	Clocks
XOR reg,reg	XOR CL,DL XOR AX,DX XOR CH,CL XOR EAX,EBX XOR ESI,EDI	8086	3
		8088	3
		80286	2
		80386	2
		80486	1
		Pentium–Pentium 4	1 or 2
XOR mem,reg	XOR DATAJ,CL XOR BYTES,CX XOR NUMBS,ECX XOR [EAX],CX	8086	16 + ea
		8088	24 + ea
		80286	7
		80386	6
		80486	3
		Pentium–Pentium 4	1 or 3
XOR reg,mem	XOR CL,DATAL XOR CX,BYTES XOR ECX,NUMBS XOR DX,[EBX+EDI]	8086	9 + ea
		8088	13 + ea
		80286	7
		80386	7
		80486	2
		Pentium–Pentium 4	1 or 2

100000sw oo110mmm disp data

Format	Examples	Microprocessor	Clocks
XOR reg,imm	XOR CX,3 XOR DI,1AH XOR DL,34H XOR EDX,1345H XOR CX,1834H	8086	4
		8088	4
		80286	3
		80386	2
		80486	1
		Pentium–Pentium 4	1 or 3

Format	Examples	Microprocessor	Clocks
XOR mem,imm	XOR DATAS,3 XOR BYTE PTR[EDI],1AH XOR DADDY,34H XOR LIST,'A' XOR TOAD,1834H	8086	17 + ea
		8088	25 + ea
		80286	7
		80386	7
		80486	3
		Pentium–Pentium 4	1 or 3

0010101w data

Format	Examples	Microprocessor	Clocks
XOR acc,imm	XOR AL,3 XOR AX,1AH XOR EAX,34H	8086	4
		8088	4
		80286	3
		80386	2
		80486	1
		Pentium–Pentium 4	1

SIMD INSTRUCTION SET SUMMARY

The SIMD (single-instruction, multiple data) instructions add a new dimension to the use of the microprocessor for performing multimedia and other operations. The XMM registers are numbered from XMM_0 to XMM_7 and are each 128 bits in width. Data formats stored in the XMM registers and used by the SIMD instructions appear in Figure B–1.

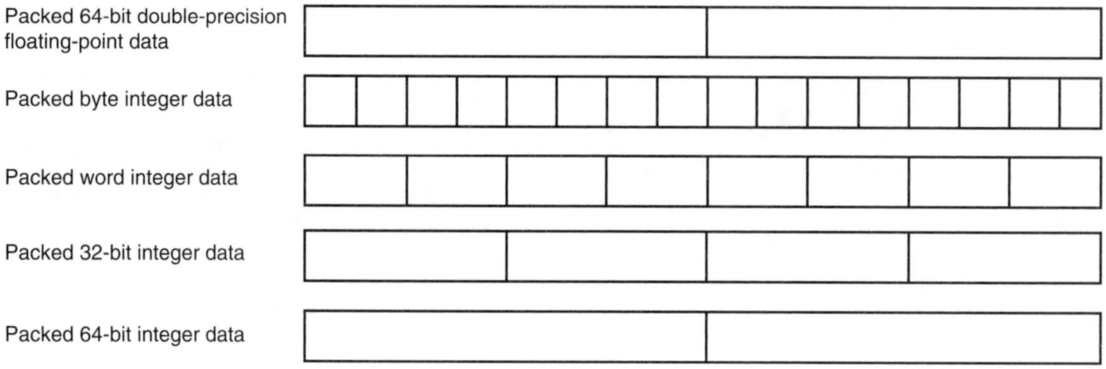

Packed 64-bit double-precision floating-point data

Packed byte integer data

Packed word integer data

Packed 32-bit integer data

Packed 64-bit integer data

FIGURE B–1 Data formats for the 128-bit-wide XMM registers in the Pentium III and Pentium 4 microprocessors.

Data stored in the memory must be stored as 16-byte-long data in a series of memory locations accessed by using the OWORD PTR override when addressed by an instruction. The OWORD PTR override is used to address an octalword of data or 16 bytes. The SIMD instructions allow operations on packed and scalar double-precision floating-point numbers. The operation of both forms is illustrated in Figure B–2, which shows both packed and scalar multiplication. Notice that scalar only copies the leftmost double-precision number into the destination register and does not use the leftmost number in the source. The scalar instruction are meant to be compatible with the floating-point coprocessor instructions.

This section of the appendix details many of the SIMD instructions and provides examples of their usage.

FIGURE B–2 Packed and scalar double-precision floating-point operation.

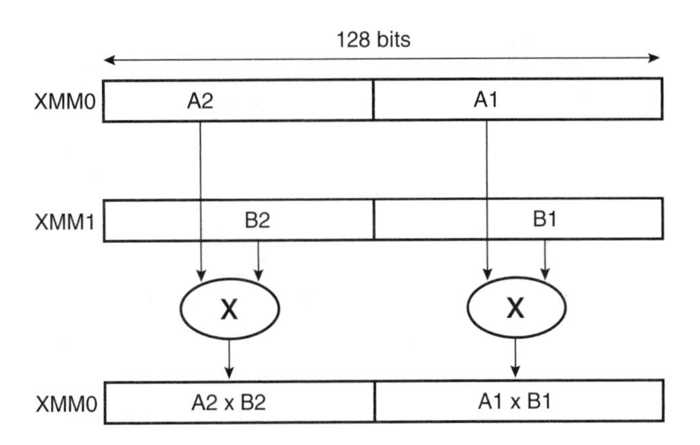

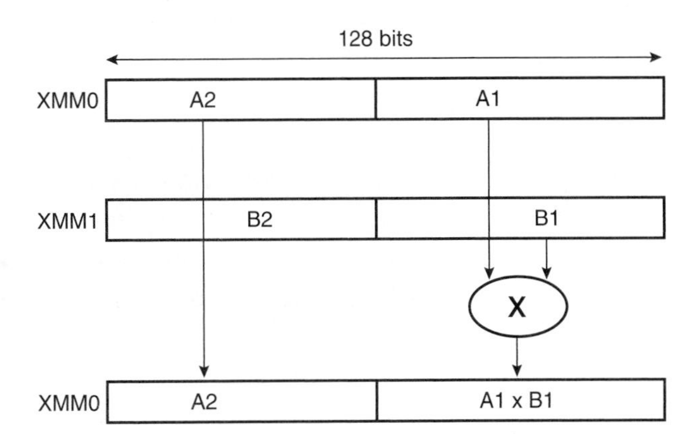

DATA MOVEMENT INSTRUCTIONS

MOVAPD Move aligned packed double-precision data, data must be aligned on 16-byte boundaries

Examples

```
MOVAPD  XMM0, OWORD DATA3        ;copies DATA3 to XMM0
MOVAPD  OWORD PTR DATA4, XMM2 ;copies XMM4 to DATA4
```

MOVUPD Move unaligned packed double-precision data

Examples

```
MOVUPD  XMM0, OWORD DATA3        ;copies DATA3 to XMM0
MOVUPD  OWORD PTR DATA4, XMM2 ;copies XMM4 to DATA4
```

MOVSD Move scalar packed double-precision data to low quadword

Examples

```
MOVSD  XMM0, DWORD DATA3         ;copies DATA3 to XMM0
MOVSD  DWORD PTR DATA4, XMM2  ;copies XMM4 to DATA4
```

MOVHPD Move packed double-precision data to high quadword

Examples

```
MOVHPD  XMM0, DWORD DATA3        ;copies DATA3 to XMM0
MOVHPD  DWORD PTR DATA4, XMM2 ;copies XMM4 to DATA4
```

MOVLPD Move packed double-precision data into low quadword

Examples

```
MOVLPD  XMM0, DWORD DATA3        ;copies DATA3 to XMM0
MOVLPD  DWORD PTR DATA4, XMM2 ;copies XMM4 to DATA4
```

MOVMSKPD Move packed double-precision mask

Examples

```
MOVMSKPD    EAX, XMM1    ;copies 2 sign bits to general-purpose register
```

MOVAPS Move 4 aligned packed single-precision data, data must be aligned on 16-byte boundaries

Examples

```
MOVAPS  XMM0, OWORD DATA3        ;copies DATA3 to XMM0
MOVAPS  OWORD PTR DATA4, XMM2 ;copies XMM4 to DATA4
```

MOVUPS Move 4 unaligned packed single-precision data

Examples

```
MOVUPS  XMM0, OWORD DATA3        ;copies DATA3 to XMM0
MOVUPS  OWORD PTR DATA4, XMM2 ;copies XMM4 to DATA4
```

MOVLPS Move 2 packed single-precision numbers to low-order quadword

Examples

```
MOVLPS  XMM0, OWORD DATA3        ;copies DATA3 to XMM0
MOVLPS  OWORD PTR DATA4, XMM2 ;copies XMM4 to DATA4
```

MOVHPS Move packed single-precision numbers to high-order quadword

Examples

```
MOVHPS  XMM0, OWORD DATA3        ;copies DATA3 to XMM0
MOVHPS  OWORD PTR DATA4, XMM2 ;copies XMM4 to DATA4
```

MOVAPD Move aligned packed double-precision data, data must be aligned on 16-byte boundaries

Examples

```
MOVAPD  XMM0, OWORD DATA3        ;copies DATA3 to XMM0
MOVAPD  OWORD PTR DATA4, XMM2 ;copies XMM4 to DATA4
```

MOVLHPS Move 2 packed single-precision numbers from the low-order quadword to the high-order quadword

Examples

```
MOVLHPS  XMM0, XMM1          ;copies XMM1 low to XMM0 high
MOVLHPS  XMM3, XMM2          ;copies XMM2 low to XMM3 high
```

MOVHLPS	Move 2 packed single-precision numbers from high-order quadword to low-order quadword

Examples

```
MOVHLPS  XMM0, XMM2          ;copies high XMM2 to low XMM0
MOVHLPS  XMM4, XMM5          ;copies high XMM5 to low XMM4
```

MOVMSKPS	Move 4-sign bits of 4 packed single-precision numbers to general-purpose register

Examples

```
MOVMSKPS  EBX, XMM0          ;copies sign bits of XMM0 to EBX
MOVMSKPS  EDX, XMM2          ;copies sign bits of XMM2 to EDX
```

ARITHMETIC INSTRUCTIONS

ADDPD	Adds packed double-precision data

Examples

```
ADDPD    XMM0, OWORD DATA3    ;adds DATA3 to XMM0
ADDPD    XMM2, XMM3           ;adds XMM3 to XMM2
```

ADDSD	Adds scalar double-precision data

Examples

```
ADDSD    XMM0, OWORD DATA3    ;adds DATA3 to XMM0
ADDSD    XMM4, XMM2           ;adds XMM2 to XMM4
```

ADDPS	Adds 2 packed single-precision numbers

Examples

```
ADDPS    XMM0, QWORD DATA3    ;adds DATA3 to XMM0
ADDPS    XMM3, XMM2           ;adds XMM2 to XMM3
```

ADDLS	Adds scalar single-precision data	
Examples		
ADDLS	XMM0, DWORD DATA3	;adds DATA3 to XMM0
ADDLS	XMM7, XMM2	;adds XMM2 to XMM7

SUBPD	Subtracts packed double-precision data	
Examples		
SUBPD	XMM0, OWORD DATA3	;subtracts DATA3 from XMM0
SUBPD	XMM2, XMM3	;subtracts XMM3 from XMM2

SUBSD	Subtracts scalar double-precision data	
Examples		
SUBSD	XMM0, OWORD DATA3	;subtracts DATA3 from XMM0
SUBSD	XMM4, XMM2	;subtracts XMM2 from XMM4

SUBPS	Subtracts 2 packed single-precision numbers	
Examples		
SUBPS	XMM0, QWORD DATA3	;subtracts DATA3 from XMM0
SUBPS	XMM3, XMM2	;subtracts XMM2 from XMM3

SUBLS	Subtracts scalar single-precision data	
Examples		
SUBLS	XMM0, DWORD DATA3	;subtracts DATA3 from XMM0
SUBLS	XMM7, XMM2	;subtracts XMM2 from XMM7

MULPD	Multiplies packed double-precision data	
Examples		
MULPD	XMM0, OWORD DATA3	;multiplies DATA3 times XMM0
MULPD	XMM3, XMM2	;multiplies XMM2 times XMM3

MULSD	Multiplies scalar double-precision data	
Examples		
MULSD	XMM0, OWORD DATA3	;multiplies DATA3 times XMM0
MULSD	XMM3, XMM6	;multiplies XMM6 times XMM3

MULPS	Multiplies 2 packed single-precision numbers	
Examples		
MULPS	XMM0, QWORD DATA3	;multiplies DATA3 times XMM0
MULPS	XMM0, XMM2	;multiplies XMM2 times XMM0

MULSS	Multiplies a single-precision number	
Examples		
MULSS	XMM0, DWORD DATA3	;multiplies DATA3 times XMM0
MULSS	XMM1, XMM2	;multiplies XMM2 times XMM1

DIVPD	Divides packed double-precision data	
Examples		
DIVPD	XMM0, OWORD DATA3	;divides XMM0 by DATA3
DIVPD	XMM3, XMM2	;divides XMM3 by XMM2

DIVSD	Divides scalar double-precision data	
Examples		
DIVSD	XMM0, OWORD DATA3	;divides XMM0 by DATA3
DIVSD	XMM3, XMM6	;divides XMM3 by XMM6

DIVPS	Divides 2 packed single-precision numbers	
Examples		
DIVPS	XMM0, QWORD DATA3	;divides XMM0 by DATA3
DIVPS	XMM0, XMM2	;divides XMM0 by XMM2

DIVSS	Divides a single-precision number	
Examples		
DIVSS	XMM0, DWORD DATA3	;divides XMM0 by DATA3
DIVSS	XMM1, XMM2	;divides XMM1 by XMM2

SQRTPD	Finds the square root of packed double-precision data	
Examples		
SQRTPD	XMM0, OWORD DATA3	;finds square root of DATA3, result to XMM0
SQRTPD	XMM3, XMM2	;finds square root of XMM2, result to XMM3

SQRTSD	Finds the square root of scalar double-precision data	
Examples		
SQRTSD	XMM0, OWORD DATA3	;finds square root of DATA3, result to XMM0
SQRTSD	XMM3, XMM6	;finds square root of XMM6, result to XMM3

SQRTPS	Finds the square root of 2 packed single-precision numbers	
Examples		
SQRTPS	XMM0, QWORD DATA3	;finds square root of DATA3, result to XMM0
SQRTPS	XMM0, XMM2	;finds square root of XMM2, result to XMM0

SQRTSS	Finds the square root of a single-precision number	
Examples		
SQRTSS	XMM0, DWORD DATA3	;finds the square root of DATA3, result to XMM0
SQRTSS	XMM1, XMM2	;finds the square root of XMM2, result to XMM1

RCPPS	Finds the reciprocal of a packed single-precision number	
Examples		
RCPPS	XMM0, OWORD DATA3	;finds the reciprocal of DATA3, result to XMM0
RCPPS	XMM3, XMM2	;finds the reciprocal of XMM2, result to XMM3

RCPSS	Finds the reciprocal of a single-precision number	
Examples		
RCPSS	XMM0, OWORD DATA3	;finds the reciprocal of DATA3, result to XMM0
RCPSS	XMM3, XMM6	;finds the reciprocal of XMM6, result to XMM3

RSQRTPS	Finds reciprocals of packed single-precision data	
Examples		
RSQRTPS	XMM0, OWORD DATA3	;finds reciprocal of square root of DATA3
RSQRTPS	XMM3, XMM2	;finds reciprocal of square root of XMM2

RSQRTSS	Finds the reciprocal of square root of a scalar single-precision number	
Examples		
RSQRTSS	XMM0, OWORD DATA3	;finds reciprocal of square root of DATA3
RSQRTSS	XMM3, XMM6	;finds reciprocal of square root of XMM6

MAXPD	Compares and returns the maximum packed double-precision floating-point number	
Examples		
MAXPD	XMM0, OWORD DATA3	;compares numbers in DATA3, largest to XMM0
MAXPD	XMM3, XMM2	;compares numbers in XMM2, largest to XMM3

MAXSD	Compares scalar double-precision data and returns the largest	
Examples		
MAXSD	XMM0, OWORD DATA3	;compares numbers in DATA3, largest to XMM0
MAXSD	XMM3, XMM6	;compares numbers in XMM6, largest to XMM3

MAXPS	Compares and returns the largest packed single-precision number	
Examples		
MAXPS	XMM0, QWORD DATA3	;compares numbers in DATA3, largest to XMM0
MAXPS	XMM0, XMM2	;compares numbers in XMM2, largest to XMM0

MAXSS Compares scalar single-precision numbers and returns the largest

Examples

```
MAXSS  XMM0, DWORD DATA3 ;compares numbers in DATA3, largest to XMM0
MAXSS  XMM1, XMM2        ;compares numbers in XMM2, largest to XMM1
```

MINPD Compares and returns the minimum packed double-precision floating-point number

Examples

```
MINPD  XMM0, OWORD DATA3 ;compares numbers in DATA3, least to XMM0
MINPD  XMM3, XMM2        ;compares numbers in XMM2, least to XMM3
```

MINSD Compares scalar double-precision data and returns the smallest

Examples

```
MINSD  XMM0, OWORD DATA3 ;compares numbers in DATA3, least to XMM0
MINSD  XMM3, XMM6        ;compares numbers in XMM6, least to XMM3
```

MINPS Compares and returns the smallest packed single-precision number

Examples

```
MINPS  XMM0, QWORD DATA3 ;compares numbers in DATA3, least to XMM0
MINPS  XMM0, XMM2        ;compares numbers in XMM2, least to XMM0
```

MINSS Compares scalar single-precision numbers and returns the smallest

Examples

```
MINSS  XMM0, DWORD DATA3 ;compares numbers in DATA3, least to XMM0
MINSS  XMM1, XMM2        ;compares numbers in XMM2, least to XMM1
```

LOGIC INSTRUCTIONS

ANDPD	ANDs packed double-precision data

Examples

```
ANDPD    XMM0, OWORD DATA3    ;ands DATA3 to XMM0
ANDPD    XMM2, XMM3           ;ands XMM3 to XMM2
```

ANDNPD	NANDs packed double-precision data

Examples

```
ANDNPD   XMM0, OWORD DATA3    ;Nands DATA3 to XMM0
ANDNPD   XMM4, XMM2           ;Nands XMM2 to XMM4
```

ANDPS	ANDs 2 packed single-precision data

Examples

```
ANDPS    XMM0, QWORD DATA3    ;ands DATA3 to XMM0
ANDPS    XMM3, XMM2           ;ands XMM2 to XMM3
```

ANDNPS	NANDs 2 packed single-precision data

Examples

```
ANDNPS   XMM0, DWORD DATA3    ;Nands DATA3 to XMM0
ANDNPS   XMM7, XMM2           ;Nands XMM2 to XMM7
```

ORPD	ORs packed double-precision data

Examples

```
ORPD     XMM0, OWORD DATA3    ;ors DATA3 to XMM0
ORPD     XMM2, XMM3           ;ors XMM3 to XMM2
```

ORPS	ORs 2 packed single-precision numbers

Examples

```
ORPS     XMM0, OWORD DATA3    ;ors DATA3 to XMM0
ORPS     XMM3, XMM2           ;ors XMM2 to XMM3
```

XORPD	Exclusive-ORs packed double-precision data

Examples

```
XORPD    XMM0, OWORD DATA3    ;exclusive-ors DATA3 to XMM0
XORPD    XMM2, XMM3           ;exclusive-ors XMM3 to XMM2
```

XORPS	Exclusive-ORs packed double-precision data

Examples

```
XORPS    XMM0, OWORD DATA3    ;exclusive-ors DATA3 to XMM0
XORPS    XMM2, XMM3           ;exclusive-ors XMM3 to XMM2
```

COMPARISON INSTRUCTIONS

CMPPD	Compares packed double-precision numbers

Examples

```
CMPPD    XMM0, OWORD DATA3    ;compares DATA3 with XMM0
CMPPD    XMM2, XMM3           ;compares XMM3 with XMM2
```

CMPSD	Compares scalar double-precision data

Examples

```
CMPSD    XMM0, QWORD DATA3    ;compares DATA3 with XMM0
CMPSD    XMM3, XMM2           ;compares XMM2 with XMM3
```

CMPISD	Compares scalar double-precision data and sets EFAGS

Examples

```
CMPISD   XMM0, OWORD DATA3    ;compares DATA3 with XMM0
CMPISD   XMM2, XMM3           ;compares XMM3 with XMM2
```

UCOMISD	Compares scalar unordered double-precision numbers and changes EFLAGS

Examples

```
UCOMISD   XMM0, QWORD DATA3    ;compares DATA3 with XMM0
UCOMISD   XMM3, XMM2           ;compares XMM2 with XMM3
```

CMPPS	Compares packed single-precision data

Examples

```
CMPPS     XMM0, OWORD DATA3    ;compares DATA3 with XMM0
CMPPS     XMM2, XMM3           ;compares XMM3 with XMM2
```

CMPSS	Compares 2 packed single-precision numbers

Examples

```
CMPSS     XMM0, QWORD DATA3    ;compares DATA3 with XMM0
CMPSS     XMM3, XMM2           ;compares XMM2 with XMM3
```

COMISS	Compares scalar single-precision data and changes EFLAGS

Examples

```
COMISS    XMM0, OWORD DATA3    ;compares DATA3 with XMM0
COMISS    XMM2, XMM3           ;compares XMM3 with XMM2
```

UCOMISS	Compares unordered single-precision numbers and changes EFLAGS

Examples

```
UCOMISS   XMM0, QWORD DATA3    ;compares DATA3 with XMM0
UCOMISS   XMM3, XMM2           ;compares XMM2 with XMM3
```

DATA CONVERSION INSTRUCTIONS

SHUFPD	Shuffles packed double-precision numbers

Examples

```
SHUFPD    XMM0, OWORD DATA3    ;shuffles DATA3 with XMM0
SHUFPD    XMM2, XMM2           ;swaps upper and lower quadword in XMM2
```

UNPCKHPD	Unpacks the upper double-precision number

Examples

```
UNPCKHPD  XMM0, OWORD DATA3    ;unpacks DATA3 into XMM0
UNPCKHPD  XXM3, XMM2           ;unpacks XMM2 into XMM3
```

UNPCKLPD	Unpacks the lower double-precision number

Examples

```
UNPCKLPD  XMM0, OWORD DATA3    ;unpacks DATA3 into XMM0
UNPCKLPD  XMM3, XMM2           ;unpacks XMM2 into XMM3
```

SHUFPS	Shuffles packed single-precision numbers

Examples

```
SHUFPS    XMM0, QWORD DATA3    ;shuffles DATA3 with XMM0
SHUFPS    XMM2, XMM2           ;swaps upper and lower quadword in XMM2
```

UNPCKHPS	Unpacks the lower double-precision number

Examples

```
UNPCKHPS  XMM0, QWORD DATA3    ;unpacks DATA3 into XMM0
UNPCKHPS  XMM3, XMM2           ;unpacks XMM2 into XMM3
```

UNPCKLPSD	Unpacks the lower double-precision number

Examples

```
UNPCKLPSD XMM0, QWORD DATA3    ;unpacks DATA3 into XMM0
UNPCKLPSD XXM3, XMM2           ;unpacks XMM2 into XMM3
```

APPENDIX C

Flag-Bit Changes

This appendix shows only the instructions that actually change the flag bits. Any instruction not listed does not affect any of the flag bits.

Instruction	*O*	*D*	*I*	*T*	*S*	*Z*	*A*	*P*	*C*
AAA	?				?	?	*	?	*
AAD	?				*	*	?	*	?
AAM	?				*	*	?	*	?
AAS	?				?	?	*	?	*
ADC	*				*	*	*	*	*
ADD	*				*	*	*	*	*
AND	0				*	*	?	*	0
ARPL						*			
BSF						*			
BSR						*			
BT									*
BTC									*
BTR									*
BTS									*
CLC									0
CLD		0							
CLI			0						
CMC									*
CMP	*				*	*	*	*	*
CMPS	*				*	*	*	*	*
CMPXCHG	*				*	*	*	*	*
CMPXCHG8B						*			
DAA	?				*	*	*	*	*
DAS	?				*	*	*	*	*
DEC	*				*	*	*	*	
DIV	?				?	?	?	?	?
IDIV	?				?	?	?	?	?
IMUL	*				?	?	?	?	*
INC	*				*	*	*	*	

Instruction	*O*	*D*	*I*	*T*	*S*	*Z*	*A*	*P*	*C*
IRET	*	*	*	*	*	*	*	*	*
LAR						*			
LSL						*			
MUL	*				?	?	?	?	*
NEG	*				*	*	*	*	*
OR	0				*	*	?	*	0
POPF	*	*	*	*	*	*	*	*	*
RCL/RCR	*								*
REPE/REPNE						*			
ROL/ROR	*								*
SAHF					*	*	*	*	*
SAL/SAR	*				*	*	?	*	*
SHL/SHR	*				*	*	?	*	*
SBB	*				*	*	*	*	*
SCAS	*				*	*	*	*	*
SHLD/SHRD	?				*	*	?	*	*
STC									1
STD		1							
STI			1						
SUB	*				*	*	*	*	*
TEST	0				*	*	?	*	0
VERR/VERW						*			
XADD	*				*	*	*	*	*
XOR	0				*	*	?	*	0

APPENDIX D

Answers to Selected Even-Numbered Questions and Problems

CHAPTER 1

2. Herman Hollerith
4. Konrad Zuse
6. ENIAC
8. Augusta Ada Byron
10. A machine that stores the instructions of a program in the memory system.
12. 200 million
14. 16M bytes
16. 1993
18. 2000
20. Millions of instructions per second
22. A binary bit stores a 1 or a 0.
24. 1024K
26. 1,000,000
28. 2G bytes
30. 2G bytes
32. 4G bytes or if 36-bit addressing is enabled, 64G bytes
34. XMS or extended memory system
36. The disk operating system
38. A now-defunct 32-bit-wide bus designed mainly for video and hard disk drives.
40. Universal serial bus
42. Extended memory system
44. 64K bytes
46. See Figure 1–6.
48. Address, control, and data buses
50. $\overline{MRDC}$
52. Memory read operation
54. (a) 8-bit signed number (b) 16-bit signed number (c) 32-bit signed number (d) 32-bit floating-point number (e) 64-bit floating-point number
56. (a) 156.625 (b) 18.375 (c) 4087.109375 (d) 83.578125 (e) 58.90625
58. (a) 10111_2, 27_8, and 17_{16} (b) 1101011_2, 153_8, and 6B (c) 10011010110_2, 2326_8, and $4D6_{16}$ (d) 1011100_2, 134_8, and $5C_{16}$ (e) 10101101_2, 255_8, and AD
60. (a) 0010 0011 (b) 1010 1101 0100 (c) 0011 0100. 1010 1101 (d) 1011 1101 0011 0010 (e) 0010 0011 0100. 0011
62. (a) 0111 0111 (b) 1010 0101 (c) 1000 1000 (d) 0111 1111
64. Byte is an 8-bit binary number, word is a 16-bit binary number, doubleword is a 32-bit binary number.
66. Enter is a 0DH and it is used to return the cursor/print head to the left margin of the screen or page of paper.
68. LINE1 DB ‘What time is it?’
70. (a) 0000 0011 1110 1000 (b) 1111 1111 1000 1000 (c) 0000 0011 0010 0000 (d) 1111 0011 0111 0100
72. char Fred1 = –34
74. Little endian numbers are stored so the least significant portion is in the lowest numbered memory location and big endian numbers are stored so the most significant part is stored in the lowest numbered memory location.
76. (a) packed = 00000001 00000010 and unpacked 00000001 00000000 00000010 (b) packed = 01000100 and unpacked 00000100 00000100 (c) packed = 00000011 00000001 and unpacked 00000011 00000000 00000001 (d) packed = 00010000 00000000 and unpacked 00000001 00000000 00000000 00000000
78. (a) 89 (b) 9 (c) 32 (d) 1
80. (a) +3.5 (b) –1.0 (c) +12.5

CHAPTER 2

2. 16
4. EBX

6. Holds the offset address of the next step in the program.
8. No, if you add +1 and –1 you have zero, which is a valid number.
10. The I-flag
12. The segment register addresses the lowest address in a 64K memory segment.
14. (a) 12000H (b) 21000H (c) 24A00H (d) 25000H (e) 3F12DH
16. DI
18. SS plus either SP or ESP
20. (a) 12000H (b) 21002H (c) 26200H (d) A1000H (e) 2CA00H
22. All 16M bytes
24. The segment register is a selector that selects the descriptor from a descriptor table. It also sets the privilege level of the request and chooses either the global or local table.
26. A00000H–A01000H
28. 00280000H–00290FFFH
30. 3
32. 64K
34.

0000 0011	1101 0000
1001 0010	0000 0000
0000 0000	0000 0000
0010 1111	1111 1111

36. Through a descriptor stored in the global table
38. The program invisible registers are the cache portions of the segment registers and also the GDTR, LDTR, and IDTR registers.
40. 4K
42. 1024
44. Entry zero or the first entry
46. The TLB caches the most recent memory accesses through the paging mechanism.

CHAPTER 3

2. AL, AH, BL, BH, CL, CH, DL, and DH
4. EAX, EBX, ECX, EDX, ESP, EBP, EDI, and ESI
6. You may not specify mixed register sizes.
8. (a) MOV EDX,EBX (b) MOV CL,BL (c) MOV BX,SI (d) MOV AX,DS (e) MOV AH,AL
10. #
12. .CODE
14. Opcode
16. The .EXIT statement returns control to DOS at the end of a program.
18. The .Startup directive indicates the start of a program and loads the DS register with the location of the data segment as well as the SS and SP registers with the location of the stack.
20. The [ ] symbols indicate indirect addressing.
22. Memory-to-memory data transfers are not allowed.
24. MOV WORD PTR DATA1, 5
26. The MOV BX,DATA instruction copies the word from memory location data into the BX register where the MOV BX,OFFSET DATA instruction copies the offset address of DATA into BX.
28. Nothing wrong with the instruction; it just uses an alternative addressing style.
30. (a) 11750H (b) 11950H (c) 11700H
32. BP or as an extended version EBP
34.

```
FIELDS      STRUC
F1          DW   ?
F2          DW   ?
F3          DW   ?
F4          DW   ?
F5          DW   ?
FIELDS      ENDS
```

36. Direct, relative, and indirect
38. The intersegment jump allows jumps between segments or to anywhere in the memory system while the intrasegment jump allows a jump to any location within the current code segment.
40. 32
42. Short
44. JMP BX
46. 2
48. AX, CX, DX, BX, SP, BP, DI and SI in the same order as listed
50. PUSHFD

CHAPTER 4

2. The D-bit indicates the direction of flow for the data (REG to R/M or R/M to REG) and the W-bit indicates the size of the data (byte or word/doubleword).
4. DL
6. DS:[BX+DI]
8. MOV AL,[BX]
10. 8B 77 02
12. The instruction will assemble, but if you only change the code segment value without changing the instruction pointer value, the outcome will cause problems in most cases.
14. 16
16. All the 16-bit registers are moved to the stack with the PUSHA instruction.
18. (a) AX is copied to the stack. (b) A 32-bit number is retrieved from the stack and placed into ESI. (c) The word contents of the data segment memory location addressed by BX is pushed onto the stack.

(d) EFLAGS are pushed onto the stack. (e) A word is retrieved from the stack and placed into DS. (f) A 32-bit number 4 is pushed onto the stack.

20. Bits 24–31 of EAX are stored in location 020FFH, bits 16–23 of EAX are stored into location 020FEH, bits 8–15 of EAX are stored into location 020FDH, and bits 0–7 of EAX are stored into location 020FCH. SP is then decremented by 4 to a value of 00FCH.
22. There are many possible locations, but SP = 0200H and SS = 0200H is one of them.
24. Both instruction load the address of NUMB into DI. The difference is that MOV DI,OFFSET NUMB assembles as a move immediate and LEA DI,NUMB assembles as an LEA instruction.
26. The LDS BX,NUMB instruction loads BX with the word stored at data segment memory location NUMB and DS is loaded from the data segment memory location addressed by NUMB+2.
28.
```
MOV   BX,NUMB
MOV   DX,BX
MOV   SI,DX
```
30. CLD clears the direction flag and STD sets the direction flag.
32. The LODSB instruction copies a byte of data from the data segment memory location addressed by SI into the AL register and then increments SI by one if the direction flag is cleared.
34. The OUTSB instruction sends the contents of the data segment memory location addressed by SI to the I/O port addressed by DX, then SI is incremented by one if the direction flag is cleared.
36.
```
MOV   SI,OFFSET SOURCE
MOV   DI,OFFSET DEST
MOV   CX,12
REP   MOVSB
```
38. XCHG EBX,ESI
40. The XLAT instruction adds the contents of AL to BX to form a data segment memory address whose content is copied into AL.
42. The contents of I/O port 12H are copied into AL.
44. A segment override prefix is a one-byte instruction that is added to the front of almost any instruction to change the default segment addressed by the instruction.
46.
```
XCHG  AX,BX
XCHG  ECX,EDX
XCHG  SI,DI
```
48. An assembly language directive is a special command to the assembler that may or may not generate code or data for the memory.
50. LIST1 DB 30 DUP(?)
52. The .686 direction informs the assembler to use the instruction set for the Pentium Pro through the Pentium 4.
54. Models
56. The program exits to DOS.
58. The USES statement allows the programmer to automatically push and pop registers in procedures.
60.
```
STORE  PROC    NEAR
       MOV     [DI],AL
       MOV     [DI+1],AL
       MOV     [DI+2],AL
       MOV     [DI+3],AL
       RET
STORE  ENDP
```

CHAPTER 5

2. You cannot use mixed-size registers.
4. AX = 3100H, C = 0, A = 1, S = 0, Z = 0, and O = 0
6.
```
ADD   AX,BX
ADD   AX,CX
ADD   AX,DX
ADD   AX,SP
MOV   DI,AX
```
8. ADC DX,BX
10. The assembler cannot determine the size of the memory location.
12. BH = 7FH, C = 1, A = ?, S = 0, Z = 0, and O = 0 (A is undefined)
14. DEC EBX
16. The only difference between SUB and CMP is that the answer is lost with CMP.
18. DX (most-significant) and AX (least-significant)
20. EDX–EAX
22.
```
MOV   DL,5
MOV   AL,DL
MUL   DL
MUL   DL
```
24. AL
26. Divide by zero and divide overflow
28. AH
30. DAA and DAS
32. It divides AL by 10. This causes numbers between 0 and 99 decimal to be converted to unpacked BCD in AH (quotient) and AL (remainder).
34.
```
XCHG   AX,BX
ADD    AL,DL
DAA
MOV    DL,AL
XCHG   AH,AL
ADD    AL,DH
DAA
MOV    DH,AL
XCHG   AX,BX
ADD    AL,CL
DAA
MOV    CL,AL
XCHG   AL,AH
ADD    AL,CH
DAA
MOV    CH,AL
```
36.
```
MOV    BH,DH
AND    BH,00011111B
```
38.
```
MOV    SI,DI
OR     SI,1FH
```

40.
```
OR    AX,0FH
AND   AX,1FFFH
XOR   AX,0380H
```

42. TEST CH,4
44. (a) SHR DI,3 (b) SHL AL,1 (c) ROL AL,3 (d) RCR EDX,1 (e) SAR DH,1
46. Extra
48. The SCASB instruction is repeated while the condition is equal as long as CX is not zero.
50. CMPSB compares the byte contents of the byte in the data segment addressed by SI with the byte in the extra segment addressed by DI.
52. In DOS the letter C is displayed.

CHAPTER 6

2. A near JMP instruction
4. A far jump
6. (a) near (b) short (c) far
8. The IP or EIP register
10. The JMP AX instruction jumps to the offset address stored in AX. This can only be a near jump.
12. The JMP [DI] instruction jumps to the memory location addressed by the offset address stored in the data segment memory location addressed by DI. The JMP FAR PTR[DI] instruction jumps to the new offset address stored in the data segment memory location addressed by DI and the new segment addressed by the data segment memory location address by DI+2. JMP [DI] is a near jump and JMP FAR PTR [DI] is a far jump.
14. JA tests the condition of an arithmetic or logic instruction to determine if the outcome is above. If the outcome is above a jump occurs, otherwise no jump occurs.
16. JNE, JE, JG, JGE, JL, or JLE
18. JA and JBE
20. SETZ or SETE
22. ECX

24.
```
        MOV  DI,OFFSET DATAZ
        MOV  CX,150H
        CLD
        MOV  AL,00H
L1:     STOSB
        LOOP L1
```

26.
```
        CMP  AL,3
        JNE  @C0001
        ADD  AL,2
@C0001:
```

28.
```
        MOV  SI,OFFSET BLOCKA
        MOV  DI,OFFSET BLOCKB
        CLD
        .REPEAT
              LODSB
              STOSB
        .UNTIL AL == 0
```

30.
```
        MOV  AL,0
        MOV  SI,OFFSET BLOCKA
        MOV  DI,OFFSET BLOCKB
        CLD
        .WHILE  AL != 12H
                LODSB
                ADD  AL,[DI]
                MOV  [DI],AL
                INC  DI
        .ENDW
```

32. The far CALL pushes IP and CS onto the stack. Next, the two bytes following the opcode are moved into IP, the two bytes following that are moved into CS, and the jump occurs.
34. RET
36. By using NEAR or FAR to the right of the PROC directive

38.
```
CUBE  PROC  NEAR USES AX DX
      MOV   AX,CX
      MUL   CX
      MUL   CX
      RET
CUBE  ENDP
```

40.
```
SUMS  PROC  NEAR
      MOV   EDI,0
      ADD   EAX,EBX
      ADD   EAX,ECX
      ADD   EAX,EDX
      ADC   EDI,0
      RET
SUMS  ENDP
```

42. INT 0 through INT 255
44. The interrupt vector is used to detect and respond to divide errors.
46. The RET instruction pops the return address from the stack, while the IRET pops the flags and the return address from the stack.
48. When overflow is a 1
50. CLI and STI
52. If the value in the register or memory location under test in the destination operand is below or above the boundaries stored in the memory address by the source operand
54. BP

CHAPTER 7

2. No, bytes must be defined in C++ using char or _int8.
4. EAX, EBX, ECX, EDX, and ES
6. Floating-point coprocessor stack
8. Data are accessed in array string1 using register SI to index the string element.
10. If no headers are used for a C++ program it will be much smaller.
12. No. INT 21H is a 16-bit DOS call that cannot be used in the Windows 32-bit environment.

14.
```
#include "stdafx.h"
#include <conio.h>

int _tmain(int argc, _TCHAR* argv[])
```

```
{
        char a = 0;
        while ( a != '@')
        {
                a = _getche();
                _putch(a);
        }
        return 0;
}
```

16. The _putch(10) instruction displays the new line function and the _putch(13) returns the cursor to the left margin of the display.
18. Separate assembly modules are the most flexible.
20. The flat model must be used with the C prototype as in .MODEL FLAT,C and the function that is linked to C++ must be made public.
22. A 16-bit word is defined with the short directive or _int16.
24. Examples of events are mouse move, key down, etc. Event handlers catch these events so they can be used in a program.
26. Yes. The C++ editor can be used to edit an assembly language module, but the module must use the .TXT extension instead of .ASM.
28. `#define RDTSC _asm _emit 0x0f _asm _emit 0x31`
30.

```
;
;External function rotates a byte 3 places
 left
;
.586                    ;select Pentium and 32-
                        bit model
.model flat, C          ;select flat model with
                        C/C++ linkage
.stack 1024             ;allocate stack space
.code                   ;start code segment

public RotateLeft3      ;define RotateLeft3 as a
                        public function

RotateLeft3 proc        ;define procedure
Rotatedata:byte         ;define byte

        mov   al,Rotatedata
        rol   al,3
        ret

RotateLeft3 endp
```

32.

```
;Function that converts
;
.model flat,c
.stack 1024
.code

Public Upper

Upper proc
Char:byte

       mov  al,Char
       .if al >= 'a' && a; <= 'z'
              sub al,30h
       .endif
       Ret
Upper endp
```

34. Properties contains information about an object such as the foreground and background colors, etc.
36. `_asm inc ptr;`

CHAPTER 8

2. The TEST.ASM file, when assembled, generates the TEST.OBJ file and the TEST.EXE file if no switches appear on the command line.
4. PUBLIC indicates that a label is available to other program modules.
6. EXTRN
8. MACRO and ENDM
10. Parameters are passed to a macro through a parameter list that follows the MACRO keyword (on the same line).
12. The LOCAL directive defines local labels and must be on the line immediately following the MACRO line.
14.

```
ADDM  MACRO   LIST,LENGTH
      PUSH CX
      PUSH SI
      MOV   CX,LENGTH
      MOV   SI,LIST
      MOV   AX,0
      .REPEAT
           ADD   AX,[SI]
           INC   SI
      .UNTILCXZ
      POP   SI
      POP   CX
      ENDM
```

16.

```
BOOL CssssDlg::PreTranslateMessage(MSG* pMsg)
{
        if ( pMsg->message == WM_CHAR )
        {
              unsigned int key = pMsg->wParam;
              if ( key < '0' || key > '9' )
                    return true;
              pMsg->wParam = key;
        }
        return CDialog::PreTranslateMessage
        (pMsg);
}
```

18.

```
BOOL CssssDlg::PreTranslateMessage(MSG* pMsg)
{
     {
          RandomNumber++;  //a global variable
          if ( RandomNumber == 63 )
               RandomNumber = 9;
     }
     return CDialog::PreTranslateMessage
     (pMsg);
}
```

20.

```
void CshiftrotateDlg::OnBnClickedButton1()
{
       Label1.put_Caption("Shift Left = ");
       shift = true;
       data1 = 1;
       Label2.put_Caption("00000001");
       SetTimer(1,500,0);
}

void CshiftrotateDlg::OnBnClickedButton2()
{
       Label1.put_Caption("Rotate Left = ");
       shift = false;
       data1 = 1;
       Label2.put_Caption("00000001");
       SetTimer(1,500,0);
}
```

```
void CshiftrotateDlg::OnBnClickedButton3()
                    //new Left/Right button
{
       direction = !direction; //new bool
                                   variable
       if ( direction )
       {
              if ( shift )
                     Label1.put_Caption("Shift
                                   Left = ");
              else
                     Label1.put_Caption("Rotate
                                   Left = ");
       }
       else
       {
              if ( shift )
                     Label1.put_Caption("Shift
                                   Right = ");
              else
                     Label1.put_Caption("Rotate
                                   Left = ");
       }
}

void CshiftrotateDlg::OnTimer(UINT nIDEvent)
{
       if ( nIDEvent == 1 )
       {
             CString temp = "";
             char temp1 = data1;
             if ( shift )
                    if ( direction )
                           _asm shl temp1,1;
                    else
                           _asm shr temp1,1;
             else
                    if ( direction )
                           _asm rol temp1,1;
                    else
                           _asm ror temp1,1;
       data1 = temp1;
       for (int a = 128; a > 0; a>>=1)
       {
                    if ( ( temp1 & a ) == a )
                           temp += "1";
                    else
                           temp += "0";
       }
       Label2.put_Caption(temp);
  }
  CDialog::OnTimer(nIDEvent);
}
```

22. A handler for the WM_RBUTTONDOWN message is inserted into the program to intercept the right mouse button in a program.

24.
```
void CBubSortDlg::OnLButtonDown(UINT nFlags,
CPoint point)
{
       if (nFlags == 3)
       {
              //left - then right

       }
       CDialog::OnLButtonDown(nFlags, point);
}

void CBubSortDlg::OnRButtonDown(UINT nFlags,
CPoint point)
{
       if (nFlags == MK_RBUTTON | MK_LBUTTON)
       {
             //right - then left

       }
             CDialog::OnRButtonDown(nFlags, point);
}
```

26. The color of the shading on the edge of command buttons.

28. A large number is converted by repeated divisions by the number 10. After each digit the remainder is saved as a significant digit of the BCD result.

30. 30H

32.
```
int GetOct(void)
{
       CString temp;
       int result = 0;
       char temp1;
       GetDlgItemText(IDC_EDIT1, temp);
       for ( int a = 0; a < temp.GetLength();
       a++ )
       {
              temp1 = temp.GetAt(a);
              _asm
              {
                     shl result,3
                     mov eax,0
                     mov al,temp1
                     sub al,30h
                     or result,eax
              }
       }
       return result;
}
```

34.
```
char Up(char temp)
{
       if ( temp >= 'a' && temp <= 'z' )
              _asm sub temp,20h;
       Return temp;
}
```

36. The boot sector is where a bootstrap loader program is located that loads the operating system. The FAT is a table that contains numbers that indicate whether a cluster is free, bad, or occupied. If occupied, an FFFFH indicates the end of a file chain or the next cluster number in the file chain. The director holds information about a file or a folder.

38. Sectors

40. A cluster is a grouping of sectors.

42. 4G bytes

44. 8

46. 256

48. `CFile::Rename( "TEST.LST", "TEST.LIS" );`

50. An ActiveX control is a common object that can be used in a visual programming language.

52. See Figure D–1 for the output (the stock ListBox contains the output).

```
//code placed in the OnInitDlg function
int tempval = 1;
for ( int a = 0; a < 8; a++ )
{
       CString temp = "2^ = ";
       temp.SetAt(2, a + 0x30);
       temp += GetNumb(tempval);
       List.InsertString(a, temp);
       tempval <<= 1;
}
String CPowersDlg::GetNumb(int temp)
```

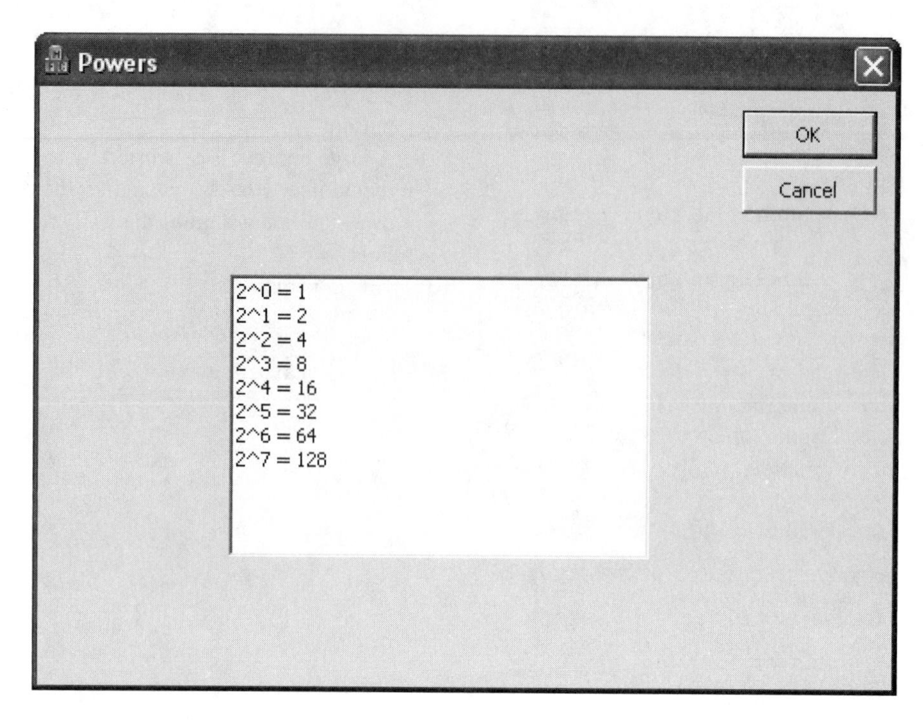

FIGURE D–1

```
{
        char numb[10];
        _asm
        {
                mov   eax,temp
                mov   ebx,10
                push  ebx
                mov   ecx,0
loop1:
                mov   edx,0
                div   ebx
                push  edx
                cmp   eax,0
                jnz   loop1
loop2:
                pop   edx
                cmp   edx,ebx
                je    loop3
                add   dl,30h
                mov   numb[ecx],dl
                inc   ecx
                jmp   loop2
loop3:
                mov   byte ptr numb[ecx],0
        }
        return numb;
}
```

54.
```
BOOL CDisplayDlg::PreTranslateMessage(MSG* pMsg)
{
        char lookup[] = {0x3f, 6, 0x5b, 0x4f,
        0x66, 0x6d, 0x7d, 7, 0x7f, 0x6f, 0x77,
        0x7c, 0x39, 0x5e, 0x79, 0x71};
        char temp;
        if ( pMsg->message == WM_KEYDOWN )
        {
                if ( pMsg->wParam >= '0' &&
                pMsg->wParam <= '9'
                        || pMsg->wParam >= 'A' &&
                        pMsg->wParam <= 'F'
                        || pMsg->wParam >= 'a' &&
                        pMsg->wParam <= 'f')
                {
                        temp = pMsg->wParam - 0x30;
                        if ( temp > 9 )
                                temp -= 7;
                        _asm        //lookup 7-segment code
                        {
                                lea  ebx,lookup
                                mov  al,temp
                                xlat
                                mov  temp,al
                        }
                        ShowDigit(temp);  //display the
                                            digit
                }
                return true;             //finished with
                                           keystroke
        }
        return CDialog::PreTranslateMessage(pMsg);
}
```

CHAPTER 9

2. Yes and no. The current drive of a logic zero is reduced to 2.0 mA and the noise immunity is reduced to 350 mV.
4. Address bits A_0–A_7
6. A read operation
8. The duty cycle must be 33%.
10. A write is occurring.
12. The data bus is sending data to the memory or I/O.
14. $IO/\overline{M}$, $DT/\overline{R}$, and $\overline{SS0}$
16. Signals to the coprocessor, which indicate what the microprocessor queue is doing

18. 3
20. 14 MHz/6 = 2.33 MHz
22. Address bus connections A_0–A_{15}
24. 74LS373 transparent latch
26. If too many memory and/or I/O devices are attached to a system the buses must be buffered.
28. 4
30. Fetch and execute
32. (a) The address is output along with ALE. (b) Time is allowed for memory access and the READY input is sampled. (c) The read or write signal is issued. (d) Data are transferred and read or write is deactivated. (e) Wait allows additional time for memory access.
36. Selects one or two stages of synchronization for READY
38. Minimum mode operation is most often used in embedded applications and maximum mode operation was most often used in early personal computers.

CHAPTER 10

2. (a) 256 (b) 2K (c) 4K (d) 8K
4. Select the memory device
6. Cause a write to occur
8. The microprocessor allows 460 ns for memory at 5 MHz, but because there is a small delay in the connections to the memory it would be best not to use a 450 ns memory device in such a system without one wait state.
10. Static random access memory
12. 250 ns
14. The address inputs to many DRAMs are multiplexed so one address input accepts two different address bits, reducing the number of pins required to address memory in a DRAM.
16. Generally the amount of time is equal to a read cycle and represents only a small amount of time in a modern memory system.
18. See Figure D–2.
20. One of the eight outputs becomes a logic zero as dictated by the address inputs.
22. See Figure D–3.
24. Verilog hardware description language
26. The architecture block between begin and end
28. $\overline{\text{MRDC}}$ and $\overline{\text{MWTC}}$
30. See Figure D–4.
32. 5
34. 1
36. $\overline{\text{BHE}}$ selects the upper memory bank and A_0 selects the lower memory bank.
38. Separate decoders and separate write signals
40. Lower memory bank
42.

```
library ieee;
use ieee.std_logic_1164.all;
entity DECODER_10_28 is

port (
        A23, A22, A21, A20, A19, A18, A17,
        A16, A0, BHE, MWTC: in STD_LOGIC;
        SEL, LWR, HWR: out STD_LOGIC
);

end;
```

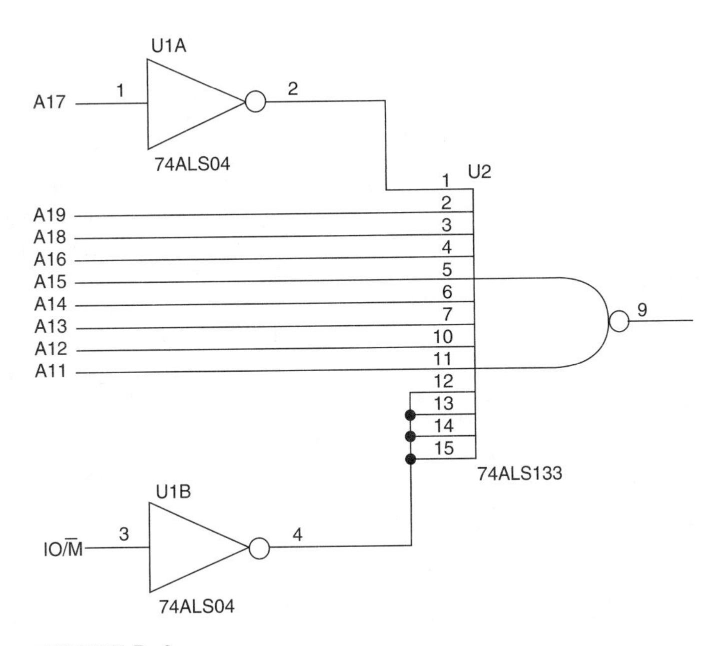

FIGURE D–2

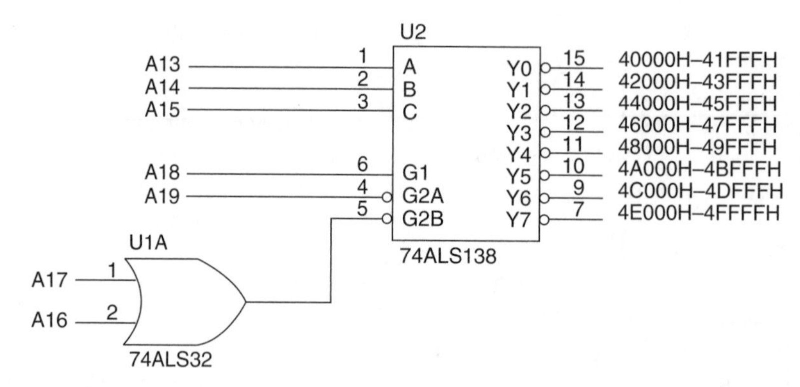

FIGURE D–3

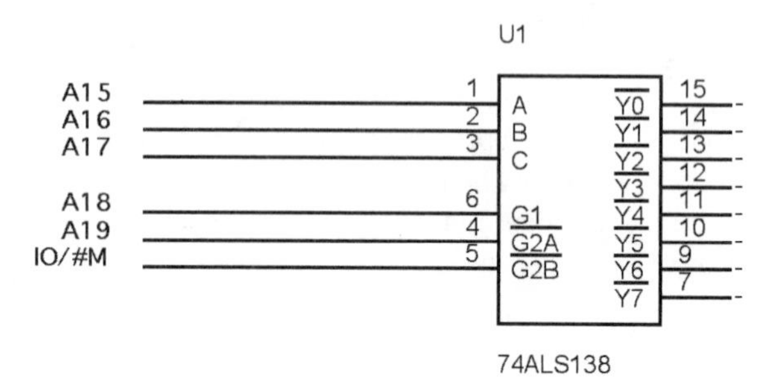

FIGURE D–4

```
architecture V1 of DECODER_10_28 is

begin

     SEL <= A23 or A22 or A21 or A20 or A19
     or A18 or (not A17) or (not A16);
     LWR <= A0 or MWTC;
     HWR <= BHE or MWTC;

end V1;
```

44. See Figure D–5.
48. Yes, as long as a memory location on the DRAM is not accessed.

CHAPTER 11

2. The I/O address is stored in the second byte of the instruction.
4. DX
6. The OUTSB instruction transfers the data segment byte addressed by SI to the I/O port addressed by DX, then SI is incremented by one.
8. Memory-mapped I/O uses any instruction that transfers data to or from the memory for I/O, while isolated I/O requires the use of the IN or OUT instruction.
10. The basic output interface is a latch that captures output data and holds it for the output device.
12. Lower
14. 4
16. It removes mechanical bounces from a switch.
18. See Figure D–6.
20. See Figure D–7.
22. See Figure D–8.
24. If the port is 16 bits wide, there is no need to enable either the low or high half.
26. D_{47}–D_{40}
28. Group A is port A and PC_4–PC_7, while group B is port B and PC_3–PC_0.
30. $\overline{RD}$
32. Inputs
34. The strobe input latches the input data and sets the buffer full flag and interrupt request.
36.

```
DELAY   PROC   NEAR USES ECX

        MOV   ECX, 7272727
D1:
        LOOPD D1
        RET

DELAY   ENDP
```

38. The strobe signal ($\overline{STB}$)
40. The INTR pin is enabled by setting the INTE bit in PC_4 (port A) or PC_2 (port B).
42. When data are output to the port $\overline{OBF}$ becomes a 0 and when $\overline{ACK}$ is sent to the port $\overline{OBF}$ becomes a 1.
44. Group or Port A contains the bidirectional data.
46. The 01H command is sent to the LCD display.
48.

```
;Displays the null terminated string
addressed by DS:BX
;uses a macro called SEND to send data to
the display
;
DISP        PROC  NEAR  USES BX
            SEND  86H,2,1       ;move cursor to
                                position 6
            .WHILE BYTE PTR [BX] != 0
                   SEND [BX],0,1
                   INC  BX
            .ENDW
            RET
DISP        ENDP
```

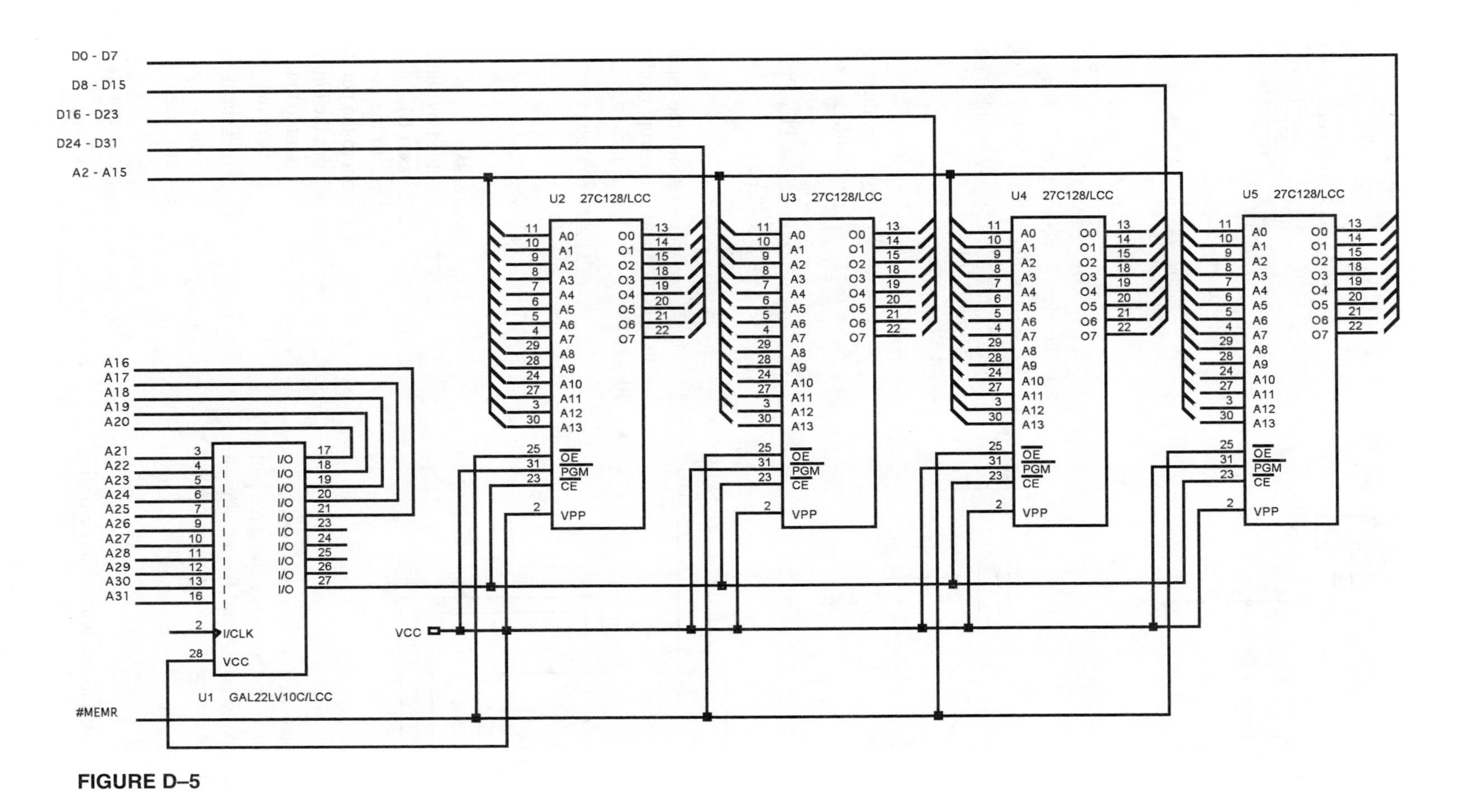
D0 - D7
D8 - D15
D16 - D23
D24 - D31
A2 - A15
A16
A17
A18
A19
A20
A21
A22
A23
A24
A25
A26
A27
A28
A29
A30
A31
#MEMR
U1 GAL22LV10C/LCC
I/CLK
VCC
I/O
VCC
U2 27C128/LCC
U3 27C128/LCC
U4 27C128/LCC
U5 27C128/LCC
A0
A1
A2
A3
A4
A5
A6
A7
A8
A9
A10
A11
A12
A13
OE
PGM
CE
VPP
O0
O1
O2
O3
O4
O5
O6
O7

FIGURE D–5

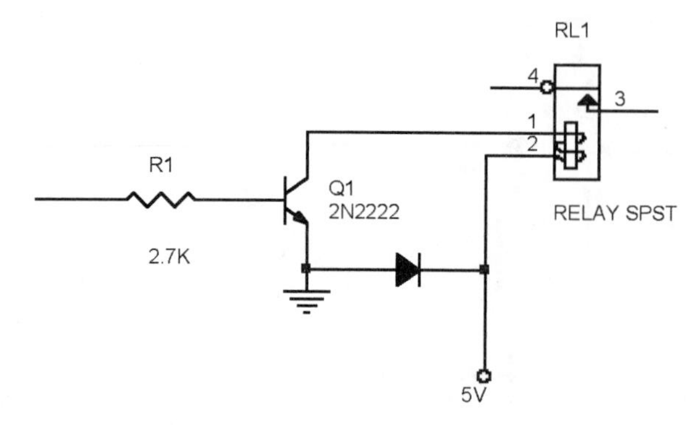

FIGURE D–6

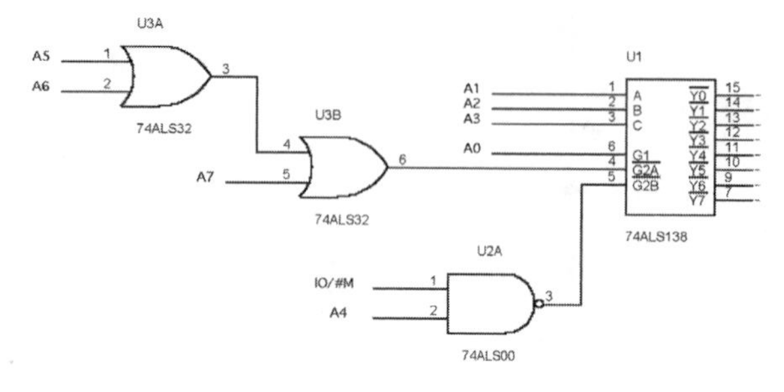

FIGURE D–7

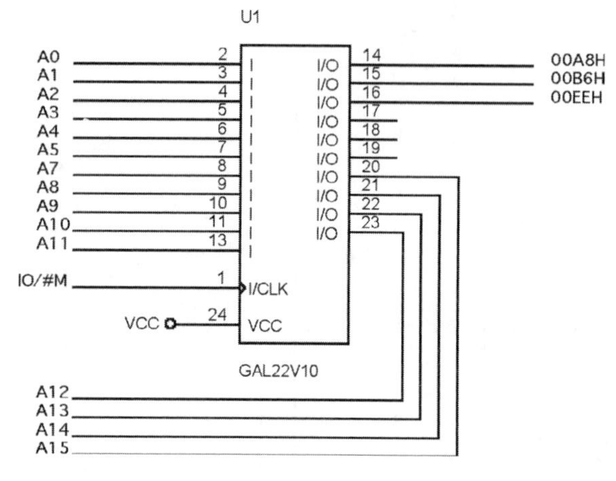

FIGURE D–8

50. The only changes that need to be made are that instead of four rows there are three rows and three pull-up resisters connected to port A and five columns to connect to Port B. Of course, the software also needs some minor changes.

54. 6

58. Least significant

62. Data that are sent a bit at a time without any clocking pulses

64.
```
LINE    EQU   023H
LSB     EQU   020H
MSB     EQU   021H
FIFO    EQU   022H

        MOV   AL,10001010B ;enable Baud divisor
        OUT   LINE,AL

        MOV   AL,60        ;program Baud rate
        OUT   LSB,AL
        MOV   AL,0
        OUT   MSB,AL

        MOV   AL,00011001B ;program 7-data, odd
        OUT   LINE,AL      ;parity, one stop

        MOV   AL,00000111B ;enable transmitter
                           and
        OUT   FIFO,AL      ;and receiver
```

66. Simplex = receiving or sending data; half-duplex = receiving and sending data; but only one direction at a time; and full-duplex = receiving and sending data at the same time.

68.
```
SENDS   PROC   NEAR

MOV     CX,16
        .REPEAT
           .REPEAT
                IN     AL,LSTAT ;get line
                                 status register
                TEST  AL,20H    ;test TH bit
           .UNTIL !ZERO?
           LODSB                ;get data
           OUT   DATA,AL        ;transmit data
        .UNTILCXZ

    RET

SENDS ENDP
```

70. 0.01V

72.
```
        .MODEL   TINY
        .CODE
        .STARTUP
                MOV       DX,400H
                .WHILE   1
                MOV       CX,256
                MOV       AL,0
                .REPEAT
                    OUT       DX,AL
                    INC       AL
                    CALL      DELAY
                .UNTILCXZ
                MOV    CX,256
                .REPEAT
                    OUT       DX,AL
                    DEC       AL
                    CALL      DELAY
                .UNTILCXZ
                .ENDW
DELAY           PROC    NEAR

; 39 microsecond time delay

DELAY   ENDP
        END
```

74. INTR indicates that the converter has completed a conversion.

76. See Figure D–9.

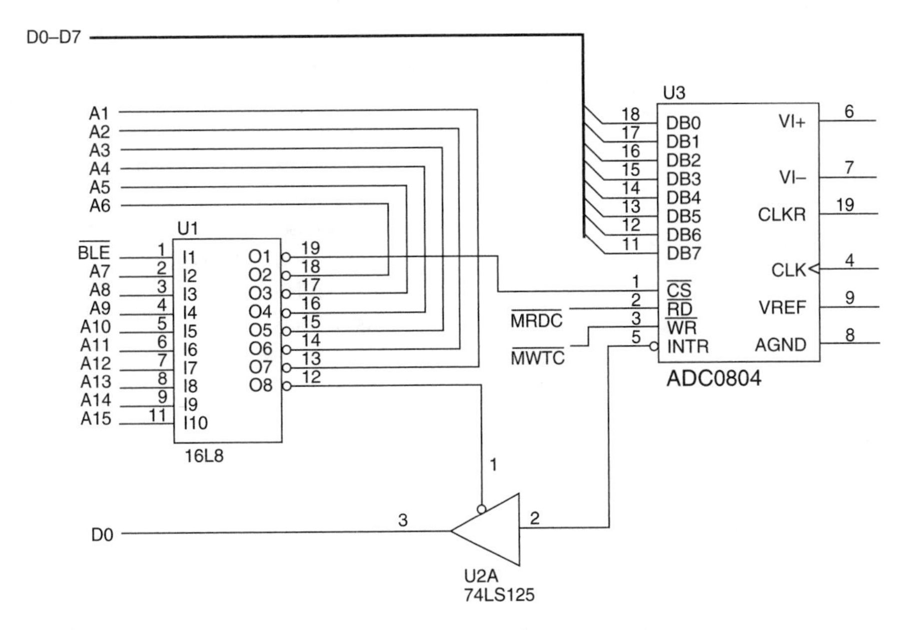

FIGURE D–9

CHAPTER 12

2. An interrupt is a hardware- or software-initiated subroutine call.
4. An interrupt only uses computer time when the interrupt is activated.
6. INT, INT_3, INTO, CLI, and STI
8. The first 1K byte of the memory system in real mode and anywhere in protected mode
10. 00H through 1FH
12. Anywhere in the memory system
14. A real mode interrupt pushes CS, IP, and the FLAGS onto the stack, while a protected mode interrupt pushes CS, EIP, and the EFLAGS onto the stack.
16. The INTO occurs if overflow is set.
18. The IRET instruction pops the flags and the return address from the stack.
20. The state of the interrupt structure is stored on the stack so when the return occurs it is restored. Both the interrupt and trace flags are cleared.
22. The T flag controls whether tracing is enabled or disabled.
24. The T flag is enabled or disabled by manipulating the flag bits directly because there are no instructions to control the trace flag.
26. No
28. Edge and level sensitive
30. A FIFO is a first-in, first-out memory structure.
32. See Figure D–10.
34. A daisy-chain is when signals are ORed together to generate a signal.
36. A programmable interrupt controller
38. These are the eight interrupt request inputs.
40. To an interrupt request input
42. An OCW is an operational control word for the 8259.
44. ICW_2
46. Program sensitivity and single or multiple 8259s
48. The most recent interrupt request level becomes the lowest level interrupt after being serviced.
50. INT 8 through INT 0FH

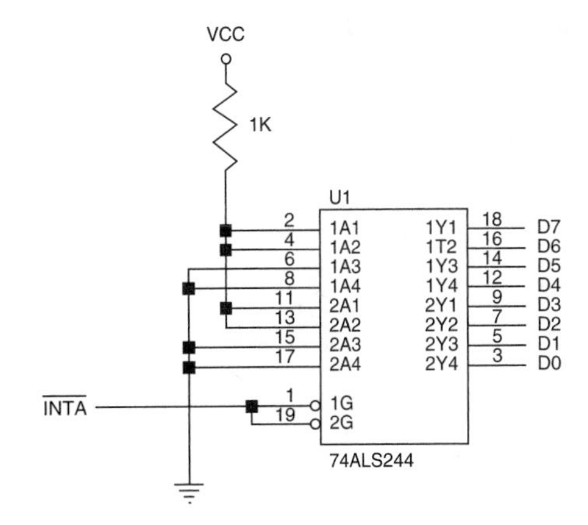

FIGURE D–10

CHAPTER 13

2. When a 1 is placed on HOLD, the program stops executing and the address, data, and control buses go to their high-impedance state.
4. I/O to memory
6. DACK
8. The microprocessor is in its hold state and the DMA controller has control of the buses.
10. 4
12. The command register
16. A pen drive is a USB device that acts as storage device using a Flash memory.
18. Tracks
20. Cylinder
22. See Figure D–11.
24. The heads in a hard disk drive are aerodynamically designed to ride on a cushion of air as the disk spins and are therefore called flying heads.
26. The stepper motor positioning mechanism is noisy and not very precise, while the voice coil positioning mechanism is silent and very accurate because its placement can be continuously adjusted.
28. A CD-ROM is an optical device for storing music or digital data and has a capacity of about 660M or 700M (80 minute) bytes.
30. A TTL monitor uses TTL signals to generate a display and an analog monitor uses analog signals.
32. Cyan, magenta, and yellow
34. 600 lines with 800 horizontal elements
36. The latest style of digital video input connector for all types of video equipment
38. 16 million colors

CHAPTER 14

2. Word (16 bits, ±32K), doubleword (32 bits, ±2G), and quadword (64 bits, $\pm 9 \times 10^{18}$)
4. Single-precision (32 bits), double-precision (64 bits), and temporary-precision (80 bits).
6. (a) −7.75 (b) .5625 (c) 76.5 (d) 2.0 (e) 10.0 (f) 0.0
8. The microprocessor continues executing microprocessor (integer) instructions while the coprocessor executes a floating-point instruction.
10. It copies the coprocessor status register to AX.
12. By comparing the two registers and then by transferring the status word to the AX register. If the SAHF instruction is next executed a JZ instruction can be used to test the outcome of the coprocessor compare instruction.
14. FSTSW AX
16. Data are always stored as an 80-bit temporary-precision number.
18. 0
20. Affine allows positive and negative infinity, while projective assumes infinity is unsigned.
22. Extended (temporary) precision
24. The contents of the top of the stack are copied into memory location DATA as a floating-point number.
26. FADD ST,ST(3)
28. FSUB ST(2),ST
30. Forward division divides the top of the stack by the contents of a memory location and returns the quotient to the top of the stack. Reverse division divides the top of the stack into the contents of the memory location and returns the result to the top of the stack. If no operand exists, then forward division divides ST(1) by ST and reverse division divides ST by ST(1).
32. It performs a MOV to ST if the condition is below.
34.
```
RECIP  PROC    NEAR
       MOV     TEMP,EAX
       FLD     TEMP
       FLD1
       FDIVR
       FSTP TEMP
       MOV   EAX,TEMP
RECIP  ENDP
TEMP   DD  ?
```
36. Finds the function $2^X - 1$.
38. FLDPI
40. It indicates that register ST(2) is free.
42. The state of the machine

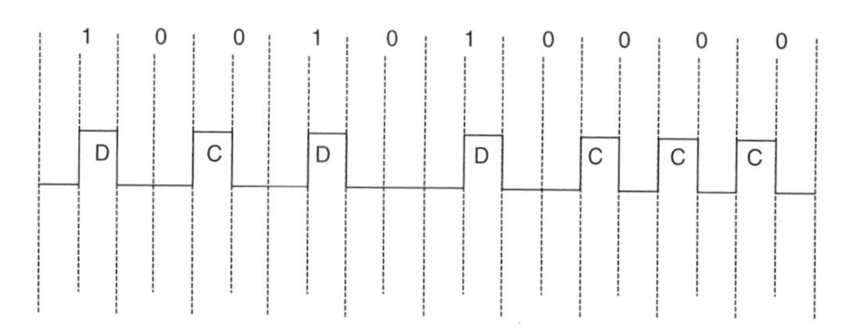

FIGURE D–11

44.
```
CAPR    PROC    NEAR
        FLDPI
        FADD ST,ST(1)
        FMUL F
        FMUL C1
        FLD1
        FDIVR
        FTSP XC
        RET
CAPR    ENDP
```

46. In modern software it is never used.

48.
```
TOT     PROC    NEAR
        FLD   R2
        FLD1
        FDIVR
        FLD   R3
        FLD1
        FDIVR
        FLD   R4
        FLD1
        FDIVR
        FADD
        FADD
        FLD1
        FDIV
        FADD  R1
        FSTP  RT
        RET
TOT     ENDP
```

50.
```
PROD    PROC    NEAR
        MOV   ECX,100
        .REPEAT
              FLD    ARRAY1[ECX*8-8]
              FUML   ARRAY2[ECX*8-8]
              FSTP   ARRAY3[ECX+8-8]
        .UNTILCXZ
        RET
PROD    ENDP
```

52.
```
POW     PROC    NEAR
        MOV   TEMP,EBX
        FLD   TEMP
        F2XM1
        FLD1
        FADD
        MOV   TEMP,EAX
        FLD   TEMP
        FYL2X
        FSTP  TEMP
        MOV   ECX,TEMP
        RET
POW     ENDP
```

54.
```
GAIN    PROC    NEAR
        MOV   ECX,100
        .REPEAT
              FLD   DWORD PTR VOUT[ECX*4-4]
              FDIV DWORD PTR VIN[ECX*4-4]
              CALL LOG10
              FIMUL TWENTY
              FSTP DWORD PTR DBG[ECX*4-4]
        .UNTILCXZ
        RET
TWENTY DW     20
GAIN    ENDP
```

56. The EMMS instruction clears the coprocessor stack to indicate that the MMX unit has completed using the stack.

58. Signed saturation occurs when byte-sized numbers are added and have values of 7FH for an overflow and 80H for an underflow.

60. The FSAVE instruction stores all the MMX registers in memory.

62. Single-instruction, multiple-data instructions

64. 128 bits

66. 16

68. Yes

CHAPTER 15

2. 8- or 16-bit depending on the socket configuration

4. See Figure D–12.

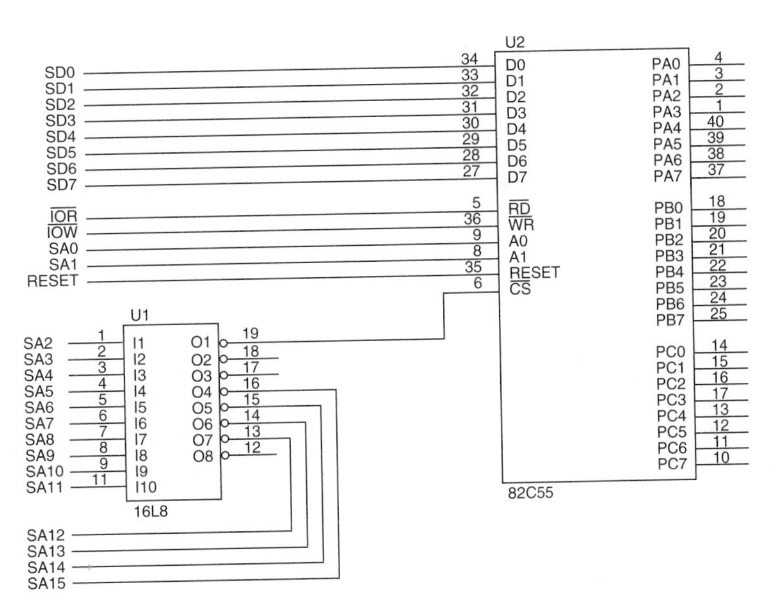

FIGURE D–12

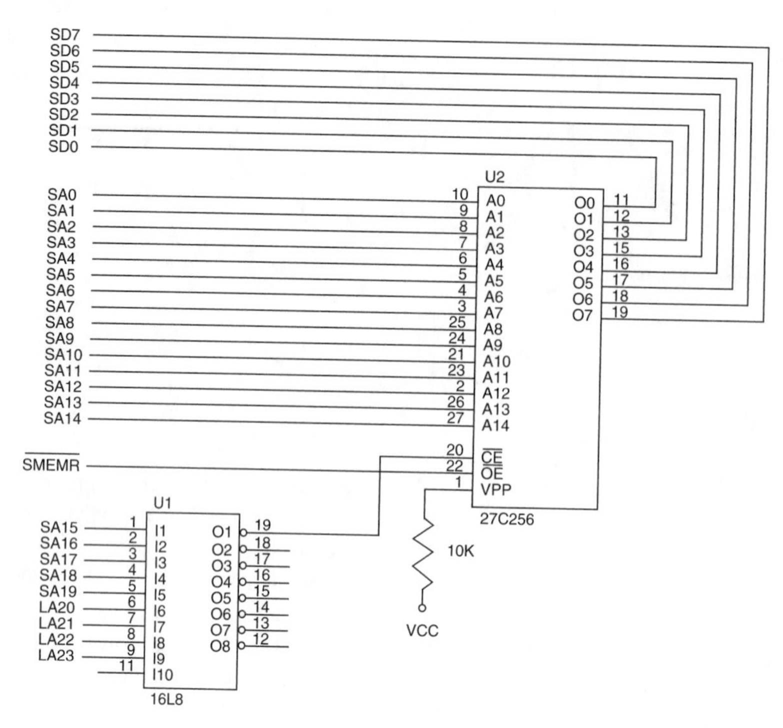

FIGURE D–13

6. See Figure D–13.
8. See Figure D–14.
12. 16 bits
14. The configuration memory identifies the vendor and also information about the interrupts.
16. This is the command/bus enable signal that is high to indicate the PCI bus contains a command and low for data.
18.
```
MOV  AX,0B108H
MOV  BX,0
MOV  DI,8
INT  1AH
```
20. 2.5 GHz
22. Yes
24. COM_1
28. 1.5 Mbps, 12 Mbps, and 480 Mbps
30. 127
32. 5 meters
34. An extra bit that is thrown in the data stream if more than six ones are sent in a row.
36. 1 to 1023 bytes
38. The PCI transfers data at 33 MBs, while AGP transfers data at 2 GBps (8×).

CHAPTER 16

2. The hardware enhancements include internal timers, additional interrupt inputs, chip selection logic, serial communications ports, parallel pins, DMA controller, and an interrupt controller.
4. 10 MHz
6. 3 mA
8. The point at which the address appears
10. 260 ns for the 16 MHz version operated at 10 MHz
12.
```
MOV  AX,1000H
MOV  DX,0FFFEH
OUT  DX,AX
```
14. 10 on most versions of the 80186/80188 including the internal interrupts.
16. The interrupt control registers control a single interrupt.
18. The interrupt poll register acknowledges the interrupt, while the interrupt poll status register does not acknowledge the interrupt.
20. 3
22. Timer 2
24. It determines whether the enable counter bit functions.
26. The ALT bit selects both compare registers so the duration of the logic 1 and logic 0 output times can be programmed.
28.
```
MOV  AX,123
MOV  DX,0FF5AH
OUT  DX,AX
MOV  AX,23
ADD  DX,2
OUT  DX,AX
MOV  AX,0C007H
MOV  DX,0FF58H
OUT  DX,AX
```

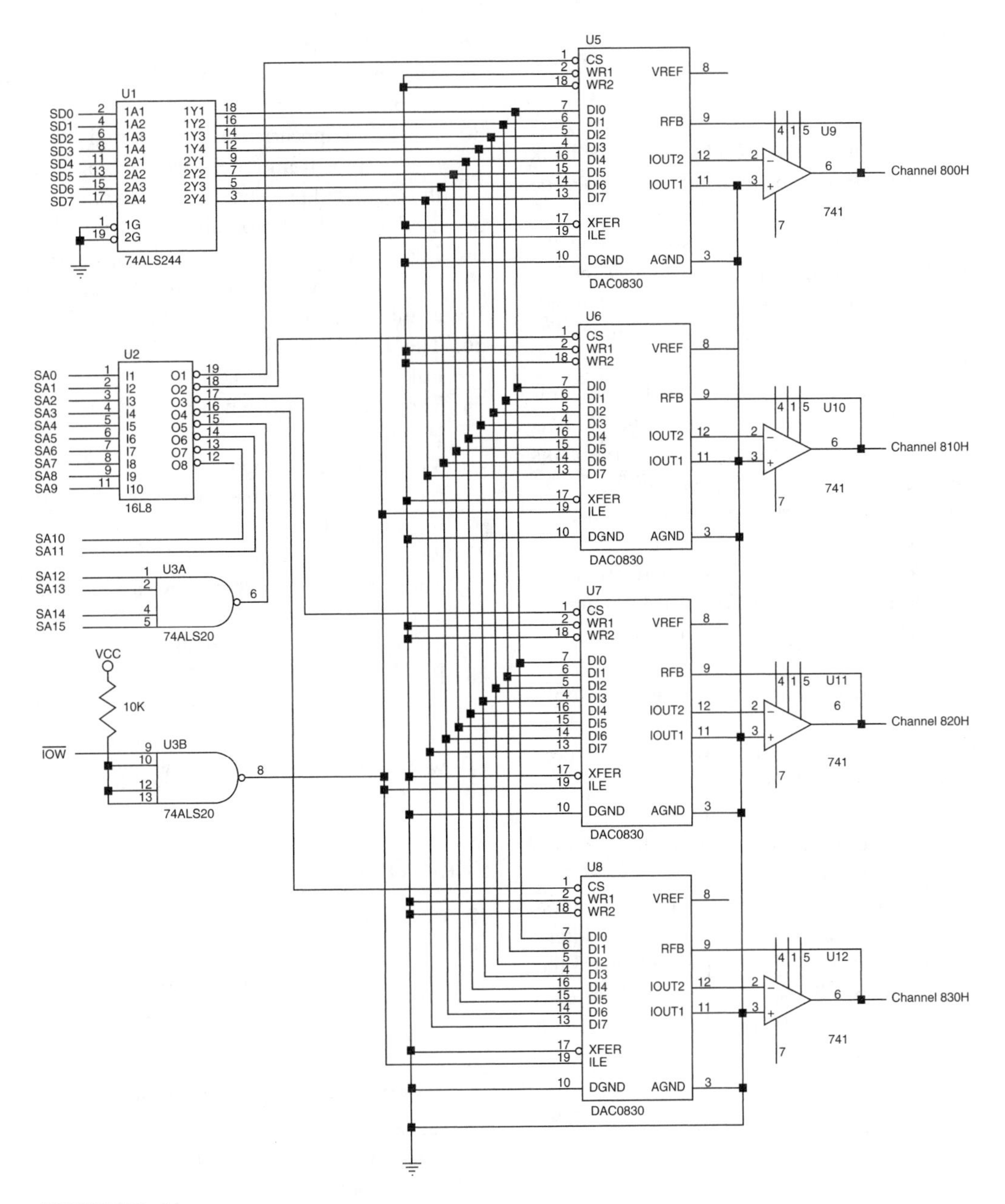

FIGURE D–14

30. 2
32. Place a logic 1 in both the CHG/$\overline{\text{NOCHG}}$ and START/$\overline{\text{STOP}}$ bits of the control register.
34. 7
36. Chip
38. 15
40. It determines the operation of the $\overline{\text{PCS5}}$ and $\overline{\text{PCS6}}$ pins.
42.

```
MOV   AX,1001H
MOV   DX,0FF90H
OUT   DX,AX
MOV   AX,1048H
OUT   DX,AX
```

44. 1G
46. Verify for read access.
48. An RTOS is a real-time operating system that has a predictable and guaranteed time for threads access.

CHAPTER 17

2. 64T
4. See Figure D–15.
6. The memory system has up to 4G bytes and the bank enable signals select one or more of the 8-bit-wide banks of memory.
8. The pipeline allows the microprocessor to send the address of the next memory location while it fetches the data from the prior memory operation. This allows the memory additional time to access the data.
10. 0000H–FFFFH
12. I/O has the same address as earlier models of the microprocessor. The difference is that the I/O is arranged as a 32-bit-wide space with four 8-bit banks that are selected by the bank enable signals.
14. The $\overline{BS16}$ pin causes the microprocessor to function with an 8-bit-wide data bus.
16. The first four debug registers (DR_0–DR_3) contain breakpoint addresses; registers DR_4 and DR_5 are reserved for Intel's use; DR_6 and DR_7 are used to control debugging.
18. The test registers are used to test the translation look-aside buffer.
20. The PE bit switches the microprocessor into protected mode if set and real mode if cleared.
22. Scaled-index addressing used a scaling factor of 1, 2, 4, or 8 times to scale addressing from byte, word, doubleword, or quadword.
24. (a) the address in the data segment at the location pointed to by EBX times 8 plus ECX (b) the address in the data segment array DATA pointed to by the sum of EAX plus EBX (c) the address at data segment location DATA (d) the address in the data segment pointed to by EBX

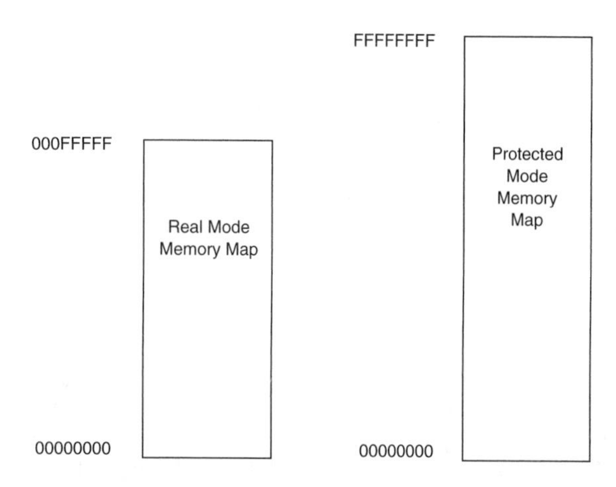

FIGURE D–15

26. Type 13 (0DH)
28. The interrupt descriptor table and its interrupt descriptors.
30. A selector appears in a segment register and it selects a descriptor from a descriptor table. It also contains the requested privilege level of the request.
32. The global descriptor table register
34. Because a descriptor addresses up to 4G of memory and there are 8K local and 8K global descriptors available at a time, 4G times 16K = 64T.
36. The TSS holds linkages and registers of a task so tasks can be switched efficiently.
38. The switch occurs when a logic 1 is placed into the PE bit of CR_0.
40. Virtual mode, which simulates DOS in protected mode, sets up 1M memory spans that can operate in the real mode.
42. 4K
44. The 80486 has an internal 8K cache and also contains a coprocessor.
46. The register sets are virtually identical.
48. $\overline{PCHK}$ and DP_0–DP_3
50. 8K
52. A burst is when four 32-bit numbers are read or written between the cache and memory.
54. Built-in self-test

CHAPTER 18

2. 64G bytes
4. These pins generate and check parity.
6. The burst ready pin is used to insert a wait state into the bus cycle.
8. 18.5 ns
10. T_2
12. An 8K-byte data cache and an 8K-byte instruction cache.
14. Yes, if one is a coprocessor instruction and the integer instructions are not dependent.
16. The SSM mode is used for power management in most systems.
18. 38000H
20. The CMPXCH8B instruction compares the 64-bit number in EDX:EAX with a 64-bit number stored in memory. If they are equal ECX:EBX is stored in memory. If they are not equal the contents of memory are moved into EDX:EAX.
22. ID, VIP, VIF, and AC
24. To access 4M pages, the page tables are dropped and only the page directory is used with a 22-bit offset address.

26. The Pentium Pro is an improved version of the Pentium that contains three integer units, an MMX unit, and a 36-bit address bus.
28. 36 address bits on A_3 through A_{35} (A_0–A_2 are encoded in the bank selection signals)
30. The access time in a 66 MHz Pentium is 18.5 ns and in the Pentium Pro at 66 MHz access time is 17 ns.
32. SDRAM that is 72 bits wide is purchased for ECC memory applications instead of 64-bit-wide memory.

CHAPTER 19

2. 512K, 1M, or 2M
4. The Pentium Pro cache is on the main board and the Pentium 2 cache is in the cartridge and operates at a high speed.
6. 64G bytes
8. 242
10. The read and write signals are developed by the chip set instead of the microprocessor.
12. 8 ns after the first quadword is accessed. The first quadword still requires 60 ns for access.
14. Model-specific registers have been added for SYSENTER_CS, STSENTER_SS, and SYSENTER _ESP.
16. The ECX register addresses the MSR number when the RDMSR instruction executes. After execution EDX:EAX contains the contents of the register.
18.

```
TESTS   PROC     NEAR
        CPUID
        BT     EDX,800H
        RET
TESTS   ENDP
```

20. EDX to the EIP register and the value in ECX to the ESP register
22. Ring 3
24. Pentium Pro
26. The Pentium 4 requires a power supply with an addition 12V connector for the main board. A Pentium 4–compliant supply must be used.
28.

```
bool Hyper()
{
     _asm
     {
          bool State = true;
          mov  eax,1
          cpuid
          mov  temp1,31h
          bt   edx,28        ;check for hyper-
                              threading
          jc   Hyper1
          mov  State, 0
Hyper1:
     }
     return State;
}
```

INDEX